SAXON MATH™

Texas Edition

Course 2

Student Edition

Stephen Hake

SAXON™
A Harcourt Achieve Imprint
www.SaxonPublishers.com
1-800-284-7019

ACKNOWLEDGEMENTS

This book was made possible by the significant contributions of many individuals and the dedicated efforts of a talented team at Harcourt Achieve.

Special thanks to:

- Melody Simmons and Chris Braun for suggestions and explanations for problem solving in Courses 1–3,
- Elizabeth Rivas and Bryon Hake for their extensive contributions to lessons and practice in Course 3,
- Sue Ellen Fealko for suggested application problems in Course 3.

The long hours and technical assistance of John and James Hake on Courses 1–3, Robert Hake on Course 3, Tom Curtis on Course 3, and Roger Phan on Course 3 were invaluable in meeting publishing deadlines. The saintly patience and unwavering support of Mary is most appreciated.

– Stephen Hake

Staff Credits

Editorial: Jean Armstrong, Shelley Farrar-Coleman, Marc Connolly, Hirva Raj, Brooke Butner, Robin Adams, Roxanne Picou, Cecilia Colome, Michael Ota

Design: Alison Klassen, Joan Cunningham, Deborah Diver, Alan Klemp, Andy Hendrix, Rhonda Holcomb

Production: Mychael Ferris-Pacheco, Heather Jernt, Greg Gaspard, Donna Brawley, John-Paxton Gremillion

Manufacturing: Cathy Voltaggio

Marketing: Marilyn Trow, Kimberly Sadler

E-Learning: Layne Hedrick, Karen Stitt

ISBN 1-5914-1999-9

2 3 4 5 6 7 8 9 048 12 11 10 09 08 07

TEXAS EDITION

ABOUT THE AUTHOR

Stephen Hake has authored five books in the Saxon Math series. He writes from 17 years of classroom experience as a teacher in grades 5 through 12 and as a math specialist in El Monte, California. As a math coach, his students won honors and recognition in local, regional, and statewide competitions.

Stephen has been writing math curriculum since 1975 and for Saxon since 1985. He has also authored several math contests including Los Angeles County's first Math Field Day contest. Stephen contributed to the 1999 National Academy of Science publication on the Nature and Teaching of Algebra in the Middle Grades.

Stephen is a member of the National Council of Teachers of Mathematics and the California Mathematics Council. He earned his BA from United States International University and his MA from Chapman College.

EDUCATIONAL CONSULTANTS

Nicole Hamilton
Consultant Manager
Richardson, TX

Joquita McKibben
Consultant Manager
Pensacola, FL

John Anderson
Lowell, IN

Beckie Fulcher
Gulf Breeze, FL

Heidi Graviette
Stockton, CA

Brenda Halulka
Atlanta, GA

Marilyn Lance
East Greenbush, NY

Ann Norris
Wichita Falls, TX

Melody Simmons
Nogales, AZ

Benjamin Swagerty
Moore, OK

Kristyn Warren
Macedonia, OH

Mary Warrington
East Wenatchee, WA

CONTENTS OVERVIEW

TEXAS EDITION

Integrated and Distributed Units of Instruction

Section 1 *Lessons 1–10, Investigation 1*

Math Focus:
Number & Operations • Algebra

Distributed Strands:
Number & Operations • Algebra • Geometry • Measurement • Problem Solving

Maintaining & Extending

Power Up

Facts pp. 6, 13, 20, 26, 34, 40, 45, 53, 60, 66

Mental Math Strategies pp. 6, 13, 20, 26, 34, 40, 45, 53, 60, 66

Problem Solving Strategies pp. 6, 13, 20, 26, 34, 40, 45, 53, 60, 66

Enrichment

Early Finishers pp. 12, 25, 33, 52

Extensions p. 74

TEKS

7.1 (A), 7.1 (B), 7.1 (C), 7.2 (A), 7.2 (B), 7.2 (D), 7.2 (G), 7.4 (C), 7.5 (B), 7.13 (A), 7.13 (B), 7.13 (C), 7.14 (A), 7.14 (B), 7.15 (A), 7.15 (B)

TAKS *Objectives*
1, 2, 6

Section 2 Lessons 11–20, Investigation 2

Math Focus:
Number & Operations • Problem Solving

Distributed Strands:
Number & Operations • Algebra • Geometry • Measurement • Problem Solving

Maintaining & Extending

Power Up
Facts pp. 75, 82, 88, 93, 100, 107, 114, 120, 128, 134

Mental Math Strategies pp. 75, 82, 88, 93, 100, 107, 114, 120, 128, 134

Problem Solving Strategies pp. 75, 82, 88, 93, 100, 107, 114, 120, 128, 134

Enrichment
Early Finishers pp. 81, 87, 92, 113, 127, 133, 142

TEKS

7.1 (B), 7.1 (C), 7.2 (E), 7.4 (A), 7.4 (C), 7.5 (B), 7.6 (D), 7.9 (A), 7.10 (B), 7.13 (A), 7.13 (B), 7.13 (C), 7.14 (A)

TAKS Objectives
1, 2, 3, 4, 5, 6

Section 3 Lessons 21–30, Investigation 3

Math Focus:
Number & Operations • Problem Solving

Distributed Strands:
Number & Operations • Geometry • Problem Solving

TEKS

7.1 (A), 7.1 (B), 7.2 (A), 7.2 (B), 7.2 (F), 7.5 (A), 7.5 (B), 7.7 (A), 7.10 (A), 7.10 (B), 7.12 (A), 7.13 (A), 7.13 (B), 7.13 (C), 7.13 (D), 7.14 (A), 7.14 (B)

TAKS Objectives

1, 2, 3, 5, 6

Section 4 Lessons 31–40, Investigation 4

Math Focus:
Number & Operations • Data Analysis & Probability

Distributed Strands:
Number & Operations • Geometry • Measurement • Data Analysis & Probability

Maintaining & Extending

Power Up

Enrichment

TEKS

7.1 (A), 7.1 (B), 7.1 (C), 7.2 (A), 7.2 (B), 7.2 (D), 7.3 (B), 7.3 (C), 7.4 (A), 7.5 (B), 7.6 (A), 7.8 (C), 7.9 (A), 7.10 (A), 7.10 (B), 7.11 (A), 7.11 (B), 7.12 (A), 7.13 (B), 7.13 (C), 7.14 (A), 7.14 (B)

TAKS *Objectives*
1, 2, 3, 4, 5, 6

Section 5 Lessons 41–50, Investigation 5

Math Focus:
Number & Operations

Distributed Strands:
Number & Operations • Measurement • Data Analysis & Probability

Maintaining & Extending

TEKS

7.1 (A), 7.1 (B), 7.2 (A), 7.2 (B), 7.2 (D), 7.2 (E), 7.2 (F), 7.2 (G), 7.4 (A), 7.9 (A), 7.10 (B), 7.11 (A), 7.11 (B), 7.13 (A), 7.13 (B), 7.13 (C), 7.13 (D), 7.14 (A), 7.14 (B), 7.15 (A)

TAKS ***Objectives***
1, 2, 4, 5, 6

Section 6 Lessons 51–60, Investigation 6

Math Focus:
Number & Operations • Problem Solving

Distributed Strands:
Number & Operations • Algebra • Geometry • Problem Solving

Maintaining & Extending

Power Up

Facts pp. 363, 369, 375, 380, 386, 393, 400, 406, 413, 420

Mental Math Strategies pp. 363, 369, 375, 380, 386, 393, 400, 406, 413, 420

Problem Solving Strategies pp. 363, 369, 375, 380, 386, 393, 400, 406, 413, 420

Enrichment

Early Finishers pp. 374, 379, 385, 392, 412, 419, 426

TEKS

7.1 (A), 7.2 (A), 7.2 (C), 7.2 (D), 7.2 (E), 7.2 (F), 7.3 (A), 7.3 (B), 7.4 (C), 7.5 (B), 7.6 (B), 7.7 (A), 7.9 (A), 7.10 (A), 7.13 (A), 7.13 (B), 7.13 (C), 7.13 (D), 7.14 (A), 7.15 (A)

TAKS *Objectives*
1, 2, 3, 4, 5, 6

Section 7 Lessons 61–70, Investigation 7

Math Focus:
Number & Operations • Geometry

Distributed Strands:
Number & Operations • Algebra • Geometry • Measurement

Maintaining & Extending

Power Up

Enrichment

TEKS

7.2 (C), 7.2 (D), 7.2 (E), 7.2 (F), 7.3 (B), 7.4 (A), 7.5 (A), 7.5 (B), 7.6 (A), 7.6 (B), 7.6 (C), 7.8 (A), 7.8 (B), 7.8 (C), 7.9 (A), 7.9 (B), 7.10 (A), 7.10 (B), 7.13 (A), 7.13 (B), 7.13 (C), 7.14 (A), 7.15 (A), 7.15 (B)

TAKS *Objectives*
1, 2, 3, 4, 5, 6

Section 8 — Lessons 71–80, Investigation 8

Math Focus:
Number & Operations • Algebra

Distributed Strands:
Number & Operations • Algebra • Geometry • Measurement

TEKS

7.1 (B), 7.2 (A), 7.2 (B), 7.2 (C), 7.2 (F), 7.2 (G), 7.3 (A), 7.3 (B), 7.4 (A), 7.5 (A), 7.5 (B), 7.7 (A), 7.7 (B), 7.8 (A), 7.9 (A), 7.10 (A), 7.10 (B), 7.10 (C), 7.11 (A), 7.11 (B), 7.13 (A), 7.13 (B), 7.13 (C), 7.14 (A), 7.14 (B), 7.15 (A), 7.15 (B)

TAKS Objectives

1, 2, 3, 4, 5, 6

Section 9 *Lessons 81–90, Investigation 9*

Math Focus:
Algebra

Distributed Strands:
Number & Operations • Algebra • Geometry • Measurement

Maintaining & Extending

Power Up

Facts pp. 562, 569, 575, 580, 586, 592, 598, 604, 610, 618

Mental Math Strategies pp. 562, 569, 575, 580, 586, 592, 598, 604, 610, 618

Problem Solving Strategies pp. 562, 569, 575, 580, 586, 592, 598, 604, 610, 618

Enrichment

Early Finishers pp. 568, 574, 585, 603, 609, 617, 623

TEKS

7.1 (B), 7.2 (E), 7.3 (A), 7.4 (A), 7.4 (B), 7.5 (A), 7.5 (B), 7.6 (A), 7.7 (A), 7.7 (B), 7.9 (A), 7.13 (A), 7.13 (B), 7.13 (C), 7.13 (D), 7.14 (A), 7.14 (B), 7.15 (B)

TAKS Objectives

1, 2, 3, 4, 6

Section 10 Lessons 91–100, Investigation 10

Math Focus:
Algebra • Geometry

Distributed Strands:
Number & Operations • Algebra • Geometry • Measurement • Data Analysis & Probability

TEKS

7.1 (A), 7.1 (C), 7.2 (D), 7.2 (E), 7.2 (F), 7.2 (G), 7.3 (A), 7.3 (B), 7.4 (A), 7.4 (B), 7.5 (A), 7.5 (B), 7.6 (D), 7.8 (A), 7.9 (A), 7.9 (B), 7.9 (C), 7.10 (B), 7.13 (A), 7.13 (B), 7.13 (C), 7.13 (D), 7.14 (A), 7.14 (B), 7.15 (B)

TAKS Objectives
1, 2, 3, 4, 5, 6

Section 11 — Lessons 101–110, Investigation 11

Math Focus:
Algebra

Distributed Strands:
Number & Operations • Algebra • Geometry • Measurement • Problem Solving

Maintaining & Extending

Power Up

Enrichment

TEKS

7.1 (C), 7.2 (E), 7.2 (F), 7.3 (A), 7.3 (B), 7.4 (A), 7.4 (C), 7.5 (A), 7.5 (B), 7.6 (A), 7.8 (B), 7.8 (C), 7.9 (A), 7.13 (A), 7.13 (B), 7.13 (C), 7.13 (D), 7.14 (A), 7.15 (A)

TAKS *Objectives*
1, 2, 3, 4, 6

Section 12

Lessons 111–120, Investigation 12

Math Focus:
Algebra • Measurement • Problem Solving

Distributed Strands:
Number & Operations • Algebra • Geometry • Measurement • Problem Solving

Maintaining & Extending

Power Up

Enrichment

TEKS

7.2 (A), 7.2 (B), 7.2 (E), 7.4 (A), 7.4 (B), 7.4 (C), 7.5 (A), 7.5 (B), 7.6 (B), 7.7 (A), 7.8 (B), 7.8 (C), 7.9 (A), 7.9 (B), 7.9 (C), 7.13 (A), 7.13 (B), 7.13 (C), 7.13 (D), 7.14 (A), 7.15 (A), 7.15 (B)

TAKS *Objectives*

1, 2, 3, 4, 6

Dear Student,

We study mathematics because of its importance to our lives. Our school schedule, our trip to the store, the preparation of our meals, and many of the games we play involve mathematics. You will find that the word problems in this book are often drawn from everyday experiences.

As you grow into adulthood, mathematics will become even more important. In fact, your future in the adult world may depend on the mathematics you have learned. This book was written to help you learn mathematics and to learn it well. For this to happen, you must use the book properly. As you work through the pages, you will see that similar problems are presented over and over again.
Solving each problem day after day is the secret to success.

Your book is made up of daily lessons and investigations. Each lesson has three parts.

1. The first part is a Power Up that includes practice of basic facts and mental math. These exercises improve your speed, accuracy, and ability to do math "in your head." The Power Up also includes a problem-solving exercise to familiarize you with strategies for solving complicated problems.

2. The second part of the lesson is the New Concept. This section introduces a new mathematical concept and presents examples that use the concept. The Practice Set provides a chance to solve problems involving the new concept. The problems are lettered a, b, c, and so on.

3. The final part of the lesson is the Written Practice. This problem set reviews previously taught concepts and prepares you for concepts that will be taught in later lessons. Solving these problems helps you remember skills and concepts for a long time.

Investigations are variations of the daily lesson. The investigations in this book often involve activities that fill an entire class period. Investigations contain their own set of questions instead of a problem set.

Remember, solve every problem in every practice set, written practice set, and investigation. Do not skip problems. With honest effort, you will experience success and true learning that will stay with you and serve you well in the future.

Temple City, California

Saxon Math Course 2 is unlike any math book you have used! It doesn't have colorful photos to distract you from learning. The Saxon approach lets you see the beauty and structure within math itself. You will understand more mathematics, become more confident in doing math, and will be well prepared when you take high school math classes.

Power Yourself Up!

Start off each lesson by practicing your basic skills and concepts, mental math, and problem solving. Make your math brain stronger by exercising it every day. Soon you'll know these facts by memory!

Learn Something New!

Each day brings you a new concept, but you'll only have to learn a small part of it now. You'll be building on this concept throughout the year so that you understand and remember it by test time.

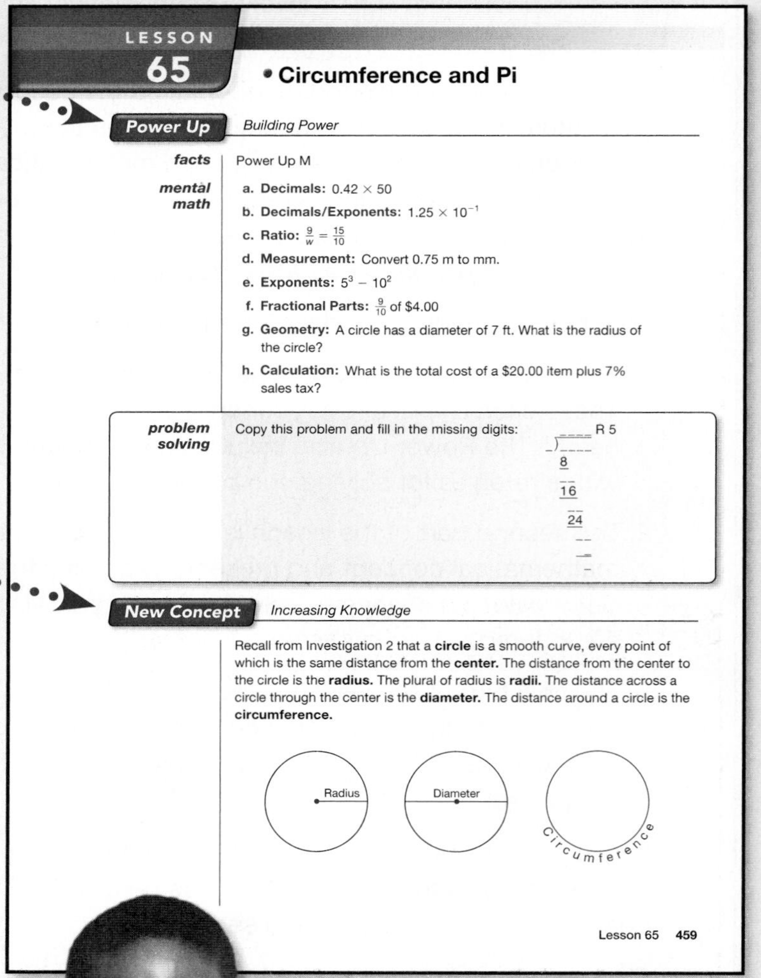
LESSON 65

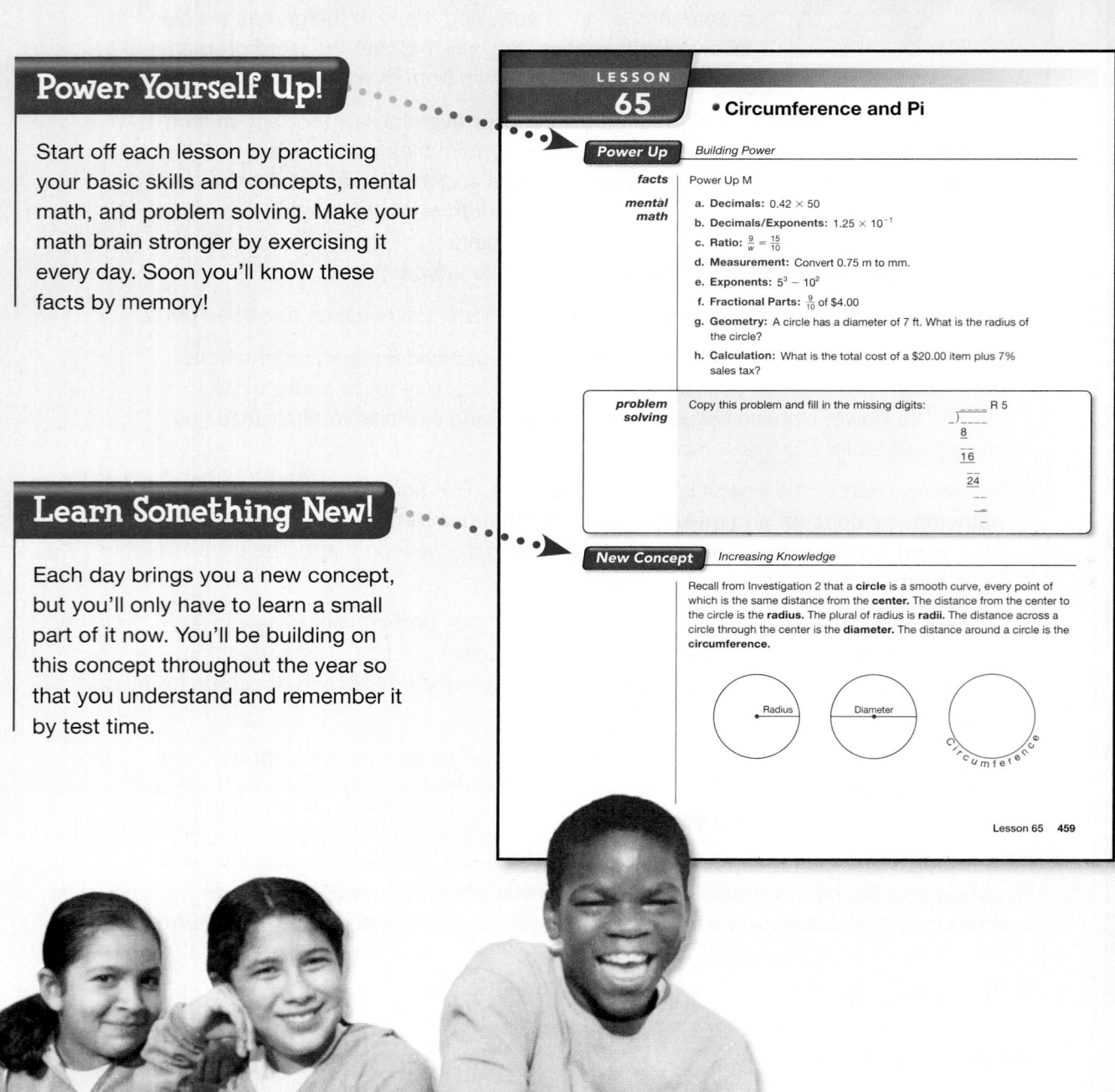

Circumference and Pi

Power Up *Building Power*

facts Power Up M

mental math

a. **Decimals:** 0.42×50

b. **Decimals/Exponents:** 1.25×10^{-1}

c. **Ratio:** $\frac{9}{w} = \frac{15}{10}$

d. **Measurement:** Convert 0.75 m to mm.

e. **Exponents:** $5^3 - 10^2$

f. **Fractional Parts:** $\frac{9}{10}$ of \$4.00

g. **Geometry:** A circle has a diameter of 7 ft. What is the radius of the circle?

h. **Calculation:** What is the total cost of a \$20.00 item plus 7% sales tax?

problem solving Copy this problem and fill in the missing digits:

```
    ___ R 5
_)____
  8
  __
  16
  __
   24
   __
    _
```

New Concept *Increasing Knowledge*

Recall from Investigation 2 that a **circle** is a smooth curve, every point of which is the same distance from the **center.** The distance from the center to the circle is the **radius.** The plural of radius is **radii.** The distance across a circle through the center is the **diameter.** The distance around a circle is the **circumference.**

Lesson 65 459

We see that the diameter of a circle is twice the radius of the circle. In the following activity we investigate the relationship between the diameter and the circumference.

Activity

Investigating Circumference and Diameter

Materials needed:

- Metric tape measure (or string and meter stick)
- Circular objects of various sizes
- Calculator (optional)

Working in a small group, select a circular object and measure its circumference and its diameter as precisely as you can. Then calculate the number of diameters that equal the circumference by dividing the circumference by the diameter. Round the quotient to two decimal places. Repeat the activity with another circular object of a different size. Record your results on a table similar to the one shown below. Extend the table by including the results of other students in the class. Are your ratios of circumference to diameter basically the same?

Sample Table

Object	Circumference	Diameter	$\frac{\text{Circumference}}{\text{Diameter}}$
Waste basket	94 cm	30 cm	3.13
Plastic cup	22 cm	7 cm	3.14

How many diameters equal a circumference? Mathematicians investigated this question for thousands of years. They found that the answer did not depend on the size of the circle. The circumference of every circle is slightly more than three diameters.

Another way to illustrate this fact is to cut a length of string equal to the diameter of a particular circle and find how many of these lengths are needed to reach around the circle. No matter what the size of the circle, it takes three diameters plus a little extra to equal the circumference.

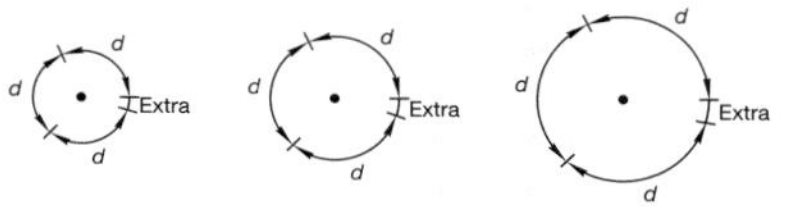

The extra amount needed is about, but not exactly, one seventh of a diameter. Thus the number of diameters needed to equal the circumference of a circle is about

$$3\frac{1}{7} \quad \text{or} \quad \frac{22}{7} \quad \text{or} \quad 3.14$$

Get Active!

Dig into math with a hands-on activity. Explore a math concept with your friends as you work together and use manipulatives to see new connections in mathematics.

Check It Out!

The Practice Set lets you check to see if you understand today's new concept.

Exercise Your Mind!

When you work the Written Practice exercises, you will review both today's new concept and also math you learned in earlier lessons. Each exercise will be on a different concept — you never know what you're going to get! It's like a mystery game — unpredictable and challenging.

As you review concepts from earlier in the book, you'll be asked to use higher-order thinking skills to show what you know and why the math works.

The mixed set of Written Practice is just like the mixed format of your state test. You'll be practicing for the "big" test every day!

Solution

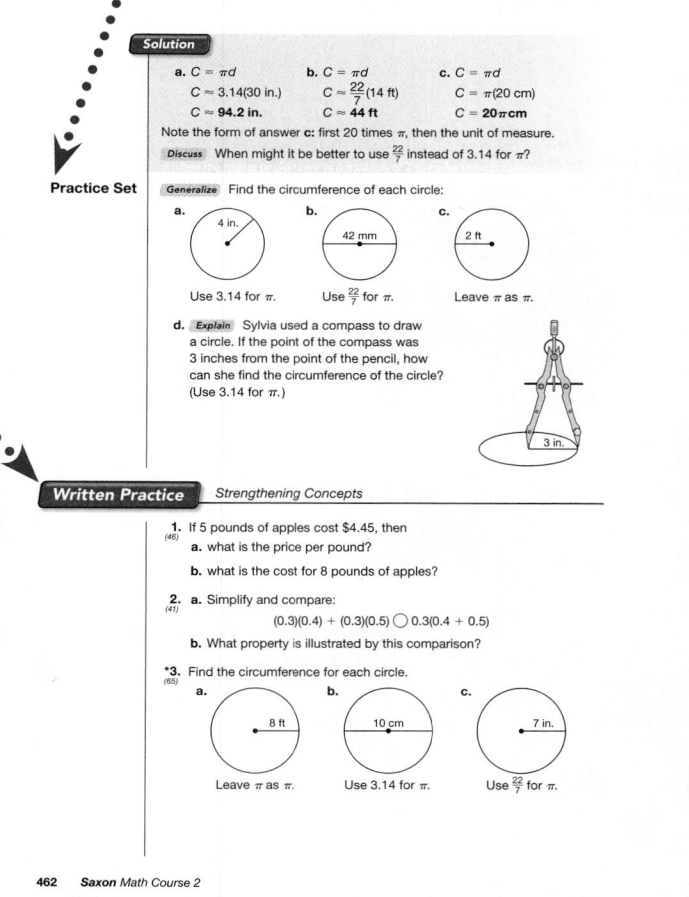

a. $C = \pi d$
$C \approx 3.14(30 \text{ in.})$
$C \approx$ **94.2 in.**

b. $C = \pi d$
$C \approx \frac{22}{7}(14 \text{ ft})$
$C \approx$ **44 ft**

c. $C = \pi d$
$C = \pi(20 \text{ cm})$
$C =$ **20πcm**

Note the form of answer **c:** first 20 times π, then the unit of measure.

Discuss When might it be better to use $\frac{22}{7}$ instead of 3.14 for π?

Practice Set

Generalize Find the circumference of each circle:

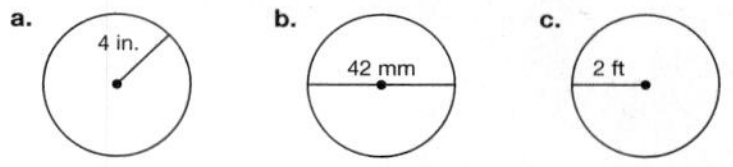

Use 3.14 for π. Use $\frac{22}{7}$ for π. Leave π as π.

d. **Explain** Sylvia used a compass to draw a circle. If the point of the compass was 3 inches from the point of the pencil, how can she find the circumference of the circle? (Use 3.14 for π.)

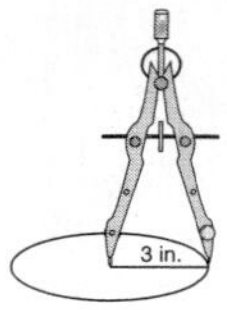

Written Practice *Strengthening Concepts*

1. (46) If 5 pounds of apples cost $4.45, then
a. what is the price per pound?
b. what is the cost for 8 pounds of apples?

2. (41) **a.** Simplify and compare:
$$(0.3)(0.4) + (0.3)(0.5) \bigcirc 0.3(0.4 + 0.5)$$
b. What property is illustrated by this comparison?

***3.** (65) Find the circumference for each circle.

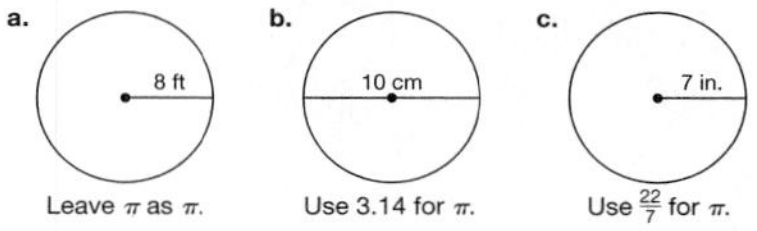

Leave π as π. Use 3.14 for π. Use $\frac{22}{7}$ for π.

Become an Investigator!

Dive into math concepts and explore the depths of math connections in the Investigations.

Continue to develop your mathematical thinking through applications, activities, and extensions.

INVESTIGATION 1

Focus on

• Investigating Fractions and Percents with Manipulatives

In this investigation, you will make a set of fraction manipulatives to use in solving problems with fractions.

Materials needed:

- **Investigation Activities 4–9**
  Halves, Thirds, Fourths, Sixths, Eighths, Twelfths
- Scissors
- Envelopes or locking plastic bags (optional)

Activity

Using Fraction Manipulatives

Discuss Working in groups of two or three, use your fraction manipulatives to help you model the following problems. Discuss how to solve each problem with your group.

Thinking Skill
Generalize
What fraction is one fourth of one half? Show how you found your answer.

1. What fraction is half of $\frac{1}{2}$?
2. What fraction is half of $\frac{1}{3}$?
3. What fraction is half of $\frac{1}{6}$?
4. A quart is one fourth of a gallon. Dan drank half a quart of milk. What fraction of a gallon did Dan drink?
5. Luz exercised for a half hour. Luz jogged for $\frac{1}{3}$ of her exercise time. For what fraction of an hour did Luz jog?
6. Four friends equally divided a whole pizza for lunch. Binli ate $\frac{1}{3}$ of his share and took the rest of his share home. What fraction of the whole pizza did Binli eat for lunch?
7. How many twelfths equal $\frac{1}{2}$?
8. Find a single fraction piece that equals $\frac{3}{12}$.
9. Find a single fraction piece that equals $\frac{4}{8}$.
10. Find a single fraction piece that equals $\frac{4}{12}$.
11. Benito is running a six-mile race, so each mile is $\frac{1}{6}$ of the race. Benito has run $\frac{2}{3}$ of the race. How many miles has he run?
12. One egg is $\frac{1}{12}$ of a dozen. How many eggs is $\frac{3}{4}$ of a dozen?

72 **Saxon** *Math Course 2*

Focus on

• Problem Solving

As we study mathematics we learn how to use tools that help us solve problems. We face mathematical problems in our daily lives, in our careers, and in our efforts to advance our technological society. We can become powerful problem solvers by improving our ability to use the tools we store in our minds. In this book we will practice solving problems every day.

This lesson has three parts:

Problem-Solving Process The four steps we follow when solving problems.

Problem-Solving Strategies Some strategies that can help us solve problems.

Writing and Problem Solving Describing how we solved a problem or formulating a problem.

four-step problem-solving process

Solving a problem is like arriving at a destination, so the process of solving a problem is similar to the process of taking a trip. Suppose we are on the mainland and want to reach a nearby island.

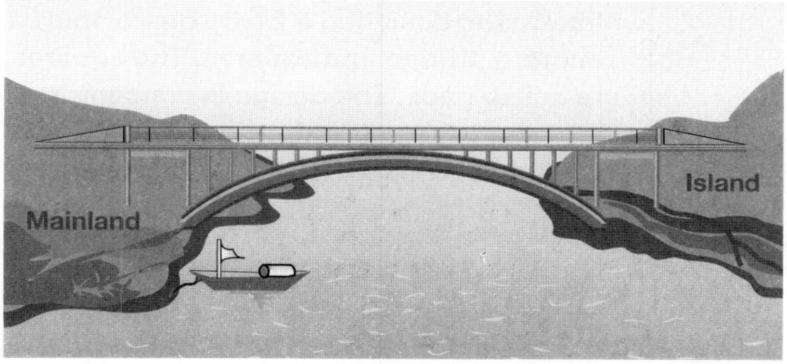

Problem-Solving Process	Taking a Trip
Step 1: *Understand* Know where you are and where you want to go.	We are on the mainland and want to go to the island.
Step 2: *Plan* Plan your route.	We might use the bridge, the boat, or swim.
Step 3: *Solve* Follow the plan.	Take the journey to the island
Step 4: *Check* Check that you have reached the right place.	Verify that you have reached your desired destination.

When we solve a problem, it helps to ask ourselves some questions along the way.

Follow the Process	Ask Yourself Questions
Step 1: *Understand*	What information am I given? What am I asked to find or do?
Step 2: *Plan*	How can I use the given information to solve the problem? What strategy can I use to solve the problem?
Step 3: *Solve*	Am I following the plan? Is my math correct?
Step 4: *Check* ***(Look Back)***	Does my solution answer the question that was asked? Is my answer reasonable?

Below we show how we follow these steps to solve a word problem.

Example

Mrs. Chang designed a 24-ft square courtyard using both gray and tan concrete divided into squares. The squares are 4 ft on each side and are called pads. The design is concentric squares in alternating colors. How many pads of each color are needed to make the design?

Solution

Step 1: Understand the problem. The problem gives the following information:

- The courtyard is a 24-ft square.
- The pads are 4 ft on each side.
- The pads are gray or tan.
- The design is concentric squares in alternating colors.

We need to find the number of pads for each color that are needed to make the design.

Step 2: Make a plan. We see that we cannot get to the answer in one step. We plan how to use the given information in a manner that will lead us toward the solution.

- Determine how many pads will cover the courtyard.
- Arrange the pads to model the design and decide how many pads are needed to show the alternating concentric squares.

Step 3: Solve the problem. (Follow the plan.) One side of the courtyard is 24 ft. One side of a pad is 4 ft.

24 ft ÷ 4 ft = 6 There are 6 pads on each side of the courtyard.

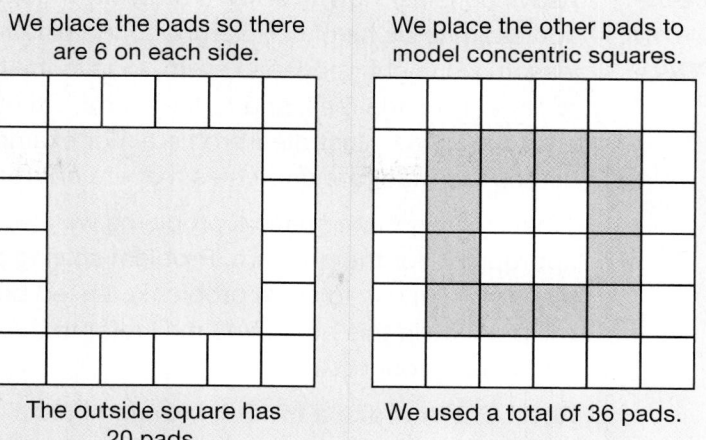

We place the pads so there are 6 on each side.

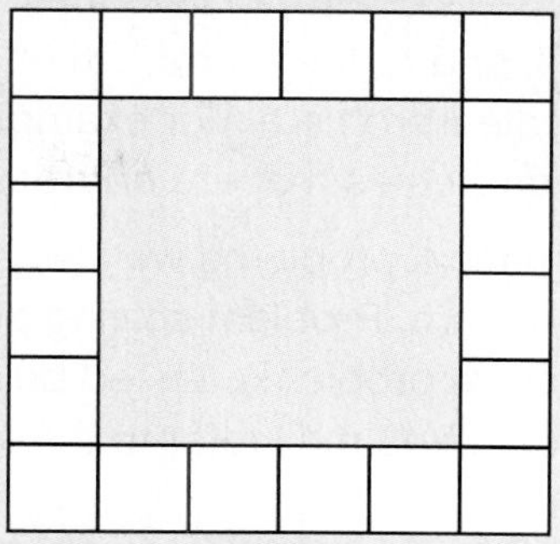

The outside square has 20 pads.

We place the other pads to model concentric squares.

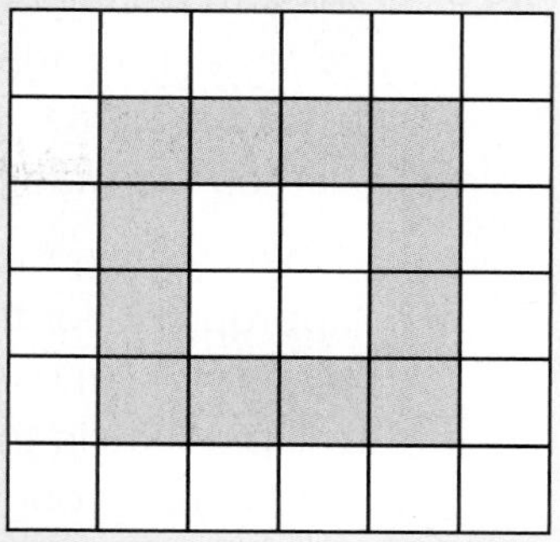

We used a total of 36 pads.

This design will need 24 tan pads and 12 gray pads.

Step 4: Check your answer. (Look back.) We read the problem again to see if our solution answers the question. We decide if our answer is reasonable.

The problem asks for the number of pads of each color to show concentric squares. Our solution shows 24 tan pads and 12 gray pads.

Conclude How would our answer change if the outside squares were gray rather than tan?

1. List in order the four steps in the problem solving process.

2. What two questions do we answer to understand a problem?

Refer to the following problem to answer questions **3–8.**

Aaron wants to buy a baseball glove that costs $86.37 including tax. Aaron has saved $28.50. He earns $15 each weekend mowing lawns. How many weekends will he need to work so that he can earn enough money to buy the baseball glove?

3. What information are we given?

4. What are we asked to find?

5. Which step of the four-step problem-solving process have you completed when you have answered questions 3 and 4?

6. Describe your plan for solving the problem. Besides the arithmetic, is there anything else you can draw or do that will help you solve the problem?

7. Solve the problem by following your plan. Show your work and any tables or diagrams you used. Write your solution to the problem in a way someone else will understand.

8. Check your work and your answer. Look back to the problem. Be sure you use the information correctly. Be sure you found what you were asked to find. Is your answer reasonable?

problem-solving strategies

As we consider how to solve a problem we choose one or more strategies that seem to be helpful. Referring to the picture at the beginning of this lesson, we might choose to swim, to take the boat, or to cross the bridge to travel from the mainland to the island. Other strategies might not be as effective for the illustrated problem. For example, choosing to walk or bike across the water are strategies that are not reasonable for this situation.

When solving mathematical problems we also select strategies that are appropriate for the problem. Problem solving **strategies** are types of plans we can use to solve problems. Listed below are ten strategies we will practice in this book. You may refer to these descriptions as you solve problems throughout the year.

Act it out or make a model. Moving objects or people can help us visualize the problem and lead us to the solution.

Use logical reasoning. All problems require logical reasoning, but for some problems we use given information to eliminate choices so that we can close in on the solution. Usually a chart, diagram, or picture can be used to organize the given information and to make the solution more apparent.

Draw a picture or diagram. Sketching a picture or a diagram can help us understand and solve problems, especially problems about graphs or maps or shapes.

Write a number sentence or equation. We can solve many story problems by fitting the given numbers into equations or number sentences and then finding the missing numbers.

Make it simpler. We can make some complicated problems easier by using smaller numbers or fewer items. Solving the simpler problem might help us see a pattern or method that can help us solve the complex problem.

Find a pattern. Identifying a pattern that helps you to predict what will come next as the pattern continues might lead to the solution.

Make an organized list. Making a list can help us organize our thinking about a problem.

Guess and check. Guessing the answer and trying the guess in the problem might start a process that leads to the answer. If the guess is not correct, use the information from the guess to make a better guess. Continue to improve your guesses until you find the answer.

Make or use a table, chart, or graph. Arranging information in a table, chart, or graph can help us organize and keep track of data needed and might reveal patterns or relationships that can help us solve the problem.

Work backwards. Finding a route through a maze is often easier by beginning at the end and tracing a path back to the start. Likewise, some problems are easier to solve by working back from information that is given toward the end of the problem to information that is missing near the beginning of the problem.

9. Name some strategies used in this lesson.

The chart below shows where each strategy is first introduced in this textbook.

Strategy	Lesson
Act It Out or Make a Model	Problem Solving Overview Example
Use Logical Reasoning	Lesson 3
Draw a Picture or Diagram	Lesson 17
Write a Number Sentence or Equation	Lesson 3
Make It Simpler	Lesson 2
Find a Pattern	Lesson 1
Make an Organized List	Lesson 8
Guess and Check	Lesson 5
Make or Use a Table, Chart, or Graph	Lesson 18
Work Backwards	Lesson 84

writing and problem solving

Sometimes, a problem will ask us to explain our thinking.

This helps us measure our understanding of math and it is easy to do.

- Explain how you solved the problem.
- Explain why your answer is reasonable.

For these situations, we can describe the way we followed our plan.

This is a description of the way we solved the problem about Aaron.

> *Subtract \$28.50 from \$86.37 to find out how much more money Aaron needs. \$86.37 − \$28.50 = \$57.87 Make a table and count by 15s to determine that Aaron needs to work 4 weekends so he can earn enough money for the basketball glove.*

10. Write a description of how we solved the problem in the Example.

Other times, we will be asked to write a problem for a given equation. Be sure to include the correct numbers and operations to represent the equation.

11. Write a word problem for $(2 \times 42) + 18$.

LESSON 1

• Arithmetic with Whole Numbers and Money
• Variables and Evaluation

Power Up[1]

Building Power

facts

Power Up A

mental math

A score is 20. Two score and 4 is 44. How many is

a. **Measurement:** 3 score

b. **Measurement:** 4 score

c. **Measurement:** 4 score and 7

d. **Measurement:** Half a dozen

e. **Measurement:** 2 dozen

f. **Measurement:** 4 dozen

g. **Probability:** What is the probability of rolling a 3 on a number cube?

h. **Calculation/Measurement:** Start with a score. Add a dozen; divide by 4; add 2; then divide by 2. What is the answer?

problem solving

A ***sequence*** is a list of terms arranged according to a certain rule. We must find the rule of a sequence in order to extend it. Finding a sequence's rule is also called *finding a pattern.* The first four triangular numbers can be shown as a sequence of diagrams or as a sequence of numbers:

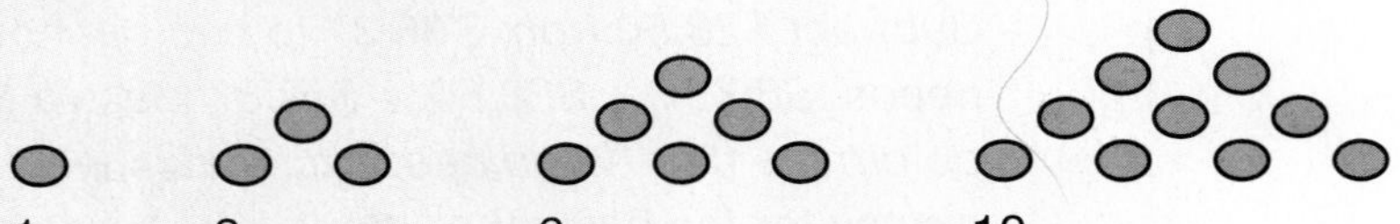

1, 3, 6, 10,

Problem: What are the next three **terms** in the sequence of triangular numbers?

Understand We are given 1, 3, 6, and 10 as the first four triangular numbers. We are asked to extend the sequence an additional three terms.

Plan We will *find the pattern* in the first four terms of the sequence, then use the pattern to extend the sequence an additional three terms.

Solve We examine the differences between the terms to see if we notice a pattern:

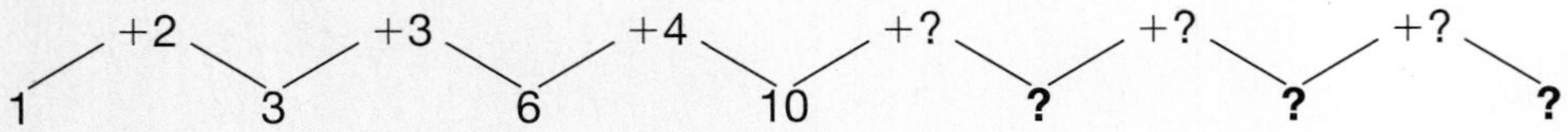

[1] For Instructions on how to use the Power-Up activities, please consult the preface.

We notice that each term in the sequence can be found by adding the term's position in the sequence to the previous term. So, to find the fifth term, we add 5 to the fourth term. To find the sixth term, we add 6 to the fifth term, and so on.

1 (+2) 3 (+3) 6 (+4) 10 (**+5**) **15** (**+6**) **21** (**+7**) **28**

Check We found the next three terms in the sequence of triangular numbers: 15, 21, and 28. Our answers are reasonable, and we can verify them by drawing the next three diagrams in the list shown at the beginning of the problem and counting the dots in each diagram.

New Concepts

Increasing Knowledge

arithmetic with whole numbers and money

Reading Math

Use braces, { }, to enclose items in a set.
Use an ellipsis, ..., to indicate a list that is infinite (goes on without end).

The numbers we say when we count are called **counting numbers** or **natural numbers.** We can show the set of counting numbers this way:

$$\{1, 2, 3, 4, 5, \ldots\}$$

Including zero with the set of counting numbers forms the set of **whole numbers.**

$$\{0, 1, 2, 3, 4, \ldots\}$$

The set of whole numbers does not include any numbers less than zero, between 0 and 1, or between any **consecutive** counting numbers.

The four fundamental **operations of arithmetic** are addition, subtraction, multiplication, and division. In this lesson we will review the operations of arithmetic with whole numbers and with money. Amounts of money are sometimes indicated with a dollar sign ($) or with a cent sign (¢), but not both. We can show 50 cents either of these two ways:

$0.50 or 50¢

Occasionally we will see a dollar sign or cent sign used incorrectly.

This sign is incorrect because it uses a **decimal point** with a cent sign. This incorrect sign literally means that soft drinks cost not half a dollar but half a cent! Take care to express amounts of money in the proper form when performing arithmetic with money.

Numbers that are added are called **addends,** and the result of their addition is the **sum.**

addend + addend = sum

Example 1

Add:

a. 36 + 472 + 3614

b. $1.45 + $6 + 8¢

Solution

a. We align the digits in the ones place and add in columns. Looking for combinations of digits that total 10 may speed the work.

$$\begin{array}{r} {}^{111}36 \\ 472 \\ +\ 3614 \\ \hline \mathbf{4122} \end{array}$$

b. We write each amount of money with a dollar sign and two places to the right of the decimal point. We align the decimal points and add.

$$\begin{array}{r} \$1.\overset{1}{4}5 \\ \$6.00 \\ +\ \$0.08 \\ \hline \mathbf{\$7.53} \end{array}$$

In subtraction the **subtrahend** is taken from the **minuend.** The result is the **difference.**

minuend − subtrahend = difference

Example 2

Subtract:

a. 5207 − 948

b. $5 − 25¢

Solution

a. We align the digits in the ones place. We must follow the correct order of subtraction by writing the minuend (first number) above the subtrahend (second number).

$$\begin{array}{r} \overset{4}{\cancel{5}}\,\overset{1}{2}\,\overset{9}{\cancel{0}}\,{}^{1}7 \\ -\quad 9\,4\,8 \\ \hline \mathbf{4\,2\,5\,9} \end{array}$$

b. We write each amount in dollar form. We align decimal points and subtract.

$$\begin{array}{r} \$\,\overset{4}{\cancel{5}}.\overset{9}{\cancel{0}}\,{}^{1}0 \\ -\ \$\,0.2\,5 \\ \hline \mathbf{\$\,4.7\,5} \end{array}$$

Numbers that are multiplied are called **factors.** The result of their multiplication is the **product.**

factor × factor = product

We can indicate the multiplication of two factors with a times sign, with a center dot, or by writing the factors next to each other with no sign between them.

$4 \times 5 \qquad 4 \cdot 5 \qquad 4(5) \qquad ab$

The parentheses in 4(5) clarify that 5 is a quantity separate from 4 and that the two digits do not represent the number 45. The expression *ab* means "*a* times *b*."

Example 3

Multiply:

a. 164 · 23 **b. $4.68 × 20** **c. 5(29¢)**

Solution

a. We usually write the number with the most digits on top. We first multiply by the 3 of 23. Then we multiply by the 20 of 23. We add the partial products to find the final product.

$$\begin{array}{r} 164 \\ \times \ \ 23 \\ \hline 492 \\ 328\ \\ \hline \mathbf{3772} \end{array}$$

b. We can let the zero in 20 "hang out" to the right. We write 0 below the line and then multiply by the 2 of 20. We write the product with a dollar sign and two decimal places.

$$\begin{array}{r} \$4.68\ \ \\ \times \quad\ 20 \\ \hline \mathbf{\$93.60} \end{array}$$

c. We can multiply 29¢ by 5 or write 29¢ as $0.29 first. Since the product is greater than $1, we use a dollar sign to write the answer.

$$\begin{array}{r} 29¢ \\ \times\ 5\ \ \\ \hline 145¢ \end{array} = \mathbf{\$1.45}$$

In division the **dividend** is divided by the **divisor.** The result is the **quotient.** We can indicate division with a division sign (÷), a division box ($\overline{)\ \ }$), or a division bar (—).

$$\text{dividend} \div \text{divisor} = \text{quotient}$$

$$\text{divisor}\overset{\text{quotient}}{\overline{)\text{dividend}}} \qquad \frac{\text{dividend}}{\text{divisor}} = \text{quotient}$$

Example 4

Divide:

a. 1234 ÷ 56

b. $\frac{\$12.60}{5}$

Solution

a. In this division there is a remainder. Other methods for dealing with a remainder will be considered later.

Analyze Should the remainder be greater than, equal to, or less than the divisor? Why?

$$\begin{array}{r} \mathbf{22\ R\ 2} \\ 56\overline{)1234}\ \ \ \ \\ \underline{112}\ \ \ \ \ \ \\ 114\ \ \ \ \\ \underline{112}\ \ \ \ \\ 2\ \ \ \ \end{array}$$

b. We write the quotient with a dollar sign. The decimal point in the quotient is directly above the decimal point in the dividend.

$$\begin{array}{r} \mathbf{\$2.52} \\ 5\overline{)\$12.60} \\ \underline{10}\ \ \ \ \ \ \\ 2\ 6\ \ \ \\ \underline{2\ 5}\ \ \ \\ 10 \\ \underline{10} \\ 0 \end{array}$$

variables and evaluation

In mathematics, letters are often used to represent numbers—in formulas and expressions, for example. The letters are called **variables** because their values are not constant; rather, they vary. We **evaluate** an expression by calculating its value when the variables are assigned specific numbers.

Example 5

Evaluate each expression for $x = 10$ and $y = 5$:

a. $x + y$ **b.** $x - y$

c. xy **d.** $\frac{x}{y}$

Solution

We substitute 10 for x and 5 for y in each expression. Then we perform the calculation.

a. $10 + 5 =$ **15** **b.** $10 - 5 =$ **5**

c. $10 \cdot 5 =$ **50** **d.** $\frac{10}{5} =$ **2**

Practice Set

a. This sign is incorrect. Show two ways to correct the sign.

b. Name a whole number that is not a counting number.

c. *Justify* When the product of 4 and 4 is divided by the sum of 4 and 4, what is the quotient? Explain how you found the answer.

Simplify by adding, subtracting, multiplying, or dividing as indicated:

d. \$1.75 + 60¢ + \$3 **e.** \$2 − 47¢

f. 5(65¢) **g.** $250 \cdot 24$

h. \$24.00 ÷ 5 **i.** $\frac{234}{18}$

Evaluate each expression for $a = 20$ and $b = 4$:

j. $a + b$ **k.** $a - b$

l. ab **m.** $\frac{a}{b}$

Connect Write equations using the number 15, where 15 is:

n. an addend **o.** the product

p. the quotient **q.** the subtrahend

r. the dividend **s.** the minuend

Written Practice *Strengthening Concepts*

* **1.** When the sum of 5 and 6 is subtracted from the product of 5 and 6, what is the difference?

* **2.** If the subtrahend is 9 and the difference is 8, what is the minuend?

3. If the divisor is 4 and the quotient is 8, what is the dividend?

* **4.** **Justify** When the product of 6 and 6 is divided by the sum of 6 and 6, what is the quotient? Explain.

* **5.** Name the four fundamental operations of arithmetic.

* **6.** Evaluate each expression for $n = 12$ and $m = 4$:

a. $n + m$ **b.** $n - m$

c. nm **d.** $\frac{n}{m}$

Simplify by adding, subtracting, multiplying, or dividing, as indicated:

7. $$\begin{array}{r} \$43.74 \\ -\ \$16.59 \\ \hline \end{array}$$

8. $$\begin{array}{r} 64 \\ \times\ 37 \\ \hline \end{array}$$

9. $$\begin{array}{r} 7 \\ 8 \\ 4 \\ 6 \\ 9 \\ 3 \\ 5 \\ +\ 7 \\ \hline \end{array}$$

10. 364 + 52 + 867 + 9

11. 4000 − 3625

12. (316)(18)

13. \$43.60 ÷ 20

14. 300 · 40

15. 8 · 12 · 0

16. 3708 ÷ 12

17. 365 × 20

18. $25\overline{)767}$

19. 30(40)

20. \$10 − \$2.34

21. 4017 − 3952

22. \$2.50 × 80

23. 20(\$2.50)

24. $\frac{560}{14}$

25. $\frac{\$10.00}{8}$

* **26.** What is another name for *counting numbers?*

27. Write 25 cents twice,

a. with a dollar sign, and

b. with a cent sign.

* Asterisks indicate exercises that should be completed in class with teacher support as needed.

* **28.** **Conclude** Which counting numbers are also whole numbers?

* **29.** What is the name for the answer to a division problem?

* **30.** **Connect** The equation below shows the relationship of addends and their sum.

$$\text{addend} + \text{addend} = \text{sum}$$

Using the vocabulary we have learned, write an equation to show the relationships in subtraction.

Early Finishers
Real-World Application

Sasha had $500.00. She purchased four shirts that cost a total of $134.00.

a. If each shirt cost the same amount, what is the cost of one shirt? Show your work.

b. The next day, Sasha returned one of the shirts. After she returned it, how much of the $500 did she have left? Show your work.

LESSON 2

• Properties of Operations

Power Up

Building Power

facts Power Up A

mental math

a. Measurement: 2 score and 8

b. Measurement: $1\frac{1}{2}$ dozen

c. Number Sense: Half of 100

d. Number Sense: 400 + 500

e. Number Sense: 9000 − 3000

f. Number Sense: 20 × 30

g. Probability: What is the probability of rolling an odd number on a number cube?

h. Calculation/Measurement: Start with a dozen. Divide by 2; multiply by 4; add 1; divide by 5; then subtract 5. What is the answer?

problem solving

German mathematician Karl Friedrich Gauss (1777–1855) developed a method for quickly adding a sequence of numbers when he was a boy. Like Gauss, we can sometimes solve difficult problems by *making the problem simpler.*

Problem: What is the sum of the first ten natural numbers?

Understand We are asked to find the sum of the first ten natural numbers.

Plan We will begin by *making the problem simpler.* If the assignment had been to add the first *four* natural numbers, we could simply add 1 + 2 + 3 + 4. However, adding columns of numbers can be time consuming. We will try to *find a pattern* that will help add the natural numbers 1–10 more quickly.

Solve We can find pairs of addends in the sequence that have the same sum and multiply by the number of pairs. We try this pairing technique on the sequence given in the problem:

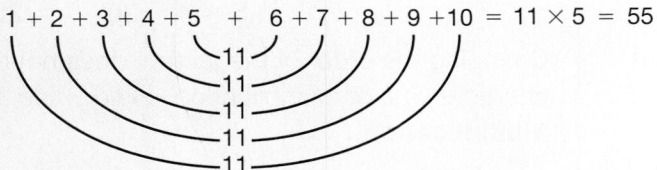

$$1 + 2 + 3 + 4 + 5 \quad + \quad 6 + 7 + 8 + 9 + 10 = 11 \times 5 = 55$$

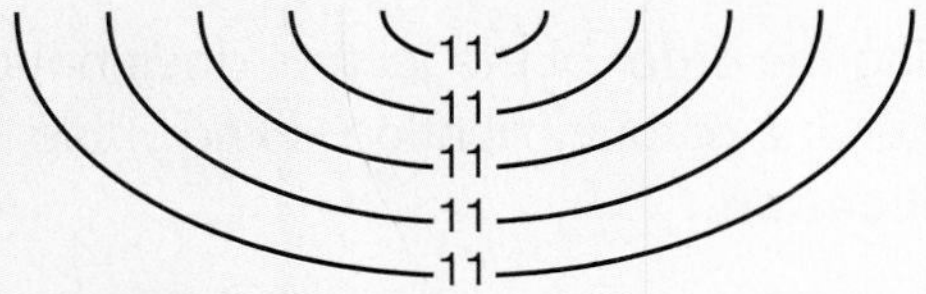

Check We found the sum of the first ten natural numbers by pairing the addends and multiplying. We can verify our solution by adding the numbers one-by-one with pencil and paper or a calculator.

New Concept

Increasing Knowledge

Addition and subtraction are **inverse operations.** We can "undo" an addition by subtracting one addend from the sum.

$$2 + 3 = 5 \qquad 5 - 3 = 2$$

Together, the numbers 2, 3, and 5 form an addition-subtraction **fact family.** With them, we can write two addition facts and two subtraction facts.

$$2 + 3 = 5 \qquad 5 - 3 = 2$$

$$3 + 2 = 5 \qquad 5 - 2 = 3$$

We see that both 2 + 3 and 3 + 2 equal 5. Changing the order of addends does not change the sum. This characteristic of addition is known as the **Commutative Property of Addition** and is often stated in equation form using variables.

$$a + b = b + a$$

Since changing the order of numbers in subtraction may change the result, subtraction is not commutative.

Addition is commutative.

$$2 + 3 = 3 + 2$$

Subtraction is not commutative.

$$5 - 3 \neq 3 - 5$$

Reading Math

The symbol ≠ means *is not equal to.*

The **Identity Property of Addition** states that when zero is added to a given number, the sum is equal to the given number.

$$a + 0 = a$$

Thus, zero is the **additive identity.**

Multiplication and division are also inverse operations. Dividing a product by one of its factors "undoes" the multiplication.

$$4 \times 5 = 20 \qquad 20 \div 5 = 4$$

Together, the numbers 4, 5, and 20 form a multiplication-division fact family that can be arranged into two multiplication facts and two division facts.

$$4 \times 5 = 20 \qquad 20 \div 5 = 4$$

$$5 \times 4 = 20 \qquad 20 \div 4 = 5$$

Thinking Skill

Conclude

How are the Commutative Properties of addition and multiplication alike?

Changing the order of the factors does not change the product. This characteristic of multiplication is known as the **Commutative Property of Multiplication.**

$$a \times b = b \times a$$

Since changing the order of division may change the quotient, division is not commutative.

Multiplication is commutative.

$$4 \times 5 = 5 \times 4$$

Division is not commutative.

$$20 \div 5 \neq 5 \div 20$$

The **Identity Property of Multiplication** states that when a given number is multiplied by 1, the result equals the given number. Thus, 1 is the **multiplicative identity.**

$$a \times 1 = a$$

The **Property of Zero for Multiplication** states that when a number is multiplied by zero, the product is zero.

$$a \times 0 = 0$$

The operations of arithmetic are **binary,** which means that we only work with two numbers in one step. If we wish to add

$$2 + 3 + 4$$

we can add two of the numbers and then add the other number. The parentheses around 2 + 3 in the expression below show that 2 + 3 should be treated as a single quantity. Therefore, 2 and 3 should be added, and then 4 should be added to the sum.

$(2 + 3) + 4$	add 2 and 3 first
$5 + 4$	then add 5 and 4
9	sum

In the expression below, the parentheses indicate that 3 and 4 are to be added first.

$2 + (3 + 4)$	add 3 and 4 first
$2 + 7$	then add 2 and 7
9	sum

Notice that the sum is the same whichever way we group the addends.

$$(2 + 3) + 4 = 2 + (3 + 4)$$

The **Associative Property of Addition** states that the grouping of addends does not change the sum.

$$(a + b) + c = a + (b + c)$$

There is a similar property for multiplication. The **Associative Property of Multiplication** states that the grouping of factors does not change the product.

$$(a \times b) \times c = a \times (b \times c)$$

Verify Show that this equation represents the Associative Property: $(2 \times 4) \times 5 = 2 \times (4 \times 5)$.

The grouping of numbers in subtraction and division does affect the result, as we see in the following expressions. Thus, there is no associative property of subtraction, and there is no associative property of division.

$$(8 - 4) - 2 \neq 8 - (4 - 2)$$

$$(8 \div 4) \div 2 \neq 8 \div (4 \div 2)$$

Discuss How do the equations above prove that the Associative Property cannot be applied to subtraction and division?

We summarize these properties in the following table:

Properties of Operations
Commutative Properties $a + b = b + a$ $a \times b = b \times a$
Associative Properties $(a + b) + c = a + (b + c)$ $(a \times b) \times c = a \times (b \times c)$
Identity Properties $a + 0 = a$ $a \times 1 = a$
Property of Zero for Multiplication $a \times 0 = 0$

Example 1

Name each property illustrated:

a. $5 \cdot 3 = 3 \cdot 5$

b. $(3 + 4) + 5 = 3 + (4 + 5)$

c. $6 + 0 = 6$

d. $6 \cdot 0 = 0$

Solution

a. Commutative Property of Multiplication

b. Associative Property of Addition

c. Identity Property of Addition

d. Property of Zero for Multiplication

Generalize Use the numbers 5, 7, and 9 to show the Associative Property of Multiplication.

Example 2

Which property can we use to find each unknown number?

a. $8 + ? = 8$

b. $1 \times ? = 9$

c. $10 \times ? = 0$

Solution

a. Identity Property of Addition

b. Identity Property of Multiplication

c. Property of Zero for Multiplication

We can use properties of operations to simplify expressions and to solve equations. In example 3 we show a way to simplify $4 \times (15 \times 25)$. We list and describe each step.

Example 3

Dena simplified $4 \times (15 \times 25)$ in four steps. Justify each step.

	Step:	Justification:
	$4 \times (15 \times 25)$	**The given expression**
Step 1:	**$4 \times (25 \times 15)$**	______________
Step 2:	**$(4 \times 25) \times 15$**	______________
Step 3:	**100×15**	______________
Step 4:	**1500**	______________

Solution

Thinking Skill

Discuss

How can the properties help us use mental math to simplify the expression?

We study how Dena changed the expression from one step to the next. To the right of each step, we justify that step by stating the property or operation Dena used.

	Step:	Justification:
	$4 \times (15 \times 25)$	The given expression
Step 1:	$4 \times (25 \times 15)$	**Commutative Property of Multiplication**
Step 2:	$(4 \times 25) \times 15$	**Associative Property of Multiplication**
Step 3:	100×15	**Multiplied 4 and 25**
Step 4:	1500	**Multiplied 100 and 15**

Practice Set

a. Which number is known as the additive identity? Which number is the multiplicative identity?

b. Which operation is the inverse of multiplication?

c. Use the letters *x*, *y*, and *z* to write an equation that illustrates the Associative Property of Addition. Then write an example using counting numbers of your choosing.

d. Name the property we can use to find the missing number in this equation:

$$5 \times ? = 8 \times 5$$

Add, subtract, multiply, or divide as indicated to simplify each expression. Remember to work within the parentheses first.

e. $(5 + 4) + 3$

f. $5 + (4 + 3)$

g. $(10 - 5) - 3$

h. $10 - (5 - 3)$

i. $(6 \cdot 2) \cdot 5$

j. $6 \cdot (2 \cdot 5)$

k. $(12 \div 6) \div 2$

l. $12 \div (6 \div 2)$

m. *Justify* List the properties used in each step to simplify the expression $5 \times (14 \times 2)$.

	Step:	Justification:
	$5 \times (14 \times 2)$	Given expression
Step 1:	$5 \times (2 \times 14)$	________
Step 2:	$(5 \times 2) \times 14$	________
Step 3:	10×14	________
Step 4:	140	________

Written Practice

Strengthening Concepts

[1]* **1.** *(1)* When the product of 2 and 3 is subtracted from the sum of 4 and 5, what is the difference?

2. *(1)* Write 4 cents twice, once with a dollar sign and once with a cent sign.

3. *(1)* The sign shown is incorrect. Show two ways to correct the sign.

* **4.** *(2)* Which operation of arithmetic is the inverse of addition?

* **5.** *(1)* If the dividend is 60 and the divisor is 4, what is the quotient?

6. *(2)* *Connect* For the fact family 3, 4, and 7, we can write two addition facts and two subtraction facts.

$$3 + 4 = 7 \qquad 7 - 4 = 3$$
$$4 + 3 = 7 \qquad 7 - 3 = 4$$

For the fact family 3, 5, and 15, write two multiplication facts and two division facts.

7. *(2)* *Justify* List the properties used in the first and second steps to simplify the expression $5 + (27 + 35)$.

	Step:	Justification:
	$5 + (27 + 35)$	Given expression
Step 1:	$(27 + 35) + 5$	**a.** ________
Step 2:	$27 + (35 + 5)$	**b.** ________
Step 3:	$27 + 40$	Added 5 and 35
Step 4:	67	Added 27 and 40

[1] The italicized numbers within parentheses underneath each problem number are called *lesson reference numbers.* These numbers refer to the lesson(s) in which the major concept of that particular problem is introduced. If additional assistance is needed, refer to the discussion, examples, or practice problems of that lesson.

Simplify:

8. (1)
$$\begin{array}{r} \$20.00 \\ -\ \$14.79 \\ \hline \end{array}$$

9. (1)
$$\begin{array}{r} \$1.54 \\ \times\quad 7 \\ \hline \end{array}$$

10. (1) $\dfrac{\$30.00}{8}$

* **11.** (1) \$4.36 + 75¢ + \$12 + 6¢

* **12.** (2) **Analyze** \$10.00 − (\$4.89 + 74¢)

13. (1)
$$\begin{array}{r} 8 \\ 5 \\ 4 \\ 6 \\ 5 \\ 4 \\ 3 \\ 7 \\ 2 \\ 4 \\ 1 \\ +\ 8 \\ \hline \end{array}$$

14. (1) 3105 ÷ 15

15. (1) $40\overline{)1630}$

16. (2) 81 ÷ (9 ÷ 3)

17. (2) (81 ÷ 9) ÷ 3

18. (1) (10)(\$3.75)

* **19.** (2) 3167 − (450 − 78)

* **20.** (2) (3167 − 450) − 78

21. (1) \$20.00 ÷ 16

22. (1) 70 · 800

23. (1) 3714 + 268 + 47 + 9

24. (1) 5 · 4 · 3 · 2 · 1

* **25.** (2) \$20 − (\$1.47 + \$8)

26. (1) 30 × 45¢

* **27.** (2) Which property can we use to find each missing number?

a. $10x = 0$ **b.** $10y = 10$

* **28.** (1) Evaluate each expression for $x = 18$ and $y = 3$:

a. $x - y$ **b.** xy

c. $\dfrac{x}{y}$ **d.** $x + y$

* **29.** (2) **Explain** Why is zero called the additive identity?

* **30.** (1) **Connect** The equation below shows the relationship of factors and their product

factor × factor = product

Using the vocabulary we have learned, write a similar equation to show the relationships in division.

LESSON 3

Unknown Numbers in Addition, Subtraction, Multiplication, and Division

Power Up

Building Power

facts Power Up A

mental math

a. **Measurement:** 3 score and 6

b. **Measurement:** $2\frac{1}{2}$ dozen

c. **Number Sense:** Half of 1000

d. **Number Sense:** 1200 + 300

e. **Number Sense:** 750 − 500

f. **Number Sense:** 30 × 30

g. **Probability:** What is the probability of rolling an even number on a number cube?

h. **Calculation:** Start with the number of minutes in an hour. Divide by 2; subtract 5; double that number; subtract 1; then divide by 7. What is the answer?

problem solving

Some math problems require us to "think through" a lot of information before arriving at the solution. We can *use logical reasoning* to interpret and apply the information given in the problem to help us find the solution.

Problem: Simon held a number cube so that he could see the dots on three of the faces. Simon said he could see 7 dots. How many dots could he not see?

Understand We must first establish a base of knowledge about **standard number cubes.** The faces of a standard number cube are numbered with 1, 2, 3, 4, 5, or 6 dots. The number of dots on opposite faces of a number cube always total 7 (1 dot is opposite 6 dots, 2 dots are opposite 5 dots, and 3 dots are opposite 4 dots). Simon sees seven dots on three faces of a standard number cube. We are asked to find the number of dots on the faces he cannot see.

Plan We will *use logical reasoning* about a number cube and *write an equation* to determine the number of unseen dots.

Solve First, we find the total number of dots on a number cube: 1 + 2 + 3 + 4 + 5 + 6 = 21 dots. Then, we write an equation to solve for the number of unseen dots: 21 total dots − 7 seen dots = 14 unseen dots.

Check We found that Sam could not see 14 dots. Our answer makes sense, because the total number of dots on the number cube is 21.

New Concept *Increasing Knowledge*

Math Language

A **variable** is a letter used to represent a number that is not given.

An **equation** is a statement that two quantities are equal. Here we show two equations:

$$3 + 4 = 7 \qquad 5 + a = 9$$

The equation on the right contains a variable. In this lesson we will practice finding the value of variables in addition, subtraction, multiplication, and division equations.

Sometimes we encounter **addition equations** in which the sum is unknown. Sometimes we encounter addition equations in which an addend is unknown. We can use a letter to represent an unknown number. The letter may be uppercase or lowercase.

Unknown Sum	Unknown Addend	Unknown Addend
$2 + 3 = N$	$2 + a = 5$	$b + 3 = 5$

Thinking Skill

Verify

Why can we use subtraction to find the missing variable in an addition equation?

If we know two of the three numbers, we can find the unknown number. We can find an unknown addend by subtracting the known addend from the sum. If there are more than two addends, we subtract all the known addends from the sum. For example, to find n in the equation

$$3 + 4 + n + 7 + 8 = 40$$

we subtract 3, 4, 7, and 8 from 40. To do this, we can add the known addends and then subtract their sum from 40.

Example 1

Find the unknown number in each equation:

a. $\mathbf{n + 53 = 75}$

b. $\mathbf{26 + a = 61}$

c. $\mathbf{3 + 4 + n + 7 + 8 = 40}$

Solution

In both **a** and **b** we can find each unknown addend by subtracting the known addend from the sum. Then we check.

a. Subtract. Try it.

$$\begin{array}{r} 75 \\ -\ 53 \\ \hline 22 \end{array} \qquad \begin{array}{r} 22 \\ +\ 53 \\ \hline 75 \end{array} \text{ check}$$

In **a,** n is **22.**

b. Subtract. Try it.

$$\begin{array}{r} 61 \\ -\ 26 \\ \hline 35 \end{array} \qquad \begin{array}{r} 26 \\ +\ 35 \\ \hline 61 \end{array} \text{ check}$$

In **b,** a is **35.**

c. We add the known addends.

$$3 + 4 + 7 + 8 = 22$$

Then we subtract their sum, 22, from 40.

$$40 - 22 = 18$$

$$n = \mathbf{18}$$

We use the answer in the original equation for a check.

$$3 + 4 + 18 + 7 + 8 = 40 \quad \text{check}$$

There are three numbers in a **subtraction equation.** If one of the three numbers is unknown, we can find the unknown number.

Unknown Minuend	Unknown Subtrahend	Unknown Difference
$a - 3 = 2$	$5 - x = 2$	$5 - 3 = m$

To find an unknown minuend, we add the other two numbers. To find an unknown subtrahend or difference, we subtract.

Example 2

Find the unknown number in each equation:

a. $p - 24 = 17$ **b.** $32 - x = 14$

Solution

a. To find the minuend in a subtraction equation, we add the other two numbers. We find that the unknown number p is **41.**

Add.	Try it.	
$\begin{array}{r} 17 \\ +\ 24 \\ \hline 41 \end{array}$	$\begin{array}{r} 41 \\ -\ 24 \\ \hline 17 \end{array}$	check

b. To find a subtrahend, we subtract the difference from the minuend. So the unknown number x is **18.**

Subtract.	Try it.	
$\begin{array}{r} 32 \\ -\ 14 \\ \hline 18 \end{array}$	$\begin{array}{r} 32 \\ -\ 18 \\ \hline 14 \end{array}$	check

A **multiplication equation** is composed of factors and a product. If any one of the numbers is unknown, we can figure out what it is.

Unknown Product	Unknown Factor	Unknown Factor
$3 \cdot 2 = p$	$3f = 6$	$r \times 2 = 6$

To find an unknown product, we multiply the factors. To find an unknown factor, we divide the product by the known factor(s).

Example 3

Find the unknown number in each equation:

a. $12n = 168$ **b.** $7k = 105$

Solution

In both **a** and **b** the unknown number is one of the two factors. Notice that $7k$ means "7 times k." We can find an unknown factor by dividing the product by the known factor.

a. Divide. Try it.

$$\begin{array}{r} 14 \\ 12\overline{)168} \\ \underline{12} \\ 48 \\ \underline{48} \\ 0 \end{array} \qquad \begin{array}{r} 12 \\ \underline{\times\ 14} \\ 48 \\ \underline{12} \\ 168 \end{array}\ \text{check}$$

In **a,** n is **14.**

b. Divide. Try it.

$$\begin{array}{r} 15 \\ 7\overline{)105} \\ \underline{7} \\ 35 \\ \underline{35} \\ 0 \end{array} \qquad \begin{array}{r} 15 \\ \underline{\times\ 7} \\ 105 \end{array}\ \text{check}$$

In **b,** k is **15.**

Model Draw a picture to help you find the unknown number in the equation $3y = 12$. Hint: How many groups of 3 can you make?

If we know two of the three numbers in a **division equation,** we can figure out the unknown number.

Unknown Quotient	Unknown Divisor	Unknown Dividend
$\frac{24}{3} = n$	$\frac{24}{m} = 8$	$\frac{p}{3} = 8$

To find an unknown quotient, we simply *divide* the dividend by the divisor. To find an unknown divisor, we *divide* the dividend by the quotient. To find an unknown dividend, we *multiply* the quotient by the divisor.

Example 4

Find the unknown number in each equation:

a. $\frac{a}{3} = 15$ **b.** $\frac{64}{b} = 4$

Solution

a. To find an unknown dividend, multiply the quotient and divisor.

$3 \times 15 = \mathbf{45}$ try it $45 \div 3 = 15$ check

b. To find an unknown divisor, divide the dividend by the quotient.

$\begin{array}{r} \mathbf{16} \\ 4\overline{)64} \end{array}$ try it $\frac{64}{16} = 4$ check

Explain How can we find the value of s in the equation $\frac{s}{14} = 7$?

Practice Set

Find the unknown number in each equation:

a. $a + 12 = 31$ **b.** $b - 24 = 15$ **c.** $15c = 180$

d. $\frac{r}{8} = 12$ **e.** $14e = 420$ **f.** $26 + f = 43$

g. $51 - g = 20$ **h.** $\frac{364}{h} = 7$ **i.** $4n = 2 \cdot 12$

j. $3 + 6 + m + 12 + 5 = 30$

k. **Represent** Write an equation using 2, 6, and the variable x. Explain how to solve the equation and check the answer.

Written Practice — *Strengthening Concepts*

1. (1) When the product of 4 and 4 is divided by the sum of 4 and 4, what is the quotient?

*** 2.** (1, 3) **Summarize** If you know the subtrahend and the difference, how can you find the minuend? Write a complete sentence to answer the question.

*** 3.** (2) **Analyze** Which property of addition is stated by this equation?

$$(a + b) + c = a + (b + c)$$

*** 4.** (3) **Analyze** If one addend is 7 and the sum is 21, what is the other addend?

*** 5.** (2) **Generalize** Use the numbers 3 and 4 to illustrate the Commutative Property of Multiplication. Use a center dot to indicate multiplication.

*** 6.** (2) **Justify** List the properties used in the first and second steps to simplify the expression $5 + (x + 7)$.

Step:	Justification:
$5 + (x + 7)$	Given expression
$5 + (7 + x)$	**a.** ________
$(5 + 7) + x$	**b.** ________
$12 + x$	$5 + 7 = 12$

Find the value of each variable.

*** 7.** (3) $x + 83 = 112$

*** 8.** (3) $96 - r = 27$

*** 9.** (3) $7k = 119$

*** 10.** (3) $127 + z = 300$

*** 11.** (3) $m - 137 = 731$

*** 12.** (3) $25n = 400$

*** 13.** (3) $\frac{625}{w} = 25$

*** 14.** (3) $\frac{x}{60} = 700$

*** 15.** (1) Evaluate each expression for $a = 20$ and $b = 5$:

a. $\frac{a}{b}$

b. $a - b$

c. ab

d. $a + b$

Simplify:

*** 16.** (2) **Analyze** $96 \div (16 \div 2)$

17. (2) $(96 \div 16) \div 2$

18. (2) $16.47 + $15 + 63¢

19. (2) $50.00 − ($6.48 + $31.75)

20. (1)
$$\begin{array}{r} 47 \\ \times\ 39 \\ \hline \end{array}$$

21. (1)
$$\begin{array}{r} \$8.79 \\ \times\ \quad 80 \\ \hline \end{array}$$

22. 1100 − (374 − 87)
(2)

23. (1100 − 374) − 87
(2)

24. 4736 + 271 + 9 + 88
(1)

25. 30,145 − 4299
(1)

26. $\frac{4740}{30}$
(1)

27. $\frac{\$40.00}{32}$
(1)

28. $35\overline{)2104}$
(1)

29. $\begin{array}{r} \$0.48 \\ \times \quad 40 \\ \hline \end{array}$
(1)

* **30.** **Verify** Why is 1 called the multiplicative identity?
(2)

Early Finishers

Real-World Application

Jessica and her friend went to the local high school football game. The price of admission was $4.50. They each bought a bottle of water for $1.25 and a pickle. Together they spent a total of $13.50.

a. Find the cost of each pickle.

b. Jessica had $10 to spend at the game. How much should she have left after buying her ticket, the water, and a pickle?

LESSON 4

• Number Line
• Sequences

Power Up — *Building Power*

facts Power Up A

mental math

a. Measurement: Five score

b. Measurement: Ten dozen

c. Number Sense: Half of 500

d. Number Sense: 350 + 400

e. Number Sense: 50×50

f. Number Sense: $400 \div 10$

g. Measurement: Convert 2 gallons to quarts

h. Calculation: Start with the number of feet in a yard. Multiply by 12; divide by 6; add 4; double that number; add 5; double that number; then double that number. What is the answer?

problem solving Lawrence drew a closed shape with a wide gray marker. Then Marta drew a closed shape with a thin black pen. Marta's shape crosses Lawrence's shape exactly eight times. Is it possible for Marta to draw a closed shape that crosses Lawrence's shape only seven times?

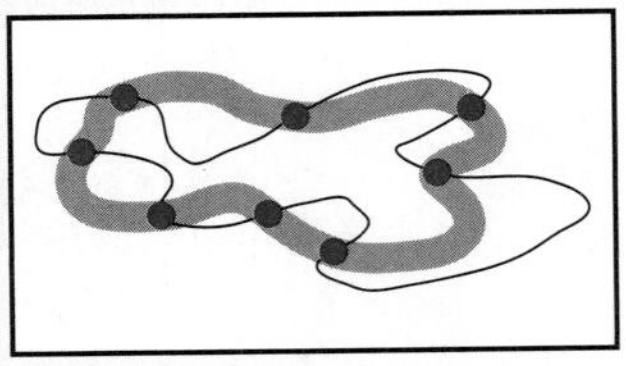

New Concepts — *Increasing Knowledge*

number line

A **number line** can be used to help us arrange numbers in order. Each number corresponds to a unique point on the number line. The zero point of a number line is called the **origin.** The numbers to the right of the origin are called **positive numbers,** and they are all **greater than zero.** Every positive number has an **opposite** that is the same distance to the left of the origin. The numbers to the left of the origin are called **negative numbers.** The negative numbers are all **less than zero.** Zero is neither positive nor negative.

Thinking Skill
Connect

Which are the counting numbers? Which are their opposites?

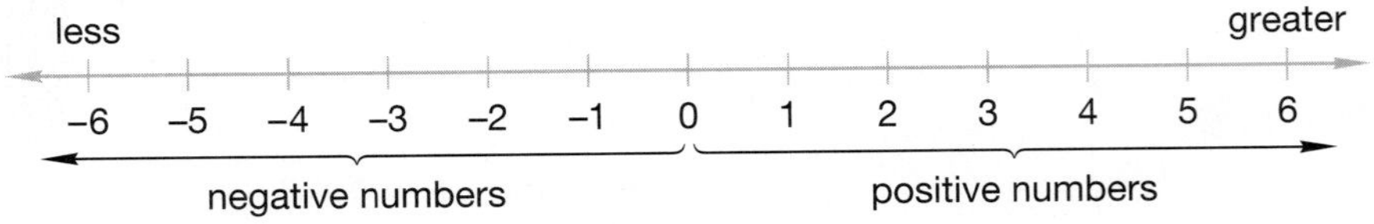

On this number line the **tick marks** indicate the location of **integers.** Integers include all of the counting numbers as well as their opposites—their negatives—and the number zero. Integers do not include fractions or any other numbers between consecutive tick marks on the number line.

Integers

$$\{\ldots, -3, -2, -1, 0, 1, 2, 3, \ldots\}$$

The ellipses to the left and the right indicate that the number of negative and positive integers is infinite. Notice that the negative numbers are written with a negative sign. For -5 we say "negative five." Positive numbers may be written with or without a positive sign. Both $+5$ and 5 are positive and equal to each other.

As we move to the right on a number line, the numbers become greater and greater. As we move to the left on a number line, the numbers become less and less. A number is greater than another number if it is farther to the right on a number line.

Thinking Skill

Generalize

What will always be true when you compare a positive and a negative number?

We **compare** two numbers by determining whether one is greater than the other or whether the numbers are equal. We place a **comparison symbol** between two numbers to show the comparison. The comparison symbols are the equal sign (=) and the greater than/less than symbols (> or <). We write the greater than and less than symbols so that the smaller end (the point) points to the number that is less. Below we show three comparisons.

$-5 < 4$	$3 + 2 = 5$	$5 > -6$
-5 is less than 4	3 plus 2 equals 5	5 is greater than -6

Example 1

Arrange these numbers in order from least to greatest:

0, 1, −2

Solution

We arrange the numbers in the order in which they appear on a number line.

−2, 0, 1

Example 2

Rewrite the expression below by replacing the circle with the correct comparison symbol. Then use words to write the comparison.

−5 ◯ 3

Solution

Since -5 is less than 3, we write

$$-5 < 3$$

Negative five is less than three.

We can use a number line to help us add and subtract. We will use arrows to show addition and subtraction. To add, we let the arrow point to the right. To subtract, we let the arrow point to the left.

Example 3

Show this addition problem on a number line: 3 + 2

Solution

We start at the origin (at zero) and draw an arrow 3 units long that points to the right. From this arrowhead we draw a second arrow 2 units long that points to the right.

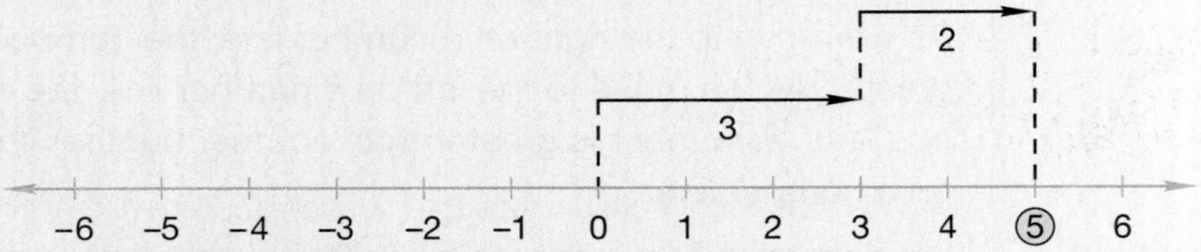

The second arrow ends at 5; 3 + 2 = **5.**

Example 4

Show this subtraction problem on a number line: 5 − 3

Solution

Starting at the origin, we draw an arrow 5 units long that points to the right. To subtract, we draw a second arrow 3 units long that points to the left. Remember to draw the second arrow from the first arrowhead.

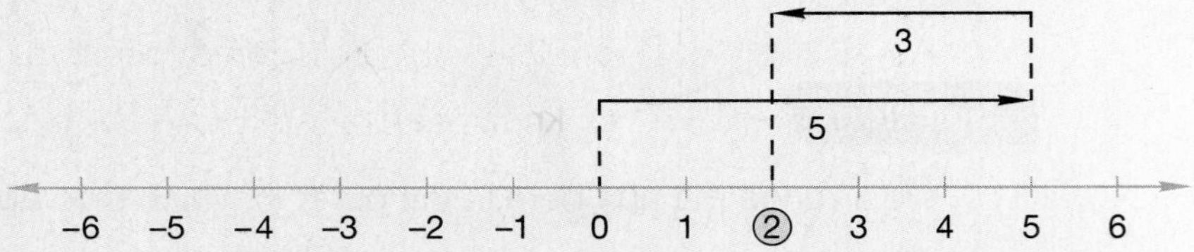

The second arrow ends at 2. This shows that 5 − 3 = **2.**

Example 5

Show this subtraction problem on a number line: 3 − 5

Solution

We take the numbers in the order given. Starting from the origin, we draw an arrow 3 units long that points to the right. From this arrowhead we draw a second arrow 5 units long that points to the left. The second arrow ends to the left of zero, which illustrates that the result of this subtraction is a negative number.

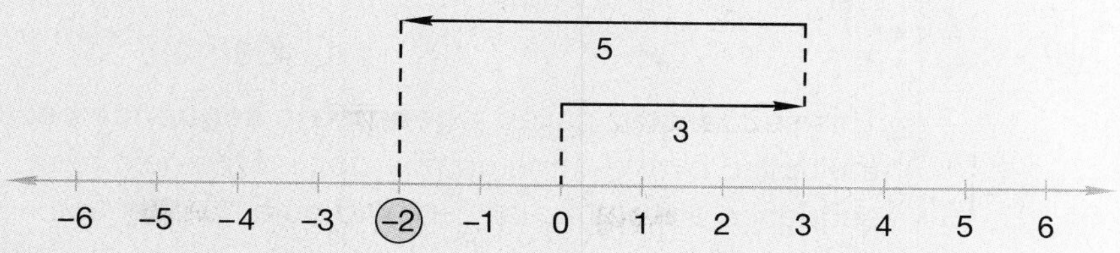

The second arrow ends at -2. This shows that $3 - 5 =$ **-2.**

Together, examples 4 and 5 show graphically that subtraction is not commutative; that is, the order of the numbers affects the outcome in subtraction. In fact, notice that reversing the order of subtraction results in opposite differences.

$$5 - 3 = 2$$
$$3 - 5 = -2$$

We can use this characteristic of subtraction to help us with subtraction problems like the next example.

Example 6

Simplify: 376 − 840

Solution

We see that the result will be negative. We reverse the order of the numbers to perform the subtraction.

$$\begin{array}{r} 840 \\ -\ 376 \\ \hline 464 \end{array}$$

The answer to the original problem is the opposite of 464, which is **−464.**

Discuss How do we know that the answer is the opposite of 464, and not 464?

sequences

A **sequence** is an ordered list of **terms** that follows a certain pattern or rule. A list of the whole numbers is an example of a sequence.

0, 1, 2, 3, 4, ...

If we wish to list the **even** or **odd** whole numbers, we could write the following sequences:

Evens: 0, 2, 4, 6, 8, ...   Odds: 1, 3, 5, 7, 9, ...

These sequences are called **arithmetic sequences** because the same number is added to each term to find the next term. Thus the numbers in the sequence are equally spaced on a number line.

Even Whole Numbers

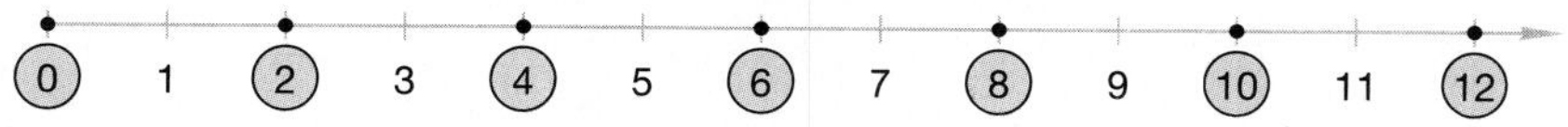

Odd Whole Numbers

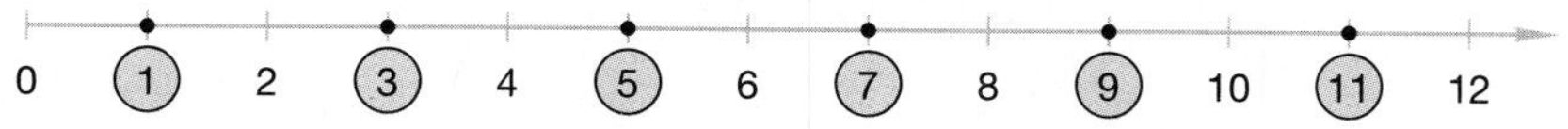

Here is a different sequence:

1, 2, 4, 8, 16, . . .

This sequence is called a **geometric sequence** because each term is multiplied by the same number to find the next term. Terms in a geometric sequence are not equally spaced on a number line.

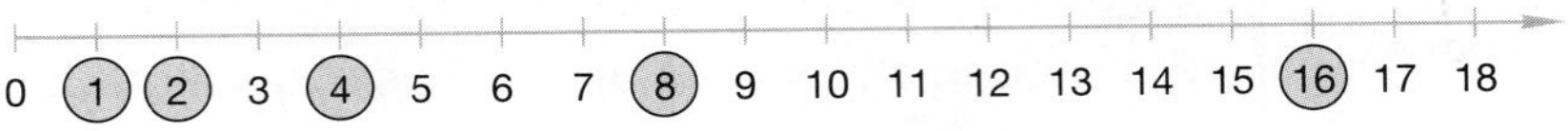

To continue a sequence, we study the sequence to understand its pattern or rule; then we apply the rule to find additional terms in the sequence.

Example 7

The first four terms of a sequence are shown below. Find the next three terms in the sequence.

1, 4, 9, 16, . . .

Solution

This sequence is neither arithmetic nor geometric. We will describe two solutions. First we see that the terms increase in size by a larger amount as we move to the right in the sequence.

+3 +5 +7

1, 4, 9, 16, . . .

The increase itself forms a sequence we may recognize: 3, 5, 7, 9, 11, ... We will continue the sequence by adding successively larger odd numbers.

+3 +5 +7 +9 +11 +13

1, 4, 9, 16, 25, 36, 49, . . .

We find that the next three numbers in the sequence are **25, 36,** and **49.**

Another solution to the problem is to recognize the sequence as a list of **perfect squares.** When we multiply a counting number by itself, the product is a perfect square.

$1 \cdot 1 = 1 \qquad 2 \cdot 2 = 4 \qquad 3 \cdot 3 = 9 \qquad 4 \cdot 4 = 16$

Here we use figures to illustrate perfect squares.

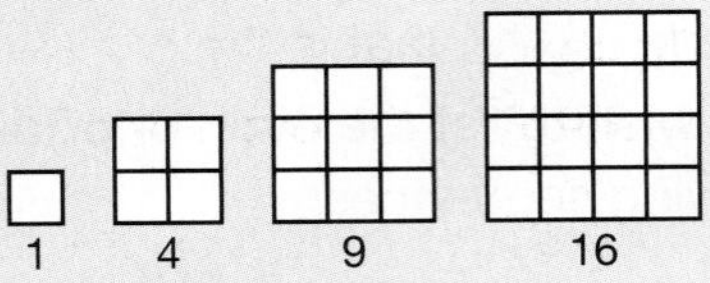

Continuing this pattern, the next three terms are

$5 \times 5 = \mathbf{25} \qquad 6 \times 6 = \mathbf{36} \qquad 7 \times 7 = \mathbf{49}$

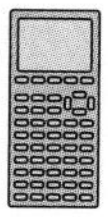

Visit www.SaxonPublishers.com/ActivitiesC2 *for a graphing calculator activity.*

The rule of a sequence might be expressed as a formula. A formula for the sequence of a perfect square is

$$k = n \cdot n$$

The variable n represents the position of the term (first, second, third, ...) and k represents the value of the term (1, 4, 9, ...). To find the eighth term of the sequence, we write 8 in place of n and find k.

$$k = 8 \cdot 8$$

Since $8 \cdot 8$ is 64, the eighth term is 64.

Example 8

The rule of a certain sequence is $k = 2n$. Find the first four terms of the sequence.

Solution

We substitute 1, 2, 3, and 4 for n to find the first four terms.

First term	Second term	Third term	Fourth term
$k = 2(1)$	$k = 2(2)$	$k = 2(3)$	$k = 2(4)$
$= 2$	$= 4$	$= 6$	$= 8$

The first four terms of the sequence are **2, 4, 6,** and **8.**

Practice Set

Model For problems **a–c,** draw a number line and use arrows to represent the addition or subtraction.

a. $4 + 2$ **b.** $4 - 2$ **c.** $2 - 4$

d. Arrange these numbers in order from least to greatest:

$$0, -1, -2, -3$$

e. Use digits and other symbols to write "The sum of 2 and 3 is less than the product of 2 and 3."

Replace each circle with the proper comparison symbol:

f. $3 - 4 \bigcirc 4 - 3$ **g.** $2 \cdot 2 \bigcirc 2 + 2$

h. **Justify** What is the first step you take before comparing the expressions in exercises **f** and **g?**

i. Where is the origin on the number line?

j. Simplify: $436 - 630$

k. **Predict** Find the next three numbers in this sequence:

$$\ldots, 3, 2, 1, 0, -1, \ldots$$

l. Find the next three terms of this sequence:

$$1, 4, 9, 16, 25, 36, 49, \ldots$$

m. Use words to describe the rule of the following sequence. Then find the next three terms.

$$1, 2, 4, 8, \ldots$$

n. The rule of a certain sequence is $k = (2n) - 1$. Find the first four terms of the sequence.

Written Practice

Strengthening Concepts

1. (1) What is the difference when the sum of 5 and 4 is subtracted from the product of 3 and 3?

*** 2.** (1, 3) If the minuend is 27 and the difference is 9, what is the subtrahend?

*** 3.** (4) What is the name for numbers that are greater than zero?

*** 4.** (1) Evaluate each expression for $n = 6$ and $m = 24$:

a. $m - n$ **b.** $n - m$

c. $\frac{m}{n}$ **d.** mn

*** 5.** (4) Use digits and other symbols to write "The product of 5 and 2 is greater than the sum of 5 and 2."

*** 6.** (4) **Analyze** Arrange these numbers in order from least to greatest:

$$-2, 1, 0, -1$$

*** 7.** (3) **Explain** If you know the divisor and the quotient, how can you find the dividend? Write a complete sentence to answer the question.

*** 8.** (4) Show this subtraction problem on a number line: $2 - 3$

Find the value of each variable.

*** 9.** (3) $12x = 12$

*** 10.** (3) $4 + 8 + n + 6 = 30$

*** 11.** (3) $z - 123 = 654$

*** 12.** (3) $1000 - m = 101$

13. (3) $p + \$1.45 = \$4.95$

*** 14.** (3) $32k = 224$

*** 15.** (3) $\frac{r}{8} = 24$

*** 16.** (2, 3) **Justify** Look at the first and second steps of solving this equation. Justify each of these steps.

Step:	Justification:
$4 + (n + 9) = 20$	Given equation
$4 + (9 + n) = 20$	**a.** ________
$(4 + 9) + n = 20$	**b.** ________
$13 + n = 20$	$4 + 9 = 13$
$n = 7$	$13 + 7 = 20$

* **17.** (4) Replace each circle with the proper comparison symbol:

a. $3 \cdot 4 \bigcirc 2(6)$

b. $-3 \bigcirc -2$

c. $3 - 5 \bigcirc 5 - 3$

d. $xy \bigcirc yx$

Simplify:

18. (1) $100.00 − $36.49

19. (1) 48(36¢)

20. (1) $5 \cdot 6 \cdot 7$

21. (1) $9900 \div 18$

22. (1) 30(20)(40)

23. (2) $(130 - 57) + 9$

24. (1) $1987 - 2014$

25. (1) $68.60 ÷ 7

26. (1) 46¢ + 64¢

27. (1) $\dfrac{4640}{80}$

28. (1)
$$\begin{array}{r} \$3.75 \\ \times \quad 30 \\ \hline \end{array}$$

* **29.** (2) **Represent** Use the numbers 2, 3, and 6 to illustrate the Associative Property of Multiplication.

* **30.** (4) **Generalize** Use words to describe the rule of the following sequence. Then find the next two terms.

1, 10, 100, . . .

Early Finishers
Real-World Application

The supply budget for the school play is $150. The manager has already spent $34.73 on costumes and $68.98 on furniture and backdrops.

a. How much money is left in the supply budget?

b. Is there enough money left to buy two lamps for $16.88 each and a vase for $12.25? Support your answer.

LESSON 5

• Place Value Through Hundred Trillions
• Reading and Writing Whole Numbers

Power Up

Building Power

facts Power Up A

mental math

a. **Measurement:** Half a score

b. **Measurement:** Twelve dozen

c. **Number Sense:** Ten hundreds

d. **Number Sense:** $475 - 200$

e. **Number Sense:** 25×20

f. **Number Sense:** $5000 \div 10$

g. **Measurement:** Convert 24 inches to feet

h. **Calculation:** Start with the number of years in a century. Subtract 1; divide by 9; add 1; multiply by 3; subtract 1; divide by 5; multiply by 4; add 2; then find half of that number. What is the answer?

problem solving

Find each missing digit:

$$\begin{array}{r} 75\_0 \\ -\ \_60\_ \\ \hline 4\_13 \end{array}$$

Understand We are shown a subtraction problem with several digits missing. We are asked to find the missing digits.

Plan We will make intelligent *guesses* for each missing digit, then *check* our guesses using arithmetic. We will begin with the ones digits and then look at each place-value column separately.

Solve **Step 1:** We find that seven is the missing digit in the ones place of the subtrahend, because $10 - 7 = 3$. We also record that a ten was borrowed from the tens digit of the minuend.

$$\begin{array}{r} 75\_\,{}^{1}0 \\ -\ \_60\underline{7} \\ \hline 4\_13 \end{array}$$

Step 2: We can find the missing digit in the tens column by "adding up" (remember to add the ten borrowed in regrouping): $1 + 0 + 1 = 2$.

$$\begin{array}{r} 75\underline{2}\,{}^{1}0 \\ -\ \_60\underline{7} \\ \hline 4\_13 \end{array}$$

Step 3: To complete the subtraction in the hundreds column we must borrow from the thousands column. Regrouping gives us $15 - 6 = 9$. We write a 9 in the hundreds place of the difference and change the 7 in the thousands place of the minuend to a 6.

$$\begin{array}{r} \overset{6}{\cancel{7}}\,{}^{1}5\underline{2}\,{}^{1}0 \\ -\ \_60\underline{7} \\ \hline 4\underline{9}13 \end{array}$$

Step 4: To find the digit missing from the thousands place of the subtrahend, we subtract: $6 - 4 = 2$. We write a 2 in the thousands place of the subtrahend.

$$\begin{array}{r} \overset{6}{\cancel{7}}\,{}^{1}5\underline{2}\,{}^{1}0 \\ -\ \underline{2}60\underline{7} \\ \hline 4\underline{9}13 \end{array}$$

Check We can check our work by verifying the arithmetic. We add up: 4913 + 2607 = 7520. The missing digits we found are correct.

New Concepts *Increasing Knowledge*

place value through hundred trillions

In our number system the value of a digit depends upon its position within a number. The value of each position is its **place value.** The chart below shows place values from the ones place to the hundred-trillions place.

Whole Number Place Values

hundred trillions	ten trillions	trillions		hundred billions	ten billions	billions		hundred millions	ten millions	millions		hundred thousands	ten thousands	thousands		hundreds	tens	ones	decimal point
___	___	___	,	___	___	___	,	___	___	___	,	___	___	___	,	___	___	___	.

Example 1

a. Which digit is in the trillions place in the number 32,567,890,000,000?

b. In 12,457,697,380,000, what is the place value of the digit 4?

Solution

a. The digit in the trillions place is **2.**

b. The place value of the digit 4 is **hundred billions.**

We write a number in **expanded notation** by writing each nonzero digit times its place value. For example, we write 5280 in expanded notation this way:

$$(5 \times 1000) + (2 \times 100) + (8 \times 10)$$
$$= 5000 + 200 + 80$$
$$= 5280$$

Example 2

Write 25,000 in expanded notation.

Solution

(2 × 10,000) + (5 × 1000)

Justify Why do we multiply 10,000 by 2 and 1000 by 5?

reading and writing whole numbers

Whole numbers with more than three digits may be written with commas to make the numbers easier to read. Commas help us read large numbers by separating the trillions, billions, millions, and thousands places. We need only to read the three-digit number in front of each comma and then say either "trillion," "billion," "million," or "thousand" when we reach the comma.

_ _ _ , _ _ _ , _ _ _ , _ _ _ , _ _ _

↑ "trillion" ↑ "billion" ↑ "million" ↑ "thousand"

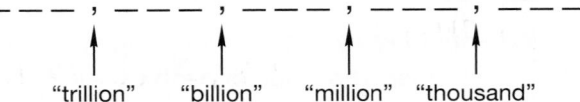

We will use the following guidelines when writing numbers as words:

1. Put commas after the words *trillion, billion, million,* and *thousand.*
2. Hyphenate numbers between 20 and 100 that do not end in zero. For example, 52, 76, and 95 are written "fifty-two," "seventy-six," and "ninety-five."

Example 3

Use words to write 1,380,000,050,200.

Solution

One trillion, three hundred eighty billion, fifty thousand, two hundred.

Explain Why do we not say the millions when we read this number aloud?

Example 4

Use words to write 3406521.

Solution

We start on the right and insert a comma every third place as we move to the left:

3,406,521

Three million, four hundred six thousand, five hundred twenty-one.

Note: We do not write "... five hundred *and* twenty-one." We never include "and" when saying or writing whole numbers.

Example 5

Use digits to write twenty trillion, five hundred ten million.

Solution

It may be helpful to first draw a "skeleton" of the number with places through the trillions. We use abbreviations for "trillion," "billion," "million," and "thousand". We will read the number until we reach a comma and then write what we have read. We read "twenty trillion," so we write 20 before the trillions comma.

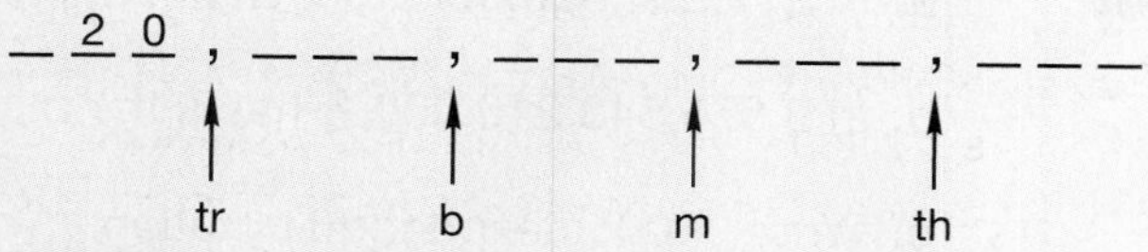

Next we read "five hundred ten million." We write 510 before the millions comma.

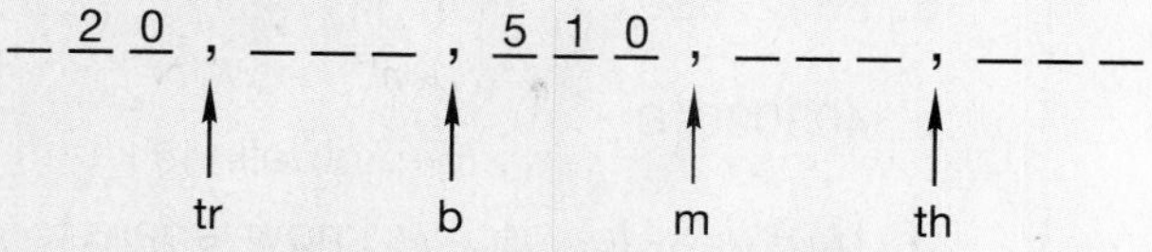

Since there are no billions, we write zeros in the three places before the billions comma.

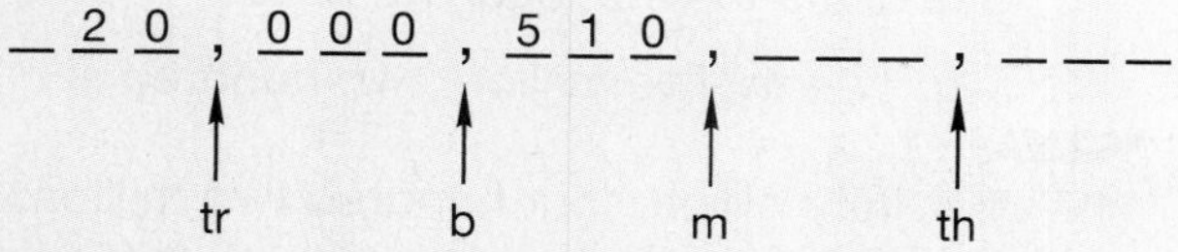

To hold place values, we write zeros in the remaining places. Now we omit the dashes and write the number.

20,000,510,000,000

Large numbers that end with many zeros are often named using a combination of digits and words, such as $3 billion for $3,000,000,000.

Example 6

Use only digits and commas to write 25 million.

Solution

25,000,000

Example 7

Terrell said he and his family traveled twenty-four hundred miles on a summer driving trip. Use digits to write the number of miles they traveled.

Solution

Reading Math

We can read the number 1200 as "twelve hundred" or as "one thousand, two hundred."

Counting up by hundreds, some people say "eight hundred," "nine hundred," "ten hundred," "eleven hundred," and so on for 800, 900, 1000, 1100, ...

In this example Terrell said "twenty-four hundred" for 2400, which is actually two thousand, four hundred. Four-digit whole numbers are often written without commas, so either of these forms is correct: **2400** or **2,400.**

Practice Set

a. In 217,534,896,000,000, which digit is in the ten-billions place?

b. In 9,876,543,210,000, what is the place value of the digit 6?

c. Write 2500 in expanded notation.

Use words to write each number:

d. 36427580

e. 40302010

f. **Explain** How do we know where to place commas when writing the numbers in **d** and **e** as words?

Use digits to write each number:

g. twenty-five million, two hundred six thousand, forty

h. fifty billion, four hundred two million, one hundred thousand

i. $15 billion

j. **Explain** What is a shorter, easier way to write $15,000,000?

Written Practice

Strengthening Concepts

* **1.** (5) What is the sum of six hundred seven and two thousand, three hundred ninety-three?

* **2.** (4, 5) Use digits and other symbols to write "One hundred one thousand is greater than one thousand, one hundred."

* **3.** (5) Use words to write 50,574,006.

* **4.** (5) Which digit is in the trillions place in the number 12,345,678,900,000?

* **5.** (5) Use digits to write two hundred fifty million, five thousand, seventy.

* **6.** (4) Replace the circle with the proper comparison symbol. Then write the comparison as a complete sentence, using words to write the numbers.

$$-12 \bigcirc -15$$

7. (4) Arrange these numbers in order from least to greatest:

$$-1, 4, -7, 0, 5, 7$$

* **8.** (4) **Model** Show this subtraction problem on a number line: $5 - 4$.

9. (4) The rule of a certain sequence is $k = 3n$. Find the first four terms of the sequence.

Find the value of each variable.

* **10.** (3) $2 \cdot 3 \cdot 5 \cdot n = 960$

* **11.** (3) $a - 1367 = 2500$

12. (3) $b + 5 + 17 = 50$

* **13.** (3) $\$25.00 - k = \$18.70$

14. (3) $6400 + d = 10{,}000$

* **15.** (3) $\frac{144}{f} = 8$

16. (5) Write 750,000 in expanded notation.

Simplify:

17. (1)
$$\begin{array}{r} 37{,}428 \\ +\ 59{,}775 \\ \hline \end{array}$$

18. (1)
$$\begin{array}{r} 31{,}014 \\ -\ 24{,}767 \\ \hline \end{array}$$

19. (1) $45 + 362 + 7 + 4319$

20. (1) $64.59 + $124 + $6.30 + 37¢

21. (2) $144 \div (12 \div 3)$

22. (2) $(144 \div 12) \div 3$

23. (1) $40(500)$

24. (1) $8505 \div 21$

25. (2) $10 − ($4.60 − 39¢)

26. (1) 29¢ × 36

* **27.** (2) Which property can we use to find each unknown number?

a. $365n = 365$

b. $52 \cdot 7 = 7m$

* **28.** (4) **Generalize** Use words to describe the rule of the following sequence. Then find the next three terms of the sequence.

$$\ldots, 10, 8, 6, 4, 2, \ldots$$

* **29.** (1, 4) **Classify** Name each set of numbers illustrated:

a. $\{1, 2, 3, 4, \ldots\}$

b. $\{0, 1, 2, 3, \ldots\}$

c. $\{\ldots, -2, -1, 0, 1, 2, \ldots\}$

* **30.** (1, 4) **Represent** Use braces, an ellipsis, and digits to illustrate the set of negative even numbers.

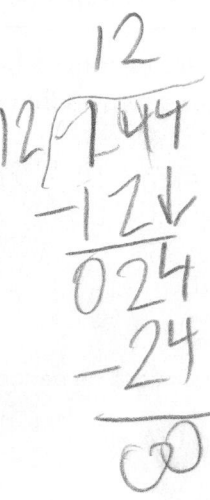

LESSON 6

• Factors
• Divisibility

Power Up

Building Power

facts

Power Up B

mental math

a. **Calculation:** \$5.00 + \$2.50

b. **Decimals:** \$1.50 × 10

c. **Calculation:** \$1.00 − \$0.45

d. **Calculation:** 450 + 35

e. **Number Sense:** 675 − 50

f. **Number Sense:** 750 ÷ 10

g. **Probability:** What is the probability of rolling a number greater than 6 on a number cube?

h. **Calculation:** 9 × 5, − 1, ÷ 4, + 1, ÷ 4, × 5, + 1, ÷ 4[1]

problem solving

The sum of five different single-digit natural numbers is 30. The product of the same five numbers is 2520. Two of the numbers are 1 and 8. What are the other three numbers?

New Concepts

Increasing Knowledge

factors

Recall that factors are the numbers multiplied to form a product.

$3 \times 5 = 15$ both 3 and 5 are factors of 15

$1 \times 15 = 15$ both 1 and 15 are factors of 15

Therefore, each of the numbers 1, 3, 5, and 15 can serve as a factor of 15.

Thinking Skill

Generalize

For any given number, what is the number's least positive factor? Its greatest positive factor?

Notice that 15 can be divided by 1, 3, 5, or 15 without a remainder. This leads us to another definition of factor.

> The **factors** of a number are the whole numbers that divide the number without a remainder.

For example, the numbers 1, 2, 5, and 10 are factors of 10 because each divides 10 without a remainder (that is, with a remainder of zero).

$$\begin{array}{r} 10 \\ \mathbf{1})\overline{10} \\ \underline{10} \\ 0 \end{array} \qquad \begin{array}{r} 5 \\ \mathbf{2})\overline{10} \\ \underline{10} \\ 0 \end{array} \qquad \begin{array}{r} 2 \\ \mathbf{5})\overline{10} \\ \underline{10} \\ 0 \end{array} \qquad \begin{array}{r} 1 \\ \mathbf{10})\overline{10} \\ \underline{10} \\ 0 \end{array}$$

[1] As a shorthand, we will use commas to separate operations to be performed sequentially from left to right. This is not standard mathematical notation.

Example 1

List the whole numbers that are factors of 12.

Solution

The factors of 12 are the whole numbers that divide 12 with no remainder. We find the factors quickly by writing factor pairs.

12 divided by **1** is **12**

12 divided by **2** is **6**

12 divided by **3** is **4**

Below we show the factor pairs arranged in order.

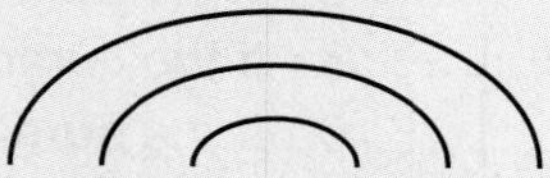

1, 2, 3, 4, 6, 12

Example 2

List the factors of 51.

Solution

As we try to think of whole numbers that divide 51 with no remainder, we may think that 51 has only two factors, 1 and 51. However, there are actually four factors of 51. Notice that 3 and 17 are also factors of 51.

$$3\overline{)51}\text{ with quotient } 17$$

17 ← 17 is a factor of 51

3 is a factor of 51 → 3)51

Thus, the four factors of 51 are **1, 3, 17,** and **51.**

From the first two examples we see that 12 and 51 have two **common factors,** 1 and 3. The **greatest common factor (GCF)** of 12 and 51 is 3, because it is the largest common factor of both numbers.

Example 3

Find the greatest common factor of 18 and 30.

Solution

We are asked to find the largest factor (divisor) of both 18 and 30. Here we list the factors of both numbers, circling the common factors.

Factors of 18: ①,②,③,⑥, 9, 18

Factors of 30: ①,②,③, 5,⑥, 10, 15, 30

The greatest common factor of 18 and 30 is **6.**

divisibility

As we saw in example 2, the number 51 **can be divided** by 3. The capability of a whole number to be divided by another whole number with no remainder is called **divisibility.** Thus, 51 is **divisible** by 3.

There are several methods for testing the divisibility of a number without actually dividing. Listed below are methods for testing whether a number is divisible by 2, 3, 4, 5, 6, 8, 9, or 10.

Tests for Divisibility
A number is divisible by ...
2 if the last digit is even.
4 if the last two digits can be divided by 4.
8 if the last three digits can be divided by 8.
5 if the last digit is 0 or 5.
10 if the last digit is 0.
3 if the **sum of the digits** can be divided by 3.
6 if the number can be divided by 2 **and** by 3.
9 if the **sum of the digits** can be divided by 9.
A number ending in ...
one zero is divisible by 2.
two zeros is divisible by 2 and 4.
three zeros is divisible by 2, 4, and 8.

Discuss Why might we want to test the divisibility of a number without dividing? Why not just divide?

Explain Why does the divisibility test for 2 work?

Example 4

Which whole numbers from 1 to 10 are divisors of 9060?

Solution

In the sense used in this problem, a **divisor** is a **factor.** The number 1 is a divisor of any whole number. As we apply the tests for divisibility, we find that 9060 passes the tests for 2, 4, 5, and 10, but not for 8. The sum of its digits (9 + 0 + 6 + 0) is 15, which can be divided by 3 but not by 9. Since 9060 is divisible by both 2 and 3, it is also divisible by 6. The only whole number from 1 to 10 we have not tried is 7, for which we have no simple test. We divide 9060 by 7 to find that 7 is not a divisor. We find that the numbers from 1 to 10 that are divisors of 9060 are **1, 2, 3, 4, 5, 6,** and **10.**

Practice Set

List the whole numbers that are factors of each number:

a. 25 **b.** 23

c. List the factor pairs of 24.

Analyze List the whole numbers from 1 to 10 that are factors of each number:

d. 1260 **e.** 73,500 **f.** 3600

g. List the single-digit divisors of 1356.

h. The number 7000 is divisible by which single-digit numbers?

i. List all the common factors of 12 and 20.

j. Find the greatest common factor (GCF) of 24 and 40.

k. **Explain** How did you find your answer to exercise **j?**

Written Practice *Strengthening Concepts*

1. (1) **Analyze** If the product of 10 and 20 is divided by the sum of 20 and 30, what is the quotient?

* **2.** (6) **a.** List all the common factors of 30 and 40.

b. Find the greatest common factor of 30 and 40.

* **3.** (4) **Connect** Use braces, an ellipsis, and digits to illustrate the set of negative odd numbers.

* **4.** (5) Use digits to write four hundred seven million, six thousand, nine hundred sixty-two.

* **5.** (6) List the whole numbers from 1 to 10 that are divisors of 12,300.

* **6.** (4) Replace the circle with the proper comparison symbol. Then write the comparison as a complete sentence using words to write the numbers.

$$-7 \bigcirc -11$$

* **7.** (6) The number 3456 is divisible by which single-digit numbers?

8. (4) **Model** Show this subtraction problem on a number line: 2 – 5

* **9.** (5) Write 6400 in expanded notation.

Find the value of each variable:

10. (3) $x + \$4.60 = \$10.00$

* **11.** (3) $p - 3850 = 4500$

* **12.** (3) $8z = \$50.00$

13. (3)

$$\begin{array}{r} 7 \\ 4 \\ 8 \\ 6 \\ 2 \\ 1 \\ 6 \\ 8 \\ 9 \\ +\ n \\ \hline 60 \end{array}$$

* **14.** (3) $1426 - k = 87$

* **15.** (3) $\dfrac{990}{p} = 45$

* **16.** (3) $\dfrac{z}{8} = 32$

Simplify:

17. $\dfrac{1225}{35}$ (1)

18. $\begin{array}{r} 800 \\ \times \ \ 50 \\ \hline \end{array}$ (1)

19. $\begin{array}{r} \$100.00 \\ -\ \$48.37 \\ \hline \end{array}$ (1)

20. $\begin{array}{r} 46{,}302 \\ +\ 49{,}998 \\ \hline \end{array}$ (1)

21. $45.00 ÷ 20 (1)

22. $7 \cdot 11 \cdot 13$ (1)

23. $9\overline{)43{,}271}$ (1)

24. 48¢ + $8.49 + $14 (1)

25. 1000 − (430 − 58) (2)

26. 140(16) (1)

27. $\begin{array}{r} 25¢ \\ \times\ 24 \\ \hline \end{array}$ (1)

28. $\dfrac{\$43.50}{10}$ (1)

* **29.** (2) **a.** **Analyze** Name the property illustrated by the following equation.

$$x \cdot 5 = 5x$$

b. **Summarize** In your own words explain the meaning of this property.

* **30.** (2) **Justify** List the properties used in the first three steps to simplify the expression $(8 \times 7) \times 5$.

$(8 \times 7) \times 5$	Given expression
$8 \times (7 \times 5)$	**a.** ________
$\mathbf{8 \times (5 \times 7)}$	**b.** ________
$\mathbf{(8 \times 5) \times 7}$	**c.** ________
40×7	$8 \times 5 = 40$
280	$40 \times 7 = 280$

7
77
×13
231
+770
1001

LESSON 7

Lines, Angles and Planes

Power Up

Building Power

facts Power Up B

mental math

a. **Positive/Negative:** 5 − 10

b. **Decimals:** $2.50 × 10

c. **Calculation:** $1.00 − 35¢

d. **Calculation:** 340 + 25

e. **Number Sense:** 565 − 300

f. **Number Sense:** 480 ÷ 10

g. **Probability:** What is the probability of rolling a number less than 3 on a number cube?

h. **Calculation:** Start with the number of years in a decade, × 7, + 5, ÷ 3, − 1, ÷ 4.

problem solving

The two pulleys on the left are both in *equilibrium.* A pulley is in equilibrium when the total weight suspended from the left side is equal to the total weight suspended from the right side. Will the pulley on the far right be in equilibrium, or will one side be heavier than the other?

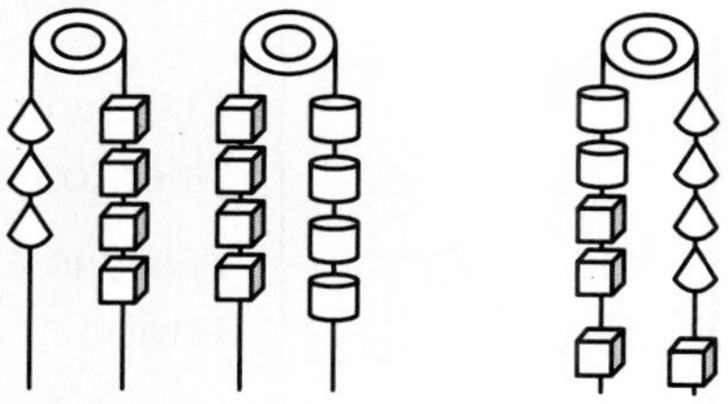

Understand We are shown three pulleys on which three kinds of weights are suspended. The first two pulleys are in equilibrium. We are asked to determine if the third pulley is in equilibrium or if one side is heavier than the other.

Plan We will use logical reasoning to help us understand the relative weights of the objects.

Solve From the first pulley we see that three cones are equal in weight to four cubes, which means that cones are heavier than cubes. The second pulley shows that four cubes weigh the same as four cylinders, which means that cubes and cylinders weigh the same. On the third pulley, we can mentally remove the bottom cubes on either side. We are left with two cylinders and two cubes on one side, and four cones on the other side. Because we know that cones are heavier than cubes or cylinders (which weigh the same), the pulley is not in equilibrium, and will pull to the right.

Check We found that the pulley was not in equilibrium. Another way to verify our solution is to compare the third pulley to the first pulley. Once we remove the bottom cubes, we are left with two cylinders and two cubes on the left side of the third pulley, which is equal to four cubes. The first pulley shows that four cubes equals three cones, so four cones will be heavier.

New Concept *Increasing Knowledge*

We live in a world of three dimensions called **space.** We can measure the length, width, and depth of objects that occupy space. We can imagine a two-dimensional world called a **plane,** a flat world having length and width but not depth. Occupants of a two-dimensional world could not pass over or under other objects because, without depth, "over" and "under" would not exist. A one-dimensional world, a **line,** has length but neither width nor depth. Occupants of a one-dimensional world could not pass over, under, or to either side of other objects. They could only move back and forth on their line.

In **geometry** we study figures that have one dimension, two dimensions, and three dimensions, but we begin with a **point,** which has no dimensions. A point is an exact location in space and is unmeasurably small. We represent points with dots and usually name them with uppercase letters. Here we show point *A:*

A **line** contains an infinite number of points extending in opposite directions without end. A line has one dimension, length. A line has no thickness. We can represent a line by sketching part of a line with two arrowheads. We identify a line by naming two points on the line in either order. Here we show line *AB* (or line *BA*):

Line *AB* or line *BA*

The symbols $\overleftrightarrow{AB}$ and $\overleftrightarrow{BA}$ (read "line *AB*" and "line *BA*") also can be used to refer to the line above.

A **ray** is a part of a line with one endpoint. We identify a ray by naming the endpoint and then one other point on the ray. Here we show ray *AB* ($\overrightarrow{AB}$):

Ray *AB*

A **segment** is a part of a line with two endpoints. We identify a segment by naming the two endpoints in either order. Here we show segment *AB* ($\overline{AB}$):

Segment *AB* or segment *BA*

Thinking Skills

Generalize

A line segment has a specific length. Why doesn't a ray or a line have a specific length?

A segment has a specific length. We may refer to the length of segment *AB* by writing $m\overline{AB}$, which means "the measure of segment *AB*," or by writing the letters *AB* without an overbar. Thus, both *AB* and $m\overline{AB}$ refer to the distance from point *A* to point *B*. We use this notation in the figure below to state that the sum of the lengths of the shorter segments equals the length of the longest segment.

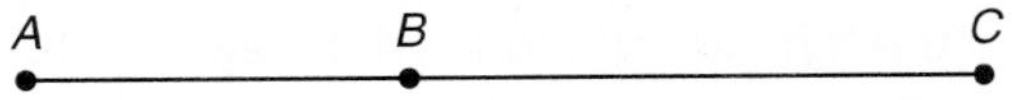

$$AB + BC = AC$$

$$m\overline{AB} + m\overline{BC} = m\overline{AC}$$

Example 1

Use symbols to name a line, two rays, and a segment in the figure at right.

Solution

The line is $\overleftrightarrow{AB}$ (or $\overleftrightarrow{BA}$). The rays are $\overrightarrow{AB}$ and $\overrightarrow{BA}$. The segment is $\overline{AB}$ (or $\overline{BA}$).

Example 2

In the figure below, *AB* is 3 cm and *AC* is 7 cm. Find *BC*.

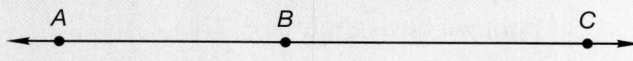

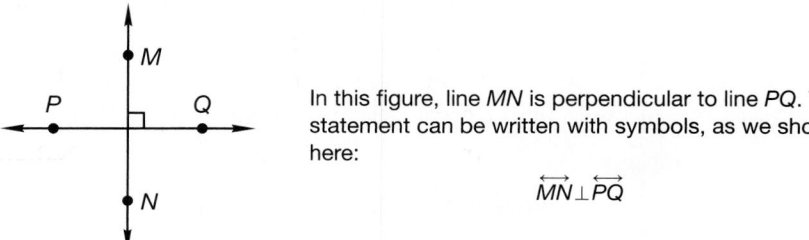

Solution

BC represents the length of segment *BC*. We are given that *AB* is 3 cm and *AC* is 7 cm. From the figure above, we see that $AB + BC = AC$. Therefore, we find that *BC* is **4 cm.**

Formulate Write an equation using numbers and variables to illustrate the example. Then show the solution to your equation.

A **plane** is a flat surface that extends without end. It has two dimensions, length and width. A desktop occupies a part of a plane.

Two lines in the same plane either cross once or do not cross at all. If two lines cross, we say that they **intersect** at one point.

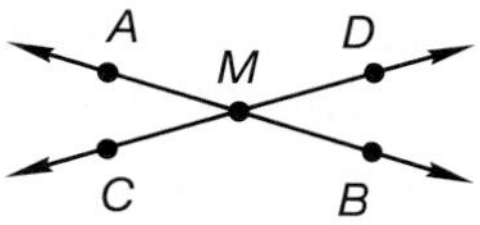

Line *AB* intersects line *CD* at point *M*.

If two lines in a plane do not intersect, they remain the same distance apart and are called **parallel lines.**

Reading Math

The symbol $\|$ means *is parallel to.* The symbol $\perp$ means *is perpendicular to.*

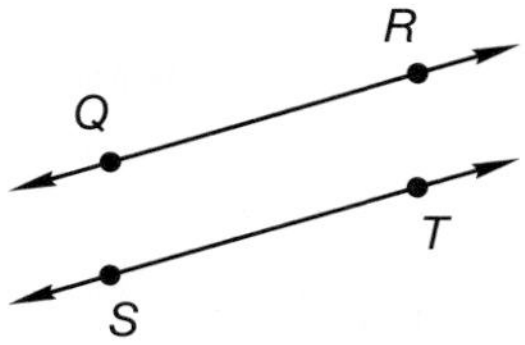

In this figure, line *QR* is parallel to line *ST*. This statement can be written with symbols, as we show here:

$$\overleftrightarrow{QR} \| \overleftrightarrow{ST}$$

Lines that intersect and form "square corners" are **perpendicular lines.** The small square in the figure below indicates a "square corner."

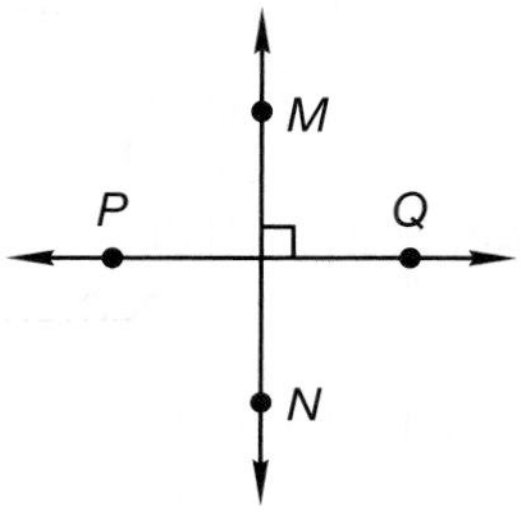

In this figure, line *MN* is perpendicular to line *PQ*. This statement can be written with symbols, as we show here:

$$\overleftrightarrow{MN} \perp \overleftrightarrow{PQ}$$

Lines in a plane that are neither parallel nor perpendicular are **oblique.** In our figure showing intersecting lines, lines *AB* and *CD* are oblique.

An **angle** is formed by two rays that have a common endpoint. The angle at right is formed by the two rays $\overrightarrow{MD}$ and $\overrightarrow{MB}$. The common endpoint is *M*. Point *M* is the **vertex** of the angle. Ray *MD* and ray *MB* are the **sides** of the angle. Angles may be named by listing the following points in order: a point on one ray, the vertex, and then a point on the other ray. So our angle may be named either angle *DMB* or angle *BMD*.

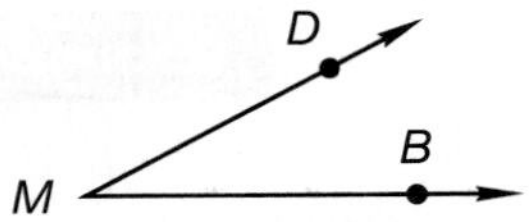

Angle *DMB* or angle *BMD*

When there is no chance of confusion, an angle may be named by only one point, the vertex. At right we have angle *A*.

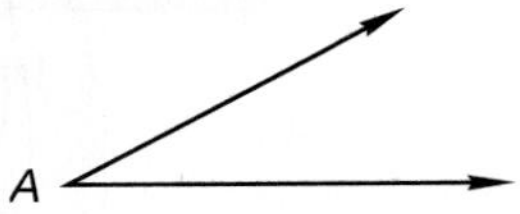

Angle *A*

An angle may also be named by placing a small letter or number near the vertex and between the rays (in the interior of the angle). Here we see angle 1.

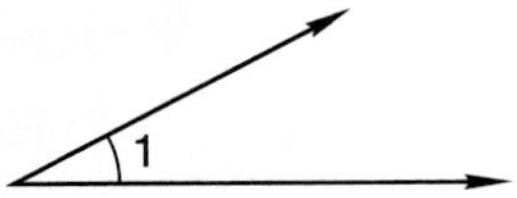

Angle 1

The symbol ∠ is often used instead of the word *angle*. Thus, the three angles just named could be referred to as:

$\angle DMB$	read as "angle *DMB*"
$\angle A$	read as "angle *A*"
$\angle 1$	read as "angle 1"

Angles are classified by their size. An angle formed by perpendicular rays is a **right angle** and is commonly marked with a small square at the vertex. An angle smaller than a right angle is an **acute angle.** An angle that forms a straight line is a **straight angle.** An angle smaller than a straight angle but larger than a right angle is an **obtuse angle.**

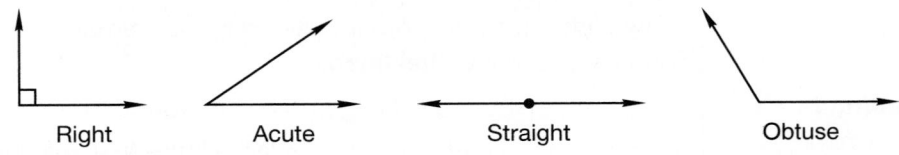

Right

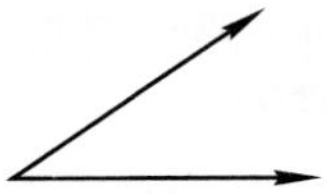
Acute

Straight

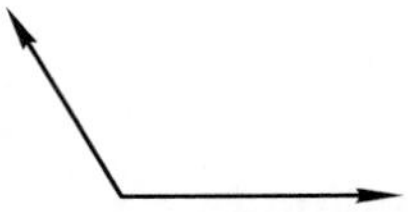
Obtuse

Example 3

a. Which line is parallel to line *AB*?

b. Which line is perpendicular to line *AB*?

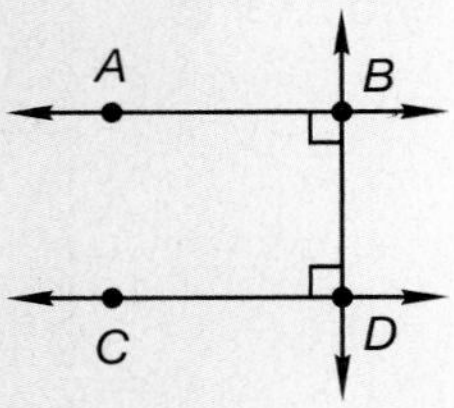

Solution

a. Line *CD* (or $\overleftrightarrow{DC}$) is parallel to line *AB*.

b. Line *BD* (or $\overleftrightarrow{DB}$) is perpendicular to line *AB*.

Conclude How many right angles are formed by two perpendicular lines?

Example 4

There are several angles in this figure.

a. Name the straight angle.

b. Name the obtuse angle.

c. Name two right angles.

d. Name two acute angles.

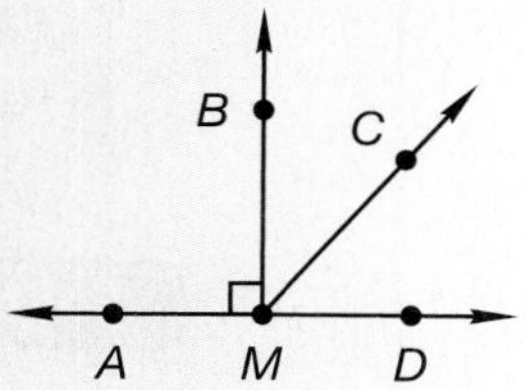

Solution

a. ∠***AMD*** (or ∠***DMA***)

b. ∠***AMC*** (or ∠***CMA***)

c. 1. ∠***AMB*** (or ∠***BMA***)

2. ∠***BMD*** (or ∠***DMB***)

d. 1. ∠***BMC*** (or ∠***CMB***)

2. ∠***CMD*** (or ∠***DMC***)

On earth we refer to objects aligned with the force of gravity as **vertical** and objects aligned with the horizon as **horizontal.**

Example 5

A power pole with two cross pieces can be represented by three segments.

a. Name a vertical segment.

b. Name a horizontal segment.

c. Name a segment perpendicular to $\overline{CD}$.

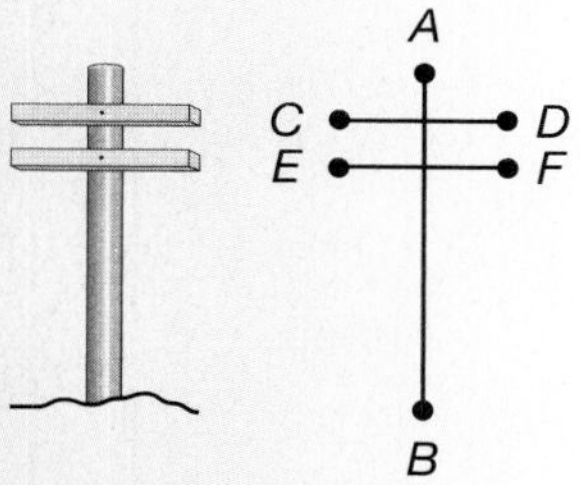

Solution

a. $\overline{AB}$ (or $\overline{BA}$)

b. $\overline{CD}$ (or $\overline{DC}$) or $\overline{EF}$ (or $\overline{FE}$)

c. $\overline{AB}$ (or $\overline{BA}$)

The wall, floor, and ceiling surfaces of your classroom are portions of planes. Planes may be parallel, like opposite walls in a classroom, or they may intersect, like adjoining walls.

We may draw parallelograms to represent planes. Below we sketch how the planes of the floor and two walls appear to intersect.

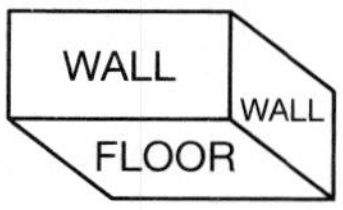

Although the walls and floor have boundaries, the planes of which they are a part do not have boundaries.

Example 6

Sketch two intersecting planes. Which word below best describes the location where the planes intersect?

A Point **B Line** **C Segment**

Solution

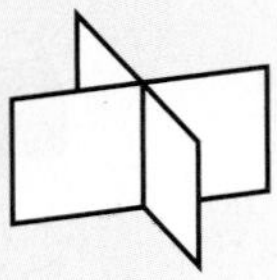

We draw parallelograms through each other to illustrate the planes. In our sketch the intersection appears to be a segment. However, the actual planes extend without boundary, so the intersection continues without end and is a **line.**

Lines in the same plane that do not intersect are parallel. Lines in different planes that do not intersect are **skew lines.**

Connect Can you identify where planes intersect in your classroom?

Practice Set

a. Name a point on this figure that is not on ray *BC:*

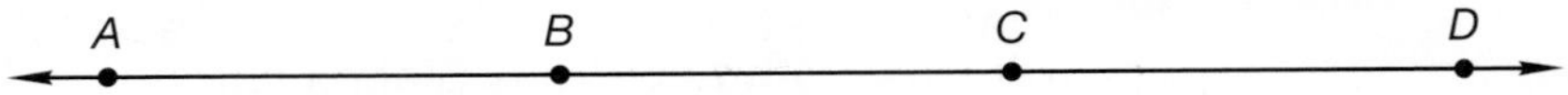

b. In this figure *XZ* is 10 cm, and *YZ* is 6 cm. Find *XY*.

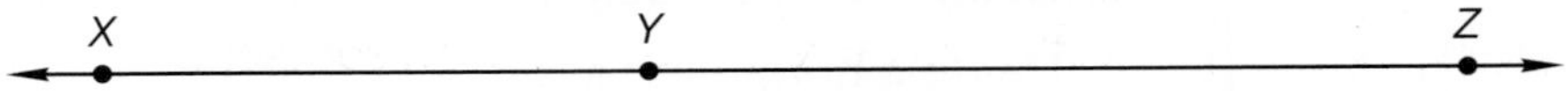

c. Draw two parallel lines.

d. Draw two perpendicular lines.

e. Draw two lines that intersect but are not perpendicular. What word describes the relationship of these lines?

f. Draw a right angle.

g. Draw an acute angle.

h. Draw an obtuse angle.

i. Two intersecting segments are drawn on the board. One segment is vertical and the other is horizontal. Are the segments parallel or perpendicular?

j. Describe a physical example of parallel planes.

k. Describe a physical example of intersecting planes.

l. Lines intersect at a point and planes intersect in a ______.

m. **Connect** If a power pole represents one line and a paint stripe in the middle of the road represents another line, then the two lines are

A parallel **B** intersecting **C** skewed

n. **Model** Sketch a part of the classroom where three planes intersect, such as two adjacent walls and the ceiling.

Written Practice *Strengthening Concepts*

* **1.** (3) If the product of two one-digit whole numbers is 35, what is the sum of the same two numbers?

* **2.** (2) **Analyze** Name the property illustrated by this equation:

$$-5 \cdot 1 = -5$$

* **3.** (6) List the factor pairs of 50.

* **4.** (4) Use digits and symbols to write "Two minus five equals negative three."

5. (5) Use only digits and commas to write 90 million.

6. (6) List the single-digit factors of 924.

7. (4) Arrange these numbers in order from least to greatest:

$$-10, 5, -7, 8, 0, -2$$

* **8.** (4) **Generalize** Use words to describe the following sequence. Then find the next three numbers in the sequence.

$$\ldots, 49, 64, 81, 100, \ldots$$

* **9.** (7) To build a fence, Megan dug holes in the ground to hold the posts upright. Then she attached rails to connect the posts. Which fence parts were vertical, the posts or the rails?

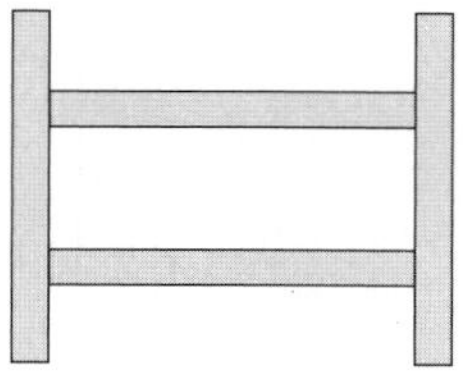

* **10.** (6) **a.** List the common factors of 24 and 32.

b. Find the greatest common factor of 24 and 32.

11. (4) **Connect** The temperature at noon was 3°C. The temperature at 5:00 p.m. was −4°C. Did the temperature rise or fall between noon and 5:00 p.m.? By how many degrees?

Find the value of each variable.

* **12.** (3) $6 \cdot 6 \cdot z = 1224$

13. (3) $\$100.00 - k = \$17.54$

14. (3) $w - 98 = 432$

15. (3) $20x = \$36.00$

* **16.** (3) $\frac{w}{20} = 200$

* **17.** (3) $\frac{300}{x} = 30$

18. (6) **Explain** Does the quotient of 4554 ÷ 9 have a remainder? How can you tell without dividing?

Simplify:

19. (1)
$$\begin{array}{r} 36{,}475 \\ +\ 55{,}984 \\ \hline \end{array}$$

20. (1)
$$\begin{array}{r} 476 \\ \times\ \ 38 \\ \hline \end{array}$$

21. (1) $80.00 − $72.45

22. (1) $68.00 ÷ 40

* **23.** (2) **Justify** Show the steps and the properties that make this multiplication easier to perform mentally: $8 \cdot 7 \cdot 5$

24. (2, 4) Compare: $4000 \div (200 \div 10) \bigcirc (4000 \div 200) \div 10$

25. (1) Evaluate each expression for $a = 200$ and $b = 400$:

a. ab **b.** $a - b$ **c.** $\frac{b}{a}$

* **26.** (7) Refer to the figure at right to answer **a** and **b.**

a. Which angle is an acute angle?

b. Which angle is a straight angle?

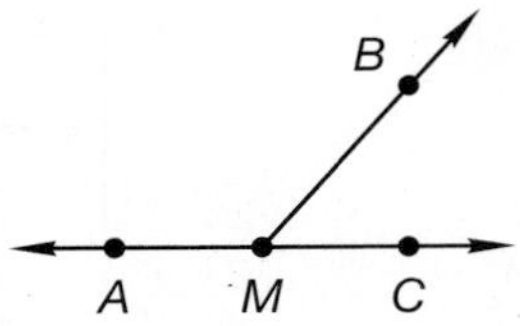

* **27.** (7) What type of angle is formed by perpendicular lines?

Refer to the figure below to answer problems **28** and **29.**

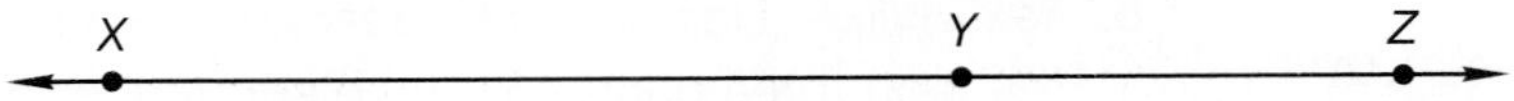

* **28.** (7) Name three segments in this figure.

* **29.** (7) **Conclude** If you knew $m\overline{XY}$ and $m\overline{YZ}$, describe how you would find $m\overline{XZ}$.

* **30.** (7) **Model** Sketch two intersecting planes.

Early Finishers
Real-World Application

Lindy and seven friends played miniature golf at a new course. Par (average score for a good player) for the course is 63. The players scores were recorded as numbers above or below par. For example a score of 62 was recorded as –1 (one under par). The recorded scores were:

1 −1 3 −3 5 0 −2 3

a. What were the seven scores?

b. Arrange the scores in order from least to greatest.

c. How many of the scores are par or under par? List the scores.

LESSON 8

• Fractions and Percents
• Inch Ruler

Power Up

Building Power

facts

Power Up A

mental math

a. **Positive/Negative:** 4 − 10

b. **Decimals:** \$0.25 × 10

c. **Calculation:** \$1.00 − 65¢

d. **Number Sense:** 325 + 50

e. **Number Sense:** 347 − 30

f. **Number Sense:** 200 × 10

g. **Measurement:** Convert 2 hours into minutes

h. **Calculation:** Start with a score, + 1, ÷ 3, × 5, + 1, ÷ 4, + 1, ÷ 2, × 6, + 3, ÷ 3.

problem solving

The number 325 contains the digits 2, 3, and 5. These three digits can be ordered in other ways to make different numbers. Each order is called a *permutation* of the three digits. The smallest permutation of 2, 3, and 5 is 235. How many permutations of the three digits are possible? Which number is the greatest permutation of 2, 3, and 5?

Understand We are told that digits can be arranged in different permutations. We are asked to find how many permutations of the digits 2, 3, and 5 are possible, and to find the greatest permutation of the three digits.

Plan We will make an organized list by working from least to greatest: first we will list the permutations that begin with 2, then with 3, then with 5.

Solve First, we write each permutation that begins with 2. Then we write each permutation that begins with 3. Finally, we write each permutation that begins with 5.

235, 253

325, 352

523, 532

There are six possible permutations of the three digits. The greatest permutation is 532.

Check We found the number of possible permutations and the greatest permutation of the three digits. We kept an organized list to ensure we did not accidentally forget any permutations.

New Concepts *Increasing Knowledge*

fractions and percents

Fractions and **percents** are commonly used to name parts of a whole or parts of a group.

At right we use a whole circle to represent 1. The circle is divided into four equal parts with one part shaded. One fourth of the circle is shaded, and $\frac{3}{4}$ of the circle is not shaded.

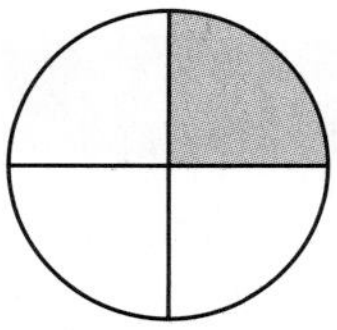

Since the whole circle also represents 100% of the circle, we may divide 100% by 4 to find the percent of the circle that is shaded.

$$100\% \div 4 = 25\%$$

We find that 25% of the circle is shaded, so 75% of the circle is not shaded.

A common fraction is written with two numbers and a division bar. The number below the bar is the **denominator** and shows how many equal parts are in the whole. The number above the bar is the **numerator** and shows how many of the parts have been selected.

numerator → 1
denominator → 4 ← division bar

A percent describes a whole as though there were 100 parts, even though the whole may not actually contain 100 parts. Thus the "denominator" of a percent is always 100.

Math Language
Instead of writing the denominator 100, we can use the word *percent* or the percent symbol, %.

$$25 \text{ percent means } \frac{25}{100}$$

A **mixed number** such as $2\frac{3}{4}$ includes an integer and a fraction. The shaded circles below show that $2\frac{3}{4}$ means $2 + \frac{3}{4}$. To read a mixed number, we first say the integer, then we say "and"; then we say the fraction.

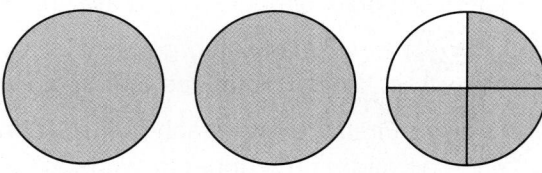

Two and three fourths

It is possible to have percents greater than 100%. When we write $2\frac{3}{4}$ as a percent, we write 275%.

Connect How do we write $3\frac{1}{4}$ as a percent?

Example 1

Name the shaded part of the circle as a fraction and as a percent.

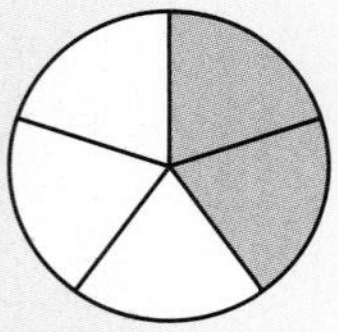

Solution

Two of the five equal parts are shaded, so the fraction that is shaded is $\frac{2}{5}$. Since the whole circle (100%) is divided into five equal parts, each part is 20%.

$$100\% \div 5 = 20\%$$

Two parts are shaded. So **2 × 20%,** or **40%,** is shaded.

Example 2

Which of the following could describe the part of this rectangle that is shaded?

A $\frac{1}{2}$ **B 40%** **C 60%**

Solution

There is a shaded and an unshaded part of this rectangle, but the parts are not equal. More than $\frac{1}{2}$ of the rectangle is shaded, so the answer is not **A.** Half of a whole is 50%.

$$100\% \div 2 = 50\%$$

Since more than 50% of the rectangle is shaded, the correct choice must be **C 60%.**

Between the points on a number line that represent integers are many points that represent fractions and mixed numbers. To identify the fraction or mixed number associated with a point on a number line, it is first necessary to discover the number of segments into which each length has been divided.

Example 3

Point *A* represents what mixed number on this number line?

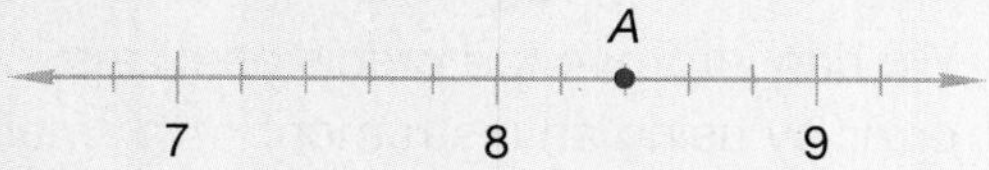

Solution

We see that point *A* represents a number greater than 8 but less than 9. It represents 8 plus a fraction. To find the fraction, we first notice that the segment from 8 to 9 has been divided into five smaller segments. The distance from 8 to point *A* crosses two of the five segments. Thus, **point *A* represents the mixed number $8\frac{2}{5}$.**

Note: It is important to focus on the *number of segments* and not on the number of vertical tick marks. The four tick marks divide the space between 8 and 9 into five segments, just as four cuts divide a strip of paper into five pieces.

inch ruler

A ruler is a practical application of a number line. The units on a ruler are of a standard length and are often divided successively in half. That is, inches are divided in half to show half inches. Then half inches are divided in half to show quarter inches. The divisions may continue in order to show eighths, sixteenths, thirty-seconds, and even sixty-fourths of an inch. In this book we will practice measuring and drawing segments to the nearest sixteenth of an inch.

Here we show a magnified view of an inch ruler with divisions to one sixteenth of an inch.

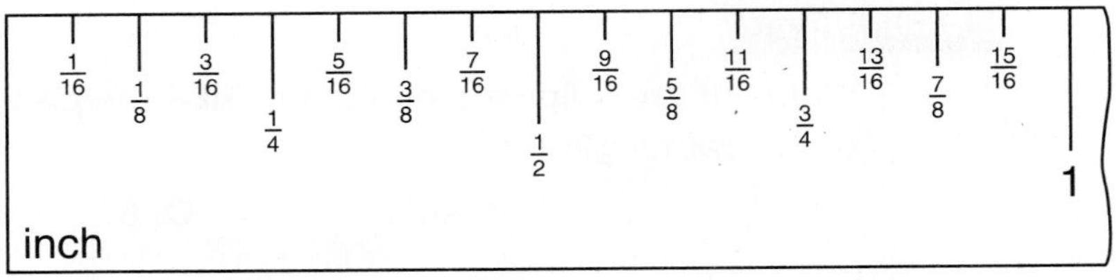

Bear in mind that all measurements are approximate. The quality of a measurement depends upon many conditions, including the care taken in performing the measurement and the precision of the measuring instrument. The finer the gradations are on the instrument, the more precise the measurement can be.

For example, if we measure segments *AB* and *CD* below with an undivided inch ruler, we would describe both segments as being about 3 inches long.

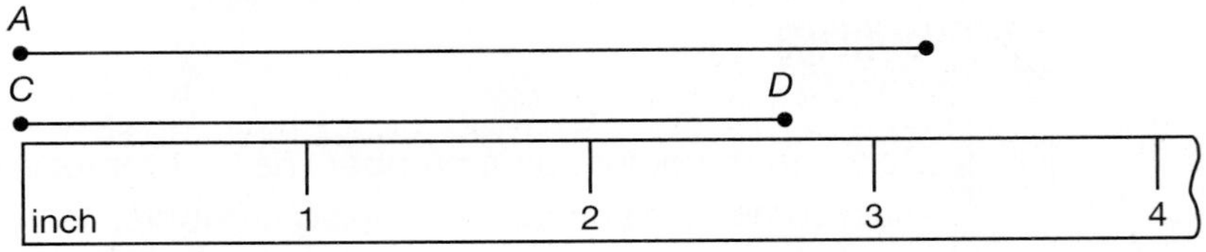

We can say that the measure of each segment is 3 inches $\pm \frac{1}{2}$ inch ("three inches plus or minus one half inch"). This means each segment is within $\frac{1}{2}$ inch of being 3 inches long. In fact, for any measuring instrument, the greatest possible error due to the instrument is one half of the unit that marks the instrument.

We can improve the precision of measurement and reduce the possible error by using an instrument with smaller units. Below we use a ruler divided into quarter inches. We see that *AB* is about $3\frac{1}{4}$ inches and *CD* is about $2\frac{3}{4}$ inches. These measures are precise to the nearest quarter inch. The greatest possible error due to the measuring instrument is one eighth of an inch, which is half of the unit used for the measure.

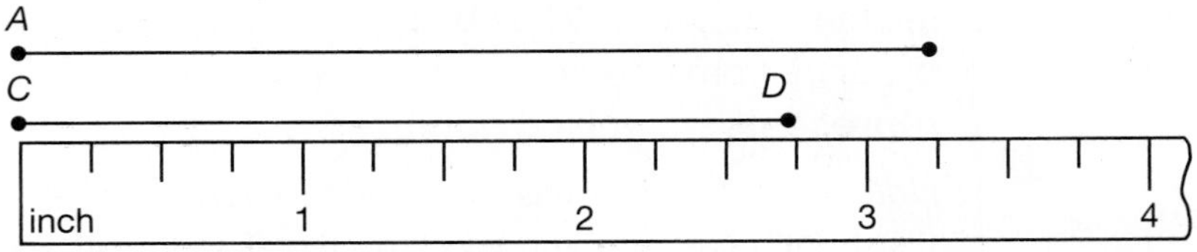

Example 4

Use an inch ruler to find *AB*, *BC*, and *AC* to the nearest sixteenth of an inch.

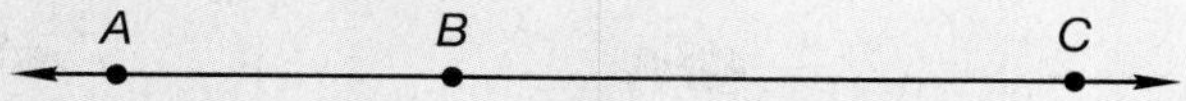

Solution

From point *A* we find *AB* and *AC*. ***AB* is about $\frac{7}{8}$ inches,** and ***AC* is about $2\frac{1}{2}$ inches.**

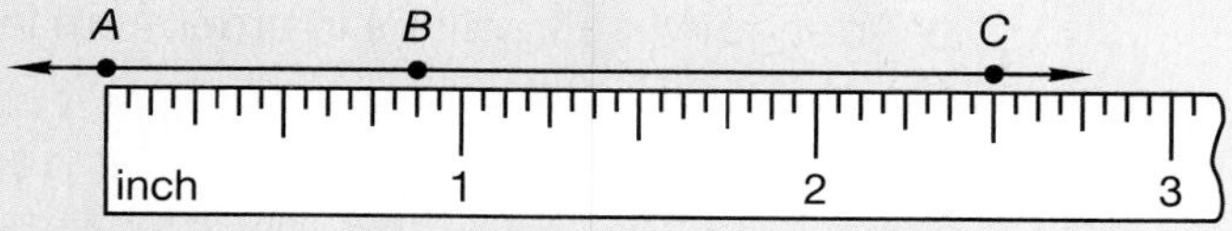

We move the zero mark on the ruler to point *B* to measure *BC*. We find ***BC* is about $1\frac{5}{8}$ inches.**

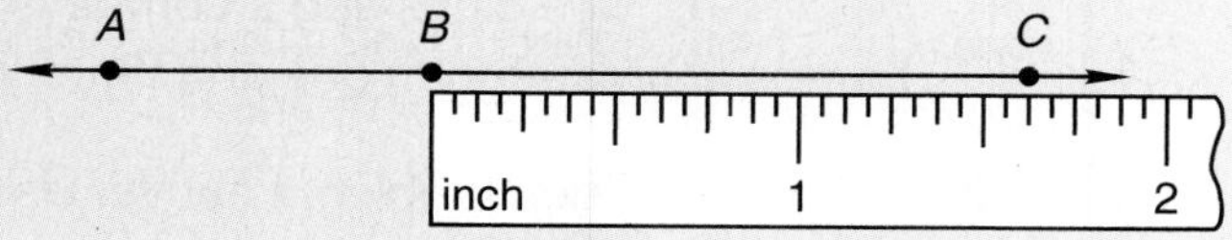

Just as we have used a number line to order integers, we may use a number line to help us order fractions.

Example 5

Arrange these fractions in order from least to greatest:

$$\frac{1}{2}, \frac{1}{4}, \frac{5}{8}, \frac{7}{16}$$

Solution

The illustrated inch ruler can help us order these fractions.

$$\frac{1}{4}, \frac{7}{16}, \frac{1}{2}, \frac{5}{8}$$

Practice Set

a. What fraction of this circle is not shaded?

b. What percent of this circle is not shaded?

c. Half of a whole is what percent of the whole?

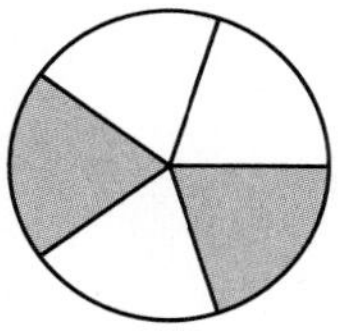

Model Draw and shade circles to illustrate each fraction, mixed number, or percent:

d. $\frac{2}{3}$ **e.** 75% **f.** $2\frac{3}{4}$

Evaluate Points **g** and **h** represent what mixed numbers on these number lines?

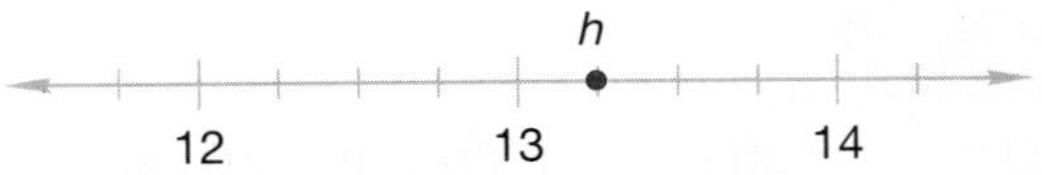

i. Find XZ to the nearest sixteenth of an inch.

j. Jack's ruler is divided into eighths of an inch. Assuming the ruler is used correctly, what is the greatest possible measurement error that can be made with Jack's ruler? Express your answer as a fraction of an inch.

k. Arrange these fractions in order from least to greatest:

$$\frac{1}{4}, \frac{1}{2}, \frac{1}{8}, \frac{1}{16}$$

Written Practice — *Strengthening Concepts*

* **1.** (4, 8) **Represent** Use digits and a comparison symbol to write "One and three fourths is greater than one and three fifths."

* **2.** (8) Refer to practice problem **i** above. Use a ruler to find XY and YZ.

3. (1) What is the quotient when the product of 20 and 20 is divided by the sum of 10 and 10?

* **4.** (6) **Analyze** List the single-digit divisors of 1680.

* **5.** (8) **Evaluate** Point A represents what mixed number on this number line?

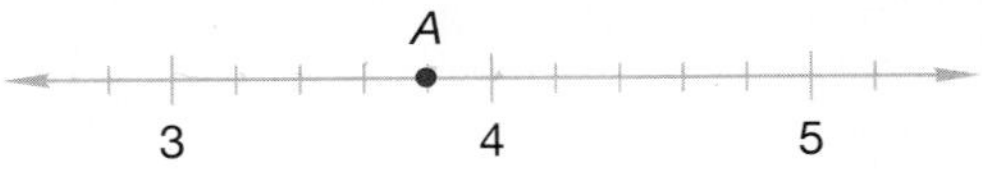

6. (2, 4) **a.** Replace the circle with the proper comparison symbol.

$$3 + 2 \bigcirc 2 + 3$$

b. **Analyze** What property of addition is illustrated by this comparison?

7. (5) Use words to write 32500000000.

* **8.** (8) **a.** What fraction of the circle is shaded?

b. What fraction of the circle is not shaded?

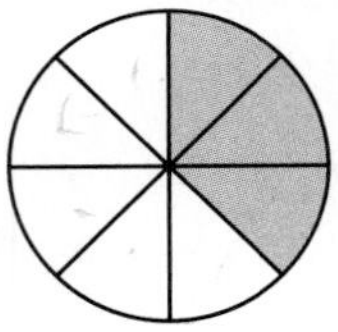

* **9.** (8) **a.** **Analyze** What percent of the rectangle is shaded?

b. What percent of the rectangle is not shaded?

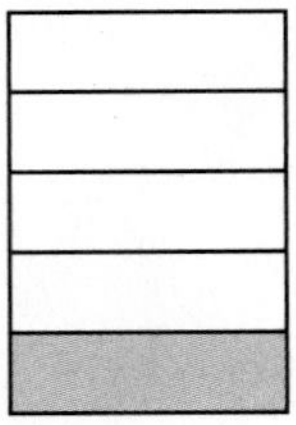

* **10.** (8) What is the name of the part of a fraction that indicates the number of equal parts in the whole?

Find the value of each variable.

11. (3) $a - \$4.70 = \$2.35$

12. (3) $b + \$25.48 = \$60.00$

13. (3) $8c = \$60.00$

14. (3) $10{,}000 - d = 5420$

* **15.** (3) $\frac{e}{15} = 15$

* **16.** (3) $\frac{196}{f} = 14$

17. (2, 3) **Justify** Give a reason for each of the first two steps taken to solve the equation $9 + (n + 8) = 20$.

$9 + (n + 8) = 20$	Given equation
$9 + (8 + n) = 20$	**a.** ____________
$(9 + 8) + n = 20$	**b.** ____________
$17 + n = 20$	$9 + 8 = 17$
$n = 3$	$17 + 3 = 20$

Simplify:

18. (1) $\begin{array}{r} 400 \\ \times\ 500 \\ \hline \end{array}$

19. (1) $\begin{array}{r} 79¢ \\ \times\ 30 \\ \hline \end{array}$

20. (1) $3625 + 431 + 687$

21. (1) $6000 \div 50$

22. (1) $20 \cdot 10 \cdot 5$

23. (1) $\frac{\$27.00}{18}$

24. (1) $\frac{3456}{6}$

25. (1) **Analyze** If t is 1000 and v is 11, find

a. $t - v$

b. $v - t$

* **26.** (4) **a.** The rule of the following sequence is $k = 3n - 1$. What is the tenth term of the sequence?

2, 5, 8, 11, ...

b. **Analyze** What pattern do you recognize in this sequence?

27. (2, 4) Compare: $416 - (86 + 119)$ ◯ $(416 - 86) + 119$

Refer to the figure at right to answer problems **28** and **29**.

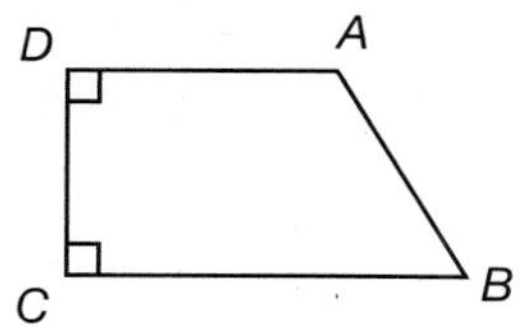

* **28.** (7) Name the acute, obtuse, and right angles.

29. (7) **a.** Name a segment parallel to $\overline{DA}$.

b. Name a segment perpendicular to $\overline{DA}$.

30. (7) **Explain** Referring to the figure below, what is the difference in meaning between the notations $\overline{QR}$ and QR?

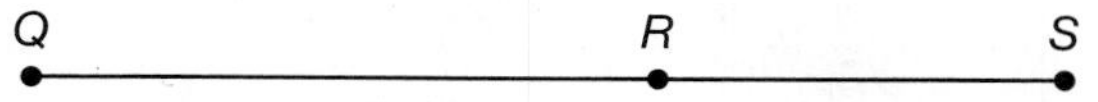

LESSON 9

• Adding, Subtracting, and Multiplying Fractions
• Reciprocals

Power Up — *Building Power*

facts Power Up A

mental math

a. **Positive/Negative:** $3 - 5$

b. **Decimals:** $\$0.39 \times 10$

c. **Calculation:** $\$1.00 - 29¢$

d. **Number Sense:** $342 + 200$

e. **Number Sense:** $580 - 40$

f. **Number Sense:** $500 \div 50$

g. **Measurement:** Convert 6 pints into quarts

h. **Calculation:** Start with half a dozen, $+ 1$, $\times 6$, $- 2$, $\div 2$, $+ 4$, $\div 4$, $- 5$, $\times 15$

problem solving

The diameter of a circle or a circular object is the distance across the circle through its center. Find the approximate diameter of the penny shown at right.

New Concepts — *Increasing Knowledge*

adding fractions

On the line below, *AB* is $1\frac{3}{8}$ in. and *BC* is $1\frac{4}{8}$ in. We can find *AC* by measuring or by adding $1\frac{3}{8}$ in. and $1\frac{4}{8}$ in.

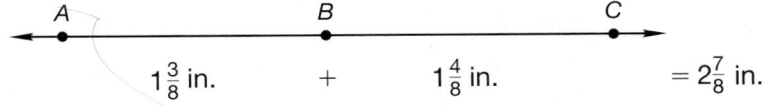

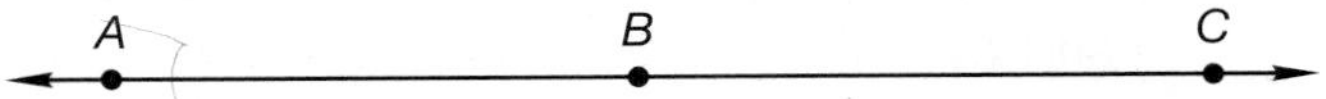

$1\frac{3}{8}$ in. $+$ $1\frac{4}{8}$ in. $= 2\frac{7}{8}$ in.

Math Language
Here, the word *common* means "shared by two or more."

When adding fractions that have the same denominators, we add the numerators and write the sum over the **common denominator**.

Example 1

Find each sum:

a. $\frac{1}{7} + \frac{2}{7} + \frac{3}{7}$

b. $33\frac{1}{3}\% + 33\frac{1}{3}\%$

Solution

Thinking Skill

Model

Draw a number line to show the solution to Example 1a.

a. $\frac{1}{7} + \frac{2}{7} + \frac{3}{7} = \mathbf{\frac{6}{7}}$

b. $33\frac{1}{3}\% + 33\frac{1}{3}\% = \mathbf{66\frac{2}{3}}\%$

Example 2

How much money is $\frac{1}{4}$ of a dollar plus $\frac{3}{4}$ of a dollar?

Solution

$\frac{1}{4} + \frac{3}{4} = \frac{4}{4} = \mathbf{1}$ The sum is **one dollar.**

When the numerator and denominator of a fraction are equal (but not zero), the fraction is equal to 1. The illustration shows $\frac{4}{4}$ of a circle, which is one whole circle.

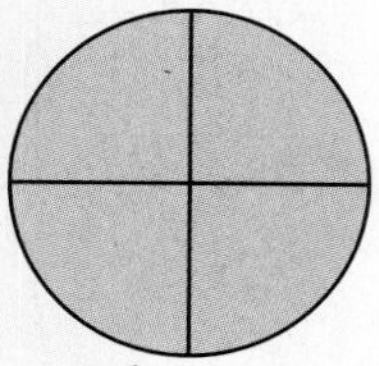

$\frac{4}{4} = 1$

subtracting fractions

To subtract a fraction from a fraction with the same denominator, we write the difference of the numerators over the common denominator.

Example 3

Find each difference:

a. $\mathbf{3\frac{5}{9} - 1\frac{1}{9}}$

b. $\mathbf{\frac{3}{5} - \frac{3}{5}}$

Solution

a. $3\frac{5}{9} - 1\frac{1}{9} = \mathbf{2\frac{4}{9}}$

b. $\frac{3}{5} - \frac{3}{5} = \frac{0}{5} = \mathbf{0}$

multiplying fractions

The first illustration shows $\frac{1}{2}$ of a circle. The second illustration shows $\frac{1}{2}$ of $\frac{1}{2}$ of a circle. We see that $\frac{1}{2}$ of $\frac{1}{2}$ is $\frac{1}{4}$. We often translate the word *of* into a multiplication symbol. We find $\frac{1}{2}$ of $\frac{1}{2}$ by multiplying:

$\frac{1}{2}$

$\frac{1}{2}$ of $\frac{1}{2}$

$$\frac{1}{2} \text{ of } \frac{1}{2} \text{ becomes } \frac{1}{2} \times \frac{1}{2} = \frac{1}{4}$$

To multiply fractions, we multiply the numerators to find the numerator of the product, and we multiply the denominators to find the denominator of the product. Notice that the product of two positive fractions less than 1 is less than either fraction.

Model Draw and shade a rectangle to show $\frac{1}{4}$. Then show $\frac{1}{2}$ of $\frac{1}{4}$ on your rectangle. What is $\frac{1}{2} \times \frac{1}{4}$?

Example 4

Find each product:

a. $\frac{1}{2}$ **of** $\frac{1}{3}$

b. $\frac{1}{2} \cdot \frac{3}{4} \cdot \frac{1}{5}$

Solution

a. $\frac{1}{2} \times \frac{1}{3} = \mathbf{\frac{1}{6}}$

b. $\frac{1}{2} \cdot \frac{3}{4} \cdot \frac{1}{5} = \mathbf{\frac{3}{40}}$

reciprocals

If we **invert** a fraction by switching the numerator and denominator, we form the **reciprocal** of the fraction.

The reciprocal of $\frac{4}{3}$ is $\frac{3}{4}$.

The reciprocal of $\frac{3}{4}$ is $\frac{4}{3}$.

The reciprocal of $\frac{1}{4}$ is $\frac{4}{1}$, which is 4.

The reciprocal of 4 (or $\frac{4}{1}$) is $\frac{1}{4}$.

Note the following relationship between a number and its reciprocal:

The product of a number and its reciprocal is 1.

Here we show two examples of multiplying a number and its reciprocal.

$$\frac{4}{3} \cdot \frac{3}{4} = \frac{12}{12} = 1$$

$$\frac{1}{4} \cdot \frac{4}{1} = \frac{4}{4} = 1$$

Math Language
A **real number** is a number that can be represented by a point on a number line.

This relationship between a number and its reciprocal applies to all real numbers except zero and is called the **Inverse Property of Multiplication.**

Inverse Property of Multiplication

$$a \cdot \frac{1}{a} = 1$$

if a is not 0.[1]

Example 5

Find the reciprocal of each number below. Then multiply the number and its reciprocal.

a. $\frac{3}{5}$

b. 3

Solution

a. The reciprocal of $\frac{3}{5}$ is $\mathbf{\frac{5}{3}}$. $\frac{3}{5} \cdot \frac{5}{3} = \frac{15}{15} = 1$

b. The reciprocal of 3, which is 3 “wholes” or $\frac{3}{1}$, is $\mathbf{\frac{1}{3}}$. $\frac{3}{1} \cdot \frac{1}{3} = \frac{3}{3} = 1$

[1] The exclusion of zero from being a divisor is presented in detail in Lesson 119.

Example 6

Find the missing number: $\frac{3}{4}n = 1$

Solution

The expression $\frac{3}{4}n$ means "$\frac{3}{4}$ times n." Since the product of $\frac{3}{4}$ and n is 1, the missing number must be the reciprocal of $\frac{3}{4}$, which is $\frac{4}{3}$.

$$\frac{3}{4} \cdot \frac{4}{3} = \frac{12}{12} = 1 \quad \text{check}$$

Generalize Give another example to show that the product of a number and its reciprocal is always 1.

Example 7

How many $\frac{3}{4}$s are in 1?

Solution

The answer is the reciprocal of $\frac{3}{4}$, which is $\frac{4}{3}$.

In Lesson 2 we noted that although multiplication is commutative ($6 \times 3 = 3 \times 6$), division is not commutative ($6 \div 3 \neq 3 \div 6$). Now we can say that reversing the order of division results in the reciprocal quotient.

$$6 \div 3 = 2$$

$$3 \div 6 = \frac{1}{2}$$

Practice Set

Simplify:

a. $\frac{5}{6} + \frac{1}{6}$

b. $\frac{4}{5} - \frac{3}{5}$

c. $\frac{3}{5} \times \frac{1}{2} \times \frac{3}{4}$

d. $\frac{3}{3} + \frac{3}{3} + \frac{2}{3}$

e. $\frac{4}{7} \times \frac{2}{3}$

f. $\frac{5}{8} - \frac{5}{8}$

g. $14\frac{2}{7}\% + 14\frac{2}{7}\%$

h. $87\frac{1}{2}\% - 12\frac{1}{2}\%$

Explain Write the reciprocal of each number. Tell what your answer shows about the product of a number and its reciprocal.

i. $\frac{4}{5}$

j. $\frac{8}{7}$

k. 5

Find each unknown number:

l. $\frac{5}{8}a = 1$

m. $6m = 1$

n. Gia's ruler is divided into tenths of an inch. What fraction of an inch represents the greatest possible measurement error due to Gia's ruler? Why?

o. How many $\frac{2}{3}$s are in 1?

p. If a ÷ b equals 4, what does $b \div a$ equal?

q. What property of multiplication is illustrated by this equation?

$$\frac{2}{3} \cdot \frac{3}{2} = 1$$

Written Practice *Strengthening Concepts*

1. (1) What is the quotient when the sum of 1, 2, and 3 is divided by the product of 1, 2, and 3?

2. (1) **Represent** The sign shown is incorrect. Show two ways to correct the sign.

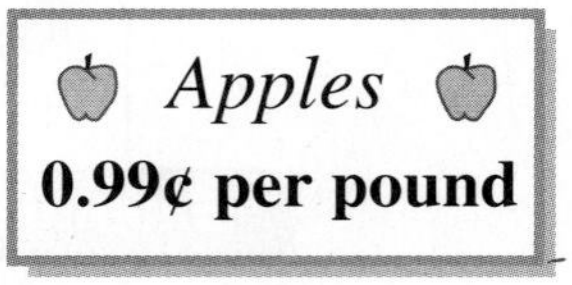

*** 3.** (4, 9) Replace each circle with the proper comparison symbol. Then write the comparison as a complete sentence, using words to write the numbers.

a. $\frac{1}{2} \bigcirc \frac{1}{2} \cdot \frac{1}{2}$ **b.** $-2 \bigcirc -4$

*** 4.** (5) Write twenty-six thousand in expanded notation.

*** 5.** (8) **a.** A dime is what fraction of a dollar?

b. A dime is what percent of a dollar?

*** 6.** (8) The flag to the right is a nautical flag that stands for the letter J.

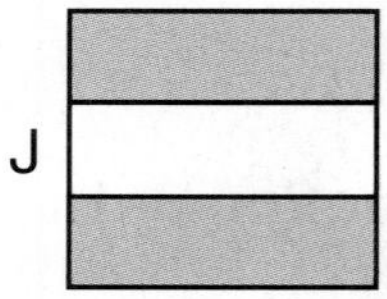

a. What fraction of the flag is shaded?

b. What fraction of the flag is not shaded?

*** 7.** (7) **Classify** Is an imaginary "line" from the Earth to the Moon a line, a ray, or a segment? Why?

*** 8.** (8) Use an inch ruler to find *LM, MN,* and *LN* to the nearest sixteenth of an inch.

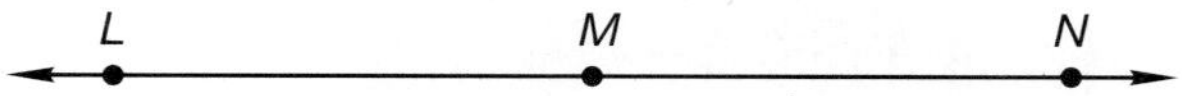

*** 9.** (6) **a.** List the factors of 18.

b. List the factors of 24.

c. Which numbers are factors of both 18 and 24?

d. Which number is the GCF of 18 and 24?

*** 10.** (1, 9) If n is $\frac{2}{5}$, find

a. $n + n$ **b.** $n - n$

Evaluate Find the value of each variable:

11. (3) $85{,}000 + b = 200{,}000$

12. (3) $900 \div c = 60$

13. $d + \$5.60 = \$20.00$
(3)

14. $e \times 12 = \$30.00$
(3)

15. $f - \$98.03 = \$12.47$
(3)

16. $5 + 7 + 5 + 7 + 6 + n + 1 + 2 + 3 + 4 = 40$
(3)

Simplify:

* **17.** $3\frac{11}{15} - 1\frac{3}{15}$
(9)

* **18.** $1\frac{3}{8} + 1\frac{4}{8}$
(9)

* **19.** $\frac{3}{4} \times \frac{1}{4}$
(9)

20. $\frac{1802}{17}$
(1)

21. $\$60.00 - \$49.49$
(1)

22. 607×78
(1)

23. $\frac{4}{5} \times \frac{2}{3} \times \frac{1}{3}$
(9)

24. $\frac{1}{9} + \frac{2}{9} + \frac{4}{9}$
(9)

* **25.** Write the steps and properties that make this multiplication easier to perform mentally: $50 \times 36 \times 20$
(2)

26. What property of multiplication is illustrated by this equation?
(9)

$$\frac{4}{5} \times \frac{5}{4} = 1$$

* **27.** **Classify** Lines *AB* and *XY* lie in different planes. Which word best describes their relationship?
(7)

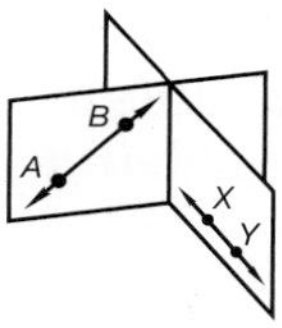

A Intersecting **B** Skew **C** Parallel

* **28.** Refer to the figure at right to answer **a** and **b.**
(7)

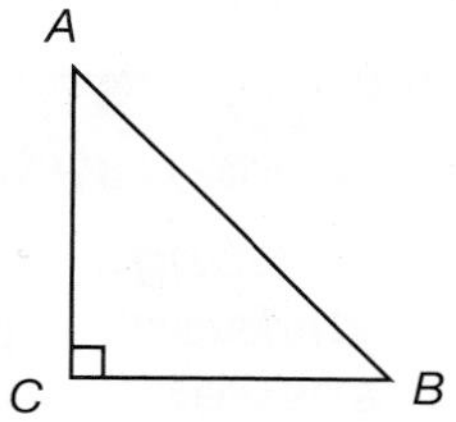

a. Which angles are acute?

b. Which segment is perpendicular to $\overline{CB}$?

* **29.** **Generalize** Describe the following sequence. Then find the next number in the sequence.
(4, 8)

$$\frac{1}{2}, \frac{1}{4}, \frac{1}{8}, \ldots$$

* **30.** How many $\frac{2}{5}$s are in 1?
(9)

LESSON

10

• Writing Division Answers as Mixed Numbers
• Improper Fractions

Power Up

Building Power

facts Power Up A

mental math

a. Positive/Negative: 7 − 10

b. Decimals: $1.25 × 10

c. Calculation: $1.00 − 82¢

d. Number Sense: 384 + 110

e. Number Sense: 649 − 200

f. Number Sense: 300 ÷ 30

g. Measurement: Convert 5 yd into feet

h. Calculation: 3 × 6, ÷ 2, × 5, + 3, ÷ 6, − 3, × 4, + 1, ÷ 3

problem solving In one section of a theater there are twelve rows of seats. In the first row there are 6 seats, in the second row there are 9 seats, and in the third row there are 12 seats. If the pattern continues, how many seats are in the twelfth row?

New Concepts

Increasing Knowledge

writing division answers as mixed numbers

Alexis cut a 25-inch ribbon into four equal lengths. How long was each piece?

To find the answer to this question, we divide. However, expressing the answer with a remainder does not answer the question.

$$\begin{array}{r} 6 \text{ R } 1 \\ 4\overline{)25} \\ \underline{24} \\ 1 \end{array}$$

Interpret What unit of measure does the answer 6 R 1 stand for?

The answer 6 R 1 means that each of the four pieces of ribbon was 6 inches long and that a piece remained that was 1 inch long. But that would make five pieces of ribbon!

Instead of writing the answer with a remainder, we will write the answer as a mixed number. The remainder becomes the numerator of the fraction, and we use the divisor as the denominator.

$$\begin{array}{r} 6\frac{1}{4} \\ 4\overline{)25} \\ \underline{24} \\ 1 \end{array}$$

Thinking Skill

Analyze

What given information shows that the answer is $6\frac{1}{4}$, rather than 6R1?

This answer means that each piece of ribbon was $6\frac{1}{4}$ inches long, which is the correct answer to the question.

Example 1

What percent of the circle is shaded?

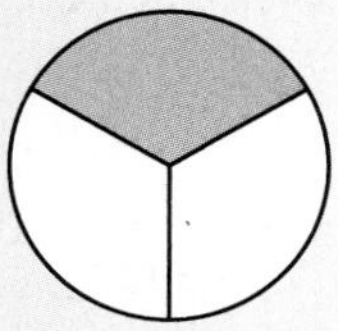

Solution

Thinking Skill

Analyze

In this problem, what does the 100% represent?

One third of the circle is shaded, so we divide 100% by 3.

$$\begin{array}{r} 33\frac{1}{3}\% \\ 3\overline{)100\%} \\ \underline{9} \\ 10 \\ \underline{9} \\ 1 \end{array}$$

We find that **$33\frac{1}{3}$%** of the circle is shaded.

improper fractions

A fraction is equal to 1 if the numerator and denominator are equal (and are not zero). Here we show four fractions equal to 1.

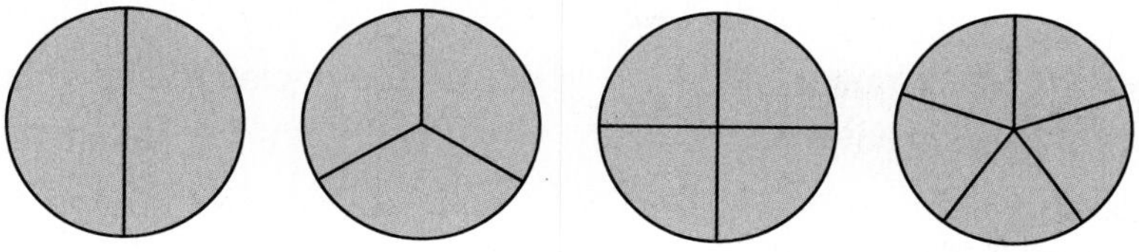

$\frac{2}{2}$ $\frac{3}{3}$ $\frac{4}{4}$ $\frac{5}{5}$

A fraction that is equal to 1 or is greater than 1 is called an **improper fraction.** Improper fractions can be rewritten either as whole numbers or as mixed numbers.

Example 2

Draw and shade circles to illustrate that $\frac{5}{3}$ equals $1\frac{2}{3}$.

Solution

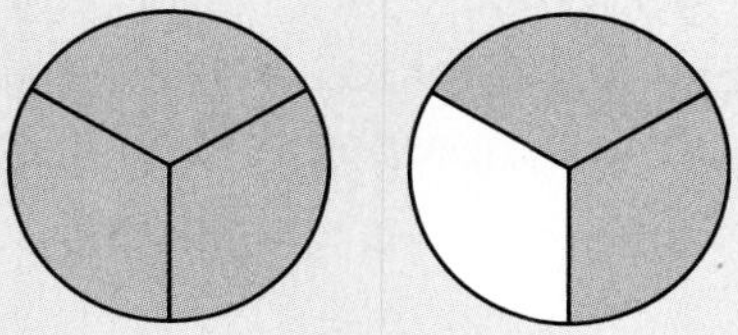

$\frac{5}{3} = 1\frac{2}{3}$

Example 3

Convert each improper fraction to either a whole number or a mixed number:

a. $\frac{5}{3}$ **b.** $\frac{6}{3}$

Solution

a. Since $\frac{3}{3}$ equals 1, the fraction $\frac{5}{3}$ is greater than 1.

$$\frac{5}{3} = \frac{3}{3} + \frac{2}{3}$$
$$= 1 + \frac{2}{3}$$
$$= \mathbf{1\frac{2}{3}}$$

b. Likewise, $\frac{6}{3}$ is greater than 1.

$$\frac{6}{3} = \frac{3}{3} + \frac{3}{3}$$
$$= 1 + 1$$
$$= \mathbf{2}$$

Math Language

A fraction bar indicates division. For example, $\frac{5}{3}$ means 5 *divided by 3.*

We can find the whole number within an improper fraction by performing the division indicated by the fraction bar. If there is a remainder, it becomes the numerator of a fraction whose denominator is the same as the denominator in the original improper fraction.

a. $\frac{5}{3} \longrightarrow 3\overline{)5}$ (quotient 1; $5 - 3 = 2$) $\longrightarrow \mathbf{1\frac{2}{3}}$ **b.** $\frac{6}{3} \longrightarrow 3\overline{)6}$ (quotient **2**)

Model How could you shade circles to show that $\frac{6}{3} = 2$? Draw your answer.

Example 4

Rewrite $3\frac{7}{5}$ with a proper fraction.

Solution

The mixed number $3\frac{7}{5}$ means $3 + \frac{7}{5}$. The fraction $\frac{7}{5}$ converts to $1\frac{2}{5}$.

$$\frac{7}{5} = 1\frac{2}{5}$$

Now we combine 3 and $1\frac{2}{5}$.

$$3 + 1\frac{2}{5} = \mathbf{4\frac{2}{5}}$$

When the answer to an arithmetic problem is an improper fraction, we may convert the improper fraction to a mixed number.

Example 5

Simplify:

a. $\frac{4}{5} + \frac{4}{5}$ **b.** $\frac{5}{2} \times \frac{3}{4}$ **c.** $1\frac{3}{5} + 1\frac{3}{5}$

Solution

a. $\frac{4}{5} + \frac{4}{5} = \frac{8}{5} = \mathbf{1\frac{3}{5}}$

b. $\frac{5}{2} \times \frac{3}{4} = \frac{15}{8} = \mathbf{1\frac{7}{8}}$

c. $1\frac{3}{5} + 1\frac{3}{5} = 2\frac{6}{5} = \mathbf{3\frac{1}{5}}$

Sometimes we need to convert a mixed number to an improper fraction. The illustration below shows $3\frac{1}{4}$ converted to the improper fraction $\frac{13}{4}$.

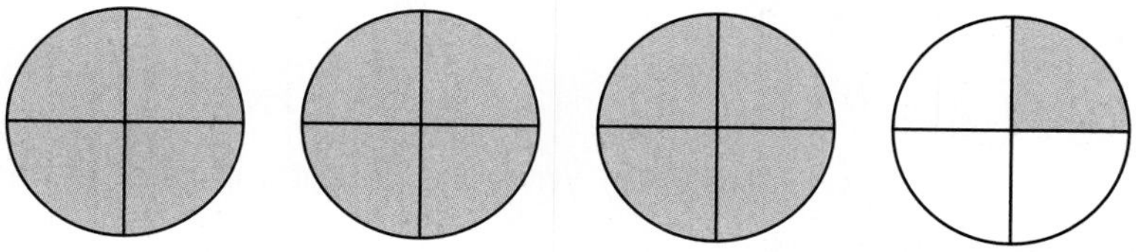

$3\frac{1}{4} = \frac{13}{4}$

We see that every whole circle equals $\frac{4}{4}$. So three whole circles is $\frac{4}{4} + \frac{4}{4} + \frac{4}{4}$, which equals $\frac{12}{4}$. Adding $\frac{1}{4}$ more totals $\frac{13}{4}$.

Example 6

Write each mixed number as an improper fraction:

a. $3\frac{1}{3}$ **b.** $2\frac{3}{4}$ **c.** $12\frac{1}{2}$

Solution

a. The denominator is 3, so we use $\frac{3}{3}$ for 1. Thus $3\frac{1}{3}$ is

$$\frac{3}{3} + \frac{3}{3} + \frac{3}{3} + \frac{1}{3} = \mathbf{\frac{10}{3}}$$

b. The denominator is 4, so we use $\frac{4}{4}$ for 1. Thus $2\frac{3}{4}$ is

$$\frac{4}{4} + \frac{4}{4} + \frac{3}{4} = \mathbf{\frac{11}{4}}$$

c. The denominator is 2, so we use $\frac{2}{2}$ for 1. If we multiply 12 by $\frac{2}{2}$, we find that 12 equals $\frac{24}{2}$. Thus, $12\frac{1}{2}$ is

$$12\left(\frac{2}{2}\right) + \frac{1}{2} = \frac{24}{2} + \frac{1}{2} = \mathbf{\frac{25}{2}}$$

The solution to example 6c suggests a quick way to convert a mixed number to an improper fraction.

$$12\frac{1}{2} = 12 \overset{+}{\curvearrowright} \underset{\times}{\frac{1}{2}} = \frac{2 \times 12 + 1}{2} = \frac{24 + 1}{2} = \frac{25}{2}$$

Practice Set

a. Alexis cut a 35-inch ribbon into four equal lengths. How long was each piece?

b. One day is what percent of one week?

Convert each improper fraction to either a whole number or a mixed number:

c. $\frac{12}{5}$ **d.** $\frac{12}{6}$ **e.** $2\frac{12}{7}$

f. Model Draw and shade circles to illustrate that $2\frac{1}{4} = \frac{9}{4}$.

Simplify:

g. $\frac{2}{3} + \frac{2}{3} + \frac{2}{3}$ **h.** $\frac{7}{3} \times \frac{2}{3}$ **i.** $1\frac{2}{3} + 1\frac{2}{3}$

Convert each mixed number to an improper fraction:

j. $1\frac{2}{3}$ **k.** $3\frac{5}{6}$ **l.** $4\frac{3}{4}$

m. $5\frac{1}{2}$ **n.** $6\frac{3}{4}$ **o.** $10\frac{2}{5}$

p. Generalize Write 3 different improper fractions for the number 4.

Written Practice

Strengthening Concepts

* **1.** (2, 9) Represent Use the fractions $\frac{1}{2}$, $\frac{1}{3}$, and $\frac{1}{6}$ to write an equation that illustrates the Associative Property of Multiplication.

* **2.** (7) Use the words *perpendicular* and *parallel* to complete the following sentence:

In a rectangle, opposite sides are **a.** _______ *and adjacent sides are* **b.** _______.

3. (1) What is the difference when the sum of 2, 3, and 4 is subtracted from the product of 2, 3, and 4?

* **4.** (8) **a.** What percent of the rectangle is shaded?

b. What percent of the rectangle is not shaded?

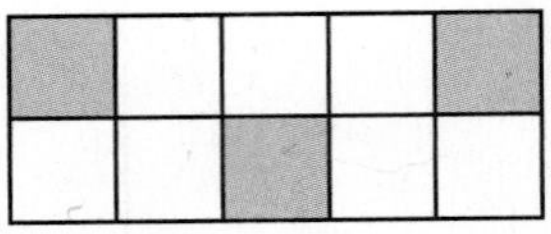

* **5.** (10) Write $3\frac{2}{3}$ as an improper fraction.

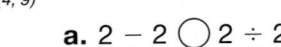

* **6.** (4, 9) Replace each circle with the proper comparison symbol:

a. $2 - 2 \bigcirc 2 \div 2$ **b.** $\frac{1}{2} + \frac{1}{2} \bigcirc \frac{1}{2} \times \frac{1}{2}$

* **7.** (8) Connect Point *M* represents what mixed number on this number line?

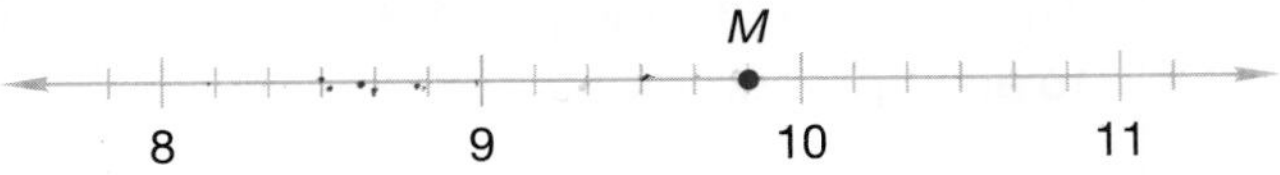

* **8.** (10) **Model** Draw and shade circles to show that $1\frac{3}{5} = \frac{8}{5}$.

* **9.** (6) List the single-digit numbers that are divisors of 420.

Find the value of each variable.

10. (3) $12{,}500 + x = 36{,}275$

11. (3) $18y = 396$

12. (3) $77{,}000 - z = 39{,}400$

13. (3) $\frac{a}{8} = \$1.25$

14. (3) $b - \$16.25 = \$8.75$

15. (3) $c + \$37.50 = \$75.00$

* **16.** (8) Arrange these fractions in order from least to greatest:

$$\frac{1}{2}, \frac{3}{8}, \frac{3}{4}, \frac{1}{16}$$

Simplify:

17. (10) $\frac{5}{2} \times \frac{5}{4}$

18. (9) $\frac{5}{8} - \frac{5}{8}$

19. (10) $\frac{11}{20} + \frac{18}{20}$

20. (2) $2000 - (680 - 59)$

21. (10) $100\% \div 9$

22. (1) 89¢ + 57¢ + \$15.74

23. (1) 800×300

* **24.** (10) $2\frac{2}{3} + 2\frac{2}{3}$

* **25.** (9) $\frac{2}{3} \cdot \frac{2}{3} \cdot \frac{2}{3}$

* **26.** (7) Describe each figure as a line, ray, or segment. Then use a symbol and letters to name each figure.

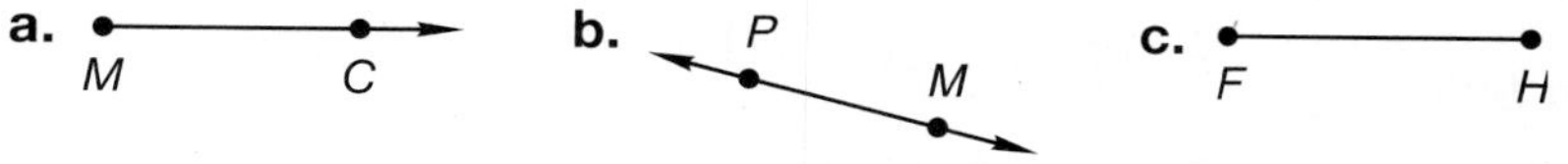

* **27.** (9) How many $\frac{5}{9}$s are in 1?

* **28.** (4, 8) **Generalize** What are the next three numbers in this sequence?

$$\ldots, 32, 16, 8, 4, 2, \ldots$$

* **29.** (4) Which of these numbers is not an integer?

A -1 **B** 0 **C** $\frac{1}{2}$ **D** 1

* **30.** (4, 9) **a.** **Conclude** If $a - b = 5$, what does $b - a$ equal?

b. If $\frac{w}{x} = 3$, what does $\frac{x}{w}$ equal?

c. **Conclude** How are $\frac{w}{x}$ and $\frac{x}{w}$ related?

INVESTIGATION 1

Focus on

• Investigating Fractions and Percents with Manipulatives

In this investigation, you will make a set of fraction manipulatives to use in solving problems with fractions.

Materials needed:

- **Investigation Activities 4–9**
  Halves, Thirds, Fourths, Sixths, Eighths, Twelfths
- Scissors
- Envelopes or locking plastic bags (optional)

Activity

Using Fraction Manipulatives

Discuss Working in groups of two or three, use your fraction manipulatives to help you model the following problems. Discuss how to solve each problem with your group.

Thinking Skill

Generalize

What fraction is one fourth of one half? Show how you found your answer.

1. What fraction is half of $\frac{1}{2}$?
2. What fraction is half of $\frac{1}{3}$?
3. What fraction is half of $\frac{1}{6}$?
4. A quart is one fourth of a gallon. Dan drank half a quart of milk. What fraction of a gallon did Dan drink?
5. Luz exercised for a half hour. Luz jogged for $\frac{1}{3}$ of her exercise time. For what fraction of an hour did Luz jog?
6. Four friends equally divided a whole pizza for lunch. Binli ate $\frac{1}{3}$ of his share and took the rest of his share home. What fraction of the whole pizza did Binli eat for lunch?
7. How many twelfths equal $\frac{1}{2}$?
8. Find a single fraction piece that equals $\frac{3}{12}$.
9. Find a single fraction piece that equals $\frac{4}{8}$.
10. Find a single fraction piece that equals $\frac{4}{12}$.
11. Benito is running a six-mile race, so each mile is $\frac{1}{6}$ of the race. Benito has run $\frac{2}{3}$ of the race. How many miles has he run?
12. One egg is $\frac{1}{12}$ of a dozen. How many eggs is $\frac{3}{4}$ of a dozen?

13. **Model** With a partner, assemble five $\frac{1}{3}$ pieces to illustrate a mixed number. Draw a picture of your work. Then write an equation that relates the improper fraction to the mixed number.

14. Find a single fraction piece that equals $\frac{3}{6}$.

15. **Model** With a partner, assemble nine $\frac{1}{6}$ pieces to form $\frac{5}{6}$ of a circle and $\frac{4}{6}$ of a circle. Then demonstrate the addition of $\frac{5}{6}$ and $\frac{4}{6}$ by recombining the pieces to make $1\frac{1}{2}$ circles. Draw a picture to illustrate your work.

16. Two $\frac{1}{4}$ pieces form half of a circle. Which two different manipulative pieces also form half of a circle?

Find a fraction to complete each equation:

17. $\frac{1}{2} + \frac{1}{3} + a = 1$

18. $\frac{1}{6} + b = \frac{1}{4}$

19. $\frac{1}{2} + c = \frac{3}{4}$

20. $\frac{1}{4} + d = \frac{1}{3}$

Find each percent:

21. What percent of a circle is $\frac{2}{3}$ of a circle?

22. What percent of a circle is $\frac{3}{12}$ of a circle?

23. What percent of a circle is $\frac{3}{8}$ of a circle?

24. **a.** What percent of a circle is $\frac{3}{6}$ of a circle?

    b. **Generalize** Name two other fractions that are the same percent of a circle as $\frac{3}{6}$ of a circle.

25. What percent of a circle is $\frac{1}{4} + \frac{1}{12}$?

26. Use four $\frac{1}{4}$ pieces to demonstrate the subtraction $1 - \frac{1}{4}$, and write the answer.

27. What fraction piece, when used twice, will cover $\frac{4}{6}$ of a circle?

28. What fraction piece, when used three times, will cover $\frac{6}{8}$ of a circle?

29. If you subtract $\frac{1}{12}$ of a circle from $\frac{1}{3}$ of a circle, what fraction of the circle is left?

30. **Represent** Find as many ways as you can to make half of a circle using two or more of the fraction manipulative pieces. Write an equation for each way you find. For example, $\frac{2}{4} = \frac{1}{2}$ and $\frac{1}{3} + \frac{1}{6} = \frac{1}{2}$.

31. Use your fraction pieces to arrange the following fractions in order from least to greatest:

$$\frac{1}{2}, \frac{1}{4}, \frac{1}{3}, \frac{1}{6}$$

32. Use your fraction manipulatives to form the following fractions and to arrange them in order from least to greatest:

$$\frac{2}{3}, \frac{3}{6}, \frac{3}{8}$$

33. **Represent** Select two $\frac{1}{4}$ and one $\frac{1}{2}$ pieces and look at the decimal numbers on those pieces. These pieces show that $\frac{1}{4} + \frac{1}{4} = \frac{1}{2}$. Write a decimal addition number sentence for these pieces.

34. Select two $\frac{1}{2}$ pieces. These pieces show that $\frac{1}{2} + \frac{1}{2} = 1$. Write a similar number sentence using the decimal numbers on these pieces.

---

extension

a. **Formulate** Write new problems for other groups to answer.

b. **Justify** Choose between mental math, manipulatives, or paper and pencil to complete the comparisons. Explain why you chose a particular method.

$$\frac{3}{5} \bigcirc 50\%$$

$$\frac{1}{2} + \frac{2}{6} \bigcirc \frac{1}{4} + \frac{1}{8}$$

$$\frac{1}{2} + \frac{1}{3} \bigcirc 1$$

c. **Infer** Look at the two sets of figures. The figures in Set 2 are nonexamples of Set 1. Based on the attributes of members of both sets, sketch another figure that would fit in Set 1. Support your choice.

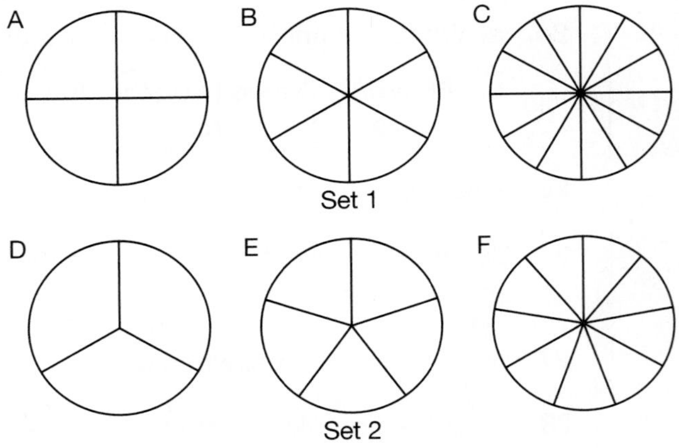

LESSON 11

• Problems About Combining
• Problems About Separating

Power Up

Building Power

facts

Power Up C

mental math

a. **Number Sense:** \$7.50 + 75¢

b. **Decimals:** \$40.00 ÷ 10

c. **Calculation:** \$10.00 − \$5.50

d. **Order of Operations:** (3 × 20) + (3 × 5)

e. **Positive/Negative:** 250 − 1000

f. **Fractional Parts:** $\frac{1}{2}$ of 28

g. **Measurement:** Convert 2 lb. into ounces

h. **Calculation:** Start with the number of hours in a day, ÷ 2, × 3, ÷ 4, × 5, + 4, ÷ 7.

problem solving

If there are twelve glubs in a lorn and four lorns in a dort, then how many glubs are in half a dort?

New Concepts

Increasing Knowledge

problems about combining

In this lesson we will begin solving one-step word problems by writing and solving equations for the problems. To write an equation, it is helpful to understand the **plot** of the word problem. Problems with the same plot can be modeled with the same equation or formula. In this lesson we consider two common plots.

One common idea in word problems is that of **combining.** Here is an example.

Albert has \$12. Betty has \$15. Together they have \$27.

Math Language
A **formula** is an equation that represents a rule or method for doing something.

Problems about combining have an addition thought pattern that we show with this formula:

some + some more = total

$$s + m = t$$

There are three numbers in this description. In a word problem, one of the numbers is unknown. To write an equation for a problem, we write the numbers we are given in the formula and use a letter to stand for the unknown number. If the total is unknown, we add to find the unknown number. If an addend is unknown, we subtract the known addend from the sum to find the unknown addend.

Although we sometimes use subtraction to solve the problem, it is important to recognize that word problems about combining have addition thought patterns.

We follow the four-step problem solving process when solving word problems.

Example 1

In the morning, the trip odometer in Mr. Chin's car read 47 miles. At the end of the day the trip odometer read 114 miles. How many miles did Mr. Chin drive that day?

Solution

Understand We recognize that this problem has an addition pattern. Mr. Chin drove some miles and then he drove some more miles.

$$s + m = t$$

Plan We write an equation for the given information. The trip odometer read 47 miles. Mr. Chin drove some more miles. Then there was a total of 114 miles on the trip odometer.

$$47 + m = 114$$

Solve To find m, an unknown addend, we subtract. We confirm our arithmetic is correct by substituting the answer into the original equation.

$$\begin{array}{r} 114 \\ -\ 47 \\ \hline 67 \end{array} \quad \longrightarrow \quad \begin{array}{r} 47 \text{ miles} \\ +\ 67 \text{ miles} \\ \hline 114 \text{ miles} \end{array} \text{ verify}$$

Check Now we review the question and write the answer. During that day Mr. Chin drove **67 miles.**

Example 2

The first scout troop encamped in the ravine. A second troop of 137 scouts joined them, making a total of 312 scouts. How many scouts were in the first troop?

Solution

Understand We recognize that this problem is about **combining.** There were some scouts. Then some more scouts came. We use the addition formula.

Plan We can solve this problem by writing an equation using s to stand for the number of scouts in the first troop.

$$s + m = t$$

$$s + 137 = 312$$

Solve To find s, we subtract. Then we verify the arithmetic by substituting into the original equation.

$$\begin{array}{r} 312 \\ -\ 137 \\ \hline 175 \end{array} \quad \longrightarrow \quad \begin{array}{r} 175 \text{ scouts} \\ +\ 137 \text{ scouts} \\ \hline 312 \text{ scouts} \end{array} \text{ verify}$$

Check Now we review the question and write the answer. There were **175 scouts** in the first troop.

Formulate Write a word problem that could be solved using this equation:

$$56 + m = 195$$

problems about separating

Another common idea in word problems is **separating** an amount into two parts. Often problems about separating involve something "going away." Here is an example:

> *Mr. Smith wrote a check to Mr. Rodriguez for $37.50. If $824.00 was available in Mr. Smith's account before he wrote the check, how much was available after he gave the check to Mr. Rodriguez?*

Problems about separating have a subtraction thought pattern that we show with this formula:

beginning amount − some went away = what remains

$$b - a = r$$

In a word problem one of the three numbers is unknown. To write an equation, we write the numbers we are given into the formula and use a letter to represent the unknown number. Then we find the unknown number and answer the question in the problem.

Example 3

Tim baked 4 dozen muffins. He made a platter with some of the muffins and gave them away to the school bake sale. He had 32 muffins left which he packed in freezer bags to store in the freezer. How many muffins did Tim give away to the bake sale?

Solution

Understand We recognize that this problem is about **separating.** Tim had some muffins. Then some went away. We use the subtraction formula.

Plan The strategy we choose is to write an equation using 48 for 4 dozen muffins and *a* for the number of muffins that went away.

$$b - a = r$$

$$48 - a = 32$$

Math Language
A **subtrahend** is a number that is subtracted.

Solve We find the unknown number. To find the subtrahend in a subtraction pattern, we subtract. Then we verify the solution by substituting the answer into the original equation.

$$\begin{array}{r} 48 \\ -\ 32 \\ \hline 16 \end{array} \quad \longrightarrow \quad \begin{array}{r} 48 \text{ muffins} \\ -\ 16 \text{ muffins} \\ \hline 32 \text{ muffins} \end{array} \ \text{verify}$$

Check Now we review the question and write the answer. Tim gave away **16 muffins** to the school bake sale.

Example 4

The room was full of boxes when Sharon began. Then she shipped out 56 boxes. Only 88 boxes were left. How many boxes were in the room when Sharon began?

Solution

Understand We recognize that this problem is about **separating.** There were boxes in a room. Then Sharon shipped some away.

Plan We write an equation using *b* to stand for the number of boxes in the room when Sharon began.

$$b - a = r$$
$$b - 56 = 88$$

Solve We find the unknown number. To find the minuend in a subtraction pattern, we add the subtrahend and the difference.

$$\begin{array}{r} 88 \\ +\ 56 \\ \hline 144 \end{array} \quad \rightarrow \quad \begin{array}{r} 144 \text{ boxes} \\ -\ 56 \text{ boxes} \\ \hline 88 \text{ boxes} \end{array} \quad \text{verify}$$

Check Now we review the question and write the answer. There were **144 boxes** in the room when Sharon began.

Formulate Use the numbers 56 and 29 to write a word problem about separating.

Practice Set

Generalize Follow the four-step method shown in this lesson for each problem. Along with each answer, include the equation you used to solve the problem.

a. Rover, a St. Bernard, and Spot, an English Sheepdog, together weighed 213 pounds. Rover weighs 118 pounds. How much did Spot weigh?

b. Dawn cranked for a number of turns. Then Tim gave the crank 216 turns. If the total number of turns was 400, how many turns did Dawn give the crank?

c. There were 254 horses in the north corral yesterday. The rancher moved some of the horses to the south corral. Only 126 horses remained in the north corral . How many horses were moved to the south corral?

d. Cynthia had a lot of paper. After using 36 sheets for a report, only 164 sheets remained. How many sheets of paper did she have at first?

e. Write a word problem about combining that fits this equation:

$$\$15.00 + t = \$16.13$$

f. Write a word problem about separating that fits this equation:

$$32 - s = 25$$

Written Practice *Strengthening Concepts*

* **1.** (11) As the day of the festival drew near, there were 200,000 people in the city. If the usual population of the city was 85,000, how many visitors had come to the city?

* **2.** (11) Syd returned from the store with $12.47. He had spent $98.03 on groceries. How much money did he have when he went to the store?

* **3.** (11) Exactly 17,926 runners began the 2004 Boston Marathon. If only 16,733 runners finished the marathon, how many dropped out along the way?

* **4.** (8, 10) **a.** What fraction of the group is shaded?

b. What fraction of the group is not shaded?

c. What percent of the group is not shaded?

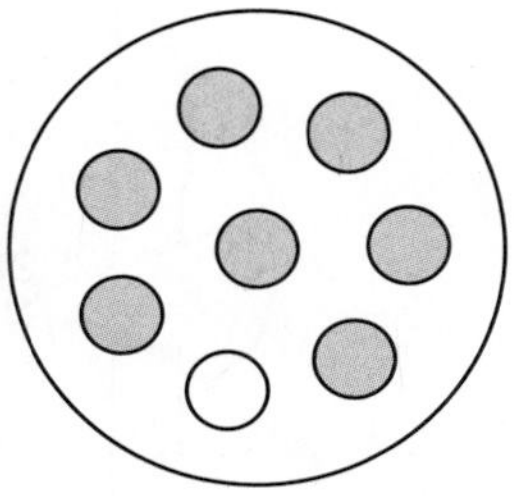

* **5.** (4, 8) **a.** Arrange these numbers in order from least to greatest:

$$\frac{1}{2}, 0, -2, 1$$

b. *Classify* Which of these numbers is not an integer?

* **6.** (10) *Explain* A 35-inch ribbon was cut into 8 equal lengths. How long was each piece? Describe how the ribbon could be cut into eighths.

7. (4) Use digits and symbols to write, "The product of one and two is less than the sum of one and two."

8. (5) Subtract 89 million from 100 million. Use words to write the difference.

9. (6) **a.** List the factors of 16.

b. List the factors of 24.

c. Which numbers are factors of both 16 and 24?

d. What is the GCF of 16 and 24?

Reading Math

Read the term GCF as "greatest common factor." The GCF is the greatest whole number that is a factor of two or more numbers.

* **10.** (11) Write and solve a word problem for this equation:

$$\$20.00 - k = \$12.50$$

11. (7) **Model** Sketch two intersecting planes.

Find the unknown number in each equation.

12. (3) $4 \cdot 9 \cdot n = 720$

13. (3) $\$126 + r = \$375$

14. (3) $\frac{169}{s} = 13$

15. (3) $\frac{t}{40} = \$25.00$

16. (2, 4) Compare: $100 - (5 \times 20) \bigcirc (100 - 5) \times 20$

Simplify:

* **17.** (10) $1\frac{5}{9} + 1\frac{5}{9}$

* **18.** (10) $\frac{5}{3} \times \frac{2}{3}$

19. (1) 135×72

20. (1) $\frac{1000}{40}$

21. (1) 30($1.49)

22. (1) $140.70 ÷ 35

* **23.** (9) $\frac{5}{9} \cdot \frac{1}{3} \cdot \frac{1}{2}$

* **24.** (9) $\frac{5}{8} + \left(\frac{3}{8} - \frac{1}{8}\right)$

* **25.** (10) **a.** Write $3\frac{3}{4}$ as an improper fraction.

b. Write the reciprocal of answer **a.**

c. Find the product of answers **a** and **b.**

* **26.** (8) **Estimate** Which choice below is the best estimate of the portion of the rectangle that is shaded?

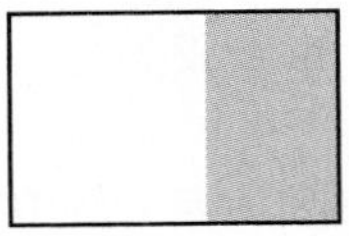

A $\frac{1}{4}$ **B** $\frac{1}{2}$ **C** 40% **D** 60%

* **27.** (4, 8) What are the next four numbers in this sequence?

$$\frac{1}{8}, \frac{1}{4}, \frac{3}{8}, \frac{1}{2}, \ldots$$

* **28.** (7) **Classify** Refer to the figure at right to answer **a** and **b.**

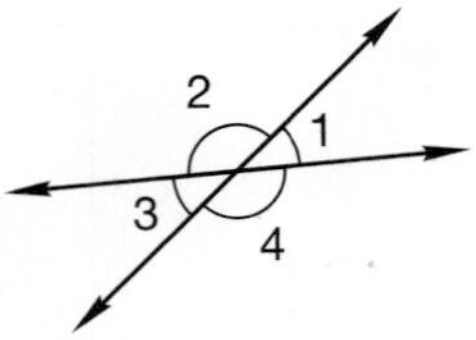

a. Which angles appear to be acute angles?

b. Which angles appear to be obtuse angles?

* **29.** (8) **Evaluate** Use an inch ruler to draw $\overline{AC}$ $3\frac{1}{2}$ inches long. On $\overline{AC}$ mark point B so that AB is $1\frac{7}{8}$ inches. Now find BC.

* **30.** (9) If $n \div m$ equals $\frac{7}{8}$, what does $m \div n$ equal?

Early Finishers

Real-World Application

Mrs. Chen purchased 30 packets of #2 pencils. Each packet contains 12 pencils. She has 50 students and wants to share the pencils equally among the students.

a. How many pencils should each student get? Show your work.

b. Will there be any pencils left over? If so, how many?

c. Express the relationship of the pencils left over to the total number of pencils as a fraction in simplest form.

LESSON

12

• Problems About Comparing
• Elapsed-Time Problems

Power Up

Building Power

facts Power Up C

mental math

a. **Number Sense:** \$6.50 + 60¢

b. **Decimals:** \$1.29 × 10

c. **Number Sense:** \$10.00 − \$2.50

d. **Order of Operations:** (4 × 20) + (4 × 3)

e. **Positive/Negative:** 500 − 2000

f. **Fractional Parts:** $\frac{1}{2}$ of 64

g. **Probability:** How many different outfits will you have with 4 shirts and 2 pair of pants?

h. **Calculation:** Start with three score, ÷ 2, + 2, ÷ 2, + 2, ÷ 2, + 2, × 2.

problem solving

What is the sum of the first ten even numbers?

Understand We are asked to find the sum of the first ten even numbers.

Plan We will begin by *making the problem simpler.* If the assignment had been to add the first *four* even numbers, we could simply add 2 + 4 + 6 + 8. However, adding columns of numbers can be time-consuming. We will try to *find a pattern* that will help add the even numbers 2–20 more quickly.

Solve We can find pairs of addends in the sequence that have the same sum and multiply by the number of pairs. We try this pairing technique on the sequence given in the problem:

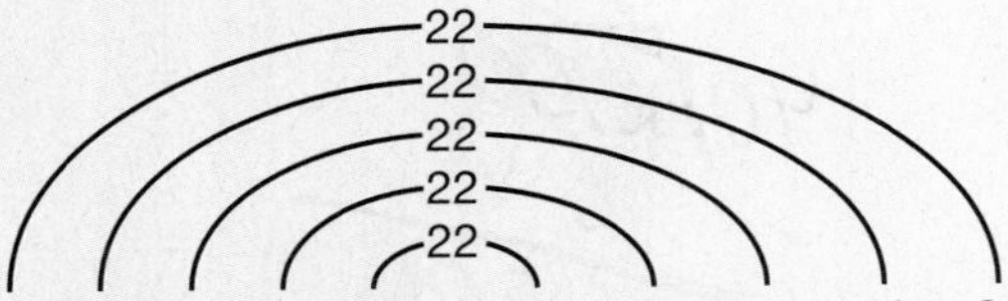

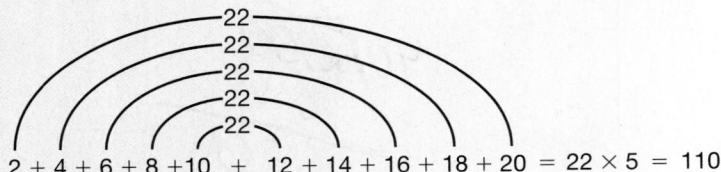

2 + 4 + 6 + 8 + 10 + 12 + 14 + 16 + 18 + 20 = 22 × 5 = 110

Check We found the sum of the first ten even numbers by pairing the addends and multiplying. We can verify our solution by adding the numbers one-by-one with pencil and paper or a calculator.

New Concepts *Increasing Knowledge*

problems about comparing

Problems about comparing often ask questions that contain words like "how much greater" or "how much less." The number that describes how much greater or how much less is called the *difference.* We find the difference by subtracting the lesser number from the greater number. Here is the formula:

$$\text{greater} - \text{lesser} = \text{difference}$$

$$g - l = d$$

Example 1

During the day 1320 employees work at the toy factory. At night 897 employees work there. How many more employees work at the factory during the day than at night?

Solution

Understand Questions such as "How many more?" or "How many fewer?" indicate a **comparison** problem. We use the greater-lesser-difference formula.

Plan We write an equation for the given information.

$$g - l = d$$

$$1320 - 897 = d$$

Solve We find the unknown number in the pattern by subtracting.

$$\begin{array}{r} 1320 \text{ employees} \\ -\ \ 897 \text{ employees} \\ \hline 423 \text{ employees} \end{array}$$

As expected, the difference is less than the greater of the two given numbers.

Check We review the question and write the answer. There are **423 more employees** who work at the factory during the day than work there at night.

Thinking Skill

Estimate

How can we use estimation to see if the answer is reasonable?

Example 2

The number 620,000 is how much less than 1,000,000?

Solution

Understand The words *how much less* indicate that this is a **comparison** problem. We use the *g-l-d* formula.

Plan We write an equation using *d* to stand for the difference between the two numbers.

$$g - l = d$$

$$1{,}000{,}000 - 620{,}000 = d$$

Solve We subtract to find the unknown number.

$$\begin{array}{r} 1{,}000{,}000 \\ -\quad 620{,}000 \\ \hline 380{,}000 \end{array}$$

Check We review the question and write the answer. Six hundred twenty thousand is **380,000** less than 1,000,000.

elapsed-time problems

Elapsed time is the length of time between two points in time. Here we use points on a ray to illustrate elapsed time. Timelines are often used to indicate the sequence of important dates in history. The distance between two dates on a timeline indicates the elapsed time.

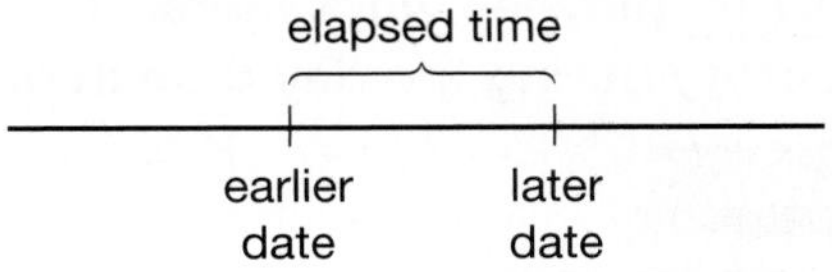

A person's age is an example of elapsed time. Your age is the time that has elapsed since you were born until this present moment. By subtracting the date you were born from today's date you can find your age.

$$\begin{array}{rl} \text{Today's date} & \text{(later)} \\ -\ \text{Your birth date} & \text{(earlier)} \\ \hline \text{Your age} & \text{(difference)} \end{array}$$

Elapsed-time problems are like comparison problems. We can use a later-earlier-difference formula.

$$\text{later} - \text{earlier} = \text{difference}$$

$$l - e = d$$

Example 3

How many years were there from 1492 to 1776? (Unless otherwise specified, years are A.D.)

Solution

Understand We recognize that this is an **elapsed-time problem.** We use the later-earlier-difference formula.

Plan The year 1776 is later than 1492. The difference is the number of years between 1492 and 1776.

$$l \quad - \quad e \quad = d$$

$$1776 - 1492 = d$$

Solve We subtract to find the difference. We confirm our arithmetic is correct.

$$\begin{array}{r} 1776 \\ -\ 1492 \\ \hline 284 \end{array} \longrightarrow \begin{array}{r} 1492 \\ +284 \\ \hline 1776 \end{array} \quad \text{verify}$$

Check Now we review the question and write the answer. There were **284 years** from 1492 to 1776.

Model Represent the solution to this problem on a timeline.

Example 4

Dr. Martin Luther King, Jr. was 34 years old in 1963 when he delivered his "I Have A Dream" speech. In what year was he born?

Solution

Understand This is an **elapsed-time problem.** We use the *l-e-d* formula. The age at which Dr. King gave his speech is the difference of the year of the speech and the year of his birth.

Plan We write an equation using *e* to stand for the year of Dr. King's birth.

$$l - e = d$$

$$1963 - e = 34$$

Solve To find the subtrahend in a subtraction problem, we subtract the difference from the minuend. Then we verify the solution by substituting into the original equation.

$$\begin{array}{r} 1963 \\ -\quad 34 \\ \hline 1929 \end{array} \longrightarrow \begin{array}{r} 1963 \\ -\ 1929 \\ \hline 34 \end{array} \text{ verify}$$

Check Now we review the question and write the answer. Dr. Martin Luther King Jr. was born in **1929.**

Practice Set

Generalize Follow the four-step method to solve each problem. Along with each answer, include the equation you used to solve the problem.

a. **Classify** The number 1,000,000,000 is how much greater than 25,000,000? What type of problem is this? Explain your reasoning.

b. How many years were there from 1215 to 1791?

c. John F. Kennedy was elected president in 1960 at the age of 43. In what year was he born?

d. **Formulate** Write a word problem about comparing that fits this equation:

$$58 \text{ in.} - 55 \text{ in.} = d$$

e. **Formulate** Write a word problem about elapsed time that fits this equation:

$$2003 - b = 14$$

Written Practice

Strengthening Concepts

* **1.** (11) In 2003, the U.S. imported one million, eight hundred seventy thousand barrels of crude oil per day. In 1988, the U.S. imported only nine hundred eleven thousand barrels per day. How many fewer barrels per day did the U.S. import in 1988?

* **2.** (11) West Street Middle School received 18 new computers for the media center. Now there are 31 computers in the media center. How many computers were there before they received the new computers?

* **3.** (12) William and the Normans conquered England in 1066. The Magna Carta was signed by King John in 1215. How many years were there from 1066 to 1215?

* **4.** (12) The Coliseum, built by the Romans in the first century A.D., could seat about 50,000 spectators. The Los Angeles Coliseum, built for the 1932 Olympics, could seat about 105,000 spectators at the time. How many fewer spectators could the Roman Coliseum seat?

* **5.** (12) **Formulate** Write a word problem about separating that fits this equation:

$$\$20.00 - c = \$7.13$$

6. (2, 9) Which properties are illustrated by these equations?

a. $\frac{1}{2} \times 1 = \frac{1}{2}$ **b.** $\frac{1}{2} \times \frac{2}{1} = 1$

* **7.** (5, 12) Twenty-three thousand is how much less than one million? Use words to write the answer.

8. (4, 8) Replace each circle with the proper comparison symbol:

a. $2 - 3 \bigcirc -1$ **b.** $\frac{1}{2} \bigcirc \frac{1}{3}$

9. (7) Name three segments in the figure below in order of length from shortest to longest.

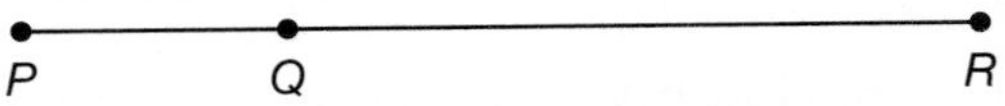

* **10.** (10) **Model** Draw and shade circles to show that $2\frac{3}{4}$ equals $\frac{11}{4}$.

* **11.** (8) **a.** What fraction of the triangle is shaded?

b. What percent of the triangle is not shaded?

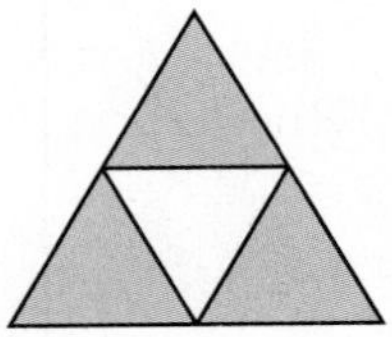

12. (6) The number 100 is divisible by which whole numbers?

Generalize Solve.

13. (3) $15x = 630$

14. (3) $y - 2714 = 3601$

15. (3) $2900 - p = 64$

16. (3) $\$1.53 + q = \$5.00$

17. (3) $20r = 1200$

18. (3) $\frac{m}{14} = 16$

Simplify:

19. (1) $\begin{array}{r} 72{,}112 \\ -\ 64{,}309 \\ \hline \end{array}$

20. (1) $\begin{array}{r} 453{,}978 \\ +\ 386{,}864 \\ \hline \end{array}$

* **21.** $\frac{8}{9} - \left(\frac{3}{9} + \frac{5}{9}\right)$ *(9)*

***22.** $\left(\frac{8}{9} - \frac{3}{9}\right) + \frac{5}{9}$ *(10)*

* **23.** $\frac{9}{2} \times \frac{3}{5}$ *(10)*

24. $37.20 ÷ 15 *(1)*

* **25.** Divide 42,847 by 9 and express the quotient as a mixed number. *(10)*

26. Justify the steps taken to simplify 25 · 36 · 4. *(2)*

25 · 36 · 4	Given
25 · 4 · 36	**a.** ________
(25 · 4) · 36	**b.** ________
100 · 36	**c.** ________
3600	**d.** ________

* **27.** **Generalize** Find the next three numbers in this sequence: *(4, 8)*

$$\frac{1}{4}, \frac{1}{2}, \frac{3}{4}, \ldots$$

* **28.** How many $\frac{2}{3}$s are in 1? *(9)*

* **29.** Write $1\frac{2}{3}$ as an improper fraction, and multiply the improper fraction by $\frac{1}{2}$. What is the product? *(10)*

* **30.** Using a ruler, draw a triangle that has two perpendicular sides, one that is $\frac{3}{4}$ in. long and one that is 1 in. long. What is the measure of the third side? *(7, 8)*

Early Finishers
Real-World Application

Patricia's grandmother was born in 1948. In 2004, her grandmother's age in years equaled her cat's age in months. What was the cat's age in years and months? Explain your thinking.

LESSON

13

• Problems About Equal Groups

Power Up

Building Power

facts Power Up C

mental math

a. **Number Sense:** \$8.00 − \$0.80

b. **Decimals:** \$25.00 ÷ 10

c. **Calculation:** \$10.00 − \$6.75

d. **Order of Operations:** (5 × 30) + (5 × 3)

e. **Positive/Negative:** 250 − 500

f. **Fractional Parts:** $\frac{1}{2}$ of 86

g. **Probability:** How many different one-topping pizzas can you make with 2 types of crust and 5 types of toppings?

h. **Calculation:** 7 × 8, + 4, ÷ 3, + 1, ÷ 3, + 8, × 2, − 3, ÷ 3

problem solving

Nelson held a standard number cube between two fingers so that he covered parallel (opposite) faces of the cube. On two of the faces Nelson could see there were 3 dots and 5 dots. How many dots were on each of the two faces his fingers covered?

New Concept

Increasing Knowledge

In this lesson we will solve word problems that have a multiplication thought pattern.

Juanita packed 25 marbles in each box. If she filled 32 boxes, how many marbles did she pack in all?

This is a problem about **equal groups.** This is the equal groups formula:

number of groups × number in each group = total

$$n \times g = t$$

To find the total, we multiply. To find an unknown factor, we divide. We will consider three examples.

Example 1

Juanita packed 25 marbles in each box. If she filled 32 boxes, how many marbles did she pack in all?

Solution

We use the four-step procedure to solve word problems.

Understand Since each box contains the same number of marbles, this problem is about **equal groups.** We use the equal groups formula.

Plan We write an equation using t for the total number of marbles. There were 32 groups with 25 marbles in each group.

$$n \times g = t$$

$$32 \times 25 = t$$

Solve To find the unknown product, we multiply the factors.

$$\begin{array}{r} 32 \\ \times\ 25 \\ \hline 160 \\ 64\ \ \\ \hline 800 \end{array}$$

Check We review the question and write the answer. Juanita packed **800 marbles** in all.

Example 2

Movie tickets were $8 each. The total ticket sales were $960. How many tickets were sold?

Solution

Thinking Skill

Generalize

Suppose some of the tickets sold for $8 and some sold for $5? What were the total ticket sales? Write an equation to solve this problem.

Understand Each ticket sold for the same price. This problem is about **equal groups of money.**

Plan We write an equation. In the equation we use n for the number of tickets. Each ticket cost $8 and the total was $960.

$$n \times g = t$$

$$n \times \$8 = \$960$$

Solve To find an unknown factor, we divide the product by the known factor. We can verify that our arithmetic is correct by substituting the answer into the original equation.

$$8\overline{)960}^{\,120} \qquad 120 \times \$8 = \$960 \quad \text{verify}$$

Check We review the question and write the answer: **120 tickets** were sold.

Example 3

Six hundred new cars were delivered to the dealer by 40 trucks. Each truck carried the same number of cars. How many cars were delivered by each truck?

Solution

Understand An equal number of cars were grouped on each truck. We can tell this problem is about equal groups because the problem states that each truck carried the same or equal number of cars.

Plan We write an equation using g to stand for the number of cars on each truck.

$$n \times g = t$$

$$40 \times g = 600$$

Solve To find an unknown factor, we divide. We test the solution by substituting into the original equation.

$$40\overline{)600}\text{ quotient } 15 \qquad 40 \times 15 = 600 \quad \text{verify}$$

Check We review the question and write the answer: **15 cars** were delivered by each truck.

Formulate Give an example of a word problem in which you would need to find g in the equation $n \times g = t$.

Practice Set

Generalize Follow the four-step method to solve each problem. Along with each answer, include the equation you use to solve the problem.

a. Beverly bought two dozen juice bars for 32¢ each. How much did she pay for all the juice bars?

b. Workers in the orchard planted 375 trees. There were 25 trees in each row they planted. How many rows of trees did the workers plant?

c. Every day Arnold did the same number of push-ups. If he did 1225 push-ups in one week, then how many push-ups did he do each day?

d. Write a word problem about equal groups that fits this equation:

$$12x = \$3.00$$

Written Practice

Strengthening Concepts

*** 1.** (12) In 1990, the population of Alabaster, Alabama was 14,619. By 2002, the population had increased to 24,877. How much greater was the population of Alabaster in 2002?

*** 2.** (10) Write $2\frac{1}{2}$ as an improper fraction. Then multiply the improper fraction by $\frac{1}{3}$. What is the product?

*** 3.** (12) President Franklin D. Roosevelt was 50 years old when he was elected in 1932. In what year was he born?

*** 4.** (13) The beach balls were packed 12 in each case. If 75 cases were delivered, how many beach balls were there in all?

*** 5.** (13) One hundred twenty poles were needed to construct the new pier. If each truckload contained eight poles, how many truckloads were needed?

*** 6.** (13) **Formulate** Write a word problem about equal groups that fits this equation. Then answer the problem.

$$5t = \$63.75$$

*** 7.** (1, 12) **Analyze** The product of 5 and 8 is how much greater than the sum of 5 and 8?

8. (8) **a.** Three quarters make up what fraction of a dollar?

b. Three quarters make up what percent of a dollar?

9. (4) How many units is it from -5 to $+5$ on the number line?

10. (7) **Classify** Describe each figure as a line, ray, or segment. Then use a symbol and letters to name each figure.

a. 

b. 

c. 

11. (6) **a.** What whole numbers are factors of both 24 and 36?

b. What is the GCF of 24 and 36?

*** 12.** (8) **a.** **Evaluate** What fractions or mixed numbers are represented by points A and B on this number line?

b. Find AB.

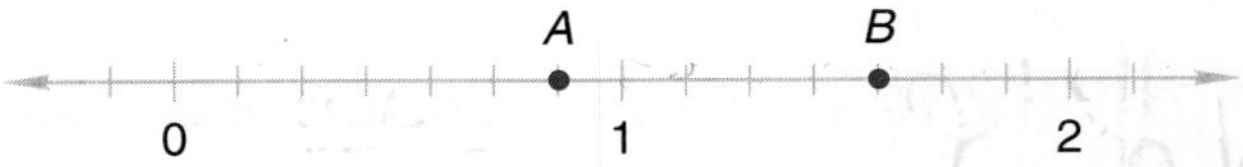

Solve.

13. (3) $36c = 1800$

14. (3) $f - \$1.64 = \$3.77$

15. (3) $\frac{d}{7} = 28$

16. (3) $\frac{4500}{e} = 30$

17. (3) $4 + 7 + 6 + 8 + 4 + 5 + 5 + 7 + 9 + 6 + n + 8 = 75$

18. (3) $3674 - a = 2159$

19. (3) $4610 + b = 5179$

Simplify:

20. (1) $363 + 4579 + 86 + 7$

21. (2) $(5 \cdot 4) \div (3 + 2)$

* **22.** (10) $\frac{5}{3} \cdot \frac{5}{2}$

* **23.** (9) $3\frac{4}{5} - \left(\frac{2}{5} + 1\frac{1}{5}\right)$

24. (1) $\frac{600}{25}$

25. (1) $\begin{array}{r} 600 \\ \times\ \ 25 \\ \hline \end{array}$

26. (2, 4) Compare: $1000 \div (100 \div 10) \bigcirc (1000 \div 100) \div 10$

* **27.** (11) **Explain** Mr. Lim bought 6 dozen eggs to make sandwiches for the school picnic. He used all but 17 eggs. How many eggs did he use? Explain how you found the answer.

* **28.** (9) **a.** What is the product of $\frac{11}{12}$ and its reciprocal?

b. What property is illustrated by the multiplication in part **a?**

Refer to the figure at right to answer exercises **29** and **30.**

* **29.** (7) **Classify** Name the obtuse, acute, and right angles.

* **30.** (7) **a.** $\overline{AB} \parallel$ ___?___

b. $\overline{AB} \perp$ ___?___

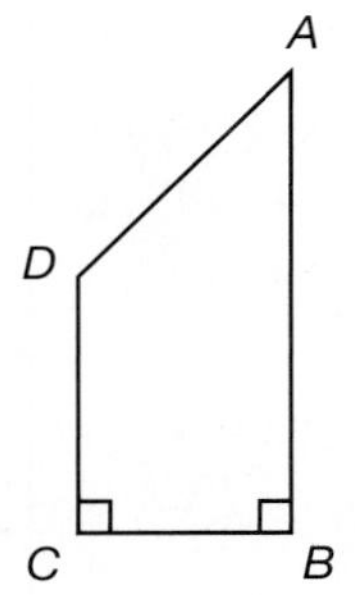

Early Finishers
Real-World Application

A small bakery wants to purchase a new oven that costs $2,530.00 plus 8.75% sales tax.

a. How much will the oven cost including tax?

b. A four-year service agreement before taxes, cost $455.40. If the bakery purchases a service agreement which is also taxed, what will be their total cost for the oven and service agreement?

LESSON 14

• Problems About Parts of a Whole
• Simple Probability

Power Up — *Building Power*

facts Power Up B

mental math

a. **Number Sense:** $7.50 − 75¢

b. **Decimals:** $0.63 × 10

c. **Calculation:** $10.00 − $8.25

d. **Order of Operations:** (6 × 20) + (6 × 4)

e. **Number Sense:** 625 − 500

f. **Fractional Parts:** $\frac{1}{2}$ of 36

g. **Probability:** How many different ways can you arrange the numbers 3, 5, 7?

h. **Calculation:** Start with three dozen, ÷ 2, + 2, ÷ 2, + 2, ÷ 2, + 2, ÷ 2, + 2, ÷ 2.

problem solving A local hardware store charges $2.50 per straight cut. How much will it cost to cut a plank of hardwood into two pieces? ... into four pieces? ... into six pieces? Each cut must be perpendicular to the length of the plank.

New Concepts — *Increasing Knowledge*

problems about parts of a whole

Problems about **parts of a whole** have an addition thought pattern.

$$\text{part} + \text{part} = \text{whole}$$

$$a + b = w$$

Sometimes the parts are expressed as fractions or percents.

Example 1

One third of the students attended the game. What fraction of the students did not attend the game?

Solution

We are not given the number of students. We are given only the fraction of students in the whole class who attended the game. The following model can help us visualize the problem:

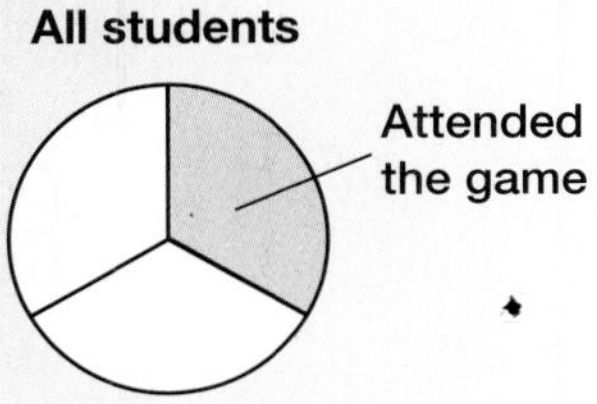

Understand This problem is about **part of a whole.** We are given the size of one part and we are asked to find the remaining part.

Plan We write an equation for the given information. It may seem as though we are given only one number, $\frac{1}{3}$, but the model reminds us that $\frac{3}{3}$ is all the students.

$$a + b = w$$

$$\frac{1}{3} + b = \frac{3}{3}$$

Thinking Skill

Justify

Why do we subtract to find the unknown number?

Solve We find the unknown number, b, by subtracting. We test our answer in the original equation.

$$\begin{array}{r} \frac{3}{3} \\ -\frac{1}{3} \\ \hline \frac{2}{3} \end{array} \qquad \rightarrow \qquad \begin{array}{rl} \frac{1}{3} & \text{attended the game} \\ +\frac{2}{3} & \text{did not attend the game} \\ \hline \frac{3}{3} & \text{total students} \end{array} \qquad \text{verify}$$

Check We review the question and write the answer. **Two thirds of the students did not attend the game.**

Example 2

Melisenda's science beaker is 61% full. What percent of Melisenda's beaker is empty?

Solution

Understand Part of Melisenda's science beaker is full, and part of it is empty. This problem is about **part of a whole.**

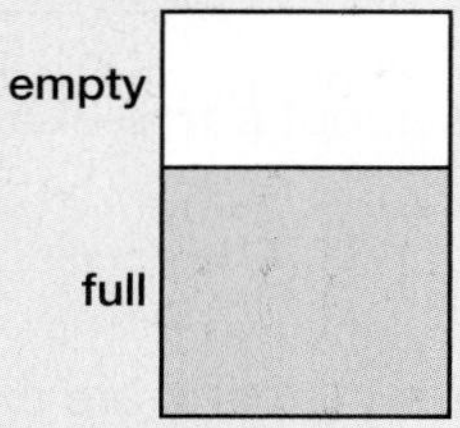

Plan We write an equation. The whole beaker is represented by 100%. We use b to stand for the percent of the beaker that is empty.

$$61\% + b = 100\%$$

Solve We can find the missing number, b, by subtracting.

$$100\% - 61\% = 39\%$$

$$61\% + 39\% = 100\% \quad \text{verify}$$

Check We review the question and write the answer. **39% of Melisenda's beaker is empty.**

Discuss How can we check our work?

simple probability

Thinking Skill

Connect

How do you write 0, $\frac{1}{2}$, and 1 as a decimal and percent?

Probability is the likelihood that a particular event will occur. We express the probability of an event occurring using the numbers from 0 through 1. The numbers between 0 and 1 can be written as fractions, decimals, or percents.

- A probability of 0 represents an event that cannot occur or is *impossible*.
- A probability of 1 represents an event that is *certain* to occur.
- A probability of $\frac{1}{2}$ represents an event that is *equally likely* to occur as to not occur.
- A probability less than $\frac{1}{2}$ means the event is *unlikely* to occur.
- A probability greater than $\frac{1}{2}$ means the event is *likely* to occur.

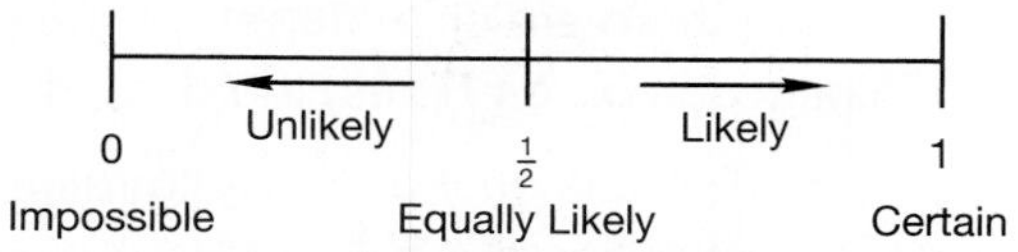

We can use this formula to find the probability of an event occurring.

$$\text{Probability (Event)} = \frac{\text{number of favorable outcomes}}{\text{total number of possible outcomes}}$$

Suppose we have a bag that contains 4 red marbles and 5 blue marbles. We want to find the probability of picking one marble of a specific color from the bag without looking.

Reading Math

The symbols P(Red) and P(R) are both read as "the probability of red."

Generalize How can we use the formula to find the probability of picking each color?

$$P(\text{Red}) = \frac{\text{number of red marbles}}{\text{total number of marbles}}$$

$$P(\text{R}) = \frac{4}{9}$$

We find that the probability of picking red is $\frac{4}{9}$.

$$P(\text{Blue}) = \frac{\text{number of blue marbles}}{\text{total number of marbles}}$$

$$P(\text{B}) = \frac{5}{9}$$

We find that the probability of picking blue is $\frac{5}{9}$.

We can also write the probability of picking a green marble from this bag of marbles.

$$P(\text{Green}) = \frac{\text{number of green marbles}}{\text{total number of marbles}} = \frac{0}{9} = 0$$

The probability of picking green is 0.

Conclude Why is P(G) equal to 0?

Example 3

This number cube has 1 through 6 dots on the faces of the cube. If the number cube is rolled once, what is the probability of each of these outcomes?

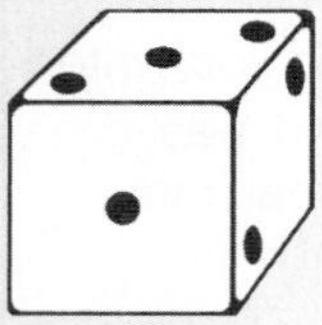

a. rolling a 4

b. rolling a number greater than 4

c. rolling a number greater than 6

d. rolling a number less than 7

Solution

Since there are six different faces on the number cube, there are six *equally* likely outcomes. Thus, there are six possible outcomes.

a. There is only one way to roll a 4 with the number cube. The probability of rolling a 4 is $\frac{1}{6}$.

b. The numbers greater than 4 on the number cube are 5 and 6, so there are two ways to roll a number greater than 4. The probability of rolling a number greater than 4 is $\frac{2}{6}$.

c. There are no numbers greater than 6 on the number cube. So it is impossible to roll a number greater than 6. The probability of rolling a number greater than 6 is $\frac{0}{6}$ **or 0.**

d. There are six numbers less than 7 on the number cube. So there are six ways to roll a number less than 7. The probability of rolling a number less than 7 is $\frac{6}{6}$ **or 1.**

Example 4

This spinner is divided into five equal sectors and is numbered 1 through 5. The arrow is spun once.

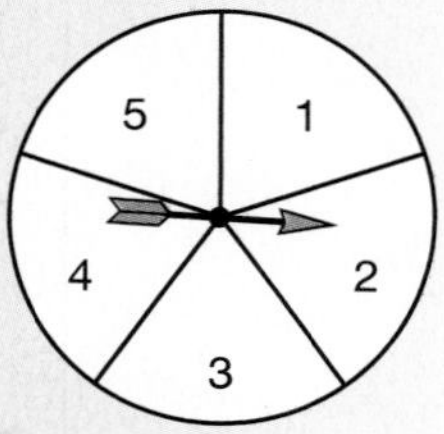

a. How many different outcomes are possible?

b. What is the probability of spinning a 3?

c. What is the probability of *not* spinning a 3?

d. What is the probability of spinning a 3 on the first spin and a 2 on the second spin?

Solution

The probability that the spinner will stop in a given sector is equal to the fraction of the spinner's face occupied by that sector.

a. There are **five** equally likely outcomes when spinning this spinner.

b. Spinning a 3 is one of five equally likely outcomes. We can use the formula to find the probability of spinning a 3.

$$P(3) = \frac{\text{number of favorable outcomes}}{\text{total number of possibles outcomes}}$$

$$P(3) = \frac{1}{5}$$

The probability of spinning a 3 is $\frac{1}{5}$.

c. We can also use the formula to find the probability of not spinning a 3. There are four ways for the spinner *not* to stop on 3.

$$P(\text{not } 3) = \frac{\text{number of favorable outcomes}}{\text{total number of possibles outcomes}}$$

$$P(\text{not } 3) = \frac{4}{5}$$

The probability of not spinning a 3 is $\frac{4}{5}$.

Notice that the sum of the probability of an event occurring plus the probability of the event *not* occurring is 1.

$$P(3) + P(\text{not } 3) = 1$$

$$\frac{1}{5} + \frac{4}{5} = \frac{5}{5} \text{ or } 1$$

When the sum of the probabilities of two events is equal to 1, they are called **complementary events.**

d. We begin by writing all the possible outcomes of two spins. The number on the left is the first spin and the number on the right is the second spin:

1, 1	2, 1	3, 1	4, 1	5, 1
1, 2	2, 2	3, 2	4, 2	5, 2
1, 3	2, 3	3, 3	4, 3	5, 3
1, 4	2, 4	3, 4	4, 4	5, 4
1, 5	2, 5	3, 5	4, 5	5, 5

There are 25 total possible outcomes and one favorable outcome (3, 2), so the probability of spinning a 3 then a 2 is $\frac{1}{25}$.

Example 5

The spinner at the right is divided into one half and two fourths. What is the probability of the spinner stopping on 3?

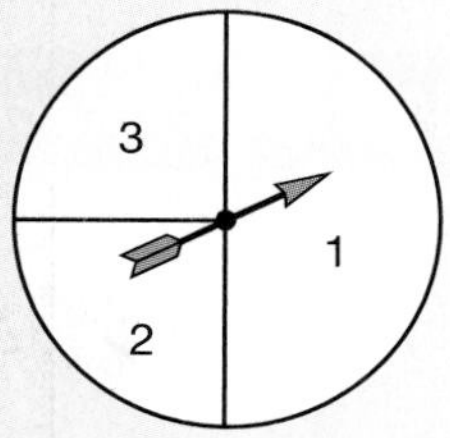

Solution

There are three possible outcomes, but the outcomes are *not* equally likely, because the sizes of the regions are not all equal.

- Since Region 1 is one half of the whole area, the probability of the spinner stopping on 1 is $\frac{1}{2}$.

- Regions 2 and 3 each represent $\frac{1}{4}$ of the whole area. The probability of the spinner stopping on 2 is $\frac{1}{4}$, and the probability of it stopping on 3 is also $\frac{1}{4}$.

Generalize The probability of the spinner stopping on 2 or 3 is $\frac{1}{2}$. What is the complement of this event?

Practice Set

Along with each answer, include the equation you used to solve the problem.

a. Only 39% of the lights were on. What percent of the lights were off?

b. Two fifths of the students did not go to the museum. What fraction of the students did go to the museum?

c. Write a word problem about parts of a whole that fits this equation:

$$45\% + g = 100\%$$

d. Rolling a number cube once, what is the probability of rolling a number less than 4?

Analyze This spinner is divided into four equal sections.

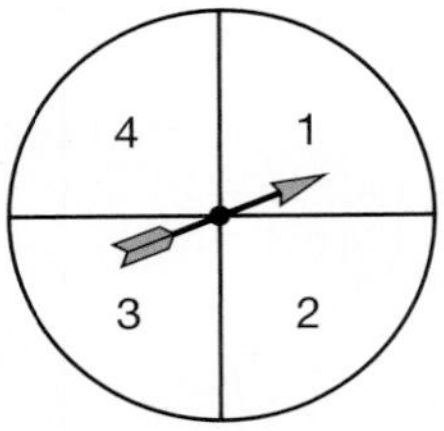

e. What is the probability of this spinner stopping on 3 and then on 1 on the second spin?

f. What is the probability of this spinner stopping on 5?

g. What is the probability of this spinner stopping on a number less than 6?

This spinner is divided into one half and two fourths.

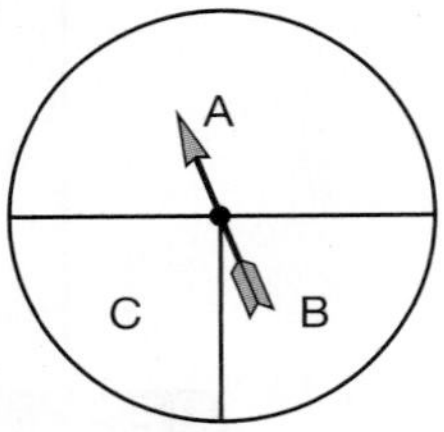

h. What is the probability of this spinner stopping on A?

i. What is the probability of this spinner not stopping on B?

Written Practice

Strengthening Concepts

* **1.** (11) The USDA recommends that adults eat at least 85 grams of whole grain products each day. Ryan ate 63 grams of whole-grain cereal. How many more grams of whole grain products should he eat?

* **2.** (14) Seven tenths of the new recruits did not like their first haircut. What fraction of the new recruits did like their first haircut?

* **3.** (12) The Declaration of Independence was signed in 1776. The U.S. Constitution was ratified in 1789. How many years passed between these two events?

* **4.** (13) **Formulate** Write a word problem that fits this equation:

$$12p = \$2.40$$

* **5.** (14) In 2000, nearly 18% of cars sold in North America were silver. What percent of cars sold were not silver?

* **6.** (10) **Model** Draw and shade circles to show that $3\frac{1}{3} = \frac{10}{3}$.

7. (5) Use digits to write four hundred seven million, forty-two thousand, six hundred three.

8. (2) **Analyze** What property is illustrated by this equation?

$$3 \cdot 2 \cdot 1 \cdot 0 = 0$$

9. (6) **a.** List the common factors of 40 and 72.

b. What is the greatest common factor of 40 and 72?

10. (7) Name three segments in the figure below in order of length from shortest to longest.

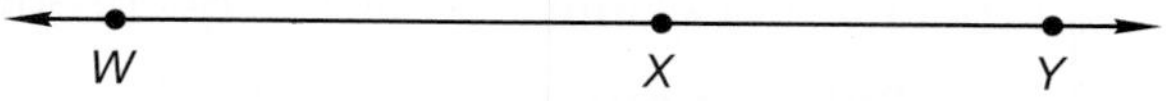

11. (8) Describe how to find the fraction of the group that is shaded.

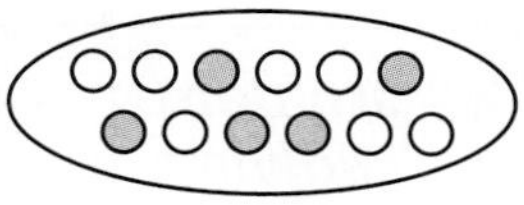

Solve:

12. (3) $b - 407 = 623$

13. (3) $\$20 - e = \$3.47$

14. (3) $7 \cdot 5f = 7070$

15. (3) $\frac{m}{25} = 25$

16. (3)
$$\begin{array}{r} 5 \\ 8 \\ 7 \\ 6 \\ 5 \\ 9 \\ 4 \\ 3 \\ 6 \\ 4 \\ 7 \\ 8 \\ 5 \\ n \\ +\ 6 \\ \hline 89 \end{array}$$

17. (3) $a + 295 = 1000$

Simplify:

* **18.** (10) $3\frac{3}{5} + 2\frac{4}{5}$

* **19.** (10) $\frac{5}{2} \cdot \frac{3}{2}$

20. (1) $3.63 + $0.87 + 96¢

21. (2) $5 \cdot 4 \cdot 3 \cdot 2 \cdot 1$

* **22.** (9) $\frac{2}{3} \cdot \frac{2}{3} \cdot \frac{2}{3}$

23. (1) $\frac{900}{20}$

24. (1)
$$\begin{array}{r} 145 \\ \times\ 74 \\ \hline \end{array}$$

25. (1) 30(65¢)

26. (2) $(5)(5 + 5)$

27. (4) $9714 - 13{,}456$

28. (7) **Classify** Name each type of angle illustrated:

a.
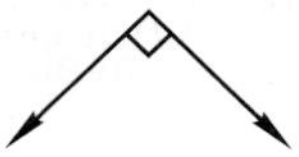

b.

c.
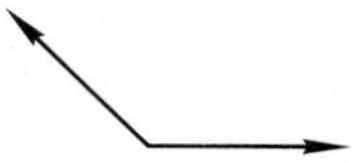

* **29.** (9) How many $\frac{4}{5}$s are in 1?

30. (14) Rolling a number cube once, what is the probability of rolling a number greater than 4?

LESSON 15

• Equivalent Fractions
• Reducing Fractions, Part 1

Power Up — *Building Power*

facts Power Up C

mental math

a. Calculation: \$3.50 + \$1.75

b. Decimals: \$4.00 ÷ 10

c. Calculation: \$10.00 − \$4.98

d. Order of Operations: (7 × 30) + (7 × 2)

e. Number Sense: 125 − 50

f. Fractional Parts: $\frac{1}{2}$ of 52

g. Measurement: Convert 8 quarts into gallons

h. Calculation: 10, − 9, + 8, − 7, + 6, − 5, + 4, − 3, + 2, − 1

problem solving

Copy the problem and fill in the missing digits:

```
  _37_
− 2_65
  59_7
```

New Concepts — *Increasing Knowledge*

equivalent fractions

Different fractions that name the same number are called **equivalent fractions.** Here we show four equivalent fractions:

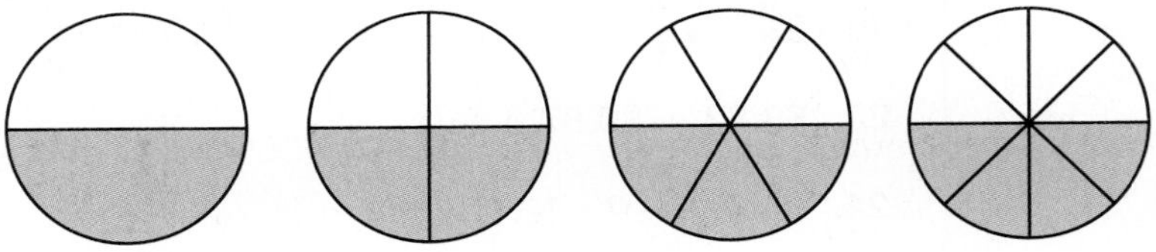

$\frac{1}{2}$ $\frac{2}{4}$ $\frac{3}{6}$ $\frac{4}{8}$

As we can see from the models, **equivalent fractions have the same value.**

$$\frac{1}{2} = \frac{2}{4} = \frac{3}{6} = \frac{4}{8}$$

Thinking Skill

Summarize

In your own words, state the Identity Property of Multiplication.

Recall the Identity Property of Multiplication, $a \times 1 = a$. We can form equivalent fractions by multiplying a fraction by fractions equal to 1. Here we multiply $\frac{1}{2}$ by $\frac{2}{2}$, $\frac{3}{3}$, and $\frac{4}{4}$ to form fractions equivalent to $\frac{1}{2}$:

$$\frac{1}{2} \times \frac{2}{2} = \frac{2}{4} \qquad \frac{1}{2} \times \frac{3}{3} = \frac{3}{6} \qquad \frac{1}{2} \times \frac{4}{4} = \frac{4}{8}$$

Model Use fraction manipulatives to model three other fractions that are equivalent to $\frac{1}{2}$. How did you form them?

Example 1

Find an equivalent fraction for $\frac{2}{3}$ that has a denominator of 12.

Solution

Thinking Skill

Justify

How do we know that we should multiply $\frac{2}{3}$ by $\frac{4}{4}$?

The denominator of $\frac{2}{3}$ is 3. To make an equivalent fraction with a denominator of 12, we multiply by $\frac{4}{4}$, which is a name for 1.

$$\frac{2}{3} \times \frac{4}{4} = \mathbf{\frac{8}{12}}$$

Example 2

Find a fraction equivalent to $\frac{1}{3}$ that has a denominator of 6. Next find a fraction equivalent to $\frac{1}{2}$ with a denominator of 6. Then add the two fractions you found.

Solution

We multiply $\frac{1}{3}$ by $\frac{2}{2}$ and $\frac{1}{2}$ by $\frac{3}{3}$ to find the fractions equivalent to $\frac{1}{3}$ and $\frac{1}{2}$ that have denominators of 6. Then we add.

$$\begin{array}{r} \frac{1}{3} \times \frac{2}{2} = \mathbf{\frac{2}{6}} \\ + \frac{1}{2} \times \frac{3}{3} = \mathbf{\frac{3}{6}} \\ \hline \mathbf{\frac{5}{6}} \end{array}$$

Explain Why do we need to find equivalent fractions with denominators of 6 before we can add $\frac{1}{3}$ and $\frac{1}{2}$?

reducing fractions, part 1

An inch ruler provides another example of equivalent fractions. The segment in the figure below is $\frac{1}{2}$ inch long. By counting the tick marks on the ruler, we see that there are several equivalent names for $\frac{1}{2}$ inch.

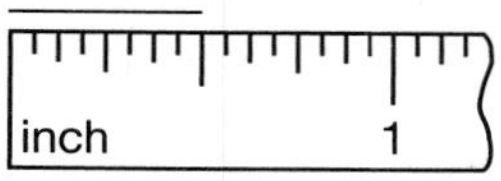

$$\frac{1}{2} \text{ in.} = \frac{2}{4} \text{ in.} = \frac{4}{8} \text{ in.} = \frac{8}{16} \text{ in.}$$

Math Language

The term **reduce** means *to rewrite a fraction in lowest terms.* Reducing a fraction does not change its value.

Math Language

A **factor** is a whole number that divides another whole number without a remainder.

We say that the fractions $\frac{2}{4}$, $\frac{4}{8}$, and $\frac{8}{16}$ each **reduce** to $\frac{1}{2}$. We can reduce some fractions by dividing the fraction to be reduced by a fraction equal to 1.

$$\frac{4}{8} \div \frac{4}{4} = \frac{1}{2} \qquad \begin{matrix}(4 \div 4 = 1)\\(8 \div 4 = 2)\end{matrix}$$

By dividing $\frac{4}{8}$ by $\frac{4}{4}$, we have reduced $\frac{4}{8}$ to $\frac{1}{2}$.

The numbers we use when we write a fraction are called the **terms** of the fraction. To reduce a fraction, we divide both terms of the fraction by a factor of both terms.

$$\frac{4 \div 2}{8 \div 2} = \frac{2}{4} \qquad \frac{4 \div 4}{8 \div 4} = \frac{1}{2}$$

Dividing each term of $\frac{4}{8}$ by 4 instead of by 2 results in a fraction with lower terms, since the terms of $\frac{1}{2}$ are lower than the terms of $\frac{2}{4}$. It is customary to reduce fractions to **lowest terms.** As we see in the next example, fractions can be reduced to lowest terms in one step by dividing the terms of the fraction by the greatest common factor of the terms.

Example 3

Reduce $\frac{18}{24}$ to lowest terms.

Solution

Both 18 and 24 are divisible by 2, so we divide both terms by 2.

$$\frac{18}{24} = \frac{18 \div 2}{24 \div 2} = \frac{9}{12}$$

This is not in lowest terms, because 9 and 12 are divisible by 3.

$$\frac{9}{12} = \frac{9 \div 3}{12 \div 3} = \mathbf{\frac{3}{4}}$$

We could have used just one step had we noticed that the greatest common factor of 18 and 24 is 6.

$$\frac{18}{24} = \frac{18 \div 6}{24 \div 6} = \mathbf{\frac{3}{4}}$$

Both methods are correct. One method took two steps, and the other took just one step.

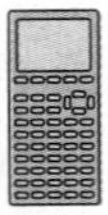

Visit www.SaxonPublishers.com/ActivitiesC2 *for a graphing calculator activity.*

Example 4

Reduce $3\frac{8}{12}$ to lowest terms.

Solution

To reduce a mixed number, we reduce the fraction and leave the whole number unchanged.

$$\frac{8}{12} = \frac{8 \div 4}{12 \div 4} = \frac{2}{3}$$

$$3\frac{8}{12} = \mathbf{3\frac{2}{3}}$$

Example 5

Write $\frac{12}{9}$ as a mixed number with the fraction reduced.

Solution

There are two steps to reduce and convert to a mixed number. Either step may be taken first.

Reduce First	**Convert First**
Reduce: $\frac{12}{9} = \frac{4}{3}$	Convert: $\frac{12}{9} = 1\frac{3}{9}$
Convert: $\frac{4}{3} = \mathbf{1\frac{1}{3}}$	Reduce: $1\frac{3}{9} = \mathbf{1\frac{1}{3}}$

Discuss Which method, reduce first or convert first, do you prefer in the example above and why? Can you think of an example for which you would prefer the other method?

Example 6

Simplify: $\frac{7}{9} - \frac{1}{9}$

Solution

First we subtract. Then we reduce.

Subtract	**Reduce**
$\frac{7}{9} - \frac{1}{9} = \frac{6}{9}$	$\frac{6 \div 3}{9 \div 3} = \mathbf{\frac{2}{3}}$

Example 7

Write 70% as a reduced fraction.

Solution

Recall that a percent is a fraction with a denominator of 100.

$$70\% = \frac{70}{100}$$

We can reduce the fraction by dividing each term by 10.

$$\frac{70}{100} \div \frac{10}{10} = \mathbf{\frac{7}{10}}$$

Example 8

With one spin, what is the probability that the spinner will stop on 4?

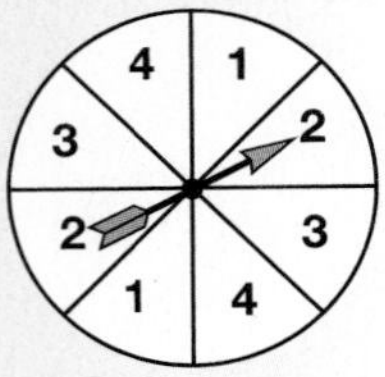

Solution

Two of the eight equally likely outcomes are 4. So the probability of 4 is $\frac{2}{8}$. We reduce $\frac{2}{8}$ to $\frac{1}{4}$.

Analyze What information do we need to know to find the probability of the spinner stopping on a specific part of an equally divided circle?

Practice Set

a. Form three equivalent fractions for $\frac{3}{4}$ by multiplying by $\frac{5}{5}$, $\frac{7}{7}$, and $\frac{3}{3}$.

b. Find an equivalent fraction for $\frac{3}{4}$ that has a denominator of 16.

Find the number that makes the two fractions equivalent.

c. $\frac{4}{5} = \frac{?}{20}$ **d.** $\frac{3}{8} = \frac{9}{?}$

e. Find a fraction equivalent to $\frac{3}{5}$ that has a denominator of 10. Next find a fraction equivalent to $\frac{1}{2}$ with a denominator of 10. Then subtract the second fraction you found from the first fraction.

Justify Reduce each fraction to lowest terms. Tell if you used the greatest common factor to reduce each fraction and how you know it is the greatest common factor.

f. $\frac{3}{6}$ **g.** $\frac{8}{10}$ **h.** $\frac{8}{16}$ **i.** $\frac{12}{16}$

j. $4\frac{4}{8}$ **k.** $6\frac{9}{12}$ **l.** $12\frac{8}{15}$ **m.** $8\frac{16}{24}$

Generalize Perform each indicated operation and reduce the result:

n. $\frac{5}{12} + \frac{5}{12}$ **o.** $3\frac{7}{10} - 1\frac{1}{10}$ **p.** $\frac{5}{8} \cdot \frac{2}{3}$

Write each percent as a reduced fraction:

q. 90% **r.** 75% **s.** 5%

t. Find a fraction equivalent to $\frac{2}{3}$ that has a denominator of 6. Subtract $\frac{1}{6}$ from the fraction you found and reduce the answer.

u. What is the probability of rolling an even number with one roll of a 1–6 number cube?

Written Practice *Strengthening Concepts*

* **1.** (12) **Connect** Mr. Chong celebrated his seventy-fifth birthday in 1998. In what year was he born?

* **2.** (14) **a.** What is the probability of not spinning a 4 with one spin?

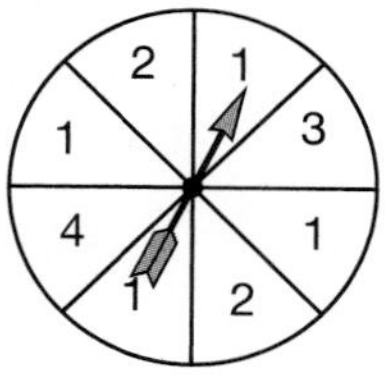

b. **Explain** Describe how you found the answer to part **a.**

* **3.** (15) If 40% of all the citizens voted "No" on the ballot, what fraction of all of the citizens voted "No"?

* **4.** (13) The farmer harvested 9000 bushels of grain from 60 acres. The crop produced an average of how many bushels of grain for each acre?

5. (8) **Evaluate** With a ruler, draw a segment $2\frac{1}{2}$ inches long. Draw a second segment $1\frac{7}{8}$ inches long. The first segment is how much longer than the second segment?

6. (4) **Represent** Use digits and symbols to write "The product of three and five is greater than the sum of three and five."

7. (6) List the single-digit divisors of 2100.

* **8.** (15) Reduce each fraction or mixed number:

a. $\frac{6}{8}$ **b.** $2\frac{6}{10}$

* **9.** (15) **Analyze** Find three equivalent fractions for $\frac{2}{3}$ by multiplying by $\frac{3}{3}$, $\frac{5}{5}$, and $\frac{6}{6}$. What property of multiplication do we use to find equivalent fractions?

* **10.** (15) For each fraction, find an equivalent fraction that has a denominator of 20:

a. $\frac{3}{5}$ **b.** $\frac{1}{2}$ **c.** $\frac{3}{4}$

11. (7) Refer to this figure to answer **a–c:**

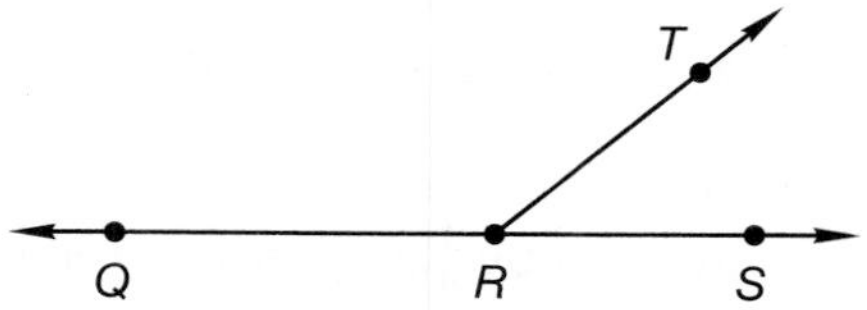

a. Name the line.

b. Name three rays originating at point *R*.

c. Name an acute angle.

12. (10) Convert each fraction to either a whole number or a mixed number:

a. $\frac{11}{3}$ **b.** $\frac{12}{3}$ **c.** $\frac{13}{3}$

13. Compare: (11)(6 + 7) ◯ 66 + 77
(4)

Solve:

14. $39 + b = 50$
(3)

15. $6a = 300$
(3)

16. $c - \$5 = 5¢$
(3)

17. $\frac{w}{35} = 35$
(3)

* **18.** Write each percent as a reduced fraction:
(15)

a. 80%

b. 35%

c. **Model** Sketch and shade a circle to show your answer to **a.**

19. How many $\frac{1}{8}$s are in 1?
(9)

* **20.** **Justify** Name the four properties used to simplify the expression.
(2, 9)

$\frac{3}{4} \cdot \frac{5}{6} \cdot \frac{4}{3}$ Given

$\frac{3}{4} \cdot \frac{4}{3} \cdot \frac{5}{6}$ **a.** ____________

$\left(\frac{3}{4} \cdot \frac{4}{3}\right) \cdot \frac{5}{6}$ **b.** ____________

$1 \cdot \frac{5}{6}$ **c.** ____________

$\frac{5}{6}$ **d.** ____________

Generalize Simplify:

21. $\frac{2}{5} + \frac{3}{5} + \frac{4}{5}$
(10)

22. $3\frac{5}{8} - 1\frac{3}{8}$
(15)

23. $\frac{4}{3} \cdot \frac{3}{4}$
(9)

24. $\frac{3}{4} + \frac{3}{4}$
(15)

25. $\frac{7}{5} + \frac{8}{5}$
(10)

* **26.** $\frac{11}{12} - \frac{1}{12}$
(15)

* **27.** $\frac{5}{6} \cdot \frac{2}{3}$
(15)

28. Evaluate each expression for $a = 4$ and $b = 8$:
(1, 9)

a. $\frac{a}{b} + \frac{a}{b}$

b. $\frac{a}{b} - \frac{a}{b}$

29. Find a fraction equal to $\frac{1}{3}$ that has a denominator of 6. Add the fraction to $\frac{1}{6}$ and reduce the answer.
(15)

* **30.** **Evaluate** Write $2\frac{2}{3}$ as an improper fraction. Then multiply the improper fraction by $\frac{1}{4}$ and reduce the product.
(15)

LESSON 16

• U.S. Customary System
• Function Tables

Power Up — *Building Power*

facts Power Up D

mental math

a. **Positive/Negative:** $10 - 20$

b. **Decimals:** $15¢ \times 10$

c. **Number Sense:** $\$1.00 - 18¢$

d. **Calculation:** 4×23

e. **Number Sense:** $875 - 750$

f. **Fractional Parts:** $\frac{1}{2}$ of $\frac{1}{3}$

g. **Algebra:** If $x = 3$, what does $2x$ equal?

h. **Calculation:** Start with 2 score and 10, $\div 2$, $\times 3$, -3, $\div 9$, $+2$, $\div 5$.

problem solving Find the next four numbers in this sequence: $\frac{1}{16}, \frac{1}{8}, \frac{3}{16}, \frac{1}{4}, \ldots$

New Concepts — *Increasing Knowledge*

U.S. customary system

In this lesson we will consider units of the **U.S. Customary System.** We can measure an object's dimensions, weight, volume, or temperature. Each type of measurement has a set of units. We should remember common equivalent measures and have a "feel" for the units so that we can estimate measurements reasonably.

The following table shows the common weight equivalences in the U.S. Customary System:

Units of Weight
16 ounces (oz) = 1 pound (lb)
2000 pounds = 1 ton (tn)

Example 1

Suppose a pickup truck can carry a load of $\frac{1}{2}$ of a ton. How many pounds can the pickup truck carry?

Solution

One ton is 2000 pounds, so $\frac{1}{2}$ of a ton is **1000 pounds.**

The following table shows the common length equivalences in the U.S. Customary System:

Units of Length		
12 inches (in.)	=	1 foot (ft)
3 feet	=	1 yard (yd)
1760 yards	=	1 mile (mi)
5280 feet	=	1 mile

Example 2

One yard is equal to how many inches?

Solution

One yard equals 3 feet. One foot equals 12 inches. Thus 1 yard is equal to 36 inches.

$$1 \text{ yard} = 3 \times 12 \text{ inches} = \mathbf{36 \text{ inches}}$$

Example 3

A mountain bicycle is about how many feet long?

Solution

We should develop a feel for various units of measure. Most mountain bicycles are about $5\frac{1}{2}$ feet long, so a good estimate would be **about 5 or 6 feet.**

Validate Without measuring a mountain bicycle, how could we determine that this estimate is reasonable?

Just as an inch ruler is divided successively in half, so units of liquid measure are divided successively in half. Half of a gallon is a half gallon. Half of a half gallon is a quart. Half of a quart is a pint. Half of a pint is a cup.

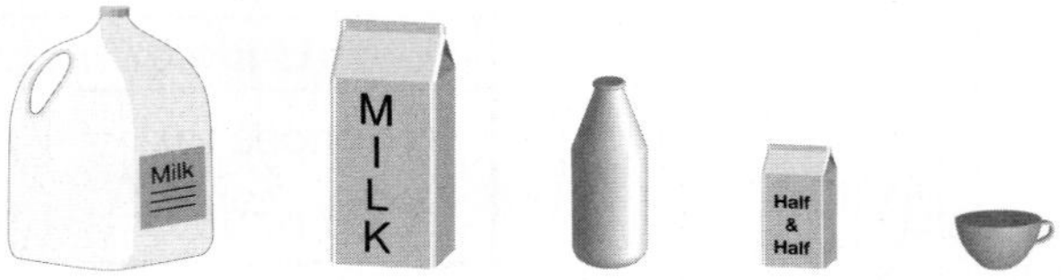

1 gallon | $\frac{1}{2}$ gallon | 1 quart | 1 pint | 1 cup

The capacity of each container above is half the capacity of the next larger container (the container to its left).

The following chart shows some equivalent liquid measures in the U.S. Customary System.

Units of Liquid Measure		
8 ounces (oz)	=	1 cup (c)
2 cups	=	1 pint (pt)
2 pints	=	1 quart (qt)
4 quarts	=	1 gallon (gal)

Example 4

Steve drinks at least 8 cups of water every day. How many quarts of water does he drink a day?

Solution

Two cups is a pint, so 8 cups is 4 pints. Two pints is a quart, so 4 pints is 2 quarts. Steve drinks at least **2 quarts** of water every day.

Reading Math

The symbol ° means *degrees.* Read 32°F as "thirty-two degrees Fahrenheit."

The following diagram shows important benchmark temperatures in the U.S. Customary System.

Fahrenheit Temperature Scale

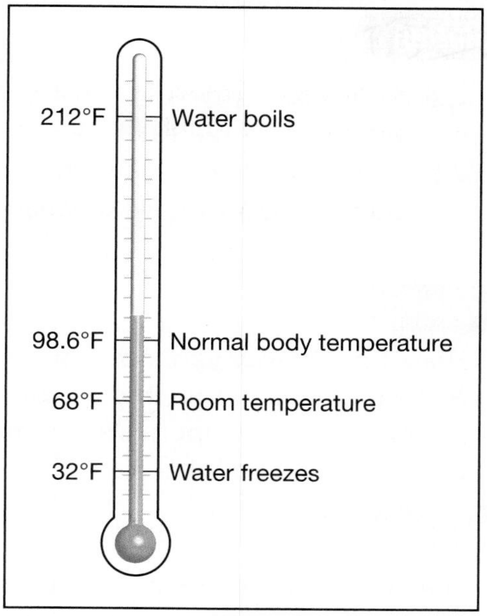

Example 5

How many Fahrenheit degrees are between the freezing and boiling temperatures of water?

Solution

212°F − 32°F = **180°F**

function tables

A **function** is a mathematical rule that identifies the relationship between two sets of numbers. The rule uses an input number to generate an output number. For each input number, there is one and only one output number.

The rule for a function may be described with words or in an equation, and the relationship between the sets of numbers may be illustrated in tables or graphs.

In this lesson we will use function tables to help us solve problems. We will also study function tables to discover the rules of functions.

Example 6

This table shows the weight in ounces for a given weight in pounds.

INPUT Pounds	OUTPUT Ounces
1	16
2	32
3	48
4	64
5	80

Thinking Skill

Connect

How is the rule for a function similar to a rule for a number pattern?

a. Describe the rule of this function.

b. Mattie weighed 7 pounds when she was born. Use the function rule to find how many ounces Mattie weighed when she was born.

Solution

a. To find the number of ounces (the output), **multiply the number of pounds (the input) by 16.**

b. To find Mattie's birth weight in ounces, we multiply 7 (her birth weight in pounds) by 16. Mattie weighed **112 ounces** at birth.

Example 7

Hana bought four yards of fabric to make a costume for a school play. Make a function table that shows the number of feet (output) for a given number of yards (input). Use the function table to find the number of feet of fabric Hana bought.

Solution

The number of feet Hana bought depends on the numbers of yards. We can say that the number of feet is a *function* of the number of yards. To solve the problem, we can set up a table to record input and output numbers for this function. Each yard equals three feet. So the input is yards and the output is feet.

INPUT Yards	OUTPUT Feet
1	3
2	6
3	9
4	12

In the function table we see 12 feet paired with 4 yards. Hana bought **12 feet** of fabric.

Conclude What is the rule for this function?

Practice Set

a. A typical door may be about how many feet tall?

b. How many quarts are in a half-gallon?

c. *Estimate* When Alberto was born, he weighed 8 lb. 7 oz. Is that weight closer to 8 lb or 9 lb?

d. How many ounces are in a 2-cup measure?

e. Both pots are filled with water. What is the temperature difference, in degrees Fahrenheit, between the two pots of water?

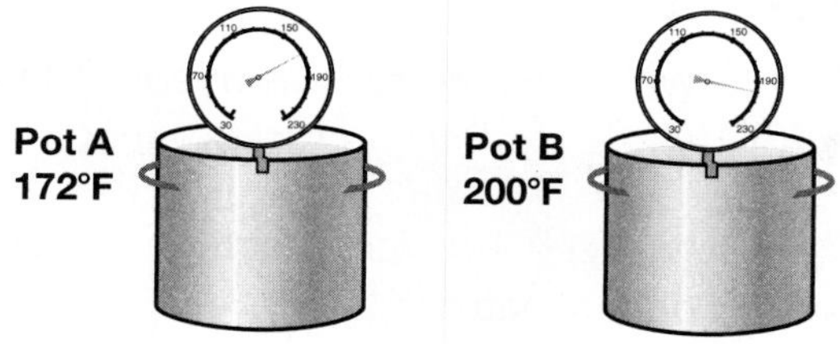

Simplify.

f. $\frac{3}{8}$ in. + $\frac{5}{8}$ in.

g. 32°F + 180°F

h. 2(3 ft + 4 ft)

i. 1 ton − 1000 pounds

j. *Analyze* A sheet of plywood is 4 feet wide. Copy and complete the function table to determine the width, in inches, of a sheet of plywood.

How wide is a sheet of plywood?

What is the rule for this function?

INPUT Feet	OUTPUT Inches
1	12
2	
3	
4	

Written Practice

Strengthening Concepts

* **1.** (14) Forty-four of the one hundred ninety-three flags of countries around the world do not feature the color red. How many flags do feature red?

* **2.** (13) At Henry's egg ranch 18 eggs are packaged in each carton. How many cartons would be needed to package 4500 eggs?

* **3.** (16) *Represent* Make a function table for the relationship between cartons and eggs described in Problem 2. In the table show the number of eggs (output) in 1, 2, 3, 4, and 5 cartons (input).

* **4.** (14) **Explain** If a coin is flipped once, which outcome is more likely, heads or tails? Explain your answer.

* **5.** (10, 15) Replace each circle with the proper comparison symbol:

a. $\frac{8}{10} \bigcirc \frac{4}{5}$ **b.** $\frac{8}{5} \bigcirc 1\frac{2}{5}$

6. (8) Use an inch ruler to find *AB, CB,* and *CA* to the nearest sixteenth of an inch.

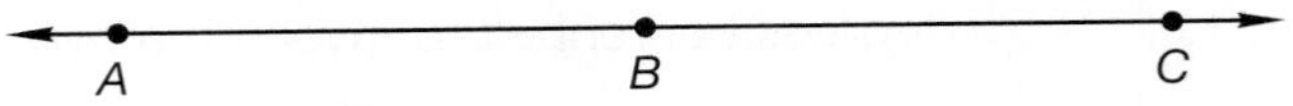

* **7.** (15) Write each number as a reduced fraction or mixed number:

a. $\frac{8}{12}$ **b.** 40% **c.** $6\frac{10}{12}$

* **8.** (15) **Analyze** For each fraction, find an equivalent fraction that has a denominator of 24:

a. $\frac{5}{6}$ **b.** $\frac{3}{8}$ **c.** $\frac{1}{4}$

* **9.** (8) **Analyze** Arrange these fractions in order from least to greatest. Explain the steps you used to order the fractions.

$$\frac{5}{6}, \frac{5}{8}, \frac{3}{4}$$

* **10.** (16) **a.** What percent of a yard is a foot?

b. What fraction of a gallon is a quart?

11. (6) The number 630 is divisible by which single-digit numbers?

12. (10) Convert each improper fraction to either a whole number or a mixed number:

a. $\frac{16}{7}$ **b.** $3\frac{16}{8}$ **c.** $2\frac{16}{9}$

13. (2, 9) Which properties are illustrated by these equations?

a. $\frac{1}{2} \cdot \frac{2}{2} = \frac{2}{4}$ **b.** $\frac{1}{3} \cdot \frac{3}{1} = 1$

Find each unknown number.

14. (3) $m - 1776 = 87$

15. (3) $\$16.25 - b = \$10.15$

16. (3) $\frac{1001}{n} = 13$

17. (3) $42d = 1764$

Simplify:

* **18.** (15) $3\frac{3}{4} - 1\frac{1}{4}$

* **19.** (10, 16) $\frac{3}{10}$ in. + $\frac{8}{10}$ in.

20. (15) $\frac{3}{4} \times \frac{1}{3}$

21. (10) $\frac{4}{3} \cdot \frac{3}{2}$

22. (1) $\frac{10,000}{16}$

* **23.** (10, 6) $\frac{100\%}{8}$

24. (10) $9\overline{)70,000}$

25. (1) $45 \cdot 45$

26. (4, 8) **Generalize** Describe the rule of this sequence, and find the next three terms:

$$\frac{1}{16}, \frac{1}{8}, \frac{3}{16}, \ldots$$

27. (7) If two intersecting lines are not perpendicular, then they form which two types of angles?

28. (7) Two walls representing planes meet at a corner of the room. The intersection of two planes is a

A point. **B** line. **C** plane.

* **29.** (15) Find a fraction equivalent to $\frac{2}{3}$ that has a denominator of 6. Then add that fraction to $\frac{1}{6}$. What is the sum?

30. (9) How many $\frac{3}{8}$s are in 1?

Early Finishers

Real-World Application

The number of miles a car can travel on one gallon of gas is expressed as "miles per gallon" (mpg). Petro's mother drives an average of 15,000 miles a year. Her car averages 25 mpg.

a. How many gallons of gasoline will Petro's mother buy every year? Show your work.

b. If the average cost of a gallon of gasoline is $2.39, how much will she spend on gasoline?

LESSON
17 • Measuring Angles with a Protractor

Power Up

Building Power

facts Power Up D

mental math

a. **Number Sense:** \$3.50 + \$1.50

b. **Decimals:** \$3.60 ÷ 10

c. **Number Sense:** \$10.00 − \$6.40

d. **Calculation:** 5 × 33

e. **Number Sense:** 250 − 125

f. **Fractional Parts:** $\frac{1}{2}$ of 32

g. **Patterns:** What is the next number in the pattern: 2, 3, 5, 8, ______

h. **Calculation:** Start with 3 score and 15, ÷ 3, × 2, ÷ 5, × 10, ÷ 2, − 25, ÷ 5.

problem solving

If a 9 in.-by-9 in. dish of casserole serves nine people, how many 12 in.-by-12 in. dishes of casserole should we make to serve 70 people? (*Hint*: You may have "leftovers.")

Understand A nine-inch-square dish of casserole will serve 9 people. We are asked to find how many 12-inch-square dishes of casserole are needed to feed 70 people.

Plan We will *draw a diagram* to help us visualize the problem. Then we will *write* an *equation* to find the number of 12-inch-square dishes of casserole needed.

Solve First, we find the size of each serving by "cutting" the 9-inch-square dish into nine pieces. Then we see how many pieces of the same size can be made from the 12-inch-square dish:

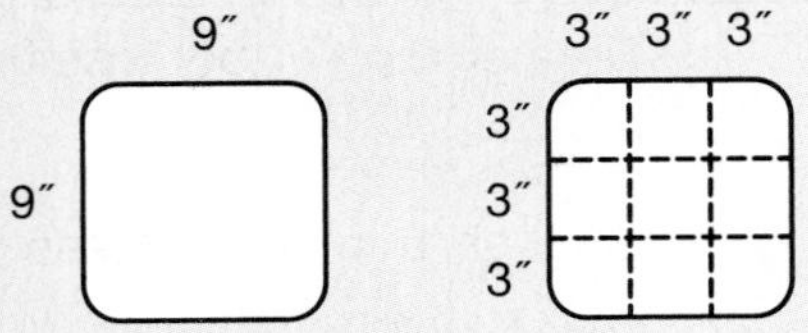

One 3-inch-square piece is one serving.

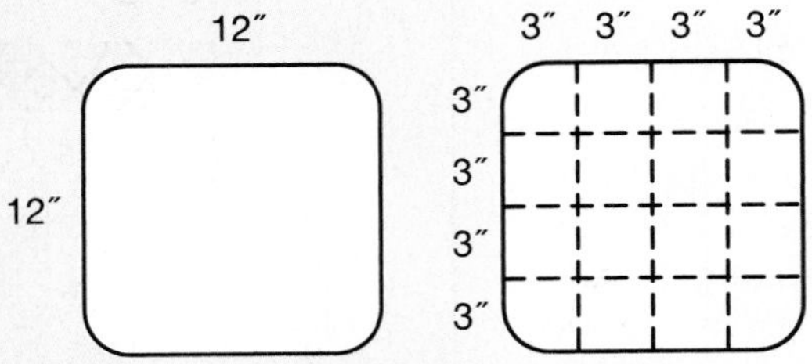

One 12-inch-square pan of casserole can be cut into sixteen 3-inch-square pieces.

One 12-inch-square dish of casserole can serve 16 people. We use this information to write an equation: $N \times 16 = 70$. We divide to find that $N = 4$ R6. Four dishes of casserole would only serve 64 people, so we must make five dishes of casserole in order to serve 70 people.

Check We found that we need to make five 12-inch dishes of casserole to feed 70 people. It would take 8 nine-inch-square dishes of casserole to feed 70 people. One 12-inch dish feeds almost twice as many people as one 9-inch dish, so our answer should be between 4 and 8. Our answer is reasonable.

New Concept *Increasing Knowledge*

In Lesson 7 we discussed angles and classified them as acute, right, obtuse, or straight. In this lesson we will begin measuring angles.

Angles are commonly measured in units called **degrees.** The abbreviation for *degrees* is a small circle written above and to the right of the number. One full rotation, a full circle, measures 360 degrees.

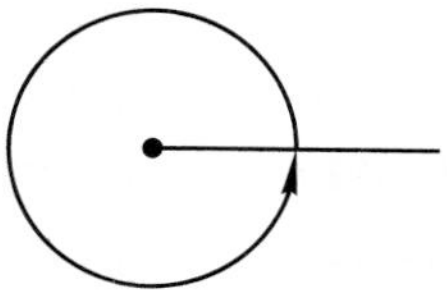

A full circle measures 360°.

A half circle measures half of 360°, which is 180°.

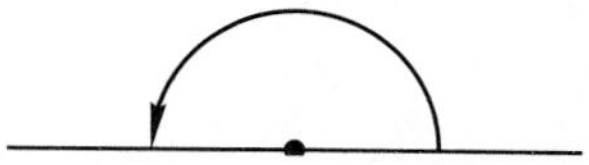

A half circle measures 180°.

One fourth of a full rotation is a right angle. A right angle measures one fourth of 360°, which is 90°.

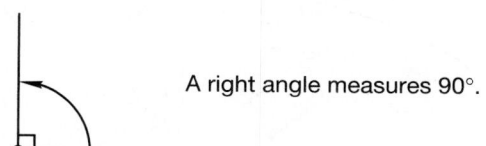

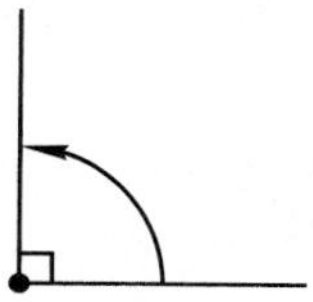

A right angle measures 90°.

Thus, the measure of an acute angle is less than 90°, and the measure of an obtuse angle is greater than 90° but less than 180°. An angle that measures 180° is a straight angle. The chart below summarizes the types of angles and their measures.

Thinking Skill
Analyze
How can we find the measure of an angle that is halfway between 0° and a right angle?

Angle Type	Measure
Acute	Greater than 0° but less than 90°
Right	Exactly 90°
Obtuse	Greater than 90° but less than 180°
Straight	Exactly 180°

A **protractor** can be used to measure angles. As shown on the following page, the protractor is placed on the angle to be measured so the vertex is under the dot, circle, or crossmark of the protractor, and one side of the angle is under the zero mark at either end of the scale of the protractor.

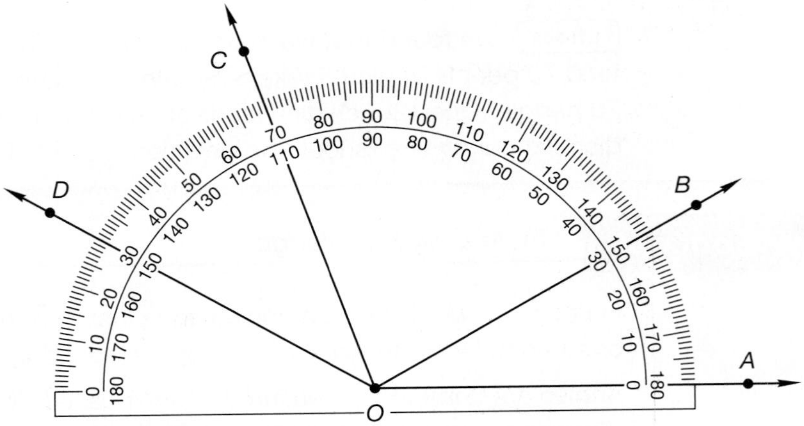

The measures of three angles shown are as follows:

$\angle AOB = 30°$ $\angle AOC = 110°$ $\angle AOD = 150°$

Notice there are two scales on a protractor, one starting from the left side, the other from the right. One way to check whether you are reading from the correct scale is to consider whether you are measuring an acute angle or an obtuse angle.

Verify Explain how you know that $\angle AOD$ does not have a measure of 30°?

Example 1

Find the measure of each angle.

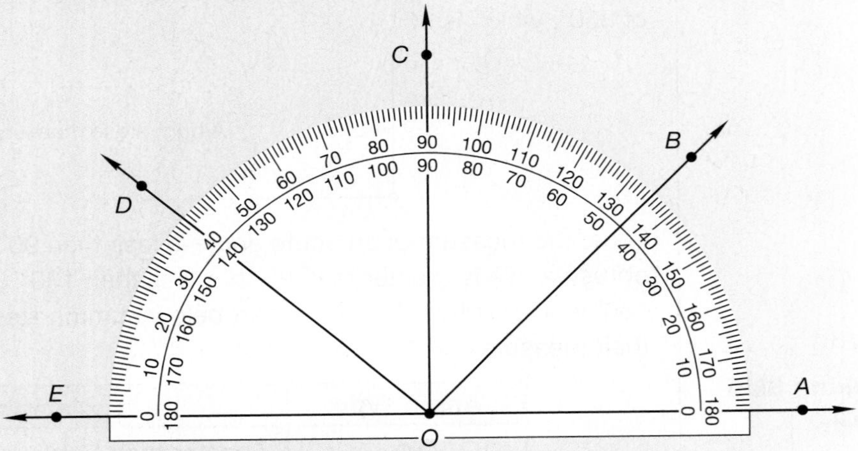

a. $\angle AOB$ **b.** $\angle AOD$ **c.** $\angle EOD$

Solution

a. Since $\angle AOB$ is acute, we read the numbers less than 90. Ray *OB* passes through the mark halfway between 40 and 50. Thus, the measure of $\angle AOB$ is **45°.**

b. Since $\angle AOD$ is obtuse, we read the numbers greater than 90. The measure of $\angle AOD$ is **140°.**

c. Angle *EOD* is acute. The measure of $\angle EOD$ is **40°.**

Example 2

Use a protractor to draw a 60° angle.

Solution

We use a ruler or the straight edge of the protractor to draw a ray. Our sketch of the ray should be longer than half the diameter of the protractor. Then we carefully position the protractor so it is centered over the endpoint of the ray, with the ray extending through either the left or right 0° mark.

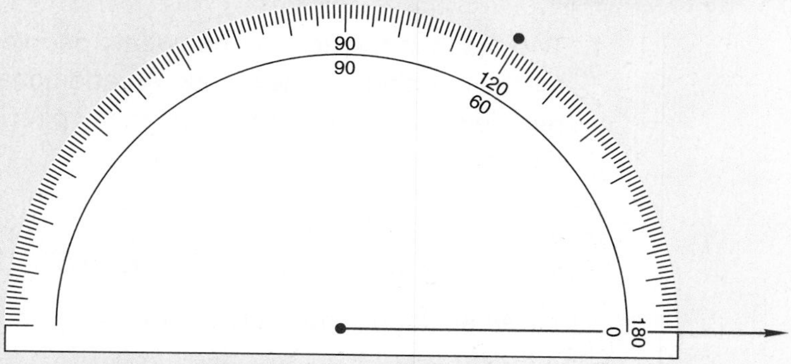

From the 0° mark we follow the curve of the protractor to the 60° mark and make a dot on the paper. Then we remove the protractor and use a straightedge or ruler to draw the second ray of the angle from the endpoint of the first ray through the dot. This completes the 60° angle.

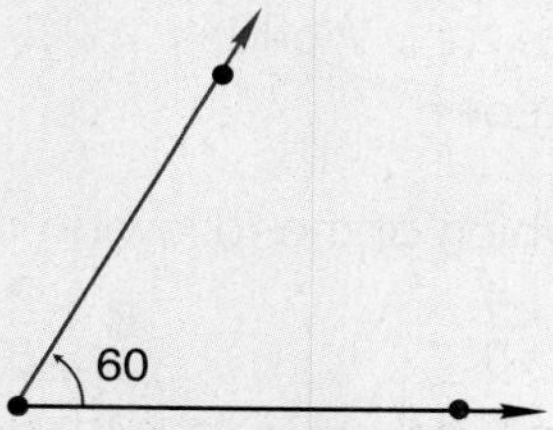

Thinking Skill

Model

Use your ruler and protractor to draw two more angles, a 75° and a 150° angle.

Classify Is this an acute, obtuse, right, or straight angle? How do you know?

Practice Set

Analyze Find the measure of each angle named in problems **a–f**.

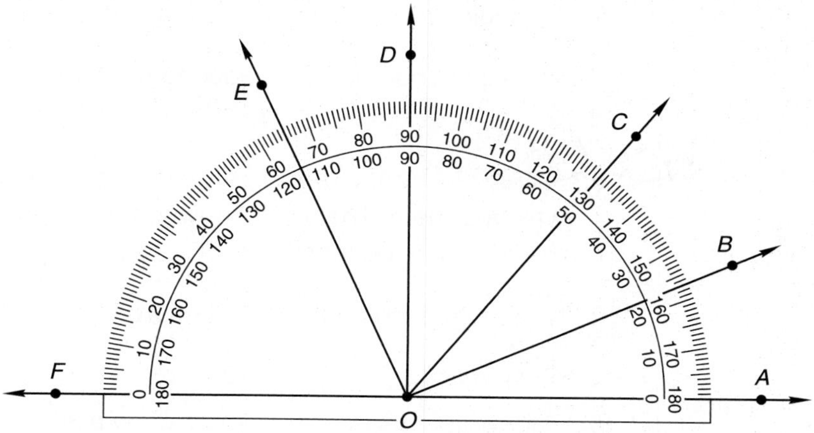

a. $\angle AOD$ **b.** $\angle AOC$ **c.** $\angle AOE$

d. $\angle FOE$ **e.** $\angle FOC$ **f.** $\angle AOB$

Represent Use your protractor to draw each of these angles:

g. 45° **h.** 120° **i.** 100° **j.** 80°

k. **Analyze** Perry's protractor is marked at each degree. Assuming the protractor is used correctly, what is the greatest possible error that can be made with Perry's protractor? Express your answer as a fraction of a degree.

Written Practice

Strengthening Concepts

1. (11) Two thousand, four hundred twenty people gathered before noon for the opening of a new park. An additional five thousand, ninety people arrived after noon. How many people were at the opening of the new park?

* **2.** (14) A number cube is rolled once.

a. What is the probability of rolling a 5?

b. What is the probability of not rolling a 5?

* **3.** (13) There are 210 students in the first-year physical education class. If they are equally divided into 15 squads, how many students will be in each squad?

4. (12) Columbus set sail for the New World in 1492. The Pilgrims set sail for the New World in 1620. How many years are there between the two dates?

* **5.** (10, 15) Which of the following does not equal $1\frac{1}{3}$?

A $\frac{4}{3}$ **B** $1\frac{2}{6}$ **C** $\frac{5}{3}$ **D** $1\frac{4}{12}$

* **6.** (7, 17) **Analyze** Refer to the figure at right to answer **a–c**:

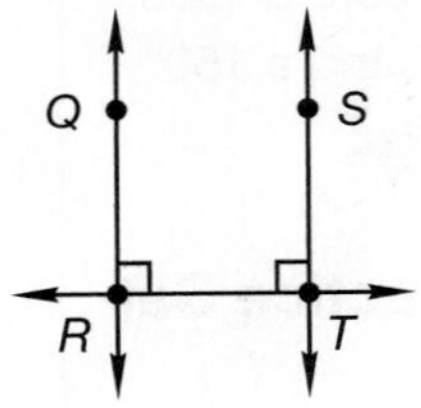

a. Which line is parallel to $\overleftrightarrow{ST}$?

b. Which line is perpendicular to $\overleftrightarrow{ST}$?

c. Angle *QRT* measures how many degrees?

* **7.** (15) Write each number as a reduced fraction or mixed number:

a. $\frac{12}{16}$ **b.** $3\frac{12}{18}$ **c.** 25%

* **8.** (16) **Generalize** At the grocery store Len put five apples in a bag and weighed them. The scale showed the weight to be 2 lb 8 oz. What is this weight in ounces?

* **9.** (15) Complete each equivalent fraction:

a. $\frac{2}{9} = \frac{?}{18}$ **b.** $\frac{1}{3} = \frac{?}{18}$ **c.** $\frac{5}{6} = \frac{?}{18}$

* **10.** (17) **Represent** Use a protractor to draw a 30° angle.

* **11.** (6) **a.** What factors of 20 are also factors of 50?

b. What is the GCF of 20 and 50?

12. (7, 8) **Represent** Draw $\overline{RS}$ $1\frac{3}{4}$ in. long. Then draw $\overrightarrow{ST}$ perpendicular to $\overline{RS}$.

13. (1, 9) If $x = 4$ and $y = 8$, find

a. $\frac{y}{x} - \frac{x}{y}$

b. $x - \frac{x}{y}$

Find the value of each variable. Check your work.

14. (3) $x - 231 = 141$

15. (3) $\$6.30 + y = \$25$

16. (3) $8w = \$30.00$

17. (3) $\frac{100\%}{m} = 20\%$

Simplify:

* **18.** (15) $3\frac{5}{6} - 1\frac{1}{6}$

* **19.** (15) $\frac{1}{2} \cdot \frac{2}{3}$

20. (1) $\frac{\$100.00}{40}$

21. (1) $55 \cdot 55$

* **22.** (1, 16) 2(8 in. + 6 in.)

* **23.** (15, 16) $\frac{3}{4}$ in. + $\frac{3}{4}$ in.

* **24.** (15, 16) $\frac{15}{16}$ in. − $\frac{3}{16}$ in.

25. (10) $\frac{1}{2} \cdot \frac{4}{3} \cdot \frac{9}{2}$

26. (1) **Analyze** The cost of the meal was $15.17. Loretha gave the cashier a $20 bill and a quarter. Name the fewest possible number of bills and coins she could have received in change.

27. (2, 9) **a.** Compare: $\left(\frac{1}{2} \cdot \frac{3}{4}\right) \cdot \frac{2}{3} \bigcirc \frac{1}{2}\left(\frac{3}{4} \cdot \frac{2}{3}\right)$

b. What property is illustrated by the comparison?

* **28.** (14) **Formulate** Write a word problem about parts of a whole that fits this equation:

$$85\% + w = 100\%$$

29. (9, 10) Write $3\frac{3}{4}$ as an improper fraction. Then write its reciprocal.

* **30.** (15) Find a fraction equal to $\frac{3}{4}$ with a denominator of 8. Add the fraction to $\frac{5}{8}$. Write the sum as a mixed number.

LESSON 18

• Polygons
• Similar and Congruent

Power Up

Building Power

facts

Power Up D

mental math

a. Calculation: \$3.75 + \$1.75

b. Decimals: \$1.65 × 10

c. Number Sense: \$20.00 − \$12.50

d. Calculation: 6 × 24

e. Number Sense: 375 − 250

f. Fractional Parts: $\frac{1}{2}$ of $\frac{1}{4}$

g. Patterns: What is the next number in the pattern: 3, 3, 6, 18, …

h. Calculation: Start with two score, × 2, + 1, ÷ 9, × 3, + 1, ÷ 4.

problem solving

Letha has 7 coins in her hand totaling 50 cents. What are the coins?

Understand Letha has a combination of coins in her hand totaling 50 cents. We have been asked to determine what seven coins Letha has.

Plan We will *use logical reasoning* to eliminate coins that are not possible. Then, we will *guess and check* until we find the combination of coins in Letha's hand. We will *make a table* to keep track of the combinations we try and to ensure we do not miss any combinations.

Solve Letha cannot have a half dollar, because one half dollar equals 50 cents. If one of Letha's coins were a quarter, she would have to have six other coins that total 25 cents, which is impossible. Letha cannot have pennies, either. In order to have exactly 50 cents, she would need either 5 pennies (which would require two additional coins totaling 45 cents, which is impossible) or 10 pennies (which is too many). We recognize that we only need to consider nickels and dimes to determine Letha's coins.

Now we are ready to make a table. We begin by considering 5 dimes and then add nickels to bring the total to 50 cents:

Dimes	Nickels	Total
5 = 50¢	0 = 0¢	5 coins = 50 cents
4 = 40¢	2 = 10¢	6 coins = 50 cents
3 = 30¢	**4** = 20¢	**7 coins = 50 cents**

Check We found that Letha has three dimes and four nickels in her hand. Our table helped us keep track of each guess and helped us work through the problem in an organized way.

New Concepts

Increasing Knowledge

polygons

When three or more line segments are connected to enclose a portion of a plane, a **polygon** is formed. The word *polygon* comes from the ancient Greeks and means "many angles." The name of a polygon tells how many angles and sides the polygon has.

Names of Polygons

Name of Polygon	Number of Sides	Name of Polygon	Number of Sides
Triangle	3	Octagon	8
Quadrilateral	4	Nonagon	9
Pentagon	5	Decagon	10
Hexagon	6	Undecagon	11
Heptagon	7	Dodecagon	12

Note: For polygons with more than 12 sides, we use the term *n*-gon, with *n* being the number of sides. Thus, a polygon with 15 sides is a 15-gon.

Math Language
The plural of vertex is **vertices.** Polygon *STVU* has four vertices.

The point where two sides of a polygon meet is called a **vertex.** A particular polygon may be identified by naming the letters of its vertices in order. Any letter may be first. The rest of the letters can be named clockwise or counterclockwise. The polygon below has eight names, as shown.

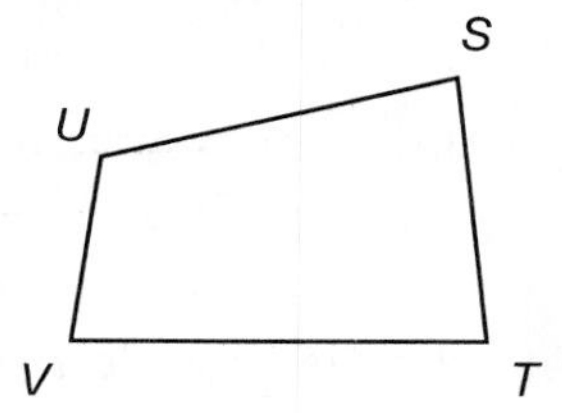

Named Clockwise	Named Counterclockwise
USTV	*UVTS*
STVU	*VTSU*
TVUS	*TSUV*
VUST	*SUVT*

If all the sides of a polygon have the same length and all the angles have the same measure, then the polygon is a **regular polygon.**

Regular and Irregular Polygons

Type	Regular	Irregular
Triangle	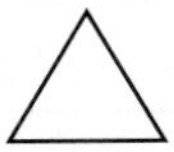	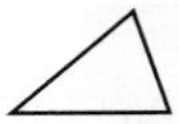
Quadrilateral		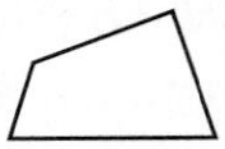
Pentagon	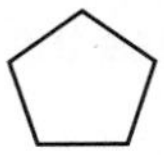	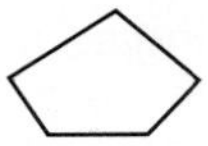
Hexagon	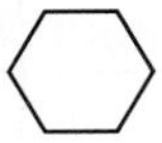	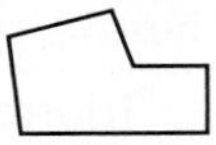

Example 1

a. Name this polygon.

b. Is the polygon regular or irregular?

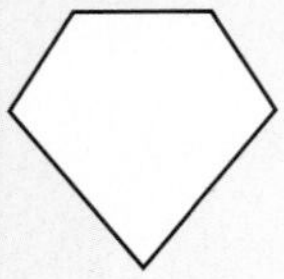

Solution

a. pentagon **b. irregular**

Example 2

Which of these figures is a polygon?

A  **B**  **C** 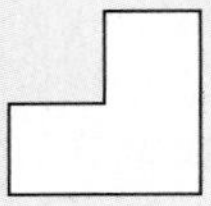 **D** 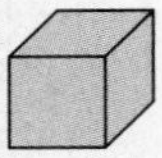

Solution

Figure A is not a polygon because its sides are not all segments. Figure B is not a polygon because it is not closed. **Figure C** is a polygon. Figure D is not a polygon because it is not a plane (2-dimensional) figure, although its faces are polygons.

similar and congruent

Two figures are **similar** if they have the same shape even though they may vary in size. In the illustration below, triangles I, II, and III are similar. To see this, we can imagine enlarging (dilating) triangle II as though we were looking through a magnifying glass. By enlarging triangle II, we could make it the same size as triangle I or triangle III. Likewise, we could reduce triangle III to the same size as triangle I or triangle II.

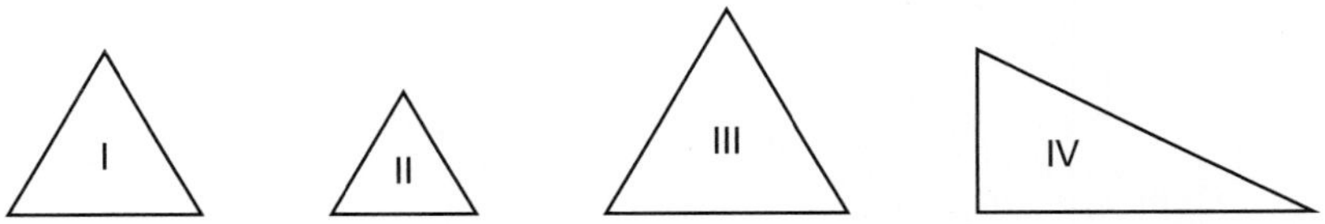

Although triangle IV is a triangle, it is not similar to the other three triangles, because its shape is different. Viewing triangle IV through a reducing or enlarging lens will change its size but not its shape.

Discuss Ask students how we might draw a triangle that is similar to triangle IV.

Figures that are the same shape and size are not only similar, they are also **congruent.** All three of the triangles below are similar, but only triangles *ABC* and *DEF* are congruent. Note that figures may be reflected (flipped) or rotated (turned) without affecting their similarity or congruence.

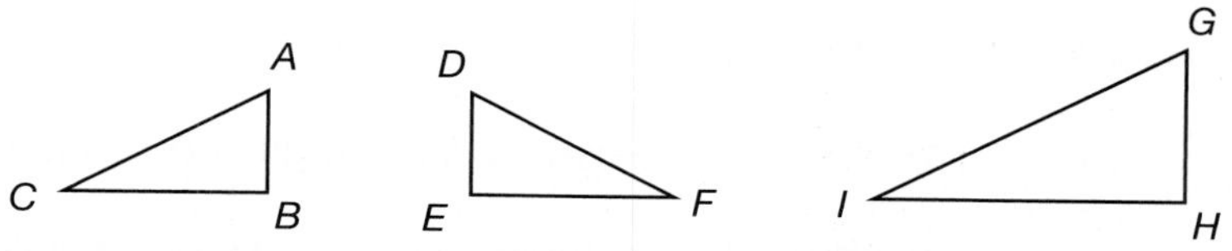

Math Language

We can use small lines to indicate corresponding parts of congruent figures.

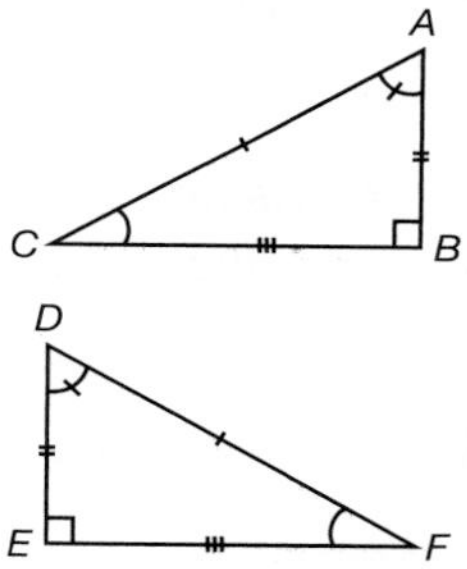

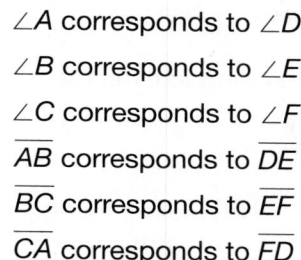

When inspecting polygons to determine whether they are similar or congruent, we compare their **corresponding parts.** A triangle has six parts—three sides and three angles. If the six parts of one triangle have the same measures as the six corresponding parts of another triangle, the triangles are congruent. Referring back to the illustration of triangle *ABC* ($\triangle ABC$) and triangle *DEF* ($\triangle DEF$), we identify the following corresponding parts:

$\angle A$ corresponds to $\angle D$

$\angle B$ corresponds to $\angle E$

$\angle C$ corresponds to $\angle F$

$\overline{AB}$ corresponds to $\overline{DE}$

$\overline{BC}$ corresponds to $\overline{EF}$

$\overline{CA}$ corresponds to $\overline{FD}$

Notice that the corresponding angles of similar figures have the same measure even though the corresponding sides may be different lengths.

Example 3

a. Which of these quadrilaterals appear to be similar?

b. Which of these quadrilaterals appear to be congruent?

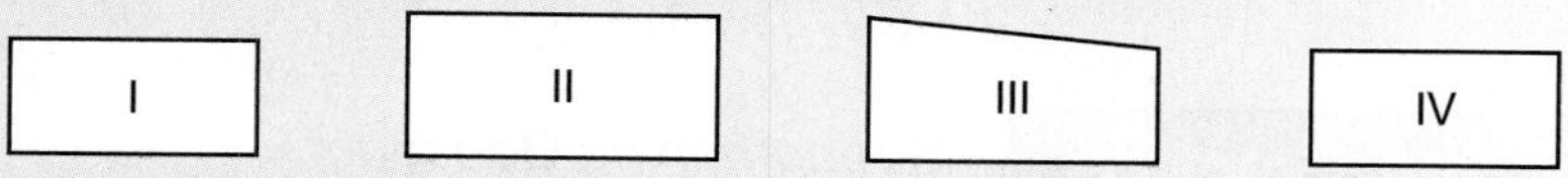

Solution

a. I, II, IV **b. I, IV**

Discuss What is the difference between similar figures and congruent figures?

Example 4

a. **Which angle in $\triangle XYZ$ corresponds to $\angle A$ in $\triangle ABC$?**

b. **Which side in $\triangle XYZ$ corresponds to $\overline{BC}$ in $\triangle ABC$?**

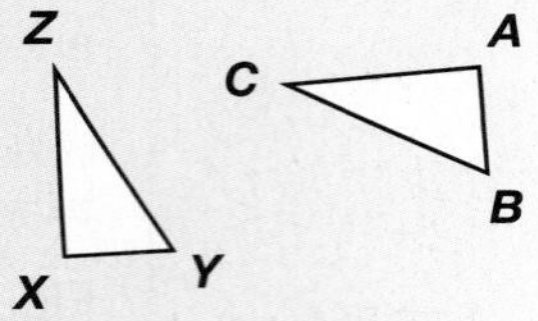

Solution

a. $\angle X$

b. $\overline{YZ}$

Practice Set

a. What is the shape of a stop sign?

b. What do we usually call a regular quadrilateral?

c. What kind of angle is each angle of a regular triangle?

d. **Explain** Are all squares similar? How do you know?

e. Are all squares congruent?

f. Sketch a pentagon, a hexagon, and a heptagon.

g. **Analyze** Referring to example 4, which angle in $\triangle ABC$ corresponds to $\angle Y$ in $\triangle XYZ$?

h. Referring to example 3, are the angles in figure II larger in measure, smaller in measure, or equal in measure to the corresponding angles in figure I?

i. **Model** Draw a triangle that contains a right angle. Label the vertices A, B, C so that the right angle is at vertex C.

j. Which of these figures is not a polygon? Write the reason for your choice.

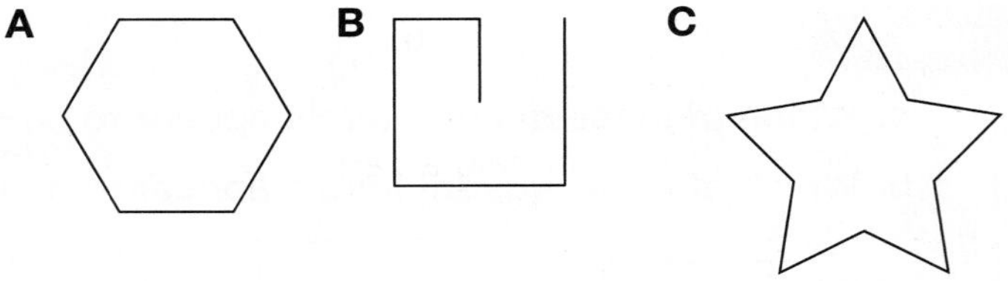

Written Practice

Strengthening Concepts

1. (13) The Collins family drove the 2825-mile, coast-to-coast drive from New York, New York to Los Angeles, California in 6 days. They drove about the same distance each day. What was the average number of miles they traveled each day?

* **2.** (16) **a.** This function table shows the number of gallons (output) in a given number of quarts (input). Copy and extend the table to include 5, 6, 7, and 8 quarts.

Quarts	Gallons
1	$\frac{1}{4}$
2	$\frac{1}{2}$
3	$\frac{3}{4}$
4	1

b. *Generalize* What is the rule for this function table? Justify your answer.

3. (14) Albert ran 3977 meters of the 5000-meter race but walked the rest of the way. How many meters of the race did Albert walk?

4. (5, 12) One billion is how much greater than ten million? Use words to write the answer.

5. (4, 10) **a.** Arrange these numbers in order from least to greatest:

$$\frac{5}{3}, -1, \frac{3}{4}, 0, 1$$

b. Which of these numbers are not positive?

6. (7) In rectangle *ABCD*, which side is parallel to side *BC*?

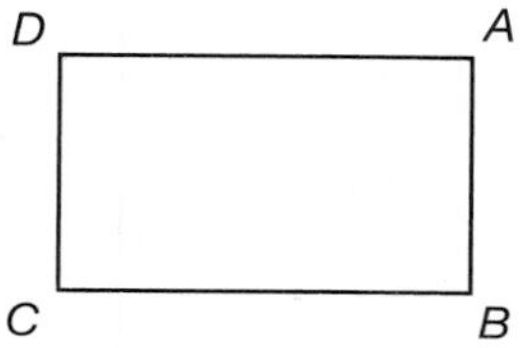

7. (4) Refer to this number line to answer **a** and **b**:

a. What integer is two units to the left of the origin?

b. What integer is seven units to the right of −3?

* **8.** (15) Write each number as a reduced fraction or mixed number:

a. 2%　　**b.** $\frac{12}{20}$　　**c.** $6\frac{15}{20}$

* **9.** (15) For each fraction, find an equivalent fraction that has a denominator of 30:

a. $\frac{4}{5}$　　**b.** $\frac{2}{3}$　　**c.** $\frac{1}{6}$

10. (18) Name each of these figures. Then tell which is not a polygon. Give the reason for your choice.

a.
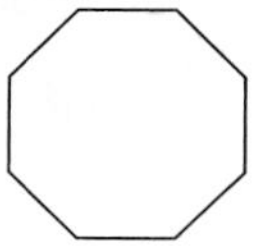

b.
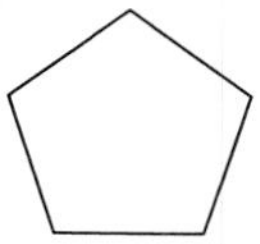

c.
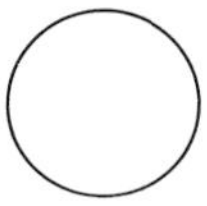

* **11.** (7, 18) **a.** *Model* Draw a triangle that has one obtuse angle.

b. What kind of angles are the other two angles of the triangle?

* **12.** **a.** **Predict** What is the probability of spinning a 3?
(14, 15)

b. What is the probability of not spinning a 3?

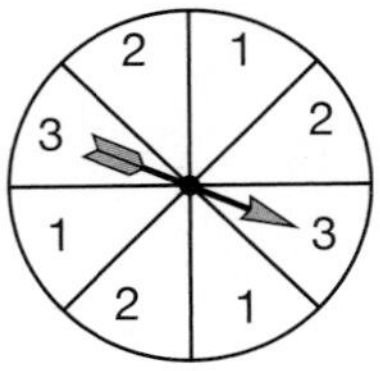

* **13.** Which properties are illustrated by these equations?
(2, 9, 15)

a. $\frac{1}{2} \times \frac{3}{3} = \frac{3}{6}$ **b.** $\frac{2}{3} \times \frac{3}{2} = 1$

Find each unknown number:

14. $x - \frac{3}{8} = \frac{5}{8}$
(9)

* **15.** $y + \frac{3}{10} = \frac{7}{10}$
(9, 15)

* **16.** $\frac{5}{6} - m = \frac{1}{6}$
(9, 15)

17. $\frac{3}{4}x = 1$
(9)

Simplify:

* **18.** $5\frac{7}{10} - \frac{3}{10}$
(15)

* **19.** $\frac{3}{2} \cdot \frac{2}{4}$
(15)

20. $\frac{2025}{45}$
(1)

21. 750×80
(1)

22. $21 \cdot 21$
(1)

23. Use and identify the commutative, associative, inverse, and identity properties of multiplication to simplify the following expression. (Hint: See problem **20** in Lesson 15.)
(2, 9)

$$\frac{5}{8} \cdot \frac{4}{9} \cdot \frac{8}{5}$$

* **24.** **Analyze** What percent of a pound is 8 ounces?
(16)

* **25.** **a.** How many degrees is $\frac{1}{4}$ of a circle or $\frac{1}{4}$ of a full turn?
(17)

b. How many degrees is $\frac{1}{6}$ of a circle or $\frac{1}{6}$ of a full turn?

* **26.** **a.** Use a protractor to draw a 135° angle.
(17)

b. A 135° angle is how many degrees less than a straight angle?

* **27.** Refer to the triangles below to answer **a–c.**
(18)

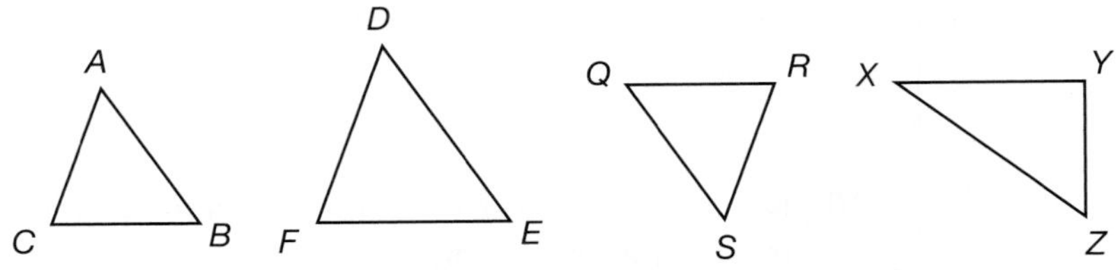

a. Which triangle appears to be congruent to $\triangle ABC$?

b. Which triangle is not similar to $\triangle ABC$?

c. Which angle in $\triangle DEF$ corresponds to $\angle R$ in $\triangle SQR$?

*** 28.** (15) Write a fraction equal to $\frac{1}{2}$ with a denominator of 6 and a fraction equal to $\frac{1}{3}$ with a denominator of 6. Then add the fractions.

29. (9, 10) Write $2\frac{1}{4}$ as an improper fraction, and multiply the improper fraction by the reciprocal of $\frac{3}{4}$.

*** 30.** (8, 18) **Connect** Different shaped figures are used for various traffic signs. A triangle with one downward vertex is used for the Yield sign. Use a ruler to draw a yield sign with each side 1 inch long. Is the triangle regular or irregular?

Early Finishers
Real-World Application

Ramona needs 10 pieces of wood that are each $1\frac{1}{4}$ feet in length. The lumber yard sells 6-foot long boards and 8-foot long boards.

a. How many feet of board does Ramona need? Show your work.

b. Ramona wants to have as little wood leftover as possible. What length boards should she choose? Justify your choices.

LESSON
19

• Perimeter

Power Up

Building Power

facts Power Up C

mental math

a. **Number Sense:** \$8.25 + \$1.75

b. **Decimals:** \$12.00 ÷ 10

c. **Number Sense:** \$1.00 – 76¢

d. **Calculation:** 7 × 32

e. **Number Sense:** 625 – 250

f. **Fractional Parts:** $\frac{1}{2}$ of 120

g. **Measurement:** Convert 36 inches to yards

h. **Calculation:** Start with 4 dozen, ÷ 6, × 5, + 2, ÷ 6, × 7, + 1, ÷ 2, – 1, ÷ 2.

problem solving The product of 10 × 10 × 10 is 1000. Find three prime numbers whose product is 1001.

New Concept

Increasing Knowledge

The distance around a polygon is the **perimeter** of the polygon. To find the perimeter of a polygon, we add the lengths of its sides.

Example 1

This figure is a rectangle. The two dimensions of a rectangle are called *length* and *width.* This rectangle is 3 cm long and 2 cm wide. What is the perimeter of this rectangle?

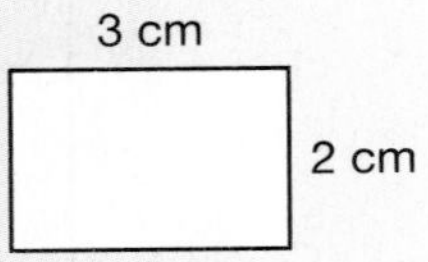

Solution

The opposite sides of a rectangle are equal in length. Tracing around the rectangle, our pencil travels 3 cm, then 2 cm, then 3 cm, then 2 cm. Thus, the perimeter is

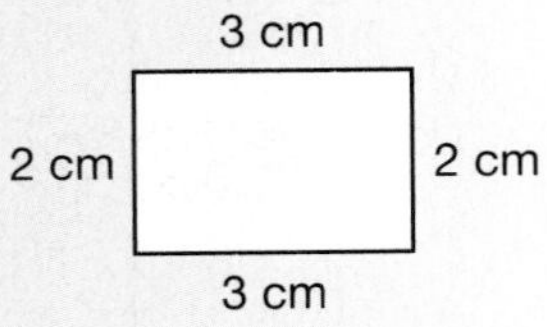

3 cm + 2 cm + 3 cm + 2 cm = **10 cm**

Example 2

Math Language
The sides of a **regular polygon** are equal in length.

What is the perimeter of this regular hexagon?

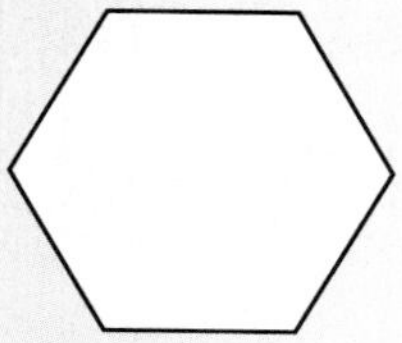

Solution

The perimeter of this hexagon is

$$8 \text{ mm} + 8 \text{ mm} + 8 \text{ mm} + 8 \text{ mm} + 8 \text{ mm} + 8 \text{ mm} = \mathbf{48 \text{ mm}}$$

or

$$6 \times 8 \text{ mm} = \mathbf{48 \text{ mm}}$$

Example 3

Find the perimeter of this polygon. All angles are right angles. Dimensions are in feet.

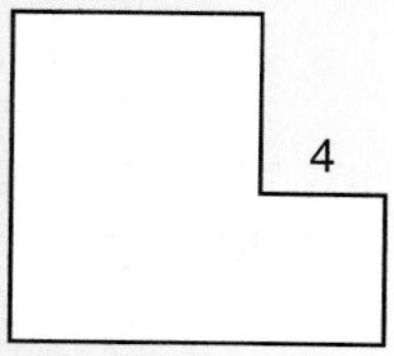

Solution

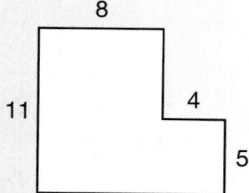

We will use the letters *a* and *b* to refer to the unmarked sides. Notice that the lengths of side *a* and the side marked 5 total 11 feet.

$a + 5 = 11$ So side *a* is 6 ft.

Also notice that the length of side *b* equals the total lengths of the sides marked 8 and 4.

$8 + 4 = b$ So side *b* is 12 ft.

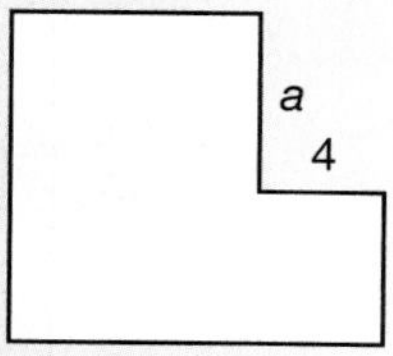

The perimeter of the figure in feet is

$$8 \text{ ft} + 6 \text{ ft} + 4 \text{ ft} + 5 \text{ ft} + 12 \text{ ft} + 11 \text{ ft} = \mathbf{46 \text{ ft}}$$

Example 4

The perimeter of a square is 48 ft. How long is each side of the square?

Solution

A square has four sides whose lengths are equal. The sum of the four lengths is 48 ft. Here are two ways to think about this problem:

1. The sum of what four identical addends is 48?

____ + ____ + ____ + ____ = 48 ft

2. What number multiplied by 4 equals 48?

$$4 \times \_\_\_\_ = 48 \text{ ft}$$

As we think about the problem the second way, we see that we can divide 48 ft by 4 to find the length of each side.

$$4\overline{)48}\quad 12$$

The length of each side of the square is **12 ft.**

Example 5

Isabel wants to fence some grazing land for her sheep. She made this sketch of her pasture. How many feet of wire fence does she need?

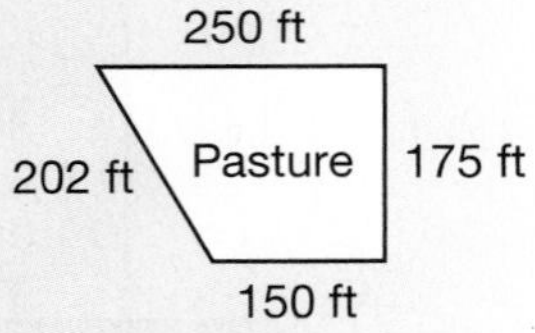

Solution

We add the lengths of the sides to find how many feet of fence Isabel needs.

$$250 \text{ ft} + 175 \text{ ft} + 150 \text{ ft} + 202 \text{ ft} = 777 \text{ ft}$$

We see that Isabel needs **777 ft** of wire fence.

Discuss In Example 4, we were given the perimeter of a square and asked to find the length of each side. In this example, if we were given only the perimeter of the grazing land and the diagram of its shape, would it be possible for us to find the length of each side?

Activity

Creating Formulas for the Perimeters of Polygons

1. Here is a triangle. The lengths of its sides are *a, b,* and *c.* What is the perimeter of the triangle? Begin your answer this way: $P =$

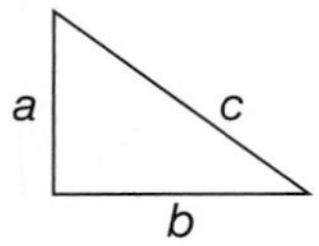

2. Here is a rectangle. Its length is L and its width is W. What is its perimeter? Begin your answer this way: $P =$

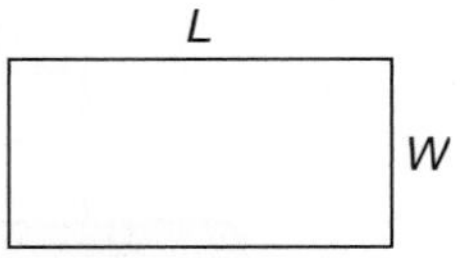

Analyze Is there another way to write your answer?

3. Here is a square. The length of each side is s. What is the perimeter of the square? Begin your answer this way: $P =$

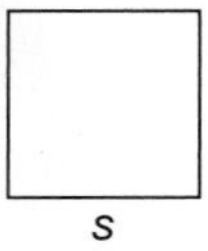

Is there another way to write your answer?

Practice Set

Thinking Skill

Connect

Use your formula to find the perimeter of some rectangular objects in your classroom.

a. What is the perimeter of this quadrilateral?

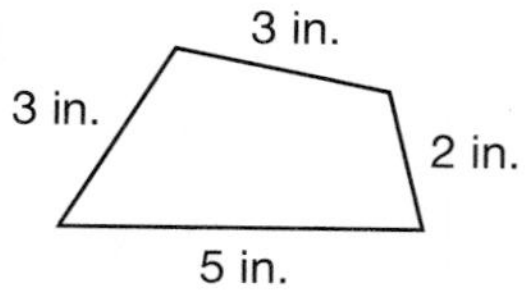

b. What is the perimeter of this regular pentagon?

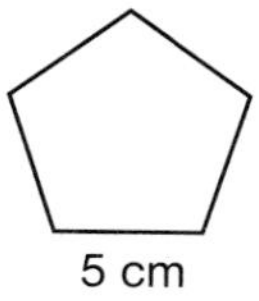

c. If each side of a regular octagon measures 12 inches, what is its perimeter?

d. What is the perimeter of this hexagon?

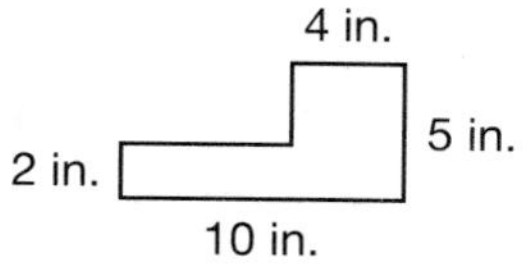

e. MacGregor has 100 feet of wire fence that he plans to use to enclose a square garden. Each side of his garden will be how many feet long?

f. Draw a quadrilateral with each side $\frac{3}{4}$ inch long. What is the perimeter of the quadrilateral?

g. *Represent* The lengths of the sides of this polygon are *a, b, c,* and *d.* What is the perimeter of the polygon? Begin your answer this way: $P =$

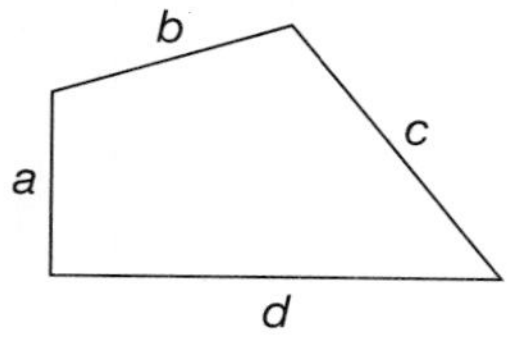

Written Practice

Strengthening Concepts

1. (14) One eighth of the students in the class are in the school band. What fraction of the total number of students are not in the band?

2. (11) The theater was full when the movie began. Seventy-six people left before the movie ended. One hundred twenty-four people remained. How many people were in the theater when it was full?

3. (13) All ants have 6 legs. A scientist studying ants observes each leg on every ant in her sample. If her sample contains 84 ants, how many legs does she observe?

* **4.** (16, 19) The perimeter of a square is a function of the length of its sides. Make a function table that shows the perimeter of squares with side lengths of 1, 2, 3, 4, and 5 units.

5. (5) **a.** Use words to write 18700000.

b. Write 874 in expanded notation.

6. (4) Use digits and other symbols to write "Three minus seven equals negative four."

7. At what temperatures on the Fahrenheit scale does water freeze and boil?
(16)

*** 8.** Write a formula for the perimeter of a rectangle. Then find the perimeter of this rectangle:
(19)

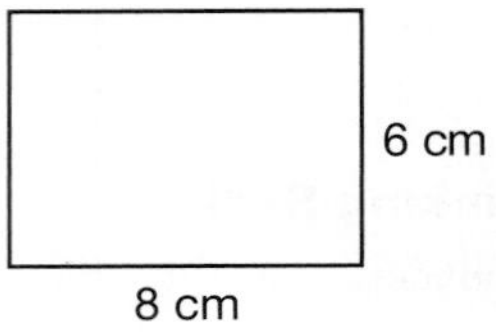

*** 9.** *Generalize* Write each number as a reduced fraction or mixed number:
(15)

a. $3\frac{16}{24}$ **b.** $\frac{15}{24}$ **c.** 4%

*** 10.** *Analyze* Find *a* and *b* to complete each equivalent fraction:
(15)

a. $\frac{3}{4} = \frac{a}{36}$ **b.** $\frac{4}{9} = \frac{b}{36}$

*** 11.** *Classify* Which of these figures is not a polygon? Write the reason for your choice.
(18)

A 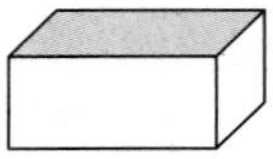 **B** 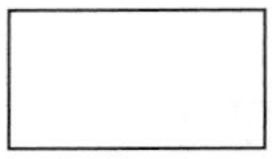 **C** 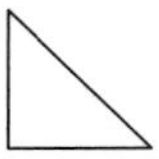

*** 12.** What is the name of a polygon that has twice as many sides as a quadrilateral?
(18)

*** 13.** **a.** Each angle of a rectangle measures how many degrees?
(17)

b. The four angles of a rectangle total how many degrees?

14. The rule of this sequence is $k = \frac{1}{8}n$. Find the eighth term of the sequence.
(4, 9)

$$\frac{1}{8}, \frac{1}{4}, \frac{3}{8}, \frac{1}{2}, \ldots$$

Find the value of each variable.

15. $a + 1547 = 8998$
(3)

16. $30b = \$41.10$
(3)

17. $\$0.32c = \$7.36$
(3)

18. $\$26.57 + d = \$30.10$
(3)

Simplify:

19. $\frac{2}{3} + \frac{2}{3} + \frac{2}{3}$
(10)

20. $3\frac{7}{8} - \frac{5}{8}$
(15)

21. $\frac{2}{3} \cdot \frac{3}{7}$
(15)

22. $3\frac{7}{8} + \frac{5}{8}$
(15)

23. $50 \cdot 50$
(1)

24. $\frac{100{,}100}{11}$
(1)

25. **a.** How many $\frac{1}{2}$s are in 1?
(9)

b. Use the answer to **a** to find the number of $\frac{1}{2}$s in 5.

26. Use your ruler to draw $\overline{AB}$ $1\frac{1}{2}$ in. long. Then draw $\overline{BC}$ perpendicular to $\overline{AB}$ 2 in. long. Draw a segment from point *A* to point *C* to complete $\triangle ABC$. What is the length of $\overline{AC}$?
(7, 8)

27. (9, 10) Write $3\frac{1}{3}$ as an improper fraction, and multiply it by the reciprocal of $\frac{2}{3}$.

* **28.** (15) **Evaluate** Find a fraction equal to $\frac{1}{2}$ that has a denominator of 10. Subtract this fraction from $\frac{9}{10}$. Write the difference as a reduced fraction.

* **29.** (8, 16) What percent of a yard is a foot?

* **30.** (19) **a.** What is the perimeter of this hexagon? All angles are right angles.

b. **Relate** How can you tell that the figure at right is a hexagon?

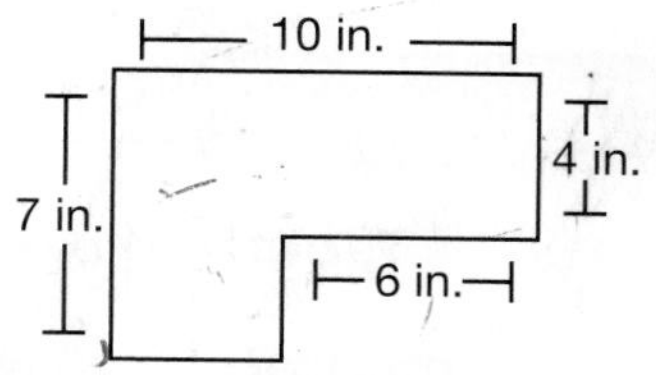

Early Finishers
Real-World Application

Magali needs to fill a five-gallon cooler with sports drink for the soccer game. The label on the drink mix says that each packet will make 10 quarts.

a. Will one packet be enough to make a full cooler? If not, how many packets will Magali need? Show your work.

b. A package of 8 packets cost \$11.95. Estimate the cost of 2 packets. Show your work.

LESSON

20

• Exponents
• Rectangular Area, Part 1
• Square Root

Power Up

Building Power

facts Power Up D

mental math

a. Number Sense: \$4.75 + \$2.50

b. Decimals: 36¢ × 10

c. Number Sense: \$5.00 − \$4.32

d. Calculation: 5 × 43

e. Number Sense: 625 − 125

f. Fractional Parts: $\frac{1}{2}$ of $\frac{3}{4}$

g. Algebra: If $r = 6$, what does $7r$ equal?

h. Calculation: 10 × 10, − 10, ÷ 10, + 1, − 10, × 10, + 10, ÷ 10

problem solving

The card shown is rotated 90° clockwise three times. Draw the card with the triangle in the correct position after each turn.

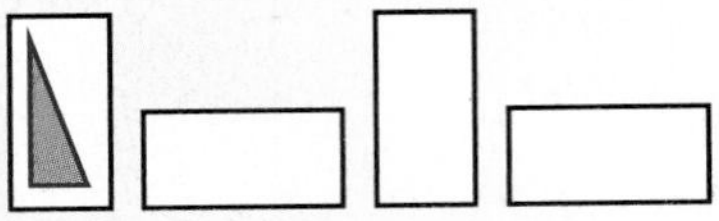

New Concepts

Increasing Knowledge

exponents

We remember that we can show repeated addition by using multiplication.

$$5 + 5 + 5 + 5 \quad \text{has the same value as} \quad 4 \times 5$$

There is also a way to show repeated multiplication. We can show repeated multiplication by using an **exponent.**

$$5 \cdot 5 \cdot 5 \cdot 5 = 5^4$$

In the expression 5^4, the exponent is 4 and the **base** is 5. The exponent shows how many times the base is to be used as a factor.

base ⟶ 5^4 ⟵ exponent

The following examples show how we read expressions with exponents, which we call **exponential expressions.**

4^2 "four squared" or "four to the second power"

2^3 "two cubed" or "two to the third power"

5^4 "five to the fourth power"

10^5 "ten to the fifth power"

To find the value of an expression with an exponent, we use the base as a factor the number of times shown by the exponent.

$$5^4 = 5 \cdot 5 \cdot 5 \cdot 5 = 625$$

Example 1

Thinking Skill

Conclude

When the base of an exponential expression is a fraction, why is the fraction in parentheses?

Simplify:

a. 4^2 **b.** 2^3 **c.** 10^5 **d.** $\left(\frac{2}{3}\right)^2$

Solution

a. $4^2 = 4 \cdot 4 = \mathbf{16}$

b. $2^3 = 2 \cdot 2 \cdot 2 = \mathbf{8}$

c. $10^5 = 10 \cdot 10 \cdot 10 \cdot 10 \cdot 10 = \mathbf{100{,}000}$

d. $\left(\frac{2}{3}\right)^2 = \frac{2}{3} \cdot \frac{2}{3} = \mathbf{\frac{4}{9}}$

Example 2

Simplify: $4^2 - 2^3$

Solution

We first find the value of each exponential expression. Then we subtract.

$$4^2 - 2^3$$

$$16 - 8 = \mathbf{8}$$

Example 3

Find the missing exponent in each equation:

a. $2^3 \cdot 2^3 = 2^n$ **b.** $\frac{2^6}{2^3} = 2^n$

Solution

a. We are asked to find the missing exponent of the product. Consider the meaning of each exponent.

$$\overbrace{2 \cdot 2 \cdot 2}^{2^3} \cdot \overbrace{2 \cdot 2 \cdot 2}^{2^3} = 2^n$$

We see that 2 appears as a factor 6 times. So the missing exponent is **6.**

b. We are asked to find the missing exponent of the quotient.

$$\frac{2^6}{2^3} = \frac{\overset{1}{\cancel{2}} \cdot \overset{1}{\cancel{2}} \cdot \overset{1}{\cancel{2}} \cdot 2 \cdot 2 \cdot 2}{\underset{1}{\cancel{2}} \cdot \underset{1}{\cancel{2}} \cdot \underset{1}{\cancel{2}}} = 2 \cdot 2 \cdot 2 = 2^3$$

We expand each expression and then reduce. We see that the missing exponent is **3.**

We can use exponents to indicate units that have been multiplied. Recall that when we add or subtract measures with like units, the units do not change.

$$4\text{ ft} + 8\text{ ft} = 12\text{ ft}$$

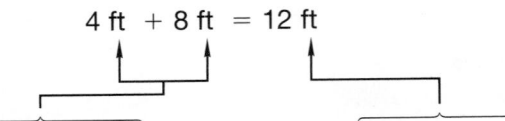

The units of the addends are the same as the units of the sum.

However, when we multiply or divide measures, the units do change.

$$4\text{ ft} \cdot 8\text{ ft} = 32\text{ ft} \cdot \text{ft}$$
$$= 32\text{ ft}^2$$

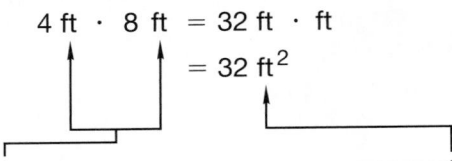

The units of the factors are not the same as the units of the product.

The result of multiplying feet by feet is **square feet,** which we can abbreviate sq. ft or ft^2. Square feet are units used to measure area, as we see in the next section of this lesson.

rectangular area, part 1

The diagram below represents the floor of a hallway that has been covered with square tiles that are 1 foot on each side. How many 1-ft square tiles does it take to cover the floor of the hallway?

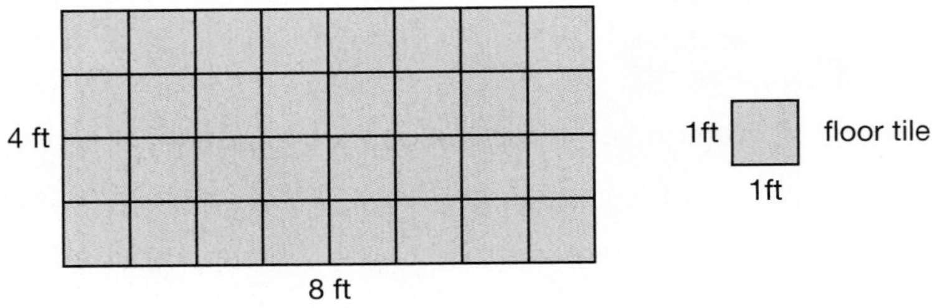

We see that there are 4 rows and 8 floor tiles in each row. So there are 32 1-ft square tiles.

The floor tiles cover the **area** of the hallway. Area is an amount of surface. Floors, ceilings, walls, sheets of paper, and polygons all have areas. If a square is 1 foot on each side, it is a **square foot.** Thus the area of the hallway is 32 square feet. Other standard square units in the U.S. system include square inches, square yards, and square miles.

It is important to distinguish between a unit of length and a unit of area. Units of length, such as inches or feet, are used for measuring distances, not for measuring areas. To measure area, we use units that occupy area. Square inches and square feet occupy area and are used to measure area.

Explain When finding the perimeter of the classroom floor, would you use feet or square feet? Explain your answer.

Discuss Name some examples of things that would be measured in inches, feet, or yards. Name some things that would be measured in square inches, square feet, or square yards.

Unit of Length **Unit of Area**

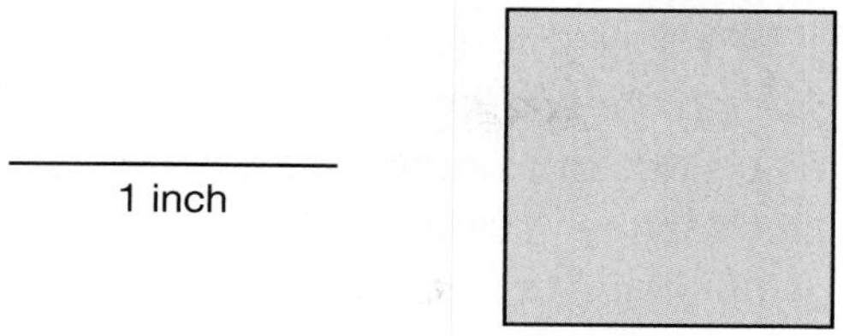

1 inch

1 square inch
1 sq. in.
1 in.^{2}

One way to find the area of the rectangular hallway is to count each tile. What is another way to find the area?

Activity

Creating Formulas for the Areas of Rectangles and Squares

1. The length of this rectangle is L and its width is W. What is its area? Begin your answer this way: $A =$

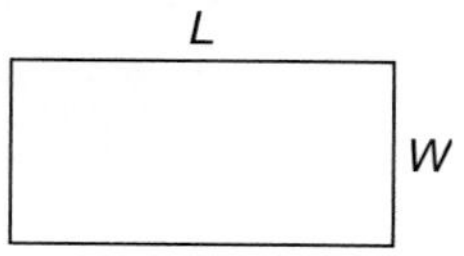

2. The length of each side of a square is s. What is the area of the square? Begin your answer this way: $A =$

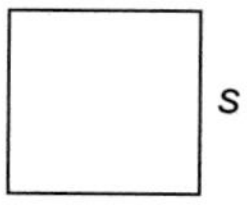

Is there another way to write the answer?

Example 4

What is the area of this rectangle?

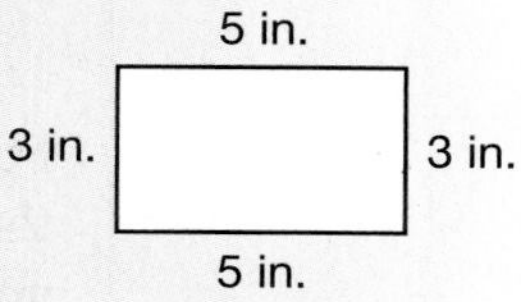

Solution

The area of the rectangle is the number of square inches needed to cover the rectangle.

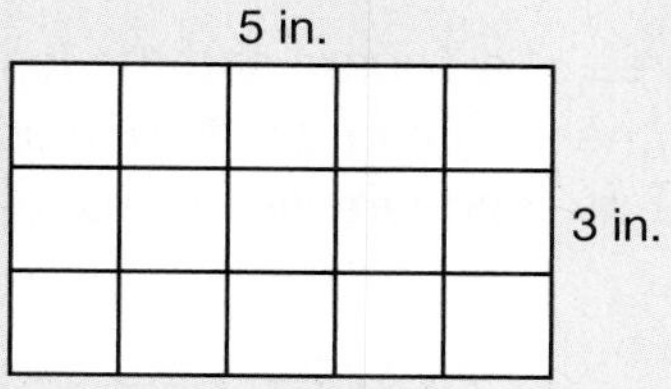

We can find this number by multiplying the length (5 in.) by the width (3 in.).

$$\text{Area of rectangle} = 5 \text{ in.} \cdot 3 \text{ in.}$$
$$= \mathbf{15 \text{ in.}^2}$$

Example 5

The perimeter of a certain square is 12 inches. What is the area of the square?

Solution

To find the area of the square, we first need to know the length of the sides. The sides of a square are equal in length, so we divide 12 inches by 4 and find that each side is 3 inches. Then we multiply the length (3 in.) by the width (3 in.) to find the area.

$$\text{Area} = 3 \text{ in.} \times 3 \text{ in.}$$
$$= \mathbf{9\ in.^2}$$

Example 6

Dickerson Ranch is a level plot of land 4 miles square. The area of Dickerson Ranch is how many square miles?

Solution

"Four miles square" does not mean "4 square miles." A plot of land that is 4 miles square is square and has sides 4 miles long. So the area is

$$4 \text{ mi} \times 4 \text{ mi} = \mathbf{16\ mi^2}$$

To summarize, if we multiply two perpendicular lengths, the product is the area of a rectangle.

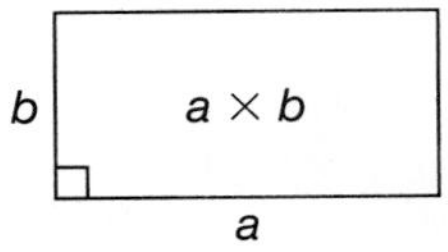

For squares the perpendicular lengths are called *sides.* For other rectangles they are called *length* and *width.* Thus, we get these formulas for the areas of squares and rectangles.

Area of a square = side × side	$A = s^2$
Area of a rectangle = length × width	$A = lw$

square root

The area of a square and the length of its side are related by "squaring." If we know the length of a side of a square, we square the length to find the area. This 3-by-3 square illustrates that $3^2 = 9$.

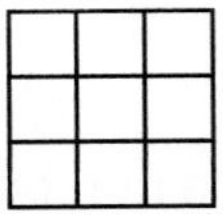

3 units squared is 9 square units.

If we know the area of a square, we can find the length of a side by finding the **square root** of the area. We often indicate square root with the radical symbol, $\sqrt{}$. This square also illustrates $\sqrt{9} = 3$ which we read as, "The square root of 9 equals 3."

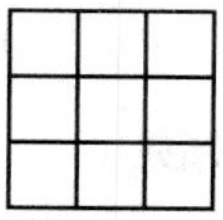

The square root of 9 square units is 3 units.

Example 7

Simplify:

a. $\sqrt{121}$ **b.** $\sqrt{8^2}$

Solution

a. To find the square root of 121 we may ask, "What number multiplied by itself equals 121?" Since $10 \times 10 = 100$, we try 11×11 and find that $11^2 = 121$. Therefore, $\sqrt{121}$ equals **11.**

b. Squaring and finding a square root are inverse operations, so one operation "undoes" the other operation.

$$\sqrt{8^2} = \sqrt{64} = \mathbf{8}$$

Practice Set

Use words to show how each exponential expression is read. Then find the value of each expression.

a. 4^3

b. $\left(\frac{1}{2}\right)^2$

c. 10^6

d. *Predict* Suppose you exchanged the base and the exponent in the expression 10^3 to get the expression 3^{10}. Would you predict that the values of the two expressions would be the same or different? Explain.

Analyze Find each missing exponent:

e. $2^3 \cdot 2^2 = 2^n$ **f.** $\dfrac{2^6}{2^2} = 2^m$

Evaluate Find each square root:

g. $\sqrt{100}$ **h.** $\sqrt{400}$ **i.** $\sqrt{15^2}$

Find the area of each rectangle:

j.

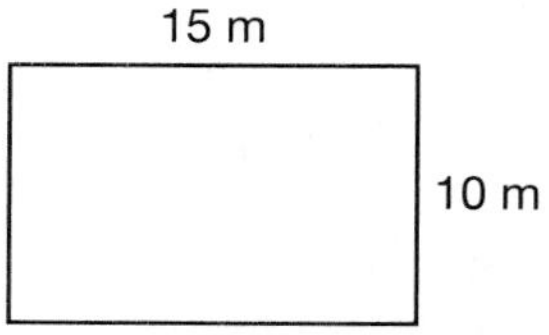

k.

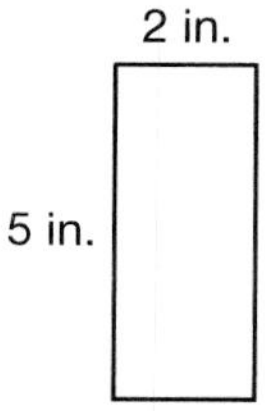

l.

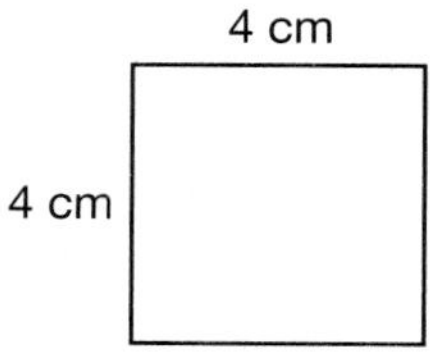

m. If the perimeter of a square is 20 cm, what is its area?

n. What is the area of a park that is 100 yards square?

o. Write a squaring fact and a square root fact illustrated by this 4-by-4 square.

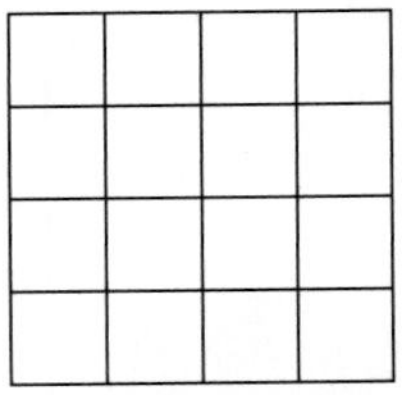

Written Practice

Strengthening Concepts

1. (13) There were 628 students in 4 college dormitories. Each dormitory housed the same number of students. How many students were housed in each dormitory?

2. (11) A candidate for the U.S. Senate from Arkansas wants to visit every county in Arkansas. She has already traveled to 36 counties. She has 39 left to visit. How many counties are there in Arkansas?

3. (16, 20) The area of a square is a function of the length of its sides. Make a function table that shows the area of a square in square units for side lengths of 1, 2, 3, 4, and 5 units.

*** 4.** (19, 20) Choose the formula for the area of a square.

A $P = 2L + 2W$

B $P = 4s$

C $A = s^2$

*** 5.** (4, 20) **Predict** The rule of the following sequence is $k = 2^n$. Find the sixth term of the sequence.

2, 4, 8, 16, . . .

6. (4, 8) **a.** Arrange these numbers in order from least to greatest:

$$\frac{1}{3}, -2, 1, -\frac{1}{2}, 0$$

b. **Justify** Draw a number line with the numbers from **a** to show that your answer to **a** is correct.

c. Which of these numbers are not integers?

7. (8) Which is the best estimate of how much of this rectangle is shaded?

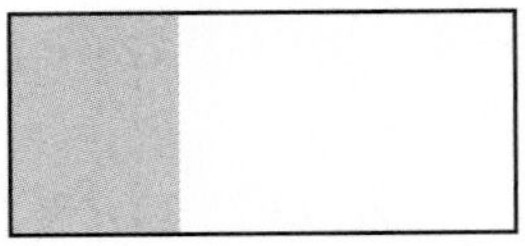

A 50%

B $33\frac{1}{3}\%$

C 25%

D 60%

8. (7) Each angle of a rectangle is a right angle. Which two sides are perpendicular to side *BC*?

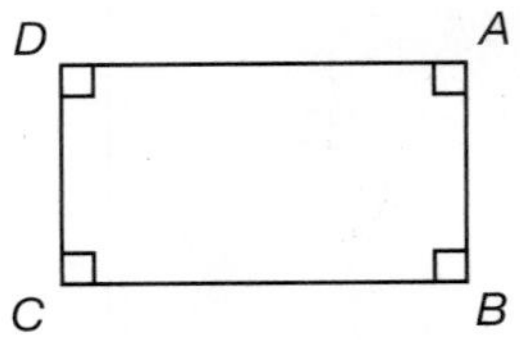

* **9.** (20) **Evaluate** Simplify:

a. $\left(\frac{1}{3}\right)^3$ **b.** 10^4 **c.** $\sqrt{12^2}$

10. (15) For each fraction, find an equivalent fraction that has a denominator of 36:

a. $\frac{2}{9}$ **b.** $\frac{3}{4}$

c. **Explain** Name one reason we might want to find equivalent fractions.

11. (6) List the factors of each number:

a. 10 **b.** 7 **c.** 1

* **12.** (16, 18) The perimeter of a certain square is 2 feet. How many inches long is each side of the square?

* **13.** (20) **Formulate** Write a squaring fact and a square root fact illustrated by this 6-by-6 square.

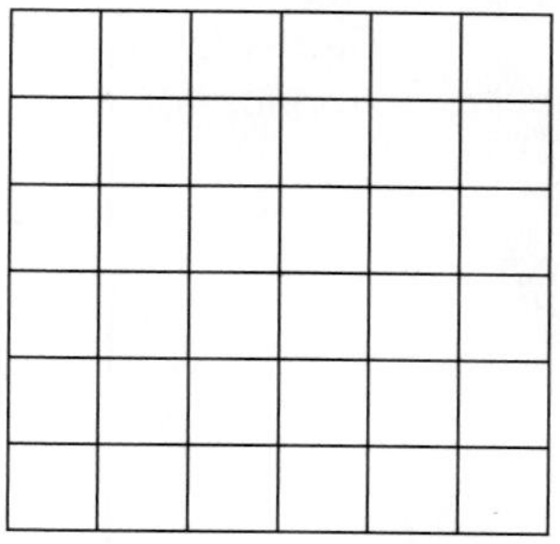

* **14.** (14) **Generalize** What is the probability of rolling an even number with one roll of a number cube?

Solve each equation:

15. (3) $5x = 60$

16. (3) $100 = m + 64$

* **17.** (20) $5^4 \cdot 5^2 = 5^n$

18. (3) $\frac{60}{y} = 4$

Simplify:

19. (10, 15) $1\frac{8}{9} + 1\frac{7}{9}$

20. (10) $\frac{5}{2} \cdot \frac{5}{6}$

21. (1) $\frac{6345}{9}$

22. (1) $\begin{array}{r} 360 \\ \times\ 25 \\ \hline \end{array}$

23. (9) $\frac{3}{4} - \left(\frac{1}{4} + \frac{2}{4}\right)$

24. (9) $\left(\frac{3}{4} - \frac{1}{4}\right) + \frac{2}{4}$

25. (1, 9) Evaluate the following expressions for $m = 3$ and $n = 10$:

a. $\frac{m}{n} + \frac{m}{n}$ **b.** $\frac{m}{n} \cdot \frac{m}{n}$

* **26.** (15) **Evaluate** Find a fraction equivalent to $\frac{1}{2}$ that has a denominator of 10. Add $\frac{3}{10}$ to that fraction and reduce the sum.

27. (10, 15) Write $1\frac{4}{5}$ as an improper fraction. Then multiply the improper fraction by $\frac{1}{3}$ and reduce the product.

*** 28.** Which properties are illustrated by these equations?
(2, 9, 15)

a. $\frac{3}{4} \times \frac{4}{3} = 1$ **b.** $\frac{1}{3} \cdot \frac{2}{2} = \frac{2}{6}$

*** 29.** **Generalize** A common floor tile is 12 inches square.
(19, 20)

a. What is the perimeter of a common floor tile?

b. What is the area of a common floor tile?

*** 30.** What is the perimeter of this hexagon?
(19)

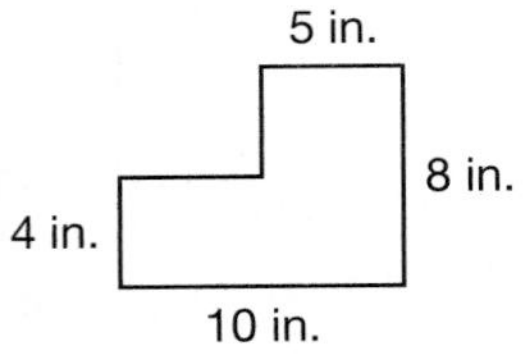

Early Finishers
Real-World Application

Baseboards line the perimeter of a room, covering the joint formed by the wall and the floor. A new house requiring baseboards has rooms with the following dimensions (in feet):

12 by 12	6 by 8
10 by 12	15 by 10
24 by 10	12 by 11

a. How many total feet of baseboards is this? Show your work.

b. If the baseboard comes in 8-foot sections, how many sections must be bought? Show your work. (Assume that sections of baseboard can be joined together as needed.)

Focus on

• Using a Compass and Straightedge, Part 1

Activity

Drawing Concentric Circles

Materials needed:

- Compass
- Ruler or straightedge
- Protractor

A **compass** is a tool used to draw **circles** and portions of circles called **arcs.** Compasses are manufactured in various forms. Here we show two forms:

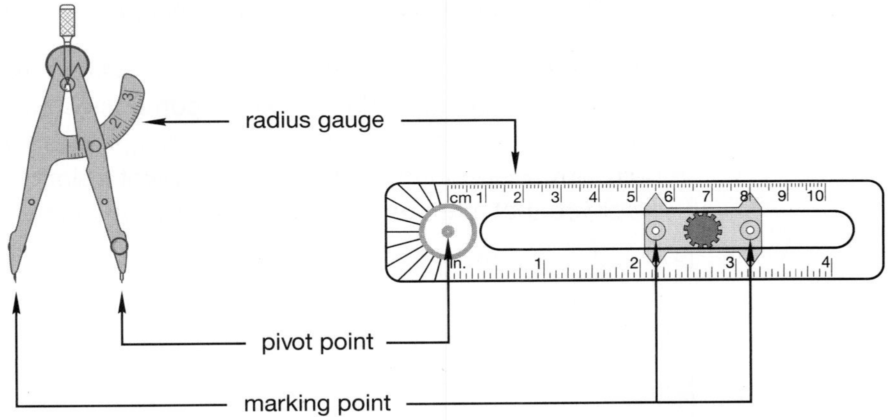

The **marking point** of a compass is the pencil point that draws circles and arcs. The marking point rotates around the **pivot point,** which is placed at the **center** of the desired circle or arc. The **radius** (plural, **radii**) of the circle, which is the distance from every point on the circle to the center of the circle, is set by the **radius gauge.** The radius gauge identifies the distance between the pivot point and the marking point of the compass.

concentric circles

Concentric circles are two or more circles with a common center. When a pebble is dropped into a quiet pool of water, waves forming concentric circles can be seen. A bull's-eye target is another example of concentric circles.

To draw concentric circles with a compass, we begin by swinging the compass a full turn to make one circle. Then we make additional circles using the same center, changing the radius for each new circle.

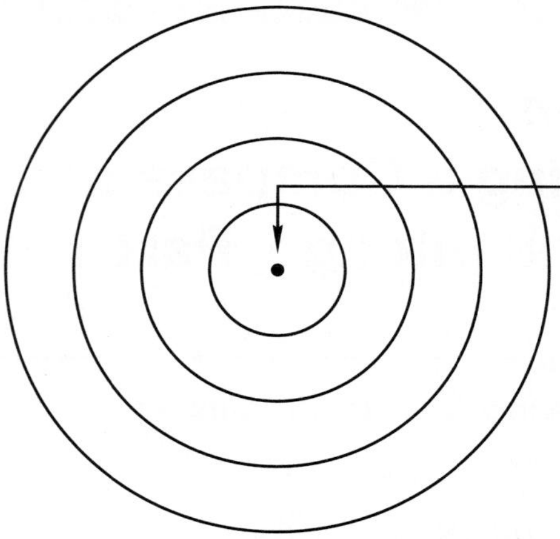

1. Practice drawing several concentric circles.

regular hexagon and regular triangle

Recall that all the sides of a regular polygon are equal in length and all the angles are equal in measure. Due to their uniform shape, regular polygons can be **inscribed** in circles. A polygon is inscribed in a circle if all of its vertices are on the circle and all of the other points of the polygon are within the circle. We will inscribe a regular hexagon and a regular triangle.

First we fix the compass at a comfortable setting that will not change until the project is finished. We swing the compass a full turn to make a circle. Then we lift the compass without changing the radius and place the pivot point anywhere on the circle. With the pivot point on the circle, we swing a small arc that intersects the circle, as shown below.

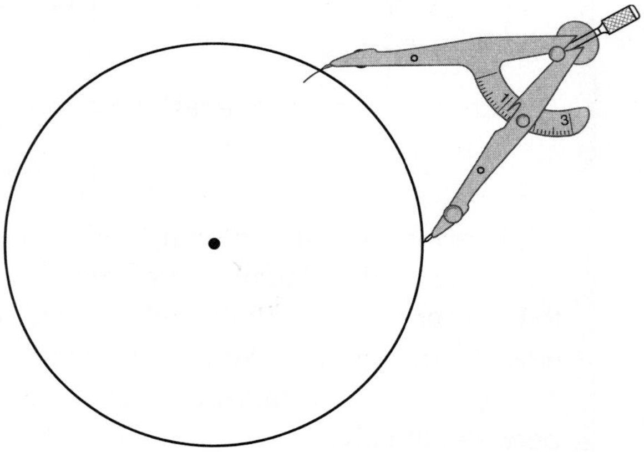

Thinking Skill

Predict

What will happen if we change the distance between the marking point and the pivot point as we make marks on the circle?

Again we lift the compass without changing the radius and place the pivot point at the point where the arc intersects the circle. From this location we swing another small arc that intersects the circle. We continue by moving the pivot point to where each new arc intersects the circle, until six small arcs are drawn on the circle. We find that the six small arcs are equally spaced around the circle.

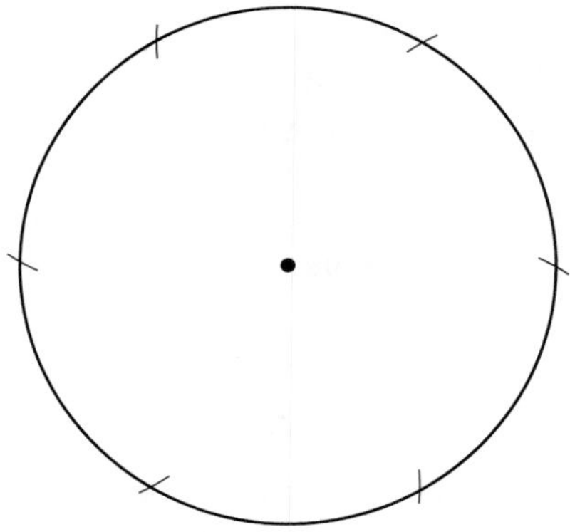

Now, to inscribe a regular hexagon, we draw line segments connecting each point where an arc intersects the circle to the next point where an arc intersects the circle.

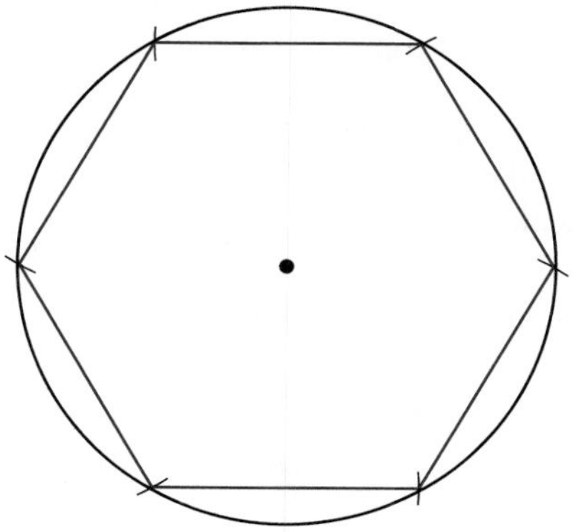

2. *Model* Use a compass and straightedge to inscribe a regular hexagon in a circle.

To inscribe a regular triangle, we will start the process over again. We swing the compass a full turn to make a circle. Then, without resetting the radius, we swing six small arcs around the circle. A triangle has three vertices, but there are six points around the circle where the small arcs intersect the circle. Therefore, to inscribe a regular triangle, we draw segments between *every other* point of intersection. In other words, we skip one point of intersection for each side of the triangle.

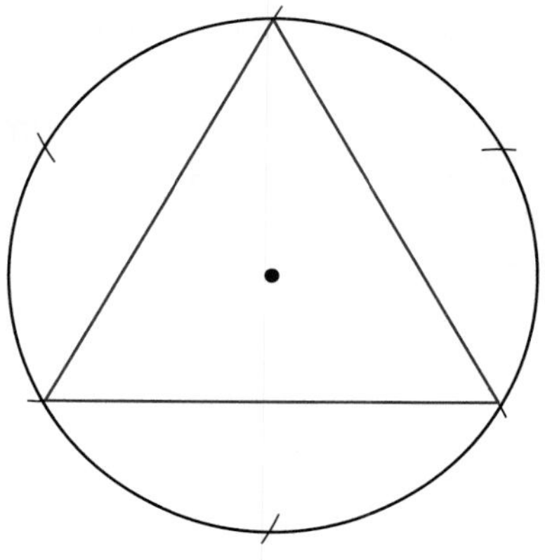

3. Use your tools to inscribe a regular triangle in a circle.

With a protractor we can measure each angle of the triangle. Since the vertex of each angle is on the circle and the angle opens to the interior of the circle, the angle is called an **inscribed angle.**

4. What is the measure of each inscribed angle? (If necessary, extend the rays of each angle to perform the measurements.)

5. What is the sum of the measures of all three angles of the triangle?

6. *Predict* What shape will we make if we now draw segments between the remaining three points of intersection?

dividing a circle into sectors

We can use a compass and straightedge to divide a circle into equal parts. First we swing the compass a full turn to make a circle. Next we draw a segment across the circle through the center of the circle. A segment with both endpoints on a circle is a **chord.** The longest chord of a circle passes through the center and is called a **diameter** of the circle.

Discuss What is the relationship between a chord and a diameter? Explain.

Math Language
The prefix *circum-* in **circumference** means "around." The prefix *semi-* in **semicircle** means "half."

Notice that a diameter equals two radii. Thus the length of a diameter of a circle is twice the length of a radius of the circle. The **circumference** is the distance around the circle and is determined by the length of the radius and diameter, as we will see in a later lesson.

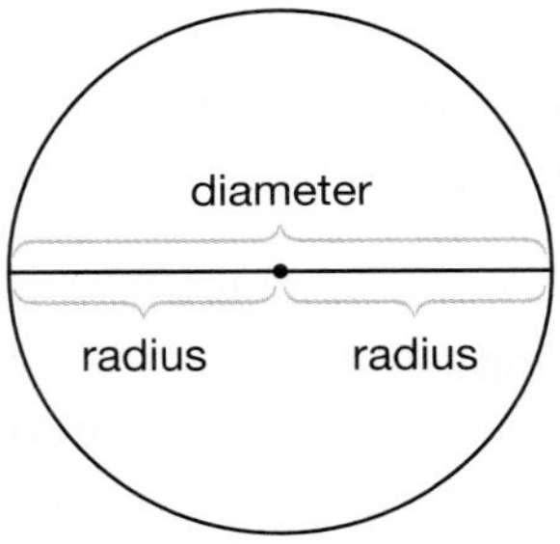

A diameter divides a circle into two half circles called **semicircles.**

To divide a circle into thirds, we begin with the process we used to inscribe a hexagon. We draw a circle and swing six small arcs. Then we draw three segments from the center of the circle to *every other* point where an arc intersects the circle. These segments divide the circle into three congruent **sectors.** A sector of a circle is a region bounded by an arc of the circle and two of its radii. A model of a sector is a slice of pizza.

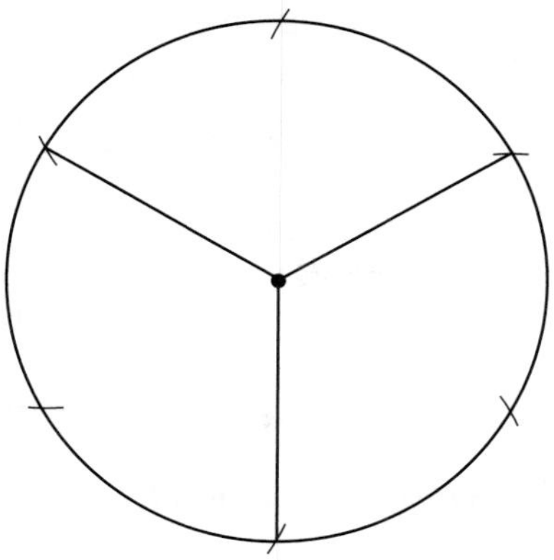

7. Use a compass and straightedge to draw a circle and to divide the circle into thirds.

The segments we drew from the center to the circle formed angles. Each angle that has its vertex at the center of the circle is a **central angle.** We can measure a central angle with a protractor. We may extend the rays of the central angle if necessary in order to use the protractor.

8. What is the measure of each central angle of a circle divided into thirds?

9. Each sector of a circle divided into thirds occupies what percent of the area of the whole circle?

To divide a circle into sixths, we again begin with the process we used to inscribe a hexagon. We divide the circle by drawing a segment from the center of the circle to the point of intersection of each small arc.

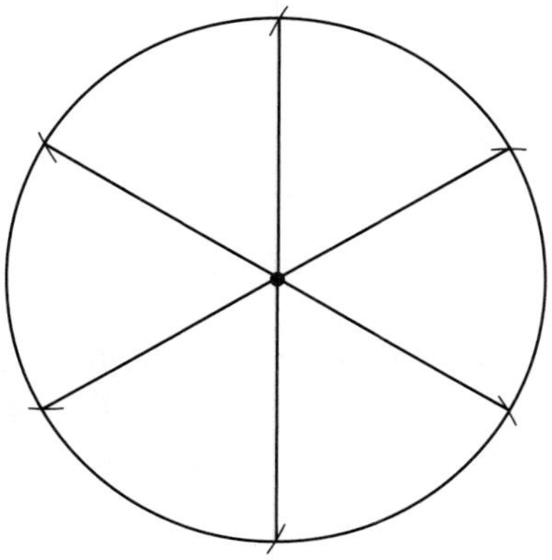

10. **a.** What is the measure of each central angle of a circle that has been divided into sixths?

    b. *Conclude* What is the sum of the angle measures of the entire circle?

11. Each sector of a circle divided into sixths occupies what percent of the area of the whole circle?

In problems **12–24** we provide definitions of terms presented in this investigation. Find the term for each definition:

12. The distance around a circle

13. The distance across a circle through its center

14. The distance from the center of a circle to any point on the circle

15. Part of the circumference of a circle

16. A region bounded by an arc of a circle and two radii

17. Two or more circles with the same center

18. A segment that passes through the interior of a circle and has both endpoints on the circle

19. A polygon whose vertices are on a circle and whose other points are inside the circle

20. A half circle

21. An angle whose vertex is the center of a circle

22. The distance between the pivot point and the marking point of a compass when drawing a circle

23. The point that is the same distance from any point on a circle

24. An angle that opens to the interior of the circle from a vertex on the circle

25. What professions might require the use of compasses, circles, and arcs?

The following paragraphs summarize important facts about circles.

The distance around a circle is its *circumference.* Every point on the circle is the same distance from the center of the circle. The distance from the center to a point on the circle is the *radius.* The distance across the circle through its center is the *diameter,* which equals two radii. A diameter divides a circle into two half circles called *semicircles.* A diameter, as well as any other segment between two points on a circle, is a *chord* of the circle. Two or more circles with the same center are concentric circles.

An angle formed by two radii of a circle is called a *central angle.* A central angle opens to a portion of a circle called an *arc,* which is part of the circumference of a circle. The region enclosed by an arc and its central angle is called a *sector.*

An angle whose vertex is on the circumference of a circle and whose sides are chords of the circle is an *inscribed angle.* A polygon is inscribed in a circle if all of its vertices are on the circumference of the circle.

LESSON 21

• Prime and Composite Numbers
• Prime Factorization

Power Up

Building Power

facts Power Up E

mental math

a. Number Sense: \$1.25 + 99¢

b. Decimals: \$6.50 ÷ 10

c. Number Sense: \$20.00 − \$15.75

d. Calculation: 6 × 34

e. Calculation: $1\frac{2}{3} + 2\frac{1}{3}$

f. Fractional Parts: $\frac{1}{3}$ of 36

g. Measurement: Which is greater 3 pints or 1 quart?

h. Calculation: Start with the number of sides of a hexagon, × 5, + 2, ÷ 8, + 1, ÷ 5.

problem solving

The first even counting number is 2; the sum of the first two even counting numbers is 6; the sum of the first three even counting numbers is 12. Add to this list the sums of first four, five, and six even counting numbers. Does this list of the sums of even counting numbers have a pattern? Can you describe a rule for continuing the sequence?

Understand We are given the sums of the first one, two, and three even counting numbers. We are asked to find the sums of the first four, five, and six even counting numbers and to find a pattern in the sums.

Plan We will *make a chart* to help us record our work in an organized way. Then we will use our chart to *find a pattern* in the sums of sequences of even counting numbers.

Solve We write the first six sequences, the number of terms in each sequence, and the sum of each sequence on our chart:

Sequence	Number of Terms	Sum
2	1	2
2 + 4	2	6
2 + 4 + 6	3	12
2 + 4 + 6 + 8	4	20
2 + 4 + 6 + 8 + 10	5	30
2 + 4 + 6 + 8 + 10 + 12	6	42

When we look at the number of terms and the resulting sums, we see several numbers that belong to the same fact families: 1 is a factor of 2, 2 is a factor of 6, 3 is a factor of 12, etc. We rewrite each sum as a multiplication problem using the number of terms as one of the factors:

Number of Terms		Sum
1	$1 \times 2 =$	2
2	$2 \times 3 =$	6
3	$3 \times 4 =$	12
4	$4 \times 5 =$	20
5	$5 \times 6 =$	30
6	$6 \times 7 =$	42

To find each sum, we can multiply the number of terms in the sequence by the next whole number.

Check We found a pattern in the sums of the sequences of even counting numbers. We can verify our solution by finding the sum of the seventh sequence using our multiplication method, then check the sum by adding the numbers one-by-one.

New Concepts *Increasing Knowledge*

prime and composite numbers

We remember that the counting numbers (or natural numbers) are the numbers we use to count. They are

1, 2, 3, 4, 5, 6, 7, 8, 9, 10, ...

Counting numbers greater than 1 are either **prime numbers** or **composite numbers.** A prime number has exactly two different factors, and a composite number has three or more factors. In the following table, we list the factors of the first ten counting numbers. The numbers 2, 3, 5, and 7 each have exactly two factors, so they are prime numbers.

Factors of Counting Numbers 1–10

Number	Factors
1	1
2	1, 2
3	1, 3
4	1, 2, 4
5	1, 5
6	1, 2, 3, 6
7	1, 7
8	1, 2, 4, 8
9	1, 3, 9
10	1, 2, 5, 10

We see that the factors of each of the prime numbers are 1 and the number itself. So we define a prime number as follows:

> A **prime number** is a counting number greater than 1 whose only factors are 1 and the number itself.

From the table we can also see that 4, 6, 8, 9, and 10 each have three or more factors, so they are composite numbers. Each composite number is divisible by a number other than 1 and itself.

Discuss The number 1 is neither a prime number nor a composite number. Why do you think that is true?

Example 1

Make a list of the prime numbers that are less than 16.

Solution

First we list the counting numbers from 1 to 15.

1, 2, 3, 4, 5, 6, 7, 8, 9, 10, 11, 12, 13, 14, 15

A prime number must be greater than 1, so we cross out 1. The next number, 2, has only two divisors (factors), so 2 is a prime number. However, all the even numbers greater than 2 are divisible by 2, so they are not prime. We cross these out.

~~1~~, 2, 3, ~~4~~, 5, ~~6~~, 7, ~~8~~, 9, ~~10~~, 11, ~~12~~, 13, ~~14~~, 15

The numbers that are left are

2, 3, 5, 7, 9, 11, 13, 15

The numbers 9 and 15 are divisible by 3, so we cross them out.

2, 3, 5, 7, ~~9~~, 11, 13, ~~15~~

The only divisors of each remaining number are 1 and the number itself. So the prime numbers less than 16 are **2, 3, 5, 7, 11,** and **13.**

Example 2

List the factor pairs for each of these numbers:

16 17 18

Classify **Which of these numbers is prime?**

Solution

The factor pairs for 16 are **1 and 16, 2 and 8, 4 and 4.**

The factor pair for 17 is **1 and 17.**

The factor pairs for 18 are **1 and 18, 2 and 9, 3 and 6.**

Note that perfect squares have one pair of identical factors. Therefore they have an odd number of different factors. Also note that prime numbers have only one factor pair since they have only two factors.

Example 3

List the composite numbers between 40 and 50.

Solution

First we write the counting numbers between 40 and 50.

41, 42, 43, 44, 45, 46, 47, 48, 49

Any number that is divisible by a number besides 1 and itself is composite. All the even numbers in this list are composite since they are divisible by 2. That leaves the odd numbers to consider. We quickly see that 45 is divisible by 5, and 49 is divisible by 7. So both 45 and 49 are composite. The remaining numbers, 41, 43, and 47, are prime. So the composite numbers between 40 and 50 are **42, 44, 45, 46, 48,** and **49.**

prime factorization

Thinking Skill

Generalize

The prime factorization of a number is: $2 \cdot 3 \cdot 3$. What is the number?

Every composite number can be *composed* (formed) by multiplying two or more prime numbers. Here we show each of the first nine composite numbers written as a product of **prime** factors,

$4 = 2 \cdot 2$	$6 = 2 \cdot 3$	$8 = 2 \cdot 2 \cdot 2$
$9 = 3 \cdot 3$	$10 = 2 \cdot 5$	$12 = 2 \cdot 2 \cdot 3$
$14 = 2 \cdot 7$	$15 = 3 \cdot 5$	$16 = 2 \cdot 2 \cdot 2 \cdot 2$

Notice that we **factor** 8 as $2 \cdot 2 \cdot 2$ and not $2 \cdot 4$, because 4 is not prime.

When we write a composite number as a product of prime numbers, we are writing the **prime factorization** of the number.

Example 4

Write the prime factorization of each number.

a. **30** **b.** **81** **c.** **420**

Solution

We will write each number as the product of two or more prime numbers.

a. $30 = \mathbf{2 \cdot 3 \cdot 5}$ We do not use $5 \cdot 6$ or $3 \cdot 10$, because neither 6 nor 10 is prime.

b. $81 = \mathbf{3 \cdot 3 \cdot 3 \cdot 3}$ We do not use $9 \cdot 9$, because 9 is not prime.

c. $420 = \mathbf{2 \cdot 2 \cdot 3 \cdot 5 \cdot 7}$ Two methods for finding this are shown after example 5.

Explain How can you quickly tell that these three numbers are composite without writing the complete prime factorization of each number?

Example 5

Write the prime factorization of 100 and of $\sqrt{100}$.

Solution

The prime factorization of 100 is **2 · 2 · 5 · 5.** We find that $\sqrt{100}$ is 10, and the prime factorization of 10 is **2 · 5.** Notice that 100 and $\sqrt{100}$ have the same prime factors, 2 and 5, but that each factor appears half as often in the prime factorization of $\sqrt{100}$.

There are two commonly used methods for factoring composite numbers. One method uses a factor tree. The other method uses division by primes. We will factor 420 using both methods.

Thinking Skill

Represent

You can begin a factor tree with any pair of factors. Find the prime factorization of 420 starting with 2 and 210. Then find it starting with 6 and 70. Is the result the same?

To factor a number using a **factor tree,** we first write the number. Below the number we write any two whole numbers greater than 1 that multiply to equal the number. If these numbers are not prime, we continue the process until there is a prime number at the end of each "branch" of the factor tree. These numbers are the prime factors of the original number. We write them in order from least to greatest.

Factor Tree

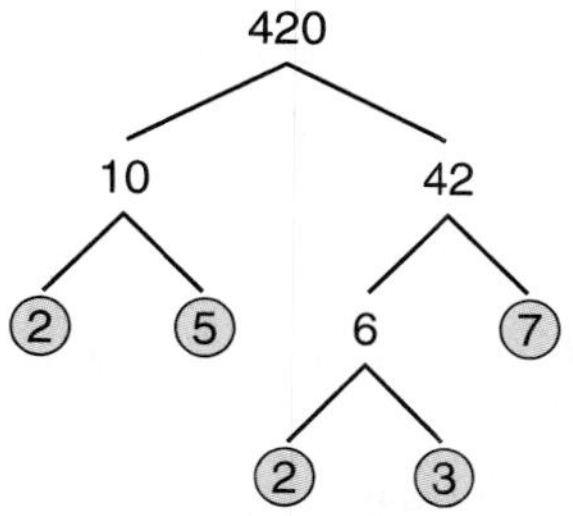

$$420 = 2 \cdot 2 \cdot 3 \cdot 5 \cdot 7$$

To factor a number using **division by primes,** we write the number in a division box and divide by the smallest prime number that is a factor. Then we divide the resulting quotient by the smallest prime number that is a factor. We repeat this process until the quotient is 1.[1] The divisors are the prime factors of the number.

Division by Primes

1
7)7
5)35
3)105
2)210
2)420

$$420 = 2 \cdot 2 \cdot 3 \cdot 5 \cdot 7$$

We can use prime factorization to help us find the greatest common factor (GCF) of two or more numbers.

Step 1: List the prime factors for each number.

Step 2: Identify the shared factors.

Step 3: Multiply the shared factors to find the GCF.

[1] Some people prefer to divide until the quotient is a prime number. In this case, the final quotient is included in the list of prime factors.

Example 6

Write the prime factorization of 36 and 60. Use the results to find the greatest common factor of 36 and 60.

Solution

1. Using a factor tree or division by primes, we find the prime factorization of 36 and 60.
2. Identify the shared factors.

$$36 = 2 \cdot 2 \cdot 3 \cdot 3$$
$$60 = 2 \cdot 2 \cdot 3 \cdot 5$$

We see 36 and 60 share two 2s and one 3.

3. We multiply the shared factors: $2 \cdot 2 \cdot 3 = 12$. The GCF of 36 and 60 is **12.**

Practice Set

a. List the first ten prime numbers.

b. *Classify* If a whole number greater than 1 is not prime, then what kind of number is it?

c. Write the prime factorization of 81 using a factor tree.

d. Write the prime factorization of 360 using division by primes.

e. *Generalize* Write the prime factorization of 64 and of $\sqrt{64}$.

f. Use prime factorization to find the GCF of 18 and 81.

Written Practice

Strengthening Concepts

1. (14) Two thirds of the students wore green on St. Patrick's Day. What fraction of the students did not wear green on St. Patrick's Day?

2. (13) Three hundred forty-three quills were carefully placed into 7 compartments. If each compartment held the same number of quills, how many quills were in each compartment?

3. (19, 20) Choose the formula for the perimeter of a rectangle.

   A $P = 2L + 2W$ **B** $P = 4s$

   C $A = LW$ **D** $A = s^2$

4. (20) Write a squaring fact and a square root fact illustrated by this square.

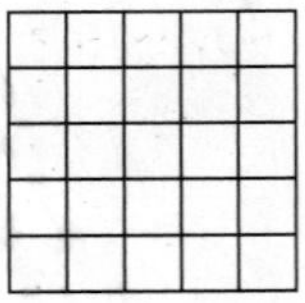

5. Write each number as a reduced fraction or mixed number:
(15)
a. $3\frac{12}{21}$ b. $\frac{12}{48}$ c. 12%

* 6. List the prime numbers between 50 and 60.
(21)

* 7. Write the prime factorization of each number:
(21)
a. 50 b. 60 c. 300

8. **Justify** Which point could represent 1610 on this number line? How did you decide?
(4)

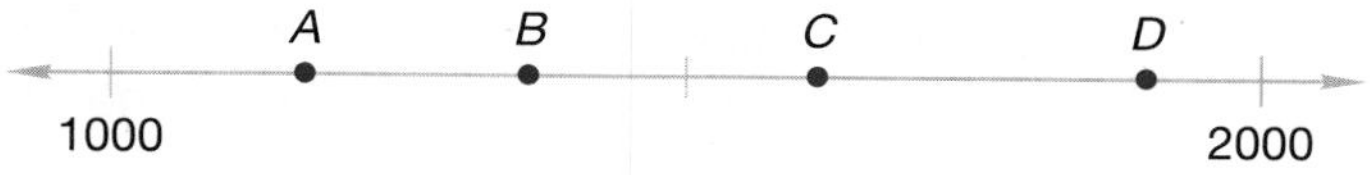

9. Complete each equivalent fraction:
(15)
a. $\frac{2}{3} = \frac{?}{15}$ b. $\frac{3}{5} = \frac{?}{15}$ c. $\frac{?}{3} = \frac{8}{12}$

d. What property of multiplication do we use to rename fractions?

10. a. How many $\frac{1}{3}$s are in 1?
(9)
b. How many $\frac{1}{3}$s are in 3?

* 11. The perimeter of a regular quadrilateral is 12 inches. What is its area?
(20)

* 12. **Represent** Use a ruler to draw a rectangle that is $\frac{3}{4}$ in. wide and twice as long as it is wide.
(8, 19)
a. How long is the rectangle?
b. What is the perimeter of the rectangle?

* 13. Find the perimeter of this hexagon:
(19)

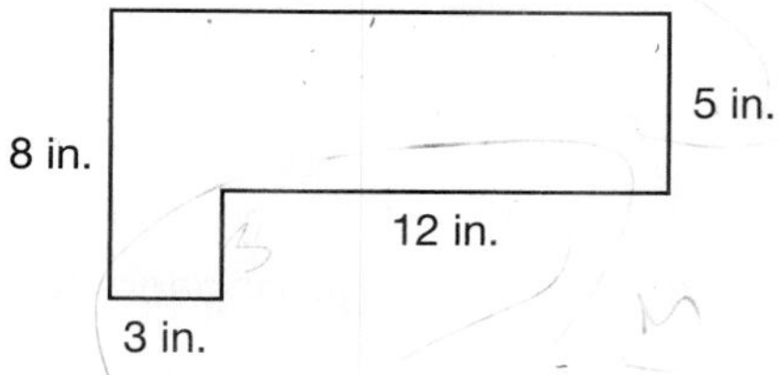

14. A number cube is rolled once. What is the probability of getting an odd number greater than 5?
(14)

Solve:

15. $p + \frac{3}{5} = 1$ (9)

16. $\frac{3}{5}q = 1$ (9)

17. $\frac{w}{25} = 50$ (3)

18. $\frac{1}{6} + f = \frac{5}{6}$ (9, 15)

19. $m - 3\frac{2}{3} = 1\frac{2}{3}$ (10)

20. $51 = 3c$ (3)

Simplify:

21. $\frac{2}{3} + \frac{2}{3} + \frac{2}{3}$ (9)

* 22. $\left(\frac{2}{3}\right)^3$ (20)

* **23.** (21) **a.** Write the prime factorization of 225.

b. **Generalize** Find $\sqrt{225}$ and write its prime factorization.

24. (15) Describe how finding the greatest common factor of the numerator and denominator of a fraction can help reduce the fraction.

25. (17) Draw $\overline{AB}$ $2\frac{1}{2}$ inches long. Then draw $\overline{BC}$ $2\frac{1}{2}$ inches long perpendicular to $\overline{AB}$. Complete the triangle by drawing $\overline{AC}$. Use a protractor to find the measure of $\angle A$.

26. (9, 10) Write $1\frac{3}{4}$ as an improper fraction. Multiply the improper fraction by the reciprocal of $\frac{2}{3}$. Then write the product as a mixed number.

* **27.** (Inv. 2) **Classify** Refer to the circle at right with center at point M to answer **a–d.**

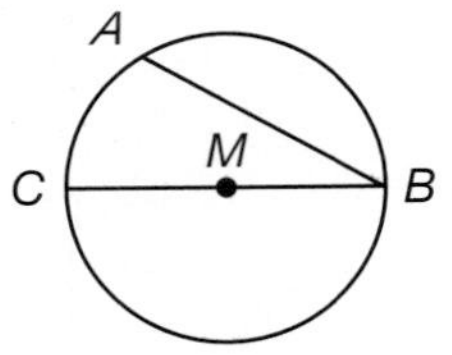

a. Which segment is a diameter?

b. Which segment is a chord but not a diameter?

c. Which two segments are radii?

d. Which angle is an inscribed angle?

28. (16) Alicia's father asked her to buy a gallon of milk at the store. The store had milk only in quart-sized containers. What percent of a gallon is a quart? How many quart containers did Alicia have to buy?

29. (2) **a.** Compare: $a + b \bigcirc b + a$

b. What property of operations applies to part **a** of this problem?

* **30.** (18) **Analyze** Refer to the triangles below to answer **a–c.**

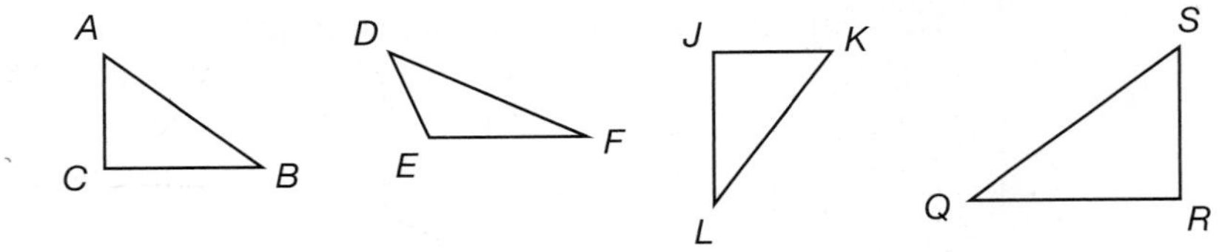

a. Which triangle appears to be congruent to $\triangle ABC$?

b. Which triangle is not similar to $\triangle ABC$?

c. Which angle in $\triangle QRS$ corresponds to $\angle A$ in $\triangle ABC$?

LESSON
22

• Problems About a Fraction of a Group

Power Up

Building Power

facts Power Up E

mental math

a. Number Sense: \$1.54 + 99¢

b. Decimals: 8¢ × 100

c. Calculation: \$10.00 − \$7.89

d. Calculation: 7 × 53

e. Calculation: $3\frac{3}{4} + 1\frac{1}{4}$

f. Fractional Parts: $\frac{1}{4}$ of 24

g. Measurement: Which is greater a gallon or 2 quarts?

h. Calculation: Start with the number of years in half a century. Add the number of inches in half a foot; then divide by the number of days in a week. What is the name of the polygon with this number of sides?

problem solving

Yin has 25 tickets, Bobby has 12 tickets, and Mary has 8 tickets. How many tickets should Yin give to Bobby and to Mary so that they all have the same number of tickets?

New Concept

Increasing Knowledge

In Lesson 13 we looked at problems about equal groups. In Lesson 14 we considered problems about parts of a whole. In this lesson we will solve problems that involve both equal groups and parts of a whole. Many of the problems will require two or more steps to solve.

Consider the following statement:

Two thirds of the students in the class wore sneakers on Monday.

We can draw a diagram for this statement. We use a rectangle to represent all the students in the class. Next we divide the rectangle into three equal parts. Then we describe the parts.

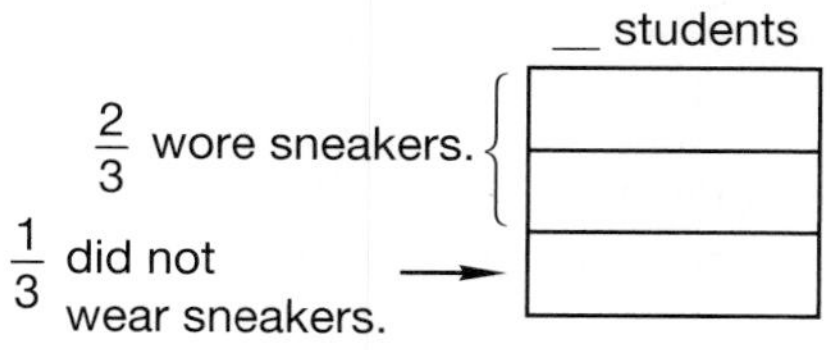

If we know how many students are in the class, we can figure out how many students are in each part.

Two thirds of the 27 students in the class wore sneakers on Monday.

There are 27 students in all. If we divide the group of 27 students into three equal parts, there will be 9 students in each part. We write these numbers on our diagram.

Analyze Why do we divide the rectangle into 3 equal parts rather than any other number of equal parts?

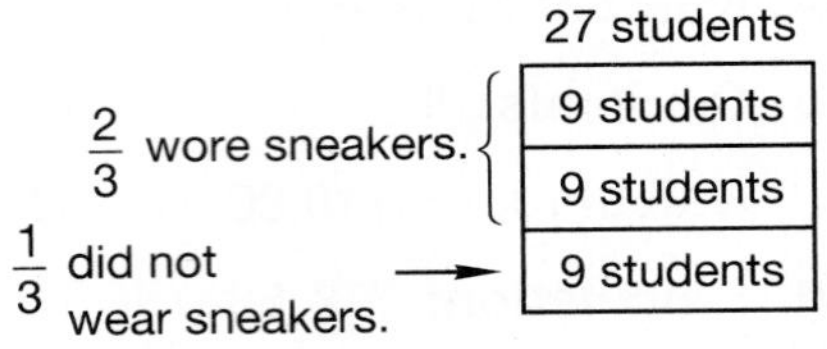

Since $\frac{2}{3}$ of the students wore sneakers, we add two of the parts and find that 18 students wore sneakers. Since $\frac{1}{3}$ of the students did not wear sneakers, we find that 9 students did not wear sneakers.

Example 1

Diagram this statement. Then answer the questions that follow.

Two fifths of the 30 students in the class are boys.

a. How many boys are in the class?

b. How many girls are in the class?

Solution

Thinking Skill

Model

How could you use colored counters to model the solution to example 1?

We draw a rectangle to represent all 30 students. Since the statement uses fifths to describe a part of the class, we divide the class of 30 students into five equal parts. Since 30 ÷ 5 is 6, there are 6 students in each part.

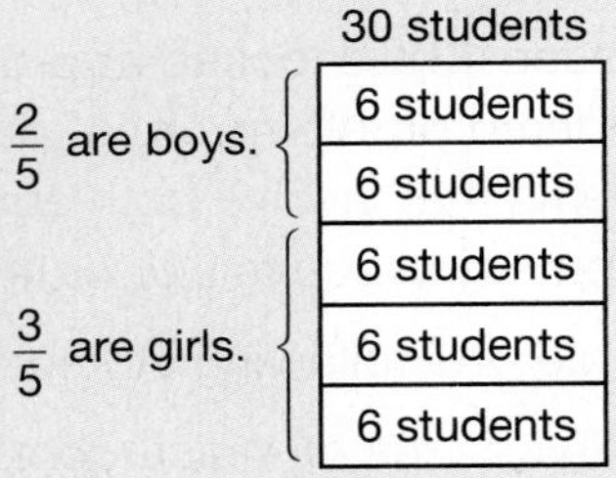

Now we can answer the questions.

a. Two of the five parts are boys. Since there are 6 students in each part, there are **12 boys.**

b. Since two of the five parts are boys, three of the five parts must be girls. Thus there are **18 girls.**

Another way to find the answer to **b** after finding the answer to **a** is to subtract. Since 12 of the 30 students are boys, the rest of the students (30 − 12 = 18) are girls.

Predict How many girls would there be in the class if $\frac{1}{5}$ were boys?

Example 2

In the following statement, change the percent to a fraction. Then diagram the statement and answer the questions.

Britt read 80% of a 40-page book in one day.

a. What fraction of the book did Britt read in one day?

b. How many pages did Britt read in one day?

Math Language
Remember that a **percent** can be expressed as a fraction with a denominator of 100.

Solution

This problem is about a fraction of a group, but the fraction is expressed as a percent. We write 80% as 80 over 100 and reduce.

$$\frac{80}{100} \div \frac{20}{20} = \frac{4}{5}$$

So 80% is equivalent to the fraction $\frac{4}{5}$.

Now we draw a rectangle to represent all 40 pages, dividing the rectangle into five equal parts. Since 40 ÷ 5 is 8, there are 8 pages in each part.

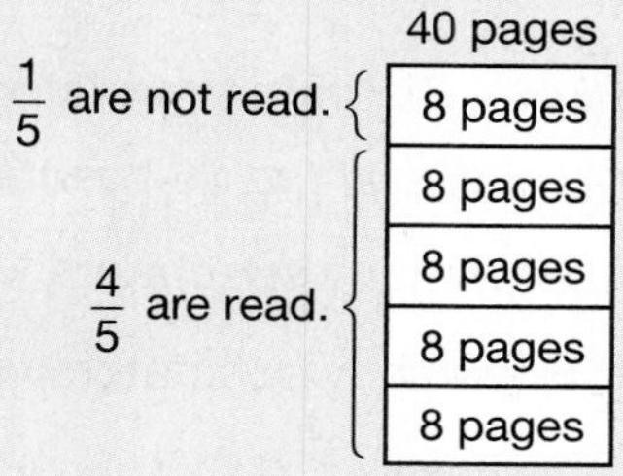

Now we can answer the questions.

a. Britt read $\frac{4}{5}$ **of the book** in one day.

b. Britt read 4 × 8 pages, which is **32 pages in one day.**

Represent Write an equation you could use to find the number of pages Britt did not read yet. Use the answer to **b** above to help you write the equation. Then solve the equation.

Practice Set

Model Diagram each statement. Then answer the questions.

First statement: *Three fourths of the 60 marbles in the bag were red.*

a. How many marbles were red?

b. How many marbles were not red?

Second statement: *Sixty percent of the 20 tomatoes were green.*

c. What fraction of the tomatoes were not green?

d. How many tomatoes were green?

e. **Formulate** For the following statement, write and answer two questions: *Three fifths of the thirty students were girls.*

Written Practice *Strengthening Concepts*

1. (11) In Room 7 there are 28 students. In Room 9 there are 30 students. In Room 11 there are 23 students. How many students are in all three rooms?

2. (13) If the total number of students in problem 1 were equally divided among three rooms, how many students would be in each room?

3. (11) The largest state is Alaska. It has an area of about 663,000 square miles. The smallest state, Rhode Island, has an area of about 1,500 square miles. About how many more square miles is Alaska than Rhode Island?

*** 4.** (19, 20)
a. Write the formula for the perimeter of a square.

b. A landscape planner designed a square garden that is 24 feet long per side. How many feet of border are needed to surround the garden?

*** 5.** (22) **Model** Diagram this statement. Then answer the questions that follow.

Five ninths of the 36 spectators were happy with the outcome.

a. How many spectators were happy with the outcome?

b. How many spectators were not happy with the outcome?

*** 6.** (22) In the following statement, change the percent to a reduced fraction. Then diagram the statement and answer the questions.

Twenty-five percent of three dozen plants are blooming.

a. What fraction of the total number of plants are not blooming?

b. How many plants are not blooming?

7. (14)
a. What fraction of the rectangle is shaded?

b. What percent of the rectangle is not shaded?

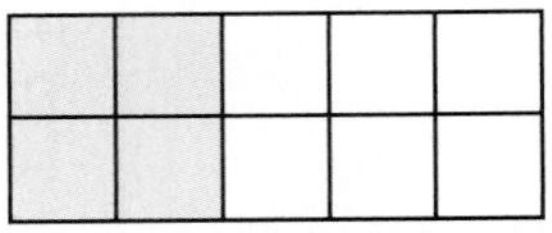

8. (9)
a. How many $\frac{1}{4}$s are in 1?

b. **Explain** Tell how you can use the answer to part **a** to find the number of $\frac{1}{4}$s in 3.

9. (2)
a. Multiply: $6 \cdot 5 \cdot 4 \cdot 3 \cdot 2 \cdot 1 \cdot 0$

b. **Analyze** What property is illustrated by the multiplication in part **a?**

10. (9) **Analyze** Simplify and compare: $\frac{3}{3} - \left(\frac{1}{3} \cdot \frac{3}{1}\right) \bigcirc \left(\frac{3}{3} - \frac{1}{3}\right) \cdot \frac{3}{1}$

* 11. (19, 20) Draw a rectangle $ABCD$ so that AB is 2 in. and BC is 1 in.

a. What is the perimeter of rectangle $ABCD$?

b. What is the area of the rectangle?

c. What is the sum of the measures of all four angles of the rectangle?

* 12. (21) **Generalize** Write the prime factorization of each number:

a. 32 b. 900 c. $\sqrt{900}$

13. (15) For each fraction, write an equivalent fraction that has a denominator of 60.

a. $\frac{5}{6}$ b. $\frac{3}{5}$ c. $\frac{7}{12}$

14. (10) Add the three fractions with denominators of 60 from problem **13,** and write their sum as a mixed number.

15. (4, 10) a. Arrange these numbers in order from least to greatest:

$$0, -\frac{2}{3}, 1, \frac{3}{2}, -2$$

b. Which of these numbers are positive?

* 16. (14, 15) **Predict** If one card is drawn from a regular deck of cards, what is the probability the card will be a heart?

Evaluate Find the value of each variable.

17. (9, 15) $\frac{5}{12} + a = \frac{11}{12}$

18. (3) $121 = 11x$

19. (10) $2\frac{2}{3} = y - 1\frac{1}{3}$

* 20. (20) $10^2 \cdot 10^5 = 10^n$

Simplify:

21. (15) $\frac{5}{6} + \frac{5}{6} + \frac{5}{6}$

22. (10) $\frac{15}{2} \cdot \frac{10}{3}$

* 23. (20) $\left(\frac{5}{6}\right)^2$

* 24. (20) $\sqrt{30^2}$

* 25. (2,9) **Justify** Give reasons for the steps used to simplify the following expression by using the commutative, associative, inverse, and identity properties of multiplication.

$$\frac{3}{1} \cdot \frac{2}{3} \cdot \frac{1}{3}$$

26. (10,15) Write $1\frac{1}{2}$ and $1\frac{2}{3}$ as improper fractions. Then multiply the improper fractions, and write the product as a mixed number.

* 27. (16) A package that weighs 1 lb 5 oz weighs how many ounces?

* 28. (17) Use a protractor to draw a 45° angle.

29. (4, 9) **Justify** Find the next number in this sequence and explain how you found your answer.

$$\ldots, 100, 10, 1, \frac{1}{10}, \ldots$$

30. (4) Write an odd negative integer greater than -3.

Early Finishers
Real-World Application

This weekend workers will cut the grass on the high school football field and repaint the white outline around the field.

a. The field is 360 feet long and 160 feet wide. Find the perimeter and area of the field.

b. One quart of paint is enough to paint a 200 ft stripe. How many quarts of paint should be purchased to paint a stripe around the entire field? Show your work.

c. If it takes a large mower 25 seconds to mow 800 ft^2, how long will it take to mow the whole field? Show your work. (Assume the paths are cut with no overlap.)

LESSON
23

Subtracting Mixed Numbers with Regrouping

Power Up

Building Power

facts Power Up E

mental math

a. **Number Sense:** \$3.65 + 98¢

b. **Decimals:** \$25.00 ÷ 100

c. **Positive/Negative:** 449 − 500

d. **Calculation:** 8 × 62

e. **Calculation:** $1\frac{1}{2} + 2\frac{1}{2}$

f. **Fractional Parts:** $\frac{1}{2}$ of 76

g. **Measurement:** What fraction of a minute is one second?

h. **Calculation:** 8 × 8, − 1, ÷ 9, × 4, − 1, ÷ 3, × 2, + 2, ÷ 4

problem solving

Altogether, how many dots on the six number cubes are not visible in the illustration?

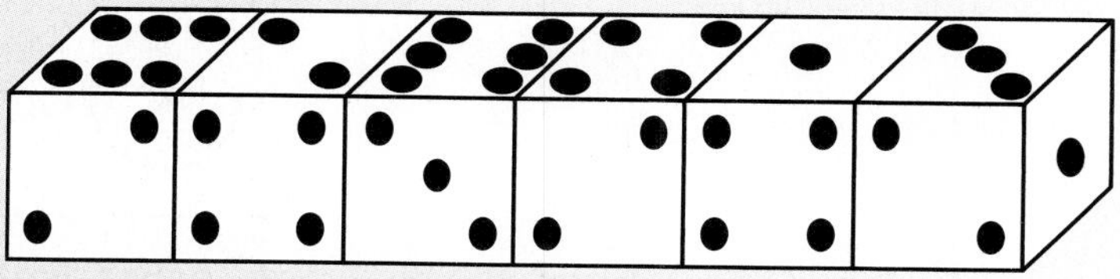

New Concept

Increasing Knowledge

Math Language
In *regrouping,* we exchange a value for an equal amount. For example, 1 ten for 10 ones, or 1 whole for $\frac{4}{4}$.

In this lesson we will practice subtracting mixed numbers that require regrouping. Regrouping that involves fractions differs from regrouping with whole numbers. When regrouping with whole numbers, we know that each unit equals ten of the next-smaller unit. However, when regrouping from a whole number to a fraction, we need to focus on the denominator of the fraction to determine how to regroup. We will use illustrations to help explain the process.

Example 1

There are $3\frac{1}{5}$ pies on the shelf. If the baker takes away $1\frac{2}{5}$ pies, how many pies will be on the shelf?

Solution

To answer this question, we subtract $1\frac{2}{5}$ from $3\frac{1}{5}$. Before we subtract, however, we will draw a picture to see how the baker solves the problem.

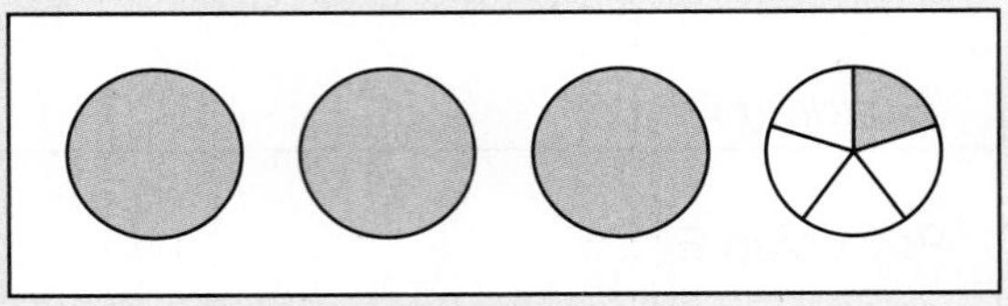

In order for the baker to remove $1\frac{2}{5}$ pies, it will be necessary to slice one of the whole pies into fifths. After cutting one pie into fifths, there are 2 whole pies plus $\frac{5}{5}$ plus $\frac{1}{5}$, which is $2\frac{6}{5}$ pies. Then the baker can remove $1\frac{2}{5}$ pies, as we illustrate.

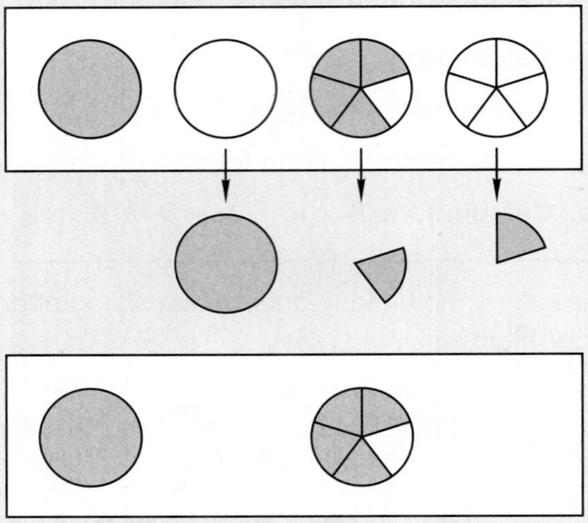

As we can see from the picture, $\mathbf{1\frac{4}{5}}$ **pies** will be left on the shelf.

To perform the subtraction on paper, we first rename $3\frac{1}{5}$ as $2\frac{6}{5}$, as shown below. Then we can subtract.

$$\begin{array}{r} 3\frac{1}{5} \\ -\ 1\frac{2}{5} \\ \hline \end{array} \xrightarrow{2 + \frac{5}{5} + \frac{1}{5}} \begin{array}{r} 2\frac{6}{5} \\ -\ 1\frac{2}{5} \\ \hline 1\frac{4}{5} \end{array}$$

Example 2

Thinking Skill

Analyze

How can we tell we will need to regroup just by looking at the example?

Simplify: $\mathbf{3\frac{5}{8} - 1\frac{7}{8}}$

Solution

We need to regroup in order to subtract. The mixed number $3\frac{5}{8}$ equals $2 + 1 + \frac{5}{8}$, which equals $2 + \frac{8}{8} + \frac{5}{8}$. Combining $\frac{8}{8}$ and $\frac{5}{8}$ gives us $\frac{13}{8}$, so we use $2\frac{13}{8}$. Now we can subtract and reduce.

$$\begin{array}{r} 3\frac{5}{8} \\ -\ 1\frac{7}{8} \\ \hline \end{array} \xrightarrow{2 + \frac{8}{8} + \frac{5}{8}} \begin{array}{r} 2\frac{13}{8} \\ -\ 1\frac{7}{8} \\ \hline 1\frac{6}{8} = \mathbf{1\frac{3}{4}} \end{array}$$

Example 3

Simplify: $83\frac{1}{3}\% - 41\frac{2}{3}\%$

Solution

The fraction in the subtrahend is greater than the fraction in the minuend, so we rename $83\frac{1}{3}\%$.

$$\begin{array}{r} 83\frac{1}{3}\% \\ -\ 41\frac{2}{3}\% \\ \hline \end{array} \xrightarrow{\left(82 + \frac{3}{3} + \frac{1}{3}\right)\%} \begin{array}{r} 82\frac{4}{3}\% \\ -\ 41\frac{2}{3}\% \\ \hline \mathbf{41\frac{2}{3}\%} \end{array}$$

Example 4

Simplify: $6 - 1\frac{3}{4}$

Solution

We rewrite 6 as a mixed number with a denominator of 4. Then we subtract.

$$\begin{array}{r} 6 \\ -\ 1\frac{3}{4} \\ \hline \end{array} \longrightarrow \begin{array}{r} 5\frac{4}{4} \\ -\ 1\frac{3}{4} \\ \hline \mathbf{4\frac{1}{4}} \end{array}$$

Example 5

Simplify: $100\% - 16\frac{2}{3}\%$

Solution

Thinking Skill

Explain

How has 100% regrouped to $99\frac{3}{3}\%$?

We rename 100% as $99\frac{3}{3}\%$ and subtract.

$$\begin{array}{r} 100\ \% \\ -\ 16\frac{2}{3}\% \\ \hline \end{array} \longrightarrow \begin{array}{r} 99\frac{3}{3}\% \\ -\ 16\frac{2}{3}\% \\ \hline \mathbf{83\frac{1}{3}\%} \end{array}$$

Practice Set

Simplify:

a. There were seven pies on the shelf. If the server removes $2\frac{1}{3}$ pies, how many pies will be on the shelf?

b. $6\frac{2}{5} - 1\frac{4}{5}$

c. $5\frac{1}{6} - 1\frac{5}{6}$

d. $100\% - 12\frac{1}{2}\%$

e. $83\frac{1}{3}\% - 16\frac{2}{3}\%$

Written Practice

Strengthening Concepts

1. (13) Willie shot eighteen rolls of film for the school annual. If there were thirty-six exposures in each roll, how many exposures were there in all?

2. (16) Carpeting is usually sold by the square yard. The number of square feet of area a carpet will cover is a function of the number of square yards installed. Create a function table that shows the number of square feet in 1, 2, 3, 4, and 5 square yards.

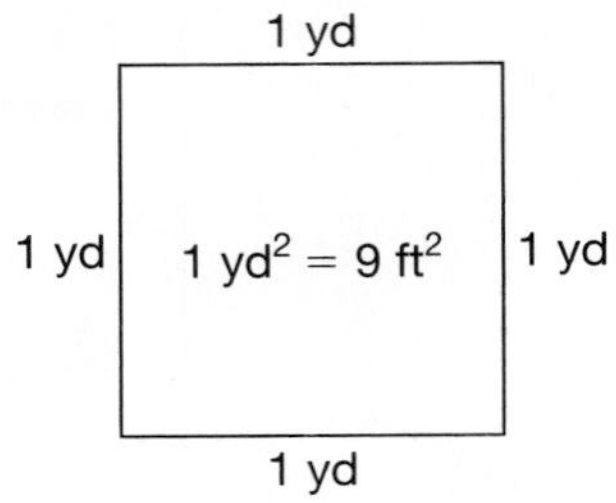

*** 3.** (20) **a.** Write the formula for the area of a rectangle.

b. A professional basketball court is a rectangle that is 50 feet wide and 94 feet long. What is the area of the basketball court?

4. (13) The 16-pound turkey cost $14.24. What was the price per pound?

*** 5.** (22) **Model** Draw a diagram of the statement. Then answer the questions that follow.

Three eighths of the 56 restaurants in town were closed on Monday.

a. How many of the restaurants in town were closed on Monday?

b. How many of the restaurants in town were open on Monday?

*** 6.** (22) **Evaluate** In the following statement, write the percent as a reduced fraction. Then draw a diagram of the statement and answer the questions.

Forty percent of the 30 students in the class were boys.

a. How many boys were in the class?

b. How many girls were in the class?

Math Language
A **spheroid** is a three-dimensional object with the shape of a sphere, such as a ball.

7. (16) After contact was made, the spheroid sailed four thousand, one hundred forty inches. How many yards did the spheroid sail after contact was made?

8. (9) **a.** How many $\frac{1}{5}$s are in 1?

b. How many $\frac{1}{5}$s are in 3?

c. **Explain** How can you use the answer to **a** to solve **b?**

9. (9, 10) **Explain** Describe how to find the reciprocal of a mixed number.

10. (15) Replace each circle with the proper comparison symbol:

a. $\frac{2}{3} \cdot \frac{3}{2} \bigcirc \frac{5}{5}$ **b.** $\frac{12}{36} \bigcirc \frac{12}{24}$

11. (10, 15) Write $2\frac{1}{4}$ and $3\frac{1}{3}$ as improper fractions. Then multiply the improper fractions, and write the product as a reduced mixed number.

12. (15) Complete each equivalent fraction:

a. $\frac{3}{4} = \frac{?}{40}$ **b.** $\frac{2}{5} = \frac{?}{40}$ **c.** $\frac{?}{8} = \frac{15}{40}$

* **13.** (21) **Generalize** The prime factorization of 100 is $2 \cdot 2 \cdot 5 \cdot 5$. We can write the prime factorization of 100 using exponents this way:

$$2^2 \cdot 5^2$$

a. Write the prime factorization of 400 using exponents.

b. Write the prime factorization of $\sqrt{400}$ using exponents.

14. (7) Refer to this figure to answer **a–d:**

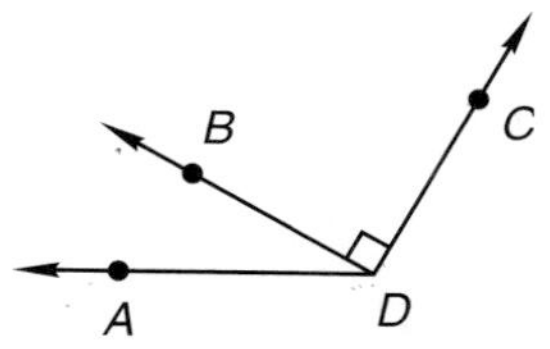

a. What type of angle is $\angle ADB$?

b. What type of angle is $\angle BDC$?

c. What type of angle is $\angle ADC$?

d. Which ray is perpendicular to $\overrightarrow{DB}$?

15. (15) Find fractions equivalent to $\frac{3}{4}$ and $\frac{2}{3}$ with denominators of 12. Then subtract the smaller fraction from the larger fraction.

Solve:

16. (3) $\frac{105}{w} = 7$

* **17.** (3, 20) $2x = 10^2$

18. (9, 15) $x + 1\frac{1}{4} = 6\frac{3}{4}$

19. (9, 15) $m - 4\frac{1}{8} = 1\frac{5}{8}$

* **20.** (23) **Analyze** There were five yards of fabric on the bolt of cloth. Fairchild bought $3\frac{1}{3}$ yards of the fabric. Then how many yards of fabric remained on the bolt?

Simplify:

* **21.** (23) $83\frac{1}{3}\% - 66\frac{2}{3}\%$

22. (9, 15) $\frac{7}{12} + \left(\frac{1}{4} \cdot \frac{1}{3}\right)$

23. (9, 15) $\frac{7}{8} - \left(\frac{3}{4} \cdot \frac{1}{2}\right)$

* **24.** (19) Draw $\overline{AB}$ $1\frac{3}{4}$ inches long. Then draw $\overline{BC}$ 1 inch long perpendicular to $\overline{AB}$. Complete the triangle by drawing $\overline{AC}$. Use a ruler to find the approximate length of $\overline{AC}$. Use that length to find the perimeter of $\triangle ABC$.

25. (17) Use a protractor to find the measure of $\angle A$ in problem 24. If necessary, extend the sides to measure the angle.

* **26.** (19) **Evaluate** Mary wants to apply a strip of wallpaper along the walls of the dining room just below the ceiling. If the room is a 14-by-12-ft rectangle, then the strip of wallpaper needs to be at least how long?

27. (9, 15) Multiply $\frac{3}{4}$ by the reciprocal of 3 and reduce the product.

28. (18) **Model** Draw an octagon. (A stop sign is a physical example of an octagon.)

* **29.** (20) **Predict** A sequence of perfect cubes ($k = n^3$) may be written as in **a** or as in **b.** Find the next two terms of both sequences.

a. $1^3, 2^3, 3^3, \ldots$

b. 1, 8, 27, ...

* **30.** (Inv. 2) The figure shows a circle with the center at point M.

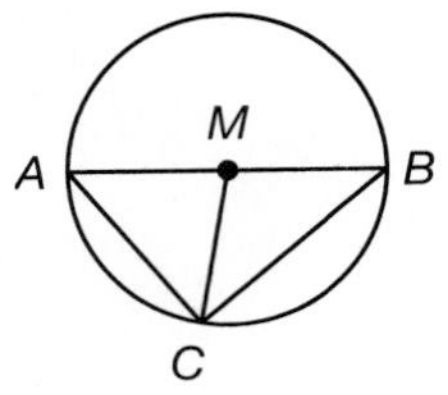

a. Which chord is a diameter?

b. Which central angle appears to be obtuse?

c. Name an inscribed angle that appears to be a right angle.

Early Finishers
Real-World Application

Zachary surveyed the 30 students in his class to find out how they get home. He found that 60% of the students ride the bus.

a. How many students ride the bus?

b. Half the students who do not ride the bus walk home. How many students walk home?

c. Based on his survey, what fraction of the students in the school might Zachary conclude walk to school?

LESSON 24

• Reducing Fractions, Part 2

Power Up

Building Power

facts Power Up D

mental math

a. **Number Sense:** $5.74 + 98¢

b. **Decimals:** $1.50 × 10

c. **Number Sense:** $1.00 − 36¢

d. **Calculation:** 4 × 65

e. **Calculation:** $3\frac{1}{3} + 1\frac{2}{3}$

f. **Fractional Parts:** $\frac{1}{3}$ of 24

g. **Measurement:** What fraction represents 15 minutes of an hour?

h. **Calculation:** What number is 3 more than half the product of 4 and 6?

problem solving

Huck followed the directions on the treasure map. Starting at the big tree, he walked six paces north, turned left, and walked seven more paces. He turned left and walked five paces, turned left again, and walked four more paces. He then turned right, and took one pace. In which direction was Huck facing, and how many paces was he from the big tree?

New Concepts

Increasing Knowledge

using prime factorization to reduce

We have been practicing reducing fractions by dividing the numerator and the denominator by a common factor. In this lesson we will practice a method of reducing that uses prime factorization to find the common factors of the terms. If we write the prime factorization of the numerator and of the the denominator, we can see how to reduce a fraction easily.

Math Language

The **greatest common factor** of two numbers is the greatest whole number that divides both numbers evenly.

Example 1

a. Use prime factorization to reduce $\frac{420}{1050}$.

b. Find the greatest common factor of 420 and 1050.

Solution

a. We rewrite the numerator and the denominator as products of prime numbers.

$$\frac{420}{1050} = \frac{2 \cdot 2 \cdot 3 \cdot 5 \cdot 7}{2 \cdot 3 \cdot 5 \cdot 5 \cdot 7}$$

Next we look for pairs of factors that form a fraction equal to 1. A fraction equals 1 if the numerator and denominator are equal. In this fraction there are four pairs of numerators and denominators that equal 1. They are $\frac{2}{2}$, $\frac{3}{3}$, $\frac{5}{5}$, and $\frac{7}{7}$. Below we have indicated each of these pairs.

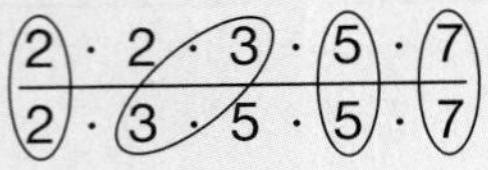

Each pair reduces to $\frac{1}{1}$.

$$\frac{\overset{1}{\cancel{2}} \cdot 2 \cdot \overset{1}{\cancel{3}} \cdot \overset{1}{\cancel{5}} \cdot \overset{1}{\cancel{7}}}{\underset{1}{\cancel{2}} \cdot \underset{1}{\cancel{3}} \cdot 5 \cdot \underset{1}{\cancel{5}} \cdot \underset{1}{\cancel{7}}}$$

The reduced fraction equals $1 \cdot 1 \cdot 1 \cdot 1 \cdot \frac{2}{5}$, which is $\mathbf{\frac{2}{5}}$.

Math Language

A **prime factor** is a factor that is a prime number.

b. In **a** we found the common prime factors of 420 and 1050. The common prime factors are 2, 3, 5, and 7. The product of these prime factors is the greatest common factor of 420 and 1050.

$$2 \cdot 3 \cdot 5 \cdot 7 = \mathbf{210}$$

Explain How could you have used this greatest common factor to reduce $\frac{420}{1050}$?

Example 2

A set of alphabet cards includes one card for each letter of the alphabet. If one card is drawn from the set of cards, what is the probability of drawing a vowel, including y?

Solution

The vowels are a, e, i, o, u, and we are told to include y. So the probability of drawing a vowel card is 6 in 26, which we can reduce.

$$\frac{6}{26} = \frac{\overset{1}{\cancel{2}} \cdot 3}{\underset{1}{\cancel{2}} \cdot 13} = \mathbf{\frac{3}{13}}$$

reducing before multiplying

When multiplying fractions, we often get a product that can be reduced even though the individual factors could not be reduced. Consider this multiplication:

$$\frac{3}{8} \cdot \frac{2}{3} = \frac{6}{24} \qquad \frac{6}{24} \text{ reduces to } \frac{1}{4}$$

We see that neither $\frac{3}{8}$ nor $\frac{2}{3}$ can be reduced. The product, $\frac{6}{24}$, can be reduced. We can avoid reducing after we multiply by reducing before we multiply. Reducing before multiplying is also known as **canceling.** To reduce, any numerator may be paired with any denominator. Below we have paired the 3 with 3 and the 2 with 8.

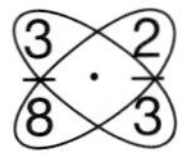

Then we reduce these pairs: $\frac{3}{3}$ reduces to $\frac{1}{1}$, and $\frac{2}{8}$ reduces to $\frac{1}{4}$, as we show below. Then we multiply the reduced terms.

$$\frac{\overset{1}{\cancel{3}}}{\underset{4}{\cancel{8}}} \cdot \frac{\overset{1}{\cancel{2}}}{\underset{1}{\cancel{3}}} = \frac{1}{4}$$

Discuss Why did we pair the 3 with the 3 and the 2 with the 8?

Example 3

Simplify: $\frac{9}{16} \cdot \frac{2}{3}$

Solution

Before multiplying, we pair 9 with 3 and 2 with 16 and reduce these pairs. Then we multiply the reduced terms.

$$\frac{\overset{3}{\cancel{9}}}{\underset{8}{\cancel{16}}} \cdot \frac{\overset{1}{\cancel{2}}}{\underset{1}{\cancel{3}}} = \mathbf{\frac{3}{8}}$$

Example 4

Simplify: $\frac{8}{9} \cdot \frac{3}{10} \cdot \frac{5}{4}$

Solution

We mentally pair 8 with 4, 3 with 9, and 5 with 10 and reduce.

$$\frac{\overset{2}{\cancel{8}}}{\underset{3}{\cancel{9}}} \cdot \frac{\overset{1}{\cancel{3}}}{\underset{2}{\cancel{10}}} \cdot \frac{\overset{1}{\cancel{5}}}{\underset{1}{\cancel{4}}}$$

We can still reduce by pairing 2 with 2. Then we multiply.

$$\frac{\overset{\overset{1}{\cancel{2}}}{\cancel{8}}}{\underset{3}{\cancel{9}}} \cdot \frac{\overset{1}{\cancel{3}}}{\underset{\underset{1}{\cancel{2}}}{\cancel{10}}} \cdot \frac{\overset{1}{\cancel{5}}}{\underset{1}{\cancel{4}}} = \mathbf{\frac{1}{3}}$$

Example 5

Simplify: $\frac{27}{32} \cdot \frac{20}{63}$

Solution

To give us easier numbers to work with, we factor the terms of the fractions before we reduce and multiply.

$$\frac{3 \cdot \overset{1}{\cancel{3}} \cdot \overset{1}{\cancel{3}}}{2 \cdot 2 \cdot 2 \cdot \underset{1}{\cancel{2}} \cdot \underset{1}{\cancel{2}}} \cdot \frac{\overset{1}{\cancel{2}} \cdot \overset{1}{\cancel{2}} \cdot 5}{\underset{1}{\cancel{3}} \cdot \underset{1}{\cancel{3}} \cdot 7} = \mathbf{\frac{15}{56}}$$

Explain What is another method we could have used to simplify the expression in Example 5? Which method do you prefer and why?

Practice Set

Generalize Use prime factorization to reduce each fraction:

a. $\frac{48}{144}$ **b.** $\frac{90}{324}$

c. Find the greatest common factor of 90 and 324.

Reduce before multiplying:

d. $\frac{5}{8} \cdot \frac{3}{10}$ **e.** $\frac{8}{15} \cdot \frac{5}{12} \cdot \frac{9}{10}$ **f.** $\frac{8}{3} \cdot \frac{6}{7} \cdot \frac{5}{16}$

Math Language

To **factor** a number means to write it as the product of factors. For example: $\frac{24}{28} = \frac{4 \cdot 6}{4 \cdot 7}$.

g. Factor and reduce before multiplying: $\frac{36}{45} \cdot \frac{25}{24}$

h. Of the 900 students at Columbia Middle School, there are 324 seventh graders. If one student is chosen at random to lead the pledge at a school assembly, what is the probability the person chosen will be a seventh grader? Show how to use prime factorization to reduce the answer.

i. **Justify** How can you use reducing fractions to demonstrate that the product of a fraction and its reciprocal is 1?

Written Practice *Strengthening Concepts*

1. (12) From Hartford to Los Angeles is two thousand, eight hundred ninety-five miles. From Hartford to Portland is three thousand, twenty-six miles. The distance from Hartford to Portland is how much greater than the distance from Hartford to Los Angeles?

2. (13) Hal ordered 15 boxes of microprocessors. If each box contained two dozen microprocessors, how many microprocessors did Hal order?

*** 3.** (22) In the following statement, write the percent as a fraction. Then draw a diagram and answer the questions.

Ashanti went to the store with $30.00 and spent 75% of the money.

a. What fraction of the money did she spend?

b. How much money did she spend?

*** 4.** (16, Inv. 2) **a.** If the diameter of a wheel is one yard, then its radius is how many inches?

b. **Justify** Write a statement telling how you know your solution to **a** is correct.

*** 5.** (22) Nancy descended the 30 steps that led to the floor of the cellar. One third of the way down she paused. How many more steps were there to the cellar floor?

6. (9) **a.** How many $\frac{1}{8}$s are in 1?

b. How many $\frac{1}{8}$s are in 3?

7. (9) **a.** Write the reciprocal of 3.

b. What fraction of 3 is 1?

*** 8.** (24) **a.** Use prime factorization to reduce $\frac{540}{600}$.

b. What is the greatest common factor of 540 and 600?

9. (17) What type of angle is formed by the hands of a clock at

a. 2 o'clock? **b.** 3 o'clock? **c.** 4 o'clock?

10. (15) **a.** *Explain* Describe how to complete this equivalent fraction.

$$\frac{3}{5} = \frac{?}{30}$$

b. *Justify* Name the property we use to find equivalent fractions.

*** 11.** (21) *Generalize* The prime factorization of 1000 using exponents is $2^3 \cdot 5^3$.

a. Write the prime factorization of 10,000 using exponents.

b. Write the prime factorization of $\sqrt{10{,}000}$ using exponents.

12. (7) **a.** *Model* Draw two parallel lines that are intersected by a third line perpendicular to the parallel lines.

b. What type of angles are formed?

*** 13.** (20) The perimeter of a square is one yard.

a. How many inches long is each side of the square?

b. What is the area of the square in square inches?

14. (2) This equation illustrates that which property does not apply to division?

$$10 \div 5 \neq 5 \div 10$$

15. (7) The front and back covers of a closed book represent two planes that are

A parallel **B** skew **C** intersecting **D** perpendicular

16. (13) *Formulate* Write and solve a word problem about equal groups that fits this equation.

$$12p = \$3.36$$

Solve:

17. (9, 15) $4\frac{7}{12} = x + 1\frac{1}{12}$

18. (9, 15) $w - 3\frac{3}{4} = 2\frac{3}{4}$

Simplify:

*** 19.** (20) $10^5 \div 10^2$

*** 20.** (20) $\sqrt{9} - \sqrt{4^2}$

*** 21.** (23) $100\% - 66\frac{2}{3}\%$

*** 22.** (23) $5\frac{1}{8} - 1\frac{7}{8}$

* **23.** $\left(\frac{5}{6}\right)^2$
(20)

* **24.** $\frac{3}{4} \cdot \frac{1}{2} \cdot \frac{8}{9}$
(24)

25. Kevin clipped three corners from a square sheet of paper. What is the name of the polygon that was formed?
(18)

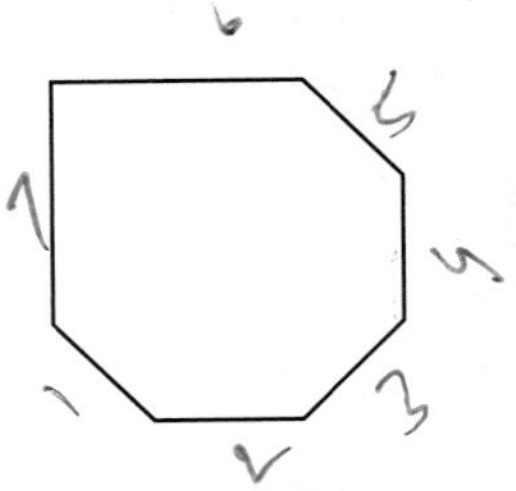

26. **Evaluate** Evaluate the following expressions for $a = 10$ and $b = 100$:
(1, 4)

a. ab **b.** $a - b$ **c.** $\frac{a}{b}$

* **27.** **a.** **Generalize** Find the perimeter of the figure at right. Dimensions are in yards. All angles are right angles.
(19)

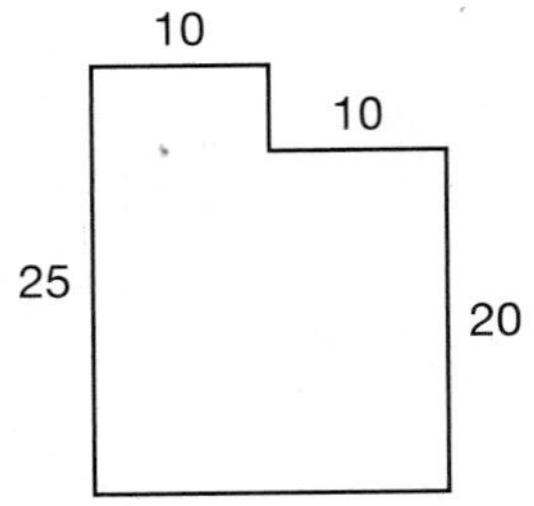

b. **Justify** How did you find the missing measures of two sides of the figure?

28. Find equivalent fractions for $\frac{1}{4}$ and $\frac{1}{6}$ that have denominators of 12. Then add them.
(15)

* **29.** **Generalize** Segment *AC* divides rectangle *ABCD* into two congruent triangles. Angle *ADC* corresponds to $\angle CBA$. Name two more pairs of corresponding angles.
(18)

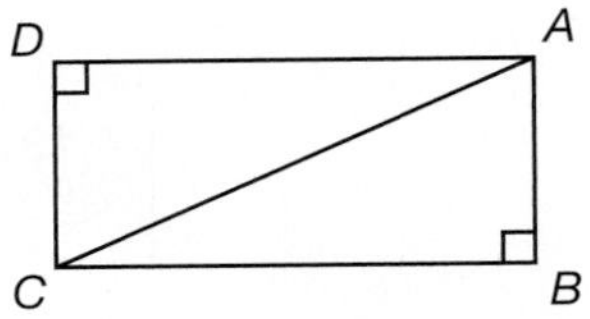

30. **a.** Arrange these numbers in order from least to greatest:
(4, 9)

$$0, 1, -1, \frac{1}{2}, -\frac{1}{2}$$

b. **Predict** The ordered numbers in the answer to part **a** form a sequence. What are the next three positive numbers in the sequence?

LESSON 25

Dividing Fractions

Power Up

Building Power

facts Power Up F

mental math

a. **Number Sense:** \$2.65 + \$1.99

b. **Decimals:** \$60.00 ÷ 10

c. **Number Sense:** \$2.00 − \$1.24

d. **Calculation:** 7 × 36

e. **Calculation:** $1\frac{3}{4} + 4\frac{1}{4}$

f. **Fractional Parts:** $\frac{1}{4}$ of 36

g. **Measurement:** What fraction represents 30 seconds of a minute?

h. **Calculation:** What number is 3 less than half the sum of 8 and 12?

problem solving Copy the problem and fill in the missing digits:

```
    3 6
  × _ _
  _ _ 0
+ _ 6
  _ _ _
```

New Concept

Increasing Knowledge

"How many quarters are in a dollar?" is a way to ask, "How many $\frac{1}{4}$s are in 1?" This question is a division question:

$$1 \div \frac{1}{4}$$

We can model the question with the fraction manipulatives we used in Investigation 1.

How many 

are in 

?

We see that the answer is 4. Recall that 4 (or $\frac{4}{1}$) is the reciprocal of $\frac{1}{4}$. Likewise, when we ask the question, "How many quarters ($\frac{1}{4}$s) are in three dollars (3)?" we are again asking a division question.

$$3 \div \frac{1}{4}$$

How many 

are in 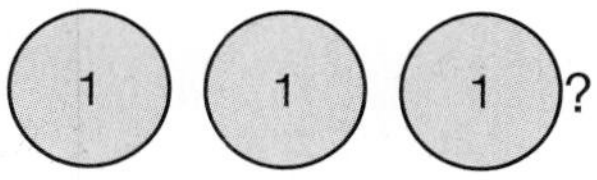

?

We can use the answer to the first question to help us answer the second question. There are four $\frac{1}{4}$s in 1, so there must be *three times as many* $\frac{1}{4}$s in 3. Thus, there are twelve $\frac{1}{4}$s in 3. We found the answer to the second question by multiplying 3 by 4, the answer to the first question. We will follow this same line of thinking in the next few examples.

Analyze your own thinking about this question: How many quarters are in five dollars? Our thinking probably takes two steps: (1) There are 4 quarters in a dollar, (2) So there are $5 \times 4 = 20$ quarters in five dollars.

Summarize How could you use this same thinking to find out how many dimes are in 5 dollars?

Example 1

a. How many $\frac{2}{3}$s are in 1? $(1 \div \frac{2}{3})$

b. How many $\frac{2}{3}$s are in 3? $(3 \div \frac{2}{3})$

Solution

a. We may model the question with manipulatives.

How many 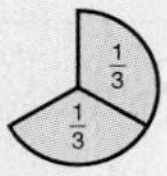 are in  ?

We see from the manipulatives that the answer is more than 1 but less than 2. If we think of the two $\frac{1}{3}$ pieces as one piece, we see that another *half* of the $\frac{2}{3}$ piece would make a whole. Thus there are $\mathbf{\frac{3}{2}}$ (or $\mathbf{1\frac{1}{2}}$) $\frac{2}{3}$s in 1. Notice that the answer to question **a** is the reciprocal of $\frac{2}{3}$.

Thinking Skill

Summarize

What rule can we state about the division of the number 1 by a fraction?

b. We use the answer to **a** to help us answer **b.** There are $\frac{3}{2}$ (or $1\frac{1}{2}$) $\frac{2}{3}$s in 1, so there are three times as many $\frac{2}{3}$s in 3. Thus, we answer the question by multiplying 3 by $\frac{3}{2}$ (or 3 by $1\frac{1}{2}$).

$$3 \times \frac{3}{2} = \frac{9}{2} \qquad 3 \times 1\frac{1}{2} = 1\frac{1}{2} + 1\frac{1}{2} + 1\frac{1}{2}$$

$$= 4\frac{1}{2} \qquad = 4\frac{1}{2}$$

The number of $\frac{2}{3}$s in 3 is $\mathbf{4\frac{1}{2}}$. We found the answer by multiplying 3 by the reciprocal of $\frac{2}{3}$.

Example 2

a. $\mathbf{1 \div \frac{2}{5}}$ **b.** $\mathbf{\frac{3}{4} \div \frac{2}{5}}$

Solution

a. The problem $1 \div \frac{2}{5}$ means, "How many $\frac{2}{5}$s are in 1?" The answer is the reciprocal of $\frac{2}{5}$, which is $\mathbf{\frac{5}{2}}$.

$$1 \div \frac{2}{5} = \frac{5}{2}$$

b. We use the answer to **a** to help us answer **b.** There are $\frac{5}{2}$ (or $2\frac{1}{2}$) $\frac{2}{5}$s in 1, so there are $\frac{3}{4}$ times as many $\frac{2}{5}$s in $\frac{3}{4}$. Thus we multiply $\frac{3}{4}$ by $\frac{5}{2}$.

$$\frac{3}{4} \times \frac{5}{2} = \frac{15}{8}$$

$$= 1\frac{7}{8}$$

The number of $\frac{2}{5}$s in $\frac{3}{4}$ is $\mathbf{1\frac{7}{8}}$. We found the answer by multiplying $\frac{3}{4}$ by the reciprocal of $\frac{2}{5}$.

Example 3

$$\mathbf{\frac{2}{3} \div \frac{3}{4}}$$

Solution

To find how many $\frac{3}{4}$s are in $\frac{2}{3}$, we take two steps. First we find how many $\frac{3}{4}$s are in 1. The answer is the reciprocal of $\frac{3}{4}$.

$$1 \div \frac{3}{4} = \frac{4}{3}$$

Then we use this reciprocal to find the number of $\frac{3}{4}$s in $\frac{2}{3}$. The number of $\frac{3}{4}$s in $\frac{2}{3}$ is $\frac{2}{3}$ times as many $\frac{3}{4}$s as are in 1. So we multiply $\frac{2}{3}$ by $\frac{4}{3}$.

$$\frac{2}{3} \times \frac{4}{3} = \mathbf{\frac{8}{9}}$$

This means there is slightly less than one $\frac{3}{4}$ in $\frac{2}{3}$. We found the answer by multiplying $\frac{2}{3}$ by the reciprocal of $\frac{3}{4}$.

Conclude Complete this sentence to create a rule for dividing fractions: To find the quotient of two fractions, ______ the dividend by the ______ of the ______.

Example 4

Sam walks $\frac{9}{10}$ of a mile to school. On his way to school he passes a bank which is $\frac{3}{4}$ of a mile from his home. What fraction of his walk has Sam completed when he reaches the bank?

Solution

The whole walk is the distance to school. The part is the distance to the bank. The fraction is "part over whole."

$$\frac{\text{part}}{\text{whole}} \quad \frac{\frac{3}{4}}{\frac{9}{10}}$$

Math Language

A **compound fraction** is a fraction whose numerator and/or denominator are also fractions.

This **compound fraction** means $\frac{3}{4}$ divided by $\frac{9}{10}$. We perform the division to find the answer.

$$\frac{3}{4} \div \frac{9}{10}$$

$$\frac{3}{4} \times \frac{10}{9} = \frac{5}{6}$$

When Sam reaches the bank he has completed $\mathbf{\frac{5}{6}}$ of his walk.

Find the number of $\frac{9}{10}$s in 1.

$$1 \div \frac{9}{10} = \frac{10}{9}$$

Use the number of $\frac{9}{10}$s in 1 to find the number of $\frac{9}{10}$s in $\frac{3}{4}$.

$$\frac{\overset{1}{\cancel{3}}}{\underset{2}{\cancel{4}}} \times \frac{\overset{5}{\cancel{10}}}{\underset{3}{\cancel{9}}} = \mathbf{\frac{5}{6}}$$

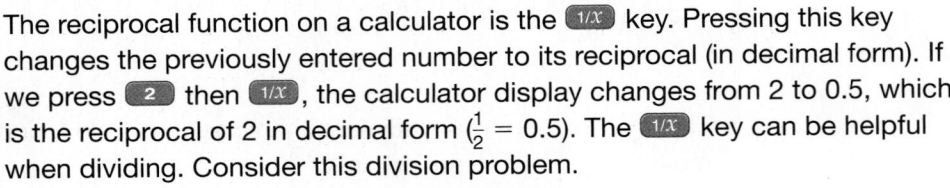

Working on paper, we often move from the original problem directly to step 2 by multiplying the dividend, the first fraction, by the reciprocal of the divisor, the second fraction.

$$\frac{3}{4} \div \frac{9}{10}$$

$$\downarrow$$

$$\frac{\overset{1}{\cancel{3}}}{\underset{2}{\cancel{4}}} \times \frac{\overset{5}{\cancel{10}}}{\underset{3}{\cancel{9}}} = \mathbf{\frac{5}{6}}$$

When Sam reaches the bank, he is $\mathbf{\frac{5}{6}}$ of the way to school.

Extend How far is the bank from Sam's school?

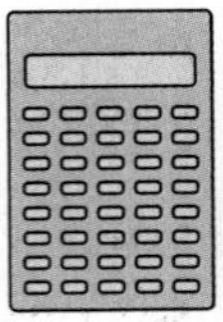

The reciprocal function on a calculator is the [1/x] key. Pressing this key changes the previously entered number to its reciprocal (in decimal form). If we press [2] then [1/x], the calculator display changes from 2 to 0.5, which is the reciprocal of 2 in decimal form ($\frac{1}{2} = 0.5$). The [1/x] key can be helpful when dividing. Consider this division problem.

$$144\overline{)\$10,461.60}$$

The divisor is 144. You could choose to divide $10,461.60 by 144 or to multiply $10,461.60 by the reciprocal of 144. Since multiplication is commutative, using the reciprocal allows you to enter the numbers in either order. The following multiplication yields the answer even though the entry begins with the divisor. Notice that we drop the terminal zero from $10,461.60, since it does not affect the value.

[1] [4] [4] [1/x] [×] [1] [0] [4] [6] [1] [.] [6] [=]

Whether we choose to divide $10,461.60 by 144 or to multiply by the reciprocal of 144, the answer is $72.65.

Practice Set

a. How many $\frac{2}{3}$s are in 1? How many $\frac{2}{3}$s are in $\frac{3}{4}$?

b. How many $\frac{3}{4}$s are in 3?

c. Describe the two steps for finding the number of quarters in six dollars.

d. **Explain** Tell how to use the reciprocal of the divisor to find the answer to a division problem.

e. Describe the function of the 1/X key on a calculator.

Generalize Use the two-step method described in this lesson to find each quotient:

f. $\frac{3}{5} \div \frac{2}{3}$ **g.** $\frac{7}{8} \div \frac{1}{4}$ **h.** $\frac{5}{6} \div \frac{2}{3}$

i. Amanda has a ribbon $\frac{3}{4}$ of a yard long. She used $\frac{1}{2}$ of a yard of ribbon for a small package. What fraction of her ribbon did Amanda use? Write a fraction division problem for this story and show the steps. Then write the answer in a sentence.

Written Practice *Strengthening Concepts*

1. (13) Three hundred twenty-four students were given individual boxes of apple juice at lunch in the school cafeteria. If each pack of apple juice contained a half dozen individual boxes of juice, how many packages of juice were used?

2. (17, 19) Use a ruler to draw square *ABCD* with sides $2\frac{1}{2}$ in. long. Then divide the square into two congruent triangles by drawing $\overline{AC}$.

a. What is the perimeter of square *ABCD*?

b. What is the measure of each angle of the square?

c. What is the measure of each acute angle in $\triangle ABC$?

d. What is the sum of the measures of the three angles in $\triangle ABC$?

*** 3.** (11, 13, 22) **Evaluate** Use this information to answer questions **a–c.**

The family reunion was a success, as 56 relatives attended. Half of those who attended played in the big game. However, the number of players on the two teams was not equal since one team had only 10 players.

a. How many relatives played in the game?

b. If one team had 10 players, how many players did the other team have?

c. If the teams were rearranged so that the number of players on each team was equal, how many players would be on each team?

*** 4.** (16, Inv. 2) **Represent** The diameter of a circle is a function of the radius of the circle. Make a function table that shows the diameters of circles with radii that are $\frac{1}{4}$, $\frac{1}{2}$, $\frac{3}{4}$ and 1 unit long.

5. (21) Use prime factorization to find the greatest common factor of 72 and 54.

*** 6.** (22) In the following statement, write the percent as a reduced fraction. Then diagram the statement and answer the questions.

Jason has read 75% of the 320 pages in the book.

a. How many pages has Jason read?

b. How many pages has Jason not read?

* **7.** (25) **a.** How many $\frac{3}{4}$s are in 1?

b. How many $\frac{3}{4}$s are in $\frac{7}{8}$?

8. (8, 14) **Estimate** Which is the best estimate of the probability of spinning a 3? Why?

A $\frac{2}{3}$ **B** $\frac{2}{4}$ **C** $\frac{2}{5}$

* **9.** (24) **a.** Write 84 and 210 as products of prime numbers. Then reduce $\frac{84}{210}$.

b. What is the greatest common factor of 84 and 210?

10. (9, 10) Write the reciprocal of each number:

a. $\frac{9}{10}$ **b.** 8 **c.** $2\frac{3}{8}$

d. What is the product of $2\frac{3}{8}$ and its reciprocal?

e. **Generalize** What rule do you know about reciprocals that could have helped you answer **d?**

11. (15) Find fractions equivalent to $\frac{3}{4}$ and $\frac{4}{5}$ with denominators of 20. Then add the two fractions you found, and write the sum as a mixed number.

* **12.** (21) **a.** The prime factorization of 40 is $2^3 \cdot 5$. Write the prime factorization of 640 using exponents.

b. Tell how you can use a calculator to verify your answer to **12a.** Then follow your procedure.

* **13.** (10, 24) Write $2\frac{2}{3}$ and $2\frac{1}{4}$ as improper fractions. Then find the product of the improper fractions.

14. (8, 15) **a.** Points *A* and *B* represent what mixed numbers on this number line?

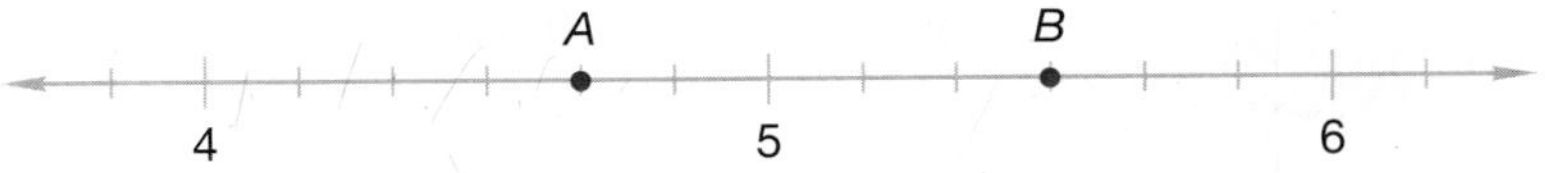

b. Find the difference between the numbers represented by points *A* and *B*.

15. (17) **a.** Draw line *AB*. Then draw ray *BC* so that angle *ABC* measures 30°. Use a protractor.

b. What type of angle is angle *ABC*?

Solve:

* **16.** (23) $1\frac{7}{12} + y = 3$

17. (9, 15) $5\frac{7}{8} = x - 4\frac{5}{8}$

18. (3) $8n = 360°$

19. (9, 20) $\frac{4}{3}m = 1^3$

Simplify:

20. (10) $6\frac{1}{6} + 1\frac{5}{6}$

***21.** (24) $\frac{3}{4} \cdot \frac{5}{9} \cdot \frac{8}{15}$

* **22.** $\frac{4}{5} \div \frac{2}{1}$
(25)

* **23.** $\frac{8}{5} \div \frac{6}{5}$
(25)

24. Veronica is using a recipe that calls for $\frac{3}{4}$ cup milk. She only needs half of the recipe. How much milk will she need?
(25)

25. $\frac{100\%}{8}$
(10)

* **26.** **Generalize** In the division $5 \div \frac{3}{5}$, instead of dividing 5 by $\frac{3}{5}$, we can find the answer by multiplying 5 by what number?
(26)

27. **a.** Simplify and compare: $2^2 \cdot 2^3 \bigcirc 2^3 \cdot 2^2$
(20)

b. Simplify: $\sqrt{2^2}$

28. A regular hexagon is inscribed in a circle. If one side of the hexagon is 6 inches long, then the perimeter of the hexagon is how many feet?
(19)

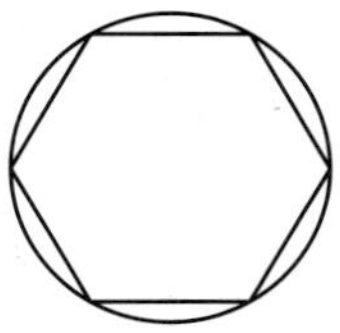

29. A 2-in. square was cut from a 4-in. square as shown in the figure. What is the perimeter of the resulting polygon?
(19)

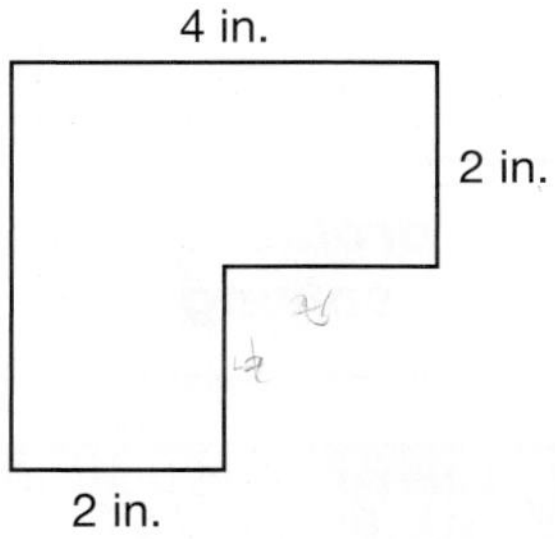

* **30.** Which negative integer is the opposite of the third prime number?
(4, 21)

LESSON

26

• Multiplying and Dividing Mixed Numbers

Power Up

Building Power

facts

Power Up F

mental math

a. Number Sense: \$8.56 + 98¢

b. Decimals: 30¢ × 100

c. Number Sense: \$1.00 − 7¢

d. Calculation: 3 × 74

e. Calculation: $\frac{2}{3} + \frac{2}{3}$

f. Fractional Parts: $\frac{2}{3}$ of 24

g. Patterns: What number comes next in the pattern: 5, 11, 15, 21, ______

h. Calculation: 7 × 7, + 1, × 2, ÷ 5, + 5, ÷ 5, − 5, × 5

problem solving

The sum of two whole numbers is 17 and their product is 60. Find the two numbers.

New Concept

Increasing Knowledge

Math Language

An **improper fraction** is a fraction whose numerator is equal to or greater than its denominator.

One way to multiply or divide mixed numbers is to first rewrite the mixed numbers as improper fractions. Then we multiply or divide the improper fractions as indicated.

Example 1

Sergio used three lengths of ribbon $2\frac{1}{2}$ feet long to wrap packages. How many feet of ribbon did he use?

Solution

This is an equal groups problem. We want to find the total.

$$3 \times 2\frac{1}{2} = T$$

We will show two ways to find the answer. One way is to recognize that $3 \times 2\frac{1}{2}$ equals three $2\frac{1}{2}$s, which we add.

$$3 \times 2\frac{1}{2} = 2\frac{1}{2} + 2\frac{1}{2} + 2\frac{1}{2} = \mathbf{7\frac{1}{2}}$$

Another way to find the product is to write 3 and $2\frac{1}{2}$ as improper fractions and multiply.

Explain How can we write 3 as an improper fraction?

$$3 \times 2\frac{1}{2}$$

$$\downarrow \quad \downarrow$$

$$\frac{3}{1} \times \frac{5}{2} = \frac{15}{2} = 7\frac{1}{2}$$

Sergio used **$7\frac{1}{2}$ feet of ribbon.**

Example 2

Simplify:

a. $3\frac{2}{3} \times 1\frac{1}{2}$

b. $\left(1\frac{1}{2}\right)^2$

Solution

Thinking Skill

Predict

If we multiply example **a** without canceling, will we get the same answer? If so, will the answer be in simplest form?

a. We first rewrite $3\frac{2}{3}$ as $\frac{11}{3}$ and $1\frac{1}{2}$ as $\frac{3}{2}$. Then we multiply and simplify.

$$\frac{11}{\cancel{3}_1} \times \frac{\cancel{3}^1}{2} = \frac{11}{2} = \mathbf{5\frac{1}{2}}$$

b. The expression $\left(1\frac{1}{2}\right)^2$ means $1\frac{1}{2} \times 1\frac{1}{2}$. We write each factor as an improper fraction and multiply.

$$1\frac{1}{2} \times 1\frac{1}{2}$$

$$\downarrow \quad \downarrow$$

$$\frac{3}{2} \times \frac{3}{2} = \frac{9}{4} = \mathbf{2\frac{1}{4}}$$

Example 3

Find the area of a square with sides $2\frac{1}{2}$ inches long.

Solution

If we draw the square on a grid, we see a physical representation of the area of the square. We see four whole square inches, four half square inches, and one quarter square inch within the shaded figure. We can calculate the area by adding.

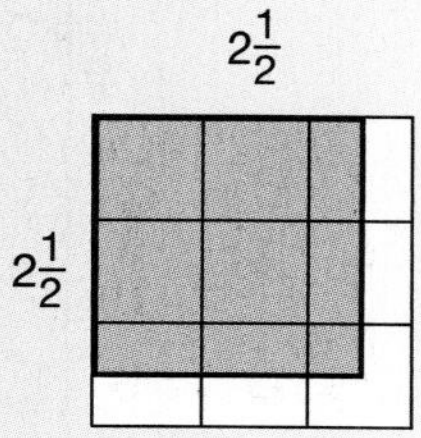

$$4 \text{ in.}^2 + \frac{4}{2} \text{ in.}^2 + \frac{1}{4} \text{ in.}^2 = \mathbf{6\frac{1}{4} \text{ in.}^2}$$

If we multiply $2\frac{1}{2}$ inches by $2\frac{1}{2}$ inches, we obtain the same result.

$$2\frac{1}{2} \text{ in.} \times 2\frac{1}{2} \text{ in.}$$

$$= \frac{5}{2} \text{ in.} \times \frac{5}{2} \text{ in.}$$

$$= \frac{25}{4} \text{ in.}^2 = \mathbf{6\frac{1}{4} \text{ in.}^2}$$

Formulate What multiplication expression can we write to show the perimeter of the same square? What is the perimeter?

Example 4

The biscuit recipe called for $3\frac{2}{3}$ cups of flour. To make half a batch, Greg divided the amount of each ingredient by 2. How many cups of flour should he use?

Solution

As we think about the problem, we see that by dividing $3\frac{2}{3}$ by 2, we will be finding *half of* $3\frac{2}{3}$. We can find half of a number either by dividing by 2 or by multiplying by $\frac{1}{2}$. In other words, the following are equivalent expressions:

$$3\frac{2}{3} \div 2 \qquad 3\frac{2}{3} \times \frac{1}{2}$$

Notice that multiplying by $\frac{1}{2}$ can be thought of as multiplying by the *reciprocal* of 2. We will write $3\frac{2}{3}$ as an improper fraction and multiply by $\frac{1}{2}$.

$$3\frac{2}{3} \times \frac{1}{2}$$

$$\downarrow$$

$$\frac{11}{3} \times \frac{1}{2} = \frac{11}{6} = \mathbf{1\frac{5}{6}}$$

Greg should use **$1\frac{5}{6}$ cups of flour.**

Example 5

Simplify: $3\frac{1}{3} \div 2\frac{1}{2}$

Solution

First we write $3\frac{1}{3}$ and $2\frac{1}{2}$ as improper fractions. Then we multiply by the reciprocal of the divisor and simplify.

$$3\frac{1}{3} \div 2\frac{1}{2} \quad \text{original problem}$$

$$\downarrow \quad \downarrow$$

$$\frac{10}{3} \div \frac{5}{2} \quad \text{changed mixed numbers to improper fractions}$$

$$\downarrow \quad \downarrow$$

$$\frac{\overset{2}{\cancel{10}}}{3} \times \frac{2}{\underset{1}{\cancel{5}}} = \frac{4}{3} \quad \text{multiplied by reciprocal of the divisor}$$

$$= \mathbf{1\frac{1}{3}} \quad \text{simplified}$$

Practice Set

a. **Model** Find the area of a rectangle that is $1\frac{1}{2}$ in. wide and $2\frac{1}{2}$ in. long. Illustrate the problem by drawing a 2 by 3 grid and sketching a $1\frac{1}{2}$ by $2\frac{1}{2}$ unit rectangle on the grid. Explain how the area of the rectangle can be found by using the sketch.

Evaluate Simplify:

b. $6\frac{2}{3} \times \frac{3}{5}$ **c.** $2\frac{1}{3} \times 3\frac{1}{2}$ **d.** $3 \times 3\frac{3}{4}$

e. $1\frac{2}{3} \div 3$ **f.** $2\frac{1}{2} \div 3\frac{1}{3}$ **g.** $5 \div \frac{2}{3}$

h. $2\frac{2}{3} \div 1\frac{1}{3}$ **i.** $1\frac{1}{3} \div 2\frac{2}{3}$ **j.** $4\frac{1}{2} \times 1\frac{2}{3}$

Written Practice *Strengthening Concepts*

1. (11) After the first hour of the monsoon, 23 millimeters of precipitation had fallen. After the second hour a total of 61 millimeters of precipitation had fallen. How many millimeters of precipitation fell during the second hour?

2. (13) Each photograph enlargement cost 85¢ and Willie needed 26 enlargements. What was the total cost of the enlargements Willie needed?

3. (12) **Connect** The Byzantine Empire can be said to have begun in 330 when the city of Byzantium was renamed Constantinople and became the capital of the Roman Empire. The Byzantine Empire came to an end in 1453 when the city of Constantinople was renamed Istanbul and became the capital of the Ottoman Empire. About how many years did the Byzantine Empire last?

4. (11) At the movie theater, Dolores gave $20 to the ticket seller and got $10.25 back in change. How much did her movie ticket cost?

5. (13) A gross is a dozen dozens. A gross of pencils is how many pencils?

*** 6.** (22) **Model** Diagram this statement and answer the questions that follow. Begin by changing the percent to a reduced fraction.

Forty percent of the 60 marbles in the bag were blue.

a. How many of the marbles in the bag were blue?

b. How many of the marbles in the bag were not blue?

*** 7.** (16) **a.** Roan estimated that the weight of the water in a full bathtub is a quarter ton. How many pounds is a quarter of a ton?

b. **Explain** Describe how you got your answer.

8. (8, 9) The figure shows a one-inch square. A smaller square that is $\frac{7}{10}$ of an inch on each side is shaded.

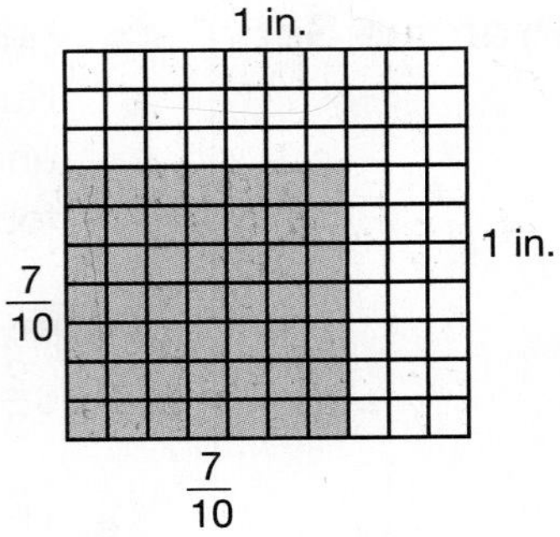

a. What fraction of the square inch is shaded?

b. What percent of the square is not shaded?

* **9.** (24) **a.** Write 210 and 252 as products of prime numbers. Then reduce $\frac{210}{252}$.

b. Find the GCF of 210 and 252.

10. (9, 10) Write the reciprocal of each number:

a. $\frac{5}{9}$ **b.** $5\frac{3}{4}$ **c.** 7

11. (15) Find the number that makes the two fractions equivalent.

a. $\frac{5}{8} = \frac{?}{24}$ **b.** $\frac{5}{12} = \frac{?}{24}$

c. Add the fractions you found in **a** and **b**.

* **12.** (18) **Represent** Draw a 2-by-2 grid. On the grid sketch a $1\frac{1}{2}$ by $1\frac{1}{2}$ square. Assume that the sketch illustrates a square with sides $1\frac{1}{2}$ inches long. What is the area of the square? Explain how the sketch illustrates the area of the square.

13. (8) Draw $\overline{AB}$ 2 in. long. Then draw $\overline{BC}$ $1\frac{1}{2}$ in. long perpendicular to $\overline{AB}$. Complete $\triangle ABC$ by drawing $\overline{AC}$. How long is $\overline{AC}$?

14. (4, 10) **a.** Arrange these numbers in order from least to greatest:

$$1, -3, \frac{5}{6}, 0, \frac{4}{3}$$

b. Which of these numbers are whole numbers?

Solve:

15. (10, 15) $x - 8\frac{11}{12} = 6\frac{5}{12}$

16. (3) $180 - y = 75$

17. (3) $12w = 360°$

* **18.** (23) $w + 58\frac{1}{3} = 100$

19. (20) **a.** Find the area of the square.

b. Find the area of the shaded part of the square.

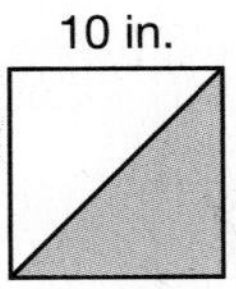

Simplify:

* **20.** (23) $9\frac{1}{9} - 4\frac{4}{9}$

* **21.** (24) $\frac{5}{8} \cdot \frac{3}{10} \cdot \frac{1}{6}$

* **22.** (20, 26) $\left(2\frac{1}{2}\right)^2$

* **23.** (26) $1\frac{3}{5} \div 2\frac{2}{3}$

* **24.** (26) $3\frac{1}{3} \div 4$

* **25.** (26) $5 \cdot 1\frac{3}{4}$

* **26.** (2, 9) **Justify** Name each property used to simplify this equation.

$$\frac{3}{2}\left(\frac{2}{3}x\right) = \frac{5}{6} \qquad \text{Given}$$

$$\left(\frac{3}{2} \cdot \frac{2}{3}\right)x = \frac{5}{6} \qquad \textbf{a.}\ \_\_\_\_\_\_\_\_\_\_$$

$$1x = \frac{5}{6} \qquad \textbf{b.}\ \_\_\_\_\_\_\_\_\_\_$$

$$x = \frac{5}{6} \qquad \textbf{c.}\ \_\_\_\_\_\_\_\_\_\_$$

27. (14, 15) Max is thinking of a counting number from 1 to 10. Deb guesses 7. What is the probability Deb's guess is correct?

28. (1, 9) **Analyze** Evaluate the following expressions for $x = 3$ and $y = 6$:

a. $x - \frac{y}{x}$ **b.** $\frac{xy}{y}$ **c.** $\frac{x}{y} \cdot \frac{y}{x}$

d. Which property is illustrated by **c**?

29. (4) **Predict** The rule of the following sequence is $k = 3n - 2$. Find the ninth term.

1, 4, 7, 10, ...

* **30.** (Inv. 2) **Conclude** The central angle of a half circle is 180°. The central angle of a quarter circle is 90°. How many degrees is the central angle of an eighth of a circle?

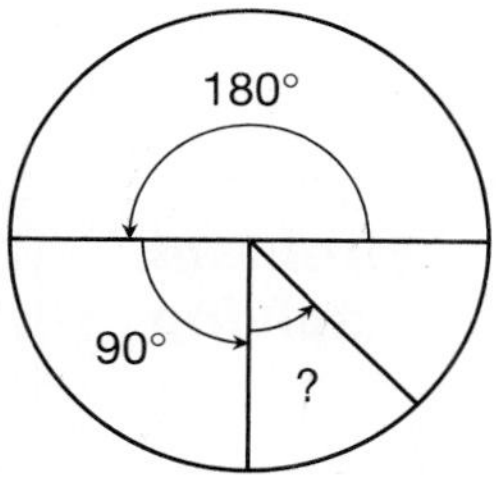

Early Finishers

Real-World Application

You and some friends volunteered to paint the concession stand for your local ball team. The back and side walls (both inside and outside) need painting.

The two side walls measure $9\frac{3}{5}$ by $8\frac{1}{3}$ feet each, while the back wall measures 39 feet by $8\frac{1}{3}$ feet. Find the total area that must be painted (in square feet). Show each step of your work.

LESSON 27

• Multiples
• Least Common Multiple
• Equivalent Division Problems

Power Up — *Building Power*

facts Power Up E

mental math

a. **Number Sense:** $3.75 + $1.98

b. **Decimals:** $125.00 ÷ 10

c. **Number Sense:** 10 × 42

d. **Calculation:** 5 × 42

e. **Calculation:** $\frac{3}{4} + \frac{3}{4}$

f. **Fractional Parts:** $\frac{3}{4}$ of 24

g. **Algebra:** If $m = 9$, what does $3m$ equal?

h. **Measurement:** Start with a score. Add a dozen; then add the number of feet in a yard. Divide by half the number of years in a decade; then subtract the number of days in a week. What is the answer?

problem solving

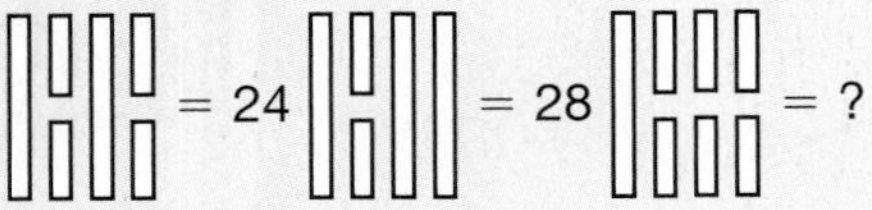

Each bar shown above has a value. All long bars are worth the same amount, and all small bars are worth the same amount. How much is one long bar worth? How much is one short bar worth? What is the value of the third arrangement?

New Concepts — *Increasing Knowledge*

multiples

The **multiples** of a number are produced by multiplying the number by 1, by 2, by 3, by 4, and so on. Thus the multiples of 4 are

4, 8, 12, 16, 20, 24, 28, 32, 36, ...

The multiples of 6 are

6, 12, 18, 24, 30, 36, 42, 48, 54, ...

If we inspect these two lists, we see that some of the numbers in both lists are the same. A number appearing in both of these lists is a **common multiple** of 4 and 6. Below we have circled some of the common multiples of 4 and 6.

Multiples of 4: 4, 8, (12), 16, 20, (24), 28, 32, (36), ...

Multiples of 6: 6, (12), 18, (24), 30, (36), 42, 48, 54, ...

We see that 12, 24, and 36 are common multiples of 4 and 6. If we continued both lists, we would find many more common multiples.

least common multiple

Of particular interest is the least (smallest) of the common multiples. The **least common multiple** of 4 and 6 is 12. Twelve is the smallest number that is a multiple of both 4 and 6. The term *least common multiple* is often abbreviated **LCM.**

Example 1

Find the least common multiple of 6 and 8.

Solution

We will list some multiples of 6 and of 8 and circle common multiples.

Multiples of 6: 6, 12, 18, (24), 30, 36, 42, (48), ...

Multiples of 8: 8, 16, (24), 32, 40, (48), 56, 64, ...

We find that the least common multiple of 6 and 8 is **24.**

It is unnecessary to list multiples each time. Often the search for the least common multiple can be conducted mentally.

Example 2

Find the LCM of 3, 4, and 6.

Solution

To find the least common multiple of 3, 4, and 6, we can mentally search for the smallest number divisible by 3, 4, and 6. We can conduct the search by first thinking of multiples of the largest number, 6.

6, 12, 18, 24, ...

Then we mentally test these multiples for divisibility by 3 and by 4. We find that 6 is divisible by 3 but not by 4, while 12 is divisible by both 3 and 4. Thus the LCM of 3, 4, and 6 is **12.**

We can use prime factorization to help us find the least common multiple of a set of numbers. The LCM of a set of numbers is the product of *all the prime factors necessary to form any number in the set.*

Example 3

Math Language

A **prime factorization** is the expression of a composite number as a product of its prime factors.

Use prime factorization to help you find the LCM of 18 and 24.

Solution

We write the prime factorization of 18 and of 24.

$$18 = 2 \cdot 3 \cdot 3 \qquad 24 = 2 \cdot 2 \cdot 2 \cdot 3$$

The prime factors of 18 and 24 are 2's and 3's. From a pool of three 2's and two 3's, we can form either 18 or 24. So the LCM of 18 and 24 is the product of three 2's and two 3's.

$$\text{LCM of 18 and 24} = 2 \cdot 2 \cdot 2 \cdot 3 \cdot 3$$
$$= \mathbf{72}$$

equivalent division problems

Tricia's teacher asked this question:

If sixteen health snacks cost $4.00, what was the price for each health snack?

Tricia quickly gave the correct answer, 25¢, and then explained how she found the answer.

I knew I had to divide $4.00 by 16, but I did not know the answer. So I mentally found half of each number, which made the problem $2.00 ÷ 8. I still couldn't think of the answer, so I found half of each of those numbers. That made the problem $1.00 ÷ 4, and I knew the answer was 25¢.

How did Tricia's mental technique work? She used the identity property of multiplication. Recall from Lesson 15 that we can form equivalent fractions by multiplying or dividing a fraction by a fraction equal to 1.

$$\frac{3}{4} \times \frac{10}{10} = \frac{30}{40} \qquad \frac{6}{9} \div \frac{3}{3} = \frac{2}{3}$$

We can form equivalent division problems in a similar way. We multiply (or divide) the dividend and divisor by the same number to form a new division problem that is easier to calculate mentally. The new division problem will produce the same quotient, as we show below.

$$\frac{\$4.00 \div 2}{16 \div 2} = \frac{\$2.00}{8} = \frac{\$2.00 \div 2}{8 \div 2} = \frac{\$1.00}{4} = \$0.25$$

Example 4

Thinking Skill

Explain

How does doubling the number and dividing by 10 make this problem easier?

Instead of dividing 220 by 5, double both numbers and mentally calculate the quotient.

Solution

We double the two numbers in 220 ÷ 5 and get 440 ÷ 10. We mentally calculate the new quotient to be **44.** Since 220 ÷ 5 and 440 ÷ 10 are equivalent division problems, we know that 44 is the quotient for both problems.

Example 5

Instead of dividing 6000 by 200, divide both numbers by 100, and then mentally calculate the quotient.

Solution

We mentally divide by 100 by removing two places (two zeros) from each number. This forms the equivalent division problem 60 ÷ 2. We mentally calculate the quotient as **30.**

Represent Show how the equivalent division problem was formed.

Practice Set Find the least common multiple (LCM) of each pair or group of numbers:

a. 8 and 10 **b.** 4, 6, and 10

Use prime factorization to help you find the LCM of these pairs of numbers:

c. 24 and 40 **d.** 30 and 75

e. Instead of dividing $7\frac{1}{2}$ by $1\frac{1}{2}$, double each number and mentally calculate the quotient.

Mentally calculate each quotient by finding an equivalent division problem. What strategy did you use and why?

f. 24,000 ÷ 400 **g.** $6.00 ÷ 12 **h.** 140 ÷ 5

Written Practice *Strengthening Concepts*

1. (11) Octavio was writing a report on New Hampshire. He found that, in 2002, the population of Hanover, NH was 11,123. The population of Hollis, NH was 7416. The population of Newmarket, NH was 8449. What was the total population of these three places?

2. (13, 16) Rebecca and her mother built a shelf that was six feet long. How many inches long is this shelf?

*** 3.** (27) **Generalize** If the cost of one dozen eggs was $1.80, what was the cost per egg? Write an equivalent division problem that is easy to calculate mentally. Then find the quotient.

4. (5, 20) Which of the following equals one billion?

A 10^3 **B** 10^6 **C** 10^9 **D** 10^{12}

*** 5.** (22) Read this statement and answer the questions that follow.

Three eighths of the 712 students bought their lunch.

a. How many students bought their lunch?

b. How many students did not buy their lunch?

6. (19, 20) The width of this rectangle is 6 inches and its perimeter is 30 inches.

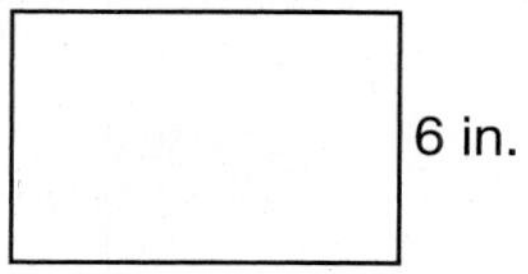

a. What is the length of the rectangle?

b. What is the area of the rectangle?

*** 7.** (27) Use prime factorization to find the least common multiple of 25 and 45.

8. (4) What number is halfway between 3000 and 4000?

*** 9.** (15, 24) **a.** Write 24% as a reduced fraction.

b. Use prime factorization to reduce $\frac{36}{180}$.

10. It was a very hot day. The temperature was 102°F in the shade.
(16)

a. The temperature was how many degrees above the freezing point of water?

b. The temperature was how many degrees below the boiling point of water?

c. *Discuss* What additional information did we need to know to answer **a** and **b?**

11. For each fraction, write an equivalent fraction that has a denominator of 36.
(2, 15)

a. $\frac{5}{12}$ **b.** $\frac{1}{6}$ **c.** $\frac{7}{9}$

d. *Analyze* What property do we use when we find equivalent fractions?

* **12.** **a.** *Generalize* Write the prime factorization of 576 using exponents.
(21)

b. Find $\sqrt{576}$.

* **13.** Write $5\frac{5}{6}$ and $6\frac{6}{7}$ as improper fractions and find their product.
(26)

In the figure below, quadrilaterals *ABCF* and *FCDE* are squares. Refer to the figure to answer problems **14–16.**

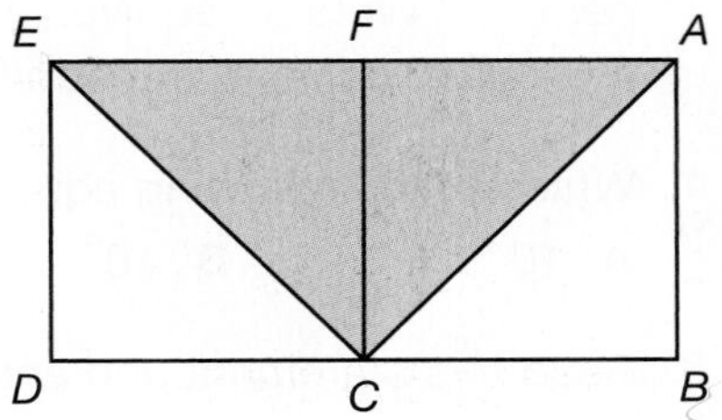

14. **a.** What kind of angle is $\angle ACD$?
(7)

b. Name two segments parallel to $\overline{FC}$.

15. **a.** What fraction of square *CDEF* is shaded?
(8)

b. What fraction of square *ABCF* is shaded?

c. What fraction of rectangle *ABDE* is shaded?

16. If *AB* is 3 ft,
(19, 20)

a. what is the perimeter of rectangle *ABDE?*

b. what is the area of rectangle *ABDE?*

Solve:

17. $10y = 360°$
(3)

18. $p + 2^4 = 12^2$
(3, 20)

* **19.** $5\frac{1}{8} - n = 1\frac{3}{8}$
(23)

20. $m - 6\frac{2}{3} = 4\frac{1}{3}$
(10)

Simplify:

* **21.** $10 - 1\frac{3}{5}$
(23)

* **22.** $5\frac{1}{3} \cdot 1\frac{1}{2}$
(26)

* **23.** $3\frac{1}{3} \div \frac{5}{6}$
(26)

* **24.** $5\frac{1}{4} \div 3$
(26, 15)

* **25.** $\frac{5}{4} \cdot \frac{9}{8} \cdot \frac{4}{15}$
(24)

26. $\frac{8}{9} - \left(\frac{7}{9} - \frac{5}{9}\right)$
(9, 15)

27. If the diameter of a circle is half of a yard, then its radius is how many inches?
(Inv. 2)

* **28.** **Generalize** Divide $12.00 by 16 or find the quotient of an equivalent division problem.
(27)

29. A 3-by-3-in. paper square is cut from a 5-by-5-in. paper square as shown.
(19, 20)

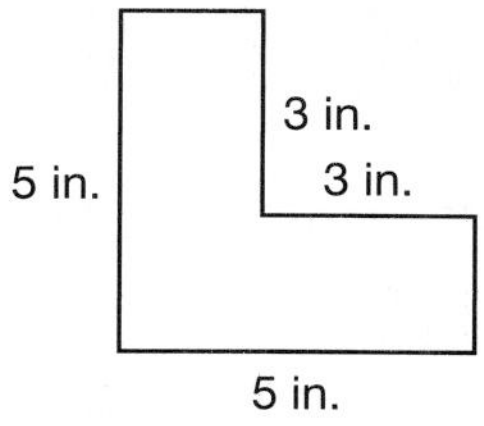

a. What is the perimeter of the resulting polygon?

b. How many square inches of the 5-by-5-in. square remain?

* **30.** **Classify** Refer to this circle with center at point *M* to answer **a–e:**
(Inv. 2)

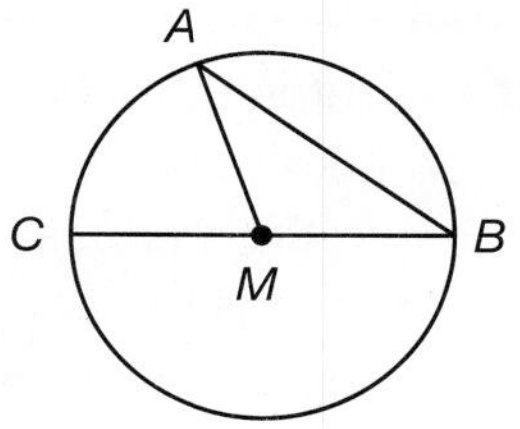

a. Which chord is a diameter?

b. Which chord is not a diameter?

c. What angle is an acute central angle?

d. Which angles are inscribed angles?

e. Which two sides of triangle *AMB* are equal in length?

LESSON 28

• Two-Step Word Problems
• Average, Part 1

Power Up

Building Power

facts Power Up F

mental math

a. **Number Sense:** \$6.23 + \$2.99

b. **Decimals:** \$1.75 × 100

c. **Calculation:** \$5.00 − \$1.29

d. **Calculation:** 8 × 53

e. **Calculation:** $\frac{5}{8} + \frac{5}{8}$

f. **Fractional Parts:** $\frac{2}{5}$ of 25

g. **Algebra:** If $w = 10$, what does $10w$ equal?

h. **Calculation:** Think of an easier equivalent division for \$56.00 ÷ 14. Then find the quotient.

problem solving

There are two routes that Imani can take to school. There are three routes Samantha can take to school. If Imani is going from her house to school and then on to Samantha's house, how many different routes can Imani take? Draw a diagram that illustrates the problem.

New Concepts

Increasing Knowledge

two-step word problems

Thus far we have considered these six one-step word-problem themes:

1. combining
2. separating
3. comparing
4. elapsed time
5. equal groups
6. parts of a whole

Word problems often require more than one step to solve. In this lesson we will continue practicing problems that require multiple steps to solve. These problems involve two or more of the themes mentioned above.

Example 1

Julie went to the store with \$20. If she bought 8 cans of dog food for 67¢ per can, how much money did she have left?

Solution

This is a two-step problem. First we find out how much Julie spent. This first step is an "equal groups" problem.

$$\begin{array}{rcr} \text{Number in group} & \longrightarrow & \$0.67 \text{ each can} \\ \text{Number of groups} & \longrightarrow & \times \quad 8 \text{ cans} \\ \hline \text{Total} & \longrightarrow & \$5.36 \end{array}$$

Now we can find out how much money Julie had left. This second step is about separating.

$$\begin{array}{r} \$20.00 \\ - \ \$5.36 \\ \hline \$14.64 \end{array}$$

After spending $5.36 of her $20 on dog food, Julie had **$14.64** left.

average, part 1

Calculating an **average** is often a two-step process. As an example, consider these five stacks of coins:

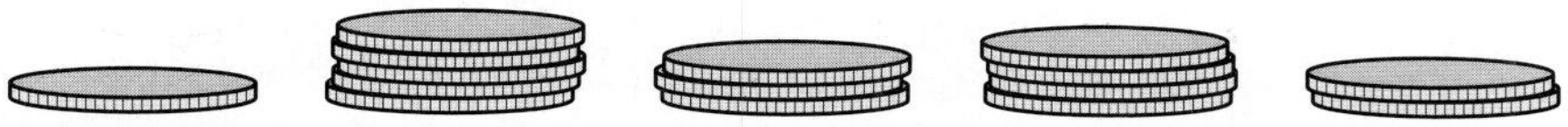

Thinking Skill

Summarize

In your own words, state a rule for finding an average?

There are 15 coins in all. If we made all the stacks the same size, there would be 3 coins in each stack.

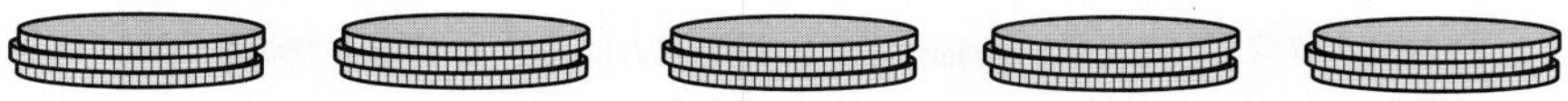

Predict If there were 20 coins in all, and we made all the stacks the same size, how many coins would be in each stack?

We say the average number of coins in each stack is 3. Now look at the following problem:

> *There are 4 squads in the physical education class. Squad A has 7 players, squad B has 9 players, squad C has 6 players, and squad D has 10 players. What is the average number of players per squad?*

The average number of players per squad is the number of players that would be on each squad if all of the squads had the same number of players. To find the average of a group of numbers, we combine the numbers by finding their sum.

$$\begin{array}{r} 7 \text{ players} \\ 6 \text{ players} \\ 9 \text{ players} \\ + \ 10 \text{ players} \\ \hline 32 \text{ players} \end{array}$$

Then we form equal groups by dividing the sum of the numbers by the number of numbers. There are 4 squads, so we divide by 4.

$$\frac{\text{sum of numbers}}{\text{number of numbers}} = \frac{32 \text{ players}}{4 \text{ squads}}$$

$$= 8 \text{ players per squad}$$

Finding the average took two steps. First we added the numbers to find the total. Then we divided the total to make equal groups.

Example 2

When people were seated, there were 3 in the first row, 8 in the second row, and 10 in the third row. What was the average number of people in each of the first three rows?

Solution

The average number of people in the first three rows is the number of people that would be in each row if the numbers were equal. First we add to find the total number of people.

$$\begin{array}{r} 3 \text{ people} \\ 8 \text{ people} \\ + \ 10 \text{ people} \\ \hline 21 \text{ people} \end{array}$$

Then we divide by 3 to separate the total into 3 equal groups.

$$\frac{21 \text{ people}}{3 \text{ rows}} = 7 \text{ people per row}$$

The average was **7 people** in each of the first 3 rows. Notice that the average of a set of numbers is *greater than the smallest number* in the set but *less than the largest number* in the set.

Another name for the average is the **mean.** We find the mean of a set of numbers by adding the numbers and then dividing the sum by the number of numbers.

Example 3

In a word game, five students in the class scored 100 points, four scored 95, six scored 90, and five scored 80. What was the mean of the scores?

Solution

First we find the total of the scores.

$$\begin{array}{rr} 5 \times 100 = & 500 \\ 4 \times 95 = & 380 \\ 6 \times 90 = & 540 \\ 5 \times 80 = & \underline{400} \\ & 1820 \end{array}$$

Visit www.SaxonPublishers.com/ActivitiesC2 *for a graphing calculator activity.*

Next we divide the total by 20 because there were 20 scores in all.

$$\frac{\text{sum of numbers}}{\text{number of numbers}} = \frac{1820}{20} = 91$$

We find that the mean of the scores was **91.**

Practice Set

Generalize Work each problem as a two-step problem:

a. Jody went to the store with $20 and returned home with $5.36. She bought 3 jars of spaghetti sauce. What was the cost of each jar of sauce?

b. Three-eighths of the 32 wild ducks feeding in the lake were wood ducks, the rest were mallards. How many mallards were in the lake?

c. In Room 1 there were 28 students, in Room 2 there were 29 students, in Room 3 there were 30 students, and in Room 4 there were 25 students. What was the average number of students per room?

d. What is the mean of 46, 37, 34, 31, 29, and 24?

e. What is the average of 40 and 70? What number is halfway between 40 and 70?

f. *Explain* The Central High School basketball team's lowest game score was 80 and highest score was 95. Which of the following could be their average score? Why?

A 80 **B** 84 **C** 95 **D** 96

Written Practice *Strengthening Concepts*

*** 1.** (28) Five volunteers collected bottles to be recycled. The number they collected were: 242, 236, 248, 268, and 226. What was the average number of bottles collected by the volunteers?

*** 2.** (28) Yori ran a mile in 5 minutes 14 seconds. How many seconds did it take Yori to run a mile?

*** 3.** (28) Luisa bought a pair of pants for $24.95 and 3 blouses for $15.99 each. Altogether, how much did she spend?

4. (12) The Italian navigator Christopher Columbus was 41 years old when he reached the Americas in 1492. In what year was he born?

5. (22) In the following statement, change the percent to a reduced fraction. Then diagram the statement and answer the questions.

Salma led for 75% of the 5000-meter race.

a. Salma led the race for how many meters?

b. Salma did not lead the race for how many meters?

6. (19, 20) This rectangle is twice as long as it is wide.

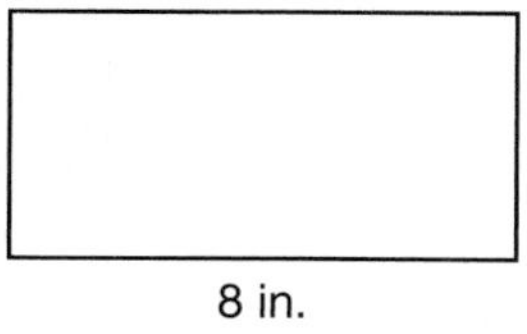

a. What is the perimeter of the rectangle?

b. What is the area of the rectangle?

*** 7.** (27) **a.** List the first six multiples of 3.

b. List the first six multiples of 4.

c. *Analyze* What is the LCM of 3 and 4?

d. Use prime factorization to find the least common multiple of 27 and 36.

*** 8.** *(27)* **Predict** On the number line below, 283 is closest to

a. which multiple of 10?

b. which multiple of 100?

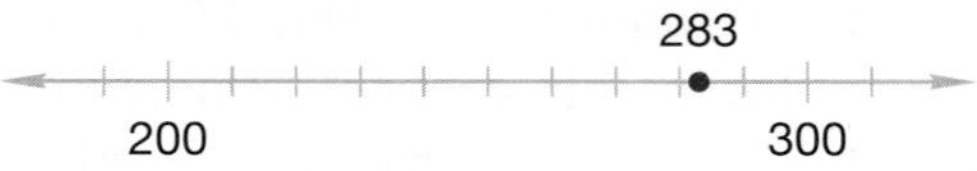

*** 9.** *(24)* **Generalize** Write 56 and 240 as products of prime numbers. Then reduce $\frac{56}{240}$.

10. *(16)* A mile is five thousand, two hundred eighty feet. Three feet equals a yard. So a mile is how many yards?

11. *(15)* For **a** and **b,** find an equivalent fraction that has a denominator of 24.

a. $\frac{7}{8}$ **b.** $\frac{11}{12}$

c. What property do we use to find equivalent fractions?

12. *(21)* **a.** Write the prime factorization of 3600 using exponents.

b. Find $\sqrt{3600}$.

*** 13.** *(28)* **Summarize** Describe how to find the mean of 45, 36, 42, 29, 16, and 24.

14. *(8, 20)* **a.** Draw square *ABCD* so that each side is 1 inch long. What is the area of the square?

b. Draw segments *AC* and *BD*. Label the point at which they intersect point *E*.

c. Shade triangle *CDE*.

d. What percent of the area of the square did you shade?

*** 15.** *(4, 10)* **a.** Arrange these numbers in order from least to greatest:

$$-1, \frac{1}{10}, 1, \frac{11}{10}, 0$$

b. **Classify** Which of these numbers are odd integers?

Solve:

16. *(3)* $12y = 360°$

17. *(3, 20)* $10^2 = m + 8^2$

18. *(3)* $\frac{180}{w} = 60$

Simplify:

19. *(9, 15)* $4\frac{5}{12} - 1\frac{1}{12}$

20. *(10, 15)* $8\frac{7}{8} + 3\frac{3}{8}$

*** 21.** *(23)* $12 - 8\frac{1}{8}$

*** 22.** *(26)* $6\frac{2}{3} \cdot 1\frac{1}{5}$

*** 23.** *(20, 26)* $\left(1\frac{1}{2}\right)^2 \div 7\frac{1}{2}$

*** 24.** *(26)* $8 \div 2\frac{2}{3}$

25. *(1)* $\frac{10,000}{80}$

*** 26.** *(25)* $\frac{3}{4} - \left(\frac{1}{2} \div \frac{2}{3}\right)$

27. Evaluate the following expressions for $x = 3$ and $y = 4$:
(1, 20)

a. x^y **b.** $x^2 + y^2$

28. Summarize What is the rule for this function?
(16)

Input	Output
1	3
2	5
4	9
5	11
10	21
12	25

29. In the figure below, the two triangles are congruent.
(18)

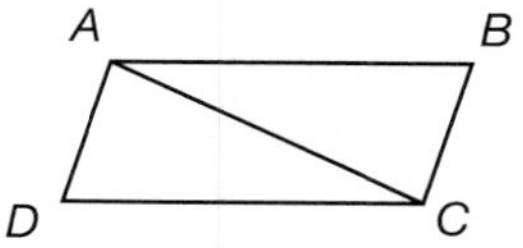

a. Which angle in $\triangle ACD$ corresponds to $\angle CAB$ in $\triangle ABC$?

b. Which segment in $\triangle ABC$ corresponds to $\overline{AD}$ in $\triangle ACD$?

c. If the area of $\triangle ABC$ is $7\frac{1}{2}$ in.^{2}, what is the area of figure $ABCD$?

30. With a ruler draw $\overline{PQ}$ $2\frac{3}{4}$ in. long. Then with a protractor draw $\overrightarrow{QR}$ so that $\angle PQR$ measures 30°. Then, from point P, draw a ray perpendicular to $\overline{PQ}$ that intersects $\overrightarrow{QR}$. (You may need to extend $\overrightarrow{QR}$ to show the intersection.) Label the point where the rays intersect point M. Use a protractor to measure $\angle PMQ$.
(17)

Early Finishers
Math and Science

On average, a heart beats about 60 to 80 times a minute when a person is at rest. Silvia wanted to know her average resting heart rate, so she recorded her resting heart rate every morning for one week, as shown below.

80, 75, 77, 66, 61, 73, 65

a. Find Silvia's average resting heart rate.

b. Is her resting heart rate normal? Support your answer.

LESSON

29

• Rounding Whole Numbers
• Rounding Mixed Numbers
• Estimating Answers

Power Up

Building Power

facts Power Up E

mental math

a. Calculation: \$4.32 + \$2.98

b. Decimals: \$12.50 ÷ 10

c. Calculation: \$10.00 − \$8.98

d. Calculation: 9×22

e. Calculation: $\frac{5}{6} + \frac{5}{6}$

f. Fractional Parts: $\frac{3}{5}$ of 20

g. Algebra: If $x = 4$, what does $4x$ equal?

h. Calculation: 6 × 6, ÷ 4, × 3, + 1, ÷ 4, × 8, − 1, ÷ 5, × 2, − 2, ÷ 2

problem solving The diameter of a penny is $\frac{3}{4}$ inch. How many pennies placed side by side would it take to make a row of pennies 1 foot long?

New Concepts

Increasing Knowledge

rounding whole numbers

The first sentence below uses an exact number to state the size of a crowd. The second sentence uses a rounded number.

There were 3947 fans at the game.

There were about 4000 fans at the game.

Rounded numbers are often used instead of exact numbers. One way to round a number is to consider its location on the number line.

Thinking Skill

Analyze

What word in the second sentence of the problem tells you that 4000 is not an exact number?

Example 1

Use a number line to

a. round 283 to the nearest hundred.

b. round 283 to the nearest ten.

Solution

a. We draw a number line showing multiples of 100 and mark the estimated location of 283.

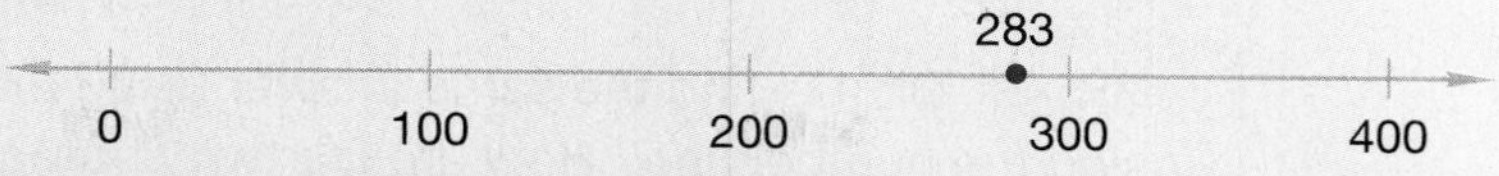

We see that 283 is between 200 and 300 and is closer to 300. To the nearest hundred, 283 rounds to **300.**

b. We draw a number line showing the tens from 200 to 300 and mark the estimated location of 283.

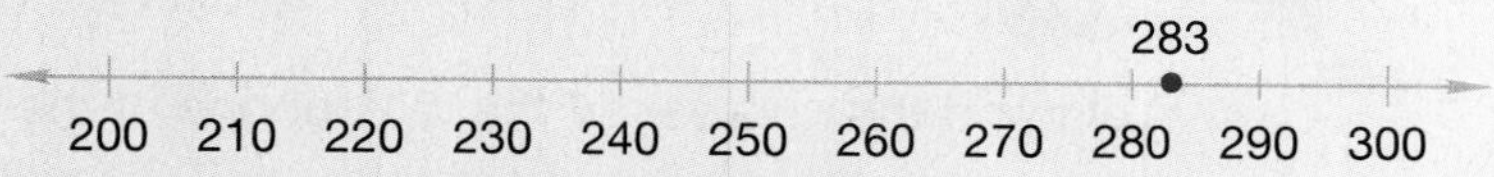

We see that 283 is between 280 and 290 and is closer to 280. To the nearest ten, 283 rounds to **280.**

Sometimes we are asked to round a number to a certain place value. We can use an underline and a circle to help us do this. We will underline the digit in the place to which we are rounding, and we will circle the next place to the right. Then we will follow these rules:

1. If the circled digit is 5 or more, we add 1 to the underlined digit. If the circled digit is less than 5, we leave the underlined digit unchanged.
2. We replace the circled digit and all digits to the right of the circled digit with zeros.

This rounding strategy is sometimes called the "4-5 split," because if the circled digit is 4 or less we round down, and if it is 5 or more we round up.

Example 2

a. Round 283 to the nearest hundred.

b. Round 283 to the nearest ten.

Solution

a. We underline the 2 since it is in the hundreds place. Then we circle the digit to its right.

2⑧3

Since the circled digit is 5 or more, we add 1 to the underlined digit, changing it from 2 to 3. Then we replace the circled digit and all digits to its right with zeros and get

300

b. Since we are rounding to the nearest ten, we underline the tens digit and circle the digit to its right.

28③

Since the circled digit is less than 5, we leave the 8 unchanged. Then we replace the 3 with a zero and get

280

Discuss Did we find the same answer using the 4–5 split strategy as we did using the number line? How are the two strategies different?

Example 3

Round 5280 so that there is one nonzero digit.

Solution

We round the number so that all but one of the digits are zeros. In this case we round to the nearest thousand, so 5280 rounds to **5000.**

Represent Write the number 5280. Underline the digit in the thousands place, and circle the digit to its right. How do we know to round down?

Example 4

Round 93,167,000 to the nearest million.

Solution

To the nearest million, 9<u>3</u>,①67,000 rounds to **93,000,000.**

rounding mixed numbers

When rounding a mixed number to a whole number, we need to determine whether the fraction part of the mixed number is greater than, equal to, or less than $\frac{1}{2}$. If the fraction is greater than or equal to $\frac{1}{2}$, the mixed number rounds up to the next whole number. If the fraction is less than $\frac{1}{2}$, the mixed number rounds down.

A fraction is greater than $\frac{1}{2}$ if the numerator of the fraction is more than half of the denominator. A fraction is less than $\frac{1}{2}$ if the numerator is less than half of the denominator.

Example 5

Round $14\frac{7}{12}$ to the nearest whole number.

Solution

The mixed number $14\frac{7}{12}$ is between the consecutive whole numbers 14 and 15. We study the fraction to decide which is nearer. The fraction $\frac{7}{12}$ is greater than $\frac{1}{2}$ because 7 is more than half of 12. So $14\frac{7}{12}$ rounds to **15.**

estimating answers

Rounding can help us **estimate** the answers to arithmetic problems. Estimating is a quick and easy way to get close to an exact answer. Sometimes a close answer is "good enough," but even when an exact answer is necessary, estimating can help us determine whether our exact answer is reasonable. One way to estimate is to round the numbers before calculating.

Example 6

Barb stopped by the store on the way home to buy two gallons of milk for \$2.79 per gallon, a loaf of bread for \$1.89, and a jar of peanut butter for \$3.15. About how much should she expect to pay for these items?

Solution

By rounding to the nearest dollar, shoppers can mentally keep a running total of the cost of items they are purchasing. Rounding the \$2.79 price per gallon of milk to \$3.00, the \$1.89 price of bread to \$2.00, and the \$3.15 price of peanut butter to \$3.00, we estimate the total to be

$$\$3 + \$3 + \$2 + \$3 = \$11$$

Barb should expect to pay about **\$11.00.**

Example 7

Mentally estimate:

a. $5\frac{7}{10} \times 3\frac{1}{3}$ **b.** 396×312 **c.** $4160 \div 19$

Solution

a. We round each mixed number to the nearest whole number before we multiply.

$$5\frac{7}{10} \times 3\frac{1}{3}$$
$$\downarrow \quad \downarrow$$
$$6 \times 3 = \mathbf{18}$$

b. When mentally estimating we often round the numbers to one nonzero digit so that the calculation is easier to perform. In this case we round to the nearest hundred.

$$400 \times 300 = \mathbf{120{,}000}$$

c. We round each number so there is one nonzero digit before we divide.

$$\frac{4160 \longrightarrow}{19 \longrightarrow} \frac{4000}{20} = \mathbf{200}$$

Performing a quick mental estimate helps us determine whether the result of a more complicated calculation is reasonable.

Example 8

Eldon calculated the area of this rectangle to be $25\frac{1}{4}$ sq. in. Is Eldon's calculation reasonable? Why or why not?

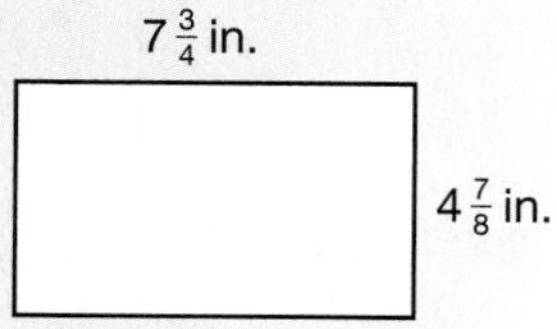

Solution

By estimating the area we can decide quickly whether Eldon's answer is reasonable. We round $7\frac{3}{4}$ up to 8 and round $4\frac{7}{8}$ up to 5 and estimate that the area of the rectangle is a little less than 40 sq. in. (8 in. × 5 in.). Based on this estimate, Eldon's calculation seems unreasonably low. Furthermore, by rounding the length and width down to 7 in. and 4 in., we see that the area of the rectangle must be more than 28 sq. in. This confirms that Eldon's calculation is not correct.

Practice Set

a. Round 1760 to the nearest hundred.

b. Round 5489 to the nearest thousand.

c. Round 186,282 to the nearest thousand.

Estimate each answer:

d. $7986 - 3074$

e. 297×31

f. $5860 \div 19$

g. $12\frac{1}{4} \div 3\frac{7}{8}$

h. Calculate the area of this rectangle. After calculating, check the reasonableness of your answer.

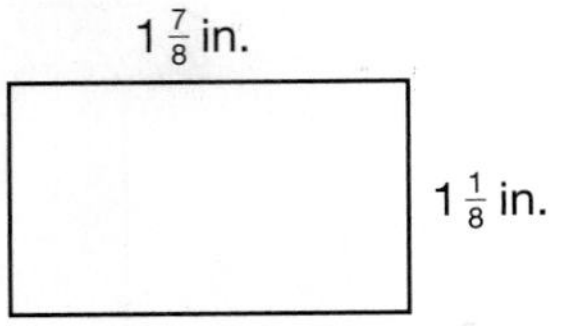

Written Practice

Strengthening Concepts

*** 1.** (16, 28) In the 1996 Summer Olympics, Charles Austin won the high jump event by jumping 7 feet 10 inches. How many inches did he jump?

2. (13) **Justify** If 8 pounds of bananas cost $5.52, what does 1 pound of bananas cost? How did you find the cost per pound?

Math Language

The **mean** of a set of numbers is the same as the average of the numbers.

*** 3.** (28) The number of fruit flies in each of Sandra's six samples were: 75, 70, 80, 80, 85, and 90. What was the mean number of fruit flies in her samples?

4. (15, 21) With one spin, what is the probability the arrow will stop on a prime number?

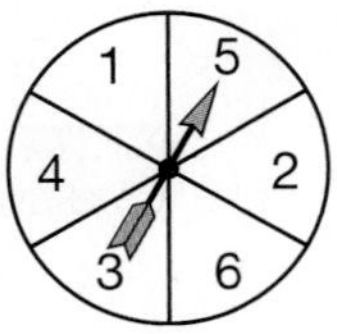

5. (22) **Evaluate** In the following statement, change the percent to a reduced fraction. Then diagram the statement and answer the questions.

Forty percent of the 80 birds were robins.

a. How many of the birds were robins?

b. How many of the birds were not robins?

*** 6.** (27) **a.** What is the least common multiple (LCM) of 4, 6, and 8?

b. **Represent** Use prime factorization to find the LCM of 16 and 36.

7. (19, 20) **a.** What is the perimeter of this square?

b. What is the area of this square?

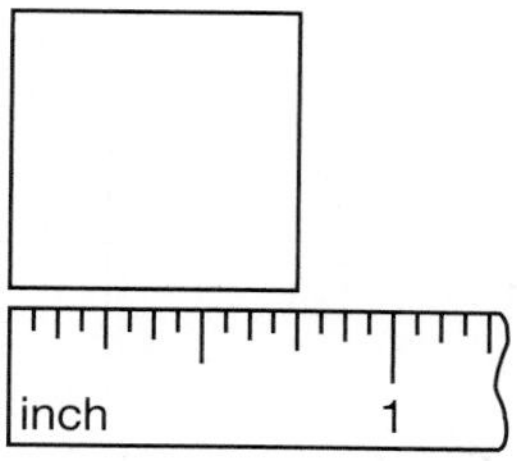

*** 8.** (29) **a.** Round 366 to the nearest hundred.

b. Round 366 to the nearest ten.

*** 9.** (29) **Estimate** Mentally estimate the sum of 6143 and 4952 by rounding each number to the nearest thousand before adding.

*** 10.** (26, 29) **a.** **Estimate** Mentally estimate the following product by rounding each number to the nearest whole number before multiplying:

$$\frac{3}{4} \cdot 5\frac{1}{3} \cdot 1\frac{1}{8}$$

b. **Estimate** Now find the exact product of these fractions and mixed numbers.

11. (15) Complete each equivalent fraction:

a. $\frac{2}{3} = \frac{?}{30}$ **b.** $\frac{?}{6} = \frac{25}{30}$

12. (20, 21) The prime factorization of 1000 is $2^3 \cdot 5^3$. Write the prime factorization of one billion using exponents.

In the figure below, quadrilaterals *ACDF*, *ABEF*, and *BCDE* are rectangles. Refer to the figure to answer problems **13–15.**

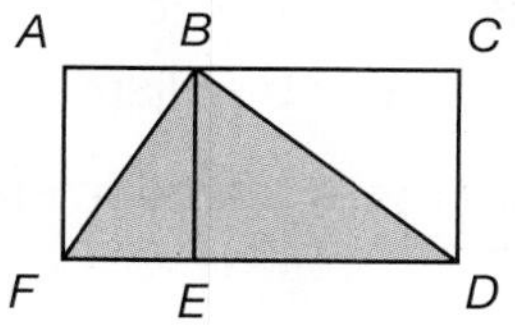

13. (8) **a.** What percent of rectangle *ABEF* is shaded?

b. What percent of rectangle *BCDE* is shaded?

c. What percent of rectangle *ACDF* is shaded?

14. (19, 20) **Infer** The relationships between the lengths of the sides of the rectangles are as follows:

$$AB + FE = BC$$
$$AF + CD = AC$$
$$AB = 2 \text{ in.}$$

a. Find the perimeter of rectangle *ABEF.*

b. Find the area of rectangle *BCDE.*

15. (18) **Infer** Triangle *ABF* is congruent to $\triangle EFB$.

a. Which angle in $\triangle ABF$ corresponds to $\angle EBF$ in $\triangle EFB$?

b. What is the measure of $\angle A$?

Solve:

16. (3, 20) $8^2 = 4m$

* **17.** (23) $x + 4\frac{4}{9} = 15$

18. (10, 15) $3\frac{5}{9} = n - 4\frac{7}{9}$

Simplify:

19. (23) $6\frac{1}{3} - 5\frac{2}{3}$

* **20.** (26) $6\frac{2}{3} \div 5$

* **21.** (26) $1\frac{2}{3} \div 3\frac{1}{2}$

22. (1) $\$7.49 \times 24$

* **23.** (29) **Explain** Describe how to estimate the product of $5\frac{1}{3}$ and $4\frac{7}{8}$.

24. (20) Find the missing exponents.

a. $10^3 \cdot 10^3 = 10^m$

b. $\frac{10^6}{10^3} = 10^n$

25. (4) The rule of the following sequence is $k = 2^n + 1$. Find the fifth term of the sequence.

3, 5, 9, 17, ...

26. (19, Inv. 2) Recall how you inscribed a regular hexagon in a circle in Investigation 2. If the radius of this circle is 1 inch,

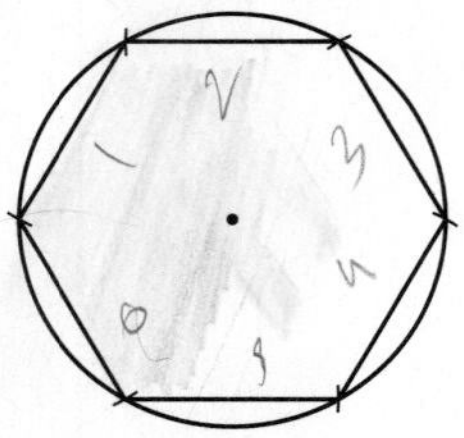

a. what is the diameter of the circle?

b. what is the perimeter of the hexagon?

27. (7) Use the figure below to identify the types of angles in **a–c**.

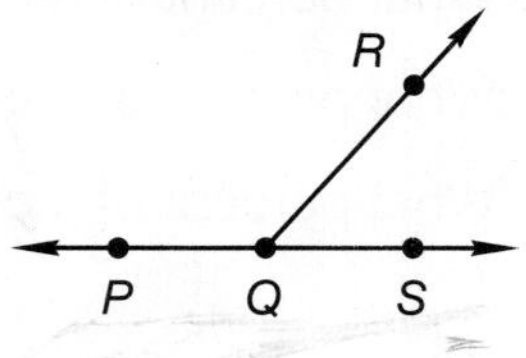

a. $\angle RQS$?

b. $\angle PQR$?

c. $\angle PQS$?

28. *(15)* Find fractions equivalent to $\frac{2}{3}$ and $\frac{1}{2}$ with denominators of 6. Subtract the smaller fraction you found from the larger fraction.

29. *(16)* Reggie and Elena each had one cup of water during a break in the soccer game. They took the water from the same 1-quart container. If they took two cups total from the full 1-quart container, how many ounces of water were left?

* **30.** *(29)* A photograph has the dimensions shown.

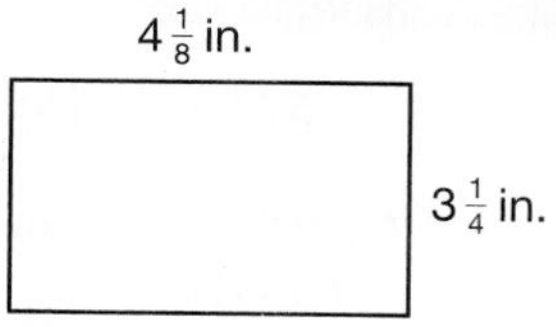

a. Estimate the area of the photograph.

b. **Verify** Is the actual area of the photograph more or less than your estimate? How do you know?

Early Finishers
Math and Science

The following list shows the average distance from the Sun to each of the nine planets in kilometers.

a. Round each distance to the nearest million.

b. Is Venus or Mars closer to Earth? Use the rounded distances to support your answers

Planet	Distance (in thousands)
Mercury	57,910
Venus	108,200
Earth	149,600
Mars	227,940
Jupiter	778,330
Saturn	1,426,940
Uranus	2,870,990
Neptune	4,497,070
Pluto	5,913,520

LESSON

30

• Common Denominators
• Adding and Subtracting Fractions with Different Denominators

Power Up

Building Power

facts Power Up F

mental math

a. **Number Sense:** \$1.99 + \$1.99

b. **Decimals:** \$0.15 × 1000

c. **Equivalent Fractions:** $\frac{3}{4} = \frac{?}{12}$

d. **Calculation:** 5 × 84

e. **Calculation:** $1\frac{2}{3} + 2\frac{2}{3}$

f. **Fractional Parts:** $\frac{3}{4}$ of 20

g. **Estimation:** Estimate the sum of 43 and 23

h. **Calculation:** Find $\frac{1}{2}$ of 88, + 4, ÷ 8, × 5, − 5, double that number, − 2, ÷ 2, ÷ 2, ÷ 2.

problem solving

Artists since the 14th century have used a geometric illusion in painting and drawing called ***one-point perspective.*** One-point perspective allows the artist to make it appear that objects in the drawing vanish into the distance, even though the drawing is two-dimensional. Follow the five steps provided to create a one-point perspective drawing.

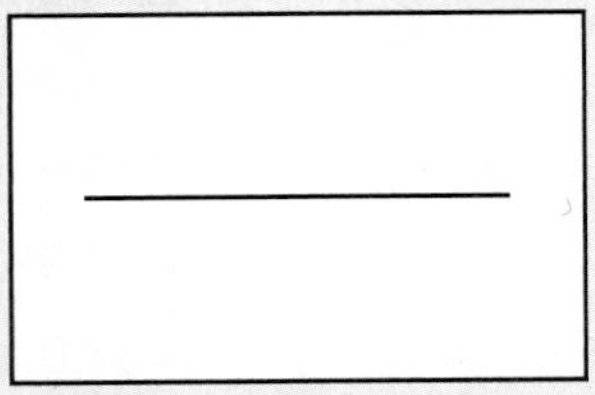

The ***horizon line*** divides the sky from the earth.

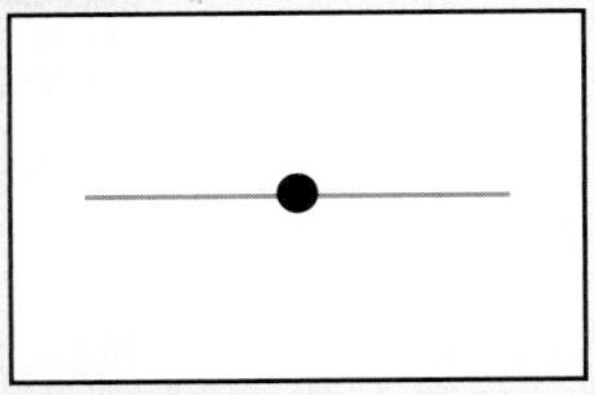

The ***vanishing point*** marks the direction in which you are looking.

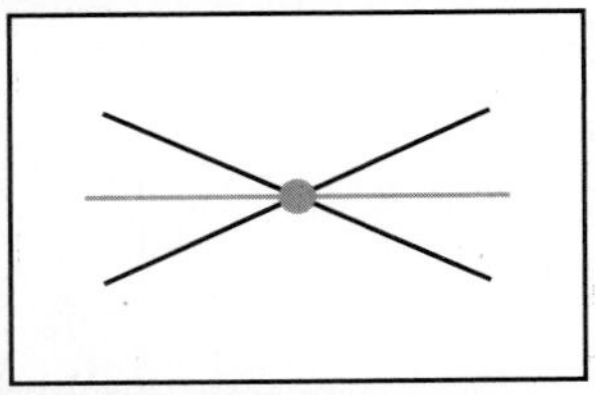

The ***construction lines*** show the tops and bottoms of the buildings.

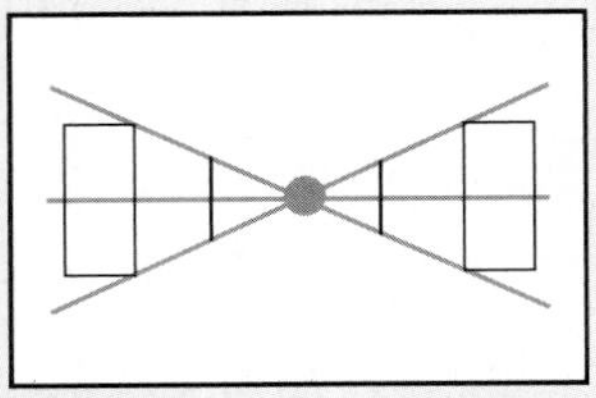

The edges of the buildings' sides will be both perpendicular and parallel to the ***horizon line.***

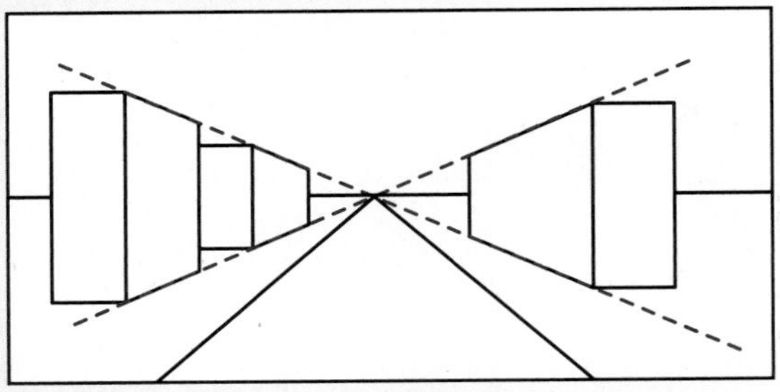

All ***receding lines*** will merge at the ***vanishing point.*** Erase ***construction lines*** and add details to complete the ***one-point perspective*** drawing.

New Concepts *Increasing Knowledge*

common denominators

If two fractions have the same denominator, we say they have **common denominators.**

$\frac{3}{8}$ $\frac{6}{8}$

These two fractions have common denominators.

$\frac{3}{8}$ $\frac{3}{4}$

These two fractions do not have common denominators.

If two fractions do not have common denominators, then one or both fractions can be renamed so both fractions do have common denominators. We remember that we can rename a fraction by multiplying it by a fraction equal to 1. Thus by multiplying by $\frac{2}{2}$, we can rename $\frac{3}{4}$ so that it has a denominator of 8.

$$\frac{3}{4} \cdot \frac{2}{2} = \frac{6}{8}$$

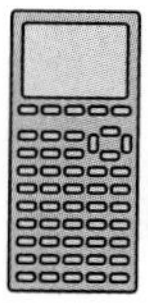

Visit www.SaxonPublishers.com/ActivitiesC2 *for a graphing calculator activity.*

Thinking Skill

Explain

Describe one way to find the least common multiple of 3 and 4.

Example 1

Rename $\frac{2}{3}$ and $\frac{1}{4}$ so that they have common denominators.

Solution

The denominators are 3 and 4. A common denominator for these two fractions would be any common multiple of 3 and 4. The **least common denominator** would be the least common multiple of 3 and 4, which is 12. We want to rename each fraction so that the denominator is 12.

$$\frac{2}{3} = \frac{\quad}{12} \qquad \frac{1}{4} = \frac{\quad}{12}$$

We multiply $\frac{2}{3}$ by $\frac{4}{4}$ and multiply $\frac{1}{4}$ by $\frac{3}{3}$.

$$\frac{2}{3} \cdot \frac{4}{4} = \frac{8}{12} \qquad \frac{1}{4} \cdot \frac{3}{3} = \frac{3}{12}$$

Thus $\frac{2}{3}$ and $\frac{1}{4}$ can be written with common denominators as

$$\mathbf{\frac{8}{12}} \quad \text{and} \quad \mathbf{\frac{3}{12}}$$

Fractions written with common denominators can be compared by simply comparing the numerators.

Explain In this example, the least common denominator is the product of the two original denominators. Is the product of the denominators always a common denominator? Is the product of the denominators always the least common denominator? Explain.

Example 2

Write these fractions with common denominators and then compare the fractions.

$$\frac{5}{6} \bigcirc \frac{7}{9}$$

Solution

Reading Math

Read the abbreviation LCM as "least common multiple."

The least common denominator for these fractions is the LCM of 6 and 9, which is 18.

$$\frac{5}{6} \cdot \frac{3}{3} = \frac{15}{18} \qquad \frac{7}{9} \cdot \frac{2}{2} = \frac{14}{18}$$

In place of $\frac{5}{6}$ we may write $\frac{15}{18}$, and in place of $\frac{7}{9}$ we may write $\frac{14}{18}$. Then we compare the renamed fractions.

$\frac{15}{18} \bigcirc \frac{14}{18}$ renamed

$\frac{15}{18} > \frac{14}{18}$ compared

Example 3

Arrange the following fractions in order from least to greatest. (You may write the fractions with common denominators to help you order them.)

$$\frac{2}{3}, \frac{5}{6}, \frac{7}{12}$$

Solution

Since $\frac{2}{3} = \frac{8}{12}$ and $\frac{5}{6} = \frac{10}{12}$, the order is $\frac{7}{12}, \frac{2}{3}, \frac{5}{6}$.

adding and subtracting fractions with different denominators

To add or subtract two fractions that do not have common denominators, we first rename one or both fractions so they do have common denominators. Then we can add or subtract.

Example 4

Add: $\frac{3}{4} + \frac{3}{8}$

Solution

First we write the fractions so they have common denominators. The denominators are 4 and 8. The least common multiple of 4 and 8 is 8. We rename $\frac{3}{4}$ so the denominator is 8 by multiplying by $\frac{2}{2}$. We do not need to rename $\frac{3}{8}$. Then we add the fractions and simplify.

$$\begin{aligned} \frac{3}{4} \cdot \frac{2}{2} &= \frac{6}{8} \quad \text{renamed } \frac{3}{4} \\ + \frac{3}{8} \quad &= \frac{3}{8} \\ \hline &\frac{9}{8} \quad \text{added} \end{aligned}$$

We finish by simplifying $\frac{9}{8}$.

$$\frac{9}{8} = \mathbf{1\frac{1}{8}}$$

Formulate Write a real world word problem involving the addition of $\frac{3}{4}$ and $\frac{3}{8}$. Then answer your problem.

Example 5

Subtract: $\mathbf{\frac{5}{6} - \frac{3}{4}}$

Solution

First we write the fractions so they have common denominators. The LCM of 6 and 4 is 12. We multiply $\frac{5}{6}$ by $\frac{2}{2}$ and multiply $\frac{3}{4}$ by $\frac{3}{3}$ so that both denominators are 12. Then we subtract the renamed fractions.

$$\begin{aligned} \frac{5}{6} \cdot \frac{2}{2} &= \frac{10}{12} \quad \text{renamed } \frac{5}{6} \\ - \frac{3}{4} \cdot \frac{3}{3} &= \frac{9}{12} \quad \text{renamed } \frac{3}{4} \\ \hline &\mathbf{\frac{1}{12}} \quad \text{subtracted} \end{aligned}$$

Example 6

Subtract: $\mathbf{8\frac{2}{3} - 5\frac{1}{6}}$

Solution

We first write the fractions so that they have common denominators. The LCM of 3 and 6 is 6. We multiply $\frac{2}{3}$ by $\frac{2}{2}$ so that the denominator is 6. Then we subtract and simplify.

$$\begin{aligned} 8\frac{2}{3} &= 8\frac{4}{6} \quad \text{renamed } 8\frac{2}{3} \\ - 5\frac{1}{6} &= 5\frac{1}{6} \\ \hline 3\frac{3}{6} &= \mathbf{3\frac{1}{2}} \quad \text{subtracted and simplified} \end{aligned}$$

Example 7

Add: $\frac{1}{2} + \frac{2}{3} + \frac{3}{4}$

Solution

The denominators are 2, 3, and 4. The LCM of 2, 3, and 4 is 12. We rename each fraction so that the denominator is 12. Then we add and simplify.

$$\begin{array}{rll} \frac{1}{2} \cdot \frac{6}{6} = & \frac{6}{12} & \text{renamed } \frac{1}{2} \\ \frac{2}{3} \cdot \frac{4}{4} = & \frac{8}{12} & \text{renamed } \frac{2}{3} \\ + \frac{3}{4} \cdot \frac{3}{3} = & \frac{9}{12} & \text{renamed } \frac{3}{4} \\ \hline & \frac{23}{12} = \mathbf{1\frac{11}{12}} & \text{added and simplified} \end{array}$$

Recall from Lesson 27 that prime factorization helps us find the least common multiple. We factor the numbers. Then we find the pool of numbers from which we can form either number. Consider 24 and 32.

$$24 = 2 \cdot 2 \cdot 2 \cdot 3$$

$$32 = 2 \cdot 2 \cdot 2 \cdot 2 \cdot 2$$

We can form either number from a pool of factors containing five 2s and one 3. Thus, the LCM of 24 and 32 is

$$2 \cdot 2 \cdot 2 \cdot 2 \cdot 2 \cdot 3 = 96$$

Example 8

Use prime factorization to help you add $\frac{5}{32} + \frac{7}{24}$.

Solution

We write the prime factorization of the denominators for both fractions.

$$\frac{5}{32} = \frac{5}{2 \cdot 2 \cdot 2 \cdot 2 \cdot 2} \qquad \frac{7}{24} = \frac{7}{2 \cdot 2 \cdot 2 \cdot 3}$$

The least common denominator of the two fractions is the least common multiple of the denominators. So the least common denominator is

$$2 \cdot 2 \cdot 2 \cdot 2 \cdot 2 \cdot 3 = 96$$

To rename the fractions with common denominators, we multiply $\frac{5}{32}$ by $\frac{3}{3}$, and we multiply $\frac{7}{24}$ by $\frac{2 \cdot 2}{2 \cdot 2}$.

$$\begin{array}{rl} \frac{5}{32} \cdot \frac{3}{3} = & \frac{15}{96} \\ + \frac{7}{24} \cdot \frac{2 \cdot 2}{2 \cdot 2} = & \frac{28}{96} \\ \hline & \mathbf{\frac{43}{96}} \end{array}$$

Practice Set

Write the fractions so that they have common denominators. Then compare the fractions.

a. $\frac{3}{5} \bigcirc \frac{7}{10}$

b. $\frac{5}{12} \bigcirc \frac{7}{15}$

c. Use common denominators to arrange these fractions in order from least to greatest:

$$\frac{1}{2}, \frac{3}{10}, \frac{2}{5}$$

Add or subtract:

d. $\frac{3}{4} + \frac{5}{6} + \frac{3}{8}$

e. $7\frac{5}{6} - 2\frac{1}{2}$

f. $4\frac{3}{4} + 5\frac{5}{8}$

g. $4\frac{1}{6} - 2\frac{5}{9}$

Use prime factorization to help you add or subtract the fractions in problems **h** and **i.**

h. $\frac{25}{36} + \frac{5}{60}$

i. $\frac{3}{25} - \frac{2}{45}$

j. *Justify* Choose one of the exercises you answered in this practice set. Explain the steps you took to find the answer.

Written Practice

Strengthening Concepts

* **1.** (28) The 5 starters on the basketball team were tall. Their heights were 76 inches, 77 inches, 77 inches, 78 inches, and 82 inches. What was the average height of the 5 starters?

* **2.** (28) Marie bought 6 pounds of carrots for $0.87 per pound and paid for them with a $10 bill. How much did she get back in change?

* **3.** (29) *Verify* While helping her father build a stone fence, Tanisha lifted 17 rocks averaging 8 pounds each. She calculated that she had lifted over 2000 pounds in all. Her father thought Tanisha's calculation was unreasonable. Do you agree or disagree with Tanisha's father? Why?

4. (14, 24) One hundred forty of the two hundred sixty students in the auditorium were not seventh graders. What fraction of the students in the auditorium were seventh graders?

5. (22) In the following statement, change the percent to a reduced fraction. Then answer the questions.

The Daltons completed 30% of their 2140-mile trip the first day.

a. How many miles did they travel the first day?

b. How many miles of their trip do they still have to travel?

6. (16, 19) If the perimeter of a square is 5 feet, how many inches long is each side of the square?

*** 7.** *(30)* ***Generalize*** Use prime factorization to subtract these fractions:

$$\frac{1}{18} - \frac{1}{30}$$

*** 8.** *(29)* Mt. Whitney in California is 14,494 ft high.

a. What is Mt. Whitney's height to the nearest thousand feet?

b. What is Mt. Whitney's height to the nearest hundred feet?

*** 9.** *(29)* ***Estimate*** Martin used a calculator to divide 28,910 by 49. The answer displayed was 59. Did Martin enter the problem correctly? (Use estimation to determine whether the displayed answer is reasonable.)

10. *(15, 24)* **a.** Write 32% as a reduced fraction.

b. Use prime factorization to reduce $\frac{48}{72}$.

11. *(30)* Write these fractions so that they have common denominators. Then compare the fractions.

$$\frac{5}{6} \bigcirc \frac{7}{8}$$

In the figure below, a 3-by-3-in. square is joined to a 4-by-4-in. square. Refer to the figure to answer problems **12** and **13**.

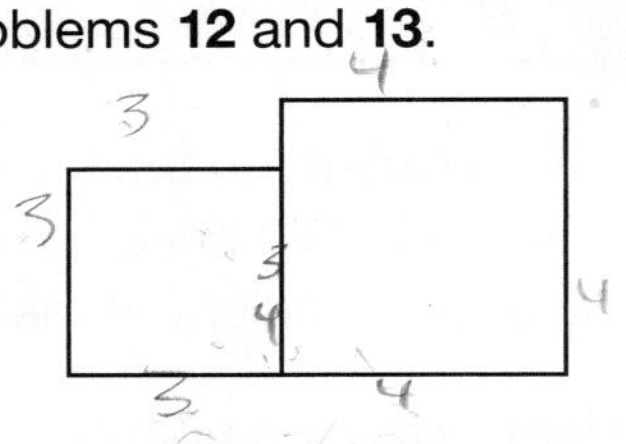

12. *(20)* **a.** What is the area of the smaller square?

b. What is the area of the larger square?

c. What is the total area of the figure?

*** 13.** *(19)* **a.** What is the perimeter of the hexagon that is formed by joining the two squares?

b. The perimeter of the hexagon is how many inches less than the combined perimeter of the two squares?

c. ***Justify*** Explain your answer to **b.**

14. *(21)* **a.** Write the prime factorization of 5184 using exponents.

b. Use the answer to **a** to find $\sqrt{5184}$.

*** 15.** *(28)* What is the mean of 5, 7, 9, 11, 12, 13, 24, 25, 26, and 28?

16. *(28)* List the single-digit divisors of 5670.

Solve:

17. *(3, 20)* $6w = 6^3$

18. *(3)* $90° + 30° + a = 180°$

19. (3, 13) **Formulate** Write an equal groups word problem for this equation and solve the problem.

$$36x = \$45.00$$

20. (16) To raise funds, the service club washed cars for $6 each. The money earned is a function of the number of cars washed. Make a function table that shows the dollars earned from washing 1, 3, 5, 10, and 20 cars.

Evaluate Simplify:

* **21.** (30) $\frac{1}{2} + \frac{1}{3}$

* **22.** (30) $\frac{3}{4} - \frac{1}{3}$

* **23.** (30) $2\frac{5}{6} - 1\frac{1}{2}$

* **24.** (26) $\frac{4}{5} \cdot 1\frac{2}{3} \cdot 1\frac{1}{8}$

* **25.** (26) $1\frac{3}{4} \div 2\frac{2}{3}$

* **26.** (26) $3 \div 1\frac{7}{8}$

Estimate For exercises **27** and **28,** record an estimated answer and the exact answer.

* **27.** (30) $3\frac{2}{3} + 1\frac{5}{6}$

* **28.** (23, 30) $5\frac{1}{8} - 1\frac{3}{4}$

29. (Inv. 2) **Represent** Draw a circle with a compass, and label the center point *O.* Draw chord *AB* through point *O.* Draw chord *CB* not through point *O.* Draw segment *CO.*

30. (Inv. 2) Refer to the figure drawn in problem **29** to answer **a–c.**

a. Which chord is a diameter?

b. Which segments are radii?

c. Which central angle is an angle of $\triangle OBC$?

Early Finishers
Real-World Application

Half the children at the park are on swings. One eighth of the children are on seesaws. One fourth of the children are on the slides. The other 6 children are playing ball.

Draw a diagram that represents the problem. Then write and solve an equation that shows how many children are in the park. Explain your work.

Focus on

• Coordinate Plane

By drawing two perpendicular number lines and extending the tick marks, we can create a grid over an entire plane called the **coordinate plane.** We can identify any point on the coordinate plane with two numbers.

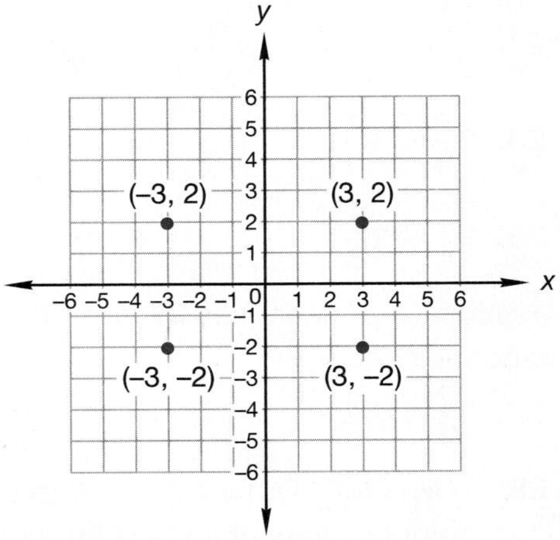

The horizontal number line is called the ***x*-axis.** The vertical number line is called the ***y*-axis.** The point at which the *x*-axis and the *y*-axis intersect is called the **origin.** The two numbers that indicate the location of a point are the **coordinates** of the point. The coordinates are written as a pair of numbers in parentheses, such as (3, 2). The first number shows the horizontal (↔) direction and distance from the origin. The second number shows the vertical (↕) direction and distance from the origin. The sign of the number indicates the direction. Positive coordinates are to the right or up. Negative coordinates are to the left or down. The origin is at point (0, 0).

The two axes divide the plane into four regions called **quadrants,** which are numbered counterclockwise, beginning with the upper right, as first, second, third, and fourth. The signs of the coordinates of each quadrant are shown below. Every point on a plane is either in a quadrant or on an axis.

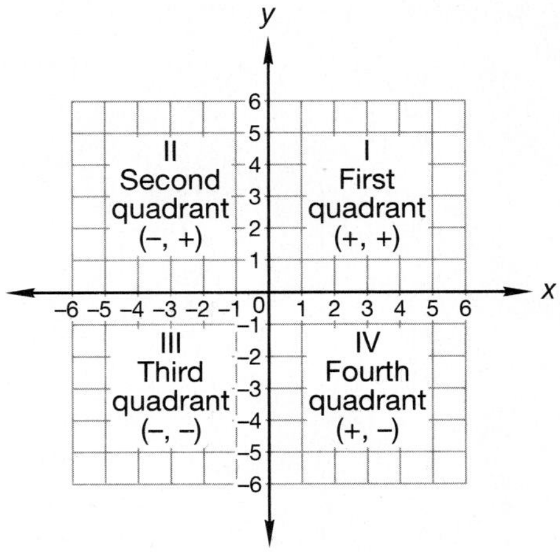

Example 1

Find the coordinates for points *A*, *B*, and *C* on this coordinate plane.

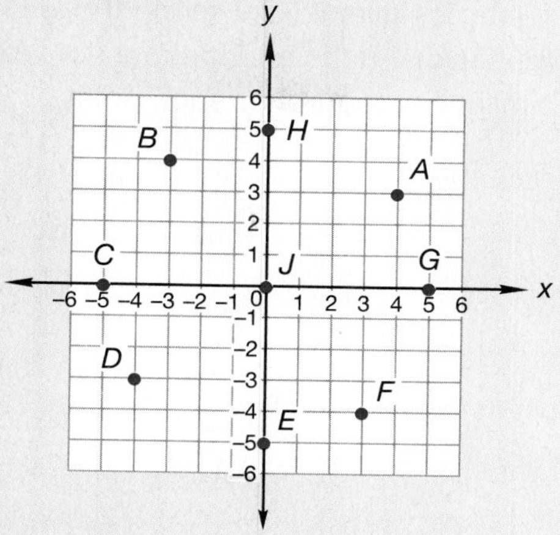

Solution

We first find the point on the *x*-axis that is directly above, below, or on the designated point. That number is the first coordinate. Then we determine how many units above or below the *x*-axis the point is. That number is the second coordinate.

Point *A* **(4, 3)**
Point *B* **(–3, 4)**
Point *C* **(–5, 0)**

Activity

Coordinate Plane

Materials needed:

- **Investigation Activity 13** Coordinate Plane (graph paper may also be used).
- Straightedge
- Protractor

Example 2

Graph the following points on a coordinate plane:

a. (3, 4) **b. (2, –3)** **c. (–1, 2)** **d. (0, –4)**

Solution

Math Language
The **origin** is the point (0, 0) on the coordinate plane.

To graph each point, we begin at the origin. To graph (3, 4), we move to the right (positive) 3 units along the *x*-axis. From there we turn and move up (positive) 4 units and make a dot. We label the location (3, 4). We follow a similar procedure for each point.

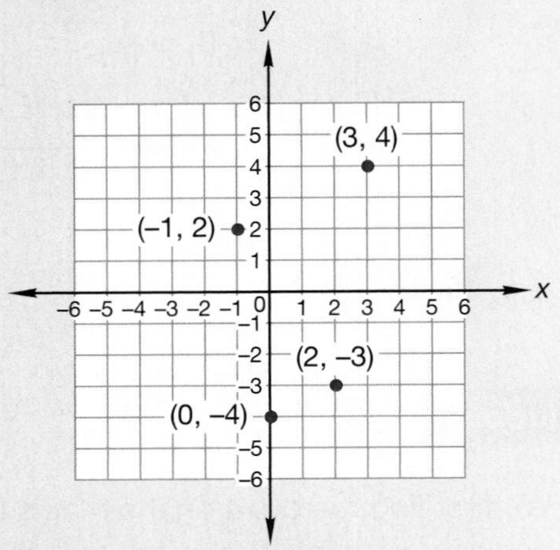

Example 3

The vertices of a square are located at (2, 2), (2, –1), (–1, –1), and (–1, 2). Draw the square and find its perimeter and area.

Solution

We graph the vertices and draw the square.

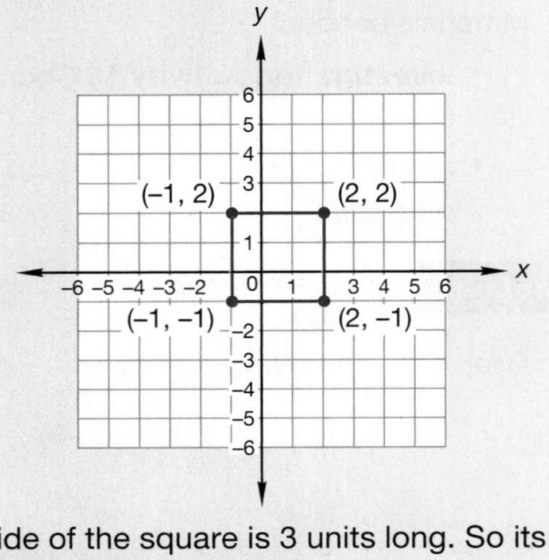

We find that each side of the square is 3 units long. So its perimeter is **12 units,** and its area is **9 square units.**

Example 4

Three vertices of a rectangle are located at (2, 1), (2, –1), and (–2, –1). Find the coordinates of the fourth vertex and the perimeter and area of the rectangle.

Solution

We graph the given coordinates.

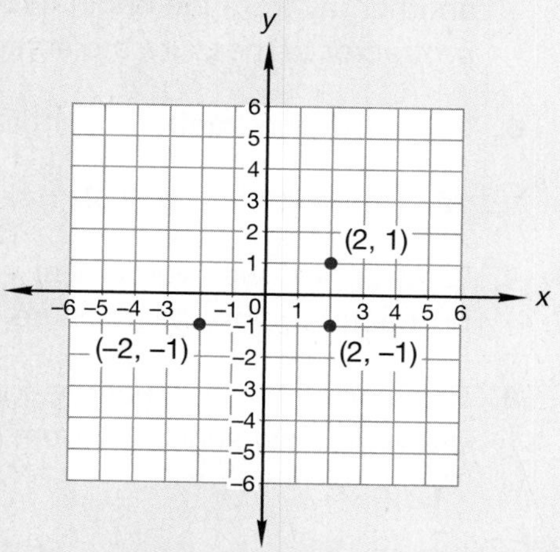

We see that the location of the fourth vertex is **(–2, 1),** which we graph.

Verify How do we know that the location of the 4th vertex must be (−2, 1)?

Then we draw the rectangle and find that it is 4 units long and 2 units wide. So its perimeter is **12 units,** and its area is **8 square units.**

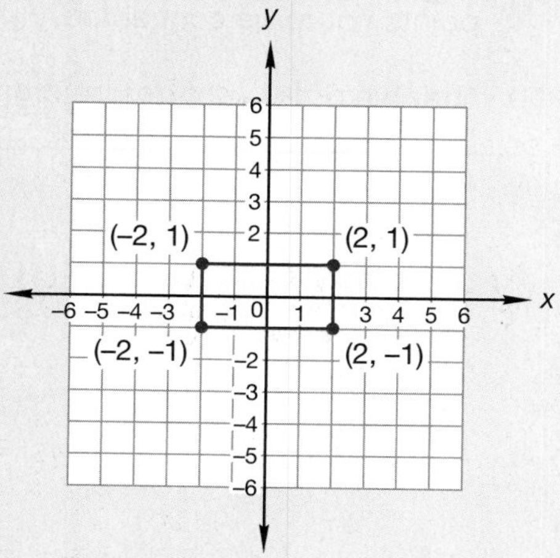

1. **Evaluate** Graph these three points: (2, 4), (0, 2), and (–3, –1). Then draw a line that passes through these points. Name a point in the second quadrant that is on the line.

2. One vertex of a square is the origin. Two other vertices are located at (−2, 0) and (0, −2). What are the coordinates of the fourth vertex?

3. Find the perimeter and area of a rectangle whose vertices are located at (3, –1), (–2, –1), (–2, – 4), and (3, – 4).

4. Points (4, 4), (4, 0), and (0, 0) are the vertices of a triangle. The triangle encloses whole squares and half squares on the grid. Determine the area of the triangle by counting the whole squares and the half squares. (Count two half squares as one square unit.)

5. *Represent* Draw a ray from the origin through the point (10, 10). Draw another ray from the origin through the point (10, 0). Then use a protractor to measure the angle.

6. Name the quadrant that contains each of these points:

   a. (–15, –20) **b.** (12, 1) **c.** (20, –20) **d.** (–3, 5)

7. Draw △ ABC with vertices at A (0, 0), B (8, –8), and C (–8, –8). Use a protractor to find the measure of each angle of the triangle.

8. Shae wrote these directions for a dot-to-dot drawing. To complete the drawing, draw segments from point to point in the order given.

   1. (0, 4)
   2. (–3, – 4)
   3. (5, 1)
   4. (–5, 1)
   5. (3, – 4)
   6. (0, 4)

9. *Model* Plan and create a straight-segment drawing on graph paper. Determine the coordinates of the vertices. Then write directions for completing the dot-to-dot drawing for other classmates to follow. Include the directions "lift pencil" between consecutive coordinates of points not to be connected.

10. Graph a dot-to-dot design created by a classmate.

LESSON 31

Reading and Writing Decimal Numbers

Power Up

Building Power

facts Power Up G

mental math

a. **Number Sense:** \$4.00 − 99¢

b. **Calculation:** $7 \times 35¢$

c. **Equivalent Fractions:** $\frac{2}{3} = \frac{?}{12}$

d. **Fractions:** Reduce $\frac{18}{24}$.

e. **Power/Roots:** $\sqrt{100} + 3^2$

f. **Fractional Parts:** $\frac{3}{4}$ of 60

g. **Estimation:** Estimate the sum of 89 and 64

h. **Calculation:** Start with the number of degrees in a right angle, ÷ 2, + 5, ÷ 5, − 1, find the square root.

problem solving

Fourteen blocks were used to build this three-layer pyramid.

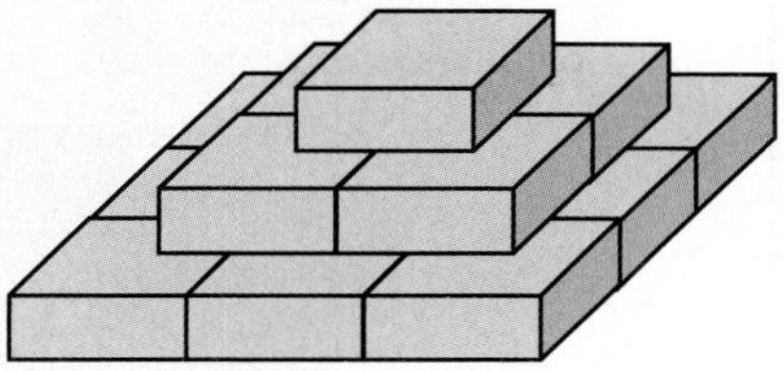

How many blocks would we need to build a five-layer pyramid?

If we build a ten-layer pyramid, how many blocks will we need for the bottom layer?

New Concept

Increasing Knowledge

We have used fractions and percents to name parts of a whole. We remember that a fraction has a numerator and a denominator. The denominator indicates the number of equal parts in the whole. The numerator indicates the number of parts that are selected.

$$\frac{\text{Number of parts selected}}{\text{Number of equal parts in the whole}} = \frac{3}{10}$$

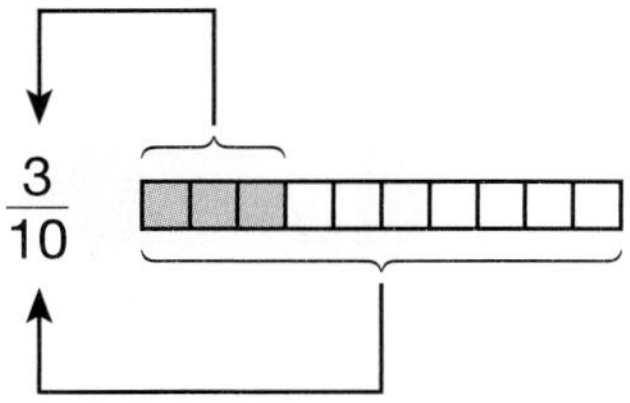

Parts of a whole can also be named by using **decimal fractions.** In a decimal fraction we can see the numerator, but we cannot see the denominator. **The denominator of a decimal fraction is indicated by place value.** On the following page is the decimal fraction three tenths.

We know the denominator is 10 because only one place is shown to the right of the decimal point.

$$0.3$$

The decimal fraction 0.3 and the common fraction $\frac{3}{10}$ are equivalent. Both are read "three tenths."

$$0.3 = \frac{3}{10} \qquad \text{three tenths}$$

A decimal fraction written with two digits after the decimal point (two decimal places) is understood to have a denominator of 100, as we show here:

$$0.03 = \frac{3}{100} \qquad \text{three hundredths}$$

$$0.21 = \frac{21}{100} \qquad \text{twenty-one hundredths}$$

A number that contains a decimal fraction is called a **decimal number** or just a **decimal.**

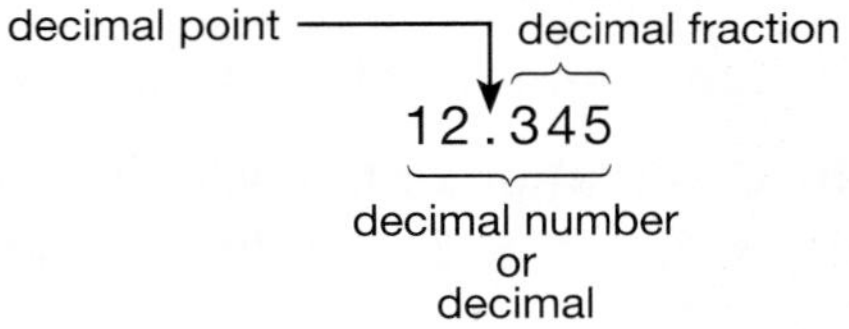

Example 1

Write seven tenths as a fraction and as a decimal.

Solution

$\frac{7}{10}$ **0.7**

Example 2

Name the shaded part of this square

a. as a fraction.

b. as a decimal.

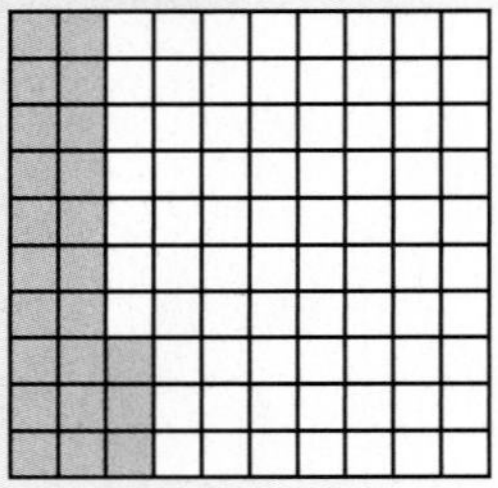

Solution

a. $\frac{23}{100}$ **b. 0.23**

In our number system the place a digit occupies has a value called **place value.** We remember that places to the left of the decimal point have values of 1, 10, 100, 1000, and so on, becoming greater and greater. Places to the right of the decimal point have values of $\frac{1}{10}$, $\frac{1}{100}$, $\frac{1}{1000}$, and so on, becoming less and less.

Reading Math

When reading a decimal number, be sure to include the ending suffix *–ths.* Read 0.6 as "6 tenths."

This chart shows decimal place values from the millions place through the millionths place:

Decimal Place Values

millions	1,000,000
,	
hundred thousands	100,000
ten thousands	10,000
thousands	1000
,	
hundreds	100
tens	10
ones	1
decimal point	.
tenths	$\frac{1}{10}$
hundredths	$\frac{1}{100}$
thousandths	$\frac{1}{1000}$
ten-thousandths	$\frac{1}{10,000}$
hundred-thousandths	$\frac{1}{100,000}$
millionths	$\frac{1}{1,000,000}$

Example 3

In the number 12.34579, which digit is in the thousandths place?

Solution

The thousandths place is the third place to the right of the decimal point and is occupied by the **5.**

Example 4

Name the place occupied by the 7 in 4.63471.

Solution

The 7 is in the fourth place to the right of the decimal point. This is the **ten-thousandths place.**

To read a decimal number, we first read the whole-number part, and then we read the fraction part. To read the fraction part of a decimal number, we read the digits to the right of the decimal point as though we were reading a whole number. This number is the numerator of the decimal fraction. Then we say the name of the last decimal place. This number is the denominator of the decimal fraction.

Example 5

Read this decimal number: 123.123

Solution

First we read the whole-number part. *When we come to the decimal point, we say "and."* Then we read the fraction part, ending with the name of the last decimal place.

We say "and" for the decimal point.

123.123

We say "thousandths" to conclude naming the number.

One hundred twenty-three and one hundred twenty-three thousandths

Discuss Which has a greater value, the 123 to the left of the decimal point, or the 123 to the right of the decimal point? How do you know?

Example 6

Use digits to write these decimal numbers:

a. Seventy-five thousandths

b. One hundred and eleven hundredths

Solution

a. The last word tells us the last place in the decimal number. "Thousandths" means there are three places to the right of the decimal point.

. ___ ___ ___

We fit the digits of 75 into the places so the 5 is in the last place. We write zero in the remaining place.

. <u>0</u> <u>7</u> <u>5</u>

Thinking Skill

Discuss

When a decimal number has zero as its whole-number part, why is it helpful to show the zero in the ones place?

Decimal numbers without a whole-number part are usually written with a zero in the ones place. Therefore, we will write the decimal number "seventy-five thousandths" as follows:

0.075

b. To write "one hundred and eleven hundredths," we remember that the word "and" separates the whole-number part from the fraction part. First we write the whole-number part followed by a decimal point for "and":

100.

Then we write the fraction part. We shift our attention to the last word to find out how many decimal places there are. "Hundredths" means there are two decimal places.

100. ___ ___

Now we fit "eleven" into the two decimal places, as follows:

100.11

Practice Set

a. **Represent** Write three hundredths as a fraction. Then write three hundredths as a decimal.

b. Name the shaded part of the circle both as a fraction and as a decimal.

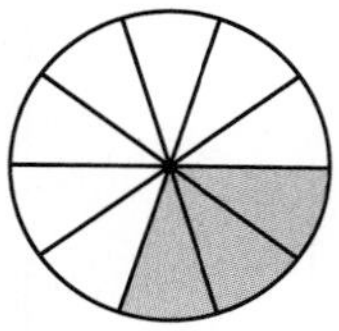

c. In the number 16.57349, which digit is in the thousandths place?

d. The number 36.4375 has how many decimal places?

Use words to write each decimal number:

e. 25.134

f. 100.01

Use digits to write each decimal number:

g. one hundred two and three tenths

h. one hundred twenty-five ten-thousandths

i. three hundred and seventy-five thousandths

j. *Conclude* What word tells you that some of the numbers in exercises **g–i** include both whole-number parts and fraction parts?

Written Practice *Strengthening Concepts*

* **1.** (28) *Evaluate* Ms. Gonzalez's class and Mr. O'Brien's class are going to use the money they raise at the school raffle to buy several new maps for their classrooms. The maps cost $89.89. Ms. Gonzalez's class has raised $26.47. Mr. O'Brien's class has raised $32.54. How much more money do they need to raise to buy the maps?

* **2.** (28) Norton read 4 books during his vacation. The first book had 326 pages, the second had 288 pages, the third had 349 pages, and the fourth had 401 pages. The 4 books he read had an average of how many pages per book?

3. (13) A one-year subscription to the monthly magazine costs $15.96. At this price, what is the cost for each issue?

4. (12) The settlement at Jamestown began in 1607. This was how many years after Columbus reached the Americas in 1492?

5. (19) *Explain* A square and a regular hexagon share a common side. The perimeter of the square is 24 in. Describe how to find the perimeter of the hexagon.

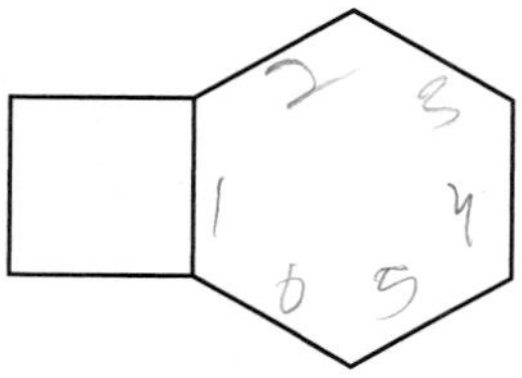

6. (22) In the following statement, change the percent to a reduced fraction and answer the questions.

Kelly bought a book of 20 stamps. She has used 80% of them.

a. How many stamps has Kelly used?

b. How many stamps does she have left?

* **7.** (29) Round 481,462

a. to the nearest hundred thousand.

b. to the nearest thousand.

* **8.** (29) **Estimate** Mentally estimate the difference between 49,623 and 20,162.

* **9.** (31) Name the shaded part of this square

a. as a fraction.

b. as a decimal.

c. as a percent.

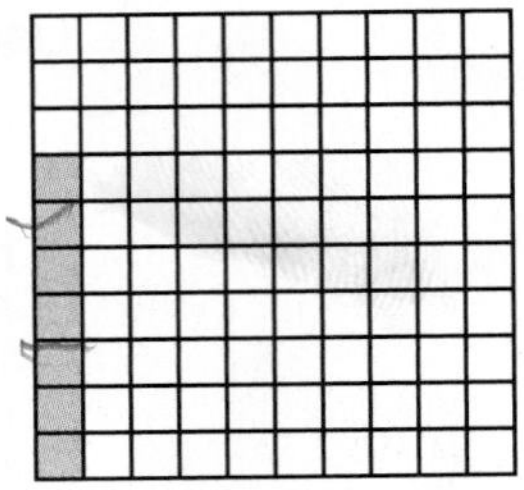

* **10.** (31) In the number 9.87654, which digit is in the hundredths place?

* **11.** (31) Replace each circle with the proper comparison symbol:

a. $\frac{3}{10} \bigcirc 0.3$ **b.** $\frac{3}{100} \bigcirc 0.3$

* **12.** (Inv. 3) The vertices of a square are located at (3, 3), (3, −3), (−3, −3), and (−3, 3).

a. **Represent** Sketch the axis. Then draw the square as described. Use your drawing to answer questions **b** and **c.**

b. What is the perimeter of the square?

c. What is the area of the square?

13. (15) Complete each equivalent fraction:

a. $\frac{5}{?} = \frac{15}{24}$ **b.** $\frac{7}{12} = \frac{?}{24}$ **c.** $\frac{?}{6} = \frac{4}{24}$

14. (21) **a.** Write the prime factorization of 2025 using exponents.

b. Find $\sqrt{2025}$.

15. (18) Draw two parallel lines. Then draw two more parallel lines that are perpendicular to the first pair of lines. Label the points of intersection *A, B, C,* and *D* consecutively in a counterclockwise direction. Draw segment *AC.* Refer to the figure to answer **a** and **b.**

a. What kind of quadrilateral is figure *ABCD?*

b. Triangles *ABC* and *CDA* are congruent. Which angle in $\triangle ABC$ corresponds to $\angle DAC$ in $\triangle CDA$?

16. (14) Twelve red marbles and 15 blue marbles are in a bag. If one marble is drawn from the bag, what is the probability the marble is not red?

17. Name the properties used to solve this equation.
(2, 9)

$$\frac{4}{3}\left(\frac{3}{4}x\right) = \frac{4}{3} \cdot \frac{1}{4} \qquad \text{Given}$$

$$\left(\frac{4}{3} \cdot \frac{3}{4}\right)x = \frac{4}{3} \cdot \frac{1}{4} \qquad \textbf{a.}\ \_\_\_\_\_\_\_\_\_\_$$

$$1x = \frac{4}{3} \cdot \frac{1}{4} \qquad \textbf{b.}\ \_\_\_\_\_\_\_\_\_\_$$

$$x = \frac{4}{3} \cdot \frac{1}{4} \qquad \textbf{c.}\ \_\_\_\_\_\_\_\_\_\_$$

$$x = \frac{1}{3} \qquad \frac{4}{3} \cdot \frac{1}{4} = \frac{1}{3}$$

Solve:

18. $9n = 6 \cdot 12$
(3)

19. $90° + 45° + b = 180°$
(3)

Generalize Simplify:

* **20.** $\frac{1}{2} + \frac{2}{3}$
(30)

21. $\frac{1}{2} - \left(\frac{3}{4} \cdot \frac{2}{3}\right)$
(9, 24)

* **22.** $3\frac{5}{6} - \frac{1}{3}$
(30)

23. $\frac{5}{8} \cdot 2\frac{2}{5} \cdot \frac{4}{9}$
(26)

24. $2\frac{2}{3} \div 1\frac{3}{4}$
(26)

25. $1\frac{7}{8} \div 3$
(26)

* **26.** $3\frac{1}{2} + 1\frac{5}{6}$
(30)

* **27.** $5\frac{1}{4} - 1\frac{5}{8}$
(23, 30)

* **28.** Evaluate this expression for $a = 3$ and $b = 4$:
(1, 30)

$$\frac{b}{a} + \frac{a}{b}$$

29. The rule of the following sequence is $k = 10^n$. Use words to name the sixth term.
(4, 20)

10, 100, 1,000, ...

30. **a.** A half circle or half turn measures how many degrees?
(17)

b. A quarter of a circle measures how many degrees?

c. An eighth of a circle measures how many degrees?

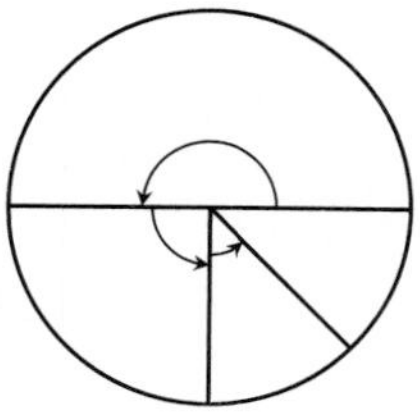

LESSON 32

• Metric System

Power Up

Building Power

facts Power Up G

mental math

a. **Number Sense:** \$3.76 − 99¢

b. **Calculation:** 8 × 25¢

c. **Equivalent Fractions:** $\frac{5}{6} = \frac{?}{24}$

d. **Fractions:** Reduce $\frac{12}{20}$.

e. **Exponents:** $3^2 + 4^2$

f. **Fractional Parts:** $\frac{2}{5}$ of 30

g. **Estimation:** Estimate the sum of 99 and 76

h. **Calculation:** Start with the number of sides of an octagon, × 5, + 2, ÷ 6, × 5, + 1, $\sqrt{\ }$, ÷ 3.

problem solving

Wes and Josh took turns multiplying numbers. Josh began by choosing the number 4. Wes multiplied Josh's number by 4 to get 16. Josh then multiplied 16 by 4 and got 64. Wes multiplied 64 by 4 and got 256. They continued this pattern until one of the brothers found the product 1,048,576. Which brother found the product 1,048,576?

New Concept

Increasing Knowledge

The system of measurement used throughout most of the world is the **metric system.** The metric system has two primary advantages over the U.S. Customary System: it is a decimal system, and the units of one category of measurement are linked to units of other categories of measurement.

The metric system is a decimal system in that units within a category of measurement differ by a factor, or power, of 10. The U.S. Customary System is not a decimal system, so converting between units is more difficult. Here we show some equivalent measures of length in the metric system:

Units of Length
10 millimeters (mm) = 1 centimeter (cm)
1000 millimeters (mm) = 1 meter (m)
100 centimeters (cm) = 1 meter (m)
1000 meters (m) = 1 kilometer (km)

Thinking Skill

Justify

A room's temperature is 22 degrees and it feels comfortably warm. Is this measurement in °C or °F? Explain.

Celsius Temperature Scale

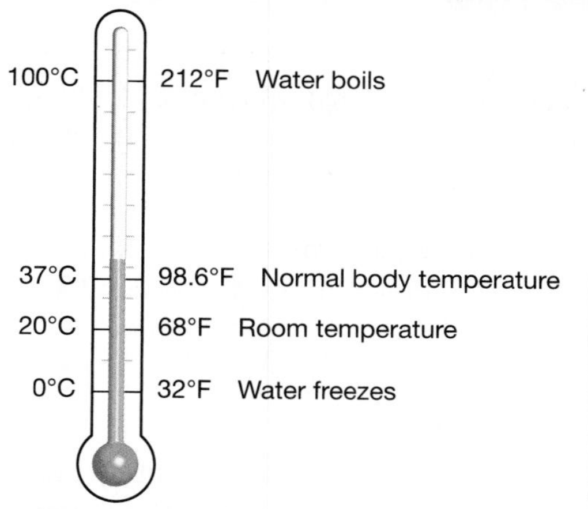

Example 4

A temperature increase of 100° on the Celsius scale is an increase of how many degrees on the Fahrenheit scale?

Solution

The Celsius and Fahrenheit scales are different scales. An increase of 1°C is not equivalent to an increase of 1°F. On the Celsius scale there are 100° between the freezing point of water and the boiling point of water. On the Fahrenheit scale water freezes at 32° and boils at 212°, a difference of 180°. So an increase of 100°C is an increase of **180°F.** Thus, a change of one degree on the Celsius scale is equivalent to a change of 1.8 degrees on the Fahrenheit scale.

Practice Set

a. The closet door is about 2 meters tall. How many centimeters is 2 meters?

b. A 1-gallon plastic jug can hold about how many liters of milk?

c. A metric ton is 1000 kilograms, so a metric ton is about how many pounds?

d. A temperature increase of 10° on the Celsius scale is equivalent to an increase of how many degrees on the Fahrenheit scale? (See example 4.)

e. After running 800 meters of a 3-kilometer race, Michelle still had how many meters to run?

f. A 30-cm ruler broke into two pieces. One piece was 120 mm long. How long was the other piece? Express your answer in millimeters.

g. *Conclude* About how many inches long was the ruler in exercise **f** before it broke?

Written Practice *Strengthening Concepts*

1. (28) There were 3 towns on the mountain. The population of Hazelhurst was 4248. The population of Baxley was 3584. The population of Jesup was 9418. What was the average population of the 3 towns on the mountain?

2. (28) The film was a long one, lasting 206 minutes. How many hours and minutes long was the film?

3. (14, 24) A mile is 1760 yards. Yolanda ran 440 yards. What fraction of a mile did she run?

*** 4.** (18, 19) **a.** A square and a regular pentagon share a common side. The perimeter of the square is 20 cm. What is the perimeter of the pentagon?

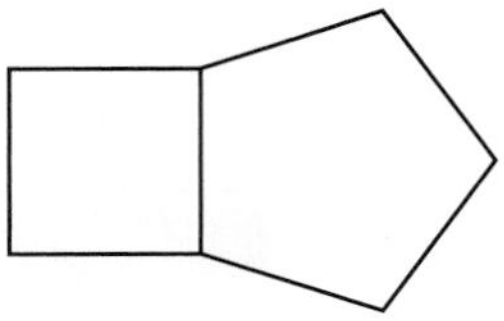

b. *Justify* Explain how you got your answer.

5. (29) Round 3,197,270

a. to the nearest million.

b. to the nearest hundred thousand.

*** 6.** (29) **Estimate** Mentally estimate the product of 313 and 489.

7. (22) Diagram this statement. Then answer the questions that follow.

Five eighths of the troubadour's 200 songs were about love and chivalry.

a. How many of the songs were about love and chivalry?

b. How many of the songs were not about love and chivalry?

*** 8.** (31) **a.** What fraction of the rectangle is not shaded?

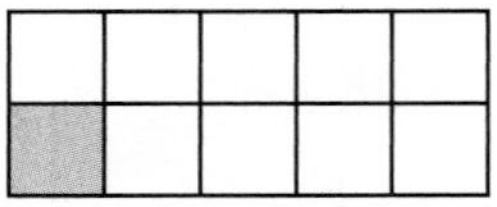

b. What decimal part of the rectangle is not shaded?

c. What percent of the rectangle is not shaded?

*** 9.** (31) Use words to write 3.025.

*** 10.** (31) In the 1988 Summer Olympics, American Jackie Joyner-Kersee won the long jump event with a jump of seven and forty hundredths meters. Write this number as a decimal.

*** 11.** (27) **Evaluate** Instead of dividing \$15.00 by $2\frac{1}{2}$, double both numbers and then find the quotient.

12. (5, 21) **a.** Write 2500 in expanded notation.

b. Write the prime factorization of 2500 using exponents.

c. Find $\sqrt{2500}$.

13. (13) If 35 liters of milk cost $28.00, what is the price per liter?

14. (17) Use a protractor to draw a triangle that has a 90° angle and a 45° angle.

In the figure below, a 6-by-6-cm square is joined to an 8-by-8-cm square. Refer to the figure to answer problems **15** and **16.**

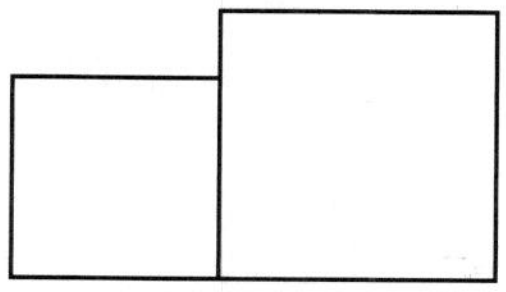

15. (20) **a.** What is the area of the smaller square?

b. What is the area of the larger square?

c. What is the total area of the figure?

16. (19) **Evaluate** What is the perimeter of the hexagon that is formed by joining the squares?

Solve:

17. (3) $10 \cdot 6 = 4w$

18. (3) $180° - s = 65°$

Generalize Simplify:

* **19.** (30) $\frac{1}{4} + \frac{3}{8} + \frac{1}{2}$

* **20.** (30) $\frac{5}{6} - \frac{3}{4}$

* **21.** (30) $\frac{5}{16} - \frac{3}{20}$

22. (26) $\frac{8}{9} \cdot 1\frac{1}{5} \cdot 10$

* **23.** (23, 30) $6\frac{1}{6} - 2\frac{1}{2}$

* **24.** (30) $4\frac{5}{8} + 1\frac{1}{2}$

25. (25) $\frac{2}{3} + \left(\frac{2}{3} \div \frac{1}{2}\right)$

26. (24) $\frac{25}{36} \cdot \frac{9}{10} \cdot \frac{8}{15}$

Estimate For problems **27** and **28,** record an estimated answer and an exact answer:

27. (26) $5\frac{2}{5} \div \frac{9}{10}$

* **28.** (10) $7\frac{3}{4} + 1\frac{7}{8}$

* **29.** (Inv. 3) **Conclude** The coordinates of three vertices of a rectangle are (−5, 3), (−5, −2), and (2, −2).

a. What are the coordinates of the fourth vertex?

b. What is the area of the rectangle?

30. Refer to the figure below to answer **a–c.**
(Inv. 2)

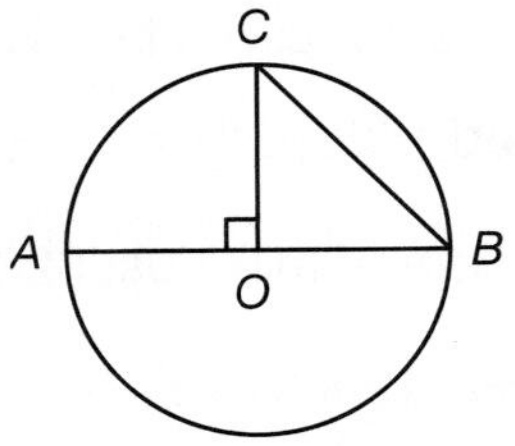

Math Language
A **chord** is a segment whose endpoints lie on a circle.

a. Which chord is not a diameter?

b. Name a central angle that is a right angle.

c. Name an inscribed angle.

Early Finishers
Real-World Application

On Saturday morning, David rode his bike to Jordan's house, which is $1\frac{3}{4}$ miles away. Then they rode to the park $2\frac{3}{5}$ miles away and played basketball. Afterwards they rode back to Jordan's house. Finally, David rode back to his house. Calculate the number of miles David rode his bike on Saturday?

LESSON

33

• Comparing Decimals
• Rounding Decimals

Power Up — *Building Power*

facts Power Up F

mental math

a. **Number Sense:** \$2.84 − 99¢

b. **Calculation:** 6 × 55¢

c. **Equivalent Fractions:** $\frac{3}{8} = \frac{?}{24} =$

d. **Fractions:** Reduce $\frac{24}{30}$.

e. **Power/Roots:** $5^2 - \sqrt{25}$

f. **Fractional Parts:** $\frac{5}{6}$ of 30

g. **Estimation:** Estimate the difference of 87 and 34.

h. **Calculation:** Think of an equivalent division problem for 600 ÷ 50. Then find the quotient.

problem solving Simon held a number cube so that he could see the dots on three adjoining faces. Simon said he could see a total of 8 dots. Could Simon be correct? Why or why not?

New Concepts — *Increasing Knowledge*

comparing decimals

When comparing decimal numbers, it is necessary to consider place value. The value of a place is determined by its position with respect to the decimal point. Aligning decimal points can help to compare decimal numbers digit by digit.

Thinking Skill

Generalize

Remember that the value of the decimal places decreases from left to right. Thus 0.1 > 0.01 > 0.001.

Example 1

Arrange these decimal numbers in order from least to greatest:

0.13 0.128 0.0475

Solution

We will align the decimal points and consider the digits column by column. First we look at the tenths place.

↓

0.13
0.128
0.0475

Two of the decimal numbers have a 1 in the tenths place, while the third number has a 0. So we can determine that 0.0475 is the least of the three numbers. Now we look at the hundredths place to compare the remaining two numbers.

$$\begin{array}{l} 0.1\overset{\downarrow}{3} \\ 0.128 \end{array}$$

Since 0.128 has a 2 in the hundredths place, it is less than 0.13, which has a 3 in the hundredths place. So from least to greatest the order is

0.0475, 0.128, 0.13

Note that terminal zeros on a decimal number add no value to the decimal number.

$$1.3 = 1.30 = 1.300 = 1.3000$$

When we compare two decimal numbers, it may be helpful to insert terminal zeros so that both numbers will have the same number of digits to the right of the decimal point. We will practice this technique in the next few examples.

Example 2

Compare: 0.12 ◯ 0.012

Solution

So that each number has the same number of decimal places, we insert a terminal zero in the number on the left.

0.120 ◯ 0.012

One hundred twenty thousandths is greater than twelve thousandths, so we write our answer this way:

0.12 > 0.012

Explain Which place value did you use to determine your answer? Explain.

Example 3

Compare: 0.4 ◯ 0.400

Solution

We can delete two terminal zeros from the number on the right.

0.4 ◯ 0.4

Or we could have added terminal zeros to the number on the left.

0.400 ◯ 0.400

We write our answer this way:

0.4 = 0.400

Example 4

Compare: 1.232 ◯ 1.23185

Solution

We insert two terminal zeros in the number on the left.

1.23200 ◯ 1.23185

Since 1.23200 is greater than 1.23185, we write

1.232 > 1.23185

rounding decimals

To round decimal numbers, we can use the same procedure that we use to round whole numbers.

Example 5

Round 3.14159 to the nearest hundredth.

Solution

Thinking Skill

Generalize

When rounding, if the circled digit is less than 5, the underlined digit does not change.

The hundredths place is two places to the right of the decimal point. We underline the digit in that place and circle the digit to its right.

3.1$\underline{4}$①59

Since the circled digit is less than 5, we leave the underlined digit unchanged. Then we replace the circled digit and all digits to the right of it with zeros.

3.14000

Terminal zeros to the right of the decimal point do not serve as placeholders as they do in whole numbers. After rounding decimal numbers, we should remove terminal zeros to the right of the decimal point.

3.14~~000~~ ⟶ **3.14**

Discuss Not every zero in a decimal is a terminal zero. Consider these numbers: 302.15, 86.2050, 900.1, and 0.65000. Which of the zeros in these numbers are terminal zeros?

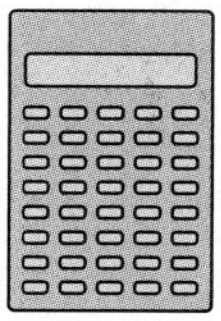

Note that a calculator simplifies decimal numbers by omitting from the display extraneous (unnecessary) zeros. For example, enter the following sequence of keystrokes:

[1] [0] [.] [2] [0] [3] [0] [0]

Notice that all entered digits are displayed. Now press the [=] key, and observe that the unnecessary zeros disappear from the display.

Example 6

Round 4396.4315 to the nearest hundred.

Solution

We are rounding to the nearest hundred, not to the nearest hundredth.

4$\underline{3}$⑨6.4315

Since the circled digit is 5 or more, we increase the underlined digit by 1. All the following digits become zeros.

4400.0000

Zeros at the end of the whole-number part are needed as placeholders. Terminal zeros to the right of the decimal point are not needed as placeholders. We remove these zeros.

4400.~~0000~~ ⟶ **4400**

Example 7

Round 38.62 to the nearest whole number.

Solution

To round a number to the nearest whole number, we round to the ones place.

3<u>8</u>.⑥2 ⟶ 39.~~00~~ ⟶ **39**

Example 8

Estimate the product of 12.21 and 4.9 by rounding each number to the nearest whole number before multiplying.

Solution

We round 12.21 to 12 and 4.9 to 5. Then we multiply 12 and 5 and find that the estimated product is **60.** (The actual product is 59.829.)

Practice Set

Compare:

a. 10.30 ◯ 10.3

b. 5.06 ◯ 5.60

c. 1.1 ◯ 1.099

Generalize For problems **d–f** underline the digit in the place that each number will be rounded to, and circle the digit to the right of that place. Then round each number.

d. Round 3.14159 to the nearest ten-thousandth.

e. Round 365.2418 to the nearest hundred.

f. Round 57.432 to the nearest whole number.

g. Simplify 10.2000 by removing extraneous zeros.

h. Estimate Estimate the sum of 8.65, 21.7, and 11.038 by rounding each decimal number to the nearest whole number before adding.

Written Practice *Strengthening Concepts*

1. (28) **Explain** The young tree was 5 feet 8 inches high. How can we find the number of inches in 5 feet 8 inches?

2. (28) During the first week of November the daily high temperatures in degrees Fahrenheit were 42°F, 43°F, 38°F, 47°F, 51°F, 52°F, and 49°F. What was the average daily high temperature during the first week of November?

3. (11) The population of Chandler, AZ increased from 89,862 in 1990 to 176,581 in 2000. How many people did the population increase by over this 10-year time span?

4. (16) To find the length in millimeters of a segment measured in centimeters, we multiply the number of centimeters by 10. Make a function table that shows the number of millimeters in 1, 3, and 5 centimeters. Then add one more pair of numbers of your choosing to the table.

5. (19) **a.** A regular hexagon and a regular octagon share a common side. If the perimeter of the hexagon is 24 cm, what is the perimeter of the octagon?

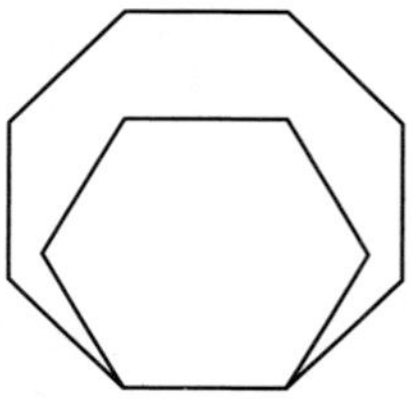

b. **Justify** How did you get your answer?

6. (22) Diagram this statement. Then answer the questions that follow.

One-third of the 60 fish were goldfish.

a. How many of the fish were goldfish?

b. How many of the fish were not goldfish?

c. What percent of the fish were goldfish?

*** 7.** (Inv. 3) **Generalize** Find the area of a square whose vertices have the coordinates (3, 6), (3, 1), (−2, 1), and (−2, 6).

*** 8.** (33) **a.** Round 15.73591 to the nearest hundredth.

b. Estimate the product of 15.73591 and 3.14 by rounding each decimal number to the nearest whole number before multiplying.

*** 9.** (31) Use words to write each of these decimal numbers:

a. 150.035

b. 0.0015

*** 10.** (31) Use digits to write each of these decimal numbers:

a. one hundred twenty-five thousandths

b. one hundred and twenty-five thousandths

*** 11.** (33) Replace each circle with the proper comparison symbol:

a. 0.128 ◯ 0.14 **b.** 0.03 ◯ 0.0015

Math Language

A **whole number** is any number that is a member of the set: {0, 1, 2, 3, 4, ...}.

* **12.** (32) Find the length of this segment

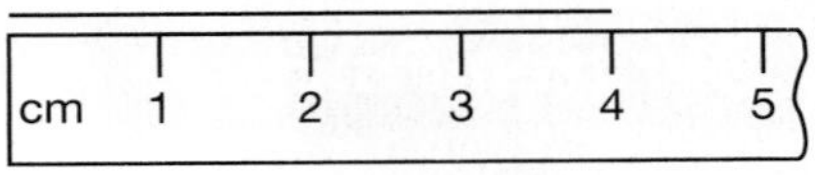

a. in centimeters.

b. in millimeters.

13. (17) Draw the straight angle *AOC*. Then use a protractor to draw ray *OD* so that angle *COD* measures 60°.

* **14.** (2) **Generalize** If we multiply one integer by another integer that is a whole number but not a counting number, what is the product?

* **15.** (27) **Generalize** Use prime factorization to find the least common denominator for the fractions $\frac{5}{27}$ and $\frac{5}{36}$.

Solve:

16. (3) $8m = 4 \cdot 18$

17. (3) $135° + a = 180°$

Simplify:

* **18.** (30) $\frac{3}{4} + \frac{5}{8} + \frac{1}{2}$

* **19.** (30) $\frac{3}{4} - \frac{1}{6}$

* **20.** (30) $4\frac{1}{2} - \frac{3}{8}$

21. (26) $\frac{3}{8} \cdot 2\frac{2}{5} \cdot 3\frac{1}{3}$

22. (26) $2\frac{7}{10} \div 5\frac{2}{5}$

23. (26) $5 \div 4\frac{1}{6}$

* **24.** (23, 30) $6\frac{1}{2} - 2\frac{5}{6}$

25. (25) $\frac{3}{4} + \left(\frac{1}{2} \div \frac{2}{3}\right)$

* **26.** (14) **Analyze** If one card is drawn from a regular deck of cards, what is the probability the card will not be an ace?

27. (2) **a.** Solve: $54 = 54 + y$

b. What property is illustrated by the equation in **a?**

* **28.** (29) **Justify** Consider the following division problem. Without dividing, decide whether the quotient will be greater than 1 or less than 1. How did you decide?

$$5 \div 4\frac{1}{6}$$

* **29.** (29) When Paulo saw the following addition problem, he knew that the sum would be greater than 13 and less than 15. How did he know?

$$8\frac{7}{8} + 5\frac{2}{3}$$

30. (17) Use a protractor to draw a triangle that has a 30° angle and a 60° angle.

LESSON 34

Decimal Numbers on the Number Line

Power Up

Building Power

facts Power Up A

mental math

a. **Number Sense:** \$6.48 − 98¢

b. **Calculation:** 5 × 48¢

c. **Equivalent Fractions:** $\frac{3}{5} = \frac{?}{30}$

d. **Fractions:** Reduce $\frac{16}{24}$.

e. **Power/Roots:** $\sqrt{36} \cdot \sqrt{49}$

f. **Fractional Parts:** $\frac{2}{3}$ of 36

g. **Estimation:** Estimate the difference of 126 and 45.

h. **Calculation:** Square the number of sides on a pentagon, double that number, − 1, $\sqrt{}$, × 4, − 1, ÷ 3, $\sqrt{}$.

problem solving

Terrance folded a square piece of paper in half diagonally to form a triangle. He folded the triangle in half two more times as shown. Then he used scissors to cut off the lower left corner (the right angle) of the resulting triangle. Which diagram will the paper look like when it is unfolded?

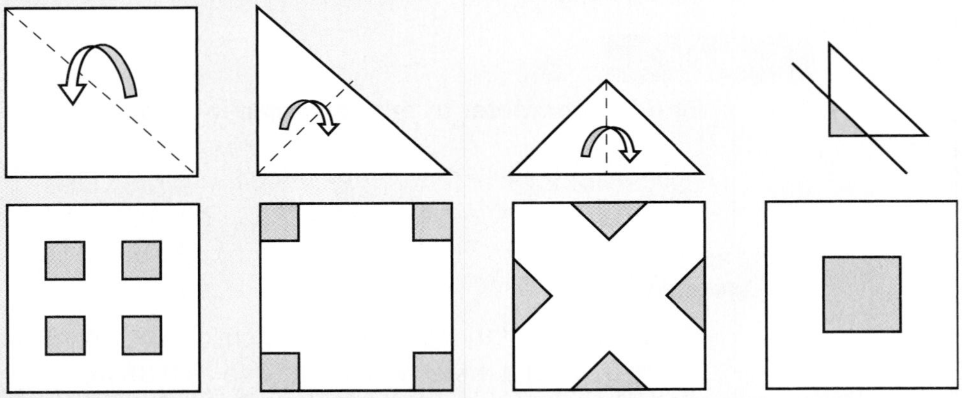

New Concept

Increasing Knowledge

If the distance between consecutive whole numbers on a number line is divided by tick marks into 10 equal units, then numbers corresponding to these marks can be named using decimal numbers with one decimal place.

If each centimeter on a centimeter scale is divided into 10 equal parts, then each part is 1 millimeter long. Each part is also one tenth of a centimeter long.

Example 1

Find the length of this segment

a. in centimeters.

b. in millimeters.

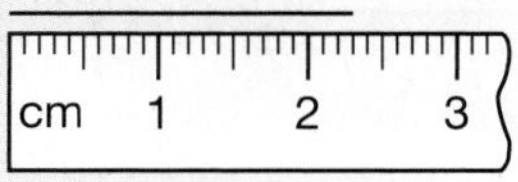

Solution

a. Each centimeter on the scale has been divided into 10 equal parts. The length of the segment is 2 centimeters plus three tenths of a centimeter. In the metric system we use decimals rather than common fractions to indicate parts of a unit. So the length of the segment is **2.3 cm.**

b. Each centimeter is 10 mm. Thus, each small segment on the scale is 1 mm. The length of the segment is **23 mm.**

Analyze What number is in the ones place in 2.3? What number is in the tenths place?

If the distance between consecutive whole numbers on a number line is divided into 100 equal units, then numbers corresponding to the marks on the number line can be named using two decimal places. For instance, a meter is 100 cm. So each centimeter segment on a meterstick is 0.01 or $\frac{1}{100}$ of the length of the meterstick. This means that an object 25 cm long is also 0.25 m long.

Example 2

Find the perimeter of this rectangle in meters.

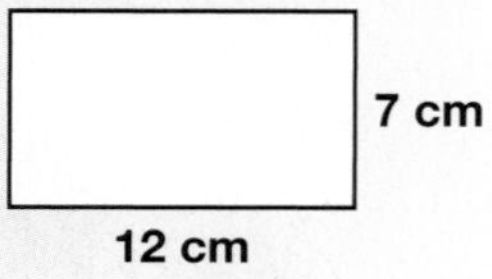

Solution

The perimeter of the rectangle is 38 cm. Each centimeter is $\frac{1}{100}$ of a meter. So 38 cm is $\frac{38}{100}$ of a meter, which we write as **0.38 m.**

Example 3

Thinking Skill

Justify

On this number line, the shorter tick marks indicate what measure? Justify your answer.

Find the number on the number line indicated by each arrow:

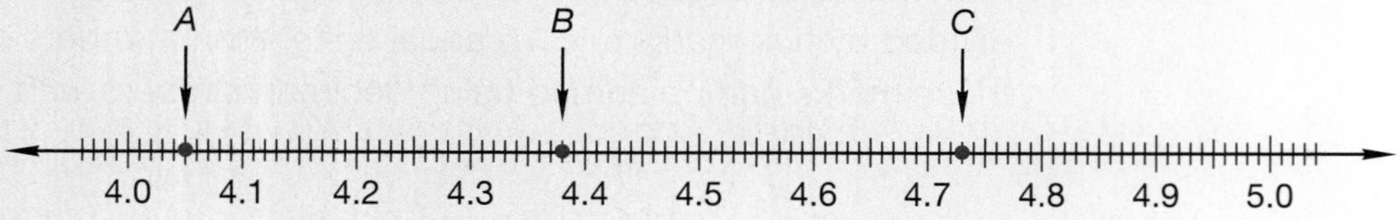

Solution

We are considering a portion of the number line from 4 to 5. The distance from 4 to 5 has been divided into 100 equal segments. Tenths have been identified. The point 4.1 is one tenth of the distance from 4 to 5. However, it is also ten hundredths of the distance from 4 to 5, so 4.1 equals 4.10.

Arrow *A* indicates **4.05.**

Arrow *B* indicates **4.38.**

Arrow *C* indicates **4.73.**

Example 4

Arrange these decimal numbers in order from least to greatest:

4.5, 4.25, 4.81

Solution

We can find points on the number line in example 3 that correspond to these numbers. The decimal 4.5 appears in the center of the scale, with 4.25 to the left and 4.81 to the right. Thus, from least to greatest, the order is

4.25, 4.5, 4.81

Activity

Decimal Numbers on a Meterstick

Use a meterstick ruled to millimeters to measure the lengths of a few objects in the classroom. Record the measures three ways, as a number of millimeters, as a number of centimeters, and as a number of meters.

For example, a one-foot ruler is about 305 mm, 30.5 cm, and 0.305 m.

Explain For each object you measured, identify the scale that was most useful to measure the object with. Explain why it is useful to have more than one scale to measure objects.

Practice Set

Refer to the figure below to answer problems **a–c.**

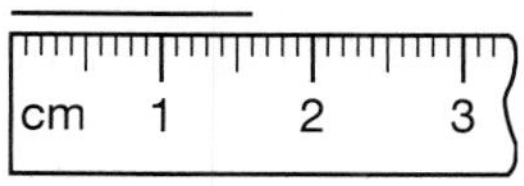

a. Find the length of the segment in centimeters.

b. Find the length of the segment to the nearest millimeter.

c. Analyze What is the greatest possible error of the measurement in problem **b?** Express your answer as a fraction of a millimeter.

d. Seventy-five centimeters is how many meters?

e. Carmen's wardrobe closet is 1.57 meters tall. How many centimeters tall is it?

f. Model What point on a number line is halfway between 2.6 and 2.7?

g. What decimal number names the point marked A on this number line?

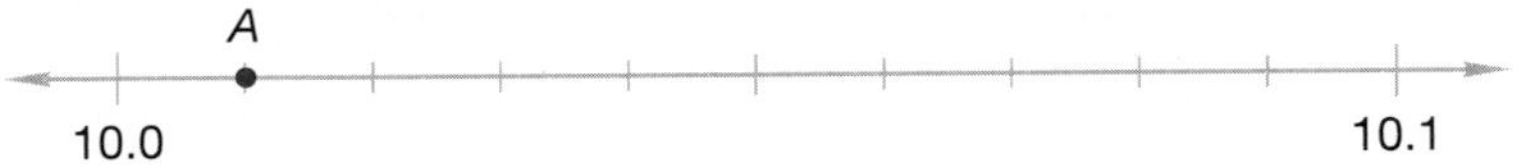

h. Estimate the length of this segment in centimeters. Then use a centimeter ruler to measure its length.

i. Compare: 4.6 ◯ 4.45

j. Compare: 2.5 cm ◯ 25 mm

Written Practice

Strengthening Concepts

* **1.** (28) Generalize In 3 boxes of cereal, Jeff counted 188 raisins, 212 raisins, and 203 raisins. What was the average number of raisins in each box of cereal?

2. (11) On April 29, 2005, the tree pollen count in Waterbury, CT was 1024 parts per cubic meter. On May 2, the count was 1698 parts per cubic meter. By how much did the pollen count increase?

3. (11) Gina spent $3.95 for lunch. She had $12.55 left. How much money did she have before she bought lunch?

* **4.** (12) Connect In 1903 the Wright brothers made the first powered airplane flight. Just 66 years later astronauts first landed on the Moon. In what year did astronauts first land on the Moon?

5. (19) The perimeter of the square equals the perimeter of the regular hexagon. If each side of the hexagon is 6 inches long, how long is each side of the square?

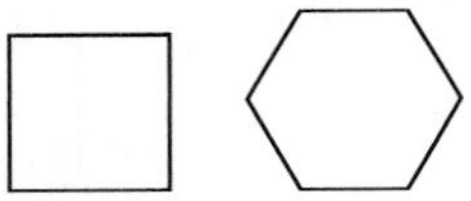

6. (22) In the following statement, write the percent as a reduced fraction. Then diagram the statement and answer the questions.

Each week Jessica earns $12 dollars doing yard work. She saves 40% of the money she makes.

a. How much money does she save each week?

b. How much money does she spend each week?

* **7.** (29) **Explain** Tom's 27-inch television screen has the dimensions shown. Describe how to estimate its area.

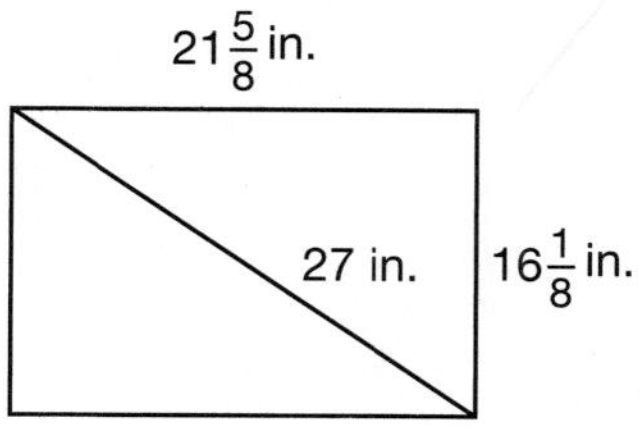

* **8.** (33) Round 7.49362 to the nearest thousandth.

* **9.** (31) Use words to write each of these decimal numbers:

a. 200.02

b. 0.001625

Reading Math

In a decimal number, the word *and* indicates the position of the decimal point.

* **10.** (31) Use digits to write each of these decimal numbers:

a. one hundred seventy-five millionths

b. three thousand, thirty and three hundredths

* **11.** (33) Replace each circle with the proper comparison symbol:

a. 6.174 ◯ 6.17401 **b.** 14.276 ◯ 1.4276

* **12.** (34) Find the length of this segment

a. in centimeters.

b. in millimeters.

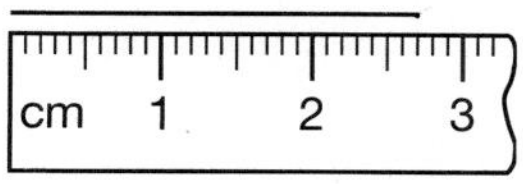

* **13.** (34) What decimal number names the point marked X on this number line?

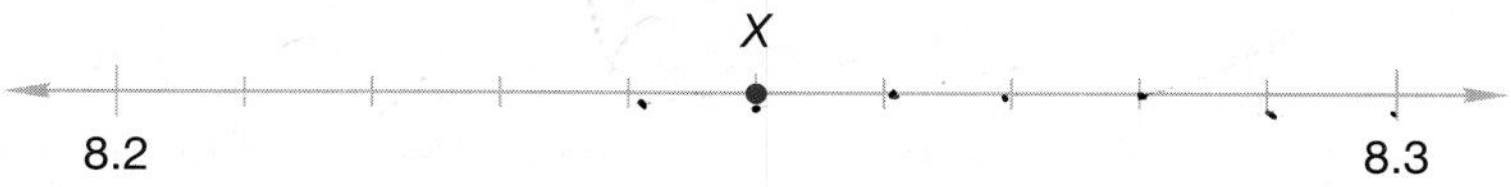

* **14.** (Inv. 3) The coordinates of three vertices of a square are (0, 0), (0, 3), and (3, 3).

a. **Represent** Draw a coordinate plane and plot the points for the coordinates given. Then draw the square.

b. What are the coordinates of the fourth vertex?

c. What is the area of the square?

* **15.** (34) **a.** What decimal number is halfway between 7 and 8?

b. What number is halfway between 0.7 and 0.8?

Solve:

16. (3) $15 \cdot 20 = 12y$

17. (3) $180° = 74° + c$

Simplify:

* **18.** (30) $\frac{5}{6} + \frac{2}{3} + \frac{1}{2}$

* **19.** (30) $\frac{5}{36} - \frac{1}{24}$

* **20.** (23, 30) $5\frac{1}{6} - 1\frac{2}{3}$

21. (26) $\frac{1}{10} \cdot 2\frac{2}{3} \cdot 3\frac{3}{4}$

22. (26) $5\frac{1}{4} \div 1\frac{2}{3}$

23. (26) $3\frac{1}{5} \div 4$

* **24.** (30) $6\frac{7}{8} + 4\frac{1}{4}$

25. (9, 24) $\frac{1}{8} + \left(\frac{5}{6} \cdot \frac{3}{4}\right)$

* **26.** (34) **Generalize** Express the following difference two ways:

3.6 cm − 24 mm

a. in centimeters

b. in millimeters

27. (20) Which is equivalent to $2^2 \cdot 2^3$?

A 2^5 **B** 2^6 **C** 12 **D** 24

* **28.** (33) Arrange these numbers in order from least to greatest:

0.365, 0.3575, 0.36

29. (1, 4) Evaluate this expression for $x = 5$ and $y = 10$:

$$\frac{y}{x} - x$$

* **30.** (14) A bag contains 3 red marbles, 4 white marbles, and 5 blue marbles. If one marble is drawn from the bag, what is the probability that the marble will be

a. red? **b.** white?

c. blue? **d.** green?

Early Finishers
Real-World Application

James's coach asked him to measure ten practice long jumps and to plot the measurements on a number line. The measurements were: 5.15 m, 5.13 m, 5.17 m, 4.99 m, 5.02 m, 5.05 m, 5.22 m, 5.09 m, 5.2 m, and 5.11 m.

a. Plot James's measurements on a number line.

b. What are some observations James can make regarding his data now that he has plotted them on a number line?

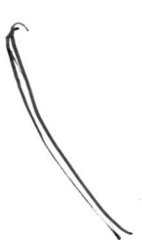

LESSON
35

• Adding, Subtracting, Multiplying, and Dividing Decimal Numbers

Power Up — *Building Power*

facts Power Up H

mental math

a. **Number Sense:** \$7.50 − \$1.99

b. **Calculation:** 5 × 64¢

c. **Equivalent Fractions:** $\frac{9}{10} = \frac{?}{30}$

d. **Fractions:** Reduce $\frac{15}{24}$.

e. **Power/Roots:** $4^2 - \sqrt{4}$

f. **Fractional Parts:** $\frac{5}{12}$ of 24

g. **Estimation:** Estimate the sum of 453 and 57

h. **Measurement:** Start with the number of inches in two feet, + 1, × 4, $\sqrt{\ }$. What do we call this many years?

problem solving Copy this problem and fill in the missing digits:

```
      ___
  8)___
    _
    __
    _
     4_
    _8
     0
```

New Concepts — *Increasing Knowledge*

adding and subtracting decimal numbers

Adding and subtracting decimal numbers is similar to adding and subtracting money. **We align the decimal points to ensure that we are adding or subtracting digits that have the same place value.**

Example 1

Add: 3.6 + 0.36 + 36

Solution

This problem is equivalent to adding \$3.60 + \$0.36 + \$36. We align the decimal points vertically. A number written without a decimal point is a whole number, so the decimal point is to the right of 36.

```
   3.6
   0.36
+ 36.
------
  39.96
```

Example 2

Find the perimeter of this rectangle.

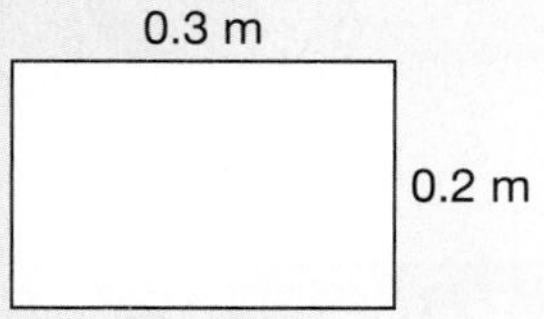

Solution

We align the decimal points vertically and add. The sum is 1.0, not 0.10. Since 1.0 equals 1, we can simplify the answer to 1. The perimeter of the rectangle is **1 meter.**

$$\begin{array}{r} 0.3 \\ 0.2 \\ 0.3 \\ +\ 0.2 \\ \hline 1.0 \end{array} = 1$$

Example 3

Subtract: 12.3 − 4.567

Solution

We write the first number above the second number, aligning the decimal points. We write zeros in the empty places and subtract.

$$\begin{array}{r} \overset{0\,11}{\not{1}}\overset{}{\not{2}}.\overset{12}{\not{3}}\overset{9}{\not{0}}\overset{1}{0} \\ -\ \ 4.567 \\ \hline \mathbf{7.733} \end{array}$$

Example 4

Subtract: 5 − 4.32

Solution

This problem is equivalent to subtracting \$4.32 from \$5. We write the whole number 5 with a decimal point and write zeros in the two empty decimal places. Then we subtract.

$$\begin{array}{r} \overset{4}{\not{5}}.\overset{9}{\not{0}}\overset{1}{0} \\ -\ 4.32 \\ \hline \mathbf{0.68} \end{array}$$

multiplying decimal numbers

If we multiply the fractions three tenths and seven tenths, the product is twenty-one hundredths.

$$\frac{3}{10} \times \frac{7}{10} = \frac{21}{100}$$

Likewise, if we multiply the decimal numbers three tenths and seven tenths, the product is twenty-one hundredths.

$$0.3 \times 0.7 = 0.21$$

Here we use an area model to illustrate this multiplication:

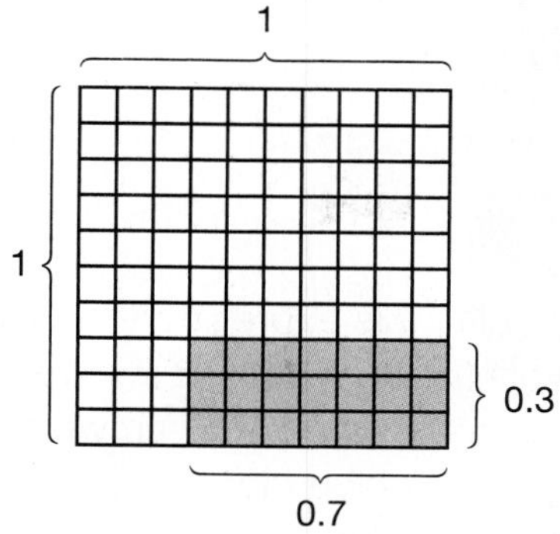

Each side of the square is one unit in length. We multiply three tenths of one side by seven tenths of a perpendicular side. The product is an area that contains twenty-one hundredths of the square.

$$0.3 \times 0.7 = 0.21$$

Notice that the factors each have one decimal place and the product has two decimal places. **When we multiply decimal numbers, the product has as many decimal places as there are in all the factors combined.**

Example 5

Find the area of this rectangle.

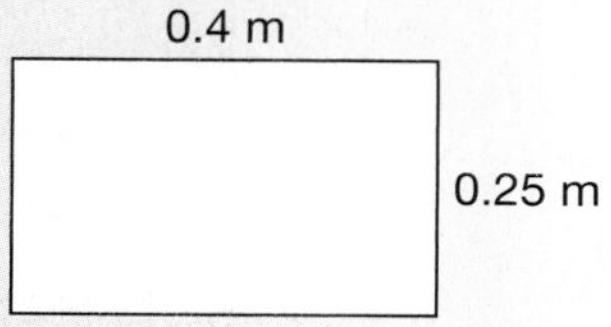

Solution

Math Language

The term *square meters* can be abbreviated in two ways: sq. m *or* m^2.

We multiply the length times the width. We need not align decimal points to multiply. We set up the problem as though we were multiplying whole numbers. After multiplying, we count the number of decimal places in both factors. There are a total of three decimal places, so we write the product with three decimal places. We remove unnecessary zeros. The area is **0.1 sq. m.**

$$\begin{array}{r} 0.25 \\ \times\ 0.4 \\ \hline 100 \end{array}$$

$$\begin{array}{rl} 0.25 & \text{2 places} \\ \times\ 0.4 & \text{1 place} \\ \hline 0.100 & \text{3 places} \end{array}$$

Example 6

A ruler that is 12 inches long is how many centimeters long? (1 in. = 2.54 cm)

Solution

Each inch is 2.54 cm, so 12 inches is 12 × 2.54 cm. We set up the problem as though we were multiplying whole numbers. After multiplying, we place the decimal point in the product. Twelve inches is **30.48 cm.**

$$\begin{array}{r} 2.54 \\ \times\ 12 \\ \hline 508 \\ 254 \\ \hline 30.48 \end{array}$$

Explain Describe the steps we use to place the decimal point in the product.

Example 7

Simplify: $(0.03)^2$

Solution

To simplify this expression, we multiply 0.03 times 0.03. We can perform the multiplication mentally. First we multiply as though we were multiplying whole numbers: $3 \times 3 = 9$. Then we count decimal places. There are four decimal places in the two factors. Starting from the right side, we count to the left four places. We write zeros in the empty places.

$$.\_\,\_\,\_\,9 \longrightarrow \mathbf{0.0009}$$

dividing decimal numbers

Dividing a decimal number by a whole number is similar to dividing dollars and cents by whole numbers. When we use long division, the decimal point in the quotient is lined up with the decimal point in the dividend.

Example 8

The perimeter of a square is 7.2 meters. How long is each side?

Solution

We divide the perimeter by 4. We place a decimal point in the quotient above the decimal point in the dividend. Then we divide as though we were dividing whole numbers. Each side is **1.8 meters** long.

$$\begin{array}{r} 1.8 \\ 4\overline{)7.2} \\ \underline{4} \\ 32 \\ \underline{32} \\ 0 \end{array}$$

Example 9

Divide: $0.0144 \div 8$

Solution

We place the decimal point in the quotient directly above the decimal point in the dividend. Then we write a digit in every place following the decimal point until the division is complete. If we cannot perform a division, we write a zero in that place. The answer is **0.0018.**

$$\begin{array}{r} 0.0018 \\ 8\overline{)0.0144} \\ \underline{8} \\ 64 \\ \underline{64} \\ 0 \end{array}$$

Example 10

Divide: $1.2 \div 5$

Solution

We do not write a decimal division answer with a remainder. Since a decimal point fixes place values, we may write a zero in the next decimal place. This zero does not change the value of the number, but it does let us continue dividing. The answer is **0.24.**

$$\begin{array}{r} 0.24 \\ 5\overline{)1.20} \\ \underline{1\,0} \\ 20 \\ \underline{20} \\ 0 \end{array}$$

Practice Set

Simplify:

a. 1.2 + 3.45 + 23.6

b. 4.5 + 0.51 + 6 + 12.4

c. What is the perimeter of this rectangle?

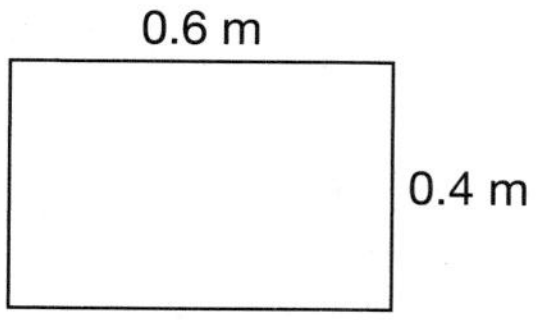

d. Drew ran 50 meters in 8.46 seconds. Mathea ran 50 meters in 8.52 seconds. Drew ran how many seconds faster than Mathea?

e. 16.7 − 1.936

f. 12 − 0.875

g. 4.2 × 0.24

h. $(0.06)^2$

i. Six inches is how many centimeters? (1 in. = 2.54 cm)

j. 0.3 × 0.2 × 0.1

k. (0.04)(10)

l. What is the area of this rectangle?

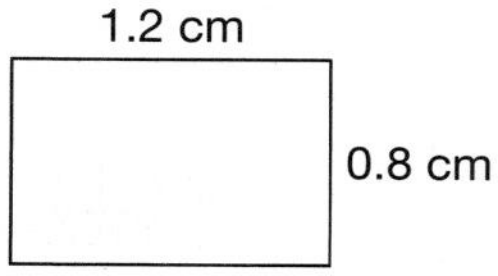

m. 14.4 ÷ 6

n. 0.048 ÷ 8

o. 3.4 ÷ 5

p. 0.3 ÷ 6

q. A loop of string 0.6 meter long is arranged to form a square. How long is each side of the square?

Written Practice

Strengthening Concepts

1. (28) **Explain** During the first six months of the year, the Montgomerys' monthly electric bills were $128.45, $131.50, $112.30, $96.25, $81.70, and $71.70. How can the Montgomerys find their average monthly electric bill for the first six months of the year?

2. (23, 30) There were $2\frac{1}{2}$ gallons of milk in the refrigerator before breakfast. There were $1\frac{3}{4}$ gallons after dinner. How many gallons of milk were consumed during the day?

3. (28) A one-year subscription to a monthly magazine costs \$15.60. The regular newsstand price is \$1.75 per issue. How much is saved per issue by paying the subscription price?

4. (28) In the 1896 Summer Olympics, Alfred Hajos of Hungary swam 100 m in about 1 minute 22 seconds. One hundred years later, Li Jinyi of China swam the same distance about 27 seconds faster than Alfred Hajos. About how many seconds did it take Li Jinyi to swim 100 m?

5. (19) The perimeter of the square equals the perimeter of the regular pentagon. Each side of the pentagon is 16 cm long. How long is each side of the square?

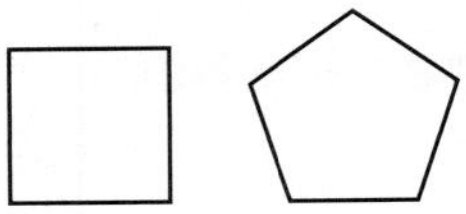

6. (22) Only $\frac{1}{11}$ of the 110 elements known to modern scientists were discovered in ancient times.

a. How many elements were discovered in ancient times?

b. How many elements were not discovered in ancient times?

7. (20) A 6-by-6-cm square is cut from a 10-by-10-cm square sheet of paper as shown below. Refer to this figure to answer **a–c:**

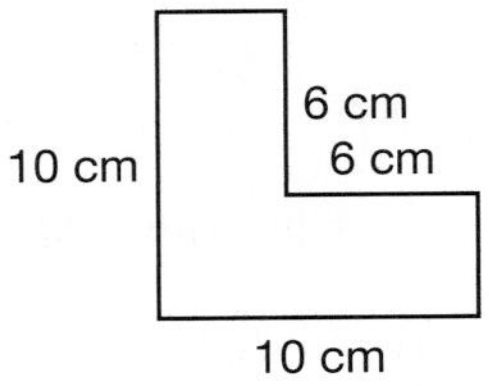

a. What was the area of the original square?

b. What was the area of the square that was cut out?

c. What is the area of the remaining figure?

8. (31) **a.** In the square at right, what fraction is not shaded?

b. What decimal part of the square is not shaded?

c. What percent of the square is not shaded?

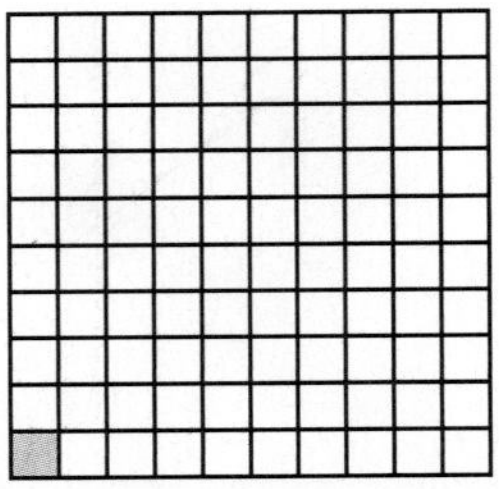

Math Language
The first number in a **coordinate** is the *x*-coordinate. The second number is the *y*-coordinate.

*** 9.** (Inv. 3) The coordinates of three vertices of a rectangle are (−3, 2), (3, −2), and (−3, −2).

a. What are the coordinates of the fourth vertex?

b. What is the area of the rectangle?

*** 10.** (31) **a.** Use words to write 100.075.

b. Use digits to write the decimal number twenty-five hundred-thousandths.

* **11.** Find the length of this segment
(34)

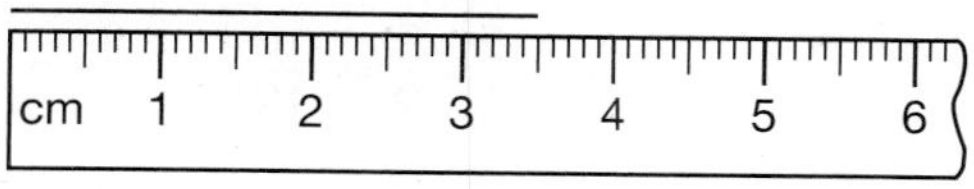

a. in centimeters.

b. in millimeters.

* **12.** Miss Gaviria bought 10.38 gallons of gasoline at $\$2.28\frac{9}{10}$ per gallon.
(33)

a. **Justify** About how much did she pay for the gasoline? Justify your answer by showing what numbers you rounded off and why.

b. **Explain** Is it possible to pay $\frac{9}{10}$ of one cent? Why do you think gas stations give their prices this way?

* **13.** What decimal number names the point marked with an arrow on this number line?
(34)

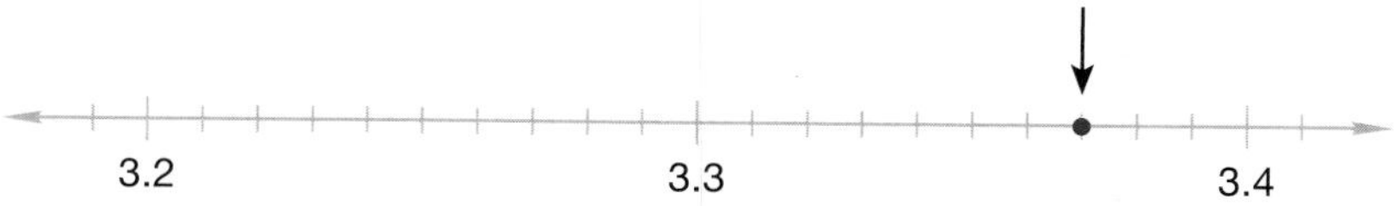

* **14.** Use the two color 10-by-10 grid transparencies to model 0.5×0.7. What is the product?
(35)

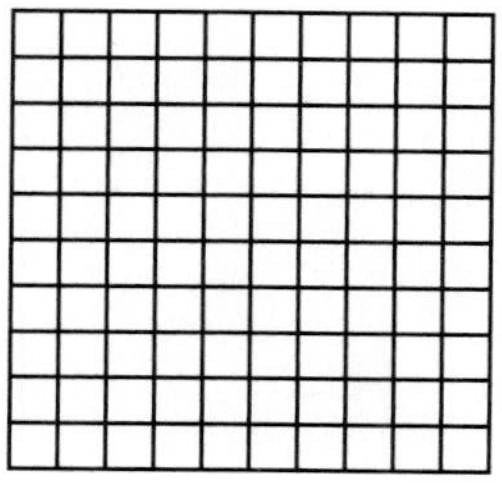

* **15.** **a.** Find the perimeter of this rectangle.
(35)

b. Find the area of this rectangle.

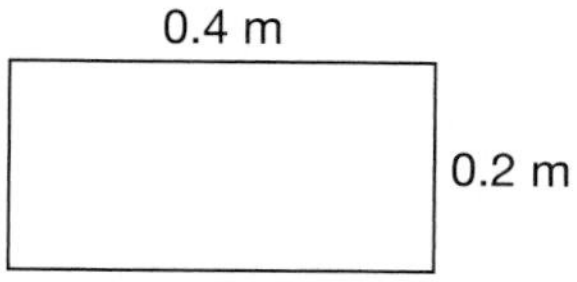

Solve:

16. $15x = 9 \cdot 10$
(3)

* **17.** $f + 4.6 = 5.83$
(3, 35)

* **18.** $8y = 46.4$
(3, 35)

* **19.** $w - 3.4 = 12$
(3, 35)

Simplify:

20. $3.65 + 0.9 + 8 + 15.23$
(35)

21. $1\frac{1}{2} + 2\frac{2}{3} + 3\frac{3}{4}$
(30)

22. $1\frac{1}{2} \cdot 2\frac{2}{3} \cdot 3\frac{3}{4}$
(26)

23. $1\frac{1}{6} - \left(\frac{1}{2} + \frac{1}{3}\right)$
(23, 30)

24. $3\frac{1}{12} - 1\frac{3}{4}$
(23, 30)

* **25.** $1.2 \div 10$
(35)

* **26.** $(0.3)(0.4)(0.5)$
(35)

27. Which property of multiplication is used to rename $\frac{5}{6}$ to $\frac{10}{12}$?
(9, 15)

$$\frac{5}{6} \cdot \frac{2}{2} = \frac{10}{12}$$

*** 28.** One inch equals 2.54 centimeters.
(16, 35)

a. Make a function table that shows the number of centimeters in 1, 2, 3, and 4 inches.

b. *Generalize* What is the rule for this function?

*** 29.** Find an estimated answer and an exact answer.
(33, 35)

a. $6.45 - 4.912$ **b.** 4.2×0.9

30. Use a protractor to draw a triangle that has two 45° angles.
(17)

LESSON

36

• Ratio
• Sample Space

Power Up

Building Power

facts

Power Up G

mental math

a. **Decimals:** $1.45 × 10

b. **Number Sense:** 4 × $1.50

c. **Equivalent Fractions:** $\frac{4}{5} = \frac{?}{20}$

d. **Decimals:** Reduce $\frac{24}{30}$.

e. **Power/Roots:** $\sqrt{144} - 3^2$

f. **Fractional Parts:** $\frac{9}{10}$ of 40

g. **Estimation:** Estimate the difference of 278 and 184.

h. **Calculation:** Find the square root of three dozen, × 5, ÷ 3, square that number, − 20, + 1, $\sqrt{\ }$.

problem solving

Silvia was thinking of a number less than 90 that she says when counting by sixes and when counting by fives, but not when counting by fours. Of what number was she thinking?

New Concepts

Increasing Knowledge

ratio

A **ratio** is a way to describe a relationship between two numbers. For example, if there are 12 boys and 16 girls in the classroom, the ratio of boys to girls is 12 to 16.

One way to write a ratio is as a fraction. The ratio 12 to 16 can be written as $\frac{12}{16}$ which reduces to $\frac{3}{4}$.

The ratio 3 to 4 can be written in the following forms:

with the word *to*	3 to 4
as a fraction	$\frac{3}{4}$
as a decimal number	0.75
with a colon	3:4

The numbers used to express a ratio are stated in the same order as the items named. If the boy-girl ratio is 3 to 4, then the girl-boy ratio is 4 to 3. Although we may reduce ratios, we do not express them as mixed numbers.

Most ratios involve three numbers, even though only two numbers may be stated. When the ratio of boys to girls is 3 to 4, the unstated number is 7, which represents the total.

$$\begin{array}{r} 3 \text{ boys} \\ + \ 4 \text{ girls} \\ \hline 7 \text{ total} \end{array}$$

Sometimes the total is given and one of the parts is unstated, as we see in examples 1 and 2.

Example 1

In a class of 28 students, there are 12 boys.

a. What is the boy-girl ratio?

b. What is the girl-boy ratio?

Solution

We will begin by stating all three numbers in the ratio. We find the missing number by subtracting, using mental math.

$$\begin{array}{r} 12 \text{ boys} \\ + \ ? \text{ girls} \\ \hline 28 \text{ total} \end{array} \qquad \begin{array}{r} 12 \text{ boys} \\ + \ 16 \text{ girls} \\ \hline 28 \text{ total} \end{array}$$

The three numbers in the ratio are: 12 boys, 16 girls, 28 total.

Now we can write the ratios.

a. The *boy-girl* ratio is $\frac{12}{16}$, which reduces to $\frac{3}{4}$, or a ratio of **3 to 4.**

b. The *girl-boy* ratio is $\frac{16}{12}$, which reduces to $\frac{4}{3}$, or a ratio of **4 to 3.** Remember, we do not change the ratio to a mixed number.

Example 2

The team won $\frac{4}{7}$ of its games and lost the rest. What was the team's win-loss ratio?

Solution

Thinking Skill

Analyze

In example 2, what does the denominator 7 represent?

We are not told the total number of games the team played, nor the number they lost. However, we are told that the team won $\frac{4}{7}$ of its games. Therefore, the team must have lost $\frac{3}{7}$ of its games. In other words, on average the team won 4 out of every 7 games it played. Now we can write the three numbers in the ratio.

$$\begin{array}{r} 4 \text{ won} \\ + \ 3 \text{ lost} \\ \hline 7 \text{ total} \end{array}$$

The team's win-loss ratio was **4 to 3** or $\frac{4}{3}$.

Sometimes the parts are given and we must find the total.

Example 3

There are red marbles and green marbles in a bag. If the ratio of red marbles to green marbles is 4 to 5, what fraction of the total number of marbles is red?

Solution

Although we are not given the actual number of marbles in the bag, we are given enough information to answer the question. This example involves a part-total relationship. We are given ratio numbers for the parts. To solve the problem, we first need to find the ratio number for the total.

4 red	4 red
+ 5 green	+ 5 green
? total	9 total

Now we choose the ratio numbers needed to state the fraction of the marbles that are red.

We choose 4 for red and 9 for total, so the fraction that are red is $\frac{4}{9}$.

So, we see that ratios can describe relationships between two parts of a total, or between one part and a total.

Conclude Does the total number of marbles in the bag have to equal 9?

sample space

Math Language
Remember that **probability** is the likelihood that a particular event will occur.

In Lesson 14 we learned that we can express **probability** as a fraction, decimal, or percent. We can also express probability as a ratio because probability involves part-total relationships. Now we can state that the probability of an event is the ratio of the number of favorable outcomes to the number of possible outcomes.

$$\text{Probability (Event)} = \frac{\text{number of favorable outcomes}}{\text{total number of possible outcomes}}$$

The probability experiments we have considered so far had outcomes that were easy to count. Many probability experiments have outcomes that are more difficult to count.

To help us count outcomes, it is often helpful to list the possible outcomes of a probability experiment. The list of all possible outcomes is called the **sample space.**

Example 4

What is the sample space for the following experiments?

a. Flip a coin once.

b. Roll a number cube once.

c. Pick one letter from the alphabet.

Solution

Sample space is often shown by listing the possible outcomes in braces separated by commas.

a. One coin toss has two possible outcomes. **Sample space = {heads, tails}**

b. There are six possible outcomes for one roll of a number cube. **Sample space = {1, 2, 3, 4, 5, 6}**

c. There are 26 possible outcomes for picking one letter of the alphabet. Rather than list all 26 letters, we may list several of the outcomes in a way that makes it clear that all unwritten outcomes are included in the list. **Sample space = {A, B, C, D, ..., W, X, Y, Z}**

For some experiments drawing a tree diagram can help us find the sample space.

Example 5

A coin is flipped and the spinner is spun. What is the sample space for one coin toss and one spin? What is the probability of heads and a number greater than 2?

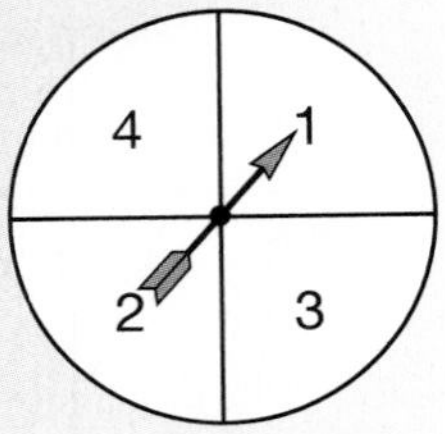

Solution

Math Language
The "branches" on a tree diagram help us visualize all the possible outcomes of an event.

The coin may end up heads or tails. The spinner may stop on 1, 2, 3, or 4. This tree diagram shows how these possible outcomes can combine:

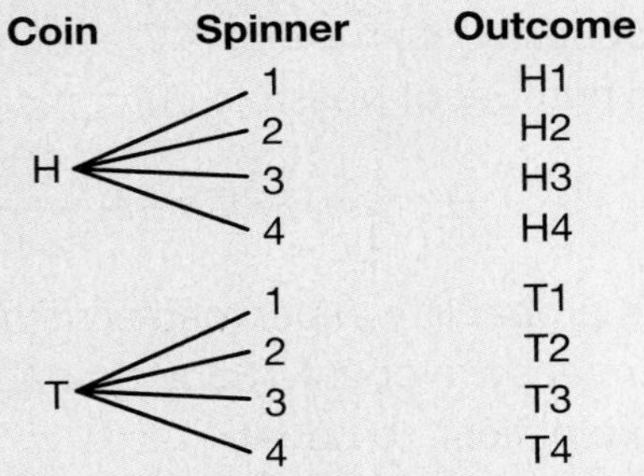

The list of outcomes from the tree diagram is the sample space. We may also show the sample space in braces.

Represent How can we represent the sample space as a list?

Since heads and tails are equally likely, and since each numbered region on the spinner is equally likely, each outcome in the sample space is equally likely.

To find the probability of heads and a number greater than 2, we inspect the sample space. We see that the outcomes H3 and H4 are the favorable outcomes, two favorable outcomes in eight.

$$P(\text{H}, >2) = \frac{2}{8} = \mathbf{\frac{1}{4}}$$

Example 6

There are two spinners shown below. Spinner A is labeled with letters and spinner B is labeled with numbers.

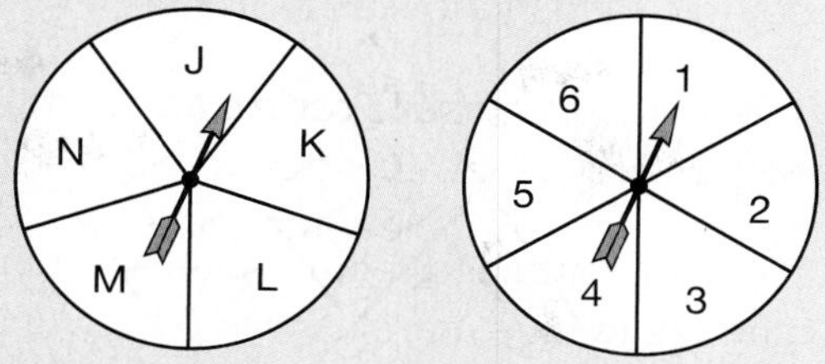

a. Find all possible outcomes when the two spinners are spun at the same time.

b. Find the probability of spinning the letter M and the number 3.

Solution

Although we can draw a tree diagram to find all the possible outcomes for the two spinners, for this problem we will make a table instead. We make a column for each possible letter and pair each possible number with each letter.

a. The table lists each outcome as a letter-number pair.

Outcomes For Spinners A and B				
J,1	K,1	L,1	M,1	N,1
J,2	K,2	L,2	M,2	N,2
J,3	K,3	L,3	M,3	N,3
J,4	K,4	L,4	M,4	N,4
J,5	K,5	L,5	M,5	N,5
J,6	K,6	L,6	M,6	N,6

The table shows the sample space when Spinners A and B are spun at the same time. In the table, we can count **30 possible outcomes.**

b. Since there are 30 possible outcomes, the probability of spinning the letter M and the number 3 is $\frac{1}{30}$, or **1 in 30.**

Thinking Skill

Discuss Give an example of when might it be easier to use the Fundamental Counting Principle than to create a tree diagram or table.

Notice in example 6 that Spinner A has 5 sectors and Spinner B has 6 sectors. Instead of drawing a tree diagram or making a table to count each outcome, we can multiply 5 by 6 to find all the possible outcomes. The total number of possible outcomes is 5×6, or 30. This method of finding the sample space is called the **Fundamental Counting Principle.**

> **Fundamental Counting Principle**
> If there are m ways for A to occur and n ways for B to occur, then there are $m \times n$ ways for A and B to occur together.

Practice Set

a. In the pond were 240 little fish and 90 big fish. What was the ratio of big fish to little fish?

b. Fourteen of the 30 students in the class were girls. What was the boy-girl ratio in the class?

c. The team won $\frac{3}{8}$ of its games and lost the rest. What was the team's win-loss ratio?

d. The bag contained red marbles and blue marbles. If the ratio of red marbles to blue marbles was 5 to 3, what fraction of the marbles were blue?

e. What is the name for the list of all possible outcomes of a probability experiment?

f. *Represent* A penny and a nickel are tossed in the air once. One possible outcome is HT (H for head, T for tails; position of penny listed first then the nickel). List all the possible outcomes. Record the sample space in braces.

g. *Represent* A coin is tossed and the spinner is spun. Use a tree diagram to find the sample space for this experiment.

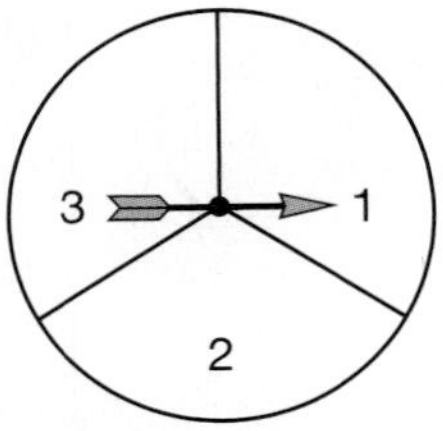

h. With one toss of a coin and one spin of the spinner in problem **g,** what is the probability of getting tails and a number less than 3?

i. *Justify* Which sample space below is the better way to show the possible outcomes of rolling two number cubes? Why? Use the better sample space to find the probability of rolling a sum of 7.

Sample space 1: {2, 3, 4, 5, 6, 7, 8, 9, 10, 11, 12}

Sample space 2:

Outcome of Second Cube (columns); Outcome of First Cube (rows)

	⚀	⚁	⚂	⚃	⚄	⚅
⚀	2	3	4	5	6	7
⚁	3	4	5	6	7	8
⚂	4	5	6	7	8	9
⚃	5	6	7	8	9	10
⚄	6	7	8	9	10	11
⚅	7	8	9	10	11	12

Use the data in the table to answer questions **1** and **2.**

Average Annual Precipitation 1971–2000 (inches)	
Mobile	66.3
San Francisco	20.1
Honolulu	18.3
Portland	45.8
San Juan	50.8
Salt Lake City	16.5

*** 1.** *(36)* **Evaluate** What is the ratio of cities with more than 20 inches of precipitation to cities with less than 20 inches?

2. *(28)* What is the average (mean) rainfall for all six cities?

3. *(13)* Darren reads 35 pages each night. At this rate, how many pages can Darren read in a week?

*** 4.** *(35)* Shannon swam 100 meters in 56.24 seconds. Rick swam 100 meters in 59.48 seconds. Rick took how many seconds longer to swim 100 meters than Shannon?

*** 5.** *(22, 36)* **Model** In the following statement, change the percent to a reduced fraction. Then diagram the statement and answer the questions.

Forty percent of the 30 players in the game had never played rugby.

a. How many of the players had never played rugby?

b. What was the ratio of those who had played rugby to those who had not played rugby?

*** 6.** *(34)* **Explain** AB is 4 cm. AC is 9.5 cm. Describe how to find BC in millimeters.

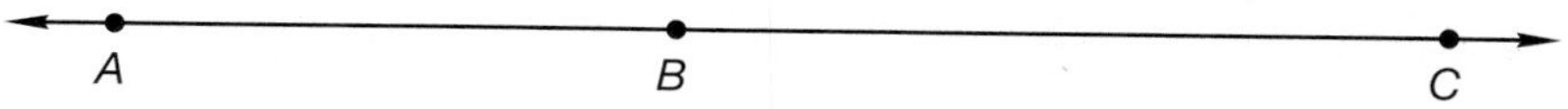

7. *(19, 20)* The length of the rectangle is 5 cm greater than its width.

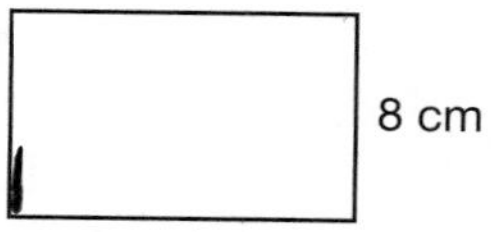

a. What is the area of the rectangle?

b. What is the perimeter of the rectangle?

*** 8.** *(29)* Estimate the perimeter of this triangle by rounding each measure to the nearest hundred millimeters before adding.

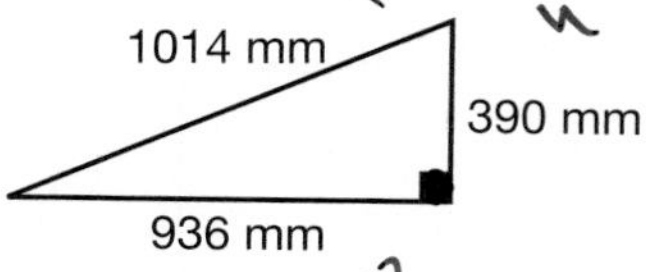

*** 9.** *(33)* **a.** Round 6.857142 to three decimal places.

b. Estimate the product of 6.8571420 and 1.9870. Round each factor to the nearest whole number before multiplying.

10. *(31)* Use digits to write each number:

a. twelve million

b. twelve millionths

*11. In a probability experiment a coin is tossed and a number cube is rolled. Make a tree diagram for the experiment, and show the sample space. Refer to the sample space to help you find the probability of getting heads and an even number.
(36)

* 12. Find the length of this segment
(34)

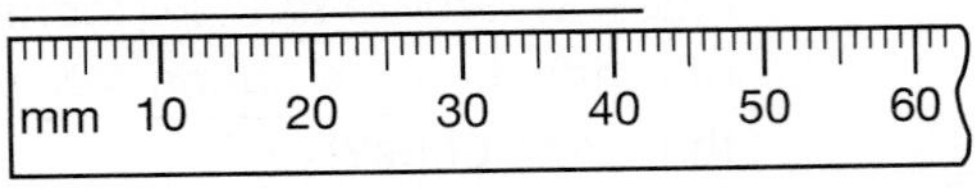

a. in centimeters.

b. in millimeters.

* 13. What decimal number names the point marked M on this number line?
(34)

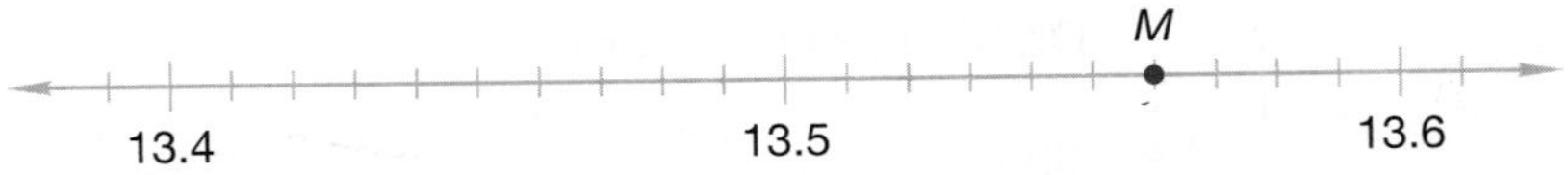

Math Language
A **prime factorization** expresses a composite number as a product of its prime factors.

14. **a.** Write 85% as a reduced fraction.
(24)

b. Write the prime factorization and reduce: $\frac{144}{600}$.

15. **Explain** Alba worked for 6 hr 45 min at $10.90 per hour. What numbers could she use to estimate how much money she earned? Estimate the amount she earned, and state whether you think the exact amount is a little more or a little less than your estimate.
(29)

16. **Classify** In this figure, which angle is
(7)

a. a right angle?

b. an acute angle?

c. an obtuse angle?

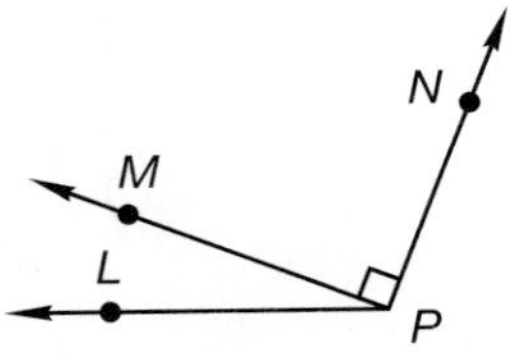

Solve:

17. $8y = 12^2$
(20)

* **18.** $\frac{W}{4} = 1.2$
(35)

Estimate each answer to the nearest whole number. Then perform the calculation.

* **19.** $4.27 + 16.3 + 10$
(35)

* **20.** $4.2 - 0.42$
(35)

Simplify:

21. $3\frac{1}{2} + 1\frac{1}{3} + 2\frac{1}{4}$
(30)

22. $3\frac{1}{2} \cdot 1\frac{1}{3} \cdot 2\frac{1}{4}$
(26)

23. $3\frac{5}{6} - \left(\frac{2}{3} - \frac{1}{2}\right)$
(30)

24. $8\frac{5}{12} - 3\frac{2}{3}$
(23, 30)

25. $2\frac{3}{4} \div 4\frac{1}{2}$
(26)

26. $5 - \left(\frac{2}{3} \div \frac{1}{2}\right)$
(23, 35)

* **27.** $1.4 \div 8$
(35)

***28.** $(0.2)(0.3)(0.4)$
(35)

* **29.** (35) **a.** 12.25×10

b. $12.25 \div 10$

30. (Inv. 3) *Model* On a coordinate plane draw a square that has an area of 25 units^2. Then write the coordinates of the vertices of the square on your paper.

Early Finishers
Real-World Application

In Rosa's class, the ratio of girls to boys is 4:3. She plans to survey her class about their favorite weekend activity.

a. If there are 9 boys in her class, how many girls are there?

b. To conduct her survey, Rosa doesn't want to ask all of her classmates. Instead she asks 8 of them while the other students are on the playground. What is the sample population for Rosa's survey?

c. The teacher suggests that Rosa survey all of the students in the class so that everyone can participate. What is the sample population for Rosa's survey now?

LESSON

37

• Area of a Triangle
• Rectangular Area, Part 2

Power Up

Building Power

facts

Power Up H

mental math

a. **Number Sense:** \$3.67 + \$0.98

b. **Calculation:** 5 × \$1.25

c. **Equivalent Fractions:** $\frac{7}{8} = \frac{?}{24}$

d. **Fractions:** Reduce $\frac{18}{30}$.

e. **Power/Roots:** $\frac{\sqrt{144}}{\sqrt{36}}$

f. **Fractional Parts:** $\frac{3}{10}$ of 60

g. **Patterns:** What is the next number in the pattern: 3, 2, 5, 4, ____

h. **Power/Roots:** What number is 5 less than the sum of 5^2 and $\sqrt{100}$?

problem solving

5 arps = 2 poms
4 poms + 2 arps = 2 dars
1 pom + 1 cob = 1 dar
1 arp + 1 cob = 1 hilp
4 hilps + 3 dars = 5 cobs + 7 poms + 1 arp

Find the value of each item if one arp is worth 2.

New Concepts

Increasing Knowledge

area of a triangle

Math Language

A small square at the vertex of an angle indicates that the angle is a right angle.

A triangle has a **base** and a **height** (or **altitude**).

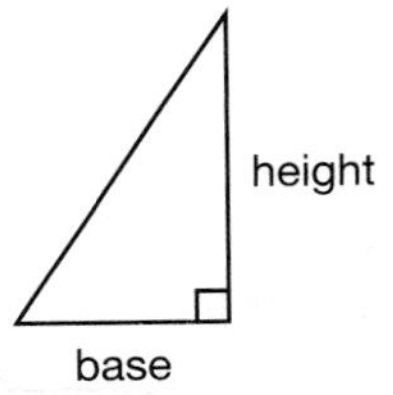

The base is one of the sides of the triangle. The height (or altitude) is the perpendicular distance between the base (or baseline) and the opposite vertex of the triangle. Since a triangle has three sides and any side can be the base, a triangle can have three base-height orientations, as we show by rotating this triangle.

One Right Triangle Rotated to Three Positions

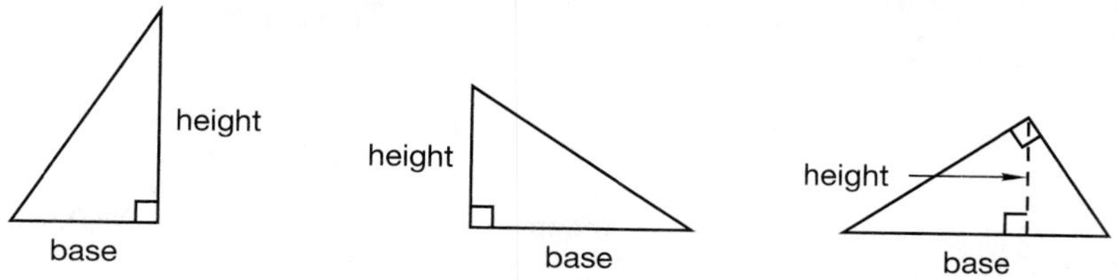

If one angle of a triangle is a right angle, the height may be a side of the triangle, as we see above. If none of the angles of a triangle are right angles, then the height will not be a side of the triangle. When the height is not a side of the triangle, a dashed line segment will represent it, as in the right-hand figure above. If one angle of a triangle is an obtuse angle, then the height is shown outside the triangle in two of the three orientations, as shown below.

One Obtuse Triangle Rotated to three Positions

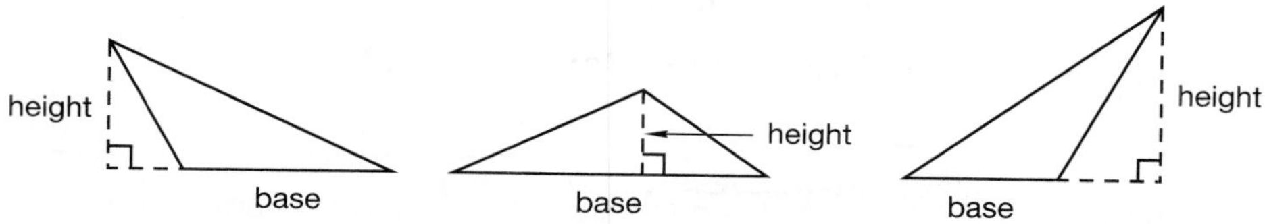

Model Using a ruler, draw a right triangle, an obtuse triangle and an acute triangle. Using the method described above, label the base and height of each triangle, adding a dashed line segment as needed.

The area of a triangle is half the area of a rectangle with the same base and height, as the following activity illustrates.

Activity

Area of a Triangle

Materials needed:

- Paper
- Ruler or straightedge
- Protractor
- Scissors

Use a ruler or straightedge to draw a triangle. Determine which side of the triangle is the longest side. The longest side will be the base of the triangle for this activity. To represent the height (altitude) of the triangle, draw a series of dashes from the topmost vertex of the triangle to the base. Make sure the dashes are perpendicular to the base, as in the figure below.

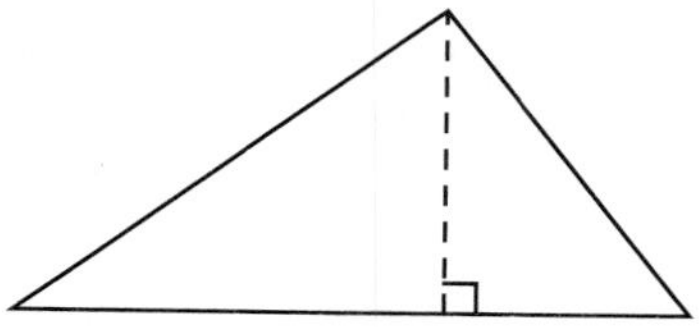

Now we draw a rectangle that contains the triangle. The base of the triangle is one side of the rectangle. The height of the triangle equals the height (width) of the rectangle.

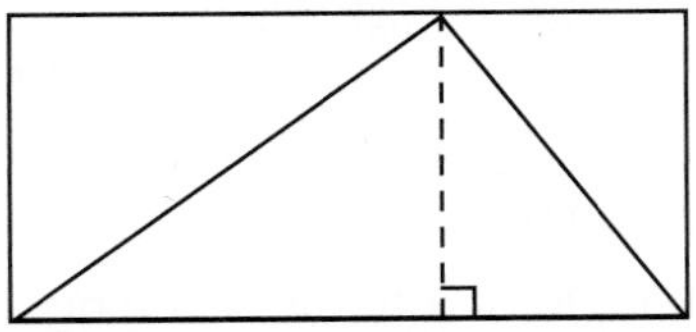

When you finish drawing the rectangle, consider this question, what fraction of the rectangle is the original triangle? The rest of this activity will answer this question.

Cut out the rectangle and set the scraps aside. Next, carefully cut out the triangle you drew from the rectangle. Save all three pieces.

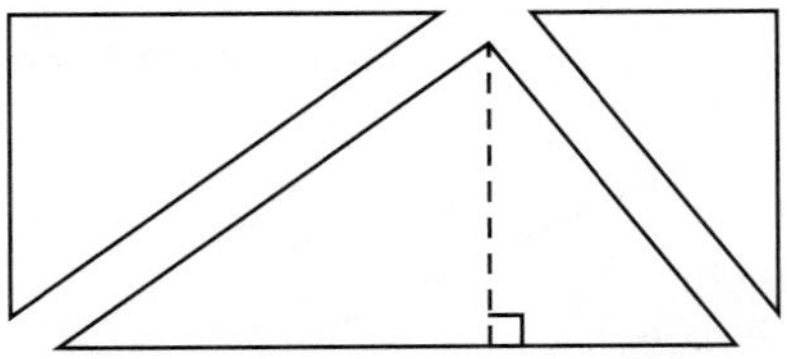

Math Language

Triangles that have the same shape *and* size are **congruent.**

Rotate the two smaller pieces, and fit them together to make a triangle congruent to the triangle you drew.

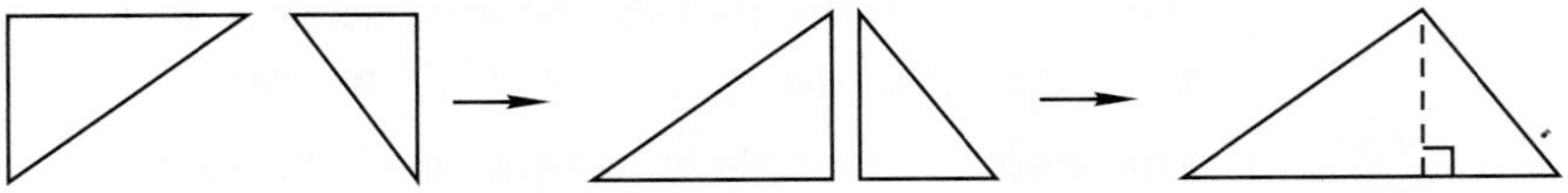

Since the two congruent triangles have equal areas, the original triangle must be half the area of the rectangle. Recall that the area of a rectangle can be found by this formula:

$$\text{Area} = \text{Length} \times \text{Width}$$
$$A = LW$$

Suggest a formula for finding the area of a triangle. Use base b and height h in place of length and width.

When we multiply two perpendicular dimensions, the product is the area of a rectangle with those dimensions.

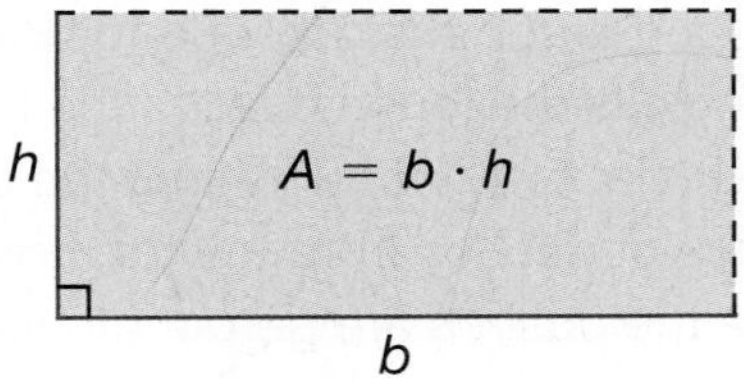

To find the area of a triangle with a base of b and a height of h, we find half of the product of b and h.

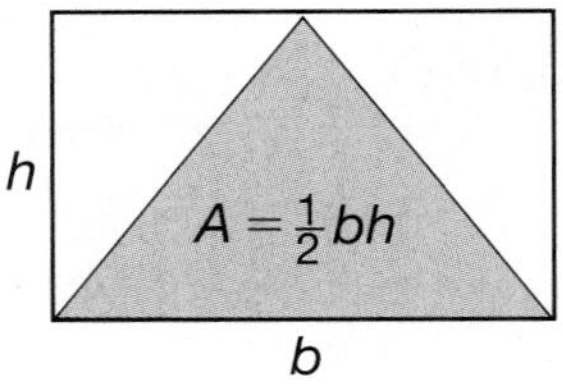

We show two formulas for finding the area of a triangle.

Discuss How are the formulas different? Why do both formulas yield the same result?

$$\text{Area of a triangle} = \frac{1}{2}bh$$

$$\text{Area of a triangle} = \frac{bh}{2}$$

Example 1

Find the area of this triangle. (Use $A = \frac{bh}{2}$.)

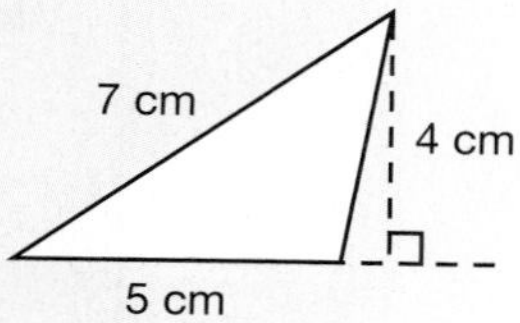

Solution

Thinking Skill

Explain

Why isn't one of the sides of this triangle its height?

We find the area of the triangle by multiplying the base by the height then dividing the product by 2. The base and height are perpendicular dimensions. In this figure the base is 5 cm, and the height is 4 cm.

$$\begin{aligned}\text{Area} &= \frac{5\text{ cm} \times 4\text{ cm}}{2} \\ &= \frac{20\text{ cm}^2}{2} \\ &= \mathbf{10\ cm^2}\end{aligned}$$

Example 2

High on the wall near the slanted ceiling was a triangular window with the dimensions shown. What is the area of the window? (Use $A = \frac{1}{2}bh$.)

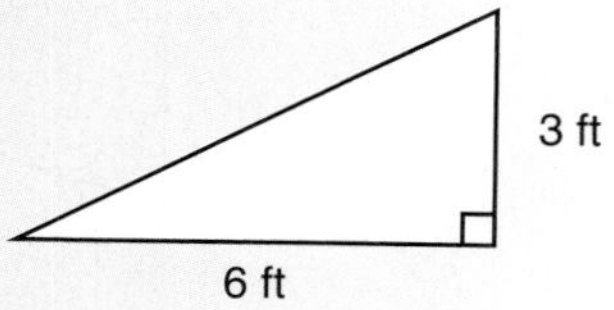

Solution

The base and height are perpendicular dimensions. Since one angle of this triangle is a right angle, the base and height are the perpendicular sides, which are 6 ft and 3 ft long.

$$\begin{aligned}\text{Area} &= \frac{1}{2} \cdot 6\text{ ft} \cdot 3\text{ ft} \\ &= \mathbf{9\ ft^2}\end{aligned}$$

The area of the window is **9 ft²**.

rectangular area, part 2

We have practiced finding the areas of rectangles. Sometimes we can find the area of a more complex shape by dividing the shape into rectangular parts. We find the area of each part and then add the areas of the parts to find the total area.

Example 3

Nate sketched this floor plan of his bedroom. Dimensions are in feet. All angles are right angles. What is the area of the room?

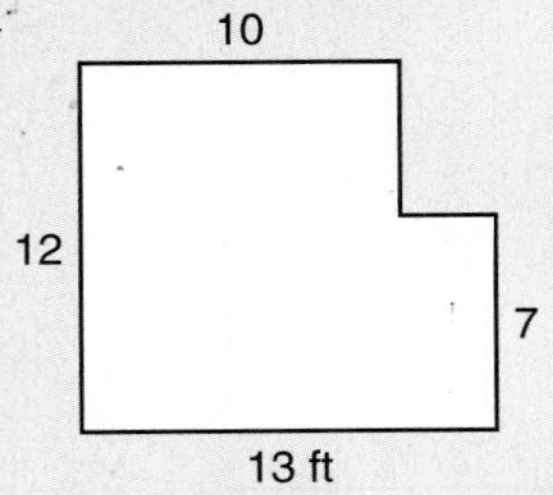

Solution

We show two ways to solve this problem.

Solution 1

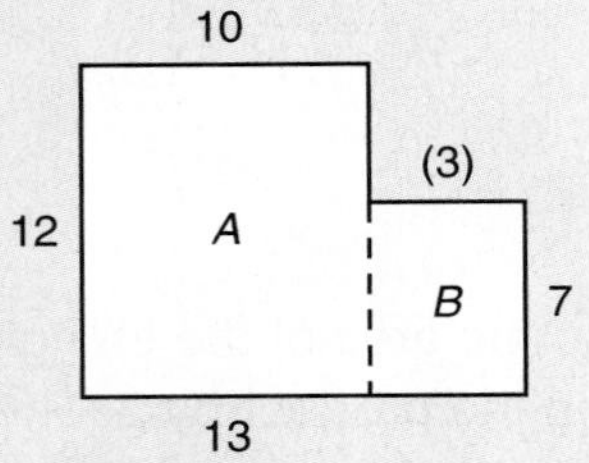

Total area = area A + area B

$$\begin{array}{lrcl} & \text{Area } A = 10 \text{ ft} \cdot 12 \text{ ft} & = & 120 \text{ ft}^2 \\ + & \text{Area } B = 3 \text{ ft} \cdot 7 \text{ ft} & = & 21 \text{ ft}^2 \\ \hline & \text{Total area} & = & \mathbf{141 \text{ ft}^2} \end{array}$$

Solution 2

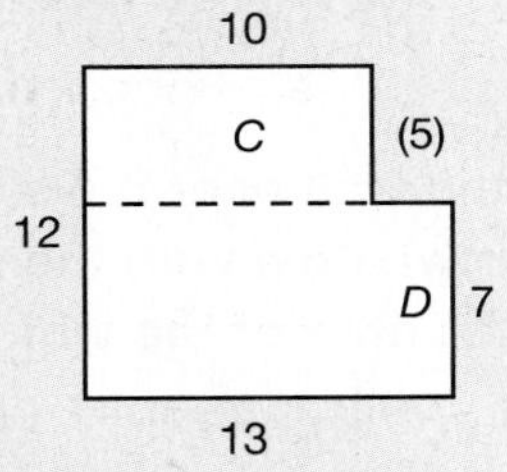

Total area = area C + area D

$$\begin{array}{lrcl} & \text{Area } C = 10 \text{ ft} \cdot 5 \text{ ft} & = & 50 \text{ ft}^2 \\ + & \text{Area } D = 13 \text{ ft} \cdot 7 \text{ ft} & = & 91 \text{ ft}^2 \\ \hline & \text{Total area} & = & \mathbf{141 \text{ ft}^2} \end{array}$$

Example 4

Find the area of the figure at right. Dimensions are in meters. All angles are right angles.

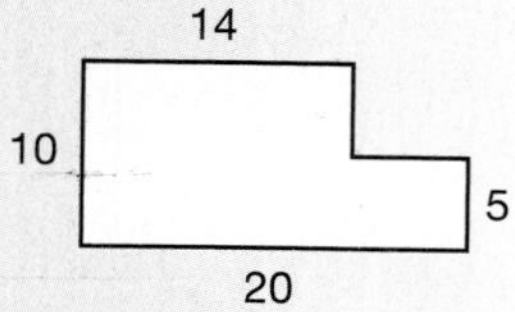

Solution

This time we will think of our figure as a large rectangle with a small rectangular piece removed. If we find the area of the large rectangle and then *subtract* the area of the small rectangle, the answer will be the area of the figure shown above.

Here we show the figure redrawn and the calculations:

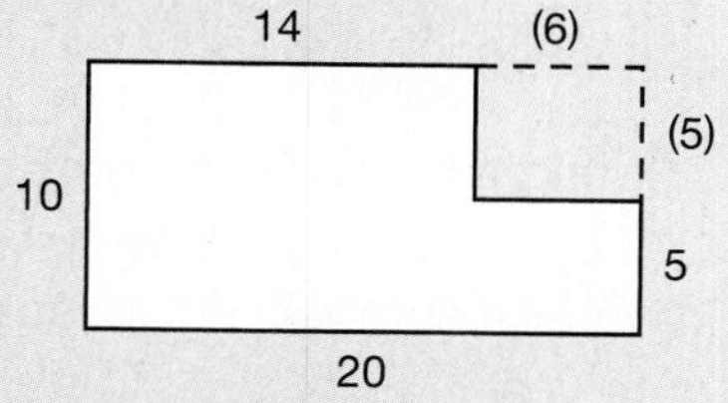

Area of figure = area of large rectangle − area of small rectangle

$$
\begin{array}{lll}
 & \text{Area of large rectangle} = 20\text{ m} \cdot 10\text{ m} = & 200\text{ m}^2 \\
- & \text{Area of small rectangle} = 6\text{ m} \cdot 5\text{ m} = & 30\text{ m}^2 \\
\hline
 & \text{Area of figure} & = \mathbf{170\text{ m}^2}
\end{array}
$$

We did not need to subtract to find the area. We could have added the areas of two smaller rectangles as we did in example 3.

Explain Which method do you think is easier? Why?

Practice Set

Find the area of each triangle. Dimensions are in centimeters.

a.

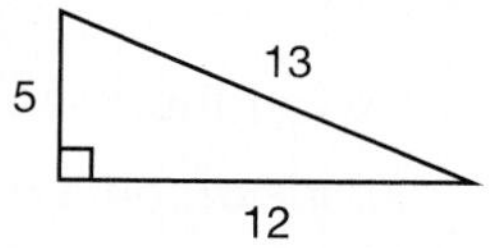

b.

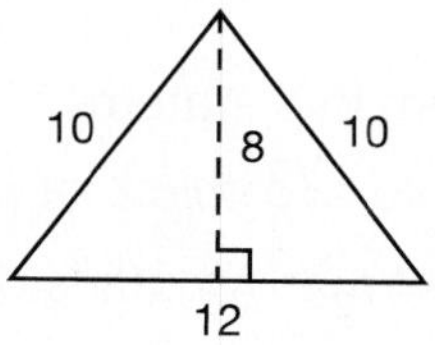

c.

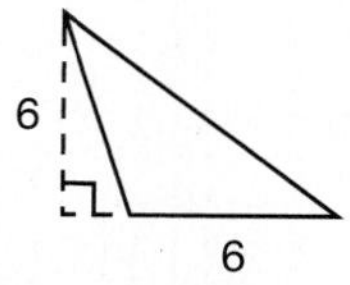

d. **Generalize** Copy the figure in example 4, and find its area by dividing the shape into two rectangles and adding the areas.

e. A 4 in.-by-4 in. square was cut from a 10 in.-by-12 in. sheet of construction paper. What is the area of the hexagon that remains? Find the area of the hexagon by subtracting the area of the square from the area of the original rectangle.

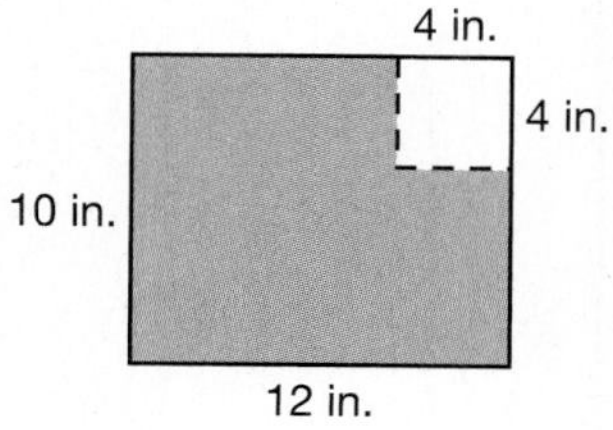

f. Find the area of the figure at right. Dimensions are in inches. All angles are right angles.

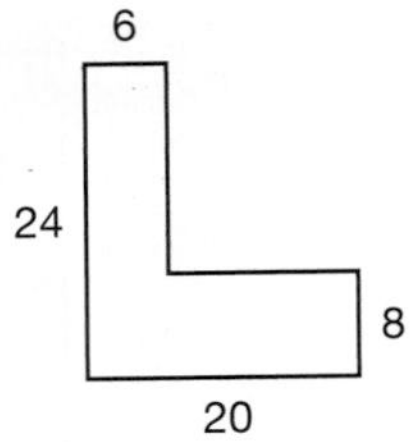

g. Justify How did you find your answer to **f?**

h. Write two formulas for finding the area of a triangle.

Written Practice *Strengthening Concepts*

1. (36) The baseball team played $\frac{2}{3}$ of its scheduled games and the rest were cancelled due to rain. What was the team's games-played to games-cancelled ratio?

2. (28) During the first six months of the year, the car dealership sold 47 cars, 53 cars, 62 cars, 56 cars, 46 cars, and 48 cars. What was the average number of cars sold per month during the first six months of the year?

*** 3.** (31, 35) Analyze The relay team carried the baton around the track. The first runner ran her leg of the relay in eleven and six tenths seconds. The second runner ran his leg in eleven and three tenths seconds. The third runner ran her leg in eleven and two tenths seconds. The fourth runner ran his leg in ten and nine tenths seconds. What was the team's total time?

4. (28) Consuela went to the store with $10 and returned home with 3 gallons of milk and $1.30 in change. How can she find the cost of each gallon of milk?

5. (22) Diagram this statement. Then answer the questions that follow.

Aziz sold two-thirds of his 18 muffins at the fundraiser.

a. How many muffins did Aziz sell?

b. How many muffins did Aziz not sell?

Copy this hexagon on your paper, and find the length of each unmarked side. Dimensions are in inches. All angles are right angles. Refer to the figure to answer problems 6 and 7.

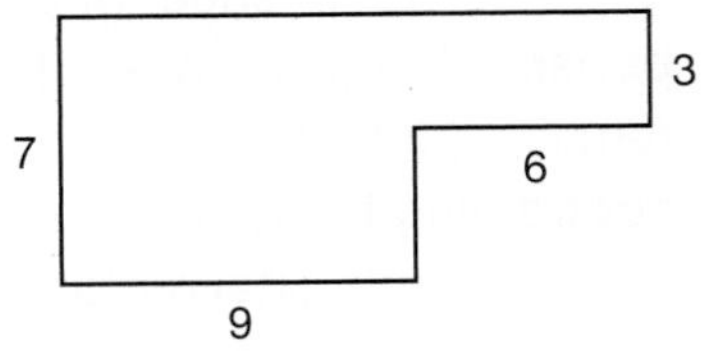

6. (19) What is the perimeter of the hexagon?

*** 7.** (37) Analyze There are two ways to find the area of the hexagon. What are they? Find the area using both methods.

8. (15) Complete each equivalent fraction:

a. $\frac{5}{6} = \frac{?}{18}$ **b.** $\frac{?}{8} = \frac{9}{24}$ **c.** $\frac{3}{4} = \frac{15}{?}$

9. (31) **a.** What decimal part of this square is shaded?

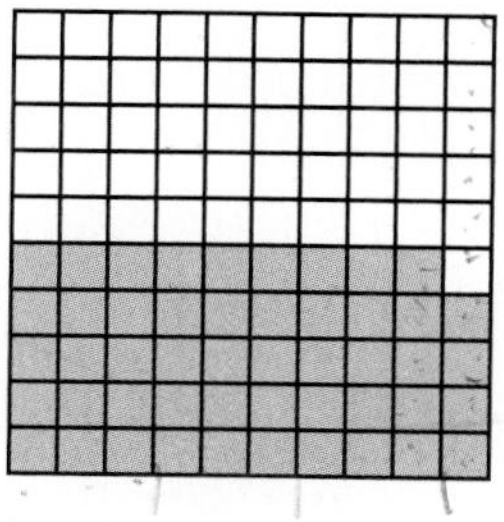

b. What decimal part of this square is not shaded?

c. What percent of this square is not shaded?

* **10.** (33) Round 3184.5641

a. to two decimal places.

b. to the nearest hundred.

11. (31) **a.** Name 0.00025.

b. Use digits to write sixty and seven hundredths.

12. (24) **a.** Write 2% as a reduced fraction.

b. Reduce $\frac{720}{1080}$.

13. (8) Find the length of segment *BC*.

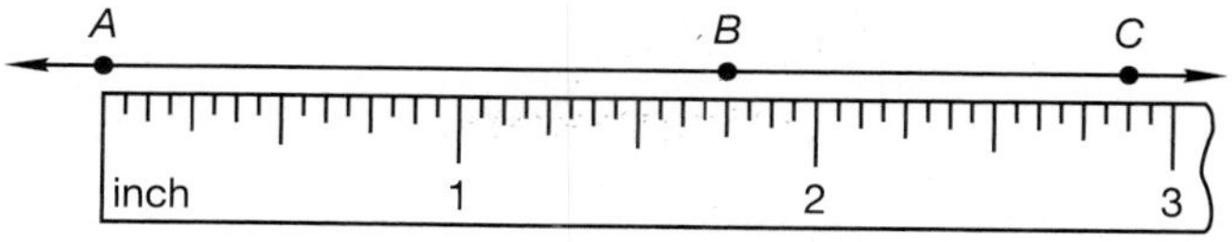

14. (7) Draw a pair of parallel lines. Next draw another pair of parallel lines that intersect the first pair of lines but are not perpendicular to them. Then shade the region enclosed by the intersecting pairs of lines.

* **15.** (37) **Generalize** Refer to the triangle below to answer **a** and **b**.

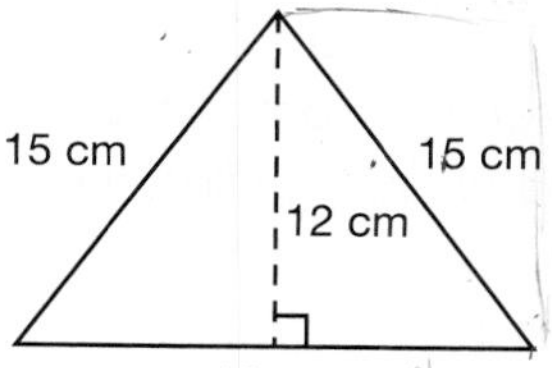

a. What is the perimeter of the triangle?

b. What is the area of the triangle?

* **16.** (33, 35) Simplify and compare: $0.2 + 0.3 \bigcirc 0.2 \times 0.3$

* **17.** (36) The face of the spinner is divided into thirds. The spinner is spun twice. One possible outcome is A then A (AA).

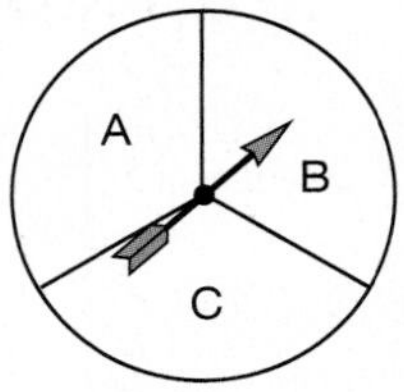

a. What is the sample space for this experiment?

b. *Analyze* Use the sample space to find the probability of spinning A at least once in two spins.

Solve:

18. (3) $7 \cdot 8 = 4x$

19. (35) $4.2 = 1.7 + y$

20. (35) $m - 3.6 = 0.45$

21. (35) $\frac{4.5}{w} = 3$

Simplify:

22. (26) $\frac{3}{5} \cdot 12 \cdot 4\frac{1}{6}$

23. (30) $\frac{5}{6} + 1\frac{3}{4} + 2\frac{1}{2}$

24. (30) $\frac{5}{8} + \left(\frac{1}{2} + \frac{3}{8}\right)$

25. (30) $3\frac{9}{20} - 1\frac{5}{12}$

26. (1, 26) Evaluate this expression for $a = 3\frac{1}{3}$ and $b = 5$:

$$\frac{a}{b}$$

27. (20) Find the missing exponent.

$$2^2 \cdot 2^2 \cdot 2^2 = 2^n$$

* **28.** (35) Simplify:

a. 0.25×10

b. $0.25 \div 10$

29. (Inv. 3) *Justify* A point with the coordinates (3, −3) lies in which quadrant of the coordinate plane? Draw a coordinate plane and plot the point to justify your answer.

30. (19) The cities of Durham, Raleigh, and Chapel Hill, North Carolina are each the home of a major research university. Thus the area between the cities is called the Research Triangle. The distance from Durham to Raleigh is about 20 miles. The distance from Raleigh to Chapel Hill is about 20 miles. The distance from Chapel Hill to Durham is about 10 miles. What is the perimeter of the triangle formed by the universities?

LESSON

38

• Interpreting Graphs

Power Up

Building Power

facts Power Up G

mental math

a. **Number Sense:** \$7.43 − \$0.99

b. **Number Sense:** 3 × \$2.50

c. **Equivalent Fractions:** $\frac{5}{6} = \frac{?}{30}$

d. **Fractions:** Reduce $\frac{18}{36}$.

e. **Power/Roots:** $\sqrt{121} + 7^2$

f. **Fractional Parts:** $\frac{7}{10}$ of 50

g. **Probability:** How many different ways can the digits 5, 6, 7 be arranged?

h. **Calculation:** 8 × 4, − 2, ÷ 3, × 4, ÷ 5, + 1, $\sqrt{\ }$, × 6, + 2, × 2, + 2, ÷ 6, × 5, + 1, $\sqrt{\ }$

problem solving

In the four class periods before lunch, Michael has math, English, science, and history, though not necessarily in that order. If each class is offered during each period, how many different permutations of the four classes are possible?

New Concept

Increasing Knowledge

Math Language
Quantitative information is data in the form of numbers. It tells us an amount of something.

We use **graphs** to help us understand quantitative information. A graph can use pictures, bars, lines, or parts of circles to help the reader visualize comparisons or changes. In this lesson we will practice interpreting graphs.

Example 1

Refer to the pictograph below to answer the questions that follow.

Adventure Tire Sales

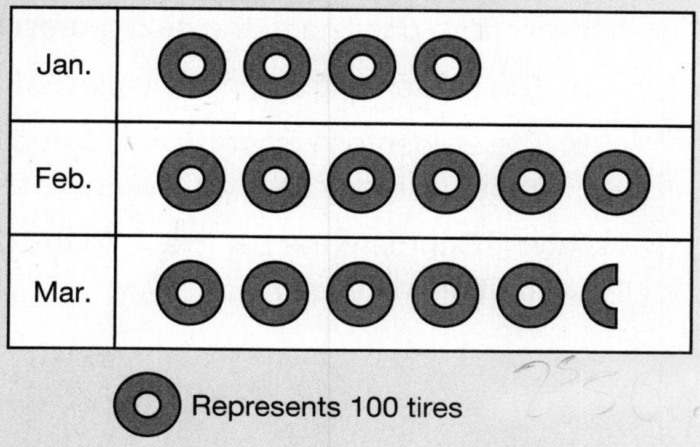

a. **Adventure sold about how many tires in March?**

b. **About how many tires were sold in the first three months of the year?**

Solution

The key at the bottom of the graph shows us that each picture of a tire represents 100 tires.

a. For March we see 5 whole tires, which represent 500 tires, and half a tire, which represents 50 tires. Thus, the $5\frac{1}{2}$ tires pictured mean that **about 550 tires** were sold in March.

b. We see a total of $15\frac{1}{2}$ tires pictured for the first three months of the year. Fifteen times 100 is 1,500. Half of 100 is 50. Thus **about 1,550 tires** were sold in the first three months of the year.

Conclude Which type of representation, a bar graph or a line graph, would be best to display the data on the pictograph? Justify your selection.

Example 2

Refer to the bar graph below to answer the questions that follow.

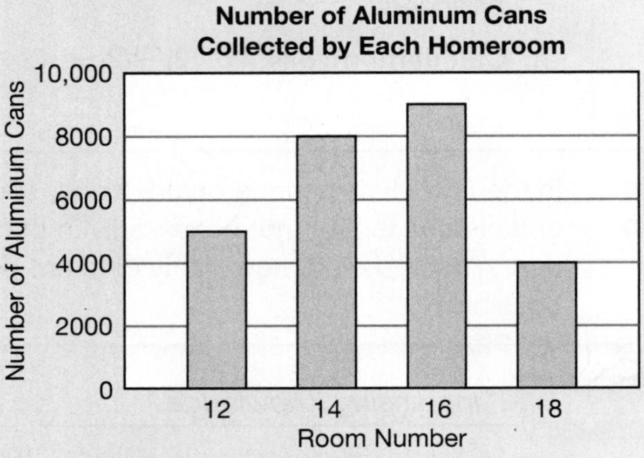

a. About how many cans were collected by the students in Room 14?

b. The students in Room 16 collected about as many cans as what other two homerooms combined?

Solution

We look at the scale on the left side of the graph. We see that the distance between two horizontal lines on the scale represents 2000 cans. Thus, halfway from one line to the next represents 1000 cans.

a. The students in Room 14 collected **about 8000 cans.**

b. The students in Room 16 collected about 9000 cans. This was about as many cans as **Room 12** and **Room 18** combined.

Discuss Compare the bar graph to the pictograph. How are they alike? How are they different?

Example 3

Reading Math

The symbol ⁓ indicates a broken scale. This graph scale is broken between 0 and 160.

This line graph shows Paul's bowling scores for the last six games he played.

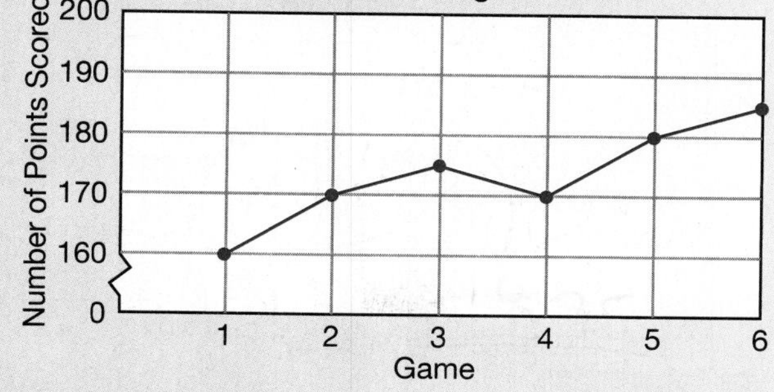

a. **What was Paul's score for game 3?**

b. **In general, were Paul's scores improving or getting worse?**

Solution

a. To find Paul's score for game 3, we look at the scale across the bottom of the graph to 3, and go up to the point that represents the score. We see that the point is halfway between 170 and 180. Thus, Paul's score for game 3 was **175.**

b. With only one exception, Paul scored higher on each succeeding game. So, in general, Paul's scores were **improving.**

Example 4

Use the information in this circle graph to answer the following questions:

a. **Altogether, how many hours are included in this graph?**

b. **What fraction of Ayisha's day is spent at school?**

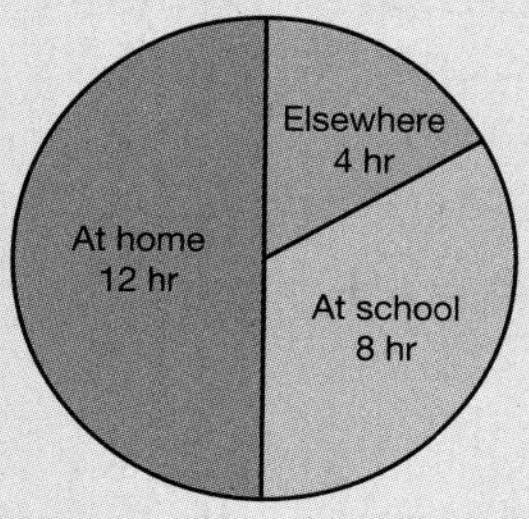

Where Ayisha Spends Her Day

Solution

A circle graph (sometimes called a pie graph) shows the relationship between parts of a whole. This graph shows parts of a whole day.

a. This graph includes **24 hours,** one whole day.

b. Ayisha spends 8 of the 24 hours at school. We reduce $\frac{8}{24}$ to $\mathbf{\frac{1}{3}}$.

Graphs should accurately and effectively display information.

Discuss How is a circle graph different from a pictograph, a bar graph, and a line graph?

Example 5

Which of these two graphs is constructed in a misleading way? What feature of the graph makes it misleading?

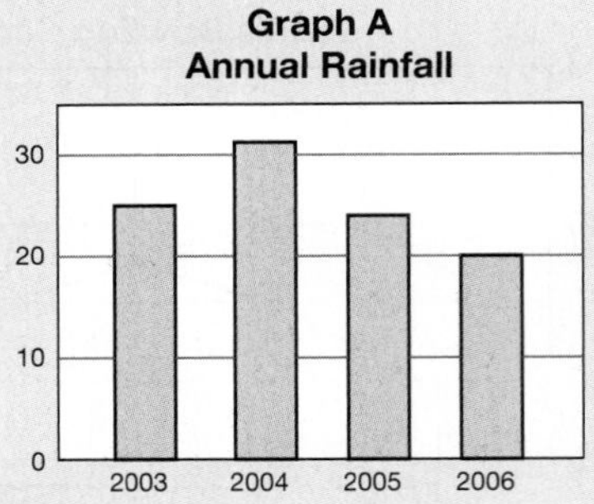

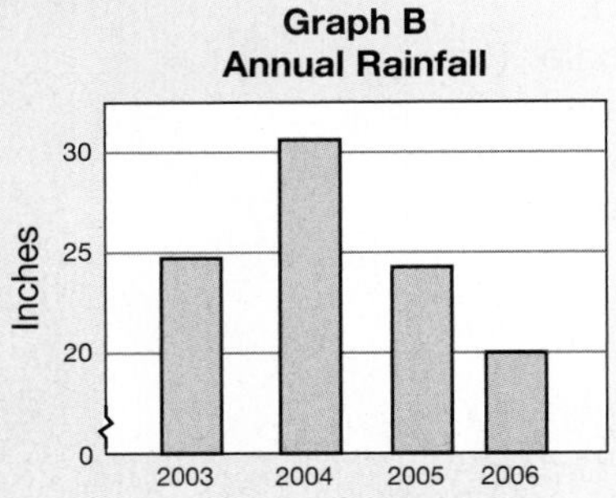

Solution

Although both bar graphs present the same information, **Graph B is visually misleading because the lengths of the bars makes it appear that three times as much rain fell in 2004 than in 2006.** Breaking the vertical scale in Graph B, which is sometimes helpful, distorted the relative lengths of the bars.

Example 6

Which of these two graphs is better to display Todd's height from age 10 to 14?

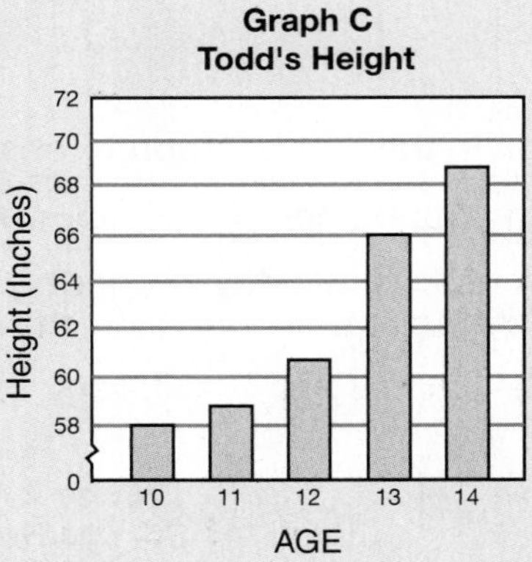

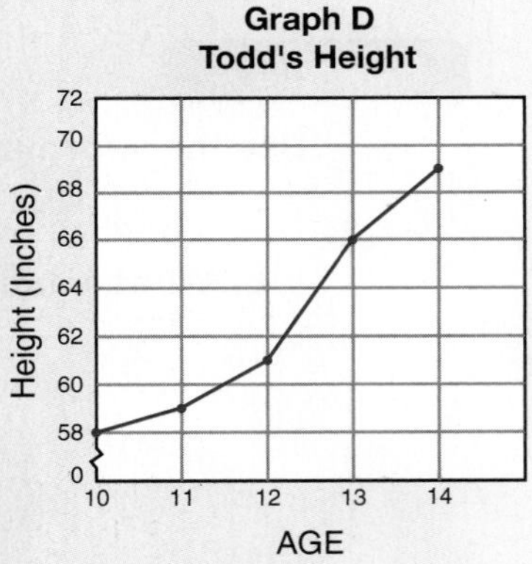

Solution

Todd's height gradually increased during these years, which is displayed better by **Graph D.** In contrast, Graph C makes it appear that Todd's height did not change until his age changed.

Practice Set

Use the information from the graphs in this lesson to answer each question.

a. How many more tires were sold in February than in January?

b. Analyze How many aluminum cans were collected by all four homerooms?

c. In which game was Paul's score lower than his score on the previous game?

d. **Evaluate** What fraction of Ayisha's day is spent somewhere other than at home or at school?

For **e, f,** and **g,** choose an answer from bar graph, line graph, and circle graph.

e. Which type of graph best displays the relationships between parts of a whole?

f. Which type of graph best displays change over time?

g. Which type of graph is best for showing comparisons?

Written Practice

Strengthening Concepts

*** 1.** (36) **Generalize** The ratio of walkers to joggers at the track was 3 to 7. What fraction of the athletes at the track were walkers?

2. (28) Denise read a 345-page book in 3 days. What was the average number of pages she read each day?

3. (28) Conner ran a mile in 5 minutes 52 seconds. How many seconds did it take Conner to run a mile?

Refer to the graphs in this lesson to answer problems **4–6.**

*** 4.** (12, 38) How many fewer cans were collected by the students in Room 18 than by the students in Room 16?

*** 5.** (28, 38) If Paul scores 185 in Game 7 (from example 3), what will be his average score for all 7 games?

*** 6.** (12, 38) **Formulate** Use the information in the graph in example 3 to write a problem about comparing.

7. (22) Diagram this statement. Then answer the questions that follow.

Mira read three eighths of the 384-page book before she could put it down.

a. How many pages did she read?

b. How many more pages does she need to read to be halfway through the book?

8. (19, 37) Refer to the figure at right to answer **a** and **b.** Dimensions are in inches. All angles are right angles.

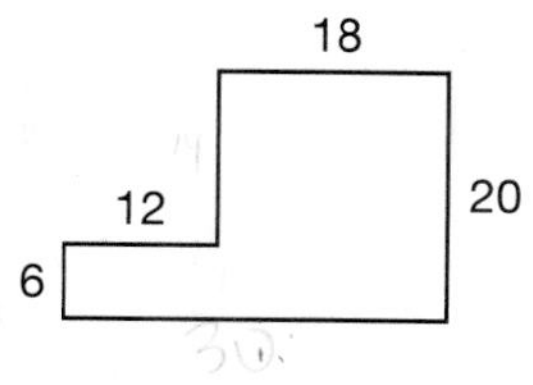

a. What is the area of the hexagon?

b. What is the perimeter of the hexagon?

9. (15) Complete each equivalent fraction:

a. $\frac{7}{9} = \frac{?}{18}$ **b.** $\frac{?}{9} = \frac{20}{36}$ **c.** $\frac{4}{5} = \frac{24}{?}$

d. What property do we use to find equivalent fractions?

10. Round 2986.34157
(33)

a. to the nearest thousand.

b. to three decimal places.

*** 11.** The face of this spinner is divided into eight congruent sectors. The spinner is spun once.
(14, 36)

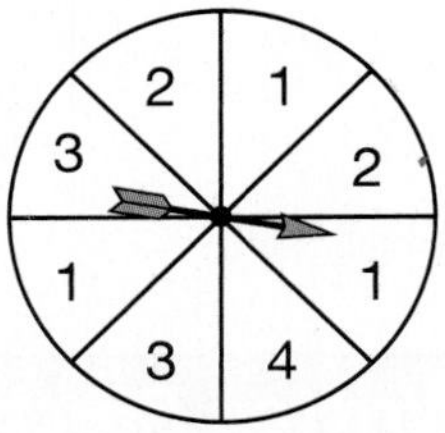

a. On which number is it most likely to stop?

b. On which number is it least likely to stop?

c. **Analyze** Which is the better sample space for this experiment and why?

Sample Space 1 = {1, 2, 3, 4}

Sample Space 2 = {1, 1, 1, 2, 2, 3, 3, 4}

12. Find the length of this segment
(34)

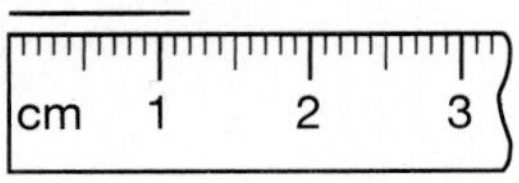

a. in centimeters.

b. in millimeters.

*** 13.** Find the perimeter of this rectangle in centimeters.
(32, 35)

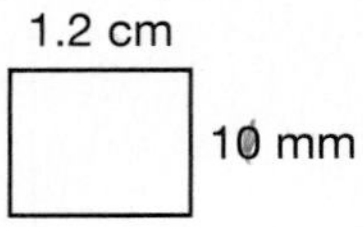

*** 14.** **Analyze** Which point marked on this number line could represent 3.4? Why?
(34)

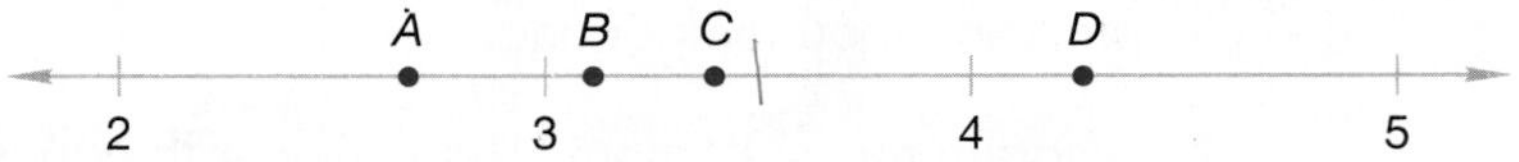

In the figure below, diagonal *AC* divides quadrilateral *ABCD* into two congruent triangles. Refer to the figure to answer problems **15** and **16.**

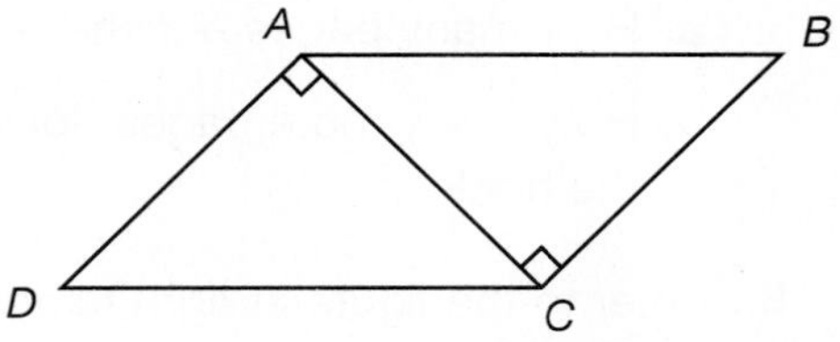

15. **a.** Which segment is perpendicular to $\overline{AD}$?
(7)

b. Which segment appears to be parallel to $\overline{AD}$?

*** 16.** The perpendicular sides of $\triangle ACD$ measure 6 cm each.
(37)

a. What is the area of $\triangle ACD$?

b. What is the area of $\triangle CAB$?

c. What is the area of the quadrilateral?

Generalize Solve:

* **17.** (35) $4.3 + a = 6.7$

* **18.** (35) $m - 3.6 = 4.7$

* **19.** (35) $10w = 4.5$

* **20.** (35) $\frac{x}{2.5} = 2.5$

Simplify:

* **21.** (35) $5.37 + 27.7 + 4$

* **22.** (35) $1.25 \div 5$

23. (26) $\frac{5}{9} \cdot 6 \cdot 2\frac{1}{10}$

24. (30) $\frac{5}{8} + \frac{3}{4} + \frac{1}{2}$

25. (26) $5 \div 3\frac{1}{3}$

26. (30) $\frac{3}{10} - \left(\frac{1}{2} - \frac{1}{5}\right)$

27. (20) Which is equivalent to $2^2 \cdot 2^4$?

A $4 \cdot 4^2$ **B** 2^8 **C** 4^8 **D** 4^6

* **28.** (32) **a.** How many milliliters of liquid are in this container?

b. The amount of liquid in this container is how much less than a liter?

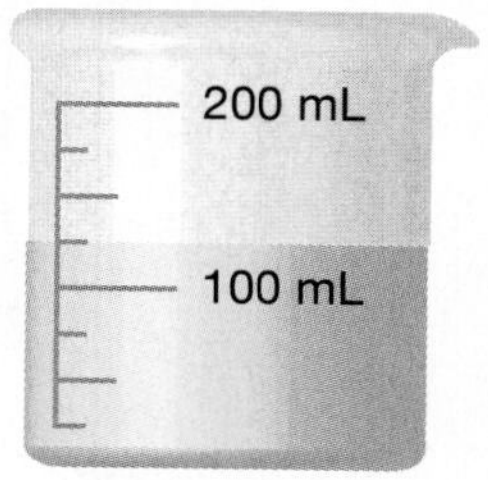

* **29.** (33) **Analyze** Five books on the library shelf were in the order shown below. Which two books should be switched so that they are arranged in the correct order?

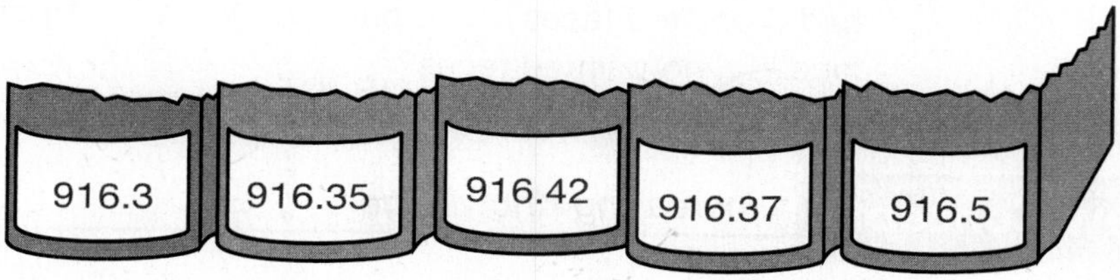

30. (Inv. 3) On graph paper draw a ray from the origin through the point (−10, −10). Then draw a ray from the origin through the point (−10, 10). Use a protractor to measure the angle formed by the two rays, and write the measure on the graph paper.

LESSON

39

• Proportions

Power Up

Building Power

facts Power Up H

mental math

a. **Decimals:** \$24.50 ÷ 10

b. **Number Sense:** $6 \times \$1.20$

c. **Equivalent Fractions:** $\frac{7}{12} = \frac{?}{60}$

d. **Fractions:** Reduce $\frac{24}{32}$.

e. **Power/Roots:** $5^2 - \sqrt{81}$

f. **Fractional Parts:** $\frac{4}{5}$ of 40

g. **Statistics:** Explain how you find the range of a set of numbers.

h. **Geometry/Measurement:** Start with the number of degrees in a straight angle. Subtract the number of years in a century; add the number of years in a decade; then subtract the number of degrees in a right angle. What is the answer?

problem solving

Jamaal glued 27 small blocks together to make this cube. Then he painted the six faces of the cube. Later the cube broke apart into 27 blocks. How many of the small blocks had 3 painted faces? ... 2 painted faces? ... 1 painted face? ... no painted faces?

New Concept

Increasing Knowledge

Reading Math

We read this proportion as "sixteen is to twenty as four is to five."

A **proportion** is a statement that two ratios are equal.

$$\frac{16}{20} = \frac{4}{5}$$

One way to test whether two ratios are equal is to compare their cross products. If we multiply the upper term of one ratio by the lower term of the other ratio, we form a **cross product.** The cross products of equal ratios are equal. We illustrate by finding the cross products of this proportion:

$$5 \cdot 16 = 80 \qquad \frac{16}{20} = \frac{4}{5} \qquad 20 \cdot 4 = 80$$

We find that both cross products equal 80. Whenever the cross products are equal, the ratios are equal as well.

We can use cross products to help us find missing terms in proportions. We will follow a two-step process.

Step 1: Find the cross products.

Step 2: Divide the known product by the known factor.

Example 1

Thinking Skill

Analyze

What does the *n* stand for in this proportion?

Mia can read 12 pages in 20 minutes. At that rate, how many pages can she read in 30 minutes? We can find the answer by solving this proportion:

$$\frac{12}{20} = \frac{n}{30}$$

Solution

We solve a proportion by finding the missing term.

Step 1: First we find the cross products. Since we are completing a proportion, the cross products are equal.

Thinking Skill

Justify

Would reducing before cross-multiplying be a helpful strategy for this problem? Why or why not?

$$\frac{12}{20} = \frac{n}{30}$$

$20 \cdot n = 30 \cdot 12$ equal cross products

$20n = 360$ simplified

Step 2: Divide the known product (360) by the known factor (20). The result is the missing term.

$n = \frac{360}{20}$ divided by 20

$n = 18$ simplified

Mia can read **18 pages** in 30 minutes. We can check our work by comparing the ratios in the completed proportion.

$$\frac{12}{20} = \frac{18}{30}$$

We see that the ratios are equal because both ratios reduce to $\frac{3}{5}$.

Example 2

Solve: $\frac{15}{x} = \frac{20}{32}$

Solution

Step 1: $20x = 480$ equal cross products

Step 2: $x = \mathbf{24}$ divided by 20

We check our work and see that both ratios reduce to $\frac{5}{8}$.

Example 3

Solve: $\frac{x}{1.2} = \frac{6}{4}$

Solution

Step 1: $4x = 7.2$ $(6 \times 1.2 = 7.2)$

Step 2: $x = 1.8$ $(7.2 \div 4 = 1.8)$

Justify Tell what was done in each step of the solution and why.

Practice Set

Solve each proportion:

a. $\frac{a}{12} = \frac{6}{8}$

b. $\frac{30}{b} = \frac{20}{16}$

c. $\frac{14}{21} = \frac{c}{15}$

d. $\frac{30}{25} = \frac{2.4}{d}$

e. $\frac{30}{100} = \frac{n}{40}$

f. $\frac{m}{100} = \frac{9}{12}$

Written Practice

Strengthening Concepts

Eduardo made a line graph to show the weight of his dog from the time the dog was 7 months old until it was 12 months old. Refer to the graph to answer problems **1** and **2.**

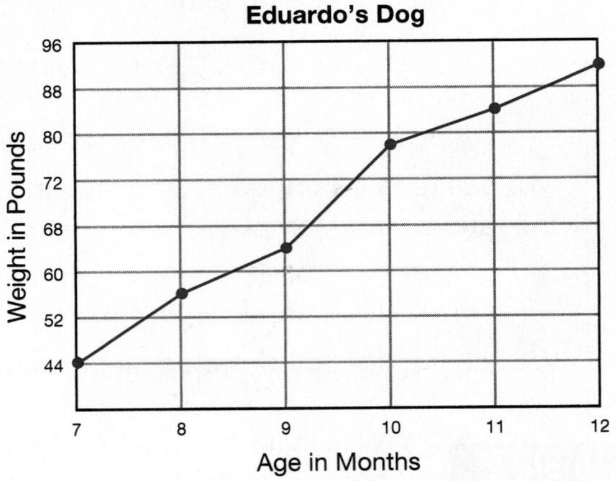

*** 1.** (38) How many pounds did Eduardo's dog gain from the time it was 9 months until it was 10 months?

*** 2.** (38) **Analyze** Between what two months did Eduardo's dog gain the most weight?

*** 3.** (36) There are 12 trumpet players and 16 flute players in the school band. What is the ratio of trumpet players to flute players in the band?

4. (28) On the first 4 days of their trip, the Curtis family drove 497 miles, 513 miles, 436 miles, and 410 miles. What was the average number of miles they drove per day on the first 4 days of their trip?

5. (16) **Represent** Don receives a weekly allowance of \$4.50. Make a function table that shows the total number of dollars received (output) in 1, 2, 3, 5, and 8 weeks.

6. (22) Diagram this statement. Then answer the questions that follow.

Three sevenths of the 105 students in the high school band have fewer than 5 years experience playing their instrument.

a. How many students have fewer than 5 years experience?

b. How many students have 5 or more years experience?

Refer to this figure to answer problems **7** and **8.** Dimensions are in millimeters. All angles are right angles.

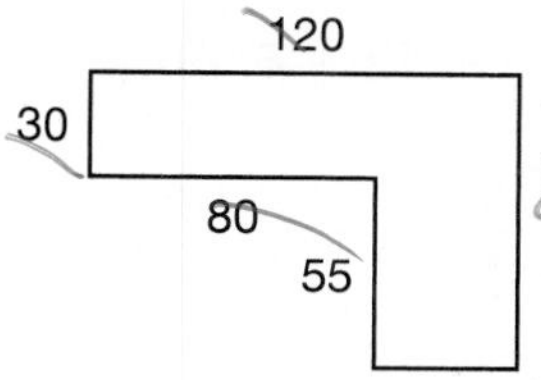

* **7.** (37) What is the area of the figure?

8. (19) What is the perimeter of the figure?

9. (31) Name the number of shaded circles

a. as a decimal number.

b. as a mixed number.

10. (33) Round 0.9166666

a. to the nearest hundredth.

b. to the nearest hundred-thousandth.

* **11.** (33) **Estimate** Sharika bought 9.16 gallons of gasoline priced at $\$1.99\frac{9}{10}$ per gallon. Estimate the total cost.

12. (31) Use digits to write each number:

a. one hundred and seventy-five thousandths

b. one hundred seventy-five thousandths

13. (7) Refer to the figure at right to name

a. an acute angle.

b. an obtuse angle.

c. a straight angle.

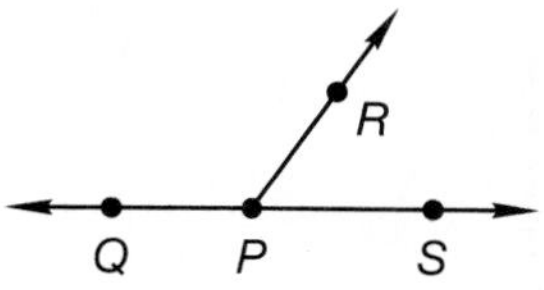

* **14.** (4, 35) **Generalize** Find the next three numbers in this sequence, and state the rule of the sequence:

$$\ldots, 100, 10, 1, 0.1, \ldots$$

Analyze Solve:

* **15.** (39) $\frac{8}{12} = \frac{6}{x}$

* **16.** (39) $\frac{16}{y} = \frac{2}{3}$

* **17.** (39) $\frac{21}{14} = \frac{n}{4}$

* **18.** (35) $m + 0.36 = 0.75$

* **19.** (35) $1.4 - w = 0.8$

* **20.** (35) $8x = 7.2$

* **21.** (35) $\frac{y}{0.4} = 1.2$

Estimate Estimate each answer to the nearest whole number. Then perform the calculation.

* **22.** (35) $9.6 + 12 + 8.59$

* **23.** (35) $3.15 - (2.1 - 0.06)$

Simplify:

24. (30) $4\frac{5}{12} + 6\frac{5}{8}$

25. (23, 30) $4\frac{1}{4} - 1\frac{3}{5}$

26. (26) $8\frac{1}{3} \cdot 1\frac{4}{5}$

27. (26) $5\frac{5}{6} \div 7$

28. (37) Refer to this triangle to answer **a** and **b:**

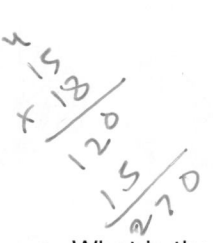

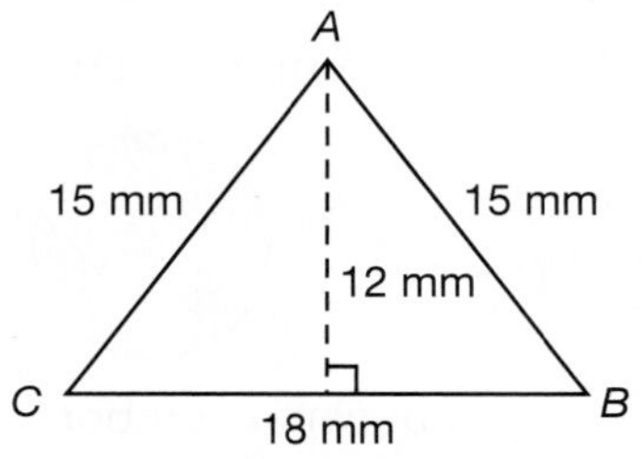

a. What is the perimeter of $\triangle ABC$?

b. What is the area of $\triangle ABC$?

29. (14, 21) What is the probability of rolling an odd prime number with one toss of a number cube?

30. (30) Use common denominators to arrange these numbers in order from least to greatest:

$$\frac{2}{3}, \frac{1}{2}, \frac{5}{6}, \frac{7}{12}$$

LESSON 40

• Sum of the Angle Measures of a Triangle
• Angle Pairs

Power Up

Building Power

facts Power Up G

mental math

a. **Decimals:** $\$0.18 \times 100$

b. **Number Sense:** $4 \times \$1.25$

c. **Equivalent Fractions:** $\frac{3}{4} = \frac{?}{24}$

d. **Fractions:** Reduce $\frac{12}{32}$.

e. **Power/Roots:** $\sqrt{144} + \sqrt{121}$

f. **Fractional Parts:** $\frac{2}{3}$ of 60

g. **Statistics:** Explain what the median is in a set of numbers.

h. **Calculation:** Start with $10.00. Divide by 4; add two quarters; multiply by 3; find half of that amount; then subtract two dimes.

problem solving

At three o'clock the hands of an analog clock form a 90° angle. What is the measure of the angle that is formed by the hands of the clock two hours after three o'clock?

Understand The hands of an analog clock form different angles depending on the time of day. At 3:00 the hands form a 90° angle. We are asked to find the angle formed by the hands two hours later, or at 5:00. We bring to this problem the knowledge that there are 360° in a circle.

Plan We will *draw a diagram* that divides the circle into 12 sections, each representing one hour. Then we will find the measure of each section and multiply to find the angle formed by the hands at 5:00.

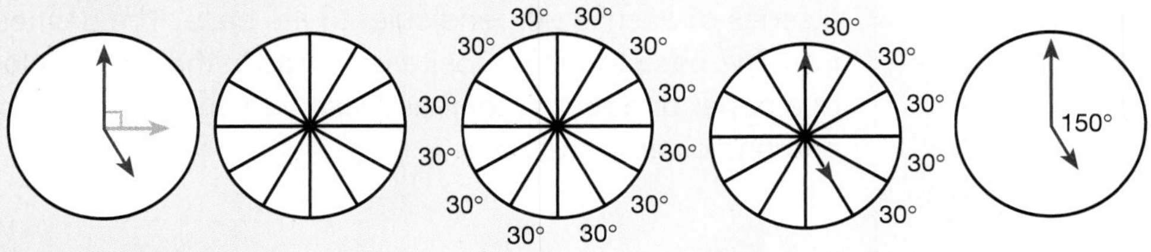

Solve We divide 360° by 12 and find that each of the 12 sections span 30°. To solve for the angle's measure, we multiply: $30° \times 5 = 150°$.

Check We found the measure of the angle that is formed by the hands of a clock at 5:00. We know that the hands form a 90° angle at 3:00 and an 180° angle at 6:00, so it makes sense that at 5:00 the angle formed would be between these two measures.

New Concepts *Increasing Knowledge*

sum of the angle measures of a triangle

Thinking Skill

Connect

Write a word sentence that states the relationship between the sum of the angle measures of a square and of a triangle.

A square has four angles that measure 90° each. If we draw a segment from one vertex of a square to the opposite vertex, we have drawn a diagonal that divides the square into two congruent triangles. We show this in the figure below.

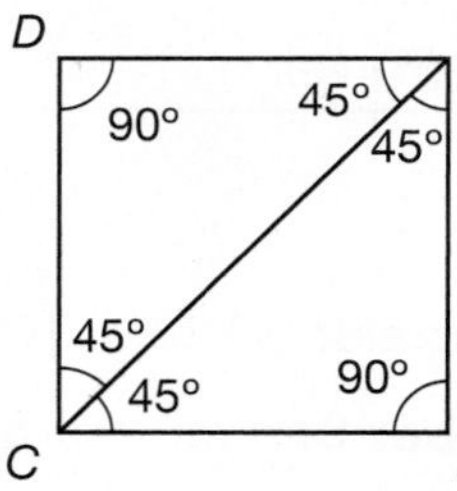

Segment AC divides the right angle at A into two angles, each measuring 45° (90° ÷ 2 = 45°). Segment AC also divides the right angle at C into two 45° angles. Each triangle has angles that measure 90°, 45°, and 45°. The sum of these angles is 180°.

$$90° + 45° + 45° = 180°$$

The three angles of every triangle have measures that total 180°. We illustrate this fact with the following activity.

Activity

Sum of the Angle Measures of a Triangle

Materials needed:

- Paper
- Ruler or straightedge
- Protractor
- Scissors

With a ruler or straightedge, draw two or three triangles of various shapes large enough to easily fold. Let the longest side of each triangle be the base of that triangle, and indicate the height (altitude) of the triangle by drawing a series of dashes perpendicular to the base. The dashes should extend from the base to the opposite vertex as in the figure below. Use the corner of a paper or a protractor if necessary to ensure the indicated height is perpendicular to the base.

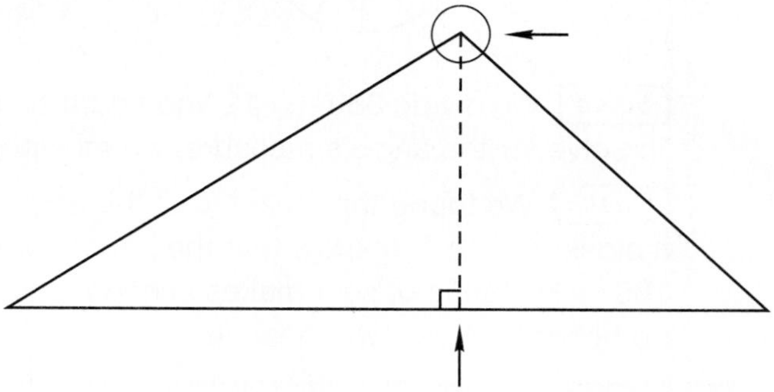

After the drawings are complete, carefully cut out each triangle. Then select a triangle for folding. First fold the vertex with the dash down to the point on the base where the row of dashes intersects the base.

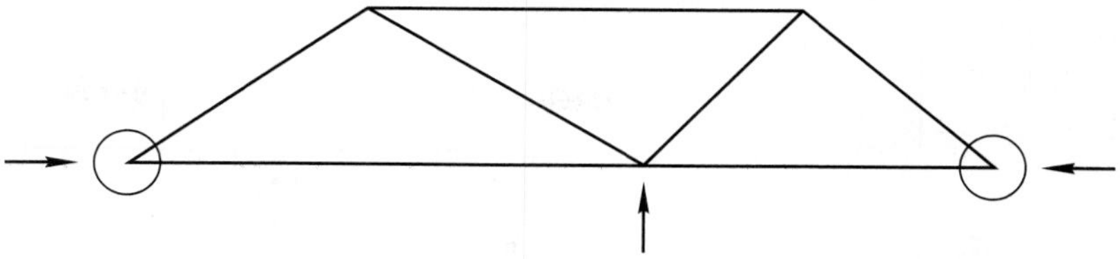

Then fold each of the other two vertices to the same point. When finished, your folded triangle should look like this:

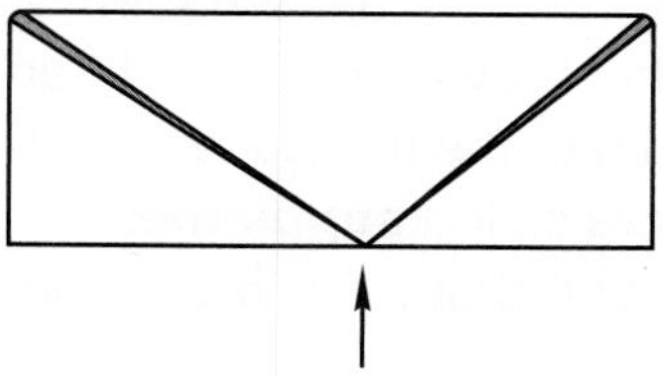

If you sketch a semicircle about the meeting point, you will see that the three angles of the triangle together form a half circle. That is, the sum of their measures is 180°.

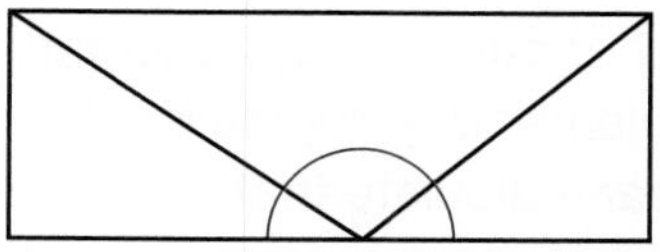

Repeat the folding activity with the other triangle(s) you drew.

Example 1

Find the measure of $\angle A$ in $\triangle ABC$.

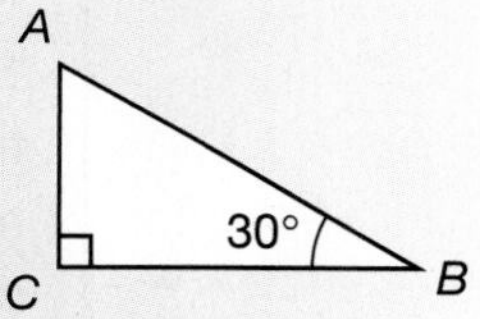

Solution

The sum of the measures of the angles is 180°. Angle *B* measures 30° and angle *C* is a right angle that measures 90°. Using this information, we can write the following equation:

$$m\angle A + 90° + 30° = 180°$$

$$m\angle A = \mathbf{60°}$$

Justify Can a triangle have more than one right angle? Explain your reasoning and support your reasoning with a sketch.

angle pairs

Two intersecting lines form four angles. We have labeled these four angles $\angle 1$, $\angle 2$, $\angle 3$, and $\angle 4$ for easy reference.

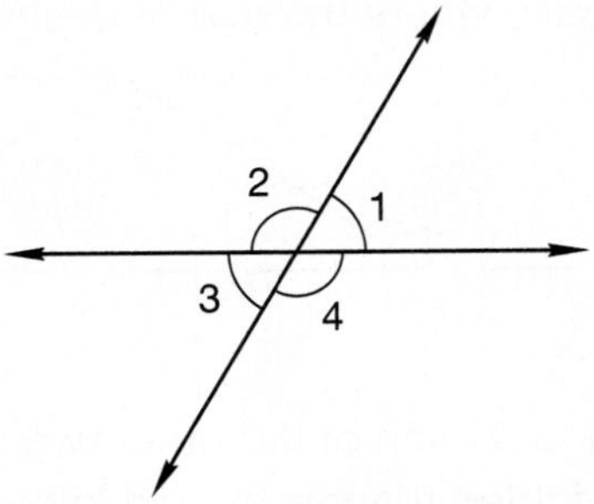

Thinking Skill

Summarize

Write a statement that summarizes the properties of each type of angles—adjacent, supplementary, vertical, and complementary.

Angle 1 and $\angle 2$ are **adjacent angles,** sharing a common side. Notice that together they form a straight angle. A straight angle measures 180°, so the sum of the measures of $\angle 1$ and $\angle 2$ is 180°. Two angles whose sum is 180° are called **supplementary angles.** We say that $\angle 1$ is a supplement of $\angle 2$ and that $\angle 2$ is a supplement of $\angle 1$.

Notice that $\angle 1$ and $\angle 4$ are also supplementary angles. If we know that $\angle 1$ measures 60°, then we can calculate that $\angle 2$ measures 120° (60° + 120° = 180°) and that $\angle 4$ measures 120°. So $\angle 2$ and $\angle 4$ have the same measure.

Another pair of supplementary angles is $\angle 2$ and $\angle 3$, and the fourth pair of supplementary angles is $\angle 3$ and $\angle 4$. Knowing that $\angle 2$ or $\angle 4$ measures 120°, we can calculate that $\angle 3$ measures 60°. So $\angle 1$ and $\angle 3$ have the same measure.

Angles 1 and 3 are not adjacent angles; they are **vertical angles.** Likewise, $\angle 2$ and $\angle 4$ are vertical angles. Vertical angles are a pair of nonadjacent angles formed by a pair of intersecting lines. Vertical angles have the same measure.

Two angles whose measures total 90° are called **complementary angles.** In the triangle below, $\angle A$ and $\angle B$ are complementary because the sum of their measures is 90°. So $\angle A$ is a complement of $\angle B$, while $\angle B$ is a complement of $\angle A$.

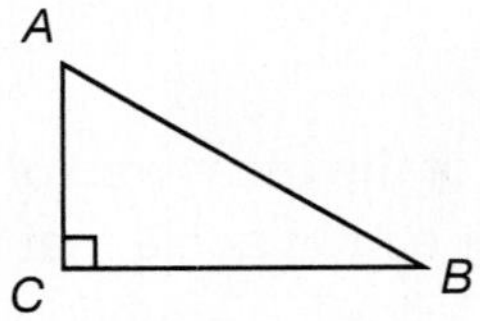

Example 2

Find the measures of $\angle x$, $\angle y$, and $\angle z$. Classify angle pairs as supplementary or complementary.

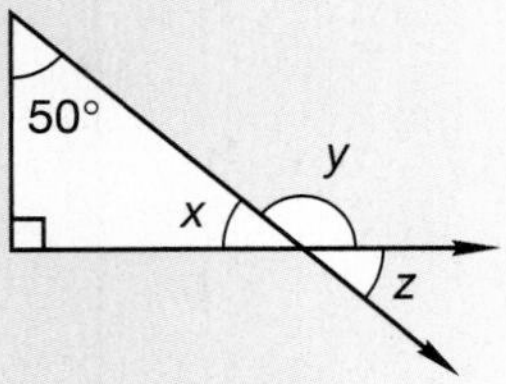

Solution

The measures of the three angles of a triangle total 180°. The right angle of the triangle measures 90°. So the two acute angles total 90°. One of the acute angles is 50°, so $\angle x$ measures **40°.**

Together $\angle x$ and $\angle y$ form a straight angle measuring 180°, so they are supplementary. Since $\angle x$ measures 40°, $\angle y$ measures the rest of 180°, which is **140°.**

Angle $\angle z$ and $\angle y$ are supplementary. Also, $\angle z$ and $\angle x$ are vertical angles. Since vertical angles have the same measure, $\angle z$ measures **40°.**

Example 3

Figure *ABCD* is a parallelogram. Segment *BD* divides the parallelogram into two congruent triangles. Angle *CBD* measures 40°. Which other angle measures 40°?

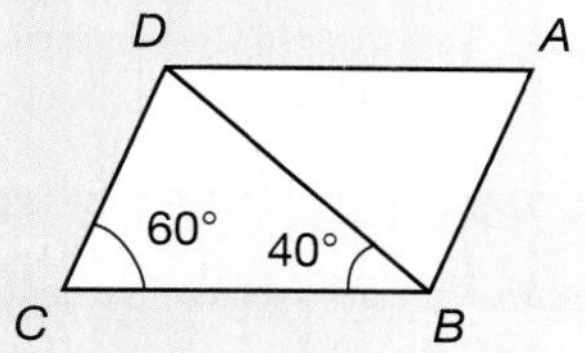

Solution

A segment that passes through a polygon to connect two vertices is a **diagonal.** A diagonal divides any parallelogram into two congruent triangles. Segment *BD* is a diagonal. Rotating (turning) one of the triangles 180° (a half turn) positions the triangle in the same orientation as the other triangle. We illustrate rotating $\triangle ABD$ 180°.

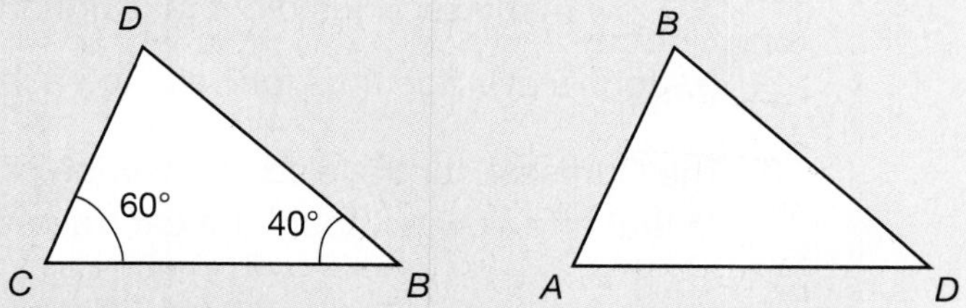

Each side and angle of $\triangle ABD$ now has the same orientation as its corresponding side or angle in $\triangle CDB$. (Notice that segments *BD* and *DB* are the same segment in both triangles. They have the same orientation in both triangles, but their vertices are reversed due to the rotation.) Angle *CBD* measures 40°. The angle in $\triangle ABD$ that corresponds to this angle is $\angle ADB$. So $\angle ADB$ **measures 40°.**

Thinking Skill

Generalize

What angle corresponds to $\angle DCB$?

Practice Set

a. Discuss The sides of a regular triangle are equal in length, and the angles are equal in measure. What is the measure of each angle of a regular triangle and why?

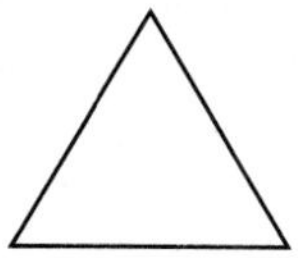

Refer to rectangle *ABCD* to answer problems **b–d.**

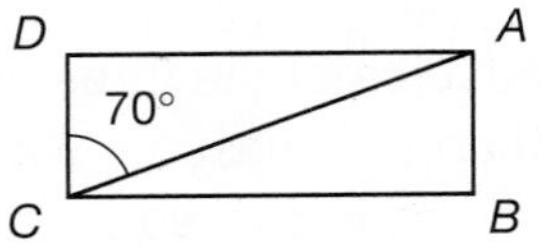

b. What is the measure of $\angle ACB$ and why?

c. What is the measure of $\angle CAB$ and why?

d. **Explain** Are angles *ACD* and *CAB* vertical angles? Why or why not?

e. Find the measures of $\angle x$, $\angle y$, and $\angle z$ in this figure.

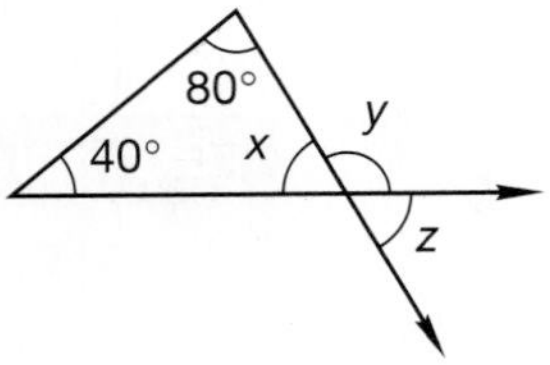

Written Practice

Strengthening Concepts

1. (14, 36) The bag contained only red marbles and white marbles in the ratio of 3 red to 2 white.

a. What fraction of the marbles were white?

b. What is the probability that a marble drawn from the bag will be white?

2. (28) John ran 4 laps of the track in 6 minutes 20 seconds.

a. How many seconds did it take John to run 4 laps?

b. John's average time for running each lap was how many seconds?

3. (13) The Curtises' car traveled an average of 24 miles per gallon of gas. At that rate how far could the car travel on a full tank of 18 gallons?

4. (12, 35) At sea level, water boils at 212°F. For every 550 feet above sea level, the boiling point of water decreases by about 1°F.

a. At 5500 feet above sea level, water boils at about 202°F. How many degrees hotter is the boiling temperature of water at sea level than at 5500 feet above sea level?

b. **Apply** At the summit of Puncak Jaya, the highest mountain in Indonesia, water boils at about 183°F. About how many feet above sea level is the summit of Puncak Jaya?

5. (19, 20) The length of the rectangle at right is twice its width.

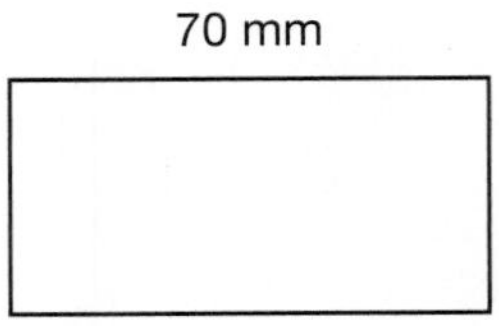

a. What is the perimeter of the rectangle?

b. What is the area of the rectangle?

6. (22) Diagram this statement. Then answer the questions that follow.

Five eighths of the 200 sheep in the flock grazed in the meadow. The rest drank from the brook.

a. How many of the sheep grazed in the meadow?

b. How many of the sheep drank from the brook?

7. (34) *AB* is 30 mm. *CD* is 45 mm. *AD* is 100 mm. Find *BC* in centimeters.

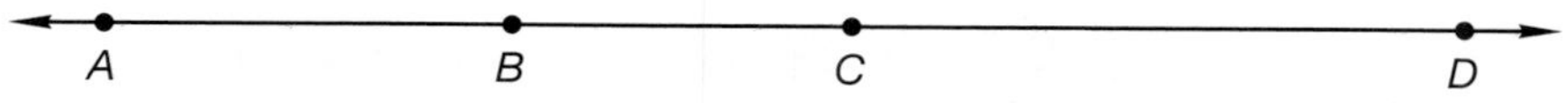

8. (33) Round 0.083333

a. to the nearest thousandth.

b. to the nearest tenth.

9. (31) Use words to write each number:

a. 12.054

b. $10\frac{11}{100}$

* **10.** (Inv. 3, 37) **Generalize** The coordinates of the three vertices of a triangle are (−2, 5), (4, 0), and (−2, 0). What is the area of the triangle?

* **11.** (34) **Analyze** What decimal number names the point marked *B* on this number line?

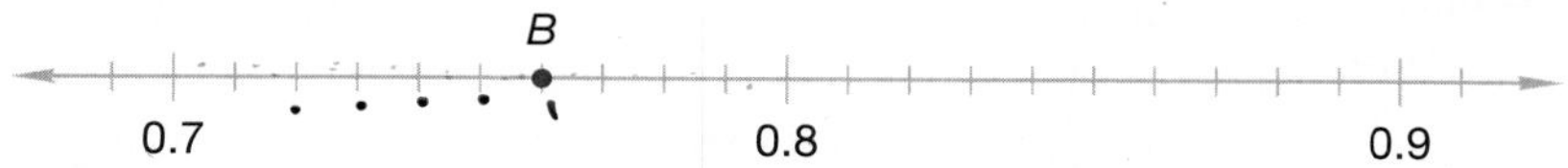

Refer to the figure below to answer problems **12** and **13.**

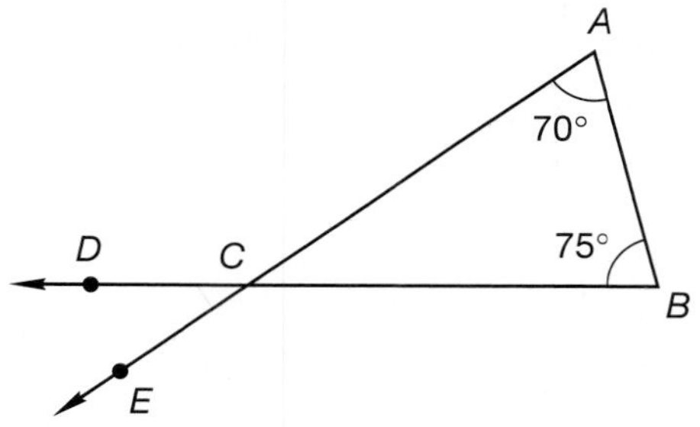

* **12.** (40) **Generalize** Find the measure of each angle:

a. $\angle ACB$

b. $\angle ACD$

c. $\angle DCE$

* **13.** (40) Angle *ACD* is supplementary to $\angle ACB$. Name another angle supplementary to $\angle ACB$.

14. (1, 2) **a.** Which property is illustrated by this equation?

$$\frac{12}{0.5} \cdot \frac{10}{10} = \frac{120}{5}$$

b. Find the quotient of $\frac{120}{5}$.

Analyze Solve:

* **15.** $\frac{8}{10} = \frac{w}{25}$ (15)

* **16.** $\frac{n}{1.5} = \frac{6}{9}$ (39)

* **17.** $\frac{9}{12} = \frac{15}{m}$ (39)

* **18.** $4 = a + 1.8$ (39)

* **19.** $3.9 = t - 0.39$ (35)

Simplify:

20. 1.2 m − 12 cm = ____ m (34)

* **21.** $(0.15)(0.05)$ (35)

* **22.** 15×1.5 (35)

* **23.** $14.4 \div 12$ (35)

* **24.** $5.6 - (4 - 1.25)$ (35)

* **25.** $5 - (3.14 + 1.2)$ (35)

26. $6\frac{1}{4} \cdot 1\frac{3}{5}$ (26)

27. $7 \div 5\frac{5}{6}$ (26)

28. $\frac{8}{15} + \frac{12}{25}$ (30)

29. $4\frac{2}{5} - 1\frac{3}{4}$ (23, 30)

* **30.** (29, 33) **Estimate** The perimeter of a square room is 15.84 m. Estimate the area of the room. How did you make your estimate?

Early Finishers

Real-World Application

To get to school, Eliza walks east to the corner of her street and then turns north 90° to the left. When she reaches the school, Eliza realizes that she forgot her lunch and turns southwest 144° to the left to face her house. She then walks straight toward her house by cutting across the soccer field. Sketch a triangle to represent Eliza's route to school and back home. Label the angles at each turn in her route.

What angle does Eliza's shortcut make with her original route to school?

INVESTIGATION 4

Focus on

• Stem-and-Leaf Plots, Box-and-Whisker Plots

A high school counselor administered a math test to eighth-grade students in local middle schools to help advise students during high school registration. The scores of one group of students are listed below.

40, 30, 43, 48, 26, 50, 55, 40, 34, 42, 47, 47,
52, 25, 32, 38, 41, 36, 32, 21, 35, 43, 51, 58,
26, 30, 41, 45, 23, 36, 41, 51, 53, 39, 28

To organize the scores, the counselor created a **stem-and-leaf plot.** Noticing that the scores ranged from a low of 21 to a high of 58, the counselor chose the initial digits of 20, 30, 40, and 50 to serve as the stem digits.

Stem
2
3
4
5

Then the counselor used the ones place digits of the scores as the leaves of the stem-and-leaf plot. (The "stem" of a stem-and-leaf plot may have more than one digit. Each "leaf" is only one digit.)

3 | 2 represents a score of 32

Stem	Leaf
2	1 3 5 6 6 8
3	0 0 2 2 4 5 6 6 8 9
4	0 0 1 1 1 2 3 3 5 7 7 8
5	0 1 1 2 3 5 8

The counselor included a key to the left to help a reader interpret the plot. The top row of leaves indicates the six scores 21, 23, 25, 26, 26, and 28.

1. **Analyze** Looking at this stem-and-leaf plot, we see that there is one score of 21, one 23, one 25, two 26's, and so on. Scanning through all of the scores, which score occurs more than twice?

Thinking Skill
Justify
Could a data set have more than one mode? Explain your reasoning.

The number that occurs most frequently in a set of numbers is the **mode.** The mode of these 35 scores is 41.

2. Looking at the plot, we immediately see that the lowest score is 21 and the highest score is 58. What is the difference of these scores?

The difference between the least and greatest numbers in a set is the **range** of the numbers. We find the range of this set of scores by subtracting 21 from 58.

3. Analyze The **median** of a set of numbers is the middle number of the set when the numbers are arranged in order. The counselor drew a vertical segment through the median on the stem-and-leaf plot. Which score was the median score?

Half of the scores are at or below the median score, and half of the scores are at or above the median score. There are 35 scores and half of 35 is $17\frac{1}{2}$. This means there are 17 whole scores below the median and 17 whole scores above the median. The $\frac{1}{2}$ means that the median is one of the scores on the list. (The median of an even number of scores is the mean—the average—of the two middle scores.) We may count 17 scores up from the lowest score or 17 scores down from the highest score. The next score is the median. We find that the median score is 40.

3 | 2 represents a score of 32

Stem	Leaf
2	1 3 5 6 6 8
3	0 0 2 2 4 5 6 6 8 9
4	0 0 1 1 1 2 3 3 5 7 7 8
5	0 1 1 2 3 5 8

median (second 0 in stem 4)

Next the counselor found the middle number of the lower 17 scores and the middle number of the upper 17 scores. The middle number of the lower half of scores is the **first quartile** or **lower quartile.** The middle number of the upper half of scores is the **third quartile** or **upper quartile.** The second quartile is the median.

4. Verify What are the first and third quartiles of these 35 scores?

There are 17 scores below the median. Half of 17 is $8\frac{1}{2}$. We count up 8 whole scores from the lowest score. The next score is the lower quartile score. Likewise, since there are also 17 scores above the median, we count down 8 whole scores from the highest score. The next score is the upper quartile score. Note that if there is an even number of numbers above and below the median, the quartiles are the mean of the two central numbers in each half.

3 | 2 represents a score of 32

Stem	Leaf
2	1 3 5 6 6 8
3	0 0 2 2 4 5 6 6 8 9
4	0 0 1 1 1 2 3 3 5 7 7 8
5	0 1 1 2 3 5 8

lower quartile (first 2 in stem 3); median (second 0 in stem 4); upper quartile (second 7 in stem 4)

We count 8 scores below the first quartile, 8 scores between the first quartile and the median, 8 scores between the median and the third quartile, and 8 scores above the third quartile. The median and quartiles have "quartered" the scores.

After locating the median and quartiles, the counselor created a **box-and-whisker plot** of the scores, which shows the location of certain scores compared to a number line. The five dots on a box-and-whisker plot show the **extremes** of the scores—the lowest score and highest score—as well as the lower quartile, median, and upper quartile. A *box* that is split at the median shows the location of the middle half of the scores. The *whiskers* show the scores below the first quartile and above the third quartile.

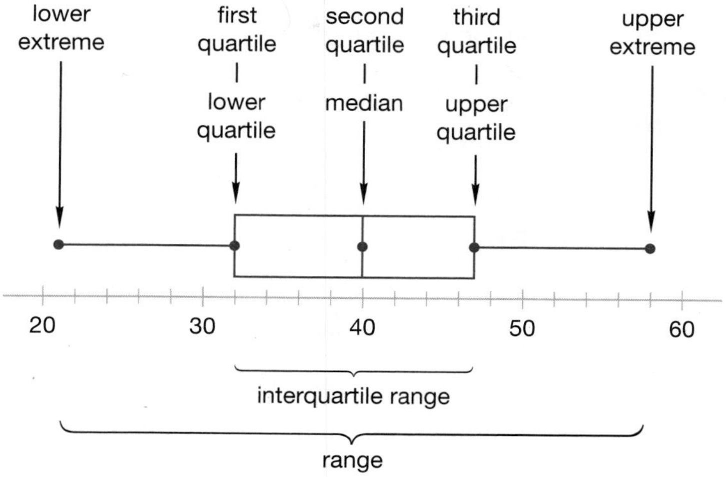

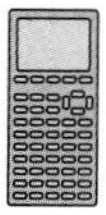

Visit www.SaxonPublishers.com/ActivitiesC2 *for a graphing calculator activity.*

5. **Represent** Create a stem-and-leaf plot for the following set of scores. Then draw vertical segments on the plot to indicate the median and the first and third quartiles.

   15, 26, 26, 27, 28, 29, 29, 30, 32, 33, 35, 36, 38, 38, 38, 38, 40, 41, 42, 43, 45, 45, 46, 47, 47, 48, 50, 52, 54, 55, 57, 58

6. **Discuss** What is the lower quartile, median, and upper quartile of this set of scores? Explain how to find the median of this set of data.

7. What is the mode of this set of scores?

8. What are the upper and lower extremes of these scores?

9. What is the range of the scores?

10. The **interquartile range** is the difference between the upper and lower quartiles. What is the interquartile range of these scores?

11. Create a box-and-whisker plot for this set of scores by using the calculations you have made for the median and the quartiles.

12. **Analyze** An **outlier** is a number in a set of numbers that is distant from the other numbers in the set. In this set of scores there is an outlier. Which score is the outlier?

extension

Create a stem-and-leaf plot and a box-and-whisker plot for the number of students in all the classes in your school.

LESSON

41

• Using Formulas
• Distributive Property

Power Up

Building Power

facts

Power Up I

mental math

a. **Number Sense:** 5×140

b. **Decimals:** 1.54×10

c. **Equivalent Fractions:** $\frac{3}{5} = \frac{15}{x}$

d. **Exponents:** $5^2 - 4^2$

e. **Estimation:** 39×29

f. **Fractional Parts:** $\frac{3}{10}$ of 70

g. **Statistics:** Explain what the mode is in a set of numbers.

h. **Calculation:** Find the sum, difference, product, and quotient of $\frac{2}{3}$ and $\frac{1}{2}$.

problem solving

Use simpler, but similar problems to find the quotient of 1 divided by 50,000,000,000,000.

New Concepts

Increasing Knowledge

using formulas

In Lesson 20 it was stated that the area (A) of a rectangle is related to the length (l) and width (w) of the rectangle by this formula:

$$A = lw$$

This formula means "the area of a rectangle equals the product of its length and width." If we are given measures for l and w, we can replace the letters in the formula with numbers and calculate the area.

Example 1

Find A in $A = lw$ when l is 8 ft and w is 4 ft.

Solution

We replace l and w in the formula with 8 ft and 4 ft respectively. Then we simplify.

$$A = lw$$

$$A = (8\text{ ft})(4\text{ ft})$$

$$A = \mathbf{32\ ft^2}$$

Notice the effect on the units when the calculation is performed. Multiplying two units of length results in a unit of area.

Example 2

Evaluate $2(l + w)$ when l is 8 cm and w is 4 cm.

Solution

In place of l and w we substitute 8 cm and 4 cm. Then we simplify.

$$2(l + w)$$

$$2(8 \text{ cm} + 4 \text{ cm})$$

$$2(12 \text{ cm})$$

$$\mathbf{24 \text{ cm}}$$

Activity

Perimeter Formulas

Kurt finds the perimeter of a rectangle by doubling the length, doubling the width, and then adding the two numbers.

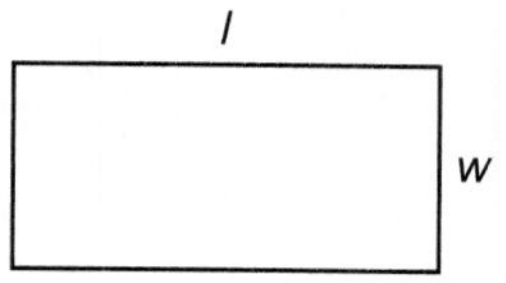

Molly finds the perimeter by adding the length and width, then doubling that number.

Write a formula for Kurt's method and another formula for Molly's method.

distributive property

There are two formulas commonly used to relate the perimeter (p) of a rectangle to its length and width.

$$p = 2(l + w)$$

$$p = 2l + 2w$$

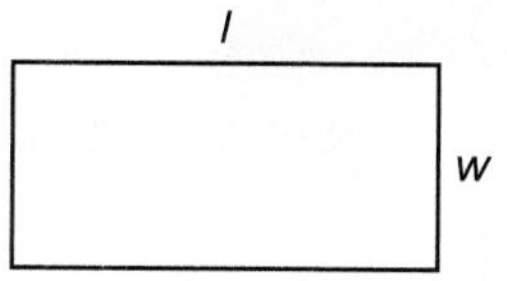

Both formulas describe how to find the perimeter of a rectangle if we are given its length and width. The first formula means "add the length and width and then double this sum." The second formula means "double the length and double the width and then add."

Example 3

Use the two perimeter formulas to find the perimeter of this rectangle.

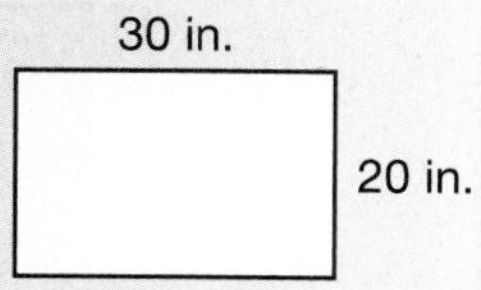

Solution

In both formulas we replace l with 30 in. and w with 20 in. Then we simplify.

$p = 2(l + w)$	$p = 2l + 2w$
$p = 2(30 \text{ in.} + 20 \text{ in.})$	$p = 2(30 \text{ in.}) + 2(20 \text{ in.})$
$p = 2(50 \text{ in.})$	$p = 60 \text{ in.} + 40 \text{ in.}$
$p =$ **100 in.**	$p =$ **100 in.**

Thinking Skill

Formulate

Write an equivalent formula for the perimeter of a rectangle using only variables.

Both formulas in example 3 yield the same result because the two formulas are equivalent.

$$2(l + w) = 2l + 2w$$

Conclude Evaluate $2(l + w)$ and $2l + 2w$ when l is 9 cm and w is 5 cm. Are the results equal?

These equivalent expressions illustrate the **Distributive Property of Multiplication Over Addition,** often called simply the **Distributive Property.** Applying the Distributive Property, we distribute, or "spread," the multiplication over the terms that are being added (or subtracted) within the parentheses. In this case we multiply l by 2, giving us $2l$, and we multiply w by 2, giving us $2w$.

$$2(l + w) = 2l + 2w$$

The Distributive Property is often expressed in equation form using variables:

$$a(b + c) = ab + ac$$

The Distributive Property also applies over subtraction.

$$a(b - c) = ab - ac$$

Example 4

Show two ways to simplify this expression:

$$\mathbf{6(20 + 5)}$$

Solution

One way is to add 20 and 5 and then multiply the sum by 6.

$$\mathbf{6(20 + 5)}$$

$$\mathbf{6(25)}$$

$$\mathbf{150}$$

Another way is to multiply 20 by 6 and multiply 5 by 6. Then add the products.

$$\mathbf{6(20 + 5)}$$

$$\mathbf{(6 \cdot 20) + (6 \cdot 5)}$$

$$\mathbf{120 + 30}$$

$$\mathbf{150}$$

Example 5

Simplify: $2(3 + n) + 4$

Solution

We show and justify each step.

$2(3 + n) + 4$	**Given**
$2 \cdot 3 + 2n + 4$	**Distributive Property**
$6 + 2n + 4$	$2 \cdot 3 = 6$
$2n + 6 + 4$	**Commutative Property**
$2n + (6 + 4)$	**Associative Property**
$2n + 10$	$6 + 4 = 10$

Practice Set

a. Connect Find A in $A = bh$ when b is 15 in. and h is 8 in.

b. Evaluate $\frac{ab}{2}$ when a is 6 ft and b is 8 ft.

c. Formulate Write an equation using the letters x, y, and z that illustrates the Distributive Property of Multiplication Over Addition.

d. Analyze Show two ways to simplify this expression:

$$6(20 - 5)$$

e. Write two formulas for finding the perimeter of a rectangle.

f. Describe two ways to simplify this expression:

$$2(6 + 4)$$

g. Simplify: $2(n + 5)$. Show and justify the steps.

Written Practice

Strengthening Concepts

1. (36) Two hundred wildebeests and 150 gazelles grazed on the savannah. What was the ratio of gazelles to wildebeests grazing on the savannah?

2. (28) In its first 5 games, the local basketball team scored 105 points, 112 points, 98 points, 113 points, and 107 points. What was the average number of points the team scored in its first 5 games?

3. (28) The crowd watched with anticipation as the pole vault bar was set to 19 feet 6 inches. How many inches is 19 feet 6 inches?

*** 4.** (2, 41) Analyze Which property is illustrated by each of these equations?

a. $(a + b) + c = a + (b + c)$

b. $a(bc) = (ab)c$

c. $a(b + c) = ab + ac$

*** 5.** (35) Model Draw a sketch to help with this problem. From Tracey's house to John's house is 0.5 kilometers. From John's house to school is 0.8 kilometer. Tracey rode from her house to John's house and then to school. Later she rode from school to John's house to her house. Altogether, how far did Tracey ride?

*** 6.** About 70% of the earth's surface is water.
(36)

a. About what fraction of the earth's surface is land?

b. On the earth's surface, what is the ratio of water area to land area?

The stem-and-leaf plot below shows the distribution of numbers of stamps collected by 20 stamp club members. Refer to the stem-and-leaf plot to answer problems **7** and **8.**

2 | 4 represents 24 stamps

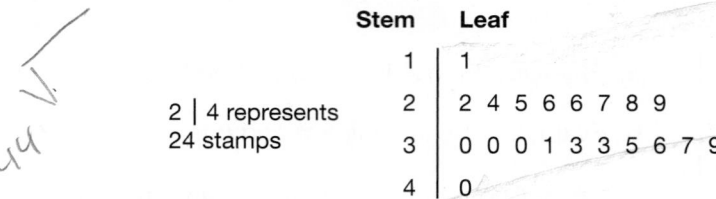

Stem	Leaf
1	1
2	2 4 5 6 6 7 8 9
3	0 0 0 1 3 3 5 6 7 9
4	0

*** 7.** For the number of stamps collected, find
(Inv. 4)

a. the median.

b. the first quartile.

c. the third quartile.

d. any outliers.

*** 8.** **Represent** Make a box-and-whisker plot of the stamps collected in the stem-and-leaf plot.
(Inv. 4)

*** 9.** **Analyze** Refer to the figure at right to answer **a** and **b.** All angles are right angles. Dimensions are in feet.
(19, 37)

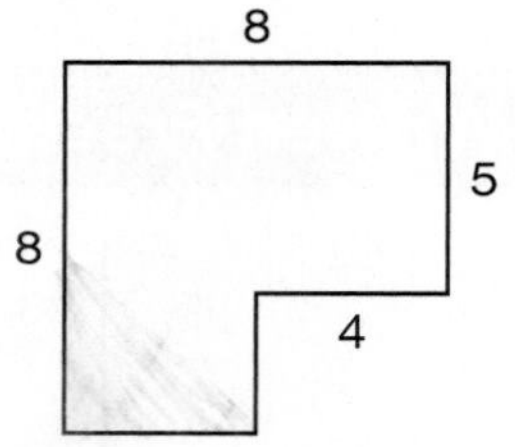

a. What is the area of the figure?

b. What is the perimeter of the figure?

10. Name the point marked M on this number line:
(34)

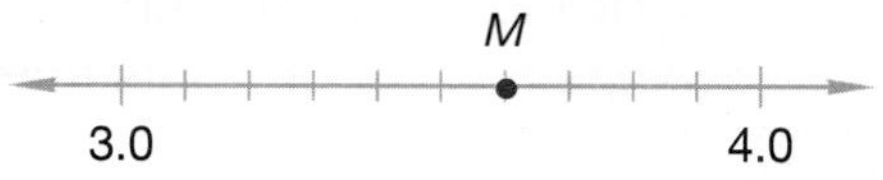

a. as a decimal number.

b. as a mixed number.

Math Language

A **prime number** is a counting number greater than 1 whose only factors are 1 and the number itself.

11. What is the sum of the first four prime numbers?
(21)

12. Dimensions of the triangle at right are in millimeters.
(37)

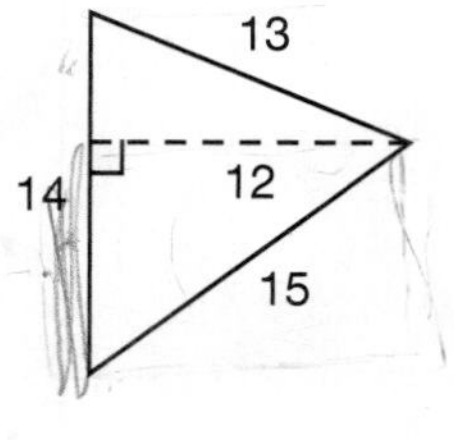

a. What is the perimeter of the triangle?

b. What is the area of the triangle?

13. **Connect** Use digits to write each number:
(31)

a. sixty-seven hundred-thousandths

b. one hundred and twenty-three thousandths

Reading Math

The symbol π stands for ***pi*** and is pronounced "pie."

* **14.** (41) **Conclude** Evaluate $2\pi r$ when π is 3.14 and r is 10.

15. (30) Write $\frac{3}{5}$, $\frac{1}{2}$, and $\frac{5}{7}$ with a common denominator, and arrange the renamed fractions in order from least to greatest.

16. (29, 35) First estimate the area of the rectangle at right. Then calculate its area.

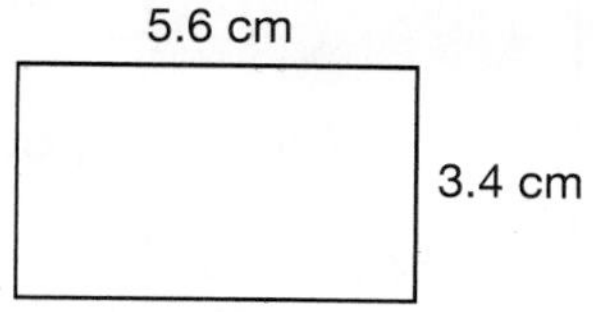

Solve:

* **17.** (39) $\frac{x}{2.4} = \frac{10}{16}$

* **18.** (39) $\frac{18}{8} = \frac{m}{20}$

19. (35) $3.45 + a = 7.6$

20. (35) $3y = 0.144$

Simplify:

21. (35) $7.4 \div 8$

22. (35) $(0.4)(0.6)(0.02)$

23. (35) $4.315 \div 5$

24. (35) $\frac{6.5}{100}$

25. (30) $3\frac{1}{3} + 1\frac{5}{6} + \frac{7}{12}$

26. (23, 30) $4\frac{1}{6} - \left(4 - 1\frac{1}{4}\right)$

27. (26) $3\frac{1}{5} \cdot 2\frac{5}{8} \cdot 1\frac{3}{7}$

28. (26) $4\frac{1}{2} \div 6$

* **29.** (41) **a.** **Discuss** Compare: $(12 \cdot 7) + (12 \cdot 13)$ ◯ $12(7 + 13)$

b. Which property of operations is illustrated by this comparison?

* **30.** (40) **Conclude** Find the measures of $\angle x$, $\angle y$, and $\angle z$ in this figure:

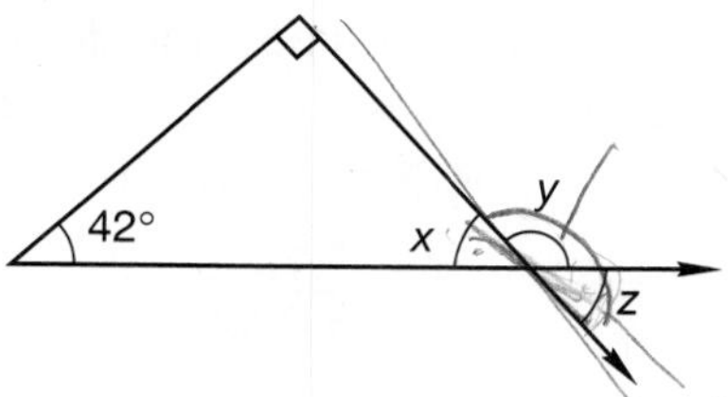

Early Finishers

Real-World Application

At an online auction site, a popular 1960's record album is listed for sale. The ten highest bids are shown below.

\$200 \$250 \$210 \$200 \$180 \$200 \$230 \$285 \$220 \$285

Which display—a stem-and-leaf plot or a line plot—is the most appropriate way to display this data? Draw the display and justify your choice.

LESSON

42

• Repeating Decimals

Power Up

Building Power

facts Power Up H

mental math

a. Calculation: 3 × 78¢

b. Decimals: 0.4 × 100

c. Equivalent Fractions: $\frac{4}{y} = \frac{20}{25}$

d. Power/Roots: $\sqrt{121} - 3^2$

e. Estimation: $1\frac{7}{8} \times 3\frac{1}{8}$

f. Fractional Parts: $\frac{4}{5}$ of 35

g. Statistics: Explain how to find the average of a set of numbers.

h. Calculation: Start with three score and 10, + 2, ÷ 8, $\sqrt{\ }$, × 5, + 1, $\sqrt{\ }$, + 1, square that number.

problem solving

Darla noticed that each of the numbers in this sequence is a prime number: 31; 331; 3,331; 33,331; 333,331; 3,333,331; 33,333,331; . . . Does the pattern of prime numbers continue?

New Concept

Increasing Knowledge

When a decimal number is divided, the division sometimes ends with a remainder of zero. The answer that results is called a **terminating decimal.** However, at other times, the division will not end with a remainder of zero. Instead the answer will have one or more digits in a pattern that repeats indefinitely. Here we show two examples:

$$\begin{array}{r} 7.1666\ldots \\ 6\overline{)43.0000\ldots} \\ \underline{42} \\ 1\,0 \\ \underline{6} \\ 40 \\ \underline{36} \\ 40 \\ \underline{36} \\ 40 \\ \underline{36} \\ 4 \end{array} \qquad \begin{array}{r} 0.31818\ldots \\ 11\overline{)3.50000\ldots} \\ \underline{3\,3} \\ 20 \\ \underline{11} \\ 90 \\ \underline{88} \\ 20 \\ \underline{11} \\ 90 \\ \underline{88} \\ 2 \end{array}$$

Discuss If we have to add terminal zeros to the dividend to carry out the division, will the answer always continue indefinitely?

The repeating digits of a decimal number are called the **repetend.** In 7.1666..., the repetend is 6. In 0.31818..., the repetend is 18 (not 81). One way to indicate that a decimal number has repeating digits is to write the number with a bar over the repetend where it first appears to the right of the decimal point. For example,

$$7.1666\ldots = 7.1\overline{6} \qquad 0.31818\ldots = 0.3\overline{18}$$

Example 1

Rewrite each of these repeating decimals with a bar over the repetend:

a. 0.0833333...

b. 5.14285714285714...

c. 454.5454545...

Solution

a. The repeating digit is 3.

$$\mathbf{0.08\overline{3}}$$

b. This is a six-digit repeating pattern.

$$\mathbf{5.\overline{142857}}$$

c. The repetend is always to the right of the decimal point. We do not write a bar over a whole number.

$$\mathbf{454.\overline{54}}$$

Example 2

Round each number to five decimal places:

a. $\mathbf{5.31\overline{6}}$ **b.** $\mathbf{25.\overline{405}}$

Solution

a. We remove the bar and write the repeating digits to the right of the desired decimal place.

$$5.31\overline{6} = 5.316666\ldots$$

Then we round to five places.

$$5.3166\underline{6}⑥\ldots \longrightarrow \mathbf{5.31667}$$

b. We remove the bar and continue the repeating pattern beyond the fifth decimal place.

$$25.\overline{405} \longrightarrow 25.405405\ldots$$

Then we round to five places.

$$25.4054\underline{0}⑤\ldots \longrightarrow \mathbf{25.40541}$$

Example 3

a. Compare: 0.3 ◯ $\mathbf{0.\overline{3}}$

b. Arrange in order from least to greatest: $\mathbf{0.\overline{6}}$, 0.6, 0.65

Solution

a. Although the digits in the tenths place are the same for both numbers, the digits in the hundredths place differ.

$$0.3 = 0.30 \qquad 0.\overline{3} = 033\ldots$$

Therefore, $\mathbf{0.3 < 0.\overline{3}}$

b. We write each number with two decimal places to make the comparison easier.

$$0.6\overline{6} \quad 0.60 \quad 0.65$$

We can now see how to arrange the numbers in order.

$$\mathbf{0.6, 0.65, 0.\overline{6}}$$

Example 4

Divide 1.5 by 11 and write the quotient

a. with a bar over the repetend.

b. rounded to the nearest hundredth.

Solution

Thinking Skill

Discuss

Why did we stop dividing at 0.13636? Would dividing again change the rounded answer?

a. Since place value is fixed by the decimal point, we can write zeros in the "empty" places to the right of the decimal point. We continue dividing until the repeating pattern is apparent. The repetend is 36 (not 63). We write the quotient with a bar over 36 where it first appears.

$$0.13636\ldots = \mathbf{0.1\overline{36}}$$

```
      0.13636 ...
11 ) 1.50000 ...
     1 1
       40
       33
        70
        66
         40
         33
          70
          66
           4
```

b. The hundredths place is the second place to the right of the decimal point.

0.13(6)36... ⟶ **0.14**

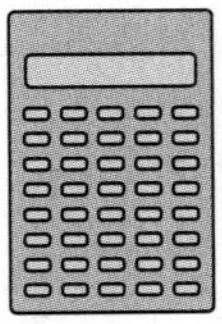

When a division problem is entered into a calculator and the display is filled with a decimal number, it is likely that the quotient is a repeating decimal. However, since a calculator either truncates (cuts off) or rounds the number displayed, the repetend may not be obvious. For example, to convert the fraction $\frac{1}{7}$ to a decimal, we enter

1 ÷ 7 =

An eight-digit display shows

0.1428571

We might wonder whether the final digit, 1, is the beginning of another 142857 pattern. We can get a peek at the next digit by shifting the digits

that are displayed one place to the left. We can do this by multiplying the numerator by 10 and dividing again. This time we divide 10 by 7.

1 0 ÷ 7 =

Math Language

Generalize

How would we write the answer to 1 ÷ 7 with a bar over the repetend?

The display shows

1.4285714

Seeing the final digit, 4, following the 1 increases the likelihood that the 142857 pattern is repeating.

Verify Why does multiplying the numerator by 10 help us get a better idea of whether the pattern repeats?

Practice Set

Write each repeating decimal with a bar over the repetend:

a. 2.72727...

b. 0.816666...

Justify Round each number to the thousandths place. Describe how you got your answer.

c. $0.\overline{6}$

d. $5.3\overline{81}$

e. Compare: $0.6 \bigcirc 0.\overline{6}$

f. Arrange in order from least to greatest: $1.3, 1.35, 1.\overline{3}$

Generalize Divide 1.7 by 12 and write the quotient

g. with a bar over the repetend.

h. rounded to four decimal places.

Written Practice

Strengthening Concepts

1. (36) Two-fifths of the photographs are black and white. The rest of the photographs are color. What is the ratio of black and white to color photographs?

2. (13, 28) Four hundred thirty-two students were assigned to 16 classrooms. What was the average number of students per classroom?

3. (13) The American Redstart, a species of bird, migrates between Central America and North America each year. Suppose that a migrating American Redstart traveled for 7 days at an average rate of 33 miles per day. How many miles did the bird travel during this period?

4. (22) Diagram this statement. Then answer the questions that follow.

Seven ninths of the 450 students in the assembly were entertained by the speaker.

a. How many students were entertained?

b. How many students were not entertained?

* **5.** (42) Round each number to four decimal places:

a. $5.1\overline{6}$

b. $5.\overline{27}$

6. (38) Refer to the circle graph below to answer **a** and **b.**

Student Movie Preferences

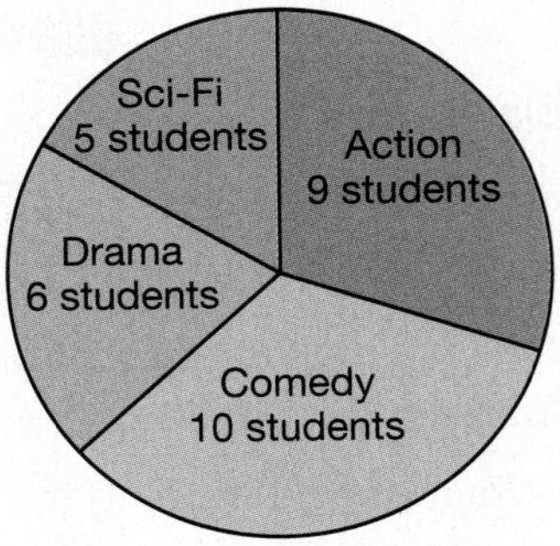

a. How many students does the graph represent in all?

b. How many more students preferred action or comedy movies than preferred drama or sci-fi movies?

c. What fraction of the students prefer comedies?

d. What fraction of the students in the class preferred action movies?

*** 7.** (Inv. 3, 37) *Analyze* The coordinates of the vertices of a triangle are (−6, 0), (0, −6), and (0, 0). What is the area of the triangle?

*** 8.** (19, 37) *Conclude* All angles in the figure are right angles. Dimensions are in inches.

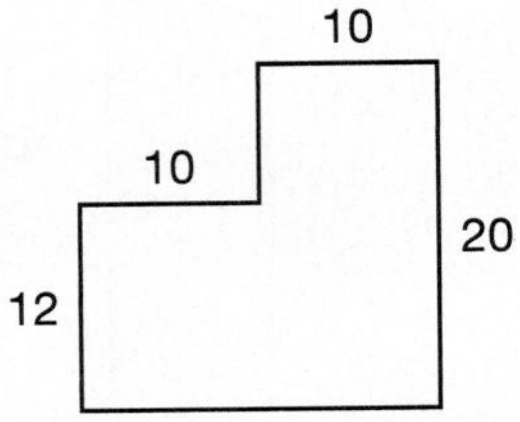

a. Find the perimeter of the figure.

b. Find the area of the figure.

*** 9.** (42) *Generalize* Divide 1.7 by 11 and write the quotient

a. with a bar over the repetend.

b. rounded to three decimal places.

c. In exercise **b,** what place value did you round to?

10. (31, 35) Use digits to write the sum of the decimal numbers twenty-seven thousandths and fifty-eight hundredths.

11. (21, 36) Ted has two spinners. He spins each spinner once.

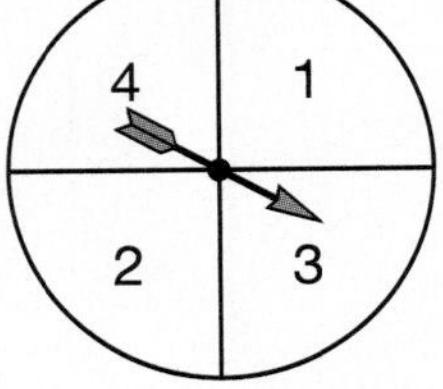

a. Draw a tree diagram to find the sample space for the experiment.

b. What is the probability of getting a consonant and a composite number?

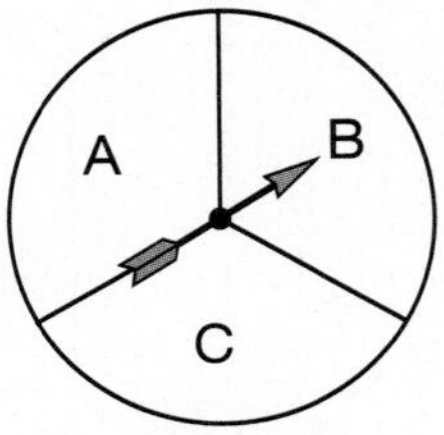

Thinking Skill

Classify

In English, the letters a, e, i, o, and u are vowels. All of the other letters are consonants.

12. a. Make a factor tree showing the prime factorization of 7200. (Start with the factors 72 and 100.)
(20, 21)

b. Write the prime factorization of 7200 using exponents.

13. Use a protractor and a ruler to draw a triangle with three 60° angles and sides 5 cm long.
(17)

14. What is the least common multiple of 12 and 15?
(27)

Analyze Solve:

* 15. $\frac{21}{24} = \frac{w}{40}$
(39)

* 16. $\frac{1.2}{x} = \frac{9}{6}$
(39)

17. $m + 9.6 = 14$
(35)

18. $n - 4.2 = 1.63$
(35)

* 19. **Conclude** Evaluate $\frac{1}{2}bh$ when $b = 12$ and $h = 10$.
(41)

* 20. **Analyze** Show two ways to simplify this expression:
(41)

$$4(5 + 6)$$

21. Justify the steps to simplify $\frac{5}{6}\left(\frac{6}{5} + \frac{6}{7}\right)$.
(9, 41)

$\frac{5}{6}\left(\frac{6}{5} + \frac{6}{7}\right)$	Given
$\frac{5}{6} \cdot \frac{6}{5} + \frac{5}{6} \cdot \frac{6}{7}$	a. ______________
$1 + \frac{5}{6} \cdot \frac{6}{7}$	b. ______________
$1 + \frac{5}{7}$	$\frac{5}{6} \cdot \frac{6}{7} = \frac{5}{7}$
$1\frac{5}{7}$	$1 + \frac{5}{7} = 1\frac{5}{7}$

22. The price of an item is $4.56 and the tax rate is 8%. Find the tax on the item by multiplying $4.56 by 0.08 and round the product to the nearest cent.
(27)

23. Estimate the quotient of 23.8 ÷ 5.975 by rounding each number to the nearest whole number before dividing.
(33)

24. What are the missing words in the following sentence?
(Inv. 2)

The longest chord of a circle is the ___a___*, which is twice the length of the* ___b___.

Simplify:

25. $7.1 \div 4$
(35)

26. $6\frac{1}{4} + 5\frac{5}{12} + \frac{2}{3}$
(30)

27. $4 - \left(4\frac{1}{6} - 1\frac{1}{4}\right)$
(23, 30)

28. $6\frac{2}{5} \cdot 2\frac{5}{8} \cdot 2\frac{6}{7}$
(26)

29. (26) Before dividing, determine whether the quotient is greater than or less than 1 and state why. Then perform the calculation.

$$6 \div 4\frac{1}{2}$$

*** 30.** (40) **Conclude** Find the measures of $\angle a$, $\angle b$, and $\angle c$ in this figure: Then classify any angle pairs as supplementary or complementary:

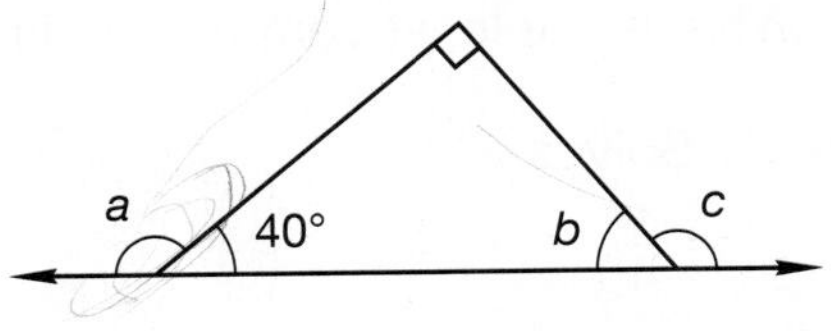

Early Finishers
Real-World Application

Mario wants to buy some running shoes that are on sale for $77.52 including tax. Mario has $135.66 to spend on clothes and shoes.

a. If he buys the shoes he wants, what fraction of his money would Mario spend?

b. If he spends $\frac{1}{3}$ of the remaining money on a jacket, how much money will he have left?

LESSON

43

• Converting Decimals to Fractions
• Converting Fractions to Decimals
• Converting Percents to Decimals

Power Up

Building Power

facts Power Up I

mental math

a. Calculation: $6 \times 48¢$

b. Decimals: $3.5 \div 100$

c. Equivalent Fractions: $\frac{n}{4} = \frac{21}{12}$

d. Power/Roots: $7^2 - \sqrt{100}$

e. Estimation: $\$9.95 \times 6$

f. Fractional Parts: $\frac{1}{5}$ of 300

g. Statistics: Find the range of the set of numbers: 34, 99, 23, 78.

h. Calculation: Find the sum, difference, product, and quotient of $\frac{2}{3}$ and $\frac{1}{4}$.

problem solving

Mavy sees three faces on each of three number cubes, for a total of nine faces. If the sum of the dots on the three faces of each number cube is different, and Mavy sees forty dots altogether, then which faces must be visible on each number cube?

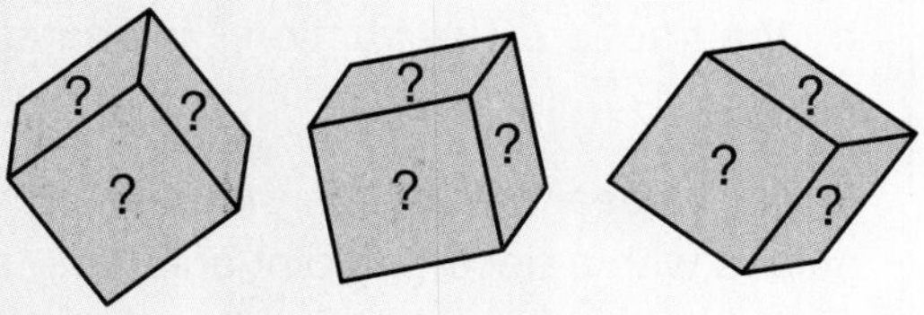

New Concepts

Increasing Knowledge

converting decimals to fractions

To write a decimal number as a fraction, we write the digits after the decimal point as the numerator of the fraction. For the denominator of the fraction, we write the place value of the last digit. Then we reduce.

Example 1

Write 0.125 as a fraction.

Solution

Thinking Skill

Generalize

What denominators would you use to express 0.3 and 0.24 as fractions? Explain why.

The digits 125 form the numerator of the fraction. The denominator of the fraction is 1000 because 5, the last digit, is in the thousandths place.

$$0.125 = \frac{125}{1000}$$

Notice that the denominator of the fraction has as many zeros as the decimal number has decimal places. Now we reduce.

$$\frac{125}{1000} = \mathbf{\frac{1}{8}}$$

Explain How can we reduce $\frac{125}{1000}$ to $\frac{1}{8}$ using the GCF of 125 and 1000?

Example 2

Mathea rode her bike 11.45 miles to the lake. Write 11.45 as a mixed number.

Solution

Visit www.SaxonPublishers.com/ActivitiesC2 *for a graphing calculator activity.*

The number 11 is the whole-number part. The numerator of the fraction is 45, and the denominator is 100 because 5 is in the hundredths place.

$$11.45 = 11\frac{45}{100}$$

Now we reduce the fraction.

$$11\frac{45}{100} = \mathbf{11\frac{9}{20}}$$

converting fractions to decimals

To change a fraction to a decimal number, we perform the division indicated by the fraction. The fraction $\frac{1}{4}$ indicates that 1 is divided by 4.

$$4\overline{)1}$$

It may appear that we cannot perform this division. However, if we fix place values with a decimal point and write zeros in the decimal places to the right of the decimal point, we can perform the division.

$$\begin{array}{r} 0.25 \\ 4\overline{)1.00} \\ \underline{8} \\ 20 \\ \underline{20} \\ 0 \end{array} \qquad \text{Thus } \frac{1}{4} = 0.25.$$

Some fractions convert to repeating decimals. We convert $\frac{1}{3}$ to a decimal by dividing 1 by 3.

$$\begin{array}{r} 0.33... \\ 3\overline{)1.00...} \\ \underline{9} \\ 10 \\ \underline{9} \\ 1 \end{array} \qquad \text{Thus } \frac{1}{3} = 0.\overline{3}.$$

Every fraction of whole numbers converts to either a terminating decimal (like 0.25) or a repeating decimal (like $0.\overline{3}$).

Example 3

Find the probability of the spinner stopping on 5. Express the probability as a decimal number.

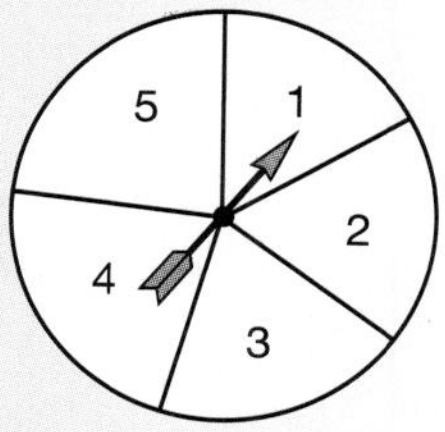

Solution

Since there are five equally likely outcomes, the probability of the spinner stopping on 5 is $\frac{1}{5}$. We convert $\frac{1}{5}$ to a decimal by dividing 1 by 5.

$$\begin{array}{r} .2 \\ 5\overline{)1.0} \\ \underline{1\,0} \\ 0 \end{array}$$

We find that P(5) is **0.2.**

Example 4

Write each of these numbers as a decimal number:

a. $\frac{23}{100}$ **b.** $\frac{7}{4}$ **c.** $3\frac{4}{5}$ **d.** $\frac{2}{3}$

Solution

a. Fractions with denominators of 10, 100, 1000, and so on can be written directly as decimal numbers, without dividing. The decimal part will have the same number of places as the number of zeros in the denominator.

$$\frac{23}{100} = \mathbf{0.23}$$

b. An improper fraction is equal to or greater than 1. When we change an improper fraction to a decimal number, the decimal number will be greater than or equal to 1.

$$\frac{7}{4} \longrightarrow \begin{array}{r} 1.75 \\ 4\overline{)7.00} \\ \underline{4} \\ 3\,0 \\ \underline{2\,8} \\ 20 \\ \underline{20} \\ 0 \end{array} \qquad \frac{7}{4} = \mathbf{1.75}$$

c. To change a mixed number to a decimal number, we can change the mixed number to an improper fraction and then divide. Another way is to separate the fraction from the whole number and change the fraction to a decimal number. Then we write the whole number and the decimal number as one number. Here we show both ways.

$$3\frac{4}{5} = \frac{19}{5} \quad \text{or} \quad 3\frac{4}{5} = 3 + \frac{4}{5}$$

$$\begin{array}{r} 3.8 \\ 5\overline{)19.0} \\ \underline{15} \\ 4\,0 \\ \underline{4\,0} \\ 0 \end{array} \qquad \begin{array}{r} 0.8 \\ 5\overline{)4.0} \\ \underline{4\,0} \\ 0 \end{array}$$

$$3\frac{4}{5} = \mathbf{3.8} \qquad 3\frac{4}{5} = \mathbf{3.8}$$

d. To change $\frac{2}{3}$ to a decimal number, we divide.

$$\frac{2}{3} \longrightarrow \begin{array}{r} 0.666... \\ 3\overline{)2.000...} \\ \underline{1\,8} \\ 20 \\ \underline{18} \\ 20 \\ \underline{18} \\ 2 \end{array} \qquad \frac{2}{3} = \mathbf{0.\overline{6}}$$

Thinking Skill

Summarize

When a fraction is equivalent to a repeating decimal, write a bar over the repetend unless directed otherwise.

Example 5

Compare: $\mathbf{\frac{3}{10} \bigcirc 0.\overline{3}}$

Solution

Since $\frac{3}{10}$ equals 0.3, we can compare

$$0.3 \text{ and } 0.\overline{3}$$

Here we show the next digits: 0.30 and $0.3\overline{3}$

Since $0.\overline{3}$ is greater than 0.3, we know that $\mathbf{\frac{3}{10} < 0.\overline{3}}$.

converting percents to decimals

Recall that *percent* means "per hundred" or "hundredths." So 75% means 75 hundredths, which can be written as a fraction or as a decimal.

$$75\% = \frac{75}{100} = 0.75$$

Likewise, 5% means 5 hundredths.

$$5\% = \frac{5}{100} = 0.05$$

We see that a percent may be written as a decimal using the same digits but with the decimal point shifted two places to the left.

Example 6

Write each percent as a decimal number:

a. **25%** **b.** **125%** **c.** **2.5%**

d. **50%** **e.** **$7\frac{1}{2}$%** **f.** **300%**

Solution

a. **0.25** **b.** **1.25** **c.** **0.025**

d. 0.50 = **0.5** **e.** 7.5% = **0.075** **f.** 300% = **3**

Many scientific calculators do not have a percent key. Designers of these calculators assume the user will mentally convert percents to decimals before entering the calculation.

If your calculator does have a percent key, you may find the decimal equivalent of a percent by entering [1] [×] the percent. For example, enter

[1] [×] [2] [5] [%]

The calculator displays the decimal equivalent 0.25.

Practice Set

Connect Change each decimal number to a reduced fraction or to a mixed number:

a. 0.24 **b.** 45.6 **c.** 2.375

Change each fraction or mixed number to a decimal number:

d. $\frac{23}{4}$ **e.** $4\frac{3}{5}$

f. $\frac{5}{8}$ **g.** $\frac{5}{6}$

Use your calculator to convert each percent to a decimal number:

h. 8% **i.** 12.5%

j. 150% **k.** $6\frac{1}{2}$%

l. *Connect* Two hundred percent (200%) is equivalent to what whole number? Describe the steps you took to find the answer.

Written Practice

Strengthening Concepts

1. (36) The ratio of weekend days to weekdays is 2 to 5. What fraction of the week consists of weekend days?

2. (28) Eric ran 8 laps in 11 minutes 44 seconds.

a. How many seconds did it take Eric to run 8 laps?

b. What is the average number of seconds it took Eric to run each lap?

3. (11, 35) Some gas was still in the tank. Mr. Wang added 13.3 gallons of gas, which filled the tank. If the tank held a total of 21.0 gallons of gas, how much gas was in the tank before Mr. Wang added the gas?

4. (5, 12) From 1750 to 1850, the estimated population of the world increased from seven hundred twenty-five million to one billion, two hundred thousand. About how many more people were living in the world in 1850 than in 1750?

5. (22, 36) Diagram this statement. Then answer the questions that follow.

The Beagles played two thirds of their 15 games in dry weather and the rest in rain.

a. How many games did the team play in dry weather?

b. What was the ratio of dry-weather games to rainy games?

The stem-and-leaf plot below shows the distribution of finish times in a 100-meter sprint. Refer to the stem-and-leaf plot to answer problems **6** and **7**.

Stem	Leaf
11	2
12	3 4 8
13	0 3 4 5 6
14	1 4 7 8
15	2 5

12 | 3 represents 12.3 seconds

*** 6.** (Inv. 4) **Analyze** For the 100-meter finish times, find the

a. median.

b. lower quartile.

c. upper quartile.

Math Language

Analyze

You know three of the values needed for your plot. What other two values do you need to know?

*** 7.** (Inv. 4) **Represent** Make a box-and-whisker plot of the 100-meter finish times in the stem-and-leaf plot.

8. (19) A square and a regular hexagon share a common side, as shown. The perimeter of the hexagon is 120 mm. What is the perimeter of the square?

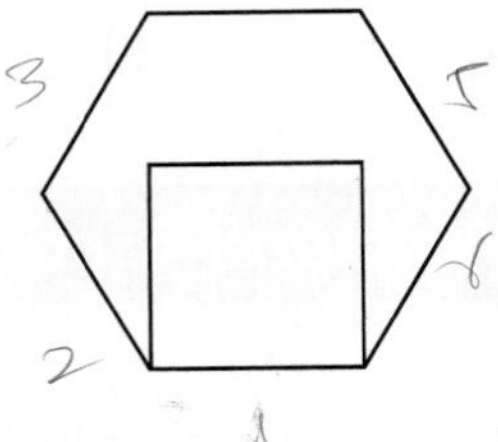

*** 9.** (43) Write each of these numbers as a reduced fraction or mixed number:

a. 0.375 **b.** 5.55 **c.** 250%

*** 10.** (43) **Connect** Write each of these numbers as a decimal number:

a. $2\frac{2}{5}$ **b.** $\frac{1}{8}$ **c.** 250%

*** 11.** (42) Round each number to the nearest thousandth:

a. $0.\overline{45}$ **b.** $3.\overline{142857}$

* **12.** Divide 1.9 by 12 and write the quotient
(42)

a. with a bar over the repetend.

b. rounded to three decimal places.

13. Four and five hundredths is how much greater than one hundred sixty-seven thousandths?
(31, 35)

14. Draw $\overline{AB}$ 1 inch long. Then draw $\overline{AC}$ $\frac{3}{4}$ inch long perpendicular to $\overline{AB}$. Complete $\triangle ABC$ by drawing $\overline{BC}$. How long is $\overline{BC}$?
(8)

* **15.** A normal deck of cards is composed of four suits (red heart, red diamond, black spade, and black club) of 13 cards each (2 through 10, jack, queen, king, and ace) for a total of 52 cards.
(14, 43)

If one card is drawn from a normal deck of cards, what is the probability that the card will be a red card? Express the probability as a reduced fraction and as a decimal number.

16. **a.** Make a factor tree showing the prime factorization of 900. (Start with the factors 30 and 30.)
(20, 21)

b. **Represent** Write the prime factorization of 900 using exponents.

c. Write the prime factorization of $\sqrt{900}$.

17. The eyedropper held 2 milliliters of liquid. How many eyedroppers of liquid would it take to fill a 1-liter container?
(32)

* **18.** **a.** **Connect** Write 8% as a decimal number.
(43)

b. Find 8% of \$8.90 by multiplying \$8.90 by the answer to **a.** Round the answer to the nearest cent.

19. Refer to the figure at right to answer **a** and **b.**
(32)

a. What is the perimeter of this triangle?

b. What is the area of this triangle?

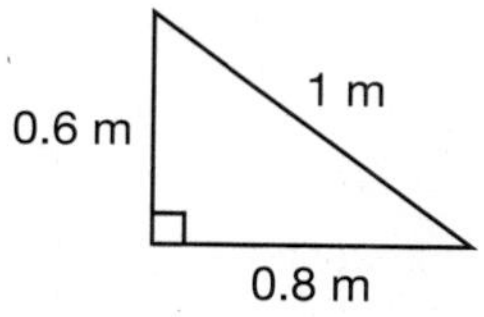

20. Compare and explain the reason for your answer:
(27)

$$\frac{32}{2} \bigcirc \frac{320}{20}$$

* **21.** **Analyze** Show two ways to evaluate $a(b + c)$ if $a = 2$, $b = 3$, and $c = 4$.
(41)

Solve:

22. $\frac{10}{18} = \frac{c}{4.5}$
(39)

23. $1.9 = w + 0.42$
(35)

Simplify:

24. $6.5 \div 4$
(35)

25. $3\frac{3}{10} - 1\frac{11}{15}$
(23, 30)

26. $5\frac{1}{2} + 6\frac{3}{10} + \frac{4}{5}$
(30)

27. $7\frac{1}{2} \cdot 3\frac{1}{3} \cdot \frac{4}{5} \div 5$
(26)

28. Find the next coordinate pair in this sequence:
(Inv. 3)

(1, 2), (2, 4), (3, 6), (4, 8), ...

* **29.** **Conclude** Find the measures of $\angle a$, $\angle b$, and $\angle c$ in the figure at right.
(40)

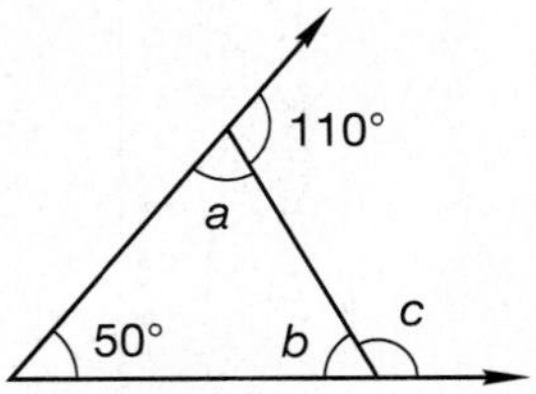

30. Refer to the figure at right to answer **a–c:**
(Inv. 2)

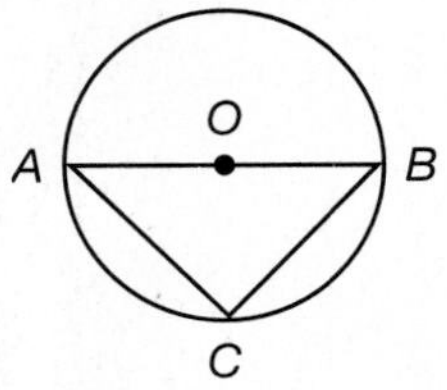

a. What is the measure of central angle *AOB*?

b. What appears to be the measure of inscribed angle *ACB*?

c. Chord *AC* is congruent to chord *BC*. What appears to be the measure of inscribed angle *ABC*?

LESSON

44

• Division Answers

Power Up

Building Power

facts Power Up H

mental math

a. **Calculation:** $5 \times 64¢$

b. **Decimals:** $0.5 \div 10$

c. **Equivalent Fractions:** $\frac{3}{m} = \frac{12}{24}$

d. **Estimation:** $596 \div 11$

e. **Power/Roots:** $\frac{\sqrt{144}}{12}$

f. **Fractional Parts:** $\frac{3}{4}$ of 200

g. **Statistics:** Find the range of the set of numbers: 56, 15, 45, 65.

h. **Calculation:** Start with the number of meters in a kilometer, $\div$ 10, $\sqrt{\ }$, $\times$ 5, $-$ 1, $\sqrt{\ }$, $\times$ 5, $+$ 1, $\sqrt{\ }$.

problem solving

Jesse has three identical cubes. Each face of each cube is painted a different color: white, yellow, blue, green, red, and purple. If the cubes are painted in the same way, what color face is parallel to the red face? ... the yellow face? ... the green face?

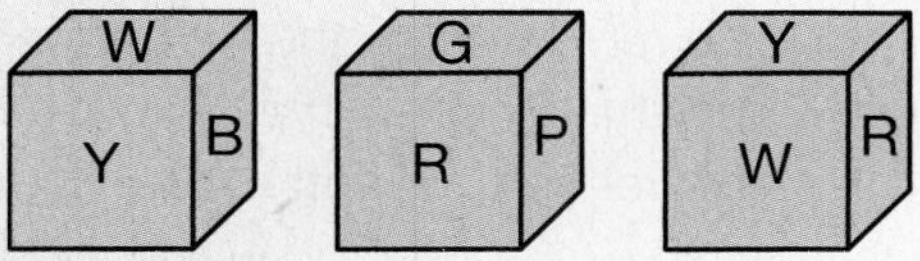

New Concept

Increasing Knowledge

We can write answers to division problems with remainders in different ways. We can write them with a remainder.

$$\begin{array}{r} 6 \text{ R } 3 \\ 4\overline{)27} \\ \underline{24} \\ 3 \end{array}$$

We can also write them as a mixed number.

$$\begin{array}{r} 6\frac{3}{4} \\ 4\overline{)27} \\ \underline{24} \\ 3 \end{array}$$

We can also write the answer as a decimal number. First, we add decimal points to the dividend and to the quotient, making sure to line them up. Then we add zeros to the right of the decimal point in the dividend, and continue dividing.

```
     6.75
  4)27.00
    24
     3 0
     2 8
       20
       20
        0
```

In some situations we might round up or round down. In this lesson we will practice writing division answers in different ways. We will also find the best way to write a division to fit the situation.

Example 1

Divide 54 by 4 and write the answer

a. with a remainder.

b. as a mixed number.

c. as a decimal.

Solution

```
    13 R 2
  4)54
    4
    14
    12
     2
```

a. We divide and find the result is **13 R 2.**

b. The remainder is the numerator of a fraction, and the divisor is the denominator. Thus this answer can be written as $13\frac{2}{4}$, which reduces to $\mathbf{13\frac{1}{2}}$.

```
    13.5
  4)54.0
    4
    14
    12
     2 0
     2 0
       0
```

c. We fix place values by placing the decimal point to the right of 54. Then we can write zeros in the following places and continue dividing until the remainder is zero. The result is **13.5.**

Sometimes a quotient written as a decimal number will be a repeating decimal. Other times it will have more decimal places than the problem requires. In this book we show the complete division of the number unless the problem states that the answer is to be rounded.

Connect Name some real-world situations in which we would not want to round a decimal number.

Example 2

Divide 37.4 by 9 and round the quotient to the nearest thousandth.

Solution

We continue dividing until the answer has four decimal places. Then we round to the nearest thousandth.

$$4.15\underline{5}⑤\ldots \longrightarrow \mathbf{4.156}$$

$$\begin{array}{r} 4.1555\ldots \\ 9\overline{)37.4000\ldots} \\ \underline{36} \\ 1\ 4 \\ \underline{9} \\ 50 \\ \underline{45} \\ 50 \\ \underline{45} \\ 50 \\ \underline{45} \\ 5 \end{array}$$

Problems involving division often require us to interpret the results of the division and express our answer in other ways. Consider the following example.

Example 3

Vans will be used to transport 27 students on a field trip. Each van can carry 6 students.

a. How many vans can be filled?

b. How many vans will be needed?

c. If all but one van will be full, then how many students will be in the van that will not be full?

Solution

Thinking Skill

Discuss

In this example, do we ignore the remainder, include it in the answer, or use it to find the answer? Explain.

The quotient when 27 is divided by 6 can be expressed in three forms. This problem involves forming groups, so we divide 27 by 6.

$$\begin{array}{r} 4 \text{ R } 3 \\ 6\overline{)27} \end{array} \qquad \begin{array}{r} 4\frac{1}{2} \\ 6\overline{)27} \end{array} \qquad \begin{array}{r} 4.5 \\ 6\overline{)27.0} \end{array}$$

The questions require us to interpret the results of the division.

a. The whole number 4 in the quotient means that **4 vans** can be filled to capacity.

b. Four vans will hold 24 students. Since 27 students are going on the field trip, another van is needed. So **5 vans** will be needed.

c. The fifth van will carry the remaining **3 students.**

Analyze What do the quotient (4), the divisor (6), and the remainder (3) represent?

Practice Set

Divide 55 by 4 and write the answer

a. with a remainder.

b. as a mixed number.

c. as a decimal number.

d. Divide 5.5 by 3 and round the answer to three decimal places.

e. Generalize Ninety-three students are assigned to four classrooms as equally as possible. How many students are in each of the four classrooms?

f. Toby bought only one pound of grapes. How much did he pay? Explain your answer.

Grapes
$2
for 3 lbs

g. *Justify* Nita is putting photos in an album. She places 6 photos on each page. If she has 74 photos, how many pages of photos are in the book? Explain how you found the answer.

Written Practice

Strengthening Concepts

1. (36) The rectangle was 24 inches long and 18 inches wide. What was the ratio of its length to its width?

2. (28) Lakeisha participates in a bowling league. In her first ten games her scores were 90, 95, 90, 85, 80, 85, 90, 80, 95, and 100. What was her mean (average) score?

3. (38) Which would be a better graph for displaying her scores, a circle graph, or a bar graph? Why?

4. (28) Rachel bought a sheet of fifty 39-cent stamps from the post office. She paid for the stamps with a $20 bill. How much money should she get back?

5. (31, 35) Ninety-seven thousandths is how much less than two and ninety-eight hundredths? Write the answer in words.

6. (22, 28) Read this statement. Then answer the questions that follow.

Five sixths of the thirty motorcycles on the lot were new.

a. How many motorcycles were used?

b. What was the ratio of new motorcycles to used motorcycles?

7. (19) Copy this figure on your paper. Find the length of each unmarked side, and find the perimeter of the polygon. Dimensions are in meters. All angles are right angles.

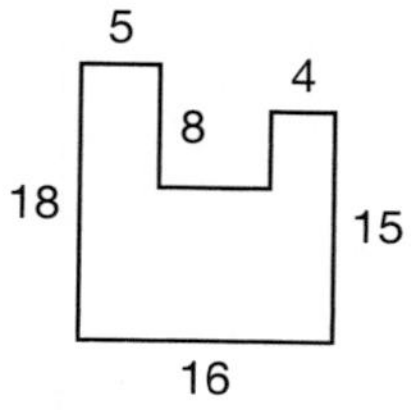

*** 8.** (43) **a.** Write 0.75 as a simplified fraction.

b. Write $\frac{5}{8}$ as a decimal number.

c. Write 125% as a decimal number.

*** 9.** (14) **Analyze** Samuel has 52 marbles in a bag. Fourteen are red, 13 are green and 25 are blue. If a marble is chosen at random, what is the probability that it is green?

*** 10.** (41) The expression 2(3 + 4) equals which of the following?

A $(2 \cdot 3) + 4$ **B** $(2 \cdot 3) + (2 \cdot 4)$

C $2 + 7$ **D** $23 + 24$

*** 11.** (4, 16) **Analyze** "Triangle numbers" are the numbers in this sequence.

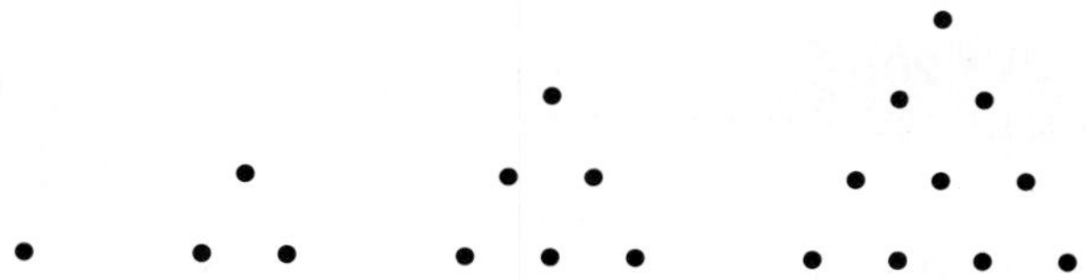

1, 3, 6, 10,

a. Find the next two terms in the sequence.

b. The rule in this function table shows how to find a term if we know its position.

Position (*n*)	1	2	3	4			
Term $\left(\frac{n^2 + n}{2}\right)$	1	3	6	10			

The fifth term is 15 because $\frac{5^2 + 5}{2} = 15$. Use the rule to find the tenth term.

*** 12.** (42, 44) Divide 5.4 by 11 and write the answer

a. with a bar over the repetend.

b. rounded to the nearest thousandth.

13. (21) What composite number is equal to the product of the first four prime numbers?

*** 14.** (33, 42) **a.** **Analyze** Arrange these numbers in order from least to greatest:

$$1.2, -12, 0.12, 0, \frac{1}{2}, 1.\overline{2}$$

b. **Classify** Which numbers in **a** are integers?

15. (26) Each math book is $1\frac{1}{2}$ inches thick.

a. A stack of 12 math books would stand how many inches tall?

b. How many math books would make a stack 1 yard tall?

16. (34, 35) What is the sum of the numbers represented by points M and N on this number line?

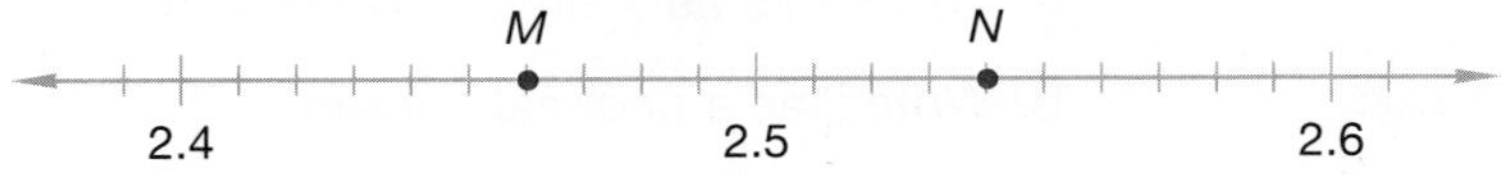

***17.** (33, 41) **Conclude** Estimate the value of πd when π is 3.14159 and d is 9.847 meters.

18. (34, 35) Draw a square with sides 2.5 cm long.

a. What is the area of the square?

b. What is the perimeter of the square?

19. (Inv. 3, 37) The coordinates of the vertices of a triangle are (−2, 0), (4, 0), and (3, 3). What is the area of the triangle?

Solve:

*** 20.** (39) $\frac{25}{15} = \frac{n}{1.2}$

*** 21.** (39) $\frac{p}{90} = \frac{4}{18}$

22. (35) $4 = 3.14 + x$

23. (35) $0.1 = 1 - z$

Simplify:

24. (35) $16.42 \div 8$

25. (35) $0.153 \div 9$

26. (30) $5\frac{3}{4} + \frac{5}{6} + 2\frac{1}{2}$

27. (23, 30) $3\frac{1}{3} - \left(5 - 1\frac{5}{6}\right)$

28. (26) $3\frac{3}{4} \cdot 3\frac{1}{3} \cdot 8$

29. (26) $7 \div 10\frac{1}{2}$

*** 30.** (40) **Justify** Figure $ABCD$ is a rectangle. The measure of $\angle ADB$ is 35°. Find the measure of each angle below. Defend how you found the answers.

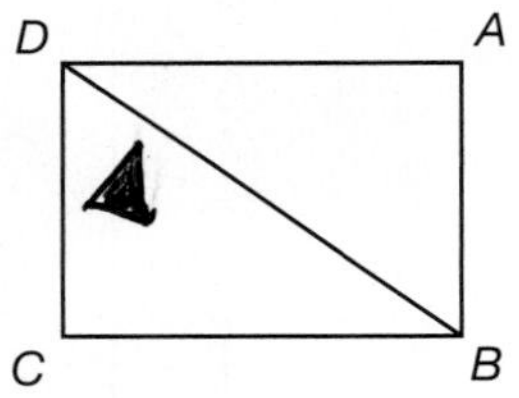

a. $\angle ABD$

b. $\angle CBD$

c. $\angle BDC$

LESSON 45

Dividing by a Decimal Number

Power Up — *Building Power*

facts Power Up I

mental math

a. **Calculation:** $7 \times \$1.50$

b. **Decimals/Exponents:** 1.25×10^2

c. **Equivalent Fractions:** $\frac{4}{6} = \frac{10}{W}$

d. **Power/Roots:** $5^2 \cdot \sqrt{16}$

e. **Estimation:** $4\frac{1}{8} \times 2\frac{6}{7}$

f. **Fractional Parts:** $\frac{2}{3}$ of 75

g. **Statistics:** Find the range of the set of numbers: 123, 321, 232, 623.

h. **Calculation:** Find the sum, difference, product, and quotient of $\frac{3}{4}$ and $\frac{2}{3}$.

problem solving

Copy the problem and fill in the missing digits:

```
   _6
 × __
   __
  __
  2_6
```

New Concept — *Increasing Knowledge*

Dan has \$2.00 and wants to buy red pens for his friends. If red pens cost \$0.25 each, how many can Dan buy?

This is a problem we may be able to solve quickly by mental calculation. Dan can buy 4 red pens priced at \$0.25 for \$1.00, so he can buy 8 red pens for \$2.00. But how do we get an answer of "8 red pens" from what seems to be a division problem?

$$\$0.25\overline{)\$2.00}$$

In this lesson we will consider how to get the "8." In a later lesson we will consider how to get the "red pens." Notice that dividing \$2.00 by \$0.25 is dividing by a decimal number (\$0.25).

$$\frac{\$2.00}{\$0.25}$$

If we wish to divide by a whole number instead of by a decimal number, we can convert to an equivalent division problem using cents rather than dollars.

$$\frac{200¢}{25¢}$$

Changing from dollars to cents shifts the decimal point two places to the right. The units (cents over cents) cancel, and 200 divided by 25 is 8.

$$\frac{200¢}{25¢} = 8$$

Recall that we can form equivalent division problems by multiplying (or dividing) the dividend and divisor by the same number. We use this method to change "division by a decimal" problems to "division by a whole number" problems.

If we want to divide 1.36 by 0.4, we have

$$\frac{1.36}{0.4}$$

Thinking Skill

Analyze

Why can we multiply by $\frac{10}{10}$ to find an equivalent division problem?

We can change the divisor to the whole number 4 by multiplying both the dividend and divisor by 10.

$$\frac{1.36}{0.4} \times \frac{10}{10} = \frac{13.6}{4}$$

The quotient of 13.6 divided by 4 is the same as the quotient of 1.36 divided by 0.4. This means that both of these division problems have the same answer.

$$0.4\overline{)1.36} \quad \text{is equivalent to} \quad 4\overline{)13.6}$$

To divide by a decimal number, we move the decimal point in the divisor to the right to make the divisor a whole number. Then we move the decimal point in the dividend the same number of places to the right.

Example 1

Divide: 3.35 ÷ 0.05

Solution

This division could answer the question, "How many nickels would make $3.35?" We use a division box and write

$$0.05\overline{)3.35}$$

First we move the decimal point in 0.05 two places to the right to make it 5.

$$0.\underbrace{05}\overline{)3.35}$$

Then we move the decimal point in 3.35 the same number of places to the right. This forms an equivalent division problem. The decimal point in the answer will be directly above the new location in the dividend.

$$0.\underbrace{05}\overline{)3.\underbrace{35}}$$

Now we divide and find that 3.35 ÷ 0.05 = **67.**

$$\begin{array}{r} 67. \\ 5\overline{)335.} \\ \underline{30} \\ 35 \\ \underline{35} \\ 0 \end{array}$$

Explain Moving the decimal point in 3.35 and 0.05 two places to the right is the same as multiplying the divisor and dividend by what fraction? Explain how you know.

Example 2

Divide: 0.144 ÷ 0.8

Solution

We want the divisor, 0.8, to be a whole number. Moving the decimal point one place to the right changes the divisor to the whole number 8. To do this, we must also move the decimal point in the dividend one place to the right.

```
       0.18
08.)1.44
      8
      64
      64
       0
```

Example 3

Solve: 0.07*x* = 5.6

Solution

We divide 5.6 by 0.07. We move both decimal points two places. This makes an empty place in the division box, which we fill with a zero. We keep dividing until we reach the decimal point.

```
          80.
007.)560.
         56
         00
          0
          0
```

Discuss Why did we need to add a zero to the end of the dividend?

Example 4

Divide: 21 ÷ 0.5

Solution

We move the decimal point in 0.5 one place to the right. The decimal point on 21 is to the right of the 1. We shift this decimal point one place to the right to form the equivalent division problem 210 ÷ 5.

```
        42.
05.)210.
       20
       10
       10
        0
```

Example 5

Divide: 1.54 ÷ 0.8

Solution

Thinking Skill

Classify

Is this quotient a repeating or terminating decimal? Explain your answer.

We do not write a remainder. We write zeros in the places to the right of the 4. We continue dividing until the remainder is zero, until the digits begin repeating, or until we have divided to the desired number of decimal places.

```
        1.925
  08.)15.400
       8
       7 4
       7 2
         20
         16
          40
          40
           0
```

Example 6

How many $0.35 erasers can be purchased with $7.00?

Solution

We record the problem as 7.00 divided by 0.35. We shift both decimal points two places and divide. The quotient is 20 and the answer to the question is **20 erasers.**

```
          20.
  035.)700.
        70
         00
          0
          0
```

Justify How can we check that our answer is correct?

Practice Set

Divide:

a. 5.16 ÷ 0.6

b. 0.144 ÷ 0.09

c. 23.8 ÷ 0.07

d. 24 ÷ 0.08

e. How many $0.75 pens can be purchased with $12.00?

f. *Explain* Why are these division problems equivalent?

$$\frac{0.25}{0.5} = \frac{2.5}{5}$$

g. *Generalize* Solve this proportion: $\frac{x}{4} = \frac{3}{0.8}$

Written Practice

Strengthening Concepts

1. (36) Raisins and nuts were mixed in a bowl. If nuts made up five eighths of the mixture, what was the ratio of raisins to nuts?

2. (28) The taxi ride cost $1 plus 80¢ more for each quarter mile traveled. What was the total cost for a 2-mile trip?

3. (31, 35) Fifty-four and five hundredths is how much greater than fifty and forty thousandths? Use words to write the answer.

4. Refer to the election tally sheet below to answer **a** and **b.**
(38)

Vote Totals	
Judy	𝍸 𝍸 𝍸 |
Carlos	𝍸 𝍸 ||||
Yolanda	𝍸 𝍸 𝍸 𝍸 ||
Khanh	𝍸 𝍸 𝍸 |||

a. The winner of the election received how many more votes than the runner-up?

b. What fraction of the votes did Carlos receive?

5. Read this statement. Then answer the questions that follow.
(22, 36)

Four sevenths of those who rode the Giant Gyro at the fair were euphoric. All the rest were vertiginous.

a. What fraction of the riders were vertiginous?

b. What was the ratio of euphoric to vertiginous riders?

6. **Analyze** Name the properties used to simplify the expression $4\left(\frac{1}{4} + 0.3\right)$.
(9, 41)

$4\left(\frac{1}{4} + 0.3\right)$	Given
$4\left(\frac{1}{4}\right) + 4(0.3)$	**a.** ______________
$1 + 4(0.3)$	**b.** ______________
$1 + 1.2$	$4(0.3) = 1.2$
2.2	$1 + 1.2 = 2.2$

7. Find the product of 5^2 and 10^2.
(20)

8. The perimeter of this rectangle is 56 cm:
(19, 20)

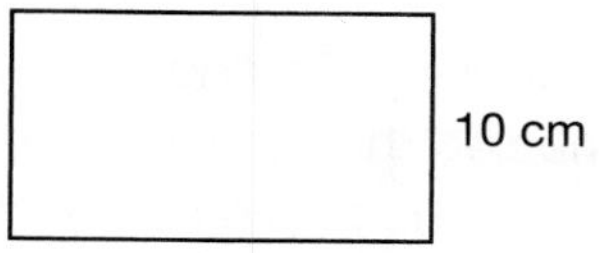

a. What is the length of the rectangle?

b. What is the area of the rectangle?

*** 9.** **a.** Write 62.5 as a mixed number.
(43)

b. Write $\frac{9}{100}$ as a decimal number.

c. Write 7.5% as a decimal number.

*** 10.** **Generalize** Round each number to five decimal places:
(42)

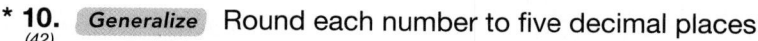

a. $23.\overline{54}$ **b.** $0.91\overline{6}$

11. (32) A 2-liter bottle of water has a mass of 2 kilograms. How many grams is that?

12. (35) Find 6.5% of \$5.00 by multiplying \$5.00 by 0.065. Round the answer to the nearest cent.

* **13.** (42, 44) **Generalize** Divide 5.1 by 9 and write the quotient

a. rounded to the nearest thousandth.

b. with a bar over the repetend.

* **14.** (14, 43) **Analyze** Letter cards *A* through *Z* are placed in a bag. If a card is chosen at random, what is the probability that it will be the card with the letter *Z*? Write the answer as a reduced fraction and as a decimal number rounded to the nearest hundredth.

15. (35) Draw $\overline{XY}$ 2 cm long. Then draw $\overline{YZ}$ 1.5 cm long perpendicular to $\overline{XY}$. Complete $\triangle XYZ$ by drawing $\overline{XZ}$. How long is $\overline{XZ}$?

16. (19, 37) Find the **a** perimeter and **b** area of the triangle drawn in problem 15.

Solve:

17. (39) $\frac{3}{w} = \frac{25}{100}$

* **18.** (39, 45) $\frac{1.2}{4.4} = \frac{3}{a}$

19. (35) $m + 0.23 = 1.2$

20. (35) $r - 1.97 = 0.65$

Simplify:

21. (35) $(0.15)(0.15)$

22. (35) $1.2 \times 2.5 \times 4$

23. (35) $14.14 \div 5$

* **24.** (45) $0.096 \div 0.12$

25. (30) $\frac{5}{8} + \frac{5}{6} + \frac{5}{12}$

26. (30) $4\frac{1}{2} - \left(2\frac{1}{3} - 1\frac{1}{4}\right)$

27. (26) $\frac{7}{15} \cdot 10 \cdot 2\frac{1}{7}$

28. (26) $6\frac{3}{5} \div 1\frac{1}{10}$

* **29.** (45) **Connect** How many \$0.21 pencils can be purchased with \$7.00?

* **30.** (40) **Conclude** Amanda cut out a triangle and labeled the corners *a*, *b*, and *c* as shown. Then she tore off the three corners and fit the pieces together to form the semicircular shape shown at right.

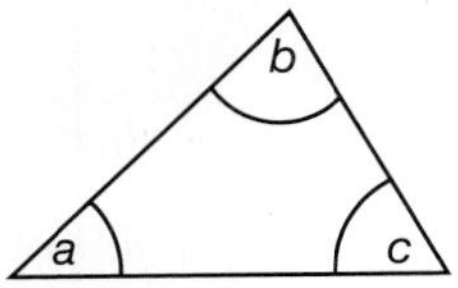

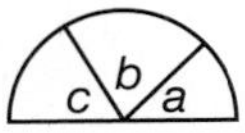

a. Try the activity described in this problem, and tape or glue the fitted corners to your paper.

b. **Infer** Describe the characteristic of triangles demonstrated by this activity.

LESSON

46

• Rates

Power Up

Building Power

facts

Power Up J

mental math

a. **Calculation:** 9 × \$0.82

b. **Decimals/Exponents:** $3.6 \div 10^2$

c. **Equivalent Fractions:** $\frac{4}{8} = \frac{a}{20}$

d. **Estimation:** 4.97 × 1.9

e. **Power/Roots:** $\sqrt{16} + 2^3$

f. **Fractional Parts:** $\frac{9}{10}$ of 80

g. **Statistics:** Find the mode of the set of numbers: 78, 87, 33, 78, 43.

h. **Measurement:** Start with the number of vertices on a quadrilateral. Add the number of years in a decade; subtract a half dozen; then multiply by the number of feet in a yard. What is the answer?

problem solving

White and black marbles were placed in four boxes as shown. From which box is the probability of choosing a black marble the greatest?

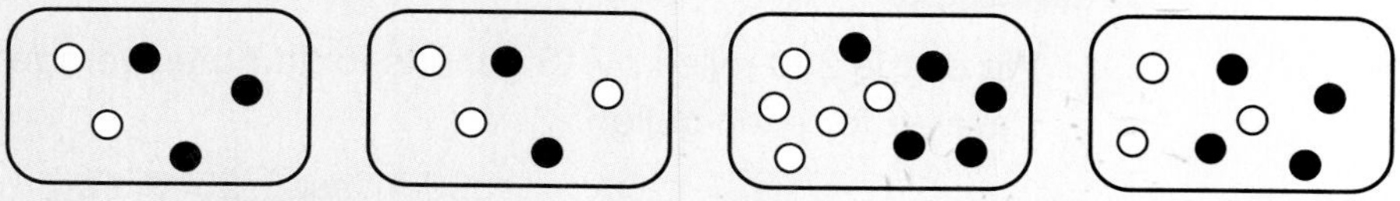

New Concept

Increasing Knowledge

A **rate** is a ratio of two measures. Consider the following examples of rates. Can you indentify the two measures in each example?

Rate of speed:	65 miles per hour
Mileage rate:	25 miles per gallon
Pay rate:	\$12 per hour

The word *per* in each rate implies division and means "in each" or "for each." To show this relationship more clearly, we can write each rate as a fraction using 1 with the unit of the denominator.

$$\frac{65 \text{ miles}}{1 \text{ hour}} \qquad \frac{25 \text{ miles}}{1 \text{ gallon}} \qquad \frac{12 \text{ dollars}}{1 \text{ hour}}$$

Thinking Skill

Summarize

Explain in your own words what the formula $r = \frac{d}{t}$ means.

Notice that a rate of speed is a distance traveled divided by the elapsed time.

$$\text{rate} = \frac{\text{distance}}{\text{time}} \qquad r = \frac{d}{t}$$

Example 1

Arnold drove 390 miles from Los Angeles to Sacramento in 6 hours. What was his average rate of speed?

Solution

We are given the total distance and the total time. To find the average speed, we divide the number of miles by the number of hours so that the number of hours is 1.

Thinking Skill

Analyze

How did we get from $\frac{390\text{ miles}}{6\text{ hours}}$ to $\frac{65\text{ miles}}{1\text{ hour}}$?

$$\text{rate} = \frac{\text{distance}}{\text{time}} \qquad \frac{390\text{ miles}}{6\text{ hours}} = \frac{65\text{ miles}}{1\text{ hour}}$$

Arnold drove at an average speed of **65 miles per hour.**

Verify How can you check that the solution is correct?

Example 2

On the 390 mile trip Arnold's car used 15 gallons of gas. His car averaged how many miles per gallon for the trip?

Solution

We divide 390 miles by 15 gallons to find the average number of miles traveled for each gallon.

$$\frac{390\text{ miles}}{15\text{ gallons}} = \frac{26\text{ miles}}{1\text{ gallon}}$$

Arnold's car averaged **26 miles per gallon.**

Example 3

Jarrod was paid $480 for 30 hours of work. What was Jarrod's hourly rate of pay?

Solution

To find the hourly rate, we divide the total pay by the number of hours worked.

$$\frac{480\text{ dollars}}{30\text{ hours}} = \frac{16\text{ dollars}}{1\text{ hour}}$$

Jarrod's hourly rate of pay is **$16 per hour.**

Unit Price is the cost for a single unit of a product. It is a ratio of price to quantity that is often posted in supermarkets to help customers identify the better buy. One of many ways to express unit price is cents per ounce.

Example 4

What is the unit price of a 24-ounce box of cereal priced at $3.60?

Solution

To find the unit price, we divide the total price by the number of units to find the price for one unit.

$$\frac{\$3.60}{24 \text{ ounces}} = \frac{\$0.15}{1 \text{ ounce}}$$

The unit price for the box of cereal is **15¢ per ounce.**

Justify How can you use the unit price to find the cost of a 32-ounce box of the cereal? Justify your procedure.

In many situations we multiply a rate or divide by a rate to find answers to questions.

Example 5

Gina wants to know how much she will earn working 40 hours at $16.00 per hour. How much will she earn?

Solution

Gina earns $16.00 in one hour, so she will earn 40 × $16.00 in 40 hours.

$$40 \times \$16.00 = \mathbf{\$640.00}$$

Example 6

Linda's car holds 12 gallons of fuel and averages 26 miles per gallon. If she starts a 280-mile trip with a full tank of gas, is she likely to reach her destination without needing to refuel?

Solution

Linda's car averages 26 miles per gallon, so it might travel 12 × 26 miles on a full tank of gas.

$$12 \times 26 \text{ miles} = 312 \text{ miles}$$

Linda is traveling 280 miles, so **she is likely to reach her destination without refueling.**

Example 7

Shelley needs to drive 180 miles from her home near Fort Worth to Austin. If Shelley begins her drive at 9:00 a.m. and averages 60 miles per hour, when can she expect to arrive in Austin?

Solution

To find when Shelley might arrive in Austin, we first need to find how many hours the trip will last. If she averages 60 miles per hour, then she will travel 180 miles in 3 hours.

$$180 \text{ miles} \div \frac{60 \text{ miles}}{1 \text{ hour}} =$$

$$\overset{3}{\cancel{180 \text{ miles}}} \times \frac{1 \text{ hour}}{\underset{1}{\cancel{60 \text{ miles}}}} = 3 \text{ hours}$$

Thinking Skill

Connect

Explain the relationship between the rate formula and the distance formula.

If Shelley can average 60 miles per hour, the trip will last 3 hours. If she begins driving at 9 a.m., **Shelley can expect to arrive in Austin at noon.**

An important formula that relates distance to rate and time is the following:

$$\text{distance} = \text{rate} \times \text{time} \qquad d = rt$$

We will use this formula in Example 8.

Example 8

Felipe is planning a bicycle tour of the state with some friends. He estimates that they can ride seven hours each day and average 12 miles per hour (mph). How far does Felipe estimate they can ride each day?

Solution

We will use the distance formula. For rate we use 12 miles per hour. For time we use 7 hours.

$$d = rt$$

$$d = \frac{12 \text{ miles}}{1 \cancel{\text{ hour}}} \cdot 7 \cancel{\text{ hours}}$$

$$d = 84 \text{ miles}$$

Using Felipe's estimates, he and his friends should be able to ride **about 84 miles per day.**

Practice Set

a. *Justify* The Chongs drove 416 miles in 8 hours. What was their average rate of speed? Explain how you arrived at your answer.

b. The Smiths' car used 16 gallons of gas to travel 416 miles, which is an average of how many miles per gallon?

c. Dillon earned $170.00 working 20 hours at the car wash. What was his hourly rate of pay?

d. Monica earns $9.25 per hour in the clothing store. How much does she earn in a 30-hour week?

e. If Marisa's hybrid car averages 45 mpg, how far can her car travel on 8 gallons of fuel?

f. What is the unit price of a 24-ounce box of cereal priced at $3.84?

g. **Connect** Use the distance formula to find how far Chris could ride in 6 hours at an average rate of 14 mph.

Written Practice *Strengthening Concepts*

*** 1.** (46) Brand X costs $2.40 for 16 ounces. Find the unit price.

*** 2.** (46) Germany's Autobahn is the world's second largest superhighway system. A car traveled 702 kilometers along the Autobahn in 6 hours. How many kilometers per hour did the car average?

*** 3.** (36) **Represent** A penny, a nickel, and a dime are flipped once. Make a tree diagram to find the possible outcomes. Then write the sample space for the experiment, recording each outcome in penny, nickel, dime order. (For example, one outcome is HTH meaning heads for the penny, tails for the nickel, and heads for the dime.)

4. (28) At four different stores the price of 1 gallon of milk was $2.86, $2.83, $2.98, and $3.09. Find the average price per gallon rounded to the nearest cent.

5. (31, 35) Two and three hundredths is how much less than three and two tenths? Write the answer in words.

6. (26) A math book is $1\frac{1}{2}$ inches thick. How many math books will fit on a shelf that is 2 feet long?

7. (22, 36) Read this statement. Then answer the questions that follow.

Three eighths of the 48 roses were red.

a. How many roses were red?

b. What was the ratio of red to not red roses?

c. What fraction of the roses were not red?

*** 8.** (33) Replace each circle with the proper comparison symbol:

a. 3.0303 ◯ 3.303 **b.** $0.\overline{6}$ ◯ 0.600

9. (16) From goal line to goal line, a football field is 100 yards long. How many feet long is a football field?

*** 10.** (43) **a.** Write 0.080 as a fraction.

b. Write $37\frac{1}{2}\%$ as a decimal.

c. Write $\frac{1}{11}$ as a decimal with a bar over the repetend.

*** 11.** (46) **Analyze** Archie earns $9.50 per hour helping a house painter. How much does Archie earn in an 8-hour workday? Tell why your answer is reasonable.

12. (Inv. 3, 37) The coordinates of three vertices of a triangle are (4, 0), (5, 3), and (0, 0). What is the area of the triangle?

13. (36) **Analyze** The numbers 1 through 52 are written on individual cards and placed face down. If you draw one of the cards, what is the probability that it will have a number less than 13 on it? Explain how you arrived at your answer.

14. (21, 28) What is the average of the first five prime numbers?

* **15.** (41) **Analyze** Show two ways to evaluate $x(y + z)$ for $x = 0.3$, $y = 0.4$, and $z = 0.5$.

* **16.** (19, 37) **Analyze** In this figure all angles are right angles. Dimensions are in inches.

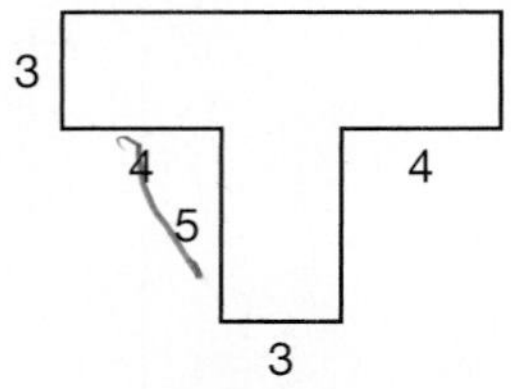

a. What is the perimeter of the figure?

b. What is the area of the figure?

17. (Inv. 2) The circle with center at point O has been divided into three sectors as shown. Find the measure of each of these central angles.

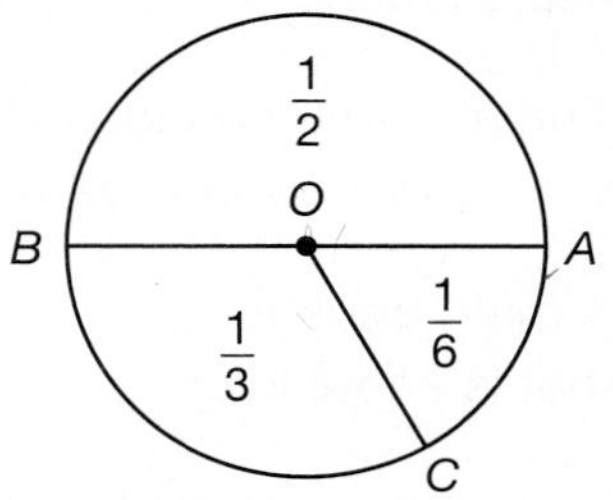

a. $\angle AOB$ **b.** $\angle BOC$ **c.** $\angle AOC$

Solve:

18. (39) $\frac{10}{12} = \frac{2.5}{a}$

19. (39) $\frac{6}{8} = \frac{b}{100}$

20. (35) $4.7 - w = 1.2$

21. (3, 20) $10x = 10^2$

Estimate each answer to the nearest whole number. Then perform the calculation.

22. (30) $1\frac{11}{18} + 2\frac{11}{24}$

23. (30) $5\frac{5}{6} - \left(3 - 1\frac{1}{3}\right)$

Simplify:

24. (26) $\frac{2}{3} \times 4 \times 1\frac{1}{8}$

25. (26) $6\frac{2}{3} \div 4$

26. (35) $3.45 + 6 + (5.2 - 0.57)$

* **27.** (45) $2.4 \div 0.016$

* **28.** (29) **Estimate** Describe how to estimate the area of this book cover.

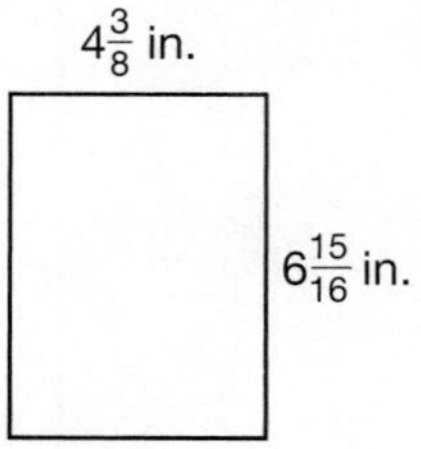

In the figure below, $\triangle ABC$ is congruent to $\triangle CDA$. Refer to the figure for problems 29 and 30.

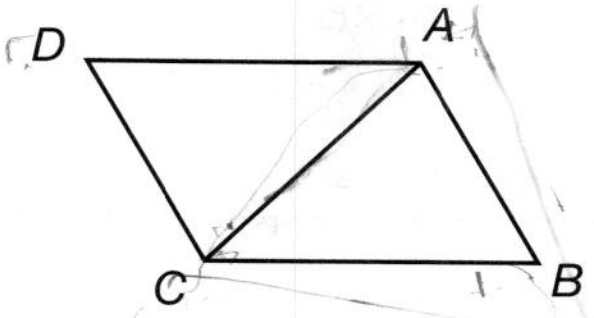

29. (18) Name the angle or side in $\triangle ABC$ that corresponds to the following angle and side from $\triangle CDA$:

a. $\angle ACD$ **b.** $\overline{DC}$

30. (40) The measure of $\angle ACB$ is 45°, and the measure of $\angle ADC$ is 60°. Find the measure of

a. $\angle B$. **b.** $\angle CAB$. **c.** $\angle CAD$.

Early Finishers
Real-World Application

Johanna, Dawn, Simone and Kelly run on a relay team. In a relay race, each runner takes a turn by running a leg that is an evenly divided portion of the total distance. Johanna, Dawn, Simone, and Kelly completed a 5-mile relay in 39 minutes.

a. If each member of the team ran at approximately the same speed, how long did it take each member to complete her leg? Write the answer as a mixed number and as a decimal of minutes and as minutes and seconds.

b. How far did each team member run? Write the answer as a mixed number and as a decimal.

LESSON 47

• Powers of 10

Power Up

Building Power

facts Power Up I

mental math

a. Calculation: $5 \times \$8.20$

b. Decimals/Exponents: 0.015×10^3

c. Equivalent Fractions: $\frac{c}{10} = \frac{9}{15}$

d. Estimation: $\$4.95 \times 19$

e. Exponents: $2^2 \cdot 2^3$

f. Fractional Parts: $\frac{5}{6}$ of 60

g. Statistics: Find the mode of the set of numbers: 123, 321, 124, 212, 321.

h. Calculation: Find the sum, difference, product, and quotient of $\frac{1}{2}$ and $\frac{2}{5}$.

problem solving

In the currency of the land a gilder is worth 6 skillings, and a skilling is worth 4 ore. Vincent offered to pay Andre 10 skillings and 2 ore for the job, but Andre wanted 2 gilders. Andre wanted how much more than Vincent's offer?

New Concept

Increasing Knowledge

Math Language

An **exponent** is a small number at the upper right corner of a base number. It shows how many times the base is to be used as a factor.

The positive powers of 10 are easy to write. The exponent matches the number of zeros in the product.

$10^2 = 10 \cdot 10 = 100$ (two zeros)

$10^3 = 10 \cdot 10 \cdot 10 = 1000$ (three zeros)

$10^4 = 10 \cdot 10 \cdot 10 \cdot 10 = 10{,}000$ (four zeros)

Observe what happens if we multiply or divide powers of 10 or find the power of a power of 10.

1. If we multiply powers of 10, the exponent of the product equals the sum of the exponents of the factors.

$$10^3 \times 10^3 = 10^6$$

$$1000 \times 1000 = 1{,}000{,}000$$

2. If we divide powers of 10, the exponent of the quotient equals the difference of the exponents of the dividend and divisor.

$$10^6 \div 10^3 = 10^3$$

$$1{,}000{,}000 \div 1000 = 1000$$

3. If we find the power of a power of 10, the result is a power of 10 that is the product of the exponents.

$$(10^3)^2 = 10^3 \cdot 10^3 = 10^6$$

Thinking Skill

Summarize

In your own words, state the three rules of exponents.

These three observations about powers of 10 apply to all powers with the same base. We summarize these rules for exponents in this table.

Rules of Exponents
$a^x \cdot a^y = a^{x+y}$
$\frac{a^x}{a^y} = a^{x-y}$
$(a^x)^y = a^{xy}$

Example 1

Find the missing exponents in a–c.

a. $2^4 \times 2^2 = 2^{\square}$

b. $2^4 \div 2^2 = 2^{\square}$

c. $(2^4)^2 = 2^{\square}$

Solution

a. $2^4 \times 2^2 = (2 \cdot 2 \cdot 2 \cdot 2) \cdot (2 \cdot 2) = \mathbf{2^6}$

b. $2^4 \div 2^2 = \frac{2 \cdot 2 \cdot 2 \cdot 2}{2 \cdot 2} = \mathbf{2^2}$

c. $(2^4)^2 = 2^4 \cdot 2^4 = \mathbf{2^8}$

We can use powers of 10 to show place value, as we see in the chart below. Notice that 10^0 equals 1.

Trillions			Billions			Millions			Thousands			Units (Ones)			
hundreds	tens	ones	hundreds	tens	ones	hundreds	tens	ones	hundreds	tens	ones	hundreds	tens	ones	Decimal point
10^{14}	10^{13}	10^{12}	10^{11}	10^{10}	10^9	10^8	10^7	10^6	10^5	10^4	10^3	10^2	10^1	10^0	.

Powers of 10 are sometimes used to write numbers in expanded notation.

Example 2

Write 5206 in expanded notation using powers of 10.

Solution

The number 5206 means 5000 + 200 + 6. We will write each number as a digit times its place value.

$$5000 \quad + \quad 200 \quad + \quad 6$$

$$\mathbf{(5 \times 10^3) \quad + \quad (2 \times 10^2) \quad + \quad (6 \times 10^0)}$$

multiplying by powers of 10

When we multiply a decimal number by a power of 10, the answer has the same digits in the same order. Only their place values are changed.

Example 3

Multiply: 46.235×10^2

Solution

Multiplying a decimal number by a power of 10 shifts each digit the number of places indicated by the exponent. For instance, the 4 in the tens place shifts two places to the thousands place.

$$46.235 \times 10^2 = \mathbf{4623.5}$$

We see that the same digits occur in the same order. Only the place values have changed. A quick way to shift place values is the shift the location of the decimal point. **To multiply a decimal number by a positive power of 10, we shift the decimal point to the right the number of places indicated by the exponent.**

Thinking Skill

Justify

Use the quick way to multiply 1.5×10^6. Explain your process.

Sometimes powers of 10 are written with words instead of with digits. For example, we might read that 1.5 million spectators lined the parade route. The expression 1.5 million means $1.5 \times 1{,}000{,}000$, which is 1,500,000.

Example 4

Write $2\frac{1}{2}$ billion in standard form.

Solution

First we write $2\frac{1}{2}$ as the decimal number 2.5. Then we multiply by one billion (10^9), which shifts the decimal point 9 places to the right.

$$2.5 \text{ billion} = 2.5 \times 10^9 = \mathbf{2{,}500{,}000{,}000}$$

dividing by powers of 10

When dividing by positive powers of 10, the quotient has the same digits as the dividend, only with smaller place values. As with multiplication, we may shift place values by shifting the location of the decimal point.

$$4.75 \div 10^3 = \mathbf{0.00475}$$

To divide a number by a positive power of 10, we shift the decimal point to the left the number of places indicated by the exponent.

Example 5

Divide: $3.5 \div 10^4$

Solution

The decimal point of the quotient is 4 places to the left of the decimal point in 3.5.

$$3.5 \div 10^4 = \mathbf{0.00035}$$

Practice Set

a. Represent Write 456 in expanded notation using powers of 10

Simplify:

b. 24.25×10^3

c. 25×10^6

d. $12.5 \div 10^3$

e. $4.8 \div 10^4$

Generalize Find each missing exponent:

f. $10^3 \cdot 10^4 = 10^{\square}$

g. $10^8 \div 10^2 = 10^{\square}$

h. $(10^4)^2 = 10^{\square}$

i. $n^3 \cdot n^4 = n^{\square}$

Write each of the following numbers in standard form:

j. $2\frac{1}{2}$ million

k. 15 billion

l. 1.6 trillion

Written Practice

Strengthening Concepts

Refer to the graph to answer problems **1–3.**

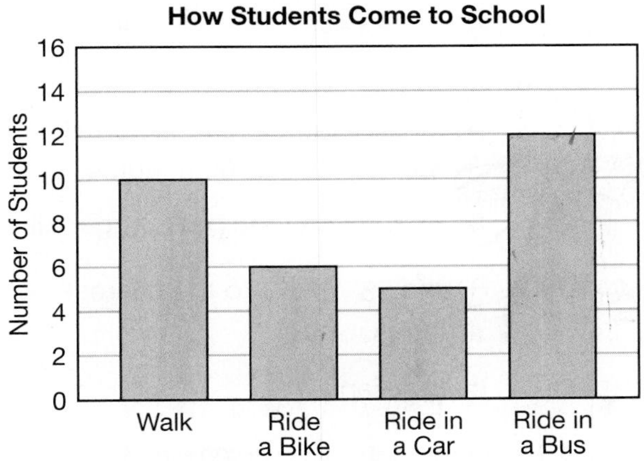

1. Answer true or false:
(38)

a. Twice as many students walk to school as ride to school in a car.

b. The majority of the students ride to school in either a bus or car.

2. What is the ratio of those who walk to school to those who ride in a bus?
(36, 38)

3. What fraction of the students ride in a bus?
(38)

4. Explain What is the mean (average) of these numbers? Describe how you found your answer and how you can check your work.
(28, 35)

1.2, 1.4, 1.5, 1.7, 2

* **5.** Connect The newspaper reported that 134.8 million viewers watched the Super Bowl. Write the number of viewers in standard form.
(47)

6. Read this statement. Then answer the questions that follow.
(22)

One eighth of the 40 paintings were abstract.

a. How many of the paintings were abstract?

b. How many of the paintings were not abstract?

*** 7.** A gallon of water (128 ounces) is poured into 12-ounce glasses.
(44)

a. How many glasses can be filled to the top?

b. How many glasses are needed to hold all of the water?

8. A cubit is an ancient unit of measure equal to the distance from the elbow to the fingertips.
(8)

a. Estimate the number of inches from your elbow to your fingertips.

b. Measure the distance from your elbow to your fingertips to the nearest inch.

*** 9.** **a.** Write 0.375 as a fraction.
(43)

b. Write $62\frac{1}{2}\%$ as a decimal.

*** 10.** **Represent** The park ranger can hike at a pace of 1 mile per 20 minutes.
(16, 46)

a. Make a function table that shows the miles hiked (output) in 10, 20, 30, 40, 50, and 60 minutes.

b. How far can the park ranger hike at that pace in two hours?

*** 11.** Round $53{,}714.\overline{54}$ to the nearest
(42)

a. thousandth.

b. thousand.

*** 12.** Find each missing exponent:
(47)

a. $10^5 \cdot 10^2 = 10^{\square}$ **b.** $10^8 \div 10^4 = 10^{\square}$

13. The point marked by the arrow represents what decimal number?
(34)

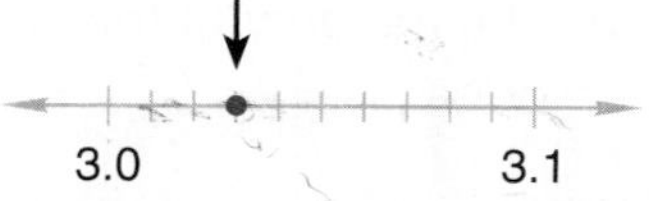

*** 14.** **Analyze** In figure *ABCDEF* all angles are right angles and $AF = AB = BC$. Segment *BC* is twice the length of $\overline{CD}$. If *CD* is 3 cm, what is the perimeter of the figure?
(19)

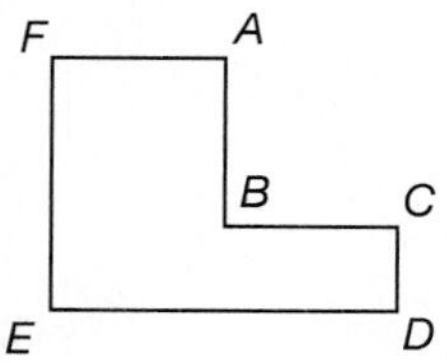

15. **Represent** Use a compass to draw a circle with a radius of 1 inch. Then inscribe a regular hexagon in the circle.
(Inv. 2)

a. What is the diameter of the circle?

b. What is the perimeter of the regular hexagon?

Solve:

16. $\frac{6}{10} = \frac{w}{100}$ (39)

17. $\frac{3.6}{x} = \frac{16}{24}$ (39)

18. $\frac{a}{1.5} = 1.5$ (35)

19. $9.8 = x + 8.9$ (35)

Estimate each answer to the nearest whole number. Then perform the calculation.

20. $4\frac{1}{5} + 5\frac{1}{3} + \frac{1}{2}$ (30)

21. $6\frac{1}{8} - \left(5 - 1\frac{2}{3}\right)$ (23,30)

Explain Simplify:

22. $\sqrt{16 \cdot 25}$ (20)

* **23.** 3.6×10^3 (47)

24. $8\frac{1}{3} \times 3\frac{3}{5} \times \frac{1}{3}$ (26)

25. $3\frac{1}{8} \div 6\frac{1}{4}$ (26)

26. $26.7 + 3.45 + 0.036 + 12 + 8.7$ (35)

27. (19, 37) The figures below illustrate one triangle rotated into three different positions. Dimensions are in inches.

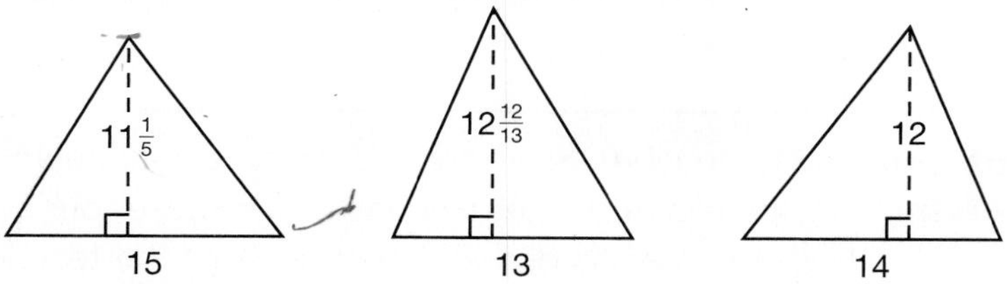

a. What is the perimeter of the triangle?

b. What is the area of the triangle?

* **28.** (47) **Connect** Simplify and compare: $125 \div 10^2 \bigcirc 0.125 \times 10^2$

29. (30) Arrange these numbers in order from least to greatest:

$$\frac{2}{3}, \frac{1}{2}, \frac{7}{12}, \frac{5}{6}$$

* **30.** (40) In this figure find the measure of

a. $\angle a$.

b. $\angle b$.

c. **Explain** Describe how to find the measure of $\angle c$.

d. Name a supplementary angle pair in the figure.

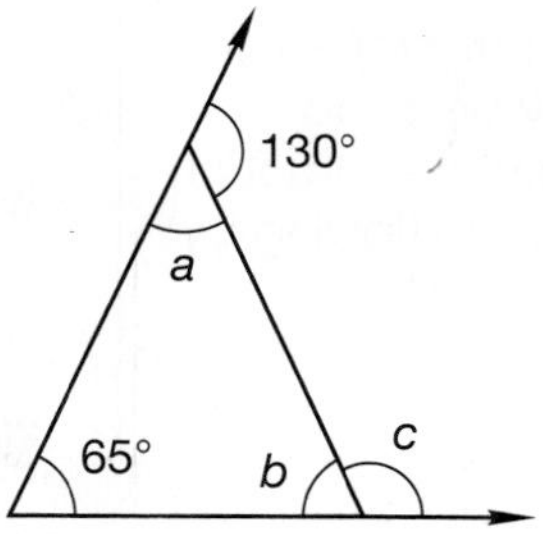

LESSON 48

• Fraction-Decimal-Percent Equivalents

Power Up

Building Power

facts Power Up J

mental math

a. Calculation: $7 \times \$35.00$

b. Decimals: $12.75 \div 10$

c. Equivalent Fractions: $\frac{6}{4} = \frac{9}{n}$

d. Exponents: $\frac{10^2}{5^2}$

e. Estimation: $6\frac{1}{6} \times 3\frac{4}{5}$

f. Fractional Parts: $\frac{3}{8}$ of 80

g. Statistics: Find the range of the set of numbers: 89, 99, 29, 89.

h. Calculation: 10×8, $+ 1$, $\sqrt{\ }$, $+ 2$, $\times 4$, $- 2$, $\div 6$, $\times 9$, $+ 1$, $\sqrt{\ }$, $\div 2$, $\div 2$, $\div 2$

problem solving

The teacher asked for two or three volunteers. Adam, Blanca, Chad and Danielle raised their hands. From these four students, how many possible combinations of two students could the teacher select? How many different combinations of three students could the teacher select?

New Concept

Increasing Knowledge

We may describe part of a whole using a fraction, a decimal, or a percent.

Thinking Skill

Identify

What property allows us to rename a fraction by multiplying by a form of 1?

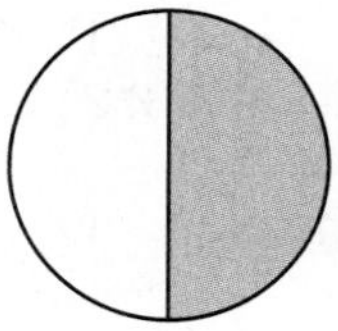

$\frac{1}{2}$ of the circle is shaded.

0.5 of the circle is shaded.

50% of the circle is shaded.

Recall that when we rename a fraction, we multiply by a form of 1 such as $\frac{2}{2}$, $\frac{5}{5}$, or $\frac{100}{100}$. Another form of 1 is 100%, so to convert a fraction or a decimal to a percent, we multiply the number by 100%.

Connect Explain how we know that 100% is a form of 1.

Example 1

Write $\frac{7}{10}$ as a percent.

Solution

To change a number to its percent equivalent, we multiply the number by 100%.

$$\frac{7}{10} \times 100\% = \frac{700\%}{10} = \mathbf{70\%}$$

Example 2

Write $\frac{2}{3}$ as a percent.

Solution

We multiply by 100 percent.

$$\frac{2}{3} \times 100\% = \frac{200\%}{3} = \mathbf{66\frac{2}{3}\%}$$

Notice the mixed-number form of the percent.

Example 3

Write 0.8 as a percent.

Solution

We multiply 0.8 by 100%.

$$0.8 \times 100\% = \mathbf{80\%}$$

Example 4

Complete the table.

Fraction	Decimal	Percent
$\frac{1}{3}$	**a.**	**b.**
c.	1.5	**d.**
e.	**f.**	60%
$\frac{2}{1}$	**g.**	**h.**

Solution

For **a** and **b** we find decimal and percent equivalents of $\frac{1}{3}$.

a. $3\overline{)1.00}$ = $\mathbf{0.\overline{3}}$

b. $\frac{1}{3} \times 100\% = \frac{100\%}{3} = \mathbf{33\frac{1}{3}\%}$

For **c** and **d** we find a fraction (or a mixed number) and a percent equivalent to 1.5.

c. $1.5 = 1\frac{5}{10} = \mathbf{1\frac{1}{2}}$

d. $1.5 \times 100\% = \mathbf{150\%}$

For **e** and **f** we find fraction and decimal equivalents of 60%.

e. $60\% = \frac{60}{100} = \mathbf{\frac{3}{5}}$

f. $60\% = \frac{60}{100} = \mathbf{0.6}$

For **g** and **h,** the whole number 2 is shown as the fraction $\frac{2}{1}$.

g. 2

h. $2 \times 100\% = \mathbf{200\%}$

Practice Set

Connect Use your calculator to complete the table.

Fraction	Decimal	Percent
$\frac{2}{3}$	**a.**	**b.**
c.	1.1	**d.**
e.	**f.**	4%
g.	3	**h.**

Written Practice

Strengthening Concepts

* **1.** (46) Bicyclist Mario Cipollini pedaled one stage of the 1999 Tour de France in about 3.9 hours. If he traveled about 195 kilometers, what was his average speed in kilometers per hour?

2. (24) Write the prime factorization of 1008 and 1323. Then reduce $\frac{1008}{1323}$.

3. (12) In 1803 the United States purchased the Louisiana Territory from France for $15 million. In 1867 the United States purchased Alaska from Russia for $7.2 million. The purchase of Alaska occurred how many years after the purchase of the Louisiana Territory?

* **4.** (36, 43) Analyze Red and blue marbles were in the bag. Five twelfths of the marbles were red.

a. What fraction of the marbles were blue?

b. What was the ratio of red marbles to blue marbles?

c. Analyze If one marble is taken from the bag, what is the probability of taking a red marble? Write the probability as a fraction and as a decimal number rounded to two places.

* **5.** (46) A 9-ounce can of peaches sells for $1.26. Find the unit price.

6. (28) The average of two numbers is the number halfway between the two numbers. What number is halfway between two thousand, five hundred fifty and two thousand, nine hundred?

7. (22) Diagram this statement. Then answer the questions that follow.

Van has read five eighths of the 336-page novel.

a. How many pages has Van read?

b. How many more pages does he have to read?

* **8.** (19) Analyze Copy this figure on your paper. Find the length of the unmarked sides, and find the perimeter of the polygon. Dimensions are in centimeters. All angles are right angles.

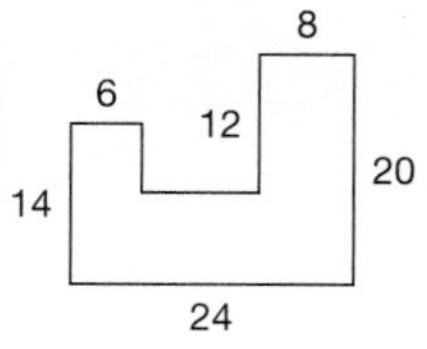

9. (38) The graph shows how one family spends their annual income. Use this graph to answer **a–d.**

How Income Is Spent

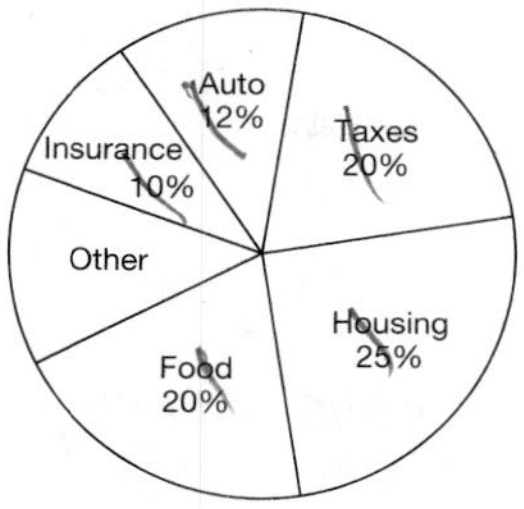

a. What percent of the family's income is spent on "other"?

b. What fraction of the family's income is spent on food?

c. If $3200 is spent on insurance, how much is spent on taxes?

d. Would this data be better displayed on a bar graph? Explain your answer.

10. (42) Write $0.\overline{54}$ as a decimal rounded to three decimal places.

11. (32) **a.** Estimate the length of $\overline{AB}$ in centimeters.

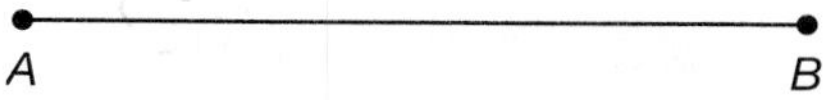

b. Use a centimeter scale to find the length of $\overline{AB}$ to the nearest centimeter.

* **12.** (20, 47) **a.** Identify the exponent and the base in the expression 5^3.

b. **Generalize** Find the value of n: $10^4 \cdot 10^4 = 10^n$

13. (18, 19) If the perimeter of a regular hexagon is 1 foot, each side is how many inches long?

* **14.** (48) Mentally complete the table.

Fraction	Decimal	Percent
$\frac{1}{2}$	**a.**	**b.**
c.	0.1	**d.**
e.	**f.**	25%
$\frac{4}{1}$	**g.**	**h.**

* **15.** (46) **Analyze** The moped traveled 78 miles on 1.2 gallons of gas. The moped averaged how many miles per gallon?

Solve:

16. (39) $\frac{6}{100} = \frac{15}{w}$

* **17.** (39) $\frac{20}{x} = \frac{15}{12}$

18. (35) $1.44 = 6m$

19. (30) $\frac{1}{2} = \frac{1}{3} + f$

Simplify:

20. (20) $2^5 + 1^4 + 3^3$

21. (20) $\sqrt{10^2 \cdot 6^2}$

22. (30) $3\frac{5}{6} - \left(1\frac{1}{4} + 1\frac{1}{6}\right)$

23. (23,30) $8\frac{3}{4} + \left(4 - \frac{2}{3}\right)$

24. (26) $\frac{15}{16} \cdot \frac{24}{25} \cdot 1\frac{1}{9}$

25. (26) $1\frac{1}{3} \div \left(2\frac{2}{3} \div 4\right)$

* **26.** (41, 45) **Conclude** Find the value of $\frac{a}{b}$ when a = \$13.93 and b = 0.07.

27. (Inv. 3, 37) The coordinates of three vertices of a triangle are $(-1, -1)$, $(5, -1)$, and $(5, -4)$. What is the area of the triangle?

28. (36, 38) Students in the class were asked how many siblings they had, and the answers were tallied. If one student from the class is selected at random, what is the probability that the selected student would have more than one sibling?

<table>
<tr><th>Number of Siblings</th><th>Number of Students</th></tr>
<tr><td>0</td><td>|||</td></tr>
<tr><td>1</td><td>𝍸 ||||</td></tr>
<tr><td>2</td><td>𝍸</td></tr>
<tr><td>3</td><td>||</td></tr>
<tr><td>4 or more</td><td>|</td></tr>
</table>

* **29.** (38) **Represent** Sketch a graph that displays the information in the table in problem **28.** Choose an appropriate type of graph and justify your selection. Make one axis the number of siblings and the other axis the number of students.

30. (40) Find the measures of $\angle a$, $\angle b$, and $\angle c$ in the figure at right.

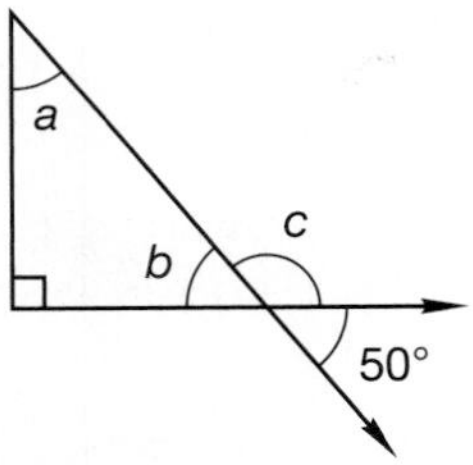

LESSON

49

Adding and Subtracting Mixed Measures

Power Up

Building Power

facts Power Up J

mental math

a. **Calculation:** $8 \times \$6.50$

b. **Decimals:** 25.75×10

c. **Equivalent Fractions:** $\frac{4}{x} = \frac{40}{100}$

d. **Estimation:** $12.11 \div 1.9$

e. **Square Roots:** $\sqrt{400}$

f. **Fractional Parts:** $\frac{3}{10}$ of 200

g. **Statistics:** Find the median of: 67, 34, 90, 23, 200.

h. **Calculation:** Find the sum, difference, product, and quotient of $\frac{3}{5}$ and $\frac{1}{3}$.

problem solving

Javier used a six-yard length of string to make a rectangle that was twice as long as it was wide. What was the area of the rectangle in square feet?

New Concept

Increasing Knowledge

A mixed measure is a measurement that includes different units from the same category (length, volume, time, etc.).

The movie was 1 hour 48 minutes long.

To add mixed measures, we align the numbers in order to add units that are the same. Then we simplify when possible.

Example 1

Add and simplify: 1 yd 2 ft 7 in. + 2 yd 2 ft 8 in.

Solution

We add like units, and then we simplify from right to left.

$$\begin{array}{rrrr} & 1\text{ yd} & 2\text{ ft} & 7\text{ in.} \\ + & 2\text{ yd} & 2\text{ ft} & 8\text{ in.} \\ \hline & 3\text{ yd} & 4\text{ ft} & 15\text{ in.} \end{array}$$

We change 15 in. to 1 ft 3 in. and add to 4 ft. Now we have

3 yd 5 ft 3 in.

Then we change 5 ft to 1 yd 2 ft and add to 3 yd. Now we have

4 yd 2 ft 3 in.

Example 2

Add and simplify:

2 hr 40 min 35 s
+ 1 hr 45 min 50 s

Solution

We add. Then we simplify from right to left.

$$\begin{array}{r} 2\text{ hr } 40\text{ min } 35\text{ s} \\ +\ 1\text{ hr } 45\text{ min } 50\text{ s} \\ \hline 3\text{ hr } 85\text{ min } 85\text{ s} \end{array}$$

Thinking Skill

Connect

What do you need to know to simplify 3 hr 85 min 85 s?

We change 85 s to 1 min 25 s and add to 85 min. Now we have

3 hr 86 min 25 s

Then we simplify 86 min to 1 hr 26 min and combine hours.

4 hr 26 min 25 s

We have practiced adding mixed measures. Now we will learn to subtract them. When subtracting mixed measures, it may be necessary to convert units.

Example 3

Subtract: **5 days 10 hr 15 min**
− 1 day 15 hr 40 min

Solution

Before we can subtract minutes, we must convert 1 hour to 60 minutes. We combine 60 minutes and 15 minutes, making 75 minutes. Then we can subtract.

$$\begin{array}{r} \scriptstyle 9 \quad (60\text{ min}) \\ 5\text{ days } \cancel{10}\text{ hr } 15\text{ min} \\ -\ 1\text{ day } \ 15\text{ hr } 40\text{ min} \\ \hline \end{array} \longrightarrow \begin{array}{r} \scriptstyle 9 \qquad 75 \\ 5\text{ days } \cancel{10}\text{ hr } \cancel{15}\text{ min} \\ -\ 1\text{ day } \ 15\text{ hr } 40\text{ min} \\ \hline 35\text{ min} \end{array}$$

Next we convert 1 day to 24 hours and complete the subtraction.

$$\longrightarrow \begin{array}{r} \scriptstyle (24\text{ hr}) \\ \scriptstyle 4 \qquad 9 \qquad 75 \\ \cancel{5}\text{ days } \cancel{10}\text{ hr } \cancel{15}\text{ min} \\ -\ 1\text{ day } \ 15\text{ hr } 40\text{ min} \\ \hline 35\text{ min} \end{array} \longrightarrow \begin{array}{r} \scriptstyle 33 \\ \scriptstyle 4 \qquad \cancel{9} \qquad 75 \\ \cancel{5}\text{ days } \cancel{10}\text{ hr } \cancel{15}\text{ min} \\ -\ 1\text{ day } \ 15\text{ hr } 40\text{ min} \\ \hline \mathbf{3\text{ days } 18\text{ hr } 35\text{ min}} \end{array}$$

Conclude How is subtracting mixed measures similar to subtracting whole numbers with the same measure? How is it different?

Example 4

Subtract: 4 yd 3 in. − 2 yd 1 ft 8 in.

Solution

We carefully align the numbers with like units. We convert 1 yard to 3 feet.

$$\begin{array}{r} \overset{3}{\cancel{4}}\text{ yd} \quad (3\text{ ft}) \quad 3\text{ in.} \\ -\ 2\text{ yd }1\text{ ft }8\text{ in.} \\ \hline \end{array}$$

Next we convert 1 foot to 12 inches. We combine 12 inches and 3 inches, making 15 inches. Then we can subtract.

$$\begin{array}{r} \overset{3}{\cancel{4}}\text{ yd }\overset{2}{\cancel{3}}\text{ ft }\overset{15}{\cancel{3}}\text{ in.} \\ -\ 2\text{ yd }1\text{ ft }8\text{ in.} \\ \hline \mathbf{1\ yd\ 1\ ft\ 7\ in.} \end{array}$$

Practice Set

a. Change 70 inches to feet and inches.

b. Change 6 feet 3 inches to inches.

c. Simplify: 5 ft 20 in.

d. Add: 2 yd 1 ft 8 in. + 1 yd 2 ft 9 in.

e. Add: 5 hr 42 min 53 s + 6 hr 17 min 27 s

Connect Subtract:

f.
$$\begin{array}{r} 3\text{ hr} \qquad\qquad\quad 3\text{ s} \\ -\ 1\text{ hr }15\text{ min }55\text{ s} \\ \hline \end{array}$$

g.
$$\begin{array}{r} 8\text{ yd }1\text{ ft }5\text{ in.} \\ -\ 3\text{ yd }2\text{ ft }7\text{ in.} \\ \hline \end{array}$$

h. 2 days 3 hr 30 min − 1 day 8 hr 45 min

i. *Justify* How can you check your answer to exercise **h?**

Written Practice

Strengthening Concepts

*** 1.** (35, 45) What is the quotient when the sum of 0.2 and 0.05 is divided by the product of 0.2 and 0.05?

2. (44) Darren carried the football 20 times and gained a total of 184 yards. What was the average number of yards he gained on each carry? Write the answer as a decimal number.

*** 3.** (46) *Analyze* At a liquidation sale Louisa bought two dozen pens for six dollars. What was the cost of each pen?

4. (18) Jeffrey counted the sides on three octagons, two hexagons, a pentagon, and two quadrilaterals. Altogether, how many sides did he count?

5. (28, 35) What is the mean of these numbers?

$$6.21, 4.38, 7.5, 6.3, 5.91, 8.04$$

6. (22, 36) Read this statement. Then answer the questions that follow.

Only two ninths of the 72 billy goats were gruff. The rest were cordial.

a. How many of the billy goats were cordial?

b. What was the ratio of gruff billy goats to cordial billy goats?

7. (42) Arrange these numbers in order from least to greatest:

$$0.\overline{5}, 0.5, 0.\overline{54}$$

8. (8) **a.** Estimate the length of segment *AB* in inches.

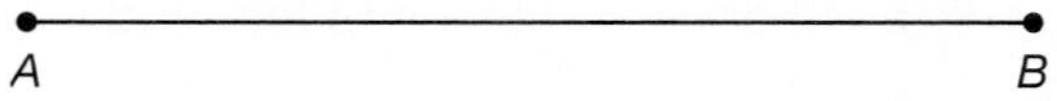

b. Measure the length of segment *AB* to the nearest eighth of an inch.

*** 9.** (48) **Connect** Write each of these numbers as a percent:

a. 0.9 **b.** $1\frac{3}{5}$ **c.** $\frac{5}{6}$

*** 10.** (48) **Connect** Complete the table.

Fraction	Decimal	Percent
a.	**b.**	75%
c.	**d.**	5%

*** 11.** (13) **Analyze** Mathea's resting heart rate is 62 beats per minute. While she is resting, about how many times will her heart beat in an hour?

*** 12.** (14, 43) What is the probability of rolling an even prime number with one roll of a number cube? Write the probability as a fraction and as a decimal number rounded to the nearest hundredth.

*** 13.** (37) A $\frac{1}{2}$-by-$\frac{1}{2}$-inch square was cut from a 1-by-1-inch square.

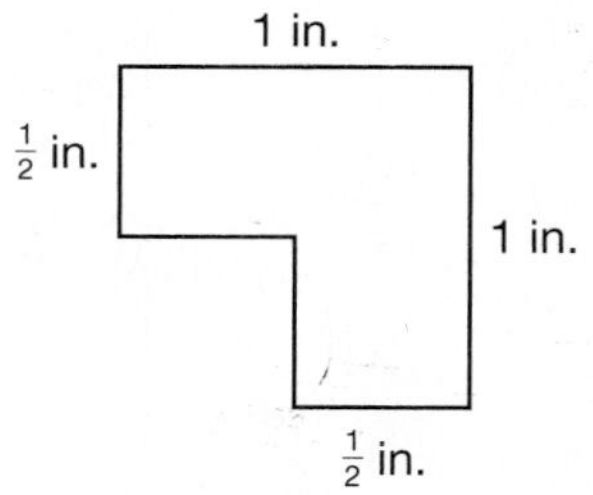

a. What was the area of the original square?

b. What is the area of the square that was removed?

c. What is the area of the remaining figure?

14. (19) What is the perimeter of the figure in problem 13?

* **15.** (37) **Generalize** The figures below show a triangle with sides 6 cm, 8 cm, and 10 cm long in three orientations. What is the height of the triangle when the base is

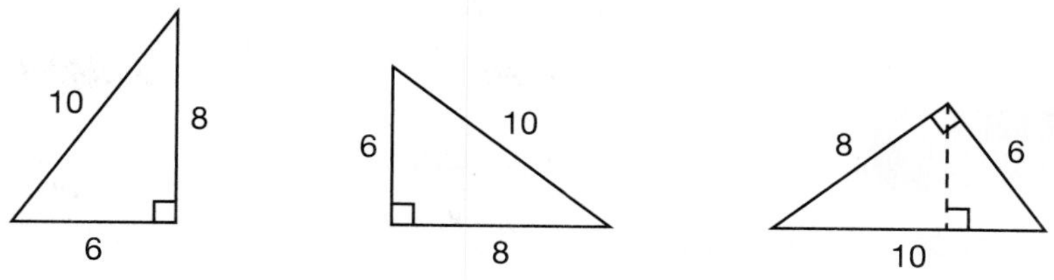

a. 6 cm? **b.** 8 cm? **c.** 10 cm?

Solve:

16. (39) $\frac{y}{100} = \frac{18}{45}$

17. (39) $\frac{35}{40} = \frac{1.4}{m}$

18. (30) $\frac{1}{2} - n = \frac{1}{6}$

19. (35) $9d = 2.61$

Simplify:

20. (20) $\sqrt{100} + 4^3$

* **21.** (47) 3.14×10^4

22. (23, 30) $3\frac{3}{4} + \left(4\frac{1}{6} - 2\frac{1}{2}\right)$

23. (26) $6\frac{2}{3} \cdot \left(3\frac{3}{4} \div 1\frac{1}{2}\right)$

* **24.** (49) **Connect**

3 days 8 hr 15 min
\+ 2 days 15 hr 45 min

* **25.** (49)

4 yd 1 ft 3 in.
− 2 yd 1 ft 9 in.

26. (45) $18.00 \div 0.06$

27. (29, 33) How would you estimate the quotient when 35.675 is divided by $2\frac{7}{8}$? What is your estimate?

* **28.** (36) **Discuss** A spinner is spun once. Derrick says the sample space is {1, 2, 3}. Jorge says the sample space is {1, 1, 1, 2, 2, 3}. Whose sample space is better? Explain why?

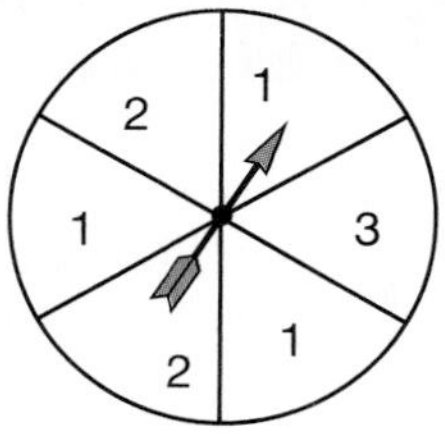

29. (41) Evaluate: LWH if $L = 0.5$, $W = 0.2$, and $H = 0.1$

30. (40) This quadrilateral is a rectangle. Find the measures of $\angle a$, $\angle b$, and $\angle c$. Name a pair of angles in the quadrilateral that are complementary.

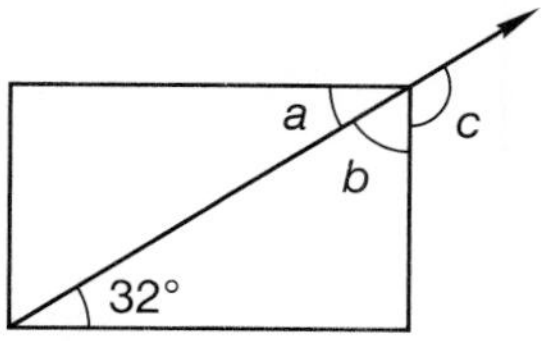

LESSON

50

• Unit Multipliers and Unit Conversion

Power Up

Building Power

facts

Power Up I

mental math

a. Calculation: $5 \times \$48.00$

b. Decimals/Exponents: 0.0125×10^2

c. Equivalent Fractions: $\frac{y}{20} = \frac{40}{100}$

d. Square Roots: $\sqrt{225}$

e. Estimation: $4\frac{3}{4} \times 1\frac{7}{8}$

f. Fractional Parts: $\frac{2}{5}$ of 40

g. Statistics: Find the median of the set of numbers: 89, 45, 47, 32.

h. Calculation: Start with a half dozen, + 4, square that number, ÷ 2, + 6, ÷ 8, × 7, + 1, ÷ 10, − 10.

problem solving

Alan wanted to form a triangle out of straws that were 5 cm, 7 cm, and 12 cm long. He threaded a piece of string through the three straws, pulled the string tight, and tied it. What was the area of the triangle formed by the three straws?

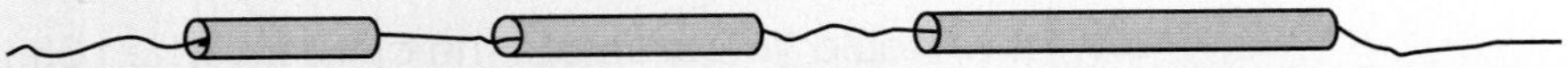

New Concept

Increasing Knowledge

Let's take a moment to review the procedure for reducing a fraction. When we reduce a fraction, we can replace factors that appear in both the numerator and denominator with 1s, since each pair reduces to 1.

$$\frac{24}{36} = \frac{\overset{1}{\cancel{2}} \cdot \overset{1}{\cancel{2}} \cdot 2 \cdot \overset{1}{\cancel{3}}}{\underset{1}{\cancel{2}} \cdot \underset{1}{\cancel{2}} \cdot 3 \cdot \underset{1}{\cancel{3}}} = \frac{2}{3}$$

Also, recall that we can reduce before we multiply. This is sometimes called **canceling.**

$$\frac{2}{\underset{1}{\cancel{3}}} \cdot \frac{\overset{1}{\cancel{3}}}{5} = \frac{2}{5}$$

Thinking Skill

Explain

Why is $\frac{12\text{ in.}}{1\text{ ft}}$ equal to 1?

We can apply this procedure to units as well. We may cancel units before we multiply.

$$\frac{5\ \cancel{\text{ft}}}{1} \cdot \frac{12\text{ in.}}{1\ \cancel{\text{ft}}} = 60\text{ in.}$$

In this instance we performed the division 5 ft ÷ 1 ft, which means, "How many feet are in 5 feet?" The answer is simply 5. Then we multiplied 5 by 12 in.

We remember that we change the name of a number by multiplying by a fraction whose value equals 1. Here we change the name of 3 to $\frac{12}{4}$ by multiplying by $\frac{4}{4}$:

$$3 \cdot \frac{4}{4} = \frac{12}{4}$$

The fraction $\frac{12}{4}$ is another name for 3 because 12 ÷ 4 = 3.

When the numerator and denominator of a fraction are equal (and are not zero), the fraction equals 1. There is an unlimited number of fractions that equal 1. A fraction equal to 1 may have units, such as

$$\frac{12 \text{ inches}}{12 \text{ inches}}$$

Since 12 inches equals 1 foot, we can write two more fractions that equal 1.

$$\frac{12 \text{ inches}}{1 \text{ foot}}$$

and,

$$\frac{1 \text{ foot}}{12 \text{ inches}}$$

Because these fractions have units and are equal to 1, we call them **unit multipliers.** Unit multipliers are very useful for converting from one unit of measure to another. For instance, if we want to convert 5 feet to inches, we can multiply 5 feet by a multiplier that has inches on top and feet on bottom. The feet units cancel and the product is 60 inches.

$$5 \cancel{\text{ft}} \cdot \frac{12 \text{ in.}}{1 \cancel{\text{ft}}} = 60 \text{ in.}$$

If we want to convert 96 inches to feet, we can multiply 96 inches by a unit multiplier that has a numerator of feet and a denominator of inches. The inch units cancel and the product is 8 feet.

$$96 \cancel{\text{in.}} \cdot \frac{1 \text{ ft}}{12 \cancel{\text{in.}}} = 8 \text{ ft}$$

Notice that we selected a unit multiplier that canceled the unit we wanted to remove and kept the unit we wanted in the answer.

When we set up unit conversion problems, we will write the numbers in this order:

Given measure	×	Unit multiplier	=	Converted measure

Connect Why is multiplying by unit multipliers the same as using the Identity Property of Multiplication.

Example 1

Write two unit multipliers for these equivalent measures:

$$\mathbf{3\ ft = 1\ yd}$$

Solution

We write one measure as the numerator and its equivalent as the denominator.

$$\frac{\mathbf{3\ ft}}{\mathbf{1\ yd}} \quad \text{and} \quad \frac{\mathbf{1\ yd}}{\mathbf{3\ ft}}$$

Example 2

Select unit multipliers from example 1 to convert

a. 240 yards to feet.

b. 240 feet to yards.

Solution

a. We are given a measure in yards. We want the answer in feet. So we write the following:

$$240\ \text{yd} \cdot \boxed{\text{Unit multiplier}} = \quad \text{ft}$$

We want to cancel the unit "yd" and keep the unit "ft," so we select the unit multiplier that has a numerator of ft and a denominator of yd. Then we multiply and cancel units.

$$240\ \cancel{\text{yd}} \cdot \frac{3\ \text{ft}}{1\ \cancel{\text{yd}}} = \mathbf{720\ ft}$$

We know our answer is reasonable because feet are shorter units than yards, and therefore it takes more feet than yards to measure the same distance.

b. We are given the measure in feet, and we want the answer in yards. We choose the unit multiplier that has a numerator of yd.

$$240\ \cancel{\text{ft}} \cdot \frac{1\ \text{yd}}{3\ \cancel{\text{ft}}} = \mathbf{80\ yd}$$

We know our answer is reasonable because yards are longer units than feet, and therefore it takes fewer yards than feet to measure the same distance.

Connect What unit multiplier would you use in **b** if you needed to convert 240 ft to inches?

Example 3

An Olympic event in track and field is the 100 meter dash. One hundred meters is about how many yards? (1 m ≈ 1.1 yd)

Solution

We can use unit multipliers to convert between the metric system and the U.S. Customary System.

$$100\ \cancel{m} \cdot \frac{1.1\ \text{yd}}{1\ \cancel{m}} \approx \mathbf{110\ yd}$$

We may also use unit multipliers to convert rates. We write the rate as a ratio and multiply the ratio by a unit multiplier to convert to the desired unit.

Example 4

Tim can sprint 9 yards per second. Convert this rate to feet per second.

Solution

We write the rate as a ratio.

$$\frac{9\ \text{yd}}{1\ \text{sec}}$$

To convert yards to feet we multiply by a unit multiplier that has yards and feet and that cancels yards. Three feet equals 1 yard.

$$\frac{9\ \cancel{\text{yd}}}{1\ \text{sec}} \cdot \frac{3\ \text{ft}}{1\ \cancel{\text{yd}}} = \frac{27\ \text{ft}}{1\ \text{sec}}$$

Tim can sprint at a rate of **27 feet per second.**

Practice Set

Write two unit multipliers for each pair of equivalent measures:

a. 1 yd = 36 in.

b. 100 cm = 1 m

c. 16 oz = 1 lb

Use unit multipliers to answer problems **d–f.**

d. Convert 10 yards to inches.

e. Twenty-four feet is how many yards (1 yd = 3 ft)?

f. Conclude Which is greater 20 inches or 50 centimeters (1 in. = 2.54 cm)?

20 in. ◯ 50 cm

Connect Use unit multipliers to convert the rates in **g** and **h.**

g. Convert 20 miles per gallon to miles per quart (1 gal = 4 qt).

h. When sleeping Diana's heart beats 60 times per minute. Convert 60 beats per minute to beats per hour.

Written Practice *Strengthening Concepts*

1. (35) When the product of 3.5 and 0.4 is subtracted from the sum of 3.5 and 0.4, what is the difference?

2. (14, 43) The face of the spinner is divided into ten congruent parts.

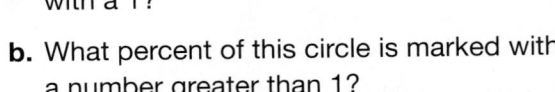

a. What fraction of this circle is marked with a 1?

b. What percent of this circle is marked with a number greater than 1?

c. If the spinner is spun, what is the probability that it will stop on a number greater than 2? Write the probability as a decimal number.

* **3.** (46) The 18-ounce box of oatmeal costs \$1.44. Find the unit price.

* **4.** (46) **Generalize** Nelson covered the first 20 miles in $2\frac{1}{2}$ hours. What was his average speed in miles per hour?

5. (28) The parking lot charges \$2 for the first hour plus 50¢ for each additional half hour or part thereof. What is the total charge for parking in the lot for 3 hours 20 minutes?

* **6.** (50) **Generalize** Monique ran one mile in 6 minutes. Her average speed was how many miles per hour? Use a unit multiplier to make the conversion. (1 hr = 60 min.)

7. (22) Read this statement. Then answer the questions that follow.

Forty percent of the 30 members of the drama club were members of the senior class.

a. How many members of the drama club were seniors?

b. What percent of the drama club were not members of the senior class?

8. (8) Which percent best identifies the shaded part of this circle?

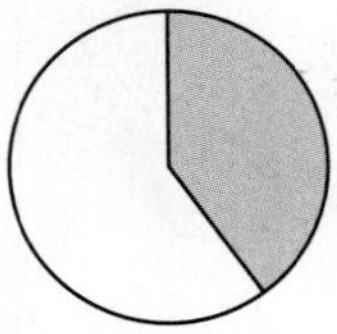

A 25% **B** 40%

C 50% **D** 60%

9. (43) Write $3\frac{5}{6}$ as a decimal number rounded to four decimal places.

* **10.** (47) **Represent** Use exponents to write 7.5 million in expanded notation.

* **11.** (48) Write each number as a percent:

a. 0.6 **b.** $\frac{1}{6}$ **c.** $1\frac{1}{2}$

* **12.** Complete the table.
(48)

Fraction	Decimal	Percent
a.	**b.**	30%
c.	**d.**	250%
e.	5	**f.**

13. List the prime numbers between 90 and 100.
(21)

14. The dashes divide this figure into a rectangle and a triangle.
(37)

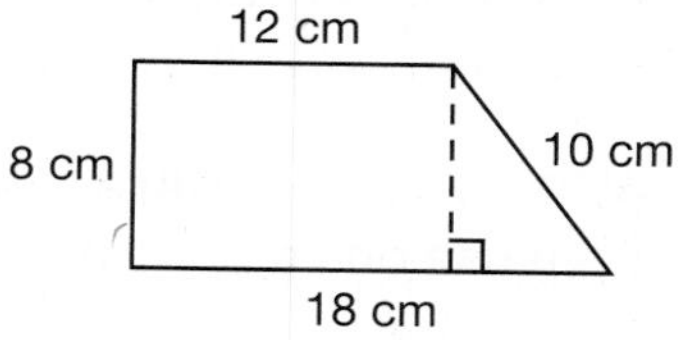

a. What is the area of the rectangle?

b. What is the area of the triangle?

c. What is the combined area of the rectangle and triangle?

15. Use a compass to draw a circle with a radius of $1\frac{1}{2}$ in. Then use a protractor to draw a central angle that measures 60°. Shade the sector that is formed by the 60° central angle.
(Inv. 2)

Explain Solve:

16. $\frac{10}{x} = \frac{7}{42}$
(39)

17. $\frac{1.5}{1} = \frac{w}{4}$
(39)

18. $3.56 = 5.6 - y$
(35)

19. $\frac{3}{20} = w + \frac{1}{15}$
(30)

20. Which property is illustrated by each of the following equations?
(2, 41)

a. $x(y + z) = xy + xz$

b. $x + y = y + x$

c. $1x = x$

* **21.** **Generalize** Which is equivalent to $\frac{10^6}{10^2}$?
(47)

A 10^3

B 10^4

C 1000

D 30

22. The coordinates of three vertices of a square are (2, 0), (0, −2), and (−2, 0).
(Inv. 3)

a. What are the coordinates of the fourth vertex?

b. Counting whole square units and half square units, find the area of the square.

* **23.** If 10 muffins are shared equally by 4 children, how many muffins will each child receive?
(44)

Thinking Skill

Summarizing

State the type of data shown at each of the five points on the box-and-whisker-plot.

* **24.** (Inv. 4) **Analyze** Below is a box-and-whisker plot of test scores. Refer to the plot to answer **a–c.**

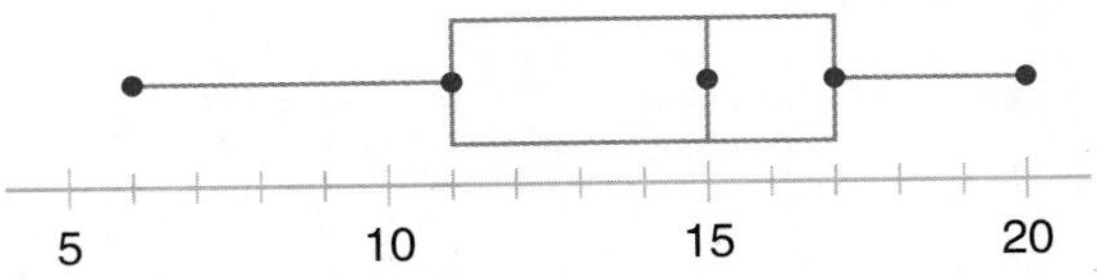

a. What is the range of scores?

b. What is the median score?

c. Write another question that can be answered by referring to the plot. Then answer the question.

* **25.** (50) Write two unit multipliers for the conversion 10 mm = 1 cm. Then use one of the unit multipliers to convert 160 mm to centimeters.

* **26.** (49) 4 yd 2 ft 7 in. + 3 yd 5 in.

27. (26, 30) $5\frac{1}{6} - \left(1\frac{3}{4} \div 2\frac{1}{3}\right)$

28. (26, 30) $3\frac{5}{7} + \left(3\frac{1}{8} \cdot 2\frac{2}{5}\right)$

29. (40) In the figure at right, $\triangle ABC$ is congruent to $\triangle DCB$. Find the measure of

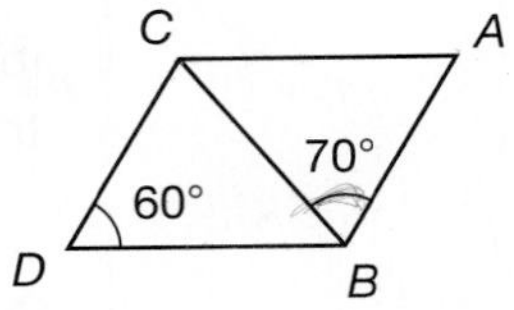

a. $\angle BAC$.

b. $\angle BCA$.

c. $\angle CBD$.

30. (41) **a.** Show two ways to evaluate $a(b - c)$ for $a = 4$, $b = 5$, and $c = 3$.

b. Do both methods simplify to the same number?

c. What property is demonstrated in **a?**

Early Finishers

Real-World Application

Brenda and her sister Brandi share a bedroom. They want to buy a CD player for their bedroom. The CD player they want costs $31.50 plus 8% sales tax.

a. What is the sales tax on the CD player?

b. If they split the cost of the CD player, how much will each girl pay?

Focus on

• Creating Graphs

Recall from Investigation 4 that we considered a stem-and-leaf plot that a counselor created to display student test scores. If we rotate that plot 90°, the display resembles a vertical bar graph, or **histogram.**

2	1 3 5 6 6 8
3	0 0 2 2 4 5 6 6 8 9
4	0 0 1 1 1 2 3 3 5 7 7 8
5	0 1 1 2 3 5 8

A histogram is a special type of bar graph that displays data in equal-sized intervals. There are no spaces between the bars. The height of the bars in this histogram show the number of test scores in each interval.

Thinking Skill

Conclude

Why is there no space between the bars on a histogram?

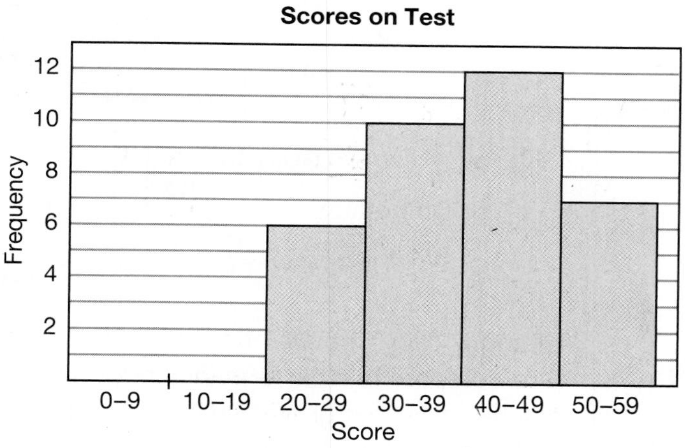

1. **Represent** Changing the intervals can change the appearance of a histogram. Create a new histogram for the test scores itemized in the stem-and-leaf plot using the intervals: 21–28, 29–36, 37– 44, 45–52, and 53–60. Draw a break in the horizontal scale (-√\^-) between 0 and 21.

Histograms and other bar graphs are useful for showing comparisons, but sometimes the visual effect can be misleading. When viewing a graph, it is important to carefully note the scale. Compare these two bar graphs that display the same information.

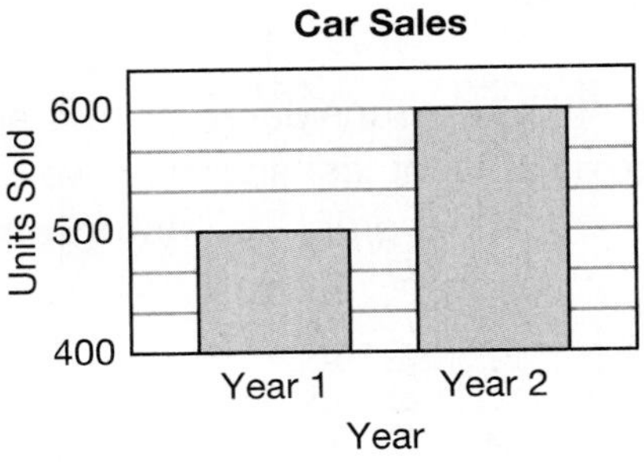

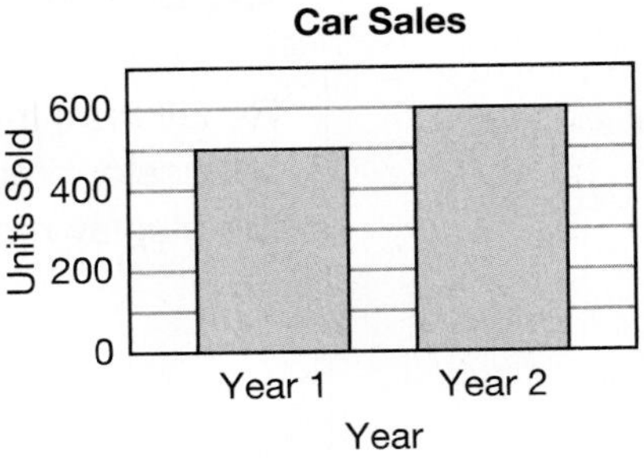

2. **Analyze** Which of the two graphs visually exaggerates the growth in sales from Year 1 to Year 2? How was the exaggerated visual effect created?

3. **Represent** Larry made a bar graph that compares his amount of reading time to Josela's. Create another bar graph that shows the same information in a less misleading way. Explain why your graph is less misleading.

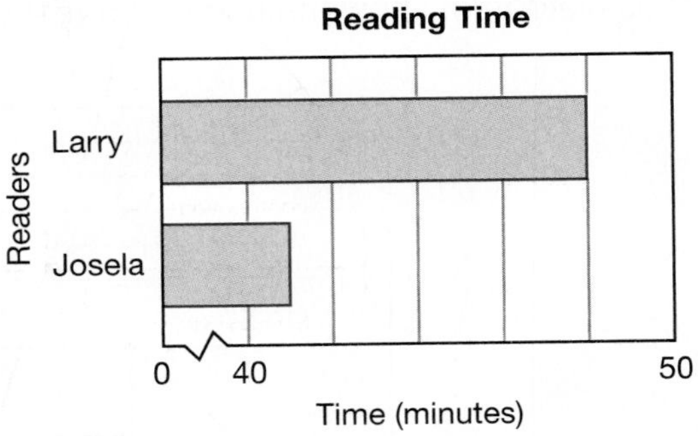

Changes over time are often displayed by line graphs. A **double-line graph** may compare two performances over time. The graph below illustrates the differences in the growing value of a $1000 investment compounded at 7% and at 10% annual interest rates.

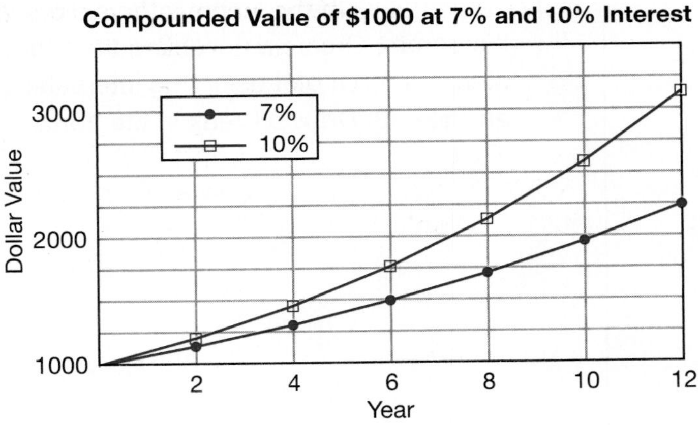

Thinking Skill

Analyze

Will this graph show an exaggerated difference if it does not include a break in the scale? Explain.

4. *Represent* Create a double-line graph using the information in the table below. Label the axes; then select and number the scales. Make a legend (or key) so that the reader can distinguish between the two graphed lines.

Stock Values ($)

First Trade	XYZ Corp	ZYX Corp
2001	30	30
2002	36	28
2003	34	36
2004	46	40
2005	50	46
2006	50	42

A **circle graph** (or pie graph) is commonly used to show components of a budget. The entire circle, 100%, may represent monthly income. The sectors of the circle show how the income is allocated.

Monthly Income Allocation

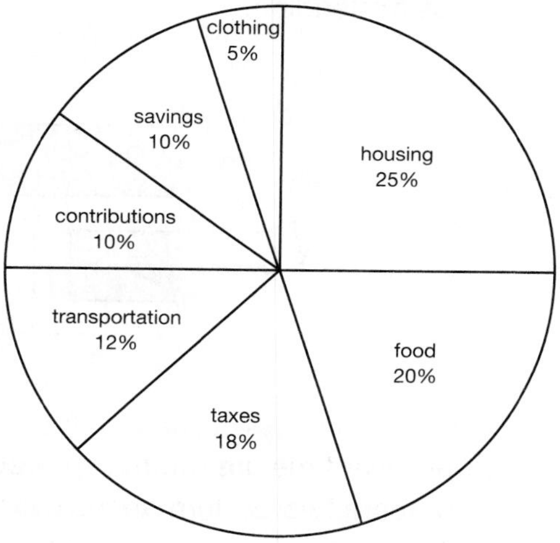

We see that the sector labeled "food" is 20% of the area of the circle, representing 20% of the income. To make a 20% sector, we could draw a central angle that measures 20% of 360°

$$20\% \text{ of } 360°$$

$$0.2 \times 360° = 72°$$

With a protractor we can draw a central angle of 72° to make a sector that is 20% of a circle.

5. *Represent* Create a circle graph for the table below to show how Kerry spends a school day. First calculate the number of degrees in the central angle for each sector of the circle graph. Next use a compass to draw a circle with a radius of about $2\frac{1}{2}$ inches. Then, with a protractor and straightedge, divide the circle into sectors of the correct size and label each sector.

How Kerry Spends a Day

Activity	% of Day
School	25%
Recreation	10%
Traveling	5%
Homework	10%
Eating	5%
Sleeping	40%
Other	5%

Explain Why is a circle graph a good way to represent this data?

extensions

a. Which other type of display—a line graph or a bar graph—would be an appropriate way to display the data about how Kerry spends her day? Draw the display and justify your choice. Then write a question a person could answer by looking at the graph.

b. During one week of a very hot summer, Dan recorded the following high temperatures in his back yard.

Mon.	Tu.	Wed.	Th.	Fri.
94	98	96	99	101

Which display—a line graph or a bar graph—is the most appropriate way to display this data if you want to emphasize the day-to-day changes in high temperature? Draw the display and justify your choice. Then write a question a person could answer by looking at the graph.

c. Develop five categories of books—mystery, science fiction, and so on. Then ask each student in the class to name which category is their favorite. Display the data in a circle graph. Then write three questions a person could answer by looking at the circle graph.

d. *Represent* Display the data you collected in **c** in a frequency table.

e. *Analyze* Read each situation below and answer the questions.

- A toothpaste company advertises that 3 out of 4 dentists recommend their toothpaste. The company surveyed 8 dentists. Do you think most dentists are recommending this toothpaste? Explain why or why not.
- A candidate for public office surveyed 1,000 people and 900 said they would vote for her. Do you think the politician has a good chance of being elected? Explain your reasoning.

LESSON

51

Scientific Notation for Large Numbers

Power Up

Building Power

facts

Power Up J

mental math

a. **Calculation:** $4 \times \$3.50$

b. **Decimals/Exponents:** 4.5×10^2

c. **Equivalent Fractions:** $\frac{5}{20} = \frac{3}{x}$

d. **Measurement:** Convert 5 km to m.

e. **Exponents:** $15^2 - 5^2$

f. **Fractional Parts:** $\frac{5}{9}$ of 45

g. **Statistics:** Find the median of the set of numbers: 567, 765, 675.

h. **Calculation:** Find the sum, difference, product, and quotient of $\frac{7}{8}$ and $\frac{1}{2}$.

problem solving

In this 3-by-3 square we see nine 1-by-1 squares, four 2-by-2 squares, and one 3-by-3 square. Find the total number of squares of any size on a standard checkerboard.

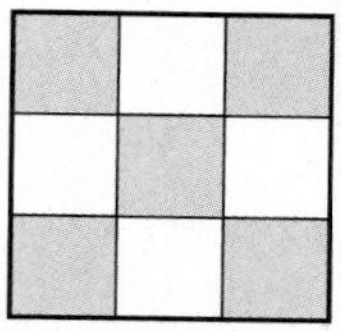

New Concept

Increasing Knowledge

The numbers used in scientific measurement are often very large or very small and occupy many places when written in standard form. For example, a light-year is about

$$9{,}461{,}000{,}000{,}000 \text{ km}$$

Scientific notation is a way of expressing numbers as a product of a decimal number and a power of 10. In scientific notation a light-year is

$$9.461 \times 10^{12} \text{ km}$$

Reading Math

Read the number 9.461×10^{12} as "Nine *point* four six one *times* ten to the twelfth."

In the table on the next page we use scientific notation to approximate some common distances. Measurements are in millimeters. Practice reading each number in scientific notation.

Thinking Skill

Justify

Why is 2.0×10^0 equivalent to 2?

Scientific Notation	Standard Form	Length
2.0×10^0 mm	2 mm	width of pencil lead
2.4×10^1 mm	24 mm	diameter of a quarter
1.6×10^2 mm	160 mm	length of a dollar bill
4.5×10^3 mm	4500 mm	length of average car
2.9×10^4 mm	29,000 mm	length of basketball court
1.1×10^5 mm	110,000 mm	length of football field
1.6×10^6 mm	1,600,000 mm	one mile
4.2×10^7 mm	42,000,000 mm	distance of runner's marathon

Example 1

There are eight point six four times ten to the fourth seconds in a day. Show how this number is written in scientific notation.

Solution

$$\mathbf{8.64 \times 10^4}$$

In scientific notation the power of 10 indicates where the decimal point is located when the number is written in standard form. Consider this number expressed in scientific notation:

$$4.62 \times 10^6$$

Multiplying 4.62 by 10^6 has the effect of shifting the decimal point six places (note the exponent in 10^6) to the right. We use zeros as placeholders.

$$4620000. \longrightarrow 4,620,000$$

Math Language

The small looped arrow below the digits shows the movement of the original decimal point to its new location.

Discuss How do we determine the number of places to shift the decimal point?

Example 2

Write 2.46×10^8 in standard form.

Solution

We shift the decimal point in 2.46 eight places to the right, using zeros as placeholders.

$$246000000. \longrightarrow \mathbf{246,000,000}$$

To write a number in scientific notation, it is customary to place the decimal point to the right of the first nonzero digit. Then we use a power of 10 to indicate the actual location of the decimal point. To write

$$405,700,000$$

in scientific notation, we begin by placing the decimal point to the right of 4 and then counting the places from the original decimal point.

$$\underbrace{4.05700000}_{\text{8 places}}$$

We see that the original decimal point was eight places to the right of where we put it. We omit the terminal zeros and write

$$4.057 \times 10^8$$

Example 3

Write 40,720,000 in scientific notation.

Solution

We begin by placing the decimal point after the 4.

$$\underbrace{4.0720000}_{\text{7 places}}$$

Now we discard the terminal zeros and write 10^7 to show that the original decimal point is really seven places to the right.

$$\mathbf{4.072 \times 10^7}$$

Example 4

Compare: $1.2 \times 10^4 \bigcirc 2.1 \times 10^3$

Solution

Since 1.2×10^4 equals 12,000 and 2.1×10^3 equals 2100, we see that

$$\mathbf{1.2 \times 10^4 > 2.1 \times 10^3}$$

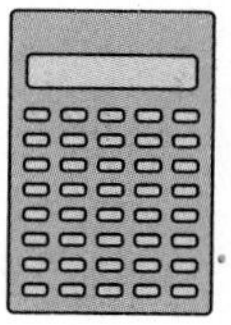

Scientific calculators will display the results of an operation in scientific notation if the number would otherwise exceed the display capabilities of the calculator. For example, to multiply one million by one million, we would enter

1 0 0 0 0 0 0 ×
1 0 0 0 0 0 0 =

The answer, one trillion, contains more digits than can be displayed by many calculators. Instead of displaying one trillion in standard form, a scientific calculator displays one trillion in some modified form of scientific notation such as

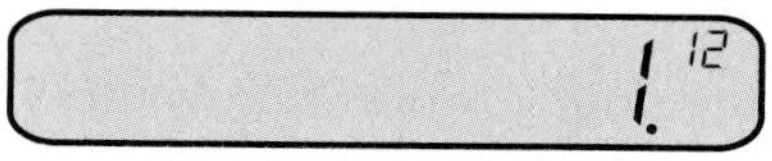 or perhaps 

Other scientific calculators will display the answer as *1e 12* or *1e + 12*. If a calculator displays an *E* or the word *Error,* it is not a scientific calculator.

Practice Set

Connect Write each number in scientific notation:

a. 15,000,000

b. 400,000,000,000

c. 5,090,000

d. two hundred fifty billion

e. two point four times ten to the fifth

Connect Write each number in standard form:

f. 3.4×10^6

g. 1×10^5

Compare:

h. $1.5 \times 10^5 \bigcirc 1.5 \times 10^6$

i. one million $\bigcirc$ 1×10^6

j. Use words to show how 9.3×10^7 is read.

Written Practice *Strengthening Concepts*

Refer to the double-line graph below to answer problems **1** and **2.**

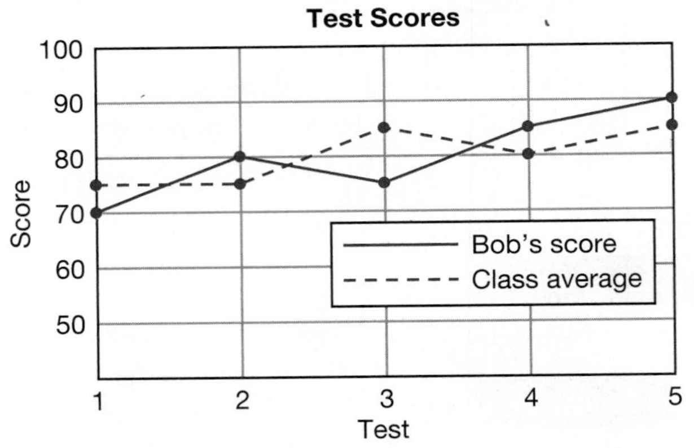

1. (38) On how many tests was Bob's score better than the class average?

2. (28, 38) What was Bob's average score on these five tests?

3. (19) In the pattern on a soccer ball, a regular hexagon and a regular pentagon share a common side. If the perimeter of the hexagon is 9 in., what is the perimeter of the pentagon?

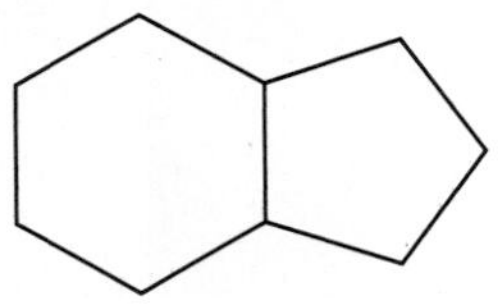

*** 4.** (46) **Analyze** The store sold juice for 65¢ per can or 6 cans for $3.36. How much money per can is saved by buying 6 cans at the 6-can price?

5. (14, 36) Five sevenths of the people who listened to the speaker were convinced. The rest were unconvinced.

a. What fraction of the people were unconvinced?

b. What was the ratio of the convinced to the unconvinced?

*** 6.** (51) **a.** Write twelve million in scientific notation.

b. Write 17,600 in scientific notation.

*** 7.** (51) **a.** Write 1.2×10^4 in standard form.

b. Write 5×10^6 in standard form.

8. (43) Write each number as a decimal:

a. $\frac{1}{8}$ **b.** $87\frac{1}{2}\%$

* **9.** (50) **a.** Connect Bernie, a Saint Bernard dog, weighs 176 pounds. What unit multiplier would you use to determine Bernie's weight in kilograms?

b. Use your answer to **a** to calculate Bernie's approximate weight in kilograms.

* **10.** (48) Complete the table.

Fraction	Decimal	Percent
a.	**b.**	40%
c.	**d.**	4%

* **11.** (Inv. 5) Analyze Find the number of degrees in the central angle of each sector of the circle shown.

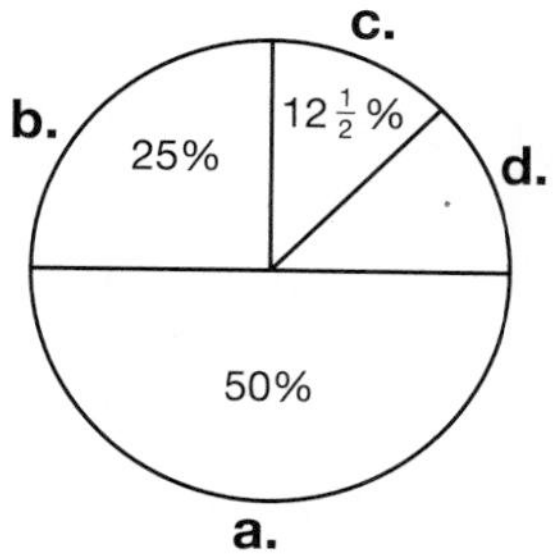

* **12.** (16, 60) Represent At a 6% sales tax rate, the sales tax is \$0.06 for each dollar.

a. Make a function table that shows the sales tax (output) for sales of \$1, \$2, \$3, \$4, \$5.

b. At the 6% rate, what is the sales tax on a \$15 purchase?

13. (14, 36) Layla is thinking of a positive, single-digit, even number. Luis guesses it is 7.

a. What is the sample space?

b. What is the probability that Luis's guess is correct?

14. (18) Quadrilaterals *ABCD* and *WXYZ* are congruent.

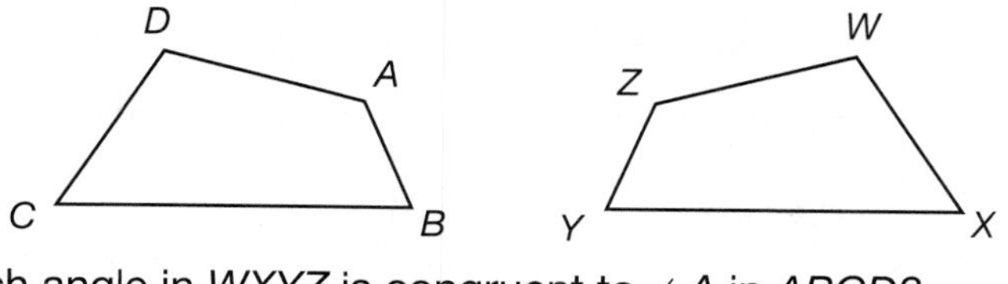

a. Which angle in *WXYZ* is congruent to $\angle A$ in *ABCD*?

b. Which segment in *ABCD* is congruent to $\overline{WX}$ in *WXYZ*?

The figure below shows the dimensions of a garden. Refer to the figure to answer problems **15** and **16.**

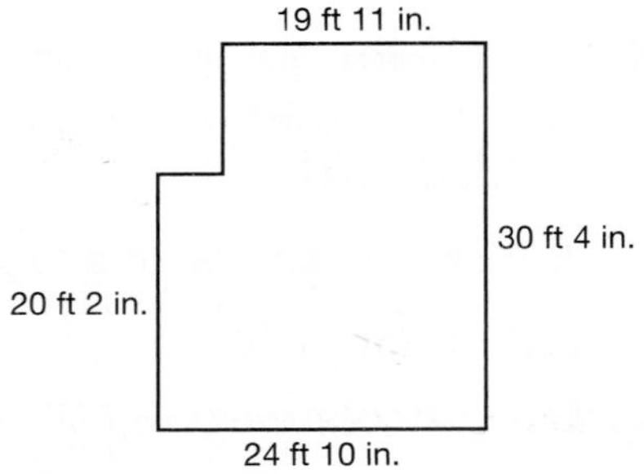

15. *(29)* Estimate the length of fencing needed for the perimeter of the garden.

16. *(29)* Estimate the area of the garden in square feet.

Solve:

17. *(39)* $\frac{24}{x} = \frac{60}{40}$

18. *(39)* $\frac{6}{4.2} = \frac{n}{7}$

19. *(35)* $5m = 8.4$

20. *(35)* $6.5 - y = 5.06$

Simplify:

21. *(20)* $5^2 + 3^3 + \sqrt{64}$

* **22.** *(50)* $16 \text{ cm} \cdot \frac{10 \text{ mm}}{1 \text{ cm}}$

* **23.** *(49)* **Connect**

$$\begin{array}{r} 8 \text{ days } \ 3 \text{ hr } \ 15 \text{ min} \\ -\ 5 \text{ days } \ 18 \text{ hr } \ 50 \text{ min} \\ \hline \end{array}$$

* **24.** *(49)*

$$\begin{array}{r} 3 \text{ yd } \ 2 \text{ ft } \ 5 \text{ in.} \\ +\ 1 \text{ yd } \qquad \ 9 \text{ in.} \\ \hline \end{array}$$

25. *(23, 30)* $6\frac{2}{3} + \left(5\frac{1}{4} - 3\frac{7}{8}\right)$

26. *(26)* $3\frac{1}{3} \cdot \left(2\frac{2}{3} \div 1\frac{1}{2}\right)$

27. *(41)* Show two ways to evaluate $x(x + y)$ for $x = 0.5$ and $y = 0.6$. What property do the results demonstrate?

The coordinates of three vertices of a triangle are $A\ (-4, 0)$, $B\ (0, -4)$, and $C\ (-8, -4)$. Graph the triangle and refer to it to answer problems **28** and **29.**

28. *(17)* Use a protractor to find the measures of $\angle A$, $\angle B$, and $\angle C$.

29. *(37)* What is the area of $\triangle ABC$?

30. *(32)* When the temperature increases from the freezing temperature of water to the boiling temperature of water, it is an increase of 100 degrees on the Celsius scale. The same increase in temperature is how many degrees on the Fahrenheit scale?

LESSON

52

• Order of Operations

Power Up

Building Power

facts Power Up K

mental math

a. **Calculation:** $6 \times 75¢$

b. **Decimals/Exponents:** $4.5 \div 10^2$

c. **Equivalent Fractions:** $\frac{15}{5} = \frac{m}{6}$

d. **Measurement:** Convert 250 cm to m.

e. **Exponents:** $10^3 - 20^2$

f. **Fractional Parts:** $\frac{9}{10}$ of 200

g. **Probability:** How many different ways can you arrange the digits 8, 4, 6, 3?

h. **Rate:** At 80 km per hour, how far will a car travel in $2\frac{1}{2}$ hours?

problem solving What is the largest 2-digit number that is divisible by three and whose digits differ by two?

New Concept

Increasing Knowledge

Recall that the four fundamental operations of arithmetic are addition, subtraction, multiplication, and division. We can also raise numbers to powers or find their roots. When more than one operation occurs in the same expression, we perform the operations in the order listed below.

Order Of Operations

1. Simplify within parentheses (or other symbols of inclusion) from innermost to outermost, before simplifying outside of the parentheses.
2. Simplify powers and roots.
3. Multiply and divide in order from left to right.
4. Add and subtract in order from left to right.

Note: **Symbols of inclusion** set apart portions of an expression so they may be evaluated first. The following are symbols of inclusion: ( ), [ ], { }, and the division bar in a fraction.

The initial letter of each word in the sentence "Please excuse my dear Aunt Sally" reminds us of the order of operations:

Please	Parenthesis (or other symbols of inclusion)
Excuse	Exponents (and roots)
My Dear	Multiplication and division (left to right)
Aunt Sally	Addition and subtraction (left to right)

Example 1

Simplify: $2 + 4 \times 3 - 4 \div 2$

Solution

We multiply and divide in order from left to right before we add or subtract.

$2 + 4 \times 3 - 4 \div 2$	problem
$2 + 12 - 2$	multiplied and divided
12	added and subtracted

Example 2

Simplify: $\dfrac{3^2 + 3 \cdot 5}{2}$

Solution

Thinking Skill

Explain

Why are the parentheses needed in the expression $4 \times (3 + 7)$?

A division bar may serve as a symbol of inclusion, like parentheses. We simplify above and below the bar before dividing.

$\dfrac{3^2 + 3 \cdot 5}{2}$	problem
$\dfrac{9 + 3 \cdot 5}{2}$	applied exponent
$\dfrac{9 + 15}{2}$	multiplied above
$\dfrac{24}{2}$	added above
12	divided

Example 3

Evaluate: $a + ab$ if $a = 3$ and $b = 4$

Solution

We will begin by writing parentheses in place of each variable. This step may seem unnecessary, but many errors can be avoided if this is always our first step.

$a + ab$	
$(\ ) + (\ )(\ )$	parentheses

Then we replace a with 3 and b with 4.

$$a + ab$$

$(3) + (3)(4)$ substituted

We follow the order of operations, multiplying before adding.

$(3) + (3)(4)$	problem
$3 + 12$	multiplied
15	added

Example 4

Evaluate: $xy - \frac{x}{2}$ **if** $x = 9$ **and** $y = \frac{2}{3}$

Solution

First we replace each variable with parentheses.

$$xy - \frac{x}{2}$$

$(\ )(\ ) - \frac{(\ )}{2}$ parentheses

Then we write 9 in place of x and $\frac{2}{3}$ in place of y.

$$xy - \frac{x}{2}$$

$(9)\left(\frac{2}{3}\right) - \frac{(9)}{2}$ substituted

We follow the order of operations, multiplying and dividing before we subtract.

$(9)\left(\frac{2}{3}\right) - \frac{(9)}{2}$	problem
$6 - 4\frac{1}{2}$	multiplied and divided
$\mathbf{1\frac{1}{2}}$	subtracted

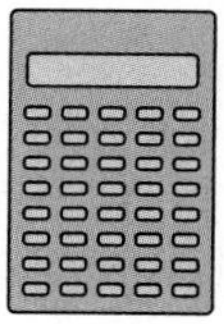

Calculators with *algebraic-logic* circuitry are designed to perform calculations according to the order of operations. Calculators without algebraic-logic circuitry perform calculations in sequence. You can test a calculator's design by selecting a problem such as that in example 1 and entering the numbers and operations from left to right, concluding with an equal sign. If the problem in example 1 is used, a displayed answer of 12 indicates an algebraic-logic design.

Practice Set

Generalize Simplify:

a. $5 + 5 \cdot 5 - 5 \div 5$

b. $50 - 8 \cdot 5 + 6 \div 3$

c. $24 - 8 - 6 \cdot 2 \div 4$

d. $\dfrac{2^3 + 3^2 + 2 \cdot 5}{3}$

Conclude Evaluate:

e. $ab - bc$ if $a = 5$, $b = 3$, and $c = 4$

f. $ab + \frac{a}{c}$ if $a = 6$, $b = 4$, and $c = 2$

g. $x - xy$ if $x = \frac{2}{3}$ and $y = \frac{3}{4}$

Written Practice *Strengthening Concepts*

1. (21) If the product of the first three prime numbers is divided by the sum of the first three prime numbers, what is the quotient?

* **2.** (50) **Connect** Compare: 40 in. ◯ 100 cm
Use a unit multiplier to convert 40 inches to centimeters to make the comparison. (1 in. = 2.54 cm)

3. (31, 35) Twenty-five and two hundred seventeen thousandths is how much less than two hundred two and two hundredths?

4. (46) Jermaine bought a 25-pack of blank CDs for $22.23. What is the price per CD to the nearest cent?

5. (28, 46) Ginger has a 330-page book. She starts to read at 4:15 PM. Suppose she reads for 4 hours and averages 35 pages per hour.

a. How many pages will she read in 4 hours?

b. After 4 hours, how many pages will she still have to read to finish the book?

c. Did we need to use all the information given in this problem? Explain.

6. (22) In the following statement, convert the percent to a reduced fraction. Then diagram the statement and answer the questions.

Seventy-five percent of the 60 passengers disembarked at the terminal.

a. How many passengers disembarked at the terminal?

b. What percent of the passengers did not disembark at the terminal?

* **7.** (51) **Connect** At its closest point, Pluto's orbit brings it to approximately 2,756,300,000 miles from the Sun. At its farthest point, Pluto is approximately 4,539,600,000 miles from the Sun.

a. Write the first distance in scientific notation.

b. Write the second distance in words.

* **8.** (51) In 2002, the automotive industry spent more than 1.6×10^{10} dollars on advertising. The oil and gas industry spent 2.4×10^{8} dollars on advertising.

a. Write the first amount in standard form.

b. Write the second amount in words.

9. (43) Write each number as a decimal:

a. $\frac{3}{8}$ **b.** 6.5%

10. (42) Write $3.\overline{27}$ as a decimal number rounded to the nearest thousandth.

* **11.** (48) Complete the table.

Fraction	Decimal	Percent
a.	**b.**	250%
c.	**d.**	25%

12. (44) Divide 70 by 9 and write the answer

a. as a mixed number.

b. as a decimal number with a bar over the repetend.

13. (34) What decimal number names the point marked by the arrow?

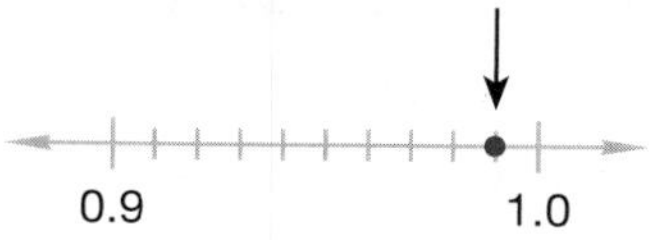

14. (32) Draw a rectangle that is 3 cm long and 2 cm wide. Then answer **a** and **b**.

a. What is the perimeter of the rectangle in millimeters?

b. What is the area of the rectangle in square centimeters?

15. (37) In quadrilateral *ABCD*, $\overline{AD}$ is parallel to $\overline{BC}$. Dimensions are in centimeters.

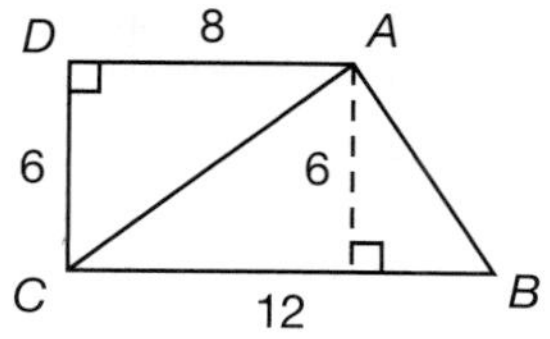

a. Find the area of $\triangle ABC$.

b. Find the area of $\triangle ACD$.

c. What is the combined area of the two triangles?

Solve:

16. (39) $\frac{8}{f} = \frac{56}{105}$

17. (39) $\frac{12}{15} = \frac{w}{2.5}$

18. (35) $p + 6.8 = 20$

19. (35) $q - 3.6 = 6.4$

Analyze Simplify:

20. (20) $5^3 - 10^2 - \sqrt{25}$

* **21.** (52) $4 + 4 \cdot 4 - 4 \div 4$

*** 22.** (35, 45) $\dfrac{4.8 - 0.24}{(0.2)(0.6)}$

*** 23.** (49)
$$\begin{array}{r} 5 \text{ hr } 45 \text{ min } 30 \text{ s} \\ + \; 2 \text{ hr } 53 \text{ min } 55 \text{ s} \\ \hline \end{array}$$

24. (26, 30) $6\frac{3}{4} + \left(5\frac{1}{3} \cdot 2\frac{1}{2}\right)$

25. (26, 30) $5\frac{1}{2} - \left(3\frac{3}{4} \div 2\right)$

26. (35) Estimate the sum to the nearest whole number. Then perform the calculation.

$$8.575 + 12.625 + 8.4 + 70.4$$

*** 27.** (47) $0.8 \times 1.25 \times 10^6$

*** 28.** (52) **Conclude** Evaluate: $ab + \dfrac{a}{b}$ if $a = 4$ and $b = 0.5$

*** 29.** (50) **Connect** Convert 1.4 meters to centimeters (1 m = 100 cm).

*** 30.** (14, 43) **Analyze** The students in a class of 30 were asked to name their favorite sport. Twelve said football, 10 said basketball, and 8 said baseball. If a student is selected at random, what is the probability that the student's favorite sport is basketball? Write the probability as a reduce fraction and as a decimal number rounded to two decimal places.

Early Finishers
Math and Science

The Sun is the largest object in our Solar System. Its diameter measures 1.390×10^6 kilometers.

a. One mile is approximately 1.6 kilometers. What is the Sun's diameter rounded to the nearest ten thousand miles?

b. Use your answer to part **a** to calculate the circumference of the Sun to the nearest hundred thousand miles.

LESSON 53

• Ratio Word Problems

Power Up

Building Power

facts

Power Up K

mental math

a. **Calculation:** $8 \times \$1.25$

b. **Decimals:** 12.75×10

c. **Algebra:** $2x + 5 = 75$

d. **Measurement:** Convert 35 cm to mm.

e. **Exponents:** $\left(\frac{1}{2}\right)^2$

f. **Fractional Parts:** $\frac{3}{5}$ of 45

g. **Probability:** How many different meals can you make with 2 types of meat, 4 types of vegetables and 2 types of bread?

h. **Calculation:** 10×6, $+ 4$, $\sqrt{\ }$, $\times 3$, double that number, $+ 1$, $\sqrt{\ }$, $\times 8$, $- 1$, $\div 5$, square that number

problem solving

Colby has two number cubes. One is a standard number cube numbered 1 through 6. The other number cube is numbered 7 through 12. If Colby rolls the two number cubes together, what totals are possible? Which total is most likely?

New Concept

Increasing Knowledge

Math Language

A **proportion** is a statement that two ratios are equal.

In this lesson we will use proportions to solve ratio word problems. Consider the following ratio word problem:

The ratio of parrots to macaws at a bird sanctuary was 3 to 5. If there were 45 parrots, how many macaws were there?

In this problem there are two kinds of numbers, ratio numbers and actual count numbers. The ratio numbers are 3 and 5. The number 45 is an actual count of parrots. We will arrange these numbers into two columns and two rows to form a ratio box. Practicing the use of ratio boxes now will pay dividends in later lessons when we extend their application to more complex problems.

	Ratio	Actual Count
Parrots	3	45
Macaws	5	*m*

We were not given the actual count of macaws, so we have used *m* to stand for the number of macaws. The numbers in this ratio box can be used to write a proportion. By solving the proportion, we find the actual count of macaws.

	Ratio	Actual Count
Parrots	3	45
Macaws	5	*m*

$$\frac{3}{5} = \frac{45}{m}$$
$$3m = 225$$
$$m = 75$$

We find that the actual count of macaws was 75.

Explain How did we form the equation $3m = 225$?

Example

In the auditorium the ratio of boys to girls was 5 to 4. If there were 200 girls in the auditorium, how many boys were there?

Solution

We begin by making a ratio box.

	Ratio	Actual Count
Boys	5	*B*
Girls	4	200

Thinking Skill

Conclude

We can use this answer to find what other information?

We use the numbers in the ratio box to write a proportion. Then we solve the proportion and answer the question.

	Ratio	Actual Count
Boys	5	*B*
Girls	4	200

$$\frac{5}{4} = \frac{B}{200}$$
$$4B = 1000$$
$$B = 250$$

There were **250 boys** in the auditorium.

Practice Set

Solve each of these ratio word problems. Begin by making a ratio box.

a. **Analyze** The girl-boy ratio was 9 to 7. If 63 girls attended, how many boys attended?

b. The ratio of sparrows to bluejays at the bird sanctuary was 5 to 3. If there were 15 bluejays in the sanctuary, how many sparrows were there?

c. The ratio of tagged fish to untagged fish was 2 to 9. Ninety fish were tagged. How many fish were untagged?

d. **Connect** Calculate the ratio of boys to girls in your classroom. Then calculate the ratio of girls to boys.

Written Practice Strengthening Concepts

Refer to this double-bar graph to answer problems **1–3:**

Distribution of Birthdays in Class

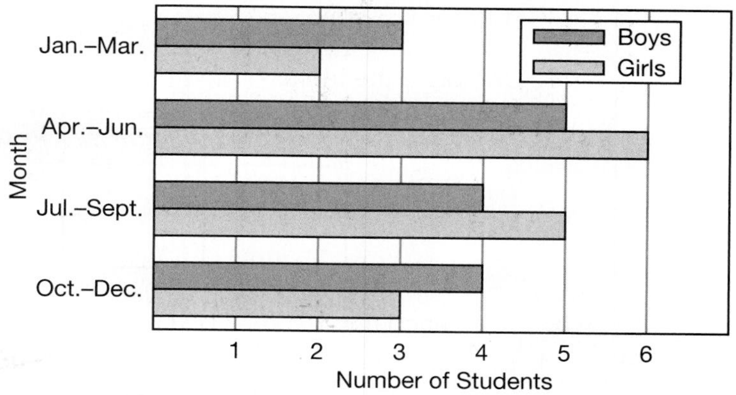

1. (38) **a.** How many boys are in the class?

b. How many girls are in the class?

2. (38) What percent of the students have a birthday in one of the months from January through June?

3. (38) What fraction of the boys have a birthday in one of the months from April through June?

4. (28) **a.** At the book fair Muhammad bought 4 books. One book cost \$3.95. Another book cost \$4.47. The other 2 books cost \$4.95 each. What was the average price per book?

b. *Predict* If Muhammed bought another book for \$4.25, would you expect the average price per book to increase or decrease? Explain your reasoning?

5. (22) Read the following and answer the questions that follow.

Seven twelfths of the 840 students attended the Spring Concert.

a. What fraction of the students did not attend the Spring Concert?

b. How many students did not attend the Spring Concert?

* **6.** (51) **a.** Write one trillion in scientific notation.

b. *Represent* Write 475,000 in scientific notation.

* **7.** (51) **a.** Write 7×10^2 in standard form.

b. *Conclude* Compare: $2.5 \times 10^6 \bigcirc 2.5 \times 10^5$

* **8.** (50) *Connect* Use unit multipliers to perform the following conversions:

a. 35 yards to feet (3 ft = 1 yd)

b. 2000 cm to m (100 cm = 1 m)

9. (27) Use prime factorization to find the least common multiple of 54 and 36.

10. (50) A car traveling 62 miles per hour is moving at a speed of about how many kilometers per hour? Use a unit multiplier to convert the rate. (1 km $\approx$ 0.62 mi)

11. (48) Complete the table.

Fraction	Decimal	Percent
a.	**b.**	150%
c.	**d.**	15%

12. (48) Write each number as a percent:

a. $\frac{4}{5}$ **b.** 0.06

13. (32) A lilac bush is 2 m tall. A rose bush is 165 cm tall. The lilac bush is how many centimeters taller than the rose bush?

14. (19, 37) Refer to this figure to answer **a** and **b.** Dimensions are in feet. All angles are right angles.

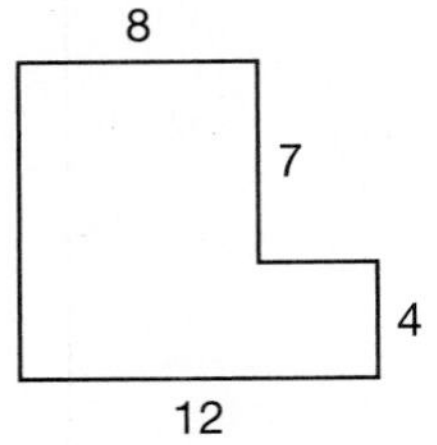

a. What is the area of the figure?

b. What is the perimeter of the figure?

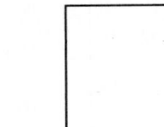

* **15.** (53) **Analyze** In the school Marching Band the ratio of trumpet players to drummers was 5 to 2. If there were six drummers in the Marching Band, how many trumpet players were there?

Solve:

16. (39) $\frac{18}{100} = \frac{90}{p}$

17. (39) $\frac{6}{9} = \frac{t}{1.5}$

18. (35) $8 = 7.25 + m$

19. (35) $1.5 = 10n$

Simplify:

20. (20) $\sqrt{81} + 9^2 - 2^5$

21. (52) $16 \div 4 \div 2 + 3 \times 4$

* **22.** (49)
$$\begin{array}{r} 3\text{ yd } 1\text{ ft } 7\tfrac{1}{2}\text{ in.} \\ +\quad 2\text{ ft } 6\tfrac{1}{2}\text{ in.} \\ \hline \end{array}$$

23. (10, 28) $12\frac{2}{3} + \left(5\frac{5}{6} \div 2\frac{1}{3}\right)$

24. (26, 30) $8\frac{3}{5} - \left(1\frac{1}{2} \cdot 3\frac{1}{5}\right)$

* **25.** (35, 47) **Analyze** $10.6 + 4.2 + 16.4 + (3.875 \times 10^1)$

26. (29, 33) Estimate: $6.85 \times 4\frac{1}{16}$

* **27.** (52) **Conclude** Find the value of $\frac{ab}{bc}$ when $a = 6$, $b = 0.9$, and $c = 5$.

* **28.** (44) **Analyze** Petersen needed to pack 1000 eggs into flats that held $2\frac{1}{2}$ dozen eggs. How many flats could he fill?

29. (14, 43) If there is one chance in five of picking a red marble, then what is the probability of not picking a red marble? Write the probablity as a fraction and as a decimal.

30. (40) Find the measures of angles *a, b,* and *c* in this figure:

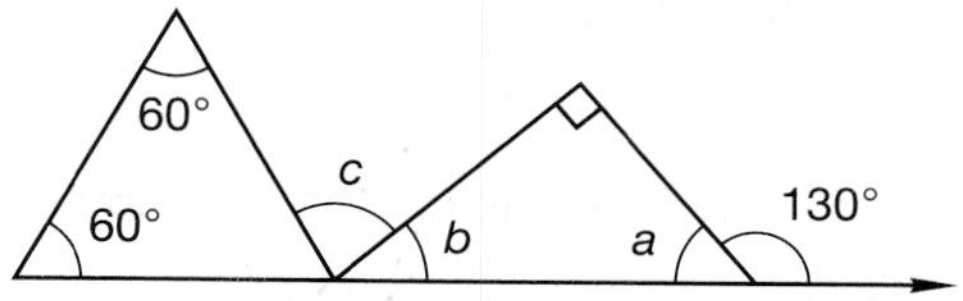

Early Finishers
Real-World Application

In the first semester 6 of the 18 students in Mrs. Eggleston's homeroom were boys. In the second semester 9 of the 21 students were boys.

a. What is the ratio of boys to boys in the class in the first to the second semester?

b. What is the ratio of girls to girls in the class in the first to the second semester?

c. Describe the changes that occurred from first semester to second semester in Mrs. Eggleston's homeroom.

LESSON
54

• Rate Word Problems

Power Up

Building Power

facts Power Up L

mental math

a. **Calculation:** 4 × \$4.50

b. **Decimals:** 12.75 ÷ 10

c. **Equivalent Fractions:** $\frac{12}{w} = \frac{9}{6}$

d. **Measurement:** Convert 1.5 m to cm.

e. **Power/Roots:** $\sqrt{900} - 3^3$

f. **Fractional Parts:** $\frac{3}{10}$ of 90

g. **Probability:** How many different ways can you arrange the digits 9, 4, 3, 7, 2?

h. **Calculation:** Mentally perform each calculation:

$$\frac{3}{4} + \frac{2}{5} \qquad \frac{3}{4} - \frac{2}{5} \qquad \frac{3}{4} \cdot \frac{2}{5} \qquad \frac{3}{4} \div \frac{2}{5}$$

problem solving

Zelda is older than Frank, but younger than Juan. Juan is younger than Celia, but older than Frank. Frank is older than Gina and Marcos. Marcos is younger than Celia and Gina. Who is the oldest? Who is the youngest?

New Concept

Increasing Knowledge

In Lesson 53, we solved ratio word problems by using proportions. We can solve rate word problems in the same way.

Consider the following problem.

Example 1

If Mr. Gomez drives his car at an average speed of 55 miles per hour, how far will he drive in 3 hours?

Solution

Thinking Skill

Summarize

How is the information in the ratio box organized?

We can make a ratio box to solve this problem. The units are miles and hours. We use the numbers 55 and 1 to write the rate. To find how far he drives in 3 hours, use the number 3 for the actual number of hours. We let *d* represent the actual distance.

	Rate	Actual Measure
Distance (mi)	55	d
Time (hr)	1	3

$\rightarrow \frac{55}{1} = \frac{d}{3}$

$$\frac{55}{1} = \frac{d}{3}$$

$$1 \times d = 3 \cdot 55$$

$$d = 165$$

Mr. Gomez will drive **165 miles.**

Discuss Why is it important to include the units of measure and time in the ratio box?

Example 2

If Mrs. Ikeda's car averages 24 miles per gallon, then about how many gallons of gas will she use on a trip of 300 miles?

Solution

The units are miles and gallons. The rate is 24 miles per gallon. The trip is 300 miles. We let g represent the number of gallons the car will use on the trip.

	Rate	Actual Measure
Miles	24	300
Gallons	1	g

$\rightarrow \frac{24}{1} = \frac{300}{g}$

$$\frac{24}{1} = \frac{300}{g}$$

$$24g = 300$$

$$g = 12.5$$

Mrs. Ikeda will use **about 12.5 gallons** of gas on the trip.

Explain Why is it appropriate to use an approximate answer for this question?

Example 3

Hana works for 8 hours at a sporting goods store and earns a total of $68.00.

a. What is her hourly rate of pay?

b. Use her hourly rate of pay to find how much she will earn if she works for 30 hours.

Solution

Thinking Skill

Discuss

In the ratio boxes for examples 1–3, the position of the unknown varies. Does this affect how we set up the proportion?

a. We will use a ratio box to solve both parts of the problem.

	Rate	Actual Measure
Amount ($)	p	68
Time (hr)	1	8

$\longrightarrow \frac{p}{1} = \frac{68}{8}$

$$\frac{p}{1} = \frac{68}{8}$$

$$8p = 68 \cdot 1$$

$$p = 8.5$$

Hana earns **$8.50 per hour.**

b. We use the rate we found in **a** to find Hana's pay for 30 hours.

	Rate	Actual Measure
Amount ($)	8.5	T
Time (hr)	1	30

$\longrightarrow \frac{8.5}{1} = \frac{T}{30}$

$$\frac{8.5}{1} = \frac{T}{30}$$

$$1 \times T = 8.5 \times 30$$

$$T = 255$$

In 30 hours Hana would earn **$255.00.**

Practice Set

Use a ratio box to help you solve these rate word problems.

On a 600-mile trip, Dixon's car averaged 50 miles per hour and 30 miles per gallon.

a. The trip took how many hours to complete?

b. During the trip the car used how many gallons of gas?

Jenna earned $68.80 working 8 hours.

c. What is Jenna's hourly rate of pay?

d. How much would Jenna earn working 20 hours?

The price of one type of cheese is $2.60 per pound.

e. What is the cost of a 2.5-pound package of cheese?

f. *Explain* How could we find the cost of a half-pound package of cheese?

Written Practice

Strengthening Concepts

1. (12, 28) Thomas Jefferson was born in 1743. He died on the fiftieth anniversary of the signing of the Declaration of Independence. The Declaration of Independence was signed in 1776. How many years did Thomas Jefferson live?

2. (28) The heights of five basketball players are given in the table below. What is the average height of the players to the nearest centimeter?

Player	Height (cm)
A	190
B	195
C	197
D	201
E	203

*** 3.** (53) **Explain** The ratio of women to men in the theater was 5 to 4. If there were 1200 women, how many men were there? Explain how you found your answer.

*** 4.** (54) What is the cost of 2.6 pounds of cheese at $6.75 per pound?

5. (6, 27) What is the quotient when the least common multiple of 4 and 6 is divided by the greatest common factor of 4 and 6?

6. (22) Draw a diagram to represent this statement. Then answer the questions that follow.

Eighty percent of the 80 seedlings were planted today.

a. How many seedlings were planted today?

b. How many seedlings remain to be planted?

*** 7.** (51) **a.** Write 405,000 in scientific notation.

b. Write 0.04×10^5 in standard form.

8. (47) **Justify** Find each missing exponent. What exponent rule did you use?

a. $10^6 \cdot 10^2 = 10^{\square}$ **b.** $10^6 \div 10^2 = 10^{\square}$

*** 9.** (50) **Connect** Use unit multipliers to perform the following conversions:

a. 5280 feet to yards (3 ft = 1 yd)

b. 300 cm to mm (1 cm = 10 mm)

10. (33) Write 3.1415926 as a decimal number rounded to four decimal places.

*** 11.** (46, 54) **Analyze** A train is traveling at a steady speed of 60 miles per hour.

a. How far will the train travel in four hours?

b. How long will it take the train to travel 300 miles?

*** 12.** (Inv. 5) **Analyze** Find the number of degrees in the central angle of each sector of the circle shown.

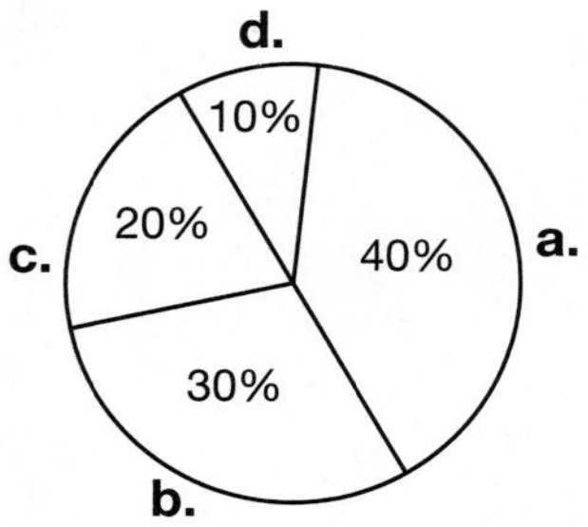

13. (20) Which expression is equivalent to $\frac{2^6}{2^2}$?

A 2^3 **B** 2^4 **C** 1^3 **D** 3

Refer to the figure below to answer problems **14** and **15.** Dimensions are in centimeters. All angles are right angles.

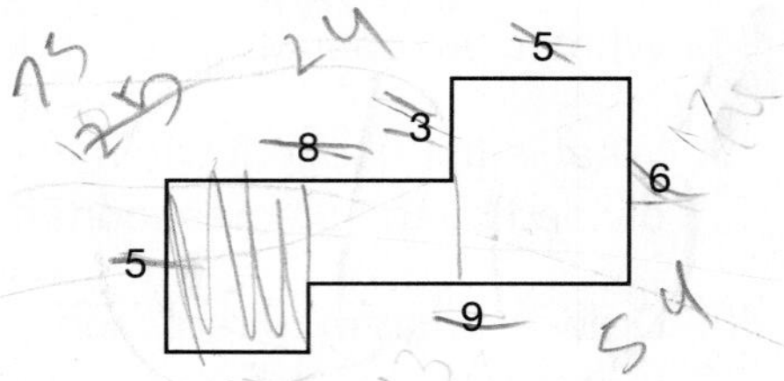

14. (19) What is the perimeter of the figure?

15. (37) What is the area of the figure?

16. (2, 9, 41) Name each property illustrated:

a. $\frac{1}{2} + 0 = \frac{1}{2}$

b. $5(6 + 7) = 30 + 35$

c. $(5 + 6) + 4 = 5 + (6 + 4)$

d. $\frac{3}{5} \cdot \frac{5}{3} = 1$

17. (34, 35) Draw a square with sides 0.5 inch long.

a. What is the perimeter of the square?

b. What is the area of the square?

*** 18.** (Inv. 4) **Predict** The box-and-whisker plot below was created from scores (number of correct out of 20) on a sports trivia quiz. Do you think that the mean (average) score is likely to be above, at, or below the median score? Explain your reasoning.

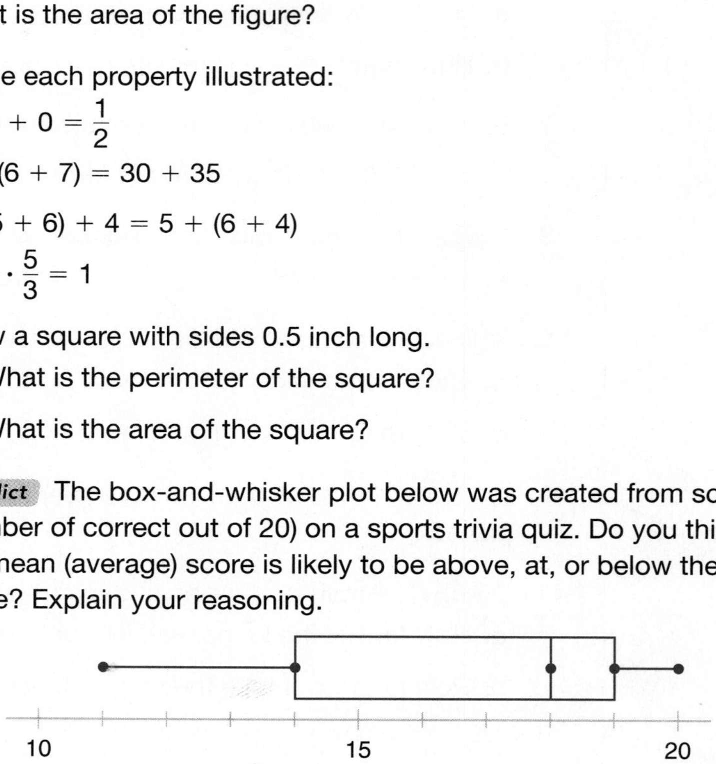

Solve:

19. (35) $6.2 = x + 4.1$

20. (35) $1.2 = y - 0.21$

21. (39) $\frac{24}{r} = \frac{36}{27}$

22. (35) $\frac{w}{0.16} = 6.25$

Analyze Simplify:

23. $11^2 + 1^3 - \sqrt{121}$
(20)

*** 24.** $24 - 4 \times 5 \div 2 + 5$
(52)

25. $\frac{(2.5)^2}{2(2.5)}$
(35)

*** 26.**
(49)

	1 week	5 days	14 hr
+	2 weeks	6 days	10 hr

27. $3\frac{5}{10} + \left(9\frac{1}{2} - 6\frac{2}{3}\right)$
(23, 30)

28. $7\frac{1}{3} \cdot \left(6 \div 3\frac{2}{3}\right)$
(26)

29. The coordinates of the vertices of $\triangle ABC$ are $A\ (-1, 3)$, $B\ (-4, 3)$, and $C\ (-4, -1)$. The coordinates of $\triangle XYZ$ are $X\ (1, 3)$, $Y\ (4, 3)$, and $Z\ (4, -1)$. Graph $\triangle ABC$ and $\triangle XYZ$.
(Inv. 3)

30. Refer to the graph drawn in problem 29 to answer **a–c.**
(18)

a. Are $\triangle ABC$ and $\triangle XYZ$ similar?

b. Are $\triangle ABC$ and $\triangle XYZ$ congruent?

c. Which angle in $\triangle ABC$ corresponds to $\angle Z$ in $\triangle XYZ$?

Early Finishers
Real-World Application

Ngo ran for student council representative and received 0.875 of the vote. Marley ran for treasurer and received $\frac{13}{65}$ of all the votes cast. Vega took 54.6% of the total votes for president.

a. Express each election result as a decimal, percent, and simplified fraction.

b. Even though we do not have the results from the other candidates, who do you think won and lost the election? Support your conclusion.

LESSON

55

Average and Rate Problems with Multiple Steps

Power Up

Building Power

facts Power Up J

mental math

a. **Calculation:** $20 \times \$0.25$

b. **Decimals/Exponents:** 0.375×10^2

c. **Algebra:** $2x - 5 = 75$

d. **Measurement:** Convert 3000 m to km.

e. **Exponents:** $\left(\frac{2}{3}\right)^2$

f. **Fractional Parts:** $\frac{3}{4}$ of 100

g. **Measurement:** One quart is what fraction of a gallon?

h. **Rate:** At 30 pages an hour, how many pages can Mike read in $2\frac{1}{2}$ hours?

problem solving

Copy this problem and fill in the missing digits:

```
        3 _
_ 3)_ 6 _ 6
    1 _ _
    -----
      _ 0 _
      _ _ _
      -----
          0
```

New Concept

Increasing Knowledge

If we know the average of a group of numbers and how many numbers are in the group, we can determine the sum of the numbers.

Example 1

The average of three numbers is 17. What is their sum?

Solution

We are not told what the numbers are, only their average.

Each set of three numbers below has an average of 17:

$$\frac{16 + 17 + 18}{3} = \frac{51}{3} = 17$$

$$\frac{10 + 11 + 30}{3} = \frac{51}{3} = 17$$

$$\frac{1 + 1 + 49}{3} = \frac{51}{3} = 17$$

Math Language

An **average** is the sum of two or more numbers divided by the number of addends.

Notice that for each set, the sum of the three numbers is 51. Since average tells us what the numbers would be if they were equalized, the sum of the three numbers is the same as if each of the numbers were 17.

$$17 + 17 + 17 = \mathbf{51}$$

Thus, multiplying the average by the quantity of numbers equals the sum of the numbers.

Example 2

The average of four numbers is 25. If three of the numbers are 16, 26, and 30, what is the fourth number?

Solution

If the average of four numbers is 25, their sum is 100.

$$4 \times 25 = 100$$

We are given three of the numbers. The sum of these three numbers plus the fourth number, n, must equal 100.

$$16 + 26 + 30 + n = 100$$

The sum of the first three numbers is 72. Since the sum of the four numbers must be 100, the fourth number is **28.**

$$16 + 26 + 30 + 28 = 100$$

Example 3

After four years, the average number of students participating in after-school sports was 89. How many students must participate in the fifth year to bring the average up to 90?

Solution

Although we do not know the specific number of students participating in each of the first four years, the total is the same as if each number were 89. Thus the total after four years is

$$4 \times 89 = 356$$

The total of the first four years is 356. However, to have an average of 90 students after five years, the total for the five years should be 450.

$$5 \times 90 = 450$$

Therefore, to raise the total from 356 to 450 in the fifth year, there needs to be 450 − 356 students participating during the fifth year.

$$\begin{array}{r} 450 \\ -\ 356 \\ \hline \mathbf{94} \end{array}$$ **students** participating in the fifth year

Discuss How do we know that 94 is a reasonable answer?

Some word problems involve many steps before a final answer can be found. When solving multiple-step problems, it is important to keep the steps and results organized.

To solve rate problems with multiple steps, we can break the problem into simpler parts. Then we can solve each part to find the final answer.

Example 4

Tyrone is buying refreshments for the next Math Club meeting. Brand A cranberry juice is on sale at $3.20 for a 64-fl oz bottle. Brand B costs $1.19 for a 17-fl oz bottle.

Which brand is the better buy?

Solution

Thinking Skill

Explain

How does finding the unit price of each brand help us solve the problem?

Break the problem into simpler parts.

Step 1: Find the unit price of Brand A.

$$\$3.20 \text{ for } 64 \text{ fl oz} \longrightarrow \frac{\$3.20}{64 \text{ fluid ounces}} = \frac{\$0.05}{1 \text{ fluid ounce}}$$

Step 2: Find the unit price of Brand B.

$$\$1.19 \text{ for } 17 \text{ fl oz} \longrightarrow \frac{\$1.19}{17 \text{ fluid ounces}} = \frac{\$0.07}{1 \text{ fluid ounce}}$$

Step 3: Compare unit prices to find the better buy.

$$\$0.05 < \$0.07$$

Brand A is the better buy.

Example 5

Brenda has two part-time jobs. During one week she worked 20 hours at the bookstore for $8.10 per hour and 16 hours as a receptionist for $9.00 per hour.

a. How much did Brenda earn at the bookstore?

b. How much did Brenda earn as a receptionist?

c. What were her total earnings for the week?

d. What was her average rate of pay for the week?

Solution

a. Brenda earned **$162.00** at the bookstore.

$$20 \cancel{\text{hrs}} \times \frac{\$8.10}{1 \cancel{\text{hr}}} = \$162.00$$

b. Brenda earned **$144.00** as a receptionist.

$$16 \cancel{\text{hrs}} \times \frac{\$9.00}{1 \cancel{\text{hr}}} = \$144.00$$

c. Brenda's total earnings were **$306.00.**

$$\$162.00 + \$144.00 = \$306.00$$

d. We find Brenda's average rate of pay by dividing her total pay by her total hours.

$$\frac{\$306.00}{36 \text{ hrs}} = \frac{\$8.50}{1 \text{ hr}}$$

Brenda's average rate of pay, for the week was **$8.50 per hour.**

Discuss Why is $8.50 a reasonable answer for Brenda's average rate of pay?

Practice Set

a. Tisha scored an average of 18 points in each of her first five basktetball games. Altogether, how many points did she score in her first five games?

b. The average of four numbers is 45. If three of the numbers are 24, 36, and 52, what is the fourth number?

c. *Analyze* After five games of bowling, Ralph's average score was 91. After six games, his average score was 89. What was his score in the sixth game?

d. *Analyze* Vin's babysitting earnings are shown in the table. If he wants to earn an average of at least $20 per month over 4 months, what is the minimum he can earn in June?

Month	Earnings
March	$18
April	$22
May	$15

e. Ray is buying cheese to make macaroni and cheese. Cheddar cheese is on sale at $3.60 for 10 oz. American cheese costs $3.04 for $\frac{1}{2}$ lb. Ray needs at least $1\frac{1}{2}$ lb cheese for the macaroni and cheese. Which type of cheese is the better buy?

16 oz = 1 lb

f. Carla drove 180 miles from her home to Houston in 3 hours. On her return home the traffic was slow and the trip took 4 hours. Find Carla's average rate of speed for her drive to Houston, for her return trip, and for the round trip to the nearest mile per hour.

g. Driving 300 miles through the prairie, the car averaged 30 miles per gallon, while driving 300 miles through the Rocky Mountains, it averaged 20 miles per gallon. What mileage did the car average for all 600 miles?

Written Practice *Strengthening Concepts*

* **1.** (53) **Connect** The ratio of white keys to black keys on a piano is 13 to 9. If there are 36 black keys, how many white keys are there? Use a ratio box to find the answer.

* **2.** (55) The average of four numbers is 85. If three of the numbers are 76, 78, and 81, what is the fourth number?

3. (46) A one-quart container of oil costs \$2.89. A case of 12 one-quart containers costs \$28.56. How much is saved per container by buying the oil by the case?

4. (8) Segment *BC* is how much longer than segment *AB*?

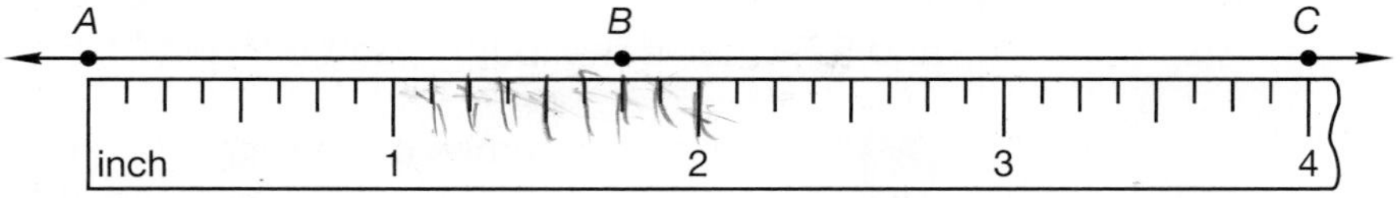

5. (22) Read this statement. Then answer the questions that follow.

Three tenths of the 30 students bought the school lunch.

a. How many students bought the school lunch?

b. What percent of the students bought the school lunch?

* **6.** (51) **a.** Write 675 million in scientific notation.

b. **Represent** The speed of light is 1.86×10^5 miles per second. Write this number in standard form.

7. (47) Find each missing exponent:

a. $10^8 \cdot 10^2 = 10^{\square}$ **b.** $10^8 \div 10^2 = 10^{\square}$

Math Language

A unit multiplier is a ratio equal to 1 that is composed of two equivalent measures.

* **8.** (50) **Analyze** Use unit multipliers to perform the following conversions:

a. 24 feet to inches

b. 60 miles per hour to miles per minute

9. (31) Use digits and other symbols to write "The product of two hundredths and twenty-five thousandths is five ten-thousandths."

* **10.** (55) One Saturday Gabe earned \$35.00 in 3 hours mowing lawns and \$7.00 an hour for 4 hours washing cars.

a. What was his total pay for the day?

b. What was his average hourly pay for the day?

11. Complete the table.
(48)

Fraction	Decimal	Percent
$\frac{1}{5}$	**a.**	**b.**
c.	0.1	**d.**
e.	**f.**	75%

Refer to the figure at right for problems **12** and **13.**

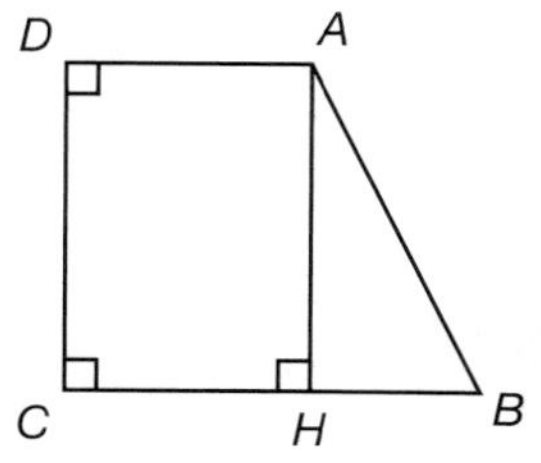

12. *(7)* **a.** Which segment is parallel to $\overline{BC}$?

b. Which two segments are perpendicular to $\overline{BC}$?

c. Angle ABC is an acute angle. Which angle is an obtuse angle?

13. *(37)* If $AD = 6$ cm, $CD = 8$ cm, and $CB = 10$ cm, then

a. what is the area of rectangle $AHCD$?

b. what is the area of triangle ABH?

c. what is the area of figure $ABCD$?

14. *(28)* Donato is 6 feet 2 inches tall. His sister is 68 inches tall. Donato is how many inches taller than his sister?

* **15.** *(54)* **Analyze** Monte swam 5 laps in 4 minutes.

a. How many laps could Monte swim in 20 minutes at this rate?

b. How long would it take for Monte to swim 20 laps at this rate?

16. *(41)* Show two ways to evaluate $b(a + b)$ for $a = \frac{1}{4}$ and $b = \frac{1}{2}$.

Solve:

17. *(39)* $\frac{30}{70} = \frac{21}{x}$

18. *(45)* $\frac{1000}{w} = 2.5$

Estimate each answer to the nearest whole number. Then perform the calculation.

19. *(30)* $2\frac{5}{12} + 6\frac{5}{6} + 4\frac{7}{8}$

20. *(23)* $6 - \left(7\frac{1}{3} - 4\frac{4}{5}\right)$

Analyze Simplify:

* **21.** (50) $10\text{ yd} \cdot \frac{36\text{ in.}}{1\text{ yd}}$

* **22.** (49)
$$\begin{array}{r} 8\text{ yd } 2\text{ ft } 7\text{ in.} \\ +\quad 5\text{ in.} \\ \hline \end{array}$$

23. (20) $12^2 - 4^3 - 2^4 - \sqrt{144}$

* **24.** (52) $50 + 30 \div 5 \cdot 2 - 6$

25. (26) $6\frac{2}{3} \cdot 5\frac{1}{4} \cdot 2\frac{1}{10}$

26. (26) $3\frac{1}{3} \div 3 \div 2\frac{1}{2}$

27. (35) $3.47 + (6 - 1.359)$

28. (45) $\$1.50 \div 0.075$

* **29.** (2, 9) **Justify** Name the property of multiplication used at each step to simplify the equation.

(4)(3.7)(0.25)	**Given**
(4)(0.25)(3.7)	**a.** ________
(1)(3.7)	**b.** ________
3.7	**c.** ________

30. (40) This quadrilateral is a rectangle. Find the measures of angles *a, b,* and *c.*

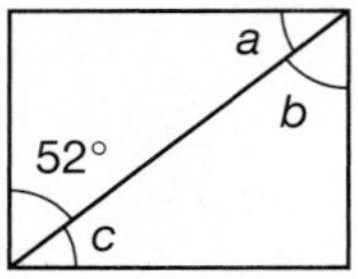

Early Finishers
Math and Art

Artists often use Leonardo da Vinci's Vitruvian ratios in order to create proportional drawings. For instance, on average a person's arm span equals his or her height.

a. If the ratio of head length to body length is 1:7, what is Marquez's height (in inches) if his head measures 9.5 inches?

b. What is Marquez's likely arm span in feet and inches?

LESSON 56

• Plotting Functions

Power Up

Building Power

facts Power Up L

mental math

a. **Calculation:** 30×2.5

b. **Decimals:** $0.25 \div 10$

c. **Algebra:** $3x + 4 = 40$

d. **Measurement:** Convert 0.5 m to cm.

e. **Exponents:** $25^2 - 15^2$

f. **Fractional Parts:** $\frac{7}{10}$ of \$50.00

g. **Measurement:** Two pints is what fraction of a gallon?

h. **Calculation:** Square 9, $-$ 1, $\div$ 2, $-$ 4, $\sqrt{\ }$, $\times$ 3, $+$ 2, $\div$ 5, $\sqrt{\ }$, $-$ 5.

problem solving

Solve the following number puzzle: If 2 is added to both the numerator and the denominator of a certain fraction, its value becomes $\frac{1}{2}$. If 2 is subtracted from both the numerator and the denominator of that same fraction, its value becomes $\frac{1}{3}$. What is the original fraction?

New Concept

Increasing Knowledge

We remember that a function is a mathematical rule defining the relationship between two sets of numbers. The numbers might be counts or they might be measures. In this lesson, we will learn to write function rules as equations, plot functions on a coordinate grid, and determine function rules from plotted points.

Example 1

Jamal is ordering chairs for the furniture store. The function table shows the number of chairs he orders for each table.

a. **Write a rule in words that shows the number of chairs Jamal orders for each table.**

b. **Complete the function table to find how many chairs Jamal orders if he has 5 tables.**

c. **Write the rule for the function table as an equation. Use the letter *y* to represent the number of chairs and *x* the number of tables.**

Input (*x*) Tables	Output (*y*) Chairs
1	4
2	8
3	12
4	16
5	

Solution

a. We can use the function table to find a rule. For every table that Jamal has, he orders 4 chairs. The rule of the function is ... ***To find the number of chairs, multiply the number of tables by 4.***

b. We can find the number of chairs Jamal orders by using our rule. The number of tables is 5. So we multiply 5 by 4 to find the number of chairs.

$$5 \times 4 = 20$$

Jamal orders **20 chairs** for **5 tables.**

Thinking Skill

Conclude

The value of *y* depends upon what?

c. Often, the rule of a function is expressed as an equation with *x* standing for the input number and *y* standing for the output number. In this example, *x* stands for the number of tables and *y* stands for the number chairs. We write the equation starting with "$y =$ ".

chairs tables

↓ ↓

$$\mathbf{y = x \cdot 4} \quad \text{or} \quad \mathbf{y = 4x}$$

Generalize Use the function rule to find the number of chairs Jamal would order if he had 7 tables.

Math Language

The term **coordinates** refers to an ordered pair of numbers used to locate a point in a coordinate plane.

Thinking Skill

Analyze

Look at the graph of the function. For each value of *x*, how many values of *y* are there?

We can use the number pairs in a function table as coordinates of points on a coordinate plane. Then we can plot the points. For the function in example 1 we can plot these points.

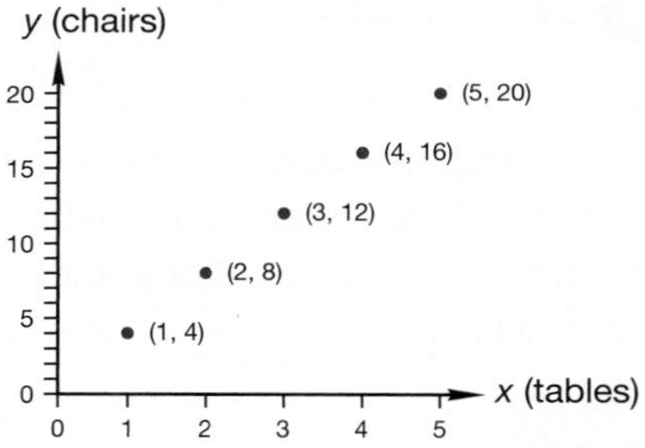

We see that the graph of this function is a set of unconnected points that are aligned. The points are unconnected because only whole tables are considered, not fractions of tables, so the input is a counting number. The points are aligned because the relationship between the number of tables and the number of chairs is constant.

The graph of some functions is a line as we see in the next example.

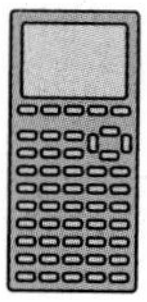

Visit www.SaxonPublishers.com/ActivitiesC2 *for a graphing calculator activity.*

Example 2

The function $y = 2x$ doubles every number that is put into it. If we input 5 the output is 10. If we input 10, the output is 20. Below is a function table that shows some input-output pairs. Graph all pairs of numbers for this function.

Input x	Output y
0	0
1	2
2	4
3	6
4	8

Solution

We begin by graphing the input-output (x, y) pairs in the function table.

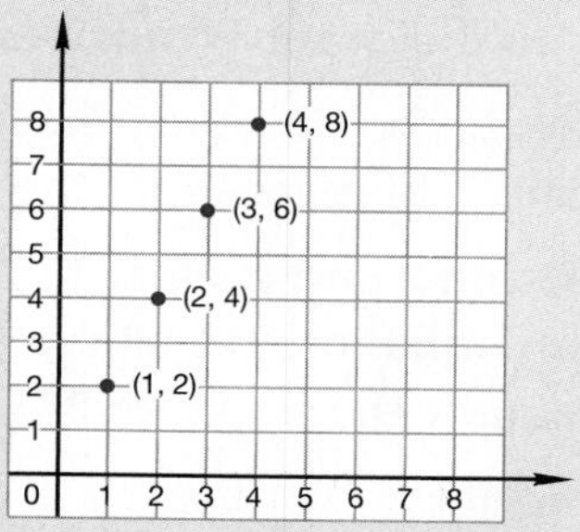

We have graphed some pairs of numbers for this function, but we are directed to graph all pairs. That means we need to graph (5, 10) and (6, 12) and so on. It also means we need to graph number pairs between these points such as $(\frac{1}{2}, 1)$ and $(\frac{1}{4}, \frac{1}{2})$ for which the output is twice the input. If we graph all such pairs the result is an uninterrupted series of points that form a line. The arrowhead shows that the line continues.

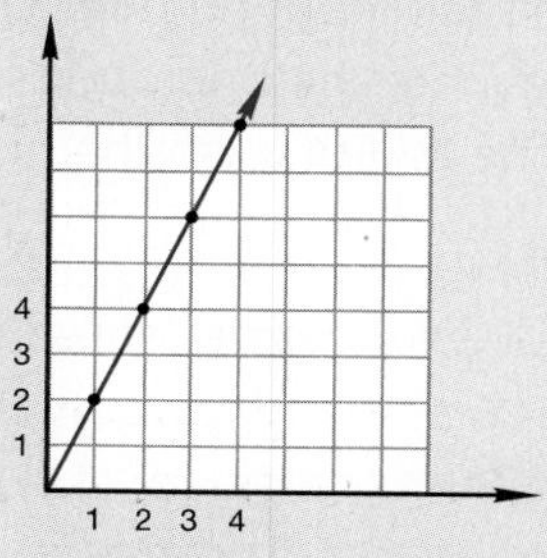

The graph of this function also continues in the opposite direction because the function applies to negative numbers as well as to positive numbers. If we input -1, the output is -2, and doubling -2 results in -4.[1] We show the pairs $(-1, -2)$ and $(-2, -4)$ and all the other pairs of negative numbers for this function by continuing the line of function pairs.

[1] Multiplication of negative numbers is taught in Lesson 73.

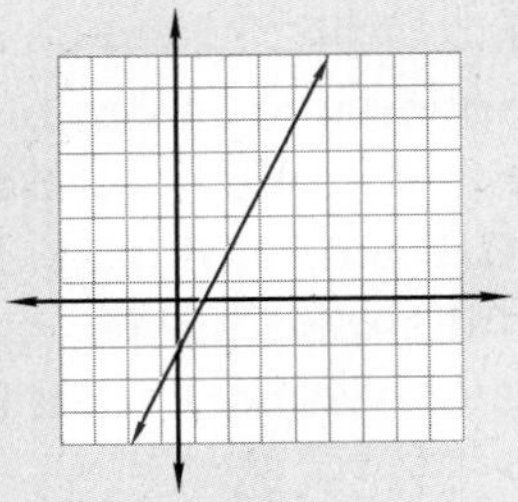

All the points on this line represent the pairs of numbers for the function $y = 2x$.

Example 3

The illustration shows the graph of a function. Refer to the graph to answer a and b.

a. Write the coordinates of three points on the line that represents the function.

b. Using x and y, write the rule of the function as an equation. Do all the (x, y) pairs from part a satisfy the equation?

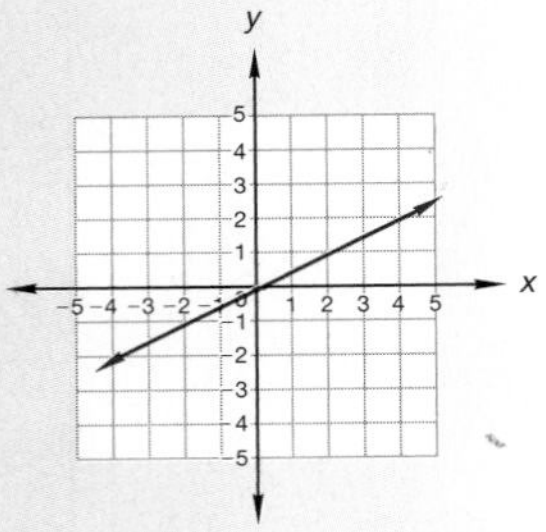

Solution

a. We look for points where the graphed line passes through intersections of the grid lines. We see that the line passes through **(0, 0), (2, 1)** and **(4, 2).** The line also passes through $(1, \frac{1}{2})$, $(-2, -1)$ and many other points, but we will use the whole number pairs.

b. By identifying points on the line we find that the output number (y) is half the input number (x). We begin the equation with "$y =$ " and then we write the operation that is performed on x to produce y. The function finds half of x, so the rule of the function can be written two ways.

$$y = \frac{1}{2}x \quad \text{or} \quad y = \frac{x}{2}$$

Now we test the equation with (x, y) pairs on the line.

For (0, 0)	**For (2, 1)**	**For (4, 2)**
$y = \frac{1}{2}x$	$y = \frac{1}{2}x$	$y = \frac{1}{2}x$
$0 = \frac{1}{2}(0)$	$1 = \frac{1}{2}(2)$	$2 = \frac{1}{2}(4)$
$0 = 0$ ✓	$1 = 1$ ✓	$2 = 2$ ✓

All tested (x, y) pairs satisfy the equation.

Practice Set

For this function table:

a. Formulate Write the rule in words.

b. Find the missing number in the table.

c. Represent Express the rule as an equation using the variables given in the table. Begin the equation with the output.

Input days (d)	Output hours (h)
1	24
2	48
5	120
10	

d. Represent On a coordinate plane plot four points that satisfy this function: the number of bicycle wheels (y) is twice the number of bicycles (x).

e. Make a function table and find four (x, y) pairs for the function $y = 3x$. (Hint: Think of small whole numbers for x and find y.) Then plot the (x, y) pairs on a coordinate plane. Then graph all the pairs of numbers (including fractions and negative numbers) by drawing a line through the plotted points as we did in example 2.

f. Write the rule for this graphed function as an equation. Begin the equation with "$y =$ ". Name another (x, y) pair not named on the graph that satisfies the function.

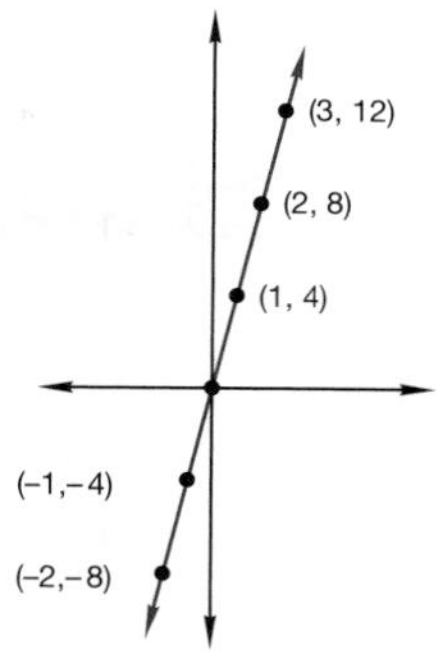

g. Discuss Which point on the graph in **f,** (4, 1) or (1, 4), satisfies the function? What are two ways we can tell?

Written Practice *Strengthening Concepts*

* **1.** (36, 43) Analyze A coin is flipped and a number cube is rolled.

a. What is the sample space of the experiment?

b. What is the probability of getting heads and a prime number? Write the probability as a reduced fraction and as a decimal number.

* **2.** (53) Model Use a ratio box to solve this problem. The ratio of the length to the width of the rectangle is 4 to 3. If the length of the rectangle is 12 feet,

a. what is its width?

b. what is its perimeter?

3. (28) The parking lot charges $2 for the first hour plus 50¢ for each additional half hour or part thereof. What is the total charge for parking a car in the lot from 11:30 a.m. until 2:15 p.m.?

* **4.** (55) After four days, Trudy's average exercise time was 45 minutes per day. If she exercises 60 minutes on the fifth day, what will be her average time in minutes per day after five days?

* **5.** (55) Twelve ounces of Brand X costs $1.50. Sixteen ounces of Brand Y costs $1.92. Find the unit price for each brand. Which brand is the better buy?

6. (36, 48) Five eighths of the rocks in the box were metamorphic. The rest were igneous.

a. What fraction of the rocks were igneous?

b. What was the ratio of igneous to metamorphic rocks?

c. What percent of the rocks were metamorphic?

7. (40) Refer to the figure at right to answer **a** and **b.**

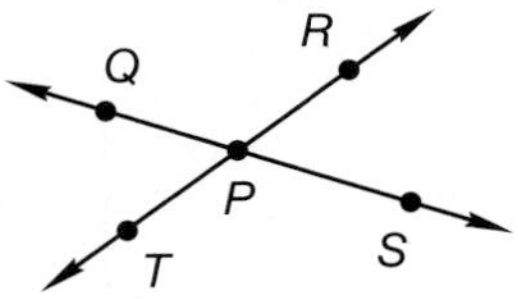

a. Name two pairs of vertical angles.

b. Name two angles that are supplemental to $\angle RPS$.

*** 8.** (51) **a.** Write six hundred ten thousand in scientific notation.

b. Write 1.5×10^4 in standard form.

*** 9.** (45, 50) **Connect** From goal line to goal line a football field is 100 yd long. One hundred yards is about how many meters? Use a unit multiplier to make the conversion and round the answer to the nearest meter. (1 meter $\approx$ 1.1 yd)

10. (43, 48) **a.** Write $\frac{1}{6}$ as a decimal number rounded to the nearest hundredth.

b. Write $\frac{1}{6}$ as a percent.

11. (51) How many pennies equal one million dollars? Write the answer in scientific notation.

12. (51) Compare: 11 million ◯ 1.1×10^6

13. (27) Which even two-digit number is a common multiple of 5 and 7?

The figure shows the dimensions of a window. Refer to the figure to answer problems **14** and **15.**

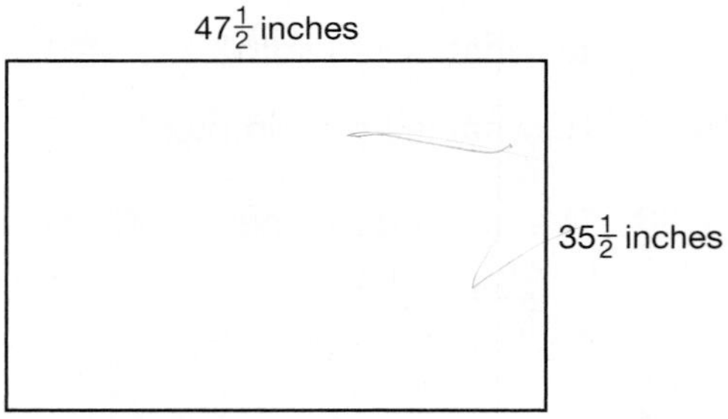

*** 14.** (29) **Estimate** What is the approximate perimeter of the window in feet?

*** 15.** (29) **Estimate** What is the approximate area of the window in square feet?

* **16.** (32, 54) **Connect** There are 100° on the Celsius scale from the freezing temperature to the boiling temperature of water. There are 180° on the Fahrenheit scale between these temperatures. So a change in temperature of 10° on the Celsius scale is equivalent to a change of how many degrees on the Fahrenheit scale?

Solve:

17. (39) $\frac{3}{2.5} = \frac{48}{c}$

18. (35) $k - 0.75 = 0.75$

Simplify:

19. (20) $15^2 - 5^3 - \sqrt{100}$

20. (52) $6 + 12 \div 3 \cdot 2 - 3 \cdot 4$

21. (49)
$$\begin{array}{r} 5 \text{ yd } 2 \text{ ft } 3 \text{ in.} \\ + \ 2 \text{ yd } 2 \text{ ft } 9 \text{ in.} \\ \hline \end{array}$$

* **22.** (49)
$$\begin{array}{r} 5 \text{ yd } 2 \text{ ft } 3 \text{ in.} \\ - \ 2 \text{ yd } 2 \text{ ft } 9 \text{ in.} \\ \hline \end{array}$$

23. (9) $\frac{18}{19} \cdot \frac{19}{18}$

24. (23, 30) $2\frac{3}{4} + \left(5\frac{1}{6} - 1\frac{1}{4}\right)$

25. (26) $3\frac{3}{4} \cdot 2\frac{1}{2} \div 3\frac{1}{8}$

26. (26) $3\frac{3}{4} \div 2\frac{1}{2} \cdot 3\frac{1}{8}$

27. (4) Describe how to find the 99th number in this sequence:

1, 4, 9, 16, 25, . . .

28. (17) Use a protractor and a straightedge to draw a triangle that has a right angle and a 30° angle. Then measure the shortest and longest sides of the triangle to the nearest millimeter. What is the relationship of the two measurements?

29. (Inv. 2, 32) If the diameter of a wheel is 0.5 meter, then the radius of the wheel is how many centimeters?

* **30.** (Inv. 3, 56) **Represent** The diameter of a circle is a function of the radius of the circle, expressed by the equation $d = 2r$.

a. Make a function table for r (input) and d (output). Record three pairs in the table using small whole numbers for the values for r.

b. On a coordinate plane, label the horizontal axis r and the vertical axis d. Then graph all the first quadrant pairs that satisfy the function.

c. **Discuss** How do we indicate that our graph includes all of the points in the first quadrant that satisfy the function?

LESSON

57

• Negative Exponents
• Scientific Notation for Small Numbers

Power Up — *Building Power*

facts Power Up K

mental math

a. Calculation: 40×3.2

b. Decimals/Exponents: 4.2×10^3

c. Ratio: $\frac{n}{20} = \frac{7}{5}$

d. Measurement: Convert 500 mL to L.

e. Exponents: $15^2 - 5^3$

f. Fractional Parts: $\frac{2}{5}$ of \$25.00

g. Statistics: Find the range of the set of numbers: 78, 56, 99, 25, 87, 12.

h. Calculation: Start with the number of pounds in a ton, ÷ 2, − 1, ÷ 9, − 11, $\sqrt{\ }$, ÷ 2, ÷ 2.

problem solving It takes 6 men 2 hours to dig a hole that is 2 m × 2 m × 2 m. How long will it take 12 men to dig a hole that is 4 m × 4 m × 4 m?

New Concepts — *Increasing Knowledge*

negative exponents

Cantara multiplied 0.000001 by 0.000001 on her scientific calculator. After she pressed [=] the display read

1. × 10 -12

The calculator displayed the product in scientific notation. Notice that the exponent is a negative number. So

$$1 \times 10^{-12} = 0.000000000001$$

Studying the pattern below may help us understand the meaning of a negative exponent.

$$\frac{10^5}{10^3} = \frac{\overset{1}{\cancel{10}} \cdot \overset{1}{\cancel{10}} \cdot \overset{1}{\cancel{10}} \cdot 10 \cdot 10}{\underset{1}{\cancel{10}} \cdot \underset{1}{\cancel{10}} \cdot \underset{1}{\cancel{10}}} = 10^2 = 100$$

$$\frac{10^4}{10^3} = \frac{\overset{1}{\cancel{10}} \cdot \overset{1}{\cancel{10}} \cdot \overset{1}{\cancel{10}} \cdot 10}{\underset{1}{\cancel{10}} \cdot \underset{1}{\cancel{10}} \cdot \underset{1}{\cancel{10}}} = 10^1 = 10$$

Recall that to divide powers of the same base we subtract (the exponent of the divisor from the exponent of the dividend). Now we will continue the pattern.

$$\frac{10^3}{10^3} = \frac{\overset{1}{\cancel{10}} \cdot \overset{1}{\cancel{10}} \cdot \overset{1}{\cancel{10}}}{\underset{1}{\cancel{10}} \cdot \underset{1}{\cancel{10}} \cdot \underset{1}{\cancel{10}}} = 10^0 \quad = 1$$

$$\frac{10^2}{10^3} = \frac{\overset{1}{\cancel{10}} \cdot \overset{1}{\cancel{10}}}{\underset{1}{\cancel{10}} \cdot \underset{1}{\cancel{10}} \cdot 10} = 10^{-1} = \frac{1}{10} = \frac{1}{10^1}$$

$$\frac{10^1}{10^3} = \frac{\overset{1}{\cancel{10}}}{\underset{1}{\cancel{10}} \cdot 10 \cdot 10} = 10^{-2} = \frac{1}{100} = \frac{1}{10^2}$$

Notice especially these results:

$$10^0 = 1$$

$$10^{-1} = \frac{1}{10^1}$$

$$10^{-2} = \frac{1}{10^2}$$

The pattern suggests two facts about exponents, which we express algebraically below.

Thinking Skill

Summarize

Write a rule that summarizes the algebraic expression $a^{-n} = \frac{1}{a^n}$.

If a number a is not zero, then
$a^0 = 1$
$a^{-n} = \frac{1}{a^n}$

Example 1

Simplify:

a. $\mathbf{2^0}$ **b.** $\mathbf{3^{-2}}$ **c.** $\mathbf{10^{-3}}$

Solution

a. The exponent is zero and the base is not zero, so 2^0 equals **1.**

b. We rewrite the expression using the reciprocal of the base with a positive exponent. Then we simplify.

$$3^{-2} = \frac{1}{3^2} = \mathbf{\frac{1}{9}}$$

c. Again we rewrite the expression with the reciprocal of the base and a positive exponent.

$$10^{-3} = \frac{1}{10^3} = \mathbf{\frac{1}{1000}} \text{ (or } \mathbf{0.001}\text{)}$$

scientific notation for small numbers

As we saw at the beginning of this lesson, negative exponents can be used to express small numbers in scientific notation. For instance, an inch is 2.54×10^{-2} (two point five four times ten to the negative two) meters. If we multiply 2.54 by 10^{-2}, the product is 0.0254.

$$2.54 \times 10^{-2} = 2.54 \times \frac{1}{10^2} = 0.0254$$

Notice the product, 0.0254, has the same digits as 2.54 but with the decimal point shifted two places to the left and with zeros used for placeholders. The two-place decimal shift to the left is indicated by the exponent −2. This is similar to the method we have used to change scientific notation to standard form. Note the sign of the exponent. If the exponent is a *positive number,* we shift the decimal point *to the right* to express the number in standard form. In the number

$$6.32 \times 10^7$$

the exponent is *positive* seven, so we shift the decimal point seven places *to the right.*

$$\underbrace{63200000.}_{\text{7 places}} \longrightarrow 63{,}200{,}000$$

If the exponent is a *negative number,* we shift the decimal point *to the left* to write the number in standard form. In the number

$$6.32 \times 10^{-7}$$

the exponent is *negative* seven, so we shift the decimal point seven places *to the left.*

$$\underbrace{.000000632}_{\text{7 places}} \longrightarrow 0.000000623$$

In either case, we use zeros as placeholders.

Example 2

Write 4.63×10^{-8} in standard notation.

Solution

The negative exponent indicates that the decimal point is eight places to the left when the number is written in standard form. We shift the decimal point and insert zeros as placeholders.

$$\underbrace{.0000000463}_{\text{8 places}} \longrightarrow \mathbf{0.0000000463}$$

Example 3

Write 0.0000033 in scientific notation.

Solution

We place the decimal point to the right of the first digit that is not a zero.

$$\underbrace{0000003.}_{\text{6 places}}3$$

In standard form the decimal point is six places to the left of where we have placed it. So we write

$$3.3 \times 10^{-6}$$

Example 4

Compare: zero ◯ 1×10^{-3}

Solution

The expression 1×10^{-3} equals 0.001. Although this number is less than 1, it is still positive, so it is greater than zero.

$$\text{zero} < 1 \times 10^{-3}$$

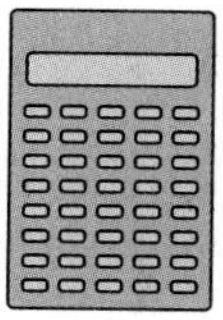

Very small numbers may exceed the display capabilities of a calculator. One millionth of one millionth is more than zero, but it is a very small number. On a calculator we enter

The product, one trillionth, contains more digits than can be displayed by many calculators. Instead of displaying one trillionth in standard form, calculators that can express numbers in scientific notation, or in a modified form of scientific notation, might display this number as shown below:

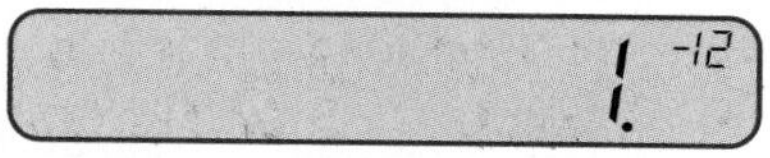

or perhaps

Practice Set

Simplify:

a. 5^{-2} **b.** 3^{0} **c.** 10^{-4}

Write each number in scientific notation:

d. 0.00000025 **e.** 0.000000001 **f.** 0.000105

Write each number in standard form:

g. 4.5×10^{-7} **h.** 1×10^{-3} **i.** 1.25×10^{-5}

j. *Explain* In exercises **g–i,** how did you shift the decimal point as you changed each number from scientific notation to standard form?

Compare:

k. 1×10^{-3} ◯ 1×10^{2} **l.** 2.5×10^{-2} ◯ 2.5×10^{-3}

m. Use digits to write "three point five times ten to the negative eight."

Written Practice *Strengthening Concepts*

* **1.** (53) Use a ratio box to solve this problem. The ratio of walkers to riders was 5 to 3. If 315 were walkers, how many were riders?

* **2.** (55) **Connect** After five games, the basketball team's average score was 88 points. After six games their average score had increased to 90. What was their score in the sixth game?

3. (28) Rico wants to mail two letters, each weighing 4 ounces. For each letter, the first ounce costs \$0.39. Each additional ounce costs \$0.25. How much does it cost Rico to mail his two letters?

4. (16) If lemonade costs \$0.52 per pint, what is the cost per cup?

5. (22) Read this statement and answer the questions that follow.

Tyrone finished his math homework in two fifths of an hour.

a. How many minutes did it take Tyrone to finish his math homework?

b. What percent of an hour did it take Tyrone to finish his math homework?

c. To answer **a** and **b,** what information that was not given did you need to know?

* **6.** (51, 57) **Represent** Write each number in scientific notation:

a. 186,000 **b.** 0.00004

* **7.** (51, 57) **Represent** Write each number in standard form:

a. 3.25×10^1 **b.** 1.5×10^{-6}

* **8.** (57) Simplify:

a. 2^{-3} **b.** 5^0 **c.** 10^{-2}

* **9.** (50) Use a unit multiplier to perform the following conversions.

a. 2000 milliliters to liters.

b. 10 liters to quarts (1 liter $\approx$ 1.06 qt)

10. (21, 43) What is the probability of rolling a composite number on one toss of a number cube? Write the probability as a fraction and as a decimal number rounded to two decimal places.

11. (46) The tickets for a dozen children to enter the amusement park cost \$330. What was the price per ticket?

* **12.** (Inv. 5) **Represent** Matt checked the reviews for the new movie and tallied the number of stars each reviewer gave the movie. Some reviewers gave half stars. Create a histogram that illustrates the data in the table.

Movie Ratings

Number of Stars	Tally	Frequency
3.5–4	𝍸 ||	7
2.5–3	𝍸 ||||	9
1.5–2	𝍸 |	6
0.5–1	|||	3

*** 13.** (57) Compare:

a. 2.5×10^{-2} ◯ $2.5 \div 10^2$

b. one millionth ◯ 1×10^{-6}

c. 3^0 ◯ 2^0

Refer to the figure below to answer problems 14 and 15. Dimensions are in yards. All angles are right angles.

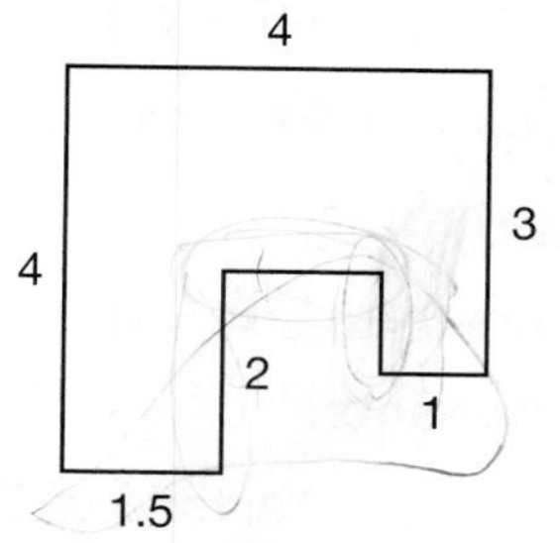

14. (19) What is the perimeter of the figure?

15. (37) What is the area of the figure? Sketch how you divided the figure to get your answer.

16. (41) Evaluate: $4ac$ if $a = 5$ and $c = 0.5$

17. (33) Estimate the quotient: $19.89 ÷ 3.987

*** 18.** (41, 56) **Analyze** The following equation is the rule of the function.

$$y = 3x + 5$$

We can use the rule to make a function table. Copy and complete this table by finding the values of y for the given values of x.

x	y
0	
1	
2	
3	

Simplify:

19. (20) $20^2 + 10^3 - \sqrt{36}$

*** 20.** (52) $48 \div 12 \div 2 + 2(3)$

*** 21.** (49) **Connect**

3 yd	2 ft	1 in.
− 1 yd	2 ft	3 in.

22. (49)

4 gal	3 qt	1 pt	6 oz
+ 1 gal	2 qt	1 pt	5 oz

23. (50) $48 \text{ oz} \cdot \frac{1 \text{ pt}}{16 \text{ oz}}$

24. (26) $5\frac{1}{3} \cdot \left(7 \div 1\frac{3}{4}\right)$

25. (30) $5\frac{1}{6} + 3\frac{5}{8} + 2\frac{7}{12}$

26. (30) $\frac{1}{20} - \frac{1}{36}$

27. (57) $(4.6 \times 10^{-2}) + 0.46$

28. (35) $10 - (2.3 - 0.575)$

29. (35) $0.24 \times 0.15 \times 0.05$

30. (45) $10 \div (0.14 \div 70)$

LESSON 58

• Symmetry

Power Up

Building Power

facts Power Up K

mental math

a. **Decimals:** 50×4.3

b. **Decimals/Exponents:** $4.2 \div 10^3$

c. **Algebra:** $3x - 5 = 40$

d. **Measurement:** Convert 1.5 kg to g.

e. **Exponents:** $10^3 \div 10^2$

f. **Fractional Parts:** $\frac{2}{3}$ of \$33.00

g. **Statistics:** Find the mode of the set of numbers: 99, 78, 28, 87, 82, 78.

h. **Calculation:** Find the sum, difference, product, and quotient of 1.2 and 0.6.

problem solving A group of four girls and two boys sat side-by-side on a bench so that every girl sat next to at least one other girl. What arrangements of girls and boys are possible?

New Concept

Increasing Knowledge

reflective symmetry

A two-dimensional figure has **reflective symmetry** or **line symmetry** if it can be divided in half so that the halves are mirror images of each other. Line *r* divides this triangle into two mirror images; so the triangle is symmetrical, and line *r* is a **line of symmetry.**

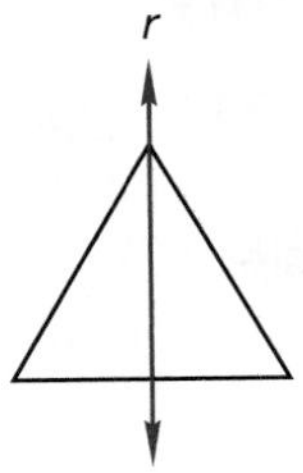

Math Language

A **regular triangle** has all sides the same length and all angles have the same measure.

Actually, the regular triangle has three lines of symmetry.

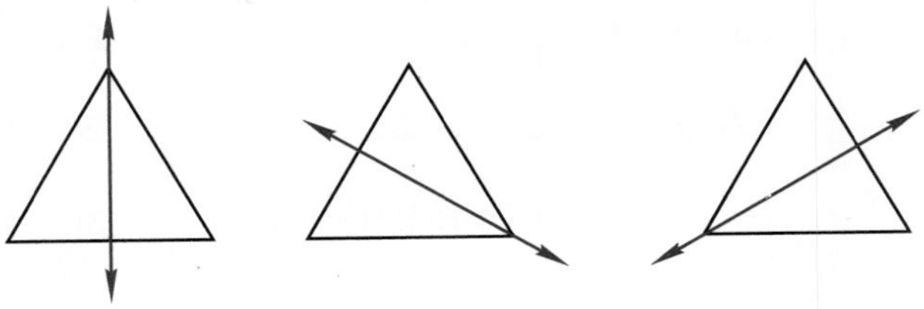

Example 1

Draw a regular quadrilateral and show all lines of symmetry.

Solution

A regular quadrilateral is a square. A square has four lines of symmetry.

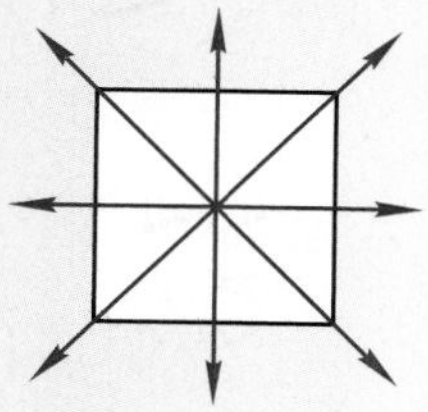

The *y*-axis is a line of symmetry for the figure below. Notice that corresponding points on the two sides of the figure are the same distance from the line of symmetry.

Thinking Skill

Analyze

Is the *x*-axis a line of symmetry? Why or why not?

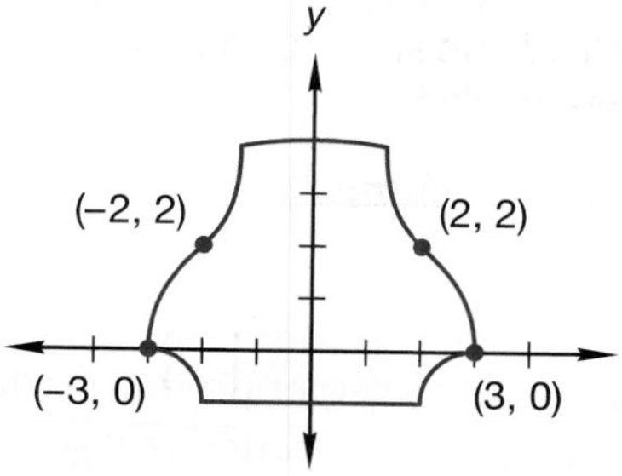

If this figure were folded along the *y*-axis, each point of the figure on one side of the *y*-axis would be folded against its corresponding point on the other side of the *y*-axis.

Connect What are some objects in your classroom that have at least one line of symmetry? What are some objects that have no lines of symmetry?

Activity

Line Symmetry

Materials needed:

- Paper and scissors

1. Fold a piece of paper in half.

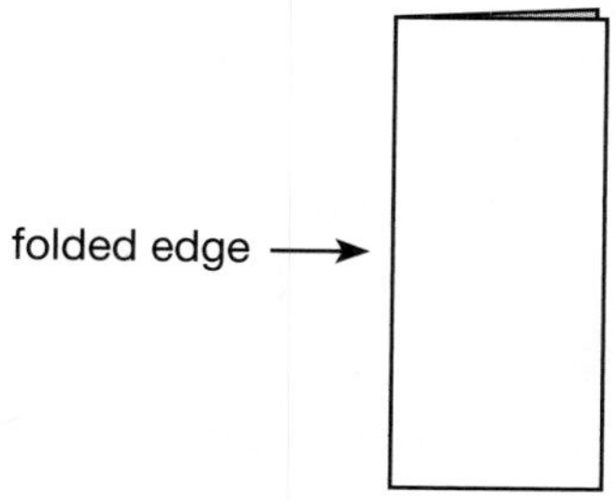

2. Beginning and ending at the folded edge, cut a pattern out of the folded paper.

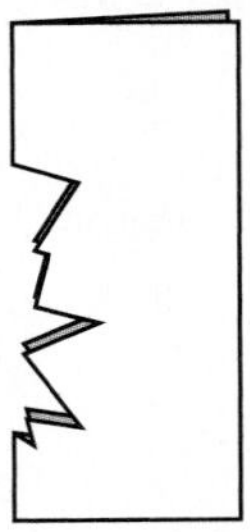

3. Open the cut-out and note its symmetry.

4. Fold a piece of paper twice as shown.

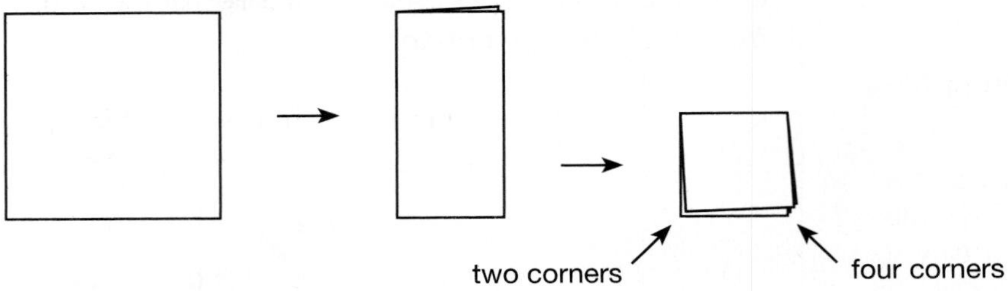

5. Hold the paper on the corner opposite the "four corners," and cut out a pattern that removes the four corners.

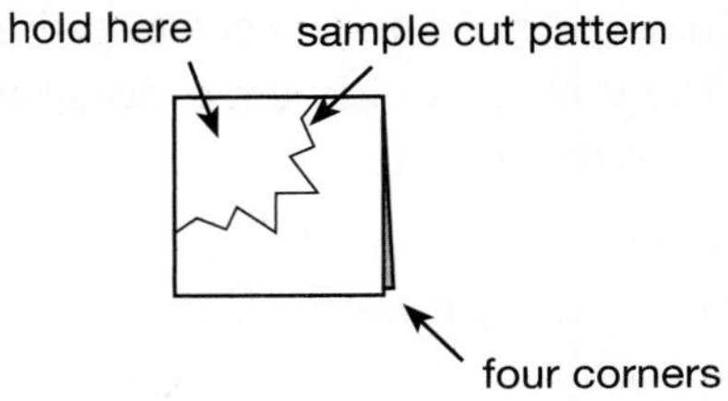

6. Unfold the cut-out. How many lines of symmetry do you see?

---

rotational symmetry

A figure has **rotational symmetry** if it re-appears in its original position more than once in a full turn.

Math Language
Rotate means to turn around a point.

For example, consider the upper case letters S and I. As these letters are rotated a full turn we see them appear in their original orientation after a half turn and after a full turn. Rotate your book to see these letters re-appear after a half turn.

S I S

Example 2

Which of these figures does not have rotational symmetry?

A  **B** 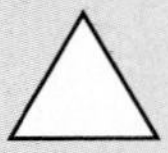 **C** 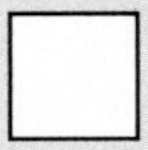 **D** 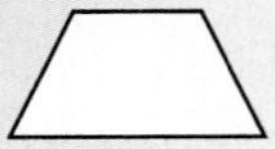

Solution

Figure A has rotational symmetry of order 2.

Figure B has rotational symmetry of order 3.

Figure C has rotational symmetry of order 4.

Figure D does not have rotational symmetry because it does not reappear in its original position in less than one full turn.

Model Draw a figure that has rotational symmetry and a figure that does not have rotational symmetry.

Practice Set

a. **Explain** Copy this rectangle on your paper, and show that it has only 2 lines of symmetry. Why aren't the diagonals of a rectangle lines of symmetry?

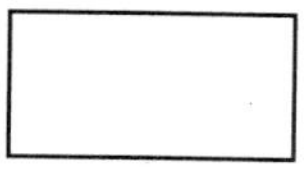

b. The y-axis is a line of symmetry for a triangle. The coordinates of two of its vertices are (0, 1) and (3, 4). What are the coordinates of the third vertex?

c. Which of these letters have rotational symmetry? Which have reflective symmetry?

V W X Y

d. Which of these polygons does not necessarily have rotational symmetry? Sketch an example of the polygon without rotational symmetry.

rectangle square

triangle parallelogram

Written Practice

Strengthening Concepts

1. (28, 35) It is 1.4 kilometers from Jim's house to school. How far does Jim walk going to and from school once every day for 5 days?

2. (28) The parking lot charges 75¢ for each half hour or part of a half hour. If Edie parks her car in the lot from 10:45 a.m. until 1:05 p.m., how much money will she pay?

3. (41) **Generalize** If the product of n and 17 is 340, what is the sum of n and 17?

4. (36) An art dealer sold 3 of the 12 paintings that were on display in her gallery.

a. What was the ratio of the number of paintings sold to the number of paintings not sold?

b. What fraction of the paintings did the dealer not sell?

c. What percent of the paintings were sold?

*** 5.** (55) Will's bowling average after 5 games was 120. In his next 3 games, Will scored 118, 124, and 142. What was Will's bowling average after 8 games?

6. (22) Diagram this statement. Then answer the questions that follow.

Three fifths of the 60 questions were multiple-choice.

a. How many of the 60 questions were multiple-choice?

b. What percent of the 60 questions were not multiple-choice?

7. (Inv. 2) Use the figure below to answer **a–d.** The center of the circle is point O and $OB = CB$.

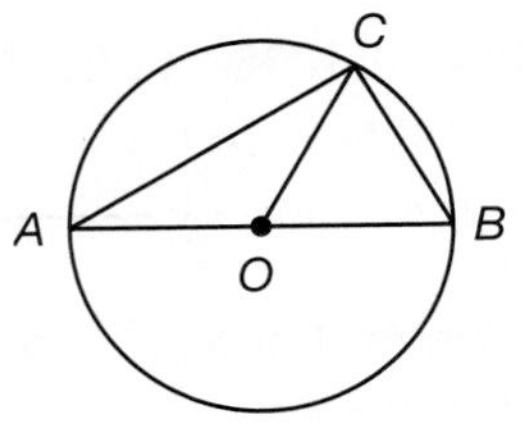

a. Name three radii.

b. Name two chords that are not diameters.

c. Estimate the measure of central angle BOC.

d. Estimate the measure of inscribed angle BAC.

*** 8.** (51, 57) Write each number in standard form:

a. 1.5×10^7

b. 2.5×10^{-4}

c. 10^{-1}

d. 10^0

*** 9.** (16, 50) Compare: 2 gal ◯ 8 liters

Use a unit multiplier to help you make the comparison.
(1 liter = 1.06 qt)(1 gal = 4 qt)

10. (33, 45) Divide 3.45 by 0.18 and write the answer rounded to the nearest whole number.

11. (4) Find the next three numbers in this sequence:

20, 15, 10, ...

12. (48) Complete the table.

Fraction	Decimal	Percent
$\frac{1}{6}$	**a.**	**b.**
c.	**d.**	16%

* **13.** (56) For this function table:

x	y
0	0
3	12
6	24
2	

a. Formulate Write the rule in words.

b. Find the value of y when x is 2.

c. Write the rule of the function as an equation.

14. (40) In the figure at right, the measure of $\angle D$ is 35° and the measure of $\angle CAB$ is 35°. Find the measure of

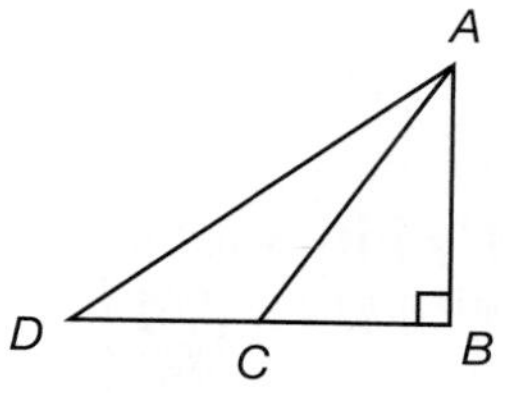

a. $\angle ACB$.

b. $\angle ACD$.

c. $\angle CAD$.

* **15.** (Inv. 3, 58) The y-axis is a line of symmetry for a triangle. The coordinates of two of its vertices are $(-3, 2)$ and $(0, 5)$.

a. What are the coordinates of the third vertex?

b. What is the area of the triangle?

* **16.** (58) **a.** A regular pentagon has how many lines of symmetry?

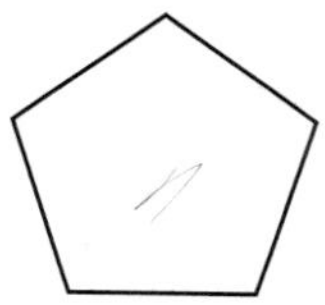

b. Justify Does a regular pentagon have rotational symmetry? How do you know?

* **17.** (46, 54) **a.** Connect Traveling at 60 miles per hour, how long would it take to travel 210 miles?

b. How long would the same trip take at 70 miles per hour?

Solve:

18. (39) $\frac{1.5}{2} = \frac{7.5}{w}$

19. (35) $1.7 - y = 0.17$

Analyze Simplify:

* **20.** (20, 57) $10^3 - 10^2 + 10^1 - 10^0$

21. (52) $6 + 3(2) - 4 - (5 + 3)$

22. (49)
$$\begin{array}{r} 1 \text{ gal } 2 \text{ qt } 1 \text{ pt} \\ +1 \text{ gal } 2 \text{ qt } 1 \text{ pt} \\ \hline \end{array}$$

* **23.** (49)
$$\begin{array}{r} 1 \text{ day } 3 \text{ hr } 15 \text{ min} \\ -\quad 8 \text{ hr } 30 \text{ min} \\ \hline \end{array}$$

24. (50) $2 \text{ mi} \cdot \frac{5280 \text{ ft}}{1 \text{ mi}}$

25. (23, 30) $10 - \left(5\frac{3}{4} - 1\frac{5}{6}\right)$

26. (26, 30) $\left(2\frac{1}{5} + 5\frac{1}{2}\right) \div 2\frac{1}{5}$

27. (26) $3\frac{3}{4} \cdot \left(6 \div 4\frac{1}{2}\right)$

* **28.** (52) **Explain** Evaluate: $b^2 - 4ac$ if $a = 3.6$, $b = 6$, and $c = 2.5$

29. (43) **a.** Arrange these numbers in order from **greatest to least:**

$$-1, \frac{3}{2}, 2.5, 0, -\frac{1}{2}, 2, 0.2$$

b. Which of the numbers in **a** are integers?

30. (27) Lindsey had the following division to perform:

$$35 \div 2\frac{1}{2}$$

Describe how Lindsey could form an equivalent division problem that would be easier to perform mentally.

Early Finishers
Math and Science

Electrons, protons, and neutrons are particles that are so small they can only be seen with a special microscope. Below are the weight in grams (g) for an electron, proton and neutron.

electron	9.1083×10^{-28}
proton	1.6726×10^{-24}
neutron	1.6750×10^{-24}

a. Which particle in this list is the lightest?

b. Which particle is the heaviest?

LESSON

59

• Adding Integers on the Number Line

Power Up

Building Power

facts Power Up L

mental math

a. **Decimals:** 60×5.4

b. **Decimals/Exponents:** 0.005×10^2

c. **Ratio:** $\frac{30}{20} = \frac{3}{t}$

d. **Measurement:** Convert 185 cm to m.

e. **Exponents:** $2 \cdot 2^3$

f. **Fractional Parts:** $\frac{7}{8}$ of \$40.00

g. **Statistics:** Find the median of the set of numbers: 78, 90, 34, 36, 55.

h. **Rate:** At \$7.50 an hour, how much money can Shelly earn in 8 hours?

problem solving

The city park has two rectangular flowerbeds. The perimeter of each flowerbed is 24 yards. However, one flowerbed has an area that is 8 square yards greater than the other. What are the dimensions of the two flowerbeds?

New Concept

Increasing Knowledge

Recall that **integers** include all the whole numbers and also the opposites of the positive integers (their negatives). All the numbers in this sequence are integers:

$$\ldots, -3, -2, -1, 0, 1, 2, 3, \ldots$$

The dots on this number line mark the integers from −5 through +5:

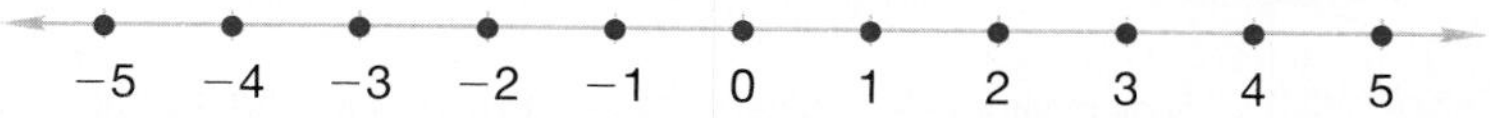

Remember that the numbers between the whole numbers, such as $3\frac{1}{2}$ and 1.3, are not integers.

All numbers on the number line except zero are **signed numbers,** either positive or negative. Zero is neither positive nor negative. Positive and negative numbers have a sign and a value, which is called **absolute value. The absolute value of a number is its distance from zero on a number line.**

Numeral	Number	Sign	Absolute Value
+3	Positive three	+	3
−3	Negative three	−	3

Thinking Skill

Generalize

What is the absolute value of +7 and −7? Explain how you know.

The absolute value of both +3 and −3 is 3 because +3 and −3 are both 3 units from zero on a number line. We may use two vertical segments to indicate absolute value.

$|3| = 3$ "The absolute value of 3 equals 3."

$|-3| = 3$ "The absolute value of −3 equals 3."

Example 1

Find each absolute value.

a. $|-2.5|$

b. $|100|$

c. $\left|-\frac{1}{2}\right|$

d. $|0|$

Solution

The absolute value of a number is its distance from zero on a number line.

a. $|-2.5| = \mathbf{2.5}$

b. $|100| = \mathbf{100}$

c. $\left|-\frac{1}{2}\right| = \mathbf{\frac{1}{2}}$

d. $|0| = \mathbf{0}$

Example 2

Simplify:

a. $|3 - 5|$

b. $|5 - 3|$

Solution

a. To find the absolute value of 3 − 5, we first subtract 5 from 3 and get −2. Then we find the absolute value of −2, which is **2.**

b. We reverse the order of subtraction. The absolute value is also **2.** Absolute value can be represented by distance, whereas the sign can be represented by direction. Thus positive and negative numbers are sometimes called **directed numbers** because the sign of the number (+ or −) can be thought of as a direction indicator.

When we add, subtract, multiply, or divide directed numbers, we need to pay attention to the signs as well as the absolute values of the numbers. In this lesson we will practice adding positive and negative numbers.

A number line can be used to illustrate the addition of signed numbers. A positive 3 is indicated by a 3-unit arrow that points to the right. A negative 3 is indicated by a 3-unit arrow that points to the left.

To show the addition of +3 and –3, we begin at zero on the number line and draw the +3 arrow. From its arrowhead we draw the −3 arrow. The sum of +3 and −3 is found at the point on the number line that corresponds to the second arrowhead.

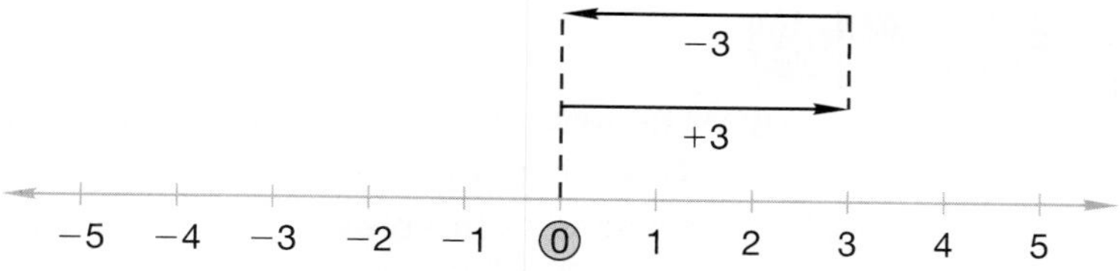

We see that the sum of +3 and −3 is 0. **The sum of two opposites is always zero.**

Example 3

Show each addition problem on a number line:

a. (−3) + (+5) **b. (−4) + (−2)**

Solution

a. We begin at zero and draw an arrow 3 units long that points to the left. From this arrowhead we draw an arrow 5 units long that points to the right. We see that the sum of −3 and +5 is **2.**

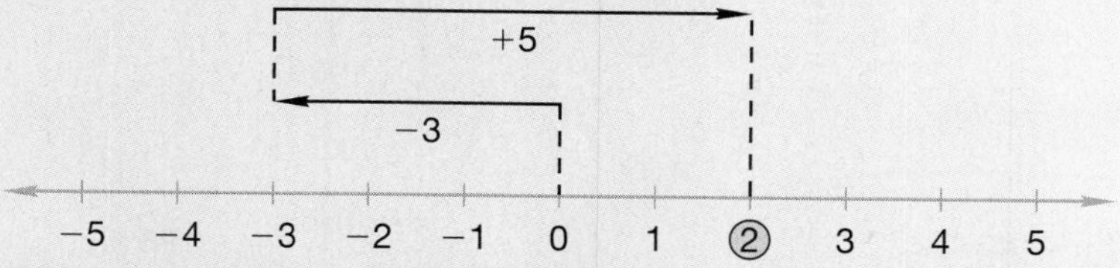

b. We use arrows to show that the sum of −4 and −2 is **−6.**

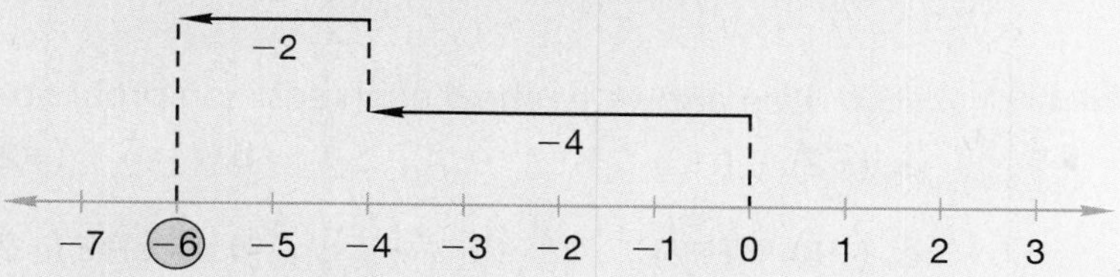

Example 4

Carmen borrowed $5 from her sister. Later Carmen received a check for $25 from her grandmother. After she repays her sister, how much money will Carmen have?

Solution

We may use negative numbers to represent debt (borrowed money). After borrowing $5, Carmen had negative five dollars. Then she received $25. We show the addition of these dollar amounts on the number line below.

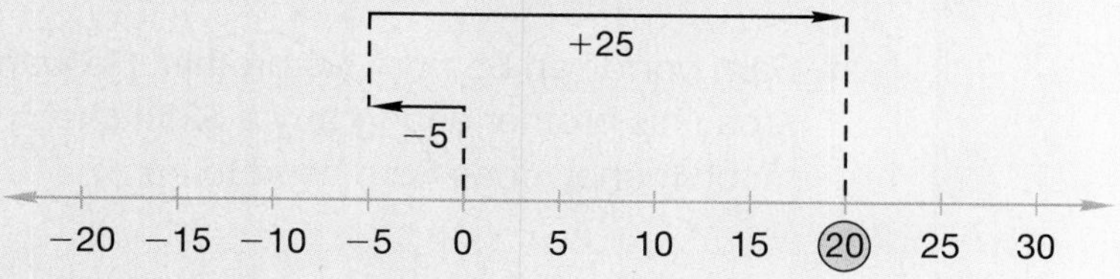

After she repays her sister, Carmen will have **$20.**

Example 5

A group of hikers began their trip at the desert floor, 126 feet below sea level. The group camped for the night on a ridge 2350 feet above sea level. What was the elevation gain from the start of the hike to the campsite?

Solution

A number line that is oriented vertically rather than horizontally is more helpful for this problem. The troop climbed 126 feet to reach sea level (zero elevation) and then climbed 2350 feet more to the campsite. We calculate the total elevation gain as shown.

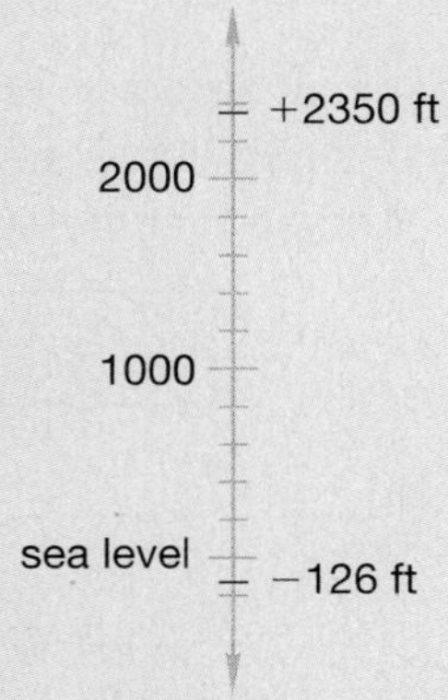

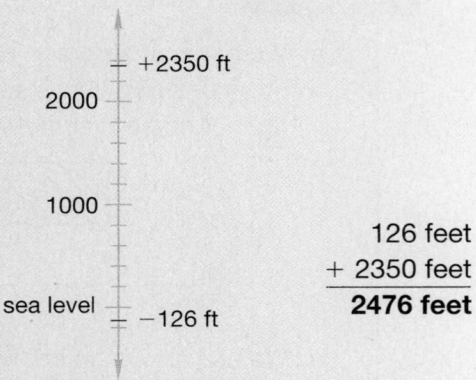

$$\begin{array}{r} 126 \text{ feet} \\ +\ 2350 \text{ feet} \\ \hline \mathbf{2476 \text{ feet}} \end{array}$$

Practice Set

Model Use arrows to show each addition problem on a number line:

a. $(-2) + (-3)$

b. $(+4) + (+2)$

c. $(-5) + (+2)$

d. $(+5) + (-2)$

e. $(-4) + (+4)$

f. $(-3) + (+6) + (-1)$

Generalize Find each absolute value.

g. $\left|-\frac{1}{4}\right|$

h. $|11|$

i. $|-0.05|$

Simplify:

j. $|-3| + |3|$

k. $|3 - 3|$

l. $|5 - 3|$

m. **Generalize** On the return trip the hikers walked down the mountain from 4362 ft above sea level to the valley floor 126 ft below sea level. What was the drop in elevation during the return trip?

n. Sam borrowed \$5 from his brother. He wants to earn enough money to repay his brother and to buy a \$25 ticket to the amusement park. How much money does Sam need to earn?

Written Practice   *Strengthening Concepts*

1. (28) School pictures cost $4.25 for an 8-by-10 print. They cost $2.35 for a 5-by-7 print and 60¢ for each wallet-size print. What is the total cost of two 5-by-7 prints and six wallet-size prints?

The double-line graph below compares the daily maximum temperatures for the first seven days of August to the average maximum temperature for the entire month of August. Refer to the graph to answer problems **2** and **3:**

Maximum Temperature Readings for Tri-City Area, August 1–7

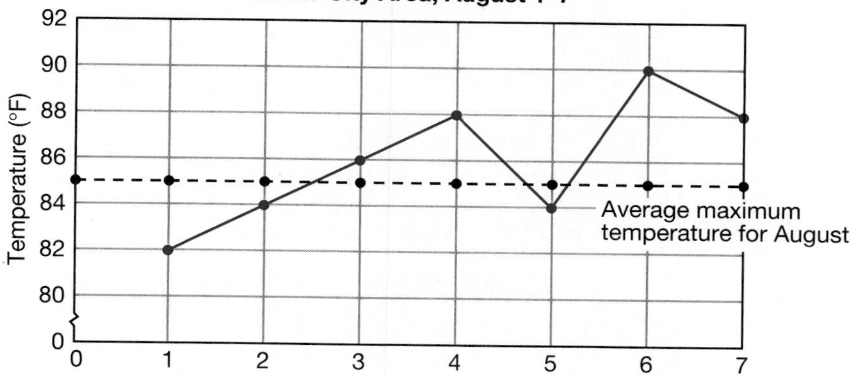

2. (38) The maximum temperature reading on August 6 was how much greater than the average maximum temperature for the month of August?

3. (28, 38) What was the average maximum temperature for the first seven days of August?

*** 4.** (59) **Generalize** On January 1 the temperature at noon was 7°F. By 10 p.m. the temperature had fallen to −9°F. The temperature dropped how many degrees from noon to 10 p.m.?

*** 5.** (53) Use a ratio box to solve this problem. The ratio of red apples to green apples was 7 to 4. If there were 56 green apples, how many red apples were there?

6. (22) Diagram this statement. Then answer the questions that follow.

The Celts won three fourths of their first 20 games.

a. How many of their first 20 games did the Celts win?

b. What percent of their first 20 games did the Celts fail to win?

*** 7.** (59) Compare: $|-3| \bigcirc |3|$

8. (51) **a.** Write 4,000,000,000,000 in scientific notation.

b. Pluto's average distance from the Sun is 3.67×10^9 miles. Write that distance in standard form.

* **9.** (57) **a.** *Represent* A micron is 1×10^{-6} meter. Write that number and unit in standard form.

b. Compare: 1 millimeter ◯ 1×10^{-3} meter

10. (50) Use a unit multiplier to convert 300 mm to m.

11. (48) Complete the table.

Fraction	Decimal	Percent
a.	**b.**	12%
$\frac{1}{3}$	**c.**	**d.**

* **12.** (59) *Model* Use arrows to show each addition problem on a number line:

a. $(+2) + (-5)$ **b.** $(-2) + (+5)$

* **13.** (56) For this function table:

a. State the value in words.

b. Find y when x is 12.

c. *Represent* Write the rule as an equation.

x	y
0	12
2	14
8	20
12	

Refer to the figure below to answer problems **14** and **15.** Dimensions are in millimeters. All angles are right angles.

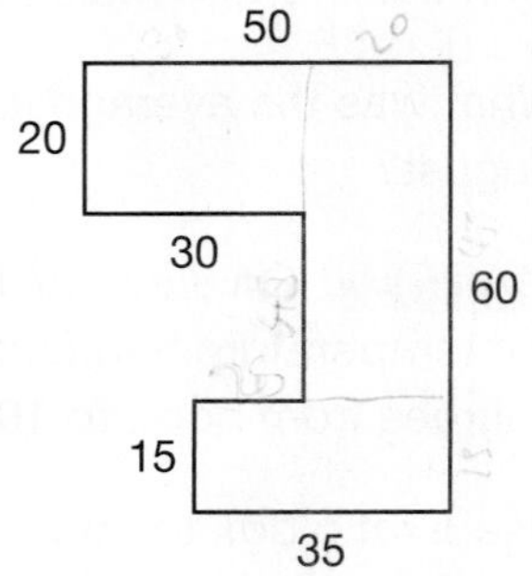

14. (19) What is the perimeter of the figure?

15. (37) What is the area of the figure?

Solve:

16. (35) $4.4 = 8w$

17. (39) $\frac{0.8}{1} = \frac{x}{1.5}$

18. (30) $n + \frac{11}{20} = \frac{17}{30}$

19. (35) $\frac{0.364}{m} = 7$

Analyze Simplify:

* **20.** (57) $2^{-1} + 2^{-1}$

* **21.** (20, 57) $\sqrt{64} - 2^3 + 4^0$

22. (49)
$$\begin{array}{r} 3 \text{ yd } 2 \text{ ft } 7\frac{1}{2} \text{ in.} \\ +\ 1 \text{ yd } \quad\ \ 5\frac{1}{2} \text{ in.} \\ \hline \end{array}$$

23. (49)
$$\begin{array}{r} 1 \text{ qt } \ 1 \text{ pt } \ \ 6 \text{ oz} \\ -\qquad 1 \text{ pt } \ 12 \text{ oz} \\ \hline \end{array}$$

24. (50) $2\frac{1}{2} \text{ hr} \cdot \frac{50 \text{ mi}}{1 \text{ hr}}$

25. (26) $\left(\frac{5}{9} \cdot 12\right) \div 6\frac{2}{3}$

Estimate each answer to the nearest whole number. Then perform the calculation.

26. (23, 30) $3\frac{5}{6} - \left(4 - 1\frac{1}{9}\right)$

27. (26, 30) $\left(5\frac{5}{8} + 6\frac{1}{4}\right) \div 6\frac{1}{4}$

* **28.** (52) **Conclude** Evaluate: $a - bc$ if $a = 0.1$, $b = 0.2$, and $c = 0.3$

* **29.** (55) **Connect** Darrin rode his bike the 30 miles out to the lake in 2 hours. The wind was against him on the return trip which took 3 hours. Find Darrin's average speed

a. to the lake ___?___.

b. on his return trip ___?___.

c. for the round trip ___?___.

30. (14, 43) This table shows the results of a class election. If one student who voted is selected at random, what is the probability that the student voted for the candidate who received the most votes? Express the probability as a reduced fraction and as a decimal.

Vote Tally

Candidate	Votes
Vasquez	卌 II
Lam	卌 卌
Enzinwa	卌 III

Early Finishers
Math Applications

A square can be constructed from congruent triangles.

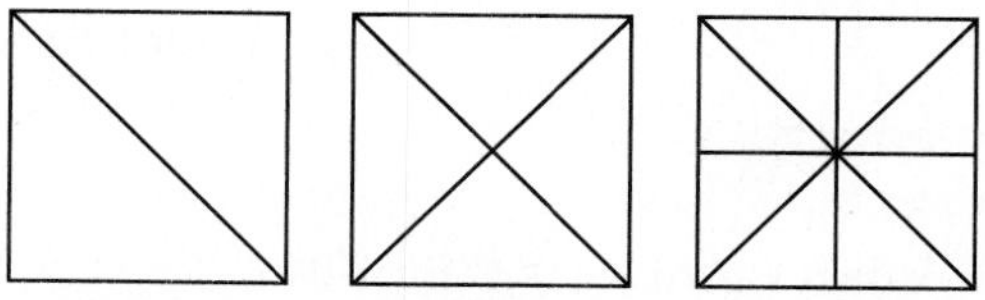

2 triangles   4 triangles   8 triangles

What other polygons can be constructed from a set of congruent triangles? Sketch examples to support your choices.

LESSON 60

• Fractional Part of a Number, Part 1
• Percent of a Number, Part 1

Power Up

Building Power

facts Power Up M

mental math

a. **Decimals:** 70×2.3

b. **Exponents:** $435 \div 10^2$

c. **Algebra:** $5x - 1 = 49$

d. **Measurement:** Convert 75 mm to cm.

e. **Square Roots:** $\sqrt{144} - \sqrt{25}$

f. **Fractional Parts:** $\frac{4}{5}$ of \$1.00

g. **Statistics:** Find the mode of the set of numbers: 23, 32, 99, 77, 23, 79.

h. **Calculation:** Start with 25¢, double that amount, double that amount, double that amount, × 5, add \$20, ÷ 10, ÷ 10.

problem solving

Amy hit a target like the one shown 4 times, earning a total score of 20. Find two sets of scores Amy could have earned. Barb earned a total score of exactly 20 in the fewest possible number of attempts. How many attempts did Barb make?

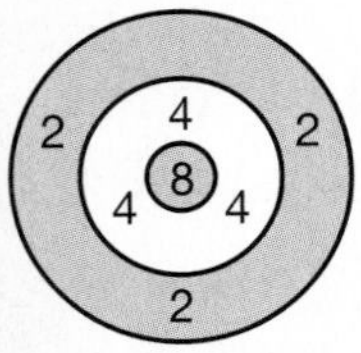

New Concepts

Increasing Knowledge

fractional part of a number, part 1

We can solve fractional-part-of-a-number problems by translating the question into an equation and then solving the equation. To translate,

we replace the word *is* with =

we replace the word *of* with ×

Example 1

What number is 0.6 of 31?

Math Language
An **equation** is a mathematical statement that two quantities are equal.

Solution

This problem uses a decimal number to ask the question. We represent *what number* with W_N. We replace *is* with an equal sign. We replace *of* with a multiplication symbol.

What number is 0.6 of 31?	question
↓ ↓ ↓ ↓ ↓	
$W_N = 0.6 \times 31$	equation

To find the answer, we multiply.

$$W_N = \mathbf{18.6} \quad \text{multiplied}$$

Example 2

Three fifths of 120 is what number?

Solution

This time the question is phrased by using a common fraction. The procedure is the same: we translate directly.

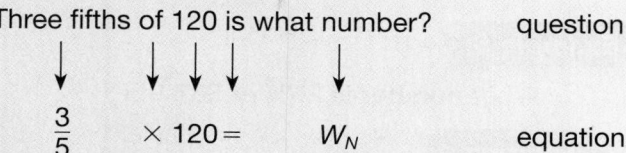

Three fifths of 120 is what number? — question

$$\frac{3}{5} \times 120 = W_N$$ — equation

To find the answer, we multiply.

$$W_N = \mathbf{72}$$

Formulate State this word problem in another way that means the same thing.

percent of a number, part 1

We can translate percent problems into equations the same way we translate fractional-part-of-a-number problems: we convert the percent to either a fraction or a decimal.

Example 3

The jacket sold for \$75. Forty percent of the selling price was profit. How much money is 40% of \$75?

Solution

We translate the question into an equation. We may convert the percent to a fraction or to a decimal. We show both ways.

Percent to Fraction	Percent to Decimal
$W_N = \frac{40}{100} \times \$75$	$W_N = 0.40 \times \$75$
$W_N = \frac{2}{5} \times \$75$	$W_N = 0.4 \times \$75$
$W_N = \mathbf{\$30}$	$W_N = \mathbf{\$30}$

Thinking Skill

Explain

How could we reduce $\frac{2}{5} \times \$75$ before multiplying?

Example 4

A certain used-car salesperson receives a commission of 8% of the selling price of a car. If the salesperson sells a car for \$10,800, how much is the salesperson's commission?

Solution

We want to find 8% of $10,800. This time we convert the percent to a decimal.

Eight percent of $10,800 is commission.

$$0.08 \times \$10{,}800 = C$$
$$\$864 = C$$

The salesperson's commission is **$864.**

Example 5

What number is 25% of 88?

Solution

This time we convert the percent to a fraction.

What number is 25% of 88?

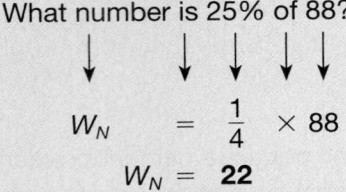

$$W_N = \frac{1}{4} \times 88$$
$$W_N = \mathbf{22}$$

Connect Show how you could have solved this problem by changing the percent to a decimal.

Whether a percent should be changed to a fraction or to a decimal is up to the person solving the problem. Often one form makes the problem easier to solve than the other form. With practice the choice of which form to use becomes more apparent.

Example 6

A bicycle is on sale for $119.95. The tax rate is 6 percent.

a. What is the tax on the bicycle?

b. What is the total price including tax?

Solution

a. To find the tax, we change 6 percent to the decimal 0.06 and multiply $119.95 by 0.06. We round the result to the nearest cent.

$$\begin{array}{r} \$119.95 \\ \times \quad 0.06 \\ \hline \$7.1970 \end{array} \longrightarrow \mathbf{\$7.20}$$

b. To find the total price, including tax, we add the tax to the initial price.

$$\begin{array}{rl} \$119.95 & \text{price} \\ + \quad \$7.20 & \text{tax} \\ \hline \mathbf{\$127.15} & \text{total} \end{array}$$

Math Language

The initial price is the original price before any sales tax is added or any discounts are taken.

Example 7

Find the total price, including tax, of an $18.95 book, a $1.89 pen, and a $2.29 pad of paper when the tax rate is 5 percent.

Solution

We begin by finding the combined price of the items.

$$\begin{array}{rl} \$18.95 & \text{book} \\ \$1.89 & \text{pen} \\ +\ \$2.29 & \text{paper} \\ \hline \$23.13 & \end{array}$$

Next we multiply the combined price by 0.05 (5 percent) and round the product to the nearest cent.

$$\begin{array}{r} \$23.13 \\ \times\ \ 0.05 \\ \hline \$1.1565 \end{array} \longrightarrow \$1.16 \text{ tax}$$

Then we add the tax to the combined price to find the total.

$$\begin{array}{rl} \$23.13 & \text{price} \\ \times\ \$1.16 & \text{tax} \\ \hline \mathbf{\$24.29} & \text{total} \end{array}$$

Practice Set

Represent For problems **a–e,** translate each question into an equation.

a. What number is $\frac{4}{5}$ of 71?

b. Seventy-five hundredths of 14.4 is what number?

c. What number is 50% of 150?

d. Three percent of $39 is how much money?

e. What number is 25% of 64?

f. If a salesperson receives a commission of 12% of sales, what is the salesperson's commission on $250,000 of sales?

g. Find the sales tax on a $36.89 radio when the tax rate is 7 percent.

h. Find the total price of the radio in problem **g,** including tax.

i. *Evaluate* Find the total price, including 6 percent tax, for a $6.95 dinner, a 95¢ beverage, and a $2.45 dessert.

Written Practice

Strengthening Concepts

1. (31, 35) Five and seven hundred eighty-four thousandths is how much less than seven and twenty-one ten-thousandths?

2. (28) Cynthia paid 20¢ per hobby magazine at a tag sale. She bought half of all the magazines that were for sale and paid $10. How many magazines were there in all?

*** 3.** (56) **Represent** In problem **2,** the amount of money Cynthia pays is a function of the number of magazines she bought.

a. Write an equation that shows what Cynthia pays in dollars (d) for the number of magazines (m) she buys.

b. Make a function table with four (m, d) pairs of your choosing that satisfy the function.

4. (36) Four fifths of the total number of students take the school bus to school.

a. What percent of the total number of students do not take the school bus to school?

b. What is the ratio of students who take the school bus to those who do not?

*** 5.** (55) **Analyze** Over a period of five days, the average time of a commuter train trip from one town to the city was 77 minutes. One trip was 71 minutes, another was 74 minutes, and two each were 78 minutes. How many minutes long was the fifth trip?

*** 6.** (51, 57) Write each number in scientific notation:

a. 0.00000008 **b.** 67.5 billion

7. (22, 48) Diagram this statement. Then answer the questions that follow.

Two thirds of the 96 members approved of the plan.

a. How many of the 96 members approved of the plan?

b. What percent of the members did not approve of the plan?

*** 8.** (59) **Model** Draw a diagram of this statement. Then answer the question.

The first stage of the rocket fell from a height of 23,000 feet and settled on the ocean floor 9000 feet below sea level. In all, how many feet did the rocket's first stage descend?

Write equations to solve problems **9** and **10.**

*** 9.** (60) What number is $\frac{3}{4}$ of 17?

*** 10.** (60) **Analyze** If 40% of the selling price of a $65 sweater is profit, then how many dollars profit does the store make when the sweater is sold?

*** 11.** (43, 59) Compare:

a. $\frac{1}{3} \bigcirc 0.33$ **b.** $|5 - 3| \bigcirc |3 - 5|$

12. (48) Complete the table.

Fraction	Decimal	Percent
$\frac{1}{8}$	**a.**	**b.**
c.	**d.**	125%

* **13.** (59) **Model** Use arrows to show each addition problem on a number line:

a. $(-3) + (-1)$ **b.** $(-3) + (+1)$

14. (21) **a.** Write the prime factorization of 3600 using exponents.

b. Write the prime factorization of $\sqrt{3600}$.

15. (Inv. 5) Find the number of degrees in the central angle of each sector of the circle at right.

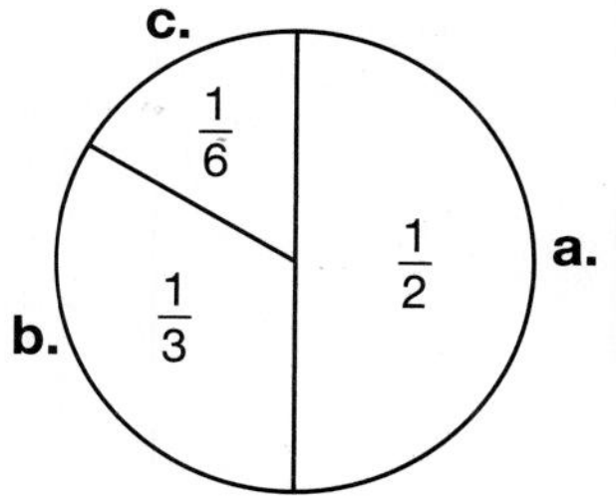

Refer to the figure below to answer problems **16–18.** Dimensions between labeled points are in feet. The measure of $\angle EDF$ equals the measure of $\angle ECA$.

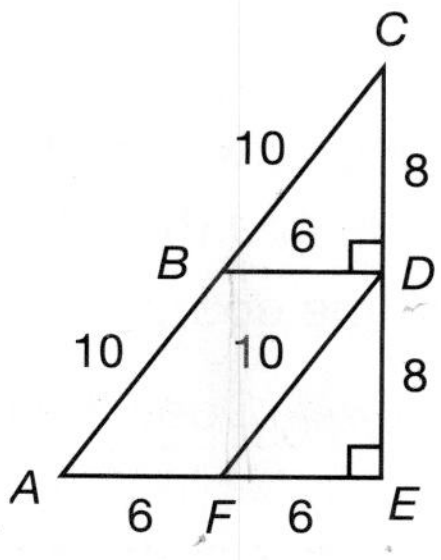

16. (18) **a.** Name a triangle congruent to triangle *DEF.*

b. Name a triangle similar to $\triangle DEF$ but not congruent to $\triangle DEF$.

17. (37) **a.** Find the area of $\triangle BCD$.

b. Find the area of $\triangle ACE$.

18. (37) By subtracting the areas of the two smaller triangles from the area of the large triangle, find the area of the quadrilateral *ABDF.*

Solve:

19. (30) $p - \frac{1}{30} = \frac{1}{20}$ **20.** (35) $9m = 0.117$

Simplify:

* **21.** (20, 52, 57) **Analyze** $3^2 + 4(3 + 2) - 2^3 \cdot 2^{-2} + \sqrt{36}$

22. (41) **a.** Compare:

$$\frac{3}{4}\left(\frac{4}{9} + \frac{2}{3}\right) \bigcirc \frac{3}{4} \cdot \frac{4}{9} + \frac{3}{4} \cdot \frac{2}{3}$$

b. Which property helps us make the comparison in **a** without performing the calculations?

23. (36) The spinner is spun twice. Two possible outcomes are *AC* and *CA*.

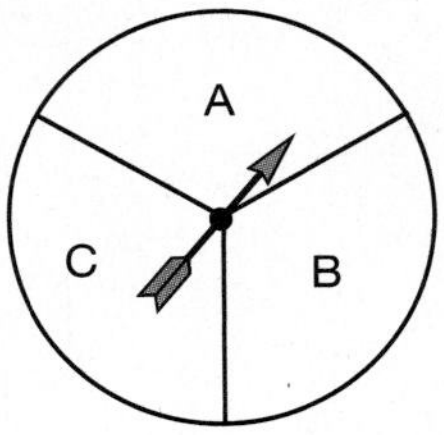

a. Create a sample space that lists all of the equally likely outcomes.

b. Refer to the sample space to find the probability of spinning A one or more times in two spins.

24. (30) $3\frac{3}{5} - \left(\frac{5}{6} \cdot 4\right)$

25. (26) $\left(1\frac{1}{4} \div \frac{5}{12}\right) \div 24$

26. (35) $6.5 - (0.65 - 0.065)$

27. (45) $0.3 \div (3 \div 0.03)$

28. (50) Use a unit multiplier to convert 3.5 centimeters to meters. (1 m = 100 cm)

29. (27, 45) Explain why these two division problems are equivalent. Then give a money example of the two divisions.

$$\frac{1.5}{0.25} = \frac{150}{25}$$

*** 30.** (Inv. 3, 58) The *x*-axis is a line of symmetry for $\triangle ABC$. The coordinates of point *A* are (3, 0), and the coordinates of point *B* are (0, −2).

a. Find the coordinates of point *C*.

b. Analyze Does the triangle have rotational symmetry?

Early Finishers
Real-World Application

Solve each problem. Choose between mental math, estimation, paper and pencil, or a calculator to solve each problem. Justify your choice.

a. The Martinez family is having a holiday dinner at a restaurant. The restaurant bill is $80.56. Mr. Martinez wants to leave a 20% tip. How much should he leave?

b. Jennifer wants to buy a coat for $95.75. She has saved $25. She earns $20 a week working at a grocery store. How many weeks must she work to buy the coat?

INVESTIGATION 6

Focus on

• Classifying Quadrilaterals

Recall from Lesson 18 that a four-sided polygon is a quadrilateral.

Analyze Refer to the quadrilaterals below to answer the problems 1–6.

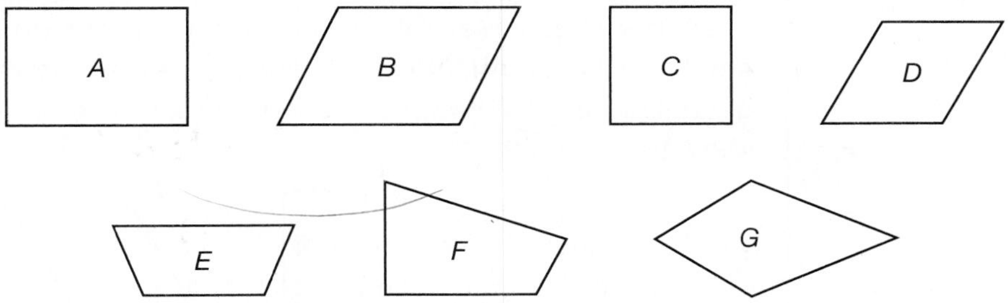

1. Which figures have four right angles?

2. Which figures have four sides of equal length?

3. Which figures have two pairs of parallel sides?

4. Which figure has just one pair of parallel sides?

5. Which figures have no pairs of parallel sides?

6. Which figures have two pairs of equal-length sides?

Math Language

Parallel lines are lines in the same plane that never intersect.

We can sort quadrilaterals by their characteristics. One way to sort is by the number of pairs of parallel sides. A quadrilateral with two pairs of parallel sides is a **parallelogram.** Here we show four parallelograms.

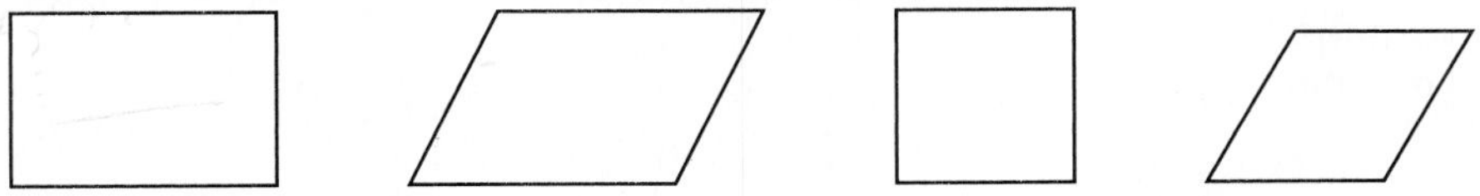

7. Which of the figures *A*–*G* are parallelograms?

A quadrilateral with just one pair of parallel sides is a **trapezoid.** The figures shown below are trapezoids. Can you find the parallel sides? (Notice that the parallel sides are not the same length.)

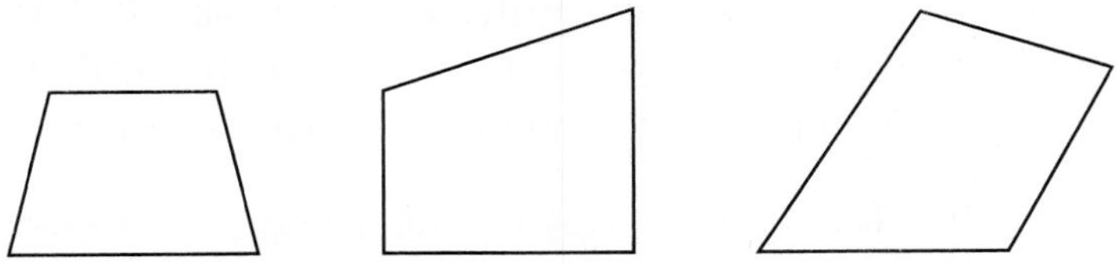

8. Which of the figures *A*–*G* is a trapezoid?

A quadrilateral with no pairs of parallel sides is a **trapezium.** Here we show two examples:

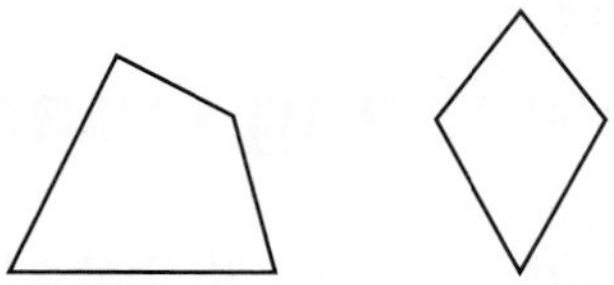

9. Which of the figures *A–G* from the previous page are trapeziums?

We can sort quadrilaterals by the lengths of their sides. If the four sides are the same length, the quadrilaterals are **equilateral.** An equilateral quadrilateral is a **rhombus.** A rhombus is a type of parallelogram. Here we show two examples.

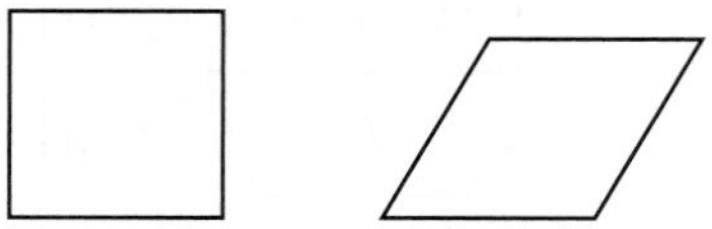

10. Which of the figures *A–G* are rhombuses?

We can sort quadrilaterals by the measures of their angles. If the four angles are of equal measure, then each angle is a right angle, and the quadrilateral is a **rectangle.** A rectangle is a type of parallelogram.

11. Which of the figures *A–G* from the previous page are rectangles?

Notice that a square is both a rectangle and a rhombus. A square is also a parallelogram. We can use a **Venn diagram** to illustrate the relationships.

Thinking Skill

Summarize

State three relationships shown in the Venn diagram.

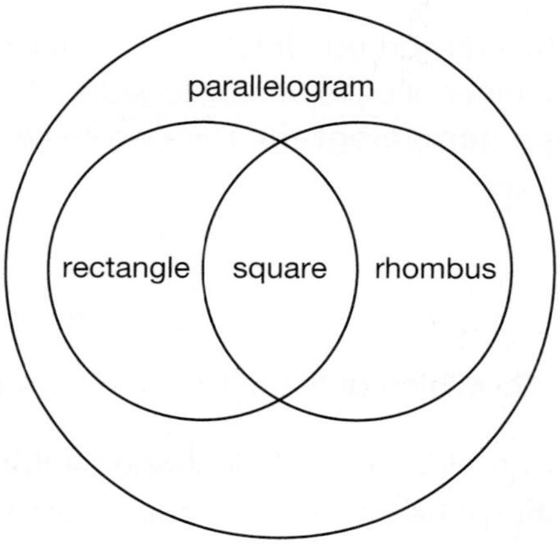

Any figure that is within the circle labeled "rectangle" is a parallelogram as well. Any figure within the circle labeled "rhombus" is also a parallelogram. A figure within both the rectangle and rhombus circles is a square.

12. *Model* Copy the Venn diagram above on your paper. Then refer to quadrilaterals *A, B, C, D,* and *E* at the beginning of this investigation. Draw each of the quadrilaterals in the Venn diagram in the proper location. (One of the figures will be outside the parallelogram category.)

A student made a model of a rectangle out of straws and pipe cleaners (Figure J). Then the student shifted the sides so that two angles became obtuse and two angles became acute (Figure K).

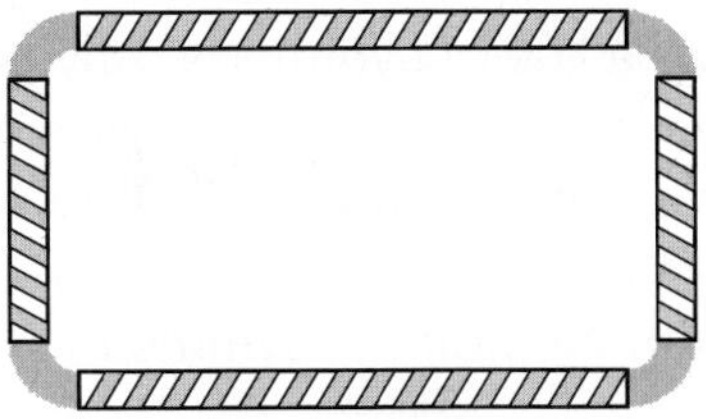

Figure J

Figure K

Analyze Refer to Figures J and K to answer problems 13–16.

13. Is Figure K a rectangle?

14. Is Figure K a parallelogram?

15. **Justify** Does the perimeter of Figure K equal the perimeter of Figure J?

16. Does the area of Figure K equal the area of Figure J?

Another student made a model of a rectangle out of straws and pipe cleaners (Figure L). Then the student reversed the positions of two of the straws so that the straws that were the same length were adjacent to each other instead of opposite each other (Figure M).

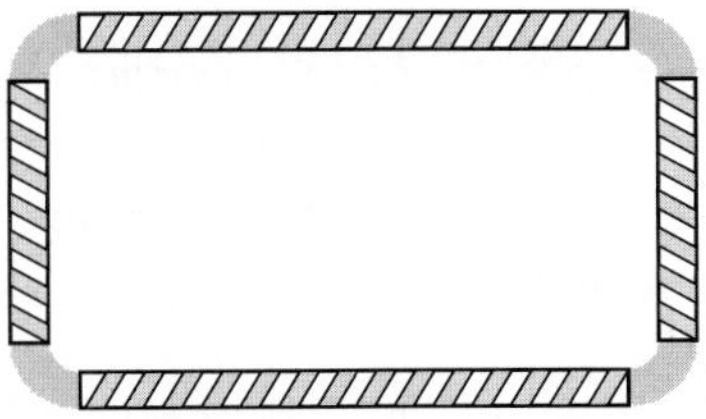

Figure L

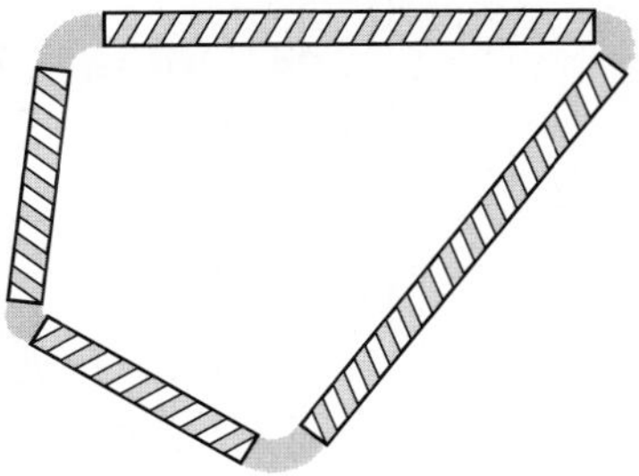

Figure M

Figure M does not have a pair of parallel sides, so it is a trapezium. However, it is a special type of trapezium called a **kite.**

17. Which of the figures *A*–*G* from the beginning of this investigation is a kite?

18. If two sides of a kite are 2 ft and 3 ft, what is the perimeter of the kite?

Notice that a kite has a line of symmetry.

19. Draw a kite and show its line of symmetry.

20. Draw a rhombus that is not a square, and show its lines of symmetry.

21. Draw a rectangle that is not a square, and show its lines of symmetry.

22. Draw a rhombus that is a rectangle, and show its lines of symmetry.

23. An **isosceles trapezoid** has a line of symmetry. The nonparallel sides of an isosceles trapezoid are the same length. Draw an isosceles trapezoid and show its line of symmetry.

Not every trapezoid has a line of symmetry. Any parallelogram that is not a rhombus or rectangle does not have line symmetry. However, every parallelogram does have **point symmetry.** A figure is symmetrical about a point if every line drawn through the point intersects the figure at points that are equal distances from the point of symmetry.

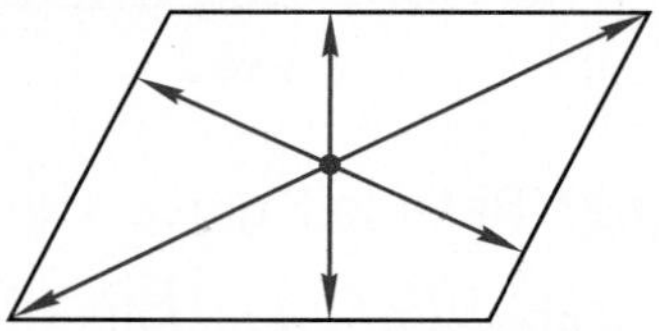

We can locate the point of symmetry of a parallelogram by finding the point where the diagonals of the parallelogram intersect. A **diagonal** of a polygon is a segment between non-consecutive vertices.

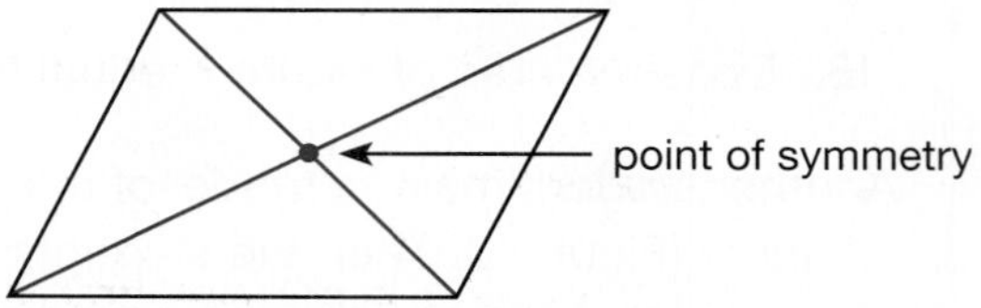

In the following problem we learn a way to test for point symmetry.

24. **Model** Draw two or three parallelograms on grid paper. Be sure that one of the parallelograms is a rectangle and one is not a rectangle. Locate and mark the point in each parallelogram where the diagonals intersect. Then carefully cut out the parallelograms.

If we rotate a figure with point symmetry a half turn (180°) about its point of symmetry, the figure will appear to be in the same position it was in before it was rotated. On one of the cut-out parallelograms, place the tip of a pencil on the point where the diagonals intersect. Then rotate the parallelogram 180°. Is the point of intersection a point of symmetry?

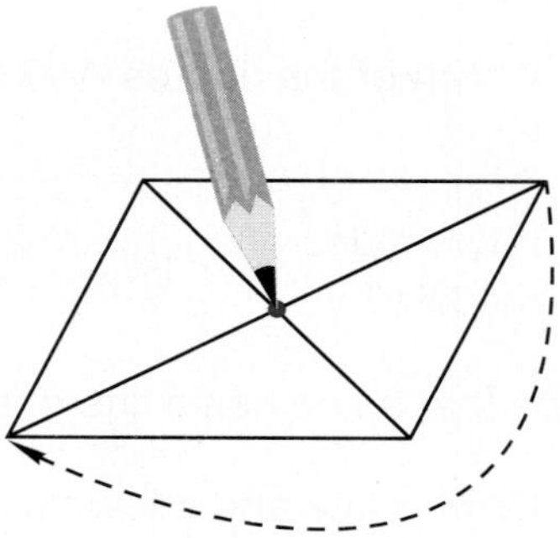

Repeat the rotation with the other parallelogram(s) you cut out.

25. **Analyze** Which of the figures *A–G* from the beginning of this investigation have point symmetry?

Below we classify the figures illustrated at the beginning of this investigation. You may refer to them as you answer the remaining problems.

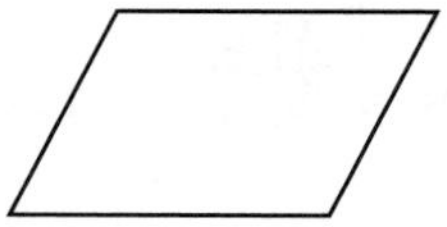

parallelogram

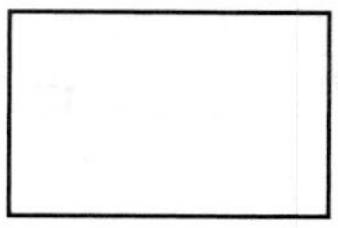

parallelogram
rectangle

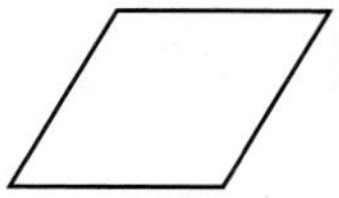

parallelogram
rhombus

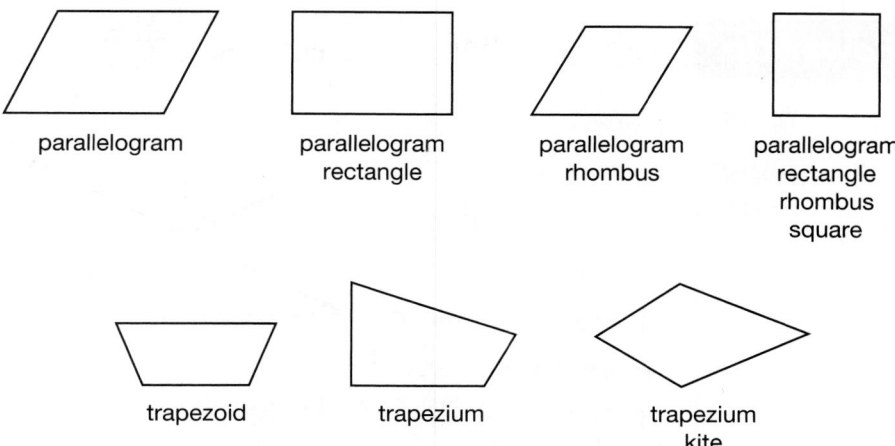

parallelogram
rectangle
rhombus
square

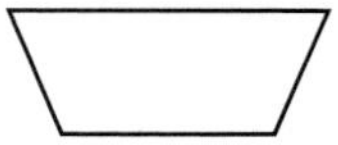

trapezoid

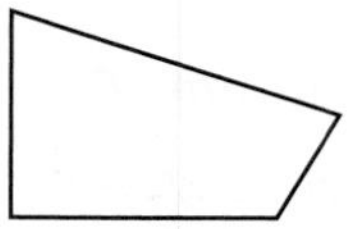

trapezium

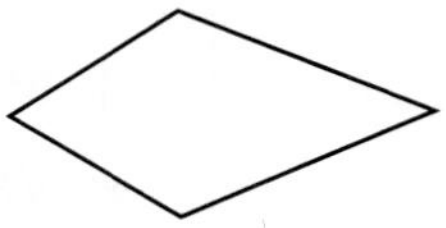

trapezium
kite

Classify Answer true or false, and state the reason(s) for your answer.

26. A square is a rectangle.

27. All rectangles are parallelograms.

28. Some squares are trapezoids.

29. Some parallelograms are rectangles.

30. Draw a Venn diagram illustrating the relationship of quadrilaterals, parallelograms, and trapezoids.

Verify Refer to the figure below to answer problems **31–33.**

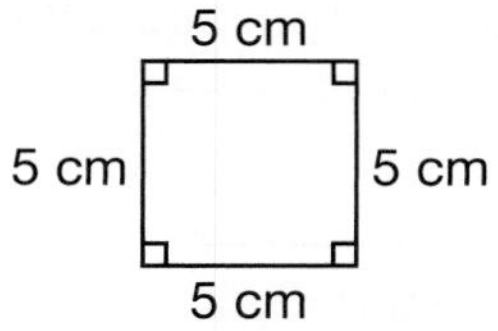

31. Explain why the figure is a parallelogram.

32. Explain why the figure is a rectangle.

33. Explain why the figure is a rhombus.

LESSON 61

• Area of a Parallelogram
• Angles of a Parallelogram

Power Up — *Building Power*

facts Power Up L

mental math

a. Decimals: 50×4.6

b. Decimals/Exponents: 2.4×10^{-1}

c. Ratio: $\frac{a}{20} = \frac{12}{8}$

d. Measurement: Convert 1.5 km to m.

e. Exponents: $3^2 - 2^3$

f. Fractional Parts: $\frac{7}{10}$ of \$3.00

g. Geometry: What type of triangle has three equal sides?

h. Calculation: What is the total cost of a \$20 item plus 6% sales tax?

problem solving

A large sheet of paper is 0.1 mm thick. If we tear it in half and stack the two pieces, the resulting stack is 2 pieces of paper, or 0.2 mm, high. If we tear that stack of paper in half and stack the remaining halves, the resulting stack is 4 pieces of paper, or 0.4 mm high. If we could continue to tear-in-half and stack, how high would our stack be after 20 times? 25 times?

New Concepts — *Increasing Knowledge*

area of a parallelogram

Thinking Skill

Discuss

Why is the figure on the right not a parallelogram?

Recall from Investigation 6 that a parallelogram is a quadrilateral in which both pairs of opposite sides are parallel.

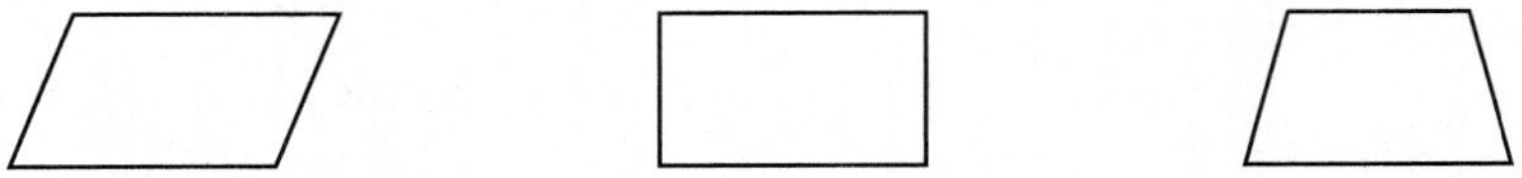

Parallelogram   Parallelogram   Not a parallelogram

In this lesson we will practice finding the areas of parallelograms, but first we will review a couple of terms. Recall that the dimensions of a rectangle are called the length and the width. When describing a parallelogram, we do not use these terms. Instead we use the terms **base** and **height.**

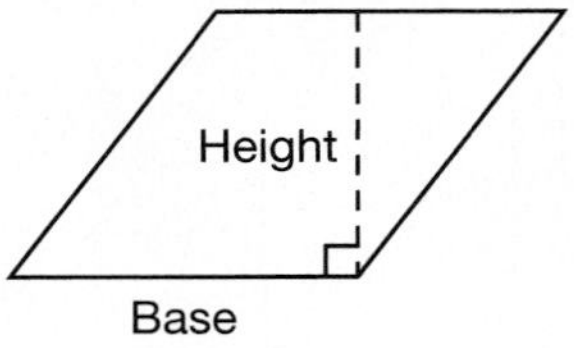

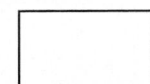

Notice that the height is not one of the sides of the parallelogram (unless the parallelogram is a rectangle). Instead, **the height is perpendicular to the base.**

Activity 1

Area of a Parallelogram

Materials needed:

- Paper
- Scissors
- Straightedge

Cut a piece of paper to form a parallelogram as shown. You may use graph paper if available.

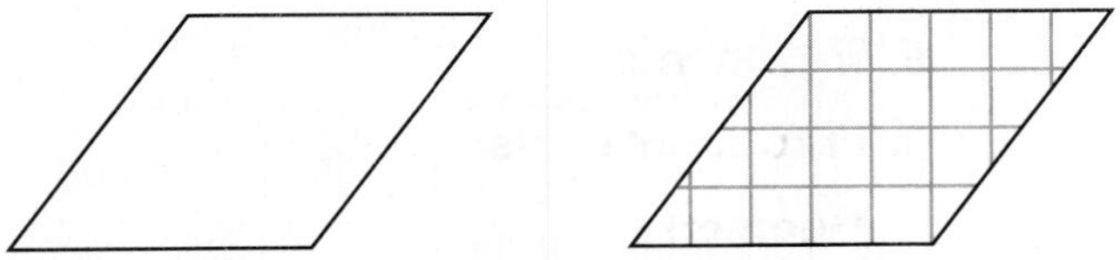

Next, sketch a segment perpendicular to two of the parallel sides of the parallelogram. The length of this segment is the height of the parallelogram. Cut the parallelogram into two pieces along the segment you drew.

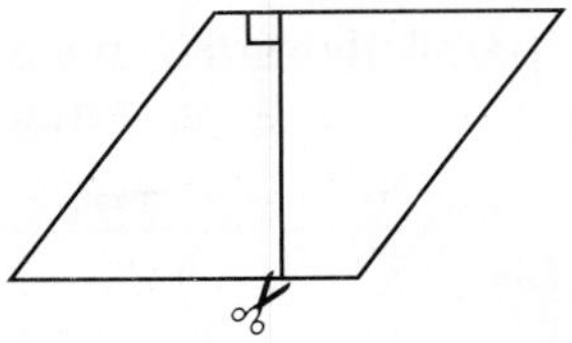

Finally, reverse the positions of the two pieces and fit them together to form a rectangle. The area of the original parallelogram equals the area of this rectangle.

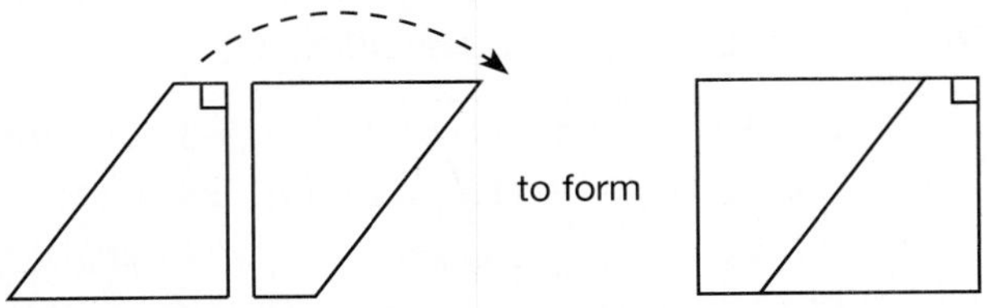

To conclude this activity, answer these questions:

1. What dimensions of the original parallelogram match the length and width of this rectangle?
2. How would you find the area of the rectangle?
3. **Discuss** How would you find the area of the original parallelogram?

As we see in Activity 1, the area of the rectangle equals the area of the parallelogram we are considering. Thus we find the area of a parallelogram by multiplying its base by its height.

Area of a parallelogram = base · height
$$A = bh$$

Example 1

Parallelogram Park is a small park in town with the dimensions shown. Find a. the perimeter of the park and b. its area.

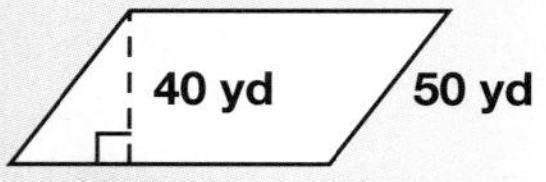

Solution

a. The perimeter is the distance around the park. We add the lengths of the sides. Opposite sides of a parallelogram are the same length.

$$50 \text{ yd} + 80 \text{ yd} + 50 \text{ yd} + 80 \text{ yd} = \mathbf{260 \text{ yd}}$$

b. We multiply the base and height to find the area.

$$80 \text{ yd} \times 40 \text{ yd} = \mathbf{3200 \text{ sq. yd}}$$

angles of a parallelogram

Figures J and K of Investigation 6 illustrated a "straw" rectangle shifted to form a parallelogram that was not a rectangle. Two of the angles became obtuse angles, and the other two angles became acute angles.

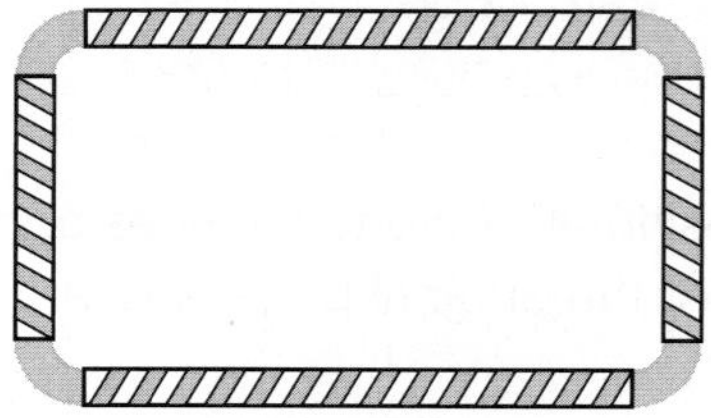
Figure J

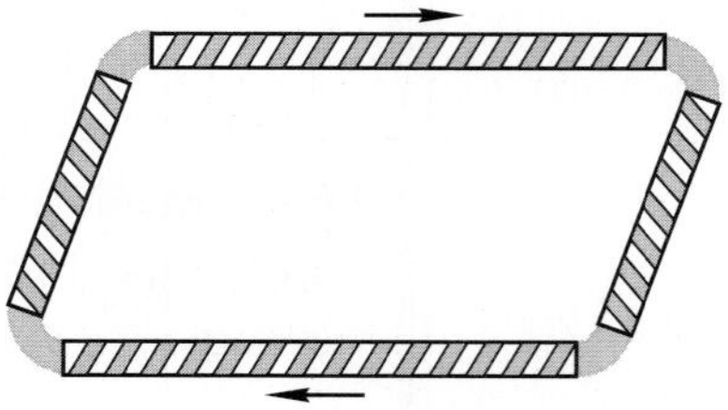
Figure K

In other words, two of the angles became more than 90°, and two of the angles became less than 90°. Each angle became greater than or less than 90° *by the same amount.* If, by shifting the sides of the straw rectangle, the obtuse angles became 10° *greater than* 90° (100° angles), then the acute angles became 10° *less than* 90° (80° angles). The following activity illustrates this relationship.

Activity 2

Angles of a Parallelogram

Materials needed:

- Protractor
- Paper
- Two pairs of plastic straws (The straws within a pair must be the same length. The two pairs may be different lengths.)
- Thread or lightweight string
- Paper clip for threading the straws (optional)

Make a "straw" parallelogram by running a string or thread through two pairs of plastic straws. If the pairs of straws are of different lengths, alternate the lengths as you thread them (long-short-long-short).

Bring the two ends of the string together, pull until the string is snug but not bowing the straws, and tie a knot.

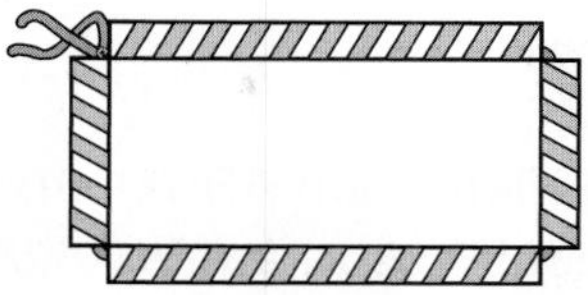

You should be able to shift the sides of the parallelogram to various positions.

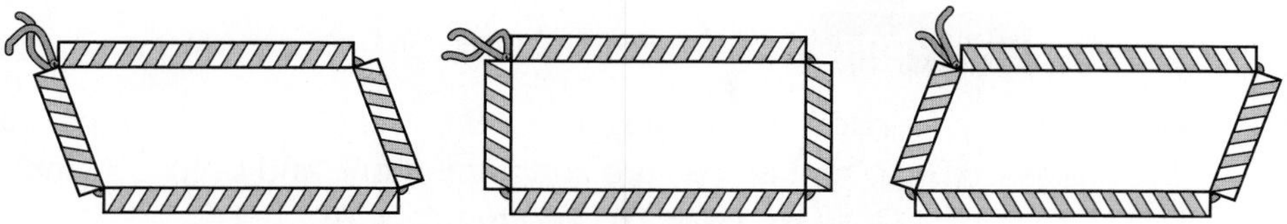

Lay the straw parallelogram on a desktop with a piece of paper under it. On the paper you will trace the parallelogram. Shift the parallelogram into a position you want to measure, hold the straws and paper still (this may require more than two hands), and carefully trace with a pencil around the *inside* of the parallelogram.

Set the straw parallelogram aside, and use a protractor to measure each angle of the traced parallelogram. Write the measure inside each angle. Some groups may wish to trace and measure the angles of a second parallelogram with a different shape before answering the questions below.

1. What were the measures of the two obtuse angles of one parallelogram?
2. What were the measures of the two acute angles of the same parallelogram?
3. What was the sum of the measures of one obtuse angle and one acute angle of the same parallelogram?

If you traced two parallelograms, answer the three questions again for the second parallelogram.

Record several groups' answers on the board.

Discuss Can any general conclusions be formed?

The quality of all types of measurement is affected by the quality of the measuring instrument, the material being measured, and the person performing the measurement. However, even rough measurements can suggest underlying relationships. The rough measurements performed in activity 2 should suggest the relationships between the angles of a parallelogram shown on the next page.

1. Nonadjacent angles (angles in opposite corners) have equal measures.
2. Adjacent angles (angles that share a common side) are supplementary—that is, their sum is 180°.

Conclude What is the sum of the measures of the angles of the parallelogram? How does this sum compare to the sum of the measures of the angles of a rectangle?

Example 2

In parallelogram *ABCD*, m∠*D* is 110°. Find the measures of angles *A*, *B*, and *C* in the parallelogram.

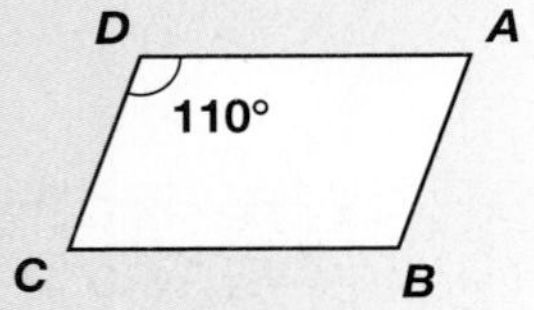

Solution

Nonadjacent angles, like ∠*B* and ∠*D*, have equal measures, so **m∠*B* = 110°.** Adjacent angles are supplementary, and both ∠*A* and ∠*C* are adjacent to ∠*D*, so **m∠*A* = 70°** and **m∠*C* = 70°.**

Practice Set

Generalize Find the perimeter and area of each parallelogram. Dimensions are in centimeters.

a.

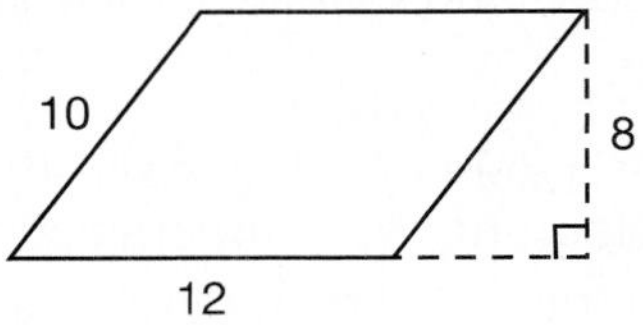

b.

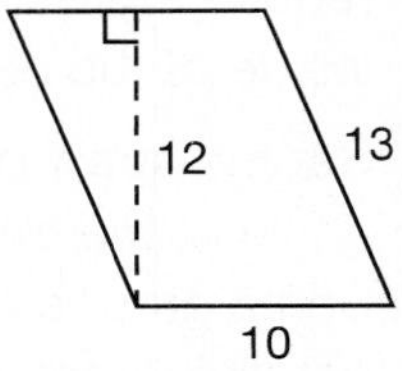

c.

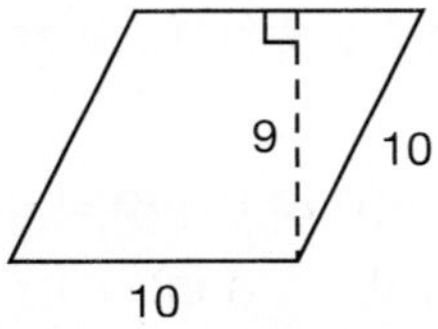

Conclude For problems **d–g,** find the measures of the angles marked *d, e, f,* and *g* in this parallelogram.

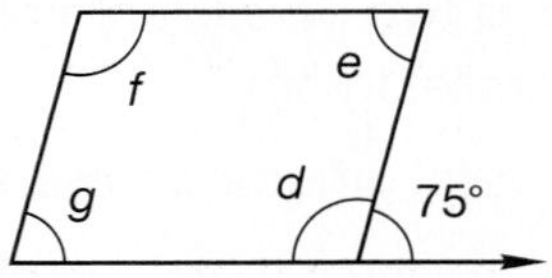

Figure *ABCD* is a parallelogram. Refer to this figure to find the measures of the angles in problems **h–j.**

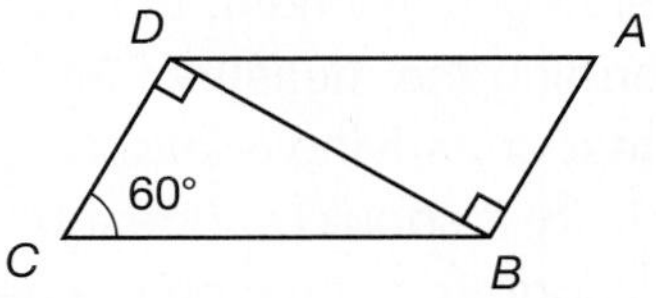

h. ∠*A* **i.** ∠*ADB* **j.** ∠*ABC*

k. Write a formula for the area of a parallelogram.

l. Sid built a rectangular door frame 80 inches tall and 30 inches wide. The frame was bumped before it was installed so that the top was only 75 inches above the base. Did the perimeter of the frame change? Did the area within the frame change? Explain your answers.

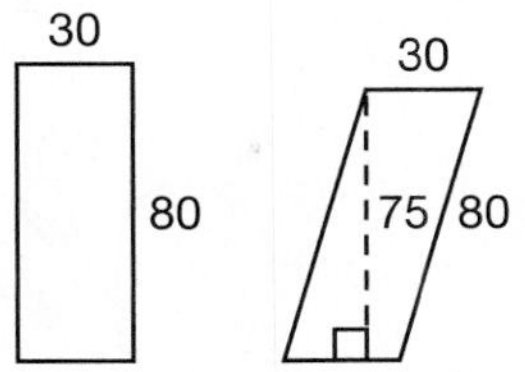

Written Practice

Strengthening Concepts

1. (16, 46) **Justify** If $\frac{1}{2}$ gallon of milk costs \$1.64, what is the cost per pint?

*** 2.** (53) Use a ratio box to solve this problem. The muffin recipe called for oatmeal and brown sugar in the ratio of 2 to 1. If 3 cups of oatmeal were called for, how many cups of brown sugar were needed?

*** 3.** (55) **Analyze** Ricardo ran the 400-meter race 3 times. His fastest time was 54.3 seconds. His slowest time was 56.1 seconds. If his average time was 55.0 seconds, what was his time for the third race?

4. (46) It is $4\frac{1}{2}$ miles to the end of the bicycle trail. If Sakari rides to the end of the trail and back in 60 minutes, what is her average speed in miles per hour?

5. (51) Sixty-three million, one hundred thousand is how much greater than seven million, sixty thousand? Write the answer in scientific notation.

6. (36, 48) Only three tenths of the print area of the newspaper carried news. The rest of the area was filled with advertisements.

a. What percent of the print area was filled with advertisements?

b. What was the ratio of news area to advertisement area?

c. If, without looking, Kali opens the newspaper and places a finger on the page, what is the probability that her finger will be on an advertisement?

*** 7.** (51, 57) **a.** **Represent** Write 0.00105 in scientific notation.

b. Write 3.02×10^5 in standard form.

8. (24) **a.** Use prime factorization to reduce $\frac{128}{192}$.

b. What is the greatest common factor of 128 and 192?

9. (50) Use a unit multiplier to convert 1760 yards to feet.

In the figure below, quadrilateral *ABDE* is a rectangle and $\overline{EC} \parallel \overline{FB}$. Refer to the figure to answer problems **10–12.**

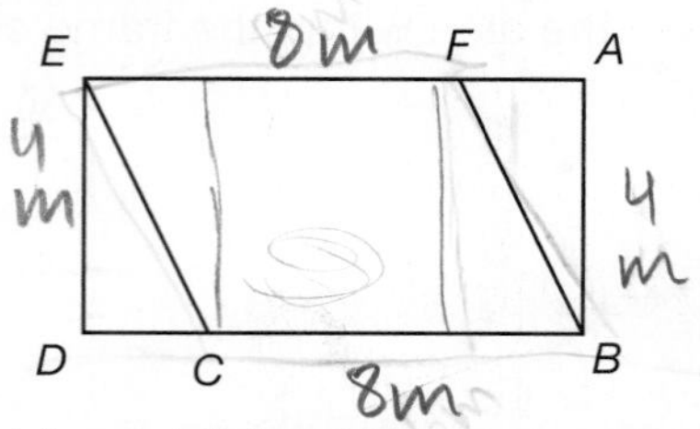

*** 10.** (Inv. 6) **Classify** What type of quadrilateral is:

a. quadrilateral *ECBF*

b. quadrilateral *ECBA*

*** 11.** (37, 61) **Analyze** If $AB = ED = 4$ m, $BC = EF = 6$ m, and $BD = AE = 8$ m, then

a. what is the area of quadrilateral *BCEF?*

b. what is the area of triangle *ABF?*

c. what is the area of quadrilateral *ECBA?*

12. (7) Classify each of the following angles as acute, right, or obtuse:

a. $\angle ECB$ **b.** $\angle EDC$ **c.** $\angle FBA$

13. (Inv. 4) The following is an ordered list of the number of used cars sold in 19 days during the month of May. Find the following for this set of numbers:

a. median **b.** first quartile

c. third quartile **d.** any outliers

2, 5, 5, 6, 6, 6, 7, 7, 7, 8, 8, 8, 8, 9, 9, 10, 10, 10, 10

*** 14.** (Inv. 6, 61) A park in the shape of a parallelogram has the dimensions shown in this figure. Refer to the figure to answer **a–c.**

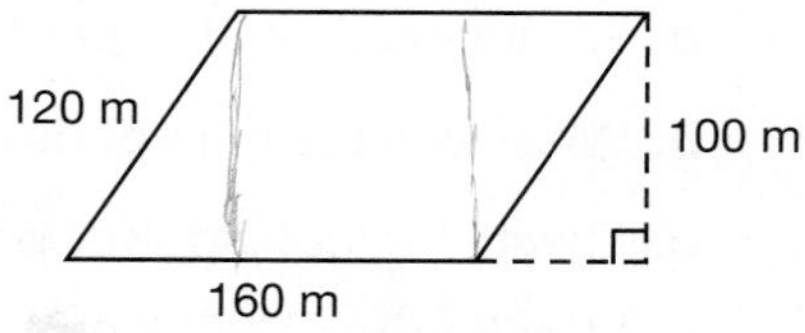

a. If Clyde walks around the park once, how far does he walk?

b. What is the area of the park?

c. **Analyze** Does the parallelogram have rotational symmetry?

*** 15.** (40, 61) The parallelogram at right is divided by a diagonal into two congruent triangles. Find the measure of

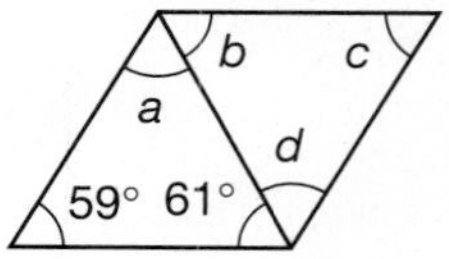

a. $\angle a$. **b.** $\angle b$.

c. $\angle c$. **d.** $\angle d$.

16. (50) Tara noticed that the tape she was using to wrap packages was 2 centimeters wide. How many meters wide was the tape?

17. (Inv. 3) A circle is drawn on a coordinate plane with its center at the origin. The circle intersects the x-axis at (5, 0) and (−5, 0).

a. At what coordinates does the circle intersect the y-axis?

b. What is the diameter of the circle?

18. (3) On one tray of a balanced scale was a 50-g mass. On the other tray were a small cube, a 10-g mass, and a 5-g mass. What was the mass of the small cube? Describe how you found your answer.

* **19.** (59) **Analyze** The trail Paula runs begins at 27 feet below sea level and ends at 164 feet above sea level. What is the gain in elevation from the beginning to the end of the trail?

Generalize Simplify:

20. (52) $10 + 10 \times 10 - 10 \div 10$

* **21.** (57) $2^0 - 2^{-3}$

22. (34) 4.5 m + 70 cm = ____ m

* **23.** (50) $2.75 \text{ L} \cdot \frac{1000 \text{ mL}}{1 \text{ L}}$

24. (23, 30) $5\frac{7}{8} + \left(3\frac{1}{3} - 1\frac{1}{2}\right)$

25. (26) $4\frac{4}{5} \cdot 1\frac{1}{9} \cdot 1\frac{7}{8}$

26. (26) $6\frac{2}{3} \div \left(3\frac{1}{5} \div 8\right)$

27. (35) $12 - (0.8 + 0.97)$

28. (35) $(2.4)(0.05)(0.005)$

29. (47) $0.2 \div (4 \times 10^2)$

30. (45) $0.36 \div (4 \div 0.25)$

Early Finishers
Real-World Application

Ida's Country Kitchen Restaurant has 6 different main courses.

Chicken and dumplings $9.00
Steak with baked potato $28.00
Pasta with vegetables $8.00
Fish and wild rice $11.00
Vegetable Stew with tofu $9.00
Spinach Salad with garlic bread $7.00

Use the mean, median, mode, and range to describe the meal prices. Which measure best describes the price of a typical meal at this restaurant? Justify your choice.

LESSON 62

• Classifying Triangles

Power Up

Building Power

facts Power Up M

mental math

a. Calculation: 5×8.6

b. Decimals/Exponents: 2.5×10^{-2}

c. Algebra: $10x + 2 = 32$

d. Measurement: Convert 2500 g to kg.

e. Exponents: $10^3 \div 10^3$

f. Fractional Parts: $\frac{2}{3}$ of \$24.00

g. Geometry: What is the name for a regular quadrilateral?

h. Calculation: 8^2, $-$ 4, $\div$ 2, $\times$ 3, $+$ 10, $\sqrt{\ }$, $\times$ 2, $+$ 5, $\sqrt{\ }$, $-$ 4, square that number, $-$ 1

problem solving

Michelle's grandfather taught her this method for converting kilometers to miles: "Divide the kilometers by 8, and then multiply by 5." Michelle's grandmother taught her this method: "Multiply the kilometers by 0.6." Use both of these methods to convert 80 km to miles. Do both methods produce the same result?

New Concept

Increasing Knowledge

Recall from Lesson 7 that we classify angles as acute angles, right angles, and obtuse angles.

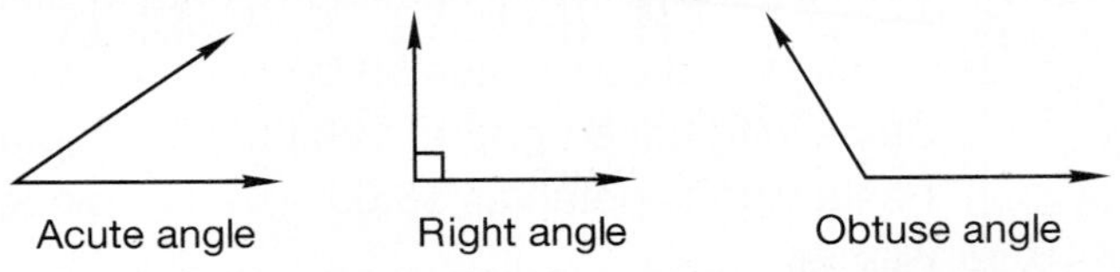

Thinking Skills

Analyze

Can a triangle have two obtuse angles? Explain.

We use the same words to describe triangles that contain these angles. If every angle of a triangle measures less than 90°, then the triangle is an **acute triangle.** If the triangle contains a 90° angle, then the triangle is a **right triangle.** An **obtuse triangle** contains one angle that measures more than 90°.

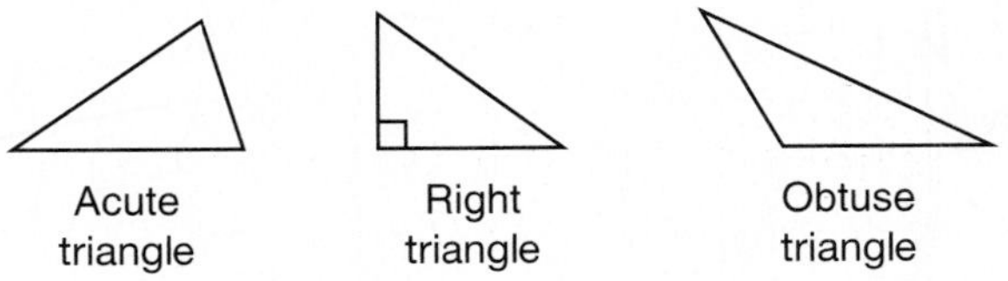

When describing triangles, we can refer to the sides and angles as "opposite" each other. For example, we might say, "The side opposite the right angle is the longest side of a right triangle." The side opposite an angle is the side the angle opens toward. In this right triangle, $\overline{AB}$ is the side opposite $\angle C$, and $\angle C$ is the angle opposite side AB.

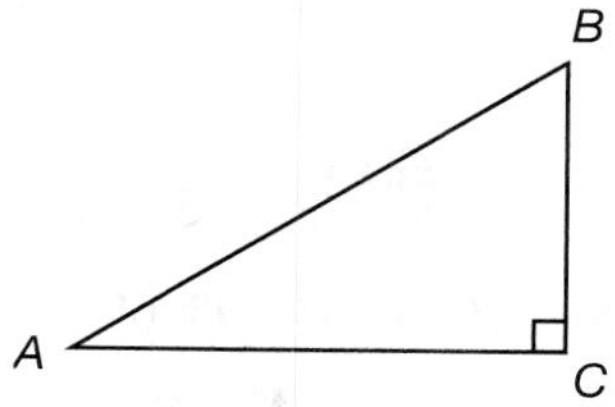

The lengths of the sides of a triangle are in the same order as the measures of their opposite angles. This means that the longest side of a triangle is opposite the largest angle, and the shortest side is opposite the smallest angle.

Example 1

Name the sides of this triangle in order from shortest to longest.

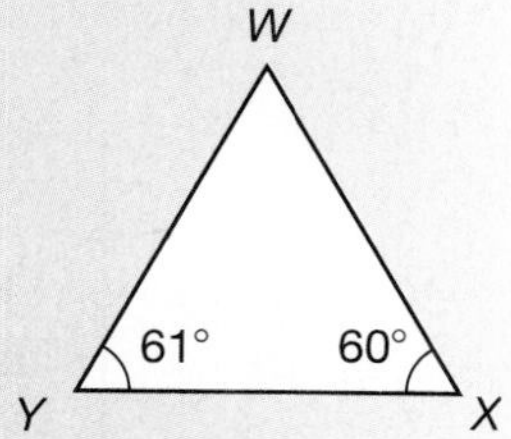

Solution

The sum of the measures of all three angles is 180°, so the measure of $\angle W$ is 59°. Since $\angle W$ is the smallest of the three angles, the side opposite $\angle W$, which is $\overline{XY}$, is the shortest side. The next angle in order of size is $\angle X$, so $\overline{YW}$ is the second longest side. The largest angle is $\angle Y$, so $\overline{WX}$ is the longest side. So the sides in order are

$$\overline{XY}, \overline{YW}, \overline{WX}$$

If two angles of a triangle are the same measure, then their opposite sides are the same length.

Example 2

Which sides of this triangle are the same length?

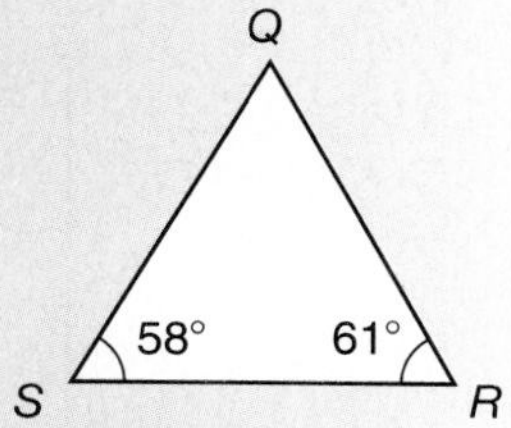

Solution

First we find that the measure of $\angle Q$ is 61°. So angles Q and R have the same measure. This means their opposite sides are the same length. The side opposite $\angle Q$ is $\overline{SR}$. The side opposite $\angle R$ is $\overline{SQ}$. So the sides that are the same length are $\overline{SR}$ and $\overline{SQ}$.

If all three angles of a triangle are the same measure, then all three sides are the same length.

Example 3

In triangle *JKL*, *JK* = *KL* = *LJ*. Find the measure of ∠*J*.

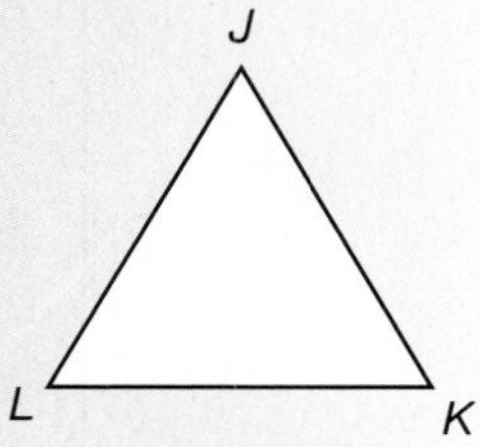

Solution

If two or more sides of a triangle are the same length, then the angles opposite those sides are equal in measure. In △*JKL* all three sides are the same length, so all three angles have the same measure. The angles equally share 180°. We find the measure of each angle by dividing 180° by 3.

$$180° \div 3 = 60°$$

We find that the measure of ∠*J* is **60°.**

The triangle in example 3 is a regular triangle. We usually call a regular triangle an **equilateral triangle.** As shown below, the three angles of an equilateral triangle each measure 60°, and the three sides are the same length. The tick marks on the sides indicate sides of equal length, while tick marks on the arcs indicate angles of equal measure.

Equilateral triangle

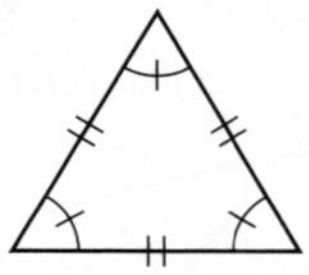

- Three equal sides
- Three equal angles

If a triangle has at least two sides of the same length (and thus two angles of the same measure), then the triangle is called an **isosceles triangle.** The triangle in example 2 is an isosceles triangle, as are each of these triangles:

Isosceles triangles

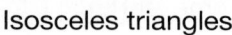

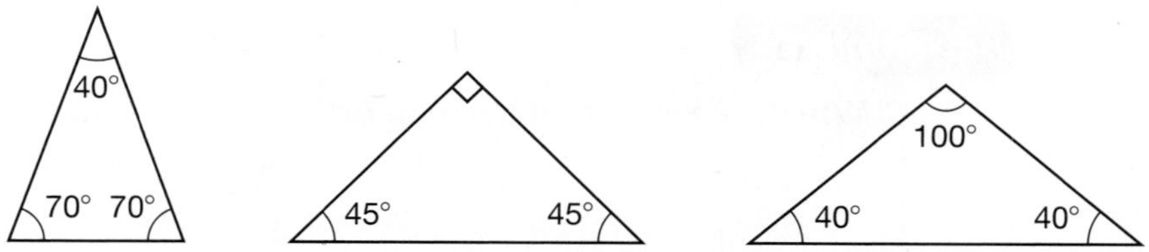

- At least two sides have the same length.
- At least two angles have the same measure.

Math Language

On geometric figures, identical tick marks mean that angles or sides are equal.

If the three sides of a triangle are all different lengths (and thus the angles are all different measures), then the triangle is called a **scalene triangle.**

Scalene triangle

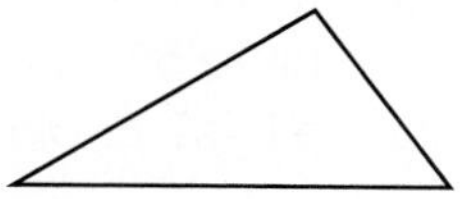

- All sides have different lengths.
- All angles have different measures.

Classify Are all equilateral triangles isosceles triangles? Are all isosceles triangles equilateral triangles? Explain.

Example 4

The perimeter of an equilateral triangle is 2 feet. How many inches long is each side?

Solution

All three sides of an equilateral triangle are equal in length. Since 2 feet equals 24 inches, we divide 24 inches by 3 and find that the length of each side is **8 inches.**

Example 5

Draw an isosceles right triangle.

Solution

Isosceles means the triangle has at least two sides that are the same length. *Right* means the triangle contains a right angle. We sketch a right angle, making both segments equal in length. Then we complete the triangle.

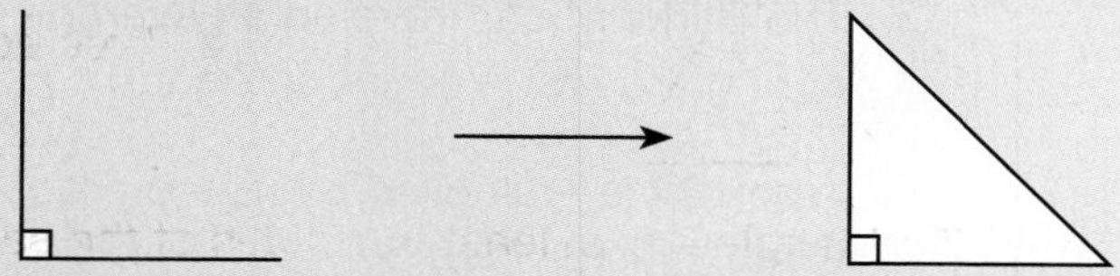

Analyze What is the measure of each of the angles opposite the sides you drew?

Practice Set

Classify Describe each triangle by its angles.

a. 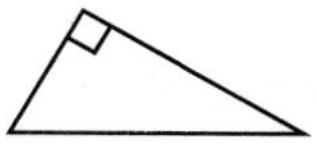

b. 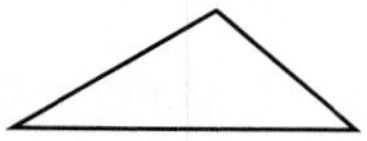

c. 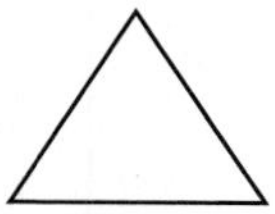

Classify Describe each triangle by its sides.

d. 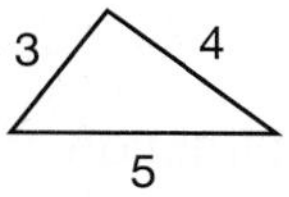

e. 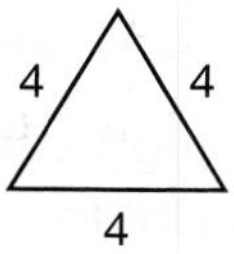

f. 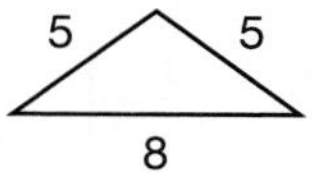

g. **Explain** If we know that two sides of an isosceles triangle are 3 cm and 4 cm and that its perimeter is not 10 cm, then what is its perimeter? Explain how you know.

h. **Classify** If the angles of a triangle measure 55°, 60°, and 65°, then what two names describe the type of triangle it is? Explain how you know.

i. Name the angles of this triangle in order from smallest to largest.

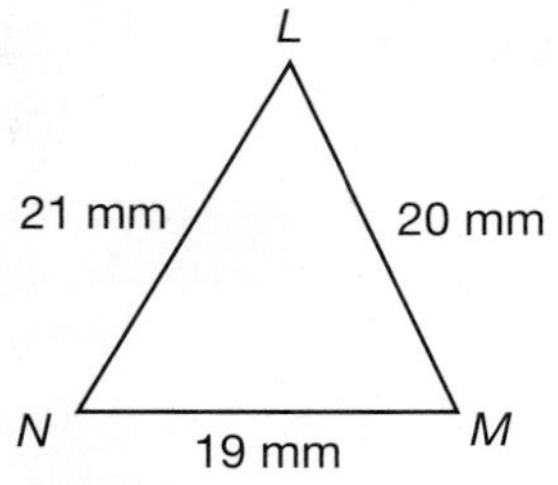

Written Practice

Strengthening Concepts

1. (28) At 1:30 p.m. Ms. Kwan found a parking meter that still had 10 minutes until it expired. She put 2 quarters into the meter and went to her meeting. If 10 cents buys 15 minutes of parking time, at what time will the meter expire?

Use the information in the paragraph below to answer problems **2** and **3.**

The Barkers started their trip with a full tank of gas and a total of 39,872 miles on their car. They stopped 4 hours later and filled the gas tank with 8.0 gallons of gas. At that time the car's total mileage was 40,092.

2. (12) How far did they travel in 4 hours?

3. (46) The Barkers' car traveled an average of how many miles per gallon during the first 4 hours of the trip?

4. (41) When 24 is multiplied by w, the product is 288. What is the quotient when 24 is divided by w?

* **5.** (53) **Represent** Use a ratio box to solve this problem. There were 144 books about mammals in the library. If the ratio of books about mammals to books about plants was 9 to 8, how many books about plants were in the library?

6. (22, 48) Read this statement. Then answer the questions that follow.

Exit polls showed that 7 out of 10 voters cast their ballot for the incumbent.

a. According to the exit polls, what percent of the voters cast their ballot for the incumbent?

b. According to the exit polls, what fraction of the voters did not cast their ballot for the incumbent?

* **7.** (60) **Formulate** Write an equation to solve this problem:

What number is $\frac{5}{6}$ of $3\frac{1}{3}$?

8. (60) What is the total price of a \$20,000 car plus 8.5% sales tax? How do you know?

9. (51) Write 1.86×10^5 in standard form. Then use words to write the number.

10. (32) Compare: 1 quart ◯ 1 liter

* **11.** (59) **Model** Show this addition problem on a number line:

$$(-3) + (+4) + (-2)$$

12. (48) Complete the table.

Fraction	Decimal	Percent
$\frac{5}{8}$	**a.**	**b.**
c.	**d.**	275%

13. (52) Evaluate: $x + \frac{x}{y} - y$ if $x = 12$ and $y = 3$

* **14.** (47) **Generalize** Find each missing exponent:

a. $2^5 \cdot 2^3 = 2^{\square}$

b. $2^5 \div 2^3 = 2^{\square}$

c. $2^3 \div 2^3 = 2^{\square}$

d. $2^3 \div 2^5 = 2^{\square}$

* **15.** (62) **Classify** In the figure below, angle *ZWX* measures 90°.

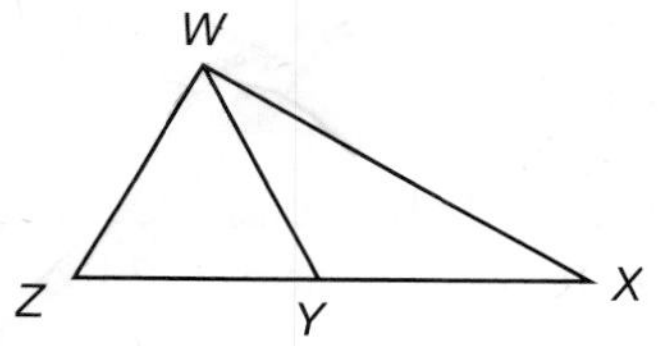

a. Which triangle is an acute triangle?

b. Which triangle is an obtuse triangle?

c. Which triangle is a right triangle?

16. (19, 37) In the figure at right, dimensions are in inches and all angles are right angles.

a. What is the perimeter of the figure?

b. What is the area of the figure?

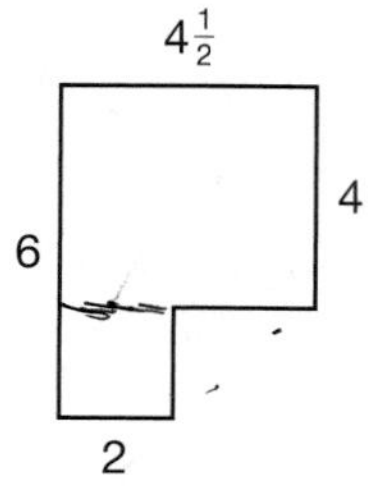

* **17.** (37, 62) **a.** Classify this triangle by its sides.

b. What is the measure of each acute angle of the triangle?

c. What is the area of the triangle?

d. The longest side of this triangle is opposite which angle?

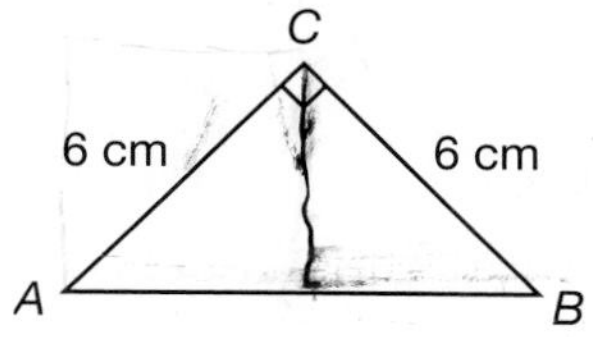

Solve:

18. (35) $7q = 1.428$

19. (39) $\frac{30}{70} = \frac{w}{\$2.10}$

Simplify:

20. (20, 52) $5^2 + 2^5 - \sqrt{49}$

21. (52) $3(8) - (5)(2) + 10 \div 2$

22. (49)
$$\begin{array}{r} 1 \text{ yd } 2 \text{ ft } 3\tfrac{3}{4} \text{ in.} \\ + \quad 2 \text{ ft } 6\tfrac{1}{2} \text{ in.} \\ \hline \end{array}$$

23. (32) 1 L − 50 mL = ___ mL

24. (50) $\frac{60 \text{ mi}}{1 \text{ hr}} \cdot \frac{1 \text{ hr}}{60 \text{ min}}$

25. (30) $2\frac{7}{24} + 3\frac{9}{32}$

26. (26) $2\frac{2}{5} \div \left(4\frac{1}{5} \div 1\frac{3}{4}\right)$

27. (23, 26) $20 - \left(7\frac{1}{2} \div \frac{2}{3}\right)$

*** 28.** (58, 62) **a.** *Model* Draw an equilateral triangle and show its lines of symmetry.

b. Does an equilateral triangle have rotational symmetry?

*** 29.** (59) *Conclude* Evaluate: $|x - y|$ if $x = 3$ and $y = 4$

30. (3, 32) On one tray of a balanced scale was a 1-kg mass. On the other tray were a box and a 250-g mass. What was the mass of the box?

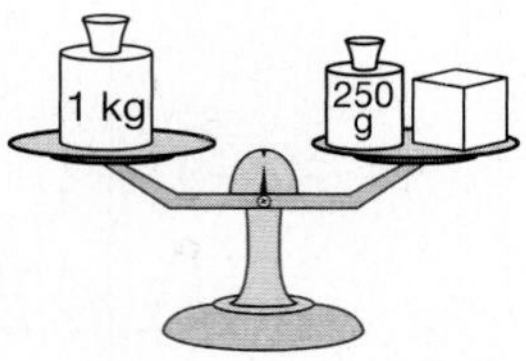

Early Finishers
Math Applications

Melody said she could use properties to make simplifying each expression easier. Is Melody correct? Use properties to validate your conclusion.

$\frac{1}{2} \cdot 41 + \frac{1}{2} \cdot 35 + \frac{1}{2} \cdot 24$

$\frac{1}{4} \cdot 20.5 + \frac{1}{4} \cdot 3.5 + \frac{1}{4} + 5.25 + \frac{3}{4}$

LESSON

63

Symbols of Inclusion

Power Up

Building Power

facts Power Up L

mental math

a. Calculation: 5×246

b. Exponents: 4×10^{-3}

c. Equivalent Fractions: $\frac{15}{20} = \frac{x}{8}$

d. Measurement: Convert 0.5 L to mL.

e. Square Roots: $\sqrt{196}$

f. Fractional Parts: $\frac{3}{8}$ of \$24.00

g. Geometry: What type of triangle has 2 equal sides?

h. Calculation: Instead of multiplying 50 and 28, double 50, find half of 28, and multiply those numbers.

problem solving

Jamillah has two standard number cubes. If she rolls the cubes together once, what totals does she have a $\frac{5}{36}$ chance of rolling?

New Concept

Increasing Knowledge

parentheses, brackets, and braces

Parentheses are called **symbols of inclusion.** We have used parentheses to show which operation to perform first. For example, to simplify the following expression, we add 5 and 7 before subtracting their sum from 15.

$$15 - (5 + 7)$$

Math Language

A *symbol of inclusion* is a symbol that groups or includes numbers together.

Brackets, [ ], and **braces,** { }, are also symbols of inclusion. When an expression contains multiple symbols of inclusion, we simplify within the innermost symbols first.

To simplify the expression

$$20 - [15 - (5 + 7)]$$

we simplify within the parentheses first.

$20 - [15 - (12)]$ simplified within parentheses

Next we simplify within the brackets.

$20 - [3]$ simplified within brackets

17 subtracted

Example 1

Simplify: 50 − [20 + (10 − 5)]

Solution

First we simplify within the parentheses.

50 − [20 + (5)]	simplified within parentheses
50 − [25]	simplified within brackets
25	subtracted

Discuss Compare the vertical arrangement of solution steps above with the same steps arranged in horizontal form:

$$50 - [20 + (10 - 5)] = 50 - [20 + (5)] = 50 - [25] = 25$$

Discuss which method is more helpful in understanding how the problem is solved.

Example 2

Simplify: 12 − (8 − |4 − 6| + 2)

Solution

Absolute value symbols may serve as symbols of inclusion. In this problem we find the absolute value of 4 − 6 as the first step of simplifying within the parentheses.

12 − (8 − 2 + 2)	found absolute value of 4 − 6
12 − (8)	simplified within parentheses
4	subtracted

division bar As we noted in Lesson 52, a division bar can serve as a symbol of inclusion. We simplify above and below the division bar before we divide. We follow the order of operations within the symbol of inclusion.

Example 3

Simplify: $\dfrac{4 + 5 \times 6 - 7}{10 - (9 - 8)}$

Solution

We simplify above and below the bar before we divide. Above the bar we multiply first. Below the bar we simplify within the parentheses first. This gives us

$$\frac{4 + 30 - 7}{10 - (1)}$$

We continue by simplifying above and below the division bar.

$$\frac{27}{9}$$

Now we divide and get

3

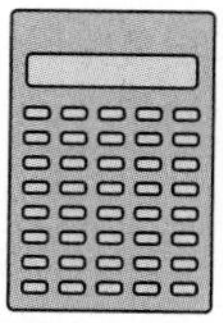

Some calculators with parenthesis keys are capable of dealing with many levels of parentheses (parentheses within parentheses within parentheses). When performing calculations such as the one in example 1, we press the "open parenthesis" key, (, for each opening parenthesis, bracket, or brace. We press the "close parenthesis" key,) , for each closing parenthesis, bracket, or brace.

For the problem in example 1, the keystrokes are

5 0 − (2 0 + (1 0 − 5)) =

To perform calculations such as the one in example 3 using a calculator, we follow one of these two procedures:

1. We perform the calculations above the bar and record the result. Then we perform the calculations below the bar and record the result. Finally, we perform the division using the two recorded numbers.
2. To perform the calculation with one uninterrupted sequence of keystrokes, we picture the problem like this:

$$\frac{4 + 5 \times 6 - 7}{10 - (9 - 8)} =$$

We press the equals key after the 7 to complete the calculations above the bar. Then we press ÷ for the division bar. We place all the operations below the division bar within a set of parentheses so that the denominator is handled by the calculator as though it were one number.

If you have a calculator with parenthesis keys and algebraic logic, perform these calculations and note the display at the indicated location in the sequence of keystrokes.

4 + 5 × 6 − 7 =

What number is displayed and what does this number represent?

Continue the calculation using the following sequence of key strokes.

÷ (1 0 − (9 − 8)) =

What number is displayed and what does this number represent?

Practice Set

Represent Simplify, then show and justify each step in your solution using vertical form.

a. $100 - 3[2(6 - 2)]$

Simplify:

b. $30 - [40 - (10 - 2)]$

c. $\dfrac{10 + 9 \cdot 8 - 7}{6 \cdot 5 - 4 - 3 + 2}$

d. $\dfrac{1 + 2(3 + 4) - 5}{10 - 9(8 - 7)}$

e. $12 + 3(8 - |-2|)$

Written Practice *Strengthening Concepts*

1. (28) Jennifer and Jason each earn $11 per hour doing yard work. On one job Jennifer worked 3 hours, and Jason worked $2\frac{1}{2}$ hours. Altogether, how much money were they paid?

2. (28, 54) When Marisol is resting, her heart beats 70 times per minute. When Marisol is jogging, her heart beats 150 times per minute. During a half hour of jogging, Marisol's heart beats how many more times than it would if she were resting?

3. (53) Use a ratio box to solve this problem. The ratio of pairs of dress shoes to pairs of running shoes sold at the Shoe-Fly Buy store last Saturday was 2 to 9. If 720 pairs of running shoes were sold, how many pairs of dress shoes were sold?

4. (55) During the first 5 days, a hiker on the Appalachian Trail averaged 17 miles a day. During the next two days the hiker traveled 13 and 19 miles. The trail is a little more than 2174 miles long. About how much farther does the hiker have to travel?

*** 5.** (59) **Analyze** Yesterday, the hiker began hiking at 2850 feet above sea level, climbed over a 5575 foot-high summit and descended 2695 feet before stopping for the night. What was the hiker's net gain or loss of elevation during the day? How did you get your answer?

6. (51) The average distance from Earth to the Sun is 1.496×10^8 km. Use words to write that distance.

7. (22, 48) Read this statement. Then answer the questions that follow.

Twelve of the 40 cars pulled by the locomotive were tankers.

a. What fraction of the cars were tankers?

b. What percent of the cars were not tankers?

*** 8.** (57) **Connect** To convert inches to meters, we can multiply the number of inches by $2.54 \cdot 10^{-2}$. Use this fact to convert 1 yard to meters.

9. (50) Use a unit multiplier to convert 1.5 km to m.

10. (45) Divide 4.36 by 0.012 and write the answer with a bar over the repetend.

*** 11.** (59) **Model** Show this addition problem on a number line:

$$(-3) + (+5) + (-2)$$

12. (48) Complete the table.

Fraction	Decimal	Percent
a.	b.	33%
$\frac{1}{3}$	c.	c.

* **13.** (56) **Formulate** Describe the rule of this function. Then write the rule as an equation using x for the input and y for the output. Begin by writing $y =$.

x Input	y Output
3	1
12	4
6	2
15	5

14. (14, 36) A deck of 26 cards has one letter of the alphabet on each card. What is the probability of drawing one of the letters in the word "MATH" from the deck?

In the figure below, $AB = AD = BD = CD = 5$ cm. The measure of angle ABC is 90°. Refer to the figure for problems **15–17.**

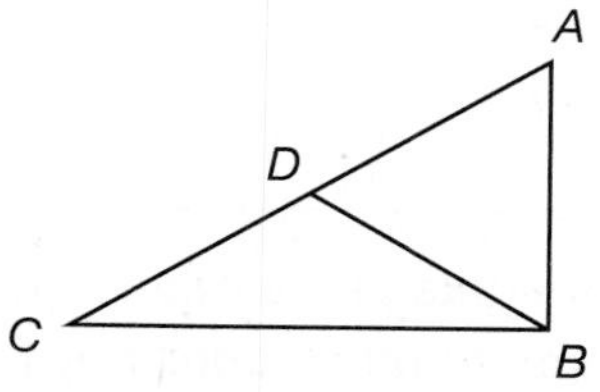

* **15.** (62) **a.** Classify $\triangle BCD$ by its sides.

b. What is the perimeter of the equilateral triangle?

c. Which triangle is a right triangle?

16. (40) Find the measure of each of the following angles:

a. $\angle BAC$ **b.** $\angle ADB$

c. $\angle BDC$ **d.** $\angle DBA$

e. $\angle DBC$ **f.** $\angle DCB$

17. (36) What is the ratio of the length of the shortest side of $\triangle ABC$ to the length of the longest side?

Solve:

18. (30) $\frac{5}{18} = x + \frac{1}{12}$

19. (45) $2 = 0.4p$

Generalize Simplify:

* **20.** (63) $3[24 - (8 + 3 \cdot 2)] - \frac{6 + 4}{|-2|}$

21. (52) $3^3 - \sqrt{3^2 + 4^2}$

22. (49)

	1 week 2 days 7 hr
−	5 days 9 hr

23. (50) $\frac{20 \text{ mi}}{1 \text{ gal}} \cdot \frac{1 \text{ gal}}{4 \text{ qt}}$

24. (30) $4\frac{2}{3} + 3\frac{5}{6} + 2\frac{5}{9}$

25. (26) $12\frac{1}{2} \cdot 4\frac{4}{5} \cdot 3\frac{1}{3}$

26. (26, 30) $6\frac{1}{3} - \left(1\frac{2}{3} \div 3\right)$

27. (52) Evaluate: $x^2 + 2xy + y^2$ if $x = 3$ and $y = 4$

* **28.** (58, 62) **Model** Draw an isosceles triangle that is not equilateral, and show its line of symmetry.

* **29.** (61) Della and Ray are planning to build a patio in the shape of a parallelogram. They sketched the patio on graph paper. Each square on the grid represents a 5 foot square. The coordinates of the four vertices are (0,0), (4,0), (1, −3), and (−3, −3).

a. **Represent** Graph the parallelogram.

b. What is the area of the patio they are planning?

30. (3) Three identical boxes are balanced on one side of a scale by a 750-g mass on the other side of the scale. What is the mass of each box?

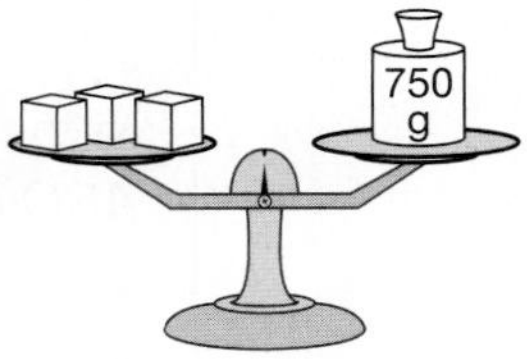

Early Finishers
Real-World Application

Dexter works at a clothing store and receives a 15% employee discount. He wants to buy a corduroy blazer that is on sale for 30% off the original price of $79.

a. Write and solve a proportion to find the price of the blazer without the employee discount.

b. Write and solve an equation to find the price Dexter would pay using his employee discount.

c. What percent of the original price did Dexter save by buying the blazer on sale and using his employee discount? Round your answer to the nearest whole percent. Show your work.

LESSON 64

Adding Positive and Negative Numbers

Power Up — *Building Power*

facts Power Up N

mental math

a. **Decimals:** 3.6×50

b. **Decimals/Exponents:** 7.5×10^2

c. **Algebra:** $4x - 5 = 35$

d. **Measurement:** Convert 20 cm to mm.

e. **Square Roots:** $\sqrt{9 + 16}$

f. **Fractional Parts:** $\frac{5}{9}$ of \$1.80

g. **Geometry:** What is the name for a regular triangle?

h. **Calculation:** $1.5 + 1, \times 2, + 3, \div 4, - 1.5$

problem solving

Candice has a red notebook, a yellow notebook, and a blue notebook. She uses one notebook for school, one notebook for sketching, and one notebook for her journal. Her yellow notebook is not used for sketching. She does not use her red notebook for school. She does not use her blue notebook for her journal. If her blue notebook is not used for school, what notebook does Candice use for each purpose? Make a table to show your work.

New Concept — *Increasing Knowledge*

From our practice on the number line, we have seen that when we add two negative numbers, the sum is a negative number. When we add two positive numbers, the sum is a positive number.

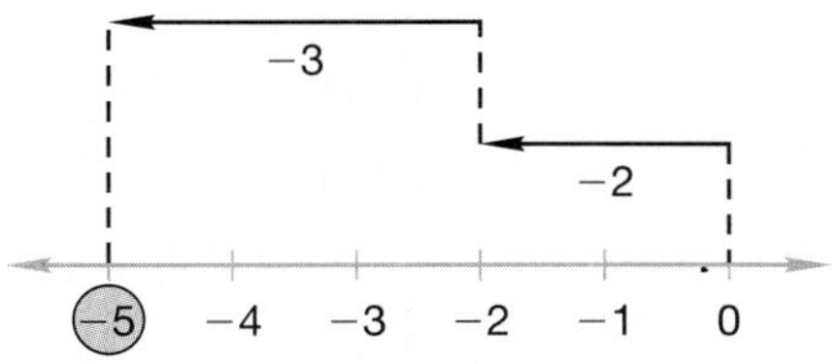

$(-2) + (-3) = -5$

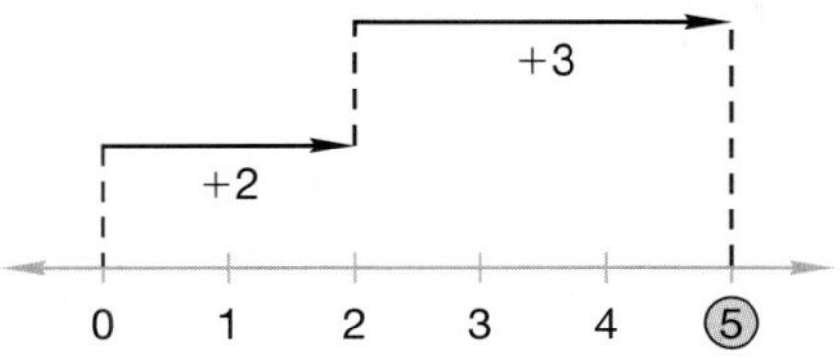

$(+2) + (+3) = +5$

We have also seen that when we add a positive number and a negative number, the sum is positive, negative, or zero depending upon which, if either, of the numbers has the greater absolute value.

Math Language
The **absolute value** of a number is the distance on the number line of that number from zero and is always positive.

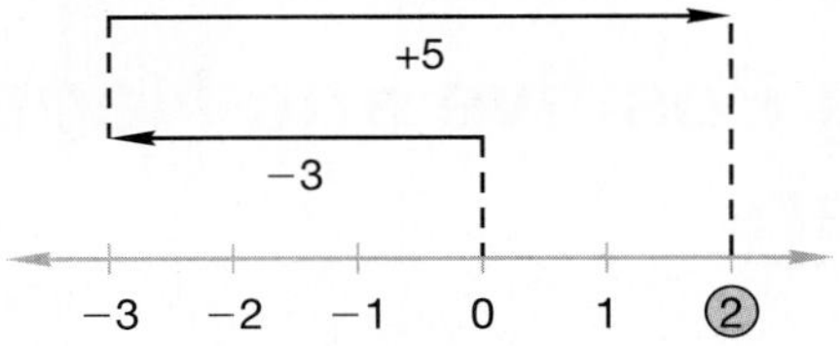

$(-3) + (+5) = +2$

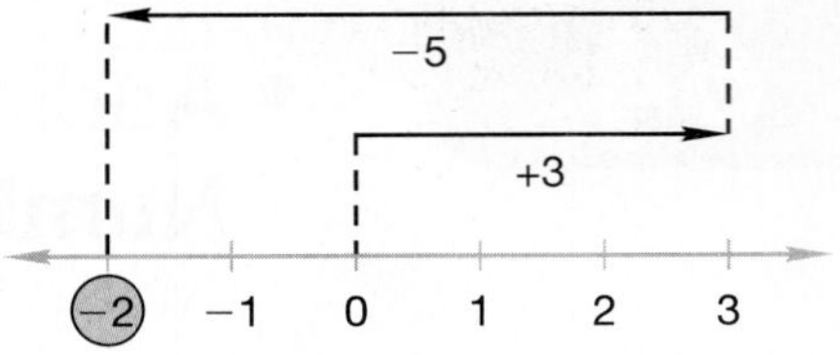

$(+3) + (-5) = -2$

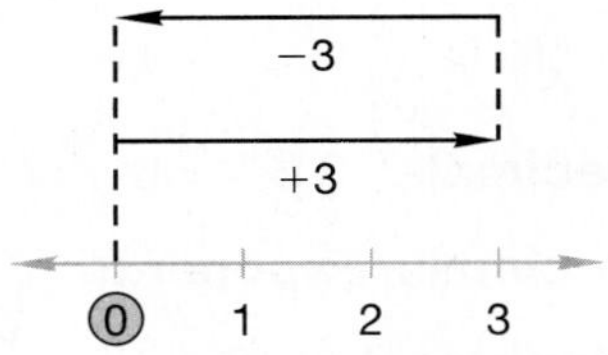

$(+3) + (-3) = 0$

We can summarize these observations with the following statements.

1. The sum of two numbers with the same sign has an absolute value equal to the sum of their absolute values. Its sign is the same as the sign of the numbers.
2. The sum of two numbers with opposite signs has an absolute value equal to the difference of their absolute values. Its sign is the same as the sign of the number with the greater absolute value.
3. The sum of two opposites is zero.

We can use these observations to help us add positive and negative numbers without drawing a number line.

Example 1

Thinking Skill

Apply

Will the sum of $(-34) + (+43)$ be positive or negative? Explain?

Find each sum:

a. $(-54) + (-78)$ **b. $(+45) + (-67)$** **c. $(-92) + (+92)$**

Solution

a. Since the signs are the same, we add the absolute values and use the same sign for the sum.

$$(-54) + (-78) = \mathbf{-132}$$

b. Since the signs are different, we find the difference of the absolute values and keep the sign of −67 because its absolute value, 67, is greater than 45.

$$(+45) + (-67) = \mathbf{-22}$$

c. The sum of two opposites is zero, a number which has no sign.

$$(-92) + (+92) = \mathbf{0}$$

Example 2

Find the sum: $(-3) + (-2) + (+7) + (-4)$

Solution

We will show two methods.

Method 1: Adding in order from left to right, add the first two numbers. Then add the third number. Then add the fourth number.

$(-3) + (-2) + (+7) + (-4)$	problem
$(-5) + (+7) + (-4)$	added -3 and -2
$(+2) + (-4)$	added -5 and $+7$
$\mathbf{-2}$	added $+2$ and -4

Method 2: Employing the commutative and associative properties, rearrange the terms and add all numbers with the same sign first.

$(-3) + (-2) + (-4) + (+7)$	rearranged
$(-9) + (+7)$	added
$\mathbf{-2}$	added

Example 3

Find each sum:

a. $\left(-2\frac{1}{2}\right) + \left(-3\frac{1}{3}\right)$ **b.** **(+4.3) + (−7.24)**

Solution

These numbers are not integers, but the method for adding these signed numbers is the same as the method for adding integers.

a. The signs are both negative. We add the absolute values and keep the same sign.

$$\left(-2\frac{1}{2}\right) + \left(-3\frac{1}{3}\right) = \mathbf{-5\frac{5}{6}}$$

$$\begin{array}{r} 2\frac{1}{2} = 2\frac{3}{6} \\ +\ 3\frac{1}{3} = 3\frac{2}{6} \\ \hline 5\frac{5}{6} \end{array}$$

b. The signs are different. We find the difference of the absolute values and keep the sign of -7.24.

$$(+4.3) + (-7.24) = \mathbf{-2.94}$$

$$\begin{array}{r} \overset{6}{\cancel{7}}.\overset{1}{2}4 \\ -\ 4.3 \\ \hline 2.94 \end{array}$$

Example 4

Thinking Skill

Conclude

What if Mr. Figuera made a third trade and lost \$350. What is the net result of the three trades?

On one stock trade Mr. Figuera lost \$450. On a second trade Mr. Figuera gained \$800. What was the net result of the two trades?

Solution

A loss may be represented by a negative number and a gain by a positive number. So the results of the two trades may be expressed this way:

$$(-450) + (+800) = \mathbf{+350}$$

Practice Set Generalize Find each sum:

a. $(-56) + (+96)$

b. $(-28) + (-145)$

c. $(-5) + (+7) + (+9) + (-3)$

d. $(-3) + (-8) + (+15)$

e. $(-12) + (-9) + (+16)$

f. $(+12) + (-18) + (+6)$

g. $\left(-3\frac{5}{6}\right) + \left(+5\frac{1}{3}\right)$

h. $(-1.6) + (-11.47)$

i. Connect On three separate stock trades Mrs. Francois gained \$250, lost \$300, and gained \$525. Write an expression that shows the results of each trade. Then find the net result of the trades.

Written Practice

Strengthening Concepts

1. (51) Two trillion is how much more than seven hundred fifty billion? Write the answer in scientific notation.

* 2. (28) Analyze The taxi cost \$2.25 for the first mile plus 15¢ for each additional tenth of a mile. For a 5.2-mile trip, Eric paid \$10 and told the driver to keep the change as a tip. How much was the driver's tip?

3. (44) Mae-Ying wanted to buy packages of crackers and cheese from the vending machine. Each package cost 35¢. Mae-Ying had 5 quarters, 3 dimes, and 2 nickels. How many packages of crackers and cheese could she buy?

4. (21) The two prime numbers p and m are between 50 and 60. Their difference is 6. What is their sum? Tell how you found your answer.

5. (28) What is the mean of 1.74, 2.8, 3.4, 0.96, 2, and 1.22?

6. (22, 48) Diagram this statement. Then answer the questions that follow.

Over the summer, Ty earned \$120 babysitting. He spent $\frac{3}{5}$ of the money to buy a new bike and saved the rest.

a. How much did he spend on the bike?

b. What percent of his earnings did he save?

* 7. (60) Formulate Write an equation to solve this problem:

What number is $\frac{5}{9}$ of 100?

* 8. (51, 57) Connect Use words to write each measure.

a. The temperature at the center of the sun is about 1.6×10^7 degrees Celsius.

b. A red blood cell is about 7×10^{-6} meter in diameter.

* **9.** (57) **Analyze** Compare:

a. $1.6 \times 10^{7} \bigcirc 7 \times 10^{-6}$

b. $7 \times 10^{-6} \bigcirc 0$

c. $2^{-3} \bigcirc 2^{-2}$

10. (44) Divide 456 by 28 and write the answer

a. as a mixed number.

b. as a decimal rounded to two decimal places.

c. rounded to the nearest whole number.

* **11.** (64) **Generalize** Find each sum:

a. $(-63) + (-14)$

b. $(-16) + (+20) + (-32)$

c. $\left(-\frac{1}{2}\right) + \left(-\frac{1}{2}\right)$

* **12.** (64) On two separate stock trades Ms. Miller lost $327 and gained $280. What was the net result of the two trades?

* **13.** (Inv. 2, 62) The figure shows an equilateral triangle inscribed in a circle.

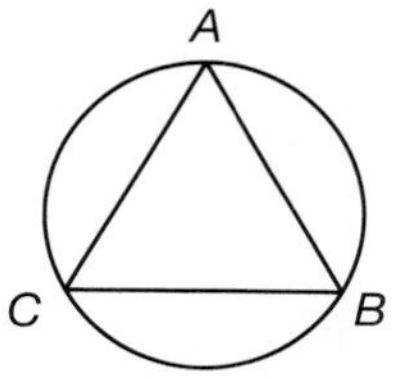

a. What is the measure of the inscribed angle *BCA*?

b. **Justify** Select a chord of this circle, and state whether the chord is longer or shorter than the diameter of the circle and why.

14. (52) Evaluate: $x + xy$ if $x = \frac{2}{3}$ and $y = \frac{3}{4}$

Math Language

Review

A **chord** is a line segment with endpoints that lie on the circle.

Refer to the hexagon below to answer problems **15** and **16.** Dimensions are in meters. All angles are right angles.

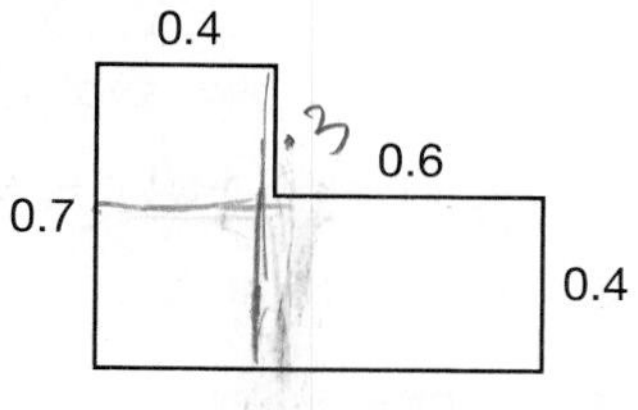

15. (19) What is the perimeter of the hexagon?

16. (37) What is the area of the hexagon?

* **17.** (36) **Analyze** What is the probability of rolling 10 with a pair of number cubes? (Hint: see the sample space in Lesson 36).

18. (Inv. 3) The center of a circle with a radius of three units is (1, 1). Which of these points is on the circle?

A (4, 4) **B** (−2, 1) **C** (−4, 1) **D** (3, 0)

Solve:

19. $\frac{4}{9} = y - \frac{2}{9}$ (15)

20. $25x = 10$ (44)

Generalize Simplify:

*** 21.** $\frac{3^2 + 4^2}{\sqrt{3^2 + 4^2}}$ (52)

22. $2\frac{4}{5} \div \left(6 \div 2\frac{1}{2}\right)$ (26)

*** 23.** $100 - [20 + 5(4) + 3(2 + 4^0)]$ (57, 63)

24. (49)

$$\begin{array}{r} 5 \text{ gal } 2 \text{ qt } 1 \text{ pt } 7 \text{ oz} \\ + \ 1 \text{ gal } 1 \text{ qt } 1 \text{ pt } 9 \text{ oz} \\ \hline \end{array}$$

25. $\left(1\frac{1}{2}\right)^2 - \left(4 - 2\frac{1}{3}\right)$ (26, 30)

26. $0.1 - (0.01 - 0.001)$ (35)

*** 27.** (56) **Formulate** Write the rule of the graphed function in an equation. Begin the equation with $y =$. Then name another (x, y) pair not named on the graph that satisfies the function.

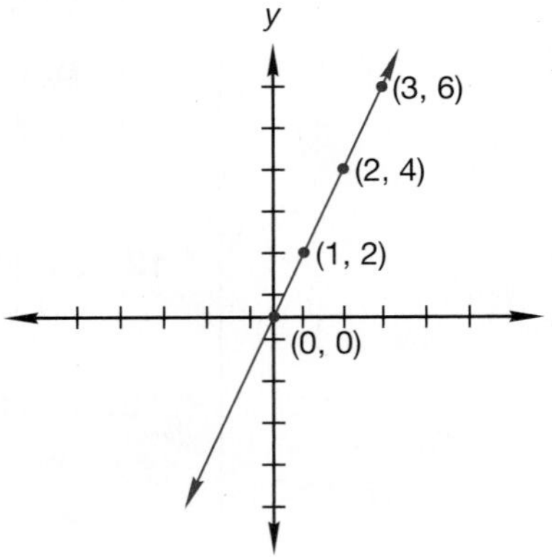

28. (43) Write $3\frac{1}{5}$ as a decimal number, and subtract it from 4.375.

29. (14) What is the probability of rolling an even prime number with one toss of a die?

***30.** (61) Figure *ABCD* is a parallelogram. Find the measure of

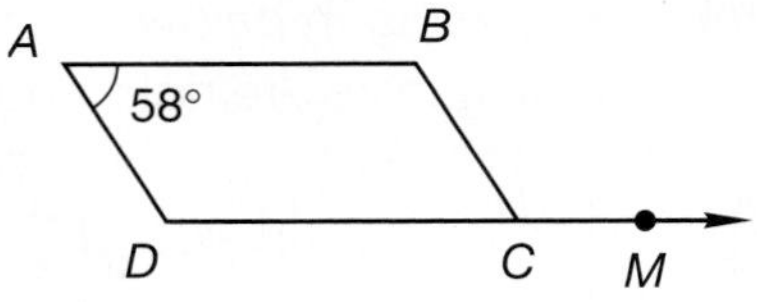

a. $\angle B$.

b. $\angle BCD$.

c. $\angle BCM$.

Circumference and Pi

Power Up — *Building Power*

facts

Power Up M

mental math

a. **Decimals:** 0.42×50

b. **Decimals/Exponents:** 1.25×10^{-1}

c. **Ratio:** $\frac{9}{w} = \frac{15}{10}$

d. **Measurement:** Convert 0.75 m to mm.

e. **Exponents:** $5^3 - 10^2$

f. **Fractional Parts:** $\frac{9}{10}$ of \$4.00

g. **Geometry:** A circle has a diameter of 7 ft. What is the radius of the circle?

h. **Calculation:** What is the total cost of a \$20.00 item plus 7% sales tax?

problem solving

Copy this problem and fill in the missing digits:

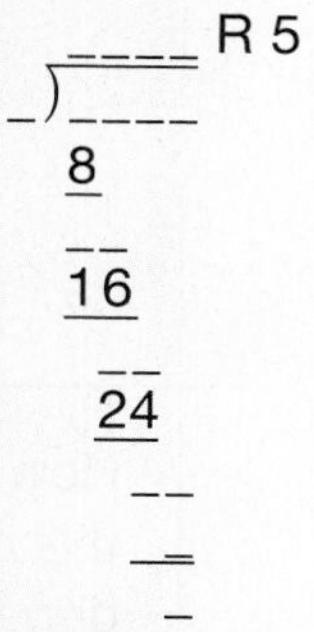

New Concept — *Increasing Knowledge*

Recall from Investigation 2 that a **circle** is a smooth curve, every point of which is the same distance from the **center.** The distance from the center to the circle is the **radius.** The plural of radius is **radii.** The distance across a circle through the center is the **diameter.** The distance around a circle is the **circumference.**

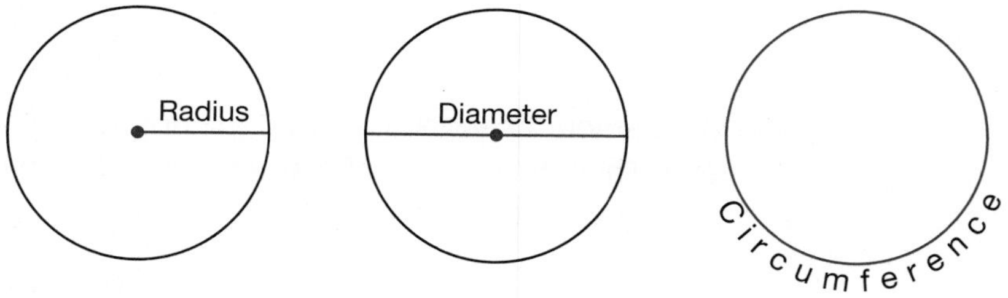

We see that the diameter of a circle is twice the radius of the circle. In the following activity we investigate the relationship between the diameter and the circumference.

Activity

Investigating Circumference and Diameter

Materials needed:

- Metric tape measure (or string and meter stick)
- Circular objects of various sizes
- Calculator (optional)

Working in a small group, select a circular object and measure its circumference and its diameter as precisely as you can. Then calculate the number of diameters that equal the circumference by dividing the circumference by the diameter. Round the quotient to two decimal places. Repeat the activity with another circular object of a different size. Record your results on a table similar to the one shown below. Extend the table by including the results of other students in the class. Are your ratios of circumference to diameter basically the same?

Sample Table

Object	Circumference	Diameter	$\frac{\text{Circumference}}{\text{Diameter}}$
Waste basket	94 cm	30 cm	3.13
Plastic cup	22 cm	7 cm	3.14

How many diameters equal a circumference? Mathematicians investigated this question for thousands of years. They found that the answer did not depend on the size of the circle. The circumference of every circle is slightly more than three diameters.

Another way to illustrate this fact is to cut a length of string equal to the diameter of a particular circle and find how many of these lengths are needed to reach around the circle. No matter what the size of the circle, it takes three diameters plus a little extra to equal the circumference.

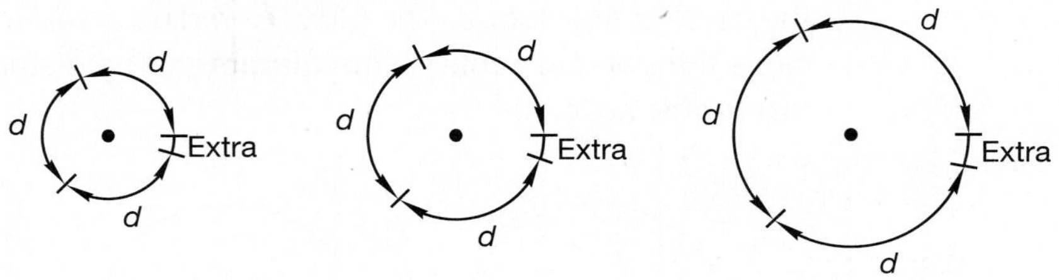

The extra amount needed is about, but not exactly, one seventh of a diameter. Thus the number of diameters needed to equal the circumference of a circle is about

$$3\frac{1}{7} \quad \text{or} \quad \frac{22}{7} \quad \text{or} \quad 3.14$$

Neither $3\frac{1}{7}$ nor 3.14 is the exact number of diameters needed. They are approximations. There is no fraction or decimal number that states the exact number of diameters in a circumference. (Some computers have calculated the number to more than 1 million decimal places.) We use the symbol π, which is the Greek letter **pi** (pronounced like "pie"), to represent this number. Note that π is not a variable. Rather, π is a **constant** because its value does not vary.

The circumference of a circle is π times the diameter of the circle. This idea is expressed by the formula

$$\boxed{C = \pi d}$$

In this formula C stands for circumference and d for diameter. To perform calculations with π, we can use an approximation. The commonly used approximations for π are

$$3.14 \qquad \text{and} \qquad \frac{22}{7}$$

For calculations that require great accuracy, more accurate approximations for π may be used, such as

$$3.14159265359$$

Sometimes π is left as π. Unless directed otherwise, we use 3.14 for π for calculations in this book.

Example 1

The radius of a circle is 10 cm. What is the circumference?

Solution

If the radius is 10 cm, the diameter is 20 cm.

$$\begin{aligned} \text{Circumference} &= \pi \cdot \text{diameter} \\ &\approx 3.14 \cdot 20 \text{ cm} \\ &\approx 62.8 \text{ cm} \end{aligned}$$

The circumference is about **62.8 cm.** In the solution we used the symbol $\approx$, which means "approximately equals," because the value for π is not exactly 3.14.

Generalize How could you write the formula for the circumference of a circle using the radius (r) instead of the diameter (d)?

Thinking Skill

Connect

For any calculation that uses an inexact value for π, the answer is approximate and we use the $\approx$ symbol.

Example 2

Find the circumference of each circle:

a.

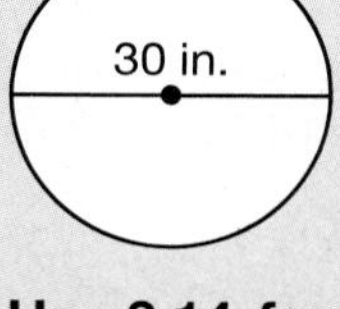

Use 3.14 for π.

b.

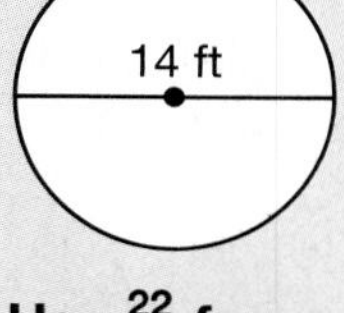

Use $\frac{22}{7}$ for π.

c.

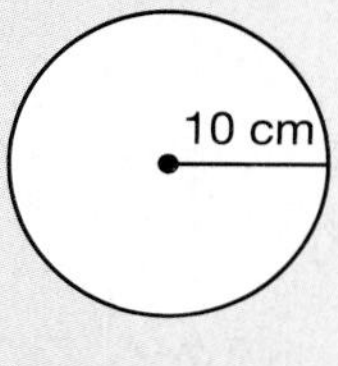

Leave π as π.

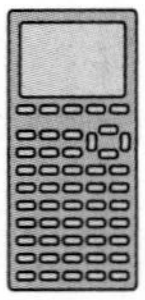

Visit www.SaxonPublishers.com/ActivitiesC2 *for a graphing calculator activity.*

Solution

a. $C = \pi d$
$C \approx 3.14(30 \text{ in.})$
$C \approx$ **94.2 in.**

b. $C = \pi d$
$C \approx \frac{22}{7}(14 \text{ ft})$
$C \approx$ **44 ft**

c. $C = \pi d$
$C = \pi(20 \text{ cm})$
$C =$ **20π cm**

Note the form of answer **c:** first 20 times π, then the unit of measure.

Discuss When might it be better to use $\frac{22}{7}$ instead of 3.14 for π?

Practice Set

Generalize Find the circumference of each circle:

a.

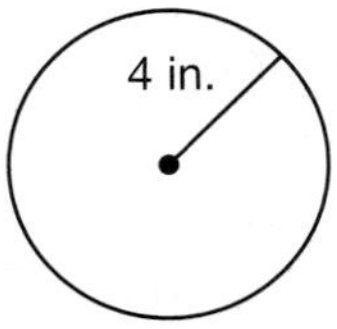

Use 3.14 for π.

b.

Use $\frac{22}{7}$ for π.

c.

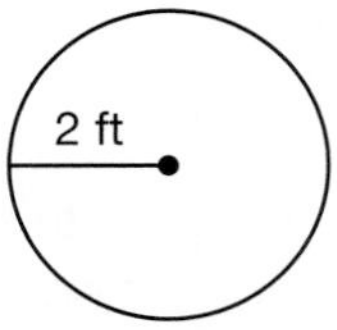

Leave π as π.

d. **Explain** Sylvia used a compass to draw a circle. If the point of the compass was 3 inches from the point of the pencil, how can she find the circumference of the circle? (Use 3.14 for π.)

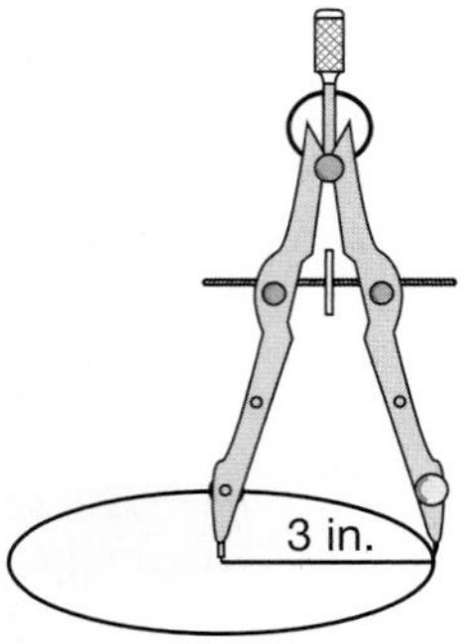

Written Practice

Strengthening Concepts

1. (46) If 5 pounds of apples cost $4.45, then

a. what is the price per pound?

b. what is the cost for 8 pounds of apples?

2. (41) **a.** Simplify and compare:

$$(0.3)(0.4) + (0.3)(0.5) \bigcirc 0.3(0.4 + 0.5)$$

b. What property is illustrated by this comparison?

***3.** (65) Find the circumference for each circle.

a.

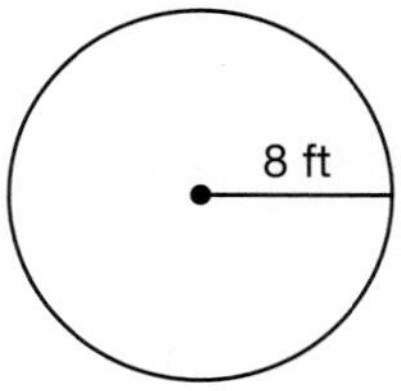

Leave π as π.

b.

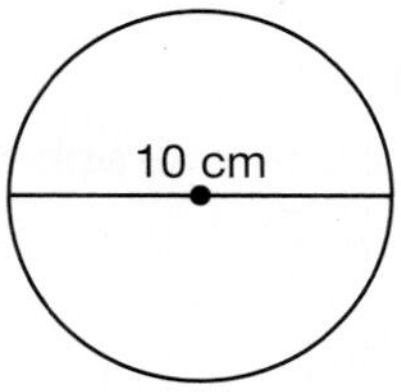

Use 3.14 for π.

c.

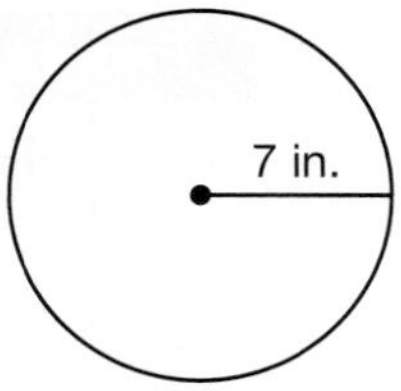

Use $\frac{22}{7}$ for π.

4. (44, 46) The car traveled 350 miles on 15 gallons of gasoline. The car averaged how many miles per gallon? Round the answer to the nearest tenth.

5. (28) The average of 2 and 4 is 3. What is the average of the reciprocals of 2 and 4?

6. (51) Write 12 billion in scientific notation.

7. (22, 36) **Model** Diagram this statement. Then answer the questions that follow.

One sixth of the five dozen computers were shipped.

a. How many computers were not shipped?

b. What was the ratio of computers shipped to those that were not shipped?

c. What percent of the computers were shipped?

*** 8.** (Inv. 6) **a.** **Represent** Draw segment AB. Draw segment DC parallel to segment AB but not the same length. Draw segments between the endpoints of segments AB and DC to form a quadrilateral.

b. **Classify** What type of quadrilateral was formed in **a?**

9. (37) Find the area of each triangle. (Dimensions are in centimeters.) Then classify each triangle as acute, obtuse, or right.

a.

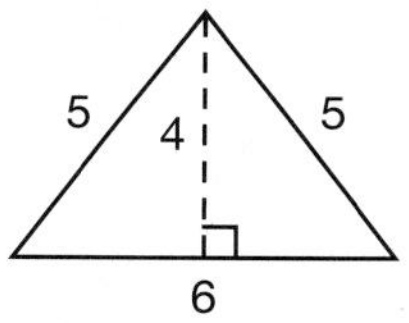

b.

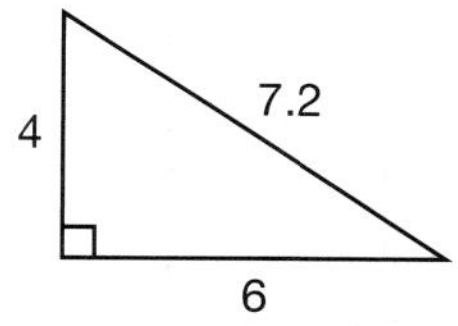

c.

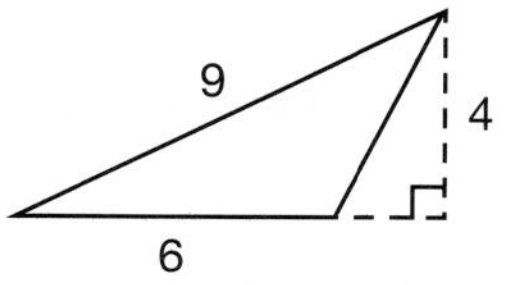

10. (28, 34) What is the average of the two numbers indicated by arrows on the number line below?

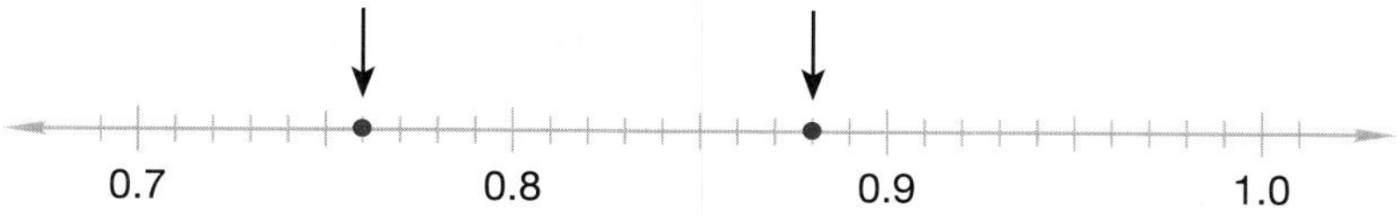

Write equations to solve problems **11** and **12.**

11. (60) What number is 75 percent of 64?

12. (60) What is the tax on a $7.40 item if the sales-tax rate is 8%?

*** 13.** (64) **Generalize** Find each sum:

a. $(-3) + (-8)$

b. $(+3) + (-8)$

c. $(-0.3) + (+0.8) + (-0.5)$

Math Language

The **origin** is the point (0, 0) on a coordinate plane.

14. (Inv. 3) A circle is drawn on a coordinate plane with its center at the origin. One point on the circle is (3, 4). Use a compass and graph paper to graph the circle. Then answer **a** and **b.**

a. What are the coordinates of the points where the circle intersects the x-axis?

b. What is the diameter of the circle?

15. (50) **Evaluate** Use a unit multiplier to convert 0.95 liters to milliliters.

16. (52) Evaluate: $ab + a + \frac{a}{b}$ if $a = 5$ and $b = 0.2$

17. (13) How many small blocks were used to build this cube?

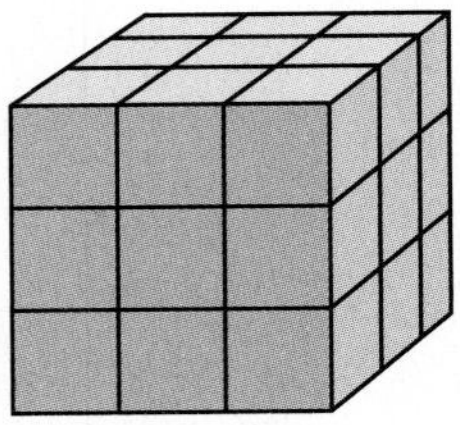

18. (40) Recall that one angle is the complement of another angle if their sum is 90°, and that one angle is the supplement of another if their sum is 180°. In this figure, **a** which angle is a complement of $\angle BOC$ and **b** which angle is a supplement of $\angle BOC$?

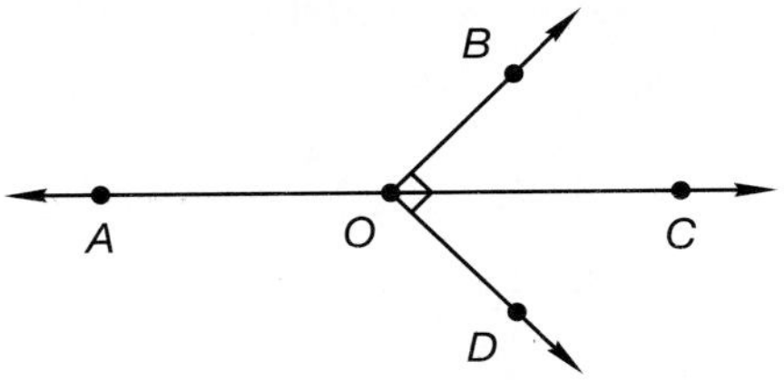

19. (29, 33) Round each number to the nearest whole number to estimate the product of 19.875 and $4\frac{7}{8}$.

*** 20.** (58, 62) **Analyze** Refer to $\triangle ABC$ at right to answer the following questions:

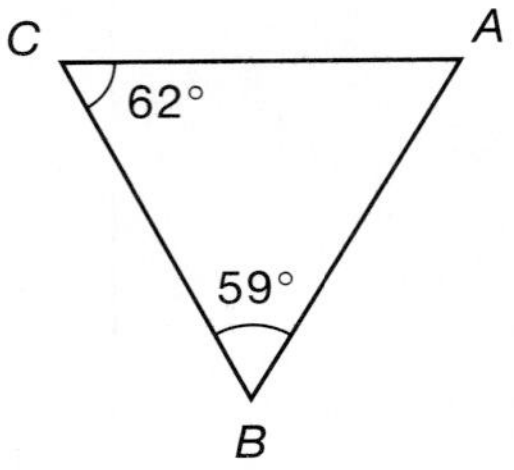

a. What is the measure of $\angle A$?

b. Which side of the triangle is the longest side?

c. Triangle ABC is an acute triangle. It is also what other type of triangle?

d. Triangle ABC's line of symmetry passes through which vertex?

21. (Inv. 4) **a.** Describe how to find the median of this set of 12 scores.

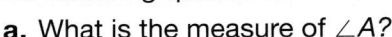

18, 17, 15, 20, 16, 14, 15, 16, 17, 18, 16, 19

b. What is the median of the set of scores?

22. (62) Answer true or false:

a. All equilateral triangles are congruent.

b. All equilateral triangles are similar.

23. (32, 34) The high jump bar was raised from 2.15 meters to 2.2 meters. How many centimeters was the bar raised?

Generalize Simplify:

24. (47) $\dfrac{10^3 \cdot 10^3}{10^2}$

25. (49)

	4 days	5 hr	15 min
−	1 day	7 hr	50 min

26. (45) $4.5 \div (0.4 + 0.5)$

27. (52) $\dfrac{3 + 0.6}{3 - 0.6}$

28. (26) $4\frac{1}{5} \div \left(1\frac{1}{6} \cdot 3\right)$

29. (52) $3^2 + \sqrt{4 \cdot 7 - 3}$

* **30.** (63) $|-3| + 4[(5 - 2)(3 + 1)]$

Early Finishers

Real-World Application

Central Park is a rectangular shaped park in the middle of New York City. The area of the park is 3.41 km^2.

a. If the width of the park is 800 meters, what is its length to the nearest meter? Show your work.

b. How many acres is the park to the nearest acre? (Note: 1 $km^2 \approx 247$ acres) Use a unit multiplier and show your work.

LESSON

66

• Ratio Problems Involving Totals

Power Up

Building Power

facts

Power Up N

mental math

a. **Decimals:** 3.65 + 1.2 + 2

b. **Decimals/Exponents:** 1.2×10^{-3}

c. **Algebra:** $9y + 3 = 75$

d. **Measurement:** Convert 20 decimeters (dm) to meters.

e. **Power/Roots:** $\sqrt{144} + 2^3$

f. **Percent:** 25% of 24

g. **Geometry:** A circle has a radius of 8 mm. What is the circumference of the circle?

h. **Estimation:** Estimate the product of 3.14 and 25.

problem solving

A palindrome is a number or word that can be read the same forward or backward, such as "42324" or "Madam, I'm Adam." Start with a 4-digit whole number, and then add to it the number formed by reversing its digits. To the sum, add the number formed by reversing the sum's digits. Continue this process, and eventually you will have a sum that is a palindrome. Why do you think this will always be the case?

New Concept

Increasing Knowledge

Some ratio problems require that we use the total to solve the problem. Consider the following problem:

The ratio of boys to girls at the assembly was 5 to 4. If there were 180 students in the assembly, how many girls were there?

We begin by making a ratio box. This time we add a third row for the total number of students.

Thinking Skills

Summarize

What is the purpose of the third row in the ratio box? How can we use it?

	Ratio	Actual Count
Boys	5	B
Girls	4	G
Total	9	180

Discuss How is this ratio box different from the ratio boxes we have been using?

In the ratio column we wrote 5 for boys and 4 for girls, then *added these to get 9 for the total ratio number*. We were given 180 as the actual count of students. This is a total. We can use two rows from this table to write a proportion. Since we were asked to find the number of girls, we will use the "girls" row.

Because we know both total numbers, we will also use the "total" row. Using these numbers, we solve the proportion.

	Ratio	Actual Count
Boys	5	*B*
Girls	4	*G*
Total	9	180

$$\frac{4}{9} = \frac{G}{180}$$
$$9G = 720$$
$$G = 80$$

We find there were 80 girls. We can use this answer to complete the ratio box.

	Ratio	Actual Count
Boys	5	100
Girls	4	80
Total	9	180

Example

The ratio of football players to soccer players in the room was 5 to 7. If the football and soccer players in the room totaled 48, how many were football players?

Solution

We use the information in the problem to form a table. We include a row for the total number of players.

Explain How do we find the number to put into the total ratio box in the table?

	Ratio	Actual Count
Football Players	5	*F*
Soccer Players	7	*S*
Total Players	12	48

$$\frac{5}{12} = \frac{F}{48}$$
$$12F = 240$$
$$F = 20$$

To find the number of football players, we write a proportion from the "football players" row and the "total players" row. We solve the proportion to find that there were **20 football players** in the room.

From this information we can complete the ratio box.

	Ratio	Actual Count
Football Players	5	20
Soccer Players	7	28
Total Players	12	48

Practice Set

Represent Solve these problems. Begin by drawing a ratio box.

a. **Explain** The bicycle store has mountain bikes and racing bikes on the showroom floor at a ratio of 3 to 5. If there are a total of 72 bicycles on the showroom floor, how can you find the number of racing bikes? Use your method to find the answer.

b. The Recycling Club collects cans and bottles for recycling. The ratio of cans to bottles collected was 8 to 9. If they collected 240 cans, how many items to be recycled did they collect in all?

c. The ratio of big fish to little fish in the pond was 4 to 11. If there were 1320 fish in the pond, how many big fish were there?

Written Practice

Strengthening Concepts

1. *(38, 60)* Use the circle graph to answer **a** and **b**.

a. What percent of Mr. Gains's income was spent on items other than food and housing?

b. If his income was $25,000, how much did he spend on food?

How Mr. Gains Spent His Income

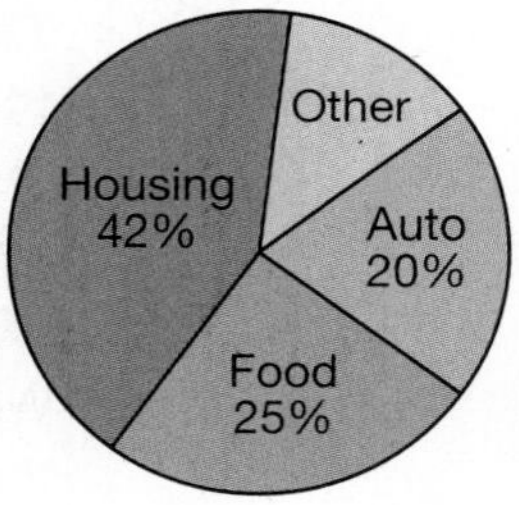

2. *(28)* It is $1\frac{1}{4}$ miles from Ahmad's house to school. How far does Ahmad travel in 5 days walking to school and back?

3. *(35)* When the sum of 1.9 and 2.2 is subtracted from the product of 1.9 and 2.2, what is the difference?

*** 4.** *(66)* **Represent** Use a ratio box to solve this problem. There were a total of 520 dimes and quarters in the bottled water vending machine.

a. If the ratio of dimes to quarters was 5 to 8, how many dimes were there?

b. Use your answer to **a** to find the total amount of money in the machine.

5. *(51)* Saturn's average distance from the Sun is about 900 million miles. Write that distance in scientific notation.

6. (22, 48) Read this statement. Then answer the questions that follow.

Three tenths of the 400 acres were planted with alfalfa.

a. What percent of the land was planted with alfalfa?

b. How many of the 400 acres were not planted with alfalfa?

7. (14, 48) Twelve of 30 students printed their name at the top of their homework sheet. The rest wrote their names in cursive.

a. What fraction of the students printed their names?

b. What percent of the students printed their names?

c. If one of the homework sheets is selected at random, which is more likely: the name at the top is printed or the name at the top is in cursive? Defend your answer.

*** 8.** (65) **Verify** Find the circumference of each circle:

a.

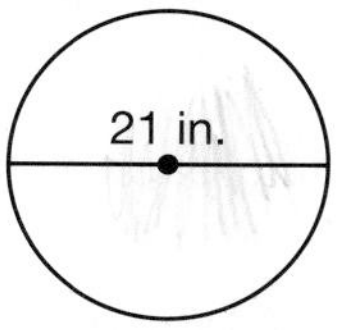

Use 3.14 for π.

b.

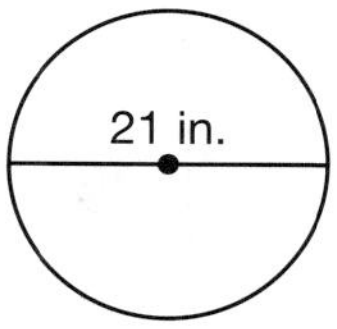

Use $\frac{22}{7}$ for π.

*** 9.** (37, 61) **Analyze** Refer to the figure at right to answer **a–c.** Dimensions are in centimeters.

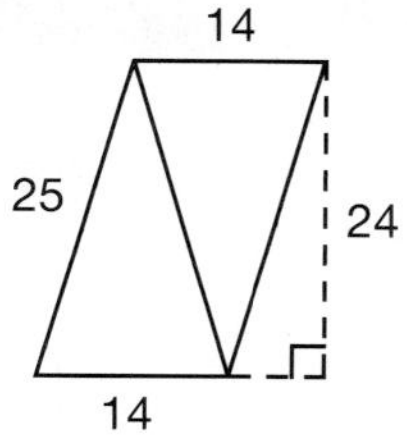

a. What is the area of the parallelogram?

b. The two triangles are congruent. What is the area of one of the triangles?

c. Each triangle is isosceles. What is the perimeter of one of the triangles?

10. (51) Write 32.5 billion in scientific notation.

Write equations to solve problems **11** and **12.**

11. (60) What number is 90 percent of 3500?

12. (60) What number is $\frac{5}{6}$ of $2\frac{2}{5}$?

13. (48) Complete the table.

Fraction	Decimal	Percent
a.	0.45	**b.**
c.	**d.**	7.5% or $7\frac{1}{2}$%

*** 14.** (64) **Generalize** Find each sum:

a. (5) + (−4) + (6) + (−1)

b. 3 + (−5) + (+4) + (−2)

c. (−0.3) + (−0.5)

15. (50) Use a unit multiplier to convert 1.4 kilograms to grams.

* **16.** (40, 61) **Conclude** Refer to the figure at right to answer **a–f.** In the figure, two sides of a parallelogram are extended to form two sides of a right triangle. The measure of ∠*M* is 35°. Find the measure of these angles.

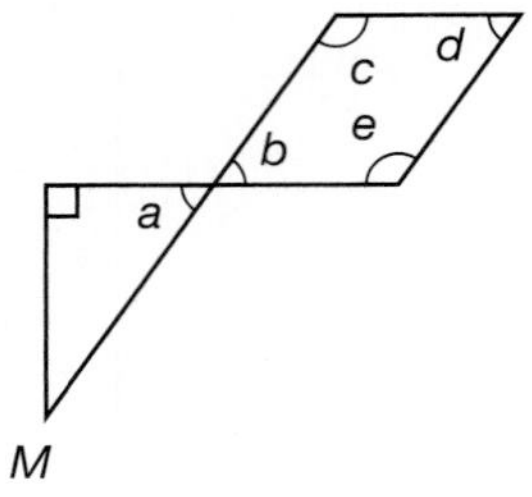

a. ∠*a*

b. ∠*b*

c. ∠*c*

d. ∠*d*

e. ∠*e*

f. **Explain** What did you have to know about a right triangle in order to solve **a?** What did you have to know about a parallelogram in order to solve **c?**

17. (29) Estimate this product by rounding each number to one nonzero digit before multiplying:

$$(2876)(513)(18)$$

18. (Inv. 3) The coordinates of the vertices of square *ABCD* are (2, 2), (2, −2), (−2, −2), and (−2, 2). The coordinates of the vertices of square *EFGH* are (2, 0), (0, −2), (−2, 0), and (0, 2). Draw both squares on the same coordinate plane and answer **a–d.**

a. What is the area of square *ABCD?*

b. What is the length of one side of square *ABCD?*

c. Counting two half squares on the grid as one square unit, what is the area of square *EFGH?*

d. Remembering that the length of the side of a square is the square root of its area, what is the length of one side of square *EFGH?*

Solve:

19. (39) $\frac{0.9}{1.5} = \frac{12}{n}$

20. (30) $\frac{11}{24} + w = \frac{11}{12}$

Generalize Simplify:

21. (57) $2^1 - 2^0 - 2^{-1}$

22. (49)
$$\begin{array}{r} 4 \text{ lb } 12 \text{ oz} \\ + \ 1 \text{ lb } \ 7 \text{ oz} \\ \hline \end{array}$$

23. (50) $\frac{3 \text{ ft}}{1 \text{ yd}} \cdot \frac{12 \text{ in.}}{1 \text{ ft}}$

24. (45) $16 \div (0.8 \div 0.04)$

* **25.** (63) $0.4[0.5 - (0.6)(0.7)]$

26. (26) $\frac{3}{8} \cdot 1\frac{2}{3} \cdot 4 \div 1\frac{2}{3}$

* **27.** (63) $30 - 5[4 + (3)(2) - 5]$

28. (13) Write a word problem for this division: $2.88 ÷ 12

29. (3) Two identical boxes balance a 9-ounce weight. What is the weight of each box?

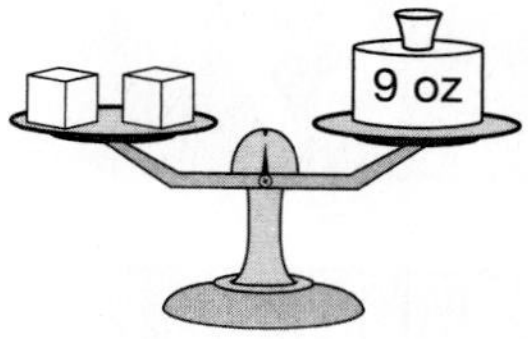

*** 30.** (Inv. 2, 62) **Analyze** Refer to the circle with center at point M to answer **a–c.**

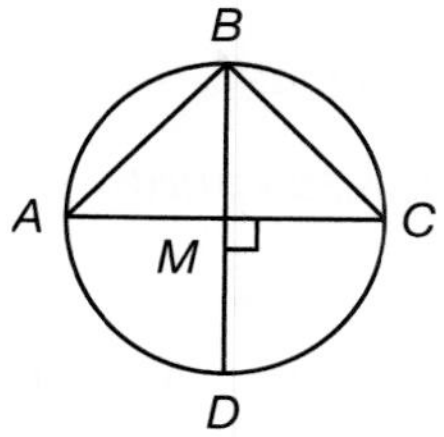

a. Name two chords that are not diameters.

b. Classify $\triangle AMB$ by its sides.

c. What is the measure of inscribed angle ABC?

Early Finishers
Real-World Application

On the first day of the month the balance in Ronda's checking account is $500. Every day for the next fourteen days, she withdraws $30. On the fifteenth day Ronda deposits two $712 checks. Over the next three days she makes three withdrawals that total $65.

a. Write an expression that represents the total amount in deposits. Write another expression that represents the total amount in withdrawals.

b. Write an expression that represents the change from the deposits and withdrawals to the original balance. What is the balance in Ronda's account after all the transactions?

LESSON 67 • Geometric Solids

Power Up

Building Power

facts Power Up M

mental math

a. **Decimals:** $43.6 - 10$

b. **Decimals/Exponents:** 3.85×10^3

c. **Ratio:** $\frac{5}{10} = \frac{2.5}{m}$

d. **Measurement:** Convert 20 m to decimeters (dm).

e. **Exponents:** $10^3 \div 10$

f. **Percent:** 75% of 24

g. **Measurement:** 4 pints is what fraction of a gallon?

h. **Mental Math:** A mental calculation technique for multiplying is to double one factor and halve the other factor. The product is the same. Use this technique to multiply 45 and 16.

$$\begin{array}{r} 35 \\ \times\ 14 \\ \hline 490 \end{array} \quad \begin{array}{c} \xrightarrow{\times 2} \\ \xrightarrow{\div 2} \\ \ \end{array} \quad \begin{array}{r} 70 \\ \times\ 7 \\ \hline 490 \end{array}$$

problem solving

Paul reads 4 pages in 3 minutes, and Art reads 3 pages in 4 minutes. If they both begin reading 120-page books at the same time, then Paul will finish how many minutes before Art?

New Concept

Increasing Knowledge

Geometric solids are shapes that take up space. Below we show a few geometric solids.

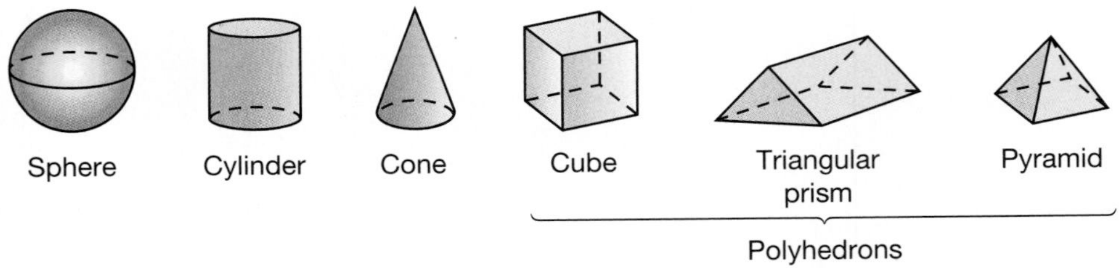

Some geometric solids, such as spheres, cylinders, and cones, have one or more curved surfaces. If a solid has only flat surfaces that are polygons, the solid is called a **polyhedron.** Cubes, triangular prisms, and pyramids are examples of polyhedrons.

When describing a polyhedron, we may refer to its faces, edges, or vertices. A **face** is one of the flat surfaces. An **edge** is formed where two faces meet. A **vertex** (plural, **vertices**) is formed where three or more edges meet.

Thinking Skill

Analyze

How many faces does a cube have? How many edges? How many vertices?

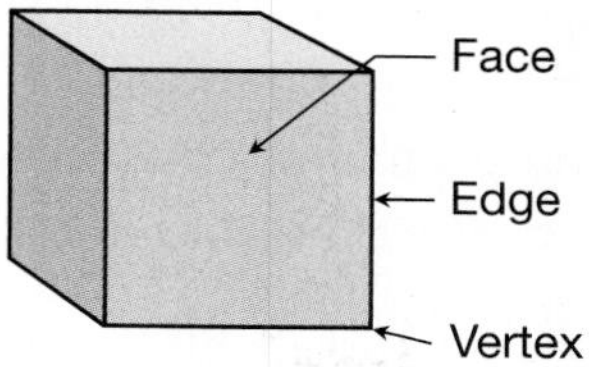

A **prism** is a special kind of polyhedron. A prism has a polygon of a constant size "running through" the prism that appears at opposite faces of the prism and determines the name of the prism. For example, the opposite faces of this prism are congruent triangles; thus this prism is called a **triangular prism.**

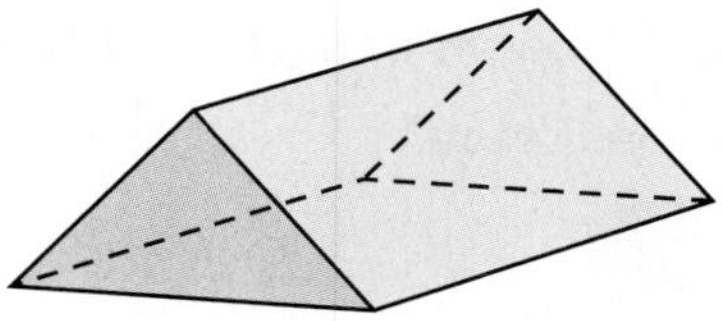

Notice that if we cut through this triangular prism perpendicular to the base, we would see the same size triangle at the cut.

To draw a prism, we draw two identical and parallel polygons, as shown below. Then we draw segments connecting corresponding vertices. We use dashes to indicate edges hidden from view.

Rectangular prism: We draw two congruent rectangles. Then we connect the corresponding vertices (using dashes for hidden edges).

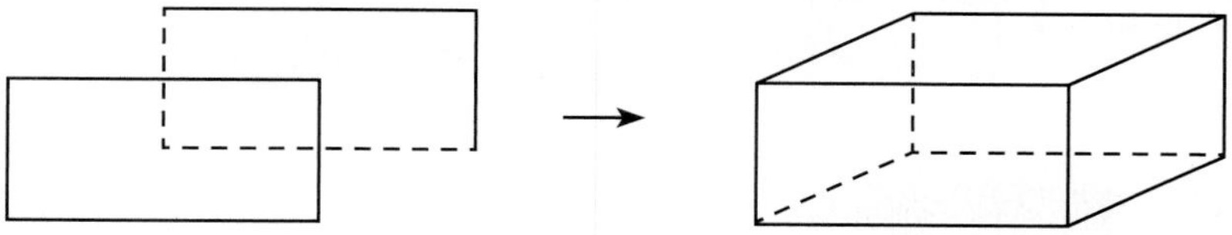

Triangular prism: We draw two congruent triangles. Then we connect corresponding vertices.

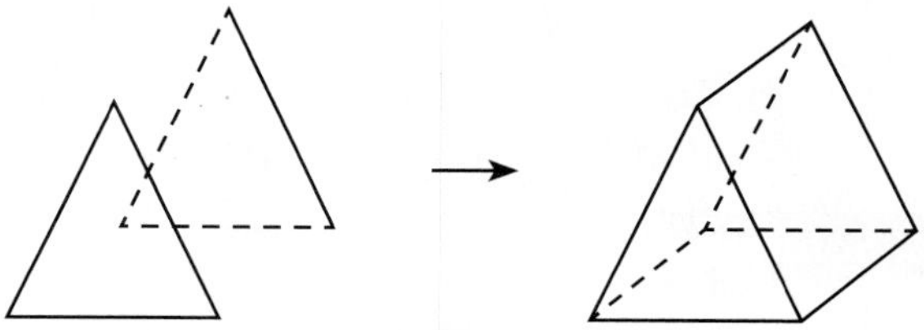

Geometric solids have three dimensions: length, width, and height (or depth). Each dimension is perpendicular to the others.

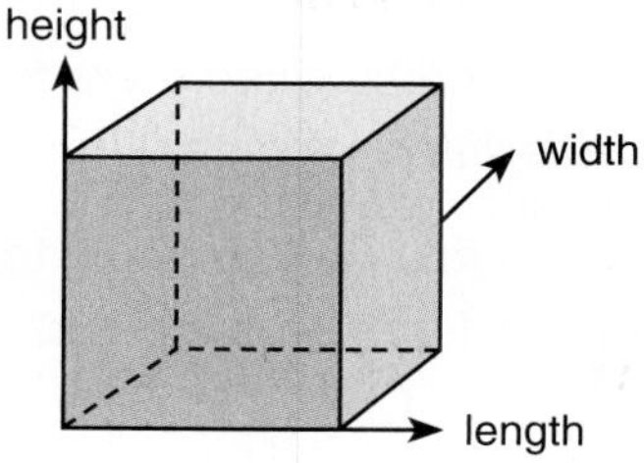

If a two-dimensional figure like a circle or polygon is moved through the third dimension, it sweeps out a three-dimensional figure. For example, if we press a quarter into soft clay, the circular shape of the quarter carves a hole the shape of a cylinder.

Example 1

Imagine pressing a flat rectangular object into clay. What would be the shape of the hole that is formed?

Solution

As the rectangular object is pressed into the clay, it would carve a hole the shape of a **rectangular prism**.

Example 2

Use the name of a geometric solid to describe the shape of each object:

a. basketball **b. can of beans** **c. shoe box**

Solution

a. sphere **b. cylinder** **c. rectangular prism**

Example 3

A cube has how many:

a. faces **b. edges** **c. vertices**

Solution

a. 6 faces **b. 12 edges** **c. 8 vertices**

Example 4

Draw a cube and answer the following questions.

a. Look at one edge of the cube. How many edges are perpendicular to that edge and how many are parallel to it?

b. Look at one face of the cube. How many faces are perpendicular to that face and how many faces are parallel to it?

Solution

A cube is a special kind of rectangular prism. All faces are squares.

a. For each edge there are **four perpendicular edges and three parallel edges.**

b. For each face there are **four perpendicular faces and one parallel face.**

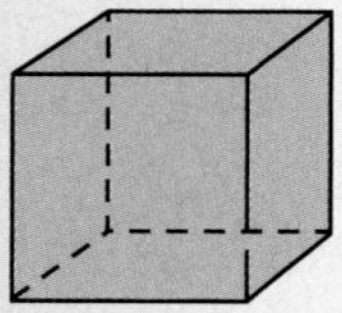

Connect How do you know that all cubes are rectangular prisms, but not all rectangular prisms are cubes?

Workers involved in the manufacturing of packaging materials make boxes and other containers out of flat sheets of cardboard or sheet metal. If we cut apart a cereal box and unfold it, we see the six rectangles that form the faces of the box.

Thinking Skill

Connect

The unfolded box is a 2-dimensional representation of a 3-dimensional figure. This type of representation is sometimes called a *net.*

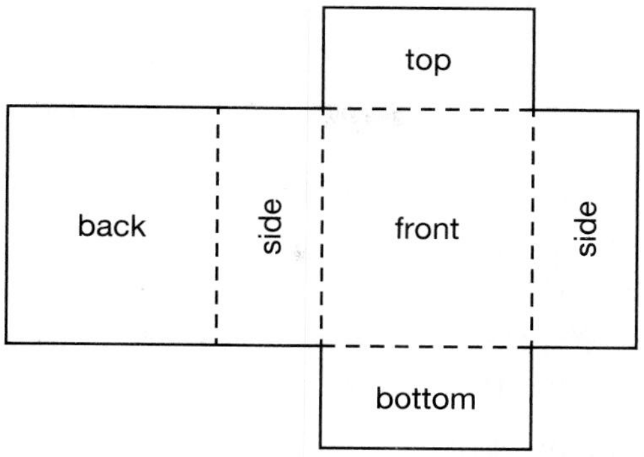

If we find the area of each rectangle and add those areas, we can calculate the **surface area** of the cereal box.

Example 5

Which of these patterns will not fold to form a cube?

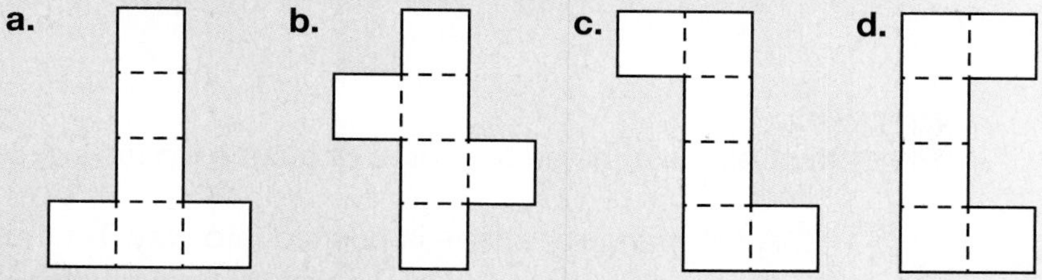

Solution

Pattern d will not fold into a cube.

Example 6

If each edge of a cube is 5 cm, what is the surface area (the combined area of all of the faces) of the cube?

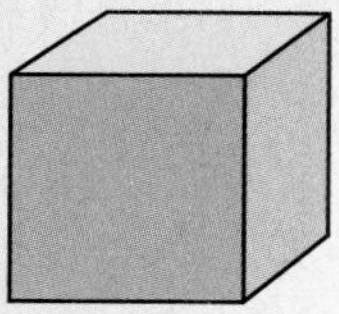

5 cm

Solution

A cube has six congruent square faces. Each face of this cube is 5 cm by 5 cm. So the area of one face is 25 cm^2, and the area of all six faces is

$$6 \times 25 \text{ cm}^2 = \mathbf{150 \text{ cm}^2}$$

Infer If the surface area of a cube is 54 in.^{2}, how can we find the dimensions of the cube?

Practice Set

Classify Use the name of a geometric solid to describe each shape:

a.

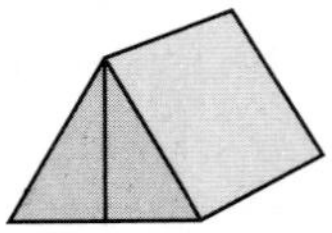

Tent

b.

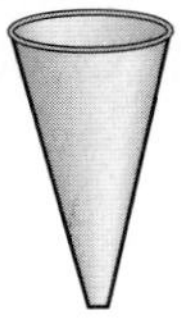

Funnel

c.

Box

A triangular prism has how many of each of the following?

d. Faces **e.** Edges **f.** Vertices

Model Draw a representation of each shape. (Refer to the representations at the beginning of this lesson.)

g. Sphere **h.** Rectangular prism

i. Cylinder

j. **Predict** What three dimensional figure could be formed by folding this pattern?

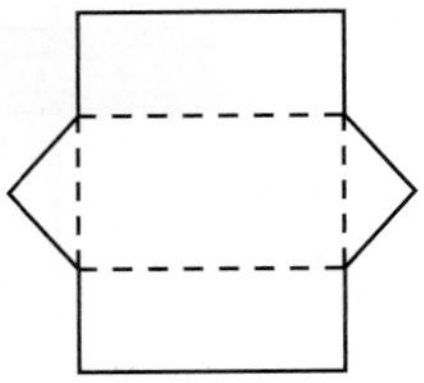

k. Calculate the surface area of a cube whose edges are 3 cm long.

l. A flat triangular shape is pushed into clay. The hole that is formed is filled with plaster. When the plaster hardens the clay is removed. What is the shape of the hardened plaster?

m. **Conclude** If we refer to the faces of a rectangular solid as front, back, top, bottom, left, and right, then which face is parallel to the top face?

Written Practice *Strengthening Concepts*

1. *(14)* The bag contains 20 red marbles, 30 white marbles, and 40 blue marbles.

a. What is the ratio of red to blue marbles?

b. What is the ratio of white to red marbles?

c. If one marble is drawn from the bag, what is the probability that the marble will not be white?

2. *(30)* When the product of $\frac{1}{3}$ and $\frac{1}{2}$ is subtracted from the sum of $\frac{1}{3}$ and $\frac{1}{2}$, what is the difference?

3. *(11, 23)* With the baby in his arms, Mr. Greer weighed 180 pounds. Without the baby, he weighed $165\frac{1}{2}$ pounds. How much did the baby weigh?

*** 4.** (28, 55) Jerome read 42 pages of a book each day for five days. On the next 3 days he read 44 pages, 35 pages, and 35 pages.

a. What was the average number of pages he read on the last 3 days?

b. What was the average number of pages he read for all 8 days?

*** 5.** (66) **Represent** Use a ratio box to solve this problem. A collection of sedimentary rocks contained sandstone and shale in the ratio of 5 to 2. If there were 210 rocks in the collection, how many were sandstone?

6. (22, 48) Read this statement. Then answer the questions that follow.

The market sold four-fifths of the 360 frozen turkeys during the second week of November.

a. How many of the turkeys were sold during the second week of November?

b. What percent of the turkeys were not sold during the second week of November?

*** 7.** (67) **Analyze** The three-dimensional figure that can be formed by folding this pattern has how many

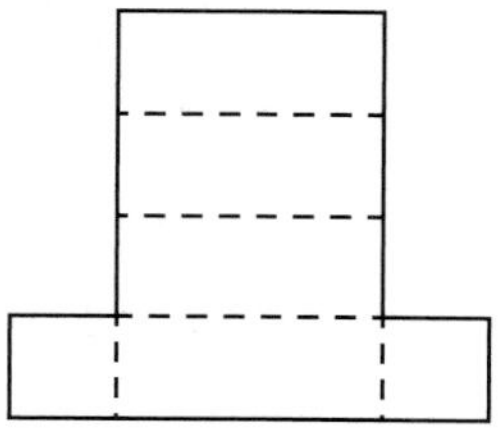

a. edges?

b. faces?

c. vertices?

*** 8.** (58, 62) **Analyze** Refer to the triangles below to answer **a–d.** Dimensions are in meters.

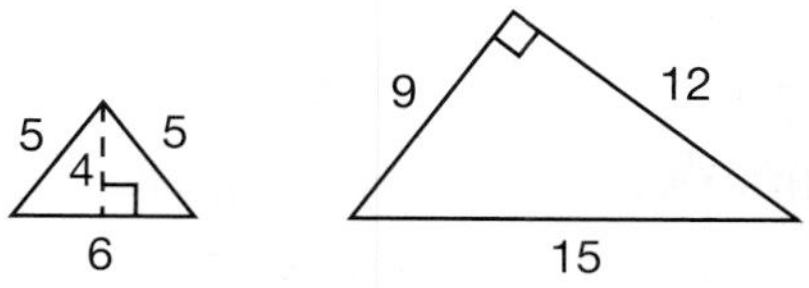

a. What is the area of the scalene triangle?

b. What is the perimeter of the isosceles triangle?

c. If one acute angle of the right triangle measures 37°, then the other acute angle measures how many degrees?

d. Which of the two triangles is not symmetrical?

9. (28, 34) What is the average of the two numbers marked by arrows on the number line below?

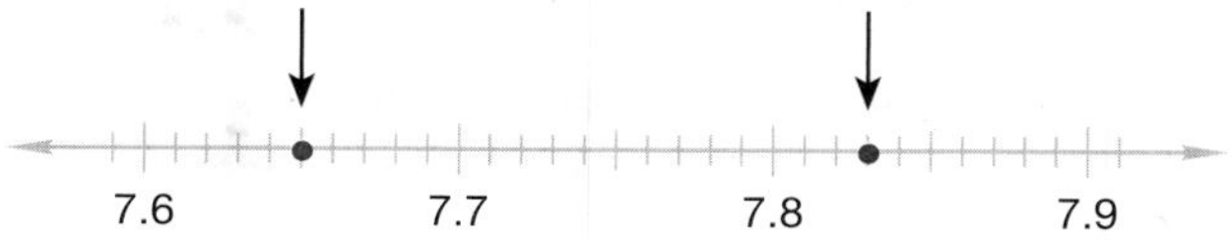

10. Write twenty-five ten-thousandths in scientific notation.
(57)

Write equations to solve problems **11** and **12.**

11. What number is 24 percent of 75?
(60)

12. What number is 120% of 12?
(60)

13. Find each sum:
(64)

a. $\left(-\frac{1}{4}\right) + \left(+\frac{1}{4}\right)$

b. $(+2) + (-3) + (+4)$

14. Complete the table.
(48)

Fraction	Decimal	Percent
a.	**b.**	4%
$\frac{7}{8}$	**c.**	**d.**

15. Use a unit multiplier to convert 700 mm to cm.
(50)

* **16.** **Analyze** In three separate stock trades Dale lost $560, gained $850, and lost $280. What was the net result of the three trades?
(64)

* **17.** **Formulate** Describe the rule of the function in words and as an equation. Then find the missing number.
(56)

Input *x*	Output *y*
7	49
0	0
11	77
1	

18. Round 7856.427
(33)

a. to the nearest hundredth.

b. to the nearest hundred.

* **19.** **Generalize** The diameter of Debby's bicycle tire is 24 inches. What is the circumference of the tire to the nearest inch?
(65)

20. Consider angles *A, B, C,* and *D* below.
(40)

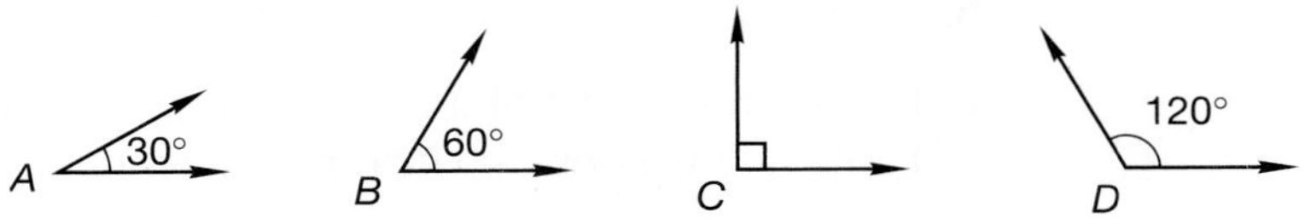

a. Which two angles are complementary?

b. Which two angles are supplementary?

21. **a.** Show two ways to simplify 2(5 ft + 3 ft).
(41)

b. Which property is illustrated in **a?**

22. Solve: $\frac{2.5}{w} = \frac{15}{12}$
(39)

Generalize Simplify:

* **23.** $9 + 8\{7 \cdot 6 - 5[4 + (3 - 2 \cdot 1)]\}$
(63)

* **24.** 1 yd − 1 ft 3 in.
(49)

25. $6.4 - (0.6 - 0.04)$
(35)

26. $\dfrac{3 + 0.6}{(3)(0.6)}$
(52)

27. $1\frac{2}{3} + 3\frac{1}{4} - 1\frac{5}{6}$
(30)

* **28.** $\frac{3}{5} \div 3\frac{1}{5} \cdot 5\frac{1}{3} \cdot |-1|$
(26, 59)

29. $3\frac{3}{4} \div \left(3 \div 1\frac{2}{3}\right)$
(26)

30. $5^2 - \sqrt{4^2} + 2^{-2}$
(52, 57)

Early Finishers

Real-World Application

The top three players on the basketball team scored a total of 1216 points this season in the ratio of 5 to 2 to 1.

a. Below is a ratio box for these ratios. How many points did each player score during the season?

Ratio	Points
5	
2	
1	
8 (total)	1216

b. Find the percents of the total that each player scored. Verify your work by showing that the percentages you find total 100%.

LESSON 68

• Algebraic Addition

Power Up

Building Power

facts Power Up N

mental math

a. Decimals: 0.75 + 0.5

b. Power/Roots: $\sqrt{1} - \left(\frac{1}{2}\right)^2$

c. Algebra: $4w - 1 = 35$

d. Mental Math: 12 × 2.5 (halve, double)

e. Measurement: 20 dm to cm

f. Percent: $33\frac{1}{3}\%$ of 24

g. Geometry: A circle has a diameter of 20 mm. What is the circumference of the circle?

h. Geometry: Find the perimeter and area of a rectangle that is 2 m long and 1.5 m wide.

problem solving

If you have five different colors of paint, how many different ways can you paint a clown mask if you make the hair, nose, and mouth each a different color?

New Concept

Increasing Knowledge

Math Language

A positive number and a negative number whose absolute values are equal are **opposites.**

Recall that the graphs of −3 and 3 are the same distance from zero on the number line. The graphs are on the opposite sides of zero.

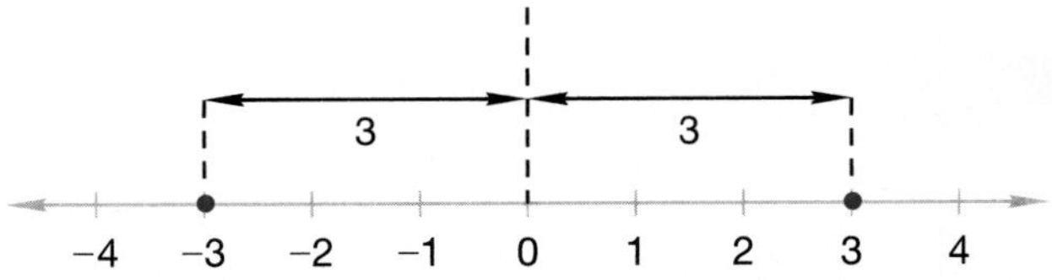

This is why we say that 3 and −3 are the opposites of each other.

3 is the opposite of −3

−3 is the opposite of 3

We can read −3 as "the opposite of 3." Furthermore, −(−3) can be read as "the opposite of the opposite of 3." This means that −(−3) is another way to write 3.

There are two ways to simplify the expression 7 − 3. The first way is to let the minus sign signify subtraction. When we subtract 3 from 7, the answer is 4.

$$7 - 3 = 4$$

The second way is to use the thought process of **algebraic addition.** To use algebraic addition, we let the minus sign mean that -3 is a negative number and we treat the problem as an addition problem.

$$7 + (-3) = 4$$

Notice that we get the same answer both ways. The only difference is in the way we think about the problem.

We can also use algebraic addition to simplify this expression:

$$7 - (-3)$$

We use an addition thought and think that 7 is added to $-(-3)$. This is what we think:

$$7 + [-(-3)]$$

But the opposite of -3 is 3, so we can write

$$7 + [3] = 10$$

We will practice using the thought process of algebraic addition because algebraic addition can be used to simplify expressions that would be very difficult to simplify if we used the thought process of subtraction.

Example 1

Thinking Skill

Generalize

What is the sum of $+7$ and -7?

Simplify: $-3 - (-2)$

Solution

We think addition. We think we are to *add* -3 and $-(-2)$. This is what we think:

$$(-3) + [-(-2)]$$

The opposite of -2 is 2 itself. So we have

$$(-3) + [2] = \mathbf{-1}$$

Example 2

Simplify: $-(-2) - 5 - (+6)$

Solution

We see three numbers. We think *addition,* so we have

$$[-(-2)] + (-5) + [-(+6)]$$

We simplify the first and third numbers and get

$$[+2] + (-5) + [-6] = \mathbf{-9}$$

Note that this time we write 2 as $+2$. Either 2 or $+2$ may be used.

Example 3

Simplify: (−1.2) − (+1.5)

Solution

We think addition.

$$(-1.2) + [-(+1.50)]$$

The opposite of positive 1.5 is negative 1.5.

$$(-1.2) + (-1.5) = \mathbf{-2.7}$$

Practice Set

Generalize Use algebraic addition to simplify each expression.

a. $(-3) - (+2)$

b. $(-3) - (-2)$

c. $(+3) - (2)$

d. $(-3) - (+2) - (-4)$

e. $(-8) + (-3) - (+2)$

f. $(-8) - (+3) + (-2)$

g. $\left(-\frac{3}{5}\right) - \left(-\frac{1}{5}\right)$

h. $(-0.2) - (+0.3)$

i. Which is greater, $3 + (-6)$ or $3 - (-6)$? Explain.

Written Practice

Strengthening Concepts

1. (11) The combined mass of the beaker and the liquid was 1037 g. The mass of the empty beaker was 350 g. What was the mass of the liquid?

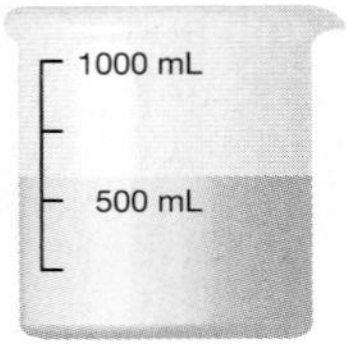

2. (53) Use a ratio box to solve this problem. Adriana's soccer ball is covered with a pattern of pentagons and hexagons in the ratio of 3 to 5. If there are 12 pentagons, how many hexagons are in the pattern?

3. (25, 30) When the sum of $\frac{1}{4}$ and $\frac{1}{2}$ is divided by the product of $\frac{1}{4}$ and $\frac{1}{2}$, what is the quotient?

4. (46) Pens were on sale 4 for \$1.24.

a. What was the price per pen?

b. How much would 100 pens cost?

5. (46) Christy rode her bike 60 miles in 5 hours.

a. What was her average speed in miles per hour?

b. What was the average number of minutes it took to ride each mile?

6. (32, 54) Sound travels through air at about $\frac{1}{3}$ of a kilometer per second. If thunder is heard 6 seconds after a flash of lightning, about how many kilometers away was the lightning?

7. Mr. Chen had the following golf scores:
(Inv. 4)

79, 81, 84, 88, 100, 88, 82

a. Which score was made most often?

b. What is the median of the scores?

c. What is the mean of the scores?

d. Choose which measure best describes the set of data. Justify your choice.

8. What is the average of the two numbers marked by arrows on the number line below?
(28, 34)

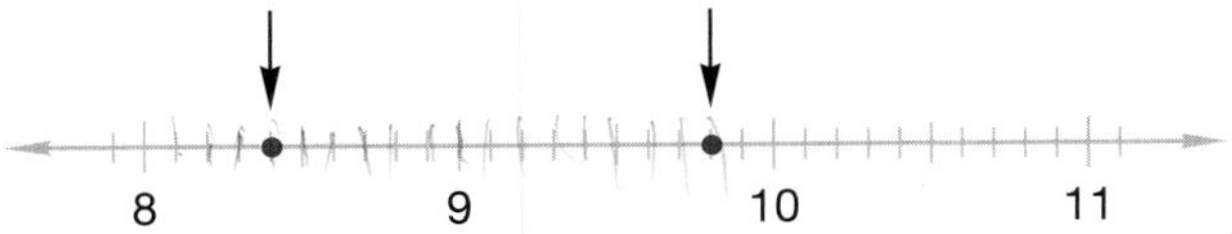

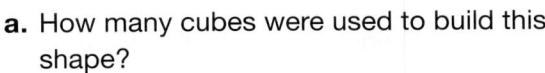

*** 9.** This rectangular shape is two cubes tall and two cubes deep.
(67)

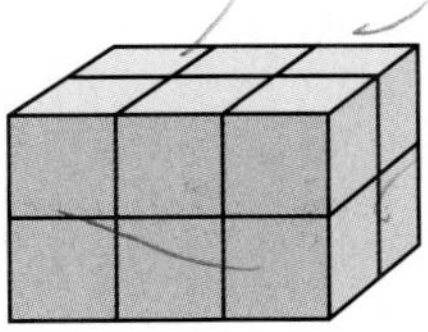

a. How many cubes were used to build this shape?

b. What is the name of this shape?

*** 10.** **Generalize** Find the circumference of each circle:
(65)

a.

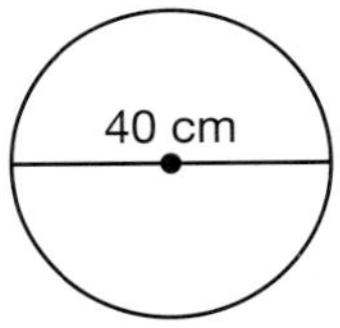

Use 3.14 for π.

b.

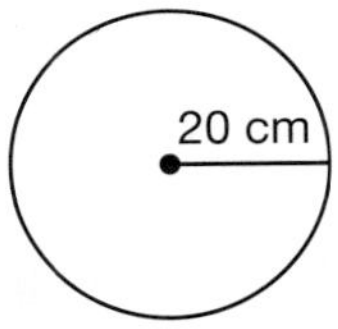

Leave π as π.

*** 11.** **Analyze** The coordinates of the vertices of △*ABC* are *A* (1, −1), *B* (−3, −1), and *C* (1, 3). Draw the triangle and answer these questions:
(58, 62)

a. What type of triangle is △*ABC* classified by angles?

b. What type of triangle is △*ABC* classified by sides?

c. Triangle *ABC*'s one line of symmetry passes through which vertex?

d. What is the measure of ∠*B*?

e. What is the area of △*ABC*?

12. Multiply twenty thousand by thirty thousand, and write the product in scientific notation.
(51)

13. What number is 75 percent of 400?
(60)

*** 14.** Simplify:
(68)

a. (−4) − (−6)

b. (−4) − (+6)

c. (−6) − (−4)

d. (+6) − (−4)

* **15.** (67) **Generalize** Find the surface area of a cube that has edges 4 inches long.

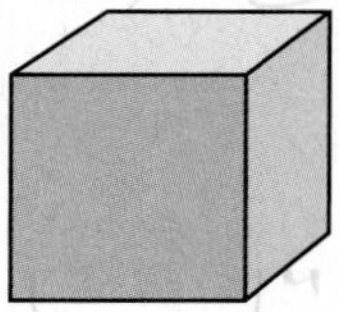

* **16.** (48) **Connect** Complete the table.

Fraction	Decimal	Percent
$\frac{3}{25}$	a.	b.
c.	d.	120%

17. (52) Evaluate: $x^2 + 2xy + y^2$ if $x = 4$ and $y = 5$

* **18.** (67) **Classify** Use the name of a geometric solid to describe each object:

a. 

b. 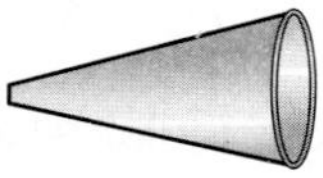

c. 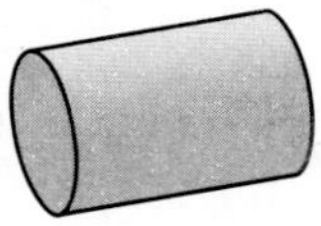

* **19.** (67) **Predict** A triangular piece of metal spins around a rod. As it spins, its path through space is shaped like which geometric solid in Problem 18?

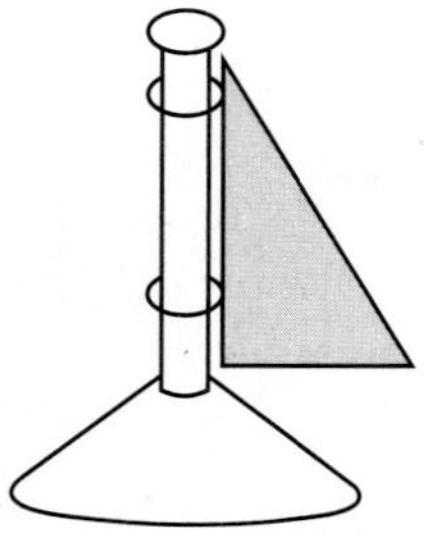

20. (40, 61) In this figure parallelogram *ABCD* is divided by a diagonal into two congruent triangles. Angle *DCA* and $\angle BAC$ have equal measures and are complementary. Find the measure of

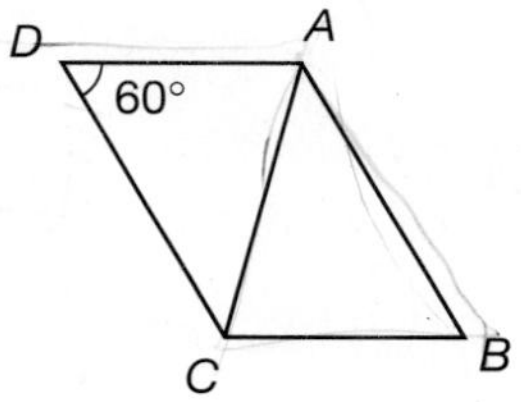

a. $\angle DCA$. **b.** $\angle DAC$.

c. $\angle CAB$. **d.** $\angle ABC$.

e. $\angle BCA$. **f.** $\angle BCD$.

21. (35) Write a word problem for this division: \$3.00 ÷ \$0.25

Solve:

22. (39) $\frac{4}{c} = \frac{3}{7\frac{1}{2}}$

23. (35) $(1.5)^2 = 15w$

Simplify:

24. (49) 1 gal − 1 qt 1 pt 1 oz

25. (45) $16 \div (0.04 \div 0.8)$

* **26.** (63) $10 - [0.1 - (0.01)(0.1)]$

27. (30, 52) $\frac{5}{8} + \frac{2}{3} \cdot \frac{3}{4} - \frac{3}{4}$

28. (26) $4\frac{1}{2} \cdot 3\frac{3}{4} \div 1\frac{2}{3}$

29. (52) $\sqrt{5^2 - 2^4}$

* **30.** (63) **Generalize** $3 + 6[10 - (3 \cdot 4 - 5)]$

LESSON
69

• Proper Form of Scientific Notation

Power Up

Building Power

facts Power Up M

mental math

a. Decimals: $4 - 1.5$

b. Exponents: 75×10^{-3}

c. Ratio: $\frac{x}{4} = \frac{1.5}{3}$

d. Mental Math: 18×35 (halve, double)

e. Measurement: 20 cm to dm

f. Percent: $66\frac{2}{3}\%$ of 24

g. Primes/Composites: Name the first 5 prime numbers.

h. Calculation: 5^2, $\times 3$, $- 3$, $\div 8$, $\sqrt{}$, $\times 7$, $- 1$, $\div 4$, $\times 10$, $- 1$, $\sqrt{}$, $\div 2$

problem solving

A rectangle has a length of 10 meters and a width of 8 meters. A second rectangle has a length of 6 meters and a width of 4 meters. The rectangles overlap as shown. What is the difference between the areas of the two non-overlapping regions of the two rectangles?

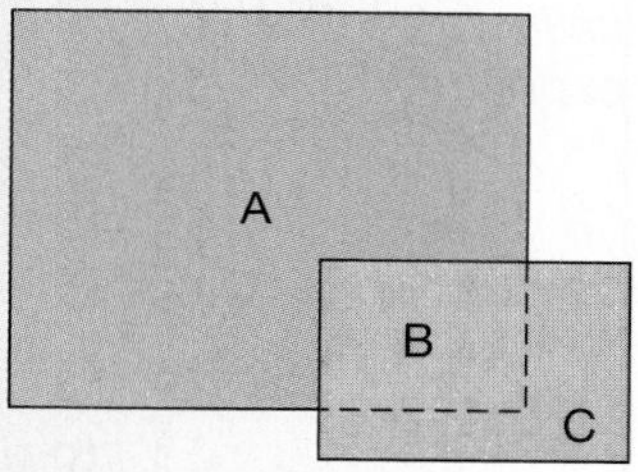

New Concept

Increasing Knowledge

When we write a number in scientific notation, we usually put the decimal point just to the right of the first digit that is not zero. To write

$$4600 \times 10^5$$

in scientific notation, we use two steps. First we write 4600 in scientific notation. In place of 4600 we write 4.6×10^3. Now we have

$$4.6 \times 10^3 \times 10^5$$

Thinking Skill

Discuss

By what number are we multiplying when we multiply by 10^3? By 10^5?

For the second step we change the two powers of 10 into one power of 10. We recall that 10^3 means the decimal point is 3 places to the right and that 10^5 means the decimal point is 5 places to the right. Since 3 places to the right and 5 places to the right is 8 places to the right, the power of 10 is 8.

$$4.6 \times 10^8$$

Discuss How can we check to see that 4600×10^5 is equal to 4.6×10^8?

Example 1

Write 25×10^{-5} in scientific notation.

Solution

First we write 25 in scientific notation.

$$2.5 \times 10^1 \times 10^{-5}$$

Then we combine the powers of 10 by remembering that 1 place to the right and 5 places to the left equals 4 places to the left.

$$\mathbf{2.5 \times 10^{-4}}$$

Example 2

Thinking Skill

Explain

When 0.25 is written in scientific notation why does it have a negative exponent?

Write 0.25×10^4 in scientific notation.

Solution

First we write 0.25 in scientific notation.

$$2.5 \times 10^{-1} \times 10^4$$

Since 1 place to the left and 4 places to the right equals 3 places to the right, we can write

$$\mathbf{2.5 \times 10^3}$$

With practice you will soon be able to perform these exercises mentally.

Practice Set

Connect Write each number in scientific notation. Show the steps you used to get each answer.

a. 0.16×10^6

b. 24×10^{-7}

c. 30×10^5

d. 0.75×10^{-8}

e. 14.4×10^8

f. 12.4×10^{-5}

Written Practice

Strengthening Concepts

1. (Inv. 4) The following is a list of scores received in a diving competition:

7.0	6.5	6.5	7.4	7.0	6.5	6.0

a. Which score was received the most often?

b. What is the median of the scores?

c. What is the mean of the scores?

d. What is the range of the scores?

e. Choose which measure best describes the set of data. Justify your choice.

*** 2.** *(66)* Use a ratio box to solve this problem. The team played 15 home games. The rest of the games were away games. If the team's ratio of home games to away games was 5 to 3, how many total games did the team play?

3. *(54)* Lucila swam 4 laps in 6 minutes. At that rate, how many minutes will it take Lucila to swim 10 laps?

*** 4.** *(69)* Write each number in scientific notation:

a. 15×10^5 **b.** 0.15×10^5

5. *(14, 60)* Refer to the following statement to answer **a–c:**

The survey found that only 2 out of 5 Lilliputians believe in giants.

a. According to the survey, what fraction of the Lilliputians do not believe in giants?

b. If 60 Lilliputians were selected for the survey, how many of them would believe in giants?

c. What is the probability that a randomly selected Lilliputian who participated in the survey would believe in giants?

*** 6.** *(65)* **Connect** The diameter of a circular tree stump was 40 cm. Find the circumference of the tree stump to the nearest centimeter.

*** 7.** *(67)* **Classify** Use the name of a geometric solid to describe the shape of these objects:

a. volleyball **b.** water pipe **c.** tepee

*** 8.** *(58, 62)* **a.** What is the perimeter of the equilateral triangle at right?

b. What is the measure of each of its angles?

c. **Represent** Trace the triangle on your paper, and show its lines of symmetry.

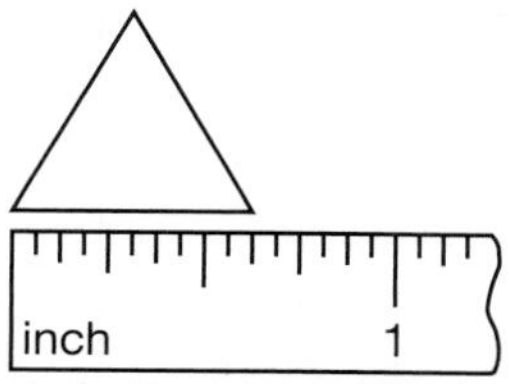

*** 9.** *(68)* **Generalize** Simplify:

a. $(-4) + (-5) - (-6)$

b. $(-2) + (-3) - (-4) - (+5)$

c. $(-0.3) - (-0.3)$

*** 10.** *(65)* Find the circumference of each circle:

a.

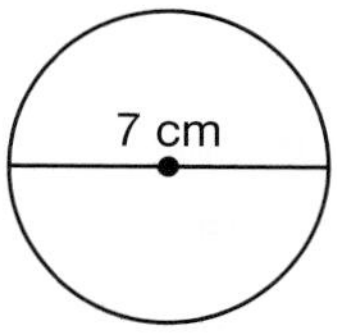

Use 3.14 for π.

b.

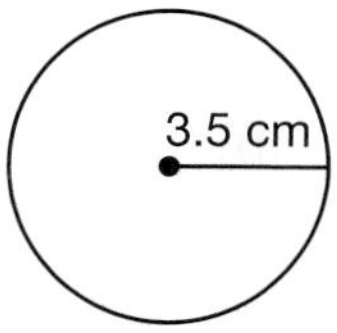

Use $\frac{22}{7}$ for π.

11. Refer to the figure to answer **a–c.**
(37)
Dimensions are in millimeters. Corners that look square are square.

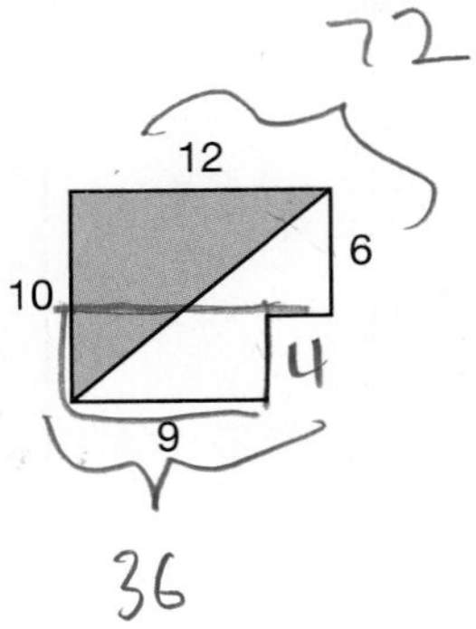

a. What is the area of the hexagon?

b. What is the area of the shaded triangle?

c. What fraction of the hexagon is shaded?

Write equations to solve problems **12** and **13.**

12. What number is 50 percent of 200?
(60)

13. What number is 250% of 4.2?
(60)

14. Complete the table.
(48)

Fraction	Decimal	Percent
$\frac{3}{20}$	**a.**	**b.**
c.	**d.**	150%

15. Refer to this figure to answer **a–c:**
(40)

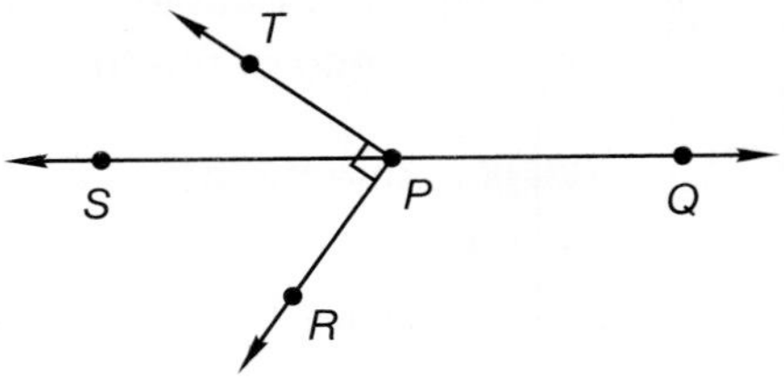

a. Which angle is supplementary to $\angle SPT$?

b. Which angle is complementary to $\angle SPT$?

c. If $\angle QPR$ measures 125°, what is the measure of $\angle QPT$?

16. Evaluate: $a^2 - \sqrt{a} + ab - a^0$ if $a = 4$ and $b = 0.5$
(52, 57)

17. Write the rule of this function with words and as an equation. Then complete the table.
(56)

Input x	Output y
8	15
6	11
10	19
4	

18. Divide 144 by 11 and write the answer
(44)

a. as a decimal with a bar over the repetend.

b. rounded to the nearest whole number.

19. (41) Anders Celsius (1701–1744) developed the Celsius scale. He used this formula to convert from degrees Celsius to degrees Fahrenheit:

$$F = 1.8\,C + 32$$

If the Celsius temperature (C) is 20°C, what is the Fahrenheit temperature (F)?

* **20.** (67) **Predict** Eva placed a quarter on its edge and gave it a spin. As it whirled, it moved through a space the shape of what geometric figure?

Solve:

21. (30) $t + \frac{5}{8} = \frac{15}{16}$

22. (39) $\frac{a}{8} = \frac{3\frac{1}{2}}{2}$

Estimate First estimate each answer to the nearest whole number. Then perform the calculation.

23. (26) $\left(3\frac{3}{4} \div 1\frac{2}{3}\right) \cdot 3$

24. (26, 30) $4\frac{1}{2} + \left(5\frac{1}{6} \div 1\frac{1}{3}\right)$

Simplify:

25. (49)
```
   5 ft 7 in.
 + 6 ft 8 in.
```

26. (50) $\frac{350 \text{ m}}{1 \text{ s}} \cdot \frac{60 \text{ s}}{1 \text{ min}} \cdot \frac{1 \text{ km}}{1000 \text{ m}}$

27. (35) $6 - (0.5 \div 4)$

28. (45) $7.50 ÷ 0.075

29. (24) **Generalize** Use prime factorization to reduce $\frac{432}{675}$.

30. (43) **a.** Convert $2\frac{1}{4}$ to a decimal and add 0.15.

b. Convert 6.5 to a mixed number and add $\frac{5}{6}$.

Early Finishers
Math Applications

Suppose you spin each of these plane figures the same way you would spin a coin. Predict the three-dimensional figures that would be formed as the two figures move through space.

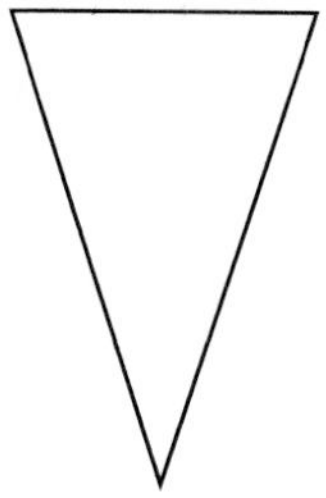

Figure A

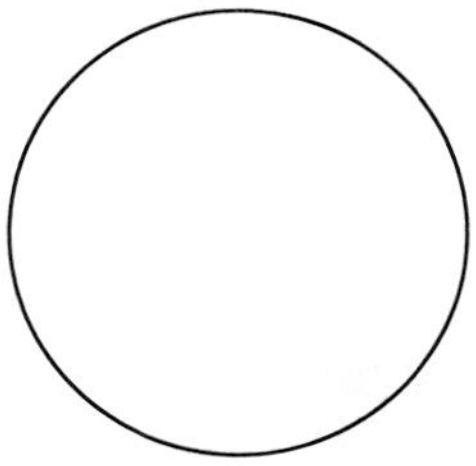

Figure B

LESSON

70

• Volume

Power Up

Building Power

facts

Power Up N

mental math

a. Decimals: $4.8 + 3 + 0.3$

b. Exponents: 25^2

c. Algebra: $5m - 3 = 27$

d. Mental Math: $\$4.80 \times 50$

e. Measurement: 20 dm to mm

f. Percent: 60% of 25

g. Primes/Composites: What is the prime factorization of 10?

h. Geometry: Find the perimeter and area of a square that has sides 0.5 m long.

problem solving

Between the whole numbers 2 and 4 is the prime number 3. Between the whole numbers 4 and 8 are the prime numbers 5 and 7. Can you always find a prime number between a number and its double?

New Concept

Increasing Knowledge

Recall from Lesson 67 that geometric solids are shapes that take up space. We use the word **volume** to describe the space occupied by a shape. To measure volume, we use units that occupy space. The units that we use to measure volume are cubes of certain sizes. We can use unit cubes to help us think of volume.

Thinking Skill

Predict

A prism has 4 layers, each containing 9 cubes. Will it have the same volume as the prism on the right? Why or why not?

Example 1

This rectangular prism was constructed of unit cubes. Its volume is how many cubes?

Solution

To find the volume of the prism, we calculate the number of cubes it contains. We see that there are 3 layers of cubes. Each layer contains 3 rows of cubes with 4 cubes in each row, or 12 cubes. Three layers with 12 cubes in each layer means that the volume of the prism is **36 cubes.**

Reading Math

Read the measure 1 cm^3 as "one cubic centimeter."

Volumes are measured by using cubes of a standard size. A cube whose edges are 1 centimeter long has a volume of 1 cubic centimeter, which we abbreviate by writing 1 cm^3.

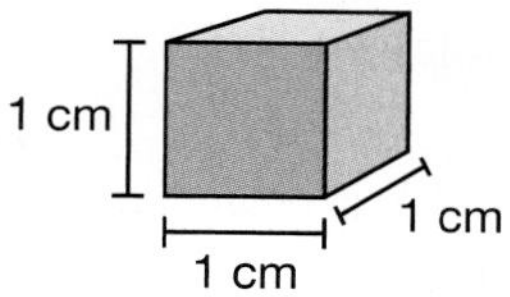

1 cubic centimeter = 1 cm^3

Similarly, if each of the edges is 1 foot long, the volume is 1 cubic foot. If each of the edges is 1 meter long, the volume is 1 cubic meter.

1 cubic foot = 1 ft^3 1 cubic meter = 1 m^3

To calculate the volume of a solid, we can imagine constructing the solid out of unit cubes of the same size. We would begin by constructing the base and then building up the layers to the specified height.

Example 2

Find the number of 1-cm cubes that can be placed inside a rectangular box with the dimensions shown.

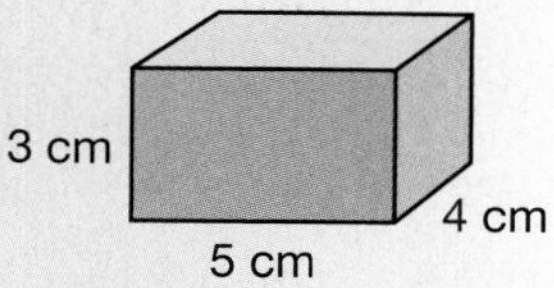

Solution

The base of the box is 5 cm by 4 cm, so we can place 5 rows of 4 cubes on the base. Thus there are 20 cubes on the first layer.

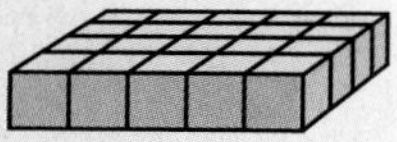

Since the box is 3 cm tall, we can fit 3 layers of cubes in the box.

$$\frac{20 \text{ cubes}}{1 \text{ }\cancel{\text{layer}}} \times 3 \text{ }\cancel{\text{layers}} = 60 \text{ cubes}$$

We find that **60 1-cm cubes** can be placed in the box.

Discuss How would we determine how many 1-cm cubes can be placed inside a rectangular box that is 4 cm high, 6 cm long, and 5 cm wide?

Example 3

What is the volume of this cube? Dimensions are in inches.

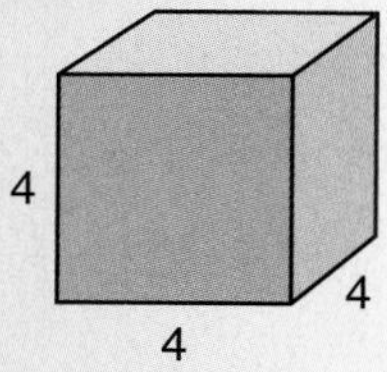

Solution

The base is 4 in. by 4 in. Thus 16 cubes can be placed on the base.

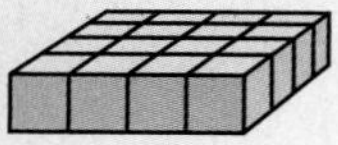

Since the big cube is 4 in. tall, there are 4 layers of small cubes.

$$\frac{16 \text{ cubes}}{1 \text{ layer}} \times 4 \text{ layers} = 64 \text{ cubes}$$

Each small cube has a volume of 1 cubic inch. Thus the volume of the big cube is **64 cubic inches (64 in.^{3}).**

We refer to the dimensions of a rectangular prism as *length, width,* and *height.*

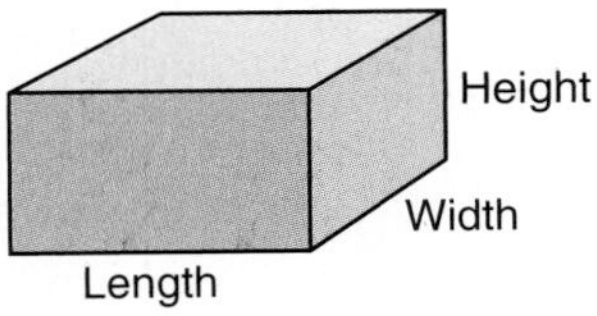

Write a formula that can be used to find the volume (V) of a rectangular prism if we know its length (l), its width (w), and its height (h).

Example 4

Find the volume of this solid.

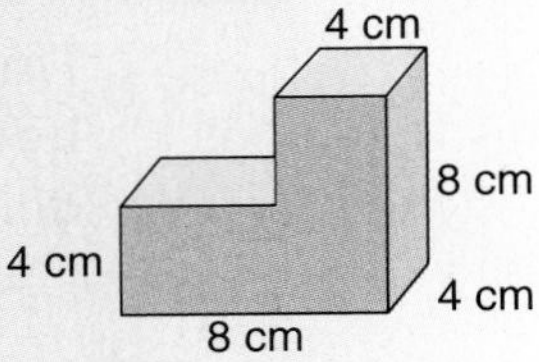

Solution

We divide the figure into two blocks and find the volume of each block.

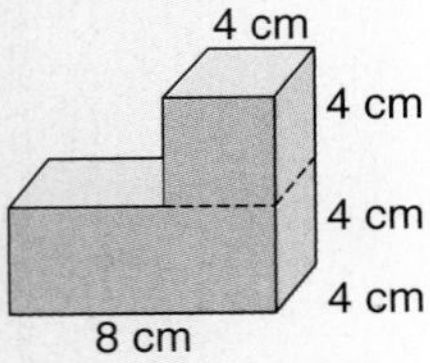

Lower:	8 cm × 4 cm × 4 cm	=	128 cm^3
+Upper:	4 cm × 4 cm × 4 cm	=	64 cm^3
	Volume of both blocks	=	**192 cm^3**

Practice Set

a. Generalize This rectangular prism was constructed of unit cubes. Its volume is how many unit cubes?

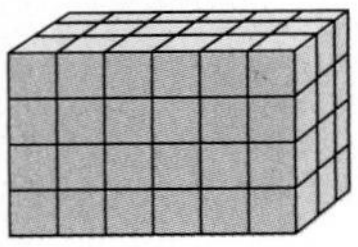

b. Find the number of 1-cm cubes that can be placed inside a box with dimensions as illustrated.

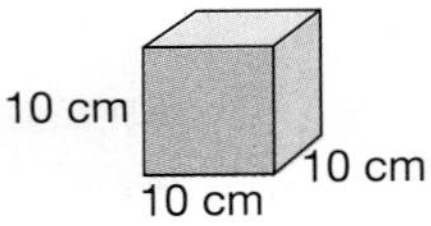

c. What is the volume of this rectangular prism? Dimensions are in feet.

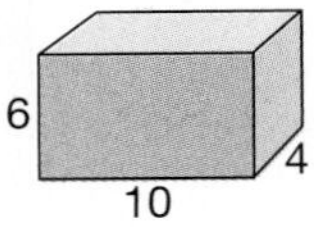

d. Estimate As a class, estimate the volume of the classroom in cubic meters. Then use a meterstick to measure the length, width, and height of the classroom to the nearest meter, and calculate the volume of the room.

e. Formulate What is the formula for the volume of a rectangular prism?

f. Generalize Find the volume of this solid.

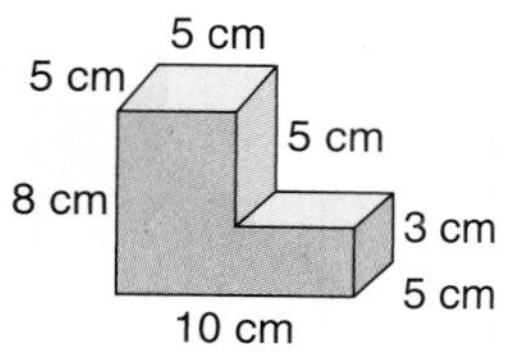

Written Practice *Strengthening Concepts*

1. (54) It was 19 kilometers from the campgrounds to the lake. Joel and Mario rode their bicycles from the campgrounds to the lake and back again. If the round trip ride took them 2 hours, what was their average speed in kilometers per hour?

2. (37, 58) The vertices of two angles of a triangle are (3, 1) and (0, −4). The *y*-axis is a line of symmetry of the triangle.

a. What are the coordinates of the third vertex of the triangle?

b. What is the area of the triangle?

* **3.** (29, 65) Verify Using a tape measure, Gretchen found that the circumference of the great oak was 600 cm. Rounding π to 3, she estimated that the tree's diameter was 200 cm. Was her estimate for the diameter a little too large or a little too small? Why?

4. (46) Grapes were priced at 3 pounds for $5.28.

a. What was the price per pound?

b. How much would 10 pounds of grapes cost?

5. (35) If the product of nine tenths and eight tenths is subtracted from the sum of seven tenths and six tenths, what is the difference?

6. (22, 48) Three fourths of the batter's 188 hits were singles.

a. How many of the batter's hits were singles?

b. What percent of the batter's hits were not singles?

7. (8) On an inch ruler, which mark is halfway between the $1\frac{1}{2}$ inch mark and the 3 inch mark?

* **8.** (70) Generalize Find the number of 1-cm cubes that can be placed in the box at right.

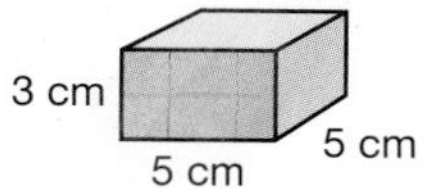

* **9.** (65) Find the circumference of each circle:

a.

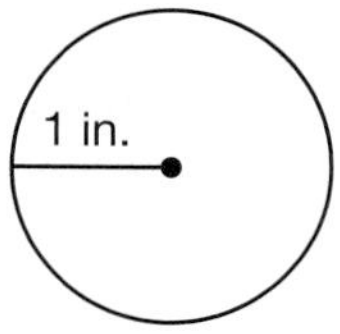

Leave π as π.

b.

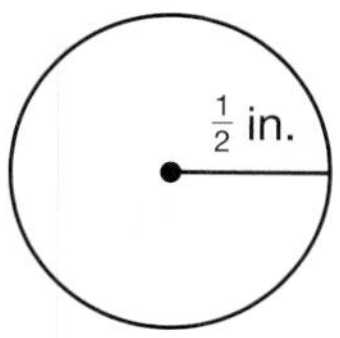

Use 3.14 for π.

* **10.** (69) Generalize Write each number in scientific notation:

a. 12×10^{-6}

b. 0.12×10^{-6}

11. What is the average of the three numbers marked by arrows on the number line below?
(28, 34)

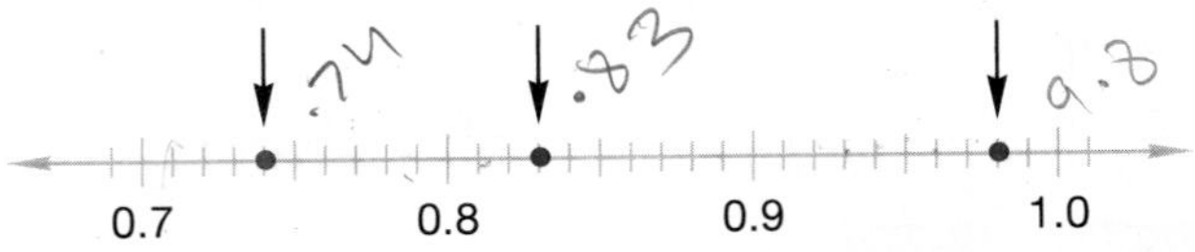

12. Use a unit multiplier to convert 1.25 kilograms to grams.
(50)

* **13.** **Generalize** Find each missing exponent:
(47)

a. $2^6 \cdot 2^3 = 2^{\square}$ **b.** $2^6 \div 2^3 = 2^{\square}$

c. $2^3 \div 2^6 = 2^{\square}$ **d.** $2^6 \div 2^6 = 2^{\square}$

14. Write an equation to solve this problem: What number is $\frac{1}{6}$ of 100?
(60)

15. Complete the table.
(48)

Fraction	Decimal	Percent
a.	b.	14%
$\frac{5}{6}$	c.	d.

* **16.** **Generalize** Simplify:
(68)

a. $(-6) - (-4) + (+2)$

b. $(-5) + (-2) - (-7) - (+9)$

c. $\left(-\frac{1}{2}\right) - \left(-\frac{1}{4}\right)$

17. Evaluate: $ab - (a - b)$ if $a = 0.4$ and $b = 0.3$
(52)

18. Round $29{,}374.\overline{65}$ to the nearest whole number.
(42)

19. Estimate the product of 6.085 and $7\frac{15}{16}$.
(29, 33)

* **20.** **Predict** What three-dimensional figure can be formed by folding this pattern? Draw the three-dimensional figure.
(67)

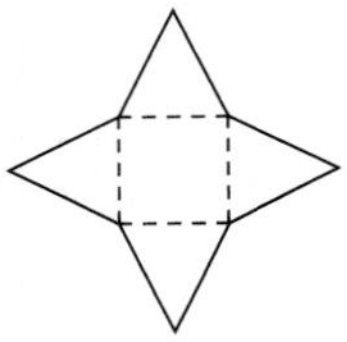

* **21.** What is the surface area of a cube with edges 2 ft long?
(67)

Solve:

22. $4.3 = x - 0.8$
(35)

23. $\frac{2}{d} = \frac{1.2}{1.5}$
(39)

Simplify:

24.
(49)
$$\begin{array}{r} 10 \text{ lb} \\ - \ 6 \text{ lb } 7 \text{ oz} \\ \hline \end{array}$$

25. $\frac{\$5.25}{1 \text{ hr}} \cdot \frac{8 \text{ hr}}{1 \text{ day}} \cdot \frac{5 \text{ days}}{1 \text{ week}}$
(55)

26. $3\frac{3}{4} \div \left(1\frac{2}{3} \cdot 3\right)$
(26)

27. $4\frac{1}{2} + 5\frac{1}{6} - 1\frac{1}{3}$
(30)

28. (45) (0.06 ÷ 5) ÷ 0.004

29. (35, 43) Write $9\frac{1}{2}$ as a decimal number, and multiply it by 9.2.

30. (36) A coin is tossed and the spinner is spun. One possible outcome is heads and A (HA).

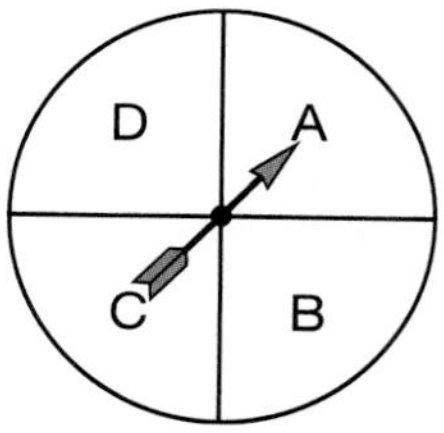

a. What is the sample space of the experiment?

b. What is the probability of getting heads and a consonant?

Early Finishers
Math and Science

Light travels at a speed of approximately 186,000 miles per second. Scientists use a unit of measurement known as a light-year for expressing distances from Earth to stars and other objects far away. A light-year is the distance light can travel in one year.

a. Express the given speed of light in scientific notation.

b. Determine the number of seconds in one year (365 days). Express this result in scientific notation.

c. Use the answers to **a** and **b** to find the distance light travels in one year. Express the answer in scientific notation.

d. Round the answer to **c** to one non-zero digit, and then use words to name that distance.

Focus on

• Balanced Equations

Equations are sometimes called **balanced equations** because the two sides of the equation "balance" each other. A balance scale can be used as a model of an equation. We replace the equal sign with a balanced scale. The left and right sides of the equation are placed on the left and right trays of the balance. For example, $x + 12 = 33$ becomes

Using a balance-scale model we think of how to simplify the equation to get the unknown number, in this case *x,* alone on one side of the scale. Using our example, we could remove 12 (subtract 12) from the left side of the scale. However, if we did that, the scale would no longer be balanced. So we make this rule for ourselves.

> Whatever operation we perform on one side of an equation, we also perform on the other side of the equation to maintain a balanced equation.

We see that there are two steps to the process.

Step 1: Select the operation that will isolate the variable.

Step 2: Perform the selected operation on both sides of the equation.

We select "subtract 12" as the operation required to isolate *x* (to "get *x* alone"). Then we perform this operation on both sides of the equation.

Select operation:

To isolate *x*, *subtract 12.*

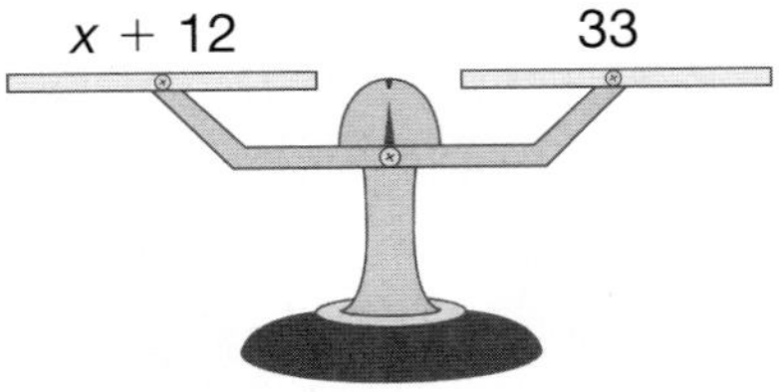

Perform operation:

To keep the scale balanced, *subtract* 12 *from both sides of the equa*

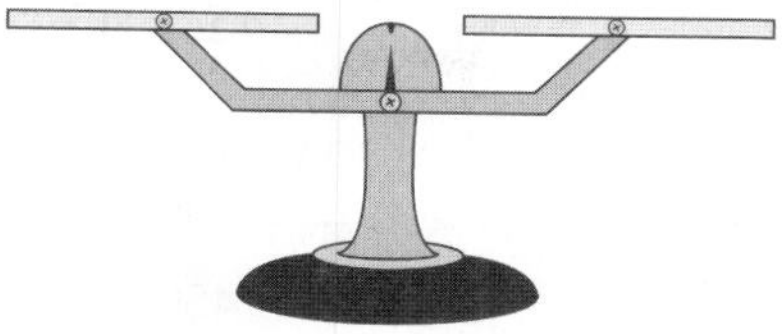

Summarize State what we do when we "isolate the variable."

After subtracting 12 from both sides of the equation, x is isolated on one side of the scale, and 21 balances x on the other side of the scale. This shows that $x = 21$. We check our solution by replacing x with 21 in the original equation.

$x + 12 = 33$	original equation
$21 + 12 = 33$	replaced x with 21
$33 = 33$	simplified left side

Both sides of the equation equal 33. This shows that the solution, $x = 21$, is correct.

Now we will illustrate a second equation $45 = x + 18$.

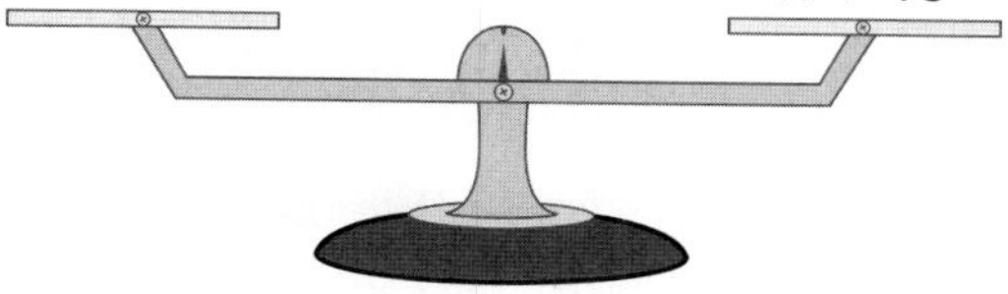

This time the unknown number is on the right side of the balance scale, added to 18.

1. **Analyze** Select the operation that will isolate the variable, and write that operation on your paper.
2. Describe how to perform the operation and keep a balanced scale.
3. **Discuss** What will remain on the left and right side of the balance scale after the operation is performed.

We show the line-by-line solution of the equation below.

$45 = x + 18$	original equation
$45 - 18 = x + 18 - 18$	subtracted 18 from both sides
$27 = x + 0$	simplified both sides
$27 = x$	$x + 0 = x$

We check the solution by replacing x with 27 in the original equation.

$45 = x + 18$	original equation
$45 = 27 + 18$	replaced x with 27
$45 = 45$	simplified right side

By checking the solution in the original equation, we see that the solution is correct. Now we revisit the equation to illustrate one more idea.

4. **Predict** Suppose the contents of the two trays of the balance scale were switched. That is, $x + 18$ was moved to the left side, and 45 was moved to the right side. Would the scale still be balanced? Write what the equation would be.

Now we will consider an equation that involves multiplication rather than addition.

$$2x = 132$$

Since $2x$ means two x's $(x + x)$, we may show this equation on a balance scale two ways.

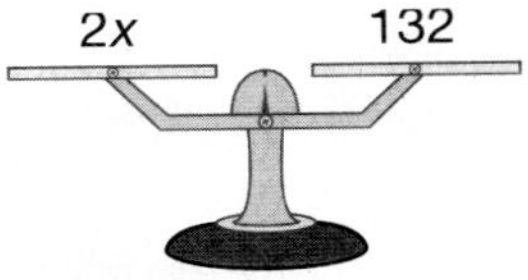

or

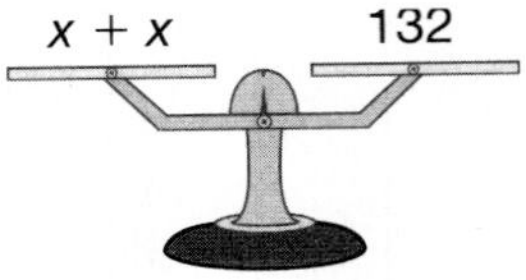

Our goal is to isolate x, that is, to have one x. We must perform the operations necessary to get one x alone on one side of the scale. We do not subtract 2, because 2 is not added to x. We do not subtract an x, because there is no x to subtract from the other side of the equation. To isolate x in this equation, we *divide by 2.* To keep the equation balanced, we *divide both sides by 2.*

Select operation:

To isolate x, *divide by 2.*

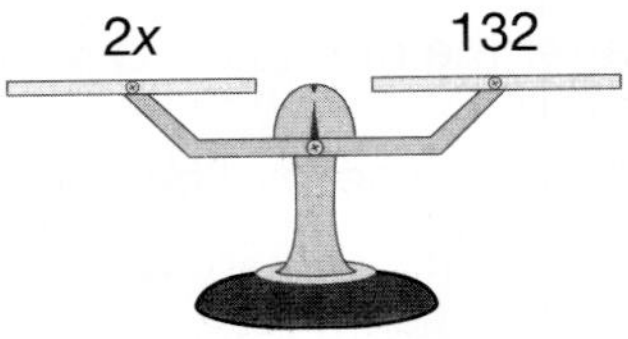

Perform operation:

To keep the equation balanced, divide *both sides by 2.*

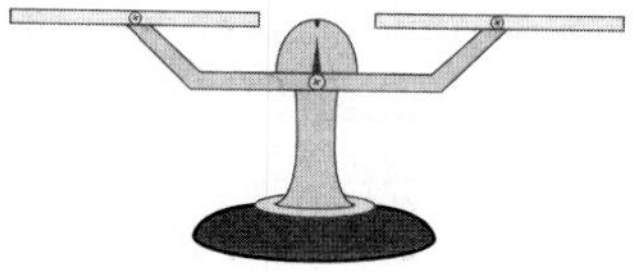

Here we show the line-by-line solution of this equation.

$2x = 132$	original equation
$\frac{2x}{2} = \frac{132}{2}$	divided both sides by 2
$1x = 66$	simplified both sides
$x = 66$	$1x = x$

Next we show the check of the solution.

$2x = 132$	original equation
$2(66) = 132$	replaced x with 66
$132 = 132$	simplified left side

This check shows that the solution, $x = 66$, is correct.

5. **Model** Draw a balance-scale model for the equation $3x = 132$.

6. **Analyze** Select the operation that will isolate the variable, and write that operation on your paper.

7. Describe how to perform the operation and keep a balanced scale.

8. **Model** Draw a balance scale and show what is on both sides of the scale after the operation is performed.

9. **Justify** Write the line-by-line solution of the equation.

10. Show the check of the solution.

Most students choose to solve the equation $3x = 132$ by dividing both sides of the equation by 3. There is another operation that could be selected that is often useful, which we will describe next. First note that the number multiplying the variable, in this case 3, is called the **coefficient** of x. Instead of dividing by the coefficient of x, we could choose to **multiply by the reciprocal** of the coefficient. In this case we could multiply by $\frac{1}{3}$. In either case the goal is to make the coefficient 1.

$$3x = 132$$

$$\frac{1}{3} \cdot 3x = \frac{1}{3} \cdot 132$$

$$1x = \frac{132}{3}$$

$$x = 44$$

Discuss Why is $\frac{1}{3}$ the reciprocal of 3?

When solving equations with whole number or decimal number coefficients, it is usually easier to think about dividing by the coefficient. However, when solving equations with fractional coefficients, it is usually easier to multiply by the reciprocal of the coefficient. Refer to the following equation for problems 11–14:

$$\frac{3}{4}x = \frac{9}{10}$$

11. **Analyze** Select the operation that will result in $\frac{3}{4}x$ becoming $1x$ in the equation. Tell why you chose the operation.

12. Describe how to perform the operation and keep the equation balanced.

13. Write a line-by-line solution of the equation.

14. Show the check of the solution.

We find that the solution to the equation is $\frac{6}{5}$ (or $1\frac{1}{5}$). In arithmetic we usually convert improper fractions to mixed numbers. In algebra we usually leave improper fractions in improper form unless the problem states or implies that a mixed number answer is preferable.

Evaluate For each of the following equations:

a. State the operation you select to isolate the variable.

b. Describe how to perform the operation and keep the equation balanced.

c. Write a line-by-line solution of the equation.

d. Show the check of the solution.

15. $x + 2.5 = 7$

16. $3.6 = y + 2$

17. $4w = 132$

18. $1.2m = 1.32$

19. $x + \frac{3}{4} = \frac{5}{6}$

20. $\frac{3}{4}x = \frac{5}{6}$

21. **Formulate** Make up your own equations. Solve and check each equation.

a. Make up an addition equation with decimal numbers.

b. Make up a multiplication equation with a fractional coefficient.

We have used some **properties of equality** to solve the equations in this investigation. The following table summarizes properties of equality involving the four operations of arithmetic.

> **1.** Addition: If $a = b$, then $a + c = b + c$.
>
> **2.** Subtraction: If $a = b$, then $a - c = b - c$.
>
> **3.** Multiplication: If $a = b$, then $ac = bc$.
>
> **4.** Division: If $a = b$, and if $c \neq 0$, then $\frac{a}{c} = \frac{b}{c}$.

The addition rule means this: If two quantities are equal and the same number is added to both quantities, then the sums are also equal.

22. **Summarize** Use your own words to describe the meaning of each of the other rules in the table.

extensions

a. **Justify** Write an equation for each problem below. Then, choose between mental math, estimation, paper and pencil, or calculator to solve the equation. Justify your choice.

Marsha hiked a total of 12 miles on two trails in a national park. One trail was twice as long as the other. How long was each trail? Let s = the shorter trail.

Jake earned $50. He put half of his money in the bank. He spent half of the money he did not put in the bank. How much money does Jake have left? Let m = the money Jake has left.

Mr. Wang's business averages $14,250 in expenses each month. Company salaries average $26,395 each month. The rent for the company is $2,825 each month. How much should Mr. Wang budget for salaries, expenses, and rent for the year? Let a = the total budget.

b. **Justify** Look at the four equations below. Which equation does not belong? Explain your reasoning.

$4x = 8 \qquad 132 = x + 130 \qquad 22x = 66 \qquad x + 12 = 14$

LESSON 71

Finding the Whole Group When a Fraction Is Known

Power Up — *Building Power*

facts Power Up O

mental math

a. **Positive/Negative:** $(-3) + (-12)$

b. **Decimals/Exponents:** 4.5×10^{-3}

c. **Ratio:** $\frac{w}{100} = \frac{24}{30}$

d. **Mental Math:** $12 \times 2\frac{1}{2}$

e. **Measurement:** 50 cm to m

f. **Percent:** 75% of $36

g. **Primes/Composites:** What is the prime factorization of 16?

h. **Calculation:** What is the total cost of a $30 item plus 8% sales tax?

problem solving

A rubber ball is dropped from a height of 12 meters. With each bounce the ball goes back up $\frac{2}{3}$ of its previous height. How many times will the ball bounce at least 12 cm high?

New Concept — *Increasing Knowledge*

Drawing diagrams of fraction problems can help us understand problems such as the following:

Three fifths of the fish in the pond are bluegills. If there are 45 bluegills in the pond, how many fish are in the pond?

The 45 bluegills are 3 of the 5 parts. We divide 45 by 3 and find there are 15 fish in each part. Since there are 15 fish in each of the 5 parts, there are 75 fish in all.

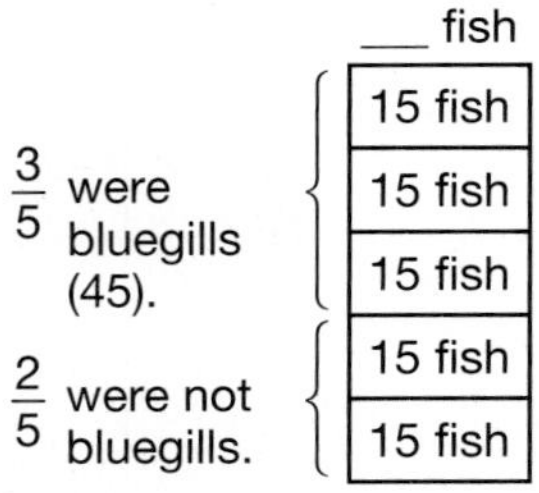

Example 1

When Juan finished page 51, he was $\frac{3}{8}$ of the way through his book. His book had how many pages?

Solution

Juan read 51 pages. This is 3 of 8 parts of the book. Since 51 ÷ 3 is 17, each part is 17 pages. Thus the whole book (8 parts) totals 8 × 17, which is **136 pages.**

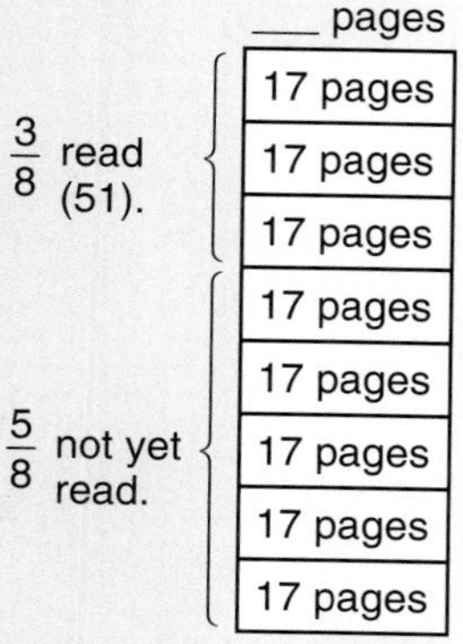

Justify How can we check that 136 pages is correct?

Example 2

The story in example 1 can be expressed with the following equation:

$$\frac{3}{8}P = 51$$

Solve the equation.

Solution

We change $\frac{3}{8}P$ to $1P$ by multiplying by $\frac{8}{3}$.

$$\frac{3}{8}P = 51$$

$$\frac{8}{3} \cdot \frac{3}{8}P = \frac{8}{3} \cdot 51$$

$$1P = \frac{408}{3} \qquad P = \mathbf{136}$$

Discuss Why do we multiply both sides of the equation by $\frac{8}{3}$?

Example 3

As Sakura went from room to room she found that $\frac{3}{5}$ of the lights were on and that 30 lights were off. How many lights were on?

Solution

Thinking Skill

Explain

How did we know that $\frac{2}{5}$ of the lights were off?

Since $\frac{3}{5}$ of the lights were on, $\frac{2}{5}$ of the lights were off. Because $\frac{2}{5}$ of the lights was 30 lights, each fifth was 15 lights. Thus **45 lights** were on.

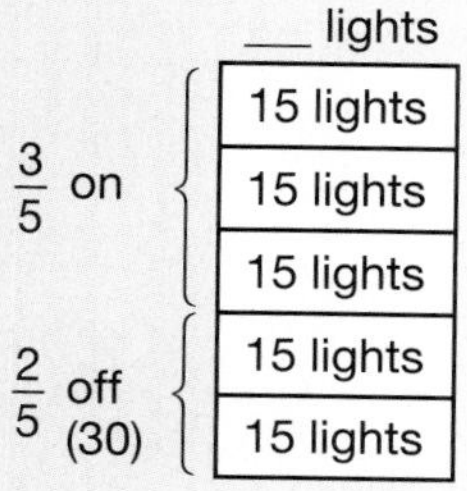

Practice Set

Model Draw a diagram to solve problems **a–c:**

a. Three fifths of the students in the class are boys. If there are 15 boys in the class, how many students are there in all?

b. Five eighths of the clowns had happy faces. If 15 clowns did not have happy faces, how many clowns were there in all?

c. Vincent was chagrined when he looked at the clock, for in $\frac{3}{4}$ of an hour he had only answered 12 homework questions. At that rate, how many questions would Vincent answer in an hour?

d. **Model** The story in problem **c** can be expressed with the equation $\frac{3}{4}H = 12$. Solve the equation.

e. **Formulate** Write and solve your own problem in which a fraction of a group is known and you must find the whole group. Use problems **a–c** as models.

Written Practice

Strengthening Concepts

1. *(32, 54)* Nine seconds elapsed from the time Abigail saw the lightning until she heard the thunder. The lightning was about how many kilometers from Abigail? Sound travels about $\frac{1}{3}$ of a kilometer per second.

2. *(28, 34)* What is the average of the three numbers marked by arrows on the number line below?

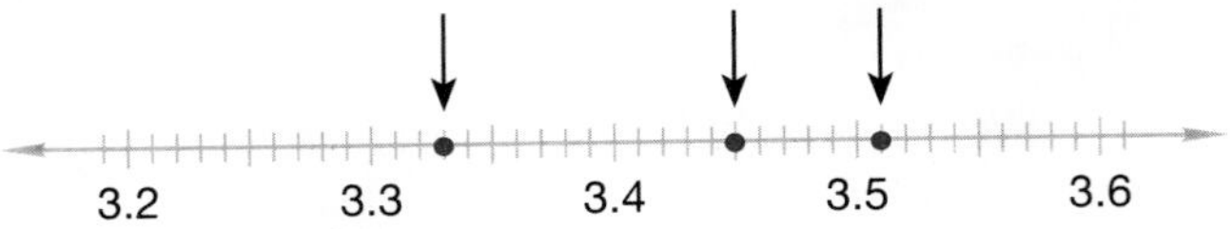

*** 3.** *(55)* **Generalize** On his first 2 tests Nate's average score was 80 percent. On his next 3 tests Nate's average score was 90 percent. What was his average score for all 5 tests?

4. *(51)* Twenty billion is how much more than nine billion? Write the answer in scientific notation.

5. *(21)* What is the sum of the first five prime numbers?

*** 6.** *(66)* **Model** Use a ratio box to solve this problem. The ratio of new pens to used pens in the box was 4 to 7. In all there were 242 new pens and used pens in the box. How many new pens were in the box?

*** 7.** *(71)* Diagram this statement. Then answer the questions that follow.

When Rosario finished page 78, she was $\frac{3}{5}$ of the way through her book.

a. How many pages are in her book?

b. How many pages does she have left to read?

*** 8.** *(67)* **Generalize** Find the surface area of the box in problem **9** and tell how you found your answer.

* **9.** *(70)* Find the number of 1-inch cubes that can be placed in this box. Dimensions are in inches.

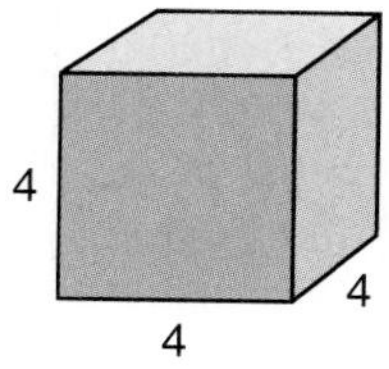

* **10.** *(65)* Find the circumference of each circle. Show the formula you used.

a.

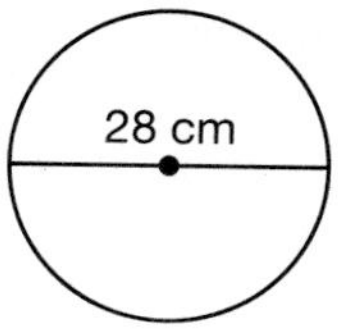

Use 3.14 for π.

b.

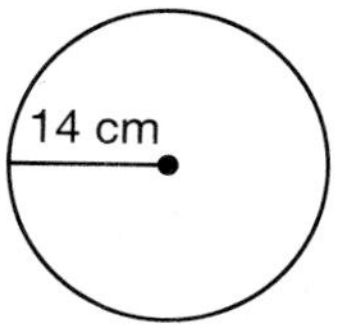

Use $\frac{22}{7}$ for π.

c. **Connect** How did your method for finding the circumference for the circle in **a** differ from your method for finding the circumference of the circle in **b?**

* **11.** *(69)* **Represent** Write each number in scientific notation:

a. 25×10^6 **b.** 25×10^{-6}

12. *(48)* Complete the table.

Fraction	Decimal	Percent
a.	0.1	**b.**
c.	**d.**	0.5%

13. *(60)* Write an equation to solve each problem:

a. What number is 35% of 80?

b. Three fourths of 24 is what number?

14. *(56)* Write the rule of this function with words and as an equation. Then find the missing number.

x	y
3	10
0	7
5	12
7	

Thinking Skill

Infer

A **vertex** is a point on a figure where two or more line segments meet. What is the meaning of *vertices?*

* **15.** *(67)* **Model** Draw a rectangular prism. A rectangular prism has how many vertices?

16. *(37, 62)* Figure *ABCD* is a trapezoid. Dimensions are in centimeters.

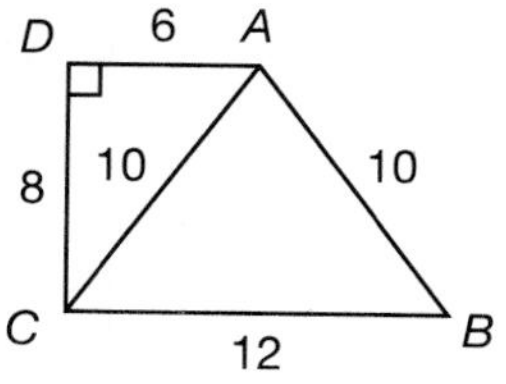

a. Find the perimeter of the trapezoid.

b. Find the area of the right triangle.

c. Find the area of the isosceles triangle.

d. Combine the areas of the triangles to find the area of the trapezoid.

e. **Analyze** What information from your solutions to problems **a–c** was not needed to solve **d?**

17. (60) The restaurant bill was \$16.50. Marcia planned to leave a tip of about 15%. After paying for the meal, she had a few dollar bills and quarters left in her purse. About how much money should she leave for the tip?

18. (Inv. 4) The principal tallied the number of middle-grade classrooms in the school that had certain numbers of students. Make a box-and-whisker plot from this information.

Distribution of Students in Middle-Grade Classrooms

Number of Students	Tally of Classes
26	|
27	|||
28	||||
29	|||
30	||
31	|

19. (Inv. 3) The coordinates of three vertices of triangle ABC are A (0, 6), B (8, −2), and C (−9, −3). Graph the triangle. Then use a protractor to find the measures of $\angle A$, $\angle B$, and $\angle C$ to the nearest degree.

20. (52) Evaluate: $y - xy$ if $x = 0.1$ and $y = 0.01$

For problems **21** and **22,** solve and check each equation. Show each step. State the property of equality you used.

21. (Inv. 7) $m + 5.75 = 26.4$

22. (Inv. 7) $\frac{3}{4}x = 48$

*** 23.** (Inv. 6) **Conclude** What is the name of a parallelogram whose sides are equal in length but whose angles are not necessarily right angles?

Simplify:

24. (63) $\dfrac{4^2 + \{20 - 2[6 - (5 - 2)]\}}{\sqrt{36}}$

25. (49)
$$\begin{array}{r} 1\text{ yd} \\ -\quad 1\text{ ft}\ \ 1\text{ in.} \\ \hline \end{array}$$

26. (50) $3.5\text{ hr} \cdot \dfrac{60\text{ min}}{1\text{ hr}} \cdot \dfrac{60\text{ s}}{1\text{ min}}$

27. (26) $6\frac{2}{3} \div \left(4\frac{1}{2} \cdot 2\frac{2}{3}\right)$

28. (30) $7\frac{1}{2} - 5\frac{1}{6} + 1\frac{1}{3}$

*** 29.** (68) **Generalize** Simplify:

a. $(-5) + (-6) - |-7|$

b. $(-15) - (-24) - (+8)$

c. $\left(-\frac{4}{5}\right) - \left(-\frac{1}{5}\right)$

30. (30, 43) Write 1.5 as a mixed number, and subtract it from $2\frac{2}{3}$.

LESSON 72

Implied Ratios

Power Up

Building Power

facts

Power Up N

mental math

a. **Positive/Negative:** $(-10) + (+17)$

b. **Exponents:** $\left(\frac{2}{3}\right)^2$

c. **Algebra:** $6x + 2 = 32$

d. **Decimals:** What decimal is 10% of 36?

e. **Measurement:** 500 g to kg

f. **Percent:** $33\frac{1}{3}\%$ of \$36

g. **Primes/Composites:** What is the prime factorization of 36?

h. **Calculation:** Find 15% of \$30 by finding 10% of \$30 plus half of 10% of \$30.

problem solving

Given that $\frac{1}{8}$ of a number is $\frac{1}{5}$, what is $\frac{5}{8}$ of that number?

New Concept

Increasing Knowledge

Thinking Skill

Predict

Would you expect the 30 books to weigh about 80, 100, or 120 pounds? Why?

Consider the following problem:

If 12 math books weigh 40 pounds, how much would 30 math books weigh?

We could solve this problem by finding the weight of one book and then multiplying that weight by 30. However, since the weight per book is constant we may write a proportion.

Notice that the problem describes two situations. In one case the number and weight of the books are given. In the other case the number is given and we are asked to find the weight. We will record the information in a ratio box. Instead of using the words "ratio" and "actual count," we will write "Case 1" and "Case 2." We will use p to stand for pounds.

	Case 1	Case 2
Books	12	30
Pounds	40	p

From the table we write a proportion and solve it.

$$\frac{12}{40} = \frac{30}{p} \qquad \text{proportion}$$

We will cross multiply to solve the proportion, as shown on the following page.

$$12p = 40 \cdot 30 \qquad \text{cross multiplied}$$

$$\frac{\overset{1}{\cancel{12}}p}{\underset{1}{\cancel{12}}} = \frac{40 \cdot 30}{12} \qquad \text{divided by 12}$$

$$p = 100 \qquad \text{simplified}$$

We find that 30 math books would weigh 100 pounds.

Example 1

If 5 pencils cost $1.20, how much would 12 pencils cost? Use a ratio box to solve the problem.

Solution

Since 12 pencils is a little more than two times 5 pencils, we estimate the cost would be a little more than two times $1.20, perhaps between $2.50 and $3.00.

Now we draw the ratio box. We use *d* for dollars.

	Case 1	Case 2
Pencils	5	12
Dollars	1.2	*d*

Now we write the proportion and solve for *d*.

Math Language
The term **cross multiply** means to multiply the numerator of one fraction in an equality by the denominator of the other fraction.

$$\frac{5}{1.2} = \frac{12}{d} \qquad \text{proportion}$$

$$5d = 12(1.2) \qquad \text{cross multiplied}$$

$$\frac{\overset{1}{\cancel{5}}d}{\underset{1}{\cancel{5}}} = \frac{12(1.2)}{5} \qquad \text{divided by 5}$$

$$d = 2.88 \qquad \text{simplified}$$

We find that 12 pencils cost **$2.88.**

Predict How could you predict the cost of 10 pencils without using a ratio box?

Example 2

Mrs. Campbell can tie 25 bows in 3 minutes. At that rate, how many bows can she tie in 1 hour?

Solution

We can use either minutes or hours but not both. *The units must be the same in both cases.* Since there are 60 minutes in 1 hour, we will use 60 minutes instead of 1 hour.

	Case 1	Case 2
Bows	25	*b*
Minutes	3	60

Next we write the proportion, cross multiply, and solve by dividing by 3.

$$\frac{25}{3} = \frac{b}{60} \quad \text{proportion}$$

$$25 \cdot 60 = 3b \quad \text{cross multiplied}$$

$$\frac{25 \cdot 60}{3} = \frac{\overset{1}{\cancel{3}}b}{\underset{1}{\cancel{3}}} \quad \text{divided by 3}$$

$$b = 500 \quad \text{simplified}$$

We see that Mrs. Campbell can tie **500 bows** in one hour.

Example 3

Six is to 15 as 9 is to what number?

Solution

Thinking Skill

Predict

Do you expect the answer will be greater than or less than 15? Why?

We can sort the numbers in this question using a case 1–case 2 ratio box.

	Case 1	Case 2
First Number	6	9
Second Number	15	n

Now we write and solve a proportion.

$$\frac{6}{15} = \frac{9}{n}$$

$$6n = 9 \cdot 15$$

$$\frac{\overset{1}{\cancel{6}}n}{\underset{1}{\cancel{6}}} = \frac{9 \cdot 15}{6}$$

$$n = \mathbf{22\tfrac{1}{2}}$$

Predict Would the solution be different if the problem read: fifteen is to 6 as 9 is to what number? Why or why not?

Practice Set

a. Estimate Kevin rode 30 km in 2 hours. At that rate, how long would it take him to ride 75 km? Estimate an answer. Then use a case 1–case 2 ratio box to solve this problem.

b. If 6 bales are needed to feed 40 head of cattle, how many bales are needed to feed 50 head of cattle?

c. Generalize Five is to 15 as 9 is to what number?

Written Practice

Strengthening Concepts

1. (12) The astronaut John Glenn first orbited Earth in 1962. In 1998, he returned to space for the last time aboard the space shuttle. How many years were there between his first and his last trip to space?

2. (55) In her first 4 basketball games Carolina averaged 4 points per game. In her next 6 games Carolina averaged 9 points per game. What was her average number of points per game after 10 games?

3. (50) Use a unit multiplier to convert 2.5 liters to milliliters.

4. (30) **a.** If the product of $\frac{1}{2}$ and $\frac{2}{5}$ is subtracted from the sum of $\frac{1}{2}$ and $\frac{2}{5}$, what is the difference?

b. **Summarize** What math language did you need to know before you could solve part **a?**

5. (53) Use a ratio box to solve this problem. The ratio of desktop to laptop computers in a school district is 7 to 2. If there are 126 desktop computers in the district, how many laptops are there?

* **6.** (72) **Model** Use a ratio box to solve this problem. If 4 books weigh 9 pounds, how many pounds would 14 books weigh?

7. (60) Write an equation to solve each problem:

a. Two fifths of 60 is what number?

b. How much money is 75% of $24?

* **8.** (65) The diameter of a bicycle tire is 20 in. Find the distance around the tire to the nearest inch. (Use 3.14 for π.)

* **9.** (71) Diagram this statement. Then answer the questions that follow.

Jasmine lives in New Jersey, which is located on the Atlantic coast. Jasmine has ridden her bike along 52 miles of the coastline. This is two-fifths of the total New Jersey coastline.

a. How many miles is the New Jersey coastline altogether?

b. How many more miles would Jasmine have to travel to ride the entire state coastline?

* **10.** (70) The volume of a block of ice with the dimensions shown is equal to how many 1-by-1-by-1-inch ice cubes?

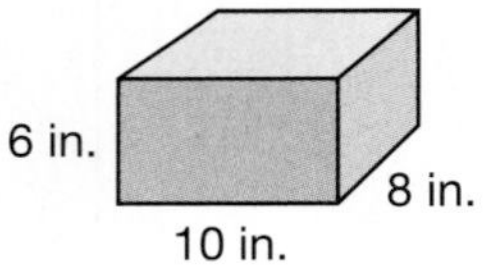

* **11.** (67) **Generalize** Find the area of each of the six surfaces of the block of ice shown in problem **10.** Then add the areas to find the total surface area of the block of ice.

* **12.** (69) **Represent** Write each number in scientific notation:

a. 0.6×10^6 **b.** 0.6×10^{-6}

13. (28, 34) What is the average of the three numbers marked by arrows on the number line below? How did you find the average?

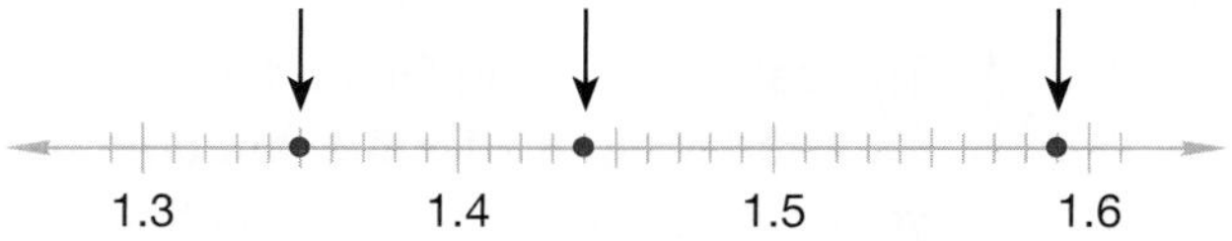

14. Complete the table.
(48)

Fraction	Decimal	Percent
$\frac{3}{5}$	a.	b.
c.	d.	2.5%

15. **a.** Write the prime factorization of 8100 using exponents.
(21)

b. Find $\sqrt{8100}$.

Thinking Skill

Conclude

What does the dashed line in the figure represent?

16. **a.** Find the area of the parallelogram.
(37, 61)

b. Find the area of the shaded triangle.

c. If each acute angle of the parallelogram measures 72°, what is the measure of each obtuse angle of the parallelogram?

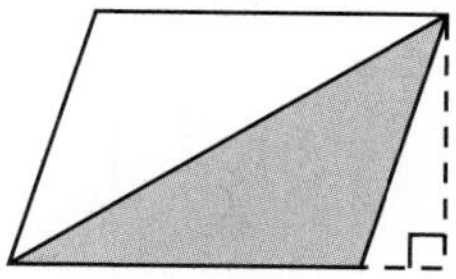

17. Name each geometric solid and tell why only one of the figures is a polyhedron:
(67)

a. 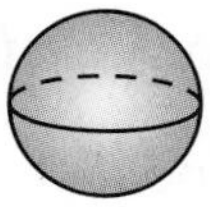

b. 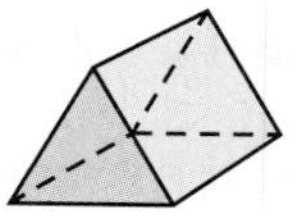

c 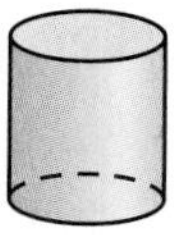

* **18.** *Connect* A rectangular piece of metal swings around a rod. As it spins, its path through space is shaded like which figure in problem 17?
(67)

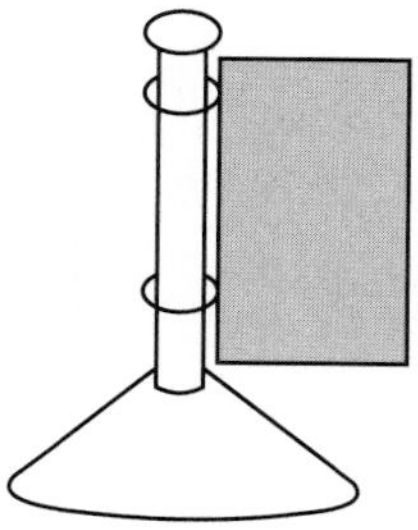

19. Find the circumference of each circle:
(65)

a.

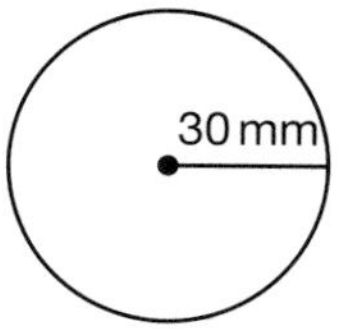

Leave π as π.

b.

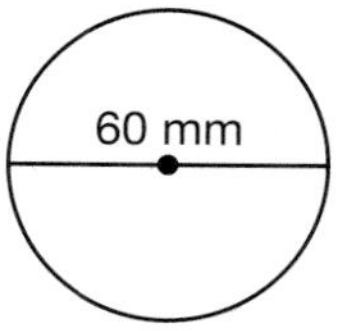

Use 3.14 for π.

20. Compare: $\frac{2}{3}$ ◯ 0.667
(43)

21. For $x = 5$ and $y = 4$, evaluate:
(52, 57)

a. $x^2 - y^2$

b. $x^0 - y^{-1}$

Analyze For problems **22** and **23,** solve and check each equation. Show each step.

* **22.** $m - \frac{2}{3} = 1\frac{3}{4}$
(Inv. 7)

* **23.** $\frac{2}{3}w = 24$
(Inv. 7)

Generalize Simplify:

* **24.** $\dfrac{[30 - 4(5 - 2)] + 5(3^3 - 5^2)}{\sqrt{9} + \sqrt{16}}$
(63)

25. $\begin{array}{r} 2 \text{ gal } 1 \text{ qt} \\ - \ 1 \text{ gal } 1 \text{ qt } 1 \text{ pt} \\ \hline \end{array}$
(49)

26. $\dfrac{1}{2} \text{ mi} \cdot \dfrac{5280 \text{ ft}}{1 \text{ mi}} \cdot \dfrac{1 \text{ yd}}{3 \text{ ft}}$
(50)

27. $\left(2\frac{1}{2}\right)^2 \div \left(4\frac{1}{2} \cdot 6\frac{2}{3}\right)$
(26)

28. $7\frac{1}{2} - \left(5\frac{1}{6} + 1\frac{1}{3}\right)$
(30)

* **29.** **a.** $(-7) + |+5| + (-9)$
(68)

b. $(16) + (-24) - (-18)$

c. $(-0.2) + (-0.3) - (-0.4)$

30. Write the sum of $5\frac{1}{4}$ and 1.9 as a decimal.
(35, 43)

Early Finishers
Real-World Application

A contractor built a lap pool for the Octave Family. The pool measures 10 ft by 40 ft by 4 ft. It can safely hold 95% of its volume with water.

a. Determine the volume of this pool in cubic feet.

b. Determine the volume of water the pool can safely hold in cubic feet.

c. If one cubic foot equals about 7.48 gallons, determine the volume of water the pool can safely hold.

d. The contractor instructed the Octaves to fill the pool for 12 hours. If the rate of water flow is 20 gallons per minute, can they safely follow these instructions?

LESSON

73

• Multiplying and Dividing Positive and Negative Numbers

Power Up

Building Power

facts Power Up O

mental math

a. Positive/Negative: $(+15) + (-25)$

b. Decimals/Exponents: 8.75×10^3

c. Ratio: $\frac{12}{x} = \frac{2.5}{7.5}$

d. Mental Math: $3\frac{1}{2} \times 18$

e. Measurement: 500 mL to L

f. Percent: $66\frac{2}{3}\%$ of \$36

g. Primes/Composites: What is the prime factorization of 48?

h. Calculation: Estimate 15% of 39 by finding 10% of 40 plus half of 10% of 40.

problem solving

Grayson is holding four cards in her hand: the ace of spades, the ace of hearts, the ace of clubs, and the ace of diamonds. Neeasha pulls two of the cards from Grayson's hand without looking. What is the probability that Neeasha has pulled at least one black ace?

New Concept

Increasing Knowledge

We can develop the rules for multiplying and dividing signed numbers if we remember that multiplication is a shorthand notation for repeated addition. Thus, 2 times −3 means (−3) + (−3), so

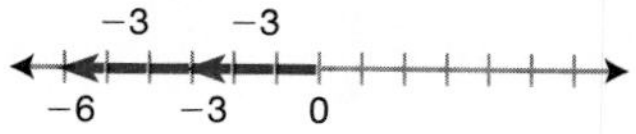

$$2(-3) = -6$$

This illustration shows that 2(−3) is −6. It also shows that if −6 is divided into two equal parts, each part is −3. We also see that the number of −3*s* in −6 is 2. We show these multiplication and division relationships below.

$$2(-3) = -6 \qquad \frac{-6}{2} = -3 \qquad \frac{-6}{-3} = +2$$

Notice that multiplying or dividing two numbers whose signs are different results in a negative number. But what happens if we multiply two negative numbers such as −2 and −3? We can read (−2) (−3) as "the opposite of 2 times −3." Since 2 times −3 equals −6, then the *opposite of* 2 times −3 equals the *opposite of* −6, which is 6.

$$(-2)(-3) = +6$$

And since division undoes multiplication, these division problems must also be true:

$$\frac{+6}{-2} = -3 \quad \text{and} \quad \frac{+6}{-3} = -2$$

These conclusions give us the rules for multiplying and dividing signed numbers.

Rules For Multiplying And Dividing Positive And Negative Numbers

1. If the two numbers in the multiplication or division problem have the same sign, the answer is a positive number.
2. If the two numbers in the multiplication or division problem have different signs, the answer is a negative number.

Thinking Skill

Summarize

Use the symbols + and − to summarize the rules for multiplying and dividing signed numbers. Example: (+)(+) = (+)

Here are some examples:

Multiplication	Division
$(+6)(+2) = +12$	$\frac{+6}{+2} = +3$
$(-6)(-2) = +12$	$\frac{-6}{-2} = +3$
$(-6)(+2) = -12$	$\frac{-6}{+2} = -3$
$(+6)(-2) = -12$	$\frac{+6}{-2} = -3$

Conclude If the result of the multiplication of two numbers is a negative number, what can we say about the signs of the factors?

Example

Divide or multiply:

a. $\frac{-12}{-4}$ **b.** $\frac{-2.4}{4}$ **c.** $(6)(-3)$ **d.** $\left(-\frac{1}{2}\right)\left(-\frac{1}{2}\right)$

Solution

We divide or multiply as indicated. If both signs are the same, the answer is positive. If one sign is positive and the other is negative, the answer is negative. Showing the positive sign is permitted but not necessary.

a. 3 **b.** −0.6 **c.** −18 **d.** $\frac{1}{4}$

Practice Set

Generalize Divide or multiply:

a. $(-7)(3)$ **b.** $(+4)(-8)$ **c.** $(8)(+5)$

d. $(-8)(-3)$ **e.** $\frac{25}{-5}$ **f.** $\frac{-27}{-3}$

g. $\frac{-28}{4}$ **h.** $\frac{+30}{6}$ **i.** $\frac{+45}{-3}$

j. $\left(-\frac{1}{2}\right)\left(\frac{1}{4}\right)$ **k.** $\frac{-1.2}{0.3}$ **l.** $(-1.2)(-2)$

m. **Model** Sketch a number line and show this multiplication: $3(-2)$.

n. Write two division facts illustrated by this number line.

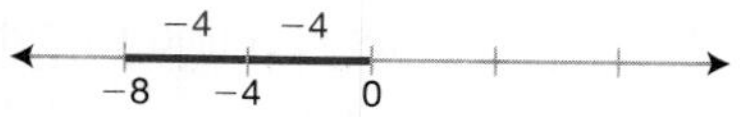

Written Practice *Strengthening Concepts*

* **1.** (72) **Model** Use a ratio box to solve this problem.

If Elena can wrap 12 packages in 5 minutes, how many packages can she wrap in 1 hour?

* **2.** (55) **Analyze** Lydia walked 30 minutes a day for 5 days. The next 3 days she walked an average of 46 minutes per day. What was the average amount of time she spent walking per day during those 8 days?

3. (35, 45) If the sum of 0.2 and 0.5 is divided by the product of 0.2 and 0.5, what is the quotient?

4. (50) Use a unit multiplier to convert 23 cm to mm.

5. (66) Use a ratio box to solve this problem. The ratio of paperback books to hardbound books in the school library was 3 to 11. If there were 9240 hardbound books in the library, how many books were there in all?

* **6.** (69) Write each number in scientific notation:

a. 24×10^{-5} **b.** 24×10^{7}

* **7.** (71) **Model** Draw a diagram of this statement:

The 30 red-tailed hawks counted during fall migration amounted to $\frac{1}{4}$ of the hawks counted.

a. How many hawks were counted?

b. How many of the hawks were not red-tails?

8. (60) Write an equation to solve each problem:

a. Five ninths of 45 is what number?

b. What number is 80% of 760?

* **9.** (73) Divide or multiply:

a. $\frac{-36}{9}$ **b.** $\frac{-3.6}{-6}$

c. $0.9(-3)$ **d.** $(+8)(+7)$

10. The face of the spinner is divided into eighths. If the spinner is spun once, what is the probability that it will stop on a composite number?
(21, 36)

11. The *x*-axis is a line of symmetry for $\triangle RST$. The coordinates of *R* and *S* are (6, 0) and (−2, −2), respectively.
(58, 62)

a. What are the coordinates of *T*?

b. What type of triangle is $\triangle RST$ classified by sides?

c. If the measure of $\angle R$ is approximately 28°, then what is the approximate measure of $\angle S$?

* **12.** **Discuss** Describe how to determine whether the product of two signed numbers is positive or negative.
(73)

* **13.** Find the number of 1-ft cubes that will fit inside a closet with dimensions as shown.
(70)

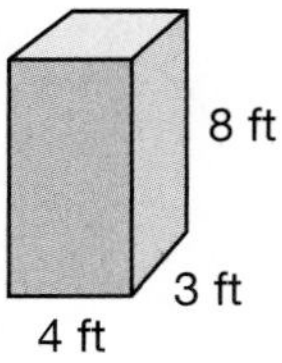

14. Find the circumference of each circle:
(65)

a.

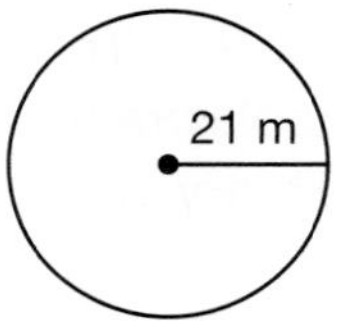

Use 3.14 for π.

b.

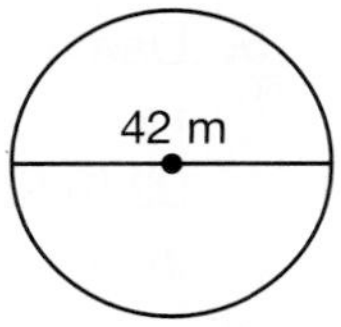

Use $\frac{22}{7}$ for π.

15. Complete the table.
(48)

Fraction	Decimal	Percent
a.	2.5	**b.**
c.	**d.**	0.2%

Classify Use the terms acute, right, or obtuse and equilateral, isosceles, or scalene to classify each triangle in problem **16–18.** Then find the area of each triangle. Dimensions are in centimeters.

* **16.**
(37, 62)

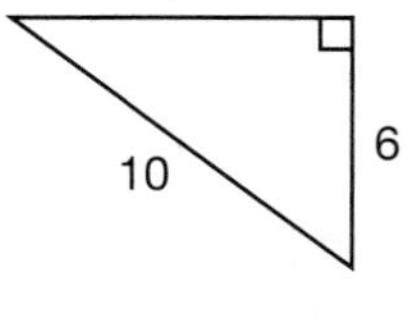

* **17.**
(37, 62)

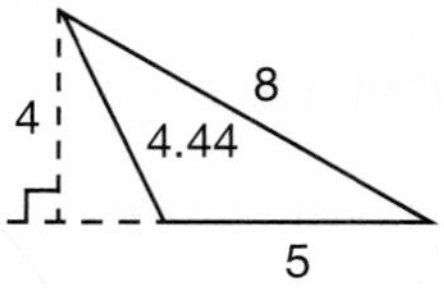

* **18.**
(37, 62)

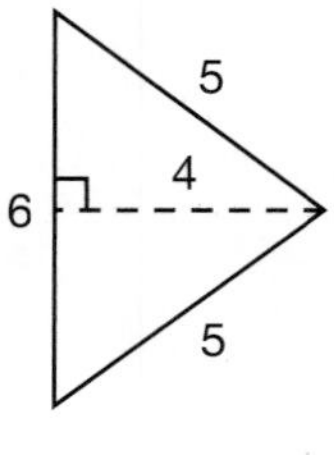

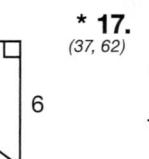

19. Name each three-dimensional figure:
(67)

a. 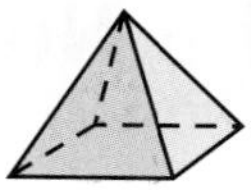

b. 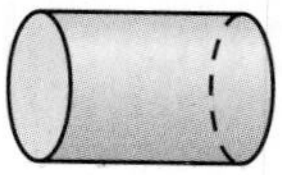

c. 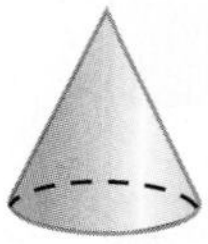

20. Compare: $\frac{2}{3}$ of 96 ◯ $\frac{5}{6}$ of 84
(60)

21. **Analyze** Find $ab - (a - b)$ if $a = \frac{5}{6}$ and $b = \frac{3}{4}$
(52)

Analyze For problems **22–24,** solve and check each equation. Show each step.

* **22.** $\frac{3}{5}w = 15$
(Inv. 7)

* **23.** $b - 1.6 = (0.4)^2$
(Inv. 7)

* **24.** $20w = 5.6$
(Inv. 7)

Simplify:

25.
(49)

$$\begin{array}{r} 2 \text{ yd } 1 \text{ ft } 7 \text{ in.} \\ + \ 1 \text{ yd } 2 \text{ ft } 8 \text{ in.} \\ \hline \end{array}$$

26. $0.5 \text{ m} \cdot \frac{100 \text{ cm}}{1 \text{ m}} \cdot \frac{10 \text{ mm}}{1 \text{ cm}}$
(50)

27. $12\frac{1}{2} \cdot 4\frac{1}{5} \cdot 2\frac{2}{3}$
(26)

28. $7\frac{1}{2} \div \left(6\frac{2}{3} \cdot 1\frac{1}{5}\right)$
(26)

* **29.** **Generalize** Simplify.
(68)

a. $(-8) + (-7) - (-15)$

b. $(-15) + (+11) - |+24|$

c. $\left(-\frac{1}{3}\right) - \left(-\frac{2}{3}\right)$

30. Find the product of 2.25 and $1\frac{1}{3}$.
(26, 43)

Early Finishers
Math Applications

Look at the three figures in problem 19 above.

a. What plane shape is formed if you make a vertical cut to divide each figure in half? Sketch the plane shape.

b. What plane shape is formed if you make a horizontal cut? Sketch the plane shape.

LESSON

74

• Fractional Part of a Number, Part 2

Power Up

Building Power

facts Power Up N

mental math

a. **Positive/Negative:** $(-8) - (-12)$

b. **Exponents:** 45×10^{-3}

c. **Algebra:** $7w + 1 = 50$

d. **Estimate/Percent:** Estimate 15% tip on a $19.81 bill.

e. **Measurement:** 400 m to km

f. **Percent:** 80% of $25

g. **Primes/Composites:** What is the prime factorization of 50?

h. **Calculation:** 10% of 80, $\times$ 2, $\sqrt{\ }$, $\times$ 7, $-$ 1, $\div$ 3, $\sqrt{\ }$, $\times$ 12, $\sqrt{\ }$, $\div$ 6

problem solving

In a class of 20 boys, 14 wear brown shoes, 12 wear watches, 11 are on a soccer team, and 10 are tall. How many tall, soccer-playing boys who wear watches and brown shoes could be in the class?

New Concept

Increasing Knowledge

In some fractional-part-of-a-number problems, the fraction is unknown. In other fractional-part-of-a-number problems, the total is unknown. As discussed in Lesson 60, we can translate these problems into equations by replacing the word "of" with a multiplication sign and the word "is" with an equal sign.

Example 1

What fraction of 56 is 42?

Solution

We translate this statement directly into an equation by replacing "what fraction" with W_F, "of" with a multiplication symbol, and "is" with an equal sign.

What fraction of 56 is 42?	question
↓ ↓ ↓ ↓ ↓	
$W_F \times 56 = 42$	equation

Thinking Skill

Explain

Why do we divide both sides of the equation by 56?

To solve, we divide both sides by 56.

$$\frac{W_F \times 56}{56} = \frac{42}{56} \quad \text{divided by 56}$$

$$W_F = \mathbf{\frac{3}{4}} \quad \text{simplified}$$

If the question had been "What decimal part of 56 is 42?" the procedure would have been the same, but as the last step we would have written $\frac{3}{4}$ as the decimal number 0.75.

$$W_D = 0.75$$

Discuss How do we change a fraction to an equivalent decimal?

Example 2

Three fourths of what number is 60?

Solution

In this problem the total is the unknown. But we can still translate directly from the question to an equation.

Three fourths	of	what number	is	60?	question
↓	↓	↓	↓	↓	
$\frac{3}{4}$	$\times$	W_N	$=$	60	equation

To solve, we multiply both sides by $\frac{4}{3}$.

$$\frac{4}{3} \times \frac{3}{4} \times W_N = 60 \times \frac{4}{3} \quad \text{multiplied by } \frac{4}{3}$$

$$W_N = \mathbf{80} \quad \text{simplified}$$

Had the question been phrased using 0.75 instead of $\frac{4}{3}$, the procedure would have been similar.

Seventy-five hundredths	of	what number	is	60?	question
↓	↓	↓	↓	↓	
0.75	$\times$	W_N	$=$	60	equation

To solve, we can divide both sides by 0.75.

$$\frac{0.75 \times W_N}{0.75} = \frac{60}{0.75} \quad \text{divided by 0.75}$$

$$W_N = 80 \quad \text{simplified}$$

Practice Set

Formulate Translate each statement into an equation and solve:

a. What fraction of 130 is 80?

b. Seventy-five is what decimal part of 300?

c. Eighty is 0.4 of what number?

d. Sixty is $\frac{5}{6}$ of what number?

e. Sixty is what fraction of 90?

f. What decimal part of 80 is 60?

g. Forty is 0.08 of what number?

h. Six fifths of what number is 60?

i. For which of the above problems were you given the fractional part and asked to find the total or whole?

Written Practice *Strengthening Concepts*

Math Language
A **unit multiplier** is a ratio equal to 1 that is composed of two equivalent measures.

1. (55) Miguel is reading the book *Call of the Wild,* by Jack London. During the first 3 days of the week he read an average of 15 pages per day. During the next 2 days, Miguel averaged 25 pages per day. For all five days, what was the average number of pages Miguel read per day?

2. (46) Twelve ounces of Brand X costs $1.14. Sixteen ounces of Brand Y costs $1.28. Brand X costs how much more per ounce than Brand Y?

3. (50) Use a unit multiplier to convert $4\frac{1}{2}$ feet to inches.

4. (66) Ira counted the number of car and food commercials during a one hour television program. The ratio was 2 to 3. If 18 of the commercials were for food, how many commercials did Ira count altogether?

*** 5.** (72) **Model** Use a ratio box to solve this problem. If 5 pounds of apples cost $4.25, how much would 8 pounds of apples cost?

6. (22, 36) Diagram this statement. Then answer the questions that follow.
Five sixths of the 300 triathletes completed the course.

a. How many triathletes completed the course?

b. What was the ratio of triathletes who completed the course to those who did not complete the course?

Formulate Write equations to solve problems **7–11.**

*** 7.** (74) Fifteen is $\frac{3}{8}$ of what number?

*** 8.** (74) Seventy is what decimal part of 200?

*** 9.** (74) Two fifths of what number is 120?

10. (60) The store made a 60% profit on the $180 selling price of the coat. What was the store's profit?

11. (60) The shoe salesperson received a 20% commission on the sale of a $35 pair of shoes. What was the salesperson's commission?

*** 12.** (67, 70) **a.** What is the volume of this cube?

b. What is its surface area?

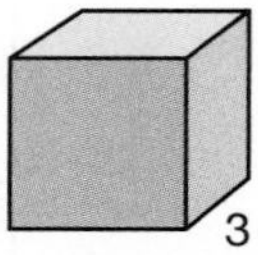
3 in.

13. Find the circumference of each circle:
(65)

a.

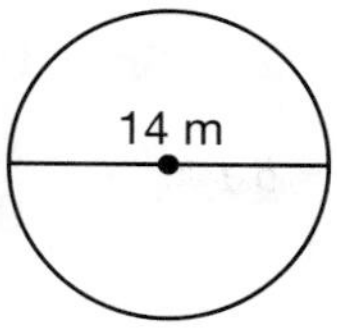

Use $\frac{22}{7}$ for π.

b.

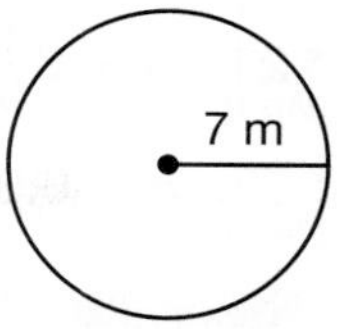

Leave π as π.

14. Complete the table.
(48)

Fraction	Decimal	Percent
$3\frac{1}{2}$	**a.**	**b.**
c.	**d.**	35%

15. Write the rule of this function with words and as an equation. Then find the missing number.
(56)

x	y
0	1
2	7
4	13
5	16
8	

16. The mean distance of the planet Mars from the Sun is 142 million miles. Write this distance in scientific notation.
(51)

Classify Refer to the figure below to answer problems **17** and **18**. In the figure, $\overline{AE} \parallel \overline{BD}$, $\overline{AB} \parallel \overline{EC}$, and $EC = ED$.

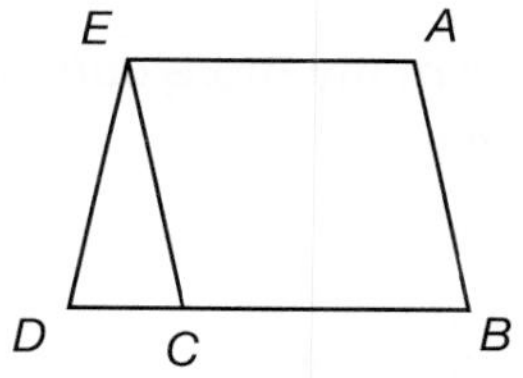

* **17.** **a.** What type of quadrilateral is figure *ABCE?*
(Inv. 6, 62)

b. What type of quadrilateral is figure *ABDE?*

c. What type of triangle is $\triangle ECD$ classified by sides?

18. If the measure of $\angle A$ is 100°, what is the measure of
(40, 61)

a. $\angle ABC$? **b.** $\angle BCE$? **c.** $\angle ECD$?

d. $\angle EDC$? **e.** $\angle DEC$? **f.** $\angle DEA$?

19. Arrange these numbers in order from least to greatest:
(33)

0.013, 0.1023, 0.0103, 0.021

20. Evaluate: $(m + n) - mn$ if $m = 1\frac{1}{2}$ and $n = 2\frac{2}{3}$
(52)

Analyze For problems **21–24**, solve and check each equation. Show each step.

* **21.** $p + 3\frac{1}{5} = 7\frac{1}{2}$ *(Inv. 7)*

* **22.** $3n = 0.138$ *(Inv. 7)*

* **23.** $n - 0.36 = 4.8$ *(Inv. 7)*

* **24.** $\frac{2}{3}x = \frac{8}{9}$ *(Inv. 7)*

Generalize Simplify:

* **25.** $\sqrt{49} + \{5[3^2 - (2^3 - \sqrt{25})] - 5^2\}$ *(63)*

26. *(49)*

$$\begin{array}{r} 4\text{ hr} \quad 5\text{ min} \quad 15\text{ s} \\ -\ 1\text{ hr} \quad 15\text{ min} \quad 30\text{ s} \\ \hline \end{array}$$

* **27.** *(68)* **a.** $(-9) + (-11) - (+14)$ **b.** $(26) + (-43) - |-36|$

* **28.** *(73)* **a.** $(-3)(1.2)$ **b.** $(-3)(-12)$

c. $\frac{-12}{3}$ **d.** $\frac{-1.2}{-3}$

29. *(30, 43)* Write the sum of $8\frac{1}{3}$ and 7.5 as a mixed number.

30. *(17)* Florence is facing north. If she turns 180°, which direction will she be facing?

Early Finishers
Real-World Application

The bowling team is having a carwash to raise money for new bowling shoes. They are charging $3 per carwash, and they have raised $366 so far.

a. There are 15 members on the team and four-fifths of them need new shoes. If each pair of shoes costs $35, how much money does the team need to cover the cost of the shoes?

b. How many more carwashes does the team need to sell to raise this money?

LESSON
75

• Area of a Complex Figure
• Area of a Trapezoid

Power Up — *Building Power*

facts Power Up O

mental math

a. **Positive/Negative:** $(-15) - (+20)$

b. **Exponents:** 15^2

c. **Ratio:** $\frac{30}{40} = \frac{g}{12}$

d. **Mental Math:** $25 \times \$2.40$

e. **Measurement:** 250 mg to g

f. **Percent:** 10% of $35

g. **Statistics:** Find the average of the set of numbers: 80, 95, 75, 70.

h. **Decimals:** What decimal is half of 10% of 36?

problem solving

Copy this problem and fill in the missing digits:

```
   _8_
 ×   _
  8__8
```

New Concepts — *Increasing Knowledge*

area of a complex figure

We have practiced finding the areas of figures that can be divided into two or more rectangles. In this lesson we will begin finding the areas of figures that include triangular regions as well.

Example 1

Find the area of this figure. Corners that look square are square. Dimensions are in millimeters.

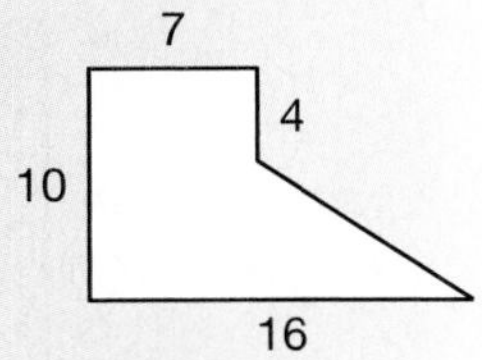

Solution

We divide the figure into smaller polygons. In this case we draw dashes that divide the figure into a rectangle and a triangle.

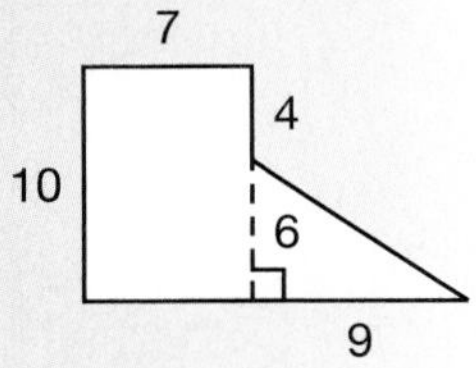

$$
\begin{aligned}
\text{Area of rectangle} &= 7 \times 10 &= 70 \text{ mm}^2 \\
+ \text{ Area of triangle} &= \frac{6 \times 9}{2} &= 27 \text{ mm}^2 \\
\hline
\text{Total area} & &= \mathbf{97\ mm^2}
\end{aligned}
$$

Math Language

To find the **area** of a triangle we use this formula: $A = \frac{1}{2}bh$.

When dividing figures, we must avoid assumptions based on appearances. Although it may appear that the figure at right is divided into two triangles, the larger "triangle" is actually a quadrilateral. The slanted "segment" bends where the solid and dashed segments intersect. The assumption that the figure is divided into two triangles leads to an incorrect calculation for the area of the figure.

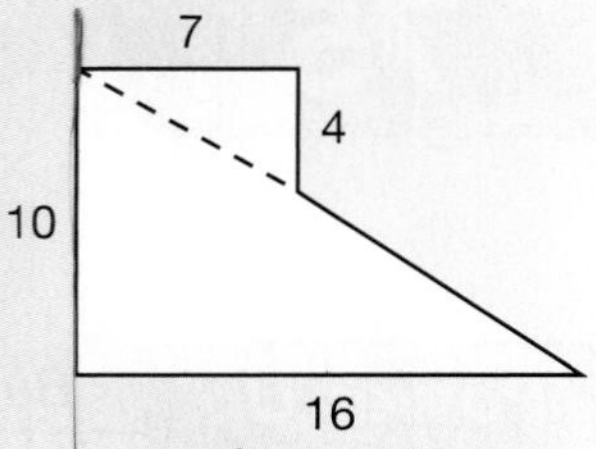

incorrect

Example 2

Find the area of this figure. Corners that look square are square. Dimensions are in centimeters.

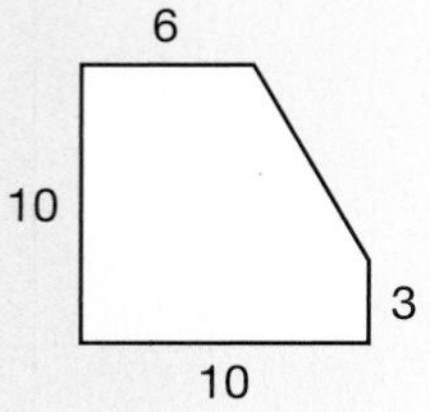

Solution

Thinking Skill

Explain

How would you find the area using method **a?**

There are many ways to divide this figure.

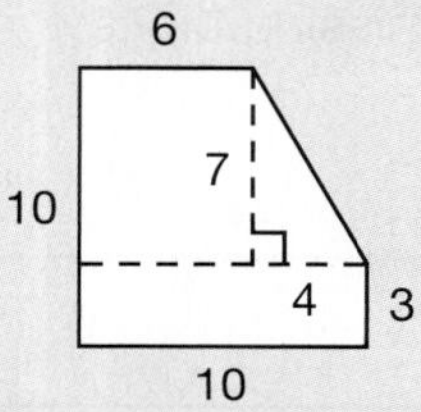

a

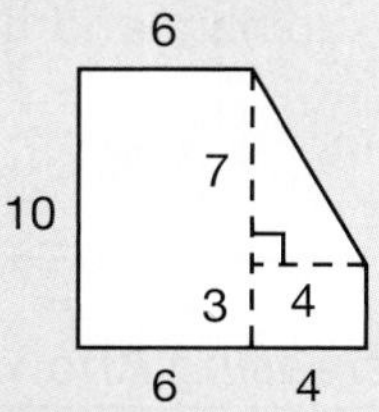

b

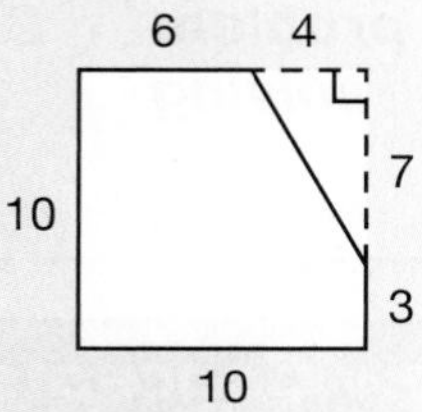

c

We decided to use **c.** We will find the area of the big rectangle and subtract from it the area of the triangle.

$$\text{Area of rectangle} = 10 \times 10 = 100 \text{ cm}^2$$

$$-\ \text{Area of triangle} = \frac{4 \times 7}{2} = 14 \text{ cm}^2$$

$$\text{Total of figure} = \mathbf{86\ cm^2}$$

area of a trapezoid

Now we will consider how to find the area of a trapezoid. The parallel sides of a trapezoid are both bases. The distance between the bases is the height. Multiplying one of the bases and the height does not give us the area of the trapezoid.

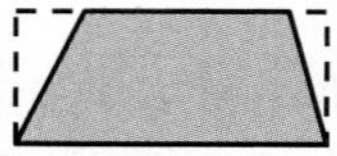
Base 2 × height
Rectangle is too big.

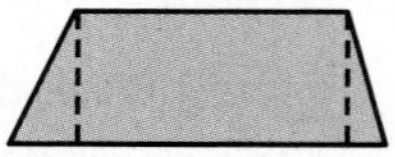
Base 1 × height
Rectangle is too small.

Recall that multiplying perpendicular lengths yields the area of a rectangle. Multiplying the base and height of a triangle results in the area of a rectangle, which we divide by 2.

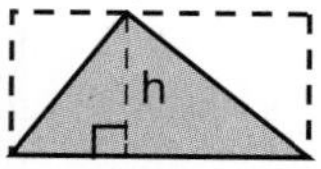

Also recall that multiplying the base and height of a parallelogram results in the area of a rectangle that equals the area of a parallelogram.

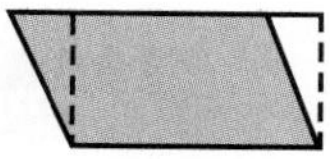

Instead of multiplying by one of the bases of a trapezoid, we multiply by the **average of the bases.** The trapezoid can be cut and rearranged to match the rectangle.

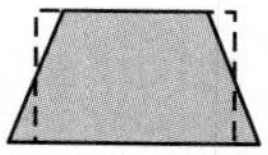

Average base $\times$ *height*
Area of rectangle equals area of trapezoid

Explain How would you find the average of the bases?

By understanding this concept we can generate a formula for area that applies to all trapezoids. We will begin with this formula:

$$\text{Area} = \text{average of the bases} \times \text{height}$$

We will label the bases of a trapezoid b_1 and b_2 and the height h as we show in this illustration:

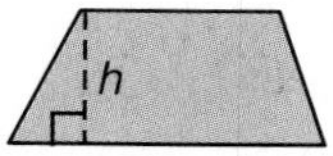

Use the labels in this illustration and the described formula to generate a formula for the area of a trapezoid.

Example 3

Find the area of this trapezoid. Dimensions are in centimeters.

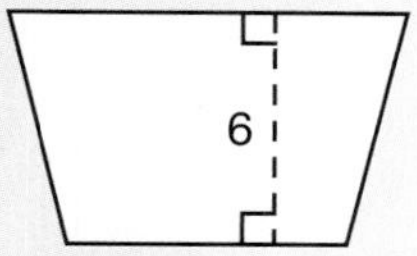

Solution

We multiply the average of the bases (7 cm and 10 cm) by the height (6 cm). We will use a formula.

$$A = \frac{1}{2}(b_1 + b_2)h$$

$$A = \frac{1}{2}(7 \text{ cm} + 10 \text{ cm})6 \text{ cm}$$

$$A = \frac{1}{2} \cdot 17 \text{ cm} \cdot 6 \text{ cm}$$

$$A = \mathbf{51 \text{ cm}^2}$$

Model Some people like to find the area of a trapezoid by dividing the trapezoid into two triangles. How could you divide the trapezoid in this example into two triangles?

Example 4

Estimate the area of this trapezoid.

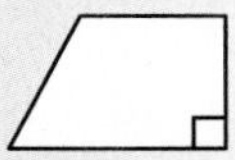

Solution

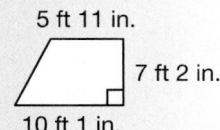

We round the two bases to 6 ft and 10 ft and the height to 7 ft. The average of the bases is 8 ft. We multiply that average by the height.

$$A = 8 \text{ ft} \cdot 7 \text{ ft} = \mathbf{56 \text{ ft}^2}$$

Practice Set

Generalize Find the area of each figure. Dimensions are in centimeters. Corners that look square are square.

a.

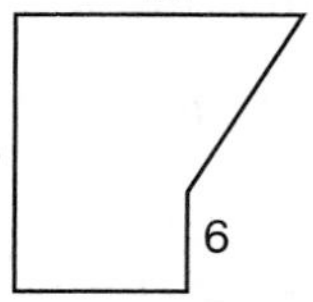

b.

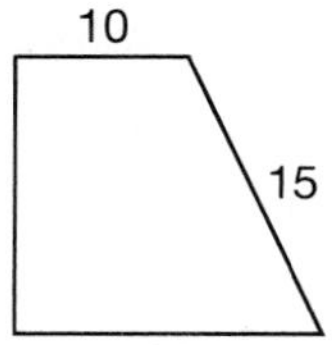

c.

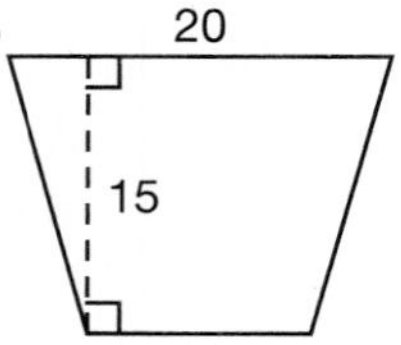

d. Write an equation for the area of a trapezoid.

e. **Estimate** In a photograph of a math book on a table, the cover appears to be a trapezoid. The parallel sides of the trapezoid are $2\frac{7}{8}$ in. and $5\frac{1}{8}$ in. The distance between the parallel sides is 3 in. Estimate the area of the photograph occupied by the cover of the book. Explain how you found the answer.

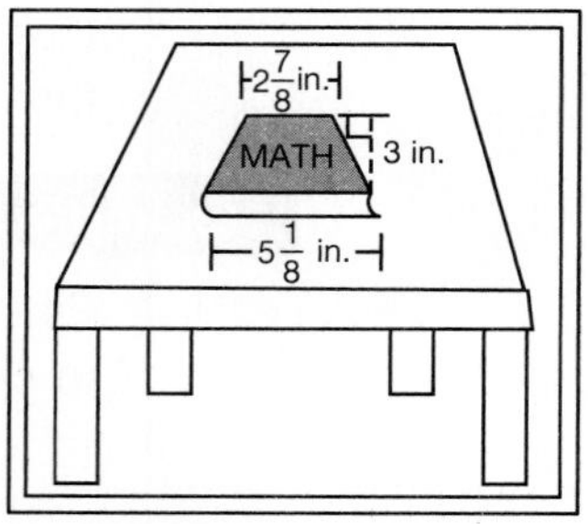

Written Practice

Strengthening Concepts

1. (55) Paula is a dog walker. She walks 8 dogs each day. She walked each of the first 6 dogs for an average of 14 minutes each. For the rest of the dogs she averaged 18 minutes each. What was the average time per walk for the 8 walks?

2. (46) If 18 ounces of cereal costs $3.69, what is the cost per ounce?

3. (32) One thousand, five hundred meters is how many kilometers?

4. (30) The sum of $\frac{1}{2}$ and $\frac{3}{5}$ is how much greater than the product of $\frac{1}{2}$ and $\frac{3}{5}$?

5. (53) The ratio of Maela's age to her niece Chelsea's age is 3 to 2. If Maela is 60 years old, how many years older than Chelsea is she?

*** 6.** (57, 69) **Analyze** Compare: $12.5 \times 10^{-4} \bigcirc 1.25 \times 10^{-3}$

*** 7.** (72) **Model** Use a ratio box to solve this problem. Maria read 40 pages in 3 hours. At this rate, how long would it take Maria to read 100 pages?

8. (22) Diagram this statement. Then answer the questions that follow.

Two fifths of the library's 21,000 books were checked out during the school year.

a. How many books were checked out?

b. How many books were not checked out?

Connect Write equations to solve problems **9–12.**

*** 9.** (74) Sixty is $\frac{5}{12}$ of what number?

10. (60) Seventy percent of $35.00 is how much money?

*** 11.** (74) Thirty-five is what fraction of 80?

*** 12.** (74) Fifty-six is what decimal part of 70?

*** 13.** (73) Simplify:

a. $\frac{-120}{4}$

b. $\left(-\frac{1}{2}\right)\left(\frac{2}{3}\right)$

c. $\frac{-120}{-5}$

d. $\left(-\frac{1}{3}\right)\left(-\frac{3}{4}\right)$

14. (70) Find the volume of this rectangular prism. Dimensions are in centimeters. What formula did you use?

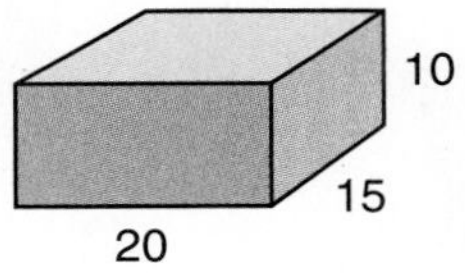

15. (65) The diameter of the plate was 11 inches. Find its circumference to the nearest half inch.

*** 16.** (75) **Estimate** Kwan's room on the second floor has a wall the shape of a trapezoid with the dimensions shown in the illustration. Estimate the area of the wall in square feet. Show how you found the area using a formula.

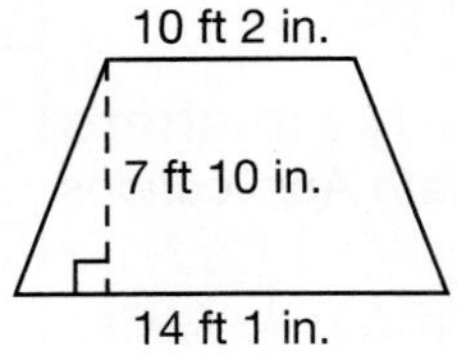

*** 17.** (75) **Analyze** A corner was trimmed from a square sheet of paper to make the shape shown. Dimensions are in centimeters.

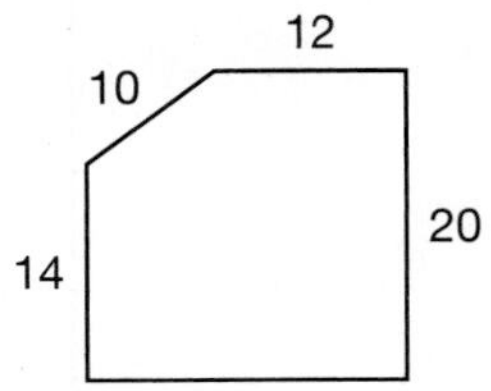

a. What was the length of each side of the square paper before the corner was trimmed?

b. Find the perimeter of the figure.

c. Find the area of the figure.

18. Complete the table.
(48)

Fraction	Decimal	Percent
a.	**b.**	125%
$\frac{1}{8}$	**c.**	**d.**

19. The taxicab bill was \$12.50. Mr. Gomez tipped the driver 20%. Altogether, how much money did Mr. Gomez pay the driver?
(60)

20. Evaluate: $x^3 - xy - \frac{x}{y}$ if $x = 2$ and $y = 0.5$
(52)

Justify For problems **21–23,** solve and check each equation. Show each step.

* **21.** $\frac{5}{8}x = 40$
(Inv. 7)

* **22.** $1.2w = 26.4$
(Inv. 7)

* **23.** $y + 3.6 = 8.47$
(Inv. 7)

Simplify:

* **24.** $9^2 - [3^3 - (9 \cdot 3 - \sqrt{9})]$
(63)

25.
(49)

$$\begin{array}{r} 2 \text{ hr } 48 \text{ min } 20 \text{ s} \\ - \ 1 \text{ hr } 23 \text{ min } 48 \text{ s} \\ \hline \end{array}$$

26. $100 \text{ yd} \cdot \frac{3 \text{ ft}}{1 \text{ yd}} \cdot \frac{12 \text{ in.}}{1 \text{ ft}}$
(50)

27. $5\frac{1}{3} \cdot \left(3 \div 1\frac{1}{3}\right)$
(26)

28. $3\frac{1}{5} + 2\frac{1}{2} - 1\frac{1}{4}$
(30)

* **29.** **Generalize**
(68)

a. $(-26) + (-15) - (-40)$

b. $(-5) + (-4) - (-3) - (+2)$

30. Find each missing exponent:
(47)

a. $5^5 \cdot 5^2 = 5^{\square}$

b. $5^5 \div 5^2 = 5^{\square}$

c. $5^2 \div 5^2 = 5^{\square}$

d. $5^2 \div 5^5 = 5^{\square}$

Early Finishers
Math Applications

Work with a partner and use 6 cubes to build a three-dimensional figure that matches all three of these views. Make a sketch of the figure you built.

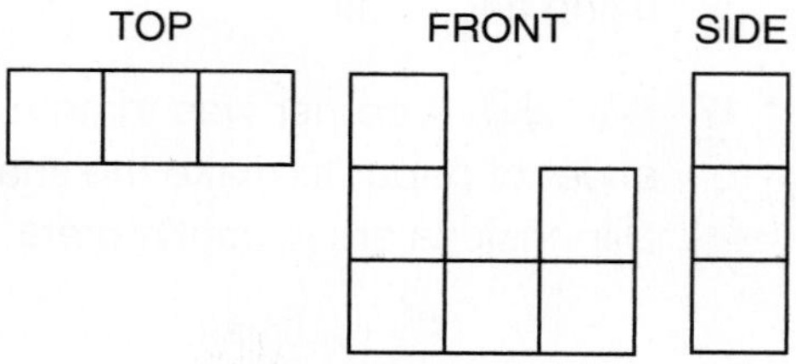

LESSON

76

• Complex Fractions

Power Up

Building Power

facts

Power Up P

mental math

a. **Positive/Negative:** $(+6) + (-18)$

b. **Decimals/Exponents:** 6.25×10^{-2}

c. **Algebra:** $9a - 4 = 32$

d. **Fractional Parts:** 12 is $\frac{2}{3}$ of what number?

e. **Measurement:** 5 mm to cm

f. **Fractional Parts:** What is $\frac{2}{3}$ of 12?

g. **Statistics:** Find the average of the set of numbers: 95, 95, 50.

h. **Calculation:** What is the total cost of a $25 video game plus 6% sales tax?

problem solving

The squares of the first nine counting numbers are each less than 100. Altogether, how many counting numbers have squares that are less than 1000?

New Concept

Increasing Knowledge

A **complex fraction** is a fraction that contains one or more fractions in the numerator or denominator. Each of the following is a complex fraction:

$$\frac{\frac{3}{5}}{\frac{2}{3}} \qquad \frac{25\frac{2}{3}}{100} \qquad \frac{15}{7\frac{1}{3}} \qquad \frac{\frac{a}{b}}{\frac{b}{c}}$$

One way to simplify a complex fraction, is to multiply the complex fraction by a fraction name for 1 that makes the denominator 1.

Example 1

Thinking Skill

Analyze

What is the relationship between fraction pairs such as $\frac{2}{3}$ and $\frac{3}{2}$? What is their product?

Simplify: $\dfrac{\frac{3}{5}}{\frac{2}{3}}$

Solution

We focus our attention on the denominator of the complex fraction. Our goal is to make the denominator 1. We multiply the denominator by the reciprocal of $\frac{2}{3}$, which is $\frac{3}{2}$, so that the new denominator is 1. We also multiply the numerator by $\frac{3}{2}$.

$$\frac{\frac{3}{5}}{\frac{2}{3}} \times \frac{\frac{3}{2}}{\frac{3}{2}} = \frac{\frac{9}{10}}{1} \text{ or } \mathbf{\frac{9}{10}}$$

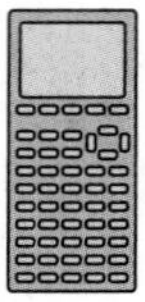

Visit www.SaxonPublishers.com/ActivitiesC2 *for a graphing calculator activity.*

We multiplied the complex fraction by a complex name for 1 to change the denominator to 1. Since $\frac{9}{10}$ divided by 1 is $\frac{9}{10}$, the complex fraction simplifies to $\frac{9}{10}$.

An alternative method for simplifying some complex fractions is to treat the fraction as a division problem. We can change the format of the division problem to a more familiar form.

$$\begin{matrix}\text{divided} \\ \text{divisor}\end{matrix} \quad \frac{\frac{3}{5}}{\frac{2}{3}} \rightarrow \frac{3}{5} \div \frac{2}{3}$$

Then we simplify the division using the method described in Lesson 25.

$$\frac{3}{5} \div \frac{2}{3} = \frac{3}{5} \cdot \frac{3}{2} = \mathbf{\frac{9}{10}}$$

Example 2

Simplify: $\mathbf{\dfrac{25\frac{2}{3}}{100}}$

Solution

First we write both numerator and denominator as fractions.

$$\frac{\frac{77}{3}}{\frac{100}{1}}$$

Now we multiply the numerator and the denominator by $\frac{1}{100}$.

$$\frac{\frac{77}{3}}{\frac{100}{1}} \cdot \frac{\frac{1}{100}}{\frac{1}{100}} = \frac{\frac{77}{300}}{1} \text{ or } \mathbf{\frac{77}{300}}$$

Example 3

Simplify: $\mathbf{\dfrac{15}{7\frac{1}{3}}}$

Solution

We begin by writing both numerator and denominator as improper fractions.

$$\frac{\frac{15}{1}}{\frac{22}{3}}$$

Now we multiply the numerator and the denominator by $\frac{3}{22}$.

$$\frac{\frac{15}{1}}{\frac{22}{3}} \cdot \frac{\frac{3}{22}}{\frac{3}{22}} = \frac{\frac{45}{22}}{1} \text{ or } \mathbf{2\frac{1}{22}}$$

Discuss Why do you think $\frac{3}{22}$ was selected as the multiplier in this example?

Example 4

Change $83\frac{1}{3}\%$ to a fraction and simplify:

Solution

Thinking Skill

Generalize

Name three complex fractions equal to 1.

A percent is a fraction that has a denominator of 100. Thus $83\frac{1}{3}\%$ is

$$\frac{83\frac{1}{3}}{100}$$

Next we write both numerator and denominator as fractions.

$$\frac{\frac{250}{3}}{\frac{100}{1}}$$

Now we multiply the numerator and the denominator by $\frac{1}{100}$.

$$\frac{\frac{250}{3}}{\frac{100}{1}} \cdot \frac{\frac{1}{100}}{\frac{1}{100}} = \frac{\frac{250}{300}}{1} = \mathbf{\frac{5}{6}}$$

Practice Set

Generalize Simplify each complex fraction:

a. $\frac{37\frac{1}{2}}{100}$ **b.** $\frac{12}{\frac{5}{6}}$ **c.** $\frac{\frac{2}{5}}{\frac{2}{3}}$

Change each percent to a fraction and simplify:

d. $66\frac{2}{3}\%$ **e.** $8\frac{1}{3}\%$ **f.** $4\frac{1}{6}\%$

g. **Justify** Which two properties of multiplication do we use to simplify complex fractions?

Written Practice

Strengthening Concepts

1. (46) **a.** Nestor finished a 42-kilometer bicycle race in 1 hour 45 minutes ($1\frac{3}{4}$ hr). What was his average speed in kilometers per hour?

b. **Explain** How did we know that 1 hour and 45 minutes is $1\frac{3}{4}$ hours?

2. (28) Akemi's scores in the diving competition were 7.9, 8.3, 8.1, 7.8, 8.4, 8.1, and 8.2. The highest and lowest scores were not counted. What was the average of the remaining scores?

3. (66) Use a ratio box to solve this problem. The school art club is having an exhibit. The ratio of oil paintings to acrylic paintings being displayed is 2 to 5. If there are 35 paintings displayed, how many of them are oil paintings?

* **4.** (50) **Connect** Use a unit multiplier to convert 3.5 grams to milligrams.

* **5.** (76) Change $16\frac{2}{3}$ percent to a fraction and simplify.

6. (17) Davie is facing north. If he turns 90° in a clockwise direction, what direction will he be facing?

7. (60) One sixth of the rock's mass was quartz. If the mass of the rock was 144 grams, what was the mass of the quartz in the rock?

8. (41, 57) For $a = 2$, evaluate

a. $\sqrt{2a^3}$ **b.** $a^{-1} \cdot a^{-2}$

* **9.** (73) **Generalize** Simplify each expression:

a. $\frac{-60}{-12}$ **b.** $\left(-\frac{1}{2}\right)\left(\frac{1}{2}\right)$

c. $\frac{40}{-8}$ **d.** $\left(-\frac{1}{4}\right)\left(-\frac{1}{4}\right)$

10. (65) What is the circumference of the circle shown?

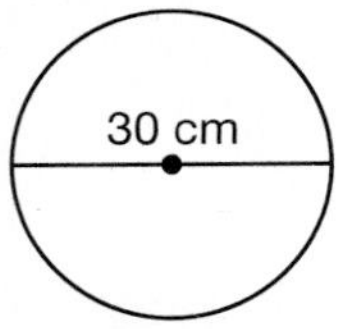

Leave π as π

11. (67) The figure at right is a pyramid with a square base. Find the number of:

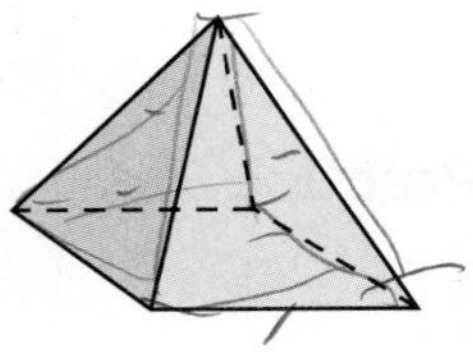

a. faces

b. edges

c. vertices

Write equations to solve problems **12–15.**

12. (60) What number is 10 percent of $37.50?

13. (60) What number is $\frac{5}{8}$ of 72?

* **14.** (74) **Analyze** Twenty-five is what fraction of 60?

* **15.** (74) **a.** Sixty is what decimal part of 80?

b. **Justify** How can you check that your answer to **a** is correct?

16. (62) In this figure $AC = AB$. Angles DCA and ACB are supplementary. Find the measure of

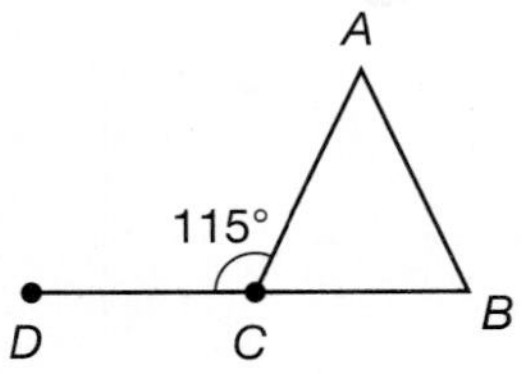

a. $\angle ACB$.

b. $\angle ABC$.

c. $\angle CAB$.

17. Complete the table.
(48)

Fraction	Decimal	Percent
$\frac{5}{6}$	**a.**	**b.**
c.	**d.**	0.1%

* **18.** **Analyze** A square sheet of paper with an area of 81 in.^{2} has a corner cut off, forming a pentagon as shown.
(75)

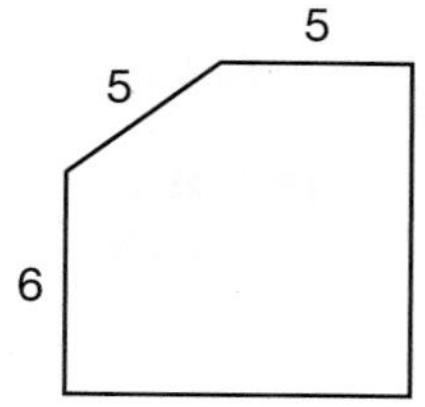

a. What is the perimeter of the pentagon?

b. What is the area of the pentagon?

c. **Explain** How did you find the area of the pentagon?

* **19.** **Conclude** What type of parallelogram has four congruent angles but not necessarily four congruent sides?
(Inv. 6)

20. When water increases in temperature from its freezing point to its boiling point, the reading on a thermometer increases from 0°C to 100°C and from 32°F to 212°F. The temperature halfway between 0°C and 100°C is 50°C. What temperature is halfway between 32°F and 212°F?
(16, 28)

Justify For problems **21–24,** solve and check each equation. Show each step.

* **21.** $x - 25 = 96$
(Inv. 7)

* **22.** $\frac{2}{3}m = 12$
(Inv.7)

* **23.** $2.5p = 6.25$
(Inv. 7)

* **24.** $10 = f + 3\frac{1}{3}$
(Inv. 7)

Simplify:

25. $\sqrt{13^2 - 5^2}$
(20)

26. 1 ton − 400 lb
(16)

27. $3\frac{3}{4} \times 4\frac{1}{6} \times (0.4)^2$ (fraction answer)
(26, 43)

28. $3\frac{1}{8} + 6.7 + 8\frac{1}{4}$ (decimal answer)
(35, 43)

* **29.** **Generalize** **a.** $(-3) + (-5) - (-3) - |+5|$
(68)

b. $(-2.4) - (+1.2)$

* **30.** **Estimate** Before dividing, determine whether the quotient is greater than or less than 1 and state why. Then perform the calculation.
(76)

$$\frac{\frac{5}{6}}{\frac{2}{3}}$$

LESSON

77

• Percent of a Number, Part 2

Power Up

Building Power

facts

Power Up P

mental math

a. **Positive/Negative:** $(+12) - (-18)$

b. **Exponents:** 4×10^6

c. **Ratio:** $\frac{100}{150} = \frac{30}{a}$

d. **Estimation:** Estimate 15% of \$61.

e. **Measurement:** 25 cm to m

f. **Fractional Parts:** 12 is $\frac{3}{4}$ of n.

g. **Statistics:** Find the average of the set of numbers: 100, 60, 90, 70.

h. **Calculation:** 10% of 50, × 6, + 2, ÷ 4, × 2, $\sqrt{\ }$, × 9, $\sqrt{\ }$, × 7, ÷ 2

problem solving

On a balanced scale are a 25-g mass, a 100-g mass, and five identical blocks marked x, which are distributed as shown. What is the mass of each block marked x? Write an equation illustrated by this balanced scale.

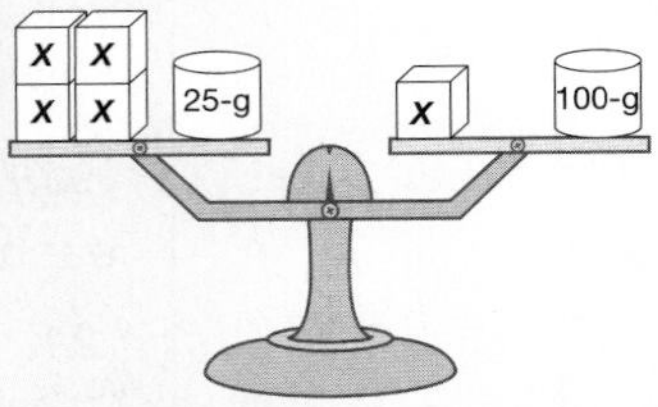

New Concept

Increasing Knowledge

In Lesson 74 we practiced fractional-part problems involving fractions and decimals. In this lesson we will practice similar problems involving percents. First we translate the problem into an equation; then we solve the equation.

Example 1

What percent of 40 is 25?

Solution

We translate the question to an equation and solve.

What percent of 40 is 25?	question
↓ ↓ ↓ ↓ ↓	
$W_P \times 40 = 25$	equation

To solve we divide both sides of the equation by 40.

$$\frac{W_P \times \overset{1}{\cancel{40}}}{\underset{1}{\cancel{40}}} = \frac{25}{40} \quad \text{divided by 40}$$

$$W_P = \frac{5}{8} \quad \text{simplified}$$

Since the question asked "what percent" and not "what fraction," we convert the fraction $\frac{5}{8}$ to a percent.

$$\frac{5}{8} \times 100\% = \mathbf{62\frac{1}{2}\%} \qquad \text{converted to a percent}$$

Verify How can you decide if this answer is reasonable?

Thinking Skills

Summarize

What steps did we follow to convert $\frac{5}{8}$ to a percent?

Example 2

What percent of \$3.50 is \$0.28?

Solution

We translate and solve.

What percent of \$3.50 is \$0.28? — question

$$W_P \times \$3.50 = \$0.28 \qquad \text{equation}$$

$$\frac{W_P \times \cancel{\$3.50}^{1}}{\cancel{\$3.50}_{1}} = \frac{\$0.28}{\$3.50} \qquad \text{divided by \$3.50}$$

$$W_P = \frac{0.28}{3.5} \qquad \text{simplified}$$

We perform the decimal division.

$$W_P = \frac{0.28}{3.5} = \frac{2.8}{35} = 0.08 \qquad \text{divided}$$

This is a decimal answer. The question asked for a percent answer so we convert the decimal 0.08 to 8%.

$$W_P = \mathbf{8\%} \qquad \text{converted to a percent}$$

Thinking Skills

Summarize

How did we convert 0.08 to 8%?

Example 3

Seventy-five percent of what number is 600?

Solution

We translate the question to an equation and solve. We can translate 75% to a fraction or to a decimal. We choose a fraction for this example.

Seventy-five percent of what number is 600? — question

$$\frac{75}{100} \times W_N = 600 \qquad \text{equation}$$

To solve, we multiply both sides by 100 over 75 as follows:

$$\frac{\cancel{100}^{1}}{\cancel{75}_{1}} \times \frac{\cancel{75}^{1}}{\cancel{100}_{1}} \times W_N = 600 \cdot \frac{100}{75} \qquad \text{multiplied by } \frac{100}{75}$$

$$W_N = \mathbf{800} \qquad \text{simplified}$$

Example 4

Fifty is what percent of 40?

Solution

Since 50 is more than 40, the answer will be greater than 100%. We translate to an equation and solve.

Fifty is what percent of 40?	question
$50 = W_P \times 40$	equation

We divide both sides by 40.

$$\frac{50}{40} = \frac{W_P \times \overset{1}{\cancel{40}}}{\underset{1}{\cancel{40}}} \quad \text{divided by 40}$$

$$\frac{5}{4} = W_P \quad \text{simplified}$$

We convert $\frac{5}{4}$ to a percent.

$$W_P = \frac{5}{4} \times 100\% = \mathbf{125\%} \quad \text{converted to a percent}$$

Example 5

Sixty is 150 percent of what number?

Solution

Predict Do you think the answer will be greater than or less than 60? Why?

We translate by writing 150% as either a decimal or a fraction. We will use the decimal form here.

Sixty is 150% of what number?	question
$60 = 1.5 \times W_N$	equation

We divide both sides of the equation by 1.5.

$$\frac{60}{1.5} = \frac{\overset{1}{\cancel{1.5}} \times W_N}{\underset{1}{\cancel{1.5}}} \quad \text{divided by 40}$$

$$\mathbf{40} = W_N \quad \text{simplified}$$

Practice Set

Generalize Find each percent or quantity in **a–f.**

a. Twenty-four is what percent of 40?

b. What percent of 6 is 2?

c. Fifteen percent of what number is 45?

d. What percent of 4 is 6?

e. Twenty-four is 120% of what number?

f. What percent of $5.00 is $0.35?

g. Rework example 5, writing 150% as a fraction instead of as a decimal.

h. **Justify** How can you show that your answer to problem **f** is correct?

Written Practice *Strengthening Concepts*

1. (66) Use a ratio box to solve this problem. Tammy saved nickels and pennies in a jar. The ratio of nickels to pennies was 2 to 5. If there were 70 nickels in the jar, how many coins were there in all?

Refer to the line graph below to answer problems **2–4.**

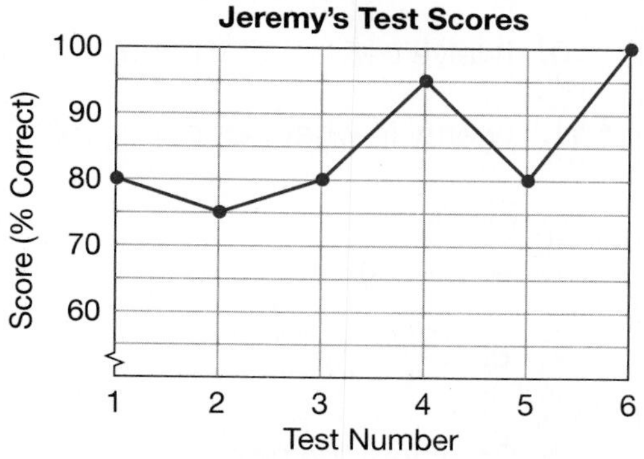

2. (38, 60) If there were 50 questions on Test 1, how many questions did Jeremy answer correctly?

3. (28, 38) What was Jeremy's average score? (What was the mean of the scores?)

4. (38, Inv. 4) **a.** Which score did Jeremy make most often? (What was the mode of the scores?)

b. What was the difference between his highest score and his lowest score? (What was the range of the scores?)

5. (67) Name the shape of each object:

a. a marble

b. a length of pipe

c. a box of tissues

*** 6.** (72) **Model** Use a Case 1/Case 2 ratio box to solve this problem. One hundred inches equals 254 centimeters. How many centimeters equals 250 inches?

*** 7.** (36, 71) **Model** Diagram this statement. Then answer the questions that follow.

Three fifths of those present agreed, but the remaining 12 disagreed.

a. What fraction of those present disagreed?

b. How many were present?

c. How many of those present agreed?

d. What was the ratio of those who agreed to those who disagreed?

e. If there were 50 people present and 30 people agreed, would the ratio of those who agreed to those who disagreed change? Why or why not?

Formulate Write equations to solve problems **8–11.**

*** 8.** (74) Forty is $\frac{4}{25}$ of what number?

9. (60) Twenty-four percent of 10,000 is what number?

*** 10.** (77) Twelve percent of what number is 240?

*** 11.** (77) Twenty is what percent of 25?

*** 12.** (73) Simplify:

a. $25(-5)$ **b.** $-15(-5)$

c. $\frac{-250}{-5}$ **d.** $\frac{-225}{15}$

13. (48) Complete the table.

Fraction	Decimal	Percent
a.	0.2	**b.**
c.	**d.**	2%

14. (60) A pair of pants costs \$21 plus 7.5% sales tax. What is the total cost of the pants? Summarize how you found your answer.

*** 15.** (76) **Generalize** Simplify:

a. $\frac{14\frac{2}{7}}{100}$ **b.** $\frac{60}{\frac{2}{3}}$

*** 16.** (75) **Analyze** Find the area of this symmetrical figure. Dimensions are in feet. Corners that look square are square.

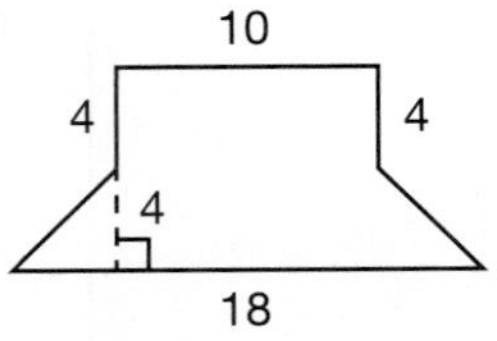

17. (67, 70) Draw a cube with edges 2 cm long.

a. What is the volume of the cube?

b. Describe how to find the surface area of a cube.

18. (51) According to the U.S. Census Bureau, in 2004 the total amount of sales on the Internet was about 69 billion dollars. Write 69 billion in scientific notation.

19. (65) Find the circumference of each circle:

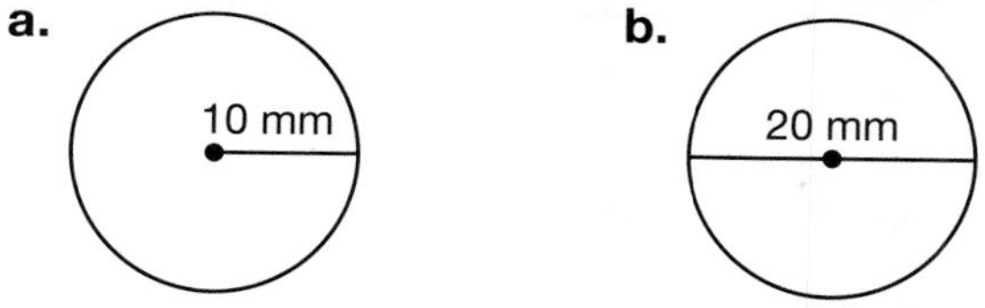

a. Leave π as π.

b. Use 3.14 for π.

Justify For problems **20–22,** solve and check each equation. Show each step.

* **20.** (Inv. 7) $3x = 26.7$

21. (Inv. 7) $y - 3\frac{1}{3} = 7$

* **22.** (Inv. 7) $\frac{2}{3}x = 48$

23. (56) Write the rule of the function with words and as an equation. Then find the missing numbers.

x	*y*
3	13
1	5
2	9
4	17
0	

Simplify:

* **24.** (63) **Generalize** $5^2 - \{2^3 + 3[4^2 - (4)(\sqrt{9})]\}$

25. (49)

$$\begin{array}{r} 4 \text{ gal } 3 \text{ qt } 1 \text{ pt} \\ +1 \text{ gal } 2 \text{ qt } 1 \text{ pt} \\ \hline \end{array}$$

26. (50) $1 \text{ ft}^2 \cdot \frac{12 \text{ in.}}{1 \text{ ft}} \cdot \frac{12 \text{ in.}}{1 \text{ ft}}$

27. (26) $5\frac{1}{3} \div \left(1\frac{1}{3} \div 3\right)$

28. (30) $3\frac{1}{5} - 2\frac{1}{2} + 1\frac{1}{4}$

29. (43) $3\frac{1}{3} \div 2.5$ (mixed-number answer)

30. (68) **a.** $(-3) + (-4) - (+5)$

b. $(-6) - (-16) - (+30)$

Early Finishers
Real-World Application

The Ortiz family had a large celebration dinner. The ages of the people at the dinner were:

1, 10, 10, 12, 16, 31, 35, 40, 65, 75, 90

Find the mean, median, mode, and range of the ages of the people at the dinner. Which best describes the ages of the people at the dinner? Justify your choice.

LESSON 78

• Graphing Inequalities

Power Up

Building Power

facts Power Up O

mental math

a. Positive/Negative: $(-8) - (-16)$

b. Exponents: $1^0 + 1^2$

c. Algebra: $10p + 3 = 63$

d. Percent: 5% of \$640.00 (double, halve)

e. Measurement: 750 g to kg

f. Fractional Parts: 12 is $\frac{1}{6}$ of m.

g. Statistics: Find the average of the set of numbers: 95, 85, 75, 65.

h. Estimation: Estimate a 15% tip on a \$31.49 bill.

problem solving

The uniform at WCMS consists of royal blue shirts with khaki pants. A light blue dress shirt and maroon tie is required for Mondays. Libosha has one dress shirt (with tie), three royal blue casual shirts, and three pairs of khaki pants. How many different combinations of the uniform can Libosha wear?

New Concept

Increasing Knowledge

We have used the symbols $>$, $<$, and $=$ to compare two numbers. In this lesson we will introduce the symbols $\geq$ and $\leq$. We will also practice graphing on the number line.

The symbols $\geq$ and $\leq$ combine the greater than/less than symbols with the equal sign. Thus, the symbol

$$\geq$$

is read, "greater than or equal to." The symbol

$$\leq$$

is read, "less than or equal to."

To graph a number on the number line, we draw a dot at the point that represents the number. Thus when we graph 4 on the number line, it looks like this:

This time we will graph *all the numbers that are greater than or equal to 4.* We might think the graph should look like this:

It is true that all the dots mark points that represent numbers greater than or equal to 4. However, we did not graph *all* the numbers that are greater than 4. For instance, we did not graph 10, 11, 12, and so on. Also, we did not graph numbers like $4\frac{1}{2}$, $5\frac{1}{3}$, $\sqrt{29}$, or 2π. If we were to graph all these numbers, the dots would be so close together that we would end up with a ray that goes on and on. Thus we graph all the numbers greater than or equal to 4 like this:

The large dot marks the 4. The blue ray marks the numbers greater than 4. The blue arrowhead shows that this ray continues without end.

Expressions such as the following are called **inequalities:**

a $x \leq 4$ **b** $x > 4$

We read **a** as "*x* is less than or equal to 4." We read **b** as "*x* is greater than 4."

We can graph inequalities on the number line by graphing all the numbers that make the inequality a true statement.

Thinking Skill

Conclude

Would a graph of the inequality $x \geq 4$ include 3? Why or why not?

Example 1

Graph on a number line: $x \leq 4$

Solution

We are told to graph all numbers that are less than or equal to 4. We draw a dot at the point that represents 4, and then we shade all the points to the left of the dot. The red arrowhead shows that the shading continues without end.

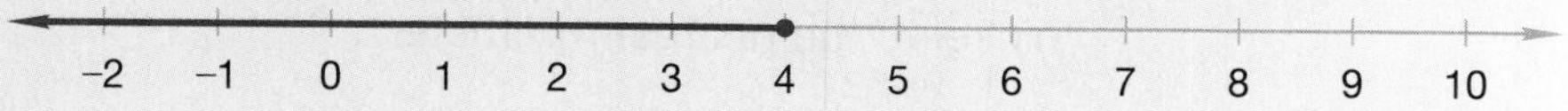

Example 2

Graph on a number line: $x > 4$

Solution

We are told to graph all numbers greater than 4 *but not including* 4. We do not start the graph at 5, because we need to graph numbers like $4\frac{1}{2}$ and 4.001. To show that the graph does not include 4, *we draw an empty circle* at 4. Then we shade the portion of the number line to the right of the circle.

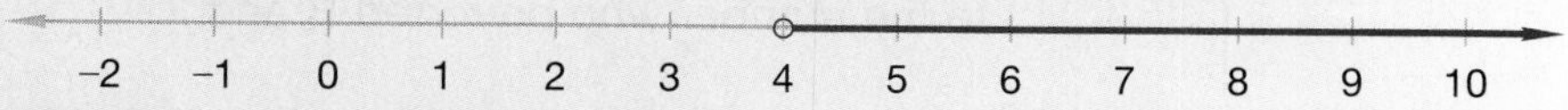

Discuss How would a graph of $x \geq 4$ differ from the graph of $x > 4$?

Practice Set

a. On a number line, graph all the numbers less than 2.

b. On a number line, graph all the numbers greater than or equal to 1.

Model Graph each inequality on a number line:

c. $x \leq -1$

d. $x > -1$

e. **Explain** What is the difference in meaning between a dot and an empty circle on a number line graph?

Written Practice

Strengthening Concepts

*** 1.** (72) **Explain** If 4 cartons are needed to pack 30 paperback books, how many cartons are needed to pack 75 paperbacks? (Assume the books are the same size.) Explain how you found your answer.

2. (55) In one northern city, the average snowfall after four months was 7 inches. What must the average snowfall be for the next two months in order to have a six-month average of 12 inches?

3. (30) If the sum of $\frac{2}{3}$ and $\frac{3}{4}$ is divided by the product of $\frac{2}{3}$ and $\frac{3}{4}$, what is the quotient?

4. (53) Use a ratio box to solve this problem. Two seed types of flowering plants are monocotyledons and dicotyledons. Suppose the ratio of monocotyledons to dicotyledons in the nursery was 3 to 4. If there were 84 dicotyledons in the nursery, how many monocotyledons were there?

5. (65) The diameter of a nickel is 21 millimeters. Find the circumference of a nickel to the nearest millimeter.

*** 6.** (78) **Model** Graph each inequality on a separate number line:

a. $x > 2$

b. $x \leq 1$

7. (50) Use a unit multiplier to convert 1.5 kg to g.

8. (65) Five sixths of the 30 people who participated in the taste test preferred Brand X. The rest preferred Brand Y.

a. How many more people preferred Brand X than preferred Brand Y?

b. What was the ratio of the number of people who preferred Brand Y to the number who preferred Brand X?

Formulate Write equations to solve problems **9–12.**

*** 9.** (74) Forty-two is seven tenths of what number?

*** 10.** (77) One hundred fifty percent of what number is 600?

11. (60) Forty percent of 50 is what number?

* **12.** Forty is what percent of 50?
(77)

13. **a.** Write 1.5×10^{-3} in standard form.
(57, 69)

b. Write 25×10^{6} in scientific notation.

* **14.** Simplify:
(73)

a. $\frac{-4.5}{9}$

b. $\frac{-2.4}{-0.6}$

c. $15(-20)$

d. $-15(-12)$

15. Complete the table.
(48)

Fraction	Decimal	Percent
a.	b.	50%
$\frac{1}{12}$	c.	d.

* **16.** Simplify: $\frac{83\frac{1}{3}}{100}$
(76)

* **17.** **Estimate** On one wall of his room Kwan has a window shaped like a trapezoid with the dimensions shown. Estimate the area of the window in square inches. Show how you used the formula.
(75)

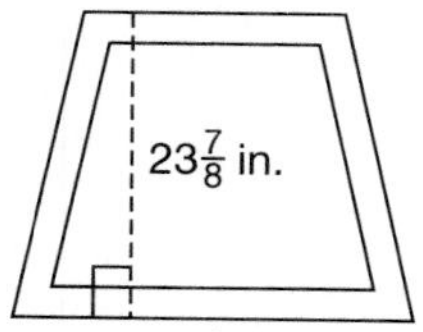

18. A box of tissues is 24 cm long, 12 cm wide, and 10 cm tall. Draw the box and find its volume.
(70)

* **19.** **Model** Draw the box in problem **18** as if it were cut open and unfolded so that the six faces are lying flat.
(70)

20. Write the rule of this function with words and as an equation. Then find the missing number.
(56)

Input x	Output y
2	9
3	14
4	
5	24

21. A merchant sold an item for $18.50. If 30% of the selling price was profit, how much profit did the merchant make on the sale?
(60)

For problems **22–24,** solve and check each equation. Show each step.

22. $m + 8.7 = 10.25$
(Inv. 7)

23. $\frac{4}{3}w = 36$
(Inv. 7)

24. $0.7y = 48.3$
(Inv. 7)

Simplify:

* **25.** (63) **Generalize** $\{4^2 + 10[2^3 - (3)(\sqrt{4})]\} - \sqrt{36}$

26. (59) $|5 - 3| - |3 - 5|$

27. (50) $1\ \text{m}^2 \cdot \dfrac{100\ \text{cm}}{1\ \text{m}} \cdot \dfrac{100\ \text{cm}}{1\ \text{m}}$

28. (26) $7\frac{1}{2} \cdot 3 \cdot \left(\frac{2}{3}\right)^2$

29. (30) $3\frac{1}{5} - \left(2\frac{1}{2} - 1\frac{1}{4}\right)$

30. (68) **a.** $(-10) - (-8) - (+6)$

b. $\left(-\frac{1}{5}\right) + \left(-\frac{2}{5}\right) - \left(-\frac{3}{5}\right)$

Early Finishers

Real-World Application

The Mendoza family needs to buy new carpet for their family room. The diagram below shows the dimensions of the family room.

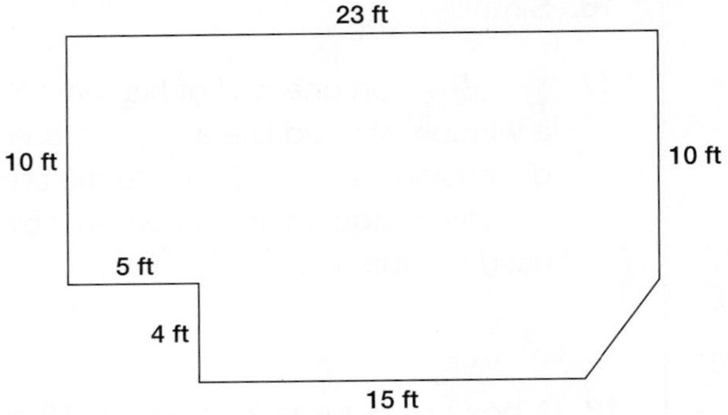

a. Calculate the area of this room.

b. If the Mendozas choose a carpet that costs \$2.95 per square foot, how much will it cost them to carpet their family room?

LESSON

79 • Estimating Areas

Power Up

Building Power

facts Power Up P

mental math

a. **Positive/Negative:** $(-25) + (-15)$

b. **Decimals/Exponents:** 3.75×10^3

c. **Ratio:** $\frac{c}{100} = \frac{25}{10}$

d. **Estimation:** Estimate 15% of \$11.95.

e. **Measurement:** 1200 mL to L

f. **Fractional Parts:** 20 is $\frac{4}{5}$ of n.

g. **Statistics:** Find the average of the set of numbers: 60, 70, 80, 90, 100.

h. **Calculation:** Square 5, $\times$ 2, $-$ 1, $\sqrt{\ }$, $\times$ 8, $-$ 1, $\div$ 5, $\times$ 3, $-$ 1, $\div$ 4, $\times$ 9, $+$ 3, $\div$ 3

problem solving

A 6-by-6 square of grid paper, ruled on the front and un-ruled on the back, is folded in half diagonally to form an isosceles right triangle. Then part of the paper is folded back as shown in the diagram. What is the area of the un-ruled part of the diagram?

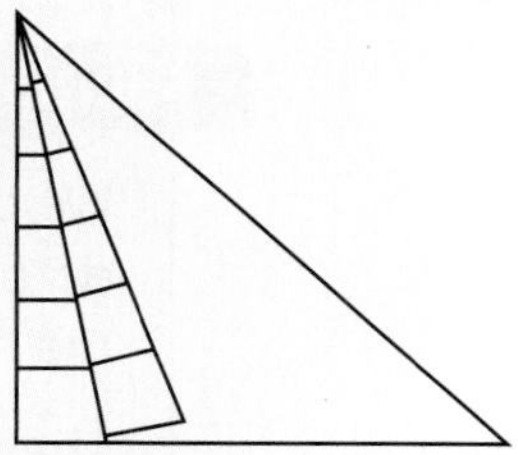

New Concept

Increasing Knowledge

Thinking Skill

Analyze

What do the dots on some of the squares mean?

As Tucker balanced one foot on the scale he wondered how many pounds per square inch the scale was supporting. Tucker traced the outline of his shoe on a one-inch square grid and counted squares to estimate the area of his shoe. Here is a reduced image of the inch grid. Can you estimate the area of Tucker's shoe print?

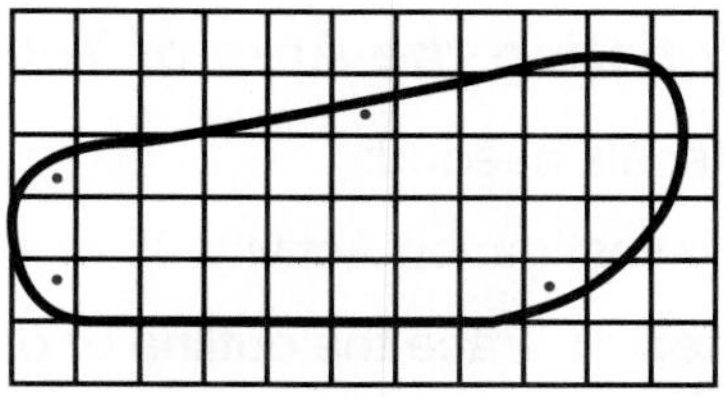

To estimate the area of the shoe print Tucker counted the complete or nearly complete squares within the shoe's outline as whole squares. Then he counted the squares that were about half within the outline as half squares. He counted 30 full or nearly full squares and four "half" squares.

Tucker estimated that the area of his shoe print was 32 square inches. Tucker weighs 116 pounds, so he divided his weight by 32 in.^{2} and found that his weight was distributed to about $3\frac{1}{2}$ pounds per square inch.

$$\frac{116 \text{ lbs}}{33 \text{ in.}^2} \approx 3.52 \text{ lbs per in.}^2$$

Discuss Would the number of pounds per square inch on each foot be greater or less than 3.52 if Tucker stood on two feet? Explain.

Example

To prevent a 500 pound piano from damaging the floor, the owner put 6-inch diameter circular castors under the three legs of the piano. Use Tucker's method to estimate the area of each castor. Then find the pounds per square inch the floor supports.

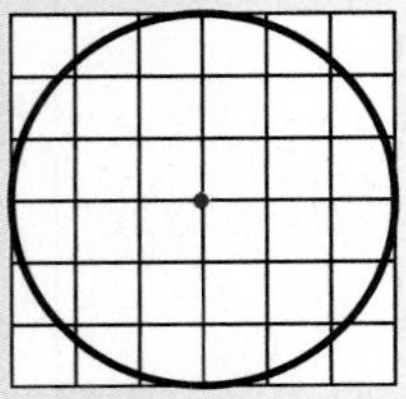

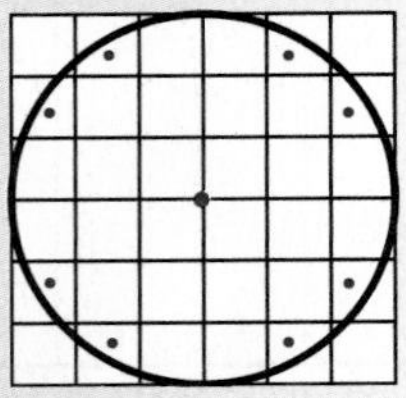

Solution

We count 24 whole or nearly whole squares. The "half squares" we mark with dots. There are 8 "half squares" that we count as 4 full squares. We estimate the total area of each castor as **28 square inches.**

There are three castors so their total area is about **84 in.^{2}.** The weight of the piano is distributed over the area of the three castors, so we divide the 500 lbs weight by 84 in.^{2} to find the pounds per square inch.

$$\frac{500 \text{ lbs}}{84 \text{ in.}^2} \approx \mathbf{6 \text{ lbs per in.}^2}$$

Infer The number of pounds per square inch may not be the same on each castor. Why is this true?

Activity

Estimating the Area of Your Handprint

Materials needed:

- Investigation Activity 25 or square centimeter grid paper

Step 1: Trace the outline of one hand with fingers together onto the grid paper.

Step 2: Then count squares to estimate the area of your handprint.

Practice Set **Estimate** Use figures A or B to estimate each area.

a. Figure A shows a map with a square kilometer grid and the outline of a lake. Estimate the area of the lake.

b. Figure B shows a square with sides 2 cm long and a circle with a radius of 2 cm. Estimate the area of the circle. The area of the circle is about how many times the area of the square?

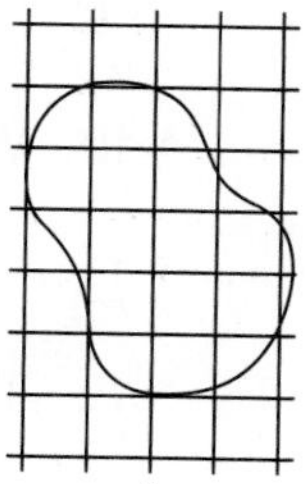

A

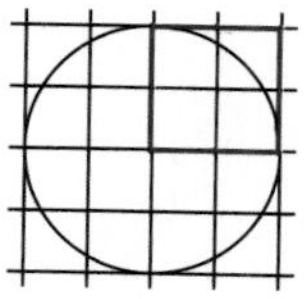

B

Written Practice *Strengthening Concepts*

1. (55) Students were grouped into 4 classrooms with an average of 33.5 students per classroom. If the students were regrouped into 5 classrooms, what would be the average number of students in each room?

2. (46) Nelda drove 315 kilometers and used 35 liters of gasoline. Her car averaged how many kilometers per liter of gas?

3. (66) The ratio of young people to adults at a contest was 7 to 5. If the total number of young people and adults was 1260, how many more young people were there than adults?

4. (69) Write each number in scientific notation:

a. 37.5×10^{-6} **b.** 37.5×10^{6}

* **5.** (79) **Analyze** A circle with a radius of one inch has an area of about how many square inches?

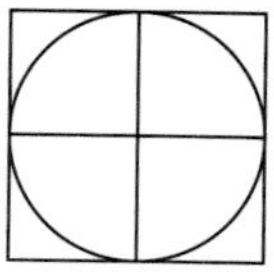

* **6.** (78) **Model** Graph each inequality on a separate number line:

a. $x < 1$ **b.** $x \geq -1$

* **7.** (72) Use a case 1–case 2 ratio box to solve this problem. Four inches of snow fell in 3 hours. At that rate, how long would it take for 1 foot of snow to fall?

8. (71) Diagram this statement. Then answer the questions that follow.

Twelve students in the school play were dancers. This was $\frac{3}{8}$ of all the students in the play.

a. How many students did not dance in the play?

b. What percent of the students did not dance in the play?

Formulate Write equations to solve problems **9–12**.

* **9.** Thirty-five is 70% of what number?
(77)

* **10.** What percent of 20 is 17?
(77)

* **11.** What percent of 20 is 25?
(77)

* **12.** Three hundred sixty is 75 percent of what number?
(77)

13. Simplify:
(73)

a. $\frac{1.44}{-8}$

b. $\frac{-14.4}{+6}$

c. $-12(1.2)$

d. $-1.6(-9)$

14. Complete the table.
(48)

Fraction	Decimal	Percent
$\frac{1}{25}$	**a.**	**b.**
c.	**d.**	8%

15. At the Stanford Used Car Dealership, a salesperson is paid a commission of 5% of the sale price for every car he or she sells. If a salesperson sells a car for $13,500, how much would he or she be paid as a commission?
(60)

* **16.** Simplify: $\frac{62\frac{1}{2}}{100}$
(76)

* **17.** **Analyze** A square sheet of paper with an area of 100 in.^{2} has a corner cut off, as shown in the figure below.
(75)

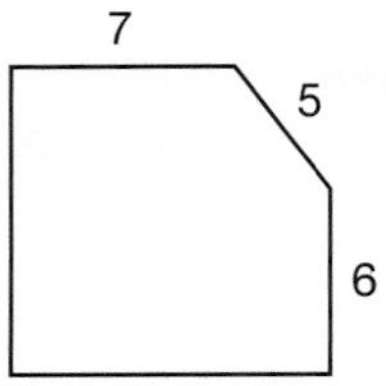

a. What is the name for this geometric figure?

b. What is the perimeter of the figure?

c. What is the area of the figure?

18. In the figure at right, each small cube has a volume of 1 cubic centimeter.
(67, 70)

a. What is the volume of this rectangular prism?

b. What is the total surface area of the rectangular prism?

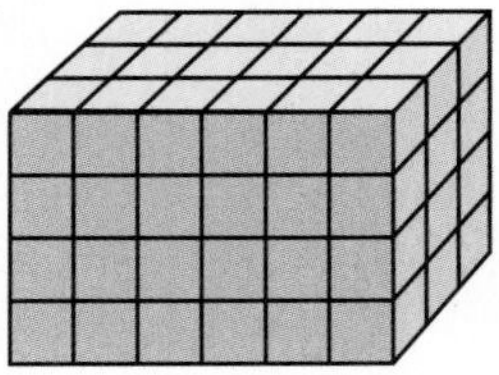

19. Find the circumference of each circle:
(65)

a.

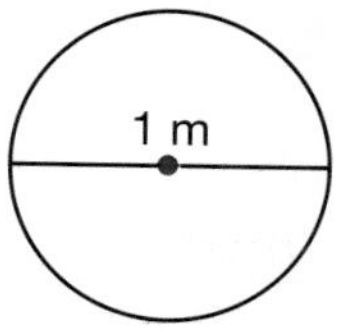

Use 3.14 for π.

b.

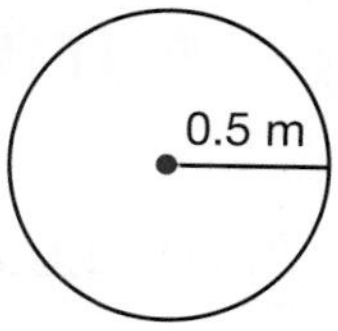

Leave π as π.

20. Classify each triangle as acute, right, or obtuse. Then classify each triangle as equilateral, isosceles, or scalene.
(62)

a.

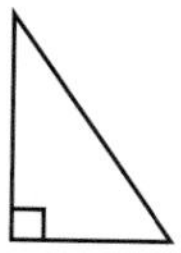

b.

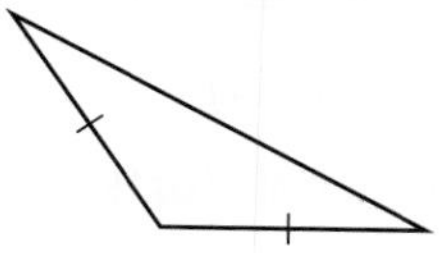

c.

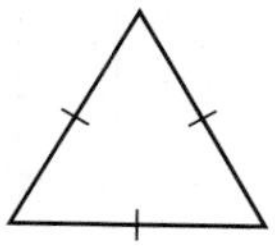

21. Which type of triangle has rotational symmetry?
(58)

Justify For problems **22–23**, solve and check each equation. Show each step.

* **22.** $1.2x = 2.88$
(Inv. 7)

* **23.** $\frac{3}{2}w = \frac{9}{10}$
(Inv. 7)

Simplify:

* **24.** ***Generalize*** $\frac{\sqrt{100} + 5[3^3 - 2(3^2 + 3)]}{5}$
(63)

25. $\begin{array}{r} 3 \text{ hr } 15 \text{ min } 24 \text{ s} \\ -\ 2 \text{ hr } 45 \text{ min } 30 \text{ s} \\ \hline \end{array}$
(49)

26. $1 \text{ yd}^2 \cdot \frac{3 \text{ ft}}{1 \text{ yd}} \cdot \frac{3 \text{ ft}}{1 \text{ yd}}$
(50)

27. $7\frac{1}{2} \cdot \left(3 \div \frac{5}{9}\right)$
(26)

28. $4\frac{5}{6} + 3\frac{1}{3} + 7\frac{1}{4}$
(30)

29. $3\frac{3}{4} \div 1.5$ (decimal answer)
(35, 43)

* **30.** ***Generalize***
(68)

a. $-0.1 - (-0.2) - (0.3)$

b. $(-10) - |(-20) - (+30)|$

Early Finishers
Real-World Application

Six houses in one community recently sold for these prices.

$130,000, $150,000, $120,000, $130,000, $210,000, $160,000

Use the mean, median, mode, and range to describe the selling prices. Which measure best describes the price of a typical house in this community? Justify your choice.

LESSON 80

• Transformations

Power Up

Building Power

facts Power Up O

mental math

a. **Positive/Negative:** $(-30) - (+45)$

b. **Exponents:** 40^2

c. **Algebra:** $5q - 4 = 36$

d. **Fractional Parts:** 15 is $\frac{3}{5}$ of n.

e. **Measurement:** 1500 m to km

f. **Fractional Parts:** What is $\frac{3}{5}$ of 15?

g. **Statistics:** Find the average of the set of numbers: 50, 60, 70.

h. **Geometry:** Find the perimeter and area of a square with sides 1.5 m long.

problem solving

Here are the front, top, and side views of an object. If this structure was constructed using 1-inch cubes, what would be the object's volume? Sketch a three-dimensional view of the structure.

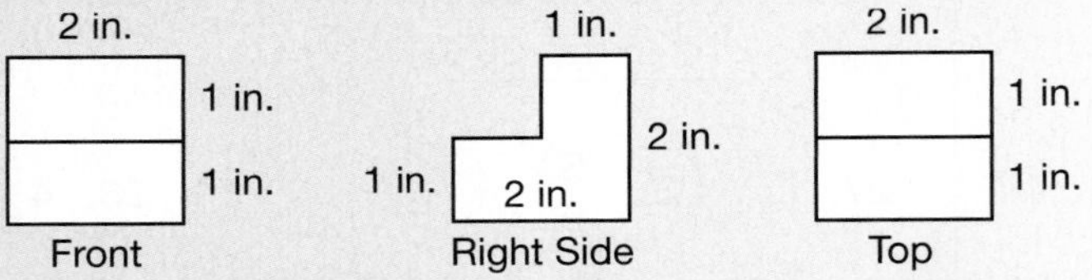

New Concept

Increasing Knowledge

Recall that two figures are congruent if they are the same shape and size. Triangles I and II below are congruent, but they are not in the same position:

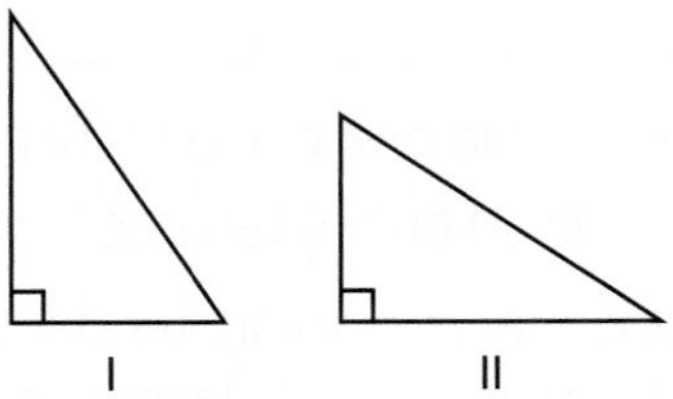

We can use three types of position change to move triangle I to the position of triangle II. One change of position is to "flip" triangle I over as though flipping a piece of paper, as shown on the following page.

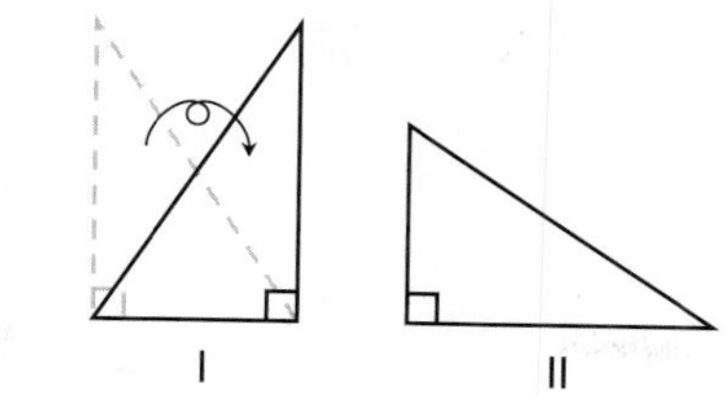

I II

A second change of position is to "slide" triangle I to the right.

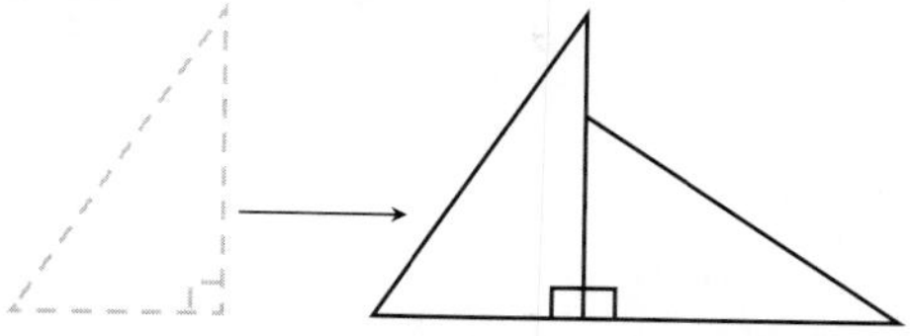

I II

The third change of position is to "turn" triangle I 90° clockwise.

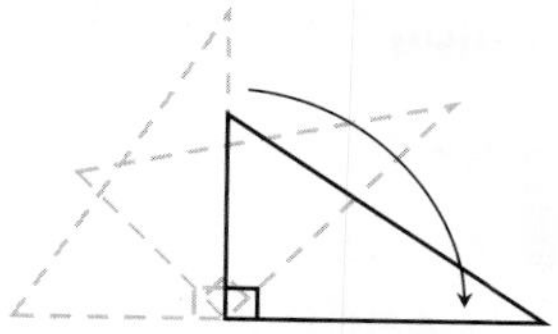

I and II

These "flips, slides, and turns" are called **transformations** and have special names, which are listed in this table.

Transformations

Movement	Name
flip	reflection
slide	transalation
turn	rotation

Math Language
A **coordinate plane** is a grid on which any point can be identified by an ordered pair of numbers.

A **reflection** of a figure in a line (a "flip") produces a mirror image of the figure that is reflected. On a coordinate plane we might reflect shapes in the lines represented by the x-axis or y-axis.

Example 1

The vertices of $\triangle RST$ are located at R (4, 3), S (4, 1), and T (1, 1). Draw $\triangle RST$ and its reflection in the x-axis.

Solution

We graph the vertices of $\triangle RST$ and draw the triangle. The triangle's reflection in the x-axis is where the triangle would appear if we flipped it across the x-axis. The reflection of each vertex is on the opposite side of the x-axis the same distance from the x-axis as the original vertex. So the reflection of R is at (4, -3). We call this vertex R', which we read as "R prime." The location of S' is (4, -1) and T' (1, -1). After we locate the vertices, we draw $\triangle R'S'T'$.

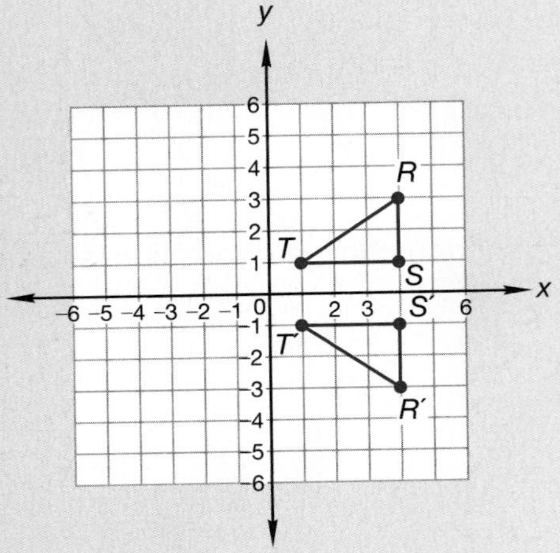

A **translation** "slides" a figure to a new position without turning or flipping the figure.

Example 2

Rectangle *ABCD* has vertices at *A* (4, 3), *B* (4, 1), *C* (1, 1), and *D* (1, 3). Draw ▭ *ABCD* and its image, ▭ *A′B′C′D′*, translated to the left 5 units and down 4 units.

Solution

We graph the vertices of ▭ *ABCD* and draw the rectangle. Then we graph its image ▭ *A′B′C′D′* by translating each vertex 5 units to the left and 4 units down. The translated vertices are ***A′*** **(–1, –1)**, ***B′*** **(–1, –3)**, ***C′*** **(–4, –3)**, and ***D′*** **(– 4, –1)**. We draw the sides of the rectangle to complete the image.

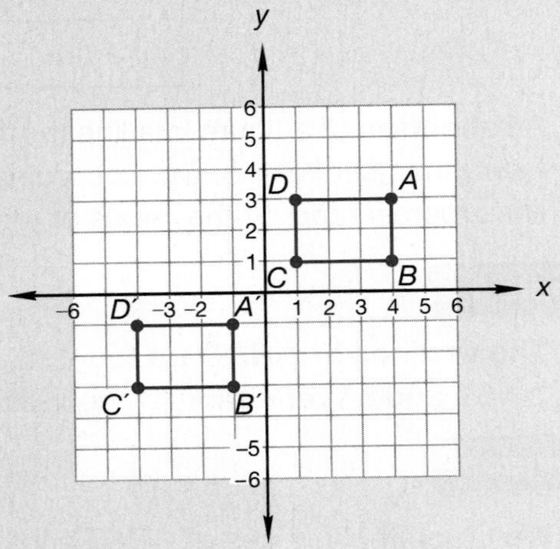

A **rotation** of a figure "turns" the figure about a specified point called the *center of rotation*. At the beginning of this lesson we rotated triangle I 90° clockwise. The center of rotation was the vertex of the right angle. In the illustration on the next page, triangle *ABC* is rotated 180° about the origin.

Thinking Skill

Conclude

What happens if you rotate a figure 180° and then rotate it 180° again around the same point?

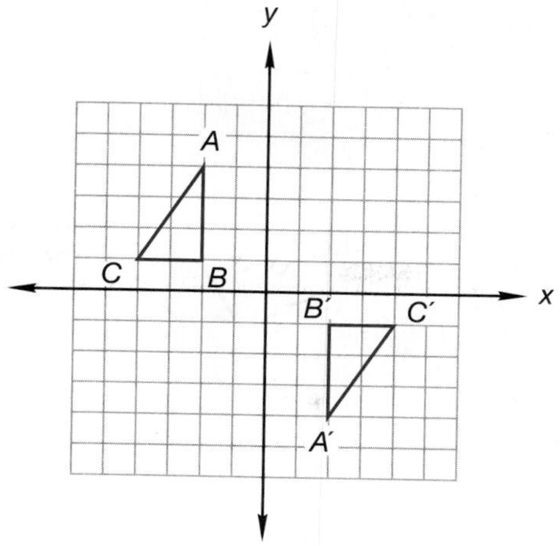

One way to view the effect of a rotation of a figure is to trace the figure on a piece of transparency film. Then place the point of a pencil on the center of rotation and turn the transparency film through the described rotation.

Example 3

The coordinates of the vertices of △*PQR* are *P* (3, 4), *Q* (3, 1), and *R* (1, 1). Draw △*PQR* and also draw its image, △*P′Q′R′*, after a counterclockwise rotation of 90° about the origin. What are the coordinates of the vertices of △*P′Q′R′*?

Solution

We graph the vertices of △*PQR* and draw the triangle.

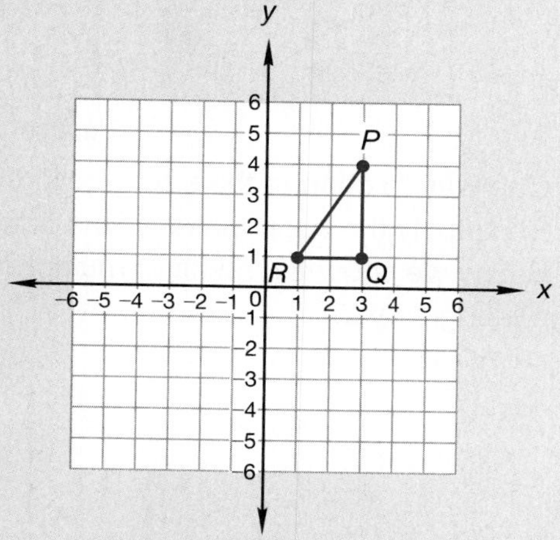

Then we place a piece of transparency film over the coordinate plane and trace the triangle. We also place a mark on the transparency aligned with the *x*-axis. This mark will align with the *y*-axis after the transparency is rotated 90°. After tracing the triangle on the transparency, we place the point of a pencil on the film over the origin, which is the center of rotation in this example. While keeping the graph paper still, we rotate the film 90° (one-quarter turn) counterclockwise. The image of the triangle rotates to the position shown, while the original triangle remains in place.

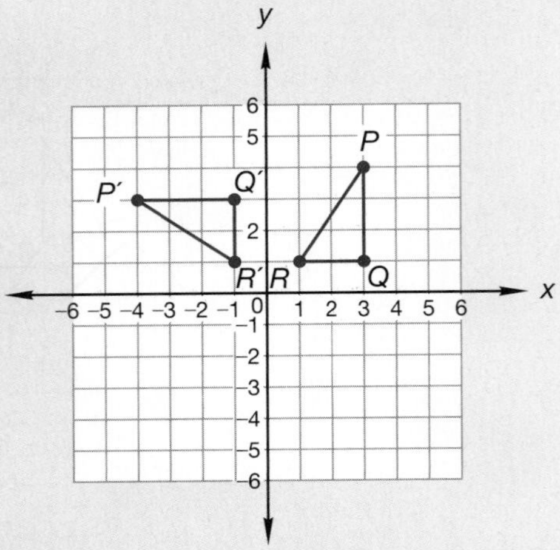

We name the rotated image $\triangle P'Q'R'$ and through the transparency see that the coordinates of the vertices are ***P′* (−4, 3), *Q′* (−1, 3),** and ***R′* (−1, 1).**

A **dilation** is a transformation that changes the size of a figure while preserving its shape. If we reduce or enlarge a figure, the resulting image is not congruent to the original figure, but the image is similar to the original figure because the shape is the same.

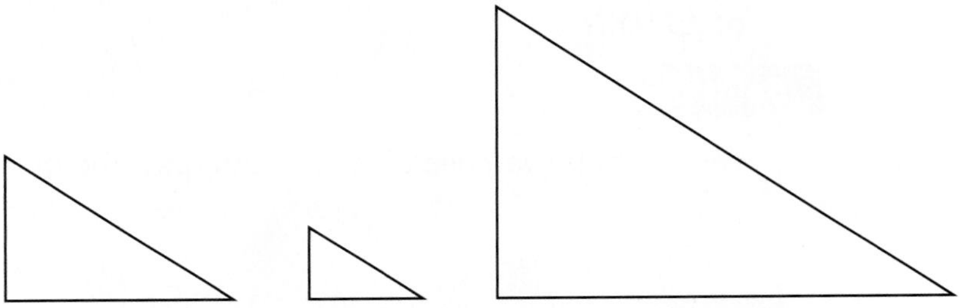

One way to change the size of a figure graphed on a coordinate plane is to multiply the coordinates of the figure's vertices by a constant number. Below we show the result of multiplying the coordinates of the vertices of a triangle by 2.

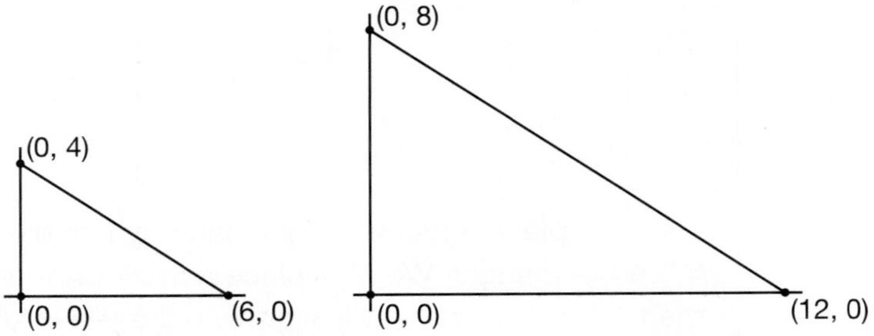

The number by which we multiply is the scale factor of the dilation. If we simply multiply the coordinates of the figure by the scale factor, then the center of the dilation is the origin.

Example 4

Triangle *XYZ* has vertices at *X* (1, 1), *Y* (2, 1), and *Z* (1, 2). Draw △*XYZ* and the dilation image of △*XYZ* with the center of the dilation at the origin and with a scale factor of 3.

Solution

We draw △*XYZ*. To locate the vertices of its image, △*X′Y′Z′*, we multiply the coordinates of the vertices of △*XYZ* by 3.

△*XYZ*	*X* (1, 1)	*Y* (2, 1)	*Z* (1, 2)
△*X′Y′Z′*	*X′* (3, 3)	*Y′* (6, 3)	*Z′* (3, 6)

We graph the vertices of the dilation and complete the triangle.

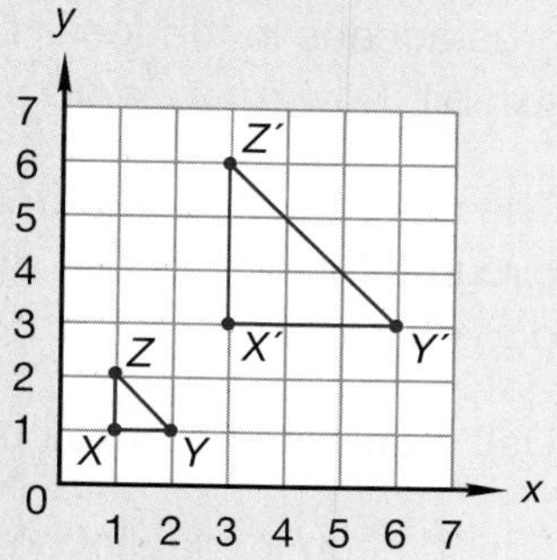

Practice Set

Perform each of the examples in this lesson before you do the Practice Set. Then draw each transformation and find the coordinates of its vertices.

a. Rectangle *WXYZ* has vertices at *W* (4, 3), *X* (4, 1), *Y* (1, 1), and *Z* (1, 3). Draw the rectangle and its image ▭*W′X′Y′Z′* after a 90° clockwise rotation about the origin. What are the coordinates of the vertices of ▭*W′X′Y′Z′*?

b. The vertices of △*JKL* are *J* (1, −1), *K* (3, −2), and *L* (1, −3). Draw the triangle and its image after reflection in the y-axis, △*J′K′L′*. What are the coordinates of the vertices of △*J′K′L′*?

c. Parallelogram *PQRS* has vertices at *P* (0, 3), *Q* (−1, 1), *R* (−4, 1), and *S* (−3, 3). Draw ▱*PQRS* and its image ▱*P′Q′R′S′* translated 6 units to the right and 3 units down. What are the coordinates of the vertices of ▱*P′Q′R′S′*?

d. Square *ABCD* has vertices at *A* (1, 1), *B* (1, −1), *C* (−1, −1), and *D* (−1, 1). Draw □*ABCD* and its dilated image □*A′B′C′D′* with the center of the dilation at the origin and with a scale factor of 2.

Written Practice — *Strengthening Concepts*

1. (55) Tina mowed lawns for 4 hours and earned \$7.00 per hour. Then she weeded flower beds for 3 hours and earned \$6.30 per hour. What was Tina's average hourly pay for the 7-hour period?

2. (52) Evaluate: $x + (x^2 - xy) - y$ if $x = 4$ and $y = 3$

3. (36) **Explain** A coin is tossed and a number cube is rolled.

a. Write the sample space for the experiment.

b. Use the sample space to find the probability of getting tails and an odd number.

4. (66) Use a ratio box to solve this problem. Nia found the ratio of trilobites to crustaceans in the fossil exhibit was 2 to 3. If there were 30 fossils in the exhibit, how many were trilobites?

5. (32, 65) The diameter of a half-dollar is 3 centimeters. What is the circumference of a half-dollar to the nearest millimeter? How do you know?

6. (50) Use a unit multiplier to convert $1\frac{1}{2}$ quarts to pints.

*** 7.** (78) Graph each inequality on a separate number line:

a. $x > -2$ **b.** $x \leq 0$

*** 8.** (72) Use a ratio box to solve this problem. In 25 minutes, 400 people entered the museum. At this rate, how many people would enter the museum in 1 hour?

9. (71) **Evaluate** Diagram this statement. Then answer the questions that follow.

Nathan found that it was 18 inches from his knee joint to his hip joint. This was $\frac{1}{4}$ of his total height.

a. What was Nathan's total height in inches?

b. What was Nathan's total height in feet?

Write equations to solve problems **10–13.**

*** 10.** (74) Six hundred is $\frac{5}{9}$ of what number?

*** 11.** (77) Two hundred eighty is what percent of 400?

12. (60) What number is 4 percent of 400?

*** 13.** (77) Sixty is 60 percent of what number?

14. (73) Simplify:

a. $\frac{600}{-15}$ **b.** $\frac{-600}{-12}$

c. $20(-30)$ **d.** $+15(40)$

15. (60) Anil is paid a commission equal to 6% of the price of each appliance he sells. If Anil sells a refrigerator for \$850, what is Anil's commission on the sale?

16. (48) Complete the table.

Fraction	Decimal	Percent
a.	0.3	**b.**
$\frac{5}{12}$	**c.**	**d.**

17. (69) Write each number in scientific notation:

a. 30×10^6 **b.** 30×10^{-6}

* **18.** (75) Find the area of the trapezoid shown. Dimensions are in meters.

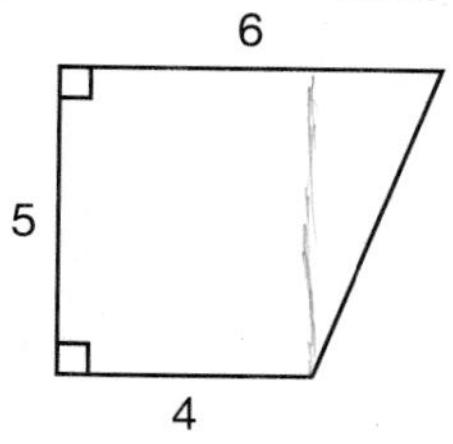

19. (67, 70) Each edge of a cube measures 5 inches.

a. What is the volume of the cube?

b. What is the surface area of the cube?

20. (14) In a bag are 100 marbles: 10 red, 20 white, 30 blue, and 40 green. If one marble is drawn from the bag, what is the probability that the marble will not be red, white, or blue?

For problems **21–23,** solve and check each equation. Show each step.

* **21.** (Inv. 7) $17a = 408$ * **22.** (Inv. 7) $\frac{3}{8}m = 48$

* **23.** (Inv. 7) $1.4 = x - 0.41$

Simplify:

24. (52) $\dfrac{2^3 + 4 \cdot 5 - 2 \cdot 3^2}{\sqrt{25} \cdot \sqrt{4}}$ **25.** (43) $7\frac{1}{7} \times 1.4$

26. (49)
$$\begin{array}{r} 10\text{ lb } 6\text{ oz} \\ -\ \ 7\text{ lb } 11\text{ oz} \\ \hline \end{array}$$

27. (50) $1\text{ cm}^2 \cdot \dfrac{10\text{ mm}}{1\text{ cm}} \cdot \dfrac{10\text{ mm}}{1\text{ cm}}$

28. (26) $7\frac{1}{2} \div \left(3 \cdot \frac{5}{9}\right)$ **29.** (20, 57) $2^{-4} + 4^{-2}$

* **30.** (80) Triangle *ABC* with vertices at *A* (0, 2), *B* (2, 2), and *C* (2, 0) is reflected in the *x*-axis. Draw $\triangle ABC$ and its image $\triangle A'B'C'$.

INVESTIGATION 8

Focus on

• Probability and Odds
• Compound Events
• Experimental Probability

probability and odds

We have defined the probability of an event as the ratio of favorable outcomes to the number of possible outcomes. The range of probability is from zero (impossible) to one (certain). We may express probability as a fraction, as a decimal, or as a percent (from 0% to 100%).

If we spin the spinner, the probability that the spinner will stop on 1 is $\frac{1}{4}$, 0.25, 25%. The **complement** of this event is the spinner not stopping on 1. The probability of the spinner not stopping on 1 is $\frac{3}{4}$, 0.75, 75%.

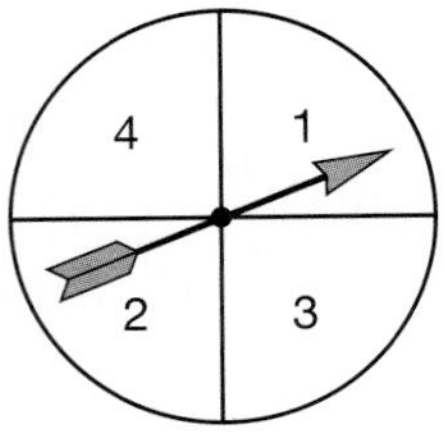

Thinking Skills

Conclude

If the chance of rain is 50%, what is the chance of "not rain?"

Sometimes the word *chance* is used to refer to probability. On a weather report we might hear that the chance of rain is 60%, which means that the complement, the chance of "not rain," is 40%.

Another way to express the likelihood of an event is as **odds.** The odds of spinning 1 on the spinner above are 1 to 3 or 1:3. Odds show the ratio of favorable to unfavorable outcomes and are written with the word "to" or with a colon (:).

Odds
favorable to unfavorable
favorable : unfavorable

In other words, odds express the relationship between the probabilities of an event and its complement. Although we do not write odds as a fraction, we do reduce the numbers used to express odds.

Analyze Refer to the following statement to answer problems **1–7.**

One marble is drawn from a bag containing 3 red marbles, 4 white marbles, and 5 blue marbles.

1. What is the probability of picking white?

2. What is the probability of not picking white?

3. What are the odds of picking white?

4. What is the probability of picking red? Express the probability as a decimal.

5. What is the probability of not picking a red marble? Express the probability as a decimal.

6. What are the odds of picking a red marble?

7. What are the odds of picking a blue marble?

The meteorologist forecast the chance of rain as 20%.

8. According to the forecast, the chance it will not rain is what percent?

9. **Analyze** According to the forecast, what are the odds of rain?

compound events

A **compound event** is composed of two or more simple events. For example, "getting heads twice" with two flips of a coin is a compound event.

To find the probability of two or more events occurring in a specific order, we multiply the probabilities of the events.

Example 1

A coin is tossed and a spinner is spun. What is the probability of getting heads and 4?

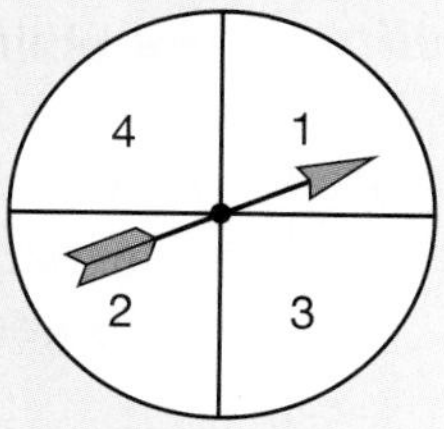

Solution

The probability of heads and 4, P (H and 4) is the probability of heads times the probability of 4.

$$
\begin{aligned}
P(\text{H and } 4) &= P(\text{H}) \cdot P(4) \\
&= \frac{1}{2} \cdot \frac{1}{4} \\
&= \frac{1}{8}
\end{aligned}
$$

Conclude What is the probability of getting tails and 4?

Refer to the illustration in example 1 to answer problems 10–12.

Math Language
A **sample space** is a list of all the possible outcomes of an event.

10. **Represent** What is the sample space for one coin toss and one spin?

11. Find the probability of heads and 2 by multiplying the probability of heads and the probability of 2. Check your answer by inspecting the sample space for the event.

12. What are the odds of getting heads and 2? Check your answer by inspecting the sample space for the event.

13. **Explain** What is the probability of flipping heads and not spinning 2? Tell how you found the answer. Check your answer by inspecting the sample space.

14. If a coin is flipped three times, what is the probability of getting heads all three times?

15. If a number cube is rolled twice, what is the probability of getting 6 both times?

16. **Analyze** If a number cube is rolled and a coin is tossed, what is the probability of getting 6 and heads?

Events like tossing a coin, spinning a spinner, and rolling a number cube are **independent events.** One outcome does not affect the next outcome. If a tossed coin happens to land heads up three times in a row, then the probability of heads on the next toss is still just $\frac{1}{2}$. The so called "law of averages" does not apply to individual events. However, over a very large number of trials, we expect a coin will land heads up $\frac{1}{2}$ of the trials, and we expect a rolled number cube will end up 6 on $\frac{1}{6}$ of the trials. This leads us to the topic of experimental probability.

experimental probability

We can distinguish between **theoretical probability** and **experimental probability.** To determine the theoretical probability of an event, we analyze the relationship between favorable and possible outcomes. However, to find the experimental probability of an event we actually conduct many trials (experiments), and we count and compare the number of favorable outcomes to the number of trials.

$$\text{Experimental probability} = \frac{\text{number of favorable outcomes}}{\text{number of trials}}$$

Example 2

A baseball coach wants to select a player to be a pinch hitter. He knows that Cruz has 80 hits in 240 at-bats. What is the experimental probability of Cruz getting a hit if he is selected to pinch-hit?

Solution

The coach uses the batter's history to determine the probability of a hit. The favorable outcomes in this situation are hits, and the number of trials are at-bats.

$$\text{Probability} = \frac{\text{hits}}{\text{at-bats}} = \frac{80}{240} \text{ or } \frac{1}{3}$$

The experimental probability of a hit is $\mathbf{\frac{1}{3}}$. (A baseball player's history of hits for at-bats is expressed as a decimal rounded to three decimal places without a leading zero. In this case, the player's batting average is .333.)

We can use probability experiments to find out information about a situation. Consider the spinner shown here. The spinner was spun 360 times. The results are recorded in the table.

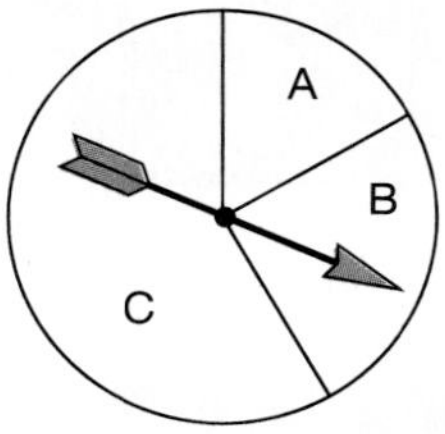

Math Language
In the table, the term *trial* means an *attempt* or *try.*

Analyze Refer to the table to answer questions 17–20.

Spinner Results	
Sector *A*	60
Sector *B*	90
Sector *C*	210
Total Trials	360

17. What is the experimental probability of spinning *A*?

18. What is the experimental probability of spinning *B*?

19. What is the experimental probability of spinning *C*?

20. Based on the results of the experiment:

a. What fraction of the spinner's face is Sector *C*?

b. About what percent of the spinner's face is Sector *B*?

Activity

Experimental Probability

- Investigation Activity 20
- Two number cubes

Section A of **Investigation Activity 20** Probability Experiment displays the 36 equally likely outcomes of tossing a pair of number cubes. Section B is the outline of a bar graph. On the graph draw bars to indicate the theoretical outcome of tossing a pair of number cubes 36 times.

After completing Section B, take turns tossing the number cubes and recording the results for 36 tosses. Record and graph the results of the tosses in Section C.

If the results of the experiment differ from the theoretical outcome, discuss why you think the results differ and write your reasons in Section D.

extension Create two classroom bar graphs to represent the results of all of the groups. First create a "Theoretical Outcomes" graph and then create an "Actual Results" graph by combining the results from every group. Does increasing the number of tosses (by counting all the tosses in the class) produce results closer to theoretical outcomes than were usually attained with just 36 rolls?

LESSON

81

Using Proportions to Solve Percent Problems

Power Up

Building Power

facts

Power Up P

mental math

a. **Positive/Negative:** $(-10)(-10)$

b. **Exponents:** 12×10^{-4}

c. **Ratio:** $\frac{40}{t} = \frac{6}{9}$

d. **Percent:** 15% of \$24.00

e. **Measurement:** 25 mm to cm

f. **Fractional Parts:** 24 is $\frac{2}{3}$ of n.

g. **Geometry:** What do the interior angles of a triangle total?

h. **Calculation:** What is the total cost of a \$25 item plus 8% sales tax?

problem solving

The Egyptians did not have symbols for numerators greater than 1, so they represented larger fractions as sums of unit fractions. For example, $\frac{6}{7}$ could be expressed $\frac{1}{7} + \frac{1}{7} + \frac{1}{7} + \frac{1}{7} + \frac{1}{7} + \frac{1}{7}$, but it could also be expressed as $\frac{1}{2} + \frac{1}{3} + \frac{1}{42}$. Can you find three ways to write $\frac{7}{8}$ as a sum of unit fractions?

New Concepts

Increasing Knowledge

A percent is a ratio in which 100 represents the total number in the group. Thus percent problems can be solved using the same method we use to solve ratio problems. Consider the following problem:

> *Thirty percent of the lunchtime customers ordered salad. If 21 customers did not order salad, how many lunchtime customers were there?*

Math Language

Percents can be written using the symbol %. You can write thirty percent as 30%.

The problem is about two parts of a group. One part of the group ordered salad for lunch; the other part did not. The whole group is 100 percent. The part that ordered salad for lunch was 30 percent. Thus 70 percent did not order salad for lunch. We record these numbers in a ratio box just as we do with ratio problems.

	Percent	Actual Count
Salad	30	
Not salad	70	
Whole group	100	

As we read the problem, we find an actual count as well. There were 21 customers who did not order salad. We record 21 in the appropriate place on the table and use letters in the remaining places.

	Percent	Actual Count
Salad	30	S
Not Salad	70	21
Whole Group	100	W

We use two of the rows in the table to write a proportion that we can use to solve the problem. **We always use the row in which both numbers are known.** Since the problem asks for the total number of lunchtime customers, we also use the third row.

	Percent	Actual Count
Salad	30	S
Not Salad	70	21
Whole Group	100	W

$$\frac{70}{100} = \frac{21}{W}$$

$$70W = 2100$$

$$W = 30$$

By solving the proportion, we find there were 30 lunchtime customers.

Conclude What other information can you find out about the lunchtime customers based on the solution to this problem?

Example 1

Thinking Skill

Explain

How is the question in example 1 different from the previous question?

Forty percent of the students at Lincoln Middle School do not buy their lunch in the school cafeteria. If 480 students buy their lunch in the cafeteria, how many of the students do not buy their lunch in the cafeteria?

Solution

We solve this problem just as we solve a ratio problem. We use the percents to fill the percent column of the table. Together, all the students in the school are 100 percent. The part that does not buy lunch is 40 percent. Therefore, the part that buys lunch is 60 percent. The number 480 is the actual count of the students who buy their lunch. We write these numbers in the table.

	Percent	Actual Count
Do Not Buy Lunch	40	N
Buy Lunch	60	480
Total	100	T

Discuss How do you think we should decide which letters to use as variables in the table? Why?

Now we use the table to write a proportion. Since we know both numbers in the second row, we will use that row in the proportion. Since the problem asks us to find the actual count of students who do not buy their lunch in the cafeteria, we also use the first row of the table in the proportion.

	Percent	Actual Count
Do Not Buy Lunch	40	N
Buy Lunch	60	480
Total	100	T

$$\frac{40}{60} = \frac{N}{480}$$

$$60N = 19{,}200$$

$$N = 320$$

We find that **320 students** do not buy their lunch in the cafeteria.

Example 2

Thinking Skill

Predict

Will the answer to question **a.** be greater than or less than 50%? Why?

In the town of Centerville, 261 of the 300 working people do not carpool.

a. What percent of the people carpool?

b. According to one national survey, approximately 13 percent of the working people in the United States carpool to work. Do the carpooling statistics in Centerville reflect the national carpooling statistics?

Solution

a. We make a ratio box and write in the numbers. The total number of working people in Centerville is 300, so 39 people carpool.

	Percent	Actual Count
Carpool	P_C	39
Do Not Carpool	P_N	261
Total	100	300

We use P_C to stand for the percent who carpool to work. We use the "carpool" row and the "total" row to write the proportion.

	Percent	Actual Count
Carpool	P_C	39
Do Not Carpool	P_N	261
Total	100	300

$$\frac{P_C}{100} = \frac{39}{300}$$

$$300P_C = 3900$$

$$P_C = 13$$

We find that **13 percent** of the people who work in Centerville carpool to work.

Conclude What percent of people do not carpool? How do you know?

b. Since survey data show that approximately 13 percent of United States workers carpool to work, we find that Centerville's statistics **do reflect the national statistics.**

Example 3

The students had a spelling test with 40 words. Six of the words had four syllables. The rest had less than four syllables. What percent of the words had less than four syllables?

Solution

We notice that a little more than one tenth of the words had four syllables.

Predict From the given information, what percent can you estimate the answer will be?

We record the given information in a ratio box.

Explain How can we find out how many words had less than four syllables?

	Percent	Actual Count
Less Than Four Syllables	P_L	34
Four Syllables	P_F	6
Total	100	40

We want to know the percent of words with less than four syllables, so we use the "Less than four" row and the "Total" row to write the proportion.

$$\frac{P_L}{100} = \frac{34}{40}$$

$$40P_L = 3400$$

$$P_L = 85$$

On this test **85%** of the words had less than four syllables.

Practice Set

Generalize Estimate each answer. Then use a ratio box to solve each problem.

a. Twenty-one of the 70 acres were planted with alfalfa. What percent of the acres was not planted in alfalfa?

b. Lori still has 60% of the book to read. If she has read 120 pages, how many pages does she still have to read?

c. Dewayne missed four of the 30 problems on the problem set. What percent of the problems did Dewayne answer correctly?

d. **Classify** Tell whether you found a part, a whole, or a percent to solve each of problems **a** through **c.**

Written Practice *Strengthening Concepts*

* **1.** (80) **Evaluate** The coordinates of the vertices of $\triangle ABC$ are A (2, −1), B (5, −1), and C (5, −3). Draw the triangle and its image $\triangle A'B'C'$ reflected in the x-axis. What are the coordinates of the vertices of $\triangle A'B'C'$?

2. (66) Use a ratio box to solve this problem. Annuals are plants that complete their life cycle in one year. Perennials are plants that live for more than one year. In a botanical garden, the ratio of annuals to perennials was 9 to 5. If the total number of plants was 2800, how many annuals were there?

Use the information below to answer problems **3** and **4**.

Rory has been learning to type on his computer. On his last 15 typing practices he typed the following number of words per minute:
70, 85, 80, 85, 90, 80, 85, 80, 90, 95, 85, 90, 100, 85, 90.

3. (Inv. 4) **a.** What was the average (mean) number of words per minute Rory typed?

b. If the number of words typed per minute were arranged in order from greatest to least, which would be the middle number (median)?

4. (Inv. 4) **a.** **Explain** How many words per minute did Rory type most often? How can you tell? Is this the mode, range, or mean of the number of words typed per minute?

b. What was the difference between the greatest and the least number of words per minute that Rory typed (range)?

5. (49) Danny is 6′1″ (6 ft 1 in.) tall. His sister is $5'6\frac{1}{2}''$ tall. Danny is how many inches taller than his sister?

6. (72) Use a ratio box to solve this problem. Carmen bought 5 pencils for 75¢. At this rate, how much would she pay for a dozen pencils?

* **7.** (78) **Represent** Graph each inequality on a separate number line:

a. $x < 4$ **b.** $x \geq -2$

8. (36, 71) Read this statement. Then answer the questions that follow.

So far Soon-Jin has sent out 48 invitations to her party. This is $\frac{4}{5}$ of all of the invitations she will send out.

a. How many invitations will Soon-Jin send out?

b. What is the ratio of invitations sent out already to those not sent out yet?

9. (8) If point B is located halfway between points A and C, what is the length of segment AB?

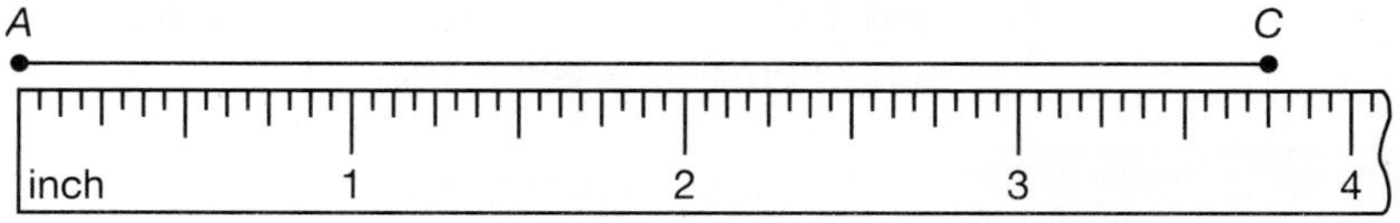

10. (41) If $x = 9$, what does $x^2 + \sqrt{x}$ equal?

* **11.** (Inv. 8) If a single number cube is rolled, what are the odds of getting a 3?

12. (48) Complete the table.

Fraction	Decimal	Percent
$2\frac{1}{4}$	**a.**	**b.**
c.	**d.**	$2\frac{1}{4}\%$

Write equations to solve problems **13** and **14.**

13. *(60)* The store owner makes a profit of 40% of the selling price of an item. If an item sells for $12, how much profit does the store owner make?

* **14.** *(77)* **Analyze** Fifty percent of what number is 0.4?

* **15.** *(76)* Simplify: $\frac{16\frac{2}{3}}{100}$

Use ratio boxes to solve problems **16** and **17.**

* **16.** *(81)* **Justify** In a restaurant, 21 of the 25 tables are round. What percent of the tables are round? Tell what you did to find out.

* **17.** *(81)* Twenty percent of the 4000 acres were plowed. How many acres were not plowed?

18. *(40)* If the measure of $\angle ABC$ is 140°, then

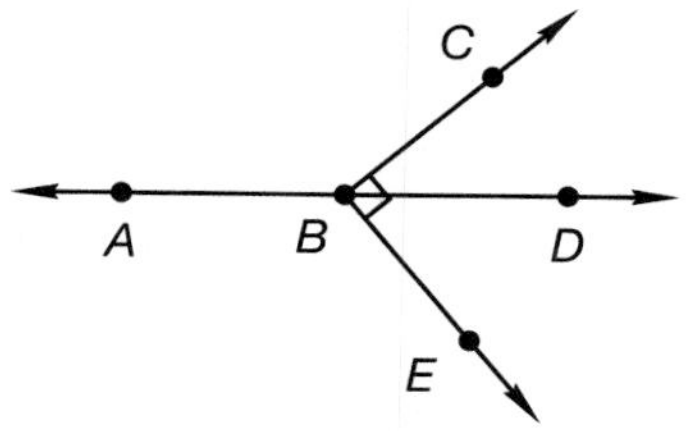

a. What is the measure of $\angle CBD$? How do you know?

b. What is the measure of $\angle DBE$? How do you know?

c. What is the measure of $\angle EBA$? How do you know?

d. What is the sum of the measures of $\angle ABC$, $\angle CBD$, $\angle DBE$, and $\angle EBA$?

19. *(24)* Write the prime factorization of the two terms of this fraction. Then reduce the fraction.

$$\frac{3000}{6300}$$

* **20.** *(58, 75)* **a.** Find the area of this isosceles trapezoid. Dimensions are in inches.

b. Trace the trapezoid on your paper. Then draw its line of symmetry.

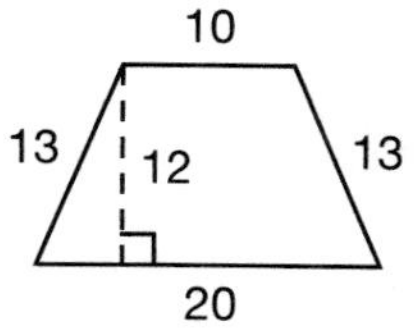

21. *(56)* Write the rule of this function with words and as an equation. Then find the missing number.

x	*y*
–1	–3
3	5
4	7
6	11
0	

22. Write each number in scientific notation:
(69)

a. 56×10^7 **b.** 56×10^{-7}

For problems **23** and **24,** solve and check the equation. Show each step.

* **23.** $5x = 16.5$ * **24.** $3\frac{1}{2} + a = 5\frac{3}{8}$
(Inv. 7) (Inv. 7)

Generalize Simplify:

* **25.** $3^2 + 5[6 - (10 - 2^3)]$
(63)

26. $\sqrt{2^2 \cdot 3^4 \cdot 5^2}$
(52)

27. $2\frac{2}{3} \times 4.5 \div 6$ (fraction answer)
(43)

28. $\left(3\frac{1}{2}\right)^2 - (5 - 3.4)$ (decimal answer)
(26, 43)

* **29.** **a.** $(-1.2)(-9)$ **b.** $(-3)(2.5)$
(73)

c. $\left(\frac{1}{2}\right)\left(\frac{-1}{2}\right)$ **d.** $\left(-\frac{1}{2}\right)\left(\frac{-1}{2}\right)$

30. **a.** $(-3) + |-4| - (-5)$
(68)

b. $(-18) - (+20) + (-7)$

c. $\frac{1}{2} - \left(-\frac{1}{2}\right)$

Early Finishers
Real-World Application

The per barrel price of crude oil, in U.S. dollars per barrel, varies from day to day. Here are the closing prices rounded to the nearest tenth for a five-day period.

Mon.	Tues.	Wed.	Th.	Fri.
65.4	66.3	66.9	67.5	66.8

Which display—a line graph or a bar graph—is the most appropriate way to display this data if you want to emphasize the changes in the price of crude oil? Draw the display and justify your choice.

LESSON

82

• Area of a Circle

Power Up

Building Power

facts Power Up Q

mental math

a. **Positive/Negative:** $(-6) - (-24)$

b. **Exponents:** 10^{-2}

c. **Algebra:** $8n + 6 = 78$

d. **Decimals/Measurement:** 3.14×10 ft

e. **Measurement:** 150 cm to m

f. **Fractional Parts:** 24 is $\frac{3}{4}$ of n.

g. **Geometry:** What do you call two adjoining angles that total 180 degrees?

h. **Calculation:** 25% of 24, $\times$ 5, $-$ 2, $\div$ 2, $+$ 1, $\div$ 3, $\times$ 7, $+$ 1, $\sqrt{\ }$, $\times$ 10

problem solving

A 60″ × 104″ rectangular tablecloth was draped over a rectangular table. Eight inches of cloth hung over the left edge of the table, 3 inches over the back, 4 inches over the right edge, and 7 inches over the front. In which directions (L, B, R, and F) and by how many inches should the tablecloth be shifted so that equal amounts of cloth hang over opposite edges of the table?

New Concept

Increasing Knowledge

Math Language

Perpendicular segments meet to form a right angle.

We can find the areas of some polygons by multiplying two perpendicular segments.

- We find the area of a rectangle by multiplying the length by the width.

$$A = lw$$

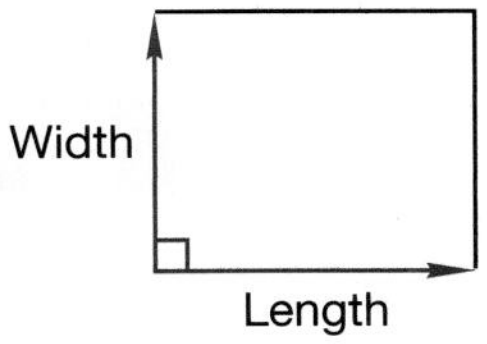

- We find the area of a parallelogram by multiplying the base by the height which gives us the area of a rectangle that equals the area of the parallelogram.

$$A = bh$$

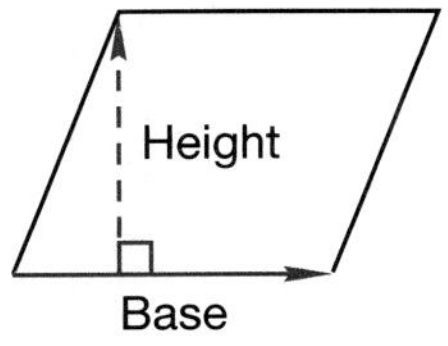

- We find the area of a triangle by multiplying the base by the height (which gives us the area of a rectangle) and then dividing by 2.

$$A = \frac{bh}{2} \quad \text{or} \quad A = \frac{1}{2}bh$$

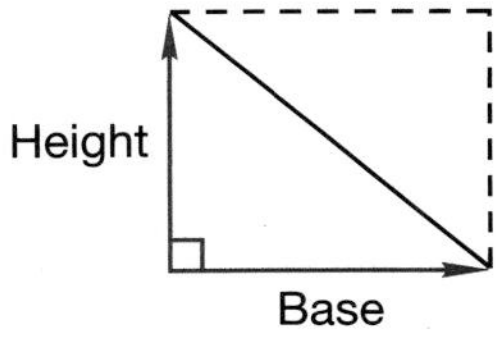

To find the area of a circle, we begin by multiplying the radius by the radius. This gives us the area of a square built on the radius.

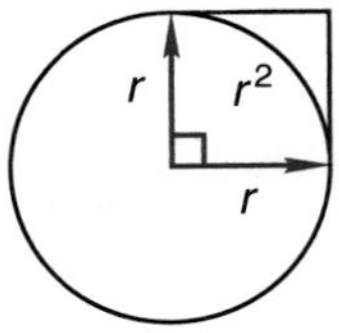

If the radius of the circle is 3, the area of the square is 3^2, which is 9. If the radius of the circle is r, the area of the square is r^2. We see that the area of the circle is less than the area of four of these squares.

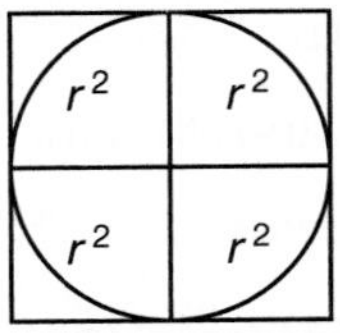

However, the area of the circle is more than the area of three squares.

The number of squares whose area exactly equals the area of the circle is between 3 and 4. The exact number is π. Thus, to find the area of the circle, we first find the area of the square built on the radius; then we multiply that area by π. This is summarized by the equation

$$\boldsymbol{A = \pi r^2}$$

Example

Find the area of each circle:

a.

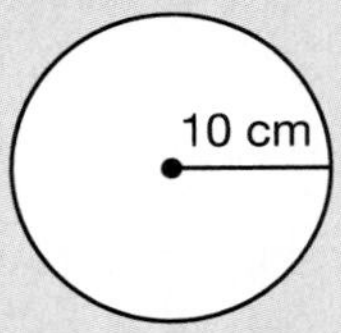

Use 3.14 for π.

b.

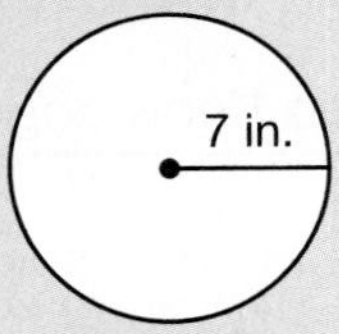

Use $\frac{22}{7}$ for π.

c.

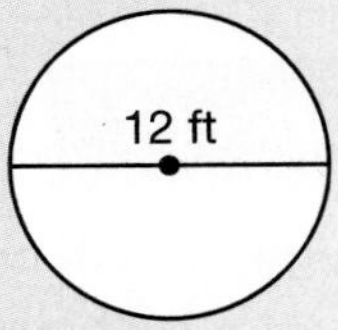

Leave π as π.

Solution

a. The area of a square built on the radius is 100 cm^2. We multiply this by π.

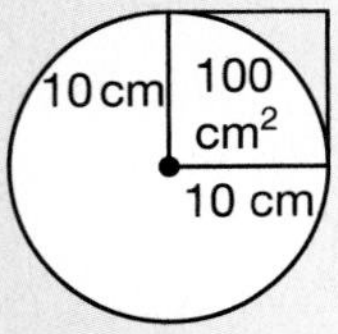

$$A = \pi r^2$$
$$A \approx (3.14)(100\text{ cm}^2)$$
$$A \approx \mathbf{314\ cm^2}$$

Explain Why is the symbol $\approx$ used instead of $=$ when 3.14 is substituted for π?

Thinking Skill

Summarize

State the relationship between the diameter and the radius of a given circle.

b. The area of a square built on the radius is 49 in.^{2}. We multiply this by π.

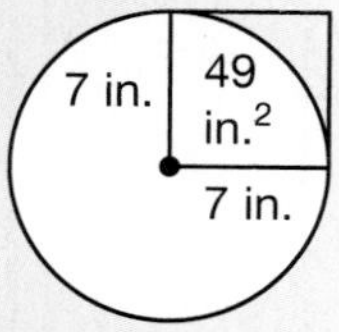

$$A = \pi r^2$$

$$A \approx \frac{\overset{}{22}}{\underset{1}{\cancel{7}}} \cdot \overset{7}{\cancel{49}} \text{ in.}^2$$

$$A \approx \mathbf{154\ in.^2}$$

c. Since the diameter is 12 ft, the radius is 6 ft. The area of a square built on the radius is 36 ft^2. We multiply this by π.

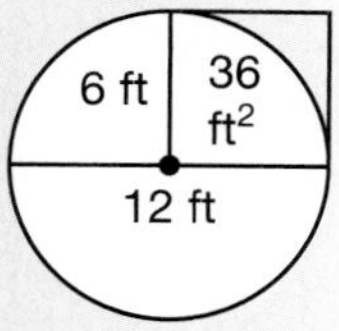

$$A = \pi r^2$$

$$A = \pi \cdot 36 \text{ ft}^2$$

$$A = \mathbf{36\pi\ ft^2}$$

Infer In the answers for **a** and **b,** we used the ≈ symbol, while in **c** we used the = sign. Why did we not use ≈ in **c?**

Practice Set

a. Using 3.14 for π, calculate to the nearest square foot the area of circle c. in this lesson's example.

Find the area of each circle:

b.

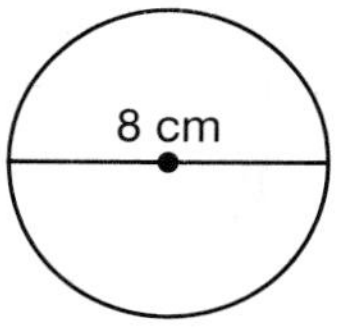

Use 3.14 for π.

c.

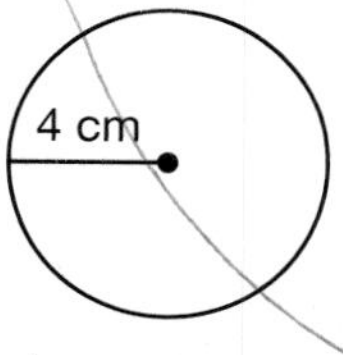

Leave π as π.

d.

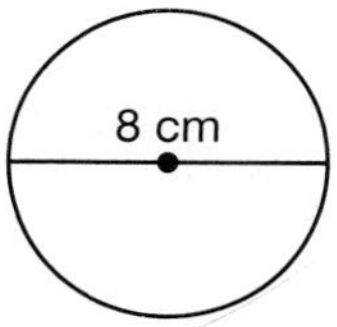

Use $\frac{22}{7}$ for π.

e. *Estimate* Make a rough estimate of the area of a circle that has a diameter of 2 meters. Tell how you made your estimate.

Written Practice

Strengthening Concepts

1. (70) Find the volume of this rectangular prism. Dimensions are in feet.

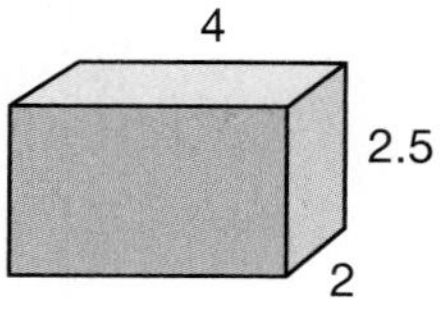

2. (28) The heights of five cherry trees are 6′3″, 6′5″, 5′11″, 6′2″, and 6′1″. Find the average height of the five trees. *Hint:* Change all measures to inches before dividing.

3. (53) Use a ratio box to solve this problem. The student-teacher ratio at the high school was 20 to 1. If there were 48 high school teachers, how many students were there?

4. (50) An inch equals 2.54 centimeters. Use a unit multiplier to convert 2.54 centimeters to meters.

* **5.** Graph each inequality on a separate number line:
(78)

a. $x < -2$ **b.** $x \geq 0$

6. Use a case 1-case 2 ratio box to solve this problem. Don's heart beats 225 times in 3 minutes. At that rate, how many times will his heart beat in 5 minutes?
(72)

7. Read this statement. Then answer the questions that follow.
(36, 71)

Two fifths of the performances were sold out. There were 15 performances that were not sold out.

a. How many performances were there?

b. What was the ratio of sold out to not sold out performances?

8. Compare: $x^2 - y^2 \bigcirc (x + y)(x - y)$ if $x = 5$ and $y = 3$
(52)

9. What percent of this circle is shaded?
(Inv. 1)

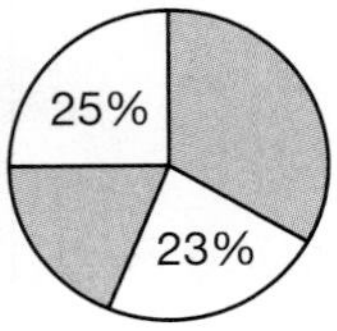

10. The meteorologist forecast the chance of rain to be 60%. Based on the forecast, what is the chance it will not rain?
(Inv. 8)

11. Find the circumference of each circle:
(65)

a.

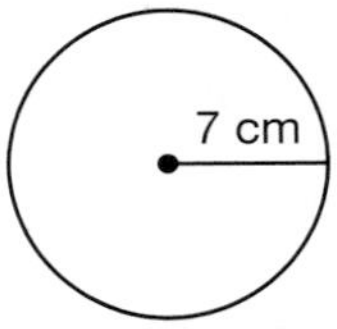

Use 3.14 for π.

b.

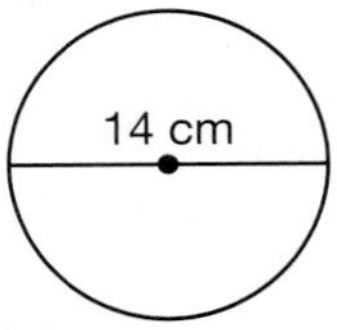

Use $\frac{22}{7}$ for π.

* **12.** Find the area of each circle in problem **11.**
(82)

13. Complete the table.
(48)

Fraction	Decimal	Percent
a.	1.6	**b.**
c.	**d.**	1.6%

14. Write an equation to solve this problem:
(60)

How much money is 6.4% of $25?

15. Write each number in scientific notation:
(69)

a. 12×10^5 **b.** 12×10^{-5}

Model Use ratio boxes to solve problems **16** and **17.**

* **16.** *(81)* Sixty-four percent of the students correctly described the process of photosynthesis. If 63 students did not correctly describe the process of photosynthesis, how many students did correctly describe the process?

* **17.** *(81)* George still has 40 percent of his book to read. If George has read 180 pages, how many pages does he still have to read? How do you know?

* **18.** *(75)* Find the area of the figure shown. Dimensions are in inches. Corners that look square are square.

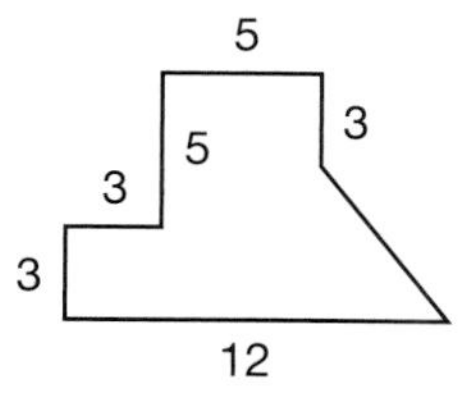

* **19.** *(80)* The coordinates of the vertices of $\triangle XYZ$ are X (4, 3), Y (4, 1), and Z (1, 1).

a. Draw $\triangle XYZ$ and its image $\triangle X'Y'Z'$ translated 5 units to the left and 3 units down.

b. What are the coordinates of the vertices of $\triangle X'Y'Z'$?

20. *(24)* Write the prime factorization of the two terms of this fraction. Then reduce the fraction.

$$\frac{240}{816}$$

21. *(Inv. 2)* The figure below illustrates regular hexagon *ABCDEF* inscribed in a circle with center at point *M*.

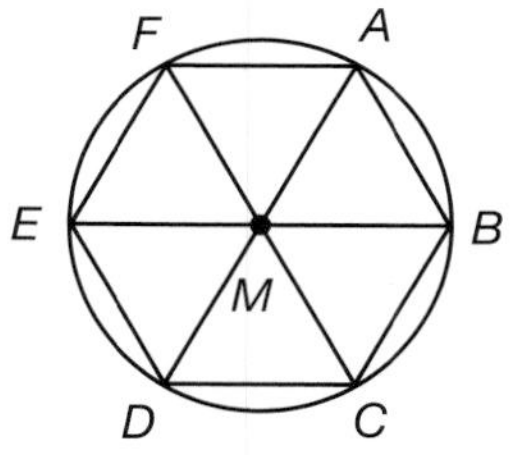

a. How many illustrated chords are diameters?

b. How many illustrated chords are not diameters?

c. What is the measure of central angle *AMB?*

d. What is the measure of inscribed angle *ABC?*

22. *(51)* Write 100 million in scientific notation.

For problems **23** and **24,** solve and check the equation. Show each step.

23. *(Inv. 7)* $\frac{3}{4}x = 36$

24. *(Inv. 7)* $3.2 + a = 3.46$

Simplify:

25. (52) $\dfrac{\sqrt{3^2 + 4^2}}{5}$

26. (52) $(8 - 3)^2 - (3 - 8)^2$

27. (43, 45) $3\frac{1}{2} \div (7 \div 0.2)$ (decimal answer)

28. (43) $4.5 + 2\frac{2}{3} - 3$ (mixed-number answer)

29. (73) **a.** $\dfrac{(-3)(-4)}{(-2)}$ **b.** $\left(-\frac{2}{3}\right)\left(-\frac{3}{4}\right)$

30. (68) **a.** $(-0.3) + (-0.4) - (-0.2)$

b. $(-20) + (+30) - |-40|$

Early Finisher
Real-World Application

Henrietta recently completed a driver's safety course, which means that her insurance premiums will decrease. Before the course, Henrietta's insurance cost $172 per month. It now costs $127 per month.

a. How much was Henrietta paying for insurance per year?

b. How much less is Henrietta now paying per year in premiums?

c. To the nearest whole number, what is the percent of decrease for her insurance premiums?

LESSON

83

Multiplying Numbers in Scientific Notation

Power Up

Building Power

facts Power Up R

mental math

a. Positive/Negative: $(-60) \div (+3)$

b. Decimals/Exponents: 6.75×10^6

c. Ratio: $\frac{100}{150} = \frac{m}{30}$

d. Percent: 15% of \$120

e. Measurement: 500 mg to g

f. Fractional Parts: 24 is $\frac{3}{8}$ of *n.*

g. Geometry: What do you call angles that share a common side and a common vertex?

h. Rate: At 60 mph, how far will a car travel in $2\frac{1}{2}$ hours?

problem solving David and Lisa are in a chess tournament. The first player to win either two consecutive games, or a total of three games, wins the match. How many different sets of games are possible in the match?

New Concept

Increasing Knowledge

From our earlier work with powers of 10, recall that

$$10^3 \text{ means } 10 \cdot 10 \cdot 10$$

and

$$10^4 \text{ means } 10 \cdot 10 \cdot 10 \cdot 10$$

Thus, $10^3 \cdot 10^4$ means 7 tens are multiplied.

$$10^3 \cdot 10^4 = 10^7$$

This multiplication illustrates an important rule of exponents learned in Lesson 47.

When we multiply powers of the same base, we add the exponents.

Explain How would you find the product of 10^5 and 10^5?

We use this rule to multiply numbers expressed in scientific notation.

To multiply numbers written in scientific notation, we multiply the decimal numbers to find the decimal-number part of the product. Then we multiply the powers of 10 to find the power-of-10 part of the product. We remember that when we multiply powers of 10, we add the exponents.

Math Language

A number expressed in **scientific notation** is written as the product of a decimal number and a power of 10.

Example 1

Multiply: $(1.2 \times 10^5)(3 \times 10^7)$

Solution

Notice that there are four factors.

$$1.2 \times 10^5 \times 3 \times 10^7$$

Using the Commutative Property, we reverse the order of two factors.

$$1.2 \times 3 \times 10^5 \times 10^7$$

Using the Associative Property, we group factors.

$$(1.2 \times 3) \times (10^5 \times 10^7)$$

Then we simplify each group.

$$\mathbf{3.6 \times 10^{12}}$$

Usually we do not show the commutative and associative steps; we simply multiply the powers of 10 separately.

Example 2

Multiply: $(4 \cdot 10^6)(3 \cdot 10^5)$

Solution

We multiply 4 by 3 and get 12. Then we multiply 10^6 by 10^5 and get 10^{11}. The product is

$$12 \times 10^{11}$$

We rewrite this expression in proper scientific notation.

$$(1.2 \times 10^1) \times 10^{11} = \mathbf{1.2 \times 10^{12}}$$

Example 3

Multiply: $(2 \cdot 10^{-5})(3 \cdot 10^{-7})$

Solution

We multiply 2 by 3 and get 6. To multiply 10^{-5} by 10^{-7}, we add the exponents and get 10^{-12}. Thus the product is

$$\mathbf{6 \times 10^{-12}}$$

Example 4

Multiply: $(5 \cdot 10^3)(7 \cdot 10^{-8})$

Solution

We multiply 5 by 7 and get 35. We multiply 10^3 by 10^{-8} and get 10^{-5}. The product is

$$35 \times 10^{-5}$$

We rewrite this expression in scientific notation.

$$(3.5 \times 10^1) \times 10^{-5} = \mathbf{3.5 \times 10^{-4}}$$

Practice Set

Generalize Multiply. Write each product in scientific notation.

a. $(4.2 \times 10^6)(1.4 \times 10^3)$

b. $(5 \times 10^5)(3 \times 10^7)$

c. $(4 \times 10^{-3})(2.1 \times 10^{-7})$

d. $(6 \times 10^{-2})(7 \times 10^{-5})$

e. **Justify** Show the commutative and associative steps to multiply 3×10^3 and 2×10^4.

Written Practice

Strengthening Concepts

1. (46) The 16-ounce box costs \$3.36. The 24-ounce box costs \$3.96. The smaller box costs how much more per ounce than the larger box?

2. (67, 70) The edges of a cube are 10 cm long.

a. What is the volume of the cube?

b. What is the surface area of the cube?

3. (55) Jan read an average of 42 pages each day for 15 days. She read an average of 50 pages for the next five days. What was the average number of pages she read for all twenty days?

4. (54) Hakim earns \$6 per hour at a part-time job. How much does he earn if he works for 2 hours 30 minutes?

5. (50) Use a unit multiplier to convert 24 shillings to pence (1 shilling = 12 pence).

*** 6.** (78) **Represent** Graph $x \le -1$ on a number line.

7. (72) Use a case 1–case 2 ratio box to solve this problem. Five is to 12 as 20 is to what number?

8. (41) If $a = 1.5$, what does $4a + 5$ equal?

9. (22) Four fifths of the football team's 30 points were scored on pass plays. How many points did the team score on pass plays?

*** 10.** (Inv. 8) **Generalize** A coin is tossed and the spinner is spun. What is the probability of heads and an even number?

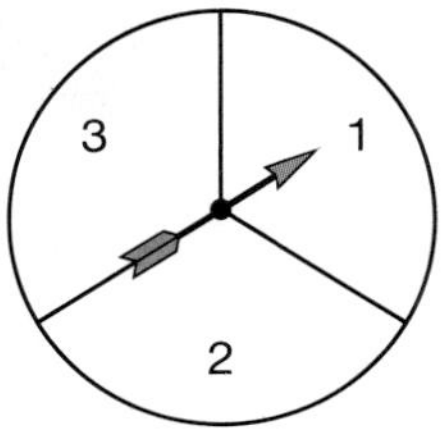

11. (65) Find the circumference of each circle:

a.

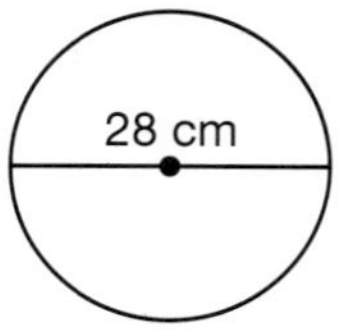

Leave π as π.

b.

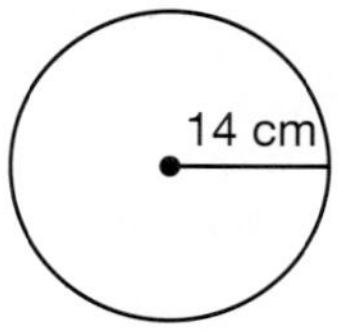

Use $\frac{22}{7}$ for π.

12. (82) Find the area of each circle in problem 11 by using the indicated values for π.

13. (66) Use a ratio box to solve this problem. The ratio of red apples to green apples in the basket was 5 to 2. If there were 70 apples in the basket, how many of them were red?

14. (48) Complete the table.

Fraction	Decimal	Percent
a.	**b.**	250%
$\frac{7}{12}$	**c.**	**d.**

15. (60) What is the sales tax on an $8.50 purchase if the sales-tax rate is $6\frac{1}{2}$%?

Model Use ratio boxes to solve problems **16** and **17.**

* **16.** (81) Monica found that 12 minutes of commercials aired during every hour of prime-time programming. Commercials were shown for what percent of each hour?

* **17.** (81) Thirty percent of the boats that traveled up the river on Monday were steam-powered. If 42 of the boats that traveled up the river were not steam-powered, how many boats were there in all?

Math Language
A **prime factorization** expresses a composite number as a product of its prime factors.

18. (24) Write the prime factorization of the numerator and denominator of this fraction. Then reduce the fraction.

$$\frac{420}{630}$$

19. (75) Find the area of the trapezoid at right.

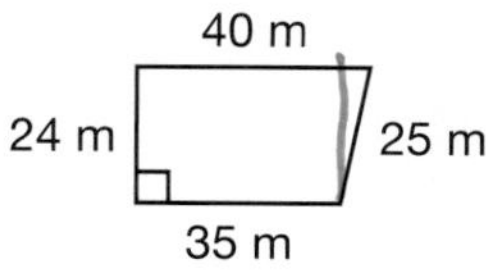

20. (40) In this figure, $\angle A$ and $\angle B$ of $\triangle ABC$ are congruent. The measure of $\angle E$ is 54°. Find the measure of

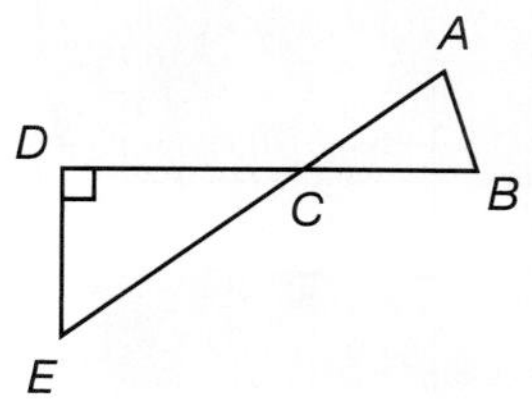

a. $\angle ECD$.

b. $\angle ECB$.

c. $\angle ACB$.

d. $\angle BAC$.

21. (56) Write the rule of this function with words and as an equation. Then find the missing number.

x	y
2	5
5	11
7	15
10	21
−5	

***22.** (83) **Generalize** Multiply. Write each product in scientific notation.

a. $(3 \times 10^4)(6 \times 10^5)$

b. $(1.2 \times 10^{-3})(4 \times 10^{-6})$

For problems **23** and **24,** solve and check the equation. Show each step.

23. (Inv. 7) $b - 1\frac{2}{3} = 4\frac{1}{2}$

24. (Inv. 7) $0.4y = 1.44$

Simplify:

25. (52, 57) $2^3 + 2^2 + 2^1 + 2^0 + 2^{-1}$

26. (43) $0.6 \times 3\frac{1}{3} \div 2$

27. (73) **a.** $\frac{(-4)(-6)}{(-2)(-3)}$

b. $(-3)(-4)(-5)$

28. (30) $\frac{5}{24} - \frac{7}{60}$

29. (68) **a.** $(-3) + (-4) - (-5)$

b. $(-1.5) - (+1.4) + (+1.0)$

*** 30.** (80) **Analyze** The coordinates of the vertices of $\triangle PQR$ are P (0, 1), Q (0, 0), and R (−2, 0). Draw the triangle and its image $\triangle P'Q'R'$ after a 180° clockwise rotation about the origin. What are the coordinates of the vertices of $\triangle P'Q'R'$?

LESSON

84

• Algebraic Terms

Power Up

Building Power

facts Power Up R

mental math

a. Positive/Negative: $(-12) - (-12)$

b. Exponents: 25^2

c. Algebra: $6m - 10 = 32$

d. Decimals/Measurement: 3.14×30 cm

e. Measurement: 1.5 cm to mm

f. Fractional Parts: 30 is $\frac{5}{6}$ of n.

g. Geometry: What four-sided figure has only one set of parallel lines?

h. Calculation: 12×12, -4, $\div 10$, $+1$, $\times 2$, $+3$, $\div 3$, $\times 5$, -1, $\div 6$, $\sqrt{}$

problem solving

Two hourglass sand timers, one of which runs for exactly 5 minutes and the other for exactly 3 minutes, were provided with a spelling game to time the players' turns. Show how to time a 7-minute turn using only the two provided timers.

Understand We need to time a 7-minute turn, but only have a five-minute and a three-minute timer to work with.

Plan We will work backwards from what we know our result needs to be. We can use logical reasoning to find a way to measure 2 minutes with the three-minute and the five-minute timers.

Solve We will turn both timers over at the same time. When the three-minute timer is empty, the players will begin their turn. When the 2 minutes remaining in the five-minute timer run out, we will immediately turn the five-minute timer back over.

Check We found a method for timing two minutes using a five-minute timer and a three-minute timer. The 2 minutes remaining in the five-minute timer + 5 minutes = a 7-minute turn.

New Concept

Increasing Knowledge

We have used the word **term** in arithmetic to refer to the numerator or denominator of a fraction. For example, we reduce a fraction to its lowest terms. In algebra, *term* refers to a part of an algebraic expression or equation. **Polynomials** are algebraic expressions that contain one, two, three, or more terms.

Some Polynomials

Type of Polynomials	Number of Terms	Example
monomial	1	$-2x$
binomial	2	$a^2 - 4b^2$
trinomial	3	$3x^2 - x - 4$

Discuss What type of polynomial is the term $m^2 + 5n^2$? Why?

Terms are separated from one another in an expression by plus or minus signs that are not within symbols of inclusion. To help us see the individual terms we have separated the terms of the binomial and trinomial examples with slashes:

$$a^2 \ / \ -4b^2 \qquad 3x^2 \ / \ -x \ / \ -4$$

Every term contains a positive or negative number and may contain one or more variables (letters). Sometimes the number is understood and not written. For instance, the understood number of a^2 is $+1$ since $a^2 = +1a^2$. **When a term is written without a number, it is understood that the number is 1. When a term is written without a sign, it is understood that the sign is positive.**

Term	$3x^2$	$-x$	-4
Number	$+3$	-1	-4
Variable	x^2	x	none

It is not necessary for a term to contain a variable. A term that does not contain a variable, like -4, is often called a **constant term,** because its value never changes.

Thinking Skill

Justify

Show how to use algebraic addition to combine the constant terms $+3$ and -1.

Constant terms can be combined by algebraic addition.

$$3x + 3 - 1 = 3x + 2 \qquad \text{added } +3 \text{ and } -1$$

Variable terms can also be combined by algebraic addition if they are **like terms.** Like terms have identical variable parts.

$$-3xy^2 + xy^2 \qquad \text{Like terms}$$

Like terms can be combined by algebraically adding the signed-number part of the terms.

$$-3xy^2 + xy^2 = -2xy^2$$

The signed number part of $+xy^2$ is $+1$. We get $-2xy^2$ by adding $-3xy^2$ and $+1xy^2$.

Example 1

Use the commutative and associative properties to collect like terms in this algebraic expression.

$$3x + y + x - y$$

Solution

There are four terms in this expression. There are two x terms and two y terms. We can use the Commutative Property to rearrange the terms.

$3x + x + y - y$ Commutative Property

We use the Associative Property to group like terms.

$(3x + x) + (y - y)$ Associative Property

Adding $+3x$ and $+1x$ we get $+4x$. Then adding $+1y$ and $-1y$ we get $0y$, which is 0.

$4x + 0$ Simplified

$\mathbf{4x}$ Zero Property of Addition

Example 2

Collect like terms in this algebraic expression:

$$3x + 2x^2 + 4 + x^2 - x - 1$$

Solution

In this expression there are three kinds of terms: x^2 terms, x terms, and constant terms. Using the Commutative Property we arrange them to put like terms next to each other.

$2x^2 + x^2 + 3x - x + 4 - 1$ Commutative Property

Now we collect like terms.

$(2x^2 + x^2) + (3x - x) + (4 - 1)$ Associative Property

$\mathbf{3x^2 + 2x + 3}$ Simplified

Thinking Skills

Conclude

Why are x^2 and x not like terms?

Notice that x^2 terms and x terms are not like terms and cannot be combined by addition. There are other possible arrangements of the collected terms, such as the following:

$$2x + 3x^2 + 3$$

Customarily, however, we arrange terms in **descending order** of exponents so that the term with the largest exponent is on the left and the constant term is on the right. An expression written without a constant term is understood to have zero as a constant term.

Generalize Rewrite the expression $4 + 3x^2 + x$ in descending order.

Practice Set

Classify Describe each of these expressions as a monomial, a binomial, or a trinomial:

a. $x^2 - y^2$

b. $3x^2 - 2x - 1$

c. $-2x^3yz^2$

d. $-2x^2y - 4xy^2$

e. State the number of terms in the expressions in exercises **a–d.**

Justify Collect like terms. Show your steps.

f. $3a + 2a^2 - a + a^2$

g. $5xy - x + xy - 2x$

h. $3 + x^2 + x - 5 + 2x^2$

i. $3\pi + 1.4 - \pi + 2.8$

Written Practice

Strengthening Concepts

1. (32) An increase in temperature of 10° on the Celsius scale corresponds to an increase of how many degrees on the Fahrenheit scale?

* **2.** (84) **Generalize** Collect like terms: $2xy + xy - 3x + x$

3. (38, 55) Refer to the graph below to answer **a–c.**

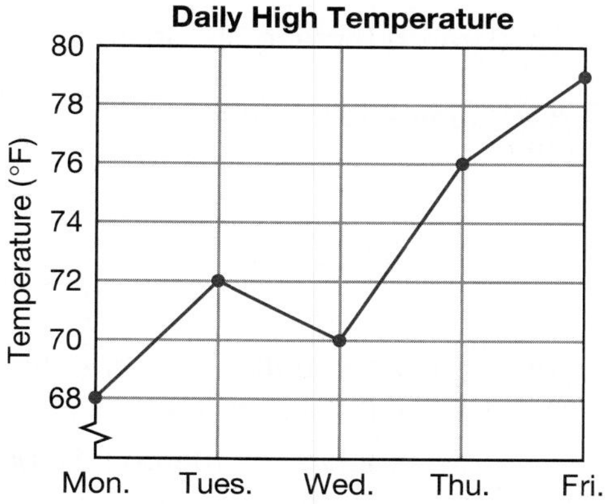

a. What was the range of the daily high temperatures from Monday to Friday?

b. Which day had the greatest increase in temperature from the previous day?

c. Wednesday's high temperature was how much lower than the average high temperature for these 5 days?

4. (Inv. 4) Frank's points in ten games were as follows:

90, 90, 100, 95, 95, 85, 100, 100, 80, 100

For this set of points, find the **a** mean, **b** median, **c** mode, and **d** range.

5. (66) Use a ratio box to solve this problem. The ratio of rowboats to sailboats in the bay was 3 to 7. If the total number of rowboats and sailboats in the bay was 210, how many sailboats were in the bay?

*** 6.** (80) **Analyze** Triangle ABC with vertices at A $(-4, 1)$, B $(-1, 3)$, and C $(-1, 1)$, is reflected in the x-axis. Draw $\triangle ABC$ and its reflection, $\triangle A'B'C'$. What are the coordinates of the vertices of $\triangle A'B'C'$?

7. (72) Write a proportion to solve this problem. If 4 cost \$1.40, how much would 10 cost?

8. (71) Five-eighths of the members supported the treaty, whereas 36 opposed the treaty. How many members supported the treaty?

9. (52) **Evaluate** Evaluate each expression for $x = 5$:

a. $x^2 - 2x + 1$ **b.** $(x - 1)^2$

10. (10) Compare: $f \bigcirc g$ if $\frac{f}{g} = 1$

11. (65, 82) **a.** Find the circumference of the circle shown.

b. Find the area of the circle.

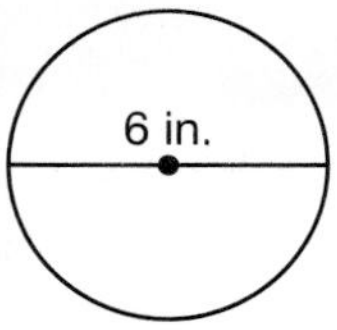

Use 3.14 for π.

12. (50) Use a unit multiplier to convert 4.8 meters to centimeters.

13. (67) Draw a rectangular prism. A rectangular prism has how many faces?

14. (48) Complete the table.

Fraction	Decimal	Percent
$1\frac{4}{5}$	**a.**	**b.**
c.	**d.**	1.8%

15. (60) Write an equation to solve this problem. A merchant priced a product so that 30% of the selling price is profit. If the product sells for \$18.00, how much is the merchant's profit?

*** 16.** (76) Simplify: $\dfrac{12\frac{1}{2}}{100}$

Model Use ratio boxes to solve problems **17** and **18.**

*** 17.** (81) When the door was left open, 36 pigeons flew the coop. If this was 40 percent of all pigeons, how many pigeons were originally in the coop?

*** 18.** (81) Sixty percent of the saplings were 3 feet tall or less. If there were 300 saplings in all, how many were more than 3 feet tall?

19. (19, 75) A square sheet of paper with a perimeter of 48 in. has a corner cut off, forming a pentagon as shown.

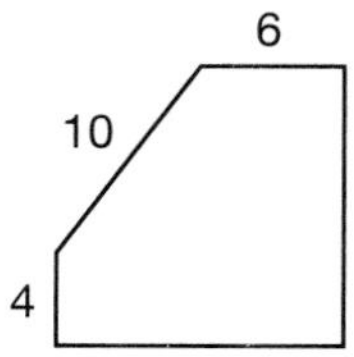

a. What is the perimeter of the pentagon?

b. What is the area of the pentagon?

20. (14) **Analyze** The face of this spinner has been divided into seven sectors, the central angles of which have the following measures:

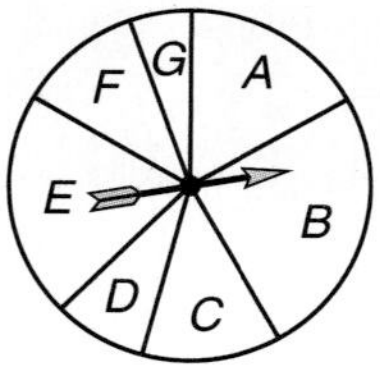

A 60° *B* 90° *C* 45° *D* 30°

E 75° *F* 40° *G* 20°

If the spinner is spun once, what is the probability that it will stop in sector

a. *A*? **b.** *C*? **c.** *E*?

21. (4) Describe the rule of this sequence. Then find the next three numbers of the sequence.

1, 3, 7, 15, 31, . . .

* **22.** (83) **Represent** Multiply. Write each product in scientific notation.

a. $(1.5 \times 10^{-3})(3 \times 10^6)$ **b.** $(3 \times 10^4)(5 \times 10^5)$

Generalize Find each missing exponent:

* **23.** (47) **a.** $10^2 \cdot 10^2 \cdot 10^2 = 10^{\square}$ **b.** $\frac{10^2}{10^6} = 10^{\square}$

For problems **24** and **25,** solve and check the equation. Show each step.

24. (Inv. 7) $b - 4.75 = 5.2$ **25.** (Inv. 7) $\frac{2}{3}y = 36$

Simplify:

26. (52) $\sqrt{5^2 - 4^2} + 2^3$ **27.** (32) 1 m − 45 mm

28. (43) $\frac{9}{10} \div 2\frac{1}{4} \cdot 24$ (decimal answer)

29. (73) **a.** $\frac{(-8)(+6)}{(-3)(+4)}$ **b.** $\left(\frac{1}{2}\right)\left(-\frac{1}{3}\right)\left(\frac{1}{4}\right)$

30. (68) **a.** $(+30) - (-50) - (+20)$

b. $(-0.3) - (-0.4) - (0.5)$

Early Finishers
Real-World Application

Twenty students guessed the number of marbles in a glass container. The results are shown below.

28	20	15	19	37	40	22	42	30	36
25	28	25	33	45	20	50	32	21	30

Which type of display—a stem-and-leaf plot or a line graph—is the most appropriate way to display this data? Draw your display and justify your choice.

LESSON 85

• Order of Operations with Positive and Negative Numbers

Power Up — *Building Power*

facts Power Up P

mental math

a. **Positive/Negative:** $(+12)(-6)$

b. **Order of Operations/Exponents:** $(4 \times 10^3)(2 \times 10^6)$

c. **Ratio:** $\frac{1}{1.5} = \frac{80}{n}$

d. **Fractional Parts:** \$12 is $\frac{1}{4}$ of how much money (m)?

e. **Measurement:** 0.8 km to m

f. **Fractional Parts:** What is $\frac{1}{4}$ of \$12?

g. **Geometry:** What face on a pyramid designates its type?

h. **Geometry:** Find the perimeter and area of a square with sides 2.5 m long.

problem solving

Copy this problem and fill in the missing digits:

$$\begin{array}{r} \_\,\_\,\_ \\ \times \quad \_ \\ \hline 1001 \end{array}$$

New Concept — *Increasing Knowledge*

Thinking Skill

Generalize

Use the sentence *<u>P</u>lease <u>e</u>xcuse <u>m</u>y <u>d</u>ear <u>A</u>unt <u>S</u>ally* to remember the correct order of operations.

To simplify expressions that involve several operations, we perform the operations in a prescribed order. We have practiced simplifying expressions with whole numbers. In this lesson we will begin simplifying expressions that contain negative integers as well.

Example 1

Simplify: $(-2) + (-2)(-2) - \frac{(-2)}{(+2)}$

Solution

First we multiply and divide in order from left to right.

$$(-2) + \underbrace{(-2)(-2)} - \underbrace{\frac{(-2)}{(+2)}}$$

$$(-2) + \quad (+4) \quad - (-1)$$

Then we add and subtract in order from left to right.

$$\underbrace{\underbrace{(-2) + (+4)}_{(+2)} - (-1)}_{\mathbf{+3}}$$

Mentally separating an expression into its terms can make an expression easier to simplify. Here is the same expression. This time we will use slashes to separate the terms.

$$(-2) \,/\, + (-2)(-2) \,/\, - \frac{(-2)}{(+2)}$$

First we simplify each term; then we combine the terms.

$$(-2) \,/\, + (-2)(-2) \,/\, - \frac{(-2)}{(+2)}$$

$$-2 \quad / \quad +4 \quad / \quad +1$$

$$\mathbf{+3}$$

Example 2

Simplify each term. Then combine the terms.

$$\mathbf{-3(2 - 4) - 4(-2)(-3) + \frac{(-3)(-4)}{2}}$$

Solution

We separate the individual terms with slashes. The slashes precede plus and minus signs that are not enclosed by parentheses or other symbols of inclusion.

$$-3(2 - 4) \,/\, - 4(-2)(-3) \,/\, + \frac{(-3)(-4)}{2}$$

Next we simplify each term.

$$-3(2 - 4) \,/\, - 4(-2)(-3) \,/\, + \frac{(-3)(-4)}{2}$$

$$-3(-2) \quad / \quad + 8(-3) \quad / \quad + \frac{12}{2}$$

$$+6 \quad / \quad - 24 \quad / \quad + 6$$

Now we combine the simplified terms.

$$+6 - 24 + 6$$

$$-18 + 6$$

$$\mathbf{-12}$$

Example 3

Simplify: (−2) − [(−3) − (−4)(−5)]

Solution

There are only two terms, −2 and the bracketed quantity. By the order of operations, we simplify within brackets first, multiplying and dividing before adding and subtracting.

$$(-2) \;/\; - [(-3) - (-4)(-5)]$$

$$(-2) \;/\; - [(-3) - (+20)]$$

$$(-2) \;/\; - (-23)$$

$$(-2) \;/\; + 23$$

$$\mathbf{+21}$$

Signed numbers are often written without parentheses. To simplify such expressions we simply add algebraically from left to right.

$$-3 + 4 - 5 = -4$$

Another way to simplify this expression is to use the commutative and associative properties to rearrange and regroup the terms by their signs.

$-3 + 4 - 5$	Given
$+4 - 3 - 5$	Commutative Property
$+4 + (-3 - 5)$	Associative Property
$+4 - 8$	$-3 - 5 = -8$
-4	$+4 - 8 = -4$

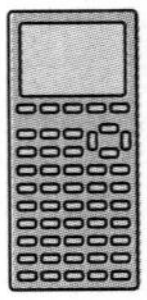

Visit www.SaxonPublishers.com/ActivitiesC2 *for a graphing calculator activity.*

Example 4

Simplify: −2 + 3(−2) − 2(+4)

Solution

To emphasize the separate terms, we first draw a slash before each plus or minus sign that is not enclosed.

$$-2 \;/\; +3(-2) \;/\; -2(+4)$$

Next we simplify each term.

$$-2 \;/\; +3(-2) \;/\; -2(+4)$$

$$-2 \qquad -6 \qquad -8$$

Then we algebraically add the terms.

$$-2 - 6 - 8 = \mathbf{-16}$$

Practice Set

Justify Simplify. Show steps and properties.

a. $(-3) + (-3)(-3) - \dfrac{(-3)}{(+3)}$

b. $(-3) - [(-4) - (-5)(-6)]$

c. $(-2)[(-3) - (-4)(-5)]$

d. $(-5) - (-5)(-5) + |-5|$

e. $-3 + 4 - 5 - 2$

f. $-2 + 3(-4) - 5(-2)$

g. $-3(-2) - 5(2) + 3(-4)$

h. $-4(-3)(-2) - 6(-4)$

Written Practice

Strengthening Concepts

1. *(Inv. 4)* Find the **a** mean, **b** median, **c** mode, and **d** range of the following set of numbers:

70, 80, 90, 80, 70, 90, 75, 95, 100, 90

Use ratio boxes to solve problems **2–4:**

2. *(66)* The ratio of fiction to nonfiction books that Sarah read was 3 to 1. If she read 24 books, how many of the books were nonfiction?

3. *(53)* Mary weeded her flower garden. She found that the ratio of dandelions to marigolds in the garden was 11 to 4. If there were 44 marigolds in the garden, how many dandelions were there?

4. *(72)* If sound travels 2 miles in 10 seconds, how far does sound travel in 1 minute?

5. *(50)* Use a unit multiplier to convert 0.98 liter to milliliters.

*** 6.** *(78)* **Represent** Graph $x > 0$ on a number line.

7. *(71)* Diagram this statement. Then answer the questions that follow.

Thirty-five thousand dollars were raised in the charity drive. This was seven tenths of the goal.

a. The goal of the charity drive was to raise how much money?

b. The drive fell short of the goal by what percent?

*** 8.** *(65, 82)* **Analyze** The radius of a circle is 4 meters. Use 3.14 for π to find the

a. circumference of the circle.

b. area of the circle.

*** 9.** *(Inv. 1)* **Generalize** What fraction of this circle is shaded?

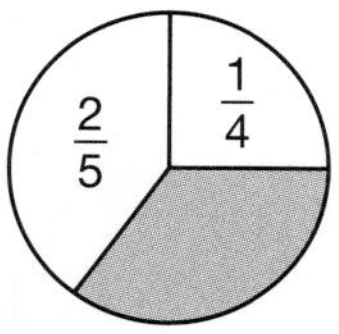

* **10.** (Inv. 8) **Analyze** If the circle in problem **9** is the face of a spinner, and if the spinner is spun twice, what is the probability of the spinner stopping in the shaded area both times?

11. (70) A certain rectangular box is 5 in. long, 4 in. wide, and 3 in. high. Draw the box and find its volume.

12. (67) Suppose that the box described in problem **11** is cut open and unfolded. Draw the unfolded pattern and find the surface area.

13. (48) Complete the table.

Fraction	Decimal	Percent
$\frac{1}{40}$	**a.**	**b.**
c.	**d.**	0.25%

14. (60) When the Nelsons sold their house, they paid the realtor a fee of 6% of the selling price. If the house sold for $180,000, how much was the realtor's fee?

15. (21) **a.** Write the prime factorization of 17,640 using exponents.

b. How can you tell by looking at the answer to **a** that $\sqrt{17,640}$ is not a whole number?

* **16.** (76) **Generalize** Simplify: $\dfrac{8\frac{1}{3}}{100}$

Use ratio boxes to solve problems **17** and **18.**

* **17.** (81) **Connect** Max was delighted when he found that 38 of the 40 seeds he planted had sprouted. What percent of the seeds sprouted?

* **18.** (81) Before the assembly began, only 35 percent of the students were seated. If 91 students were not seated, how many students were there in all?

19. (Inv. 6, 61) **a.** Classify the quadrilateral shown.

b. Find its perimeter.

c. Find its area.

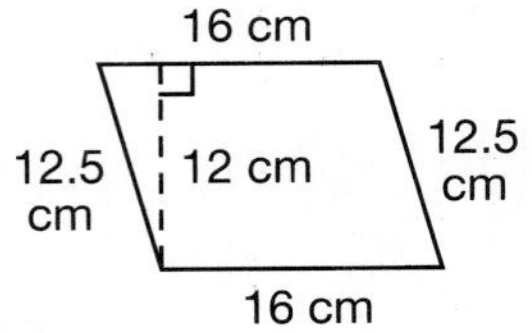

d. This figure does not have line symmetry, but it does have point symmetry. Trace the figure on your paper, and locate its point of symmetry.

20. (56) Find the missing numbers in the table by using the function rule. Then graph the x, y pairs on a coordinate plane and draw a line through the points and extending beyond the points.

$y = 2x - 1$

x	y
5	
3	
1	

21. Refer to the figure below to answer **a–d.**
(40)

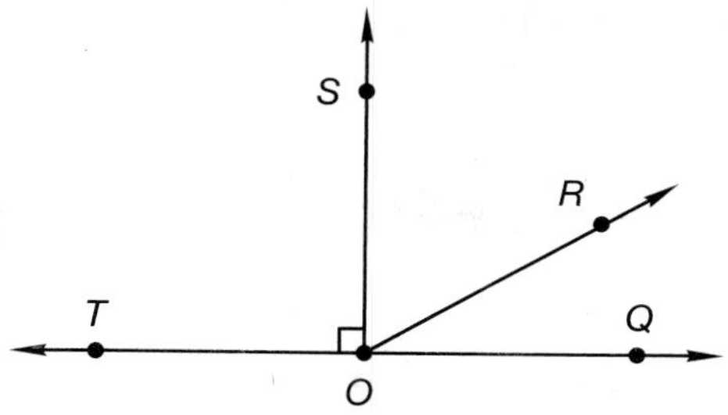

a. Find m∠*TOS.*

b. Find m∠*QOT.*

c. Angle *QOR* is one third of a right angle. Find m∠*QOR.*

d. Find m∠*TOR.*

* **22.** *(51, 83)* **Evaluate** Compare: $(5 \times 10^{-3})(6 \times 10^{8}) \bigcirc (5 \times 10^{8})(6 \times 10^{-3})$

For problems **23** and **24,** solve and check the equation. Show each step.

23. *(Inv. 7)* $13.2 = 1.2w$

24. *(Inv. 7)* $c + \frac{5}{6} = 1\frac{1}{4}$

Generalize Simplify:

25. *(63)* $3\{20 - [6^2 - 3(10 - 4)]\}$

26. *(49)*

$$\begin{array}{r} 3 \text{ hr } 15 \text{ min } 25 \text{ s} \\ - \; 2 \text{ hr } 45 \text{ min } 30 \text{ s} \\ \hline \end{array}$$

27. *(57)* $2^0 + 0.2 + 2^{-2}$ (decimal answer)

* **28.** *(85)* **a.** $(-3) - \left[(-2) - (+2) - \frac{(-2)}{(-2)}\right]$

b. $(-3) - [(-2) - (+4)(-5)]$

* **29.** *(84)* **Generalize** Collect like terms: $x^2 + 6x - 2x - 12$

* **30.** *(80)* The coordinates of three vertices of square *ABCD* are *A* (1, 2), *B* (4, 2), and *C* (4, –1).

a. Find the coordinates of *D* and draw the square.

b. Reflect square *ABCD* in the *y*-axis, and draw its image, square *A′ B′ C′ D′*. What are the coordinates of the vertices of this reflection?

LESSON 86

• Number Families

Power Up

Building Power

facts

Power Up S

mental math

a. **Positive/Negative:** $(-18) + (-40)$

b. **Order of Operations/Exponents:** $(3 \times 10^{-3})(3 \times 10^{-3})$

c. **Algebra:** $7x + 4 = 60$

d. **Percent/Estimation:** Estimate 15% of \$17.90.

e. **Measurement:** 0.2 L to mL

f. **Fractional Parts:** \$30 is $\frac{1}{3}$ of *m.*

g. **Probability:** There are 5 red marbles, 2 green marbles, and 4 blue marbles in a bag. What is the probability of choosing a red marble?

h. **Calculation:** What is the total cost of a \$200 item plus 7% sales tax?

problem solving

The expression $\sqrt[3]{8}$ means "the cube root of 8." The cube root notation, $\sqrt[3]{\ }$, is asking, "What number used as a factor three times equals the number under the symbol?" The cube root of 8 is 2 because $2 \times 2 \times 2 = 8$. Find $\sqrt[3]{64}$. Find $\sqrt[3]{125}$. Find $\sqrt[3]{1{,}000{,}000}$.

New Concept

Increasing Knowledge

Thinking Skill

Relate

If every counting number is a whole number, is every whole number a counting number? Explain.

In mathematics we give special names to certain sets of numbers. Some of these sets are the counting numbers, the whole numbers, the integers, and the rational numbers. In this lesson we will review each of these **number families** and discuss how they are related.

- **The Counting Numbers.** Counting numbers are the numbers we say when we count. The first counting number is 1, the next is 2, then 3, and so on.

  Counting numbers: 1, 2, 3, 4, 5, ...

- **The Whole Numbers.** The members of the whole-number family are the counting numbers as well as the number zero.

  Whole numbers: 0, 1, 2, 3, 4, 5, ...

  If we use a dot to mark each of the whole numbers on the number line, the graph looks like this:

Notice that there are no dots to the left of zero. This is because no whole number is a negative number. Also notice that there are no dots between consecutive whole numbers. Numbers between consecutive whole numbers are not "whole." The blue arrowhead on the right end of the number line indicates that the whole numbers increase without end.

- **The Integers.** The integer family includes all the whole numbers. It also includes the opposites (negatives) of the positive whole numbers. The list of integers goes on and on in both directions as indicated by the ellipses below.

  Integers: ..., − 4, −3, −2, −1, 0, 1, 2, 3, 4, ...

  A graph of the integers looks like this:

  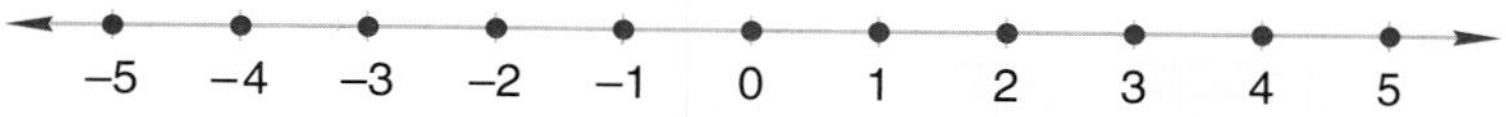

  The blue arrowheads on both ends of the number line indicate that the set of integers continues without end in both directions. Notice that integers do not include such numbers as $\frac{1}{2}$, $\frac{5}{3}$, and other fractions.

- **The Rational Numbers.** The family of **rational numbers** includes all numbers that can be written as a *ratio* (fraction) of two integers. Here are some examples of rational numbers:

  $$\frac{1}{2} \quad \frac{5}{3} \quad \frac{-3}{2} \quad \frac{-4}{1} \quad \frac{0}{2} \quad \frac{3}{1}$$

  Notice that the family of rational numbers includes all the integers, because every integer can be written as a fraction whose denominator is the number 1. For example, we can write −4 as a fraction by writing

  $$\frac{-4}{1}$$

  The set of rational numbers also includes all the positive and negative mixed numbers, because these numbers can be written as fractions. For example, we can write $4\frac{1}{5}$ as

  $$\frac{21}{5}$$

  Sometimes rational numbers are written in decimal form, in which case the decimal number will either terminate or repeat.

  $$\frac{1}{8} = 0.125 \qquad \frac{5}{6} = 0.8333\ldots = 0.8\overline{3}$$

  The diagram on the next page may be helpful in visualizing the relationships between these families of numbers. The diagram shows that the set of rational numbers includes all the other number families described in this lesson.

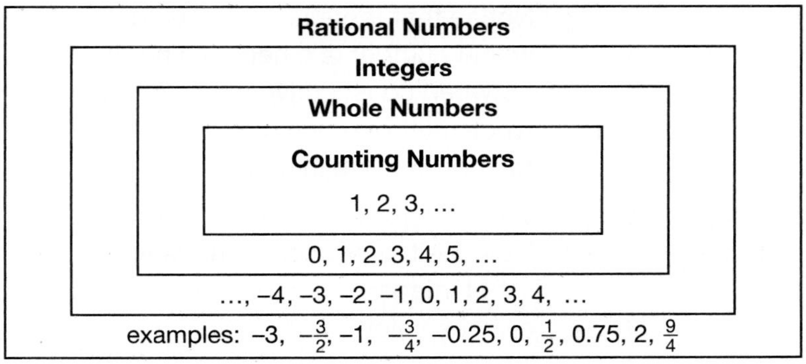

Discuss How does the diagram above visually represent the relationship between the different number families?

Example 1

Graph the integers that are less than 4.

Solution

We draw a number line and mark a dot at every integer that is less than 4. Since the set of integers includes whole numbers, we mark dots at 3, 2, 1, and 0. Since the integers also include the opposites of the positive whole numbers, we continue marking dots at −1, −2, −3, and so on. We then mark an arrowhead on the negative end of the line to indicate that the graph of integers that are less than 4 continues without end.

Verify Fractions such as $\frac{1}{2}$ are also less than 4. Why didn't we graph points for fractions such as $\frac{1}{2}$ on the number line above?

Example 2

Answer true or false:

a. All whole numbers are integers.

b. All rational numbers are integers.

Solution

a. True. Every whole number is included in the family of integers.

b. False. Although every integer is a rational number, it is not true that every rational number is an integer. Rational numbers such as $\frac{1}{2}$ and $\frac{5}{3}$ are not integers.

Practice Set

Represent Graph the integers that are:

a. greater than −4.

b. less than 4.

Justify Answer true or false. If the statement is false, explain why and give an example that proves it is false.

c. Every integer is a whole number.

d. Every integer is a rational number.

Written Practice *Strengthening Concepts*

1. (46) Fragrant Scent was priced at $28.50 for 2 ounces, while Eau de Rue cost only $4.96 for 4 ounces. Fragrant Scent cost how much more per ounce than Eau de Rue?

2. (66) Use a ratio box to solve this problem. The ratio of rookies to veterans in the camp was 2 to 7. Altogether there were 252 rookies and veterans in the camp. How many of them were rookies?

3. (Inv. 4) Seven trunks were delivered to the costume department at the theater. The trunks weighed 197 lb, 213 lb, 246 lb, 205 lb, 238 lb, 213 lb, and 207 lb. Find the **a** mode, **b** median, **c** mean, and **d** range of this group of measures.

4. (50) Use a unit multiplier to convert 12 bushels to pecks (1 bushel = 4 pecks).

5. (46) The Martins drove the 468 miles between Memphis, Tennessee and Cincinnati, Ohio in one day. If they left Memphis at 7 a.m. and arrived in Cincinnati at 4 p.m., what was the Martins' average speed?

* **6.** (86) **Represent** Illustrate on a number line, the whole numbers that are less than or equal to 3.

7. (72) Use a ratio box to solve this problem. Nine is to 6 as what number is to 30?

8. (22) Nine tenths of the school's 1800 students attended the homecoming game.

a. How many of the school's students attended the homecoming game?

b. What percent of the school's students did not attend the homecoming game?

9. (52) Evaluate: $\sqrt{b^2 - 4ac}$ if $a = 1$, $b = 5$, and $c = 4$

* **10.** (Inv. 8) **Analyze** During basketball season Heidi has made 40 free throws and missed 20. What is the experimental probability that she will make her next free throw?

* **11.** (66, 82) **a.** Find the circumference of the circle shown.

b. Find the area of the circle.

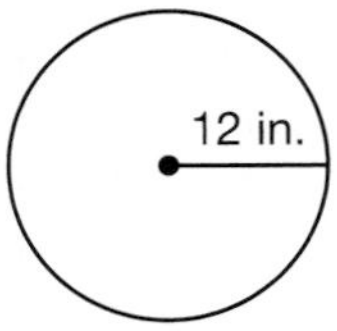

Leave π as π.

12. (47) Find each missing exponent:

a. $10^8 \cdot 10^{-3} = 10^{\square}$ **b.** $10^5 \div 10^8 = 10^{\square}$

13. (67) The figure shown is a triangular prism. Copy the figure on your paper, and find the number of its

a. faces.

b. edges.

c. vertices.

14. (48) Complete the table.

Fraction	Decimal	Percent
a.	0.9	**b.**
$\frac{11}{12}$	**c.**	**d.**

15. (17) Obi is facing north. If he turns 360° in a clockwise direction, what direction will he be facing? Does the direction he turns affect the answer? Explain.

Model Use ratio boxes to solve problems **16** and **17.**

* **16.** (81) The sale price of $24 was 60 percent of the regular price. What was the regular price?

* **17.** (81) Forty-eight corn seeds sprouted. This was 75 percent of the seeds that were planted. How many of the planted seeds did not sprout?

18. (77) Write an equation to solve this problem:

Thirty is what percent of 20?

* **19.** (Inv. 6, 75) **a.** *Classify* What type of quadrilateral is shown?

b. Find its perimeter.

c. Find its area.

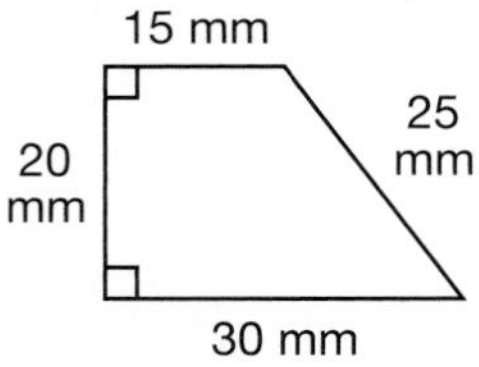

20. Find the measure of each angle.
(17)

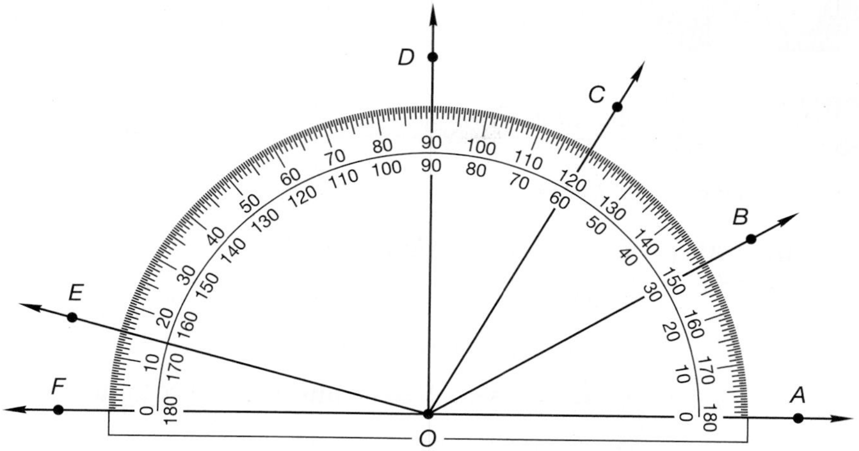

a. $\angle COF$ **b.** $\angle AOE$ **c.** $\angle BOE$

21. Find the missing numbers in the table by using the function rule. Then graph the *x*, *y* pairs on a coordinate plane and extend a line through the points. Find another *x*, *y* pair on the line and test the numbers in the equation.
(56)

$y = x + 1$

x	y
3	
5	
1	

* **22.** **Represent** Multiply. Write each product in scientific notation.
(83)

a. $(1.2 \times 10^5)(1.2 \times 10^{-8})$

b. $(6 \times 10^{-3})(7 \times 10^{-4})$

For problems **23** and **24,** solve and check the equation. Show each step.

23. $56 = \frac{7}{8}w$
(Inv. 7)

24. $4.8 + c = 7.34$
(Inv. 7)

Simplify:

25. $\sqrt{10^2 - 6^2} + \sqrt{10^2 - 8^2}$
(52)

26.
(49)

$$\begin{array}{r} 5 \text{ lb } 9 \text{ oz} \\ + \ 4 \text{ lb } 7 \text{ oz} \\ \hline \end{array}$$

27. $1.4 \div 3\frac{1}{2} \times 10^3$ (decimal answer)
(43, 45)

* **28.** Simplify each expression.
(85)

a. $(-4)(-5) - (-4)(+3)$

b. $(-2)[(-3) - (-4)(+5)]$

* **29.** Collect like terms: $x^2 + 3xy + 2x^2 - xy$
(84)

30. The factorization of $6x^2y$ is $2 \cdot 3 \cdot x \cdot x \cdot y$. Write the factorization of $9xy^2$.
(21)

LESSON

87 • Multiplying Algebraic Terms

Power Up

Building Power

facts Power Up R

mental math

a. Positive/Negative: $(-60) - (-30)$

b. Order of Operations/Exponents: $(2 \times 10^5)(4 \times 10^{-3})$

c. Equivalent Fractions: $\frac{f}{100} = \frac{10}{25}$

d. Decimals/Measurement: 3.14×20 ft

e. Measurement: 750 g to kg

f. Fractional Parts: \$50 is $\frac{2}{5}$ of m.

g. Probability: There are 9 blue marbles, 3 red marbles, and 8 purple marbles in a bag. Which color marble are you most likely to pick from the bag?

h. Calculation: $33\frac{1}{3}\%$ of 12, $\times$ 9, $\sqrt{}$, $\times$ 8, $+$ 1, $\sqrt{}$, $\times$ 3, $-$ 1, $\div$ 2, $\div$ 2, $\div$ 2

problem solving

On a balanced scale, four identical blocks marked m, a 250-g mass, and a 1000-g mass are distributed as shown. Find the mass of each block marked m. Write an equation illustrated by this balanced scale.

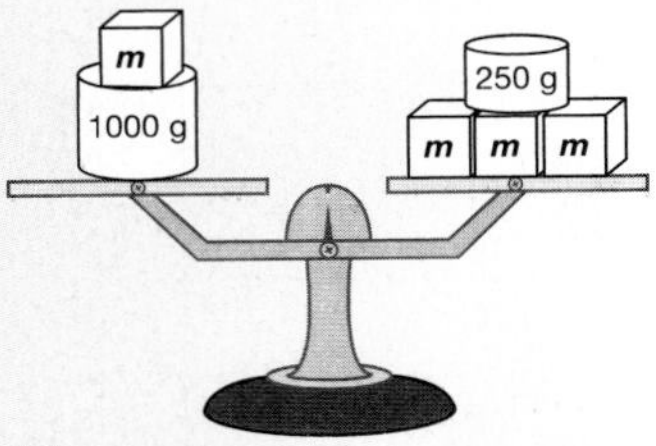

New Concept

Increasing Knowledge

Recall from Lesson 84 that like terms can be added and that adding like terms does not change the variable part of the term.

$$3x + 2x = 5x$$

However, if we multiply terms, all the factors in the multiplied terms appear in the product.

$(3x)(2x)$	$= 3 \cdot x \cdot 2 \cdot x$	
	$= 3 \cdot 2 \cdot x \cdot x$	Commutative Property
	$= (3 \cdot 2)(x \cdot x)$	Associative Property
	$= 6x^2$	Simplified

Explain What does the exponent 2 in x^2 mean?

Briefly, we multiply the numerical parts of the terms and gather variable factors with exponents. Terms may be multiplied even if they are not like terms.

$$(-2x)(-3y) = 6xy$$

Example 1

Simplify: $(-3x^2y)(2x)(-4xy)$

Solution

The minus signs on these terms indicate negative numbers, not subtraction. We will multiply the three terms to make one term. First we list all the factors:

$$(-3) \cdot x \cdot x \cdot y \cdot (+2) \cdot x \cdot (-4) \cdot x \cdot y$$

Using the Commutative Property, we rearrange the factors as shown below.

$$(-3)(+2)(-4) \cdot x \cdot x \cdot x \cdot x \cdot y \cdot y$$

Using the Associative Property, we then group the factors by multiplying the numerical factors and gathering the variable factors with exponents.

$$\mathbf{24x^4y^2}$$

Discuss How can you determine that the product will be negative or positive?

Example 2

Simplify: $(-2ab)(a^2b)(3b^3)$

Solution

Thinking Skill

Explain

What happens to the exponents when a and a^2 are multiplied?

The numerical factors are −2, 1, and 3, and their product is −6. The product of a and a^2 is a^3. The product of b, b, and b^3 is b^5.

$$(-2ab)(a^2b)(3b^3) = \mathbf{-6a^3b^5}$$

Recall that we can multiply exponents to find a power of a power.

$$(x^a)^b = x^{ab}$$

If a term with multiple factors is raised to a power, then the exponent applies to each of the factors.

$$(3xy)^2 = (3xy)(3xy)$$

$$\text{So, } (3xy)^2 = 3^2x^2y^2 = 9x^2y^2$$

If a term that includes exponents is raised to a power, then we can multiply the exponents in the term by the power.

$$(3x^2y^3)^2 = 3^2(x^2)^2(y^3)^2 = 9x^4y^6$$

Example 3

Simplify: $(-3a^2b^3)^2$

Solution

We show two methods.

Method 1: Show factors.

$$\begin{aligned}(-3a^2b^3)^2 &= (-3a^2b^3)(-3a^2b^3)\\ &= (-3)aabbb(-3)aabbb\\ &= (-3)(-3)aaaabbbbbb\\ &= \mathbf{9a^4b^6}\end{aligned}$$

Method 2: Apply exponent rules.

$$\begin{aligned}(-3a^2b^3)^2 &= (-3)^2(a^2)^2(b^3)^2\\ &= \mathbf{9a^4b^6}\end{aligned}$$

Practice Set

Justify Find the following products. Show and justify each step.

a. $(-3x)(-2xy)$

b. $3x^2(xy^3)$

c. $(2a^2)(-3ab^3)$

d. $(-4x)(-5x^2y)$

Generalize Find each product by multiplying the numerical factors and gathering variable factors with exponents.

e. $(-xy^2)(xy)(2y)$

f. $(-3m)(-2mn)(m^2n)$

g. $(4wy)(3wx)(-w^2)(x^2y)$

h. $5d(-2df)(-3d^2fg)$

i. Simplify (show two methods): $(3xy^3)^2$

Written Practice

Strengthening Concepts

1. (54) How far will a jet travel in 2 hours 30 minutes if its average speed is 450 miles per hour?

2. (50) Use a unit multiplier to convert 12.5 centimeters to meters.

3. (53) Use a ratio box to solve this problem. If 240 of the 420 students in the auditorium were girls, what was the ratio of boys to girls in the auditorium?

4. (28) Giraffes are the tallest of all land mammals. One giraffe is 18′3″ tall. Another giraffe is 17′10″ tall and a third giraffe is 17′11″ tall. What is the average height of these 3 giraffes?

5. The Xiong family traveled by car 468 miles on 18 gallons of gas. Their car averaged how many miles per gallon?
(46)

*** 6.** **Represent** On a number line, graph the whole numbers that are less than or equal to 5.
(86)

7. Use a ratio box to solve this problem. The road was steep. Every 100 yards the elevation increased 36 feet. How many feet did the elevation increase in 1500 yards?
(72)

8. The quadrilateral in this figure is a parallelogram. Find the measure of
(61)

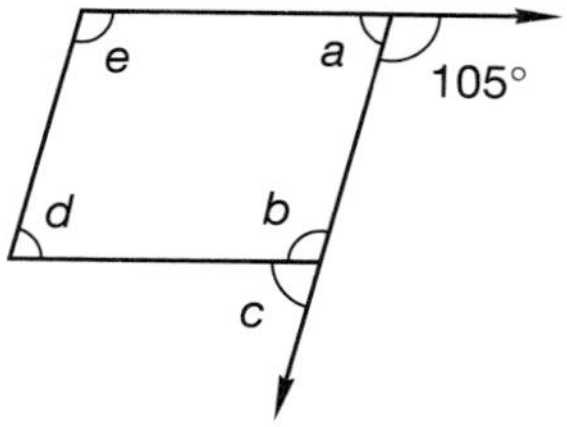

a. $\angle a$. **b.** $\angle b$.

c. $\angle c$. **d.** $\angle d$.

e. $\angle e$.

9. If $x = -4$ and $y = 3x - 1$, then y equals what number?
(41)

*** 10.** **Analyze** Find each measure for this circle. Use $\frac{22}{7}$ for π.
(65, 82)

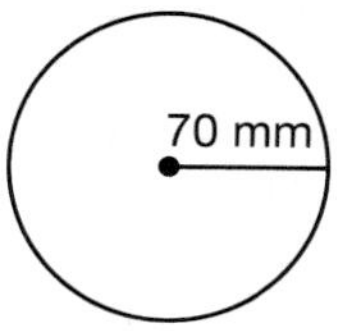

a. circumference

b. area

11. The coordinates of the vertices of parallelogram *ABCD* are *A* (5, 5), *B* (10, 5), *C* (5, 0), and *D* (0, 0).
(Inv. 3, 61)

a. On graph paper, graph the vertices. Then draw the parallelogram.

b. Find the area of the parallelogram.

c. Find the measure of each angle of the parallelogram.

12. The shape shown was built of 1-inch cubes. What is the volume of the shape?
(70)

13. Complete the table.
(48)

Fraction	Decimal	Percent
a.	**b.**	$12\frac{1}{2}\%$
$\frac{7}{8}$	**c.**	**d.**

14. Write an equation to solve this problem and find the answer.
(60)

What number is 25 percent of 4?

Model Use ratio boxes to solve problems **15** and **16.**

* **15.** (81) The sale price of $24 for a board game was 80 percent of the regular price. What was the regular price?

* **16.** (81) David had finished 60 percent of the race, but he still had 2000 meters to run. How long was his race?

17. (77) Write an equation to solve this problem and find the answer.

One hundred is what percent of 80?

* **18.** (75) **Analyze** Find the area of this figure. Dimensions are in centimeters. Corners that look square are square.

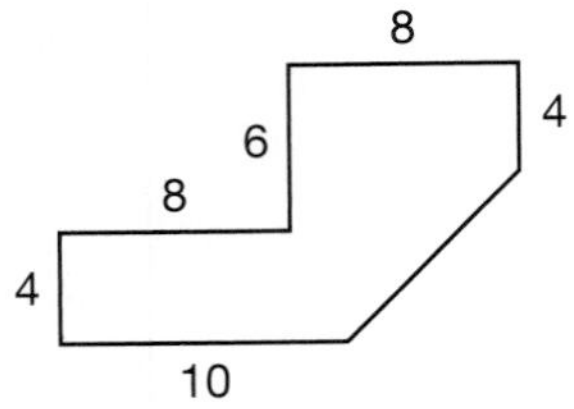

Math Reading

The small square marks a right angle. $\angle EOD$ is a right angle.

19. (40) In the figure below, angle *AOE* is a straight angle and $m\angle AOB = m\angle BOC = m\angle COD$.

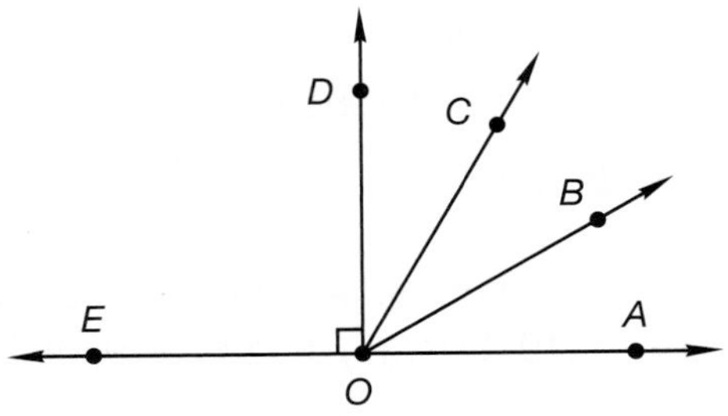

a. Find $m\angle AOB$.

b. Find $m\angle AOC$.

c. Find $m\angle EOC$.

d. Which angle is the supplement of $\angle EOC$?

* **20.** (76) Simplify: $\dfrac{66\frac{2}{3}}{100}$

* **21.** (56) **Generalize** Find the missing numbers in the table using the function rule. Then graph the *x, y* pairs on a coordinate plane and extend a line through the points. Find another *x, y* pair on the line and check the numbers in the equation.

$y = 2x - 3$

x	*y*
1	
2	
3	

* **22.** (83) Multiply. Write each product in scientific notation.

a. $(4 \times 10^{-5})(2.1 \times 10^{-7})$

b. $(4 \times 10^{5})(6 \times 10^{7})$

For problems **23** and **24,** solve and check the equation. Show each step.

23. (Inv. 7) $d - 8.47 = 9.1$

24. (Inv. 7) $0.25m = 3.6$

Generalize Simplify:

25. (52) $\dfrac{3 + 5.2 - 1}{4 - 3 + 2}$

26. (32) 1 kg − 75 g

27. (43) $3.7 + 2\frac{5}{8} + 15$ (decimal answer)

* **28.** (85) **a.** $(-5) - (-2)[(-3) - (+4)]$

b. $\dfrac{(-3) + (-3)(+4)}{(+3) + (-4)}$

* **29.** (87) **a.** $(3x)(4y)$

b. $(6m)(-4m^2n)(-mnp)$

c. $(3x^3)^2$

* **30.** (84) **Generalize** Collect like terms:

$$3ab + a - ab - 2ab + a$$

Early Finishers
Real-World Application

While on vacation, Sanjay and Aditi decide to spend an afternoon biking. They can rent the bikes from one of two companies in town. Beau's Bikes charges $10 per hour per bike. Celine's Cycles charges $4 per hour per bike plus an initial fixed fee of $15 per bike.

a. Let *t* represent the time one bike is rented (in hours) and *y* represent the total cost of renting one bike. Represent the cost from each company with an algebraic expression.

b. If Sanjay and Aditi expect to bike for 2 hours, from which bike shop should they rent? Support your answer.

c. If they want to bike for 4 hours, from which bike shop should they rent? Support your answer.

LESSON

88

• Multiple Unit Multipliers

Power Up

Building Power

facts

Power Up S

mental math

a. **Positive/Negative:** $(-15)(+5)$

b. **Order of Operations/Exponents:** $(1.5 \times 10^4)(2 \times 10^5)$

c. **Algebra:** $3t + 4 = 40$

d. **Percent/Mental Math:** $7\frac{1}{2}\% \times \$200$

e. **Measurement:** 2.54 cm to mm

f. **Fractional Parts:** \$1.50 is $\frac{3}{5}$ of m.

g. **Geometry:** When looking at a geometric shape and its mirror image what is this called?

h. **Rate:** At 500 mph, how far will an airplane fly in $2\frac{1}{2}$ hours?

problem solving

Some auto license plates take the form of one letter, followed by three numbers, followed by three letters. How many different license plates are possible?

New Concept

Increasing Knowledge

We can repeatedly multiply a number by 1 without changing the number.

$$5 \cdot 1 = 5$$

$$5 \cdot 1 \cdot 1 = 5$$

$$5 \cdot 1 \cdot 1 \cdot 1 = 5$$

Since unit multipliers are forms of 1, we can also multiply a measure by several unit multipliers without changing the measure.

Example 1

Use two unit multipliers to convert 5 hours to seconds.

Solution

We are changing units from hours to seconds.

$$\text{hours} \longrightarrow \text{seconds}$$

We will perform the conversion in two steps. We will change from hours to minutes with one unit multiplier and from minutes to seconds with a second unit multiplier. For each step we write a unit multiplier.

hours ⟶ minutes ⟶ seconds

$$5\ \cancel{\text{hr}} \cdot \frac{60\ \cancel{\text{min}}}{1\ \cancel{\text{hr}}} \cdot \frac{60\ \text{s}}{1\ \cancel{\text{min}}} = \mathbf{18{,}000\ s}$$

Discuss What is the difference between the 60 in the first unit multiplier and the 60 in the second unit multiplier?

To convert from one unit of area to another, it is helpful to use two unit multipliers.

Example 2

Diego wants to carpet a 16 ft by 9 ft room that has an area of 144 ft^2. Carpet is sold by the square yard. How many square yards of carpet does Diego need?

Solution

Thinking Skill

Explain

Diego could have converted 16 ft and 9 ft to yards. Which method would you use? Explain.

Recall that ft^2 means ft · ft. Thus, to convert from square feet to square yards, we convert from feet to yards twice. To perform the conversion we will use two unit multipliers. (1 yd = 3 ft)

$$144\ \cancel{\text{ft}} \cdot \cancel{\text{ft}} \cdot \frac{1\ \text{yd}}{3\ \cancel{\text{ft}}} \cdot \frac{1\ \text{yd}}{3\ \cancel{\text{ft}}} = \frac{144\ \text{yd}^2}{9} = 16\ \text{yd}^2$$

Diego needs **16 yd^2** of carpet

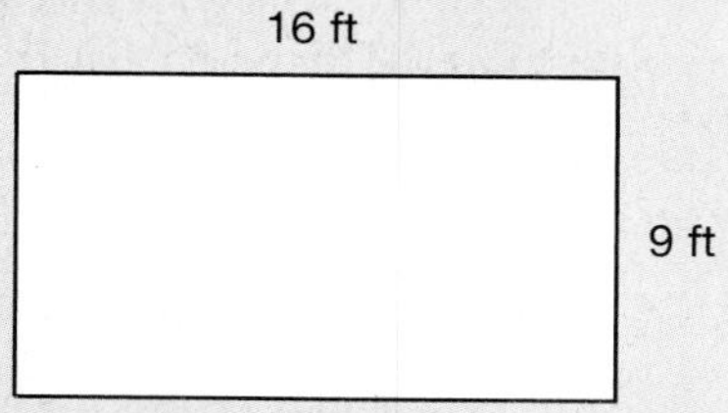

To convert units of volume we may use three unit multipliers.

Example 3

A masonry contractor needs to pour a concrete foundation for a block wall. The foundation is one foot deep, one foot wide, and 100 feet long. A cement truck will deliver the concrete. How many cubic yards of concrete should the contractor order?

Solution

The foundation is a rectangular prism. Its volume is 100 ft^3 (1 ft · 1 ft · 100 ft). We will use three unit multipliers to convert cubic feet to cubic yards.

$$100\ \cancel{\text{ft}^3} \cdot \frac{1\ \text{yd}}{3\ \cancel{\text{ft}}} \cdot \frac{1\ \text{yd}}{3\ \cancel{\text{ft}}} \cdot \frac{1\ \text{yd}}{3\ \cancel{\text{ft}}} = \frac{100\ \text{yd}^3}{27} \approx 3.7\ \text{yd}^3$$

We find that 100 ft^3 is about 3.7 yd^3. The contractor wants to be sure there is enough concrete, so the contractor should order **4 cubic yards.**

We use two or more unit multipliers to convert rates if both units are changing.

Example 4

As Tina approached the traffic light she slowed to 36 kilometers per hour. Convert 36 kilometers per hour to meters per minute.

Solution

We convert from kilometers per hour to meters per minute $\left(\frac{\text{km}}{\text{hr}} \rightarrow \frac{\text{m}}{\text{min}}\right)$. We use one unit multiplier to convert the distance, kilometers to meters (1 km = 1,000 m). We use another unit multiplier to convert the time, hours to minutes (1 hr = 60 min).

$$\text{km} \longrightarrow \text{m} \qquad \text{hr} \longrightarrow \text{min}$$

$$\frac{36\ \cancel{\text{km}}}{1\ \cancel{\text{hr}}} \cdot \frac{1000\ \text{m}}{1\ \cancel{\text{km}}} \cdot \frac{1\ \cancel{\text{hr}}}{60\ \text{min}} = \frac{36000\ \text{m}}{60\ \text{min}} = \mathbf{\frac{600\ m}{min}}$$

When using multiple unit multipliers it is often helpful to use a calculator for the arithmetic.

Example 5

Claude's car averages 12 kilometers per liter of fuel. Claude's car averages about how many miles per gallon?

Solution

We convert kilometers per liter to miles per gallon $\left(\frac{\text{km}}{\text{L}} \rightarrow \frac{\text{mi}}{\text{gal}}\right)$. We use one unit multiplier to convert kilometers to miles (1 km ≈ 0.62 mi) and another unit multiplier to convert liters to gallons (1 gal ≈ 3.781 L).

$$\text{km} \longrightarrow \text{mi} \qquad \text{L} \longrightarrow \text{gal}$$

$$\frac{12\ \cancel{\text{km}}}{1\ \cancel{\text{L}}} \cdot \frac{0.62\ \text{mi}}{1\ \cancel{\text{km}}} \cdot \frac{3.78\ \cancel{\text{L}}}{1\ \text{gal}} = \frac{(12)(0.62)(3.78)\ \text{mi}}{1\ \text{gal}}$$

Using a calculator we find the product is 28.1232 mpg, which we round to the nearest whole number. Claude's car averages about **28 miles per gallon.**

Practice Set

Connect Use two or more unit multipliers to perform each conversion:

a. 5 yards to inches

b. $1\frac{1}{2}$ hours to seconds

c. 15 yd^2 to square feet

d. 270 ft^3 to cubic yards

e. Robert ran 800 meters in two minutes. His average speed was how many kilometers per hour?

f. One cubic inch is about how many cubic centimeters? (1 in. ≈ 2.54 cm) You may use a calculator to multiply. Round your answer to the nearest cubic centimeter.

Written Practice   *Strengthening Concepts*

1. (54) Janis earns $8 per hour at a part-time job. How much does she earn working 3 hours 15 minutes?

2. (55) During a 10-day period an athlete's average resting heart rate after exercising was 58. The athlete's average heart rate during the first 4 days was 61. What was the average heart rate during the last 6 days?

*** 3.** (88) **Connect** Use unit multipliers to perform each conversion:

a. 4 yd^2 to square feet   **b.** 4 yd^3 to cubic feet

4. (53) Use a ratio box to solve this problem. The ratio of woodwinds to brass instruments in the orchestra was 3 to 2. If there were 15 woodwinds, how many brass instruments were there?

*** 5.** (86) **Represent** On a number line, graph the counting numbers that are less than 4.

6. (72) Use a ratio box to solve this problem. Oranges were on sale 8 for $3. At that price, how much would 3 dozen oranges cost?

7. (71) Diagram this statement. Then answer the questions that follow.

When Sandra walked through the house, she saw that 18 lights were on and only $\frac{1}{3}$ of the lights were off.

a. How many lights were off?

b. What percent of the lights were on?

8. (63) Evaluate: $a - [b - (a - b)]$ if $a = 5$ and $b = 3$

*** 9.** (Inv. 8) A spinner is divided into two regions as shown. The spinner was spun 60 times and stopped in region B 42 times. Based on the experiment, what is the probability of the spinner stopping in region A on the next spin? Write the probability as a decimal.

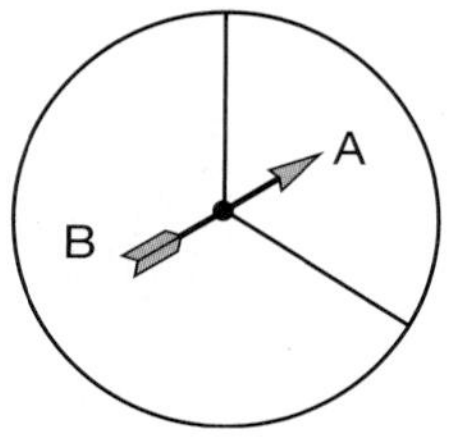

10. (66, 82) To make a large circle on the field, Nathan tied a 30-foot rope to a pole. Then he walked around the stake with the rope extended as he scraped the ground with another stake. Use 3.14 for π to find the

a. circumference of the circle.

b. area of the circle.

c. What measure of a circle does the rope represent?

11. What percent of this circle is shaded?
(Inv. 1)

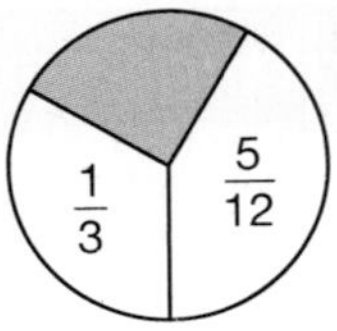

12. Draw a cube with edges 3 cm long.
(67, 70)

a. What is the volume of the cube?

b. What is the surface area of the cube?

Collect like terms:

13. $2x + 3y - 5 + x - y - 1$
(84)

14. $x^2 + 2x - x - 2$
(84)

15. Complete the table.
(48)

Fraction	Decimal	Percent
a.	0.125	**b.**
$\frac{3}{8}$	**c.**	**d.**

16. Simplify: $\dfrac{60}{1\frac{1}{4}}$
(76)

Model Use ratio boxes to solve problems **17** and **18.**

* **17.** The regular price of a shirt is $24. The sale price is $18. The sale price is what percent of the regular price?
(81)

* **18.** The auditorium seated 375, but this was enough for only 30 percent of those who wanted a seat. How many wanted a seat but could not get one?
(81)

19. Write an equation to solve this problem:
(77)

Twenty-four is 25 percent of what number? Find the answer.

* **20.** **a.** Classify this quadrilateral.
(Inv. 6, 75)

b. Find its perimeter.

c. **Analyze** Find its area.

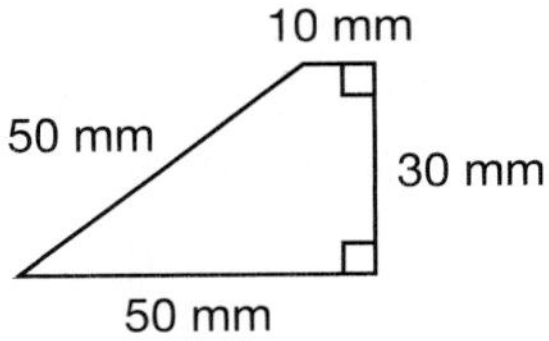

21. Find the missing numbers in the table by using the function rule. Then graph the *x*, *y* pairs and a line through those points that shows all *x*, *y* pairs that satisfy the equation. Select the *x*, *y* pair where the line crosses the *y*-axis and check the numbers in the equation.
(56)

$y = x - 5$

x	*y*
3	
7	
5	

* **22.** Multiply. Write each product in scientific notation.
(83)

a. $(9 \times 10^{-6})(4 \times 10^{-8})$

b. $(9 \times 10^{6})(4 \times 10^{8})$

For problems **23** and **24,** solve and check the equation. Show each step.

23. $8\frac{5}{6} = d - 5\frac{1}{2}$
(Inv. 7)

24. $\frac{5}{6}m = 90$
(Inv. 7)

* **25.** **Evaluate** Three vertices of rectangle *JKLM* are *J* (−4, 2), *K* (0, 2), and *L* (0, 0).
(80)

a. Find the coordinates of *M* and draw the rectangle.

b. Translate ▭ *JKLM* 4 units right, 2 down. Draw the translated image ▭ *J′K′L′M′*, and write the coordinates of its vertices.

Math Language
The symbol ▭ *JKLM* means rectangle *JKLM*.

26. Which of the following does not equal 4^3?
(20, 52)

a. 2^6 **b.** $4 \cdot 4^2$ **c.** $\frac{4^4}{4}$ **d.** $4^2 + 4$

27. **a.** Find 50% of $\frac{2}{3}$ of 0.12. Write the answer as a decimal.
(43)

b. Is $\frac{2}{3}$ of 50% of 0.12 the same as 50% of $\frac{2}{3}$ of 0.12? Explain.

Generalize Simplify:

28. $6\{5 \cdot 4 - 3[6 - (3 - 1)]\}$
(63)

* **29.** **a.** $\frac{(-3)(-4) - (-3)}{(-3) - (+4)(+3)}$
(85)

b. $(+5) + (-2)[(+3) - (-4)]$

* **30.** **a.** $(-2x)(-3x)$
(87)

b. $(ab)(2a^2b)(-3a)$

c. $(-3x)^2$

Early Finishers
Real-World Application

A group of 200 volunteers were shown television commercials A, B, C, and D. Then the volunteers were asked to identify their favorite commercial. The results are shown below.

A	B	C	D
78	35	45	42

a. Which type of display—a circle graph or a histogram—is the most appropriate way to display this data? Draw your display and justify your choice.

LESSON 89

• Diagonals
• Interior Angles
• Exterior Angles

Power Up

Building Power

facts Power Up R

mental math

a. **Positive/Negative:** $(-80) \div (-4)$

b. **Order of Operations/Exponents:** $(2.5 \times 10^{-4})(3 \times 10^{8})$

c. **Ratio:** $\frac{8}{g} = \frac{2}{2.5}$

d. **Percent/Estimation:** Estimate $7\frac{3}{4}\%$ of \$8.29.

e. **Measurement:** 1.87 m to cm

f. **Fractional Parts:** \$1.00 is $\frac{2}{5}$ of m.

g. **Geometry:** What do you call an angle that measures 90 degrees?

h. **Calculation:** 10% of 80, $\times$ 3, + 1, $\sqrt{\ }$, $\times$ 7, + 1, $\div$ 2, $\div$ 2, $\sqrt{\ }$, $\times$ 10, + 2, $\div$ 4

problem solving

Sarita rode the G-Force Ride at the fair. The cylindrical chamber spins, forcing riders against the wall while the floor drops away. If the chamber is 30 feet in diameter and if it spins around 30 times during a ride, how far do the riders travel? Do riders travel more or less than $\frac{1}{2}$ mile?

New Concepts

Increasing Knowledge

diagonals

Recall that a **diagonal** of a polygon is a line segment that passes through the polygon between two nonadjacent vertices. In the figure below, segment AC is a diagonal of quadrilateral $ABCD$.

Math Language

A **polygon** is a closed flat figure with straight sides. In a **regular polygon,** all sides and angles are equal in measure.

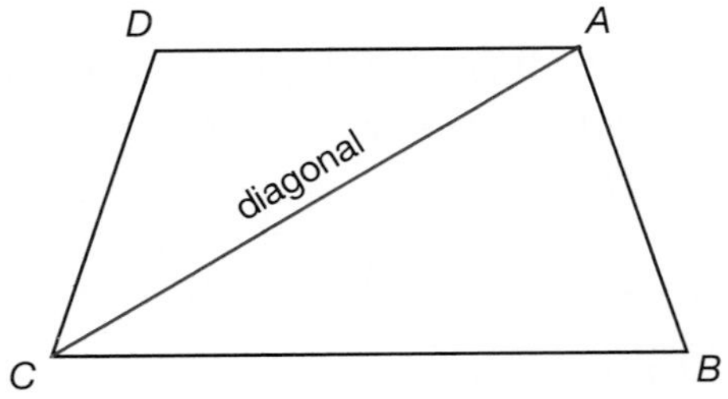

Example 1

From one vertex of regular hexagon *ABCDEF*, how many diagonals can be drawn? (Trace the hexagon and illustrate your answer.)

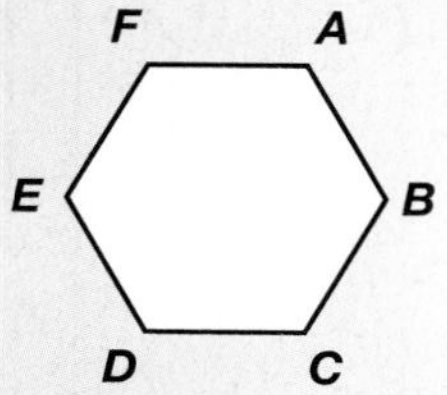

Solution

We can select any vertex from which to draw the diagonals. We choose vertex *A*. Segments *AB* and *AF* are sides of the hexagon and are not diagonals. Segments drawn from *A* to *C*, *D*, and *E* are diagonals. So **3 diagonals** can be drawn.

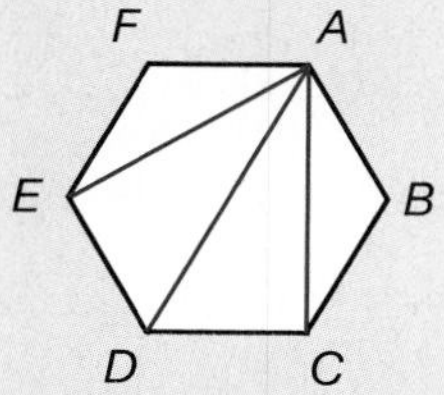

interior angles

Notice in example 1 that the three diagonals from vertex *A* divide the hexagon into four triangles. We will draw arcs to emphasize each angle of the four triangles.

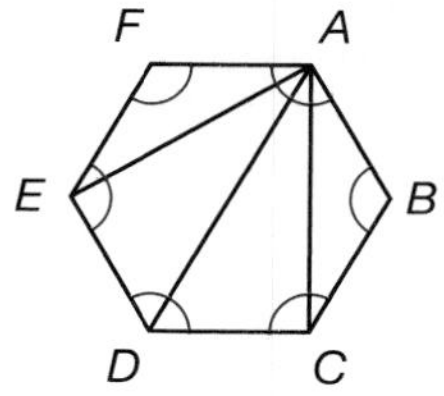

Angles that open to the interior of a polygon are called **interior angles.** We see that $\angle B$ of the hexagon is also $\angle B$ of $\triangle ABC$. Angle *C* of the hexagon includes $\angle BCA$ of $\triangle ABC$ and $\angle ACD$ of $\triangle ACD$.

Analyze Are there any angles of the four triangles that are not included in the angles of the hexagon?

Although we may not know the measure of each angle of each triangle, we nevertheless can conclude that the measures of the six angles of a hexagon have the same total as the measures of the angles of four triangles, which is $4 \times 180° = 720°$.

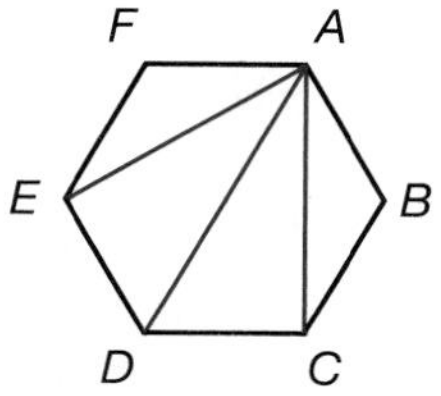

The sum of the measures of the six interior angles of a hexagon is 720°.
$4 \times 180° = 720°$

Since hexagon *ABCDEF* is a regular hexagon, we can also calculate the measure of each angle of the hexagon.

Example 2

Maura inscribed a regular hexagon in a circle. Find the measure of each angle of the regular hexagon *ABCDEF*.

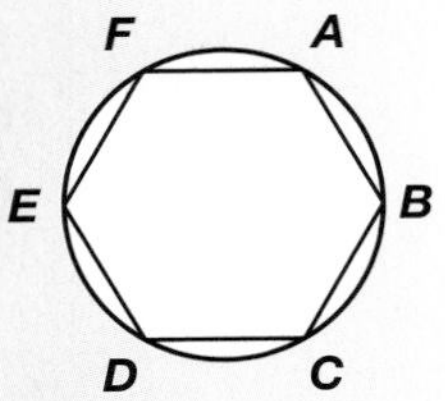

Solution

From the explanation above we know that the hexagon can be divided into four triangles. So the sum of the measures of the angles of the hexagon is 4 × 180°, which is 720°. Since the hexagon is regular, the six angles equally share the available 720°. So we divide 720° by 6 to find the measure of each angle.

$$720° \div 6 = 120°$$

We find that each angle of the hexagon measures **120°.**

Example 3

Draw a quadrilateral and one of its diagonals. What is the sum of the measures of the interior angles of the quadrilateral?

Solution

We draw a four-sided polygon and a diagonal.

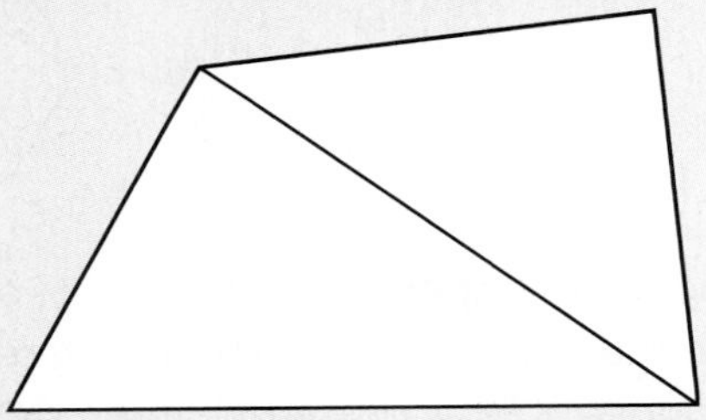

Although we do not know the measure of each angle, we can find the sum of their measures. The sum of the measures of the angles of a triangle is 180°. From the drawing above, we see that the total measure of the angles of the quadrilateral equals the total measure of the angles of two triangles. So the sum of the measures of the interior angles of the quadrilateral is

$$2 \times 180° = \mathbf{360°}$$

Discuss Will the sum of the measures of the interior angles of *all* quadrilaterals be 360°? Explain.

exterior angles

In example 2 we found that each interior angle of a regular hexagon measures 120°. By performing the following activity, we can get another perspective on the angles of a polygon.

Activity

Exterior Angles

Materials needed:

- A length of string (5 feet or more)
- Chalk
- Masking tape (optional)

Before performing this activity, lay out a regular hexagon in the classroom or on the playground. This can be done by inscribing a hexagon inside a circle as described in Investigation 2. Use the string and chalk to sweep out the circle and to mark the vertices and sides of the hexagon. If desired, mark the sides of the hexagon with masking tape.

After the hexagon has been prepared, walk the perimeter of the hexagon while making these observations:

1. Notice the direction you were facing when you started around the hexagon as well as when you finished going around the hexagon after six turns.
2. Notice how much you turned at each "corner" of the hexagon. Did you turn more than, less than, or the same as you would have turned at the corner of a square?

Each student should have the opportunity to walk the perimeter of the hexagon.

Going around the hexagon, we turned at every corner. If we did not turn, we would continue going straight.

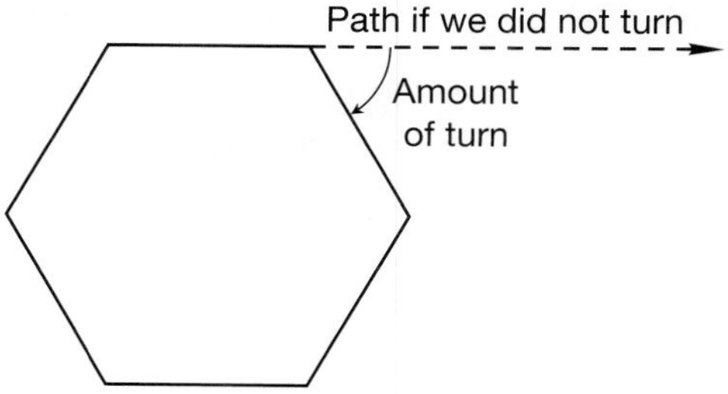

The amount we turned at the corner in order to stay on the hexagon equals the measure of the **exterior angle** of the hexagon at that vertex.

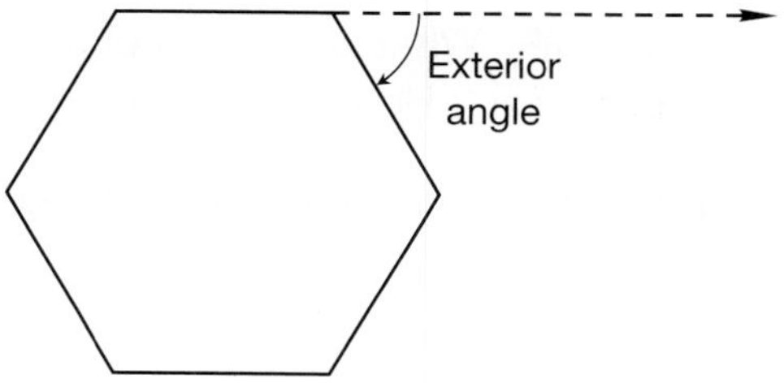

We can calculate the measure of each exterior angle of a regular hexagon by remembering how many turns were required in order to face the same direction as when we started. We remember that we made six small turns. In other words, after six turns we had completed one full turn of 360°.

If all the turns are in the same direction, the sum of the exterior angles of any polygon is 360°.

Example 4

What is the measure of each exterior angle of a regular hexagon?

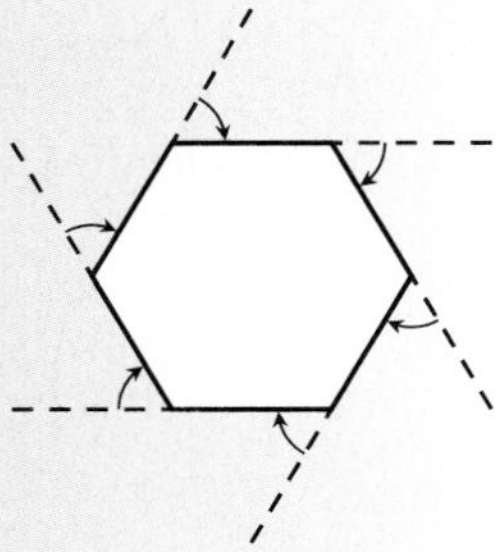

Solution

Traveling all the way around the hexagon completes one full turn of 360°. Each exterior angle of a regular hexagon has the same measure, so we can find the measure by dividing 360° by 6.

$$360° \div 6 = 60°$$

We find that each exterior angle of a regular hexagon measures **60°.**

Notice that an interior angle of a polygon and its exterior angle are supplementary, so their combined measures total 180°.

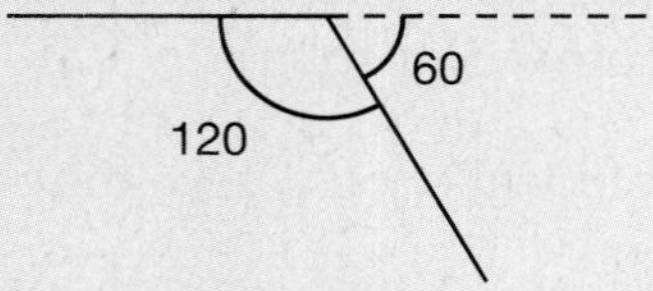

Analyze How can you determine the measure of an exterior angle if you know the measure of the interior angle?

Practice Set

Analyze Work examples 1–4. Then solve **a–f.**

a. Trace this regular pentagon. How many diagonals can be drawn from one vertex? Show your work.

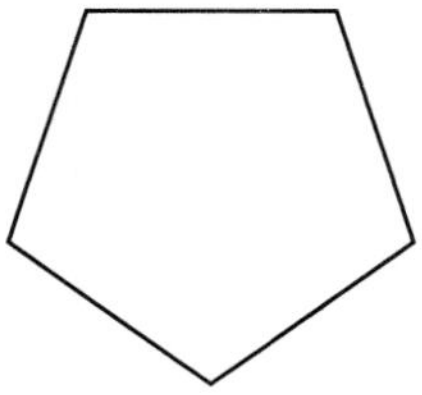

b. The diagonals drawn in problem **a** divide the pentagon into how many triangles?

c. **Explain** What is the sum of the measures of the five interior angles of a pentagon? How do you know?

d. What is the measure of each interior angle of a regular pentagon?

e. What is the measure of each exterior angle of a regular pentagon?

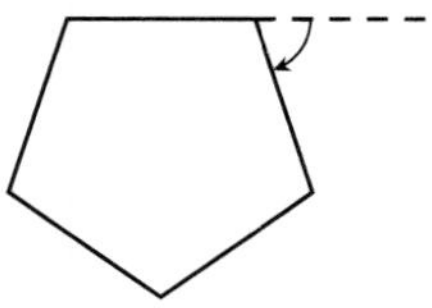

f. What is the sum of the measures of an interior and exterior angle of a regular pentagon?

Written Practice

Strengthening Concepts

1. (72) Use a ratio box to solve this problem. Jason's remote-control car traveled 440 feet in 10 seconds. At that rate, how long would it take the car to travel a mile?

2. (53) Use a ratio box to help you solve this problem. There are novels, biographies, and field guides in one bookcase of the class library. The ratio of novels to biographies is 3 to 2. The ratio of biographies to field guides is 3 to 4. If there are 18 novels, how many field guides are there? Tell how you found the answer.

3. (70) Kwame measured the shoe box and found that it was 30 cm long, 15 cm wide, and 12 cm tall. What was the volume of the shoe box?

*** 4.** (44, Inv. 8) **Generalize** A baseball player's batting average is a ratio found by dividing the number of hits by the number of at-bats and writing the result as a decimal number rounded to the nearest thousandth. If Erika had 24 hits in 61 at-bats, what was her batting average?

*** 5.** (88) **Connect** Christina ran 3,000 meters in 10 minutes. Use two unit multipliers to convert her average speed to kilometers per hour.

*** 6.** (86) **Represent** On a number line, graph the integers greater than -4.

7. (71) Draw a diagram of this statement. Then answer the questions that follow.

Jimmy bought the astronomy book for \$12. This was $\frac{3}{4}$ of the regular price.

a. What was the regular price of the astronomy book?

b. Jimmy bought the astronomy book for what percent of the regular price?

8. (40) Use the figure below to find the measure of each angle.

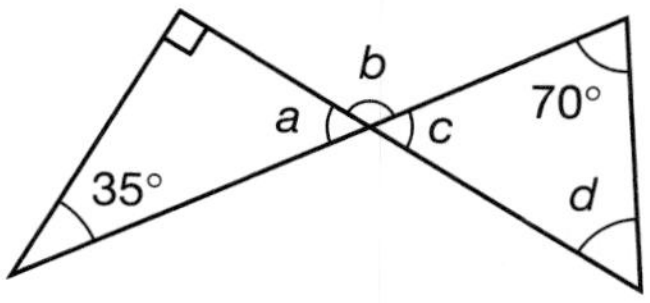

a. $\angle a$ **b.** $\angle b$ **c.** $\angle c$ **d.** $\angle d$

9. (65, 82) **a.** What is the circumference of this circle?

b. What is the area of this circle?

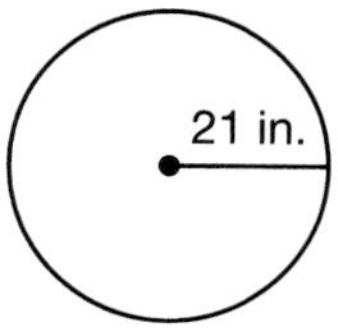

Use $\frac{22}{7}$ for π.

10. (76) Simplify: $\dfrac{91\frac{2}{3}}{100}$

11. (52) Evaluate: $\dfrac{ab + a}{a + b}$ if $a = 10$ and $b = 5$

12. (20, 33) Compare: $a^2 \bigcirc a$ if $a = 0.5$

13. (48) Complete the table.

Fraction	Decimal	Percent
$\frac{7}{8}$	**a.**	**b.**
c.	**d.**	875%

14. (17) At three o'clock and at nine o'clock, the hands of a clock form angles equal to $\frac{1}{4}$ of a circle.

a. At which two hours do the hands of a clock form angles equal to $\frac{1}{3}$ of a circle?

b. The angle described in part **a** measures how many degrees?

Model Use ratio boxes to solve problems **15** and **16.**

* **15.** (81) Forty-five percent of the 300 shoppers in a supermarket bought milk. How many of the shoppers in a supermarket bought milk?

* **16.** (81) The sale price of $24 was 75% of the regular price. The sale price was how many dollars less than the regular price?

17. (77) Write an equation to solve this problem:

Twenty is what percent of 200?

18. (58, 75) **a.** Trace this isosceles trapezoid and draw its line of symmetry.

b. Find the area of the trapezoid.

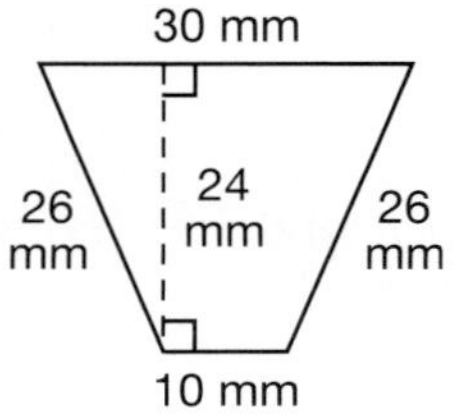

* **19.** (89) **Analyze** What is the measure of each exterior angle of a regular triangle?

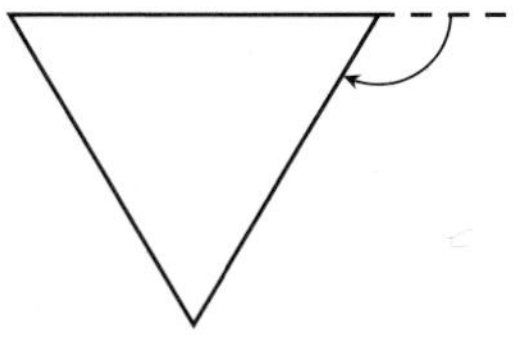

* **20.** (56) Find the missing numbers in the table by using the function rule. Then graph the x, y pairs and extend a line through the points. At what x, y pair does the graph of the equation intercept (cross) the y-axis?

$y = \frac{1}{2}x$

x	y
6	
2	
−2	

* **21.** (83) **Represent** Multiply. Write the product in scientific notation.

$$(1.25 \times 10^{-3})(8 \times 10^{-5})$$

22. (62) The lengths of two sides of an isosceles triangle are 4 cm and 10 cm.

a. Draw the triangle and find its perimeter.

b. Can there be more than one answer? Why or why not?

For problems **23** and **24,** solve and check the equation. Show each step.

23. (Inv. 7) $\frac{4}{9}p = 72$

24. (Inv. 7) $12.3 = 4.56 + f$

* **25.** (84) **Generalize** Collect like terms:

$$2x + 3y - 4 + x - 3y - 1$$

Generalize Simplify:

26. (52) $\frac{9 \cdot 8 - 7 \cdot 6}{6 \cdot 5}$

27. (57) $3.2 \times 4^{-2} \times 10^{2}$

28. (43) $13\frac{1}{3} - \left(4.75 + \frac{3}{4}\right)$ (mixed-number answer)

* **29.** (85) **a.** $\frac{(+3) + (-4)(-6)}{(-3) + (-4) - (-6)}$

b. $(-5) - (+6)(-2) + (-2)(-3)(-1)$

* **30.** (87) **a.** $(3x^2)(2x)$

b. $(-2ab)(-3b^2)(-a)$

c. $(-4ab^3)^2$

Early Finishers
Math Applications

Write *True or False* for each statement.

Some prisms have a curved surface.
No pyramids have a circular base.
All prisms are constructed from polygons.
All three-dimensional figures have at least one vertex.
Some three-dimensional figures have no edges.
All faces of a triangular prism are triangles.
One face of a pyramid can be a rectangle.

LESSON

90

• Mixed-Number Coefficients
• Negative Coefficients

Power Up

Building Power

facts Power Up Q

mental math

a. **Positive/Negative:** (–50) – (–30)

b. **Order of Operations/Exponents:** $(4.2 \times 10^{-6})(2 \times 10^{-4})$

c. **Algebra:** $4w - 8 = 36$

d. **Percent/Estimation:** Estimate 15% of $23.89.

e. **Measurement:** 800 g to kg

f. **Fractional Parts:** \$1.00 is $\frac{4}{5}$ of m.

g. **Geometry:** What do you call an angle that is greater than 90 degrees?

h. **Geometry:** A cube with edges 10 inches long has a volume of how many cubic inches?

problem solving

Here are the front, side, and top views of a structure. Construct this object using 1-inch cubes, and find the volume of the object. Then draw a three-dimensional view of the structure.

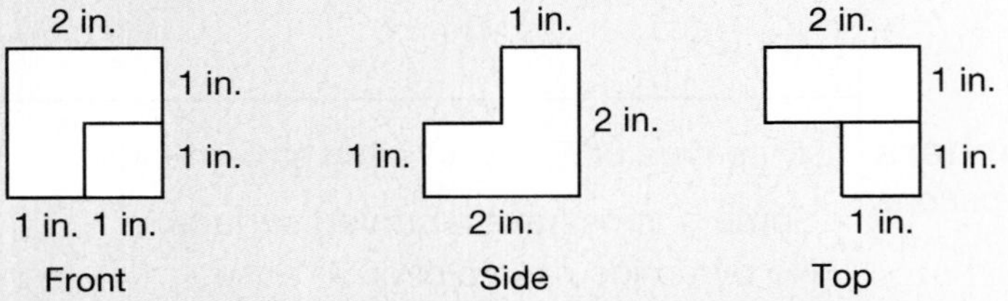

New Concepts

Increasing Knowledge

mixed-number coefficients

We have solved equations that have fractional coefficients by multiplying by the reciprocal of the coefficient.

$$\frac{4}{5}x = 7$$

Math Language

A **coefficient** is the number that is multiplied by a variable in an algebraic term.

Here the coefficient of x is $\frac{4}{5}$, so we multiply both sides by the reciprocal of $\frac{4}{5}$, which is $\frac{5}{4}$.

$$\frac{\overset{1}{\cancel{5}}}{\underset{1}{\cancel{4}}} \cdot \frac{\overset{1}{\cancel{4}}}{\underset{1}{\cancel{5}}}x = \frac{5}{4} \cdot 7 \qquad \text{multiplied both sides by } \frac{5}{4}$$

$$x = \frac{35}{4} \qquad \text{simplified}$$

To solve an equation that has a mixed-number coefficient, we convert the mixed number to an improper fraction as the first step. Then we multiply both sides by the reciprocal of the improper fraction.

Example 1

Solve: $3\frac{1}{3}x = 5$

Solution

First we write $3\frac{1}{3}$ as an improper fraction.

$$\frac{10}{3}x = 5 \quad \text{fraction form}$$

Then we multiply both sides of the equation by $\frac{3}{10}$, which is the reciprocal of $\frac{10}{3}$.

$$\frac{\overset{1}{\cancel{3}}}{\underset{1}{\cancel{10}}} \cdot \frac{\overset{1}{\cancel{10}}}{\underset{1}{\cancel{3}}}x = \frac{3}{\underset{2}{\cancel{10}}} \cdot \overset{1}{\cancel{5}} \quad \text{multiplied both sides by } \frac{3}{10}$$

$$x = \mathbf{\frac{3}{2}} \quad \text{simplified}$$

In arithmetic we usually convert an improper fraction such as $\frac{3}{2}$ to a mixed number. Recall that in algebra we usually leave improper fractions in fraction form.

Example 2

Solve: $2\frac{1}{2}y = 1\frac{7}{8}$

Solution

Since we will be multiplying both sides of the equation by a fraction, we first convert both mixed numbers to improper fractions.

$$\frac{5}{2}y = \frac{15}{8} \quad \text{fraction form}$$

Then we multiply both sides by $\frac{2}{5}$, which is the reciprocal of $\frac{5}{2}$.

$$\frac{\overset{1}{2}}{\underset{1}{\cancel{5}}} \cdot \frac{\overset{1}{\cancel{5}}}{\underset{1}{2}}y = \frac{\overset{1}{2}}{\underset{1}{\cancel{5}}} \cdot \frac{\overset{3}{\cancel{15}}}{\underset{4}{\cancel{8}}} \quad \text{multiplied both sides by } \frac{2}{5}$$

$$y = \mathbf{\frac{3}{4}} \quad \text{simplified}$$

negative coefficients

To solve an equation with a negative coefficient, we multiply (or divide) both sides of the equation by a negative number. The coefficient of x in this equation is negative.

$$-3x = 126$$

To solve this equation, we can either divide both sides by -3 or multiply both sides by $-\frac{1}{3}$. The effect of either method is to make $+1$ the coefficient of x. We show both ways on the following page.

Thinking Skill

Identify

What is the coefficient of x in the equation: $-x + 2 = 17$?

Method 1	Method 2
$-3x = 126$	$-3x = 126$
$\frac{-3x}{-3} = \frac{126}{-3}$	$\left(-\frac{1}{3}\right)(-3x) = \left(-\frac{1}{3}\right)(126)$
$x = -42$	$x = -42$

Example 3

Solve: $-\frac{2}{3}x = \frac{4}{5}$

Solution

We multiply both sides of the equation by the reciprocal of $-\frac{2}{3}$, which is $-\frac{3}{2}$.

$-\frac{2}{3}x = \frac{4}{5}$	equation
$\left(-\frac{3}{2}\right)\left(-\frac{2}{3}x\right) = \left(-\frac{3}{2}\right)\left(\frac{4}{5}\right)$	multiplied by $-\frac{3}{2}$
$x = \mathbf{-\frac{6}{5}}$	simplified

Example 4

Solve: $-5x = 0.24$

Solution

We may either multiply both sides by $-\frac{1}{5}$ or divide both sides by -5. Since the right side of the equation is a decimal number, it appears that dividing by -5 will be easier.

$-5x = 0.24$	equation
$\frac{-5x}{-5} = \frac{0.24}{-5}$	divided by -5
$x = \mathbf{-0.048}$	simplified

Practice Set

Generalize Solve:

a. $1\frac{1}{8}x = 36$

b. $3\frac{1}{2}a = 490$

c. $2\frac{3}{4}w = 6\frac{3}{5}$

d. $2\frac{2}{3}y = 1\frac{4}{5}$

e. $-3x = 0.45$

f. $-\frac{3}{4}m = \frac{2}{3}$

g. $-10y = -1.6$

h. $-2\frac{1}{2}w = 3\frac{1}{3}$

i. *Conclude* When solving an equation, how can you show that your answer is correct?

Written Practice — Strengthening Concepts

1. (31, 35) The sum of 0.8 and 0.9 is how much greater than the product of 0.8 and 0.9? Use words to write the answer.

2. (Inv. 4) For this set of scores, find the:

a. mean **b.** median

c. mode **d.** range

8, 6, 9, 10, 8, 7, 9, 10, 8, 10, 9, 8

3. (46) A 24-ounce jar of applesauce costs $1.68. A 58-ounce jar costs $2.88. Which costs more per ounce? How much more?

* **4.** (89) **Analyze** The figure at right is a regular decagon. One of the exterior angles is labeled *a*, and one of the interior angles is labeled *b*.

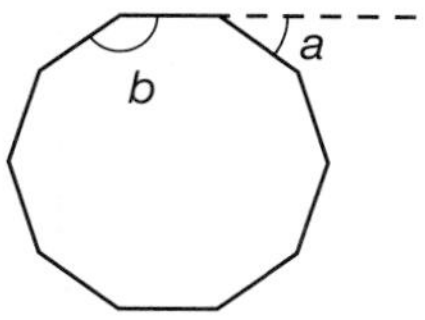

a. What is the measure of each exterior angle of the decagon?

b. What is the measure of each interior angle?

* **5.** (84) **Generalize** Collect like terms: $x^2 + 2xy + y^2 + x^2 - y^2$

Model Use ratio boxes to solve problems **6** and **7.**

* **6.** (81) A pair of binoculars were on sale for $36. This was 90% of the regular price. What was the regular price?

* **7.** (81) Seventy-five percent of the citizens voted for Graham. If there were 800 citizens, how many of them did not vote for Graham?

8. (77) **Model** Write equations to solve **a** and **b.**

a. Twenty-four is what percent of 30?

b. Thirty is what percent of 24?

* **9.** (88) **Connect** Use unit multipliers to perform these conversions:

a. 2 ft^2 to square inches

b. 1 m^3 to cubic centimeters

10. (71) Three hundred people said they prefer a vacation in a warm climate. This was $\frac{2}{3}$ of the people surveyed.

a. How many people were surveyed?

b. How many people surveyed did not choose a warm climate?

11. (41) If $x = 4.5$ and $y = 2x + 1$, then y equals what number?

12. A coin is tossed three times. One possible outcome is HHH.
(36)

a. Write the sample space for the experiment.

b. What is the probability of getting heads at least twice?

13. If the perimeter of a square is 1 foot, what is the area of the square in square inches?
(20)

14. Complete the table.
(48)

Fraction	Decimal	Percent
a.	1.75	**b.**

15. If the sales-tax rate is 6%, what is the total price of a $325 printer including sales tax?
(60)

16. Multiply. Write the product in scientific notation.
(83)

$$(6 \times 10^4)(8 \times 10^{-7})$$

17. A cereal box 8 inches long, 3 inches wide, and 12 inches tall is shown.
(67, 70)

a. What is the volume of the box?

b. What is the surface area of the box?

18. **a.** Find the circumference of the circle at right.
(65, 82)

b. Find the area of the circle.

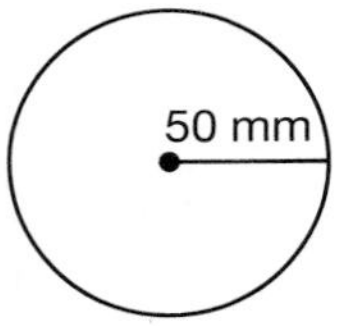

Use 3.14 for π.

*** 19.** Classify Make a list of the whole numbers that are not counting numbers.
(86)

*** 20.** Analyze The coordinates of three vertices of ▭*WXYZ* are *W* (0, 3), *X* (5, 3), and *Y* (5, 0).
(80)

a. Find the coordinates of *Z* and draw the rectangle.

b. Rotate ▭*WXYZ* 90° counterclockwise about the origin, and draw its image ▭*W′X′Y′Z′*. Write the coordinates of the vertices.

21. What mixed number is $\frac{2}{3}$ of 20?
(60)

22. On a number line graph $x \leq 4$.
(78)

For problems **23–25,** solve the equation. Show each step.

23. $x + 3.5 = 4.28$ *(Inv. 7)*

24. $2\frac{2}{3}w = 24$ *(90)*

25. $-4y = 1.4$ *(90)*

Generalize Simplify:

26. (57) $10^1 + 10^0 + 10^{-1}$ (decimal answer)

* **27.** (87) **a.** $(-2x^2)(-3xy)(-y)$

b. $(3x^2y^3)^2$

28. (30) $\dfrac{8}{75} - \dfrac{9}{100}$

* **29.** (85) **a.** $(-3) + (-4)(-5) - (-6)$

b. $\dfrac{(-2)(-4)}{(-4) - (-2)}$

30. (52) Compare for $x = 10$ and $y = 5$:

$$x^2 - y^2 \bigcirc (x + y)(x - y)$$

Early Finishers
Real-World Application

The Fernandez family is creating a semicircle driveway. In order to buy the right amount of concrete and grass, they need to calculate the area of the semicircle below. Use 3.14 for π.

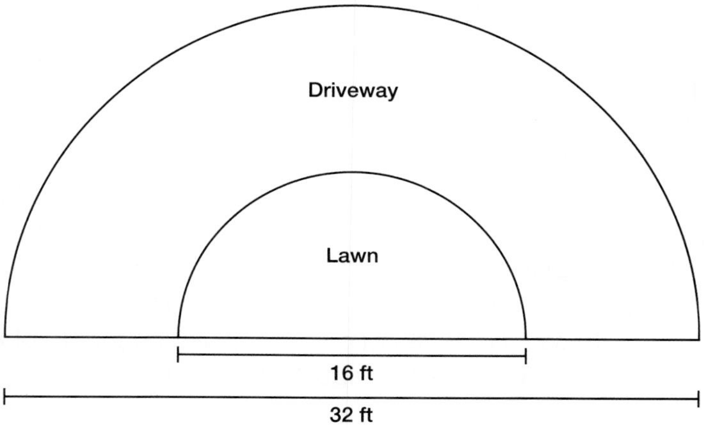

a. What is the area of the driveway?

b. Approximately how many tons of asphalt does the Fernandez family need to purchase if one ton covers 22.5 ft^2.

c. If each ton of asphalt costs $33, how much will is cost the Fernandez family to pave the driveway?

Focus on

• Graphing Functions

As we have seen, functions can be displayed as graphs on a coordinate plane. Let's review what we have learned. To graph a function, we use each (*x*, *y*) pair of numbers as coordinates of a point on the plane.

$y = 2x$

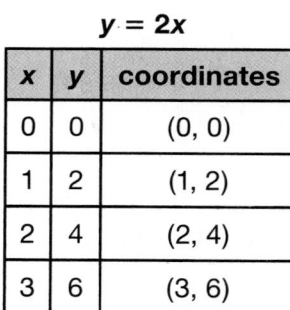

x	y	coordinates
0	0	(0, 0)
1	2	(1, 2)
2	4	(2, 4)
3	6	(3, 6)

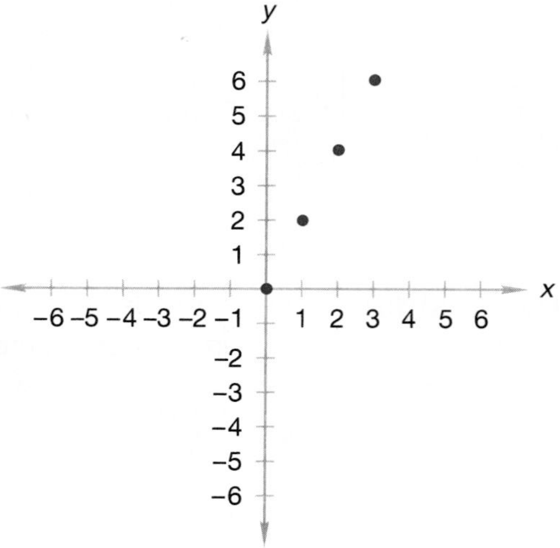

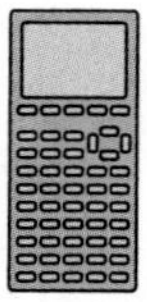

Visit www.SaxonPublishers.com/ActivitiesC2 *for a graphing calculator activity.*

On the coordinate plane above, we graphed four pairs of numbers that satisfy the equation of the function. Although the table lists only four pairs of numbers for the function, the graph of the function includes many other pairs of numbers that satisfy the equation. By extending a line through and beyond the graphed points, we graph all possible pairs of numbers that satisfy the equation. Each point on the graphed line below represents a pair of numbers that satisfies the equation of the function $y = 2x$.

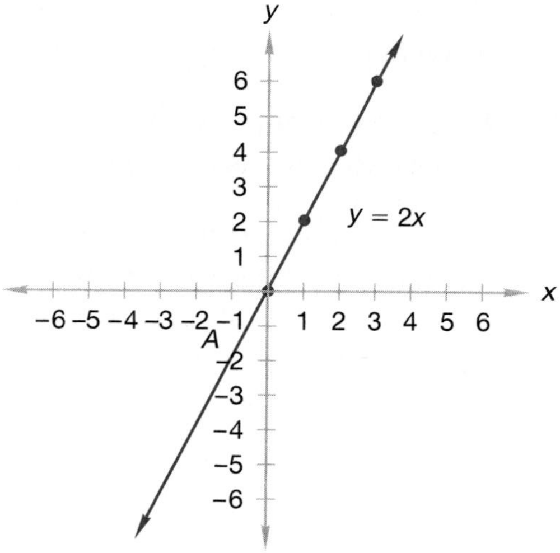

Thinking Skill

Explain

Why doesn't the graph need to indicate units?

Below we graph the relationship between the side-length and perimeter of a square. In the equation we use *s* for the length of a side and *p* for the perimeter. We only graph the numerical relationship of the measures, so we do not designate units. It is assumed that the unit used to measure the length of the side is also used to measure the perimeter.

$p = 4s$

s	p
1	4
2	8
3	12

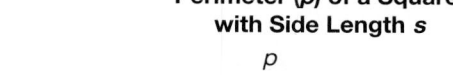

Perimeter (*p*) of a Square with Side Length *s*

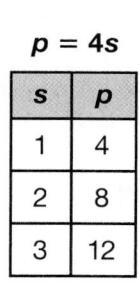

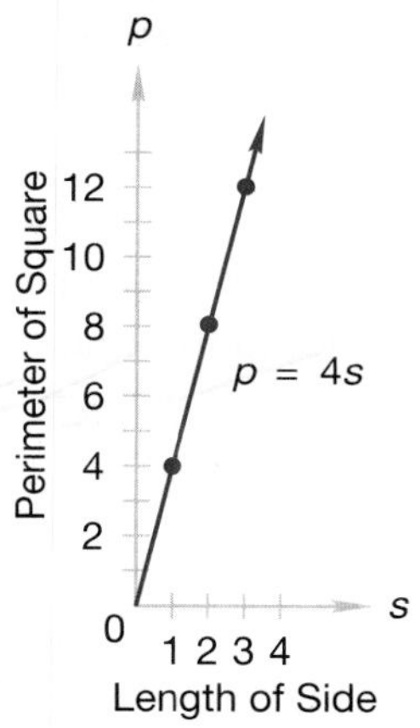

Notice that the *x*- and *y*-axes are renamed *s* and *p* for this function.

1. The graph of this function is a ray. What are the coordinates of the endpoint of the ray? Why is the graph of this function a ray and not a line?

2. *Analyze* Notice that the graph of the function $p = 4s$ is straight and not curved. Imagine two squares in which the side lengths of the larger square are twice the side lengths of the smaller square. Draw the two squares and find the perimeter of each. What effect does doubling the side length of a square have on the perimeter?

3. *Represent* The relationship between the length of a side of an equilateral triangle and the perimeter of the triangle is a function that can be graphed. Write an equation for the function using *s* for the length of a side and *p* for the perimeter. Next make a table of (*s*, *p*) pairs for side lengths 1, 2, and 3. Then draw a rectangular coordinate system with an *s*-axis and a *p*-axis, and graph the function.

Rates are functions that can be graphed. Many rates involve time as one of the variables. Speed, for example, is a function of distance and time. Suppose Sam enters a walk-a-thon and walks at a steady rate of 3 miles per hour. The distance (*d*) in miles that Sam travels is a function of the number of hours (*h*) that Sam walks. This relationship is expressed in the following equation:

$$d = 3h$$

The table below shows how far Sam walks in 1, 2, and 3 hours.

$d = 3h$

h	d
1	3
2	6
3	9

d = distance in miles
h = time in hours

A graph of the function shows how far Sam walks in any number of hours, including fractions of hours.

Distance Sam Walks at 3 mph

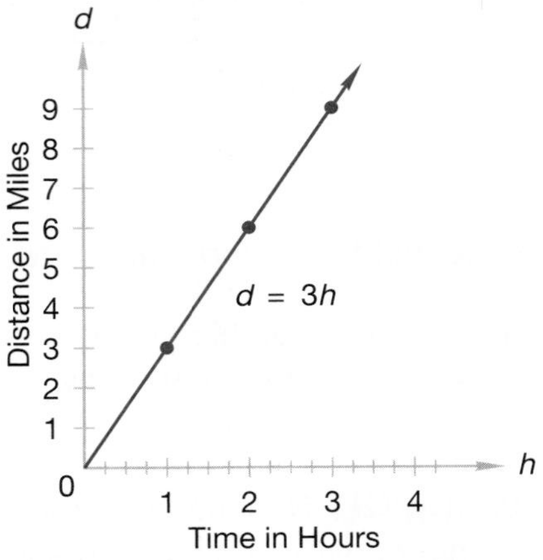

Thinking Skill

Explain

Could you have divided each hour into sixths? Why or why not?

Notice that we labeled both axes of the graph. Also notice that we adjusted the scale of the graph so that each tick mark on the horizontal scale represents one fourth of an hour (15 minutes).

4. *Represent* Suppose Sam entered a jog-a-thon and was able to jog at a steady pace of 6 miles per hour for 2 hours. Following the pattern described for the walk-a-thon, write an equation for a 6-mile-per-hour rate, and make a table that shows how far Sam had jogged after 1 hour and how far he had jogged after 2 hours. Then draw a graph of the function and label each axis. Let every tick mark on the time axis represent 10 minutes.

5. Refer to the graph drawn in problem 4 to find the distance Sam had jogged in 40 minutes.

6. Refer to the graph drawn in problem 4 to find how long it took Sam to jog 9 miles.

7. *Explain* Why is the graph of the distance Sam jogged a segment and not a ray?

The functions we have considered in this investigation so far are examples of direct variation. **Direct variation** is a relationship between two variables in which a change in one variable results in a proportional change in the other variable. In most real-world applications, if one variable increases, the other increases or if one variable decreases the other decreases. For example, if Gina is paid by the hour, then the more hours she works, the more she is paid. If she works fewer hours, her total pay decreases. We say that Gina's total pay is **directly proportional** to the number of hours she works.

This means we can form a proportion with two pairs of numbers that follow the rule of the function. A square with a side length of 3 has a perimeter of 12. A square with a side length of 5 has a perimeter of 20. These pairs of numbers form a proportion because the ratios are equal to each other.

Side length:	3	5
Perimeter:	12	20

8. *Verify* This table shows the number of dollars Gina earns working a given number of hours. Select two pairs of numbers from the table and use them to write a proportion. Are the cross products of the proportion equal?

Hours	Dollars
1	5
2	10
3	15
4	20

An equation with variables that are directly proportional often has this form.

$$y = \mathrm{k}x$$

The letter k refers to a constant, meaning a number that does not change. For example, the perimeter and side length of a square are directly proportional and the formula that relates them has this form.

$$p = 4s \text{ (The constant is 4.)}$$

Here are some more examples of direct variation and their equations. Notice that as one variable increases, the other also increases.

- The number of pounds (p) of apples purchased and the cost (c) ($c = \mathrm{k}p$)
- The time (t) spent walking and the distance (d) walked ($d = \mathrm{k}t$)

9. *Classify* Which of the following is not an example of direct variation?

A The length of the side of a regular hexagon and the perimeter of the hexagon.

B The number of books in a stack and the height of the stack.

C The speed at which you travel to school and the time it takes to get to school.

D The length of a segment in inches and its length in centimeters.

The graph of a function with direct variation is a straight line (or ray or segment or aligned points) that includes the origin.

10. *Classify* Which graphed function below is an example of direct variation? Explain your choice.

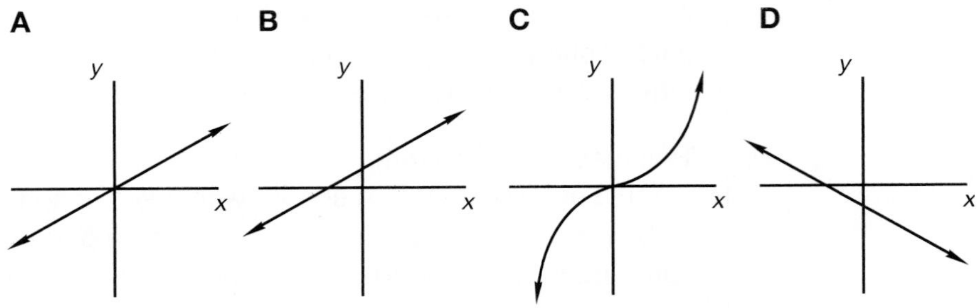

11. The large carousel at the fair makes three revolutions per minute and operates for four minutes at a time. Write an equation and create a function table that relates the input, minutes (m), to the output, revolutions (r). Then plot the numbers on a coordinate plane and draw a segment for the length of a ride.

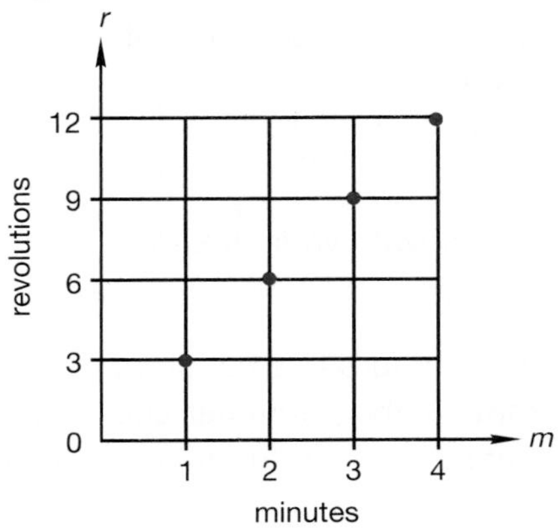

In summary, there are two requirements for a function to be a direct proportion.

- The pair (0, 0) satisfies the function. (The graph of the function includes the origin.)
- The number pairs that satisfy the function, other than (0, 0), form a constant ratio, so any two pairs form a proportion. (The graph of the function is a line or aligned points.)

Now we will consider inverse variation. **Inverse variation** is a relationship between two variables in which a change in one variable results in an inversely proportional change in the other variable. For instance, if one variable doubles (is multiplied by 2), then the other variable is halved (multiplied by the reciprocal of 2).

Choice C in problem 9 is an example of inverse variation. The variables are the speed or rate (r) of the trip and the time (t) for the trip. The product of the two variables is the distance (d) to school, which does not change. So as the speed increases, the time required to complete the trip decreases proportionately. We can express this relationship with this equation.

$$rt = d$$

In this inverse variation equation, the variables are multiplied and the product is constant. Here is another example. If a trip takes Adam 2 hours traveling at 60 miles per hour, then the trip would take Adam twice as long, 4 hours, if he traveled at half the speed, 30 miles per hour. Notice that the products of these two descriptions are equal to each other. Both equal 120 miles.

$$2 \text{ hr} \times 60 \text{ mph} = 4 \text{ hr} \times 30 \text{ mph}$$

Since the products are equal for any two pairs of numbers that satisfy the relationship, we can solve inverse proportion problems by writing an equation with two products equal to each other. Thus, to find how long it would take Adam to make the trip at 40 mph, we can write and solve this equation.

$$2 \text{ hr} \times 60 \text{ mph} = t \times 40 \text{ mph}$$

12. What is the solution to the equation?

13. **Represent** Make a function table that shows Adam's rate in miles per hour and his time in hours to make the 120 mile trip at these speeds: 60 mph, 40 mph, 30 mph, and 20 mph.

14. Multiply each pair of numbers in the table. What do you find about the products?

15. **Represent** Graph the pairs of numbers in the table from problem **13.** Notice that the points are not aligned and that 0, 0 is not a solution pair for the function.

The graph of an inverse proportion is a curve and not a line. The graph does not include the origin and does not intercept an axis. The graph may include points in the third quadrant as we see in this graph of $xy = 6$.

"Work" problems also provide examples of functions that involve measures that are products. Consider the following "work" problem.

Rafi and one friend can roof the whole house in 24 hours. How many hours would it take Rafi and two friends to roof the whole house?

"Work" problems often involve the measures "person-hours" or "person-days." A person-hour is one person working one hour. Projects can be estimated to require a certain number of person-hours of labor. To find the number of person-hours it takes to complete a job, we multiply the number of people working times the number of hours they work.

In the example on the previous page, the job is "to roof the whole house." The labor required to do the job is 2 people for 24 hours which is 48 person-hours. With three people working, the job is still 48 person-hours, but because more people are working, less time is needed. To solve the problem we can begin by writing an equation that shows the two products are equal.

$$\text{person-hours} = \text{person-hours}$$

$$(2 \text{ people})\ (24 \text{ hours}) = (3 \text{ people})\ (t \text{ hours})$$

16. What is the solution to the equation?

17. **Formulate** Solve the following "work" problem by writing and solving an equation that shows two products are equal.

If three people can harvest the field in 8 days, how many people are needed to harvest the field in two days?

We can use inverse variation to help us multiply numbers mentally. If we need to find the product of two numbers, we can decrease one factor and increase the other factor proportionately.

$$24 \times 15 = p$$

Instead of multiplying 24 and 15, we can first mentally halve 24 and double 15. This gives us 12 and 30 as factors which we can mentally multiply to find the product, 360.

18. **Explain** Mentally find the product of 25 and 18 and explain how you found the product.

Here we summarize the characteristics of inverse variation.

- As one variable changes, the other variable changes inversely.
- Zero is not a solution for either variable of the function. (The graph does not include the origin or points on either axis.)
- The products of the number pairs that satisfy the function are equal to each other.

LESSON 91

Evaluations with Positive and Negative Numbers

Power Up

Building Power

facts Power Up S

mental math

a. **Positive/Negative:** $(-84) + (-50)$

b. **Order of Operations/Exponents:** $(1.2 \times 10^3)(1.2 \times 10^3)$

c. **Ratio:** $\frac{w}{90} = \frac{80}{120}$

d. **Mental Math:** $6 \times 2\frac{1}{2}$ (*Think:* $6 \times 2 + 6 \times \frac{1}{2}$)

e. **Measurement:** 1.5 L to mL

f. **Fractional Parts:** \$20 is $\frac{1}{10}$ of m.

g. **Geometry:** What type of angle measures less than 90 degrees?

h. **Calculation:** 50% of 40, + 1, ÷ 3, × 7, + 1, × 2, $\sqrt{}$, × 5, − 1, $\sqrt{}$, × 4, + 2, ÷ 2

problem solving

Simplify the first three terms of the following sequence. Then write and simplify the next two terms.

$$\sqrt{1^3}, \sqrt{1^3 + 2^3}, \sqrt{1^3 + 2^3 + 3^3}, \ldots$$

New Concept

Increasing Knowledge

We have practiced evaluating expressions such as

$$x - xy - y$$

with positive numbers in place of x and y. In this lesson we will practice evaluating such expressions with negative numbers as well. When evaluating expressions with signed numbers, it is helpful to first replace each variable with parentheses. This will help prevent making mistakes in signs.

Thinking Skill

Analyze

What are the terms in this expression?

Example 1

Evaluate: $x - xy - y$ if $x = -2$ and $y = -3$

Solution

We write parentheses for each variable.

$$(\ ) - (\ )(\ ) - (\ ) \qquad \text{parentheses}$$

Now we write the proper numbers within the parentheses.

$$(-2) - (-2)(-3) - (-3) \qquad \text{insert numbers}$$

By the order of operations, we multiply before adding.

$(-2) - (+6) - (-3)$ multiplied

Then we add algebraically from left to right.

$(-8) - (-3)$ added -2 and -6

$\mathbf{-5}$ added -8 and $+3$

Discuss What are some values for x and y that will result in a positive solution?

Example 2

Nikki says the solution of the equation $3x + 5 = -1$ is -2. Show how Nikki can check her answer.

Solution

Nikki can replace x in the original equation with –2 and simplify.

$3x + 5 = -1$	
$3(-2) + 5 = -1$	Substituted -2
$-6 + 5 = -1$	Multiplied $3(-2)$
$-1 = -1$	Added -6 and $+5$

The two sides of the equation are equal. Nikki's solution is checked.

Practice Set

Generalize Evaluate each expression. Write parentheses as the first step.

a. $x + xy - y$ if $x = 3$ and $y = -2$

b. $-m + n - mn$ if $m = -2$ and $n = -5$

Verify Check each solution for the original equations.

c. $4x + 25 = 9$, solution: $x = -4$

d. $3x + 7 = 8$, solution: $x = -5$

Justify Solve and check these equations.

e. $4x = -20$

f. $-2x = 16$

g. **Explain** How did you check your solutions for **e** and **f?**

Written Practice

Strengthening Concepts

1. (55) In a museum gallery, the average height of the six paintings on one wall is 86 cm. The average height of the four paintings on another wall is 94 cm. What is the average height of all ten of the paintings?

2. (Inv. 4) The mean of these numbers is how much greater than the median?

3, 12, 7, 5, 18, 6, 9, 28

3. (46) The Singhs completed a 130-mile trip in $2\frac{1}{2}$ hours. What was their average speed in miles per hour?

Use ratio boxes to solve problems **4–7.**

4. (66) The ratio of workers to supervisors at the job site was 3 to 5. Of the 120 workers and supervisors at the job site, how many were workers?

5. (72) Vera bought 3 notebooks for $8.55. At this rate, how much would 5 notebooks cost?

6. (60) A software program for guitar lessons is on sale for 80% of the original price of $54. What is the sale price?

7. (81) Forty people came to the party. This was 80 percent of those who were invited. How many were invited?

8. (77) Write equations to solve **a** and **b.**

a. Twenty is 40 percent of what number?

b. Twenty is what percent of 40?

9. (88) Use two unit multipliers to convert 3600 $in.^2$ to square feet.

10. (48, 71) Read this statement. Then answer the questions that follow:

Three fourths of the exhibits at the science fair were about physical science. There were 60 physical science exhibits.

a. How many exhibits were at the science fair?

b. What percent of the exhibits were not about physical science?

*** 11.** (91) **Generalize** Evaluate: $x - y - xy$ if $x = -3$ and $y = -2$

*** 12.** (Inv. 9) **Analyze** If three painters can paint the house in 12 hours, how long would it take four painters to paint the house?

13. (19, Inv. 6, 75, 89) **a.** Classify this quadrilateral.

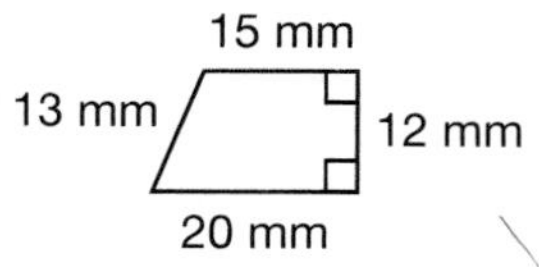

b. Find the perimeter of the figure.

c. Find the area of the figure.

d. The sum of the measures of the interior angles of a quadrilateral is 360°. If the acute interior angle of the figure measures 75°, what does the obtuse interior angle measure?

14. (2, 41) Which property is illustrated by each equation?

a. $a + (b + c) = (a + b) + c$

b. $ab = ba$

c. $a(b + c) = ab + ac$

15. (62) The lengths of two sides of an isosceles triangle are 5 inches and 1 foot. Draw the triangle and find its perimeter in inches.

* **16.** Multiply. Write the product in scientific notation.
(83)

$$(2.4 \times 10^{-4})(5 \times 10^{-7})$$

17. A pyramid with a square base has how many
(67)

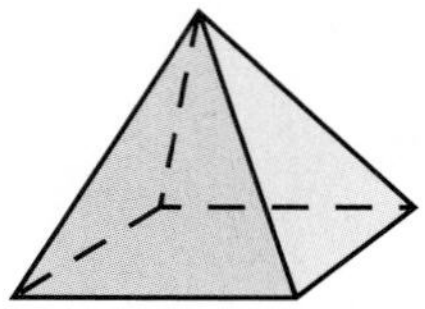

a. faces?

b. edges?

c. vertices?

18. Find each measure of the circle. (Use 3.14 for π.)
(65, 82)

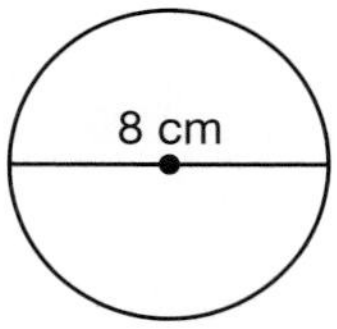

a. circumference

b. area

* **19.** **Analyze** One yard equals three feet. The table at right shows three pairs of equivalent measures.
(Inv. 9)

yd	ft
1	3
2	6
3	9

a. Plot these points on a coordinate graph, using yards for the horizontal axis and feet for the vertical axis. Then draw a ray through the points to show all pairs of equivalent measures.

b. Does the graph show direct variation? Explain your answer.

20. Refer to the figure below to answer **a–d.**
(40)

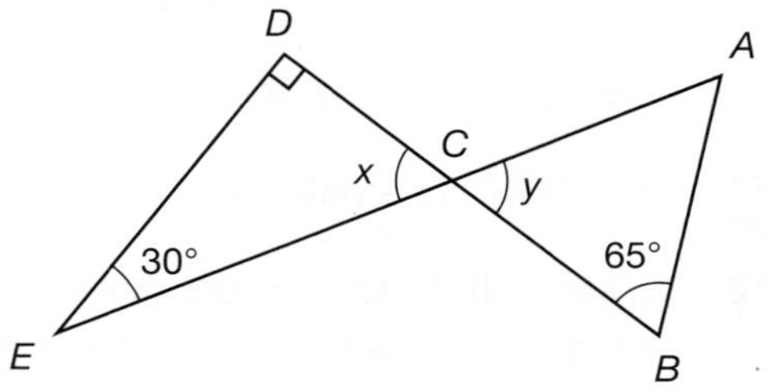

a. What is m∠x?

b. What is m∠y?

c. What is m∠A?

d. Are the two triangles similar? Why or why not?

* **21.** **a.** Add algebraically: $-3x - 3 - x - 1$
(84, 87)

b. Multiply: $(-3x)(-3)(-x)(-1)$

* **22.** **Represent** On a number line, graph all integers greater than or equal to –3.
(78, 86)

23. (32) Segment AB is how many millimeters longer than segment BC?

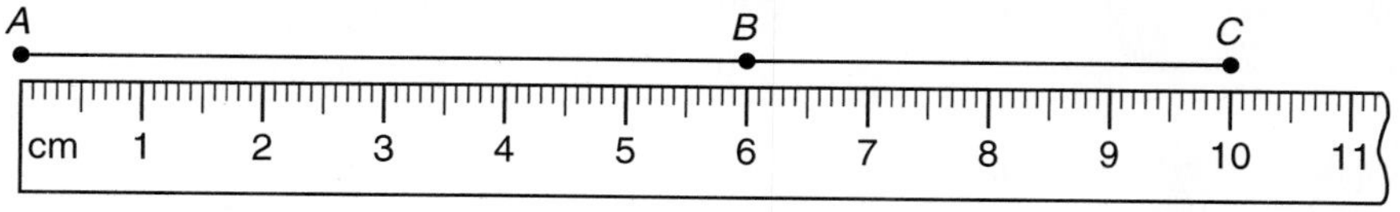

Justify For problems **24–26**, solve and check. Show each step.

24. (Inv. 7) $5 = y - 4.75$

* **25.** (90) $3\frac{1}{3}y = 7\frac{1}{2}$

* **26.** (90) $-9x = 414$

Simplify:

27. (50) $\frac{32 \text{ ft}}{1 \text{ s}} \cdot \frac{60 \text{ s}}{1 \text{ min}}$

28. (43) $5\frac{1}{3} + 2.5 + \frac{1}{6}$ (mixed-number answer)

29. (43, 45) $\frac{2\frac{3}{4} + 3.5}{2\frac{1}{2}}$ (decimal answer)

* **30.** (85) **a.** $\frac{(-3) - (-4)(+5)}{(-2)}$

b. $-3(+4) - 5(+6) - 7$

Early Finishers
Real-World Application

Tobey is using a step counter as part of a fitness program. During lunch, he walks 441 steps. Each step averages 30 inches in length.

a. Use unit multipliers to calculate Tobey's total distance during lunch (in feet).

b. Use a unit multiplier and a calculator to find his total distance during lunch (in miles). Round to the nearest hundredth of a mile.

c. In a week, will Tobey's lunchtime activity total a mile? Explain.

LESSON

92

• Percent of Change

Power Up

Building Power

facts Power Up T

mental math

a. **Positive/Negative:** (–75) – (–50)

b. **Order of Operations/Exponents:** $(1.5 \times 10^{-5})(1.5 \times 10^{-5})$

c. **Algebra:** $50 = 3m + 2$

d. **Mental Math:** $6 \times 3\frac{1}{3}$ (*Think:* $6 \times 3 + 6 \times \frac{1}{3}$)

e. **Measurement:** 0.3 m to cm

f. **Percent:** \$6 is 10% of *m*.

g. **Geometry:** What type of angle measures 180 degrees?

h. **Rate:** At \$8.00 per hour, how much money will Kenji earn working $4\frac{1}{2}$ hours?

problem solving What is the average of these fractions? $\frac{1}{12}, \frac{1}{6}, \frac{1}{4}, \frac{1}{3}, \frac{5}{12}$

New Concept

Increasing Knowledge

The percent problems that we have considered before now have used a percent to describe part of a whole. In this lesson we will consider percent problems that use a percent to describe an amount of change. The change may be an increase or a decrease. Adding sales tax to a purchase is an example of an increase. Marking down the price of an item for a sale is an example of a decrease.

Increase

original number + amount of change = new number

Decrease

original number – amount of change = new number

We can use a ratio box to help us with "increase-decrease" problems. However, there is a difference in the way we set up the ratio box. When we make a table for a "parts of a whole" problem, the bottom number in the percent column is 100 percent.

	Percent	Actual Count
Part		
Part		
Whole	100	

When we set up a ratio box for an "increase-decrease" problem, we also have three rows. The three rows represent the original number, the amount of change, and the new number. We will use the words ***original, change,*** and ***new*** on the left side of the ratio box. The difference in the setup is where we put 100 percent. Most "increase-decrease" problems consider the original amount to be 100 percent. So the top number in the percent column will be 100 percent.

	Percent	Actual Count
Original	100	
Change		
New		

If the change is an **increase,** we **add** it to the original amount to get the new amount. If the change is a **decrease,** we **subtract** it from the original amount to get the new amount.

Example 1

The county's population increased 15 percent from 1980 to 1990. If the population in 1980 was 120,000, what was the population in 1990?

Solution

First we identify the type of problem. The percent describes an amount of change. This is an increase problem. We make a ratio box and write the words "original," "change," and "new" down the side. Since the change was an increase, we write a plus sign in front of "change."

In the "percent" column we write 100 percent for the original (1980 population) and 15 percent for the change. We add to get 115 percent for the new (1990 population).

In the "actual count" column we write 120,000 for the original population and use the letters C for "change" and N for "new."

	Percent	Actual Count
Original	100	120,000
+ Change	15	C
New	115	N

$$\frac{100}{115} = \frac{120{,}000}{N}$$

We are asked for the new population. Since we know both numbers in the first row, we use the first and third rows to write the proportion.

$$\frac{100}{115} = \frac{120{,}000}{N}$$

$$100N = 13{,}800{,}000$$

$$N = 138{,}000$$

The county's population in 1990 was **138,000.**

Example 2

The price was reduced 30 percent. If the sale price was $24.50, what was the original price?

Solution

First we identify the problem. This is a decrease problem. We make a ratio box and write "original," "change," and "new" down the side, with a minus sign in front of "change." In the percent column we write 100 percent for original, 30 percent for change, and 70 percent for new. The sale price is the new actual count. We are asked to find the original price.

	Percent	Actual Count
Original	100	R
− Change	30	C
New	70	24.50

$$\frac{100}{70} = \frac{R}{24.50}$$

$$70R = 2450$$

$$R = 35$$

The original price was **$35.00.**

Example 3

A merchant bought an item at wholesale for $20 and marked the price up 75% to sell the item at retail. What was the merchant's retail price for the item?

Solution

Thinking Skill

Formulate

We can use a ratio box to solve *part-of-a-whole* and *increase-decrease* word problems. How are the problems different?

This is an increase problem. We make a table and record the given information.

	Percent	Actual Count
Original (Wholesale)	100	20
+ Change (Markup)	75	M
New (Retail)	175	R

$$\frac{100}{175} = \frac{20}{R}$$

$$100R = 3500$$

$$R = 35$$

The merchant's retail price for the item was **$35.**

Practice Set

Model Use a ratio box to solve each problem.

a. The regular price was $24.50, but the item was on sale for 30 percent off. What was the sale price?

b. The number of students taking algebra increased 20% in one year. If 60 students are taking algebra this year, how many took algebra last year?

c. Bikes were on sale for 20 percent off. Tomas bought one for \$120. How much money did he save by buying the bike at the sale price instead of at the regular price?

d. The clothing store bought shirts for \$15 each and marked up the price 80% to sell the shirts at retail. What was the retail price of each shirt?

e. **Classify** In which of the problems above is the change an increase?

Written Practice

Strengthening Concepts

1. (21) The product of the first three prime numbers is how much less than the sum of the next three prime numbers?

2. (55) Dara scored an average of 88 points per game in five crossword games. What score must she average on the next two games to have a seven-game average of 90?

3. (46) Jenna finished a 2-mile race in 15 minutes. What was her average speed in miles per hour?

Use ratio boxes to solve problems **4–7.**

4. (66) Forty-five of the 80 students walk to school. What is the ratio of students who walk to school to students who do not walk to school?

5. (72) Two dozen pencils cost \$3.60. At that rate, how much would 60 pencils cost?

*** 6.** (92) **Generalize** The population of Houston, TX increased about 25 percent from 1990 to 2000. If the population of Houston was about 1,600,000 people in 1990, what was the population in 2000?

*** 7.** (92) Because of an unexpected cold snap, the price of oranges increased 50% in one month. If the price after the increase was 75¢ each, what was the price before the increase?

8. (77) Write equations to solve **a** and **b.**

a. Sixty is what percent of 75?

b. Seventy-five is what percent of 60?

*** 9.** (88) **Connect** The snail moved one foot in 2 minutes. Use two unit multipliers to find its average speed in yards per hour.

10. (71) Diagram this statement. Then answer the questions that follow.

Five eighths of the trees in the grove were deciduous. There were 160 deciduous trees in the grove.

a. How many trees were in the grove?

b. How many of the trees in the grove were not deciduous?

* **11.** (91) *Analyze* If $x = -5$ and $y = 3x - 1$, then y equals what number?

12. (60) Compare: 30% of 20 ◯ 20% of 30

13. (58, 75) **a.** Find the area of this isosceles trapezoid.

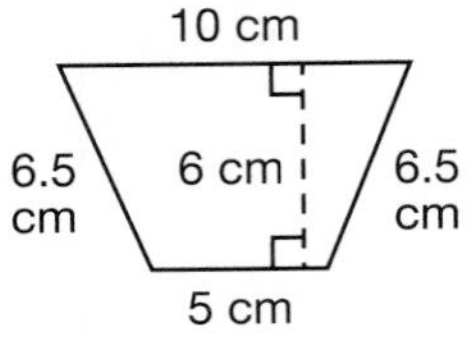

b. Trace the figure and draw its line of symmetry.

* **14.** (60, 92) *Generalize* A merchant bought a stereo at wholesale for $90.00 and marked up the price 75% to sell the stereo at retail.

a. What was the retail price of the stereo?

b. If the stereo sells at the retail price, and the sales-tax rate is 6%, what is the total price including sales tax?

15. (83) Multiply. Write the product in scientific notation.

$$(8 \times 10^{-5})(3 \times 10^{12})$$

16. (48) Complete the table.

Fraction	Decimal	Percent
$2\frac{1}{3}$	a.	b.
c.	d.	$3\frac{1}{3}\%$

* **17.** (36) One ace is missing from an otherwise normal deck of cards. How many aces are still in the deck? How many cards are still in the deck? If a card is selected at random, what is the probability that the selected card will be an ace?

18. (60) What number is 250 percent of 60?

19. (67) A triangular prism has how many

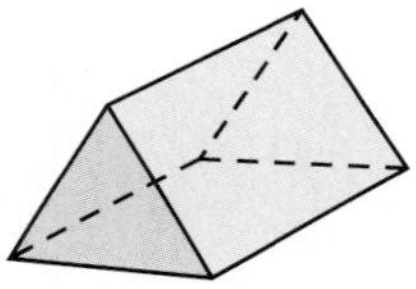

a. triangular faces?

b. rectangular faces?

20. (65) John measured the diameter of his bicycle tire and found that it was 24 inches. What is the distance around the tire to the nearest inch? (Use 3.14 for π.)

* **21.** (56, Inv. 9) **a.** Find the missing numbers in the table by using the function rule.

$y = 2x + 1$

x	y
0	
3	
−2	

b. *Represent* Plot the (x, y) pairs on a coordinate plane, and then draw a line showing all (x, y) pairs that satisfy the function.

c. *Analyze* In this function are x and y directly proportional? Explain your answer.

22. (66) The ratio of the measures of two angles was 4 to 5. If the sum of their measures was 180°, what was the measure of the smaller angle?

*** 23.** (84, 87) **Generalize** Simplify:

a. $x + y + 3 + x - y - 1$

b. $(3x)(2x) + (3x)(2)$

c. $(x^3y)^2$

24. (Inv. 6) Draw a pair of parallel lines. Draw a second pair of parallel lines that intersect but are not perpendicular to the first pair. What kind of quadrilateral is formed?

Justify For problems **25–27,** solve and check. Show each step.

*** 25.** (90) $3\frac{1}{7}x = 66$

26. (Inv. 7) $w - 0.15 = 4.9$

*** 27.** (90, 91) $-8y = 600$

Simplify:

28. (52) $(2 \cdot 3)^2 - 2(3^2)$

29. (43) $5 - \left(3\frac{1}{3} - 1.5\right)$

*** 30.** (85, 91) **a.** $\dfrac{(-8)(-6)(-5)}{(-4)(-3)(-2)}$

b. $-6 - 5(-4) - 3(-2)(-1)$

Early Finishers

Real-World Application

A toy manufacturer has 45 workers and releases 3,600 dolls a week. The manufacturer plans on increasing the number of workers by 70%, which will increase the number of dolls a week proportionally.

a. On average, how many dolls are created by each worker weekly?

b. If the manufacturer increases the workforce by 70%, how many dolls will be released weekly?

LESSON

93 • Two-Step Equations and Inequalities

Power Up

Building Power

facts Power Up T

mental math

a. Positive/Negative: $(-25)(-8)$

b. Order of Operations/Exponents: $(2.5 \times 10^8)(3 \times 10^{-4})$

c. Ratio: $\frac{100}{x} = \frac{22}{55}$

d. Measurement: 2 ft^2 equals how many square inches?

e. Calculation: $8 \times 2\frac{1}{4}$

f. Percent: 10% less than 60

g. Geometry: A slide moves triangle *ABC* to form triangle *A′B′C′*. What is the proper name of the transformation that formed $\triangle A'B'C'$?

h. Estimation: Estimate the product of 3.14 and 4.9 cm.

problem solving

Luke put 2 red pens, 7 blue pens, and 1 black pen in a bag, closed it, and shook it up. If the pens are the same size, weight, and texture, what is the probability of choosing a red pen? A red *or* black pen? What is the probability of *not* choosing a blue pen? If Luke does choose a blue pen, but gives it away, what is the probability he will choose *another* blue pen?

New Concept

Increasing Knowledge

Since Investigation 7 we have practiced solving one-step balanced equations. In this lesson we will practice solving two-step equations.

This balance scale illustrates a two-step equation.

$2x + 5 = 35$

On the left side of the equation are two terms, $2x$ and 5. To solve the equation, we first isolate the variable term by subtracting 5 from (or adding -5 to) both sides of the equation.

$2x + 5 - 5 = 35 - 5$ subtracted 5 from both sides

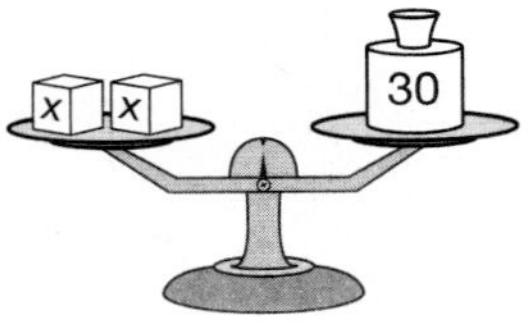

$2x = 30$
simplified

Thinking Skill

Discuss

Demonstrate that you can multiply by $\frac{1}{2}$ in order to isolate the variable. Why does this work?

We see that $2x = 30$, so we divide by 2 (or multiply by $\frac{1}{2}$) to find $1x$.

$$\frac{2x}{2} = \frac{30}{2} \quad \text{divided both sides by 2}$$

$$x = 15 \quad \text{simplified}$$

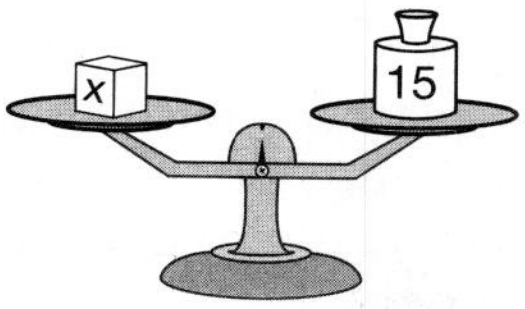

Example 1

Solve this equation. Show all steps.

$$\mathbf{0.4x + 1.2 = 6}$$

Solution

First we isolate the variable term by subtracting 1.2 from both sides of the equation. Then we divide both sides by 0.4.

$0.4x + 1.2 = 6$	equation
$0.4x + 1.2 - 1.2 = 6 - 1.2$	subtracted 1.2 from both sides
$0.4x = 4.8$	simplified
$\frac{0.4x}{0.4} = \frac{4.8}{0.4}$	divided both sides by 0.4
$x = \mathbf{12}$	simplified

Justify Show how to check the solution. Describe your process.

Example 2

Solve this equation. Show all steps.

$$\mathbf{-\frac{2}{3}x - \frac{1}{2} = \frac{1}{3}}$$

Solution

First, we isolate the variable term by adding $\frac{1}{2}$ to both sides of the equation. Then we find $1x$ by multiplying both sides of the equation by $-\frac{3}{2}$.

$-\frac{2}{3}x - \frac{1}{2} = \frac{1}{3}$	equation
$-\frac{2}{3}x - \frac{1}{2} + \frac{1}{2} = \frac{1}{3} + \frac{1}{2}$	added $\frac{1}{2}$ to both sides
$-\frac{2}{3}x = \frac{5}{6}$	simplified
$\left(-\frac{3}{2}\right)\left(-\frac{2}{3}x\right) = \left(-\frac{3}{2}\right)\left(\frac{5}{6}\right)$	multiplied both sides by $-\frac{3}{2}$
$x = \mathbf{-\frac{5}{4}}$	simplified

Example 3

Solve: $-15 = 3x + 6$

Solution

Thinking Skill

Explain

Interchange the sides of the equation. Why will the solution remain the same?

The variable term is on the right side of the equal sign. We may interchange the entire right side of the equation with the entire left side if we wish (just as we may interchange the contents of one pan of a balance scale with the contents of the other pan). However, we will solve this equation without interchanging the sides of the equation.

$-15 = 3x + 6$	equation
$-15 - 6 = 3x + 6 - 6$	subtracted 6 from both sides
$-21 = 3x$	simplified
$\frac{-21}{3} = \frac{3x}{3}$	divided both sides by 3
$\mathbf{-7} = x$	simplified

Justify Show how to check the solution.

In this lesson we have practiced procedures for solving equations. We may follow similar procedures for solving inequalities in which the variable term is positive.[1] To solve an inequality, we isolate the variable while maintaining the inequality.

Example 4

Solve this inequality and graph its solution: $2x - 5 \geq 1$

Solution

We see that the variable term ($2x$) is positive. We begin by adding 5 to both sides of the inequality. Then we divide both sides of the inequality by 2.

$2x - 5 \geq 1$	inequality
$2x - 5 + 5 \geq 1 + 5$	added 5 to both sides
$2x \geq 6$	simplified
$\frac{2x}{2} \geq \frac{6}{2}$	divided both sides by 2
$x \geq \mathbf{3}$	simplified

[1] Procedures for solving inequalities with a negative variable term will be taught in a later course.

We check the solution by replacing x in the original inequality with numbers equal to and greater than 3. We try 3 and 4 below.

$2x - 5 \geq 1$	original inequality
$2(3) - 5 \geq 1$	replaced x with 3
$1 \geq 1$	simplified and checked
$2(4) - 5 \geq 1$	replaced x with 4
$3 \geq 1$	simplified and checked

Predict Would the number 2 satisfy the inequality? Why or why not?

Now we graph the solution $x \geq 3$.

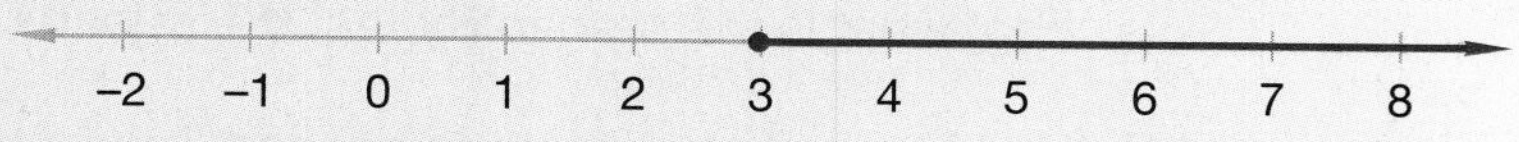

This graph indicates that all numbers greater than or equal to 3 satisfy the original inequality.

Discuss Why is the dot at 3 solid?

Practice Set

Justify Solve each equation. Show and justify all steps. Check each solution.

a. $8x - 15 = 185$

b. $0.2y + 1.5 = 3.7$

c. $\frac{3}{4}m - \frac{1}{3} = \frac{1}{2}$

d. $1\frac{1}{2}n + 3\frac{1}{2} = 14$

e. $-6p + 36 = 12$

f. $38 = 4w - 26$

Represent Solve these inequalities and graph their solutions:

g. $2x + 5 \geq 1$

h. $2x - 5 < 1$

Written Practice

Strengthening Concepts

1. (46) From Sim's house to the lake is 30 kilometers. If he completed the round trip on his bike in 2 hours 30 minutes, what was his average speed in kilometers per hour?

2. (Inv. 4) Find the **a** mean and **b** range for this set of numbers:

3, 9, 7, 5, 10, 4, 5, 8, 5, 4, 8, 40

Use ratio boxes to solve problems **3–5.**

3. (36, 66) The ratio of red marbles to blue marbles in a bag of 600 red and blue marbles was 7 to 5.

a. How many marbles were blue?

b. If one marble is drawn from the bag, what is the probability that the marble will be blue?

4. (72) The machine could punch out 500 plastic pterodactyls in 20 minutes. At that rate, how many could it punch out in $1\frac{1}{2}$ hours?

* **5.** (92) **a.** The price of a T-shirt was reduced by 25%. If the original price was \$24, what was the sale price?

b. The price of a backpack was reduced by 25%. If the sale price was \$24, what was the original price?

* **6.** (87) **Generalize** Multiply: $(-3x^2)(2xy)(-x)(3y^2)$

7. (36) A bag normally contains a dozen marbles, 4 of which are blue. However, 2 blue marbles are missing. If a marble is selected at random from the bag, what is the probability that the selected marble will be blue?

* **8.** (88) Use two unit multipliers to convert 7 days to minutes.

9. (22) Diagram this statement. Then answer the questions that follow.

Five ninths of the 45 cars pulled by the locomotive were not tanker cars.

a. How many tanker cars were pulled by the locomotive?

b. What percent of the cars pulled by the locomotive were not tanker cars?

10. (33, 48) Compare: $\frac{1}{3} \bigcirc 33\%$

* **11.** (91) Evaluate: $ab - a - b$ if $a = -3$ and $b = -1$

12. (60) Find the total price, including 5% tax, for a meal that includes a \$7.95 dish, a 90¢ beverage, and a \$2.35 dessert.

13. (19, 75) A corner was cut from a square sheet of paper to form this pentagon. Dimensions are in inches.

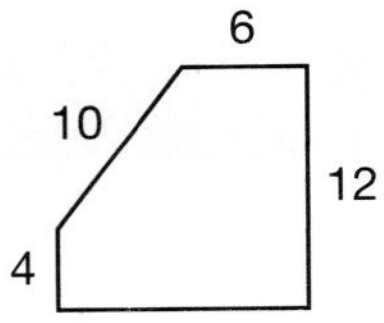

a. What is the perimeter of the pentagon?

b. What is the area of the pentagon?

14. (48) Complete the table.

Fraction	Decimal	Percent
a.	0.08	**b.**
c.	**d.**	$8\frac{1}{3}\%$

* **15.** (92) **Generalize** A retailer buys a toy for \$3.60 and marks up the price 120%. What is the retail price of the toy?

16. (83) Multiply. Write the product in scientific notation.

$$(8 \times 10^{-3})(6 \times 10^{7})$$

17. (67, 70) Each edge of a cube is 10 cm long.

a. What is the volume of the cube?

b. What is the surface area of the cube?

18. (65, 82) **a.** What is the area of the circle shown?

b. What is the circumference of the circle?

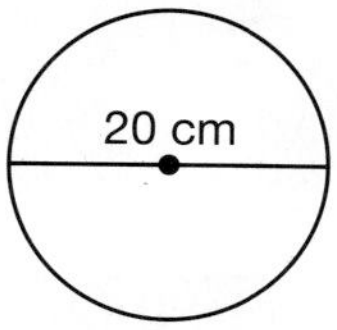

Use 3.14 for π.

19. (84) Collect like terms: $-x + 2x^2 - 1 + x - x^2$

20. (56, Inv. 9) **a.** Find the missing numbers in the table by using the function rule.

b. Represent Plot the points and draw a line.

c. Analyze Are x and y directly proportional? Why or why not?

$y = 2x + 3$

x	y
1	
0	
−2	

21. (74) Write an equation to solve this problem:

Sixty is $\frac{3}{8}$ of what number?

* **22.** (93) Represent Solve this inequality and graph its solution: $2x - 5 > -1$

23. (40) Refer to the figure below to answer **a** and **b.**

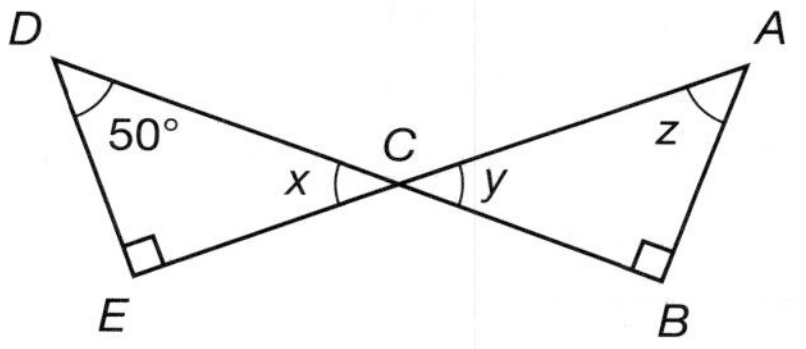

a. Find the measures of $\angle x$, $\angle y$, and $\angle z$.

b. Are the two triangles similar? Why or why not?

24. (34) Connect What is the sum of the numbers labeled A and B on the number line below?

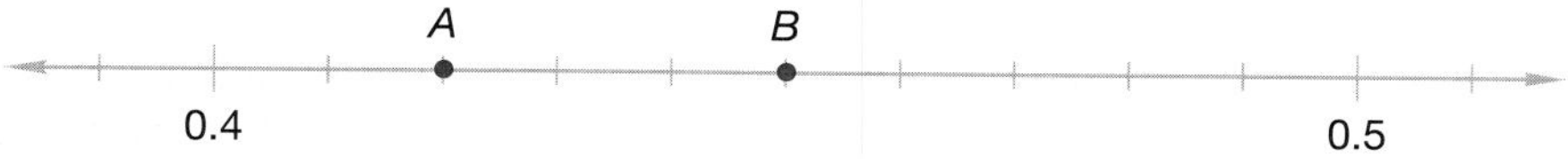

Generalize For problems **25–28,** solve and check. Show each step.

* **25.** (93) $3x + 2 = 9$

* **26.** (93) $\frac{2}{3}w + 4 = 14$

* **27.** (93) $0.2y - 1 = 7$

* **28.** (90) $-\frac{2}{3}m = 6$

Simplify:

29. (57, 63) $3(2^3 + \sqrt{16}) - 4^0 - 8 \cdot 2^{-3}$

* **30.** (85, 91) **a.** $\frac{(-9)(+6)(-5)}{(-4) - (-1)}$

b. $-3(4) + 2(3) - 1$

LESSON
94

• Probability of Dependent Events

Power Up

Building Power

facts

Power Up S

mental math

a. **Positive/Negative:** $(-144) \div (-6)$

b. **Order of Operations/Exponents:** $(1.5 \times 10^{-8})(4 \times 10^{3})$

c. **Algebra:** $5w + 1.5 = 4.5$

d. **Measurement:** Convert 30°C to degrees Fahrenheit.

e. **Calculation:** $6 \times 2\frac{2}{3}$

f. **Percent:** 10% more than 50

g. **Geometry:** Name the transformation that turns a figure around a point.

h. **Calculation:** 25% of 40, × 4, + 2, ÷ 6, × 9, + 1, $\sqrt{}$, × 3, + 1, $\sqrt{}$

problem solving

Darla designed a stained glass window that incorporates 20 squares, triangles, and circles altogether. The window has 2 more triangles than twice the number of squares. The number of circles in the window is a multiple of three. How many of each shape appears in Darla's window if there are at least two of each shape?

New Concept

Increasing Knowledge

Thinking Skills

Explain

In your own words explain the difference between independent and dependent events?

The probability experiments we have studied so far involve **independent events** which means that the probability of an event occurring is not affected by the other events. The outcome of the toss of a coin, the roll of a number cube, or the spin of a spinner does not affect future outcomes.

An event is a **dependent event** if its probability is influenced by a prior event. For example, if a marble is drawn from a bag of mixed marbles and is not replaced, then the probability for the next draw differs from the probability of the first draw because the number and mix of marbles remaining in the bag has changed.

The probability of dependent events occurring in a specified order is a product of the first event and the recalculated probabilities of each subsequent event.

Example 1

Two red marbles, three white marbles, and four blue marbles are in a bag. If one marble is drawn and not replaced, and a second marble is drawn, what is the probability that both marbles will be red?

Solution

Two of the nine marbles are red, so the probability of red on the first draw is $\frac{2}{9}$.

$$\text{1st Draw: } P(\text{Red}) = \frac{2}{9}$$

If a red marble is removed from the bag on the first draw, then only one of the remaining eight marbles is red. Thus, the probability of red on the second draw is $\frac{1}{8}$.

$$\text{2nd Draw: } P(\text{Red}) = \frac{1}{8}$$

To find the probability of red on both the first and second draw, we multiply the probabilities of the two events.

$$P(\text{Red, Red}) = \frac{2}{9} \cdot \frac{1}{8} = \mathbf{\frac{1}{36}}$$

Explain Why is the product of $\frac{2}{9} \cdot \frac{1}{8}$ shown as $\frac{1}{36}$?

Example 2

Two cards are drawn from a regular shuffled deck. What is the probability of drawing two aces?

Solution

Although two cards are drawn simultaneously we calculate the probability of each card separately and in sequence. Four of the 52 cards are aces, so the probability of the first card being an ace is $\frac{4}{52}$ which reduces to $\frac{1}{13}$.

$$\text{First card: } P(\text{Ace}) = \frac{4}{52} = \frac{1}{13}$$

Since the first card must be an ace in order for both cases to be aces, only three of the remaining 51 cards are aces. Thus the probability of the second card being an ace is $\frac{3}{51}$ which reduces to $\frac{1}{17}$.

$$\text{Second card: } P(\text{Ace}) = \frac{3}{51} = \frac{1}{17}$$

The probability that both cards drawn are aces is the product of the probabilities of each event.

$$P(Ace, Ace) = \frac{1}{13} \cdot \frac{1}{17} = \mathbf{\frac{1}{221}}$$

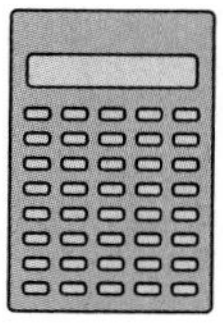

Many calculators have an **exponent key** such as y^x or x^y that can be used to calculate probabilities.

Suppose you were going to take a ten-question true-false test. Instead of reading the questions, you decide to guess every answer. What is the probability of guessing the correct answer to all ten questions?

The probability of correctly guessing the first answer is $\frac{1}{2}$. The probability of correctly guessing the first two answers is $\frac{1}{2} \cdot \frac{1}{2}$, or $(\frac{1}{2})^2$. (Since $\frac{1}{2} \cdot \frac{1}{2} = \frac{1}{4}$, we may also write $(\frac{1}{2})^2$ as $\frac{1}{2^2}$.) The probability of correctly guessing the first three answers is $(\frac{1}{2})^3$, or $\frac{1}{2^3}$. Thus, the probability of guessing the correct answer to all ten questions is $(\frac{1}{2})^{10}$, or $\frac{1}{2^{10}}$. To find 2^{10} on a calculator with a y^x or x^y key, we use these keystrokes:

2 y^x 1 0 =

The number displayed is 1024. Therefore, the probability of correctly guessing all ten true-false answers is

$$\frac{1}{1024}$$

Correctly guessing all ten answers has the same likelihood as tossing heads with a coin ten times in a row. The chance of being successful by guessing is very small.

Probabilities that are extremely unlikely may be displayed by a calculator in scientific notation.

Analyze Find the probability of correctly guessing the correct answer to every question on a twenty-question, four-option multiple-choice test.

Practice Set

a. **Analyze** In a bag are two red marbles, three white marbles, and four blue marbles. If one marble is drawn from the bag and not replaced and then a second marble is drawn from the bag, what is the probability of drawing two blue marbles?

b. If two cards are drawn from a regular shuffled deck, what is the probability that both cards will be diamonds?

Written Practice *Strengthening Concepts*

1. *(51)* Twenty-one billion is how much more than 9.8 billion? Write the answer in scientific notation.

2. *(54, 55)* The train traveled at an average speed of 68 miles per hour for the first 2 hours and at an average speed of 80 miles per hour for the next 4 hours. What was the train's average speed for the 6-hour trip?

3. *(46)* A 10-pound box of laundry detergent costs $15.45. A 15-pound box costs $20.50. Which box costs the most per pound? How much more per pound does it cost?

4. *(36, 67)* In a rectangular prism, what is the ratio of faces to edges?

Use ratio boxes to solve problems **5–8.**

5. *(66)* The team's win-loss ratio was 3 to 2. If the team won 12 games and did not tie any games, how many games did the team play?

6. *(72)* Twenty-four is to 36 as 42 is to what number?

*** 7.** *(92)* **Generalize** The number that is 20% less than 360 is what percent of 360?

*** 8.** *(92)* **Generalize** During the sale shirts were marked down 20 percent to $20. What was the regular price of the shirts (the price before the sale)?

*** 9.** *(88)* Use two unit multipliers to perform each conversion:

a. 144 ft^2 to square yards

b. 1 kilometer to millimeters

10. Read this statement. Then answer the questions that follow.
(71)

An author wrote $\frac{2}{5}$ of her new book. She has written a total of 120 pages.

a. How many pages will her book have when it is finished?

b. How many more pages does she need to write?

11. If a pair of 1–6 number generators is tossed once, what is the probability that the total number rolled will be
(36)

a. 1? **b.** 2? **c.** 3?

* **12.** **Analyze** If $y = 4x - 3$ and $x = -2$, then y equals what number?
(91)

13. The perimeter of a certain square is 4 yards. Find the area of the square in square feet.
(19, 20)

14. The sale price of the new car was $18,500. The sales-tax rate was 6.5 percent.
(60)

a. What was the sales tax on the car?

b. What was the total price including tax?

c. If the commission paid to a salesperson is 2 percent of the sale price, how much is the commission on a $18,500 sale?

15. Complete the table.
(48)

Fraction	Decimal	Percent
a.	b.	$66\frac{2}{3}\%$
$1\frac{3}{4}$	c.	d.

* **16.** **a.** What is 200 percent of $7.50?
(60, 92)

b. **Analyze** What is 200 percent more than $7.50?

17. Multiply. Write the product in scientific notation.
(83)

$$(2 \times 10^8)(8 \times 10^2)$$

18. Roberto puts 1-inch cubes in a box with inside dimensions as shown. How many cubes will fit in this box?
(70)

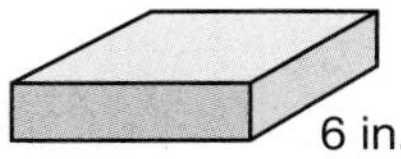

19. The length of each side of the square equals the diameter of the circle. The area of the square is how much greater than the area of the circle?
(82)

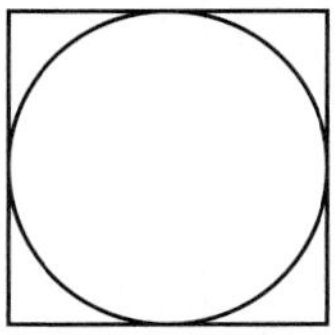

Use $\frac{22}{7}$ for π.

20. Divide 7.2 by 0.11 and write the quotient with a bar over the repetend.
(42, 45)

* **21.** (56, Inv. 9)
a. Find the missing numbers in the table by using the function rule.

b. Represent Graph the function.

c. Are x and y directly proportional? Why or why not?

$y = 3x$

x	y
1	
0	
−1	

* **22.** Represent Solve this inequality and graph its solution: $2x - 5 < -1$
(93)

23. In the figure at right, the measure of $\angle AOC$ is half the measure of $\angle AOD$. The measure of $\angle AOB$ is one third the measure of $\angle AOD$.
(40)

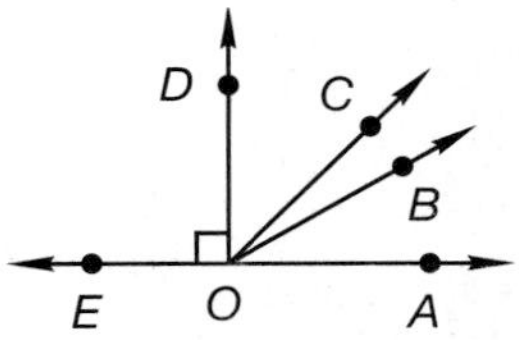

a. Find $m\angle AOB$.

b. Find $m\angle EOC$.

24. The length of segment BC is how much less than the length of segment AB?
(8)

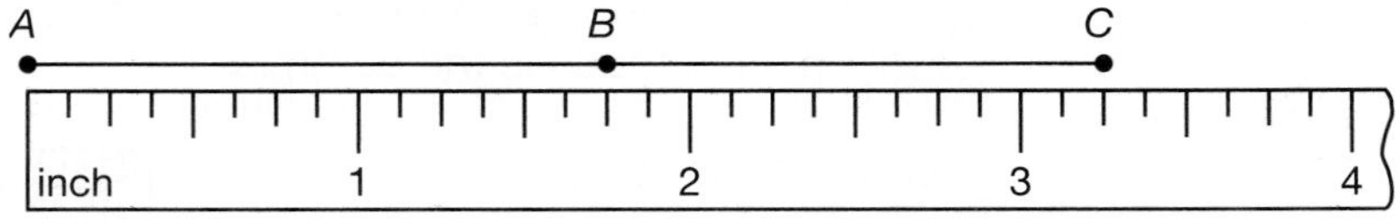

Justify For problems **25** and **26,** solve the equation. Show each step.

* **25.** $1.2p + 4 = 28$
(93)

* **26.** $-6\frac{2}{3}m = 1\frac{1}{9}$
(90)

Simplify:

27. **a.** $6x^2 + 3x - 2x - 1$
(84, 87)

b. $(5x)(3x) - (5x)(-4)$

* **28.** **a.** $\dfrac{(-8) - (-6) - (4)}{-3}$
(85, 91)

b. $-5(-4) - 3(-2) - 1$

* **29.** Evaluate: $b^2 - 4ac$ if $a = -1$, $b = -2$, and $c = 3$
(91)

* **30.** Analyze In a bag are two red marbles, three white marbles, and four blue marbles.
(94)

a. One marble is drawn and put back in the bag. Then a marble is drawn again. What is the probability of drawing a white marble on both draws? Are the events dependent or independent?

b. One marble is drawn and not replaced. Then a second marble is drawn. What is the probability of drawing a white marble on both draws? Are the events independent or dependent?

LESSON 95

Volume of a Right Solid

Power Up

Building Power

facts Power Up T

mental math

a. **Positive/Negative:** (72) + (−100)

b. **Order of Operations/Exponents:** $(2.5 \times 10^6)(2.5 \times 10^6)$

c. **Ratio:** $\frac{60}{100} = \frac{y}{1.5}$

d. **Measurement:** Convert 25°C to degrees Fahrenheit.

e. **Calculation:** $8 \times 2\frac{3}{4}$

f. **Percent:** 50% more than 60

g. **Probability:** There are 6 pairs of black socks, 3 pairs of white socks, and 2 pairs of brown socks. What is the probability of selecting a pair of white socks?

h. **Calculation/Percent:** 10% of 300 is how much more than 20% of 100?

problem solving

Copy this problem and fill in the missing digits:

$$\begin{array}{r} \_\,\_\,\_ \\ \times \quad \_ \\ \hline 1101 \end{array}$$

New Concept

Increasing Knowledge

Math Language
Volume (*V*) is the measure of cubic units of space occupied by a solid figure.

A **right solid** is a geometric solid whose sides are perpendicular to the base. **The volume of a right solid equals the area of the base times the height.** This rectangular solid is a right solid. It is 5 m long and 2 m wide, so the area of the base is 10 m^2.

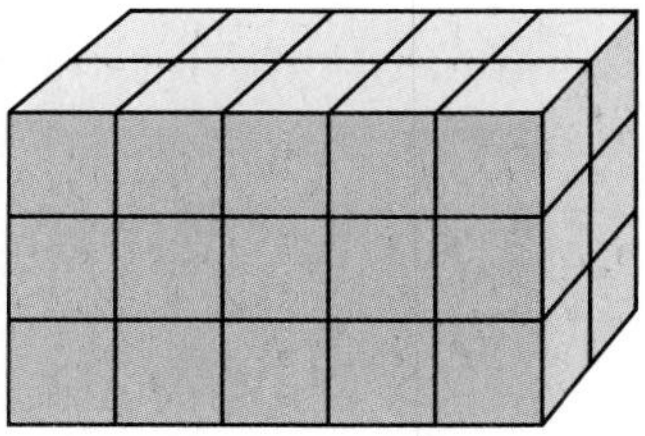

One cube will fit on each square meter of the base, and the cubes are stacked 3 m high, so

$$\begin{aligned} \text{Volume} &= \text{area of the base} \times \text{height} \\ &= 10\text{ m}^2 \times 3\text{ m} \\ &= 30\text{ m}^3 \end{aligned}$$

If the base of the solid is a polygon, the solid is called a **prism.** If the base of a right solid is a circle, the solid is called a **right circular cylinder.**

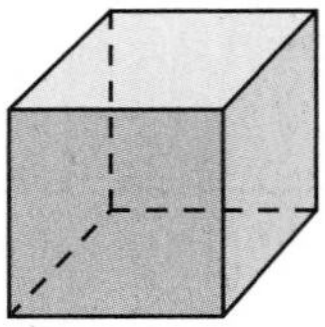

Right square prism

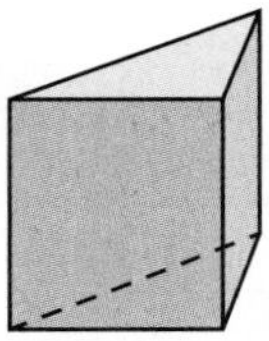

Right triangular prism

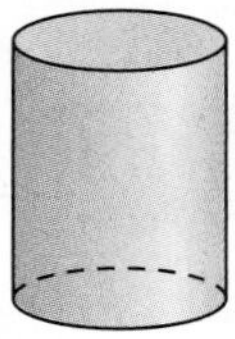

Right circular cylinder

Identify What is another name for a right square prism? What is the shape of the base of each figure shown?

Example 1

Find the volume of the right triangular prism below. Dimensions are in centimeters. We show two views of the prism.

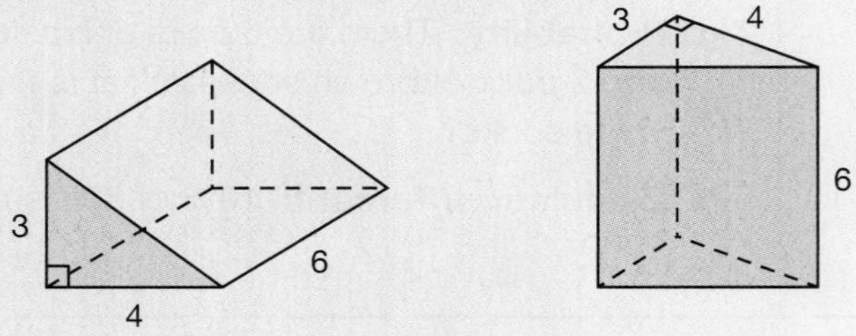

Solution

The area of the base is the area of the triangle.

$$\text{Area of base} = \frac{(4\text{ cm})(3\text{ cm})}{2} = 6\text{ cm}^2$$

The volume equals the area of the base times the height.

$$\text{Volume} = (6\text{ cm}^2)(6\text{ cm}) = \mathbf{36\ cm^3}$$

Example 2

The diameter of this right circular cylinder is 20 cm. Its height is 25 cm. What is its volume? Leave π as π.

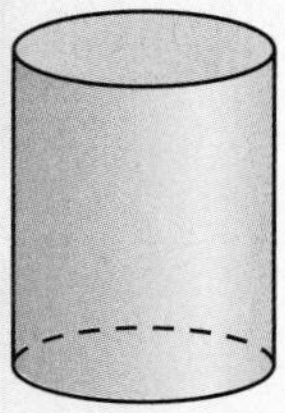

Solution

First we find the area of the base. The diameter of the circular base is 20 cm, so the radius is 10 cm.

$$\text{Area of base} = \pi r^2 = \pi(10\text{ cm})^2 = 100\pi\text{ cm}^2$$

The volume equals the area of the base times the height.

$$\text{Volume} = (100\pi\text{ cm}^2)(25\text{ cm}) = \mathbf{2500\pi\ cm^3}$$

Recall the formula for the volume of a rectangular solid.

$$V = \text{Area of base} \times \text{height } (V = Bh)$$

We can apply this formula to cylinders and prisms as well.

Formulate The base of a cylinder is a circle. Create a specific formula for the volume of a cylinder by replacing B in the formula with the formula for the area of a circle.

Formulate Create a specific formula for the volume of a triangular prism.

Many objects in the real world are composed of combinations of basic shapes like the solids in this lesson. We can calculate the volumes of such objects by dividing them into parts, finding the volume of each part, and then adding the volumes.

Example 3

Eric stores gardening tools in a shed with the dimensions shown. Find the approximate volume of the shed.

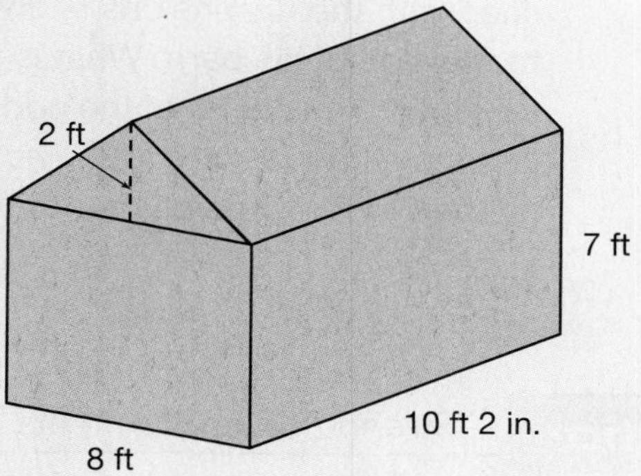

Solution

We separate the structure into a triangular prism and a rectangular prism. We round the dimensions. For example, the notation 10′2″ means 10 feet, 2 inches, which we round to 10 ft.

The volume of the triangular prism is about

$$V = \frac{1}{2}abh$$

$$V = \frac{1}{2}(2\text{ ft})(10\text{ ft})(8\text{ ft}) = 80\text{ ft}^3$$

The volume of the rectangular prism is about

$$V = lwh$$

$$V = (10\text{ ft})(8\text{ ft})(7\text{ ft}) = 560\text{ ft}^3$$

The volume of the shed is about

$$80\text{ ft}^3 + 560\text{ ft}^3 = \mathbf{640\text{ ft}^3}$$

Practice Set

Analyze Find the volume of each right solid shown. Dimensions are in centimeters.

a.

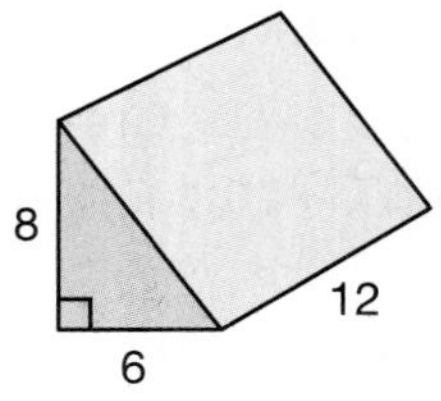

b.

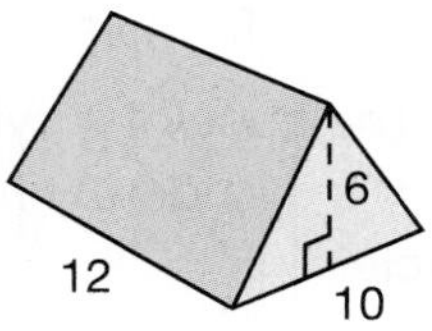

c.

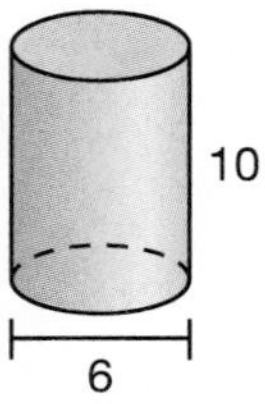

Leave π as π.

d.

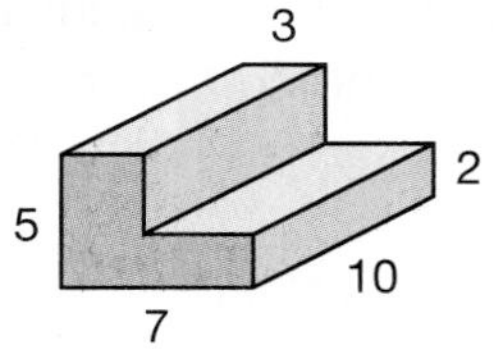

e.

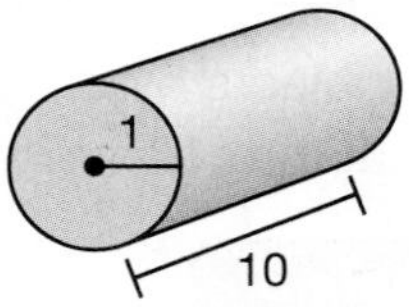

Leave π as π.

f. **List** A farmer added an attached shed with the dimensions shown to the side of his barn. What is the approximate volume of the addition? List the steps and the formulas needed to find the volume.

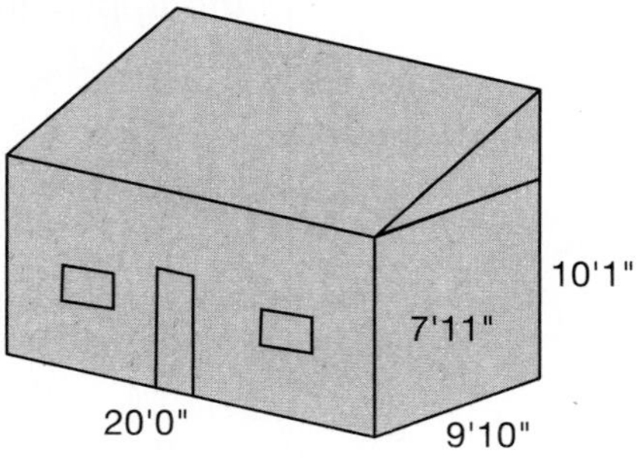

Written Practice

Strengthening Concepts

1. (55) The taxi ride cost $1.40 plus 35¢ for each tenth of a mile. What was the average cost per mile for a 4-mile taxi ride?

2. (Inv. 4) The table at right shows student responses to a survey. The students were asked to rate a television special on a scale from 1.0 to 10.0. Create a box-and-whisker plot for these ratings.

Ratings of Television Special

Rating	Number of Students
10.0	IIII
9.5	卌 I
9.0	卌 III
8.5	卌 II
8.0	III
7.5	I
7.0	I

3. (80) The coordinates of the vertices of $\triangle ABC$ are A (−1, −1), B (−1, −4), and C (−3, −2). The reflection of $\triangle ABC$ in the y-axis is its image $\triangle A'B'C'$.

a. Draw both triangles.

b. Write the coordinates of the vertices of $\triangle A'B'C'$.

4. (54) *If* Gabriela is paid $12 per hour, how much will she earn in 4 hours 20 minutes?

5. (36, 75) In this rectangle, what is the ratio of the shaded area to the unshaded area? How do you know?

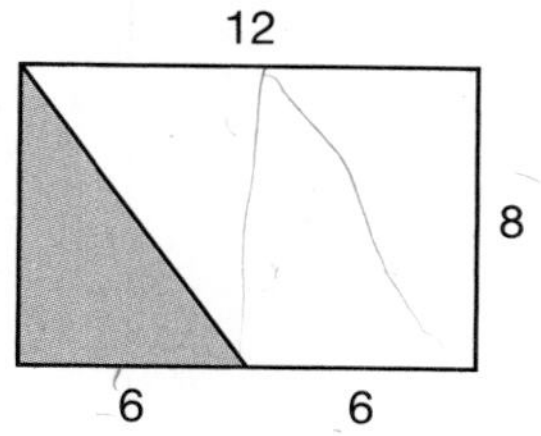

6. (72) If 600 pounds of sand costs $7.20, what would be the cost of 1 ton of sand at the same price per pound?

*** 7.** (92) **Generalize** The cost of production rose 30%. If the new cost is $3.90 per unit, what was the old cost per unit?

*** 8.** (92) **Generalize** If a grocery store marks up cereal 30%, what is the retail price of a large box of cereal that costs the store $3.90?

*** 9.** (88) Use two unit multipliers to convert 1000 mm^2 to square centimeters.

10. (71) Read this statement. Then answer the questions that follow.

Three fifths of the middle school students do not play a musical instrument. The other 60 students do play an instrument.

a. How many middle school students are there?

b. How many middle school students do not play an instrument?

*** 11.** (Inv. 9) If Jim jogs to school at 6 miles per hour the trip takes 10 minutes. If he rides his bike to school at 12 miles per hour, how long does the trip take?

*** 12.** (91) Evaluate: $m(m + n)$ if $m = -2$ and $n = -3$

13. (36) If two number cubes are tossed once, what is the probability that the total number rolled will be

a. 7?

b. a number less than 7?

*** 14.** (95) The diameter of a soup can is 6 cm. Its height is 10 cm. What is the volume of the soup can? (Use 3.14 for π.)

*** 15.** (95) What is the volume of the building shown? Show the two formulas used to find the volume. The peak of the roof is 16 feet above the ground. (Hint: Divide the figure into a rectangular prism and a triangular prism.)

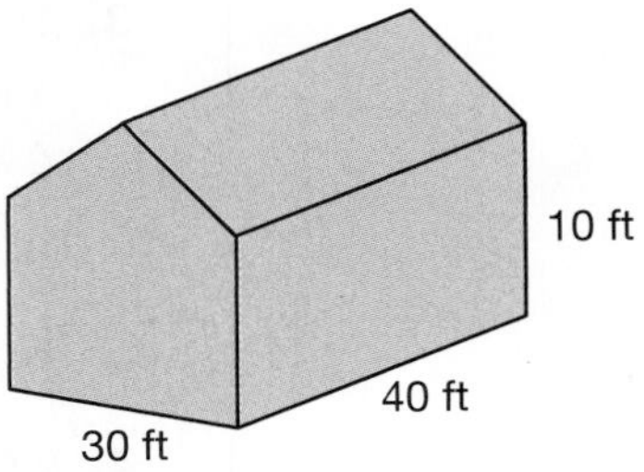

16. (60) Find the total cost, including 6 percent sales tax, of 2 tacos at $2.25 each, 2 fruit drinks at $1.20 each, and a salad at $1.30.

*** 17.** (94) **a.** **Analyze** There are two red, three white, and four blue marbles in a bag. If Chad pulls two marbles from the bag, what is the probability that one will be white and the other blue?

b. **Explain** Will the order of the draws affect the probability. Support your answer.

18. (84, 87) Simplify:

a. $(-2xy)(-2x)(x^2y)$

b. $6x - 4y + 3 - 6x - 5y - 8$

19. (83) Multiply. Write the product in scientific notation.

$$(8 \times 10^{-6})(4 \times 10^{4})$$

*** 20.** (56, Inv. 9) **a.** Use the function rule to complete the table.

b. Graph the function.

c. **Analyze** At what point does the graph of the function intersect the y axis?

$y = \frac{1}{2}x + 1$

x	y
6	
4	
−2	

21. (40) Find the measures of the following angles.

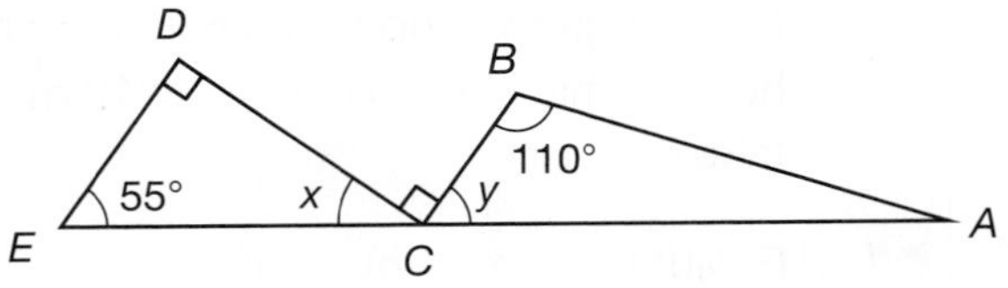

a. $\angle x$ **b.** $\angle y$ **c.** $\angle A$

22. (78) **Represent** On a number line, graph all the negative numbers that are greater than −2.

23. (34) What is the average of the numbers labeled A and B on the number line below?

Justify For problems 24 and 25, solve and check. Show each step.

* **24.** $-5w + 11 = 51$
(93)

* **25.** $\frac{4}{3}x - 2 = 14$
(93)

* **26.** Solve this inequality and graph its solution:
(93)

$$0.9x + 1.2 \leq 3$$

Generalize Simplify:

27. $\frac{10^3 \cdot 10^2}{10^5} - 10^{-1}$
(57)

28. $\sqrt{1^3 + 2^3} + (1 + 2)^3$
(63)

29. $5 - 2\frac{2}{3}\left(1\frac{3}{4}\right)$
(23, 26)

30. **a.** $\frac{(-10) + (-8) - (-6)}{(-2)(+3)}$
(85, 91)

b. $-8 + 3(-2) - 6$

Early Finishers

Real-World Application

The Li family has four children born in different years.

a. List all possible orders of boy and/or girl children that the Li family could have. Use B for boy and G for girl.

b. Find the probability that a family with four children has exactly two boys and two girls in any order.

LESSON 96

• Estimating Angle Measures
• Distributive Property with Algebraic Terms

Power Up — *Building Power*

facts Power Up U

mental math

a. **Positive/Negative:** $(-27) - (-50)$

b. **Order of Operations/Exponents:** $(5 \times 10^5)(2 \times 10^7)$

c. **Algebra:** $160 = 80 + 4y$

d. **Measurement:** Convert 15°C to degrees Fahrenheit.

e. **Calculation:** $9 \times 1\frac{2}{3}$

f. **Percent:** 25% more than $80

g. **Geometry:** A rectangular solid measures 5 in. × 3 in. × 2 in. What is the volume of the solid?

h. **Percent/Estimation:** Estimate 15% of $49.75.

problem solving At a recent banquet, every two guests shared a dish of chicken, every three guests shared a dish of rice, and every four guests shared a dish of vegetables. If there were a total of 65 dishes in all, how many guests were at the banquet?

New Concepts — *Increasing Knowledge*

estimating angle measures

In this lesson we will learn a technique for estimating the measure of an angle. To estimate angle measures, we need a mental image of a degree scale—a mental protractor. We can "build" a mental image of a protractor from a mental image we already have—the face of a clock.

The face of a clock is a full circle, which is 360°, and is divided into 12 numbered divisions that mark the hours. From one numbered division to the next is $\frac{1}{12}$ of a full circle. One twelfth of 360° is 30°. Thus the measure of the angle formed by the hands of a clock at 1 o'clock is 30°, at 2 o'clock is 60°, and at 3 o'clock is 90°.

A clock face is further divided into 60 smaller divisions that mark the minutes. From one small division to the next is $\frac{1}{60}$ of a circle, and $\frac{1}{60}$ of 360° is 6°.

Thus, **from one minute mark to the next on the face of a clock is 6°.**

Here we have drawn an angle on the face of a clock. The vertex of the angle is at the center of the clock. One side of the angle is set at 12, and the other side of the angle is at "8 minutes after."

Since each minute of separation represents 6°, the measure of this angle is 8 × 6°, which is 48°. With some practice we can usually estimate the measure of an angle to within 5° of its actual measure.

Example 1

a. Estimate the measure of ∠*BOC* in the figure below.

b. Use a protractor to find the measure of ∠*BOC*.

c. By how many degrees did your estimate differ from your measurement?

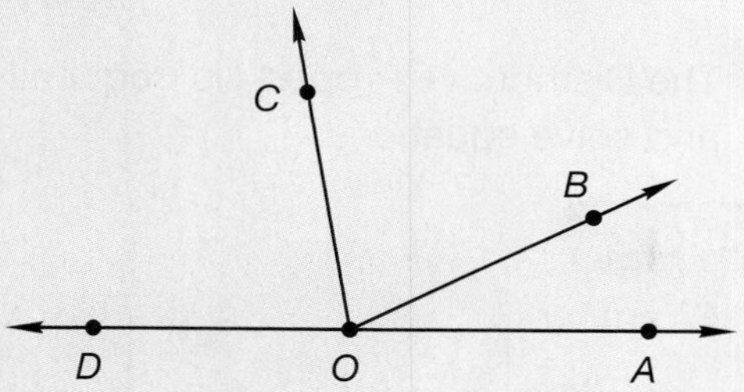

Solution

a. We use a mental image of a clock face on ∠*BOC* with $\overrightarrow{OC}$ set at 12. Mentally we see that $\overrightarrow{OB}$ falls more than 10 minutes "after." Perhaps it is 12 minutes after. Since 12 × 6° = 72°, we estimate that m∠*BOC* is **72°.**

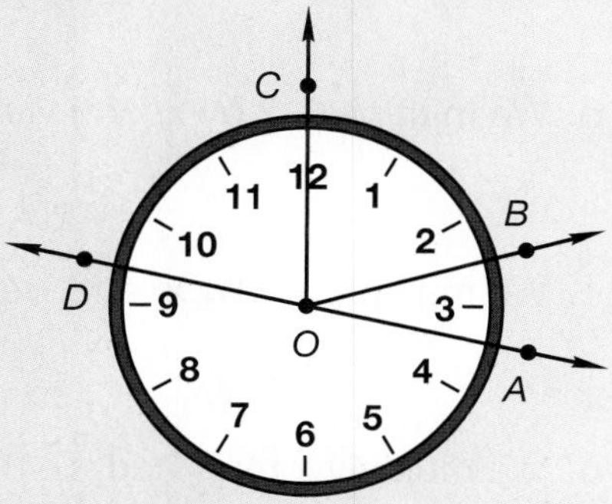

b. We trace $\angle BOC$ on our paper and extend the sides so that we can use a protractor. We find that m$\angle BOC$ is **75°.**

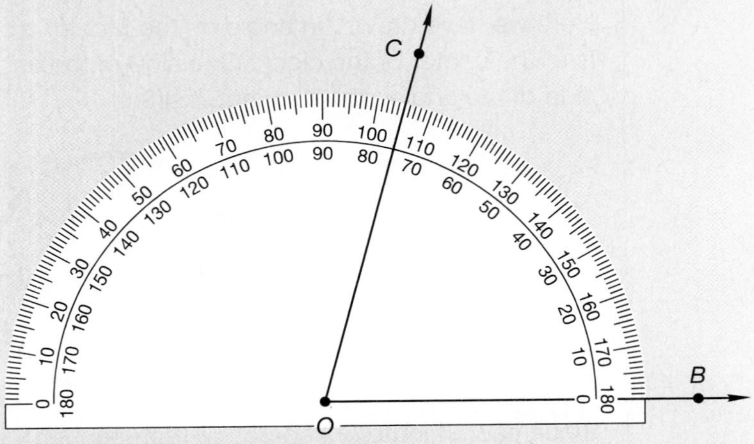

c. Our estimate of 72° differs from our measurement of 75° by **3°.**

Discuss A student estimated that m$\angle DOC$ was about 36°. Without measuring this angle, do you think that this estimate is close to the actual measure? Explain.

distributive property with algebraic terms

Recall from Lesson 41 that the Distributive Property "spreads" multiplication over terms that are algebraically added. We illustrated the Distributive Property with this equation:

$$a(b + c) = ab + ac$$

The Distributive Property is frequently used in algebra to simplify expressions and solve equations.

Example 2

Simplify:

a. $\mathbf{2(x - 3)}$ **b.** $\mathbf{-2(x + 3)}$

c. $\mathbf{-2(x - 3)}$ **d.** $\mathbf{x(x - 3)}$

Solution

Thinking Skills

Analyze

Why is the simplified expression in **b** $-2x - 6$ and not $-2x + 6$?

a. We multiply 2 by x, and we multiply 2 by -3.

$$2(x - 3) = \mathbf{2x - 6}$$

b. We multiply -2 by x, and we multiply -2 by 3.

$$-2(x + 3) = \mathbf{-2x - 6}$$

c. We multiply -2 by x, and we multiply -2 by -3.

$$-2(x - 3) = \mathbf{-2x + 6}$$

d. We multiply x by x, and we multiply x by -3.

$$x(x - 3) = \mathbf{x^2 - 3x}$$

Example 3

Simplify: $x^2 + 2x + 3(x - 2)$

Solution

Math Language
Like terms have identical variables.

We first use the Distributive Property to clear parentheses. Then we add like terms.

$x^2 + 2x + 3(x - 2)$	expression
$x^2 + 2x + 3x - 6$	Distributive Property
$x^2 + (2x + 3x) - 6$	Associative Property
$\mathbf{x^2 + 5x - 6}$	added $2x$ and $3x$

Example 4

Simplify: $x^2 + 2x - 3(x - 2)$

Solution

In this example multiply $x - 2$ by -3.

$x^2 + 2x - 3(x - 2)$	Given
$x^2 + 2x - 3x + 6$	Distributive Property
$x^2 + (2x - 3x) + 6$	Associative Property
$\mathbf{x^2 - x + 6}$	$2x - 3x = -x$

Practice Set

Analyze By counting minute marks on the clock face below, find the measure of each angle:

a. $\angle AOB$ **b.** $\angle AOC$ **c.** $\angle AOD$

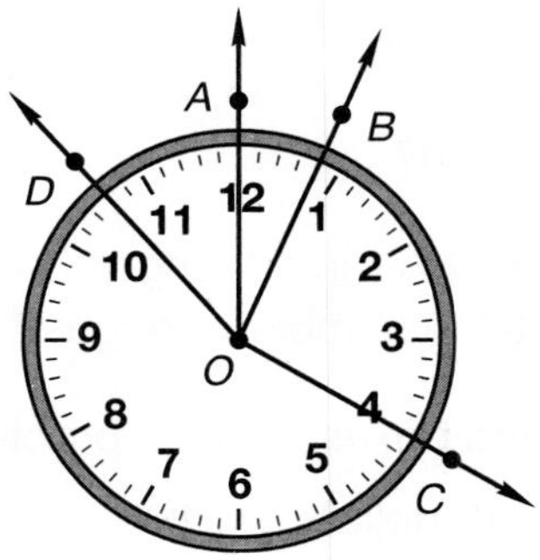

Simplify:

d. $x(x - y)$ **e.** $-3(2x - 1)$

f. $-x(x - 2)$ **g.** $-2(4 - 3x)$

Justify Simplify. Show and justify each step.

h. $x^2 + 2x - 3(x + 2)$ **i.** $x^2 - 2x - 3(x - 2)$

Estimate In practice problems **j–m,** estimate the measure of each angle. Then use a protractor to measure each angle. By how many degrees did your estimate differ from your measurement?

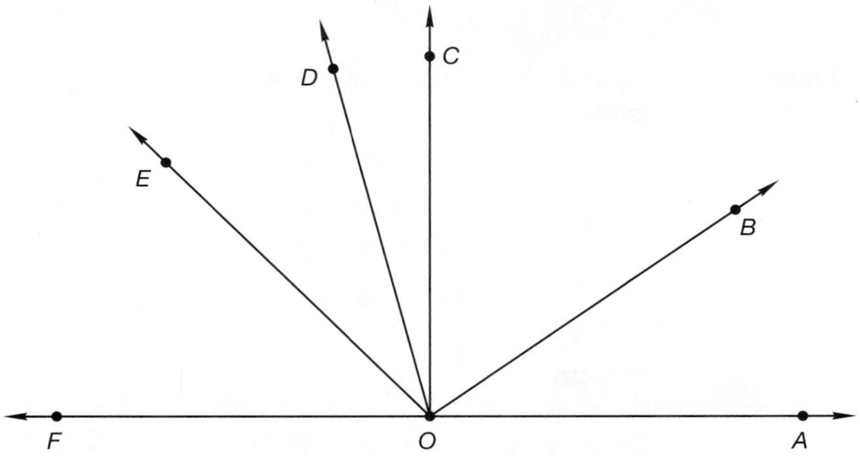

j. $\angle BOC$ **k.** $\angle DOC$

l. $\angle FOE$ **m.** $\angle FOB$

Written Practice *Strengthening Concepts*

1. (55) In May the merchant bought 3 bikes for an average price of \$280 per bike. In June the merchant bought 5 bikes at an average price of \$240 per bike. What was the average price of all the bikes the merchant bought in May and June?

2. (20) What is the quotient when 9 squared is divided by the square root of 9?

3. (28) The Adams' car has a 16-gallon gas tank. How many tanks of gas will the car use on a 2000-mile trip if the car averages 25 miles per gallon?

4. (36) **Explain** In a triangular prism, what is the ratio of the number of vertices to the number of edges? Tell how you found your answer.

Use ratio boxes to solve problems **5–7.**

5. (72) If 12 dollars can be exchanged for 100 yuan, what is the cost in dollars of an item that sells for 475 yuan?

* **6.** (92) Sixty is 20 percent less than what number?

* **7.** (92) The average number of customers per day increased 25 percent during the sale. If the average number of customers per day before the sale was 120, what was the average number of customers per day during the sale?

8. (77) Write equations to solve these problems:

a. Sixty is what percent of 50?

b. Fifty is what percent of 60?

9. (88, Inv. 9) **a.** A college student earns 12 dollars per hour at a part-time job. Use two unit multipliers to convert this pay rate into cents per min.

b. Is the student's pay directly proportional or inversely proportional to the number of hours worked?

10. (18, 40) Triangle *ABC* is similar to triangle *EDC*.

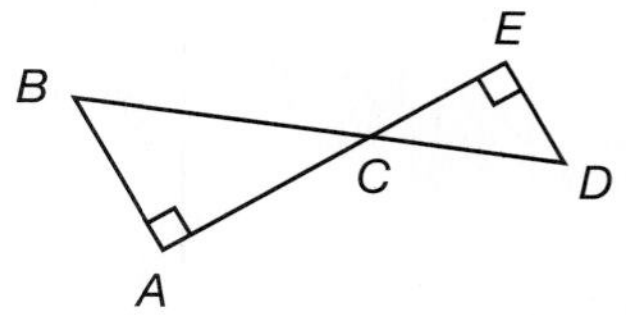

a. List three pairs of corresponding angles and three pairs of corresponding sides.

b. If $m\angle ABC = 53°$, what is $m\angle ECD$?

*** 11.** (Inv. 9) **Analyze** Ralph has 48 square floor tiles which he wants to arrange into a rectangle. Here is one arrangement that has 3 rows of 16 tiles.

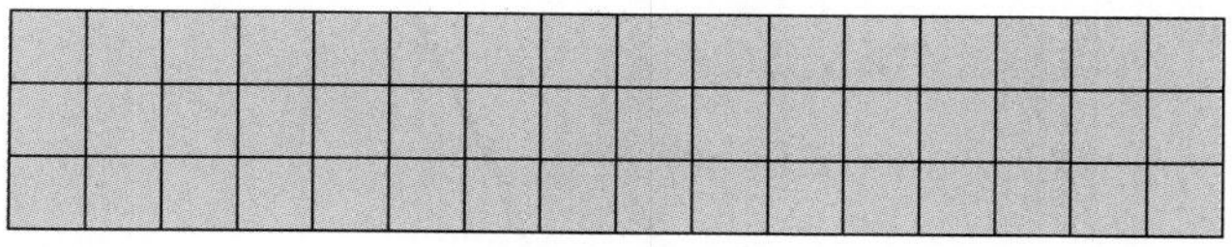

Describe the arrangement of 48 tiles that makes a rectangle with the smallest perimeter.

12. (91) Evaluate: $c(a + b)$ if $a = -4$, $b = -3$, and $c = -2$

13. (16, 20) The perimeter of a certain square is 1 yard. Find the area of the square in square inches.

14. (Inv. 8) The face of this spinner is divided into one half and two fourths.

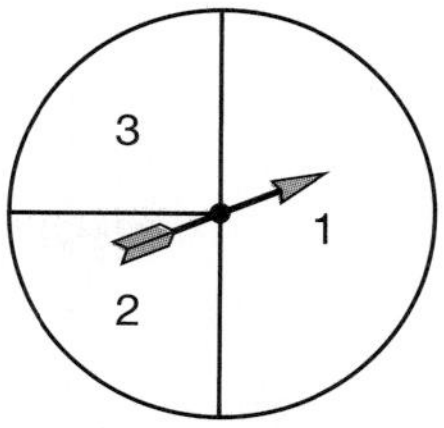

a. If the spinner is spun two times, what is the probability that it will stop on 3 both times?

b. If the spinner is spun four times, what is the probability that it will stop on 1 four times in a row?

*** 15.** (95) **Generalize** Find the volume of each solid. Dimensions are in centimeters.

a.

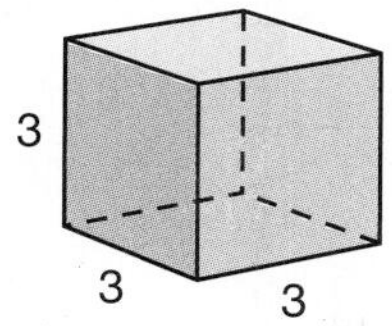

b.

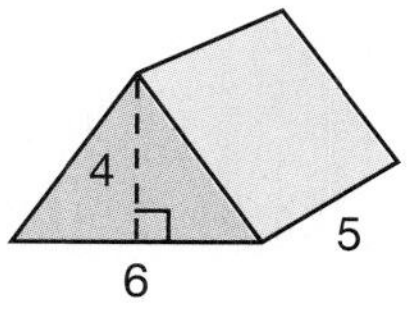

16. (60) Find the total price, including 7 percent sales tax, of 20 square yards of carpeting priced at $14.50 per square yard.

17. (48) Complete the table.

Fraction	Decimal	Percent
a.	b.	$3\frac{3}{4}\%$

*** 18.** (60, 92) Raincoats regularly priced at $24 were on sale for $33\frac{1}{3}\%$ off.

a. What is $33\frac{1}{3}\%$ of $24?

b. What is $33\frac{1}{3}\%$ less than $24?

19. Multiply. Write the product in scientific notation.
(83)

$$(3 \times 10^3)(8 \times 10^{-8})$$

20. **a.** Find the circumference of the circle at right.
(65, 82)

b. Find the area of the circle.

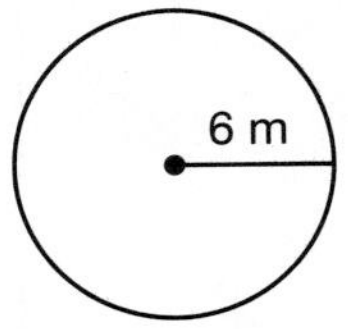

Leave π as π.

*** 21.** **Estimate** Use the clock face to estimate the measure of each angle:
(96)

a. $\angle BOC$

b. $\angle COA$

c. $\angle DOA$

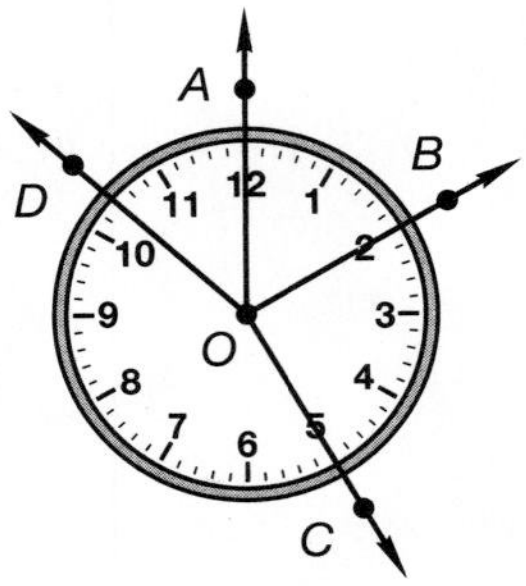

*** 22.** **a.** **Conclude** What is the measure of each exterior angle of a regular octagon?
(89)

b. What is the measure of each interior angle of a regular octagon?

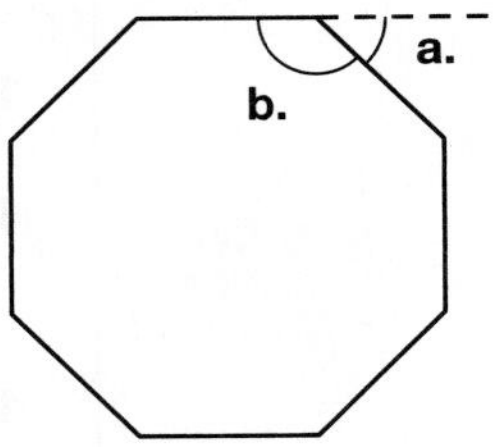

23. Find the coordinates of the vertices of □ $Q'R'S'T'$, which is the image of □ $QRST$, translated 4 units right and 4 units down.
(80)

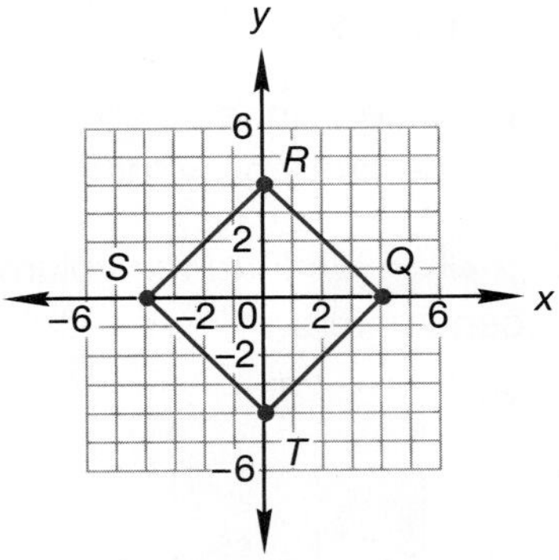

*** 24.** **Represent** Solve this inequality and graph its solution:
(93)

$$0.8x + 1.5 < 4.7$$

Justify For problems **25** and **26,** solve and check. Show each step.

* **25.** (93) $2\frac{1}{2}x - 7 = 13$

* **26.** (93) $-3x + 8 = 14$

Generalize Simplify:

* **27.** (96) **a.** $-3(x - 4)$

b. $x(x + y)$

28. (85) **a.** $\frac{(-4) - (-8)(-3)(-2)}{-2}$

b. $(-3)^2 + 3^2$

29. (87) **a.** $(-4ab^2)(-3b^2c)(5a)$

b. $a^2 + ab - ab - b^2$

* **30.** (94) **Analyze** Gloria had two sets of alphabet cards (A–Z). She mixed the two sets together to form a single stack of cards. Then Gloria drew three cards from the stack without replacing them. What is the probability that she drew either a card with an L or a card with an R all three times?

Early Finishers
Real-World Application

A fuel plant has 2 cylindrical fuel tanks that are 85 ft tall, with a diameter of 32 ft. You may use a calculator for the following questions.

a. How many cubic feet can both tanks hold? Use $\pi = 3.14$.

b. To the nearest whole gallon, how many gallons can both tanks hold? Note: one cubic foot is approximately 7.48 gallons.

LESSON **97**

• Similar Triangles
• Indirect Measure

Power Up — *Building Power*

facts Power Up T

mental math

a. **Positive/Negative:** $(-5)(-5)(-5)$

b. **Order of Operations/Exponents:** $(5 \times 10^6)(6 \times 10^5)$

c. **Ratio:** $\frac{m}{4.5} = \frac{0.6}{3}$

d. **Measurement:** Convert 5°C to degrees Fahrenheit.

e. **Calculation:** $10 \times 6\frac{1}{2}$

f. **Percent:** 25% less than \$80

g. **Measurement:** One cup is what fraction of a pint?

h. **Calculation:** $\sqrt{100}$, $\times$ 7, + 2, ÷ 8, $\times$ 4, $\sqrt{\ }$, $\times$ 5, − 2, ÷ 2, ÷ 2, ÷ 2

problem solving

Six identical blocks marked *x*, a 1.7-lb weight, and a 4.3-lb weight were balanced on a scale as shown. Write an equation to represent this balanced scale and find the weight of each block marked *x*.

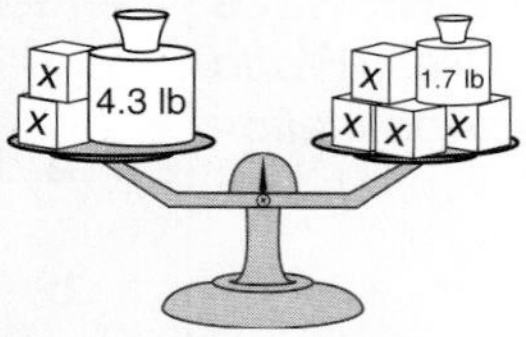

New Concepts — *Increasing Knowledge*

similar triangles

We often use tick marks to indicate congruent angles.

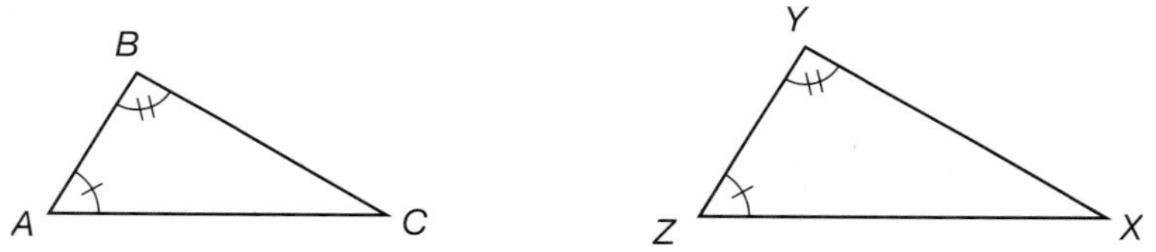

In the triangles above, the single tick marks indicate that angles *A* and *Z* have equal measures. The double tick marks indicate that angles *B* and *Y* have equal measures. Can we conclude that angles *C* and *X* also have equal measures?

Recall from Lesson 18 that corresponding angles of similar figures have the same measure. The converse is also true. If three angles in one triangle have the same measures as three angles in another triangle, the triangles are *similar triangles*. So triangles *ABC* and *ZYX* are similar. Also recall that similar triangles have three pairs of corresponding angles and three pairs of corresponding sides. On the following page, we show the corresponding angles and sides for the triangles above:

Corresponding Angles	Corresponding Sides
$\angle A$ and $\angle Z$	$\overline{AB}$ and $\overline{ZY}$
$\angle B$ and $\angle Y$	$\overline{BC}$ and $\overline{YX}$
$\angle C$ and $\angle X$	$\overline{CA}$ and $\overline{XZ}$

In this lesson we will focus our attention on the following characteristic of similar triangles:

The lengths of corresponding sides of similar triangles are proportional.

This means that ratios formed by corresponding sides are equal, as we illustrate with the two triangles below.

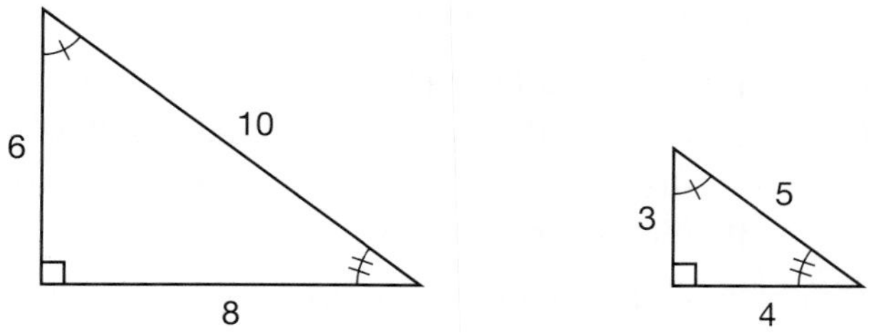

Triangle *a* Triangle *b*

The lengths of the corresponding sides of triangles *a* and *b* are 6 and 3, 8 and 4, and 10 and 5. These pairs of lengths can be written as equal ratios.

$$\frac{\text{triangle } a}{\text{triangle } b} \quad \frac{6}{3} = \frac{8}{4} = \frac{10}{5}$$

Notice that each of these ratios equals 2. If we choose to put the lengths of the sides of triangle *b* on top, we get three ratios, each equal to $\frac{1}{2}$.

$$\frac{\text{triangle } b}{\text{triangle } a} \quad \frac{3}{6} = \frac{4}{8} = \frac{5}{10}$$

Thinking Skills

Explain

Do triangles have to be the same size to be similar?

We can write proportions using equal ratios in order to find the lengths of unknown sides of similar triangles.

Example 1

Estimate the length *a*. Then use a proportion to find *a*.

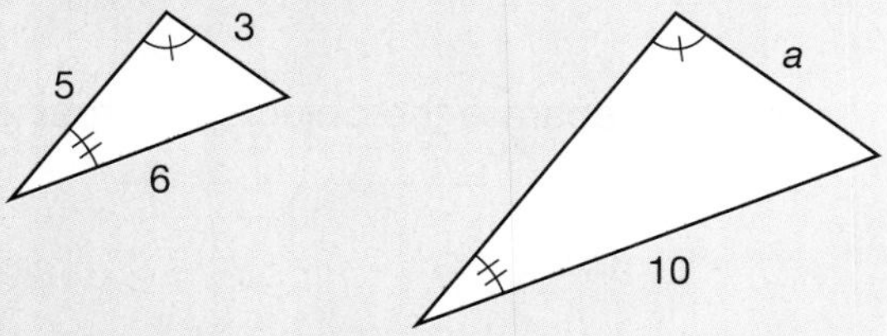

Solution

The tick marks indicate two pairs of congruent angles in the triangles. Because the sum of the interior angles of every triangle is 180°, we know that the unmarked angles are also congruent. Thus the two triangles are similar, and the lengths of the corresponding sides are proportional.

The side of length 6 in the smaller triangle corresponds to the side of length 10 in the larger triangle. Thus the side lengths of the larger triangle are not quite double the side lengths of the smaller triangle. Since the side of length *a* in the larger triangle corresponds to the side of length 3 in the smaller triangle, *a* should be a little less than 6. We estimate *a* to be 5.

We now use corresponding sides to write a proportion and solve for *a*. We decide to write the ratios so that the sides from the smaller triangle are on top.

$$\frac{6}{10} = \frac{3}{a} \quad \text{equal ratios}$$

$$6a = 30 \quad \text{cross multiplied}$$

$$a = \mathbf{5} \quad \text{solved}$$

indirect measure

Sarah looked up and said, "I wonder how tall that tree is." Beth looked down and said, "It's about 25 feet tall." Beth did not *directly* measure the height of the tree. Instead she used her knowledge of proportions to *indirectly* estimate the height of the tree.

The lengths of the shadows cast by two objects are proportional to the heights of the two objects (assuming the objects are in the same general location).

We can separate the objects and their shadows into two "triangles."

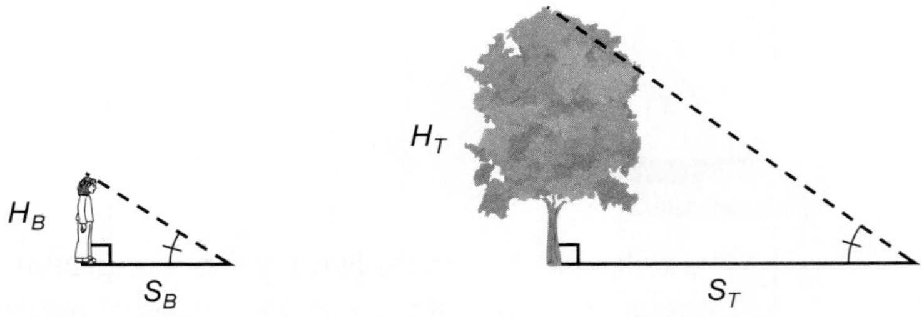

We assume that the ground is flat and level, and that both the tree and Beth are perpendicular to the ground. We also assume that the angle of the Sun's light is the same for both Beth and the tree. Thus the triangles are similar.

Thinking Skill

Explain

In indirect measurement, why must the light strike the objects at the same angle?

Discuss Why do we have to make these assumptions?

The height of Beth (H_B) and the length of Beth's shadow (S_B) are proportional to the height of the tree (H_T) and the length of the tree's shadow (S_T). We can record the relationship in a ratio box.

	Height of Object	Length of Shadow
Beth	H_B	S_B
Tree	H_T	S_T

How did Beth perform the calculation? We suggest two ways. Knowing her own height (5 ft), she may have estimated the length of her shadow (6 ft) and the length of the shadow of the tree (30 ft). She then could have solved this proportion in which the dimensions are feet:

$$\frac{5}{H_T} = \frac{6}{30}$$

$$6H_T = 5 \cdot 30$$

$$H_T = 25$$

Another way Beth may have estimated the height of the tree is by estimating that the tree's shadow was five times as long as her own shadow. If the tree's shadow was five times as long as her shadow, then the tree's height must have been five times her height.

$$5 \text{ ft} \times 5 = 25 \text{ ft}$$

Example 2

To indirectly measure the height of a telephone pole on the playground, some students measured the length of the shadow cast by the pole. The shadow measured 24 ft and the length of the shadow cast by a vertical meterstick was 40 cm. About how tall was the telephone pole?

Solution

We sketch the objects and their shadows using the given information.

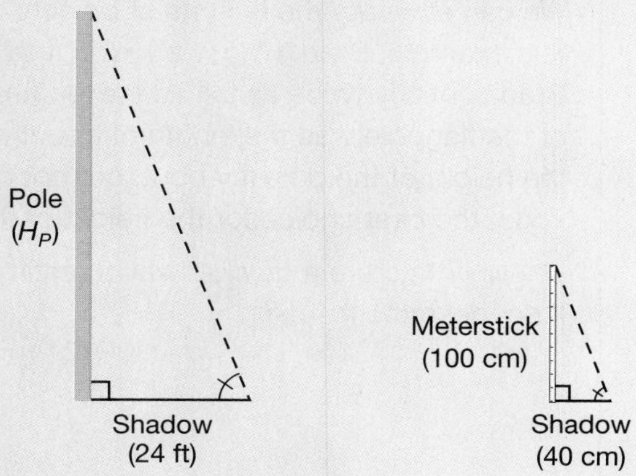

Note: Figures not drawn to scale.

We use a ratio box to record the given information. Then we write and solve a proportion.

	Height of Object	Length of Shadow
Meterstick	100 cm	40 cm
Pole	H_P	24 ft

$$\frac{100}{H_P} = \frac{40}{24}$$

$$40\,H_P = 24 \cdot 100$$

$$H_P = 60$$

We find that the height of the telephone pole was about **60 feet.** Note that the two objects were measured using different units. Exercise caution whenever mixing units in proportions to ensure the solution is expressed in the desired units. You may choose to include units in your calculation, as we have done below.

$$\frac{100 \text{ cm}}{H_P} = \frac{40 \text{ cm}}{24 \text{ ft}}$$

$$40 \text{ cm} \cdot H_P = 24 \text{ ft} \cdot 100 \text{ cm}$$

$$H_P = \frac{\overset{6}{\cancel{24}} \text{ ft} \cdot \overset{10}{\cancel{100}\,\cancel{\text{cm}}}}{\underset{1}{\cancel{\underset{\cancel{4}}{40}}}\,\cancel{\text{cm}}}$$

$$H_P = 60 \text{ ft}$$

Example 3

Brad saw that his shadow was the length of about one big step (3 ft). He then walked along the shadow of a nearby flagpole and found that it was eight big steps long. Brad is about 5 ft tall. Which is the best choice for the height of the flagpole?

A **15 ft** **B** **24 ft** **C** **40 ft** **D** **60 ft**

Solution

We can estimate the heights of objects by the lengths of their shadows. In this example, Brad, who is about 5 ft tall, cast a shadow about 3 ft long. So Brad is nearly twice as tall as the length of the shadow he cast. The shadow of the flagpole was the length of about eight big steps, or about 24 ft. So the height of the pole is about, but not quite, twice the length of its shadow. Thus, the best choice for the height of the pole is **C 40 ft.**

Discuss Is there a quicker way to estimate that the pole is about 40 feet tall? If so, describe it.

Activity

Indirect Measure

Materials: metersticks, rulers and/or tape measures.

Have small groups of students indirectly measure the height of a tree, building, pole, or other tall object as described in example 2.

Discuss If different groups measure the same object, discuss why and how answers differ.

Practice Set

a. **Analyze** Identify each pair of corresponding angles and each pair of corresponding sides in these two triangles:

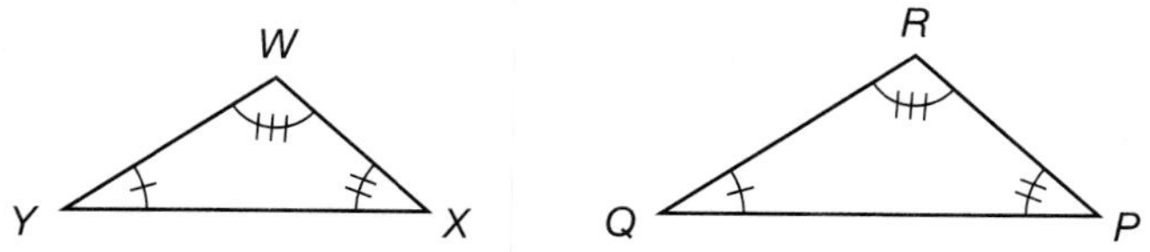

Refer to the figures shown to answer problems **b–e.**

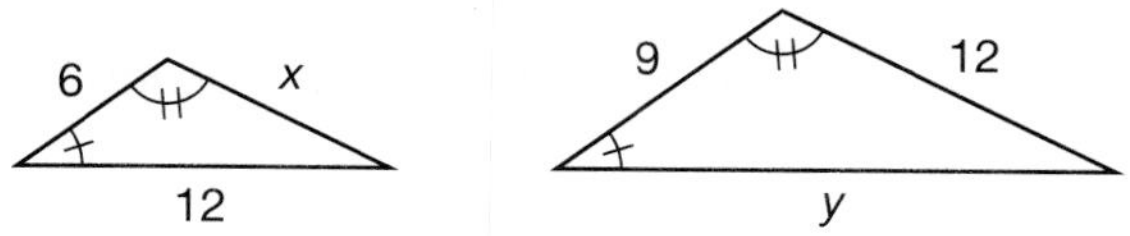

b. Estimate the length x.

c. Find the length x.

d. Estimate the length y.

e. Find the length y.

f. **Model** A tree casts a shadow 18 feet long, while a 6-foot pole casts a shadow 9 feet long. How tall is the tree? Draw a diagram to illustrate the problem.

g. **Estimate** As Donald stood next to a pole supporting a basketball hoop, he noticed that the shadow of the pole was about twice the length of his own shadow. If Donald is 5 ft 6 in. tall, what is a reasonable estimate of the height of the pole?

Written Practice

Strengthening Concepts

1. *(60)* Marta gave the clerk $10 for a CD that was on sale for $8.95 plus 6 percent tax. How much money should she get back?

2. *(51)* Three hundred billion is how much less than two trillion? Write the answer in scientific notation.

3. (Inv. 4) The table shows how much protein is in one serving of different foods made from grain.

Food	Protein (grams)
Oatmeal	6
Shredded Wheat	3
Bagel (plain)	7
White Bread	2
Whole-wheat Bread	3
Bran Flakes	4
Corn Flakes	2
Pasta	7
Bran Muffin	3

a. Find the median, mode, and range of the data.

b. Which combination would give you more protein, one serving of white bread and one serving of pasta, or one bran muffin and one bagel?

Use ratio boxes to solve problems **4** and **5.**

4. (72) Coming down a long hill on their bikes, members of a cycling club averaged 18 miles per hour. If it took them 2 minutes to come down the hill, how long was the hill?

5. (72) If Nelson biked 2640 yards in 5 minutes, how far could he bike in 8 minutes at the same rate?

* **6.** (80) **Analyze** Describe the transformation that moves $\triangle ABC$ to its image $\triangle A'B'C'$.

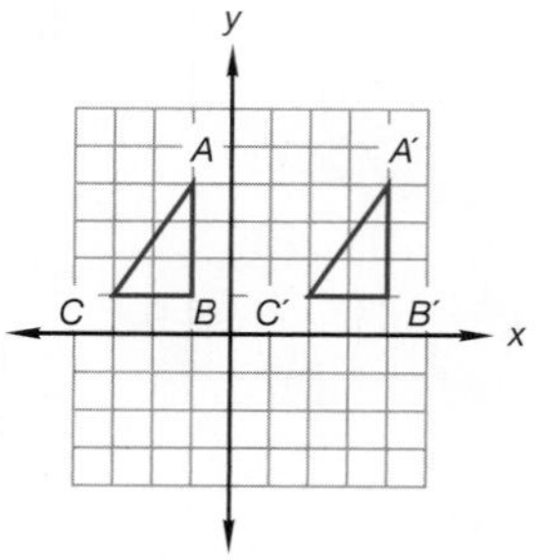

7. (16, 60) Three fourths of a yard is how many inches?

8. (53) Use a ratio box to solve this problem. The ratio of leeks to radishes growing in a garden was 5 to 7. If 420 radishes were growing in the garden, how many leeks were there?

Write equations to solve problems **9–11.**

9. (77) Forty is 250 percent of what number?

10. (77) Forty is what percent of 60?

11. (60) What decimal number is 40 percent of 6?

* **12.** (92) **Model** Use a ratio box to solve this problem. The price of one model of car increased 10 percent in one year. If the price this year is \$17,600, what was the price last year?

13. (28, 34) What is the average of the two numbers marked by arrows on the number line below?

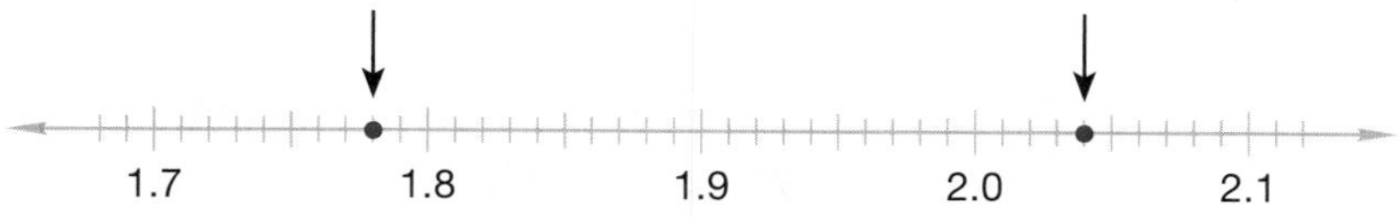

14. (48) Complete the table.

Fraction	Decimal	Percent
a.	3.25	**b.**
$\frac{1}{6}$	**c.**	**d.**

15. (Inv. 9) If Betty walks to school at 3 miles per hour the trip takes 15 minutes. How long does the trip take if she jogs to school at 5 miles per hour?

16. (83) Multiply. Write the product in scientific notation.

$$(5.4 \times 10^8)(6 \times 10^{-4})$$

17. (62, 82) **Connect** Find the **a** circumference and **b** area of a circle with a radius of 10 millimeters. (Use 3.14 for π.)

18. (75) Find the area of the trapezoid shown. Dimensions are in feet.

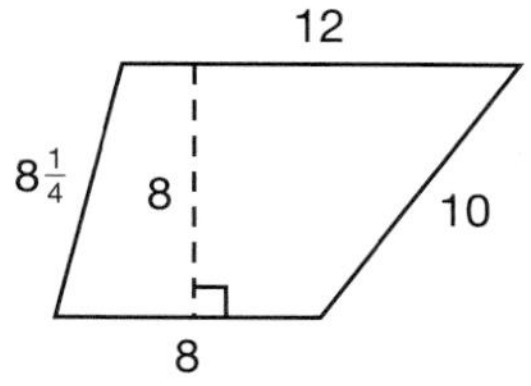

* **19.** (95) Find the volume of each of these right solids. Dimensions are in meters. (Leave π as π.)

a.

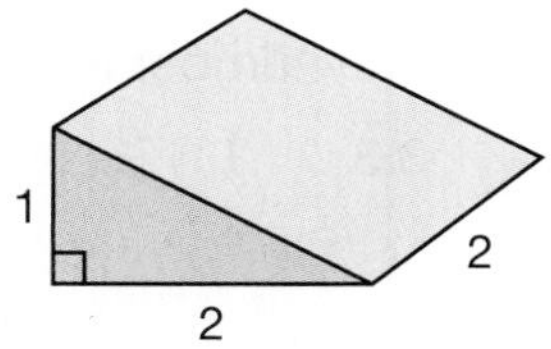

b.

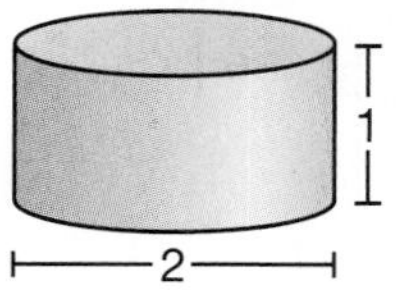

20. (40) Refer to the figure below. What are the measures of the following angles?

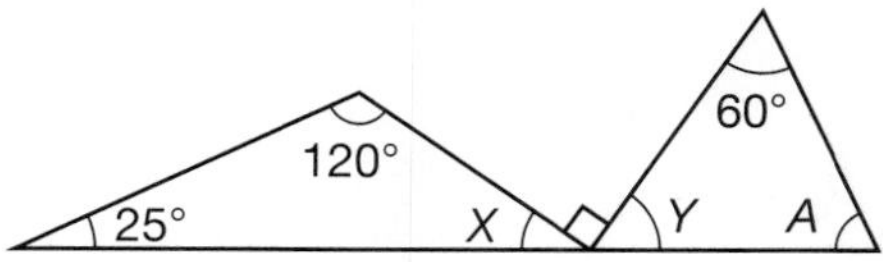

a. $\angle X$ **b.** $\angle Y$ **c.** $\angle A$

* **21.** (97) **Generalize** The triangles below are similar. Find the length of x. Dimensions are in centimeters.

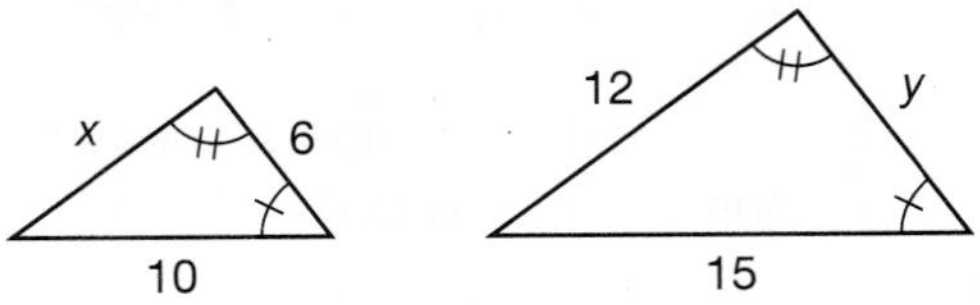

* **22.** (97) **Model** How tall is a power pole that casts a 72-foot shadow if a nearby vertical yardstick casts a 48-inch shadow? Draw a diagram to illustrate the problem.

23. (29) Estimate: $\frac{(38,470)(607)}{79}$

Generalize For problems **24** and **25**, solve and check. Show each step.

* **24.** (93) $1.2m + 0.12 = 12$

* **25.** (93) $1\frac{3}{4}y - 2 = 12$

Generalize Simplify:

26. (84) $3x - y + 8 + x + y - 2$

* **27.** (96) **a.** $3(x - y)$

b. $x(y - 3)$

28. (43) $3\frac{1}{3} \div \left(4.5 \div 1\frac{1}{8}\right)$

29. (85) $\frac{(-2) - (+3) + (-4)(-3)}{(-2) + (+3) - (+4)}$

* **30.** (94) **Analyze** Sam placed five alphabet cards face down on a table. Then he asked students to pick up two cards.

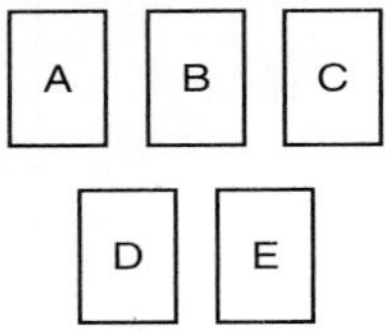

a. What is the probability that a student will pick up two vowels?

b. What is the probability the student will pick up two consonants?

Early Finishers
Math Applications

Draw a 0 to 4 number line and divide each unit into fourths. Then, estimate the position of each number on the number line.

$\frac{1}{2}$ 2.25 $3.\overline{3}$ 0.125 $3\frac{3}{4}$ $2.8\overline{3}$

LESSON
98

• Scale
• Scale Factor

Power Up

Building Power

facts

Power Up U

mental math

a. **Positive/Negative:** (–360) ÷ (8)

b. **Order of Operations/Exponents:** $(2.5 \times 10^7)(4 \times 10^{-2})$

c. **Algebra:** $2c + 1\frac{1}{2} = 6\frac{1}{2}$

d. **Measurement:** Convert 0.02 kg to g.

e. **Calculation:** $4 \times 3\frac{3}{4}$

f. **Percent:** $33\frac{1}{3}\%$ more than $60

g. **Measurement:** A centimeter is what fraction of a meter?

h. **Rate:** At 12 mph, how far can Toby ride a bike in 1 hour and 45 minutes?

problem solving

If two people shake hands, one handshake takes place. If three people all shake hands with one another, three handshakes take place. If four people all shake hands with one another, we can determine how many handshakes take place by drawing four dots (for people) and connecting the dots with segments (for handshakes). Then we count the segments (six). Use this method to count the number of handshakes that would take place between five people.

New Concepts

Increasing Knowledge

scale

In the preceding lesson we discussed similar triangles. Scale models and scale drawings are other examples of similar shapes. Scale models and scale drawings are reduced (or enlarged) renderings of actual objects. As is true of similar triangles, the lengths of corresponding parts of scale models and the objects they represent are proportional.

The **scale** of a model is stated as a ratio. For instance, if a model airplane is $\frac{1}{24}$ the size of the actual airplane, the scale is stated as $\frac{1}{24}$ or 1:24. We can use the given scale to write a proportion to find a measurement on either the model or the actual object. A ratio box helps us put the numbers in the proper places.

Example 1

A model airplane is built with a scale of 1:24. If the wingspan of the model is 18 inches, the wingspan of the actual airplane is how many feet?

Solution

Thinking Skill

Explain

If you made a model with a scale of 3:1, would the model be larger or smaller than the original object? How do you know?

The scale indicates that the dimensions of the actual airplane are 24 times the dimensions of the model, so the wingspan is 24 times 18 inches. Note that the model is measured in inches, but the question asks for the wingspan in feet. Thus we will divide the product of 18 and 24 by 12. This is the calculation:

$$\frac{18 \cdot 24}{12} = 36$$

The wingspan is 36 feet.

Another way to approach this problem is to construct a ratio box as we have done with ratio problems. In one column we write the ratio numbers, which are the scale numbers. In the other column we write the measures. The first number of the scale refers to the model. The second number refers to the object. We can use the entries in the ratio box to write a proportion.

	Scale	Measure
Model	1	18
Object	24	w

$$\frac{1}{24} = \frac{18}{w}$$

$$w = 432$$

The wingspan of the model was given in inches. Solving the proportion, we find that the full-size wingspan is 432 inches. We are asked for the wingspan in feet, so we convert units from inches to feet.

$$432 \cancel{\text{in.}} \cdot \frac{1 \text{ ft}}{12 \cancel{\text{in.}}} = 36 \text{ ft}$$

We find that the wingspan of the airplane is **36 feet.**

Discuss Could you find the actual length of the airplane if you were given the length of the model plane? Why or why not?

We can graphically portray the relationship between the measures of an object and a scale model. Below we show a graph of the measures of the airplane and its model. Notice that the scale of the model is inches while the scale of the actual airplane is feet. Since the scale is 1:24, one inch on the model corresponds to 24 inches, which is 2 feet, on the actual airplane.

Scale of Airplane Model

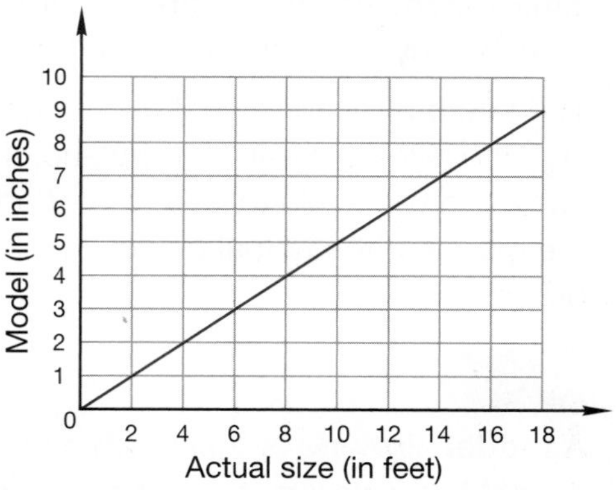

Every point on the graphed line represents a length on the airplane and its corresponding measure on the model.

Summarize Why is a graph of the relationship between the measures of a scale model and an original always a straight line?

Example 2

Sofia is molding a model of a car from clay. The scale of the model is 1:36. If the height of the car is 4 feet 6 inches, what should be the height of the model in inches?

Solution

First we convert 4 feet 6 inches to inches:

$$4 \text{ feet } 6 \text{ inches} = 4(12) + 6 = 54 \text{ inches}$$

Then we construct a ratio box using 1 and 36 as the ratio numbers, write the proportion, and solve.

	Scale	Measure
Model	1	*m*
Object	36	54

$$\frac{1}{36} = \frac{m}{54}$$

$$36m = 54$$

$$m = \frac{54}{36} = 1\frac{1}{2}$$

The height of the model car should be **$1\frac{1}{2}$ inches.**

scale factor

We have solved proportions by using cross products. Sometimes a proportion can be solved more quickly by noting the **scale factor.** The scale factor is the number of times larger (or smaller) the terms of one ratio are when compared with the terms of the other ratio. The scale factor in the proportion below is 6 because the terms of the second ratio are 6 times the terms of the first ratio.

$$\frac{3}{4} = \frac{18}{24}$$

Example 3

Solve: $\frac{3}{7} = \frac{15}{n}$

Solution

Instead of finding cross products, we note that multiplying the numerator 3 by 5 gives us the other numerator, 15. Thus the scale factor is 5. We use this scale factor to find *n*.

$$\frac{3}{7} \times \frac{5}{5} = \frac{15}{35}$$

We find that *n* is **35.**

Example 4

These two triangles are similar. Calculate the scale factor from the smaller triangle to the larger triangle.

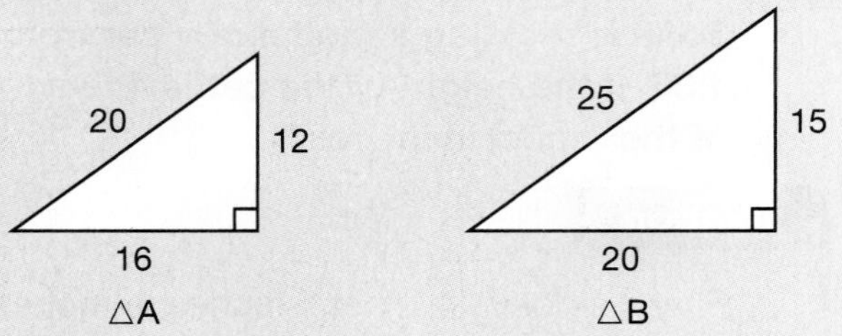

Solution

Math Language
The term *dimension* refers to the length of any side of △A or △B.

If we multiply the length of one side of the smaller triangle by the scale factor, we get the length of the corresponding side of the larger triangle.

Dimension of △A × scale factor = dimension of △B

We may select any pair of corresponding sides to calculate the scale factor. Here, we will select the longest sides. We write an equation using *f* for the scale factor and then solve for *f*.

$$20f = 25$$

$$f = \frac{25}{20}$$

$$f = \mathbf{\frac{5}{4}} \text{ or } \mathbf{1.25}$$

Generalize Find the scale factor of △A to △B using a different pair of corresponding sides.

In this book we will express the scale factor in decimal form unless otherwise directed.

Math Language
The term *linear measures* refers to measures of lengths of lines.

Note that the scale factor refers to the *linear measures* of two similar figures and not to the area or volume measures of the figures. The scale factor from cube A to cube B below is 2 because the linear measures of cube B are twice the corresponding measures of cube A.

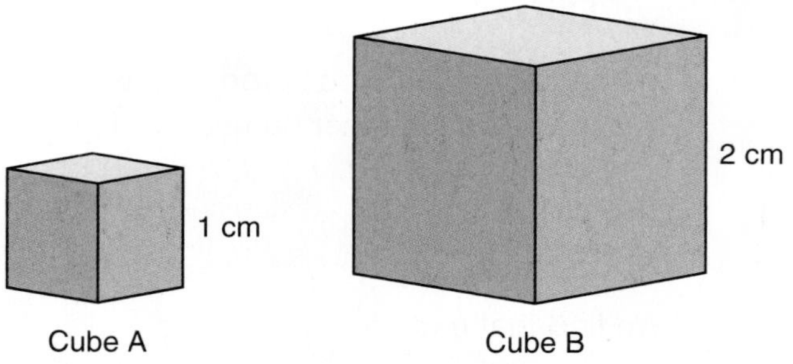

However, the surface area of cube B is 4 times the surface area of cube A, and the volume of cube B is 8 times the volume of cube A. Since we multiply two dimensions of a figure to calculate the area of the figure, the relationship between the areas of two figures is the scale factor times the scale factor; in other words, the scale factor squared. Likewise, the relationship between the volumes of two similar figures is the scale factor cubed.

Example 5

The smaller of two similar rectangular prisms has dimensions of 2 cm by 3 cm by 4 cm. The larger rectangular prism has dimensions of 6 cm by 9 cm by 12 cm.

a. **What is the scale factor from the smaller to the larger rectangular prism?**

b. **The area of any face of the larger prism is how many times the area of the corresponding face of the smaller prism?**

c. **The volume of the larger solid is how many times the volume of the smaller solid?**

Solution

Before answering the questions, we draw the two figures.

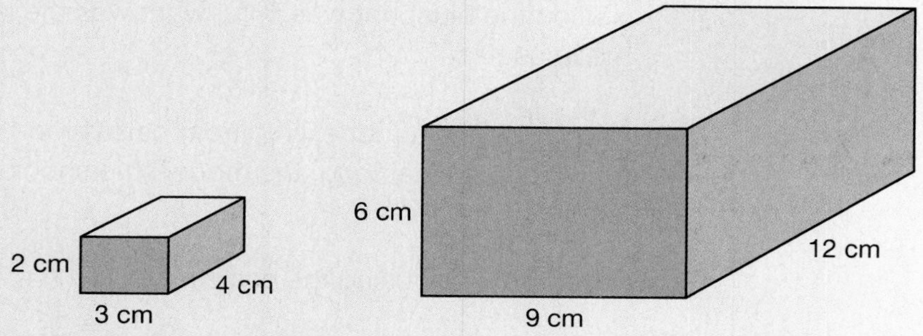

a. We select any two corresponding linear measures to calculate the scale factor. We choose the 2-cm and the 6-cm measures.

$$\text{Dimension of smaller} \times \text{scale factor} = \text{dimension of larger}$$

$$2f = 6$$

$$f = 3$$

We find that the scale factor is **3.**

b. Since the scale factor from the smaller to the larger figure is 3, the area of any face of the larger figure should be $3^2 = 9$ times the area of the corresponding face of the smaller figure. We confirm this relationship by comparing the area of a 6-by-9-cm face of the larger prism with the corresponding 2-by-3-cm face of the smaller prism.

$$\text{Area of 6-by-9-cm face} = 54 \text{ cm}^2$$

$$\text{Area of 2-by-3-cm face} = 6 \text{ cm}^2$$

We see that the area of the selected face of the larger prism is indeed **9 times** the area of the corresponding face of the smaller prism.

c. Since the scale factor of the linear dimensions of the two figures is 3, the volume of the larger prism should be $3^3 = 27$ times the volume of the smaller prism. We confirm this relationship by performing the calculations.

$$\text{Volume of 6-by-9-by-12-cm prism} = 648 \text{ cm}^3$$

$$\text{Volume of 2-by-3-by-4-cm prism} = 24 \text{ cm}^3$$

Dividing 648 cm^3 by 24 cm^3, we find that the larger volume is indeed **27 times** the smaller volume.

$$\frac{648 \text{ cm}^3}{24 \text{ cm}^3} = 27$$

Showing this calculation another way demonstrates more clearly why the larger volume is 3^3 times the smaller volume.

$$\frac{\text{Volume of larger prism}}{\text{Volume of smaller prism}} = \frac{\overset{3}{\cancel{6 \text{ cm}}} \cdot \overset{3}{\cancel{9 \text{ cm}}} \cdot \overset{3}{\cancel{12 \text{ cm}}}}{\underset{1}{\cancel{2 \text{ cm}}} \cdot \underset{1}{\cancel{3 \text{ cm}}} \cdot \underset{1}{\cancel{4 \text{ cm}}}} = 3^3$$

It is important to note that the measurements used to calculate scale factor must have the same units. If the measurements have different units, we should convert before calculating scale factor.

Practice Set

a. *Analyze* The blueprints were drawn to a scale of 1:24. If a length of a wall on the blueprint was 6 in., what was the length in feet of the wall in the house?

b. Bret is carving a model ship from balsa wood using a scale of 1:36. If the ship is 54 feet long, the model ship should be how many inches long?

Generalize Solve by using the scale factor:

c. $\frac{5}{7} = \frac{15}{w}$　　**d.** $\frac{x}{3} = \frac{42}{21}$

e. *Generalize* These two rectangles are similar. Calculate the scale factor from the smaller rectangle to the larger rectangle.

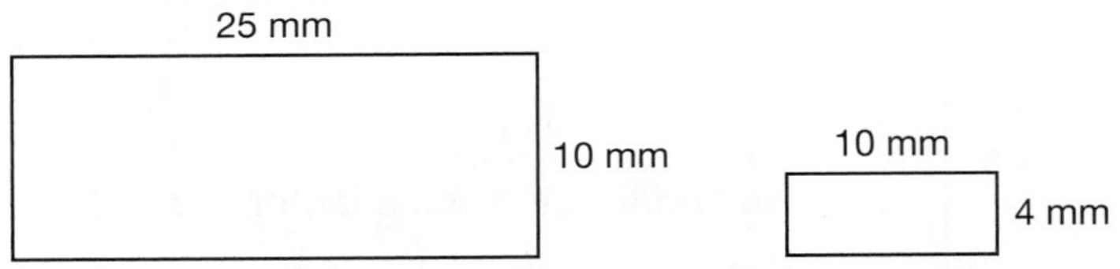

f. *Explain* The area of the larger rectangle above is how many times the area of the smaller rectangle? Show two ways to find the answer.

g. *Model* The scale of the car model in example 2 is 1:36. This means 1 inch on the model corresponds to 36 inches (that is, 3 feet) on the actual car. On grid paper make a graph that shows this relationship. Make the units of the horizontal axis feet to represent the car. On the vertical axis use inches for the model. Use the graph following example 1 as a pattern.

h. *Estimate* The statue of the standing World War II general was $1\frac{1}{2}$ times life-size. Which is the most reasonable estimate for the height of the statue?

A 4 ft　　**B** 6 ft　　**C** 9 ft　　**D** 15 ft

Written Practice *Strengthening Concepts*

* **1.** (94) **Explain** Ariana writes the letters of her first name on index cards, one letter per card. She turns the cards over and mixes them up. Then she chooses a card. If this card is an A, what is the probability that the next card will also be an A? Tell how you found your answer.

Use ratio boxes to solve problems **2–4.**

* **2.** (92) The regular price of the shoes was \$45, but they were on sale for 20 percent off. What was the sale price?

3. (72) In 2002, \$5.00 was equal to about 40 Norwegian kroner. At that time, what was the cost in dollars of an item that cost 100 kroner?

* **4.** (92) **Formulate** The number of students in chorus increased 25 percent this year. If there are 20 more students in chorus this year than there were last year, how many students are in chorus this year?

5. (84, 87) Simplify: $(3x)(x) - (x)(2x)$

6. (55) In her first 6 basketball games Ann averaged 10 points per game. In her next 9 games she averaged 15 points per game. How many points per game did Ann average during her first 15 games?

7. (46) Ingrid started her trip at 8:30 a.m. with a full tank of gas and an odometer reading of 43,764 miles. When she stopped for gas at 1:30 p.m., the odometer read 44,010 miles.

a. If it took 12 gallons to fill the tank, her car averaged how many miles per gallon?

b. Ingrid traveled at an average speed of how many miles per hour?

8. (74) Write an equation to solve this problem. Three fifths of Tyrone's favorite number is 60. What is Tyrone's favorite number?

9. (Inv. 3, 80) On a coordinate plane, graph the points $(-3, 2)$, $(3, 2)$, and $(-3, -2)$.

a. If these points designate three of the vertices of a rectangle, what are the coordinates of the fourth vertex of the rectangle? Draw the rectangle.

b. Draw the image of the rectangle in **a** after a 90° clockwise rotation about the origin. What are the coordinates of the vertices of the rotated image?

10. (36, 86) What is the ratio of counting numbers to integers in this set of numbers?

$$\{-3, -2, -1, 0, 1, 2\}$$

11. (20, 41) Find a^2 if $\sqrt{a} = 3$.

* **12.** (92) **Generalize** An antique dealer bought a chair for \$40 and sold the chair for 60% more. What was the selling price?

Write equations to solve problems **13** and **14.**

13. Forty is what percent of 250?
(77)

14. Forty percent of what number is 60?
(77)

15. **a.** Segment *BC* is how much longer than segment *AB?*
(8, 85)

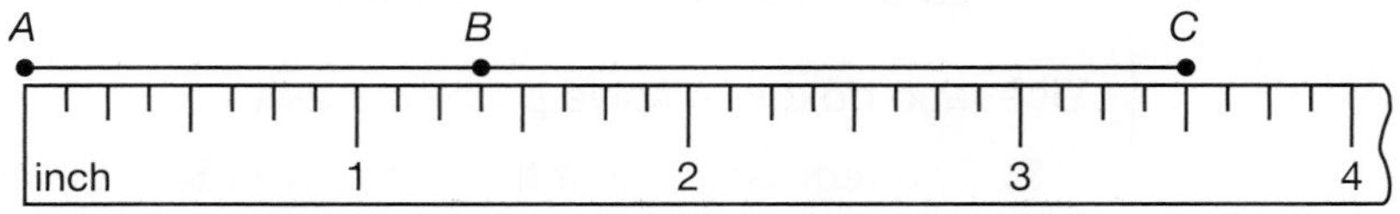

b. Convert the length of segment *AC* to centimeters.

16. Graph on a number line: $x \leq 3$
(78)

17. Complete the table.
(48)

Fraction	Decimal	Percent
a.	**b.**	1.4%

* **18.** **Represent** Find the missing numbers in the table by using the function rule. Then graph the function.
(56, Inv. 9)

$y = -2x$

x	*y*
3	
0	
−2	

a. Where does the graph of the function intersect the *y*-axis?

b. **Analyze** Are *x* and *y* directly proportional? Explain your answer.

19. Find each measure of a circle that has a diameter of 2 feet. (Use 3.14 for π.)
(65, 82)

a. circumference **b.** area

* **20.** Estimate the measure of $\angle ABC$. Then trace the angle, extend the sides, and measure the angle with a protractor.
(96)

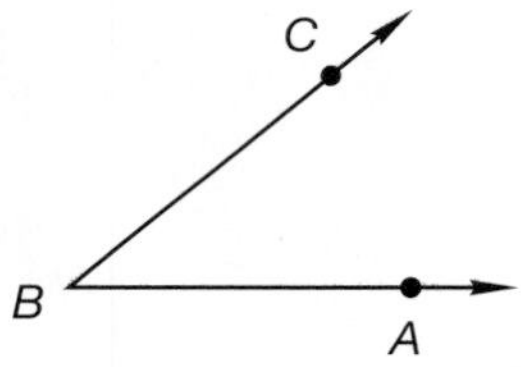

Generalize Refer to the figures below to answer problems **21** and **22.**

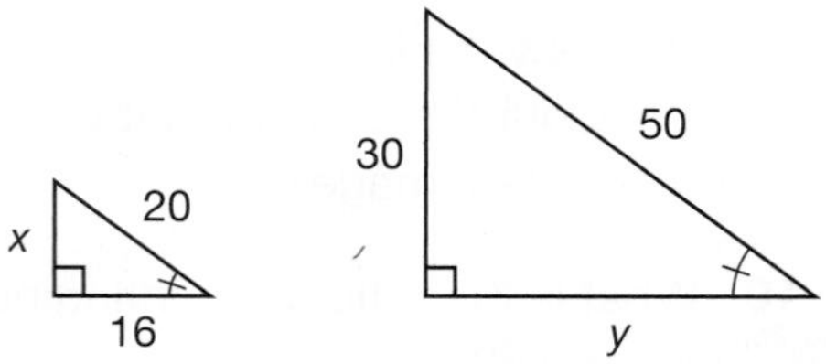

* **21.** Find *x* and *y*. Then find the area of the smaller triangle. Dimensions are in inches.
(97)

* **22.** Calculate the scale factor from the smaller triangle to the larger triangle.
(98)

23. Multiply. Write the product in scientific notation.
(83)

$$(1.4 \times 10^{-6})(5 \times 10^{4})$$

Justify For problems **24** and **25,** solve and check. Show each step.

* **24.** $-\frac{3}{5}m + 8 = 20$
(93)

* **25.** $0.3x - 2.7 = 9$
(93)

Generalize Simplify:

26. $\sqrt{5^3 - 5^2}$
(52)

27. 1 gal 1 qt − 1 qt 1 pt
(49)

28. $(0.25)\left(1\frac{1}{4} - 1.2\right)$
(43)

29. $7\frac{1}{3} - \left(1\frac{3}{4} \div 3.5\right)$
(43)

30. $\dfrac{(-2)(3) - (3)(-4)}{(-2)(-3) - (4)}$
(85)

Early Finishers
Real-World Application

A local farmer wishes to fertilize one of his fields, but he must first find its area. The field is in the shape of right triangle *ABC* shown below. After making some measurements, the farmer knows the following information:

Segment *DE* is 18 feet long and parallel to segment *BC*. *AD* is 15 feet long. The total length of *AB* is 900 feet. (Note: △*ABC* is not to scale.)

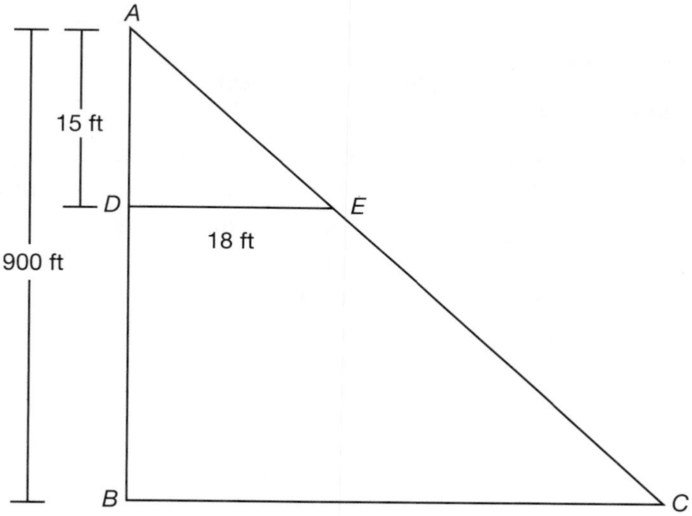

a. Use a proportion to find the length of $\overline{BC}$.

b. What is the total area of the field in square feet?

LESSON

99

• Pythagorean Theorem

Power Up

Building Power

facts

Power Up T

mental math

a. **Positive/Negative:** $(-1.5) + (4.5)$

b. **Order of Operations/Exponents:** $(8 \times 10^6)(4 \times 10^4)$

c. **Ratio:** $\frac{0.15}{30} = \frac{0.005}{n}$

d. **Measurement:** Convert -15°C to degrees Fahrenheit.

e. **Calculation:** $12 \times 2\frac{1}{3}$

f. **Percent:** $33\frac{1}{3}\%$ less than $60

g. **Geometry:** What shape(s) has 6 faces and 8 vertices?

h. **Power/Roots:** What is the square root of the sum of 6^2 and 8^2?

problem solving

Carpeting is sold by the square yard. If carpet is priced at $25 per square yard (including tax and installation), how much would it cost to carpet a classroom that is 36 feet long and 36 feet wide?

New Concept

Increasing Knowledge

Thinking Skill

Explain

Why is the hypotenuse of a right triangle always opposite the right angle?

The longest side of a right triangle is called the **hypotenuse.** The other two sides are called **legs.** Every right triangle has a property that makes right triangles very important in mathematics. **The area of the square drawn on the hypotenuse of a right triangle equals the sum of the areas of the squares drawn on the legs.**

Discuss Identify by their measure the hypotenuse and legs of the triangle shown.

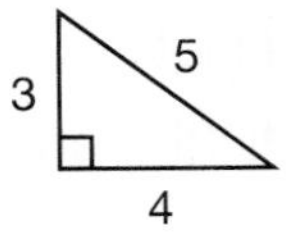

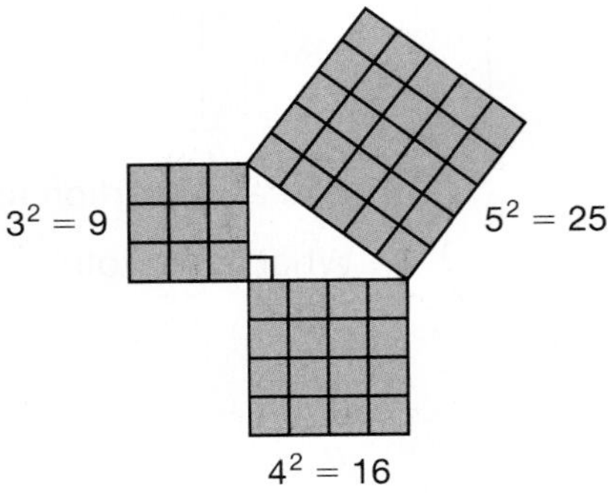

The triangle on the left is a right triangle. On the right we have drawn and shaded a square on each side of the triangle. We have divided the squares into units and can see that their areas are 9, 16, and 25. Notice that the area of the largest square equals the sum of the areas of the other two squares.

$$25 = 16 + 9$$

To solve right-triangle problems using the Pythagorean theorem, we will draw the right triangle, as well as squares on each side of the triangle.

Example 1

Copy this triangle. Draw a square on each side. Find the area of each square. Then find *c*.

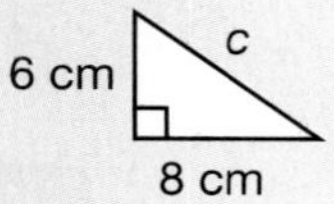

Solution

We copy the triangle and draw a square on each side of the triangle as shown.

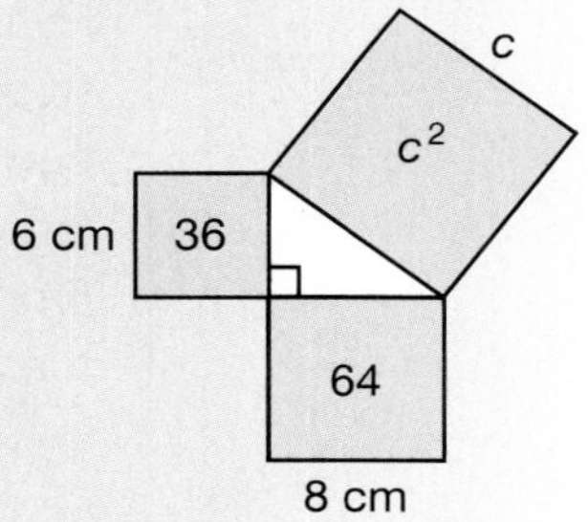

We were given the lengths of the two shorter sides. The areas of the squares on these sides are **36 cm²** and **64 cm².** The Pythagorean theorem says that the sum of the areas of the smaller squares equals the area of the largest square.

$$36 \text{ cm}^2 + 64 \text{ cm}^2 = \mathbf{100 \text{ cm}^2}$$

Math Language
Remember that $\sqrt{100} = 10$, or the square root of 100 is 10.

This means that each side of the largest square must be 10 cm long because $(10 \text{ cm})^2$ equals 100 cm^2. Thus

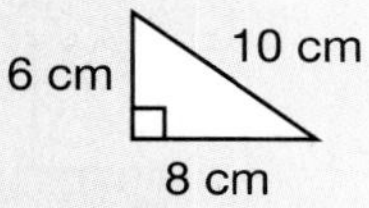

$$c = \mathbf{10 \text{ cm}}$$

Explain How is the formula for the area of a square, $A = s^2$, used when finding the lengths of the sides of a right triangle?

Example 2

In this triangle, find *a*. Dimensions are in inches.

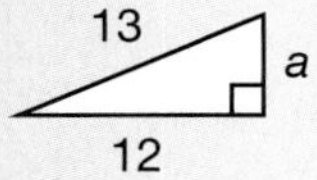

Solution

We copy the triangle and draw a square on each side. The area of the largest square is 169 in.². The areas of the smaller squares are 144 in.² and a^2. By the Pythagorean theorem, a^2 plus 144 in.² must equal 169 in.².

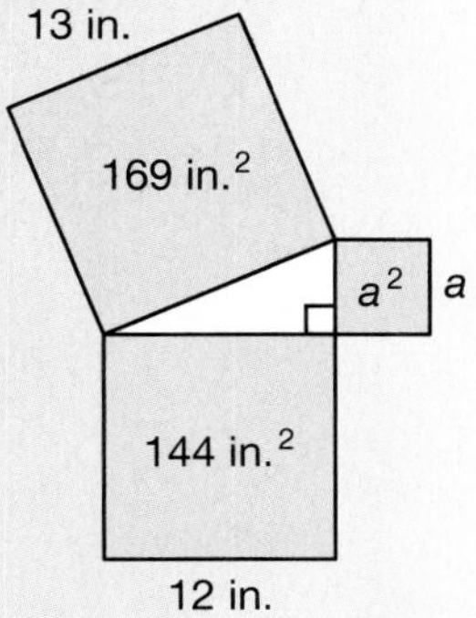

$$a^2 + 144 \text{ in.}^2 = 169 \text{ in.}^2$$

Subtracting 144 in.² from both sides, we see that

$$a^2 = 25 \text{ in.}^2$$

This means that a equals **5 in.,** because $(5 \text{ in.})^2$ is 25 in².

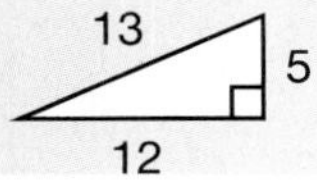

Example 3

Find the perimeter of this triangle. Dimensions are in centimeters.

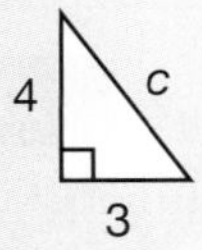

Solution

We can draw a square on each side and use the Pythagorean theorem to find *c*. The areas of the two smaller squares are 16 cm^2 and 9 cm^2. The sum of these areas is 25 cm^2, so the area of the largest square is 25 cm^2. Thus the length *c* is 5 cm. Now we add the lengths of the sides to find the perimeter.

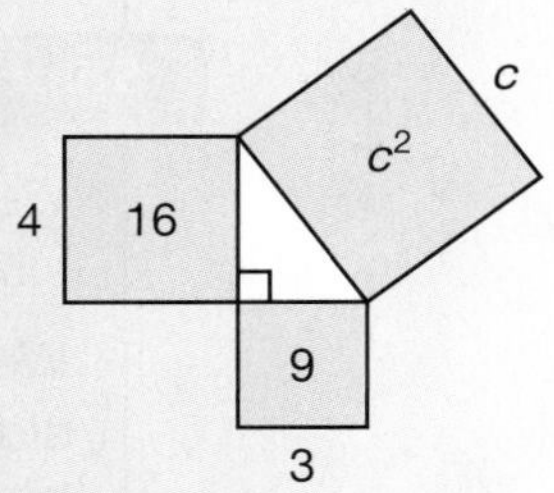

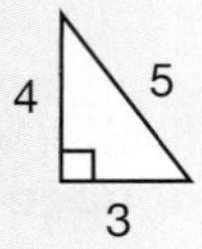

Perimeter = 4 cm + 3 cm + 5 cm

= **12 cm**

Example 4

Copy this right triangle and draw squares on the sides. Then write an equation that shows the relationship between the areas of the squares.

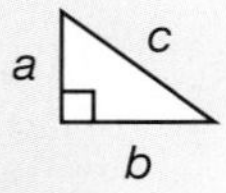

Solution

We copy the triangle and draw a square on each side. Squaring the lengths of the sides of the triangle gives us the areas of the squares: a^2, b^2, and c^2. By the Pythagorean theorem, the sum of the areas of the smaller two squares is equal to the area of the largest square.

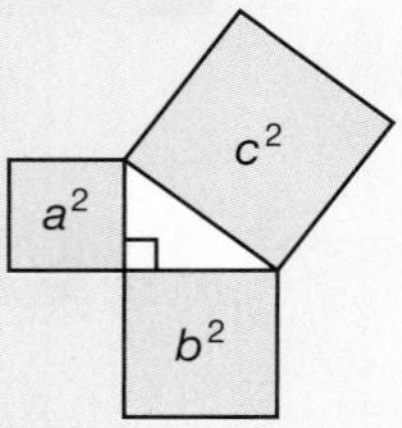

$$\boldsymbol{a^2 + b^2 = c^2}$$

This equation is commonly used to algebraically express the Pythagorean theorem: The sum of the squares of the legs equals the square of the hypotenuse. **The Pythagorean theorem applies to all right triangles and it applies only to right triangles.**

The converse of the Pythagorean theorem is true. If the sum of the squares of two sides of a triangle equals the square of the third side, then the triangle is a right triangle.

This property of right triangles was known to the Egyptians as early as 2000 b.c., but it is named for a Greek mathematician who lived about 550 b.c. The Greek's name was Pythagoras, and the property is called the **Pythagorean theorem.** The Greeks are so proud of Pythagoras that they have issued a postage stamp that illustrates the theorem. Here we show a reproduction of the stamp:

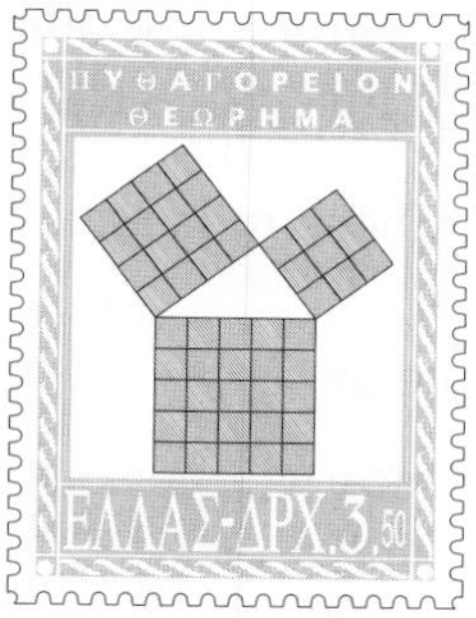

Example 5

The side lengths of three triangles are given. Which triangle is a right triangle?

A
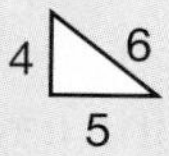

B
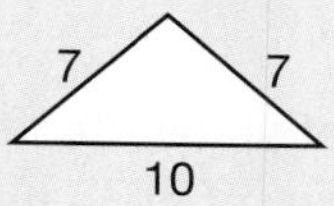

C
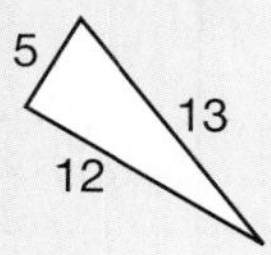

Solution

We square the lengths of the sides looking for a Pythagorean relationship.

A $4^2 \quad 5^2 \quad 6^2$
$16 + 25 \neq 36$

B $7^2 \quad 7^2 \quad 10^2$
$49 + 49 \neq 100$

C $5^2 \quad 12^2 \quad 13^2$
$25 + 144 = 169$

Only triangle **C** is a right triangle because it is the only triangle in which the sum of the squares of two sides equals the square of the third side.

Practice Set

Analyze Copy the triangles and draw the squares on the sides of the triangles as you work problems **a–c.**

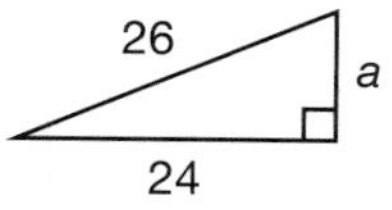

a. Use the Pythagorean theorem to find the length *a.*

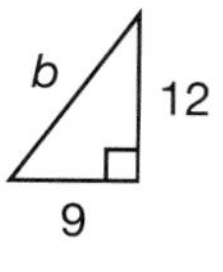

b. Use the Pythagorean theorem to find the length *b.*

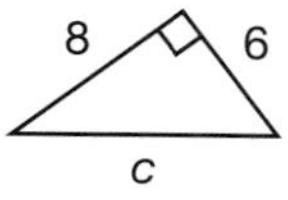

c. Find the perimeter of this triangle. Dimensions are in feet.

d. *Justify* Which triangle below is a right triangle? Show your work and defend your answer.

A
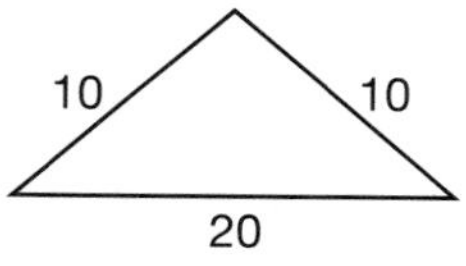

B
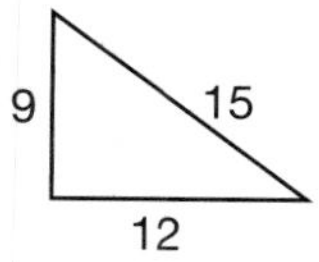

C
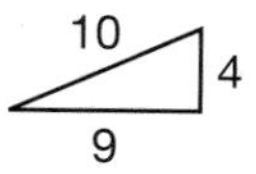

Written Practice — Strengthening Concepts

1. (60) **Explain** The meal cost $15. Christie left a tip that was 15 percent of the cost of the meal. How much money did Christie leave for a tip?

2. (57) Twenty-five ten-thousandths is how much greater than twenty millionths? Write the answer in scientific notation.

3. (Inv. 4) Find these measures for the number of days in the months of a leap year.

   a. mean **b.** median

   c. mode **d.** range

4. (46) The 2-pound box costs $2.72. The 48-ounce box costs $3.60. The smaller box costs how much more per ounce than the larger box?

Use ratio boxes to solve problems **5** and **6.**

5. (72) If 80 pounds of seed costs $96, what would be the cost of 300 pounds of seed?

6. (66) The ratio of stalactites to stalagmites in the cavern was 9 to 5. If the total number of stalactites and stalagmites was 1260, how many stalagmites were in the cavern?

7. (16, 60) Five eighths of a pound is how many ounces?

8. (77) Write equations to solve **a** and **b.**

   a. Ten percent of what number is 20?

   b. Twenty is what percent of 60?

9. (62) In this figure, central angle *BDC* measures 60°, and inscribed angle *BAC* measures 30°. Angles *ACD* and *BCD* are complementary.

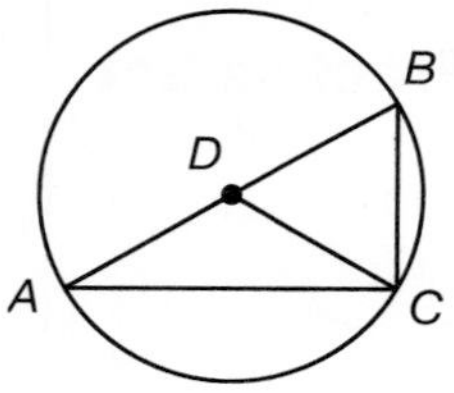

   a. Classify $\triangle ABC$ by angles.

   b. Classify $\triangle BCD$ by sides.

   c. Classify $\triangle ADC$ by sides.

* 10. (92) **Model** Use a ratio box to solve this problem. The cost of a 10-minute call to Boise decreased by 20%. If the cost before the decrease was $3.40, what was the cost after the decrease?

* 11. (92) **Explain** An item is on sale for 20% off the regular price. The sale price is what percent of the regular price? How do you know?

12. *(37)* What is the area of the shaded region of this rectangle?

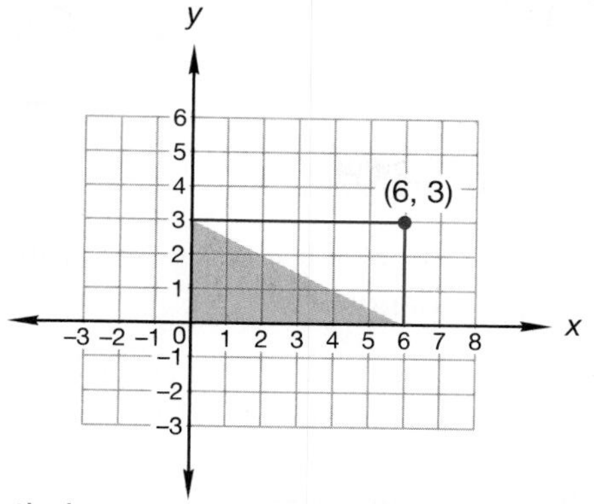

*** 13.** *(98)* **Model** Use a ratio box to solve this problem. On a 1:60 scale-model airplane, the wingspan is 8 inches. The wingspan of the actual airplane is how many inches? What is the wingspan of the actual airplane in feet?

14. *(48)* Complete the table.

Fraction	Decimal	Percent
$1\frac{1}{3}$	**a.**	**b.**
c.	**d.**	$1\frac{1}{3}\%$

15. *(84, 87)* Simplify:

a. $(ax^2)(-2ax)(-a^2)$

b. $\frac{1}{2}\pi + \frac{2}{3}\pi - \pi$

16. *(83)* Multiply. Write the product in scientific notation.

$$(8.1 \times 10^{-6})(9 \times 10^{10})$$

17. *(20, 52)* Evaluate: $\sqrt{c^2 - b^2}$ if $c = 15$ and $b = 12$

*** 18.** *(99)* **Analyze** Use the Pythagorean theorem to find c.

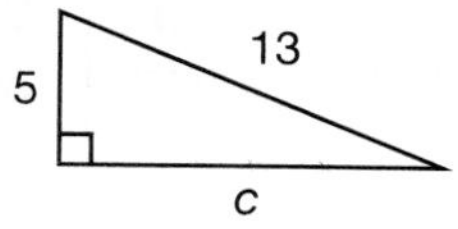

*** 19.** *(95)* **Explain** How would you would find the volume of this right solid? Show the formulas you would use to find the volume. Then find the volume. Dimensions are in centimeters. (Use 3.14 for π.)

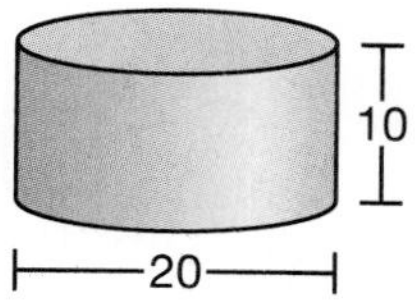

20. *(40)* Refer to the figure below to find the measures of the following angles.

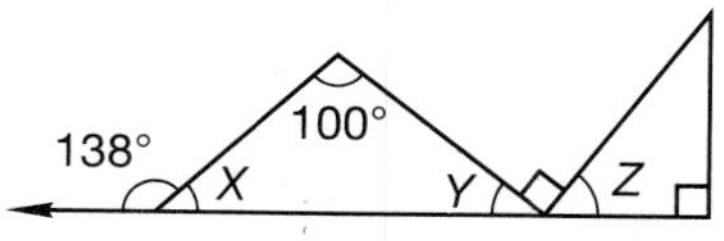

a. $\angle X$ **b.** $\angle Y$ **c.** $\angle Z$

* **21.** (97, 98) **Analyze** These triangles are similar. Dimensions are in inches.

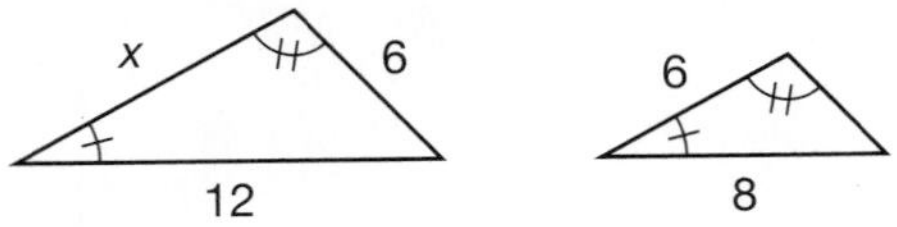

a. Find x.

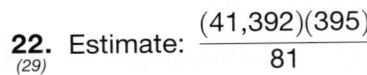

b. What is the scale factor from the smaller triangle to the larger triangle?

c. The area of the larger triangle is how many times the area of the smaller triangle? *Hint:* Use the scale factor found in **b.**

22. (29) Estimate: $\frac{(41{,}392)(395)}{81}$

Generalize For problems **23** and **24,** solve and check. Show each step.

* **23.** (93) $4n + 1.64 = 2$

* **24.** (93) $3\frac{1}{3}x - 1 = 49$

25. (39) $\frac{17}{25} = \frac{m}{75}$

Simplify:

26. (20, 52) $3^3 + 4^2 - \sqrt{225}$

27. (52, 57) $\sqrt{225} - 15^0 + 10^{-1}$

28. (43) $\left(3\frac{1}{3}\right)(0.75)(40)$

29. (85, 91) $\frac{-12 - (6)(-3)}{(-12) - (-6) + (3)}$

* **30.** (96) **Generalize** Using the Distributive Property, we know that $2(x - 4)$ equals $2x - 8$. Use the Distributive Property to multiply $3(x - 2)$.

Early Finishers
Math Applications

Which type of display—a bar graph or a Venn diagram—is the most appropriate way to display the factors and common factors of the numbers 12 and 36? Draw your display and justify your choice.

LESSON

100

• Estimating Square Roots
• Irrational Numbers

Power Up — *Building Power*

facts Power Up U

mental math

a. **Positive/Negative:** $(-1.5) - (-7.5)$

b. **Order of Operations/Exponents:** $(5 \times 10^{-5})(5 \times 10^{-5})$

c. **Algebra:** $100 = 5w - 20$

d. **Measurement:** Convert $-20°C$ to degrees Fahrenheit.

e. **Calculation:** $20 \times 3\frac{3}{4}$

f. **Percent:** $33\frac{1}{3}\%$ less than \$24

g. **Geometry:** Which shape(s) has no vertices?

h. **Calculation:** 25% of 44, $\times$ 3, $-$ 1, $\div$ 4, $\times$ 7, $-$ 1, $\div$ 5, $\times$ 9, $+$ 1, $\sqrt{\ }$, $-$ 1, $\sqrt{\ }$

problem solving

The figure represents a three-dimensional solid. Draw the top, front, back, left, and right views of the solid.

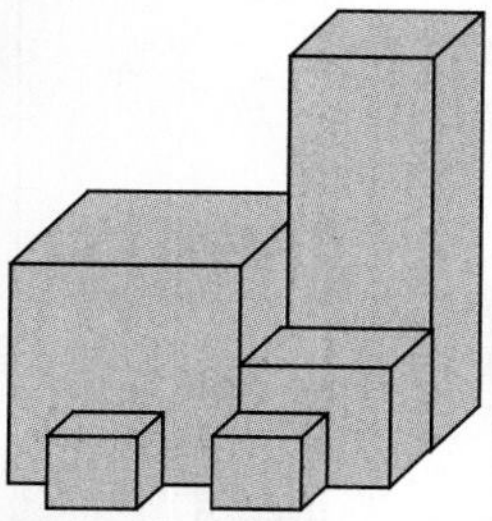

New Concepts — *Increasing Knowledge*

estimating square roots

Reading Math
Read $\sqrt{25}$ as "the square root of 25."

These counting numbers are perfect squares:

$$1, 4, 9, 16, 25, 36, 49, 64, \ldots$$

Recall that the square root of a perfect square is an integer.

$$\sqrt{25} = 5 \qquad \sqrt{36} = 6$$

The square root of a number that is between two perfect squares is not an integer but can be estimated.

$$\sqrt{29} = ?$$

Since 29 is between the perfect squares 25 and 36, we can conclude that $\sqrt{29}$ is between $\sqrt{25}$ and $\sqrt{36}$.

$$\sqrt{25} = 5 \qquad \sqrt{29} = ? \qquad \sqrt{36} = 6$$

We see that $\sqrt{29}$ is between 5 and 6. On this number line we see that $\sqrt{29}$ is between 5 and 6 but not exactly halfway between.

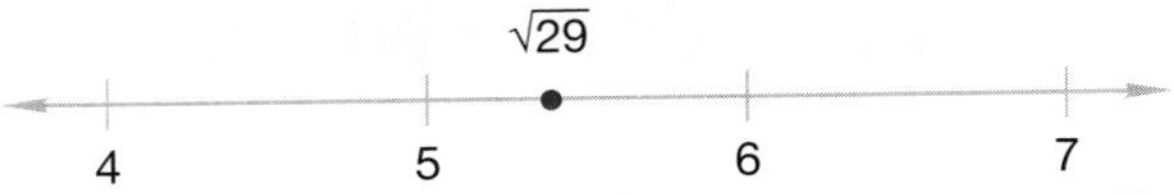

Example 1

Between which two consecutive whole numbers is $\sqrt{200}$?

Solution

We remember that $\sqrt{100}$ is 10, so $\sqrt{200}$ is more than 10. We might guess that $\sqrt{200}$ is 20. We check our guess.

$$20 \times 20 = 400 \qquad \text{too large}$$

Our guess is much too large. Next we guess 15.

$$15 \times 15 = 225 \qquad \text{too large}$$

Since 15 is still too large, we try 14.

$$14 \times 14 = 196 \qquad \text{too small}$$

We see that 14 is less than $\sqrt{200}$ and 15 is more than $\sqrt{200}$. So $\sqrt{200}$ is between the consecutive whole numbers **14** and **15.**

Generalize Using the same method, we find that $\sqrt{10}$ is between which two consecutive whole numbers?

irrational numbers

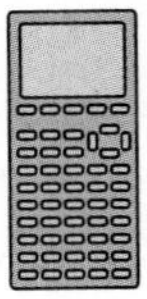

Visit www.SaxonPublishers.com/ActivitiesC2 *for a graphing calculator activity.*

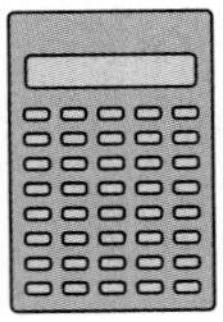

Thinking Skill

Explain

How many decimal places will be in the product of these two factors? How do you know?

At the beginning of this lesson we found that $\sqrt{29}$ is between 5 and 6. We can refine our estimate by finding a decimal (or fraction) that is closer to $\sqrt{29}$. We try 5.4.

$$5.4 \times 5.4 = 29.16 \qquad \text{too large}$$

Since 5.4 is too large, we try 5.3.

$$5.3 \times 5.3 = 28.09 \qquad \text{too small}$$

We see that $\sqrt{29}$ is between 5.3 and 5.4. We may continue refining our estimate by finding numbers closer to $\sqrt{29}$. However, no matter how many numbers we try, we will not find a decimal (or fraction) that equals $\sqrt{29}$.

If we use a calculator, we can quickly find a number close to $\sqrt{29}$. If we enter these keystrokes

[√] [2] [9] [=] or [2] [9] [√]

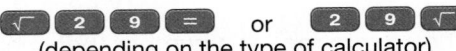

(depending on the type of calculator)

the number displayed on an 8-digit calculator is

5.3851648

This number is close to $\sqrt{29}$ but does not equal $\sqrt{29}$, as we see in the first step of checking the answer:

$$\begin{array}{r} \scriptstyle 6 \\ 5.3851648 \\ \times\ 5.3851648 \\ \hline 4 \end{array}$$

We see immediately that the product is not 29.00000000000000 because the digit in the last decimal place is 4.

Actually $\sqrt{29}$ is a number that cannot be exactly expressed as a decimal or fraction and therefore is *not a rational number*. Rather $\sqrt{29}$ is an **irrational number**—a number that cannot be expressed as a ratio of two integers.

Nevertheless, $\sqrt{29}$ is a number that has an exact value. For instance, if the legs of this right triangle are exactly 2 cm and 5 cm, we find by the Pythagorean theorem that the length of the hypotenuse is $\sqrt{29}$ cm.

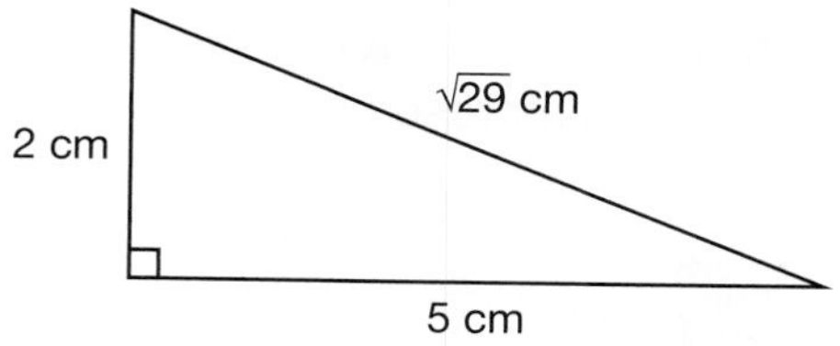

If we measure the hypotenuse with a centimeter ruler, we find that the length is about 5.4 cm, which is an approximation of $\sqrt{29}$ cm.

Other examples of irrational numbers include π (the circumference of a circle with a diameter of 1), $\sqrt{2}$ (the length of the diagonal of a square with sides of 1), and the square roots of counting numbers that are not perfect squares. The irrational numbers, together with the rational numbers, make up the set of **real numbers.**

Real Numbers

Rational Numbers	Irrational Numbers

All of the numbers represented by points on the number line are real numbers and are either rational or irrational.

Example 2

Draw a number line and show the approximate location of the points representing the following real numbers. Then describe each number as rational or irrational.

$$\pi \qquad \sqrt{2} \qquad 2.\overline{3} \qquad -\tfrac{1}{2}$$

Solution

We draw a number line and mark the location of the integers from -1 through 4. We position π ($\approx$ 3.14) between 3 and 4 but closer to 3. Since $\sqrt{2}$ is between $\sqrt{1}$ (= 1) and $\sqrt{4}$ (= 2), we position $\sqrt{2}$ between 1 and 2 but closer to 1. The repeating decimal $2.\overline{3}$ $(= 2\frac{1}{3})$ is closer to 2 than to 3. The negative fraction $-\frac{1}{2}$ is halfway between 0 and -1.

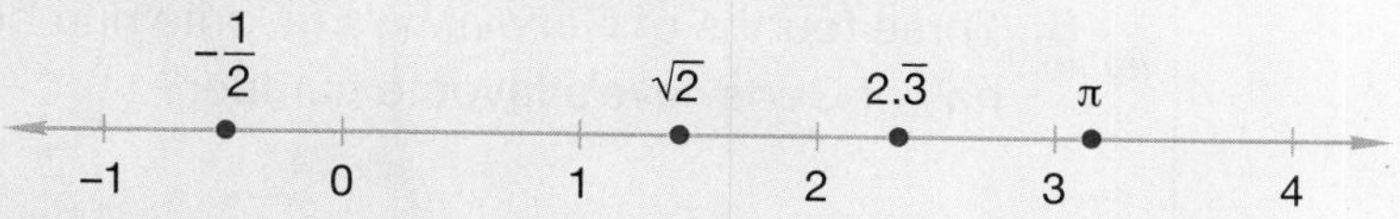

Repeating decimal numbers are rational. Thus, both **$-\frac{1}{2}$ and $2.\overline{3}$ are rational, while $\sqrt{2}$ and π are irrational.**

Classify Which of these types of decimal numbers are irrational?
repeating/terminating non-repeating/non-terminating

Practice Set

Each square root below is between which two consecutive whole numbers?

a. $\sqrt{7}$ **b.** $\sqrt{70}$ **c.** $\sqrt{700}$

d. Find x:

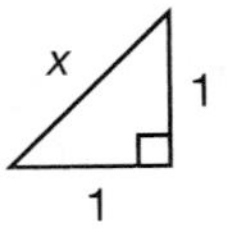

e. Find y:

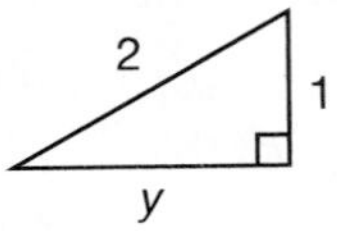

f. **Model** Draw a number line and show the approximate location of the points representing these real numbers. Which of them are irrational?

$\sqrt{3}$ $0.\overline{3}$ π $-\frac{1}{3}$

Written Practice

Strengthening Concepts

1. (35) Alberto paid \$30.00 for $2\frac{1}{2}$ pounds of cheese that cost \$6.60 per pound and 2 boxes of crackers that cost \$1.79 each. How much money should he get back?

* **2.** (21, Inv. 8) The face of this spinner is divided into fifths.

a. What is the probability that the spinner will not stop on a prime number on one spin?

b. **Explain** If the spinner is spun twice, what is the probability that it will stop on a prime number both times? Explain how you found the answer.

3. (28, 86) What is the average of the first 10 counting numbers?

4. (54) At an average speed of 50 miles per hour, how long would it take to complete a 375-mile trip?

Use ratio boxes to solve problems **5–7.**

5. (72) The Johnsons traveled 300 kilometers in 4 hours. At that rate, how long will it take them to travel 500 kilometers? Write the answer in hours and minutes.

6. (66) The ratio of children to adults at the museum was 1 to 15. If there were 800 visitors at the museum, how many children were there?

7. (92) The population of a colony of birds decreased 30 percent over one winter. If the population the next spring was 350, what was the population before winter came?

8. (60, 74) Three fourths of Genevieve's favorite number is 36. What number is one half of Genevieve's favorite number?

9. (77) Write equations to solve **a** and **b**.

a. Three hundred is 6 percent of what number?

b. Twenty is what percent of 10?

10. (60) What is the total price of a $40 item including 6.5% sales tax?

* **11.** (96) **Generalize** Using the Distributive Property, we know that $3(x + 3)$ equals $3x + 9$. Use the Distributive Property to multiply $x(x + 3)$.

12. (Inv. 9) The ordered pairs (0, 0), (−2, −4), and (2, 4) designate points that lie on the graph of the equation $y = 2x$. Graph the equation on a coordinate plane, and name another (x, y) pair from the 3rd quadrant that satisfies the equation.

* **13.** (Inv. 9, 98) Nathan used the data in the graph below to mold a scale model of a car from clay. The car is 4 feet tall, and he used the graph to find that the model should be 2 inches tall.

a. The length of the car's bumper is 5 feet. Use the graph to find the proper length of the model's bumper.

b. What is the scale factor from the car to the model? Write the scale factor as a fraction.

c. **Estimate** Nathan's completed model was 7 inches long. Estimate the length of the car.

Nathan's Model Car

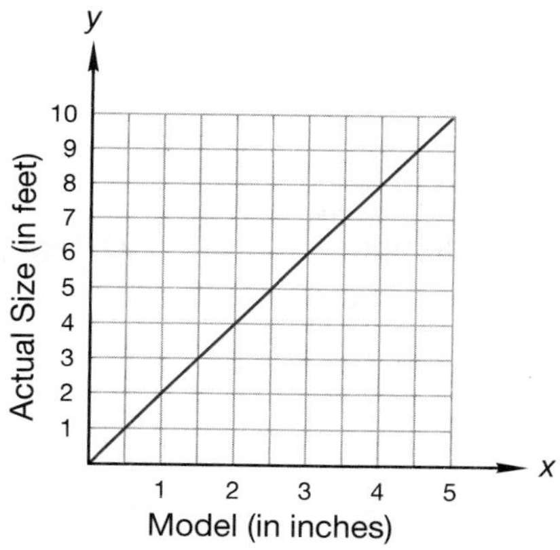

* **14.** (98) **Analyze** The edge of one cube measures 2 cm. The edge of a larger cube measures 6 cm.

a. What is the scale factor from the smaller cube to the larger cube?

b. The area of each face of the larger cube is how many times the area of a face of the smaller cube?

c. The volume of the larger cube is how many times the volume of the smaller cube?

15. (36) If the spinner is spun twice, two of the possible outcomes are C, A and A, C. Write the sample space for this experiment.

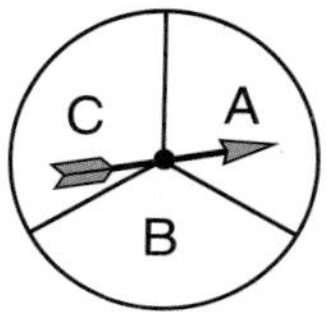

16. Complete the table.
(48)

Fraction	Decimal	Percent
a.	**b.**	72%

17. Multiply. Write the product in scientific notation.
(83)

$$(4.5 \times 10^6)(6 \times 10^3)$$

* **18.** **Analyze** Each square root is between which two consecutive whole numbers?
(100)

a. $\sqrt{40}$ **b.** $\sqrt{20}$

19. Find the **a** circumference and **b** area of a circle that has a radius of 7 inches. (Use $\frac{22}{7}$ for π.)
(65, 82)

* **20.** **Analyze** Use the Pythagorean theorem to find a. Dimensions are in centimeters.
(99)

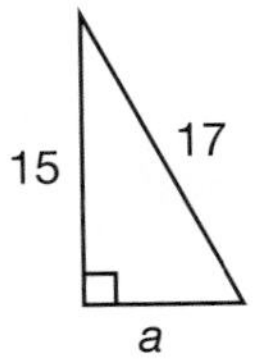

Analyze For problems **21** and **22,** find the volume of each right solid. Dimensions are in centimeters. (Use 3.14 for π.)

* **21.**
(95)

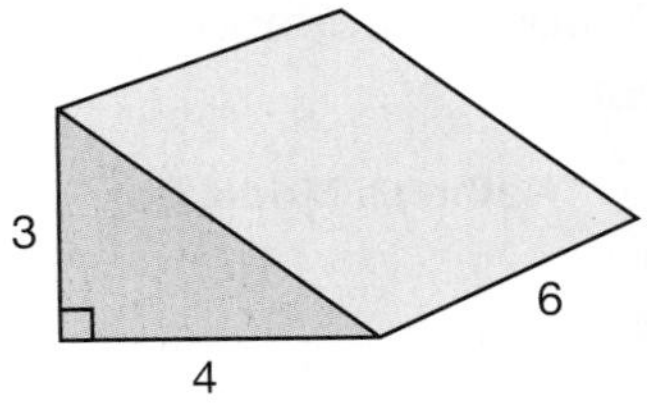

* **22.**
(95)

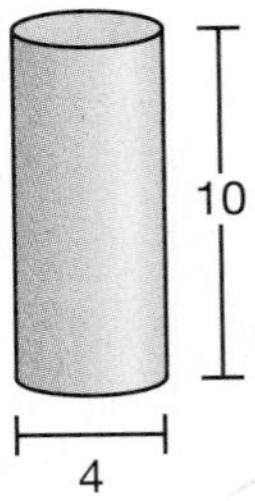

23. In the figure at right, find the measures of angles a, b, and c.
(40)

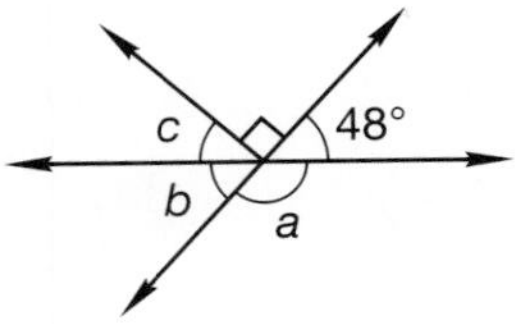

Justify For problems **24** and **25,** solve the equation. Show each step.

* **24.** $-4\frac{1}{2}x + 8^0 = 4^3$
(57, 93)

* **25.** $\frac{15}{w} = \frac{45}{3.3}$
(39)

Simplify:

26. $\sqrt{6^2 + 8^2}$
(52)

27. $3\frac{1}{3}\left(7.2 \div \frac{3}{5}\right)$
(43)

28. $8\frac{5}{6} - 2.5 - 1\frac{1}{3}$
(43)

29. $\frac{|-18| - (2)(-3)}{(-3) + (-2) - (-4)}$
(85)

* **30.** **Model** Draw a number line and show the approximate locations of 1.5, -0.5, and $\sqrt{5}$.
(100)

Focus on

• Using a Compass and Straightedge, Part 2

In Investigation 2 we used a compass to draw circles, and we used a compass and straightedge to inscribe a regular hexagon and a regular triangle in a circle. In this investigation we will use a compass and straightedge to **bisect** (divide in half) a line segment and an angle. We will also inscribe a square and a regular octagon in a circle.

Materials needed:

- Compass
- Ruler or straightedge
- Protractor

Use a metric ruler to draw a segment 6 cm long. Label the endpoints *A* and *C*.

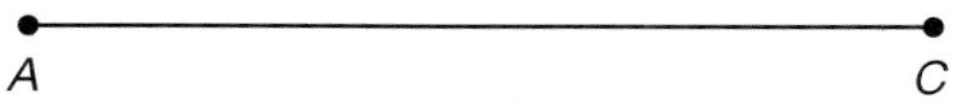

Next open a compass so that the distance between the pivot point and pencil point is more than half the length of the segment to be bisected (in this case, more than 3 cm). You will swing arcs from both endpoints of the segment, so do not change the compass radius once you have it set. Place the pivot point of the compass on one endpoint of the segment, and make a curve by swinging an arc on both sides of the segment as shown.

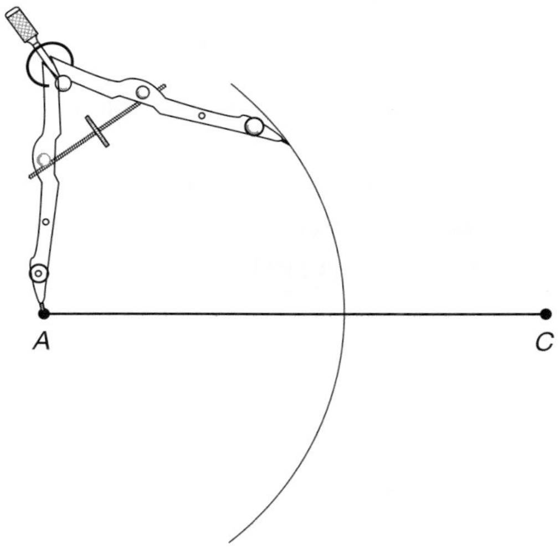

Then move the pivot point of the compass to the other endpoint of the segment, and, without resetting the compass, swing an arc that intersects the other arc on both sides of the segment. Now draw a line through the two points where the arcs intersect to divide the original segment into two parts. Label the point where the line intersects the segment point *B*.

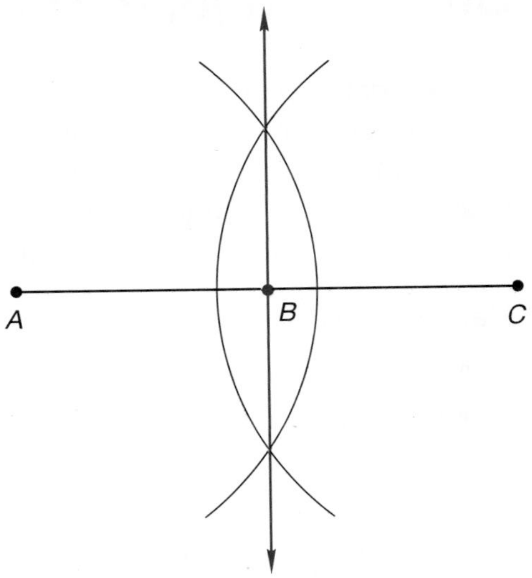

Thinking Skill

Predict

Can we predict the measures of *AB* and *BC*? Explain.

1. Use a metric ruler to find *AB* and *BC*.

2. Where the line and segment intersect, four angles are formed. What is the measure of each angle?

Using a compass and straightedge to create geometric figures is called **construction.** You just constructed the **perpendicular bisector** of a segment.

3. *Explain* Why is the line you constructed called the perpendicular bisector of the segment?

We can use a perpendicular bisector to help us **inscribe** a square in a circle. Draw a dot on your paper to be the center of a circle. Set the distance between the points of your compass to 2 cm. Then place the pivot point of the compass on the dot and draw a circle. Use a straightedge to draw a diameter of the circle.

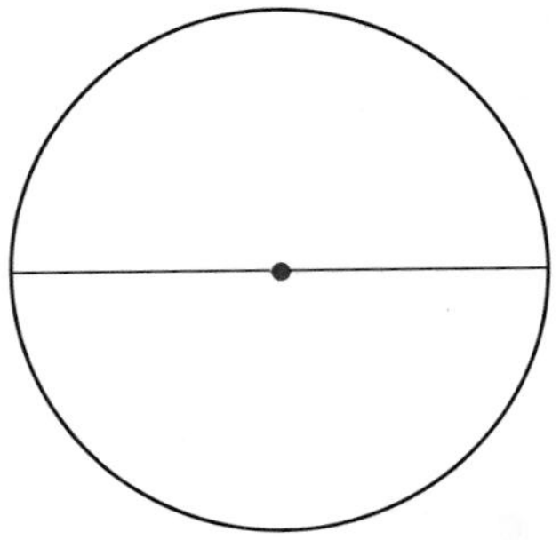

The two points where the diameter intersects the circle are the endpoints of the diameter. Open the compass a little more than the radius of the circle, and construct the perpendicular bisector of the diameter you drew.

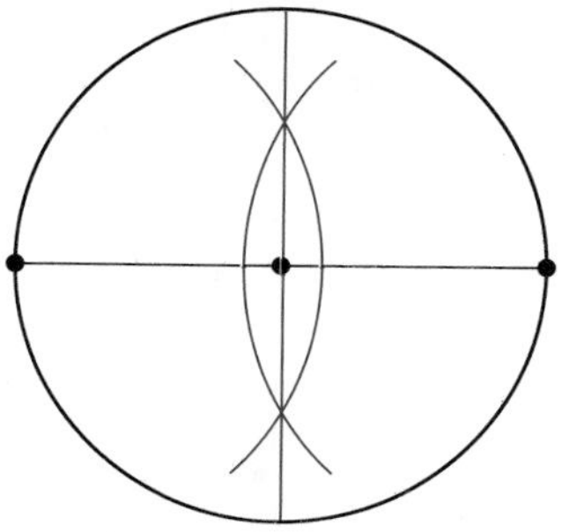

The perpendicular bisector is another diameter of the circle. The two diameters divide the circle into quarters. Draw chords between the points on the circle that are the endpoints of the two diameters.

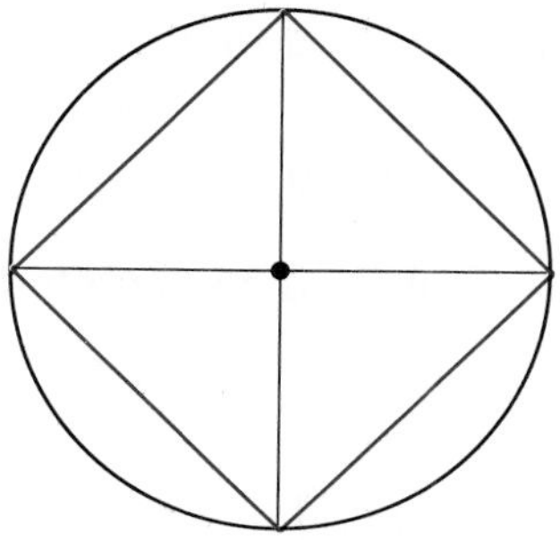

You have inscribed a square in a circle.

4. Each angle of the square is an inscribed angle of the circle. What is the measure of each angle of the square?

5. **Conclude** Notice that within the square are four small right triangles. Two sides of each small triangle are radii of the circle. If the radius of the circle is 2 cm, then

   a. what is the area of each small right triangle?

   b. how can we find the area of the inscribed square?

Use a straightedge to draw an angle. With the pivot point of the compass on the vertex of the angle, draw an arc that intersects the sides of the angle. For reference call these points R and S, and label the vertex V.

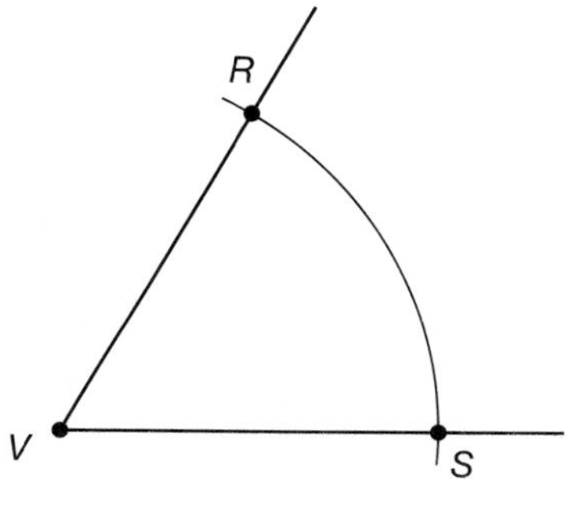

Set the compass so that it is open more than half the distance between *R* and *S*. With the pivot point on *R,* swing an arc. Then, with the pivot point on *S,* swing another arc that intersects the one centered at *R*. Label the point of intersection *T*.

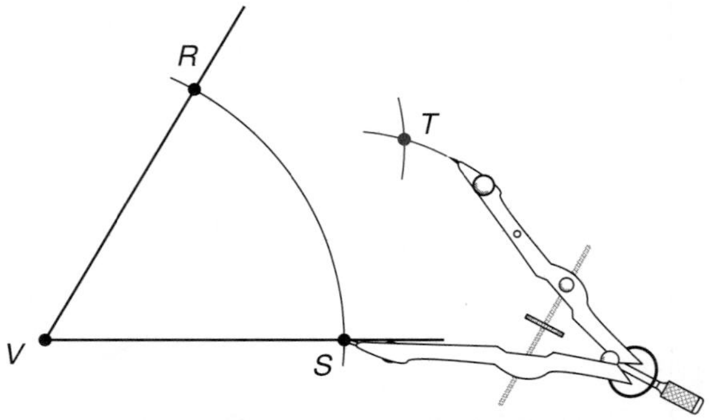

Using a straightedge, draw a ray from the vertex *V* through point *T*. Ray *VT* divides ∠*RVS* into two congruent angles.

6. Use a protractor to measure the original angle you drew and the two smaller angles formed when you constructed the ray. Record all three angle measures for your answer. What did you notice?

In this activity you constructed an **angle bisector.**

7. Discuss Why is the ray called an angle bisector?

Draw a circle and a diameter of the circle. Then construct a diameter that is a perpendicular bisector of the first diameter. Your work should look like the circle shown below. We have labeled points *M, X, Y,* and *Z* for reference.

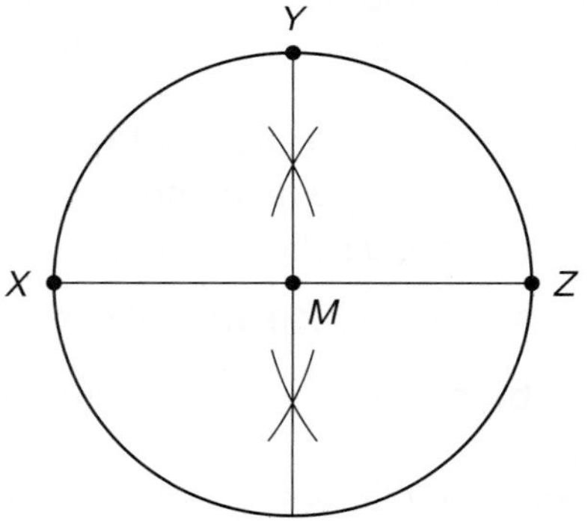

Swing intersecting arcs from points *Y* and *Z* to locate the angle bisector of ∠*YMZ*. Also swing intersecting arcs from points *X* and *Y* to locate the angle bisector of ∠*XMY*.

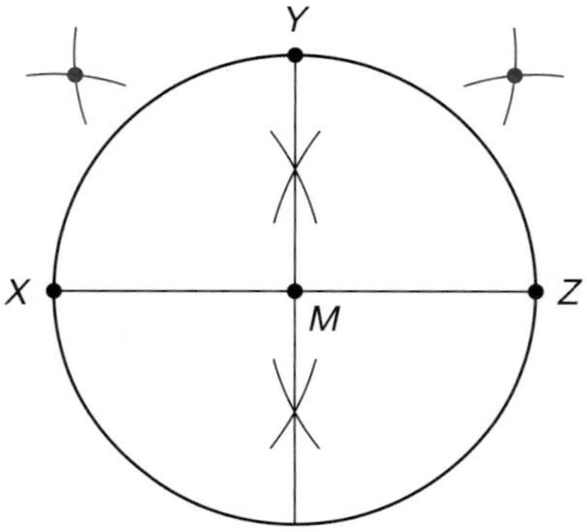

Draw two lines through the center of the circle that passes through the points where the arcs intersect. These two lines together with the two diameters divide the circle into eighths.

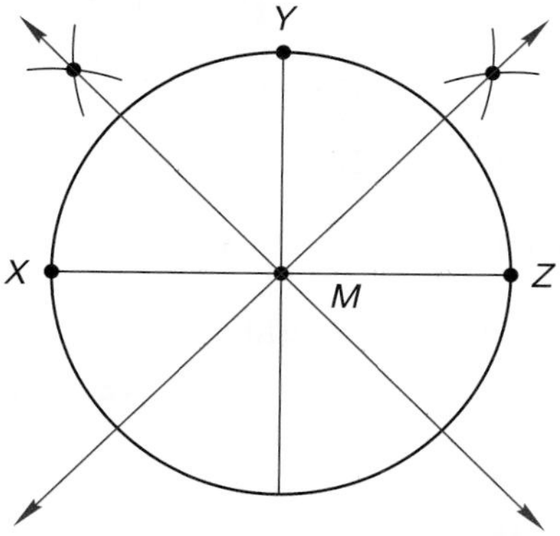

8. **Conclude** What is the measure of each small central angle that is formed?

There are 8 points of intersection around the circle. Draw chords from point to point around the circle to inscribe a regular octagon in the circle.

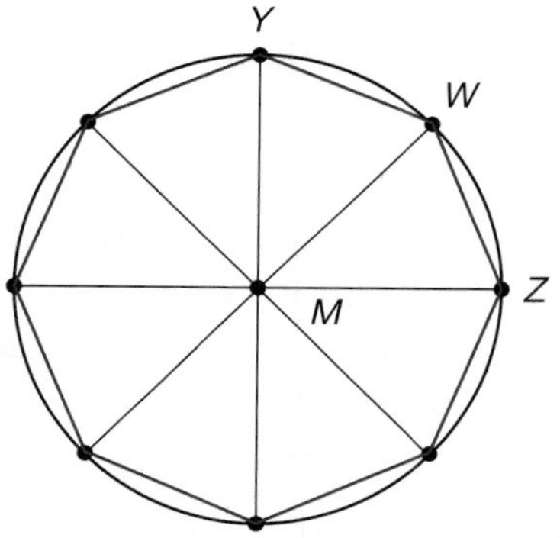

9. The octagon is divided into how many congruent triangles?

Analyze Refer to the inscribed octagon to answer problems **10–14.**

10. Segments *MW,* and *MZ* are radii of the same circle. Classify $\triangle WMZ$ by sides.

11. What is the measure of $\angle WMZ$? How do you know?

12. What are the measures of $\angle MWZ$ and $\angle MZW$? How do you know?

13. What is the measure of $\angle YWZ$? How do you know?

14. What is the measure of each inscribed angle formed by the sides of the octagon?

LESSON

101

• Translating Expressions into Equations

Power Up

Building Power

facts

Power Up Q

mental math[1]

a. **Positive/Negative:** $(-2)(-2)(-2)(-2)$

b. **Order of Operations/Exponents:** $(5 \times 10^{-5})(6 \times 10^{-6})$

c. **Ratio:** $\frac{0.2}{w} = \frac{0.4}{0.12}$

d. **Measurement:** Convert $-25°C$ to degrees Fahrenheit.

e. **Fractional Parts:** $\frac{3}{4}$ of \$80

f. **Percent:** 25% less than \$80

g. **Probability:** A spinner has 4 equal sections with blue, red, orange, and purple. What is the probability that the spinner will land on orange?

h. **Percent/Estimation:** Estimate a 15% tip on a \$29.78 bill.

problem solving

Jess works at an electronics store that is having a 30% off sale. In addition to the sale, he gets an employee discount on all purchases. The original prices of the items he buys are \$22.50, \$37.85, and \$19.23. If the total price after the sale and employee discounts are applied is \$49.02, what percentage is Jess's employee discount?

New Concept

Increasing Knowledge

An essential skill in mathematics is the ability to translate language, situations, and relationships into mathematical form. Since the earliest lessons of this book we have practiced translating word problems into equations that we then solved. In this lesson we will practice translating other common patterns of language into algebraic form. We will also use our knowledge of geometric relationships to write equations to solve geometry problems.

The table on the next page shows examples of mathematical language translated into algebraic form. Notice that the word *number* is represented by a letter in italic form. We can use any letter to represent an unknown number.

Examples of Translations

English	Symbols
twice a number	$2n$
five more than a number	$x + 5$
three less than a number	$a - 3$
half of a number	$\frac{1}{2}h$ or $\frac{h}{2}$
the product of a number and seven	$7b$
Seventeen is five more than twice a number.	$17 = 2n + 5$

The last translation in the examples above resulted in an equation that can be solved to find the unstated number. We will solve a similar equation in the next example.

Example 1

If five less than twice a number is seventeen, what is the number?

Solution

We will use the letter x to represent the unknown as we translate the sentence into an equation.

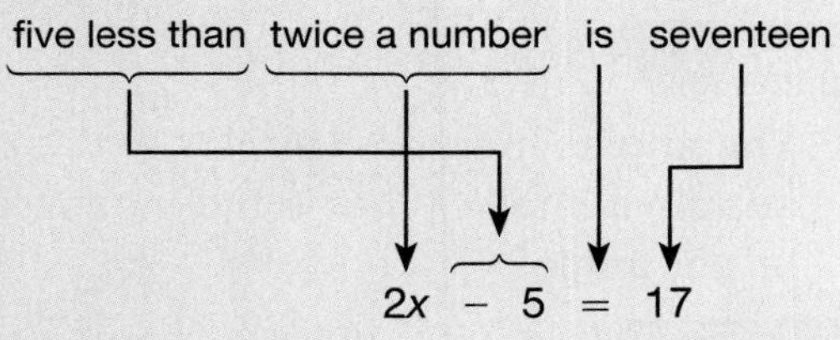

Now we solve the equation.

$$2x - 5 = 17 \quad \text{equation}$$

$$2x - 5 + 5 = 17 + 5 \quad \text{added 5 to both sides}$$

$$2x = 22 \quad \text{simplified}$$

$$\frac{2x}{2} = \frac{22}{2} \quad \text{divided both sides by 2}$$

$$x = 11 \quad \text{simplified}$$

We find that the number described is **11.** Five less than twice eleven is seventeen.

Formulate Write and simplify an equation that can be used to verify that 11 is the correct number.

In example 2 on the next page, we will translate a situation involving variables into an equation. Then we will use the equation to solve the problem.

Example 2

A taxi company charges \$1.50 plus \$3.00 per mile metered at tenths of a mile. Write an equation that relates the total fare (*f*) in dollars to the number of miles (*m*) of a taxi ride. Then use the equation to find the fare for a 6.4 mile taxi ride.

Solution

To find the fare (the price of the ride), we multiply the distance in miles (to the tenth of a mile) by \$3.00. Then we add \$1.50. Without the dollar signs the formula is:

$$f = 3m + 1.5$$

To find the fare for a 6.4 mile ride we substitute 6.4 in place of m and simplify to find f.

$f = 3(6.4) + 1.5$	Substituted 6.4
$f = 19.2 + 1.5$	Multiplied 3 and 6.4
$f = 20.7$	Added 19.2 and 1.5

Now we express the answer in dollar form. The fare for the ride is **\$20.70.**

Generalize Find the fare for a 10-mile ride.

We can also translate geometric relationships into algebraic expressions.

Example 3

The angles marked *x* and 2*x* in this figure are supplementary. What is the measure of the larger angle?

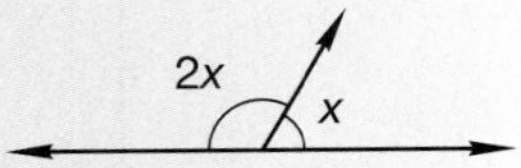

Solution

The sum of the angle measures is 180°. We write this relationship as an equation.

$$2x + x = 180°$$

Since $2x + x = 3x$, we may simplify then solve the equation.

$2x + x = 180°$	equation
$3x = 180°$	simplified
$\frac{3x}{3} = \frac{180°}{3}$	divided both sides by 3
$x = 60°$	simplified

The solution of the equation is 60°, but 60° is not the answer to the question. We were asked to find the measure of the larger angle, which in the diagram is marked $2x$. Since x is 60°, we find that the larger angle measures $2(60°) =$ **120°.**

Discuss How are angles x and 2x related in size?

Practice Set

Formulate Write and solve an equation for each of these problems:

a. Six more than the product of a number and three is 30. What is the number?

b. Ten less than half of what number is 30?

c. What is the measure of the smallest angle in this figure?

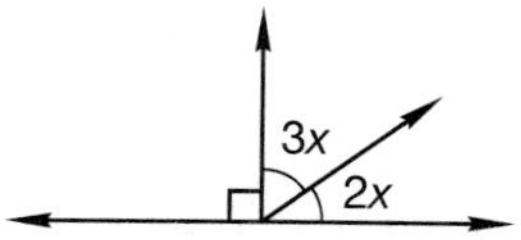

d. Find the measure of each angle of this triangle.

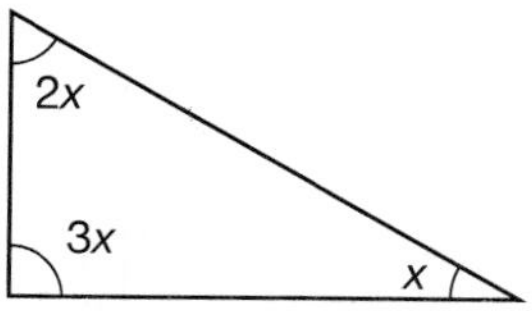

e. An online movie ticket service charges \$8 per ticket plus a \$2 transaction fee. Write an equation that relates the price (p) of buying tickets (t) through this service. Then use the equation to find the cost of 5 tickets.

Written Practice

Strengthening Concepts

1. (Inv. 4) The following numbers are Katie's 100-meter dash times, in seconds, during track season. Find the **a** median, **b** mode, and **c** range of these times.

12.3, 11.8, 11.9, 11.7, 12.0, 11.9, 12.1, 11.6, 11.8

2. (54) Jackson earns \$10 per hour at his part-time job. How much money does he earn for working 3 hours and 45 minutes?

Use ratio boxes to solve problems **3–6.**

3. (72) The recipe called for 3 cups of flour and 2 eggs to make 6 servings. If 15 cups of flour were used to make more servings, how many eggs should be used?

4. (72) Lester can type 48 words per minute. At that rate, how many words can he type in 90 seconds?

5. (81) Ten students were wearing athletic shoes. This was 40 percent of the class. How many students were in the class?

6. (92) The dress was on sale for 40 percent off the regular price. If the regular price was \$45, what was the sale price?

*** 7.** (96) **Generalize** Use the Distributive Property to clear parentheses. Then simplify by adding like terms.

$$3(x - 4) - x$$

8. (88) Use two unit multipliers to convert 3 gallons to pints.

9. (36, 71) Diagram this statement. Then answer the questions that follow.

The Trotters won $\frac{5}{6}$ of their games. They won 20 games and lost the rest.

a. How many games did they play?

b. What was the Trotters' win-loss ratio?

*** 10.** (100) **Generalize** Between which two consecutive whole numbers is $\sqrt{200}$?

*** 11.** (9) **Analyze** Compare: $w \bigcirc m$ if w is 0.5 and m is the reciprocal of w.

12. (75) Find the area of the hexagon shown at right. Dimensions are in centimeters. Corners that look square are square.

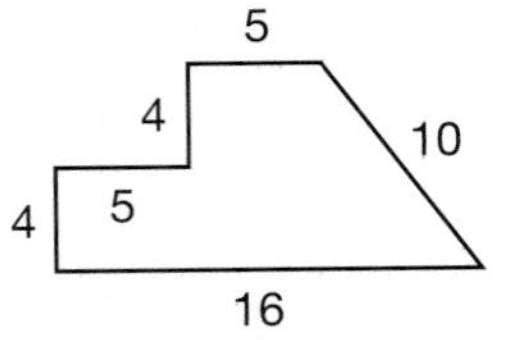

*** 13.** (101) **Formulate** Write an equation to solve this problem:

Three less than the product of six and what number is 45?

14. (83) Multiply. Write the product in scientific notation.

$$(8 \times 10^8)(4 \times 10^{-2})$$

15. (48) Complete the table.

Fraction	Decimal	Percent
a.	0.02	**b.**
c.	**d.**	0.2%

16. (56, Inv. 9) Find the missing numbers in the table by using the function rule. Then graph the function on a coordinate plane.

$y = 2x + 1$

x	y
−1	
0	
1	
2	

17. (95) Find the volume of this solid. Dimensions are in inches.

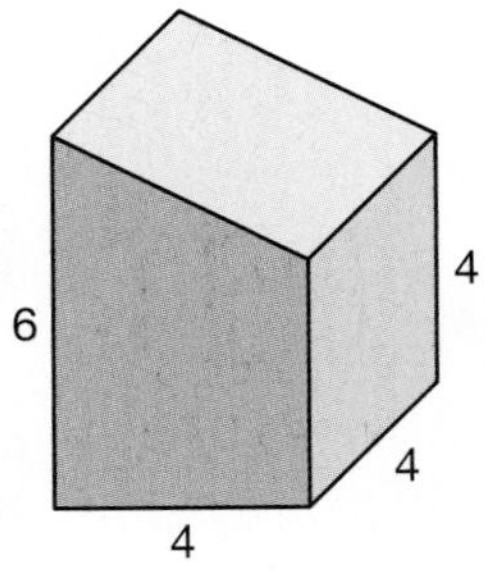

18. (65, 82) Find each measure of the circle at right. (Leave π as π.)

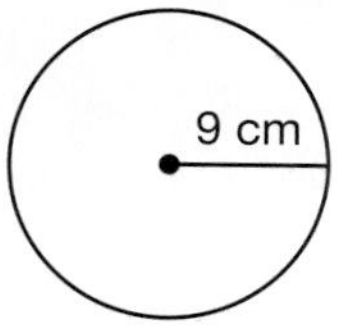

a. circumference

b. area

*** 19.** (Inv. 8) If a number cube is rolled what are the odds of not getting 6?

*** 20.** (101) **Conclude** The two acute angles in this figure are complementary. What are the measures of the two angles?

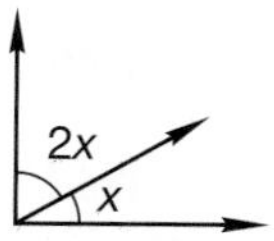

21. (44) Divide 1.23 by 9 and write the quotient

a. with a bar over the repetend.

b. rounded to three decimal places.

*** 22.** (99) **Analyze** If *BC* is 9 cm and *AC* is 12 cm, then what is *AB*?

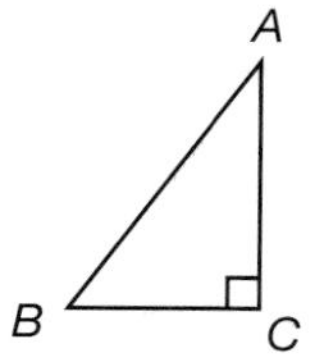

*** 23.** (98) **Model** The scale factor from $\triangle ABC$ in problem 22 to $\triangle DEF$ is 2. Draw $\triangle DEF$, then find

a. the perimeter of $\triangle DEF$.

b. the area of $\triangle DEF$.

*** 24.** (94) In a bag were 6 red marbles and 4 blue marbles. If Lily pulls a marble out of the bag with her left hand and then pulls a marble out with her right hand, what is the probability that the marble in each hand will be blue?

25. (90) Solve: $3\frac{1}{7}d = 88$

*** 26.** (93) **Model** Solve this inequality and graph its solution:

$$3x + 20 \geq 14$$

Simplify:

27. (52) $5^2 + (3^3 - \sqrt{81})$

*** 28.** (96) $3x + 2(x - 1)$

29. (43) $\left(4\frac{4}{9}\right)(2.7)\left(1\frac{1}{3}\right)$

30. (85) $(-2)(-3) - (-4)(-5)$

LESSON

102

• Transversals
• Simplifying Equations

Power Up

Building Power

facts

Power Up U

mental math

a. **Positive/Negative:** $(-0.25) + (-0.75)$

b. **Order of Operations/Exponents:** $(3 \times 10^{10})(2 \times 10^{-2})$

c. **Algebra:** $3x + 2\frac{1}{2} = 10$

d. **Measurement:** Convert 500 mL to L.

e. **Percent:** 150% of \$40

f. **Percent:** 150% more than \$40

g. **Probability:** A spinner has 3 sections with red, blue, and red. What is the probability that the spinner will land on red?

h. **Calculation:** Start with a score, $-$ 5, $\times$ 2, + 2, $\div$ 4, + 1, $\sqrt{\ }$, $\times$ 7, $-$ 1, $\div$ 10.

problem solving

The first computer can complete a payroll for 720 employees in 8 hours. The second computer can complete the same payroll in 12 hours. Working together, how long will it take the two computers to complete a payroll for 720 employees?

New Concepts

Increasing Knowledge

transversals

A **transversal** is a line that intersects one or more other lines in a plane. In this lesson we will pay particular attention to the angles formed when a transversal intersects a pair of parallel lines. Notice the eight angles that are formed.

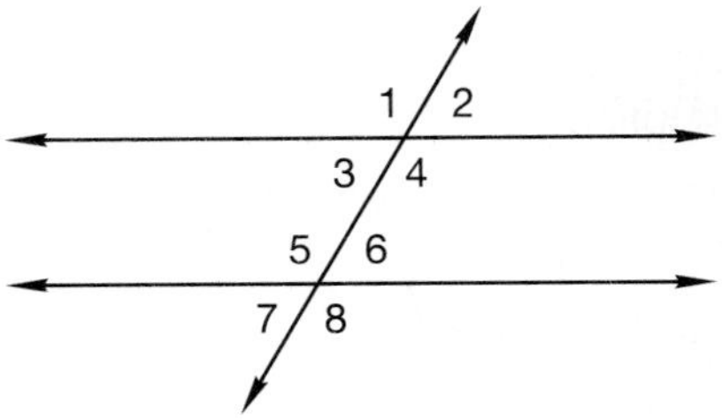

In this figure there are four acute angles numbered 2, 3, 6, and 7, and there are four obtuse angles numbered 1, 4, 5, and 8. All of the acute angles have the same measure, and all of the obtuse angles have the same measure.

Example 1

Thinking Skill

Conclude

When would all eight angles formed by a transversal intersecting two parallel lines be equal?

Transversal *t* intersects parallel lines *l* and *m* so that the measure of ∠*a* is 105°. Find the measure of angles *b–h*.

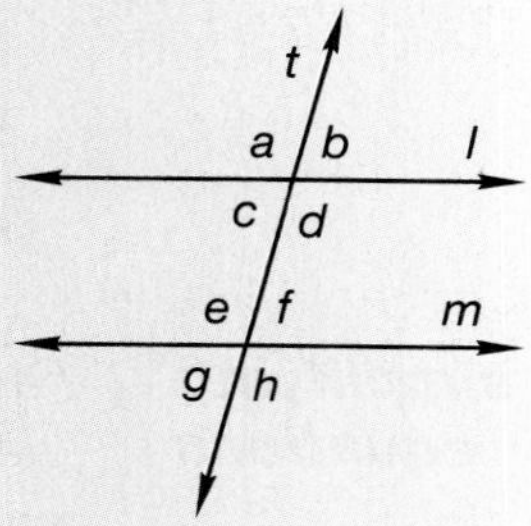

Solution

All the obtuse angles have the same measure, so ***∠a, ∠d, ∠e,*** and ***∠h* each measure 105°.** Each of the acute angles is a supplement of an obtuse angle, so each acute angle measures

$$180° - 105° = 75°$$

Thus, ***∠b, ∠c, ∠f,*** and ***∠g* each measure 75°.**

When a transversal intersects a pair of lines, special pairs of angles are formed. In example 1, ∠*a* and ∠*e* are **corresponding angles** because the position of ∠*e* corresponds to the position of ∠*a* (to the left of the transversal and above lines *m* and *l*). Name three more pairs of corresponding angles in example 1.

Angle *a* and ∠*h* in example 1 also form a special pair of angles. They are on alternate sides of the transversal and are outside of (not between) the parallel lines. So ∠*a* and ∠*h* are called **alternate exterior angles.** Name another pair of alternate exterior angles in example 1.

Angles *d* and *e* in example 1 are **alternate interior angles** because they are on alternate sides of the transversal and in the interior of (between) the parallel lines. Name another pair of alternate interior angles in example 1.

Model Draw 2 vertical parallel lines with a transversal intersecting them. Label the angles formed 1–8. List the angles that have the same measure.

Example 2

Transversal *r* intersects parallel lines *p* and *q* to form angles 1–8.

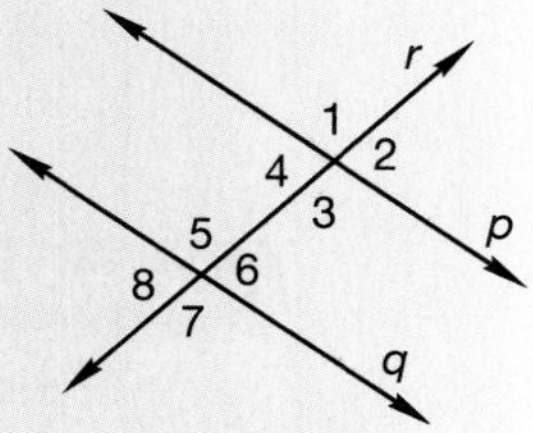

a. Name four pairs of corresponding angles.

b. Name two pairs of alternate exterior angles.

c. Name two pairs of alternate interior angles.

Solution

a. $\angle 1$ **and** $\angle 5$**,** $\angle 2$ **and** $\angle 6$**,** $\angle 3$ **and** $\angle 7$**,** $\angle 4$ **and** $\angle 8$

b. $\angle 1$ **and** $\angle 7$**,** $\angle 2$ **and** $\angle 8$

c. $\angle 4$ **and** $\angle 6$**,** $\angle 3$ **and** $\angle 5$

simplifying equations

One step to solving some equations is to collect like terms. In this equation we can collect variable terms and constant terms as the first step of the solution.

$$3x - x - 5 + 8 = 17$$
$$(3x - x) + (-5 + 8) = 17$$
$$2x + 3 = 17$$

Math Language

Here we use the term *collect* to mean "to bring together in one place."

Example 3

Simplify and then solve this equation.

$$\mathbf{3x + 5 - x = 17 + x - x}$$

Solution

We collect like terms on each side of the equal sign. On the left side, combining $3x$ and $-x$ gives us $2x$. On the right side, combining $+x$ and $-x$ gives us zero.

$3x + 5 - x = 17 + x - x$	equation
$3x - x + 5 = 17 + x - x$	Commutative Property
$(3x - x) + 5 = 17 + (x - x)$	Associative Property
$2x + 5 = 17$	simplified

Now we solve the simplified equation.

$2x + 5 = 17$	equation
$2x + 5 - 5 = 17 - 5$	subtracted 5 from both sides
$2x = 12$	simplified
$\frac{2x}{2} = \frac{12}{2}$	divided both sides by 2
$x = \mathbf{6}$	simplified

Predict If the example included the term x^2, would x^2 be collected as a like term with $3x$ and $-x$? Support your reasoning.

Example 4

Simplify this equation by removing the variable term from one side of the equation. Then solve the equation.

$$\mathbf{5x - 17 = 2x - 5}$$

Solution

We see an x-term on both sides of the equal sign. We may remove the x-term from either side. We choose to remove the variable term from the right side. We do this by subtracting $2x$ from both sides of the equation.

$5x - 17 = 2x - 5$	equation
$5x - 17 - 2x = 2x - 5 - 2x$	subtracted $2x$ from both sides
$3x - 17 = -5$	simplified

Now we solve the simplified equation.

$3x - 17 = -5$	equation
$3x - 17 + 17 = -5 + 17$	added 17 to both sides
$3x = 12$	simplified
$\frac{3x}{3} = \frac{12}{3}$	divided both sides by 3
$x = \mathbf{4}$	simplified

Example 5

Solve: $3x + 2(x - 4) = 32$

Solution

We first apply the Distributive Property to clear parentheses.

$3x + 2(x - 4) = 32$	equation
$3x + 2x - 8 = 32$	Distributive Property
$5x - 8 = 32$	added $3x$ and $2x$
$5x = 40$	added 8 to both sides
$x = \mathbf{8}$	multiplied both sides by $\frac{1}{5}$

Practice Set

Refer to this figure to answer problems **a–d.**

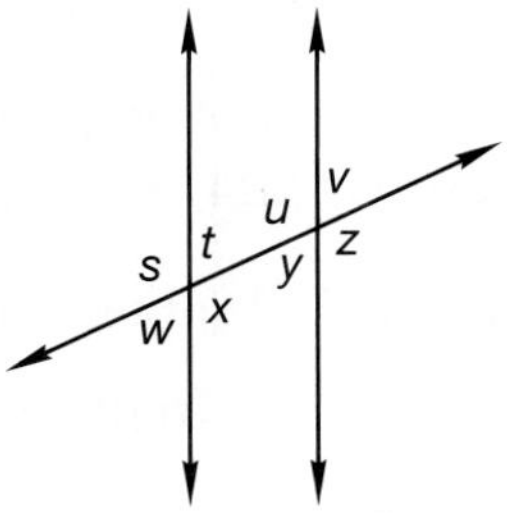

a. Name four pairs of corresponding angles.

b. Name two pairs of alternate interior angles.

c. Name two pairs of alternate exterior angles.

d. Conclude If the measure of $\angle w$ is 80°, what is the measure of each of the other angles?

Solve the following equations. Show and justify each step.

e. $3w - 10 + w = 90$

f. $x + x + 10 + 2x - 10 = 180$

g. $3y + 5 = y - 25$

h. $4n - 5 = 2n + 3$

i. $3x - 2(x - 4) = 32$

j. $3x = 2(x - 4)$

Written Practice *Strengthening Concepts*

*** 1.** (36, Inv. 8) Jorge is playing a board game and hopes to roll a sum of 5. He tosses a pair of number cubes.

a. What is the probability that the sum of the numbers tossed will be 5?

b. What are the odds that the sum of the numbers tossed will not be 5?

*** 2.** (101) **Formulate** Write an equation to solve this problem:

Twelve less than the product of what number and three is 36?

*** 3.** (89) The figure shows three sides of a regular decagon.

a. What is the measure of each exterior angle?

b. What is the measure of each interior angle?

Use ratio boxes to solve problems **4–7.**

4. (66) The ratio of youths to adults in the stadium was 3 to 7. If 4500 people were in the stadium, how many adults were present?

5. (72) Every time the knight moved over 2, he moved up 1. If the knight moved over 8, how far did he move up?

6. (81) Eighty percent of those who were invited came to the party. If 40 people were invited to the party, how many did not come?

7. (81) The dress was on sale for 60 percent of the regular price. If the sale price was $24, what was the regular price?

*** 8.** (101) **Formulate** Write an equation to solve this problem:

Three more than twice a number is -13.

*** 9.** (101) **Justify** The obtuse and acute angles in the figure at the right are supplementary. Find the measure of each angle. Show your work.

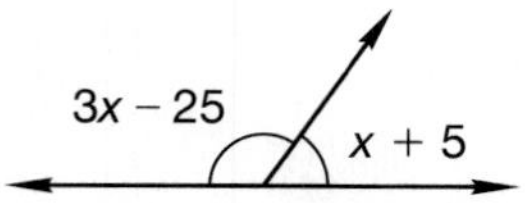

10. (71) Read this statement. Then answer the questions that follow.

Exit polls showed that 7 out of 10 voters cast their ballots for the incumbent. The incumbent received 1400 votes.

a. How many people were surveyed?

b. What percent of people did not vote for the incumbent?

* **11.** (91) **Evaluate** $x + xy - xy$ if $x = 3$ and $y = -2$

12. (99) **Verify** Is this triangle a right triangle? Explain how you know.

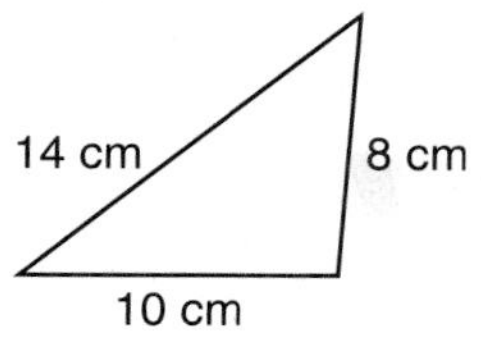

13. (32) If the perimeter of a square is 1 meter, what is the area of the square in square centimeters?

14. (60) Find the total price, including tax, of a \$12.95 bat, a \$7.85 baseball, and a \$49.50 glove. The tax rate is 7 percent.

15. (83) Multiply. Write the product in scientific notation.

$$(3.5 \times 10^5)(3 \times 10^6)$$

* **16.** (102) **Conclude** Lines *l* and *m* are parallel and are intersected by transversal *q*.

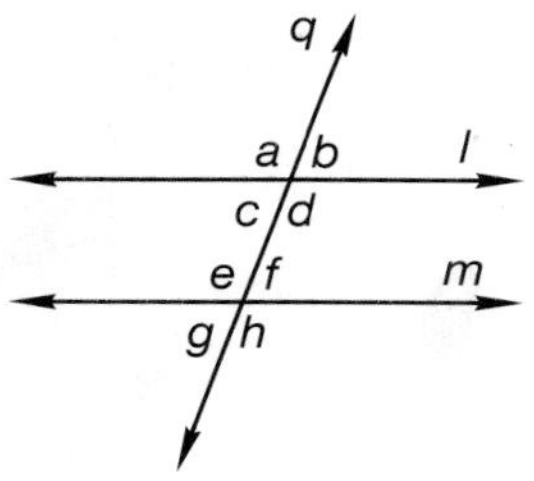

a. Which angle corresponds to $\angle c$?

b. Which angle is the alternate interior angle of $\angle e$?

c. Which angle is the alternate exterior angle of $\angle h$?

d. If $m\angle a$ is 110°, what is $m\angle f$?

17. (60, 92) **a.** What number is 125% of 84?

b. What number is 25% more than 84?

18. (Inv. 8) If the chance of rain is 40%, what are the odds that it will not rain?

19. (70) What is the volume of this rectangular prism? Dimensions are in feet.

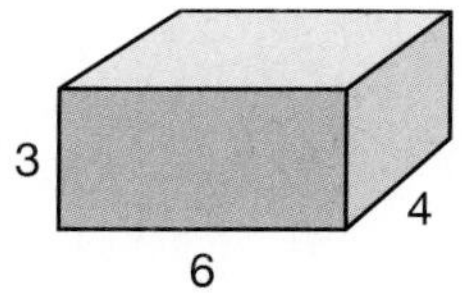

20. (65, 82) Find both measures of the circle at right. (Use $\frac{22}{7}$ for π.)

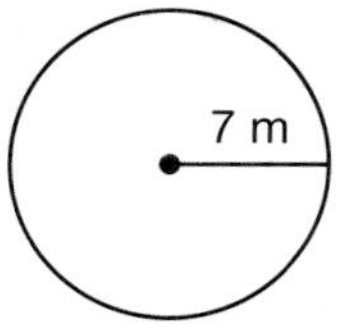

a. circumference

b. area

21. Create a table using the function rule $y = 3x$. In the table choose -1, 0, and 2 for x and find the value of y. Then graph the function on a coordinate plane.
(56, Inv. 9)

22. Polygon $ZWXY$ is a rectangle. Find the measures of the following angles:
(40)

a. $\angle a$ **b.** $\angle b$ **c.** $\angle c$

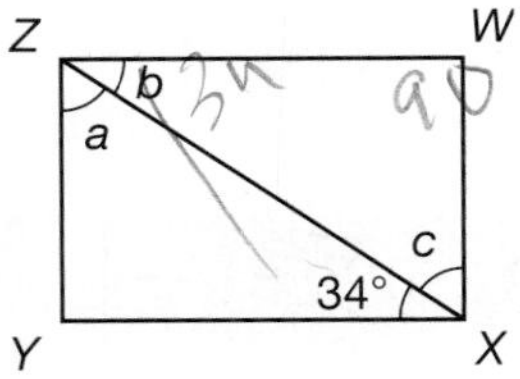

23. Graph on a number line: $x \geq -2$
(78)

* **24.** **Model** Draw a number line and show the locations of these numbers. Which are rational?
(100)

$$0.4, \frac{1}{4}, \sqrt{4}$$

Generalize Solve:

* **25.** $3x + x + 3^0 = 49$
(57, 102)

* **26.** $3y + 2 = y + 32$
(102)

* **27.** $x + 2(x + 3) = 36$
(102)

Simplify:

28. **a.** $(3x^2y)(-2x)(xy^2)$
(84, 87)

b. $-3x + 2y - x - y$

c. $(-2xy^2)^2$

29. $\left(4\frac{1}{2}\right)(0.2)(10^2)$
(43)

30. $\frac{(-4)(+3)}{(-2)} - (-1)$
(85)

Early Finishers
Choose A Strategy

The figure below is made up of a regular octagon and eight isosceles triangles. The height of each triangle is equal to the distance from the center of the octagon to the midpoint of the base of the triangle. Explain how you can determine the fractional part of the entire figure that one triangle represents.

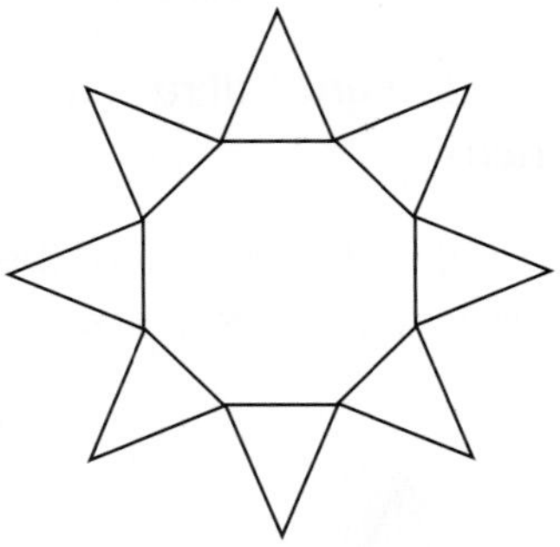

LESSON 103

• Powers of Negative Numbers
• Dividing Terms
• Square Roots of Monomials

Power Up

Building Power

facts Power Up U

mental math

a. **Positive/Negative:** $(-2.5) \div (-5)$

b. **Order of Operations/Exponents:** $(3 \times 10^{-4})(4 \times 10^{-3})$

c. **Algebra:** $2y + y = 45$

d. **Measurement:** Convert −30°C to degrees Fahrenheit.

e. **Percent:** $33\frac{1}{3}\%$ of \$600

f. **Percent:** $33\frac{1}{3}\%$ less than \$600

g. **Geometry:** If you are buying sod for your yard, what will you need to find to determine how much sod you need to cover your yard?

h. **Scientific Notation:** The expression $(3 \times 10^3)^2$ means $(3 \times 10^3)(3 \times 10^3)$. Write the product in scientific notation.

problem solving

Sheena's younger sister quickly calculated several products on a piece of scrap paper. In her hurry, she misplaced the product of 0.203 and 4.16. She determined from her notes that the product was 0.84448, 8.21218, 0.88444, or 0.04848. Without calculating, Sheena told her sister the correct answer. What is the correct answer, and how did Sheena know?

New Concepts

Increasing Knowledge

powers of negative numbers

One way to multiply three or more signed numbers is to multiply the factors in order from left to right, keeping track of the signs with each step, as we show here:

$(-3)(-4)(+5)(-2)(+3)$	problem
$(+12)(+5)(-2)(+3)$	multiplied $(-3)(-4)$
$(+60)(-2)(+3)$	multiplied $(+12)(+5)$
$(-120)(+3)$	multiplied $(+60)(-2)$
-360	multiplied $(-120)(+3)$

Another way to keep track of signs when multiplying signed numbers is to count the number of negative factors. Notice the pattern in the multiplications on the following page.

$$-1 = -1 \quad \text{odd}$$
$$(-1)(-1) = +1 \quad \text{even}$$
$$(-1)(-1)(-1) = -1 \quad \text{odd}$$
$$(-1)(-1)(-1)(-1) = +1 \quad \text{even}$$
$$(-1)(-1)(-1)(-1)(-1) = -1 \quad \text{odd}$$

If the number of negative factors is even, the product is positive. If the number of negative factors is odd, the product is negative.

Thinking Skill

Explain

Tell how to find the following product without multiplying: (–2)(4)(–5)(0)(5).

Note: If any of the multiplied factors are zero, their product is also zero (which has no sign).

Example 1

Find the product: (+3)(+4)(−5)(−2)(−3)

Solution

There are three negative factors (an odd number), so the product will be a negative number. We multiply and get

$$(+3)(+4)(-5)(-2)(-3) = \mathbf{-360}$$

We did not need to count the number of positive factors, because positive factors do not change the sign of a product.

We remember that an exponent can indicate how many times the base is used as a factor.

$$(-3)^4 \text{ means } (-3)(-3)(-3)(-3)$$

Example 2

Simplify:

a. $\mathbf{(-2)^4}$ **b.** $\mathbf{(-2)^5}$

Solution

a. The expression $(-2)^4$ means $(-2)(-2)(-2)(-2)$. There is an even number of negative factors, so the product is a positive number. Since 2^4 is 16, we find that $(-2)^4$ is **+16.**

b. The expression $(-2)^5$ means $(-2)(-2)(-2)(-2)(-2)$. This time there is an odd number of negative factors, so the product is a negative number. Since $2^5 = 32$, we find that $(-2)^5 =$ **−32.**

dividing terms

We divide terms by removing pairs of factors that equal 1.

$$\frac{4a^3b^2}{ab} = \frac{2 \cdot 2 \cdot \overset{1}{\cancel{a}} \cdot a \cdot a \cdot \overset{1}{\cancel{b}} \cdot b}{\underset{1}{\cancel{a}} \cdot \underset{1}{\cancel{b}}}$$
$$= 4a^2b$$

Example 3

Simplify: $\frac{12x^3yz^2}{3x^2y}$

Solution

We factor the two terms and remove pairs of factors that equal 1. Then we regroup the remaining factors.

$$\frac{12x^3yz^2}{3x^2y} = \frac{2 \cdot 2 \cdot \overset{1}{\cancel{3}} \cdot \overset{1}{\cancel{x}} \cdot \overset{1}{\cancel{x}} \cdot x \cdot \overset{1}{\cancel{y}} \cdot z \cdot z}{\underset{1}{\cancel{3}} \cdot \underset{1}{\cancel{x}} \cdot \underset{1}{\cancel{x}} \cdot \underset{1}{\cancel{y}}}$$

$$= \mathbf{4xz^2}$$

Summarize In $4xz^2$, what exponent do you understand x to have?

Example 4

Simplify: $\frac{10a^3bc^2}{8ab^2c}$

Solution

We factor the terms and remove common factors. Then we regroup the remaining factors.

$$\frac{10a^3bc^2}{8ab^2c} = \frac{\overset{1}{\cancel{2}} \cdot 5 \cdot \overset{1}{\cancel{a}} \cdot a \cdot a \cdot \overset{1}{\cancel{b}} \cdot \overset{1}{\cancel{c}} \cdot c}{\underset{1}{\cancel{2}} \cdot 2 \cdot 2 \cdot \underset{1}{\cancel{a}} \cdot \underset{1}{\cancel{b}} \cdot b \cdot \underset{1}{\cancel{c}}}$$

$$= \mathbf{\frac{5a^2c}{4b}}$$

square roots of monomials

A monomial is a perfect square if its prime factorization can be separated into two identical groups of factors.

$$25x^2 = 5 \cdot 5 \cdot x \cdot x$$
$$= 5x \cdot 5x$$

Since the factors of $25x^2$ can be separated into $5x \cdot 5x$, we know that $25x^2$ is a perfect square and that $5x$ is a square root of $25x^2$.

Example 5

Which monomial is a perfect square?

A $8x^2y^4$ **B** $4x^2y$ **C** $36x^2y^4$

Solution

A monomial is a perfect square if the coefficient is a perfect square (like 4 or 36) and if the exponents of the variables are all even (like x^2y^4). Thus $8x^2y^4$ is not a perfect square because 8 is not a perfect square, and $4x^2y$ is not a perfect square because the exponent of $y(1)$ is odd. The perfect square is choice **C** $\mathbf{36x^2y^4}$.

Example 6

Simplify: $\sqrt{36x^2y^4}$

Solution

We show two methods.

Method 1: Separate the factors of $36x^2y^4$ into two identical groups.

$$36x^2y^4 = 2 \cdot 2 \cdot 3 \cdot 3 \cdot x \cdot x \cdot y \cdot y \cdot y \cdot y$$
$$= (2 \cdot 3 \cdot x \cdot y \cdot y)(2 \cdot 3 \cdot x \cdot y \cdot y)$$
$$= (6xy^2)(6xy^2)$$

We see that a square root of $36x^2y^4$ is $\mathbf{6xy^2}$.

Method 2: Find the square root of 36 and half of each exponent.

$$\sqrt{36x^2y^4} = \mathbf{6xy^2}$$

Practice Set

Generalize Simplify:

a. $(-5)(-4)(-3)(-2)(-1)$

b. $(+5)(-4)(+3)(-2)(+1)$

c. $(-2)^3$

d. $(-3)^4$

e. $(-9)^2$

f. $(-1)^5$

g. $\frac{6a^2b^3c}{3ab}$

h. $\frac{8xy^3z^2}{6x^2y}$

i. $\frac{15mn^2p}{25m^2n^2}$

Classify For **j–m** state whether the monomial is a perfect square. If it is a perfect square, find a square root of the monomial.

j. $12x^2y^4$

k. $49a^2b^6c^4$

l. x^6y^2

m. $16a^2b^3$

Written Practice

Strengthening Concepts

1. (Inv. 4) The table below shows a tally of the scores earned by students on a class test. Find the **a** mode and **b** median of the 29 scores.

Class Test Scores

Score	Number of Students
100	|||
95	~~||||~~ |
90	~~||||~~ |
85	~~||||~~ ||||
80	||||
70	|

2. (Inv. 4) Draw a box-and-whisker plot for the data presented in problem **1.**

3. (60) The dinner bill totaled \$25. Mike left a 15% tip. How much money did Mike leave for a tip?

4. (46) The plane completed the flight in $2\frac{1}{2}$ hours. If the flight covered 1280 kilometers, what was the plane's average speed in kilometers per hour?

Use ratio boxes to solve problems **5–9.**

5. (72) Jeremy earned \$33 for 4 hours of work. How much would he earn for 7 hours of work at the same rate?

6. (36) If 40 percent of the lights were on, what was the ratio of lights on to lights off ?

7. (92) Lesley saved \$25 buying a coat at a sale that offered 20 percent off. What was the regular price of the coat?

8. (92) A shopkeeper bought the item for \$30 and sold it for 60 percent more. How much profit did the shopkeeper make on the item?

* **9.** (50, 98) **a.** The $\frac{1}{20}$ scale model of the rocket stood 54 inches high. What was the height of the actual rocket?

b. Find the height of the actual rocket in feet.

* **10.** (98) **Analyze** The volume of the rocket in problem 9 is how many times the volume of the model?

11. (60) A mile is about eight fifths kilometers. Eight fifths kilometers is how many meters?

* **12.** (99) **Analyze** Use the Pythagorean theorem to find the length of the longest side of a right triangle whose vertices are (3, 1), (3, −2), and (−1, −2).

13. (Inv. 8) Alina has made 35 out of 50 free throws. What is the statistical chance that Alina will make her next free throw?

14. (77) What percent of 25 is 20?

* **15.** (Inv. 10) **a.** **Generalize** The shaded sector is what fraction of the whole circle?

b. The unshaded sector is what percent of the circle?

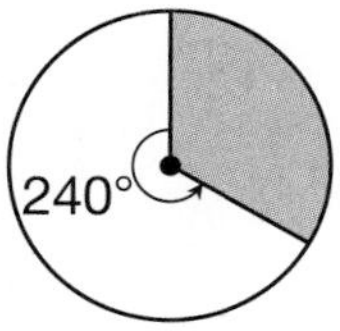

16. (56, Inv. 9) Find the missing numbers in the table by using the function rule. Then graph the function on a coordinate plane. Are x and y directly proportional? Explain your answer.

$y = -x$

x	y
2	
0	
−1	

*** 17.** (101) **Formulate** Write an equation to solve this problem:

Three less than twice what number is -7?

18. (40) Quadrilateral *ABCD* is a rectangle. The measure of $\angle ACB$ is 36°. Find the measures of the following angles:

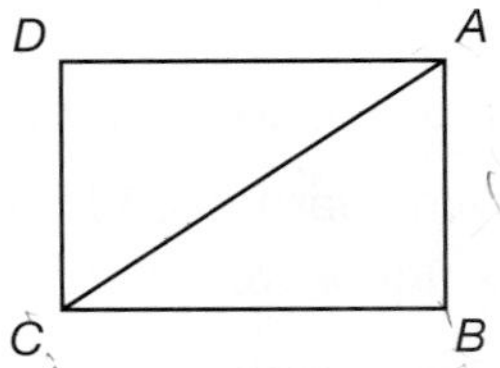

a. $\angle CAB$ **b.** $\angle CAD$ **c.** $\angle ACD$

*** 19.** (97, 98) **Generalize** The two triangles below are similar.

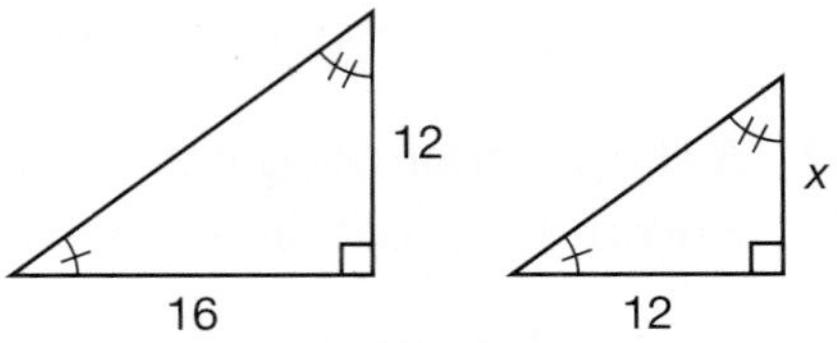

a. Estimate, then calculate, the length x.

b. Find the scale factor from the larger triangle to the smaller triangle.

*** 20.** (96) **Estimate** What is the approximate measure of $\angle AOB$. Use a protractor to verify your answer.

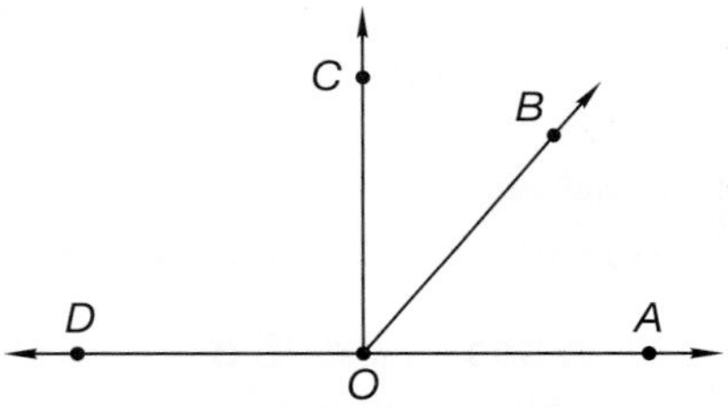

21. (65, 82) Find the **a** circumference and **b** area of a circle with a diameter of 2 feet. (Use 3.14 for π.)

*** 22.** (100) Which of these numbers is between 12 and 13?

A $\sqrt{13}$ **B** $\sqrt{130}$ **C** $\sqrt{150}$

*** 23.** (88, 95) **a.** **Justify** Find the volume of this right prism. Units are feet. Show each step of your solution.

b. Convert the volume to cubic yards.

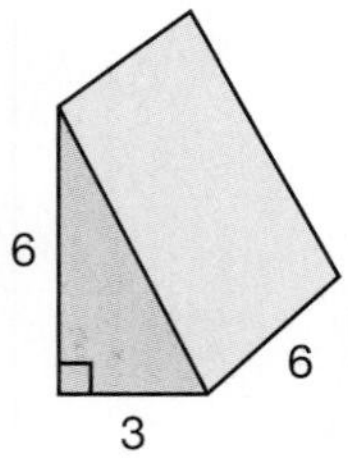

*** 24.** *(95)* Find the volume of this right circular cylinder. (Use 3.14 for π.)

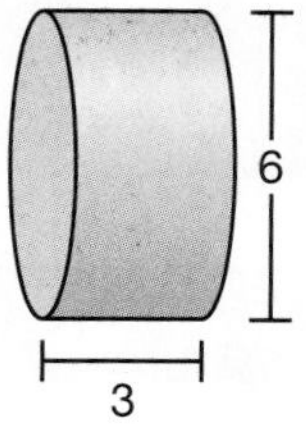

Solve:

*** 25.** *(102)* **Generalize** $3x + x - 5 = 2(x - 2)$

26. *(90, 93)* $6\frac{2}{3}f - 5 = 5$

Simplify:

27. *(26, 57)* $10\frac{1}{2} \cdot 1\frac{3}{7} \cdot 5^{-2}$

28. *(43)* $12.5 - 8\frac{1}{3} + 1\frac{1}{6}$

*** 29.** *(85, 103)* **a.** $\dfrac{(-3)(-2)(-1)}{-|(-3)(+2)|}$

b. $3^2 - (-3)^2$

*** 30.** *(103)* **a.** $\dfrac{6a^3b^2c}{2abc}$

b. $\sqrt{9x^2y^4}$

Early Finishers
Real-World Application

India is creating a map of her neighborhood as shown below. She draws a right triangle with a base of 8 cm and a height of 6 cm.

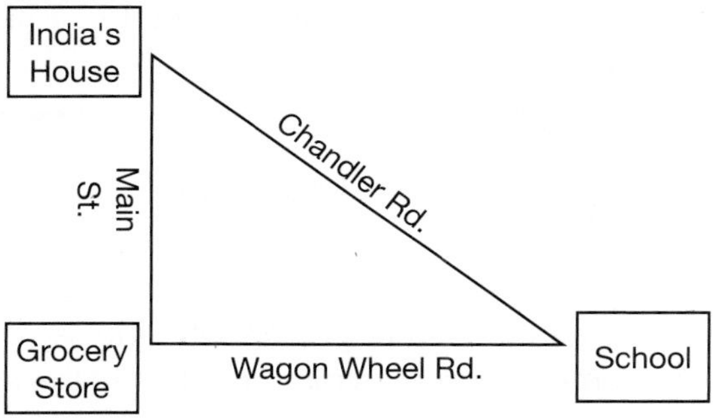

a. Redraw India's triangle on graph paper with a ruler. Measure the length of the hypotenuse.

b. Using the Pythagorean Theorem, solve for the length of the hypotenuse ($a^2 + b^2 = c^2$). Did you get the same answer as your measurement?

c. If India's scale for the map is 1 cm = 0.6 mi., how many miles is it from her house to school using Main St. then Wagon Wheel Rd.?

LESSON

104

• Semicircles, Arcs, and Sectors

Power Up

Building Power

facts Power Up V

mental math

a. **Positive/Negative:** $(-0.25) - (-0.75)$

b. **Scientific Notation:** $(5 \times 10^5)^2$

c. **Algebra:** $80 = 4m - 20$

d. **Measurement:** Convert 1 sq. yd to sq. ft.

e. **Percent:** 200% 0f $25

f. **Percent:** 200% more than $25

g. **Geometry:** A yard is shaped like an octagon and each side measures 4 ft. What is the perimeter of the yard?

h. **Rate:** At 12 mph, how far can Sherry skate in 45 minutes?

problem solving

If three standard number cubes are tossed simultaneously, what is the probability that the sum of the three number cubes will be less than or equal to 5?

New Concept

Increasing Knowledge

A **semicircle** is half of a circle. Thus the length of a semicircle is half the circumference of a circle with the same radius. The area enclosed by a semicircle and its diameter is half the area of the full circle.

Example 1

Find the perimeter of this figure. Dimensions are in meters.

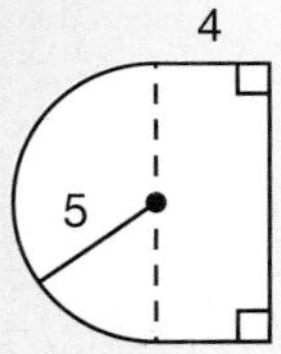

Solution

The length of the semicircle is half the circumference of a circle whose diameter is 10.

$$\text{Length of semicircle} = \frac{\pi d}{2}$$

$$\approx \frac{(3.14)(10 \text{ m})}{2}$$

$$\approx 15.7 \text{ m}$$

Now we can label all the dimensions on the figure and add them to find the perimeter.

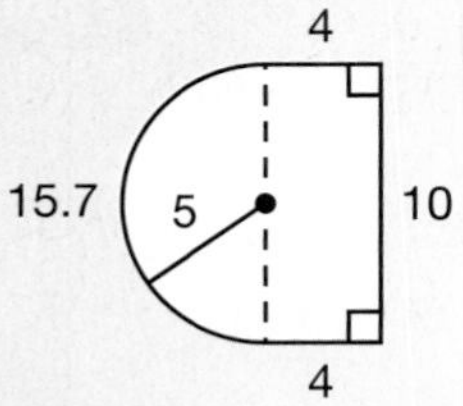

Thinking Skill

Analyze

How do we know that the height of the rectangle is 10 meters?

$$\text{Perimeter} \approx 10 \text{ m} + 4 \text{ m} + 15.7 \text{ m} + 4 \text{ m}$$
$$\approx \mathbf{33.7\ m}$$

Example 2

Find the area of this figure. Dimensions are in meters.

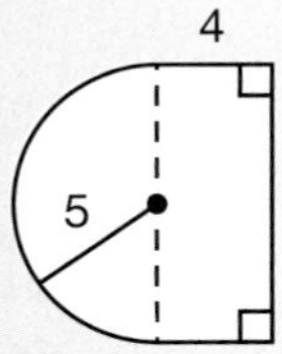

Solution

We divide the figure into two parts. Then we find the area of each part.

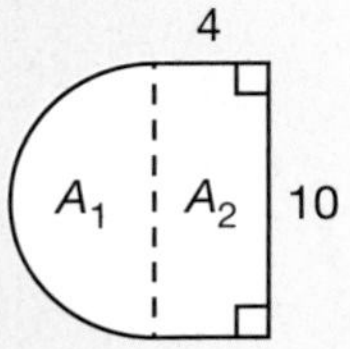

$$A_1 = \frac{\pi r^2}{2} \approx \frac{3.14(25 \text{ m}^2)}{2} \approx 39.25 \text{ m}^2$$

$$A_2 = l \times w = 4 \text{ m} \times 10 \text{ m} = 40 \text{ m}^2$$

The total area of the figure equals $A_1 + A_2$.

$$\text{Total area} = A_1 + A_2$$
$$\approx 39.25 \text{ m}^2 + 40 \text{ m}^2$$
$$\approx \mathbf{79.25\ m^2}$$

Math Language

A **sector** is a portion of a circle bound by two radii and an arc.

We can calculate the lengths of **arcs** and the areas of **sectors** by determining the fraction of a circle represented by the arc or sector.

Example 3

Find the area of the shaded sector of this circle.

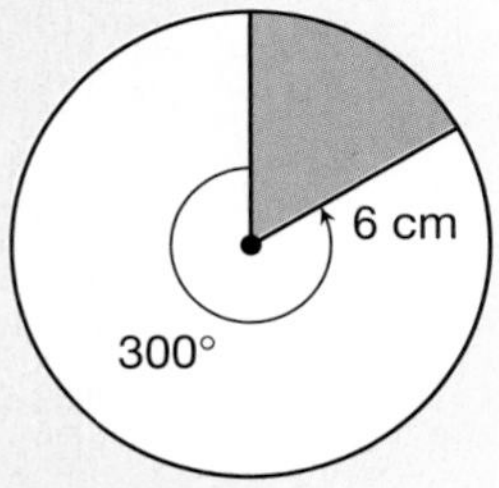

Solution

The central angle of the shaded sector measures 60° (the full circle, 360°, minus the given angle, 300°). Since 60° is $\frac{1}{6}$ of a circle ($\frac{60}{360} = \frac{1}{6}$), the area of the sector is $\frac{1}{6}$ of the area of the circle.

$$\text{Area of } 60^\circ \text{ sector} = \frac{\pi r^2}{6}$$

$$\approx \frac{3.14(\overset{1}{\cancel{6}}\text{ cm})(6\text{ cm})}{\underset{1}{\cancel{6}}}$$

$$\approx \mathbf{18.84\ cm^2}$$

As we discussed in Investigation 2, an arc is part of the circumference of a circle. In the following figure we see two arcs between point *A* and point *B*.

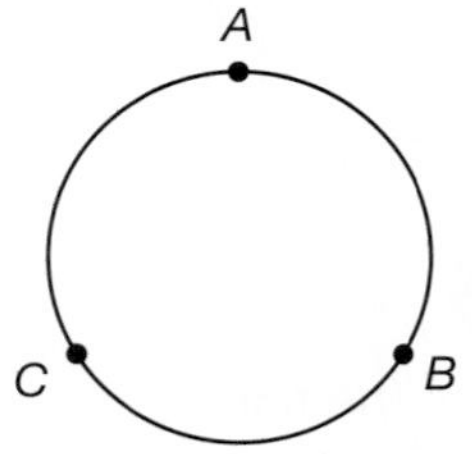

The arc from *A* clockwise to *B* is called a **minor arc** because it is less than a semicircle. We may refer to this arc as arc *AB* (abbreviated $\overset{\frown}{AB}$). The arc from *A* counterclockwise to *B* is called a **major arc** because it is greater than a semicircle. Major arcs are named with three letters. The major arc between point *A* and point *B* may be named arc *ACB*.

Analyze Name all the minor arcs found in the figure above. Then name all the major arcs.

We can measure an arc in degrees. The number of degrees in an arc equals the measure of the central angle of the arc. If minor arc *AB* in the figure above measures 120°, then the measure of major arc *ACB* is 240° because the sum of the measures of the major arc and minor arc must be 360°.

Example 4

In this figure, central angle *AOC* measures 70°. What is the measure of major arc *ADC*?

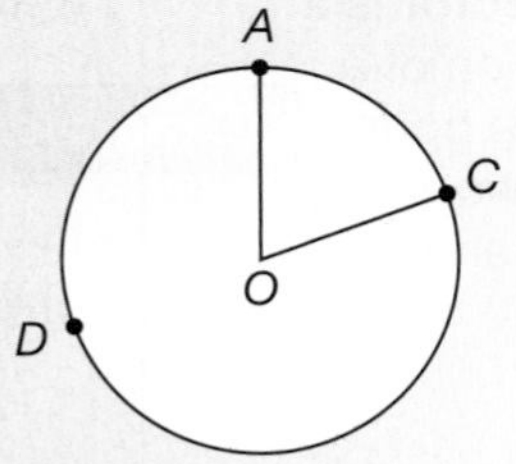

Solution

An arc may be described by the measure of its central angle. The arc in the interior of the 70° angle *AOC* is a 70° arc. However, the larger arc from point *A* counterclockwise through point *D* to point *C* measures 360° minus 70°, which is **290°.**

Example 5

A minor arc with a radius of 2, centered at the origin, is drawn from the positive *x*-axis to the positive *y*-axis. What is the length of the arc?

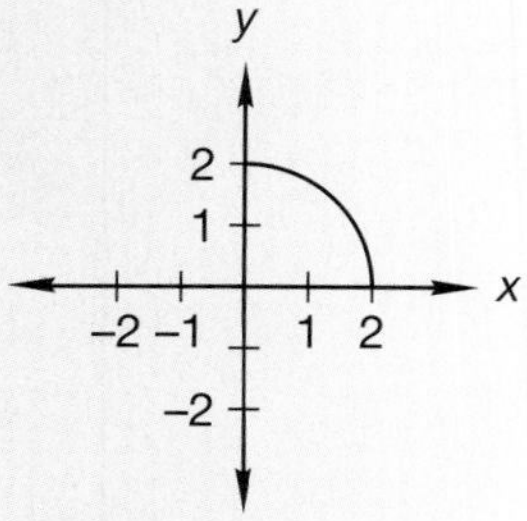

Solution

A minor arc is less than 180°. We see that the arc is $\frac{1}{4}$ of a circle, which is 90°. The length of the arc is $\frac{1}{4}$ of the circumference of a circle with a radius of 2 (and a diameter of 4).

$$\text{Length of } 90° \text{ arc} = \frac{\pi d}{4}$$

$$\approx \frac{3.14(\overset{1}{\cancel{4}} \text{ units})}{\underset{1}{\cancel{4}}}$$

$$\approx \mathbf{3.14\ units}$$

Practice Set **Analyze** Find each measure of this figure. Dimensions are in centimeters. (Use 3.14 for π.)

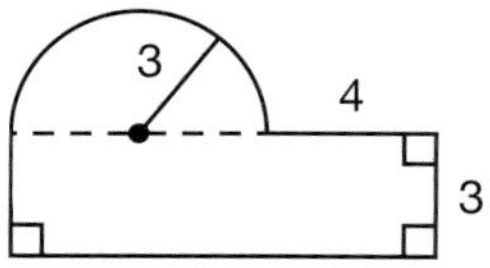

a. perimeter

b. area

c. **Explain** Describe how to find the area of this 45° sector. (Leave π as π.) Then find the area.

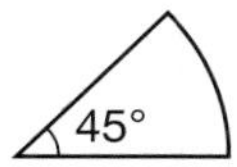

d. Find the perimeter of the figure in problem **c.** (Include the arc and two segments. Use 3.14 for π.)

Written Practice *Strengthening Concepts*

1. *(60)* The merchant sold the item for $12.50. If 40 percent of the selling price was profit, how much money did the merchant earn in profit?

*** 2.** *(94)* **Analyze** If two marbles are pulled from a bag containing eight blue marbles, seven green marbles, and six yellow marbles, what is the probability that both marbles will be green? Express the probability as a decimal number.

3. *(55)* In 10 jump-rope trials, Kiesha's average number of jumps per minute is 88. If her least number of jumps, 70, is not counted, what is her average for the remaining 9 trials?

4. *(46)* The 36-ounce container cost $3.42. The 3-pound container cost $3.84. The smaller container cost how much more per ounce than the larger container?

5. (72) Sean read 18 pages in 30 minutes. If he has finished page 128, how many hours will it take him to finish his 308-page book if he continues reading at the same rate?

6. (74) Matthew was thinking of a certain number. If $\frac{5}{6}$ of the number was 75, what was $\frac{3}{5}$ of the number?

7. (53) A naturalist collected and released 12 crayfish and 180 tadpoles from a creek. Based on this sample, what was the ratio of crayfish to tadpoles in the creek?

8. (60, 77) **Generalize** Write equations to solve **a** and **b**.

a. What percent of \$60 is \$45?

b. How much money is 45 percent of \$60?

In the figure below, $\overline{AD}$ is a diameter and $\overline{CB}$ is a radius of 12 units. Central angle *ACB* measures 60°. Refer to the figure to answer problems **9–11.** (Leave π as π.)

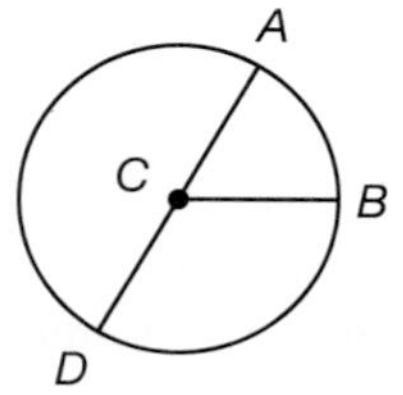

9. (65, 82) **a.** What is the area of the circle?

b. What is the circumference of the circle?

*** 10.** (104) What is the area of sector *BCD?*

*** 11.** (104) **a.** How many degrees is the major arc from *B* through *A* to *D* (arc *BAD*)?

b. How long is arc *BAD?*

12. (Inv. 9) Make a table of ordered pairs for the function $y = 2x - 1$. Use -1, 0, and 1 as x values in the table. Then graph the function on a coordinate plane.

13. (48) Complete the table on the right.

Fraction	Decimal	Percent
a.	**b.**	2.2%

*** 14.** (Inv. 9) **Analyze** The graph on the next page shows the distance a car traveled at a certain constant speed.

a. According to this graph, how far did the car travel in 1 hour 15 minutes?

b. **Analyze** Estimate the speed of the car in miles per hour as indicated by the graph.

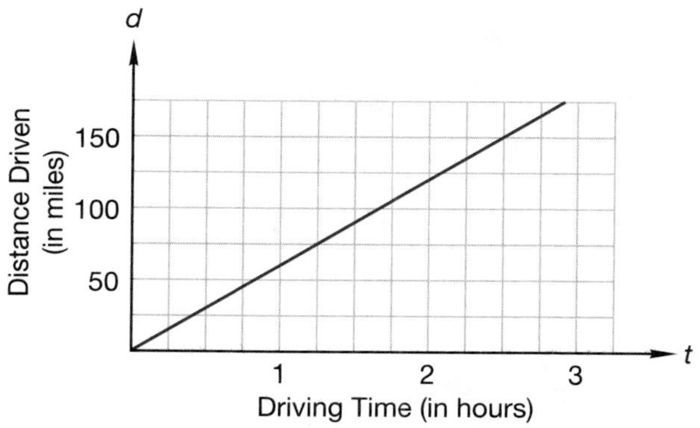

*** 15.** (91) Compare: $ab \bigcirc a - b$ if a is positive and b is negative.

16. (83) Multiply. Write the product in scientific notation.

$$(3.6 \times 10^{-4})(9 \times 10^{8})$$

*** 17.** (104) **Analyze** Find the area of the figure at right. Dimensions are in centimeters. (Use 3.14 for π.)

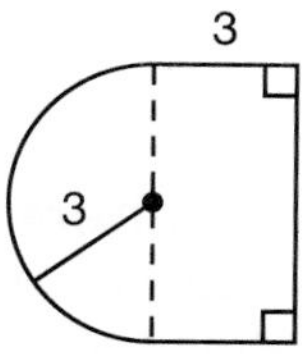

*** 18.** (104) Find the perimeter of the figure in problem **17.**

19. (67, 70) **a.** Find the volume of this solid in cubic inches. Dimensions are in feet.

b. Find the surface area of this cube in square feet.

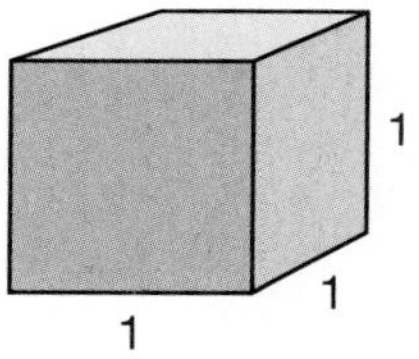

*** 20.** (96) **Predict** What angle is formed by the hands of a clock at 5:00?

21. (40) Find m$\angle x$ in the figure at right.

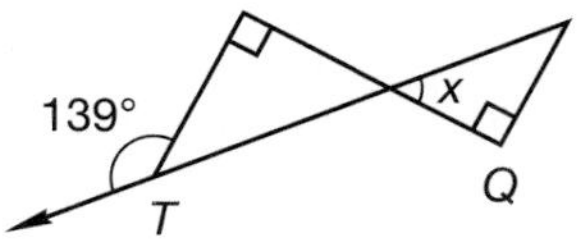

*** 22.** (97, 98) **Analyze** The triangles below are similar.

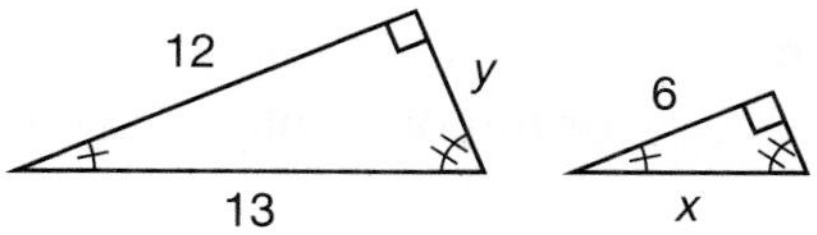

a. Find x.

b. Find the scale factor from the smaller triangle to the larger triangle.

c. The area of the larger triangle is how many times the area of the smaller triangle?

*** 23.** (99) **Justify** Find the value of y in the triangle in problem **22.** Justify your answer.

Generalize Solve and check:

24. (90, 93) $2\frac{3}{4}w + 4 = 48$

*** 25.** (102) $2.4n + 1.2n - 0.12 = 7.08$

Simplify:

26. (20) $\sqrt{(3^2)(10^2)}$

27. (102, 103) **a.** $\frac{24x^2y}{8x^3y^2}$ **b.** $2x(x - 1) - \sqrt{4x^4}$

28. (43) $12.5 - \left(8\frac{1}{3} + 1\frac{1}{6}\right)$

29. (43) $4\frac{1}{6} \div 3\frac{3}{4} \div 2.5$

*** 30.** (85, 103) **Generalize**

a. $\frac{(-3)(4)}{-2} - \frac{(-3)(-4)}{-2}$ **b.** $\frac{(-2)^3}{(-2)^2}$

Early Finishers

Real-World Application

Salim wants to create a triangular-shaped garden to grow prize roses. He plans to lay brick around the perimeter. With a tape measure Salim measured the length of the sides of the triangle to be 5 yards and 9 yards as shown below.

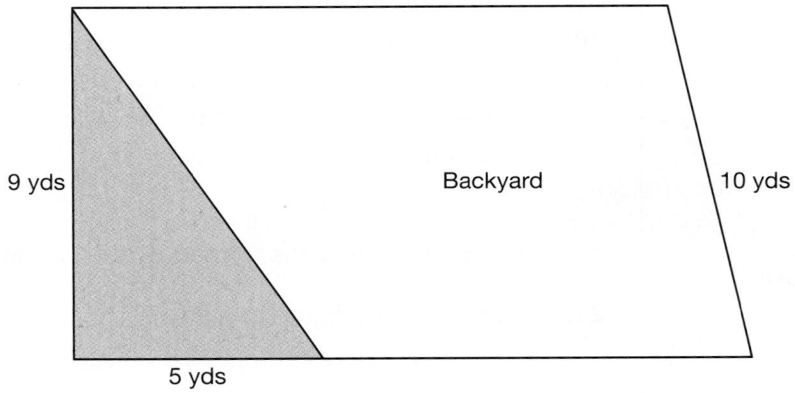

a. What is the approximate length of the third side of Salim's garden?

b. What is the approximate perimeter of the rose garden (in feet)?

c. If each brick is 7 inches long, approximately how many will he need to buy for the perimeter of the garden?

LESSON

105

• Surface Area of a Right Solid
• Surface Area of a Sphere

Power Up — *Building Power*

facts Power Up Q

mental math

a. Positive/Negative: $(-2)^4$

b. Scientific Notation: $(4 \times 10^{-4})^2$

c. Algebra: $2w + 3w = 60$

d. Measurement: Convert −35°C to degrees Fahrenheit.

e. Percent: 200% of $50

f. Percent: 100% more than $50

g. Geometry: Two triangles are similar if their corresponding sides are ____________.

h. Calculation: Square 10, − 1, ÷ 9, × 4, + 1, ÷ 9, × 10, − 1, $\sqrt{}$, × 5, + 1, $\sqrt{}$, ÷ 3.

problem solving Copy this problem and fill in the missing digits:

```
      91½
  __)___
     __
      __
      __
       _
```

New Concepts — *Increasing Knowledge*

surface area of a right solid

Recall that the total area of the surface of a geometric solid is called the **surface area** of the solid.

The block shown has six rectangular faces. We add the areas of these six faces to find the total surface area.

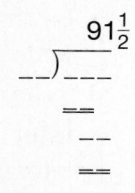

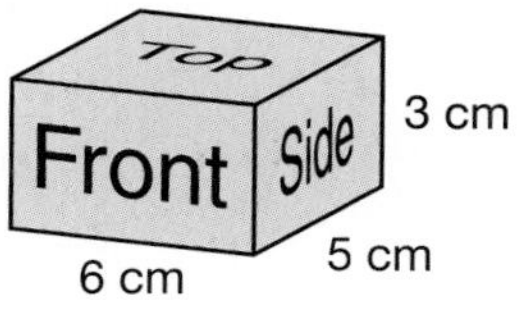

Area of top	= 5 cm × 6 cm =	30 cm^2
Area of bottom	= 5 cm × 6 cm =	30 cm^2
Area of front	= 3 cm × 6 cm =	18 cm^2
Area of back	= 3 cm × 6 cm =	18 cm^2
Area of side	= 3 cm × 5 cm =	15 cm^2
+ Area of side	= 3 cm × 5 cm =	15 cm^2
Total surface area		= 126 cm^2

Example 1

Math Language

This triangular prism is a **right solid** because its sides are perpendicular to its base(s).

Find the surface area of this triangular prism. Dimensions are in centimeters.

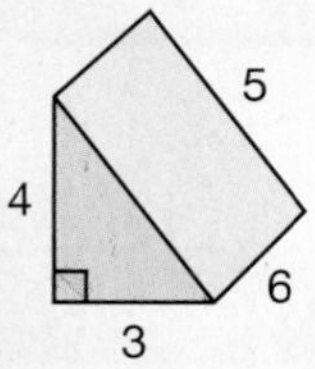

Solution

There are two triangular faces and three rectangular faces.

$$\begin{array}{rl}
\text{Area of triangle} & = \dfrac{3\text{ cm} \cdot 4\text{ cm}}{2} = 6\text{ cm}^2 \\
\text{Area of triangle} & = \dfrac{3\text{ cm} \cdot 4\text{ cm}}{2} = 6\text{ cm}^2 \\
\text{Area of rectangle} & = 6\text{ cm} \cdot 3\text{ cm} = 18\text{ cm}^2 \\
\text{Area of rectangle} & = 6\text{ cm} \cdot 4\text{ cm} = 24\text{ cm}^2 \\
+\ \text{Area of rectangle} & = 6\text{ cm} \cdot 5\text{ cm} = 30\text{ cm}^2 \\
\hline
\text{Total surface area} & = \mathbf{84\text{ cm}^2}
\end{array}$$

Discuss Would seeing every face of a right solid be helpful in finding its surface area? Is there a way that we could see every face?

The triangular prism in example 1 has two bases that are triangles and three lateral faces that are rectangles.

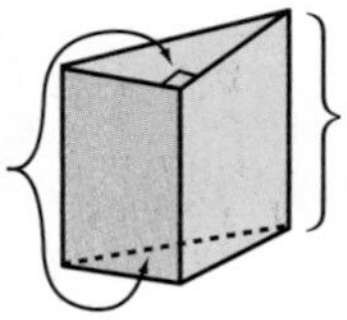

The total area of the lateral faces is called the **lateral surface area.** A quick way to find the lateral surface area of a prism is to multiply the perimeter of the base by the height (the distance between the bases).

Perimeter of base: 3 cm + 4 cm + 5 cm = 12 cm

Height: 6 cm

Lateral surface area: 12 cm · 6 cm = 72 cm^2

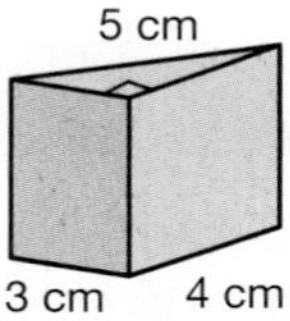

Explain When finding the perimeter of the base, how did we know which face to use?

We use the concept of lateral surface area to find the surface area of a circular cylinder.

Example 2

a. What is the area of the label on a soup can with dimensions as shown?

b. What is the total surface area of the soup can?

Use $\frac{22}{7}$ for π.

Solution

a. If we remove the label from a soup can, we see that it is a rectangle. One dimension of the rectangle is the circumference of the can, and the other dimension is the height of the can. The area of the label equals the *lateral surface area* of the soup can.

To find the area of the label, we multiply these two dimensions.

$$\begin{aligned}\text{Lateral area} &= \text{circumference} \cdot \text{height}\\ &= \pi d \cdot \text{height}\\ &= \frac{22}{\cancel{7}_1} \cdot \cancel{7}^1 \text{ cm} \cdot 10 \text{ cm}\\ &= \mathbf{220\ cm^2}\end{aligned}$$

b. The total surface area of the can consists of the lateral surface area plus the circular top and bottom of the can. Recall that the area of a circle is found by multiplying the square of the radius by π.

$$A = \pi r^2$$

We use $\frac{22}{7}$ for π and $\frac{7}{2}$ cm (or 3.5 cm) for the radius.

$$A = \frac{22}{7}\left(\frac{7}{2}\text{ cm}\right)^2$$

$$A \approx \frac{\cancel{22}^{11}}{\cancel{7}_1} \cdot \frac{\cancel{7}^1}{\cancel{2}_1} \cdot \frac{7}{2}\text{ cm}^2$$

$$A = \frac{77}{2}\text{ cm}^2 \text{ (or 38.5 cm}^2\text{)}$$

We have found the area of one circular surface. However, the can has both a top and a bottom, so we add the areas of the top, bottom, and lateral surface.

Area of top	=	38.5 cm^2
Area of bottom	=	38.5 cm^2
+ Area of lateral surface	=	220.0 cm^2
Total surface area	=	297.0 cm^2

The total surface area of the soup can is **297 cm²**.

We can use the concepts in this lesson to create formulas for surface area.

A cube has six congruent faces. If we let *s* stand for the length of each edge, then what is the area of each face? Write a formula for the total surface area (A_s) for the cube.

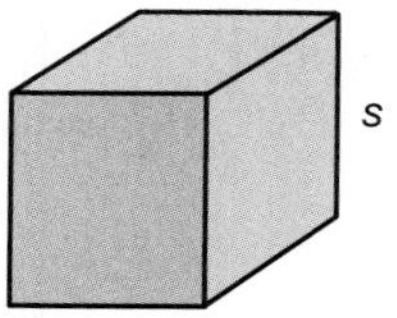

A rectangular prism has six faces. The top and bottom are congruent. The front and back are congruent. The left face and right face are congruent. Write a formula for the total surface area (A_S) for a rectangular prism that uses l, w, and h.

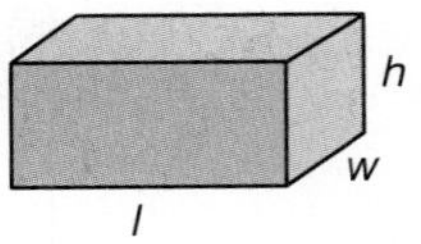

A cylinder has two bases that are circles and one curved lateral surface. Write a formula for the lateral surface area (A_L) of a cylinder using r and h. Then write a formula for the total surface area.

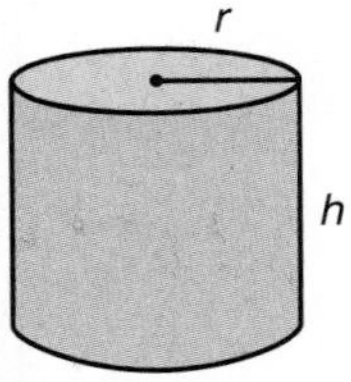

surface area of a sphere

To calculate the surface area of a sphere, we may first calculate the area of the largest cross section of the sphere. Slicing a grapefruit in half provides a visual representation of a cross section of a sphere.

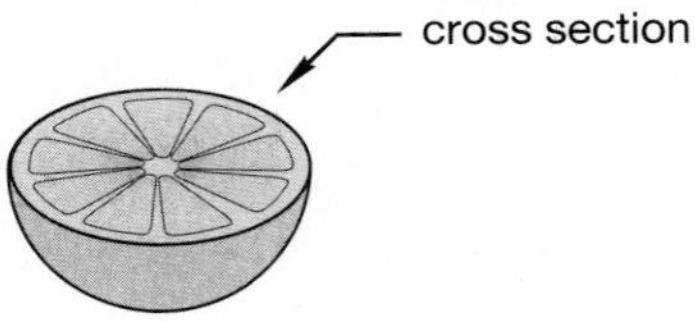

The circle formed by cutting the grapefruit in half is the cross section of the spherical grapefruit. The surface area of the entire sphere is four times the area of this circle. To find the surface area of the sphere, we calculate the area of its largest cross section ($A = \pi r^2$); then we multiply the cross sectional area by four.

Surface area of a sphere = $4\pi r^2$

Example 3

A tennis ball has a diameter of about 6 cm. Find the surface area of the tennis ball to the nearest square centimeter.

Use 3.14 for π.

Solution

A tennis ball is spherical. If its diameter is 6 cm, then its radius is 3 cm.

$$\begin{aligned} \text{Surface area} &= 4\pi r^2 \\ &\approx 4(3.14)(3\text{ cm})^2 \\ &\approx 4(3.14)(9\text{ cm}^2) \\ &\approx 113.04\text{ cm}^2 \end{aligned}$$

We round the answer to **113 cm²**.

Example 4

Find the total surface area of this figure. (Units are centimeters.)

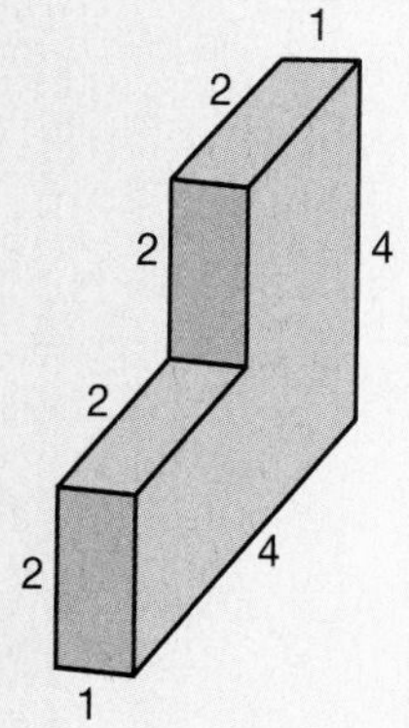

Solution

The figure has eight surfaces:

Two congruent L-shapes (each 12 cm^2)

One 1×4 back rectangle (4 cm^2)

One 1×4 bottom rectangle (4 cm^2)

Four 1×2 stairstep rectangles (each 2 cm^2)

Adding these areas we find the total surface area is **40 cm^2.**

Practice Set

a. Find the surface area of this rectangular solid. Dimensions are in inches.

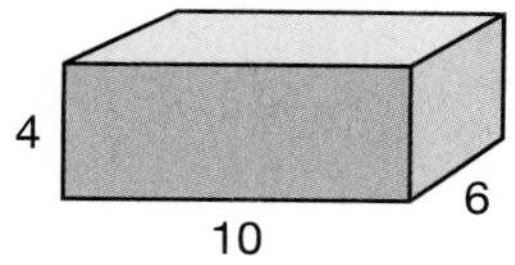

b. Analyze Find the surface area of this triangular prism. Dimensions are in inches.

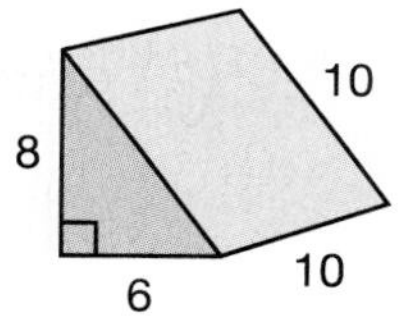

c. Analyze Find the area of the label on this can of tuna.

d. Find the total surface area of the can.

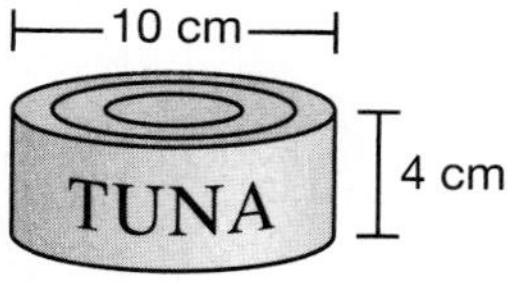

Use 3.14 for π.

e. The diameter of a golf ball is about 4 cm. Find the surface area of a golf ball to the nearest square centimeter. (Use 3.14 for π.)

f. Analyze If the figure in problem **b** is attached to the figure in problem **a** as shown, what is the total surface area of the combined figures?

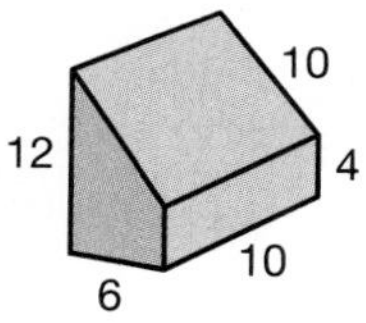

g. Explain Is the surface area of **f** equal to the sum of the surface areas of **a** and **b?** Why or why not?

Written Practice   *Strengthening Concepts*

1. (51) Twenty billion is how much greater than nine hundred million? Write the answer in scientific notation.

2. (Inv. 4) The mean of the following numbers is how much less than the median?

3.2, 4.28, 1.2, 3.1, 1.17

3. (52) Evaluate: $\sqrt{a^2 - b^2}$ if $a = 10$ and $b = 8$

4. (54) If Tyra is paid at a rate of $8.50 per hour, how much will she earn if she works $6\frac{1}{2}$ hours?

Use ratio boxes to solve problems **5–9.**

5. (72) A 5-pound bag of flour costs $1.24. What is the cost of 75 pounds of flour?

6. (92) The regular price of the dress was $30. The dress was on sale for 25% off.

a. What was the sale price?

b. What percent of the regular price was the sale price?

7. (66) The ratio of students to parents at the assembly was 7 to 3. If the total number of students and parents assembled was 210, how many parents were at the assembly?

8. (92) The original price of a hockey helmet and mask was $60. Brandon bought one on sale for 30% off. What did he pay for the mask?

* **9.** (98) **Generalize** Chen and his sister are making a model plane at a 1:36 scale. If the length of the actual plane is 60 feet, how many inches long should they make the model? Begin by converting 60 feet to inches.

* **10.** (102) **Explain** Transversal *r* intersects parallel lines *s* and *t*. If the measure of each acute angle is $4x$ and the measure of each obtuse angle is $5x$, then what is the measure of each acute and each obtuse angle?

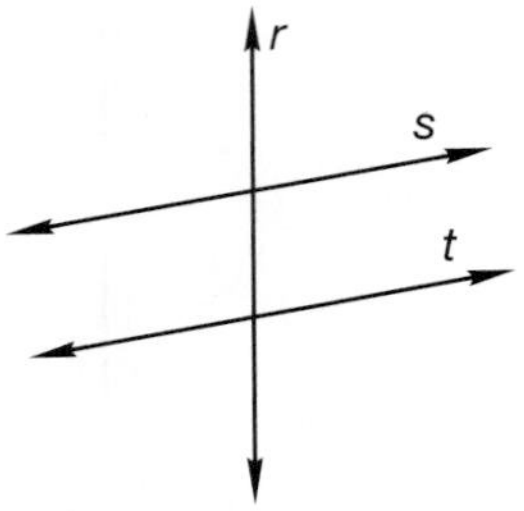

Formulate Write equations to solve problems **11–13.**

11. (77) What percent of $60 is $3?

12. (60) What fraction is 10 percent of 4?

* **13.** (101) Twelve less than twice what number is 86?

* **14.** (99) The coordinates of the vertices of a right triangle are (−2, −2), (−2, 2), and (1, −2). Find the length of the hypotenuse of this triangle.

* **15.** (103) Compare: $a^3 \bigcirc a^2$ if a is negative

* **16.** (94) Carmela has a deck of 26 alphabet cards. She mixes the cards, places them face down on the table, and turns them over one at a time. What is the probability that the first two cards she turns over are vowel cards (a, e, i, o, u)?

* **17.** (105) Alejandro bounced a big ball with a diameter of 20 inches. Using 3.14 for π, find the surface area of the ball.

18. (83) Multiply. Write the product in scientific notation.

$$(8 \times 10^{-4})(3.2 \times 10^{-10})$$

* **19.** (104) **Analyze** Find the perimeter of the figure at right. Dimensions are in meters. (Use 3.14 for π.)

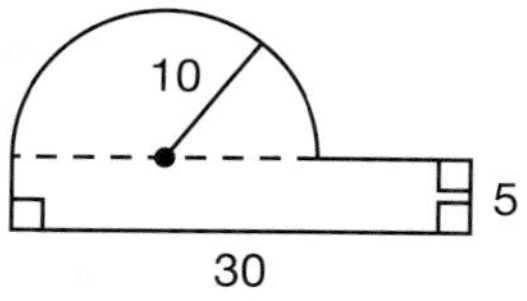

20. (56, Inv. 9) Find the missing numbers in the table by using the function rule. Then graph the function on a coordinate plane.

$y = -2x - 1$

x	y
3	
−2	
0	

21. (67, 70) Find the **a** volume and **b** surface area of this cube. Dimensions are in millimeters.

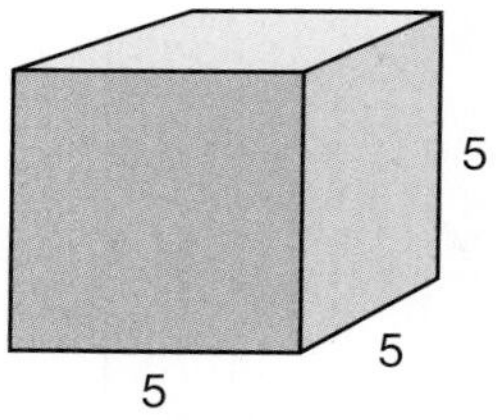

* **22.** (95) **Analyze** Find the volume of this right circular cylinder. Dimensions are in centimeters. (Use 3.14 for π.)

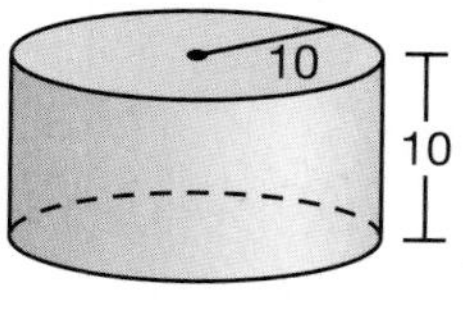

* **23.** (105) **Analyze** The total surface area of the cylinder in problem **22** includes the areas of two circles and the curved side. What is the total surface area of the cylinder?

24. (40) Find m$\angle b$ in the figure at right. Explain how you found your answer.

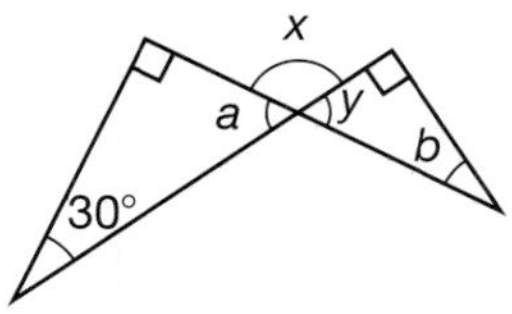

*** 25.** (97, 98) **Generalize** The triangles shown are similar.

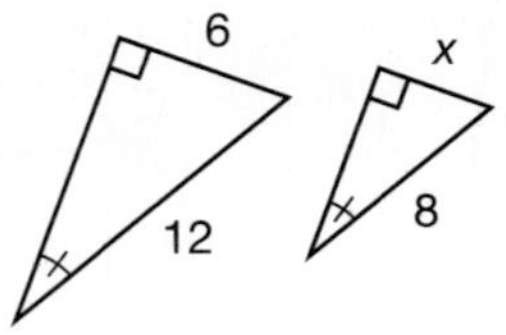

a. Estimate, then calculate, the length x.

b. Find the scale factor from the smaller triangle to the larger triangle.

c. The area of the larger triangle is how many times the area of the smaller triangle?

Generalize Solve:

*** 26.** (102) $4\frac{1}{2}x + 4 = 48 - x$

27. (98) $\frac{3.9}{75} = \frac{c}{25}$

Generalize Simplify:

28. (43) $3.2 \div \left(2\frac{1}{2} \div \frac{5}{8}\right)$

*** 29.** (96, 103) **a.** $\frac{(2xy)(4x^2y)}{8x^2y}$ **b.** $3(x+3) - \sqrt{9x^2}$

30. (57, 85) **a.** $\frac{(-10)(-4) - (3)(-2)(-1)}{(-4) - (-2)}$

b. $(-2)^4 - (-2)^2 + 2^0$

Early Finishers
Real-World Application

Whitney's family is moving across the state. The moving company will move 4000 lbs. of boxes for $1800. Forty-five pounds can be safely moved in a small moving box and 65 lbs. in a large moving box.

a. If Whitney's father filled 21 small boxes to the weight limit, how many large boxes can he fill?

b. Approximately how much is it costing Whitney's family to have the large boxes moved?

LESSON

106

• Solving Literal Equations
• Transforming Formulas
• More on Roots

Power Up

Building Power

facts

Power Up V

mental math

a. **Positive/Negative:** $(-5)^3$

b. **Scientific Notation:** $(8 \times 10^3)(5 \times 10^{-5})$

c. **Ratio:** $\frac{a}{3.6} = \frac{0.9}{1.8}$

d. **Measurement:** Convert 2 sq. yd to sq. ft.

e. **Percent:** $66\frac{2}{3}\%$ of \$45

f. **Percent:** $33\frac{1}{3}\%$ less than \$45

g. **Geometry:** When plotting an ordered pair on a coordinate grid, which axis do you start with?

h. **Percent/Estimation:** Estimate a 15% tip on a \$39.67 bill.

problem solving

Every whole number can be expressed as the sum of, *at most,* four squares. In the diagram we can see that 12 is made up of one 3 × 3 square and three 1 × 1 squares. The number sentence that represents the diagram is 12 = 9 + 1 + 1 + 1.

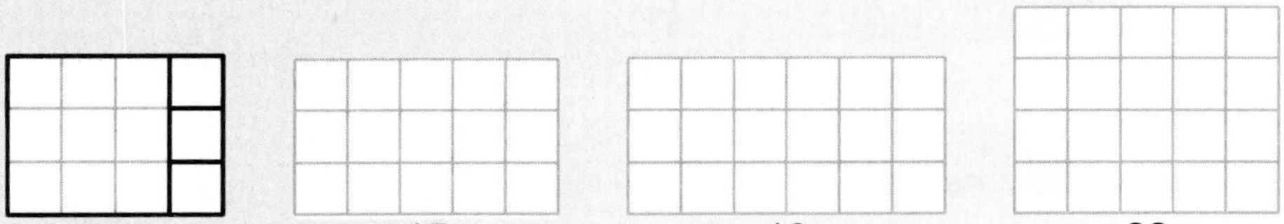

12 15 18 20

Diagram how 15, 18, and 20 are composed of—at most—four smaller squares. Write a number sentence to represent each diagram. (*Hint:* You may reposition the smaller squares if necessary.)

New Concepts

Increasing Knowledge

solving literal equations

A **literal equation** is an equation that contains letters instead of numbers. We can rearrange (transform) literal equations by using the rules we have learned.

Example 1

Solve for x: $x + a = b$

Solution

We solve for x by isolating x on one side of the equation. We do this by adding $-a$ to both sides of the equation.

$x + a = b$	equation
$x + a - a = b - a$	added $-a$ to both sides
$x = \mathbf{b - a}$	simplified

Example 2

Solve for x: $ax = b$

Solution

To solve for x, we divide both sides of the equation by a.

$ax = b$	equation
$\frac{\overset{1}{\cancel{a}}x}{\underset{1}{\cancel{a}}} = \frac{b}{a}$	divided by a
$x = \mathbf{\frac{b}{a}}$	simplified

transforming formulas

Formulas are literal equations that we can use to solve certain kinds of problems. Often it is necessary to change the way a formula is written.

Example 3

Solve for w: $A = lw$

Solution

This is a formula for finding the area of a rectangle. We see that w is to the right of the equal sign and is multiplied by l. To undo the multiplication by l, we can divide both sides of the equation by l.

Thinking Skill

Verify

How can we verify that the changed formula is true?

$A = lw$	equation
$\frac{A}{l} = \frac{\overset{1}{\cancel{l}}w}{\underset{1}{\cancel{l}}}$	divided by l
$w = \mathbf{\frac{A}{l}}$	simplified

more on roots

The perfect square 25 has both positive and negative square roots, 5 and –5.

$$5 \cdot 5 = 25 \qquad (-5)(-5) = 25$$

Thus the equation $x^2 = 25$ has two solutions, 5 and -5.

The positive square root of a number is sometimes called the *principal* square root. So the principal square root of 25 is 5. The radical symbol $\sqrt{\ }$ implies the principal root. So $\sqrt{25}$ is 5 only and does not include -5.

Example 4

What are the two square roots of 5?

Solution

The two square roots of 5 are $\mathbf{\sqrt{5}}$ and the opposite of $\sqrt{5}$, which is $\mathbf{-\sqrt{5}}$.

A radical symbol may be used to indicate other roots besides square roots. The expression below means **cube root** of 64. The small 3 is called the **index** of the root. The cube root of 64 is the number that, when used as a factor three times, yields a product of 64.

$$\sqrt[3]{64}$$

$$(?)(?)(?) = 64$$

Thus the cube root of 64 is 4 because

$$4 \cdot 4 \cdot 4 = 64$$

Discuss Is -4 a cube root of 64? Why or why not?

Example 5

Simplify:

a. $\mathbf{\sqrt[3]{1000}}$ **b.** $\mathbf{\sqrt[3]{-27}}$

Solution

a. The cube root of 1000 is **10** because $10 \cdot 10 \cdot 10 = 1000$. Notice that -10 is not a cube root of 1000, because $(-10)(-10)(-10) = -1000$.

b. The cube root of -27 is $\mathbf{-3}$ because $(-3)(-3)(-3) = -27$.

Predict Which will be a negative number, the cube root of 8 or the cube root of -8?

Practice Set

a. Solve for *a*: $a + b = c$

b. Analyze Solve for *w*: $wx = y$

c. Solve for *y*: $y - b = mx$

d. Solve the formula for the area of a parallelogram for *b*.

$$A = bh$$

e. Predict What are the two square roots of 16?

Generalize Simplify:

f. $\sqrt[3]{125}$ **g.** $\sqrt[3]{-8}$

Written Practice
Strengthening Concepts

1. (60) **Estimate** Marcos bought 2 pairs of socks priced at \$1.85 per pair and a T-shirt priced at \$12.95. The sales tax was 6%. If Marcos paid with a \$20 bill, what was his change? (round to the nearest cent)

*** 2.** (Inv. 8) The face of this spinner is divided into twelfths.

a. If the spinner is spun once, what are the odds that the spinner will land on a one-digit prime number?

b. If the spinner is spun twice, what is the chance that it will land on an even number both times?

3. (46) At \$5.60 per pound, the cheddar cheese costs how many cents per ounce?

4. (55) After 6 days at her new job, Katelyn had wrapped an average number of 90 gifts per day. She wrapped 75 of the gifts on her first day. If the first day is not counted, what is the average number of gifts Katelyn wrapped per day during the next 5 days?

*** 5.** (99) **Analyze** The ordered pairs (2, 4), (2, −1), and (0, −1) designate the vertices of a right triangle. What is the length of the hypotenuse of the triangle?

Use ratio boxes to solve problems **6–8.**

6. (72) Justin finished 3 problems in 4 minutes. At that rate, how long will it take him to finish the remaining 27 problems?

7. (66) The ratio of members to visitors in the pool was 2 to 3. If there were 60 people in the pool, how many were visitors?

8. (92) The number of students enrolled in chemistry increased 25 percent this year. If there are 80 students enrolled in chemistry this year, how many were enrolled in chemistry last year?

9. (106) **a.** What are the two square roots of 64?

b. What is the cube root of −64?

10. (60) Write an equation to solve this problem:

What number is 225 percent of 40?

*** 11.** (100) **a.** **Model** Draw a number line and show the locations of these numbers:

$$|-2|, \frac{2}{2}, \sqrt{2}, 2^2$$

b. Which of these numbers are rational numbers?

Write equations to solve problems **12** and **13**.

12. *(77)* Sixty-six is $66\frac{2}{3}$ percent of what number?

13. *(77)* Seventy-five percent of what number is 2.4?

14. *(48)* Complete the table.

Fraction	Decimal	Percent
a.	**b.**	105%

15. *(Inv. 9)* Make a table of ordered pairs for the function $y = x - 2$. Then graph the function on a coordinate plane.

16. *(42)* Divide 6.75 by 81 and write the quotient rounded to three decimal places.

17. *(83)* Multiply. Write the product in scientific notation.

$$(4.8 \times 10^{-10})(6 \times 10^{-6})$$

18. *(91)* Evaluate: $x^2 + bx + c$ if $x = -3$, $b = -5$, and $c = 6$

*** 19.** *(104)* **Explain** Find the area of this figure. Dimensions are in millimeters. Corners that look square are square. (Use 3.14 for π.) Show your work and explain how you found your answer.

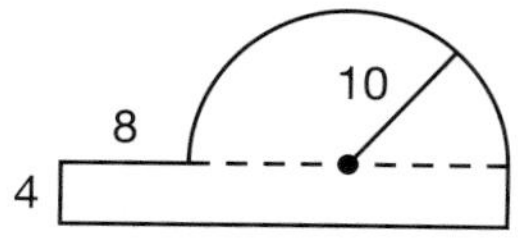

*** 20.** *(88, 105)* **a.** Find the surface area of this right triangular prism. Dimensions are in centimeters.

b. Convert the surface area to square millimeters.

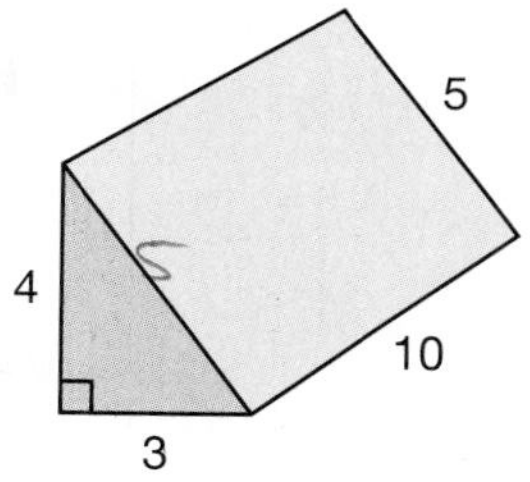

*** 21.** *(95)* **Analyze** Find the volume of this right circular cylinder. Dimensions are in inches. (Use 3.14 for π.)

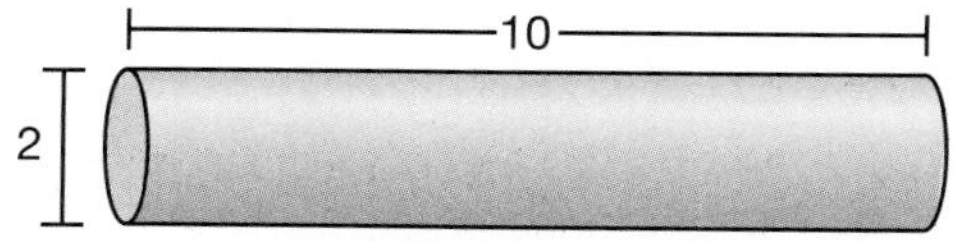

22. *(40)* Find m$\angle b$ in the figure below.

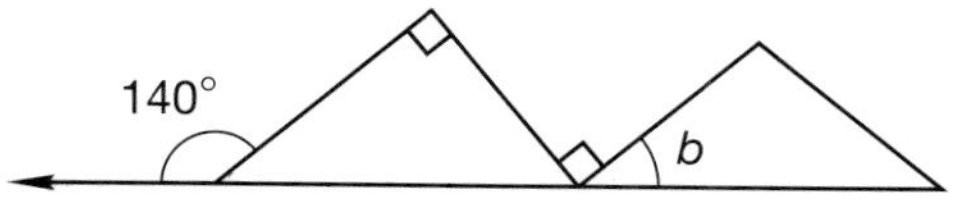

*** 23.** *(106)* **a.** Solve for x: $x + c = d$

b. Solve for n: $an = b$

*** 24.** *(102)* **Generalize** Solve: $6w - 2(4 + w) = w + 7$

25. Solve this inequality and graph its solution:
(93)

$$6x + 8 < 14$$

* **26.** Thirty-seven is five less than the product of what number and three?
(101)

Generalize Simplify:

27. $25 - [3^2 + 2(5 - 3)]$
(63)

* **28.** $\frac{6x^2 + (5x)(2x)}{4x}$
(103)

29. $4^0 + 3^{-1} + 2^{-2}$
(57)

* **30.** $(-3)(-2)(+4)(-1) + (-3)^2 + \sqrt[3]{-64} - (-2)^3$
(103, 106)

Early Finishers
Real-World Application

The illustration below represents a semicircular window with a stained glass circle in the middle.

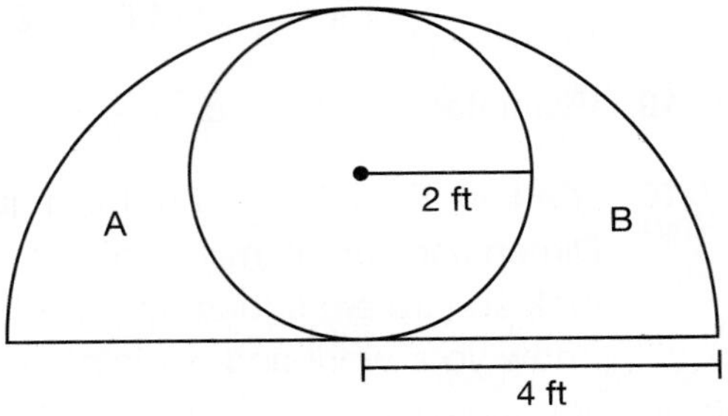

a. What is the area of the entire window?

b. What is the area of the circular stained glass window?

c. What is the combined area of sections A and B of the window? Use 3.14 for π.

d. What portion of the whole window is the stained glass?

LESSON 107

• Slope

Power Up — *Building Power*

facts Power Up Q

mental math

a. Positive/Negative: $(-2.5)(-4)$

b. Scientific Notation: $(2.5 \times 10^6)^2$

c. Algebra: $2x - 1\frac{1}{2} = 4\frac{1}{2}$

d. Measurement: Convert $-50°C$ to degrees Fahrenheit.

e. Percent: 75% of \$60

f. Percent: 75% more than \$60

g. Measurement: 32 oz. is what fraction of a gallon?

h. Calculation: 7×8, -1, $\div 5$, $\times 3$, $+2$, $\div 5$, $\times 7$, $+1$, $\times 2$, -1, $\div 3$, $+3$, $\sqrt{\ }$

problem solving Four identical blocks marked *x,* a 250-g mass, and a 500-g mass were balanced on a scale as shown. Write an equation to represent this balanced scale, and find the mass of each block.

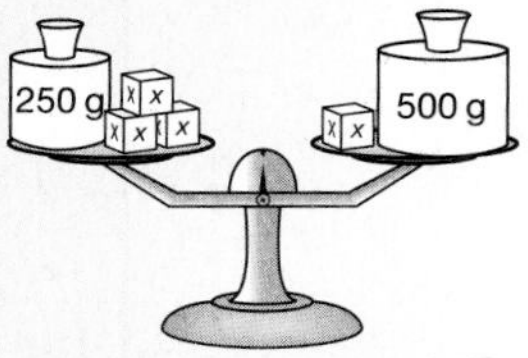

New Concept — *Increasing Knowledge*

Below are the graphs of two functions. The graph of the function on the left indicates the number of feet that equal a given number of yards. Changing the number of yards by one changes the number of feet by three. The graph of the function on the right shows the inverse relationship, the number of yards that equal a given number of feet. Changing the number of feet by one changes the number of yards by one third.

Thinking Skill

Explain

Why does changing the number of feet by 1 change the number of yards by $\frac{1}{3}$?

Yards to Feet

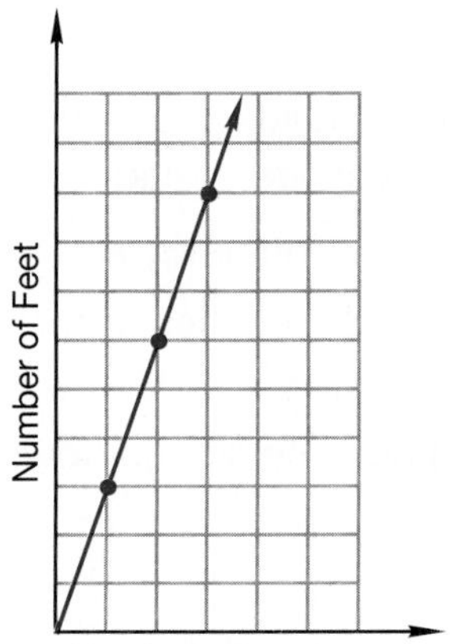

Feet to Yards

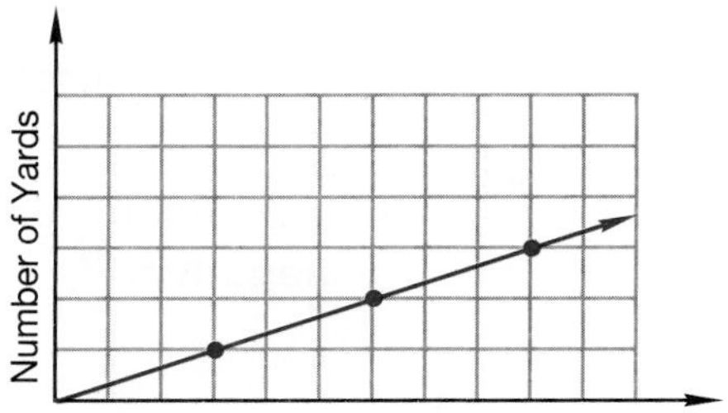

Notice that the graph of the function on the left has a steep upward slant going from left to right, while the graph of the function on the right also has an upward slant but is not as steep. The "slant" of the graph of a function is called its **slope.** We assign a number to a slope to indicate how steep the slope is and whether the slope is upward or downward. If the slope is upward, the number is positive. If the slope is downward, the number is negative. If the graph is horizontal, the slope is neither positive nor negative; it is zero. If the graph is vertical, the slope cannot be determined.

Example 1

State whether the slope of each line is positive, negative, zero, or cannot be determined.

a.

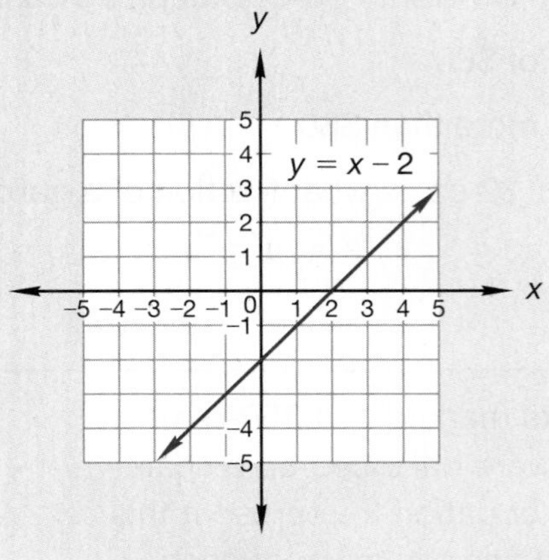

b.

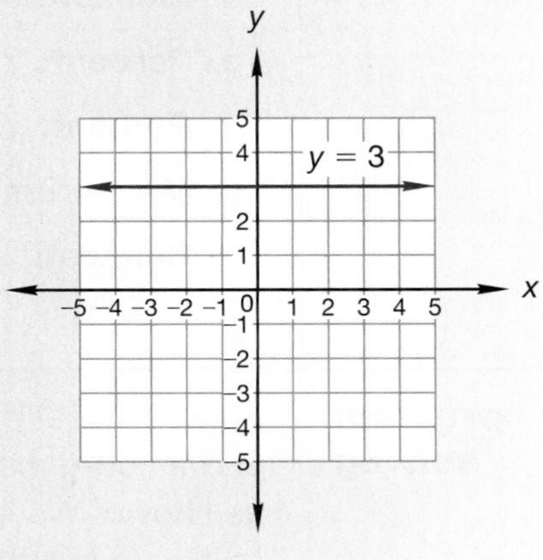

c.

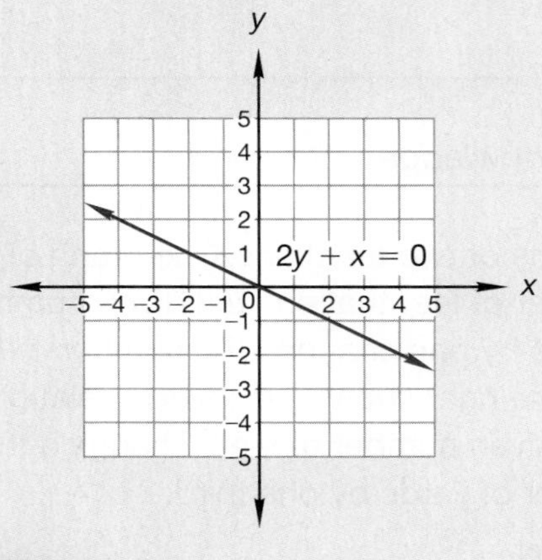

d.

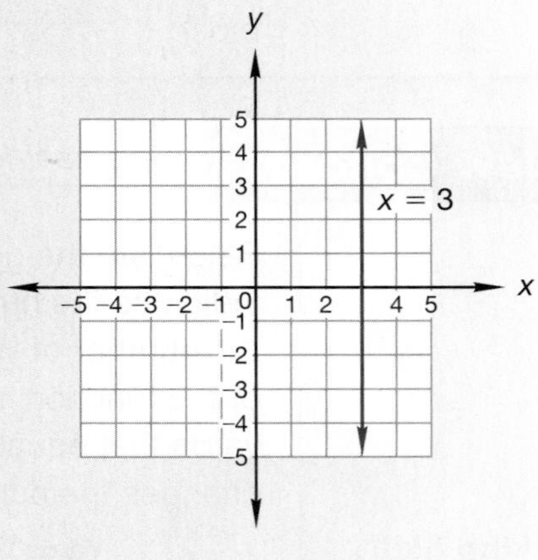

Solution

To determine the sign of the slope, follow the graph of the function with your eyes *from left to right* as though you were reading.

a. From left to right, the graphed line rises, so the slope is **positive.**

b. From left to right, the graphed line does not rise or fall, so the slope is **zero.**

c. From left to right, the graphed line slopes downward, so the slope is **negative.**

d. There is no left to right component of the graphed line, so we cannot determine if the line is rising or falling. The slope is not positive, not negative, and not zero. The slope of a vertical line **cannot be determined.**

To determine the numerical value of the slope of a line, it is helpful to draw a right triangle using the background grid of the coordinate plane and a portion of the graphed line. First we look for points where the graphed line crosses intersections of the grid. We have circled some of these points on the graphs below.

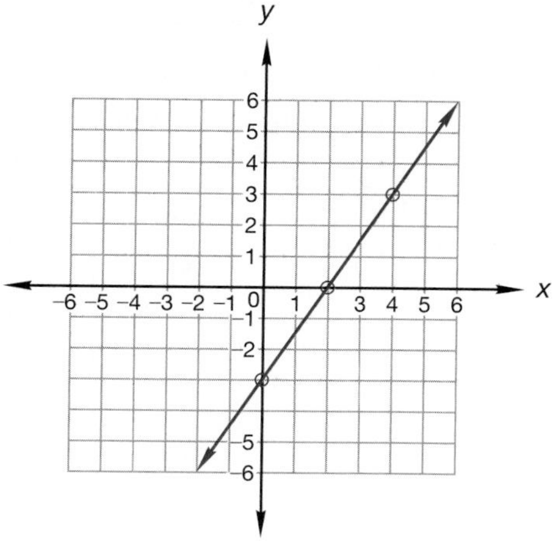

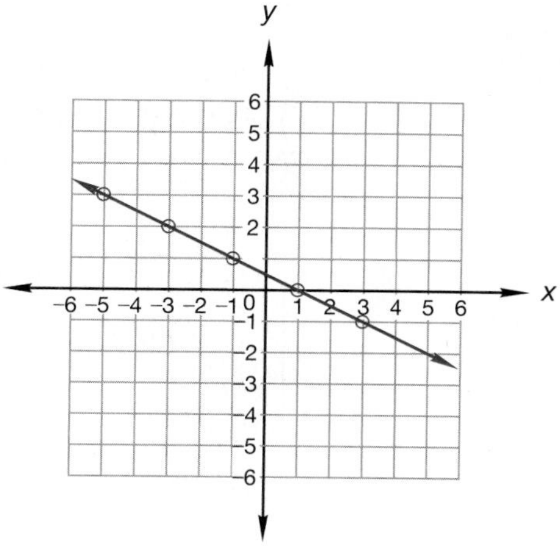

Next we select two points from the graphed line and, following the background grid, sketch the legs of a right triangle so that the legs intersect the chosen points. (It is a helpful practice to first select the point to the left and draw the horizontal leg to the right. Then draw the vertical leg to meet the line.)

Discuss Why do you think we look for points where the graphed line crosses intersections of the grid?

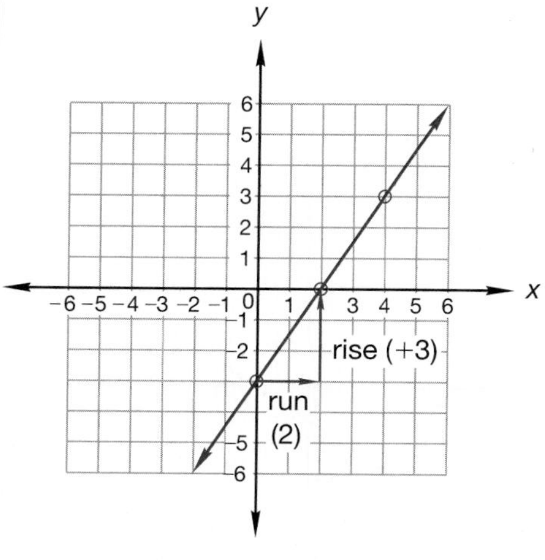

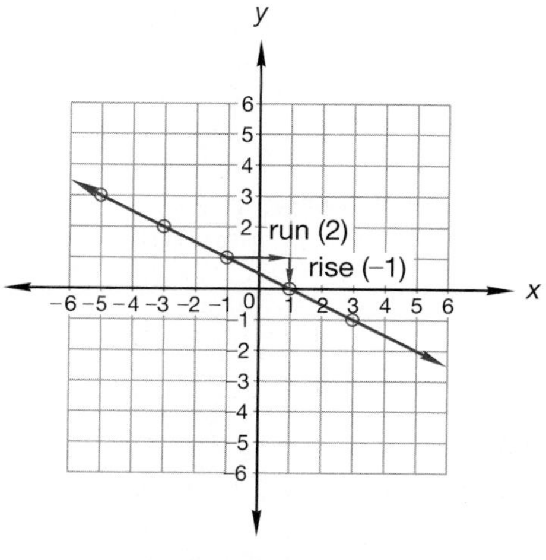

We use the words **run** and **rise** to describe the two legs of the right triangle. The *run* is the length of the horizontal leg, and the *rise* is the length of the vertical leg. We assign a positive sign to the rise if it goes up to meet the graphed line and a negative sign if it goes down to meet the graphed line. In the graph on the left, the run is 2 and the rise is +3. In the graph on the right, the run is 2 and the rise is −1. We use these numbers to write the slope of each graphed line.

So the slopes of the graphed lines are these ratios:

$$\frac{\text{rise}}{\text{run}} = \frac{+3}{2} = \frac{3}{2} \qquad \frac{\text{rise}}{\text{run}} = \frac{-1}{2} = -\frac{1}{2}$$

> The slope of a line is the ratio of its rise to its run ("rise over run").
>
> $$\textbf{slope} = \frac{\textbf{rise}}{\textbf{run}}$$

A line whose rise and run have equal values has a slope of 1. A line whose rise has the opposite value of its run has a slope of −1.

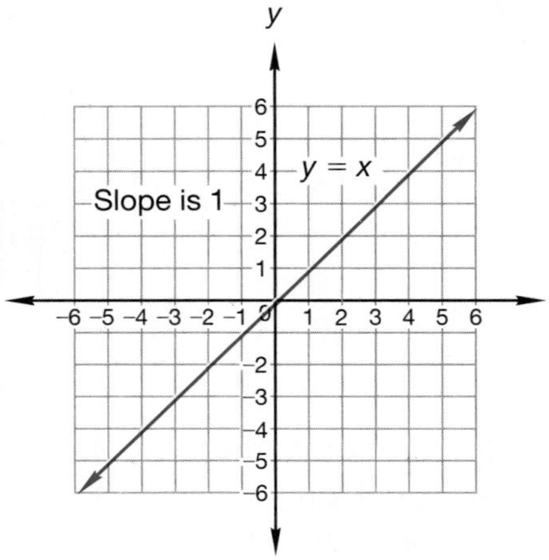

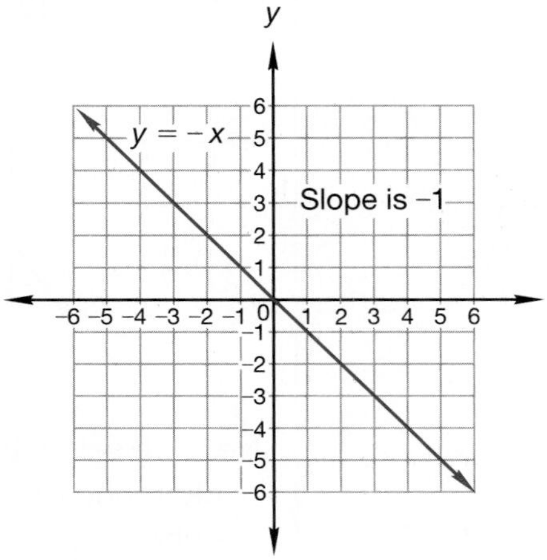

A line that is steeper than the lines above has a slope either greater than 1 or less than −1. A line that is less steep than the lines above has a slope that is between −1 and 1.

Example 2

Find the slope of the graphed line below.

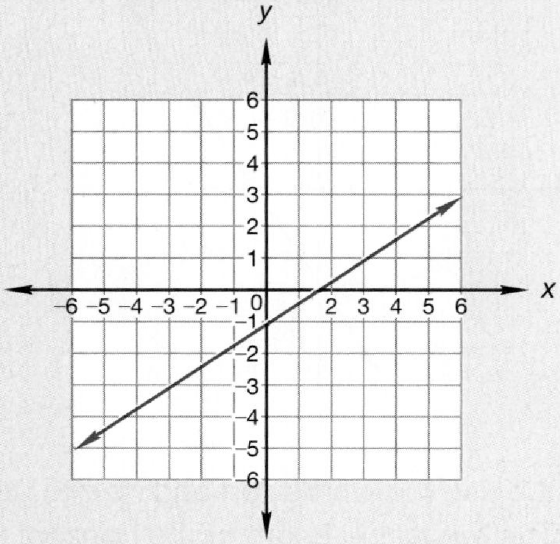

Solution

We note that the slope is positive. We locate and select two points where the graphed line passes through intersections of the grid. We choose the points (0, −1) and (3, 1). Starting from the point to the left, (0, −1), we draw the horizontal leg to the right. Then we draw the vertical leg up to (3, 1).

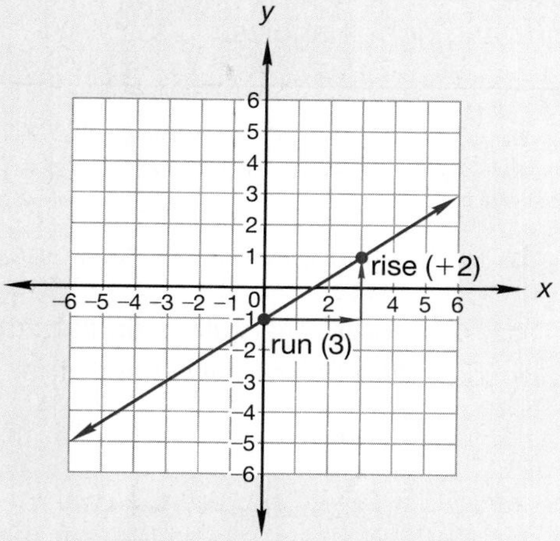

We see that the run is 3 and the rise is positive 2. We write the slope as "rise over run."

$$\text{slope} = \frac{2}{3}$$

Analyze If we had chosen the points (−3, −3) and (3, 1), the run would be 6 and the rise 4. However, the slope would be the same. Why is this true?

One way to check the calculation of a slope is to "zoom in" on the graph. When the horizontal change is one unit to the right, the vertical change will equal the slope. To illustrate this, we will zoom in on the square just below and to the right of the origin of the graph above.

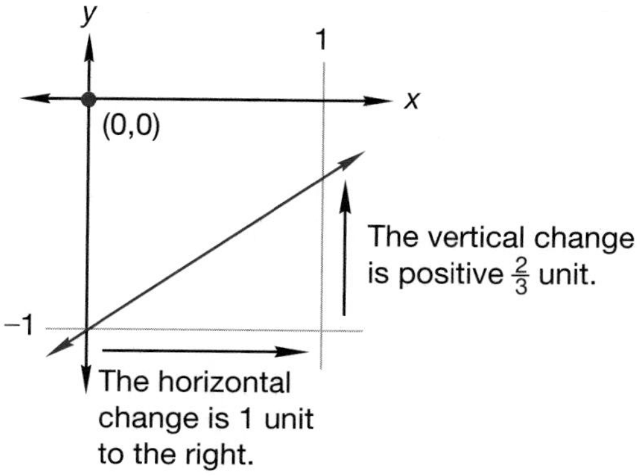

Discuss Summarize how we can *zoom in* to see if a calculated slope is reasonable.

Activity

Slope

Materials needed:

- **Lesson Activity 24**

Calculate the slope (rise over run) of each graphed line on the activity by drawing right triangles.

Practice Set

a. Generalize Find the slopes of the "Yards to Feet" and the "Feet to Yards" graphs on the first page of this lesson.

b. Find the slopes of graphs **a** and **c** in example 1.

c. Generalize Mentally calculate the slope of each graphed line below by counting the run and rise rather than by drawing right triangles.

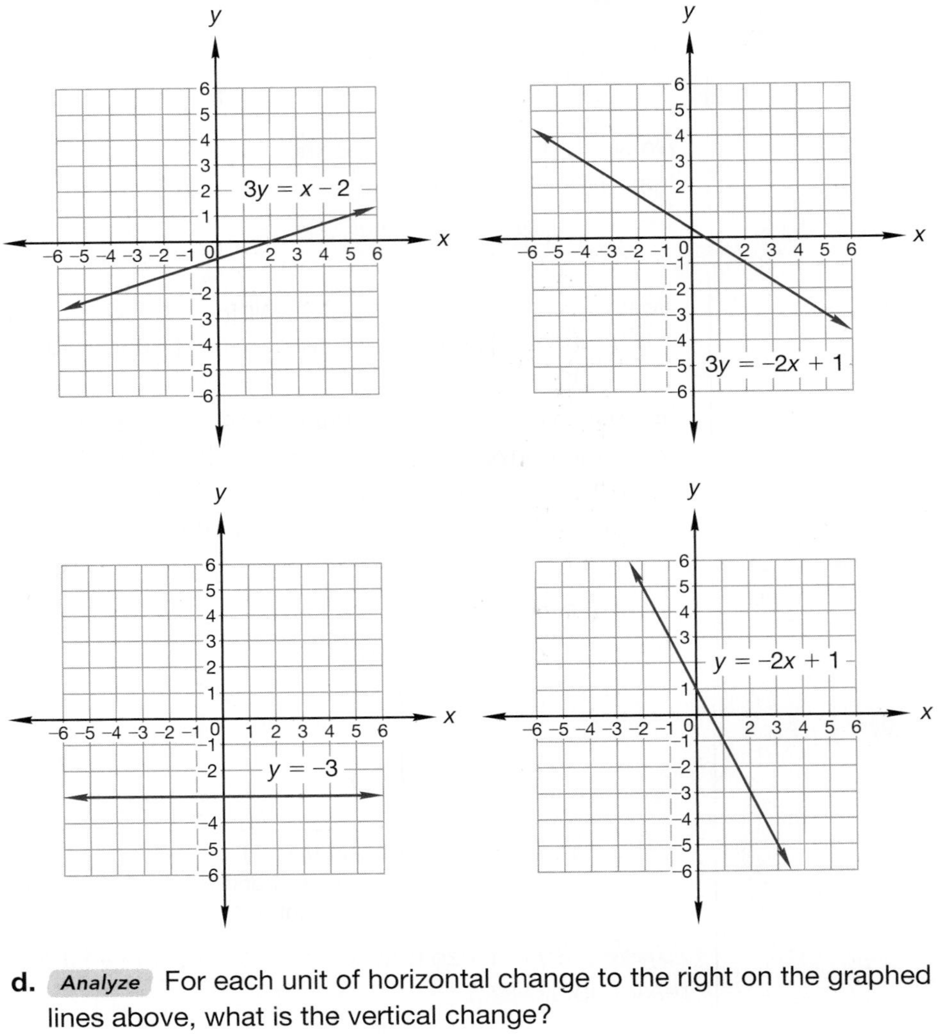

d. Analyze For each unit of horizontal change to the right on the graphed lines above, what is the vertical change?

Written Practice *Strengthening Concepts*

1. (92) The shirt regularly priced at \$27 was on sale for $\frac{1}{3}$ off. What was the sale price?

2. (51) Nine hundred seventy-five billion is how much less than one trillion? Write the answer in scientific notation.

3. (Inv. 4) What is the **a** range and **b** mode of this set of numbers?

16, 6, 8, 17, 14, 16, 12

Use ratio boxes to solve problems **4–6.**

4. (72) Sonia rode her bike from home to the lake. Her average speed was 12 miles per hour. If it took her 40 minutes to reach the lake, what was the distance from her home to the lake? Explain how you solved the problem.

5. (66) The ratio of swallowtail butterflies to skipper butterflies in the butterfly conservatory was 5 to 2. If there were 140 butterflies in the conservatory, how many of the butterflies were swallowtails?

6. (92) The average cost of a new car increased 8 percent in one year. Before the increase the average cost of a new car was \$16,550. What was the average cost of a new car after the increase?

*** 7.** (99) **Analyze** The points (3, −2), (−3, −2), and (−3, 6) are the vertices of a right triangle. Find the perimeter of the triangle.

8. (40) In this figure, $\angle ABC$ is a right angle.

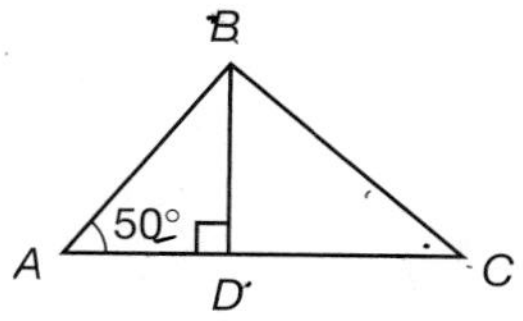

a. Find $m\angle ABD$.

b. Find $m\angle DBC$.

c. Find $m\angle BCD$.

d. Which triangles in this figure are similar?

Write equations to solve problems **9–11.**

9. (77) Sixty is 125 percent of what number?

10. (77) Sixty is what percent of 25?

*** 11.** (101) Sixty is four more than twice what number?

12. (14, 94) In a can are 100 marbles: 10 yellow, 20 red, 30 green, and 40 blue.

a. If a marble is drawn from the can, what is the chance that the marble will not be red?

b. If the first marble is not replaced and a second marble is drawn from the can, what is the probability that both marbles will be yellow?

13. (48) Complete the table.

Fraction	Decimal	Percent
$\frac{5}{6}$	**a.**	**b.**

* **14.** (19, 49) **Analyze** The Washington Monument is a little more than 555 ft tall. It has a square base that measures 55 ft, $1\frac{1}{2}$ in. on each side. What is the perimeter of the base of the monument?

15. (83) Multiply. Write the product in scientific notation.

$$(1.8 \times 10^{10})(9 \times 10^{-6})$$

* **16.** (100, 106) **a.** Between which two consecutive whole numbers is $\sqrt{600}$?

b. What are the two square roots of 10?

* **17.** (Inv. 9, 107) **Analyze** Use the function $y = x + 1$ to answer **a–c.**

a. Find three x, y pairs for the function.

b. Graph these number pairs on a coordinate plane and draw a line through the points.

c. What is the slope of the graphed line?

* **18.** (104) If the radius of this circle is 6 cm, what is the area of the shaded region?

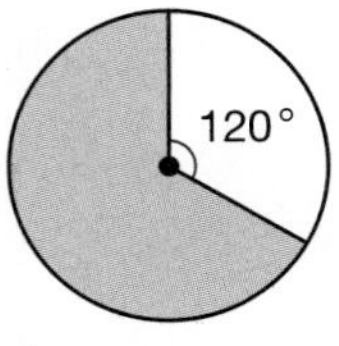

Leave π as π.

* **19.** (105) Find the surface area of this rectangular solid. Dimensions are in inches.

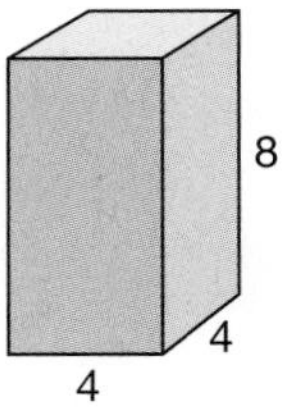

20. (95) Find the volume of this right circular cylinder. Dimensions are in centimeters. (Use 3.14 for π.)

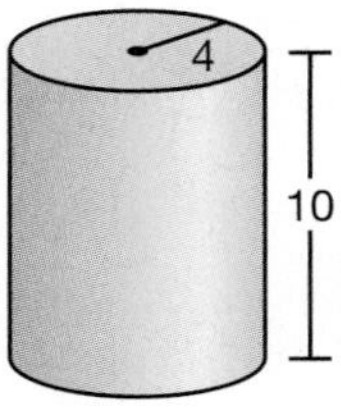

* **21.** (105) **Explain** Find the total surface area of the cylinder in problem **20.** Explain how you got your answer.

22. (40) The polygon *ABCD* is a rectangle. Find m$\angle x$.

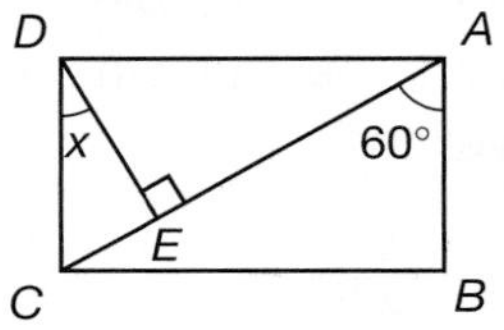

*** 23.** (107) **Analyze** Find the slope of the graphed line:

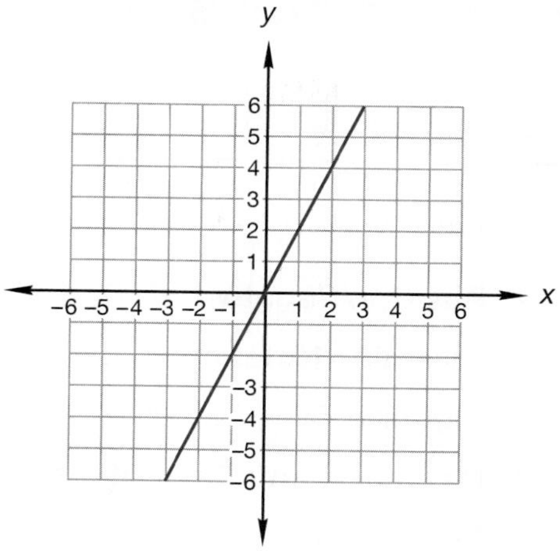

***24.** (106) **Generalize** Solve for x in each literal equation:

a. $x - y = z$ **b.** $w = xy$

Solve:

25. (98) $\frac{a}{21} = \frac{1.5}{7}$

*** 26.** (102) $6x + 5 = 7 + 2x$

Generalize Simplify:

27. (63) $62 + 5\{20 - [4^2 + 3(2 - 1)]\}$

*** 28.** (103) $\frac{(6x^2y)(2xy)}{4xy^2}$

29. (43) $5\frac{1}{6} + 3.5 - \frac{1}{3}$

30. (85) $\frac{(5)(-3)(2)(-4) + (-2)(-3)}{|-6|}$

Early Finishers
Real-World Application

Mrs. Cohen took a survey of 10 students in her class. She asked them how many school sporting events they attended last year. These are the results of the survey.

Two students attended 7 games.
Three students attended 8 games.
One student attended 9 games.
Two students attended 10 games.
One student attended 11 games.
One student attended 17 games.

a. Choose a line plot, line graph, or a Venn diagram to display this data. Justify your choice.

b. Draw your display and then find the mean, median, mode, and range. Identify any outliers.

c. Formulate questions to conduct a similar survey of 30 students in your school. Write three conclusions that can be made based on your collected data.

LESSON

108 • Formulas and Substitution

Power Up

Building Power

facts Power Up V

mental math

a. **Positive/Negative:** $(-1)^5 + (-1)^6$

b. **Scientific Notation:** $(2.5 \times 10^{-5})(4 \times 10^{-3})$

c. **Algebra:** $5y - 2y = 24$

d. **Measurement:** Convert 3 sq. yd to sq. ft.

e. **Percent:** 150% of $120

f. **Percent:** $120 increased 50%

g. **Algebra:** Complete the table by using the function rule. $y = 4x + 4$

y	20	24	28	
x	4	5	6	8

h. **Rate:** At 60 mph, how far can Freddy drive in $3\frac{1}{2}$ hours?

problem solving

Some auto license plates take the form of two letters, followed by three numbers, followed by two letters. How many different license plates are possible?

New Concept

Increasing Knowledge

A formula is a literal equation that describes a relationship between two or more variables. Formulas are used in mathematics, science, economics, the construction industry, food preparation—anywhere that measurement is used.

To use a formula, we replace the letters in the formula with measures that are known. Then we solve the equation for the measure we wish to find.

Example 1

Thinking Skill

Analyze

What measures do you know? What unknown measure do you need to find?

Use the formula $d = rt$ to find t when d is 36 and r is 9.

Solution

This formula describes the relationship between distance (d), rate (r), and time (t). We replace d with 36 and r with 9. Then we solve the equation for t.

$d = rt$	formula
$36 = 9t$	substituted
$t = \mathbf{4}$	divided by 9

Another way to find t is to first solve the formula for t.

$$d = rt \quad \text{formula}$$
$$t = \frac{d}{r} \quad \text{divided by } r$$

Then replace d and r with 36 and 9 and simplify.

$$t = \frac{36}{9} \quad \text{substituted}$$
$$t = \mathbf{4} \quad \text{divided}$$

Connect If a bike ride is 36 miles, and the rate is 9 miles per hour, how long will the trip take?

Example 2

Use the formula $F = 1.8C + 32$ to find F when C is 37.

Solution

This formula is used to convert measurements of temperature from degrees Celsius to degrees Fahrenheit. We replace C with 37 and simplify.

$$F = 1.8C + 32 \quad \text{formula}$$
$$F = 1.8(37) + 32 \quad \text{substituted}$$
$$F = 66.6 + 32 \quad \text{multiplied}$$
$$F = \mathbf{98.6} \quad \text{added}$$

Thus, 37 degrees Celsius equals 98.6 degrees Fahrenheit.

Discuss How could we use the formula to convert from degrees Fahrenheit to degrees Celsius?

Practice Set

a. Use the formula $A = bh$ to find b when A is 20 and h is 4.

b. Use the formula $A = \frac{1}{2}bh$ to find b when A is 20 and h is 4.

c. Use the formula $F = 1.8C + 32$ to find F when C is -40.

d. **Connect** The formula for converting from Fahrenheit to Celsius is often given as $F = \frac{9}{5}C + 32$. How are $\frac{9}{5}$ and 1.8 related?

Written Practice

Strengthening Concepts

1. (60) The main course cost $8.35. The beverage cost $1.25. Dessert cost $2.40. Alexi left a tip that was 15 percent of the total price of the meal. How much money did Alexi leave for a tip?

2. (57) Twelve hundred-thousandths is how much greater than twenty millionths? Write the answer in scientific notation.

3. (Inv. 4) Arrange the following numbers in order from least to greatest. Then find the median and the mode of the set of numbers.

8, 12, 9, 15, 8, 10, 9, 8, 7, 4

4. (94) There are 12 markers in a box: 4 blue, 4 red, and 4 yellow. Without looking, Lee gives one marker to Lily and takes one for himself. What is the probability that both markers will be blue?

Use ratio boxes to solve problems **5–7.**

5. (72) If Milton can exchange $200 for 300 Swiss francs, how many dollars would a 240-franc Swiss watch cost?

6. (66) The jar was filled with red beans and brown beans in the ratio of 5 to 7. If there were 175 red beans in the jar, what was the total number of beans in the jar?

7. (92) During the off-season the room rates at the resort were reduced by 35 percent. If the usual rates were $90 per day, what would be the cost of a 2-day stay during the off-season?

8. (60) Three eighths of a ton is how many pounds?

Write equations to solve problems **9–11.**

9. (60) What number is 2.5 percent of 800?

10. (77) Ten percent of what number is $2500?

* **11.** (101) Fifty-six is eight less than twice what number?

* **12.** (107) **Analyze** Find the slope of the graphed line:

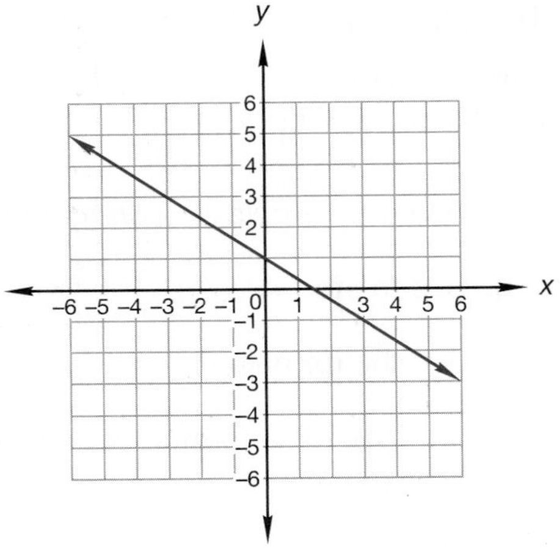

* **13.** (98) **Analyze** Liz is drawing a floor plan of her house. On the plan, 1 inch equals 2 feet.

   a. What is the floor area of a room that measures 6 inches by $7\frac{1}{2}$ inches on the plan? Use a ratio box to solve the problem.

   b. One of the walls in Liz's house is 17 feet $9\frac{1}{2}$ inches long. Estimate how long this wall would appear in Liz's floor plan, and explain how you arrived at your estimate.

* **14.** (101) **Formulate** Find the measure of each angle of this triangle by writing and solving an equation.

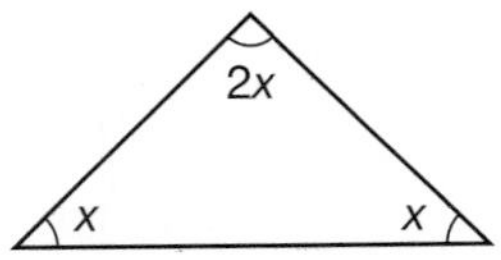

15. (83) Multiply. Write the product in scientific notation.

$$(2.8 \times 10^5)(8 \times 10^{-8})$$

* **16.** (108) The formula $c = 2.54n$ is used to convert inches (n) to centimeters (c). Find c when n is 12.

17. (Inv. 9) **a.** Make a table that shows three pairs of numbers that satisfy the function $y = 2x$.

b. Graph the number pairs on a coordinate plane, and draw a line through the points.

c. Find the slope of the line.

18. (104) Find the perimeter of this figure. Dimensions are in inches. (Use 3.14 for π.)

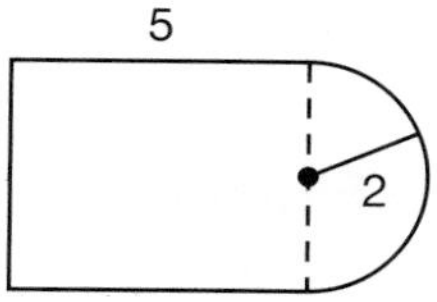

* **19.** (105) **Analyze** Find the surface area of this cube. Dimensions are in inches.

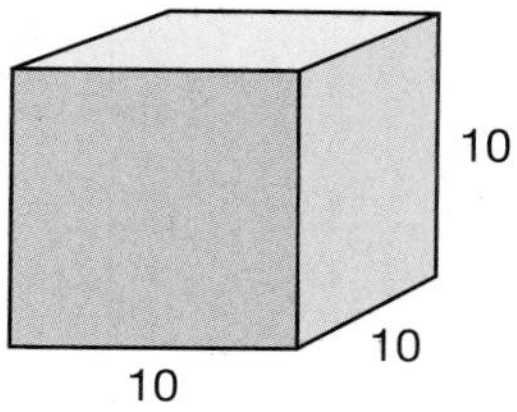

20. (95) Find the volume of this right circular cylinder. Dimensions are in centimeters. (Use 3.14 for π.)

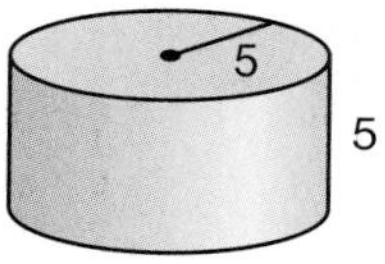

21. (40) Find m∠x in the figure below.

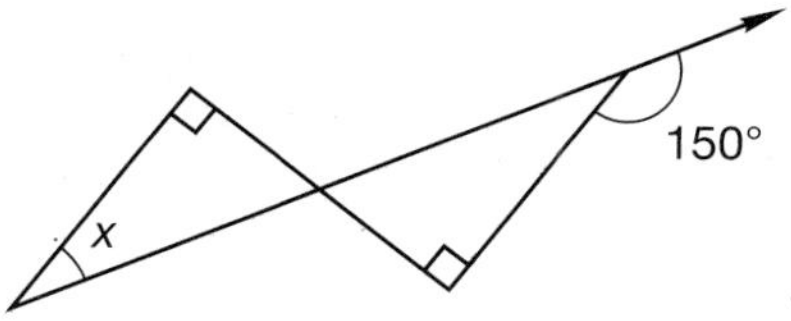

22. (97, 98) These triangles are similar. Dimensions are in centimeters.

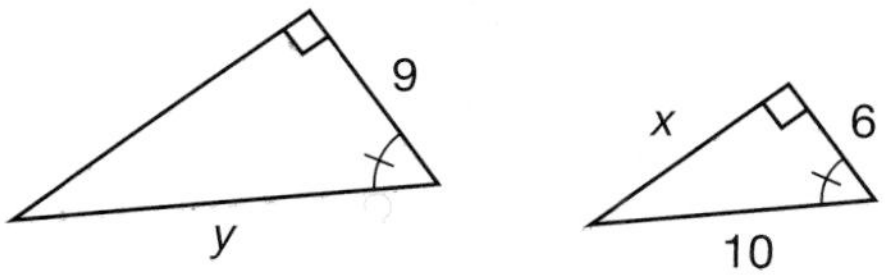

a. Find y.

b. Find the scale factor from the smaller to the larger triangle.

c. The area of the larger triangle is how many times the area of the smaller triangle?

23. (99) Use the Pythagorean theorem to find x in the smaller triangle from problem **22.**

* **24.** (105) Find the surface area of a globe that has a diameter of 10 inches. (Use 3.14 for π.)

* **25.** (102) **Generalize** Solve: $1\frac{2}{3}x = 32 - x$

Generalize Simplify:

* **26.** (96, 103) $2x(2y + 1) - \sqrt{16x^2y^2}$

* **27.** (103) $\dfrac{(-4ax)(3xy)}{-6x^2}$

28. (63) $1.1\{1.1[1.1(1000)]\}$

29. (26) $3\frac{3}{4} \cdot 2\frac{2}{3} \div 10$

* **30.** (103, 106) **a.** $(-6) - (7)(-4) + \sqrt[3]{125} + \dfrac{(-8)(-9)}{(-3)(-2)}$

b. $(-1) + (-1)^2 + (-1)^3 + (-1)^4$

Early Finishers

Math and Architecture

The Pantheon was built in 125 A.D. in ancient Rome. The structure is in the shape of a cylinder with a dome on the top. The diameter and height of the cylinder is about 43 meters. The diameter of the dome (half sphere) is also about 43 meters.

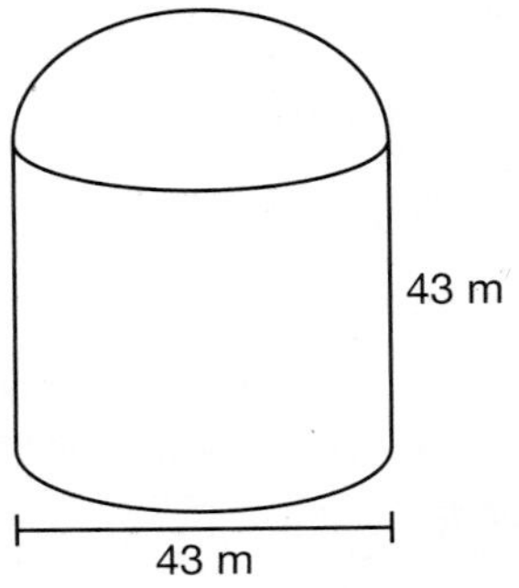

a. What is the surface area of the cylindrical portion of the Pantheon? Use 3.14 for π.

b. What is the surface area of the dome? Use 3.14 for π.

c. What is the interior surface area of the Pantheon?

LESSON

109

• Equations with Exponents

Power Up

Building Power

facts

Power Up V

mental math

a. Positive/Negative: $(-\frac{1}{2})(-\frac{1}{2})$

b. Scientific Notation: $(1.2 \times 10^{12})^2$

c. Ratio: $\frac{20}{40} = \frac{c}{2}$

d. Measurement: Convert 150 cm to m.

e. Percent: $12\frac{1}{2}\%$ of $80

f. Percent: $12\frac{1}{2}\%$ less than $80

g. Number Sense: Change $\frac{1}{3}$ to a decimal and percent.

h. Calculation: Find $\frac{1}{3}$ of 60, + 5, × 2, − 1, $\sqrt{}$, × 4, − 1, ÷ 3, square that number, − 1, ÷ 2.

problem solving

When you purchase hardwood, each board is sawn to specified widths and lengths. Hardwoods have their own unit of measure called "board foot" (bf). A board that is 1 in. thick, 12 in. wide, and 12 in. long measures 1 board foot (1 bf). Each of the following boards also measures 1 bf:

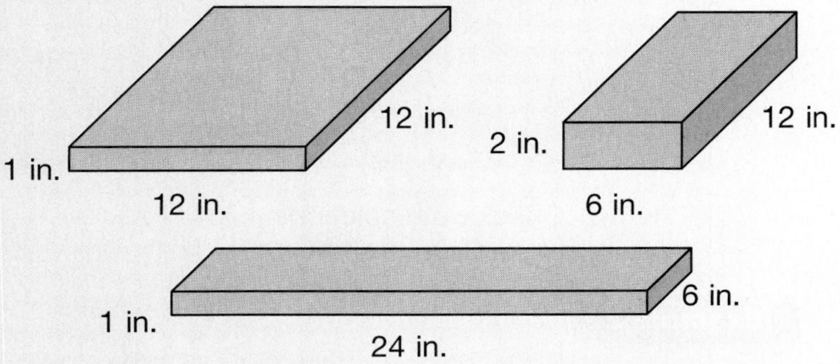

Can you find a formula for determining bf? Draw three more examples of 1 board foot. (Draw a diagram, use logical reasoning)

New Concept

Increasing Knowledge

In the equations we have solved thus far, the variables have had an exponent of 1. You have not seen the exponent, because we usually do not write the exponent when it is 1. In this lesson we will consider equations that have variables with exponents of 2, such as the following equation:

$$3x^2 + 1 = 28$$

Isolating the variable in this equation takes three steps: first we subtract 1 from both sides; next we divide both sides by 3; then we find the square root of both sides. We show the results of each step below.

Thinking Skill

Explain

Why do we subtract 1 from both sides of the equation first?

$3x^2 + 1 = 28$	equation
$3x^2 = 27$	subtracted 1 from both sides
$x^2 = 9$	divided both sides by 3
$x = 3, -3$	found the square root of both sides

Notice that there are two solutions, 3 and −3. Both solutions satisfy the equation, as we show below.

$3(3)^2 + 1 = 28$	$3(-3)^2 + 1 = 28$
$3(9) + 1 = 28$	$3(9) + 1 = 28$
$27 + 1 = 28$	$27 + 1 = 28$
$28 = 28$	$28 = 28$

When the variable of an equation has an exponent of 2, we remember to look for two solutions.

Example 1

Solve: $3x^2 - 1 = 47$

Solution

There are three steps. We show the results of each step.

$3x^2 - 1 = 47$	equation
$3x^2 = 48$	added 1 to both sides
$x^2 = 16$	divided both sides by 3
$x = \mathbf{4, -4}$	found the square root of both sides

Verify Check both solutions.

Example 2

Solve: $2x^2 = 10$

Solution

We divide both sides by 2. Then we find the square root of both sides.

$2x^2 = 10$	equation
$x^2 = 5$	divided both sides by 2
$x = \mathbf{\sqrt{5}, -\sqrt{5}}$	found the square root of both sides

Since $\sqrt{5}$ is an irrational number, we leave it in radical form. The negative of $\sqrt{5}$ is $-\sqrt{5}$ and not $\sqrt{-5}$.

Verify Check both solutions.

Example 3

Five less than what number squared is 20?

Solution

We translate the question into an equation.

$$n^2 - 5 = 20$$

We solve the equation in two steps.

$n^2 - 5 = 20$	equation
$n^2 = 25$	added 5 to both sides
$n = 5, -5$	found the square root of both sides

There are two numbers that answer the question, **5** and **−5.**

Example 4

In this figure the area of the larger square is 4 square units, which is twice the area of the smaller square. What is the length of each side of the smaller square?

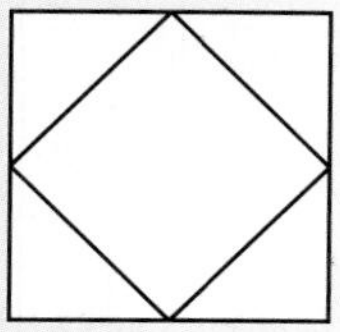

Solution

We will use the letter s to stand for the length of each side of the smaller square. So its area is s^2. Since the area of the large square (4) is twice the area of the small square, the area of the small square is 2.

$$s^2 = 2$$

We solve this equation in two steps by finding the square root of each side.

$s^2 = 2$	equation
$s = \sqrt{2}, -\sqrt{2}$	found the square root of both sides

Although there are two solutions there is only one answer because lengths are positive, not negative. Thus, each side of the smaller square is $\sqrt{2}$ **units.**

Example 5

Solve: $\frac{x}{3} = \frac{12}{x}$

Solution

Math Language

A **proportion** is a statement that shows two ratios are equal.

First we cross multiply. Then we find the square root of both sides.

$\frac{x}{3} = \frac{12}{x}$	proportion
$x^2 = 36$	cross multiplied
$x = 6, -6$	found the square root of both sides

There are two solutions to the proportion, **6** and **−6.**

Practice Set

Generalize Solve and check each equation:

a. $3x^2 - 8 = 100$ **b.** $x^2 + x^2 = 12$

c. Five less than twice what negative number squared is 157?

d. The area of a square is 3 square units. What is the length of each side?

e. $\frac{w}{4} = \frac{9}{w}$

Written Practice

Strengthening Concepts

1. (45) What is the quotient when the product of 0.2 and 0.05 is divided by the sum of 0.2 and 0.05?

2. (102) **Conclude** In the figure at right, a transversal intersects two parallel lines.

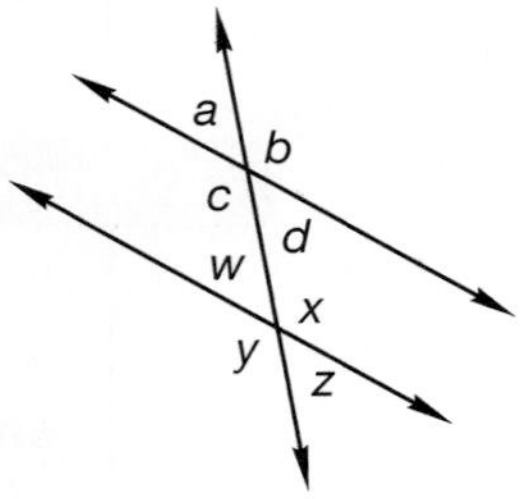

a. Which angle corresponds to $\angle d$?

b. Which angle is the alternate interior angle to $\angle d$?

c. Which angle is the alternate exterior angle to $\angle b$?

d. If the measure of $\angle a$ is m and the measure of $\angle b$ is $3m$, then each obtuse angle measures how many degrees?

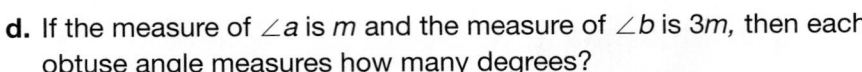

3. (101) Twenty is five more than the product of ten and what decimal number?

4. (36) **Justify** Santiago has \$5 in quarters and \$5 in dimes.

a. What is the ratio of the number of quarters to the number of dimes?

b. Explain how you found your answer.

Use ratio boxes to solve problems **5–7.**

5. (92) Sixty is 20 percent more than what number?

6. (92) The City-Wide Department Store is having a 25% off sale. What is the sale price of an item that originally cost \$36?

7. (72) Jaime ran the first 3000 meters in 9 minutes. At that rate, how long will it take Jaime to run the entire 5000 meters race?

8. (88) Use unit multipliers to convert Jaime's average speed in problem 7 to kilometers per hour.

9. (77) Write an equation to solve this problem:

Sixty is 150 percent of what number?

10. Diagram this statement. Then answer the questions that follow.
(71)

Diane kept $\frac{2}{3}$ of her baseball cards and gave the remaining 234 cards to her brother.

a. How many cards did Diane have before she gave some to her brother?

b. How many baseball cards did Diane keep?

11. **Explain** In the formula for the circumference of a circle, $C = \pi d$, are C and d directly or inversely proportional? Why?
(Inv. 9)

12. Yasmine is playing a true-false geography game. She has answered 15 of her 20 questions correctly. What is the probability that she will answer the five remaining questions correctly?
(Inv. 8)

13. Find the area of this trapezoid. Dimensions are in centimeters.
(75)

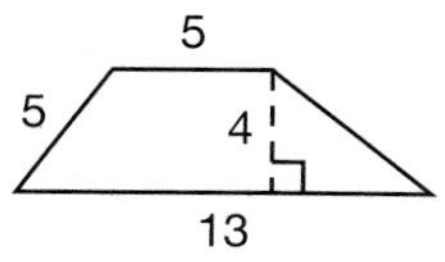

* **14.** **Analyze** Find the volume of this solid. Dimensions are in inches.
(95)

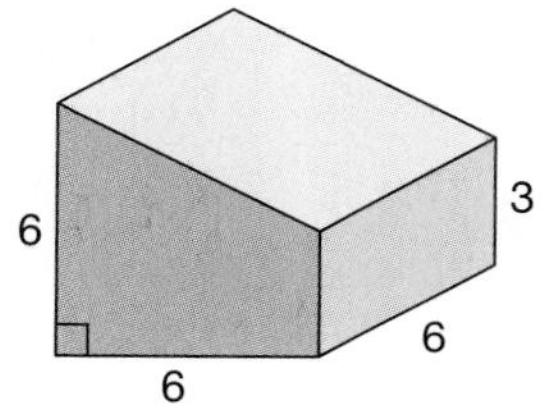

* **15.** **Analyze** A rectangular label is wrapped around a can with the dimensions shown. The label has an area of how many square inches? (Use 3.14 for π.)
(105)

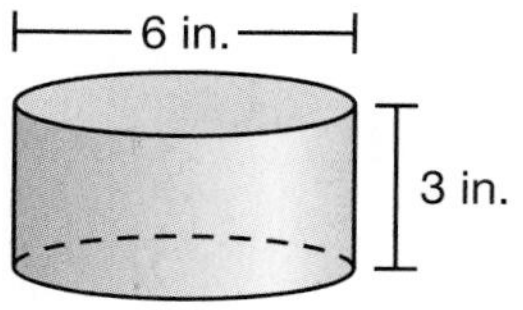

16. The skateboard costs $36. The tax rate is 6.5 percent.
(60)

a. What is the tax on the skateboard?

b. What is the total price, including tax?

17. Complete the table.
(48)

Fraction	Decimal	Percent
a.	**b.**	$\frac{1}{2}$%

18. What number is $66\frac{2}{3}$ percent more than 48?
(92)

19. Multiply. Write the product in scientific notation.
(83)

$$(6 \times 10^{-8})(8 \times 10^{4})$$

* **20.** (Inv. 9, 107) **a.** Find the missing numbers in the table by using the function rule.

b. Then graph the function on a coordinate plane.

c. What is the slope of the graphed line?

$y = \frac{2}{3}x - 1$

x	y
6	
0	
−3	

21. (66) Use a ratio box to solve this problem. The ratio of the measures of the two acute angles of the right triangle is 7 to 8. What is the measure of the smallest angle of the triangle?

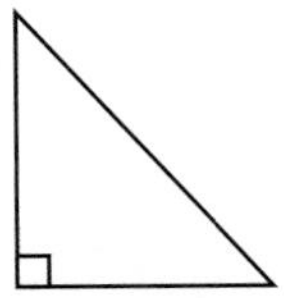

* **22.** (101) **Conclude** The relationship between the measures of four central angles of a circle is shown in this figure. What is the measure of the smallest central angle shown?

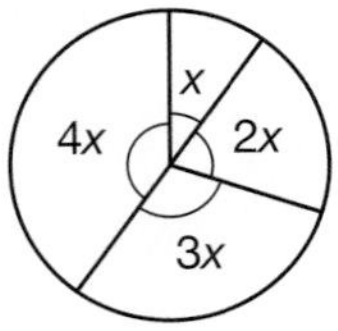

* **23.** (99) **Analyze** We can use the Pythagorean theorem to find the distance between two points on a coordinate plane. To find the distance from point M to point P, we draw a right triangle and use the lengths of the legs to find the length of the hypotenuse. What is the distance from point M to point P?

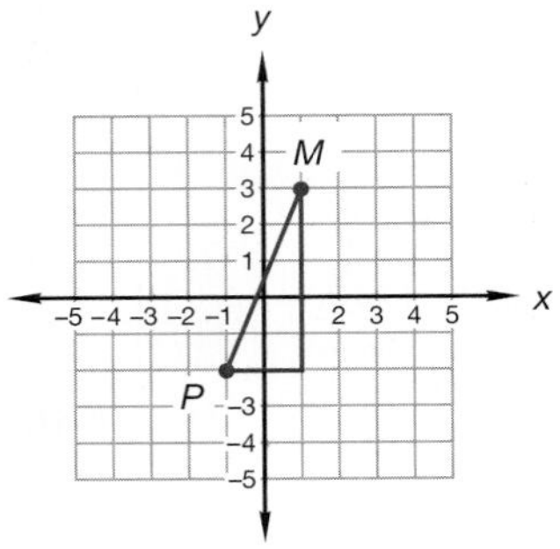

Solve and check.

24. (109) $3m^2 + 2 = 50$

* **25.** (102) $7(y - 2) = 4 - 2y$

Simplify:

26. (20) $\sqrt{144} - (\sqrt{36})(\sqrt{4})$

27. (96) $x^2y + xy^2 + x(xy - y^2)$

28. (43) $\left(1\frac{5}{9}\right)(1.5) \div 2\frac{2}{3}$

29. (43) $9.5 - \left(4\frac{1}{5} - 3.4\right)$

30. (57, 85, 106) **a.** $\dfrac{(-18) + (-12) - (-6)(3)}{-3}$

b. $\sqrt[3]{1000} - \sqrt[3]{125}$

c. $2^2 + 2^1 + 2^0 + 2^{-1}$

LESSON

110

• Simple Interest and Compound Interest
• Successive Discounts

Power Up

Building Power

facts

Power Up Q

mental math

a. Exponents: $(-\frac{1}{2})^2$

b. Scientific Notation: $(9 \times 10^6)(6 \times 10^9)$

c. Algebra: $4w - 1 = 9$

d. Measurement: 1.5 L to mL.

e. Percent: 150% of $60

f. Percent: $60 increase 50%

g. Number Sense: Change $\frac{1}{5}$ to a decimal and percent.

h. Measurement: Start with the number of minutes in half an hour. Multiply by the number of feet in a yard; add the number of years in a decade; then find the square root of that number. What is the answer?

problem solving

Sonya, Sid, and Sinead met at the gym on Monday. Sonya goes to the gym every two days. The next day she will be at the gym is Wednesday. Sid goes to the gym every three days. The next day Sid will be at the gym is Thursday. Sinead goes to the gym every four days. She will next be at the gym on Friday. What will be the next day that Sonya, Sid, and Sinead are at the gym on the same day?

New Concepts

Increasing Knowledge

simple interest and compound interest

When you deposit money in a bank, the bank does not simply hold your money for safekeeping. Instead, it spends your money in other places to make more money. For this opportunity the bank pays you a percentage of the money deposited. The amount of money you deposit is called the **principal.** The amount of money the bank pays you is called **interest.**

There is a difference between **simple interest** and **compound interest.** Simple interest is paid on the principal only and not paid on any accumulated interest. For instance, if you deposited $100 in an account that pays 6% simple interest, you would be paid 6% of $100 ($6) each year your $100 was on deposit. If you take your money out after three years, you would have a total of $118.

Simple Interest

$100.00	principal
$6.00	first-year interest
$6.00	second-year interest
+ $6.00	third-year interest
$118.00	total

Most interest-bearing accounts, however, are compound-interest accounts, not simple-interest accounts. In a compound-interest account, interest is paid on accumulated interest as well as on the principal. If you deposited $100 in an account with 6% annual percentage rate, the amount of interest you would be paid each year increases if the earned interest is left in the account. After three years you would have a total of $119.10.

Compound Interest

$100.00	principal
$6.00	first-year interest (6% of $100.00)
$106.00	total after one year
$6.36	second-year interest (6% of $106.00)
$112.36	total after two years
$6.74	third-year interest (6% of $112.36)
$119.10	total after three years

Notice that in three years, $100.00 grows to $118.00 at 6% simple interest, while it grows to $119.10 at 6% compound interest. The difference is not very large in three years, but as this table shows, the difference can become large over time.

Thinking Skill

Analyze

After 10 years, what is the difference in the interest earned through simple and compound interest? After 30 years? After 50 years?

Total Value of $100 at 6% Interest

Number of Years	Simple Interest	Compound Interest
3	$118.00	$119.10
10	$160.00	$179.08
20	$220.00	$320.71
30	$280.00	$574.35
40	$340.00	$1028.57
50	$400.00	$1842.02

Example 1

Make a table that shows the value of a $1000 investment growing at 10% compounded annually after 1, 2, 3, 4, and 5 years.

Solution

After the first year, \$1000 grows 10% to \$1100. After the second year, the value increases 10% of \$1100 (\$110) to a total of \$1210. We continue the pattern for five years in the table below.

Total Value of \$1000 at 10% Interest

Number of Years	Compound Interest
1	\$1100.00
2	\$1210.00
3	\$1331.00
4	\$1464.10
5	\$1610.51

Notice that the amount of money in the account after one year is 110% of the original deposit of \$1000. This 110% is composed of the starting amount, 100%, plus 10%, which is the interest earned in one year. Likewise, the amount of money in the account the second year is 110% of the amount in the account after one year. To find the amount of money in the account each year, we multiply the previous year's balance by 110% (or the decimal equivalent, which is 1.1).

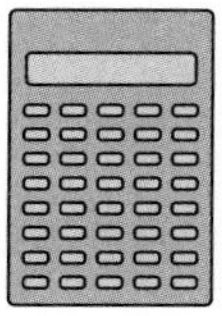

Even with a simple calculator we can calculate compound interest. To perform the calculation in example 1, we could follow this sequence:

$$1000 \times 1.1 \times 1.1 \times 1.1 \times 1.1 \times 1.1 =$$

The circuitry of some calculators permits repeating a calculation by pressing the [=] key repeatedly.[1] To make the calculations in example 1, we try this keystroke sequence:

[1] [.] [1] [×] [1] [0] [0] [0] [=]

This keystroke sequence first enters 1.1, which is the decimal form of 110% (100% principal plus 10% interest), then the times sign, then 1000 for the \$1000 investment. Pressing the [=] key once displays

1100.

which is the value (\$1100) after one year. Pressing the [=] key a second time multiplies the displayed number by 1.1, the first number we entered. The new number displayed is

1210.

representing \$1210, the value after two years. Each time the [=] key is pressed, the calculator displays the account value after a successive year. Using this method, find the value of the account after 10 years and after 20 years.

[1] This calculator function varies with make and model of calculator. See instructions for your calculator if the keystroke sequence described in this lesson does not work for you.

Try entering the factors in the reverse order.

1 0 0 0 × 1 . 1 =

Explain Are the same amounts displayed as were displayed with the prior entry when the = key is repeatedly pressed? Why or why not?

Example 2

Use a calculator to find the value after 12 years of a \$2000 investment that earns $7\frac{1}{2}$% interest compounded annually.

Solution

The interest rate is $7\frac{1}{2}$%, which is 0.075 in decimal form. We want to find the total value, including the principal. So we multiply the \$2000 investment by $107\frac{1}{2}$%, which we enter as 1.075. The keystroke sequence is

1 . 0 7 5 × 2 0 0 0 =

We then press the = key 11 more times to find the value after 12 years. We round the final display to the nearest cent, **\$4763.56.**

Example 3

Calculate the interest earned on an \$8000 deposit in 9 months if the annual interest rate is 6%.

Solution

The deposit earns 6% interest in one year, which is

$$0.06 \times \$8000 = \$480$$

In 9 months the deposit earns just $\frac{9}{12}$ of this amount.

$$\frac{9}{12} \times \$480 = \mathbf{\$360}$$

successive discounts

Thinking Skill

Connect

How is calculating compound interest related to calculating successive discounts?

Related to compound interest is **successive discount.** To calculate successive discount, we find a percent of a percent. In the following example we show two methods for finding successive discounts.

Example 4

An appliance store reduced the price of a \$400 washing machine 25%. When the washing machine did not sell at the sale price, the store reduced the sale price 20% to its clearance price. What was the clearance price of the washing machine?

Solution

One way to find the answer is to first find the sale price and then find the clearance price. We will use a ratio box to find the sale price.

	Percent	Actual Count
Original	100	400
– Change	25	D
New (Sale)	75	S

$$\frac{100}{75} = \frac{400}{S}$$
$$100S = 30{,}000$$
$$S = 300$$

We find that the sale price was $300. The second discount, 20%, was applied to the sale price, not to the original price. So for the next calculation we consider the sale price to be 100% and the clearance price to be what remains after the discount.

	Percent	Actual Count
Original	100	300
– Change	20	D
New (Clearance)	80	C

$$\frac{100}{80} = \frac{300}{C}$$
$$100C = 24{,}000$$
$$C = 240$$

We find that the clearance price of the washing machine was **$240.**

Another way to look at this problem is to consider what percent of the original price is represented by the sale price. Since the original price was discounted 25%, the sale price represents 75% of the original price.

Sale price = 75% of the original price

Furthermore, since the sale price was discounted 20%, the clearance price was 80% of the sale price.

Clearance price = 80% of the sale price

So the clearance price was 80% of the sale price, which was 75% of the original price.

$$\begin{aligned}\text{Clearance price} &= 80\% \text{ of } 75\% \text{ of } \$400\\ &= 0.8 \times 0.75 \times \$400\\ &= 0.6 \times \$400\\ &= \mathbf{\$240}\end{aligned}$$

Evaluate Which method, the first or the second, is a more efficient way to find a clearance price after a successive discount?

Practice Set

a. **Generalize** When Mai turned 21, she invested $2000 in an Individual Retirement Account (IRA) that has grown at a rate of 10% compounded annually. If the account continues to grow at that rate, what will be its value when Mai turns 65? (Use the calculator method taught in this lesson.)

b. Mrs. Rojas deposited $6000 into an account paying 4% interest annually. After 8 months Mrs. Rojas withdrew the $6000 plus interest. Altogether, how much money did Mrs. Rojas withdraw? What fraction of a year's interest was earned?

c. A television regularly priced at \$300 was placed on sale for 20% off. When the television still did not sell, the sale price was reduced 20% for clearance. What was the clearance price of the television?

d. **Formulate** Write and solve an original word problem that involves compound or simple interest.

Written Practice *Strengthening Concepts*

1. (60) Rosita bought 3 paperback books for \$5.95 each. The sales-tax rate was 6 percent. If she paid for the purchase with a \$20 bill, how much money did she get back?

*** 2.** (110) **Generalize** Hector has a coupon for 10% off the price of any item in the store. He decides to buy a shirt regularly priced at \$24 that is on sale for 25% off. If he uses his coupon, how much will Hector pay for the shirt before sales tax is applied?

3. (80) Triangle *ABC* with vertices *A* (3, 0), *B* (3, 4), and *C* (0, 0) is rotated 90° clockwise about the origin to its image $\triangle A'B'C'$. Graph both triangles. What are the coordinates of the vertices of $\triangle A'B'C'$?

4. (72) Jorge burned 370 calories by hiking for one hour.

a. How many calories would Jorge burn if he hiked for 2 hours 30 minutes?

b. Suppose Jorge's dinner consisted of a 6 oz steak (480 calories), a baked potato (145 calories), 4 spears of asparagus (15 calories), and one cup of strawberries (45 calories). How many hours would Jorge need to hike to burn the calories he ate at dinner? Round your answer to the nearest minute.

5. (72) If a dozen roses costs \$20.90, what is the cost of 30 roses?

*** 6.** (104) **Analyze** A semicircle was cut out of a square. What is the perimeter of the resulting figure shown? (Use $\frac{22}{7}$ for π.)

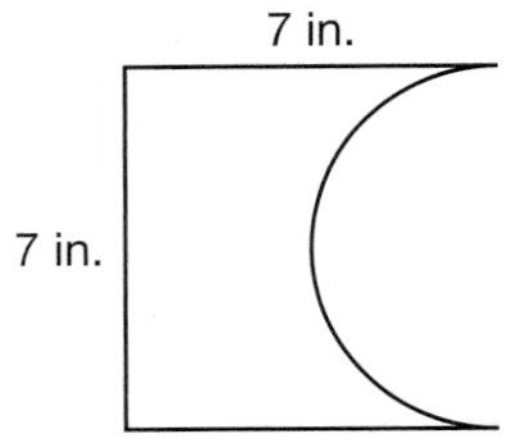

7. (55) The average of four numbers is 8. Three of the numbers are 2, 4, and 6. What is the fourth number?

Write equations to solve problems **8–10.**

8. (77) One hundred fifty is what percent of 60?

9. (77) Sixty percent of what number is 150?

*** 10.** (109) **Generalize** Six more than the square of what negative number is 150?

11. (99) Graph the points (3, 1) and (–1, –2) on a coordinate plane. Then draw a right triangle, and use the Pythagorean theorem to find the distance between the points.

Use ratio boxes to solve problems **12** and **13.**

12. (92) The price of the dress was reduced by 40 percent. If the sale price was \$48, what was the regular price?

Reading Math

Read the term *1:36* as "one to thirty-six."

* **13.** (98) **Generalize** The car model was built on a 1:36 scale. If the length of the car is 180 inches, how many inches long is the model?

14. (100) The positive square root of 80 is between which two consecutive whole numbers?

* **15.** (Inv. 9, 107) **Analyze** Make a table of ordered pairs for the function $y = -x + 1$. Then graph the function. What is the slope of the graphed line?

16. (93) Solve this inequality and graph its solution:

$$5x + 12 \geq 2$$

17. (83) Multiply. Write the product in scientific notation.

$$(6.3 \times 10^7)(9 \times 10^{-3})$$

* **18.** (102, 106) **Generalize** Solve for y: $\frac{1}{2}y = x + 2$

* **19.** (110) **Generalize** What is the total account value after 3 years on a deposit of \$4000 at 9% interest compounded annually?

20. (97, 98) The triangles below are similar. Dimensions are in inches.

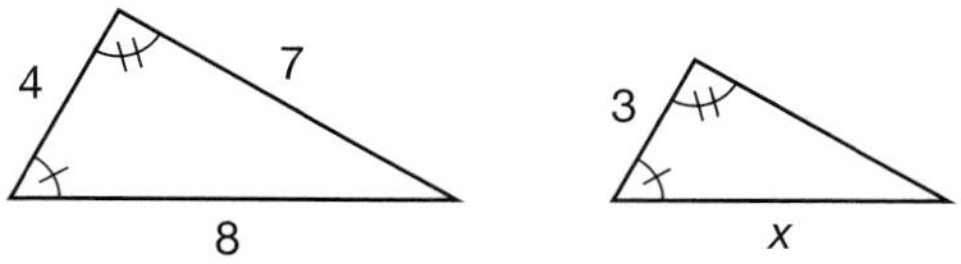

a. Estimate, then calculate, the length x.

b. Find the scale factor from the larger to the smaller triangle.

21. (95) Find the volume of this triangular prism. Dimensions are in inches.

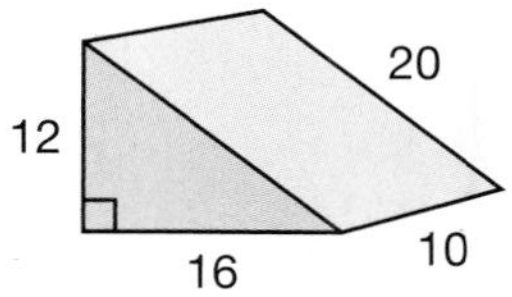

* **22.** (105) **Generalize** Find the total surface area of the triangular prism in problem **21.**

23. (40) Find m∠x in the figure at right.

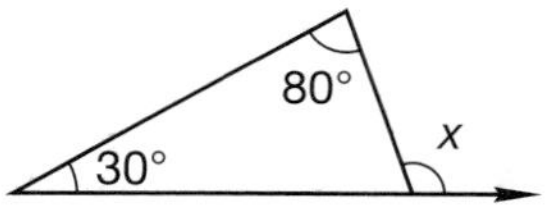

Generalize Solve:

* **24.** (109) $\frac{w}{2} = \frac{18}{w}$

* **25.** (109) $3\frac{1}{3}w^2 - 4 = 26$

Generalize Simplify:

26. (63) $16 - \{27 - 3[8 - (3^2 - 2^3)]\}$

* **27.** (103) $\frac{(6ab^2)(8ab)}{12a^2b^2}$

28. (43) $3\frac{1}{3} + 1.5 + 4\frac{5}{6}$

29. (26) $20 \div \left(3\frac{1}{3} \div 1\frac{1}{5}\right)$

* **30.** (103) $(-3)^2 + (-2)^3$

Early Finishers

Real-World Application

A local town wants to paint the entire surface of its water tank. The tank is in the shape of a cylinder. Because it is elevated, the bottom surface needs to be painted. The diameter of the tank is 30 feet and the height is 20 feet.

a. What is the surface area of the tank? Use 3.14 for π.

b. Paint costs $22 per gallon and is only sold in whole gallons. The tank only needs one coat of paint. If one gallon covers approximately 250 square feet, how many gallons of paint should the town purchase?

Focus on

Scale Factor in Surface Area and Volume

In this investigation we will study the relationship between length, surface area, and volume of three-dimensional shapes. We begin by comparing the measures of cubes of different sizes.

Activity

Scale Factor in Surface Area and Volume

Materials needed:

- 3 photocopies of **Investigation Activity 25** Square Centimeter Grid or 3 sheets of 1-cm grid paper
- Scissors
- Tape

Use the provided materials to build models of four cubes with edges 1 cm, 2 cm, 3 cm, and 4 cm long. Mark, cut, fold, and tape the grid paper so that the *grid is visible* when each model is finished.

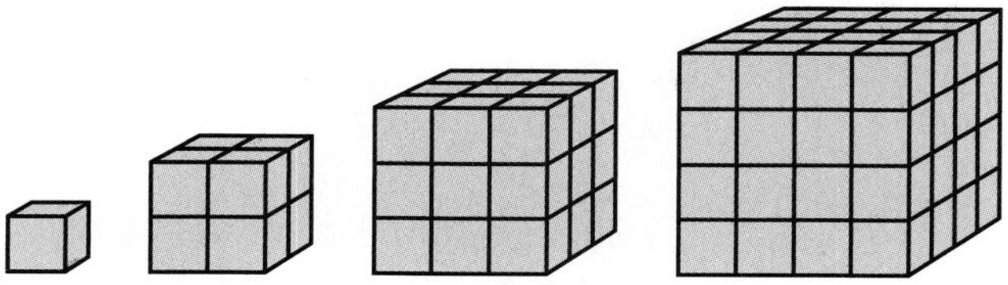

One pattern that folds to form a model of a cube is shown below. Several other patterns also work.

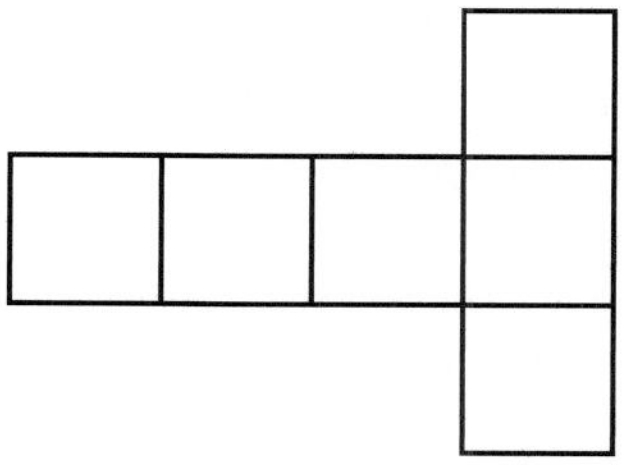

Copy this table on your paper and record the measures for each cube.

Thinking Skill

Model

Draw a different pattern with 6 squares that folds to form a cube.

Measures of Four Cubes

	1-cm cube	2-cm cube	3-cm cube	4-cm cube
Edge length (cm)				
Surface area (cm^2)				
Volume (cm^3)				

Refer to the table on the previous page to answer the following questions:

Compare the 2-cm cube to the 1-cm cube.

1. The edge length of the 2-cm cube is how many times the edge length of the 1-cm cube?

2. The surface area of the 2-cm cube is how many times the surface area of the 1-cm cube?

3. The volume of the 2-cm cube is how many times the volume of the 1-cm cube?

Compare the 4-cm cube to the 2-cm cube.

4. The edge length of the 4-cm cube is how many times the edge length of the 2-cm cube?

5. The surface area of the 4-cm cube is how many times the surface area of the 2-cm cube?

6. The volume of the 4-cm cube is how many times the volume of the 2-cm cube?

Analyze Compare the 3-cm cube to the 1-cm cube.

7. The edge length of the 3-cm cube is how many times the edge length of the 1-cm cube?

8. The surface area of the 3-cm cube is how many times the surface area of the 1-cm cube?

9. The volume of the 3-cm cube is how many times the volume of the 1-cm cube?

Predict Use the patterns that can be found in answers 1–9 to predict the comparison of a 6-cm cube to a 2-cm cube.

10. The edge length of a 6-cm cube is how many times the edge length of a 2-cm cube?

11. The surface area of a 6-cm cube is how many times the surface area of a 2-cm cube?

12. The volume of a 6-cm cube is how many times the volume of a 2-cm cube?

13. Calculate **a** the surface area and **b** the volume of a 6-cm cube.

14. The calculated surface area of a 6-cm cube is how many times the surface area of a 2-cm cube?

15. The calculated volume of a 6-cm cube is how many times the volume of a 2-cm cube?

In problems 1–6 we compared the measures of a 2-cm cube to a 1-cm cube and the measures of a 4-cm cube to a 2-cm cube. In both sets of comparisons, the scale factors from the smaller cube to the larger cube were calculated. The dimensions of the larger square were twice the dimensions of the smaller square. That means that the scale factor from the smaller square to the larger square was 2. If we know the scale factor we can find the multiplier for area and volume as we see in this table.

Length, Area, and Volume Multipliers for Scale Factor 2

Measurement	Multiplier
Edge Length	2
Surface Area	$2^2 = 4$
Volume	$2^3 = 8$

Likewise, in problems **7–15** we compared the measures of a 3-cm cube to a 1-cm cube and a 6-cm cube to a 2-cm cube. We calculated the following multipliers for scale factor 3:

Length, Area, and Volume Multipliers for Scale Factor 3

Measurement	Multiplier
Edge Length	3
Surface Area	$3^2 = 9$
Volume	$3^3 = 27$

Represent Make a table similar to the one above where the edge length scale factor is 4.

Generalize Refer to the above description of scale factors to answer problems **16–20.**

16. Mia calculated the scale factors from a 6-cm cube to a 30-cm cube.

a. From the smaller cube to the larger cube, what is the scale factor?

b. *Conclude* Make a table like the two above, identifying the multiplier for the edge length, surface area, and volume of a figure with this scale factor.

17. *Generalize* Bethany noticed that the scale factor relationships for cubes also applies to spheres. She found the approximate diameters of a table tennis ball ($1\frac{1}{2}$ in.), a baseball (3 in.), and a playground ball (9 in.). Find the multiplier for:

a. the volume of the table tennis ball to the volume of the baseball.

b. the surface area of the baseball to the surface area of the playground ball.

18. **Generalize** The photo lab makes 5-by-7-in. enlargements from $2\frac{1}{2}$-by-$3\frac{1}{2}$-in. wallet-size photos.

a. Find the scale factor from the smaller photo to the enlargement.

b. The picture area of the larger photo is what multiple of the picture area of the smaller photo?

19. **Generalize** Rommy wanted to charge the same price per square inch of cheese pizza regardless of the size of the pizza. Since all of Rommy's pizzas were the same thickness, he based his prices on the scale factor for area. If he sells a 10-inch diameter cheese pizza for $10.00, how much should he charge for a 15-inch diameter cheese pizza?

20. **Generalize** The Egyptian archaeologist knew that the scale-factor relationships for cubes also applies to similar pyramids. The archaeologist built a $\frac{1}{100}$ scale model of the Great Pyramid. Each edge of the base of the model was 2.3 meters, while each edge of the base of the Great Pyramid measured 230 meters.

a. From the smaller model to the Great Pyramid, what was the scale factor for the length of corresponding edges?

b. The area of the Great Pyramid is what multiple of the area of the model?

c. The volume of the Great Pyramid is what multiple of the volume of the model?

Notice from the chart that you completed near the beginning of this investigation that as the size of the cube becomes greater, the surface area and volume become much greater. Also notice that the volume increases at a faster rate than the surface area. The ratio of surface area to volume changes as the size of an object changes.

Ratio of Surface Area to Volume of Four Cubes

	1-cm cube	2-cm cube	3-cm cube	4-cm cube
Surface Area to Volume	6 to 1	3 to 1	2 to 1	1.5 to 1

The ratio of surface area to volume affects the size and shape of containers used to package products. The ratio of surface area to volume also affects the world of nature.

Analyze Consider the relationship between surface area and volume as you answer problems **21–25.**

21. Sixty-four 1-cm cubes were arranged to form one large cube. Austin wrapped the large cube with paper and sent the package to Betsy. The volume of the package was 64 cm^3. What was the surface area of the exposed wrapping paper?

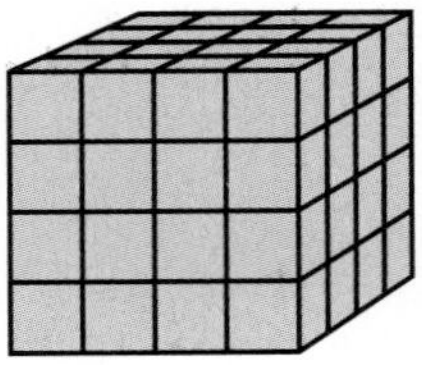

22. When Betsy received the package she divided the contents into eight smaller cubes each composed of eight 2-cm cubes. Betsy wrapped the eight cubes with paper and sent them on to Charlie. The total volume of the eight packages was still 64 cm^3. What was the total surface area of the exposed wrapping paper of the eight packages?

23. Charlie opened each of the eight packages and individually wrapped each 1-cm cube. Since there were 64 cubes, the total volume was still 64 cm^3. What was the total surface area of exposed wrapping paper for all 64 packages?

24. *Predict* After a summer picnic, the ice in two large insulated containers was emptied on the ground to melt. A large block of ice in the form of a 6-inch cube fell out of one container. An equal quantity of ice, but in the form of 1-inch cubes, fell scattered out of the other container. Which, if either, do you think will melt sooner, the large block of ice or the small scattered cubes? Explain your answer.

25. If someone does not eat very much, we might say that he or she "eats like a bird." However, birds must eat large amounts, relative to their body weights, in order to maintain their body temperature. Since mammals and birds regulate their own body temperature, there is a limit to how small a mammal or bird may be. Comparing a hawk and a sparrow in the same environment, which of the two might eat a greater percentage of its weight in food every day? Explain your answer.

extensions

a. Investigate how the weight of a bird and its wingspan are related.

b. Investigate reasons why the largest sea mammals are so much larger than the largest land mammals.

c. Brad's dad is 25% taller than Brad and weighs twice as much. Explain why you think this height-weight relationship may or may not be reasonable.

LESSON

111 • Dividing in Scientific Notation

Power Up

Building Power

facts Power Up V

mental math

a. **Positive/Negative:** $(-0.25) \div (-5)$

b. **Scientific Notation:** $(8 \times 10^{-4})^2$

c. **Algebra:** $3m + 7m = 600$

d. **Measurement:** Convert 1 ft^2 to square inches.

e. **Percent:** $33\frac{1}{3}\%$ of \$150

f. **Percent:** \$150 reduced by $33\frac{1}{3}\%$

g. **Probability:** A spinner is in 6 sections with red, blue, yellow, red, purple, and pink. What is the probability of the spinner landing on red?

h. **Percent/Estimation:** Estimate 8% tax on a \$198.75 purchase.

problem solving Find the next three numbers in this sequence: 1, 1, 2, 3, 5, 8, 13, ...

New Concept

Increasing Knowledge

One unit astronomers use to measure distances within our solar system is the **astronomical unit** (AU). An astronomical unit is the average distance between Earth and the Sun, which is roughly 150,000,000 km (or 93,000,000 mi).

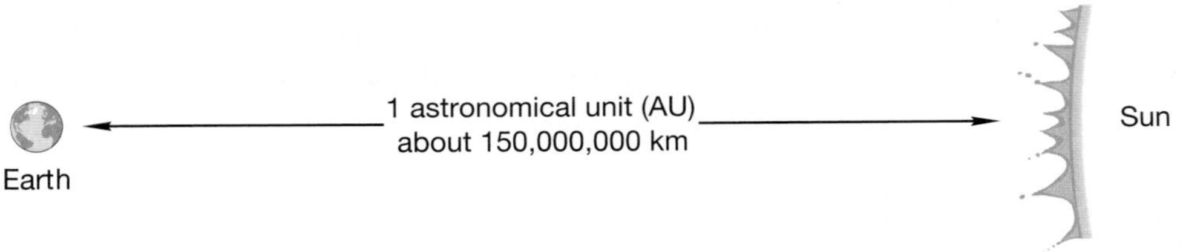

For instance, at a point in Saturn's orbit when it is 1.5 billion kilometers from the Sun, its distance from the Sun is 10 AU.

$$1{,}500{,}000{,}000 \text{ km} \cdot \frac{1 \text{ AU}}{150{,}000{,}000 \text{ km}} = 10 \text{ AU}$$

This means that the distance from Saturn to the Sun is about 10 times the average distance between Earth and the Sun.

Thinking Skill
Generalize
How many times smaller is the distance from the Sun to Earth compared to the distance from the Sun to Saturn?

When dividing very large or very small numbers, it is helpful to use scientific notation. Here we show the same calculation in scientific notation:

$$\frac{1.5 \times 10^9 \text{ km}}{1.5 \times 10^8 \text{ km/AU}} = 10 \text{ AU}$$

In this lesson we will practice dividing numbers in scientific notation.

Recall that when we multiply numbers in scientific notation, we multiply the powers of 10 by adding their exponents.

$$(6 \times 10^6)(1.5 \times 10^2) = 9 \times 10^8$$

Furthermore, we have this important rule:

When we divide numbers written in scientific notation, we divide the powers of 10 by subtracting their exponents.

$$\frac{6 \times 10^6}{1.5 \times 10^2} = 4 \times 10^4 \longleftarrow (6-2=4)$$

Example 1

Write each quotient in scientific notation:

a. $\frac{6 \times 10^8}{1.2 \times 10^6}$ **b.** $\frac{3 \times 10^3}{6 \times 10^6}$ **c.** $\frac{2 \times 10^{-2}}{8 \times 10^{-8}}$

Solution

a. To find the quotient, we divide 6 by 1.2 and 10^8 by 10^6.

$$12.\overline{)60.} = 5. \qquad 10^8 \div 10^6 = 10^2 \longleftarrow (8-6=2)$$

The quotient is $\mathbf{5 \times 10^2}$.

b. We divide 3 by 6 and 10^3 by 10^6.

$$6\overline{)3.0} = 0.5 \qquad 10^3 \div 10^6 = 10^{-3} \longleftarrow (3-6=-3)$$

The quotient is 0.5×10^{-3}. We write the quotient in scientific notation.

$$\mathbf{5 \times 10^{-4}}$$

c. We divide 2 by 8 and 10^{-2} by 10^{-8}.

$$8\overline{)2.00} = 0.25 \qquad 10^{-2} \div 10^{-8} = 10^6 \longleftarrow [-2-(-8)=6]$$

The quotient is 0.25×10^6. We write the quotient in scientific notation.

$$\mathbf{2.5 \times 10^5}$$

Example 2

The distance from the Sun to Earth is about 1.5×10^8 km. Light travels at a speed of about 3×10^5 km per second. About how many seconds does it take light to travel from the Sun to Earth?

Solution

We divide 1.5×10^8 km by 3×10^5 km/s.

$$\frac{1.5 \times 10^8 \text{ km}}{3 \times 10^5 \text{ km/s}} = 0.5 \times 10^3 \text{ s}$$

We may write the quotient in proper scientific notation, $\mathbf{5 \times 10^2}$ **s** or in standard form, **500 s.** It takes about 500 seconds.

Practice Set Generalize Write each quotient in scientific notation:

a. $\frac{3.6 \times 10^9}{2 \times 10^3}$

b. $\frac{7.5 \times 10^3}{2.5 \times 10^9}$

c. $\frac{4.5 \times 10^{-8}}{3 \times 10^{-4}}$

d. $\frac{6 \times 10^{-4}}{1.5 \times 10^{-8}}$

e. $\frac{4 \times 10^{12}}{8 \times 10^4}$

f. $\frac{1.5 \times 10^4}{3 \times 10^{12}}$

g. $\frac{3.6 \times 10^{-8}}{6 \times 10^{-2}}$

h. $\frac{1.8 \times 10^{-2}}{9 \times 10^{-8}}$

i. Justify Which rule for working with exponents did you use to complete the divisions in **a–h?**

Written Practice *Strengthening Concepts*

1. (12) The first Indian-head penny was minted in 1859. The last Indian-head penny was minted in 1909. For how many years were Indian-head pennies minted?

2. (28) Analyze The product of y and 15 is 600. What is the sum of y and 15?

3. (36, 53) Thirty percent of the class wanted to go to the Museum of Natural History. The rest of the class wanted to go to the Planetarium.

a. What fraction of the class wanted to go to the Planetarium?

b. What was the ratio of those who wanted to go to the Museum of Natural History to those who wanted to go to the Planetarium?

4. (80) Triangle ABC with vertices A (0, 3), B (0, 0), and C (4, 0) is translated one unit left, one unit down to make the image $\triangle A'B'C'$. What are the coordinates of the vertices of $\triangle A'B'C'$?

5. (21) **a.** Write the prime factorization of 1024 using exponents.

b. Find $\sqrt{1024}$.

6. (89) A portion of a regular polygon is shown at right. Each interior angle measures 150°.

a. What is the measure of each exterior angle?

b. The polygon has how many sides?

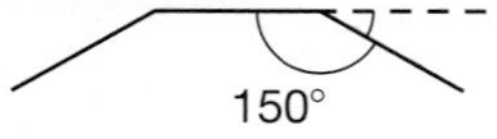

c. What is the name for a polygon with this number of sides?

7. (92) The Science Store advertised a 40% discount on all equipment. A rock tumbler's sale price was $48.00. What was the original price?

8. (94) In a bag are 12 marbles: 3 red, 4 white, and 5 blue. One marble is drawn from the bag and not replaced. A second marble is drawn and not replaced. Then a third marble is drawn.

a. What is the probability of drawing a red, a white, and a blue marble in that order?

b. What is the probability of drawing a blue, a white, and a red marble in that order?

* **9.** (101) Write an equation to solve this problem:

Six more than twice what number is 36?

* **10.** (40, 101) **a.** **Conclude** What is the measure of each acute angle of this triangle?

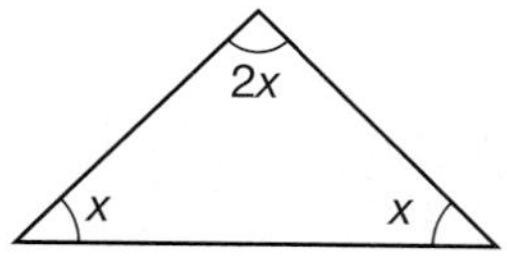

b. **Explain** What did you need to know about the measures of the angles of a triangle to solve this problem?

* **11.** (106) Solve for c^2: $c^2 - b^2 = a^2$

Math Language

The symbol $\|$ means *is parallel to*. $l \| q$ means line l is parallel to line q.

* **12.** (102) **Conclude** In the figure below, if $l \| q$ and $m\angle h = 105°$, what is the measure of

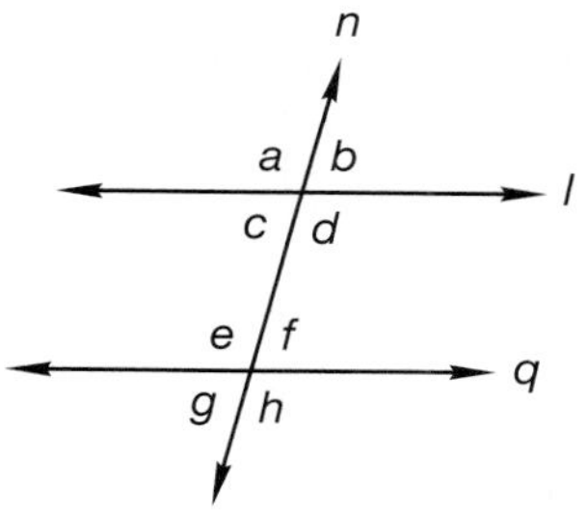

a. $\angle a$?

b. $\angle b$?

c. $\angle c$?

d. $\angle d$?

* **13.** (108) The formula below may be used to convert temperature measurements from degrees Celsius (C) to degrees Fahrenheit (F). Find F to the nearest degree when C is 17.

$$F = 1.8C + 32$$

14. (56) Make a table of ordered pairs showing three or four solutions for the equation $x + y = 1$. Then graph all possible solutions.

* **15.** (Inv. 9, 107) **Analyze** Refer to the problem you made in problem 14 to answer **a** and **b**.

a. What is the slope of the graph of $x + y = 1$?

b. Where does the graph of $x + y = 1$ intersect the y-axis?

* **16.** (104) **Analyze** What is the area of a 45° sector of a circle with a radius of 12 in.? Use 3.14 for π and round the answer to the nearest square inch.

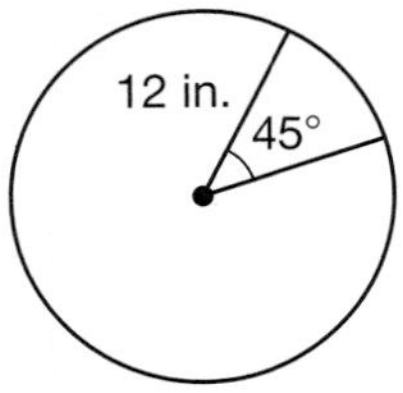

Refer to the illustration of the trash can to answer problems **17** and **18**.

* **17.** (105) **Connect** The students in Room 8 decided to wrap posters around school trash cans to encourage others to properly dispose of trash. The illustration shows the dimensions of the trash can. Converting the dimensions to feet and using 3.14 for π, find the number of square feet of paper needed to wrap around each trash can.

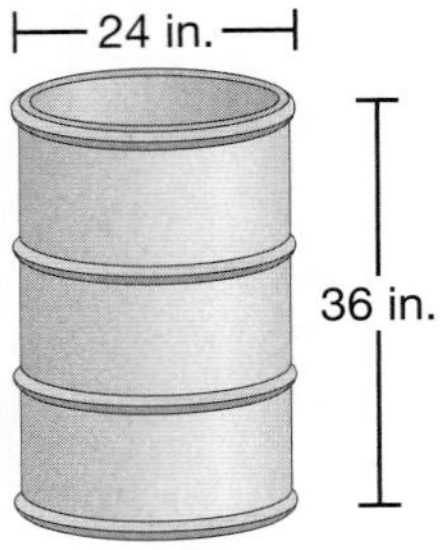

18. (95) The trash can illustrated has the capacity to hold how many cubic feet of trash?

* **19.** (109) Find two solutions to each of these equations:

a. $2x^2 + 1 = 19$ **b.** $2x^2 - 1 = 19$

20. (99) What is the perimeter of a triangle with vertices $(-1, 2)$, $(-1, -1)$, and $(3, -1)$?

* **21.** (110) **Connect** Sal deposited \$5000 in a 60-month CD that paid 5% interest compounded annually. What was the total value of the CD after 5 years?

* **22.** (97, 99) **Conclude** The figure at right shows three similar triangles. If AC is 15 cm and BC is 20 cm,

a. what is AB?

b. what is CD?

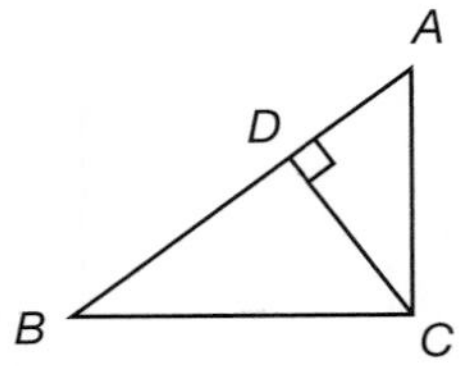

* **23.** (111) Express each quotient in scientific notation:

a. $\dfrac{3.6 \times 10^8}{6 \times 10^6}$ **b.** $\dfrac{3.6 \times 10^{-8}}{1.2 \times 10^{-6}}$

24. (40) In the figure below, if the measure of $\angle x$ is 140°, what is the measure of $\angle y$?

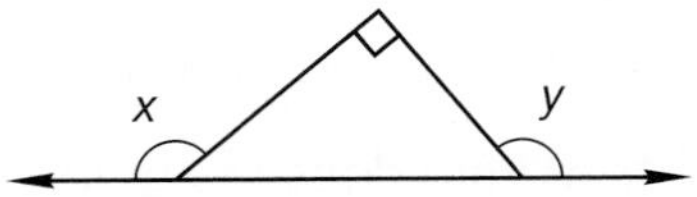

Solve:

25. (102) $5x + 3x = 18 + 2x$

26. (98) $\dfrac{3.6}{x} = \dfrac{4.5}{0.06}$

Generalize Simplify:

* **27.** (103) **a.** $(-1)^6 + (-1)^5$ **b.** $(-10)^6 \div (-10)^5$

* **28.** (96, 103) **a.** $\dfrac{(4a^2b)(9ab^2c)}{6abc}$ **b.** $x(x - c) + \sqrt{c^2x^2}$

29. (85) $(-3) + (+2)(-4) - (-6)(-2) - (-8)$

30. (43, 45) $\dfrac{3\frac{1}{3} \cdot 1\frac{4}{5} + 1.5}{0.03}$

Early Finishers

Real-World Application

You open a savings account with an initial deposit of $1500. The bank offers an annual interest rate of 4% compounded annually on its savings accounts.

a. What will your balance be after two years, if no additional deposits or withdrawals are made?

b. Three years after your initial deposit, you decide to use the money in your account to buy a $2000 car. How much more money will you need to buy the car?

LESSON

112

Applications of the Pythagorean Theorem

Power Up

Building Power

facts

Power Up W

mental math

a. **Positive/Negative:** $(-10)^2 + (-10)^3$

b. **Scientific Notation:** $(8 \times 10^6) \div (4 \times 10^3)$

c. **Algebra:** $m^2 = 100$

d. **Measurement:** Convert 50°C to degrees Fahrenheit.

e. **Percent:** 25% of $2000

f. **Percent:** $2000 increased 25%

g. **Geometry:** Which pyramid has 5 faces and 5 vertices?

h. **Calculation:** Start with 2 dozen, + 1, × 4, + 20, ÷ 3, + 2, ÷ 6, × 4, − 3, $\sqrt{\ }$, ÷ 2.

problem solving

When all the cards from a 52-card deck are dealt to three players, each player receives 17 cards, and there is one extra card. Sharla invented a new deck of cards so that any number of players up to 6 can play and there will be no extra cards. How many cards are in Sharla's deck if the number is less than 100?

New Concept

Increasing Knowledge

Workers who construct buildings need to be sure that the structures have square corners. If the corner of a 40-foot-long building is 89° or 91° instead of 90°, the other end of the building will be about 8 inches out of position.

One way construction workers can check whether a building under construction is square is by using a **Pythagorean triplet.** The numbers 3, 4, and 5 satisfy the Pythagorean theorem and are an example of a Pythagorean triplet.

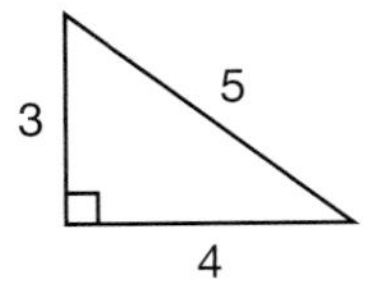

$$3^2 + 4^2 = 5^2$$

Thinking Skill

Analyze

Name two more Pythagorean triplets.

Multiples of 3-4-5 are also Pythagorean triplets.

3-4-5

6-8-10

9-12-15

12-16-20

Before pouring a concrete foundation for a building, construction workers build wooden forms to hold the concrete. Then a worker or building inspector may use a Pythagorean triplet to check that the forms make a right angle. First the perpendicular sides are marked at selected lengths. Then the diagonal distance between the marks is checked to be sure the three measures are a Pythagorean triplet.

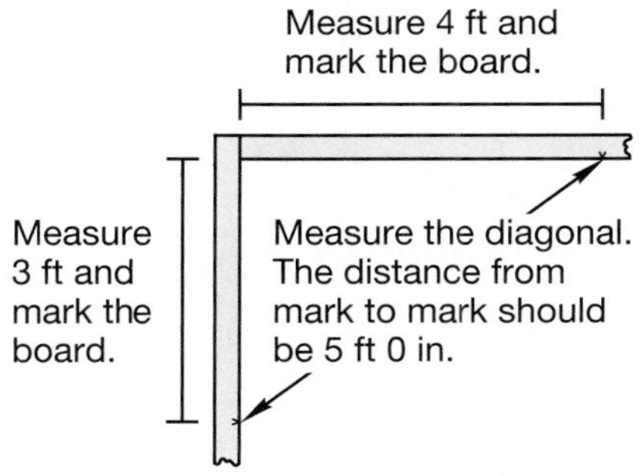
Measure 4 ft and mark the board.
Measure 3 ft and mark the board.
Measure the diagonal. The distance from mark to mark should be 5 ft 0 in.

If the three measures form a Pythagorean triplet, what can the worker conclude?

Activity

Application of the Pythagorean Theorem

Materials needed by each group of 2 or 3 students:

- Two full-length, unsharpened pencils (or other straightedges)
- Ruler
- Protractor

Position two pencils (or straightedges) so that they appear to form a right angle. Mark one pencil 3 inches from the vertex of the angle and the other pencil 4 inches from the vertex. Then measure from mark to mark to see whether the distance between the marks is 5 inches. Adjust the pencils if necessary.

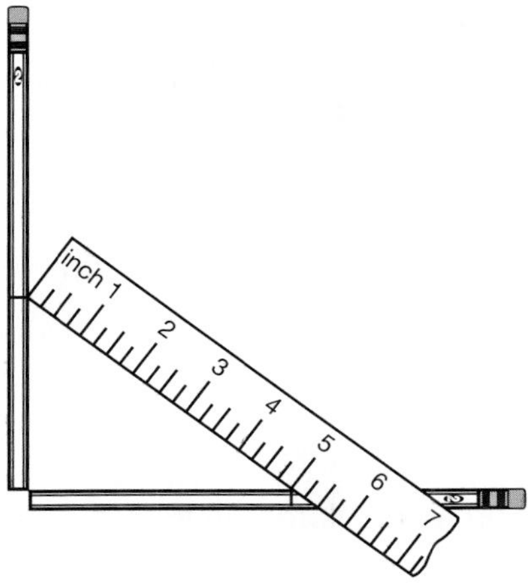
inch 1
2
3
4
5
6
7

Trace the angle formed. Then use a protractor to confirm that the angle formed by the pencils measures 90°.

Conclude Keep the pencils at a right angle. If you mark the pencils at 6 cm and 8 cm, what will be the distance between the marks?

Example 1

The numbers 5, 12, and 13 are a Pythagorean triplet because $5^2 + 12^2 = 13^2$. What are the next three multiples of this Pythagorean triplet?

Solution

To find the next three multiples of 5-12-13, we multiply each number by 2, by 3, and by 4.

10-24-26

15-36-39

20-48-52

Example 2

A roof is being built over a 24-ft-wide room. The slope of the roof is 4 in 12. Calculate the length of the rafters needed for the roof. (Include 2 ft for the rafter tail.)

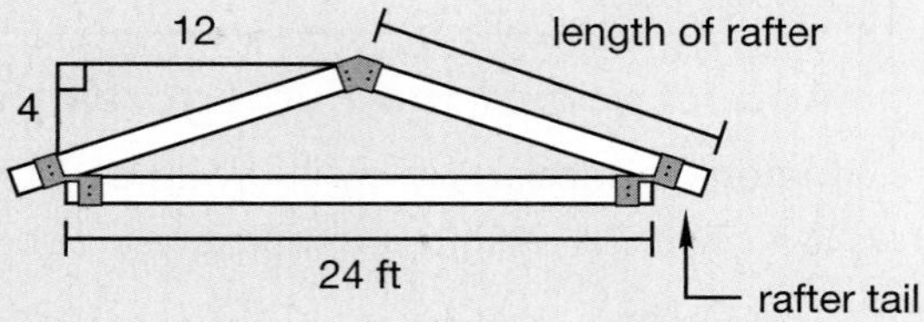

Solution

Math Language

The longest side of a right triangle is called the *hypotenuse.* The other two sides are called *legs*.

We consider the length of a rafter to be the hypotenuse of a right triangle. The width of the room is 24 ft, but a rafter spans only half the width of the room. So the base of the right triangle is 12 ft. The slope of the roof, 4 in 12, means that for every 12 horizontal units, the roof rises (or falls) 4 vertical units. Thus, since the base of the triangle is 12 ft, its height is 4 ft, as shown above.

We use the Pythagorean theorem to calculate the hypotenuse.

$$a^2 + b^2 = c^2$$
$$(4 \text{ ft})^2 + (12 \text{ ft})^2 = c^2$$
$$16 \text{ ft}^2 + 144 \text{ ft}^2 = c^2$$
$$160 \text{ ft}^2 = c^2$$
$$\sqrt{160} \text{ ft} = c$$
$$12.65 \text{ ft} \approx c$$

Using a calculator we find that the hypotenuse is about 12.65 feet. We add 2 feet for the rafter tail.

$$12.65 \text{ ft} + 2 \text{ ft} = 14.65 \text{ ft}$$

To convert 0.65 ft to inches, we multiply.

$$0.65 \text{ ft} \times \frac{12 \text{ in.}}{1 \text{ ft}} = 7.8 \text{ in.}$$

We round this up to 8 inches. So the length of each rafter is about **14 ft 8 in.**

Example 3

Serena went to a level field to fly a kite. She let out all 200 ft of string and tied it to a stake. Then she walked out on the field until she was directly under the kite, 150 feet from the stake. About how high was the kite?

Solution

We begin by sketching the problem. The length of the kite string is the hypotenuse of a right triangle, and the distance between Serena and the stake is one leg of the triangle. We use the Pythagorean theorem to find the remaining leg, which is the height of the kite.

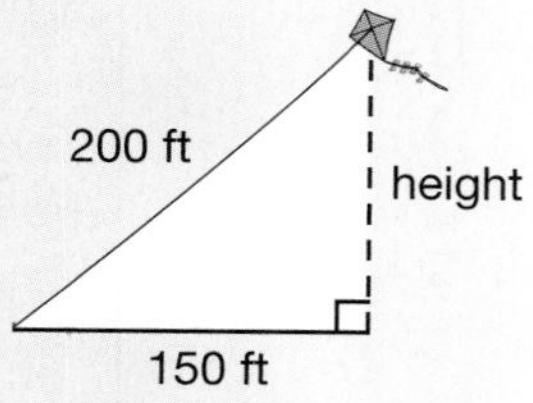

$$a^2 + b^2 = c^2$$
$$a^2 + (150 \text{ ft})^2 = (200 \text{ ft})^2$$
$$a^2 + 22{,}500 \text{ ft}^2 = 40{,}000 \text{ ft}^2$$
$$a^2 = 17{,}500 \text{ ft}^2$$
$$a = \sqrt{17{,}500} \text{ ft}$$
$$a \approx 132 \text{ ft}$$

Using a calculator, we find that the height of the kite was about **132 ft.**

Discuss How do we know we can use the Pythagorean theorem to solve this problem?

Practice Set

a. A 12-foot ladder was leaning against a building. The base of the ladder was 5 feet from the building. How high up the side of the building did the ladder reach? Write the answer in feet and inches rounded to the nearest inch.

b. Figure *ABCD* illustrates a rectangular field 400 feet long and 300 feet wide. The path from *A* to *C* is how much shorter than the path from *A* to *B* to *C*?

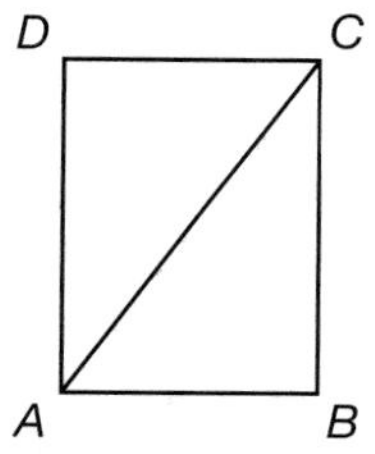

c. A contractor checks the forms for a concrete slab. He marks the forms at 8 ft and at 6 ft. Then he measures between the marks and finds that the distance is 10 feet 2 inches. Is the corner a right angle? How do you know?

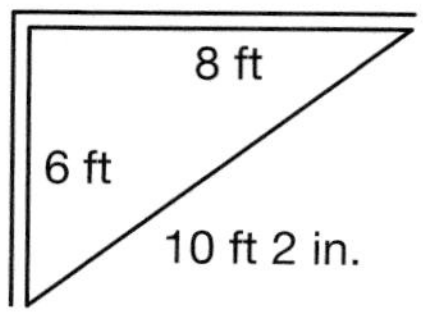

Written Practice — *Strengthening Concepts*

* **1.** (110) Mrs. Garcia deposited $3000 in an account paying 8 percent interest compounded annually. She withdrew her money with interest 3 years later. How much did she withdraw?

* **2.** (20, 28) **Connect** What is the square root of the sum of 3 squared and 4 squared?

3. (Inv. 4) Find the **a** median and **b** mode of the following quiz scores:

Class Quiz Scores

Score	Number of Students
100	2
95	7
90	6
85	6
80	3
70	3

4. (46) The driver completed the 840-kilometer trip in 10 hours 30 minutes. What was the driver's average speed in kilometers per hour?

Use ratio boxes to solve problems **5–7:**

5. (72) Barbara earned $48 for 6 hours of work. At that rate, how much would she earn for 8 hours of work?

6. (92) José paid $48 for a jacket at 25 percent off of the regular price. What was the regular price of the jacket?

7. (92) Troy bought a baseball card for $6 and sold it for 25 percent more than he paid for it. How much profit did he make on the sale?

* **8.** (110) **Analyze** At a yard sale an item marked $1.00 was reduced 50%. When the item still did not sell, the sale price was reduced 50%. What was the price of the item after the second discount?

9. (36) A test was made up of multiple-choice questions and short-answer questions. If 60% of the questions were multiple-choice, what was the ratio of multiple-choice to short-answer questions?

10. (99) The points (3, 11), (−2, −1), and (−2, 11) are the vertices of a right triangle. Use the Pythagorean theorem to find the length of the hypotenuse of this triangle.

11. (88) The final stage of the 2004 Tour de France was 163 km long. Bicyclist Tom Boonen covered that distance in about 4 hours. Find his approximate average speed in miles per hour. One km $\approx$ 0.6 mile.

12. (77) What percent of 2.5 is 2? Explain how you can determine if your answer is reasonable.

13. (Inv. 8) What are the odds of having a coin land tails up on 4 consecutive tosses of a coin?

* **14.** (110) **Generalize** How much interest is earned in 6 months on $4000 deposited at 9 percent annual simple interest?

15. (48) Complete the table.

Fraction	Decimal	Percent
$\frac{5}{8}$	**a.**	**b.**

* **16.** (111) Divide. Write each quotient in scientific notation:

a. $\frac{5 \times 10^8}{2 \times 10^4}$ **b.** $\frac{1.2 \times 10^4}{4 \times 10^8}$

* **17.** (112) **Connect** The frame of this kite is formed by two perpendicular pieces of wood whose lengths are shown in inches. A loop of string connects the four ends of the sticks. How long is the string?

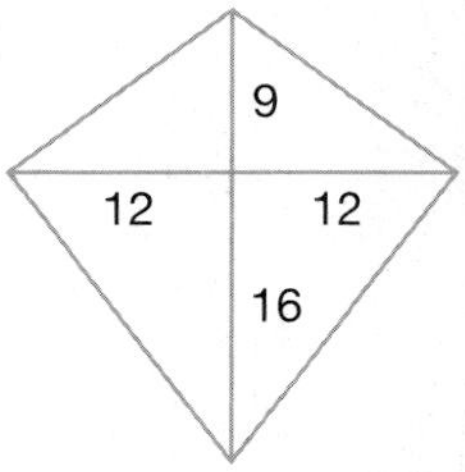

* **18.** (106) **Analyze** Solve for t: $d = rt$

19. (Inv. 9) Make a table that shows three pairs of numbers for the function $y = -x$. Then graph the number pairs on a coordinate plane, and draw a line through the points.

* **20.** (104) Find the perimeter of this figure. The arc in the figure is a semicircle. Dimensions are in centimeters. (Use 3.14 for π.)

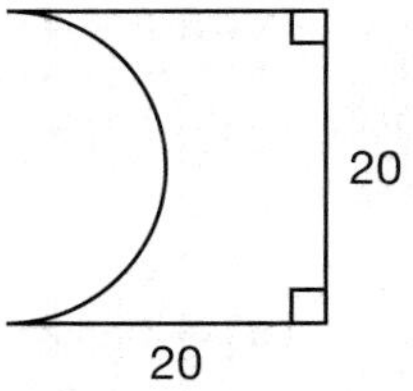

21. (97, 98) The triangles below are similar. Dimensions are in centimeters.

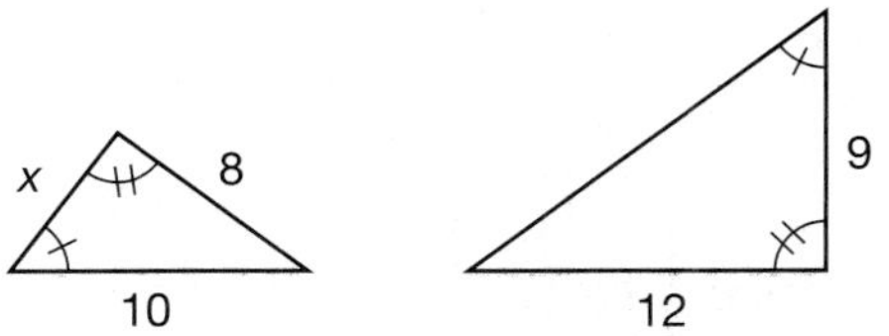

a. Find x.

b. Find the scale factor from the smaller to the larger triangle.

22. (21) **a.** Write the prime factorization of 1 trillion using exponents.

b. Find the positive square root of 1 trillion.

* **23.** (105) **Connect** Find the surface area of this right triangular prism. Dimensions are in feet.

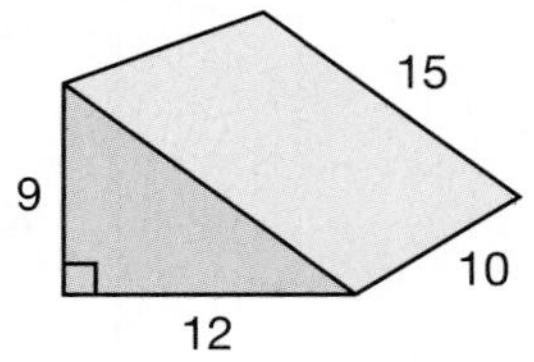

Solve:

24. (98) $\frac{16}{2.5} = \frac{48}{f}$

25. (93) $2\frac{2}{3}x - 3 = 21$

Simplify:

26. (57, 63) $10^2 - [10 - 10(10^0 - 10^{-1})]$

27. (43) $2\frac{3}{4} - \left(1.5 - \frac{1}{6}\right)$

28. (43) $3.5 \div 1\frac{2}{5} \div 3$

29. (85) $|-4| - (-3)(-2)(-1) + \frac{(-5)(4)(-3)(2)}{-1}$

* **30.** (105) The large grapefruit was nearly spherical and had a diameter of 14 cm. Using $\frac{22}{7}$ for π, find the approximate surface area of the grapefruit to the nearest hundred square centimeters.

Early Finishers
Real-World Application

A student surveyed 8 friends asking the type of pets they have. Here are the results.

Mark and Janet have both a cat and a parakeet.
Lisa has a dog and myna bird.
James has a cat.
Loretta and Jorge have a dog and a cat.
Rosa has a dog.
Rodney has a parrot.

a. Choose a line plot, line graph, or a Venn diagram to display this data. Draw your display and justify your choice.

b. Formulate questions to conduct a similar survey of 30 students in your school. Write three conclusions that can be made based on your collected data.

LESSON

113

Volume of Pyramids, Cones, and Spheres

Power Up

Building Power

facts Power Up V

mental math

a. **Positive/Negative:** $0.75 \div (-3)$

b. **Scientific Notation:** $(6 \times 10^6) \div (3 \times 10^{10})$

c. **Algebra:** $10m + 1.5 = 7.5$

d. **Measurement:** Convert 2 ft^2 to square inches.

e. **Percent:** $66\frac{2}{3}\%$ of \$2400

f. **Percent:** \$2400 reduced by $66\frac{2}{3}\%$

g. **Geometry:** Which pyramid has 6 faces and 6 vertices?

h. **Rate:** At 500 mph, how long will it take a plane to travel 1250 miles?

problem solving

Sylvia wants to pack a 9-by-14 in. rectangular picture frame that is $\frac{1}{2}$ in. thick into a rectangular box. The box has inside dimensions of 12-by-9-by-10 in. Describe why you think the frame will or will not fit into the box.

New Concept

Increasing Knowledge

A **pyramid** is a geometric solid that has three or more triangular faces and a base that is a polygon. Each of these figures is a pyramid:

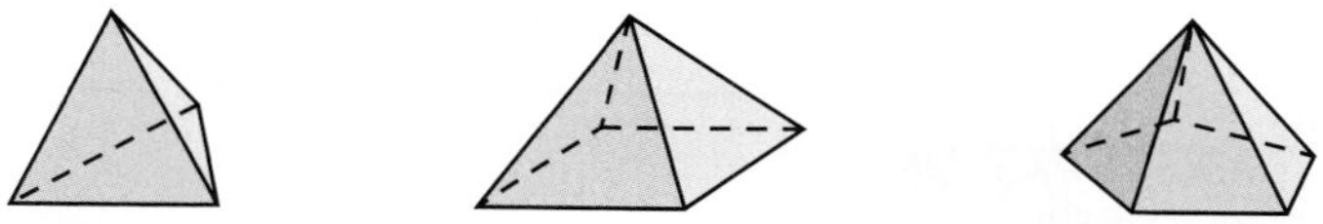

Thinking Skill

Explain

Both pyramids and triangular prisms have surfaces that are triangles. How are the two different?

The volume of a pyramid is $\frac{1}{3}$ the volume of a prism that has the same base and height. Recall that the volume of a prism is equal to the area of its base times its height.

$$\text{Volume of a prism} = \text{area of base} \cdot \text{height}$$

To find the volume of a pyramid, we first find the volume of a prism that has the same base and height. Then we divide the result by 3 (or multiply by $\frac{1}{3}$).

$$\textbf{Volume of pyramid} = \frac{1}{3} \cdot \textbf{area of base} \cdot \textbf{height}$$

Example 1

The pyramid at right has the same base and height as the cube that contains it. Each edge of the cube is 6 centimeters.

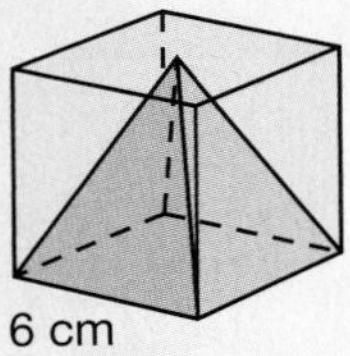

a. Find the volume of the cube.

b. Find the volume of the pyramid.

Solution

a. The volume of the cube equals the area of the base times the height.

$$\text{Area of base} = 6 \text{ cm} \times 6 \text{ cm} = 36 \text{ cm}^2$$

$$\begin{aligned}\text{Volume of cube} &= \text{area of base} \cdot \text{height} \\ &= (36 \text{ cm}^2)(6 \text{ cm}) \\ &= \mathbf{216\ cm^3}\end{aligned}$$

b. The volume of the pyramid is $\frac{1}{3}$ the volume of the cube, so we divide the volume of the cube by 3 (or multiply by $\frac{1}{3}$).

$$\begin{aligned}\text{Volume of pyramid} &= \frac{1}{3}\text{ (volume of prism)} \\ &= \frac{1}{3}\,(216 \text{ cm}^3) \\ &= \mathbf{72\ cm^3}\end{aligned}$$

Discuss In this example, if all we were given was the volume of the pyramid, could we find the length of the edge of the cube? Explain your answer.

The volume of a cone is $\frac{1}{3}$ the volume of a cylinder with the same base and height.

Math Language

The *height* of a cone is the perpendicular distance from the vertex to the base.

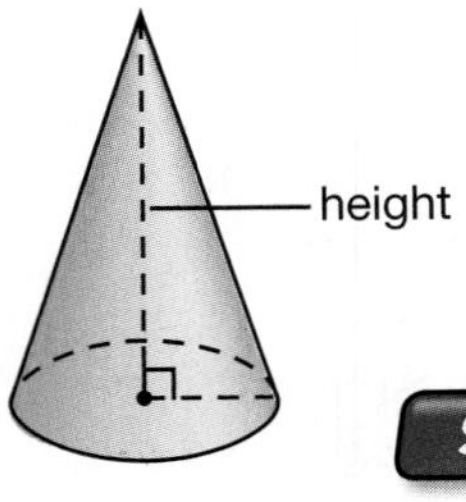

$$\textbf{Volume of cone} = \frac{\mathbf{1}}{\mathbf{3}} \cdot \textbf{area of base} \cdot \textbf{height}$$

Example 2

Find the volume of this circular cone. Dimensions are in centimeters. (Use 3.14 for π.)

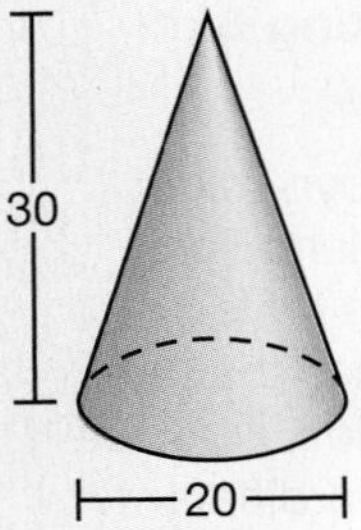

Solution

We first find the volume of a cylinder with the same base and height as the cone.

$$\begin{aligned}\text{Volume of cylinder} &= \text{area of circle} \cdot \text{height} \\ &\approx (3.14)(10 \text{ cm})^2 \cdot 30 \text{ cm} \\ &\approx 9420 \text{ cm}^3\end{aligned}$$

Then we find $\frac{1}{3}$ of this volume.

$$\text{Volume of cone} = \frac{1}{3}\,(\text{volume of cylinder})$$

$$\approx \frac{1}{3} \cdot 9420 \text{ cm}^3$$

$$\approx \mathbf{3140\ cm^3}$$

We have used the general formula for the volume of a pyramid and cone. Using an uppercase A for the area of the base, the general formula is

$$V = \frac{1}{3}Ah$$

We can develop a specific formula for various pyramids and cones by replacing A in the general formula with the formula for the area of the base.

Write a formula for the volume of a pyramid with a square base.

Write a formula for the volume of a cone with a circular base.

There is a special relationship between the volume of a cylinder, the volume of a cone, and the volume of a sphere. Picture two identical cylinders whose heights are equal to their diameters. In one cylinder is a cone with the same diameter and height as the cylinder. In the other cylinder is a sphere with the same diameter as the cylinder.

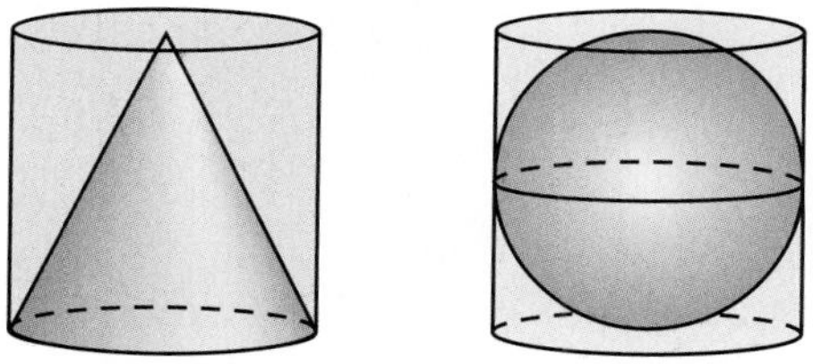

We have learned that the volume of the cone is $\frac{1}{3}$ the volume of the cylinder. Remarkably, the volume of the sphere is twice the volume of the cone, that is, $\frac{2}{3}$ the volume of the cylinder. Here we use a balance scale to provide another view of this relationship.

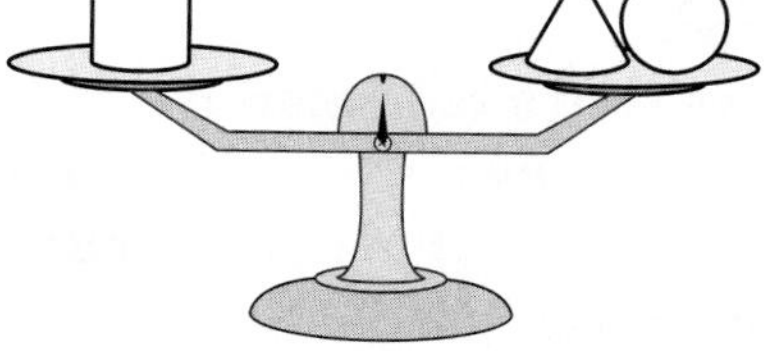

Imagine that the objects on the balance scale are solid and composed of the same material. The diameters of all three solids are equal. The heights of the cylinder and cone equal their diameters. The balance scale shows that the combined masses of the cone and sphere equal the mass of the cylinder. The cone's mass is $\frac{1}{3}$ of the cylinder's mass, and the sphere's mass is $\frac{2}{3}$ of the cylinder's mass.

The formula for the volume of a sphere can be derived from our knowledge of cylinders and their relationship to spheres. First consider the volume of any cylinder whose height is equal to its diameter. In this diagram we have labeled the diameter, d; the radius, r; and the height, h.

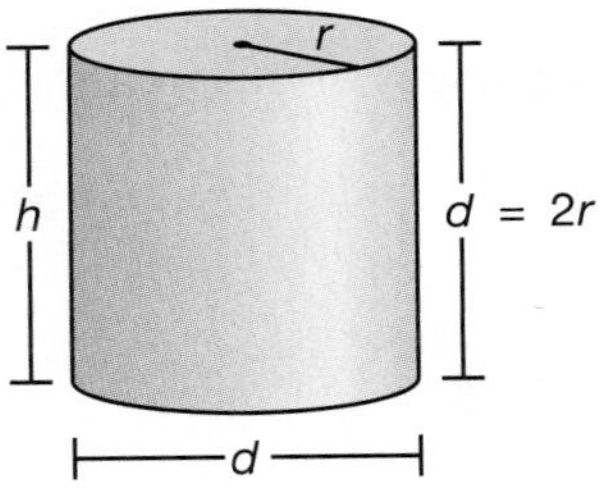

Recall that we can calculate the volume of a cylinder by multiplying the area of its circular base times its height.

$$\text{Volume of cylinder} = \text{area of circle} \cdot \text{height}$$

$$V = \pi r^2 \cdot h$$

For a cylinder whose height is equal to its diameter, we can replace *h* in the formula with *d* or with 2*r*, since two radii equal the diameter.

$V = \pi r^2 \cdot h$	formula for volume of a cylinder
$V = \pi r^2 \cdot 2r$	replaced *h* with 2*r*, which equals the height of the cylinder
$V = 2\pi r^3$	rearranged factors

This formula, $V = 2\pi r^3$, gives the volume of a cylinder whose height is equal to its diameter. The volume of a sphere is $\frac{2}{3}$ of the volume of a cylinder with the same diameter and height.

$$\text{Volume of sphere} = \frac{2}{3} \cdot \text{volume of cylinder}$$

$$= \frac{2}{3} \cdot 2\pi r^3 \quad \text{substituted}$$

$$= \frac{4}{3}\pi r^3 \quad \text{multiplied } \tfrac{2}{3} \text{ by 2}$$

We have found the formula for the volume of a sphere.

Volume of a sphere = $\frac{4}{3}\pi r^3$

Example 3

A ball with a diameter of 20 cm has a volume of how many cubic centimeters? (Round the answer to the nearest hundred cubic centimeters.)

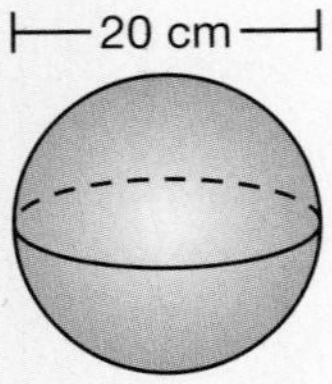

Use 3.14 for π.

Solution

The diameter of the sphere is 20 cm, so its radius is 10 cm. We use the formula for the volume of a sphere, substituting 3.14 for π and 10 cm for *r*.

$V = \frac{4}{3}\pi r^3$ formula

$\approx \frac{4}{3}(3.14)(10\text{ cm})^3$	substituted 3.14 for π and 10 cm for r
$\approx \frac{4}{3}(3.14)(1000\text{ cm}^3)$	cubed 10 cm
$\approx \frac{4}{3}(3140\text{ cm}^3)$	multiplied 3.14 by 1000 cm^3
$\approx 4186\frac{2}{3}\text{ cm}^3$	multiplied $\frac{4}{3}$ by 3140 cm^3
$\approx$ **4200 cm^3**	rounded to nearest hundred cubic centimeters

Practice Set

Pictured are two identical cylinders whose heights are equal to their diameters. In one cylinder is the largest cone it can contain. In the other cylinder is the largest sphere it can contain. Packing material is used to fill all the voids in the cylinders not occupied by the cone or sphere. Use this information to answer problems **a–e.**

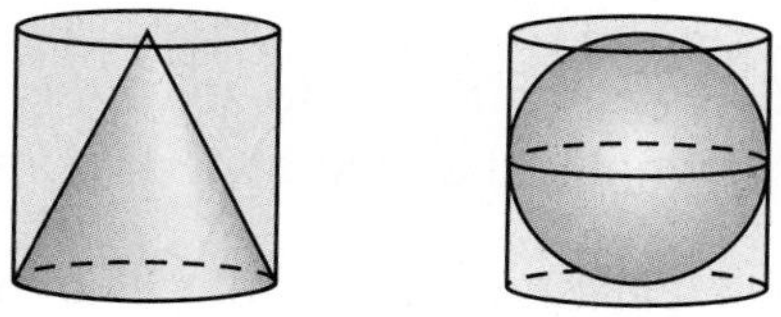

a. What fraction of the cylinder containing the cone is occupied by the cone?

b. What fraction of the cylinder containing the cone is occupied by the packing material?

c. What fraction of the cylinder containing the sphere is occupied by the sphere?

d. What fraction of the cylinder containing the sphere is occupied by the packing material?

e. *Conclude* If the cone and sphere were removed from their boxes and all the packing material from both boxes was put into one box, what portion of the box would be filled with packing material?

f. A pyramid with a height of 12 inches and a base 12 inches square is packed in the smallest cubical box that can contain it. What is the volume of the box? What is the volume of the pyramid?

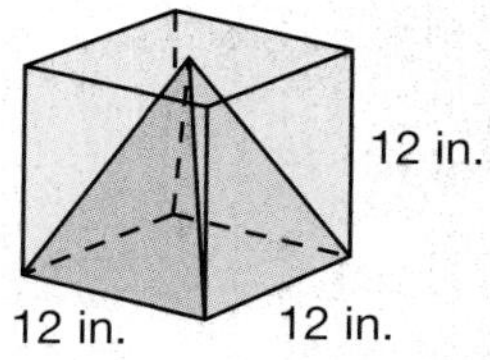

Analyze Find the volume of each figure below. For both calculations, leave π as π.

g.

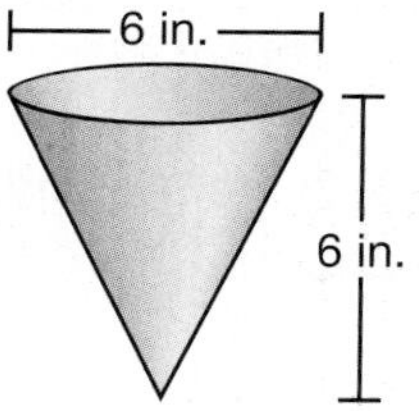

h.

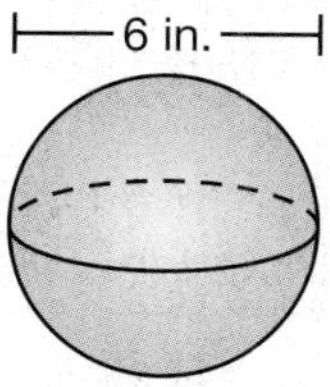

Written Practice Strengthening Concepts

* **1.** (110) **Generalize** Find the sale price of a \$24 item after successive discounts of 25% and 25%.

2. (51) Ten billion is how much greater than nine hundred eighty million? Write the answer in scientific notation.

3. (Inv. 4) The median of the following numbers is how much less than the mean?

1.4, 0.5, 0.6, 0.75, 5.2

4. (46) Nelda worked 5 hours and earned \$34. Christy worked 6 hours and earned \$45.

a. How much did Nelda earn per hour?

b. How much did Christy earn per hour?

c. Christy earned how much more per hour than Nelda?

5. (72) Use a ratio box to solve this problem. If 24 kilograms of seed costs \$31, what is the cost of 42 kilograms of seed?

6. (60) A kilometer is about $\frac{5}{8}$ of a mile. A mile is 1760 yards. A kilometer is about how many yards?

7. (Inv. 8) **a.** A card is drawn from a deck of 52 playing cards and replaced. Then another card is drawn. What is the probability of drawing a heart both times?

b. What is the probability of drawing two hearts if the first card is not replaced?

Write equations to solve problems **8** and **9.**

8. (77) What percent of \$30 is \$1.50?

9. (77) Fifty percent of what number is $2\frac{1}{2}$?

* **10.** (110) **Generalize** Mr. Rodrigo put \$5000 he had saved from his salary in an account that paid 4 percent interest compounded annually. How much interest was earned in 3 years?

Use ratio boxes to solve problems **11** and **12.**

11. (92) A merchant sold an item at a 20 percent discount from the regular price. If the regular price was \$12, what was the sale price of the item?

12. (98) After visiting the Abraham Lincoln Memorial in Washington D.C., Jessica decided to sculpt a replica of Lincoln in his chair. The statue is 19 feet tall, and Jessica will sculpt it at $\frac{1}{24}$ of its actual height. How many inches tall will her version be?

13. (37, 62) The points (0, 4), (−3, 2), and (3, 2) are the vertices of an isosceles triangle. Find the area of the triangle.

* **14.** Use the Pythagorean theorem to find the length of one of the two congruent sides of the triangle in problem 13. (*Hint:* First draw an altitude to form two right triangles.)
(112)

* **15.** **Estimate** Roughly estimate the volume of a tennis ball in cubic centimeters by using 6 cm for the diameter and 3 for π. Round the answer to the nearest ten cubic centimeters.
(113)

16. Multiply. Write the product in scientific notation.
(83)

$$(6.3 \times 10^6)(7 \times 10^{-3})$$

* **17.** **Analyze** Tim can get from point *A* to point *B* by staying on the sidewalk and turning left at the corner *C*, or he can take the shortcut and walk straight from point *A* to point *B*. How many yards can Tim save by taking the shortcut? Begin by using the Pythagorean theorem to find the length of the shortcut.
(112)

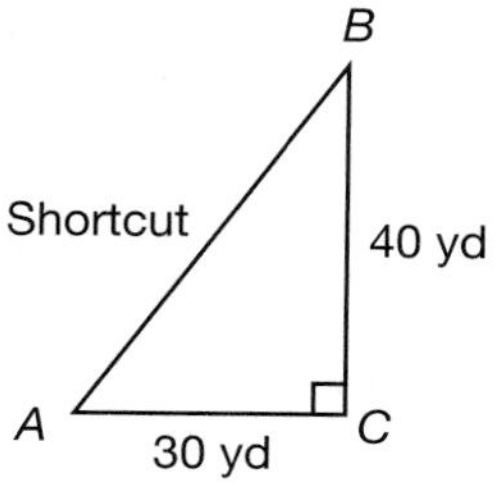

* **18.** **a.** Solve for *h*: $A = \frac{1}{2}bh$
(108)

b. **Analyze** Use the formula $A = \frac{1}{2}bh$ to find h when $A = 16$ and $b = 8$.

* **19.** **Evaluate** Make a table that shows three pairs of numbers for the function $y = -2x + 1$. Then graph the number pairs on a coordinate plane, and draw a line through the points to show other number pairs that satisfy the function. What is the slope of the graphed line?
(Inv. 9, 107)

* **20.** Find the volume of the pyramid shown. Dimensions are in meters.
(113)

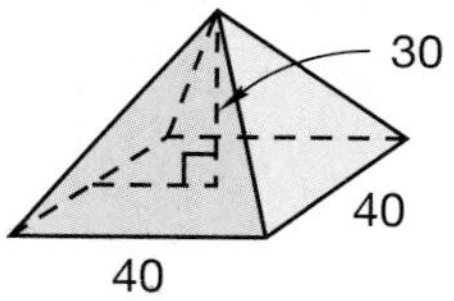

* **21.** Find the volume of the cone at right. Dimensions are in centimeters.
(113)

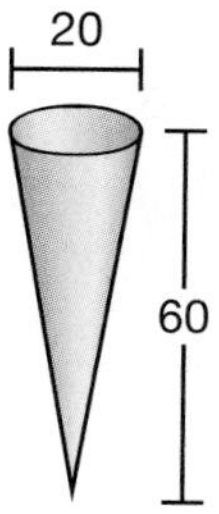

Use 3.14 for π.

* **22.** (40) **Evaluate** Refer to the figure below to find the measures of the following angles. Dimensions are in centimeters.

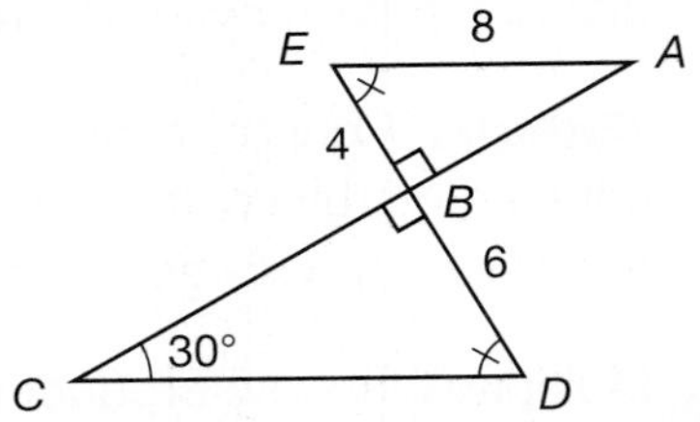

a. $\angle D$ **b.** $\angle E$ **c.** $\angle A$

23. (97) In the figure in problem **22,** what is the length of $\overline{CD}$?

Solve:

24. (98) $\frac{7.5}{d} = \frac{25}{16}$

25. (93) $1\frac{3}{5}w + 17 = 49$

Simplify:

26. (63) $5^2 - \{4^2 - [3^2 - (2^2 - 1^2)]\}$

27. (30) $1\frac{3}{4} + 2\frac{2}{3} - 3\frac{5}{6}$

28. (26) $\left(1\frac{3}{4}\right)\left(2\frac{2}{3}\right) \div 3\frac{5}{6}$

29. (85) $(-7) + |-3| - (2)(-3) + (-4) - (-3)(-2)(-1)$

* **30.** (113) **Generalize** The Great Pyramid of Khufu in Egypt, is a giant pyramid with a square base. The length of each side of the square base is about 750 feet and its height is about 480 feet. About what is the volume of the Great Pyramid of Khufu?

Early Finishers
Math Applications

Use grid paper to sketch three-dimensional figures that would match all three views shown below.

a.

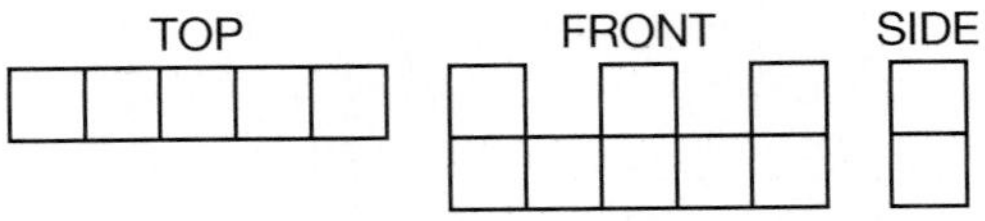

b.

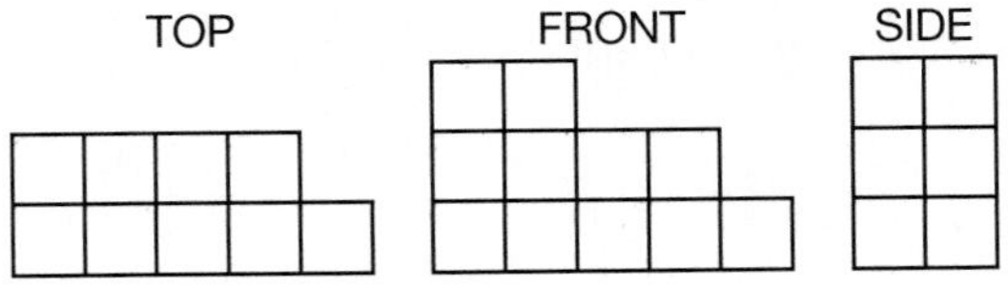

LESSON 114

Volume, Capacity, and Mass in the Metric System

Power Up

Building Power

facts Power Up W

mental math

a. **Positive/Negative:** $(-3)^3 + (-3)^2$

b. **Scientific Notation:** $(4 \times 10^8)^2$

c. **Algebra:** $10m - m = 9^2$

d. **Measurement:** Convert 60°C to degrees Fahrenheit.

e. **Percent:** 150% of \$3000

f. **Percent:** 150% more than \$3000

g. **Geometry:** Which figure has unlimited lines of symmetry?

h. **Calculation:** Find 25% of 40, − 1, × 5, − 1, ÷ 2, − 1, ÷ 3, × 10, + 2, ÷ 9, ÷ 2, $\sqrt{\ }$.

problem solving

There are three types of balls, each a different size and color, placed against a 7-foot long wall. They are orange, red, and green with surface areas of 201 in.^{2}, 314 in.^{2}, and 706.5 in.^{2}, respectively. If they take up the full length of the wall with no space left over, how many balls of each color are there?

New Concept

Increasing Knowledge

Thinking Skill

Summarize

What is the difference between volume, capacity, and mass?

Units of volume, capacity, and mass are closely related in the metric system. The relationships between these units are based on the physical characteristics of water under certain standard conditions. We state two commonly used relationships.

One milliliter of water has a volume of 1 cubic centimeter and a mass of 1 gram.

One cubic centimeter can contain 1 milliliter of water, which has a mass of 1 gram.

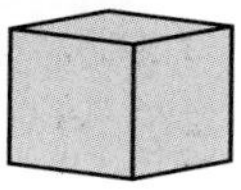

One liter of water has a volume of 1000 cubic centimeters and a mass of 1 kilogram.

One thousand cubic centimeters can contain 1 liter of water, which has a mass of 1 kilogram.

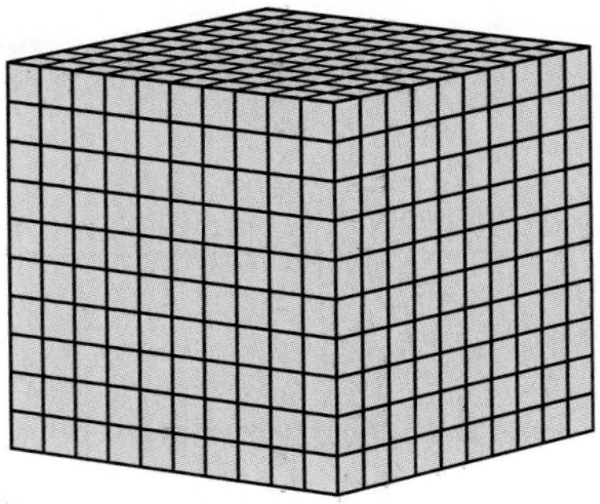

Explain How many cubic centimeters can a 2.5 liter container of water hold? Explain how you determined your answer.

Example 1

Ray has a fish aquarium that is 50 cm long and 20 cm wide. If the aquarium is filled with water to a depth of 30 cm,

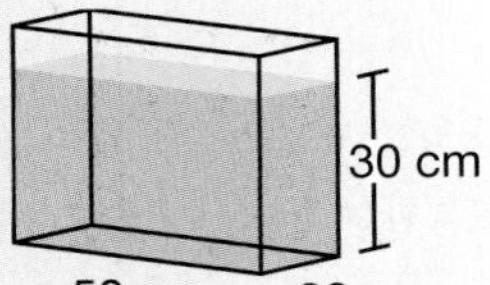

a. how many liters of water would be in the aquarium?

b. what would be the mass of the water in the aquarium?

Solution

First we find the volume of the water in the aquarium.

$$(50 \text{ cm})(20 \text{ cm})(30 \text{ cm}) = 30{,}000 \text{ cm}^3$$

a. Each cubic centimeter of water is 1 milliliter. Thirty thousand milliliters is **30 liters.**

b. Each liter of water has a mass of 1 kilogram, so the mass of the water in the aquarium is **30 kilograms.** (Since a 1-kilogram mass weighs about 2.2 pounds on Earth, the water in the aquarium weighs about 66 pounds.)

Conclude Ray put an aquarium rock in the tank. The water level increased from 30 cm to 30.2 cm. What is the volume of the rock?

Example 2

Malaika wanted to find the volume of a vase. She filled a 1-liter beaker with water and then used all but 240 milliliters to fill the vase.

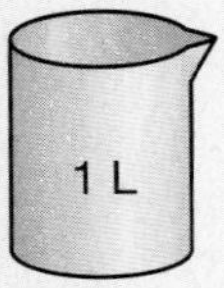

beaker

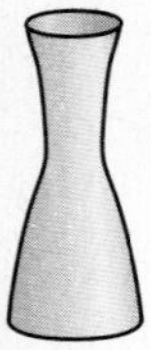
vase

a. What is the volume of the vase?

b. If the mass of the vase is 640 grams, what is the mass of the vase filled with water?

Solution

a. The 1-liter beaker contains 1000 mL of water. Since Malaika used 760 mL, (1000 mL − 240 mL), the volume of the inside of the vase is **760 cm^3.**

b. The mass of the water (760 g) plus the mass of the vase (640 g) is **1400 g.**

Practice Set

a. What is the mass of 2 liters of water?

b. What is the volume of 3 liters of water?

c. Analyze When the bottle was filled with water, the mass increased by 1 kilogram. How many milliliters of water were added?

d. Analyze A tank that is 25 cm long, 10 cm wide, and 8 cm deep can hold how many liters of water?

Written Practice

Strengthening Concepts

1. *(60, 92)* The regular price of a bowling ball was $72.50, but it was on sale for 20% off. What was the total sale price including 7% sales tax? Use a ratio box to find the sale price. Then find the sales tax and total price.

2. *(55)* A zoologist tagged 4 tortoises in a conservation area. The average weight of the 4 tortoises was 87 pounds. The zoologist then tagged 2 more tortoises. The average weight of all 6 tortoises was 90 pounds. What was the average weight of the last 2 tortoises?

3. *(Inv. 8)* There are 27 marbles in a bag: 6 red, 9 green, and 12 blue. If one marble is drawn from the bag,

a. what is the probability that the marble will be blue?

b. what is the chance that the marble will be green?

c. what are the odds that the marble will not be red?

Math Language
The **odds** of an event occurring is the ratio of the favorable outcomes to unfavorable outcomes.

4. *(46)* If a box of 12 dozen pencils costs $10.80, what is the cost per pencil?

*** 5.** *(110)* Generalize How much interest is earned in 6 months on $5000 at 8% annual simple interest?

6. *(Inv. 5)* One fourth of the students on a bus were listening to music. One third of the students were talking. The rest were reading.

a. Draw a circle graph that displays this information.

b. If six students were listening to music, how many students were reading?

7. (66) The ratio of students who ride a bus to students who walk to school is 5 to 2. If there are 1400 students in the school, how many students ride a bus?

*** 8.** (113) **Estimate** The snowball grew in size as it rolled down the hill. By the time it came to a stop, its diameter was about four feet. Using 3 for π, estimate the number of cubic feet of snow in the snowball.

Write equations to solve problems **9** and **10.**

9. (60) What is 120% of $240?

10. (77) Sixty is what percent of 150?

*** 11.** (75, 99) **Analyze** The points (3, 2), (6, −2), (−2, −2), and (−2, 2) are the vertices of a trapezoid.

a. Find the area of the trapezoid.

b. Find the perimeter of the trapezoid.

12. (100) **a.** Arrange these numbers in order from least to greatest:

$$\sqrt{6}, 6^2, -6, 0.6$$

b. Which of the numbers in **a** are rational numbers? Why?

13. (48) Complete the table.

Fraction	Decimal	Percent
$1\frac{4}{5}$	**a.**	**b.**

*** 14.** (111) **Generalize** Divide. Write each quotient in scientific notation:

a. $\dfrac{5 \times 10^{-9}}{2 \times 10^{-6}}$ **b.** $\dfrac{2 \times 10^{-6}}{5 \times 10^{-9}}$

15. (83) What is the product of answers **a** and **b** in problem 14?

16. (88) Use unit multipliers to convert one square kilometer to square meters.

*** 17.** (108) **a.** **Analyze** Solve for d: $C = \pi d$

b. Use the formula $C = \pi d$ to find d when C is 62.8. (Use 3.14 for π.)

18. (94) With one toss of a pair of number cubes, what is the probability that the total rolled will be a prime number? (Add the probabilities for each prime-number total.)

*** 19.** (104) Find the perimeter of the figure at right. Dimensions are in centimeters. (Use 3.14 for π.)

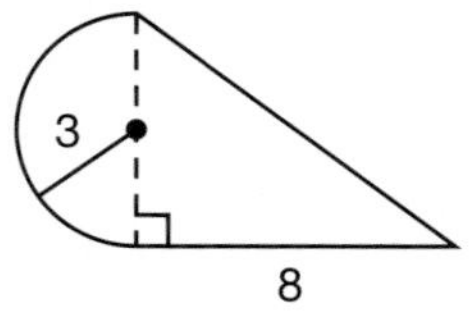

* **20.** (105, 113) **a.** Find the surface area of the cube shown. Dimensions are in feet.

b. Analyze If the cube contains the largest pyramid it can hold, what is the volume of the pyramid?

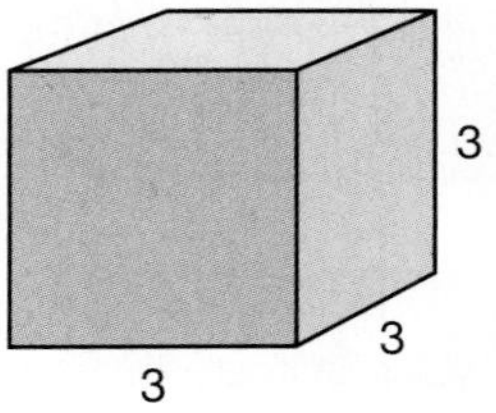

21. (95) Find the volume of this right circular cylinder. Dimensions are in meters. (Use 3.14 for π.)

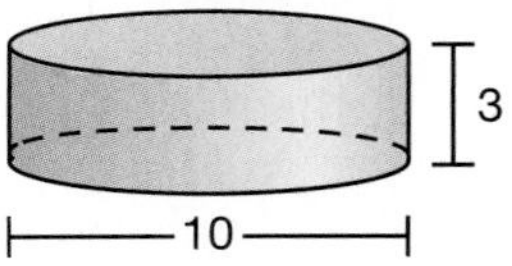

22. (40) Find the measures of the following angles:

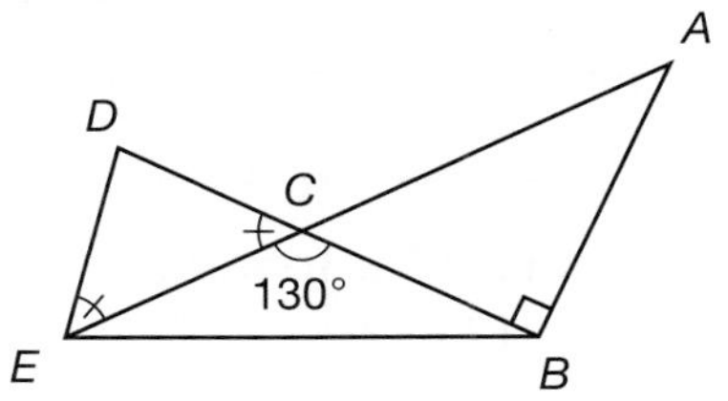

a. $\angle ACB$ **b.** $\angle CAB$ **c.** $\angle CDE$

23. (70) An aquarium that is 40 cm long, 10 cm wide, and 20 cm deep is filled with water. Find the volume of the water in the aquarium.

24. (93) Solve: $0.8m - 1.2 = 6$

25. (93) Model Solve this inequality and graph its solution:

$$3(x - 4) < x - 8$$

Simplify:

26. (57) $4^2 \cdot 2^{-3} \cdot 2^{-1}$

27. (32) 1 kilogram − 50 grams

28. (43) $(1.2)\left(3\frac{3}{4}\right) \div 4\frac{1}{2}$

29. (43) $2\frac{3}{4} - 1.5 - \frac{1}{6}$

30. (85) $(-3)(-2) - (2)(-3) - (-8) + (-2)(-3) + |-5|$

Early Finishers
Math Applications

Use graph paper to show Triangle *ABC* with these coordinates:

A (1, 4), *B* (4, 4) and *C* (1, 1).

a. Reflect Triangle *ABC* over the horizontal axis and label its image *A′B′C′*. What are the coordinates of *A′B′C′*?

b. Now reflect Triangle *ABC* over the vertical axis. What are the coordinates of the triangle formed?

LESSON

115 • Factoring Algebraic Expressions

Power Up

Building Power

facts Power Up V

mental math

a. **Positive/Negative:** 10^{-2}

b. **Scientific Notation:** $(4 \times 10^8) \div (4 \times 10^8)$

c. **Ratio:** $\frac{1.44}{1.2} = \frac{1.2}{g}$

d. **Measurement:** Convert 250 cm to m.

e. **Fractional Parts:** $\frac{2}{3}$ of \$1200

f. **Fractional Parts:** \$1200 reduced $\frac{1}{3}$

g. **Primes/Composites:** Which does not belong in the list 12, 15, 19, 21?

h. **Money:** A nickel is how many cents less than 3 dimes and 3 quarters?

problem solving

In the following three problems, each letter represents the same missing digit in each of the problems. Work as a team to determine the digit each letter represents.

$$\begin{array}{r} \text{ABCDEF} \\ \times\ \text{B} \\ \hline 248{,}712 \end{array} \qquad \begin{array}{r} \text{BCGHAB} \\ \times\ \text{D} \\ \hline 746{,}136 \end{array} \qquad \begin{array}{r} \text{HCFADF} \\ \times\ \text{C} \\ \hline 2{,}984{,}544 \end{array}$$

New Concept

Increasing Knowledge

Algebraic expressions are classified as either **monomials** or **polynomials.** Monomials are single-term expressions such as the following three examples:

$$6x^2y^3 \qquad \frac{5xy}{2w} \qquad -6$$

Polynomials are composed of two or more terms. All of the following algebraic expressions are polynomials:

$$3x^2y + 6xy^2 \qquad x^2 + 2x + 1 \qquad 3a + 4b + 5c + d$$

Polynomials may be further classified by the number of terms they contain. For example, expressions with two terms are called binomials, and expressions with three terms are called trinomials. So $3x^2y + 6xy^2$ is a binomial, and $x^2 + 2x + 1$ is a trinomial.

Recall that to factor a monomial, we express the numerical part of the term as a product of prime factors, and we express the literal (letter) part of the term as a product of factors (instead of using exponents). Here we factor $6x^2y^3$:

$$6x^2y^3 \quad \text{original form}$$

$$(2)(3)xxyyy \quad \text{factored form}$$

Thinking Skill

Analyze

Some polynomials have a GCF of 1. What is the only factor common to all terms in $6a^2 + 4b^2 + 3ab$?

Some polynomials can also be factored. To factor a polynomial we first find the greatest common factor (GCF) of the terms of the polynomial. Then we use the Distributive Property to write the expression as a product of the GCF and the remaining polynomial.

To factor $3x^2y + 6xy^2$, we first find the GCF of $3x^2y$ and $6xy^2$. With practice we may find the GCF visually. This time we will factor both terms and circle the common factors.

$$3x^2y \quad + \quad 6xy^2$$

$$\textcircled{3} \cdot \textcircled{x} \cdot x \cdot \textcircled{y} + 2 \cdot \textcircled{3} \cdot \textcircled{x} \cdot y \cdot \textcircled{y}$$

We find that the GCF of $3x^2y$ and $6xy^2$ is $3xy$. Notice that removing $3xy$ from $3x^2y$ by division leaves x. Removing $3xy$ from $6xy^2$ by division leaves $2y$.

$$\frac{3x^2y}{3xy} + \frac{6xy^2}{3xy} \quad 3xy \text{ removed by division}$$

$$x + 2y \quad \text{remaining binomial}$$

We write the factored form of $3x^2y + 6xy^2$ this way:

$$3xy(x + 2y) \quad \text{factored form}$$

Notice that we began with a binomial and ended with the GCF of its terms times a binomial.

Conclude What property can you use to mentally check that you have factored correctly?

Discuss What might indicate that you have not found the greatest common factor?

Example 1

Factor the monomial $12a^2b^3c$.

Solution

We factor 12 as (2)(2)(3), and we factor a^2b^3c as $aabbbc$.

$$12a^2b^3c = \mathbf{(2)(2)(3)}\boldsymbol{aabbbc}$$

Example 2

Factor the trinomial $6a^2b + 4ab^2 + 2ab$.

Solution

First we find the greatest common factor of the three terms. Often we can do this visually. Notice that each term has 2 as a factor, *a* as a factor, and *b* as a factor. So the GCF is $2ab$. Next we divide each term of the trinomial by $2ab$ to find what remains of each term after $2ab$ is factored out of the expression.

$$\frac{6a^2b}{2ab} + \frac{4ab^2}{2ab} + \frac{2ab}{2ab} \qquad 2ab \text{ removed by division}$$

$$3a + 2b + 1 \qquad \text{remaining trinomial}$$

Notice that the third term is 1, not zero. This is because we divided $2ab$ by $2ab$; we did not subtract.

Now we write the factored expression in this form:

GCF(remaining polynomial)

The GCF is $2ab$ and the remaining trinomial is $3a + 2b + 1$.

$$\mathbf{2ab(3a + 2b + 1)}$$

Practice Set

Factor each algebraic expression:

a. $8m^2n$

b. $12mn^2$

c. $18x^3y^2$

d. $8m^2n + 12mn^2$

e. $8xy^2 - 4xy$

f. $6a^2b^3 + 9a^3b^2 + 3a^2b^2$

g. *Classify* What type of polynomial is the expression in problem **f?**

Written Practice

Strengthening Concepts

*** 1.** (110) *Explain* How much interest is earned in 9 months on a deposit of \$7,000 at 4% simple interest? How did you find your answer?

2. (Inv. 8) With two tosses of a coin,

a. what is the probability of getting two heads?

b. what is the chance of getting two tails?

c. what are the odds of getting heads, then tails?

3. (55) On the first 4 days of their trip, the Schmidts averaged 310 miles per day. On the fifth day they traveled 500 miles. How many miles per day did they average for the first 5 days of their trip?

4. (46) An 18-ounce bottle of strawberry-melon juice costs \$2.16. The 1-quart container costs \$3.36. Which costs more per ounce, the bottle or container? How much more per ounce?

Use ratio boxes to solve problems **5** and **6.**

5. (72) The school's new laser printer printed 160 pages in 5 minutes. At this rate, how long would it take to print 800 pages?

6. (66) Volunteers for the local birding group located a roost of robins and starlings. They were able to determine that the ratio of robins to starlings was 7 to 5 in a hemlock tree that had 120 robins and starlings. How many robins were in the tree?

7. (74) Kenny was thinking of a certain number. If $\frac{3}{4}$ of the number was 48, what was $\frac{5}{8}$ of the number?

8. (92) A used car dealer bought a car for $5500 and sold the car at a 40% markup. If the purchaser paid a sales tax of 8%, what was the total price of the car including tax?

*** 9.** (110) **Generalize** What is the sale price of an $80 skateboard after successive discounts of 25% and 20%?

10. (99) The points (−3, 4), (5, −2), and (−3, −2) are the vertices of a triangle.

a. Find the area of the triangle.

b. Find the perimeter of the triangle.

*** 11.** (114) A glass aquarium with the dimensions shown has a mass of 5 kg when empty. What is the mass of the aquarium when it is half full of water?

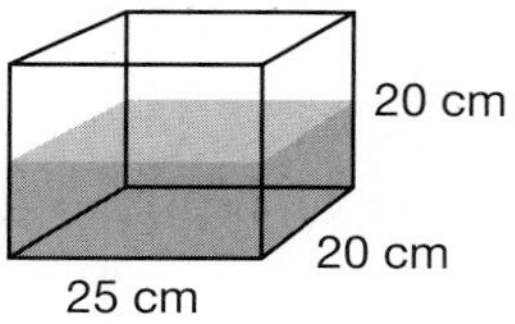

12. (48) Complete the table.

Fraction	Decimal	Percent
a.	0.875	**b.**

13. (54, 72) The nurse measured Latisha's resting heart rate by counting the number of times her heart beat over a 15 second interval. If the nurse counted 17 beats, what is Latisha's heart rate in beats per minute?

*** 14.** (83, 111) Simplify and express each answer in scientific notation:

a. $(6.4 \times 10^{6})(8 \times 10^{-8})$

b. $\dfrac{6.4 \times 10^{6}}{8 \times 10^{-8}}$

15. (50) Use a unit multiplier to convert three feet to centimeters.

*** 16.** (108) **a.** Solve for b: $A = \frac{1}{2}bh$

b. Use the formula $A = \frac{1}{2}bh$ to find b when A is 24 and h is 6.

c. **Extend** Solve for h: $A = \frac{1}{2}bh$.

*** 17.** (Inv. 9, 107) **Analyze** Find three pairs of numbers that satisfy the function $y = -2x$. Then graph the number pairs on a coordinate plane, and draw a line through the points to show other number pairs that satisfy the function. What is the slope of the graphed line?

* **18.** (104) **Analyze** Find the area of this figure. Dimensions are in millimeters. Corners that look square are square. (Use 3.14 for π.)

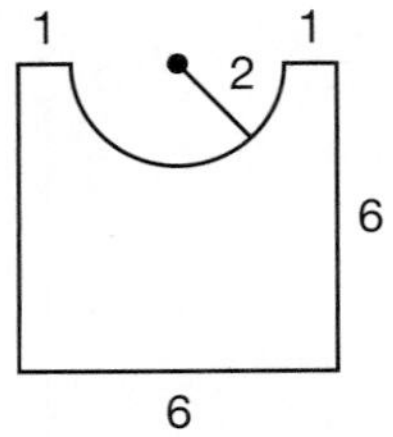

* **19.** (95, 105) **a.** Find the surface area of the cube.

b. Find the volume of the cube.

c. How many meters long is each edge of the cube?

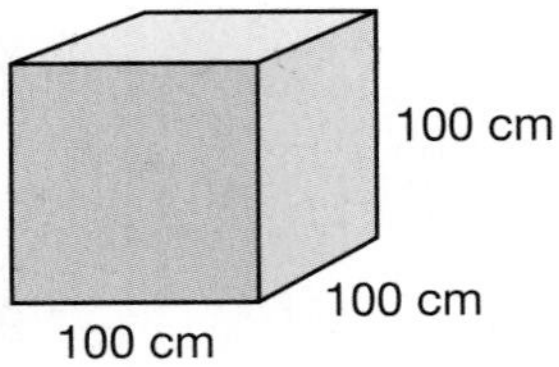

* **20.** (95, 113) **a.** Find the volume of the right circular cylinder. Dimensions are in inches.

b. **Analyze** If within the cylinder is the largest sphere it can contain, what is the volume of the sphere?

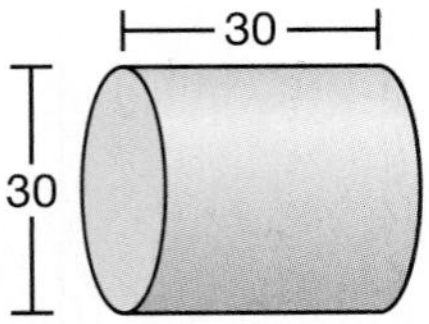

21. (40) Find the measures of the following angles:

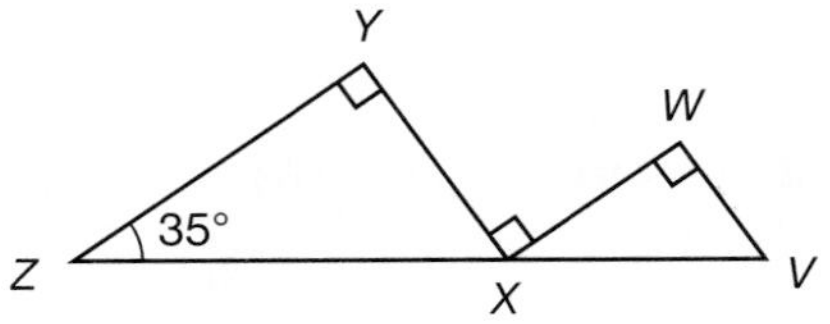

a. $\angle YXZ$ **b.** $\angle WXV$ **c.** $\angle WVX$

22. (97) In the figure in problem **21,** ZX is 21 cm, YX is 12 cm, and XV is 14 cm. Write a proportion to find WV.

* **23.** (113) **Analyze** A pyramid is cut out of a plastic cube with dimensions as shown. What is the volume of the pyramid?

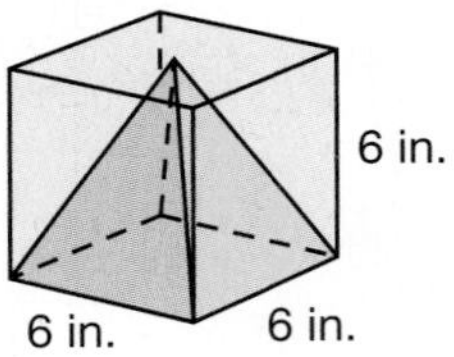

Solve:

24. (93) $0.4n + 5.2 = 12$

25. (98) $\frac{18}{y} = \frac{36}{28}$

Simplify:

26. (20) $\sqrt{5^2 - 3^2} + \sqrt{5^2 - 4^2}$

27. (49) 3 yd − 2 ft 1 in.

28. (43) $3.5 \div \left(1\frac{2}{5} \div 3\right)$

29. (57) $3.5 + 2^{-2} - 2^{-3}$

30. (85) $\frac{(3)(-2)(4)}{(-6)(2)} + (-8) + (-4)(+5) - (2)(-3)$

LESSON

116

• Slope-Intercept Form of Linear Equations

Power Up — *Building Power*

facts Power Up W

mental math

a. **Positive/Negative:** $(-2)^2 + 2^{-2}$

b. **Scientific Notation:** $(5 \times 10^5) \div (2 \times 10^2)$

c. **Algebra:** $3x + 1.2 = 2.4$

d. **Measurement:** Convert 1 m^2 to cm^2.

e. **Percent:** 125% of \$400

f. **Percent:** \$400 increased 25%

g. **Primes/Composites:** Which does not belong in the list 2, 23, 41, 48?

h. **Percent/Estimation:** Estimate $8\frac{3}{4}$% sales tax on a \$41.19 purchase.

problem solving

Three tennis balls just fit into a cylindrical container. What fraction of the volume of the container is occupied by the tennis balls?

New Concept — *Increasing Knowledge*

The three equations below are equivalent equations. Each equation has the same graph as shown.

a. $2x + y - 4 = 0$

b. $2x + y = 4$

c. $y = -2x + 4$

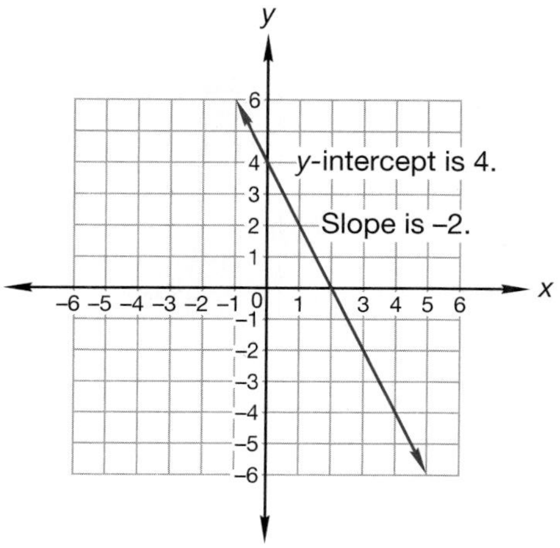

Thinking Skill

Verify

Solve equations **a** and **b** for *y* to prove the equations are equivalent. Show your work.

Equation **c** is in a special form called **slope-intercept form.** When an equation is in slope-intercept form, the coefficient of x is the slope of the graph of the equation, and the constant is the ***y*-intercept** (where the graph of the equation intercepts the *y-axis*).

slope

$$y = (-2)x\ (+4)$$

y-intercept

Notice the order of the terms in this equation. The equation is solved for y, and y is to the left of the equal sign. To the right of the equal sign is the x-term and then the constant term. The model for slope-intercept form is written this way:

Slope-Intercept Form
$\boldsymbol{y = mx + b}$

In this model, m stands for the slope and b for the y-intercept.

Represent What is the equation, in slope-intercept form, for a line whose slope is 4 and whose y-intercept is -2?

Example 1

Transform this equation so that it is in slope-intercept form.

$$\mathbf{3x + y = 6}$$

Solution

We solve the equation for y by subtracting $3x$ from both sides of the equation.

$3x + y = 6$	equation
$3x + y - 3x = 6 - 3x$	subtracted $3x$ from both sides
$y = 6 - 3x$	simplified

Next, using the Commutative Property, we rearrange the terms on the right side of the equal sign so that the x-term precedes the constant term.

$y = 6 - 3x$	equation
$\boldsymbol{y = -3x + 6}$	Commutative Property

Discuss What happens to the signs when you use the Commutative Property?

Example 2

Graph $y = -3x + 6$ using the slope and y-intercept.

Solution

The slope of the graph is the coefficient of x, which is -3, and the y-intercept is $+6$, which is located at $+6$ on the y-axis. From this point we move to the right 1 unit and down 3 units because the slope is -3. This gives us another point on the line. Continuing this pattern, we identify a series of points through which we draw the graph of the equation.

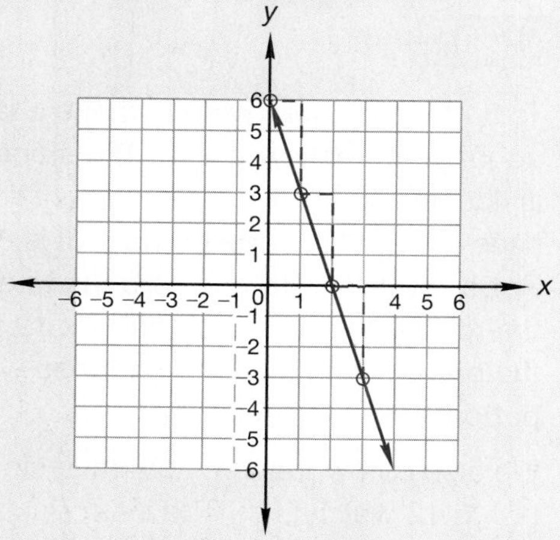

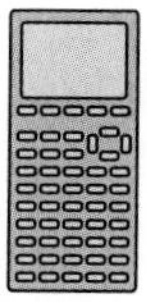

Visit www.SaxonPublishers.com/ActivitiesC2 *for a graphing calculator activity.*

Example 3

Using only slope and y-intercept, graph $y = x - 2$.

Solution

The slope is the coefficient of x, which is $+1$. The y-intercept is -2. We begin at -2 on the y-axis and sketch a line that has a slope of $+1$.

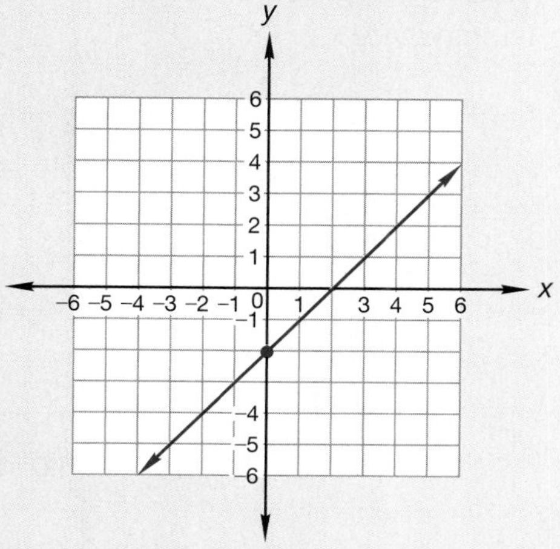

Example 4

Robert noticed a bird's nest in line with the slope of his roof. He knew that the slope of the roof was 4 in 12. He also knew that the edge of the roof was 8 feet high.

He measured 48 feet from the edge of the roof to a spot directly under the nest. Then he calculated the height of the nest. How high was the nest?

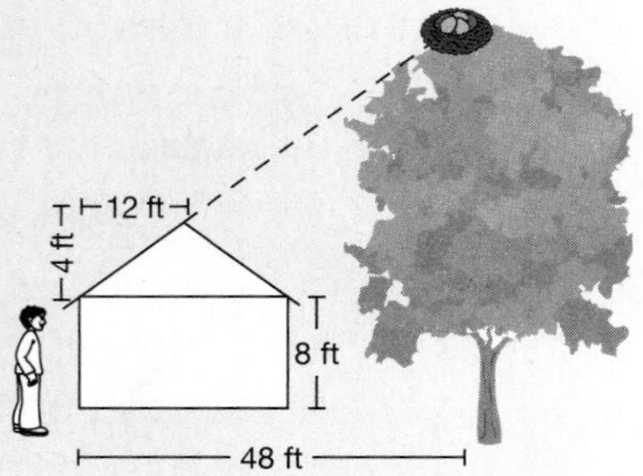

Solution

To find the answer, we can make a table, write an equation or make a graph. The slope of the roof is 4 in 12. So for every 12 horizontal feet the roof rises 4 feet. The edge of the roof is 8 feet high. As Robert walks 12 feet toward the tree the line of the roof is 4 feet higher, which is 12 feet. We begin the pattern with 0 and 8, and then we continue the pattern.

Run	Height
0	8
12	12
24	16
36	20
48	24

We also can write an equation in slope-intercept form. The slope of the roof is 4 in 12 which is $\frac{4}{12}$. The intercept is 8 feet, where the roof begins.

$$y = \frac{4}{12}x + 8$$

In this equation, y equals the height of the roof slope x feet from the edge. So the height of the roof slope 48 feet from the edge is 24 feet.

$$y = \frac{4}{12}(48) + 8$$

$$y = 24$$

We also can graph the equation $y = \frac{4}{12}x + 8$ in the first quadrant and find the height where x is 48.

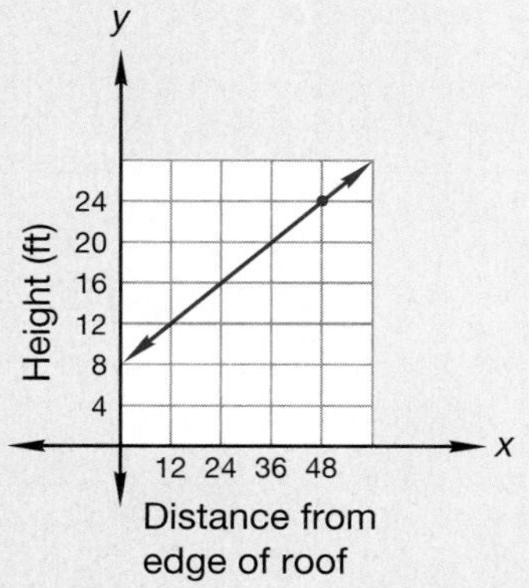

The table, the equation and the graph show that the height of the nest is **24 feet.**

Practice Set

Represent Write each equation below in slope-intercept form:

a. $2x + y = 3$

b. $y - 3 = x$

c. $2x + y - 3 = 0$

d. $x + y = 4 - x$

Connect Using only slope and *y*-intercept, graph each of these equations:

e. $y = x - 3$

f. $y = -2x + 6$

g. $y = \frac{1}{2}x - 2$

h. $y = -x + 3$

i. **Explain** How did you graph the equation in **h?**

Written Practice

Strengthening Concepts

1. (Inv. 8) A pair of number cubes is rolled once.

a. What is the probability of rolling a total of 5 (expressed as a decimal rounded to two decimal places)?

b. What is the chance of rolling a total of either 4 or 7?

c. What are the odds of rolling a total of 12?

* **2.** (20) A kilobyte of memory is 2^{10} bytes. Express the number of bytes in a kilobyte in standard form.

3. (92) Which sign seems to advertise the better sale? Explain your choice.

Sale!
40% off the regular price!

Sale!
40% of the regular price!

4. (48) Complete the table.

Fraction	Decimal	Percent
a.	b.	175%
$\frac{1}{12}$	c.	d.

5. (80) Triangle *ABC* with vertices *A* (0, 3), *B* (0, 0), and *C* (4, 0) is rotated 180° about the origin to $\triangle A'B'C'$. What are the coordinates of the vertices of $\triangle A'B'C'$?

6. (89) What is the measure of each exterior angle and each interior angle of a regular 20-gon?

7. (92) At a 30%-off sale Melba bought a jacket for $42. How much money did Melba save by buying the jacket on sale instead of paying the regular price?

* **8.** (114) The figure illustrates an aquarium with interior dimensions as shown.

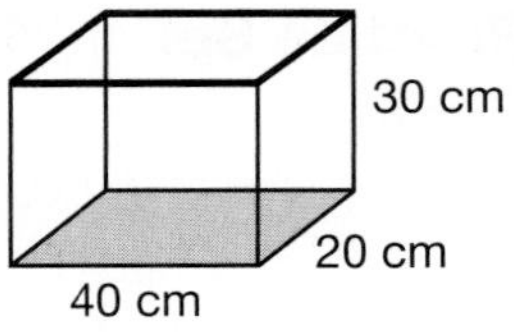

a. The aquarium has a maximum capacity of how many liters?

b. If the aquarium is filled with water, what would be the mass of the water in the aquarium?

9. (50) Use a unit multiplier to convert 24 kg to lb. (Use the approximation 1 kg $\approx$ 2.2 lb.)

10. (101) Write an equation to solve this problem:

Six less than twice what number is 48?

* **11.** (101) **Represent** Find the measure of the largest angle of the triangle shown. What equation could you write to help solve this problem?

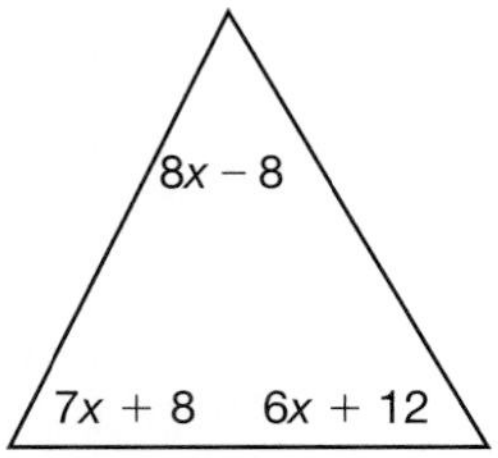

* **12.** (106) Solve for C: $F = 1.8C + 32$

* **13.** (104) **Analyze** The inside surface of this archway will be covered with a strip of wallpaper. How long must the strip of wallpaper be in order to reach from the floor on one side of the archway around to the floor on the other side of the archway? Round the answer to the nearest inch. (Use 3.14 for π.)

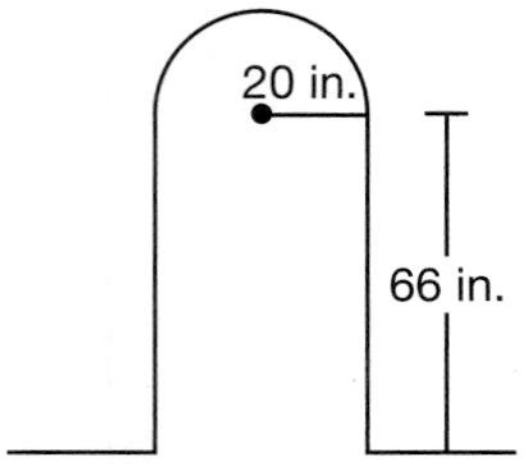

* **14.** (105) What is the total surface area of the right triangular prism below?

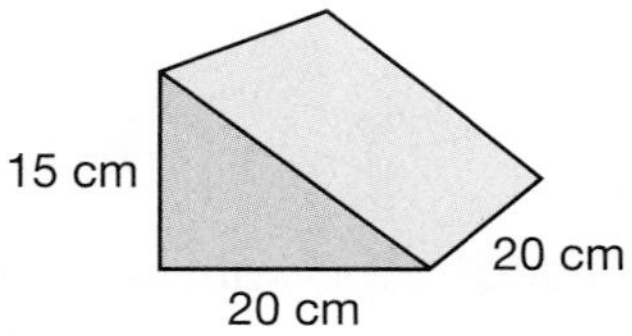

15. (95) What is the volume of the right triangular prism in problem 14?

Thinking Skill

Identify

If $a = 12$, what values will be substituted in the formula for b and h?

* **16.** (108) **Analyze** The following formula can be used to find the area, A, of a trapezoid. The lengths of the parallel sides are a and b, and the height, h, is the perpendicular distance between the parallel sides.

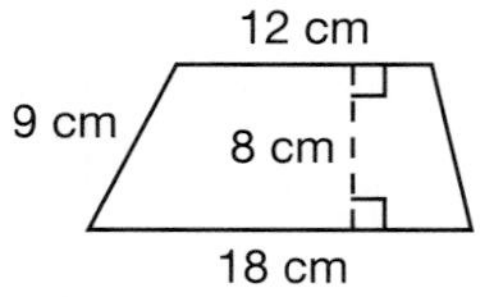

$$A = \frac{1}{2}(a + b)h$$

Use this formula to find the area of the trapezoid shown above.

* **17.** (107) **Analyze** Find the slope of each line and the point where each line intersects the y-axis:

a.

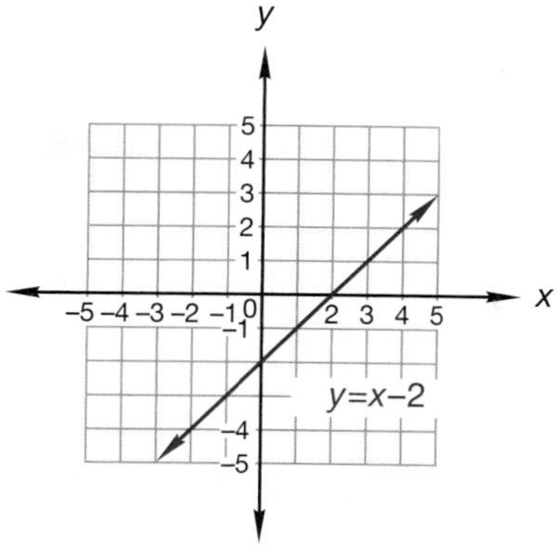

b.

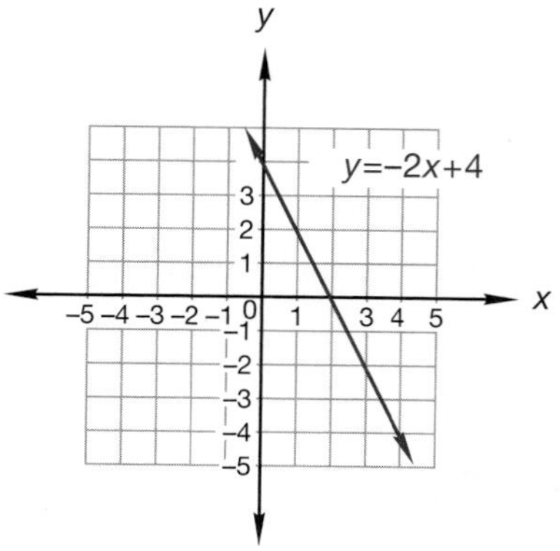

* **18.** (109) Find two solutions for $3x^2 - 5 = 40$.

* **19.** (111) **Generalize** Express each quotient in scientific notation:

a. $\dfrac{8 \times 10^{-4}}{4 \times 10^{8}}$ **b.** $\dfrac{4 \times 10^{8}}{8 \times 10^{-4}}$

20. (83) **Explain** What is the product of the two quotients in Exercise 19? Explain how you found your answer.

* **21.** (115) **Generalize** Factor each algebraic expression:

a. $9x^2y$

b. $10a^2b + 15a^2b^2 + 20abc$

* **22.** (113) **Analyze** A playground ball just fits inside a cylinder with an interior diameter of 12 in. What is the volume of the ball? Round the answer to the nearest cubic inch. (Use 3.14 for π.)

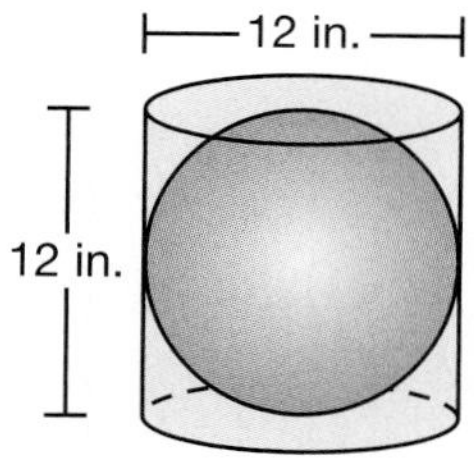

* **23.** (40) **a.** In the figure, what is m$\angle BCD$?

b. In the figure, what is m$\angle BAC$?

c. In the figure, what is m$\angle ACD$?

d. **Conclude** What can you conclude about the three triangles in the figure?

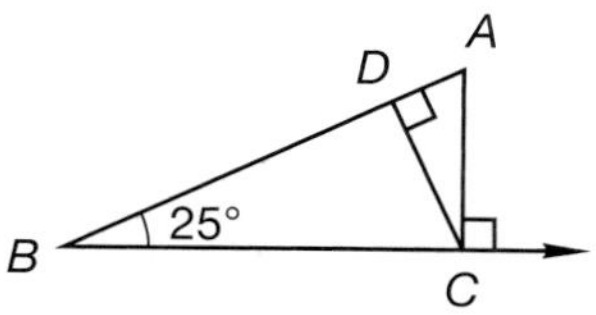

24. (97) Refer to the figure in problem **23** to complete this proportion:

$$\frac{BD}{BC} = \frac{?}{CA}$$

Solve:

25. (102) $x - 15 = x + 2x + 1$

26. (102) $0.12(m - 5) = 0.96$

Analyze Simplify:

27. (96) $a(b - c) + b(c - a)$

* 28. (103) $\sqrt{\frac{(8x^2y)(12x^3y^2)}{(4xy)(6y^2)}}$

* 29. (103, 106) a. $(-3)^2 + (-2)(-3) - (-2)^3$

b. $\sqrt[3]{-8} + \sqrt[3]{8}$

30. (7, 35) If $\overline{AB}$ is 1.2 units long and $\overline{BD}$ is 0.75 unit long, what is the length of $\overline{AD}$?

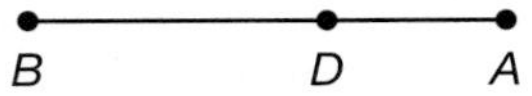

Early Finishers

Real-World Application

Hugh owns the triangular plot of land $\triangle XYZ$. He hopes to buy the adjacent plot $\triangle WXY$, seen in the figure below.

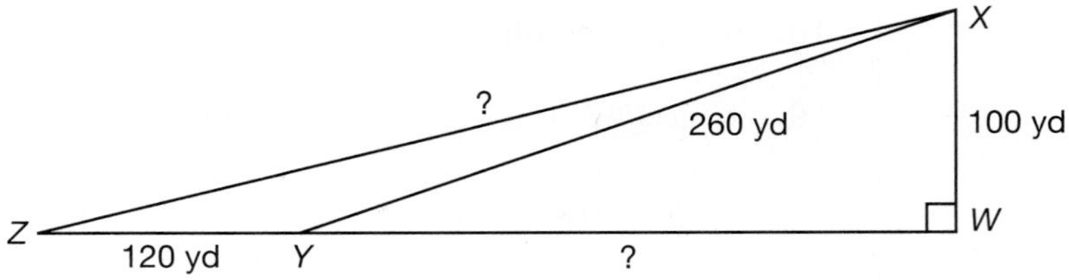

a. Find the new southern length of his land, segment WZ, if he buys the adjacent plot.

b. Find the new diagonal length of his land, segment XZ, to the nearest yard.

c. Once Hugh buys all of this land, he wants to fence the outer perimeter, $\triangle WXZ$, uniformly with a white, wooden fence. If the fence costs \$8 per linear foot, about how much will it cost for Hugh to fence in this land?

LESSON
117 • Copying Geometric Figures

Power Up
Building Power

facts Power Up V

mental math

a. **Positive/Negative:** $\frac{(-90)(-4)}{-6}$

b. **Scientific Notation:** $(7 \times 10^{-4}) \div (2 \times 10^{-6})$

c. **Algebra:** $2a^2 = 50$

d. **Measurement:** Convert 100°C to Fahrenheit.

e. **Percent:** $12\frac{1}{2}\%$ of \$4000

f. **Percent:** $12\frac{1}{2}\%$ less than \$4000

g. **Geometry:** Angles *A* and *B* are

A. complementary

B. supplementary

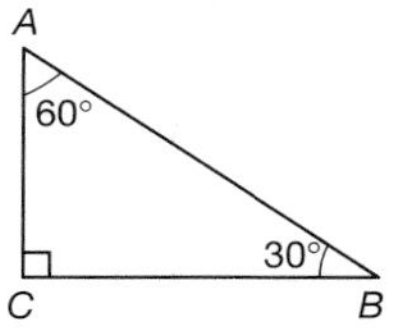

h. **Calculation:** Find 10% of 60, + 4, × 8, + 1, $\sqrt{}$, × 3, + 1, ÷ 4, × 5, + 1, $\sqrt{}$, − 7.

problem solving

There are three numbers whose sum is 180. The second number is twice the first number, and the third number is three times the first number. Create a visual representation of the equation, and then find the three numbers.

New Concept
Increasing Knowledge

Recall from Investigations 2 and 10 that we used a compass and straightedge to construct circles, regular polygons, angle bisectors, and perpendicular bisectors of segments. We may also use a compass and straightedge to copy figures.

Copying a Segment

We set the radius of the compass to match the length of the segment. Then we draw a dot to represent one endpoint of the new segment and swing an arc from that dot with the compass with the same setting.

Original segment

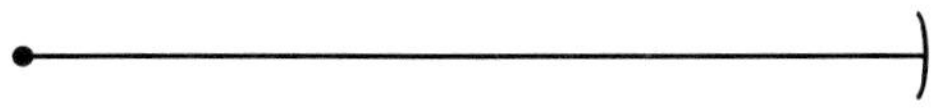

Set the radius of the compass to match the length of the segment. Then draw a dot and swing an arc from the dot using the compass setting.

Draw a segment from the dot to any point on the arc to match the length of the original segment.

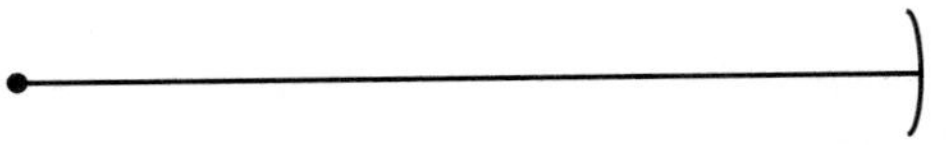

Now suppose we are asked to construct two parallel lines that are the same distance apart as another pair of parallel lines. We use the compass to find the distance between the lines and set the radius of the compass at that distance. We use that setting to mark the distance between the copied lines.

Set the compass to match the distance between the original parallel lines.

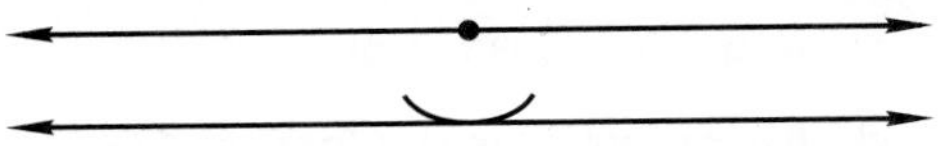

Draw a line. Then, using a compass, swing arcs from any two points on the line.

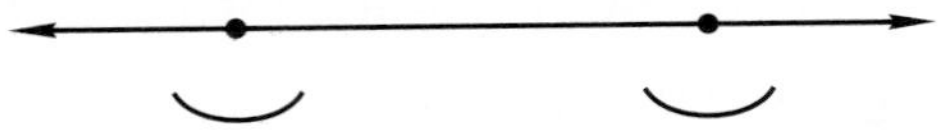

Now draw a second line that just touches the two arcs.

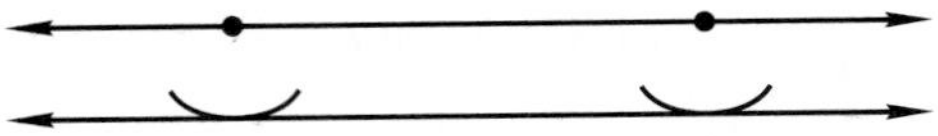

Copying an Angle

Suppose we are given this angle to copy:

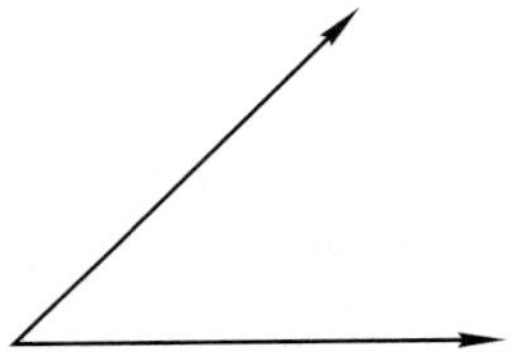

We begin by drawing a ray to form one side of the angle.

Now we need to find a point through which to draw the second ray. We find this point in two steps. First we set the compass and draw an arc across both rays of the original angle from the vertex of the angle. Without resetting the compass, we then draw an arc of the same size from the endpoint of the ray, as we show here.

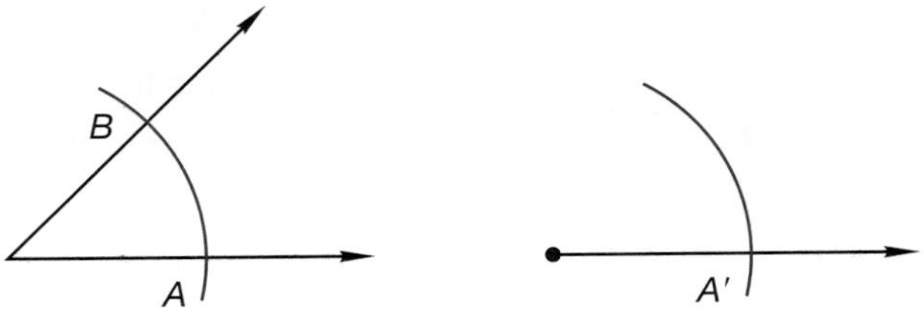

Original Angle

Copy

For the second step, we reset the compass to equal the distance from *A* to *B* on the original angle. To verify the correct setting, we swing a small arc through point *B* while the pivot point is on point *A*. With the compass at this setting, we move the pivot point to point *A′* of the copy and draw an arc that intersects the first arc we drew on the copy.

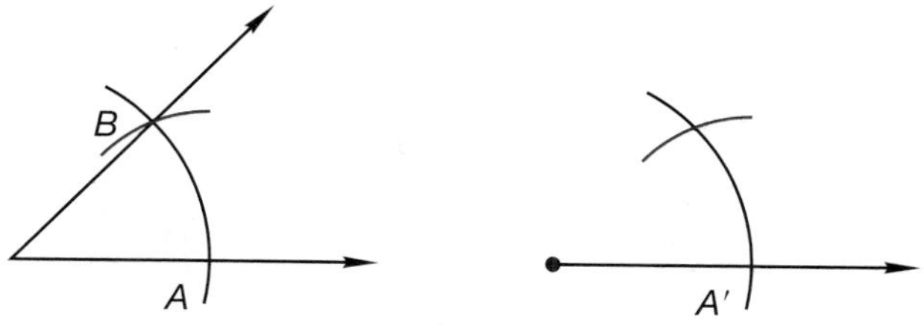

As a final step, we draw the second ray of the copied angle through the point at which the arcs intersect.

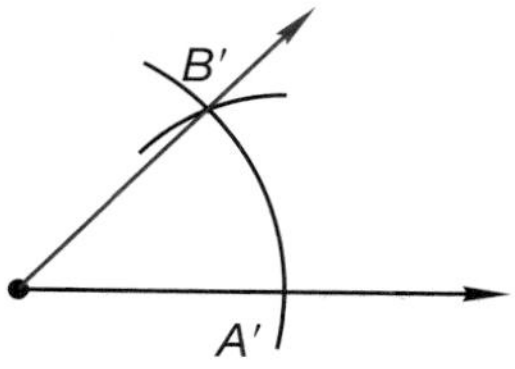

Copy

Discuss What tool can we use to check our work?

Model Using a straightedge, draw an obtuse angle. Using a compass, copy the angle.

Copying a Triangle

We use a similar method to copy a triangle. Suppose we are asked to copy $\triangle XYZ$.

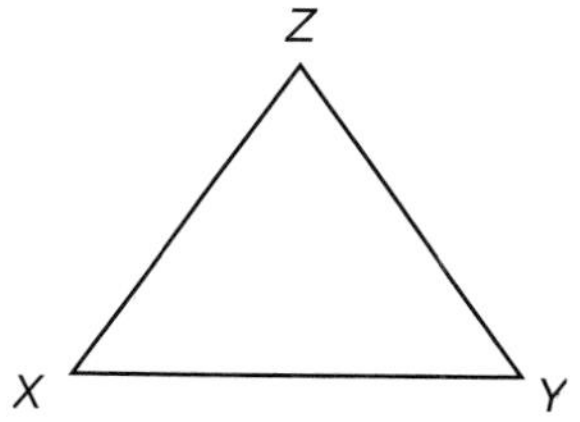

Math Language
We use an apostrophe or the word *prime* to indicate that an endpoint or vertex is related to another endpoint or vertex in a different figure.

We begin by drawing a segment equal in length to segment *XY*. We do this by setting the compass so that the pivot point is on *X* and the drawing point is on *Y*. We verify the setting by drawing a small arc through point *Y*. To copy this segment, we first sketch a ray with endpoint *X′*. Then we locate *Y′* by swinging an arc with the preset compass from point *X′*.

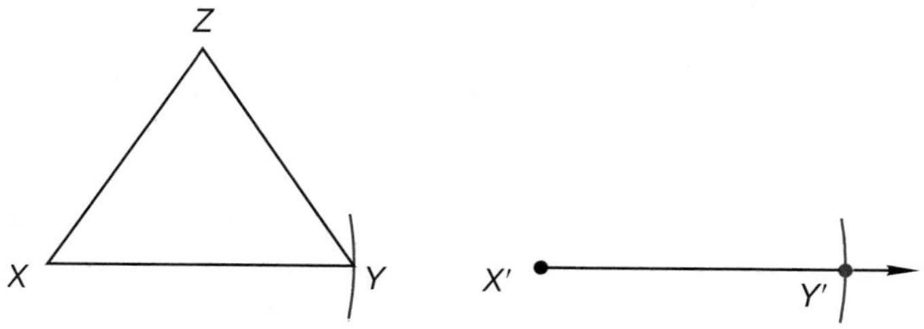

Original Triangle
Copy

To locate Z' on the copy, we will need to draw two different arcs, one from point X' and one from point Y'. We set the compass on the original triangle so that the distance between its points equals XZ. With the compass at this setting, we draw an arc from X' on the copy.

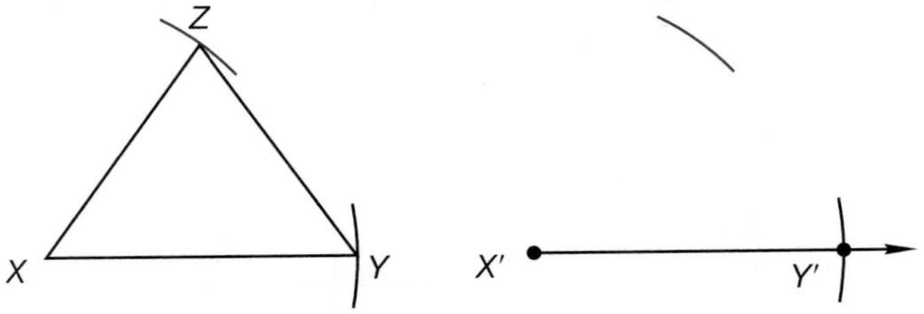

Now we change the setting of the compass to equal YZ on the original. With this setting we draw an arc from Y' that intersects the other arc.

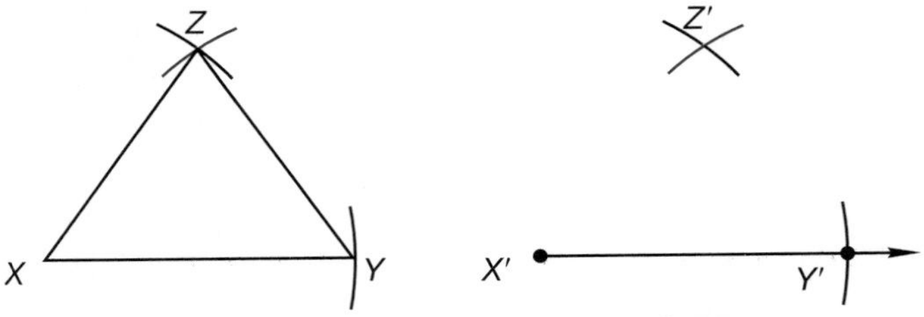

The point where the arcs intersect, which we have labeled Z', corresponds to point Z on the original triangle. To complete the copy, we draw segments $X'Z'$ and $Y'Z'$.

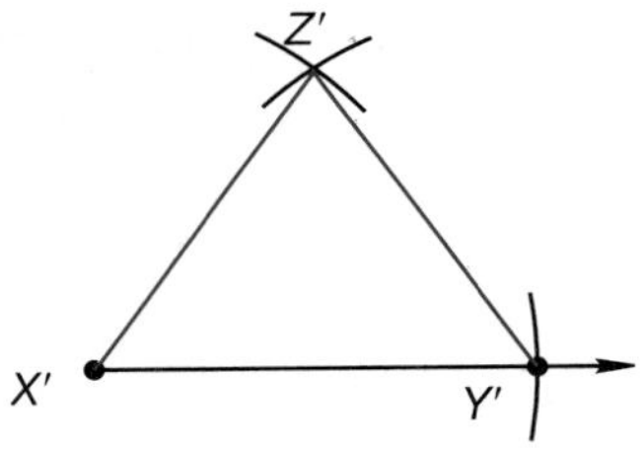

Activity

Copying Angles and Triangles

For this activity work with a partner. One partner draws an angle that the partner copies. Then switch roles. After each partner has drawn and copied an angle, repeat the process with triangles.

Practice Set

a. Use a protractor to draw an 80° angle. Then use a compass and straightedge to copy the angle.

b. With a protractor, draw a triangle with angles of 30°, 60°, and 90°. Then use a compass and straightedge to copy the triangle.

c. **Summarize** Without looking back at the lesson, explain how to copy a pair of parallel lines using a compass and a straight edge.

Summarize Without looking back at the lesson, explain how to copy a pair of parallel lines using a compass and a straight edge.

Written Practice

Strengthening Concepts

* **1.** (110) **Generalize** How much interest is earned in four years on a deposit of $10,000 if it is allowed to accumulate in an account paying 7% interest compounded annually?

2. (Inv. 8) In 240 at-bats Taneisha has 60 hits.

a. What is the statistical probability that Taneisha will get a hit in her next at-bat?

b. What are the odds of Taneisha getting a hit in her next at-bat?

3. (55) The average attendance at the first four meetings of the drama club was 75%. At the next six meetings, the average attendance was 85%. What was the average attendance at all ten meetings?

4. (48) Complete the table.

Fraction	Decimal	Percent
a.	1.4	**b.**
$\frac{11}{12}$	**c.**	**d.**

5. (80) The image of $\triangle ABC$ reflected in the y-axis is $\triangle A'B'C'$. If the coordinates of vertices A, B, and C are $(-1, 3)$, $(-3, 0)$, and $(0, -2)$, respectively, then what are the coordinates of vertices A', B', and C'?

6. (89) The figure at right shows regular octagon *ABCDEFGH*.

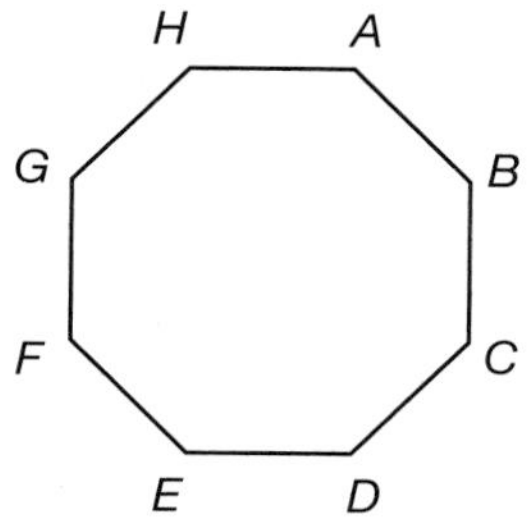

a. What is the measure of each exterior angle?

b. What is the measure of each interior angle?

c. How many diagonals can be drawn from vertex *A*?

7. (92) From 1990 to 2000, the population of Phoenix, Arizona increased from about 980,000 people to about 1,321,000 people. The population increased by what percent? Round your answer to the nearest one percent.

8. (114) A beaker is filled with water to the 500 mL level.

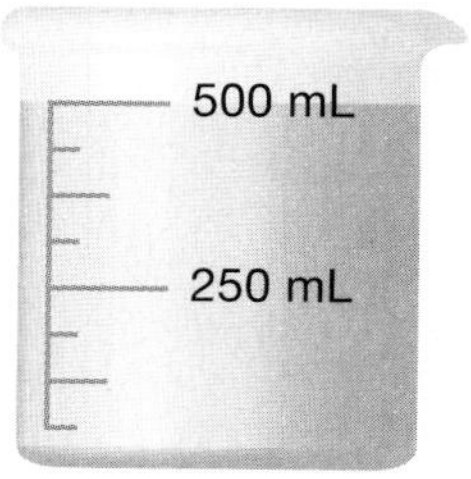

a. What is the volume of the water in cubic centimeters?

b. What is the mass of the water in kilograms?

9. Use two unit multipliers to convert 540 ft^2 to yd^2.
(88)

10. Write an equation to solve this problem:
(101)
Six more than three times what number squared is 81?

11. Solve this inequality and graph its solution:
(93)

$$\frac{3}{4}x + 12 < 15$$

* **12.** **Generalize** Solve for c^2: $c^2 - a^2 = b^2$
(106)

13. The face of this spinner is divided into four sectors. Sectors *B* and *D* are 90° sectors, and sector *C* is a 120° sector. If the arrow is spun once,
(Inv. 8)

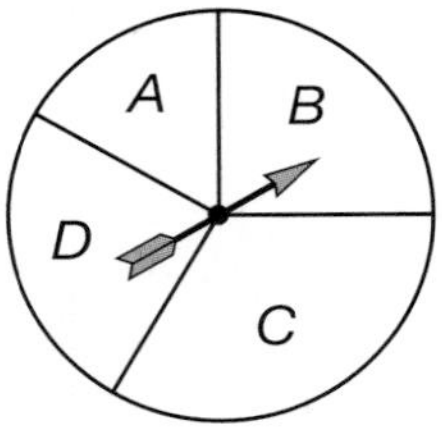

a. what is the probability (expressed as a decimal) that it will stop in sector *B*?

b. what is the chance that it will stop in sector *C*?

c. what are the odds that it will stop in sector *A*?

* **14.** **Analyze** The coordinates of the vertices of a square are (0, 4), (3, 0), (−1, −3), and (−4, 1).
(Inv. 3)

a. What is the length of each side of the square?

b. What is the perimeter of the square?

c. What is the area of the square?

* **15.** **Analyze** A right circular cylinder and a cone have an equal height and an equal diameter as shown.
(113)

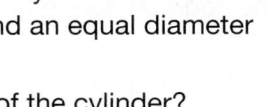

a. What is the volume of the cylinder?

b. What is the volume of the cone?

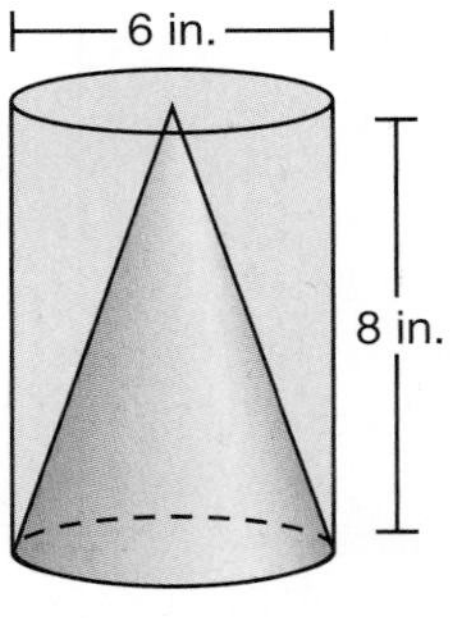

Leave π as π.

* **16.** The formula for the volume of a rectangular prism is
(106)

$$V = lwh$$

a. Transform this formula to solve for *h*.

b. Find *h* when *V* is 6000 cm^3, *l* is 20 cm, and *w* is 30 cm.

* **17.** (116) Refer to the graph shown below to answer **a–c.**

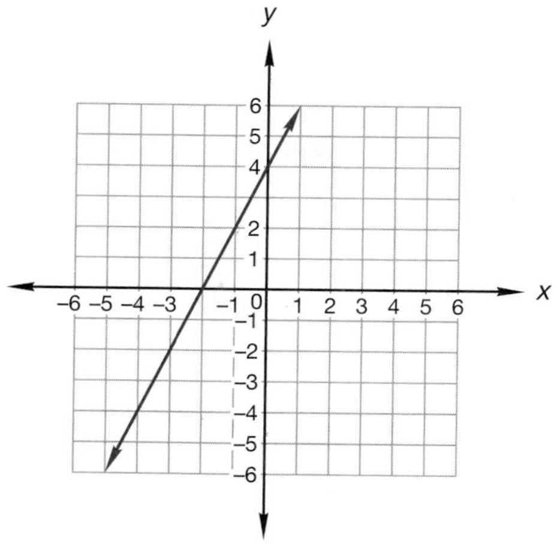

a. Analyze What is the slope of the line?

b. At what point does the line intersect the y-axis?

c. What is the equation of the line in slope-intercept form?

* **18.** (116) Connect Write each equation in slope-intercept form:

a. $y + 5 = x$ **b.** $2x + y = 4$

* **19.** (115) Generalize Factor each algebraic expression:

a. $24xy^2$ **b.** $3x^2 + 6xy - 9x$

20. (83) Find the area of a square with sides 5×10^3 mm long. Express the area

a. in scientific notation.

b. as a standard numeral.

21. (88) Use two unit multipliers to convert the answer to problem **20b** to square meters.

22. (97) Triangle ABC is a right triangle and is similar to triangles CAD and CBD.

a. Which side of $\triangle CBD$ corresponds to side BC of $\triangle ABC$?

b. Which side of $\triangle CAD$ corresponds to side AC of $\triangle ABC$?

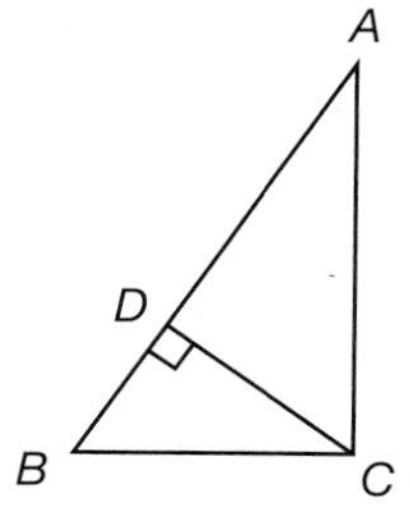

23. (7, 30) Refer to the figure below to find the length of segment BD.

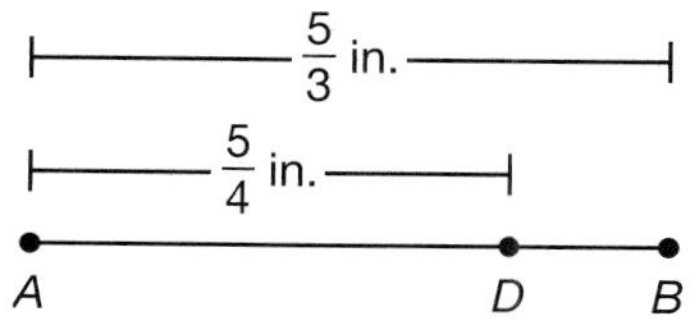

24. (101) Find the measure of the angle marked y in the figure shown.

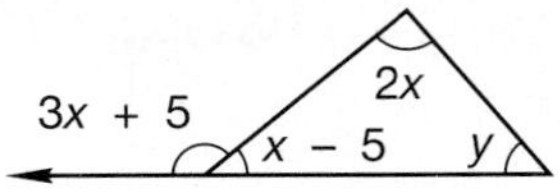

25. (102) Solve: $6w - 3w + 18 = 9(w - 4)$

Generalize Simplify:

* **26.** (96) $3x(x - 2y) + 2xy(x + 3)$

* **27.** (57, 103, 106) $2^{-2} + 4^{-1} + \sqrt[3]{27} + (-1)^3$

28. (85) $(-3) + (-2)[(-3)(-2) - (+4)] - (-3)(-4)$

* **29.** (111) $\dfrac{1.2 \times 10^{-6}}{4 \times 10^3}$

30. (103) $\dfrac{36a^2b^3c}{12ab^2c}$

Early Finishers Activity

Math and Architecture

The Great Pyramid of Giza was built by the Egyptian pharaoh Khufu in 2560 B.C. Its shape is a square pyramid with a base length of 756 feet and a height of 480 feet.

a. What is the perimeter of the pyramid's base?

b. What is the volume of the Great Pyramid?

c. Approximately how much area does the floor and outer shell of the pyramid cover?

LESSON 118

• Division by Zero

Power Up

Building Power

facts Power Up W

mental math

a. **Positive/Negative:** $(-3)^2 + 3^{-2}$

b. **Scientific Notation:** $(5 \times 10^{-6})(3 \times 10^2)$

c. **Ratio:** $\frac{k}{33} = \frac{200}{300}$

d. **Measurement:** Convert 7500 g to kg.

e. **Percent:** 150% of $4000

f. **Percent:** $4000 increased 150%

g. **Geometry:** A triangle has a base of 6 ft and a height of 2 ft. What is the area of the triangle?

h. **Rate:** At an average speed of 30 mph, how long will it take to drive 40 miles?

problem solving

A group of citizens meet to discuss improvements of a local park. If each person shakes hands with every other person at the meeting, and 28 handshakes take place altogether, how many citizens attended the meeting?

New Concept

Increasing Knowledge

When performing algebraic operations, it is necessary to guard against dividing by zero. For example, the following expression reduces to 2 only if x is not zero:

$$\frac{2x}{x} = 2 \qquad \text{if } x \neq 0$$

Thinking Skill

Analyze

What is the value of $\frac{3x}{x}$ for all values of x except 0?

What is the value of this expression if x is zero?

$\frac{2x}{x}$	expression
$\frac{2 \cdot 0}{0}$	substituted 0 for x
$\frac{0}{0}$	multiplied $2 \cdot 0$

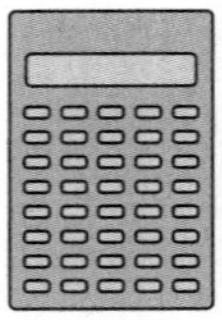

What is the value of $\frac{0}{0}$? How many zeros are in zero? Is the quotient 0? Is the quotient 1? Is the quotient some other number? Try the division with a calculator. What answer does the calculator display? Notice that the calculator displays an error message when division by zero is entered. The display is frozen and other calculations cannot be performed until the erroneous entry is cleared. In this lesson we will consider why division by zero is not possible.

Consider what happens to a quotient when a number is divided by numbers closer and closer to zero. As we know, zero lies on the number line between -1 and 1. Zero is also between -0.1 and 0.1, and between -0.01 and 0.01.

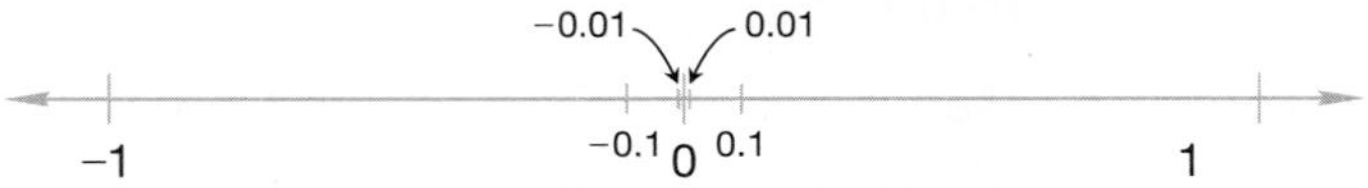

In the following example, notice the quotients we get when we divide a number by numbers closer and closer to zero.

Discuss What are two other divisors that are closer to zero than -0.01 and 0.01.

Example 1

Find each set of quotients. As the divisors become closer to zero, do the quotients become closer to zero or farther from zero?

a. $\frac{10}{1}, \frac{10}{0.1}, \frac{10}{0.01}$ **b.** $\frac{10}{-1}, \frac{10}{-0.1}, \frac{10}{-0.01}$

Solution

a. 10, 100, 1000 **b. –10, –100, –1000**

As the divisors become closer to zero, the quotients become farther from zero.

Notice from example 1a that as the divisors approach zero from the positive side, the quotients become greater and greater toward positive infinity ($+\infty$). However, in example 1b as the divisors approach zero from the negative side, the quotients become less and less toward negative infinity ($-\infty$).

In other words, as the divisors of a number approach zero from opposite sides of zero, the quotients do not become closer. Rather, the quotients grow infinitely far apart. As the divisor finally reaches zero, we might wonder whether the quotient would equal positive infinity or negative infinity! Considering this growing difference in quotients as divisors approach zero from opposite sides can help us understand why division by zero is not possible.

Another consideration is the relationship between multiplication and division. Recall that multiplication and division are inverse operations. The numbers that form a multiplication fact may be arranged to form two division facts. For the multiplication fact $4 \times 5 = 20$, we may arrange the numbers to form these two division facts:

$$\frac{20}{4} = 5 \quad \text{and} \quad \frac{20}{5} = 4$$

We see that if we divide the product of two factors by either factor, the result is the other factor.

$$\frac{\text{product}}{\text{factor}_1} = \text{factor}_2 \quad \text{and} \quad \frac{\text{product}}{\text{factor}_2} = \text{factor}_1$$

List Rearrange some other multiplication facts to form two division facts.

Predict Does the relationship between multiplication and division apply when zero is a factor? Try $2 \times 0 = 0$.

This relationship between multiplication and division breaks down when zero is one of the factors, as we see in example 2.

Example 2

The numbers in the multiplication fact $2 \times 3 = 6$ can be arranged to form two division facts.

$$\frac{6}{3} = 2 \quad \text{and} \quad \frac{6}{2} = 3$$

If we attempt to form two division facts for the multiplication fact $2 \times 0 = 0$, one of the arrangements is not a fact. Which arrangement is not a fact?

Solution

The product is 0 and the factors are 2 and 0. So the possible arrangements are these:

$$\underset{\text{fact}}{\frac{0}{2} = 0} \quad \text{and} \quad \underset{\text{not a fact}}{\frac{0}{0} = 2}$$

The arrangement **$0 \div 0 = 2$ is not a fact.**

The multiplication fact $2 \times 0 = 0$ does not imply $0 \div 0 = 2$ any more than $3 \times 0 = 0$ implies $0 \div 0 = 3$. This breakdown in the inverse relationship between multiplication and division when zero is one of the factors is another indication that division by zero is not possible.

Example 3

If we were asked to graph the following equation, what number could we not use in place of x when generating a table of ordered pairs?

$$y = \frac{12}{3 + x}$$

Solution

This equation involves division. Since division by zero is not possible, we need to guard against the divisor, $3 + x$, being zero. When x is 0, the expression $3 + x$ equals 3. So we may use 0 in place of x. However, when x is -3, the expression $3 + x$ equals zero.

$$y = \frac{12}{3 + x} \quad \text{equation}$$

$$y = \frac{12}{3 + (-3)} \quad \text{replaced } x \text{ with } -3$$

$$y = \frac{12}{0} \quad \text{not permitted}$$

Therefore, we may not use −3 in place of x in this equation. We can write our answer this way:

$$x \neq -3$$

Discuss If the numerator of the equation were $3 + x$ and the denominator were 12, could you use −3 in place of x? Why or why not?

Practice Set

a. Use a calculator to divide several different numbers of your choosing by zero. Remember to clear the calculator before entering a new problem. What answers are displayed?

b. The numbers in the multiplication fact $7 \times 8 = 56$ can be arranged to form two division facts. If we attempt to form two division facts for the multiplication fact $7 \times 0 = 0$, one of the arrangements is not a fact. Which arrangement is not a fact and why?

Analyze For the following expressions, find the number or numbers that may not be used in place of the variable.

c. $\frac{6}{w}$

d. $\frac{3}{x - 1}$

e. $\frac{4}{2w}$

f. $\frac{y + 3}{y - 3}$

g. $\frac{8}{x^2 - 4}$

h. $\frac{3ab}{c}$

i. **Justify** Explain your answer to **g.**

Written Practice

Strengthening Concepts

1. (92) The median home price in the county increased from $360,000 to $378,000 in one year. This was an increase of what percent?

2. (97) To indirectly measure the height of a power pole, Teddy compared the lengths of the shadows of a vertical meterstick and of the power pole. When the shadow of the meterstick was 40 centimeters long, the shadow of the power pole was 6 meters long. About how tall was the power pole?

*** 3.** (112) **Explain** Armando is marking off a grass field for a soccer game. He has a long tape measure and chalk for lining the field. Armando wants to be sure that the corners of the field are right angles. How can he use the tape measure to ensure that he makes right angles?

4. (50) Convert 15 meters to feet using the approximation $1 \text{ m} \approx 3.28 \text{ ft}$. Round the answer to the nearest foot.

5. The illustration below shows one room of a scale drawing of a house. One inch on the drawing represents a distance of 10 feet. Use a ruler to help calculate the actual area of the room.
(98)

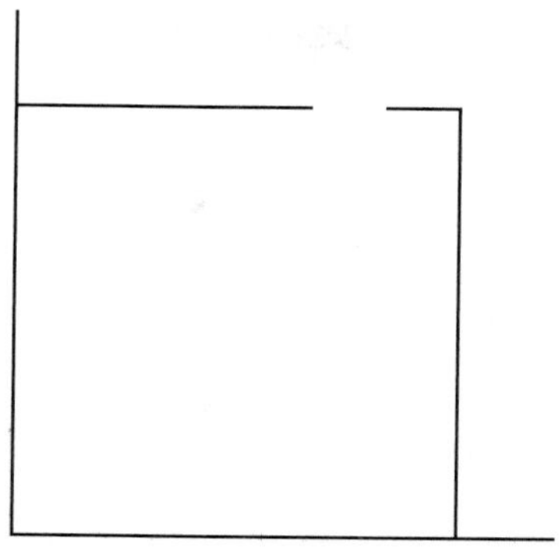

6. A pair of number cubes is rolled once.
(Inv. 8)

a. What is the probability of rolling a total of 9?

b. What is the chance of rolling a total of 10?

c. What are the odds of rolling a total of 11?

7. Use the Pythagorean theorem to find the distance from (4, 6) to (−1, −6).
(99)

*** 8.** **Analyze** A two-liter bottle filled with water
(114)

a. contains how many cubic centimeters of water?

b. has a mass of how many kilograms?

9. Write an equation to solve this problem:
(101)

Two thirds less than half of what number is five sixths?

*** 10.** **Conclude** In this figure lines *m* and *n* are parallel. If the sum of the measures of angles *a* and *e* is 200°, what is the measure of $\angle g$?
(102)

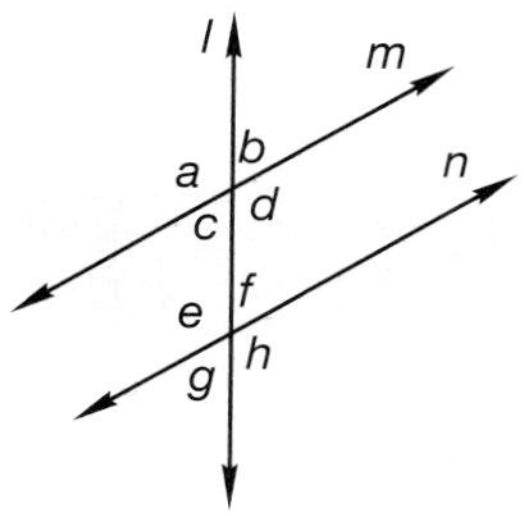

*** 11.** **Connect** Transform the equation $3x + y = 6$ into slope-intercept form. Then graph the equation on a coordinate plane.
(116)

12. **Explain** Find the measure of the smallest angle of the triangle shown. Explain how you found your answer.
(101)

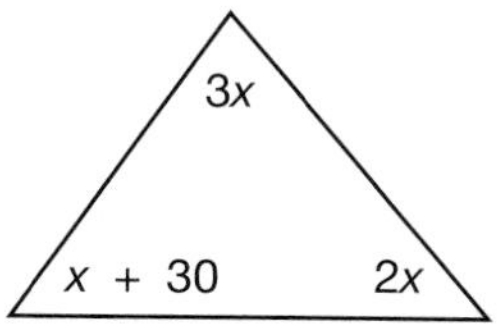

* **13.** A cube, 12 inches on edge, is topped with a pyramid so that the total height of the cube and pyramid is 20 inches. What is the total volume of the figure?
(113)

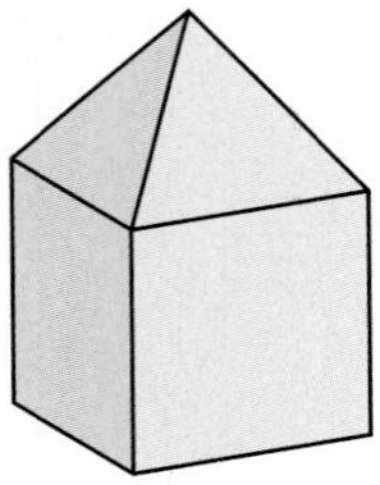

14. The length of segment BD is 12. The length of segment BA is c. Using 12 and c, write an expression that indicates the length of segment AD.
(101)

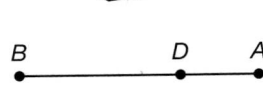

15. The three triangles in the figure shown are similar. The sum of x and y is 25. Use proportions to find x and y.
(97)

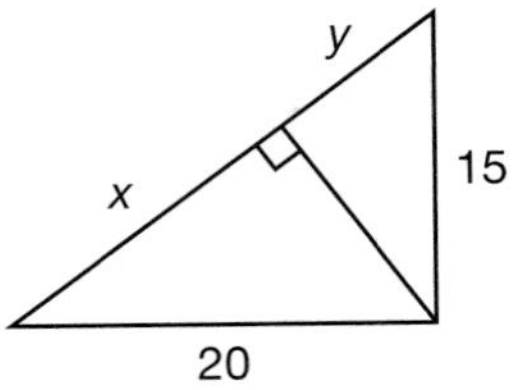

* **16.** **Estimate** Lina needs to estimate the surface area of a grapefruit. She cut a grapefruit in half. (The flat surface formed is called a cross section.)
(105)

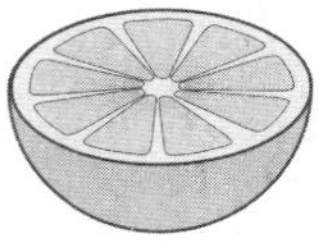

Lina estimated that the diameter of the grapefruit was 8 cm, and she used 3 in place of π. Using Lina's numbers, estimate the area of the whole grapefruit peel.

* **17.** Write the equation $y - 2x + 5 = 1$ in slope-intercept form. Then graph the equation.
(116)

* **18.** Refer to the graph shown below to answer **a–c.**
(116)

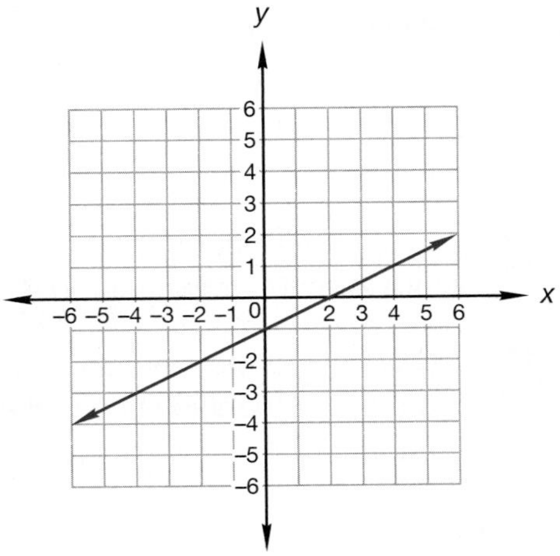

a. **Analyze** What is the slope of the line?

b. What is the y-intercept of the line?

c. **Represent** What is the equation of the line in slope-intercept form?

* **19.** (117) **Model** Draw an estimate of a 60° angle, and check your estimate with a protractor. Then set the protractor aside, and use a compass and straightedge to copy the angle.

20. (104) A semicircle with a 7-inch diameter was cut from a rectangular half sheet of paper. What is the perimeter of the resulting shape? (Use $\frac{22}{7}$ for π.)

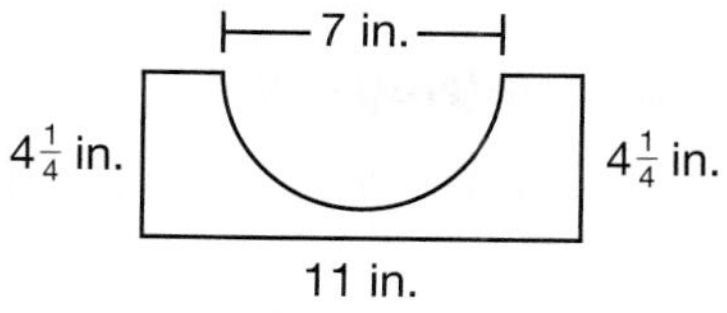

* **21.** (111) **Analyze** A dime is about 1×10^{-3} m thick. A kilometer is 1×10^{3} m. About how many dimes would be needed to make a stack of dimes one kilometer high? Express the answer in scientific notation.

* **22.** (115) **Generalize** Factor each algebraic expression:

a. $x^2 + x$

b. $12m^2n^3 + 18mn^2 - 24m^2n^2$

Solve:

23. (93) $-2\frac{2}{3}w - 1\frac{1}{3} = 4$

* **24.** (109) $5x^2 + 1 = 81$

25. (57) $\left(\frac{1}{2}\right)^2 - 2^{-2}$

26. (48) $66\frac{2}{3}\%$ of $\frac{5}{6}$ of 0.144

27. (91) $[-3 + (-4)(-5)] - [-4 - (-5)(-2)]$

Simplify:

28. (103) $\dfrac{(5x^2yz)(6xy^2z)}{10xyz}$

29. (96) $x(x + 2) + 2(x + 2)$

30. (100) The length of the hypotenuse of this right triangle is between which two consecutive whole numbers of millimeters?

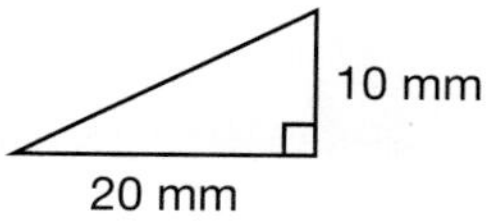

Early Finishers
Math Applications

Look at the three transformations below.

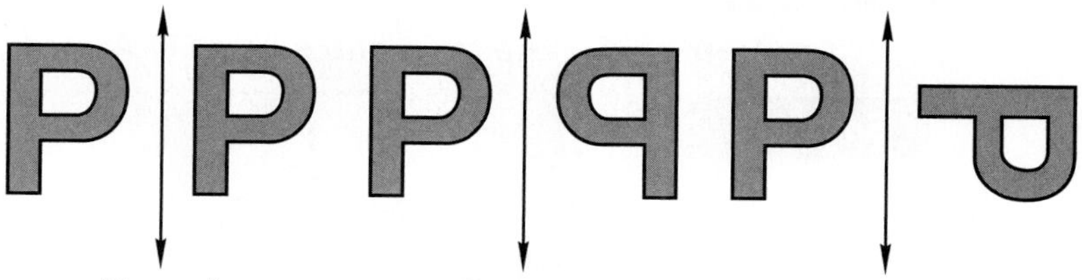

Figure 1 Figure 2 Figure 3

a. Name each transformation. Explain your reasoning for choosing each.

b. Which figure shows a transformation that has a line of symmetry? Explain your answer.

LESSON

119

• Graphing Area and Volume Formulas

Power Up

Building Power

facts Power Up V

mental math

a. Positive/Negative: $(2^{-2})(-2)^2$

b. Scientific Notation: $(1 \times 10^{-8}) \div (1 \times 10^{-4})$

c. Algebra: $2y + \frac{1}{2} = \frac{1}{2}$

d. Measurement: Convert 5 cm^2 to mm^2.

e. Percent: $66\frac{2}{3}\%$ of \$600

f. Percent: \$600 reduced $33\frac{1}{3}\%$

g. Geometry: This pair of angles is

A supplementary.

B complementary.

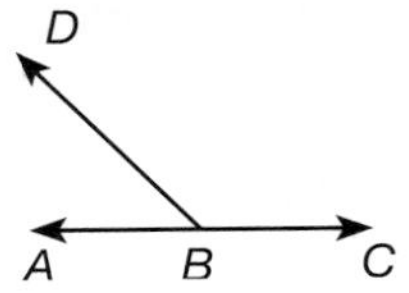

h. Fractional Parts/Measurement: What fraction of an hour is 5 minutes less than $\frac{1}{3}$ of an hour?

problem solving

A paper cone is filled with water. Then the water is poured into a cylindrical glass beaker that has the same height and diameter as the paper cone. How many cones of water are needed to fill the beaker?

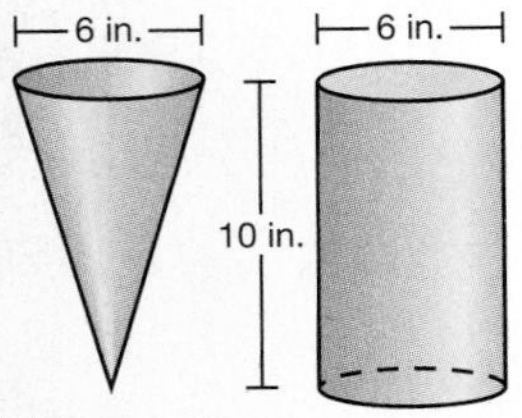

New Concept

Increasing Knowledge

The graph of a function may be a curve. In the example below we graph the relationship between the length of the side of a square and its area.

Example 1

The formula for the area of a square is $A = s^2$. Graph this function.

Solution

We use the letter *A* and *s* to represent the area in square units and the side length of the square respectively.

Thinking Skill

Predict

Is it possible to tell by looking at the table if the function is nonlinear?

$A = s^2$

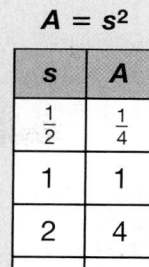

s	A
$\frac{1}{2}$	$\frac{1}{4}$
1	1
2	4
3	9

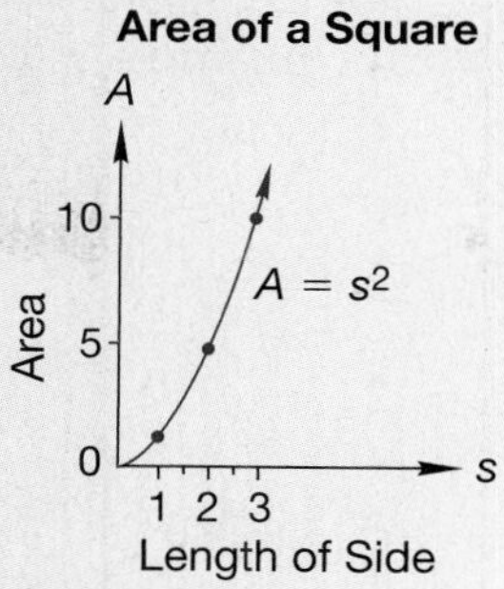

The dots on the graph of the function show the (s, A) pairs from the table. Other (s, A) pairs of numbers are represented by other points on the curve. Notice that the graph of the function rapidly becomes steeper as the side of the square becomes longer.

The graph relating the edge length to the volume of a cube grows steeper even more rapidly. In example 2, we use different vertical and horizontal scales to contain the graph for a few points that are easy to calculate.

Predict What could you do to the graph to plot more points? What would happen to the graph if you plotted more ordered pairs?

Example 2

The formula for the volume of a cube is $V = e^3$ in which V is volume and e is the length of an edge. Make a function table for this formula and describe how to graph the function.

Solution

The input is the length of the edge, the output is the volume. We use 0, 1, 2, 3 and 4 for edge lengths and calculate the volumes. To graph these points we need to space the intervals on the e-axis relatively far apart and reduce the interval length on the V-axis. Then we plot the points and draw a smooth curve.

$V = e^3$

e	V
0	0
1	1
2	8
3	27
4	64

Analyze Refer to the table. How do the values for V relate to the value for e?

Practice Set

a. Connect Half of the area of a square is shaded. As the square becomes larger, the area of the shaded region becomes greater, as indicated by the function table in which A represents the area of half of a square and s represents the length of a side of the square. Copy this table and find the missing numbers. Then graph the function.

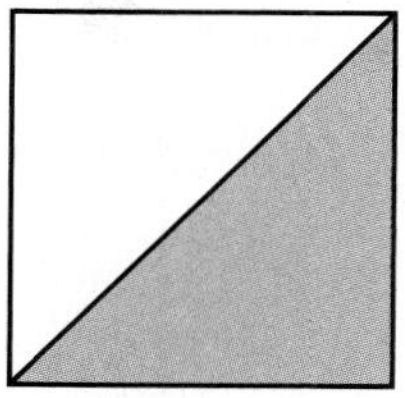

$A = \frac{1}{2}s^2$

s	A
1	$\frac{1}{2}$
2	2
3	
4	

b. Connect Graph the function $V = e^3$ using the function table from example 2. Make the scale on the axes is similar to the one shown here.

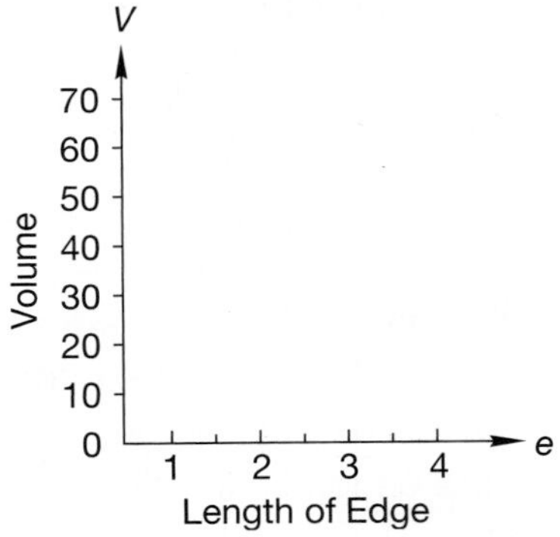

c. Analyze Refer to **a.** What effect does doubling the side length of a square have on the area of the square?

Written Practice *Strengthening Concepts*

1. (94) Lily is asked to select and hold 3 marbles from a bag of 52 marbles. Four of the marbles are green. If the first two marbles she selects are green, what is the chance that the third marble she selects will be one of the two remaining green marbles?

2. (92) If Khalid saved \$5 by purchasing a box of blank CDs at a sale price of \$15, then the regular price was reduced by what percent?

3. (78) On a number line graph all real numbers that are both greater than or equal to −3 and less than 2.

4. (89) What is the sum of the measures of the interior angles of any quadrilateral?

5. (48) Complete the table.

Fraction	Decimal	Percent
a.	**b.**	0.5%
$\frac{8}{9}$	**c.**	**d.**

6. (17) **a.** Use a centimeter ruler and a protractor to draw a right triangle with legs 10 cm long.

b. What is the measure of each acute angle?

c. Measure the length of the hotenuse to the nearest centimeter.

*** 7.** (111) Generalize Simplify. Write the answer in scientific notation.

$$\frac{(6 \times 10^5)(2 \times 10^6)}{(3 \times 10^4)}$$

*** 8.** (115) **Generalize** Factor each expression:

a. $2x^2 + x$

b. $3a^2b - 12a^2 + 9ab^2$

Analyze The figure at right was formed by stacking 1-cm cubes. Refer to the figure to answer problems **9** and **10.**

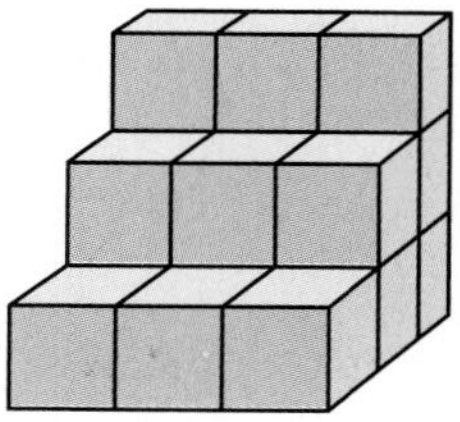

9. (70) What is the volume of the figure?

*** 10.** (105) What is the surface area of the figure?

Thinking Skill

Analyze

How many cube faces are visible from each view (top, bottom, back, front, and sides)?

*** 11.** (108) Transform the formula $A = \frac{1}{2}bh$ to solve for h. Then use the transformed formula to find h when A is 1.44 m^2 and b is 1.6 m.

12. (66) If the ratio of blue T-shirts to white T-shirts is 3 to 5, then what percent of the T-shirts are blue?

*** 13.** (112) **Analyze** If a 10-foot ladder is leaning against a wall so that the foot of the ladder is 6 feet from the base of the wall, how far up the wall will the ladder reach?

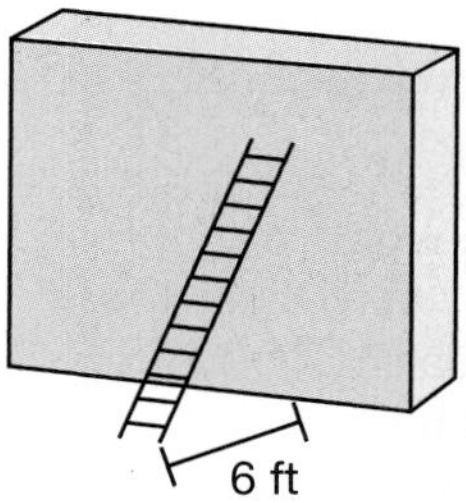

Connect The graph below shows line *l* perpendicular to line *m*. Refer to the graph to answer problems **14** and **15.**

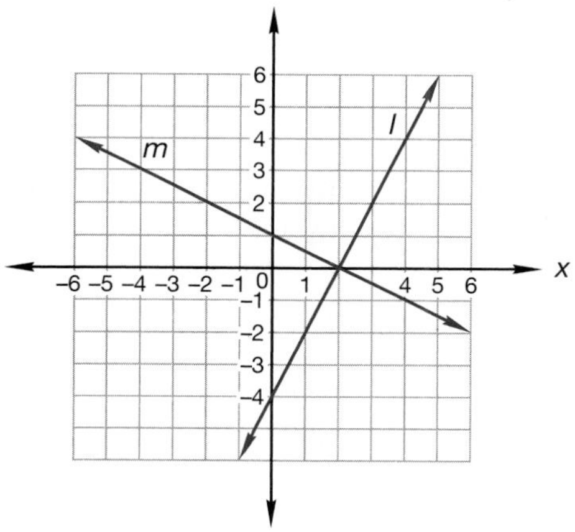

*** 14.** (116) **a.** What is the equation of line *l* in slope-intercept form?

b. What is the equation of line *m* in slope-intercept form?

*** 15.** (107) What is the product of the slopes of line *l* and line *m*? Why?

* **16.** (110) If \$8000 is deposited in an account paying 6% interest compounded annually, then what is the total value of the account after four years?

17. (88) The Joneses are planning to carpet their home. The area to be carpeted is 1250 square feet. How many square yards of carpeting must be installed? Round the answer up to the next square yard.

* **18.** (118) *Analyze* In the following expressions, what number may not be used for the variable?

a. $\frac{12}{3w}$ **b.** $\frac{12}{3 + m}$

19. (101) In the figure shown, $\overline{BD}$ is x units long, and $\overline{BA}$ is c units long. Using x and c, write an expression that indicates the length of $\overline{DA}$.

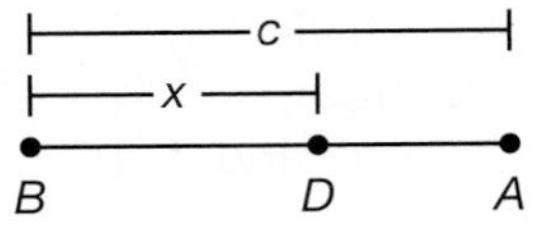

Thinking Skill

Analyze

Use what you know about similar triangles and the Pythagorean Theorem to find y and z.

20. (97) In the figure at right, the three triangles are similar. Find the area of the smallest triangle. Dimensions are in inches.

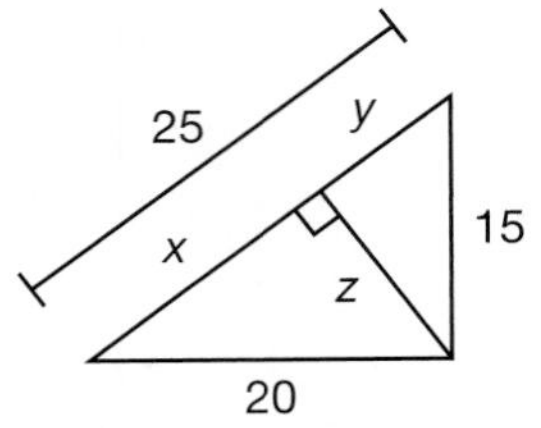

* **21.** (113) A sphere with a diameter of 30 cm has a volume of how many cubic centimeters?

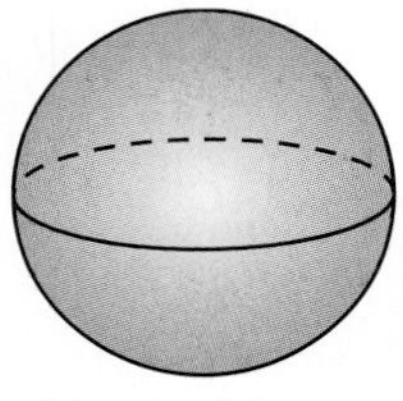

Use 3.14 for π.

22. (96, 117) Draw an estimate of a 45° angle. Then use a compass and straightedge to copy the angle.

Solve:

23. (93) $\frac{2}{3}m + \frac{1}{4} = \frac{7}{12}$

24. (102) $5(3 - x) = 55$

25. (102) $x + x + 12 = 5x$

26. (109) $10x^2 = 100$

Simplify:

27. (20) $\sqrt{90{,}000}$

28. (96) $x(x + 5) - 2(x + 5)$

29. (103) $\frac{(12xy^2z)(9x^2y^2z)}{36xyz^2}$

30. (48) $33\frac{1}{3}\%$ of 0.12 of $3\frac{1}{3}$

LESSON 120

• Graphing Nonlinear Equations

Power Up

Building Power

facts Power Up W

mental math

a. **Positive/Negative:** $(10^2)(10^{-2})$

b. **Scientific Notation:** $(5 \times 10^{-5})^2$

c. **Algebra:** $2x^2 = 32$

d. **Measurement:** Convert 0°C to Fahrenheit.

e. **Percent:** 10% of \$250

f. **Percent:** 10% more than \$250

g. **Probability:** The weather forecast stated that the chance of rain is 70%. What is the chance it will not rain?

h. **Calculation:** 2×12, $+ 1$, $\sqrt{\ }$, $\times 3$, $+ 1$, $\sqrt{\ }$, $\times 2$, $+ 1$, $\sqrt{\ }$, $+ 1$, $\sqrt{\ }$, $- 1$, $\sqrt{\ }$

problem solving

Does $0.\overline{9} = 1$? Here are two arguments that say it is:

$$\begin{array}{rl} & \frac{1}{3} = 0.\overline{3} \\ + & \frac{2}{3} = 0.\overline{6} \\ \hline & \frac{3}{3} = 0.\overline{9} \\ & 1 = 0.\overline{9} \end{array}$$

$$\begin{array}{r} 0.9999 \\ 5\overline{)5.0000} \\ \underline{4\,5} \\ 50 \\ \underline{45} \\ 50 \\ \underline{45} \\ 5 \end{array}$$

Verify or refute each argument.

New Concept

Increasing Knowledge

Equations whose graphs are lines are called **linear equations.** (Notice the word *line* in *linear*.) In Lesson 119 and in this lesson we graph equations whose graphs are not lines but are curves. These equations are called **nonlinear equations.** One way to graph nonlinear equations is to make a table of ordered pairs (a function table) and plot enough points to get an idea of the path of the curve.

Recall that the graphs in Lesson 119 were confined to the first quadrant. That is because we were graphing lengths and areas and volumes which are positive numbers. In this lesson we will graph functions whose variables may be negative, so we need to select some values for the input numbers that are negative as well as positive.

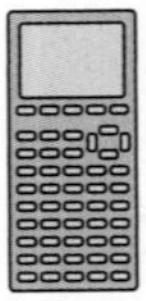

Visit www. SaxonPublishers. com/ActivitiesC2 *for a graphing calculator activity.*

Example 1

Graph: $y = x^2$

Solution

This equation is like the formula for the area of a square, except that *y* and *x* replace *A* and *S* respectively. We begin by making a table of ordered pairs. We think of numbers for *x* and then calculate *y*. We replace *x* with negative numbers as well. Remember that squaring a negative number results in a positive number.

$y = x^2$

x	y	(x, y)
0	0	(0, 0)
1	1	(1, 1)
2	4	(2, 4)
3	9	(3, 9)

x	y	(x, y)
−1	1	(−1, 1)
−2	4	(−2, 4)
−3	9	(−3, 9)

After generating several pairs of coordinates, we graph the points on a coordinate plane.

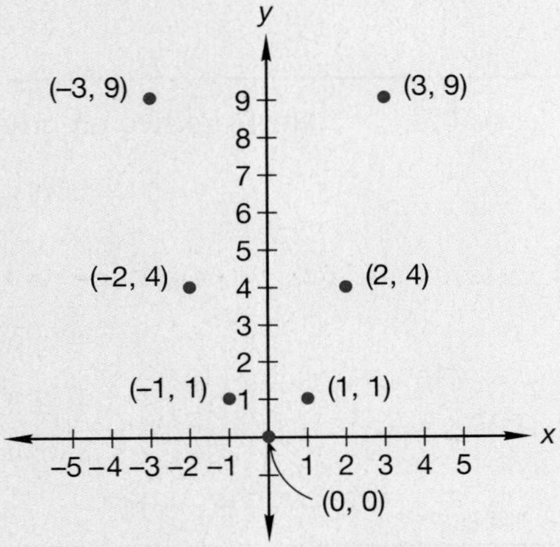

Discuss Would you expect to find an ordered pair that satisfies this equation in the third or fourth quadrant? Why or why not?

We complete the graph by drawing a smooth curve through the graphed points.

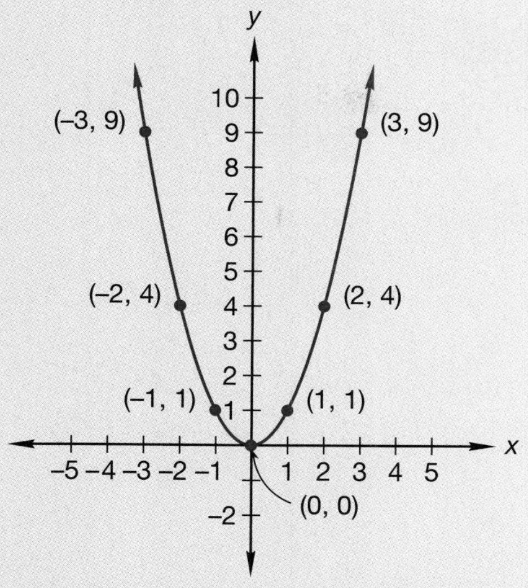

The coordinates of any point on the curve should satisfy the original equation.

Analyze What is a line of symmetry for this graph?

In example 2 we graph another nonlinear equation using a table of ordered pairs. We have used the terms *input* and *output* to refer to variables when creating function tables. Mathematicians use the words **independent variable** and **dependent variable** respectively to refer to these variables.

The dependent variable is usually isolated on one side of the equation. In example 2 the dependent variable is *y*. We freely select numbers to substitute for the independent variable, which in example 2 is *x*. The value of *y* depends upon the number we select for the independent variable.

Example 2

Graph: $y = \frac{6}{x}$

Solution

We make a table of ordered pairs. For convenience we select *x* values that are factors of 6. We remember to select negative values as well. Note that we may not select zero for *x*.

Thinking Skill

Explain

Why can't you use zero in place of the variable *x*?

$$y = \frac{6}{x}$$

x	y	(x, y)
1	6	(1, 6)
2	3	(2, 3)
3	2	(3, 2)
6	1	(6, 1)

x	y	(x, y)
−1	−6	(−1, −6)
−2	−3	(−2, −3)
−3	−2	(−3, −2)
−6	−1	(−6, −1)

On a coordinate plane we graph the x, y pairs we found that satisfy the equation.

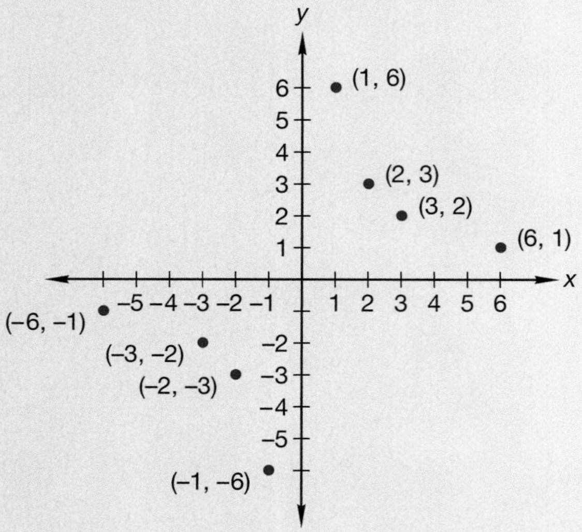

This arrangement of points on the coordinate plane suggests two curves that do not intersect.

We draw two smooth curves through the two sets of points.

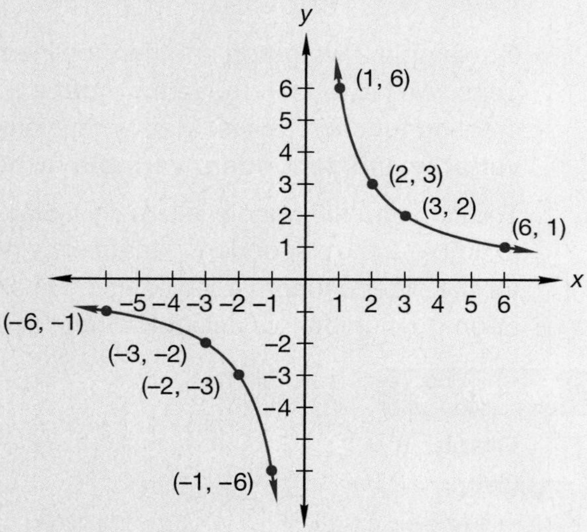

You might recall seeing this graph in Investigation 9 where we graphed $xy = 6$. Notice that $y = \frac{6}{x}$ is an alternate form of the same equation. Can you transform one equation into the other? The equation $y = \frac{6}{x}$ is an example of inverse variation.

Conclude Will either curve ever cross the x- or y-axis? Why or why not?

Practice Set

Connect Graph each equation.

a. $y = \dfrac{12}{x}$ Begin by creating a table of ordered pairs. Use 6, 4, 3, 2, −2, −3, −4, and −6 in place of x.

b. $y = x^2 - 2$ Compare your graph to the graph in example 2.

c. $y = \frac{10}{x}$ Compare your graph to the graph in example 1.

d. $y = 2x^2$ Compare your graph to the graph in example 2.

Written Practice

Strengthening Concepts

1. (Inv. 8) Schuster was playing a board game with a pair of number cubes. He rolled a 7 three times in a row. What are the odds of Schuster getting a 7 with the next roll?

2. (92) If the total cost of a package of paper clips including 8% sales tax is \$2.70, then what was the price before tax was added?

3. (77) In the year 2000, the population of the United States was about 280 million. In the election that year about 105 million people voted. About what percent of the population voted in the presidential election of 2000?

4. (58, 89) If a trapezoid has a line of symmetry and one of its angles measures 100°, what is the measure of each of its other angles?

5. (48) Complete the table.

Fraction	Decimal	Percent
a.	b.	0.1%
$\frac{8}{5}$	c.	d.

*** 6.** (111) **Generalize** Simplify. Write the answer in scientific notation.

$$\frac{(4 \times 10^{-5})(6 \times 10^{-4})}{8 \times 10^{3}}$$

*** 7.** (115) **Generalize** Factor each expression:

a. $3y^2 - y$ **b.** $6w^2 + 9wx - 12w$

The figure below shows a cylinder and a cone whose heights and diameters are equal. Refer to the figure to answer problems **8** and **9.**

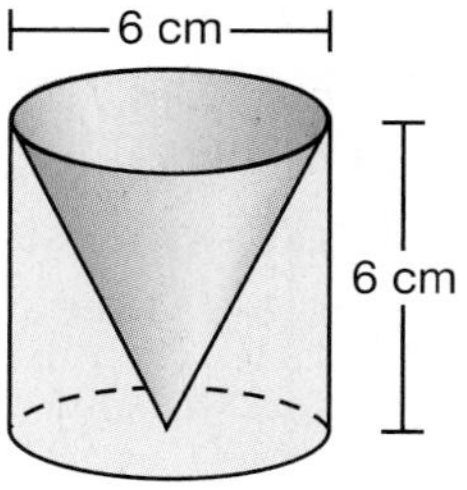

*** 8.** (113) **Analyze** What is the ratio of the volume of the cone to the volume of the cylinder?

*** 9.** (105) The lateral surface area of a cylinder is the area of the curved side and excludes the areas of the circular ends. What is the lateral surface area of the cylinder rounded to the nearest square centimeter? (Use 3.14 for π.)

10. The hypotenuse of this triangle is twice the length of the shorter leg.
(99)

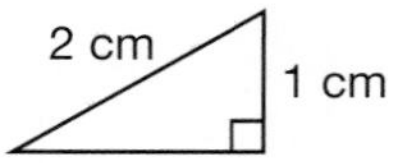

a. Use the Pythagorean theorem to find the length of the remaining side.

b. Use a centimeter ruler to find the length of the unmarked side to the nearest tenth of a centimeter.

* **11.** Transform the formula $E = mc^2$ to solve for m.
(106)

12. If 60% of the students in the assembly were girls, what was the ratio of boys to girls in the assembly?
(36)

The graph below shows $m \perp n$. Refer to the graph to answer problems **13** and **14.**

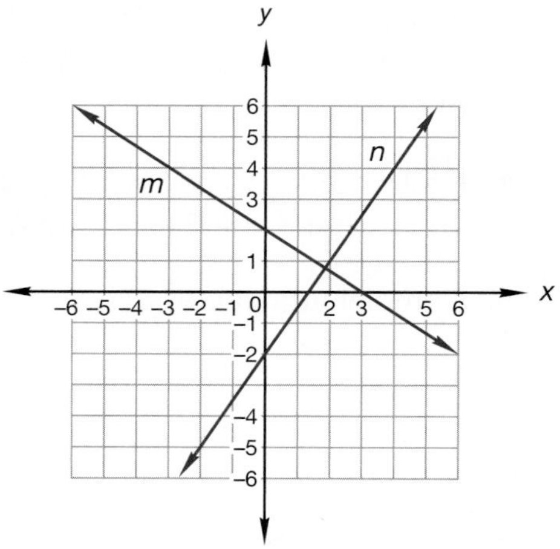

* **13.** Connect What is the equation of each line in slope-intercept form?
(116)

14. What is the product of the slopes of lines m and n? Why?
(107)

* **15.** Generalize If a $1000 investment earns 20% interest compounded annually, then the investment will double in value in how many years?
(110)

* **16.** The stated size of a TV screen or computer monitor is its diagonal measure. A screen that is 17 in. wide and 12 in. tall would be described as what size of screen? Round the answer to the nearest inch.
(112)

17. Premixed concrete is sold by the cubic yard. The Jeffersons are pouring a concrete driveway that is 36 feet long, 21 feet wide, and $\frac{1}{2}$ foot thick.
(70, 88)

a. Find the number of cubic feet of concrete needed.

b. Use three unit multipliers to convert answer **a** to cubic yards.

*** 18.** (118) **Analyze** In the following expressions, what number may not be used for the variable?

a. $\frac{12}{4 - 2m}$ **b.** $\frac{y - 5}{y + 5}$

*** 19.** (120) **Connect** Graph: $y = x^2 - 4$

20. (97) Refer to this drawing of three similar triangles to find the letter that completes the proportion below.

$$\frac{c}{a} = \frac{a}{?}$$

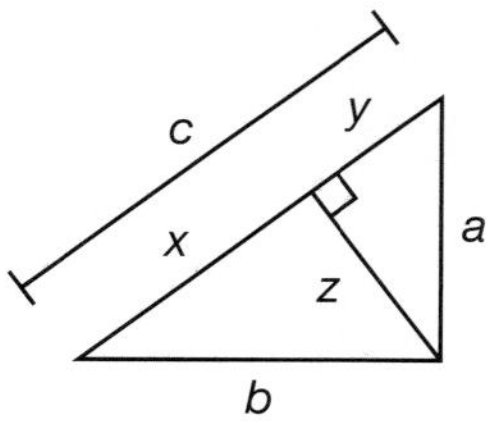

*** 21.** (105) **Analyze** Recall that the surface area of a sphere is four times the area of its largest "cross section." What is the approximate surface area of a cantaloupe that is 6 inches in diameter? Round the answer to the nearest square inch. (Use 3.14 for π.)

*** 22.** (114) **Connect** A cup containing 250 cubic centimeters of water holds how many liters of water?

Solve:

23. (102) $15 + x = 3x - 17$

24. (93) $3\frac{1}{3}x - 16 = 74$

25. (109) $\frac{m^2}{4} = 9$

26. (98) $\frac{1.2}{m} = \frac{0.04}{8}$

Simplify:

27. (96) $x(x - 5) - 2(x - 5)$

28. (103) $\frac{(3xy)(4x^2y)(5x^2y^2)}{10x^3y^3}$

29. (91) $|-8| + 3(-7) - [(-4)(-5) - 3(-2)]$

30. (43, 45) $\frac{7\frac{1}{2} - \frac{2}{3}(0.9)}{0.03}$

Focus on

• Platonic Solids

Recall that polygons are closed, two-dimensional figures with straight sides. If every face of a solid figure is a polygon, then the solid figure is called a **polyhedron.** Thus polyhedrons do not have any curved surfaces. So rectangular prisms and pyramids are polyhedrons, but spheres and circular cylinders are not.

Remember also that regular polygons have sides of equal length and angles of equal measure. Just as there are regular polygons, so there are regular polyhedrons. A cube is one example of a regular polyhedron. All the edges of a cube are of equal length, and all the angles are of equal measure, so all the faces are congruent regular polygons.

There are five regular polyhedrons. These polyhedrons are known as the **Platonic solids,** named after the ancient Greek philosopher Plato. We illustrate the five Platonic solids below.

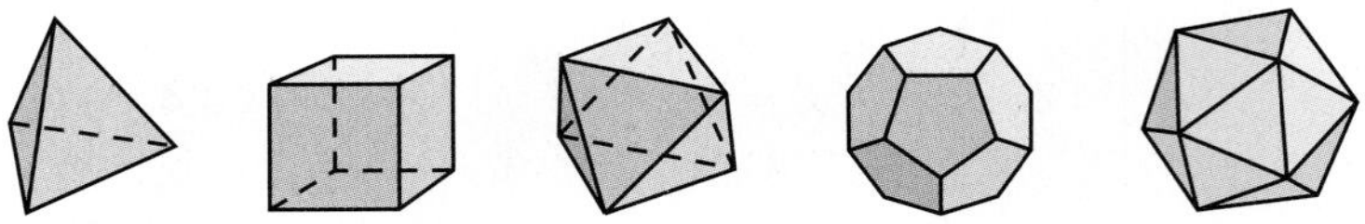

tetrahedron cube octahedron dodecahedron icosahedron

In this activity we will construct models of four of the Platonic solids.

Activity

Platonic Solids

Materials needed:

- copies of Investigation Activity 26 and Investigation Activity 27
- ruler
- scissors
- glue or tape

Working in pairs or small groups is helpful. Sometimes more than two hands are needed to fold and glue.

Beginning with the tetrahedron pattern, cut around the border of the pattern. The line segments in the pattern are fold lines. Do not cut these. The folds will become the edges of the polyhedron. The triangles marked with "T" are tabs for gluing. These tabs are tucked inside the polyhedron and are hidden from view when the polyhedron is finished.

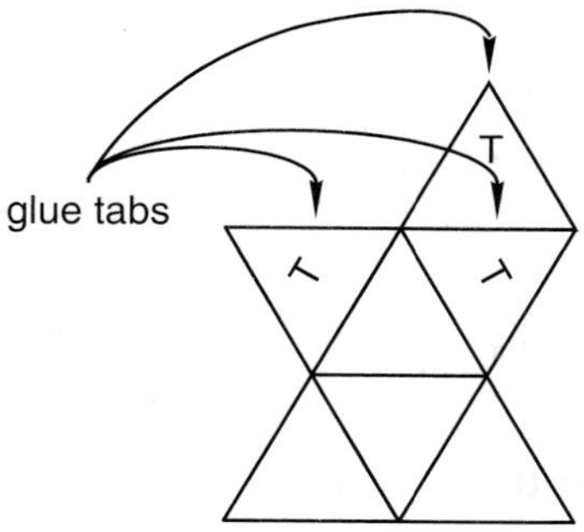

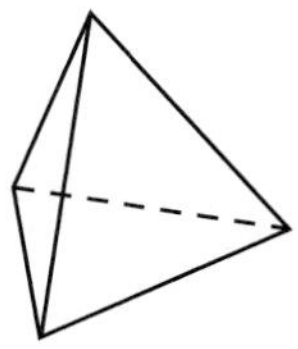

tetrahedron

Fold the pattern to make a pyramid with four faces. Glue the tabs or tape across the joining edges to hold the pattern in place.

When you have completed the tetrahedron, select another pattern to cut, fold, and form. All tabs are hidden when the pattern is properly folded, but all other polygons should be fully visible. When you have completed the models, copy this table and fill in the missing information by studying your models.

Thinking Skill

Generalize

For each polyhedron in this table, add together the number of faces (*F*) and vertices (*V*). Then find a relationship between this sum and the number of edges (*E*) in the polyhedron.

Platonic Solid	Each Face is What Polygon?	How Many Faces?	How Many Vertices?	How Many Edges?
tetrahedron	equilateral triangle	4		
cube	square	6	8	12
octahedron				
dodecahedron				

extensions

a. This arrangement of four equilateral triangles was folded to make a model of a tetrahedron. Draw another arrangement of four adjoining equilateral triangles that can be folded to make a tetrahedron model. (Omit tabs.)

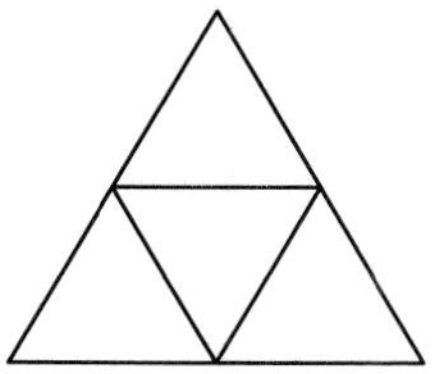

b. This arrangement of six squares was folded to make a model of a cube. How many other different patterns of six adjacent squares can you draw that can be folded to make a model of a cube? (Omit tabs.)

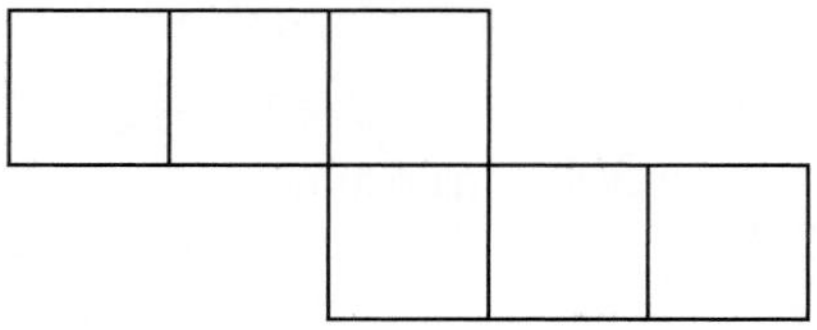

c. Using scissors and glue, cut out and construct the icosahedron model on **Instruction Activity 28** Icosahedon. Working in pairs or in small groups is helpful. We suggest pre-folding the pattern before making the cuts to separate the tabs. Remember that the triangles marked with a "T" are tabs and should be hidden from view when the model is finished.

Once you have constructed the model, hold it lightly between your thumb and forefinger. You should be able to rotate the icosahedron while it is in this position. Since your icosahedron is a regular polyhedron, you also should be able to reposition the model so that your fingers touch two different vertices but the appearance of the figure remains unchanged.

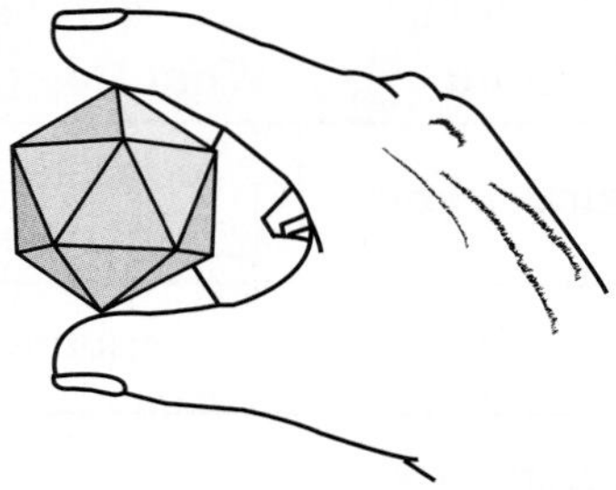

Holding the model as shown, your thumb and finger each touch a vertex. As you rotate the icosahedron, you can count the vertices. How many vertices are there in all? How many faces are there in all? What is the shape of each face?

A

absolute value
valor absoluto
(59)

The distance from the graph of a number to the number 0 on a number line. The symbol for absolute value is a vertical bar on each side of a numeral or variable, e.g., $|-x|$.

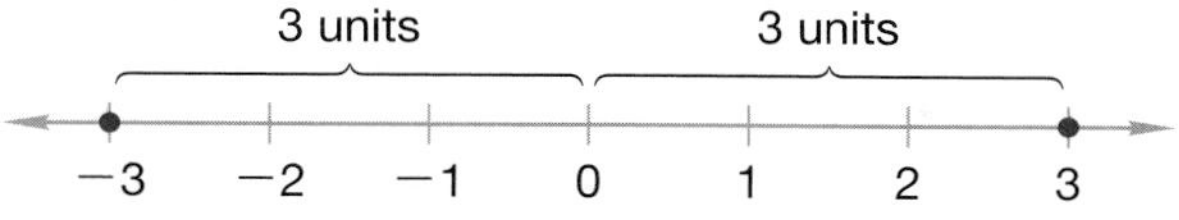

$|+3| = |-3| = 3$

Since the graphs of −3 and +3 are both 3 units from the number 0, the ***absolute value*** *of both numbers is 3.*

acute angle
ángulo agudo
(7)

An angle whose measure is between 0° and 90°.

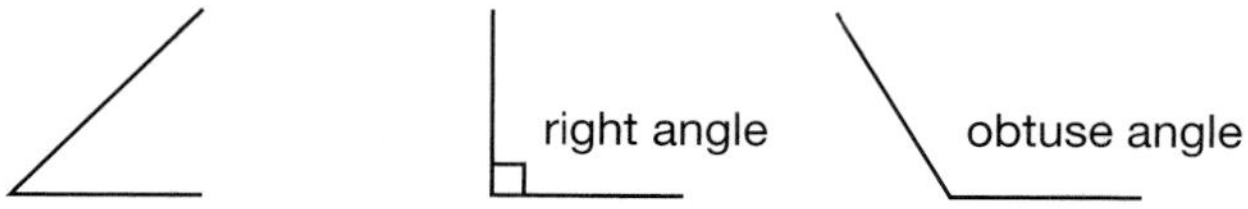

acute angle not **acute angles**

An ***acute angle*** *is smaller than both a right angle and an obtuse angle.*

acute triangle
triángulo acutángulo
(62)

A triangle whose largest angle measures less than 90°.

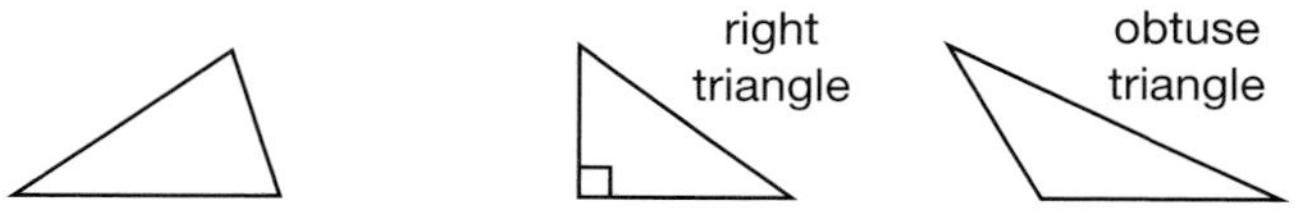

acute triangle not **acute triangles**

addend
sumando
(1)

One of two or more numbers that are added to find a sum.

$7 + 3 = 10$ *The* ***addends*** *in this problem are 7 and 3.*

additive identity
identidad aditiva
(2)

The number 0. *See also* **identity property of addition.**

$$7 + 0 = 7$$

↑

additive identity

We call zero the ***additive identity*** *because adding zero to any number does not change the number.*

adjacent angles
ángulos adyacentes
(40)

Two angles that have a common side and a common vertex. The angles lie on opposite sides of their common side.

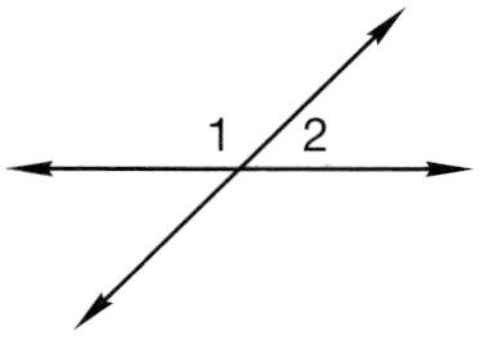

∠1 and ∠2 ***are adjacent angles.*** *They share a common side and a common vertex.*

adjacent sides
lados adyacentes
(Inv. 2)

In a polygon, two sides that intersect to form a vertex.

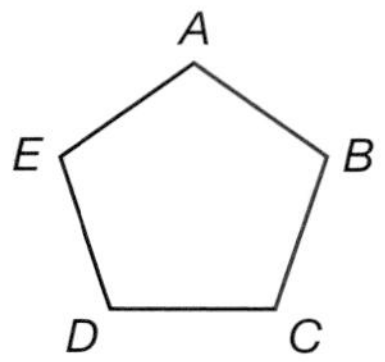

$\overline{AB}$ *and* $\overline{BC}$ *are* ***adjacent sides.*** *They form vertex B.*

algebraic addition
suma algebraica
(68)

The combining of positive and negative numbers to form a sum.

We use ***algebraic addition*** *to find the sum of* -3, $+2$, *and* -11:

$$(-3) + (+2) + (-11) = -12$$

---

alternate exterior angles
ángulos alternos externos
(102)

A special pair of angles formed when a transversal intersects two lines. Alternate exterior angles lie on opposite sides of the transversal and are outside the two intersected lines.

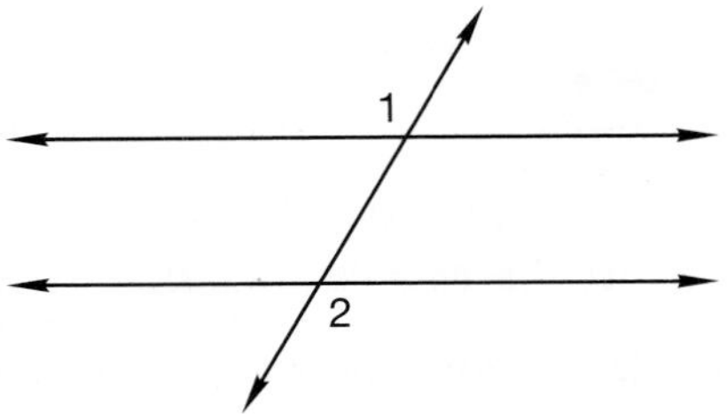

∠1 and ∠2 are ***alternate exterior angles.*** *When a transversal intersects parallel lines, as in this figure,* ***alternate exterior angles*** *have the same measure.*

---

alternate interior angles
ángulos alternos internos
(102)

A special pair of angles formed when a transversal intersects two lines. Alternate interior angles lie on opposite sides of the transversal and are inside the two intersected lines.

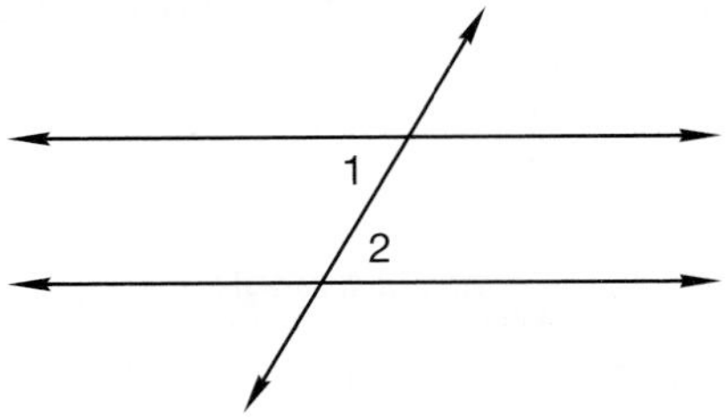

∠1 and ∠2 are ***alternate interior angles.*** *When a transversal intersects parallel lines, as in this figure,* ***alternate interior angles*** *have the same measure.*

---

altitude
altura
(37)

The perpendicular distance from the base of a triangle to the opposite vertex; also called *height.*

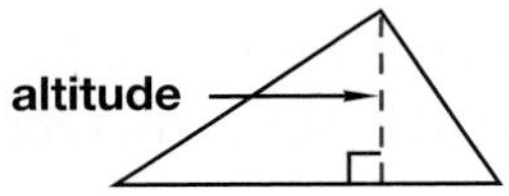

---

angle
ángulo
(7)

The opening that is formed when two lines, rays, or segments intersect.

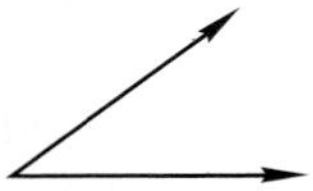

These rays form an ***angle.***

angle bisector A line, ray, or line segment that divides an angle into two equal halves.
bisectriz
(Inv. 10)

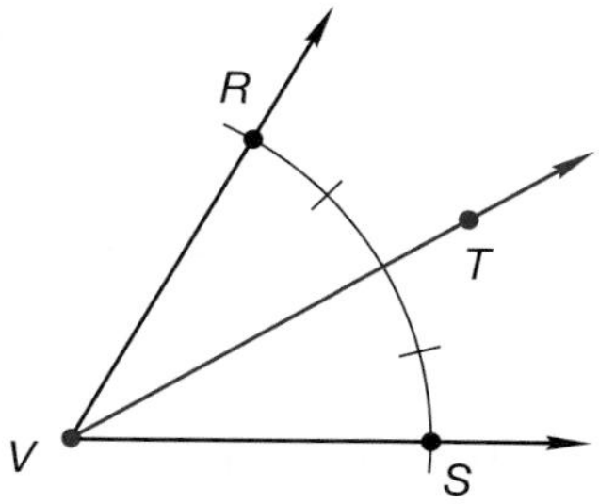

*$\overrightarrow{VT}$ is an **angle bisector.** It divides ∠RVS into two equal halves.*

arc Part of a circle.
arco
(Inv. 2)

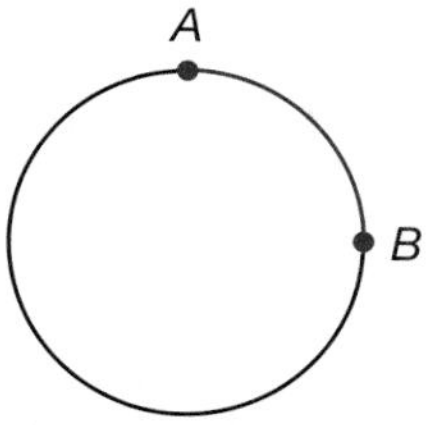

*The portion of the circle between points A and B is **arc** AB.*

area The size of the inside of a flat shape. Area is measured in square units.
área
(20)

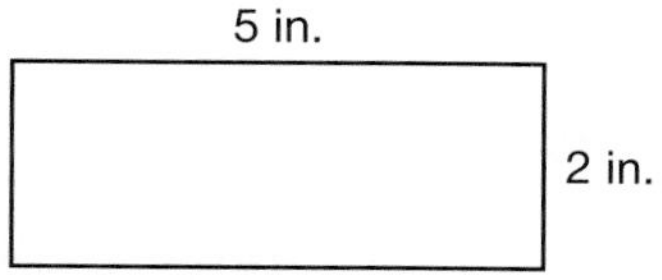

*The **area** of this rectangle is 10 square inches.*

Associative Property of Addition The grouping of addends does not affect their sum. In symbolic form, $a + (b + c) = (a + b) + c$. Unlike addition, subtraction is not associative.
propiedad asociativa de la suma
(2)

$(8 + 4) + 2 = 8 + (4 + 2)$	$(8 - 4) - 2 \neq 8 - (4 - 2)$
*Addition is **associative.***	*Subtraction is not **associative.***

Associative Property of Multiplication The grouping of factors does not affect their product. In symbolic form, $a \times (b \times c) = (a \times b) \times c$. Unlike multiplication, division is not associative.
propiedad asociativa de la multiplicación
(2)

$(8 \times 4) \times 2 = 8 \times (4 \times 2)$	$(8 \div 4) \div 2 \neq 8 \div (4 \div 2)$
*Multiplication is **associative.***	*Division is not **associative.***

average The number found when the sum of two or more numbers is divided by the number of addends in the sum; also called *mean.*
promedio
(28)

*To find the **average** of the numbers 5, 6, and 10, add.*

$$5 + 6 + 10 = 21$$

There were three addends, so divide the sum by 3.

$$21 \div 3 = 7$$

*The **average** of 5, 6, and 10 is 7.*

B

base *base* *(20, 37)*

1. A designated side (or face) of a geometric figure.

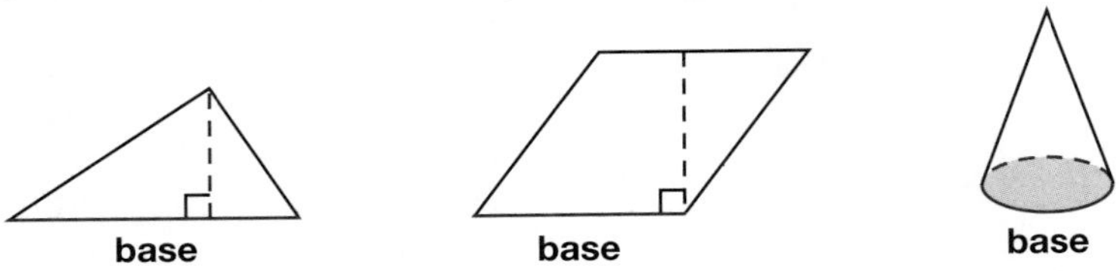

2. The lower number in an exponential expression.

base → 5^3 ← exponent

5^3 means $5 \times 5 \times 5$, and its value is 125.

---

bisect *bisecar* *(Inv. 10)*

To divide a segment or angle into two equal halves.

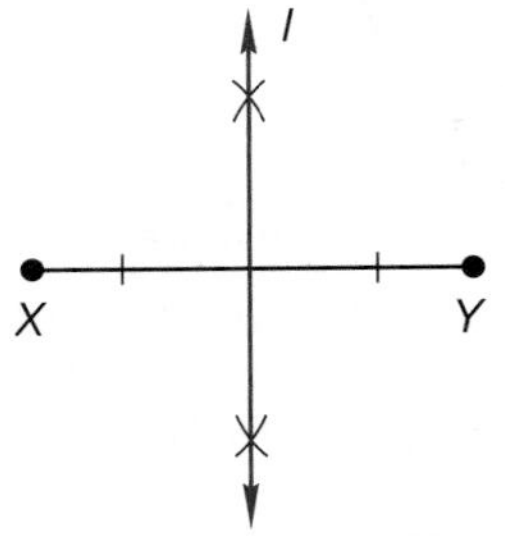

*Line l **bisects** $\overline{XY}$.*

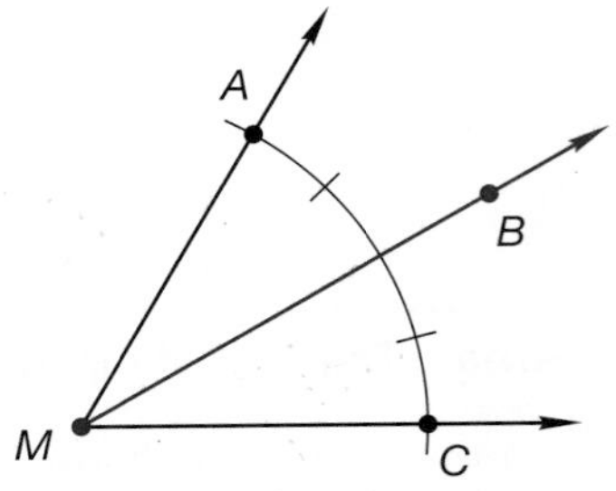

*Ray MB **bisects** ∠AMC.*

---

box-and-whisker plot *gráfica de frecuencias acumuladas* *(Inv. 4)*

A method of displaying data that involves splitting the numbers into four groups of equal size.

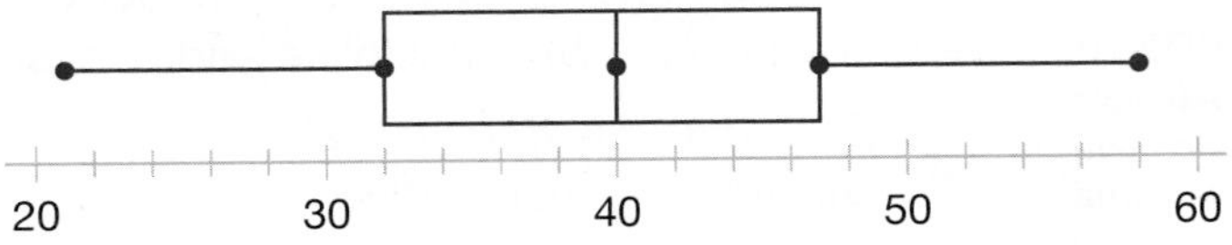

box-and-whisker plot

C

cancel (canceling) *cancelar* *(24)*

The process of reducing a fraction by matching equivalent factors from both the numerator and denominator.

$$\frac{14}{28} = \frac{\cancel{7} \cdot \cancel{2}}{\cancel{7} \cdot \cancel{2} \cdot 2} = \frac{1}{2}$$

Celsius *Celsius* *(32)*

Method of temperature measurement where 0° is the temperature for freezing water and 100° is the temperature for boiling water.

---

center *centro* *(Inv. 2)*

The point inside a circle or sphere from which all points on the circle or sphere are equally distant.

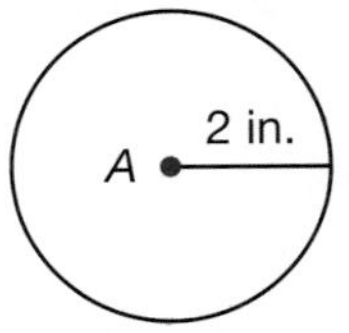

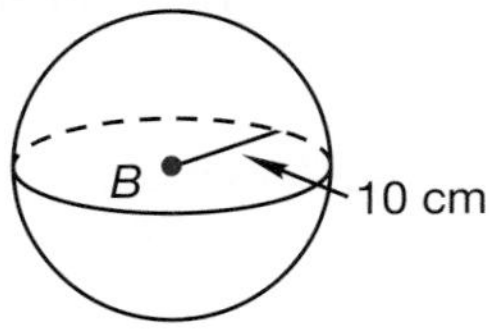

*The **center** of circle A is 2 inches from every point on the circle.*
*The **center** of sphere B is 10 centimeters from every point on the sphere.*

central angle *ángulo central* *(Inv. 2)*

An angle whose vertex is the center of a circle.

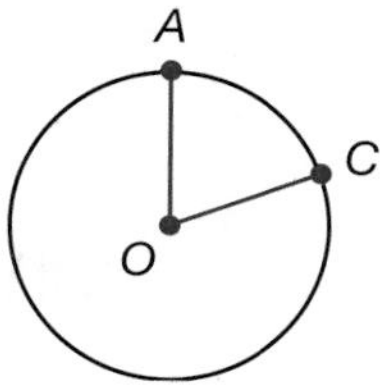

$\angle AOC$ is a ***central angle.***

chance *posibilidad* *(Inv. 8)*

A way of expressing the likelihood of an event; the probability of an event expressed as a percent.

The ***chance*** *of snow is 10%. It is not likely to snow.*
There is an 80% ***chance*** *of rain. It is likely to rain.*

chord *cuerda* *(Inv. 2)*

A segment whose endpoints lie on a circle.

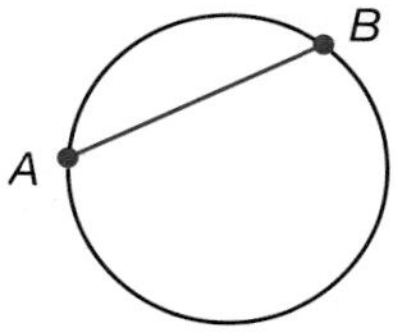

$\overline{AB}$ is a ***chord*** *of the circle.*

circle *círculo* *(Inv. 2)*

A closed, curved shape in which all points on the shape are the same distance from its center.

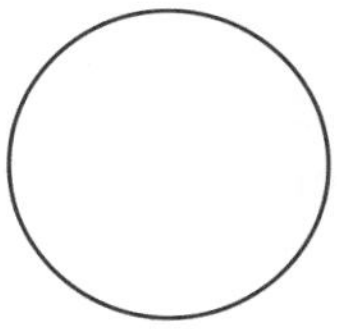

circle

circle graph *gráfica circular* *(Inv. 5)*

A method of displaying data, often used to show information about percentages or parts of a whole. A circle graph is made of a circle divided into sectors.

Class Test Grades

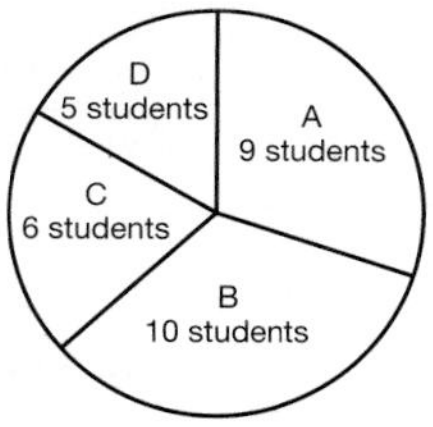

This ***circle graph*** *shows data for a class's test grades.*

circumference *circunferencia* *(Inv. 2)*

The perimeter of a circle.

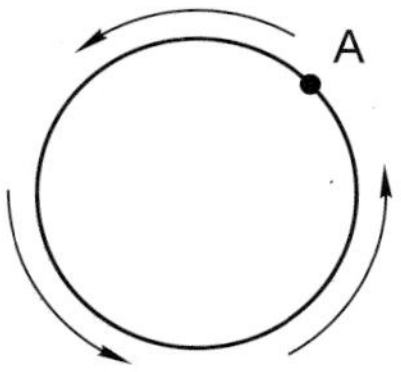

If the distance from point A around to point A is 3 inches, then the ***circumference*** *of the circle is 3 inches.*

coefficient *coeficiente* *(Inv. 7)*

In common use, the number that multiplies the variable(s) in an algebraic term. If no number is specified, the coefficient is 1.

In the term $-3x$, the ***coefficient*** *is -3.*
In the term y^2, the ***coefficient*** *is 1.*

common denominator
denominator común
(9)

A common multiple of the denominators of two or more fractions.

A common denominator of $\frac{5}{6}$ and $\frac{3}{8}$ is a common multiple of 6 and 8, such as 24, 48 and 72.

Commutative Property of Addition
propiedad conmutativa de la suma
(2)

Changing the order of addends does not change their sum. In symbolic form, $a + b = b + a$. Unlike addition, subtraction is not commutative.

$8 + 2 = 2 + 8$	$8 - 2 \neq 2 - 8$
*Addition is **commutative.***	*Subtraction is not **commutative.***

Commutative Property of Multiplication
propiedad conmutativa de la multiplicación
(2)

Changing the order of factors does not change their product. In symbolic form, $a \times b = b \times a$. Unlike multiplication, division is not commutative.

$8 \times 2 = 2 \times 8$	$8 \div 2 \neq 2 \div 8$
*Multiplication is **commutative.***	*Division is not **commutative.***

compare
comparar
(4)

Looking at two numbers to find out if one number is greater than, less than, or equal to another number. This can be done using the number line.

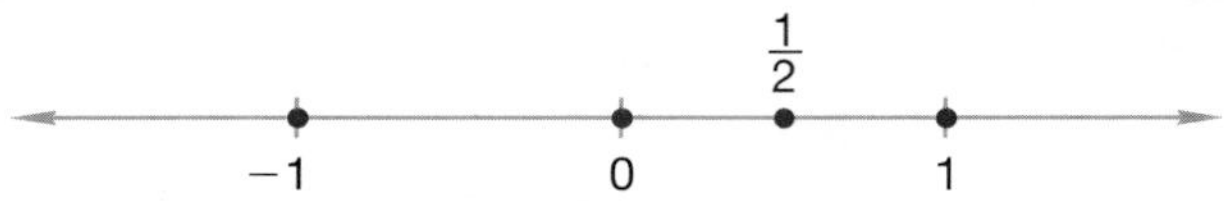

$\frac{1}{2}$ is less than 1 and 0 is greater than −1

comparison symbol
símbolo de comparación
(4)

The symbol used to show the comparison of two numbers: greater than (>), less than (<), or equal (=). The pointed end of the symbol points to the lesser number. For example, $4 < 6$ and $8 > 4$.

compass
compás
(Inv. 2)

A tool used to draw circles and arcs.

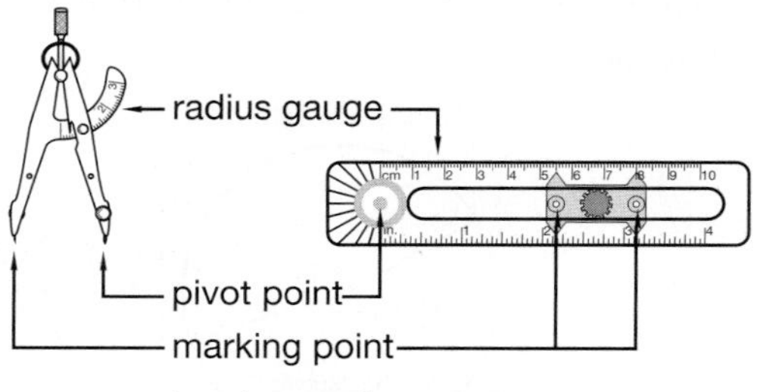

two types of **compasses**

complementary angles
ángulos complementarios
(40)

Two angles whose sum is 90°.

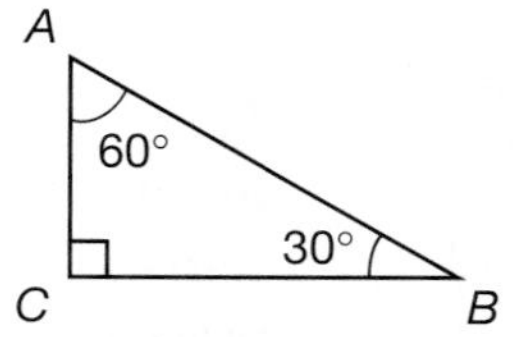

*∠A and ∠B are **complementary angles.***

complement of an event
complemento de un evento
(14, Inv. 8)

In probability, the opposite of an event. The complement of event B is "not B." The probabilities of an event and its complement total one.

complex fraction
fracción compleja
(76)

A fraction that contains one or more fractions in its numerator or denominator.

$\frac{\frac{3}{5}}{\frac{2}{3}}$ $\quad \frac{25\frac{2}{3}}{100}$ $\quad \frac{15}{7\frac{1}{3}}$ $\quad \frac{\frac{a}{b}}{\frac{b}{c}}$ $\qquad \frac{1}{2}$ $\quad \frac{12}{101}$ $\quad \frac{xy}{z}$

complex fractions — not **complex fractions**

composite number
número compuesto
(21)

A counting number greater than 1 that is divisible by a number other than itself and 1. Every composite number has three or more factors.

9 is divisible by 1, 3, and 9. It is ***composite.***
11 is divisible by 1 and 11. It is not ***composite.***

compound event
evento compuesto
(Inv. 8)

In probability, the result of combining two or more simple events.

An outcome of one coin flip is a simple event. An outcome of more than one flip is a ***compound event.***

compound interest
interés compuesto
(110)

Interest that pays on principal and previously earned interest.

Compound Interest

\$100.00	*principal*
+ \$6.00	*first-year interest (6% of \$100.00)*
\$106.00	*total after one year*
+ \$6.36	*second-year interest (6% of \$106.00)*
\$112.36	*total after two years*

Simple Interest

\$100.00	*principal*
\$6.00	*first-year interest*
+ \$6.00	*second-year interest*
\$112.00	*total after two years*

concentric circles
círculos concéntricos
(Inv. 2)

Two or more circles with a common center.

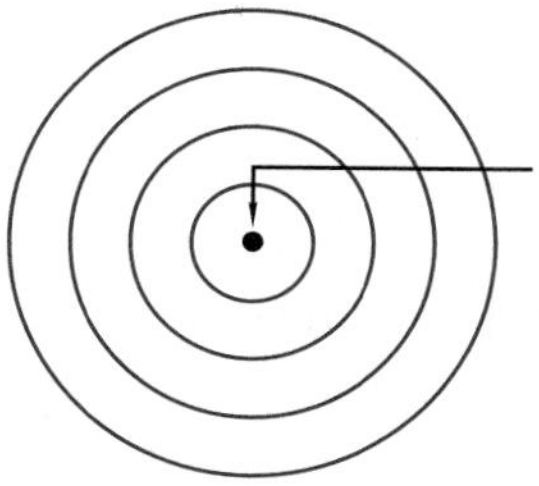

congruent
congruente
(18)

Having the same size and shape.

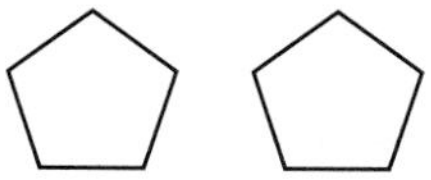

These polygons are ***congruent.*** *They have the same size and shape.*

constant
constante
(65)

A number whose value does not change.

In the expression $2\pi r$*, the numbers 2 and* π *are* ***constants,*** *while r is a variable.*

constant term
término constante
(84)

The term in a polynomial that is a number not multiplying with a variable. If there is no constant term in a polynomial, assume that it is zero.

$$3x^3 + 2x^2 + 5$$

term ↑ ($3x^3$) term ↑ ($2x^2$) **constant term** ↖ (5)

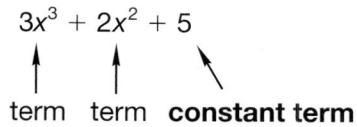

coordinate(s)
coordenada(s)
(Inv. 3)

1. A number used to locate a point on a number line.

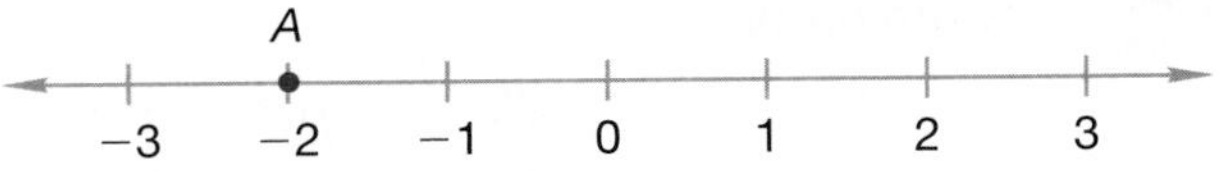

*The **coordinate** of point A is −2.*

2. An ordered pair of numbers used to locate a point in a coordinate plane.

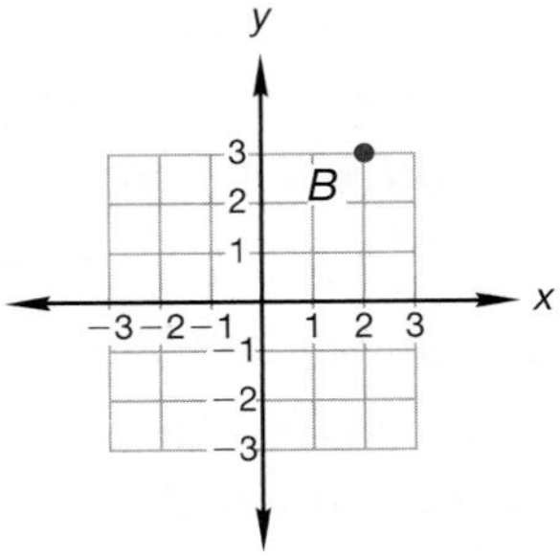

*The **coordinates** of point B are (2, 3). The x-coordinate is listed first, the y-coordinate second.*

---

coordinate plane
plano coordenado
(Inv. 3)

A grid on which any point can be identified by an ordered pair of numbers.

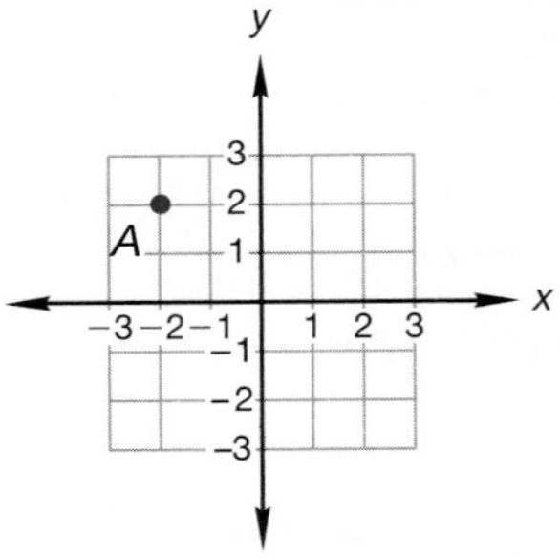

*Point A is located at (−2, 2) on this **coordinate plane.***

---

corresponding angles
ángulos correspondientes
(102)

A special pair of angles formed when a transversal intersects two lines. Corresponding angles lie on the same side of the transversal and are in the same position relative to the two intersected lines.

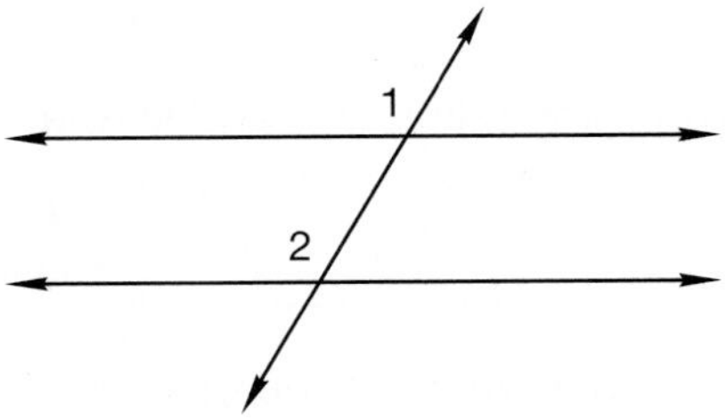

*∠1 and ∠2 are **corresponding angles.** When a transversal intersects parallel lines, as in this figure, **corresponding angles** have the same measure.*

---

corresponding parts
partes correspondientes
(18)

Sides or angles of similar polygons that occupy the same relative positions.

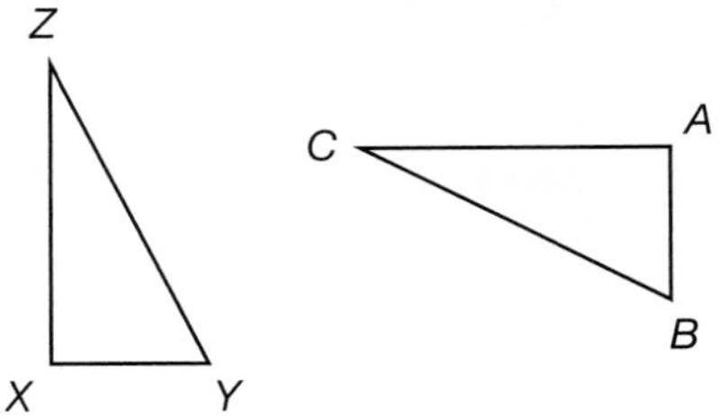

*$\overline{BC}$ **corresponds** to $\overline{YZ}$.*
*∠A **corresponds** to ∠X.*

counting numbers
números de conteo
(1)

The numbers used to count; the members of the set {1, 2, 3, 4, 5, ...}. Also called *natural numbers.*

1, 24, and 108 are ***counting numbers.***
-2, 3.14, 0, and $2\frac{7}{9}$ are not ***counting numbers.***

cross product
productos cruzados
(39)

The product of the numerator of one fraction and the denominator of another.

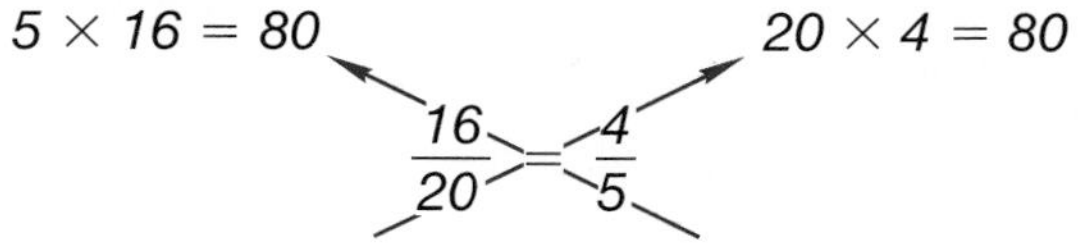

The ***cross products*** *of these two fractions are equal.*

D

decimal number
número decimal
(31)

A numeral that contains a decimal point, sometimes called a decimal fraction or a decimal.

23.94 is a ***decimal number*** *because it contains a decimal point.*

decimal point
punto decimal
(1)

The symbol in a decimal number used as a reference point for place value.

34.15
↑
decimal point

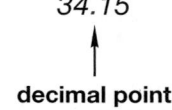

degree (°)
grado
(16, 17)

1. A unit for measuring angles.

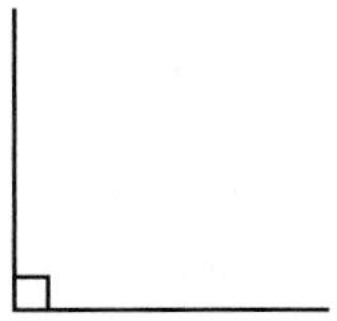

There are 90 ***degrees*** *(90°) in a right angle.*

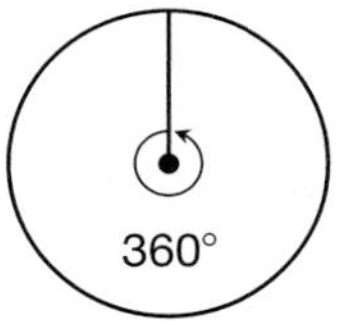

There are 360 ***degrees*** *(360°) in a circle.*

2. A unit for measuring temperature.

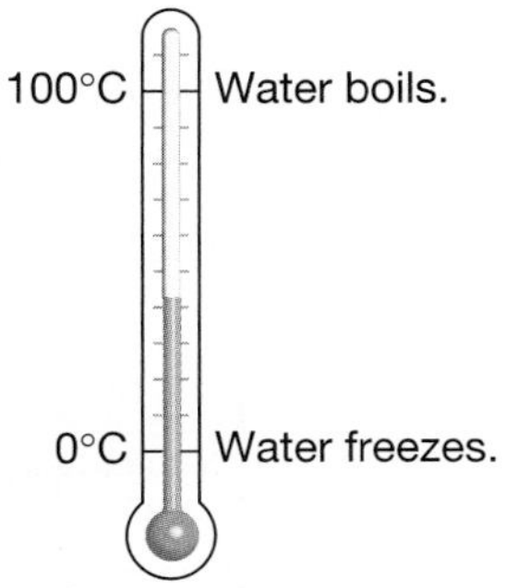

There are 100 ***degrees*** *between the freezing and boiling points of water on the Celsius scale.*

denominator
denominador
(8)

The bottom term of a fraction.

5 ← numerator
9 ← **denominator**

dependent events
eventos dependientes
(94)

In probability, events that are not independent because the outcome of one event affects the probability of the other event.

*If a bag contains 4 red marbles and 2 blue marbles and a marble is drawn from the bag twice without replacing the first draw, then the probabilities for the second draw is **dependent** upon the outcome of the first draw.*

---

dependent variable
variable dependiente
(120)

A variable whose value is determined by the value of one or more other variables.

In the equation y = 2x, *the dependent variable is* y *because its value depends upon the value chosen for* x.

---

diagonal
diagonal
(Inv. 6)

A line segment, other than a side, that connects two vertices of a polygon.

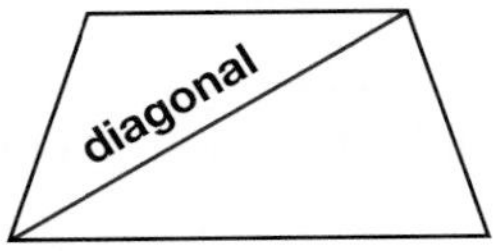

---

diameter
diámetro
(Inv. 2)

The distance across a circle through its center.

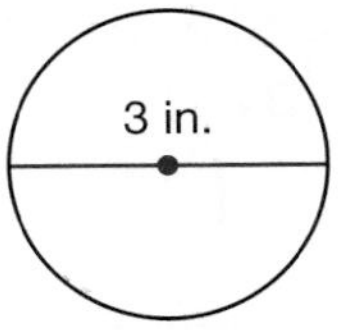

*The **diameter** of this circle is 3 inches.*

---

difference
diferencia
(1)

The result of subtraction.

*In 12 − 8 = 4, the **difference** is 4.*

---

digit
dígito
(5)

Any of the symbols used to write numbers: 0, 1, 2, 3, 4, 5, 6, 7, 8, 9.

*The last **digit** in the number 7862 is 2.*

---

direct variation
variación directa
(Inv. 9)

A relationship between two variables in which one variable is a constant multiple of the other, $y = kx$; also known as **direct proportion** because any two pairs of values form a proportion.

A graph of direct variation is a line or ray that intersects the origin.

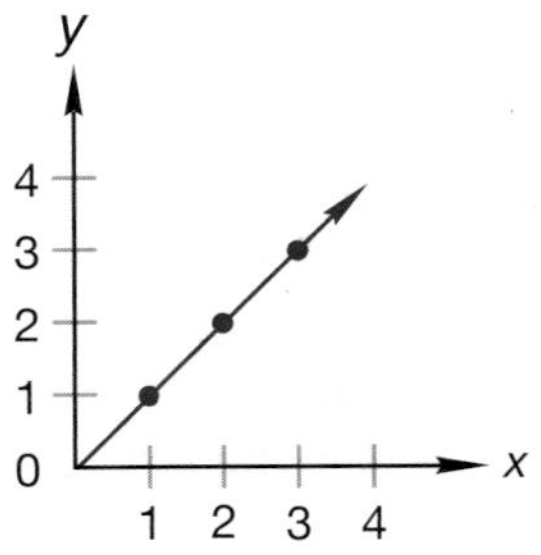

A function table displays direct variation if the x- *and* y-*values share a common ratio.*

x	y	$\frac{y}{x}$
1	2	$\frac{2}{1} = 2$
2	4	$\frac{4}{2} = 2$
3	6	$\frac{6}{3} = 2$
4	8	$\frac{8}{4} = 2$

---

direct proportion
proporción directa
(Inv. 9)

See **direct variation.**

Distributive Property
propiedad distributiva
(41)

A number times the sum of two addends is equal to the sum of that same number times each individual addend: $a \times (b + c) = (a \times b) + (a \times c)$.

$8 \times (2 + 3) = (8 \times 2) + (8 \times 3)$

*Multiplication is **distributive** over addition.*

---

dividend
dividendo
(1)

A number that is divided.

$12 \div 3 = 4$ $\qquad$ $3\overline{)12}$ with quotient 4 $\qquad$ $\frac{12}{3} = 4$

*The **dividend** is 12 in each of these problems.*

---

divisible
divisible
(6)

Able to be divided by a whole number without a remainder.

$4\overline{)20}$ with quotient 5 $\qquad$ *The number 20 is **divisible** by 4, since $20 \div 4$ has no remainder.*

$3\overline{)20}$ with quotient $6\ R\ 2$ $\qquad$ *The number 20 is not **divisible** by 3, since $20 \div 3$ has a remainder.*

---

divisor
divisor
(1)

1. A number by which another number is divided.

$12 \div 3 = 4$ $\qquad$ $3\overline{)12}$ with quotient 4 $\qquad$ $\frac{12}{3} = 4$

*The **divisor** is 3 in each of these problems.*

2. A factor of a number.

*2 and 5 are **divisors** of 10.*

---

double-line graph
gráfica de doble línea
(Inv. 5)

A method of displaying a set of data, often used to compare two performances over time.

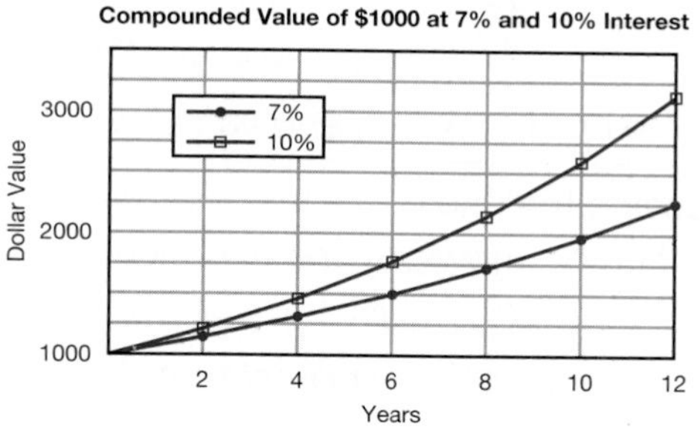

double-line graph

E

---

edge
arista
(67)

A line segment formed where two faces of a polyhedron intersect.

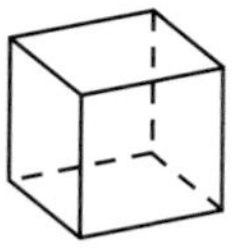

*One **edge** of this cube is blue.*
*A cube has 12 **edges.***

---

equation
ecuación
(3)

A statement that uses the symbol "=" to show that two quantities are equal.

$x = 3$	$3 + 7 = 10$	$4 + 1$	$x < 7$
equations		not **equations**	

---

equilateral triangle
triángulo equilátero
(62)

A triangle in which all sides are the same length.

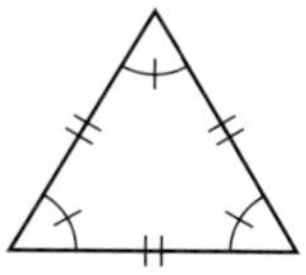

*This is an **equilateral triangle.***
All of its sides are the same length.

equivalent fractions *fracciones equivalentes* (15)
Different fractions that name the same amount.

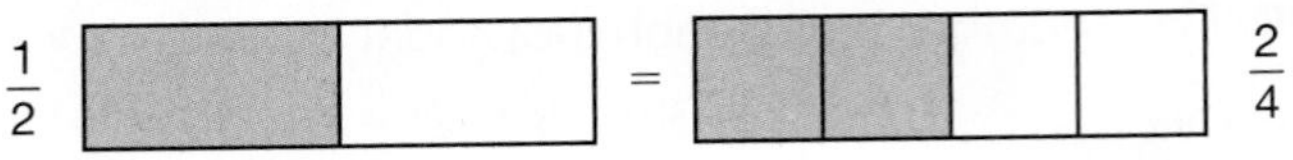

$\frac{1}{2}$ *and* $\frac{2}{4}$ *are* ***equivalent fractions.***

---

estimate *estimar* (29)
To determine an approximate value.

I ***estimate*** *that the sum of 199 and 205 is about 400.*

---

evaluate *evaluar* (1)
To find the value of an expression.

To ***evaluate*** $a + b$ *for* $a = 7$ *and* $b = 13$, *we replace* a *with 7 and* b *with 13:*

$$7 + 13 = 20$$

---

expanded notation *notación desarrollada* (5)
A way of writing a number as the sum of the products of the digits and the place values of the digits.

In ***expanded notation*** *6753 is written*

$$(6 \times 1000) + (7 \times 100) + (5 \times 10) + (3 \times 1).$$

---

experimental probability *probabilidad experimental* (Inv. 8)
The chances of an event happening as determined by repeated testing.

---

exponent *exponente* (20)
The upper number in an exponential expression that shows how many times the base is to be used as a factor.

base ⟶ 5^3 ⟵ ***exponent***

5^3 *means* $5 \times 5 \times 5$, *and its value is 125.*

---

exponential expression *expresión exponencial* (20)
An expression that indicates that the base is to be used as a factor the number of times shown by the exponent.

$$4^3 = 4 \times 4 \times 4 = 64$$

The ***exponential expression*** 4^3 *is evaluated by using 4 as a factor 3 times. Its value is 64.*

---

expression *expresión* (1)
A combination of numbers and/or variables by operations, but not including an equal or inequality sign.

	equation	inequality
$3x + 2y$ $(x - 1)^2$	$y = 3x - 1$	$x < 4$
expressions	not **expressions**	

---

exterior angle *ángulo externo* (89)
In a polygon, the supplementary angle of an interior angle.

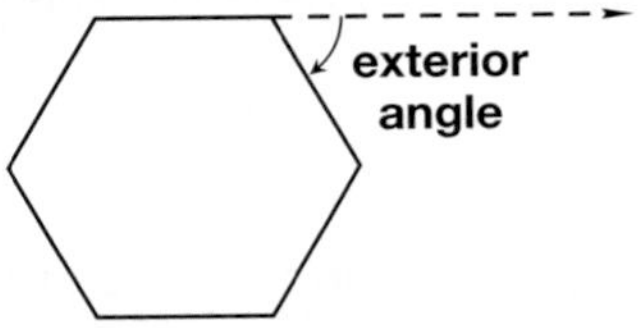

---

extremes *extremos* (Inv. 4)
The greatest value and the least value from a list of data.

F

face *cara* *(67)*

A flat surface of a geometric solid.

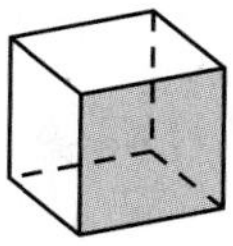

*One **face** of the cube is shaded.*
*A cube has six **faces.***

---

fact family *familia de operaciones* *(2)*

A group of three numbers related by addition and subtraction or by multiplication and division. Four mathematical fact statements can be formed using the numbers in a fact family.

*The numbers 3, 4, and 7 are a **fact family.** They make these four facts:*

$3 + 4 = 7$ $\quad 4 + 3 = 7$ $\quad 7 - 3 = 4$ $\quad 7 - 4 = 3$

---

factor *factor* *(1)*

1. Noun: One of two or more numbers that are multiplied.

$5 \times 6 = 30$ *The **factors** in this problem are 5 and 6.*

2. Noun: A whole number that divides another whole number without a remainder.

*The numbers 5 and 6 are **factors** of 30.*

3. Verb: To write as a product of factors.

*We can **factor** the number 6 by writing it as 2 × 3.*

---

factor tree *árbol de factores* *(21)*

Method for listing the prime factors of a number. Start by "branching" the original number with two factors (two numbers whose product equals the original number). Then branch those factors with two factors each. Continue branching until all of the prime factors of the original number are found.

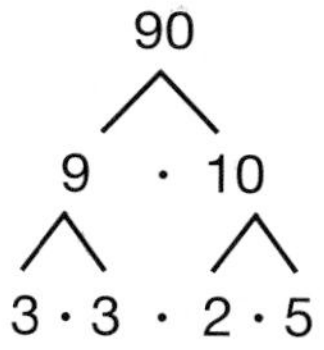

*By branching out each number into 2 factors in the **factor tree** we find the prime factorization.*

---

first quartile *primer cuartil* *(Inv. 4)*

In a set of data, the middle number of the lower half; the number below which 25% of the data lies; the 25th percentile; also called the **lower quartile.** In a box-and-whisker plot, the **first quartile** represents the left end of the box.

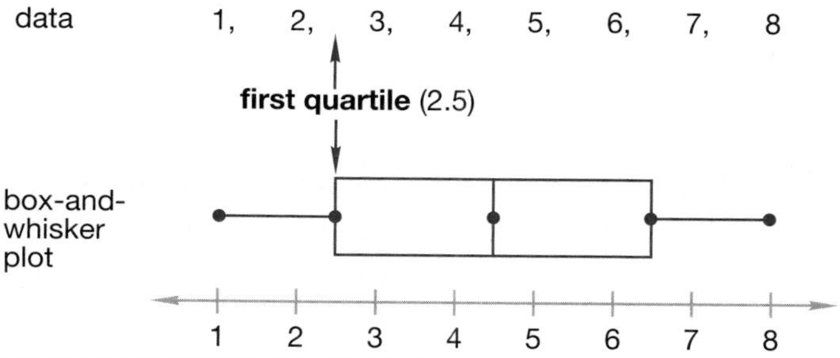

---

fraction *fracción* *(8)*

A number that names part of a whole.

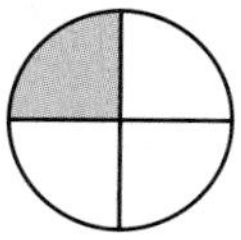

$\frac{1}{4}$ of the circle is shaded.
$\frac{1}{4}$ *is a **fraction.***

function *función* *(16)* A rule for using one number (an input) to calculate another number (an output). Each input produces only one output.

$y = 3x$

x	y
3	9
5	15
7	21
10	30

There is exactly one resulting number for every number we multiply by 3. Thus, $y = 3x$ is a ***function.***

Fundamental Counting Principle *principio fundamental de conteo* *(36)* The number of ways two or more events can occur is the product of the number of ways each event can occur.

There are 6 faces on a number cube and 2 sides of a coin. There are $6 \times 2 = 12$ outcomes of rolling a number cube and flipping a coin.

G

geometric sequence *secuencia geométrica* *(4)* A sequence whose terms share a common ratio. In the sequence {2, 4, 8, 16, 32,...} each term can be multiplied by 2 to find the next term. Thus the sequence is a **geometric sequence.**

geometric solid *sólido geométrico* *(67)* A three-dimensional geometric figure.

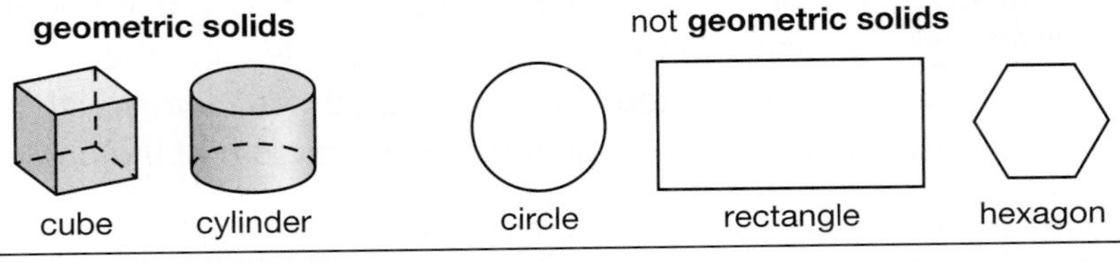

geometry *geometría* *(7)* A major branch of mathematics that deals with shapes, sizes, and other properties of figures.

Some figures we study in ***geometry*** *are angles, circles, and polygons.*

greatest common factor (GCF) *máximo común divisor (MCD)* *(6)* The largest whole number that is a factor of two or more indicated numbers.

The factors of 12 are 1, 2, 3, 4, 6, and 12.
The factors of 18 are 1, 2, 3, 6, 9, and 18.
The ***greatest common factor*** *of 12 and 18 is 6.*

H

height *altura* *(37)* The perpendicular distance from the base to the opposite side of a parallelogram or trapezoid; from the base to the opposite face of a prism or cylinder; or from the base to the opposite vertex of a triangle, pyramid, or cone. *See also* **altitude.**

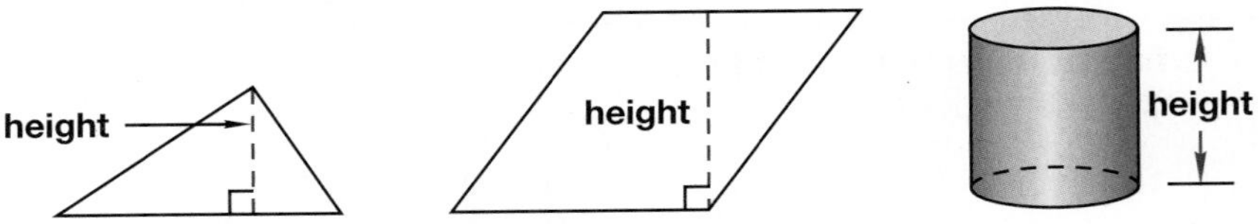

histogram
histograma
(Inv. 5)

A method of displaying a range of data. A histogram is a special type of bar graph that displays data in intervals of equal size with no space between bars.

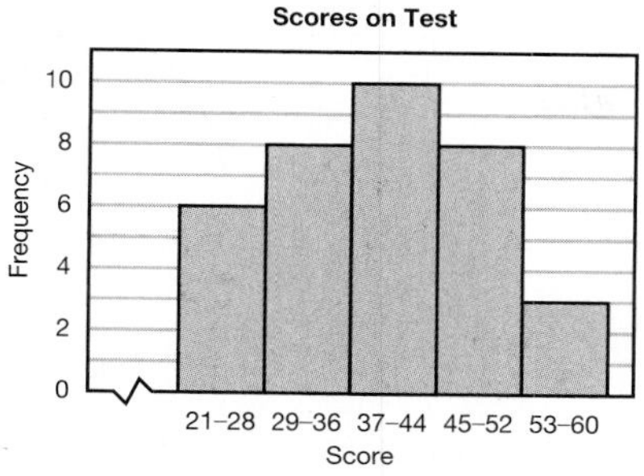

histogram

---

hypotenuse
hipotenusa
(99)

The longest side of a right triangle.

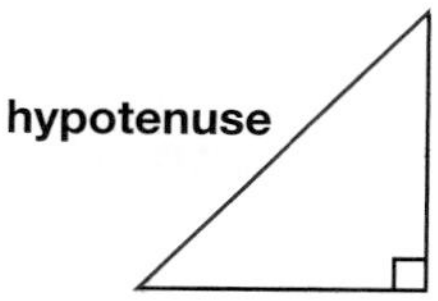

*The **hypotenuse** of a right triangle is always the side opposite the right angle.*

I

Identity Property of Addition
propiedad de identidad de la suma
(2)

The sum of any number and 0 is equal to the initial number. In symbolic form, $a + 0 = a$. The number 0 is referred to as the *additive identity.*

*The **identity property of addition** is shown by this statement:*

$$13 + 0 = 13$$

Identity Property of Multiplication
propiedad de identidad de la multiplicación
(2)

The product of any number and 1 is equal to the initial number. In symbolic form, $a \times 1 = a$. The number 1 is referred to as the *multiplicative identity.*

*The **identity property of multiplication** is shown by this statement:*

$$94 \times 1 = 94$$

---

improper fraction
fracción impropia
(10)

A fraction with a numerator equal to or greater than the denominator.

*$\frac{12}{12}$, $\frac{57}{3}$, and $2\frac{13}{2}$ are **improper fractions.***

*All **improper fractions** are greater than or equal to 1.*

---

independent events
eventos independientes
(Inv. 8)

Two events are *independent* if the outcome of one event does not affect the probability that the other event will occur.

*If a number cube is rolled twice, the outcome (1, 2, 3, 4, 5, or 6) of the first roll does not affect the probability of getting 1, 2, 3, 4, 5, or 6 on the second roll. The first and second rolls are **independent events.***

---

independent variable
variable independiente
(120)

The variable in an equation whose value can be chosen to determine the value of another variable.

*In $y = 2x$, the variable x is the **independent variable.***

---

inequalities
desigualdades
(78)

Algebraic statements that have $<$, $>$, $\le$, or $\ge$ as their symbols of comparison.

$x \le 4$	$2 < 7$	$11 \ge 10$	$x = 2$	$9 + 10$
	inequalities		not **inequalities**	

inscribed
inscrito
(Inv. 2)

A polygon is said to be *inscribed* within another shape if all points of the polygon lie within the other shape, and all of the polygon's vertices lie on the other shape.

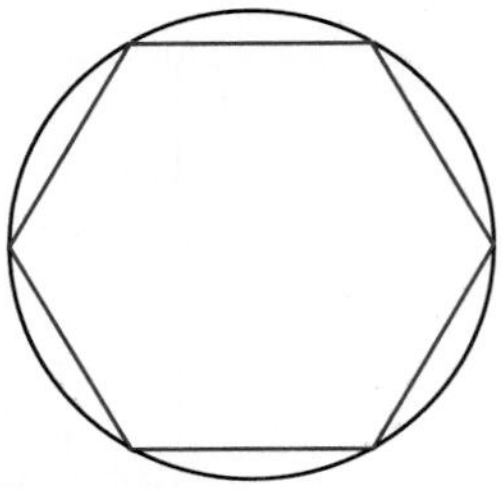

*The hexagon is **inscribed** within the circle.*

inscribed angle
ángulo inscrito
(Inv. 2)

An angle that opens to the interior of a circle and has its vertex on the circle.

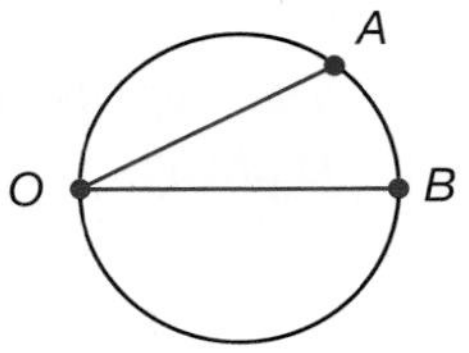

*$\angle AOB$ is an **inscribed angle.***

integers
números positivos, negativos y el cero
(4)

The set of counting numbers, their opposites, and zero; the members of the set {..., −2, −1, 0, 1, 2, ...}.

*−57 and 4 are **integers.** $\frac{15}{8}$ and −0.98 are not **integers.***

interest
interés
(110)

An additional amount added to a loan, account, or fund, usually based on a percentage of the principal; the difference between the principal and the total amount owed (with loans) or earned (with accounts and funds).

*If we borrow \$500.00 from the bank and repay the bank \$575.00 for the loan, the **interest** on the loan is \$575.00 − \$500.00 = \$75.00.*

interest rate
tasa de interés
(110)

A percent that determines the amount of interest paid on a loan over a period of time.

*If we borrow \$1000.00 and pay back \$1100.00 after one year, our **interest rate** is*

$$\frac{\$1100.00 - \$1000.00}{\$1000.00} \times 100\% = 10\% \text{ per year}$$

interior angle
ángulo interno
(89)

An angle that opens to the inside of a polygon.

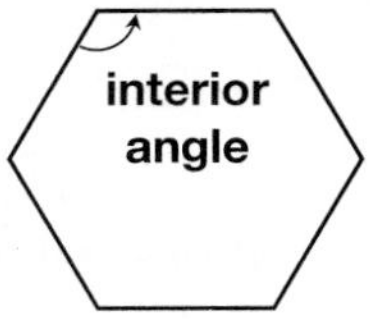

*This hexagon has six **interior angles.***

International System
Sistema internacional
(32)

See **metric system.**

interquartile range
intervalo entre cuartiles
(Inv. 4)

In a set of data, the difference between the upper and lower quartiles; the range of the middle half of the data.

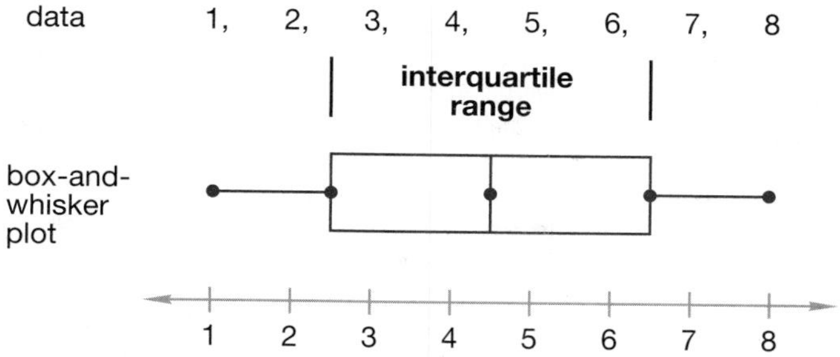

intersect
intersecar
(7)

To share a common point or points.

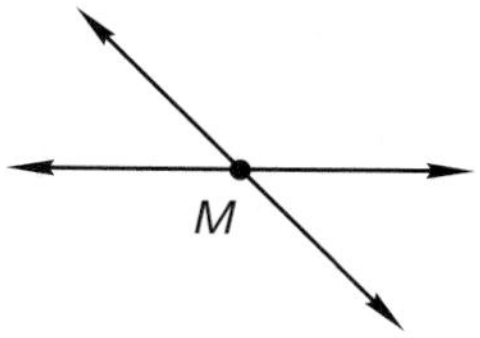

These two lines ***intersect.***
They share the common point M.

inverse operations
operaciones inversas
(2)

Operations that "undo" one another.

$a + b - b = a$
$a - b + b = a$

Addition and subtraction are ***inverse operations.***

$a \times b \div b = a \quad (b \neq 0)$
$a \div b \times b = a \quad (b \neq 0)$

Multiplication and division are ***inverse operations.***

$(\sqrt{a})^2 = a \quad (a \geq 0)$
$(\sqrt{a})^2 = a \quad (a \geq 0)$

Raising to powers and finding roots are ***inverse operations.***

Inverse Property of Multiplication
propiedad inversa de la multiplicación
(9)

The product of a number and its reciprocal equals 1. For example, $2 \times \frac{1}{2} = 1$ or $\frac{2}{3} \times \frac{3}{2} = 1$.

inverse variation
variación inversa
(Inv. 9)

A relationship between two variables such that their product is constant, $xy = k$; also known as inverse proportion.

A graph of inverse variation is a curve that does not intersect either axis.

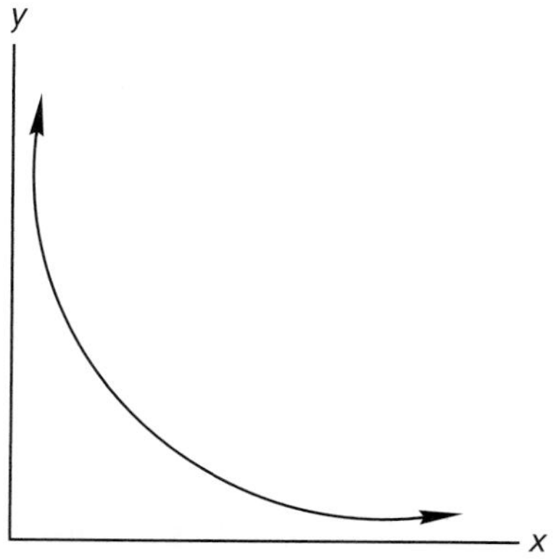

A function table displays inverse variation if the product of the variables is constant.

x	y	xy
3	20	60
4	15	60
5	12	60
6	10	60

invert
invertir
(9)

To switch the numerator and denominator of a fraction.

If we ***invert*** $\frac{7}{8}$*, we get* $\frac{8}{7}$.

irrational numbers
números irracionales
(100)

Numbers that cannot be expressed as a ratio of two integers. Their decimal expansions are nonending and nonrepeating.

π and $\sqrt{3}$ are ***irrational numbers.***

isosceles triangle
triángulo isósceles
(62)

A triangle with at least two sides of equal length.

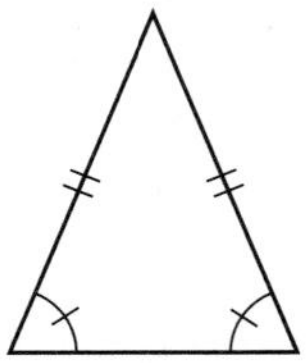

Two of the sides of this ***isosceles triangle*** *have equal lengths.*

L

least common denominator (LCD)
mínimo común denominador (mcd)
(30)

The least common multiple of the denominators of two or more fractions.

The ***least common denominator*** *of $\frac{5}{6}$ and $\frac{3}{8}$ is the least common multiple of 6 and 8, which is 24.*

least common multiple (LCM)
mínimo común múltiplo (mcm)
(27)

The smallest whole number that is a multiple of two or more given numbers.

Multiples of 6 are 6, 12, 18, 24, 30, 36, ...
Multiples of 8 are 8, 16, 24, 32, 40, 48, ...
The ***least common multiple*** *of 6 and 8 is 24.*

legs
catetos
(99)

The two shorter sides of a right triangle that form a 90° angle at their intersection.

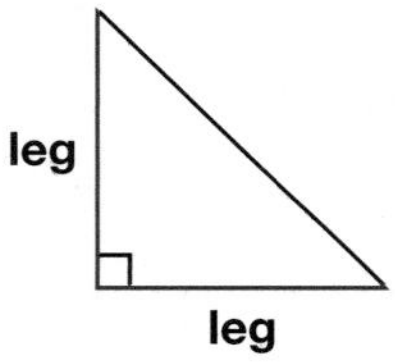

Each ***leg*** *of this right triangle is shorter than the hypotenuse.*

like terms
términos semejantes
(84)

Terms in a polynomial that share the same variable(s) and power(s).

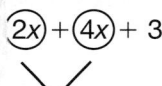

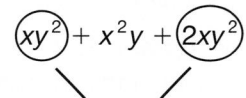

$2x + 4x + 3$ $\quad$ $xy^2 + x^2y + 2xy^2$

Like Terms $\quad$ ***Like Terms***

$x + xy + xyz$

No Like Terms

line
línea
(7)

A straight collection of points extending in opposite directions without end.

line *AB* or **line** *BA*

linear equation
ecuación lineal
(120)

An equation whose graph is a line.

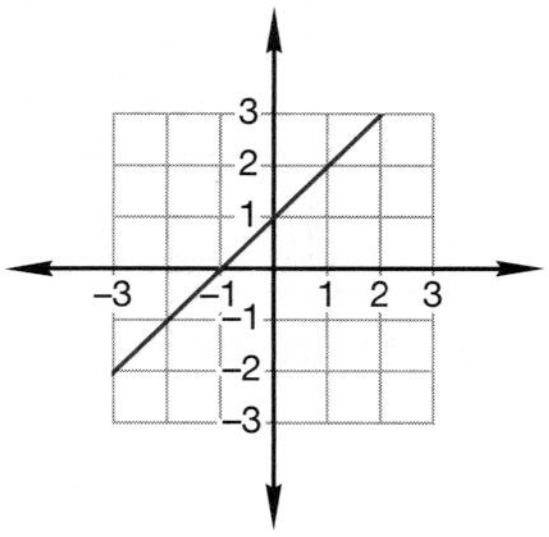

$y = x + 1$ is a ***linear equation*** *because its graph is a line.*

line of symmetry *línea de simetría* (58) — A line that divides a figure into two halves that are mirror images of each other.

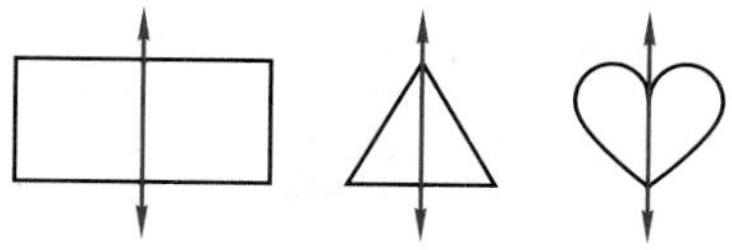

lines of symmetry

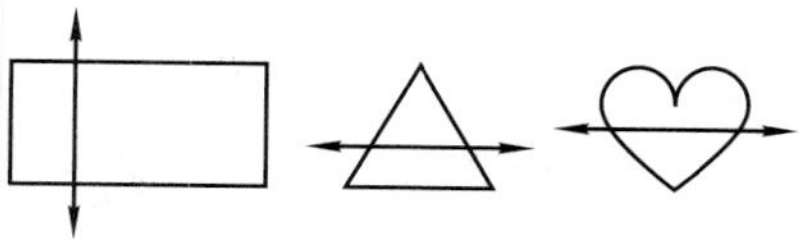

not **lines of symmetry**

lower quartile *cuartil inferior* (Inv. 4) — See **first quartile.**

lowest terms *mínima expresión* (15) — A fraction is in *lowest terms* if the only common factor of the numerator and the denominator is 1.

*When written in **lowest terms,** the fraction $\frac{8}{16}$ becomes $\frac{1}{2}$.*

M

major arc *arco mayor* (104) — An arc whose measure is between 180° and 360°.

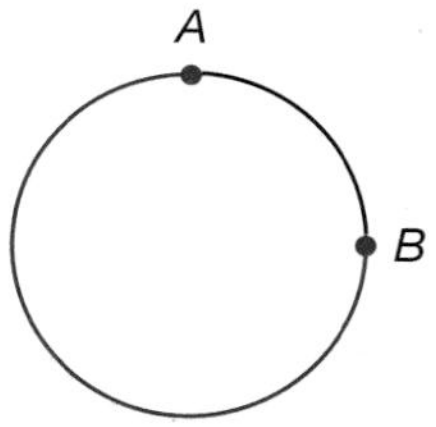

*The arc formed by moving counterclockwise from point A to point B is a **major arc.***

mean *media* (28) — See **average.**

median *mediana* (Inv. 4) — The middle number of a list of data when the numbers are arranged in order from the least to the greatest.

1, 1, 2, 5, 6, 7, 9, 15, 24, 36, 44

*In this list of data, 7 is the **median.***

metric system *sistema métrico* (32) — An international system of measurement based on multiples of ten. Also called *International System.*

*Centimeters and kilograms are units in the **metric system.***

minor arc *arco menor* (104) — An arc whose measure is between 0° and 180°.

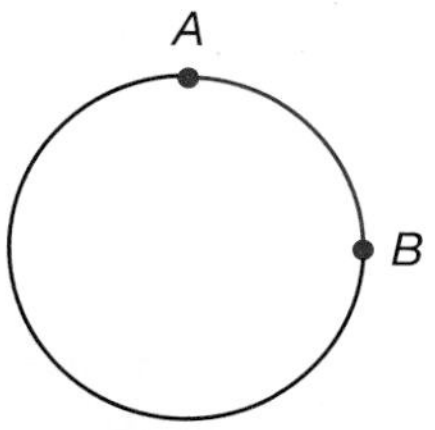

*The arc formed by moving clockwise from point A to point B is a **minor arc.***

minuend *minuendo* (1) — A number from which another number is subtracted.

*In $12 - 8 = 4$, the **minuend** is 12.*

mixed number *número mixto* (8) — A whole number and a fraction together.

*The **mixed number** $2\frac{1}{3}$ means "two and one third."*

mode
moda
(Inv. 4)

The number or numbers that appear most often in a list of data.

5, 12, 32, 5, 16, 5, 7, 12

In this list of data, the number 5 is the ***mode.***

---

monomial
monomio
(115)

An algebraic expression that contains only one term.

$3x$	$4ab$	$21mn$	$2 + a$	$x + y + z$	$2r + 3$
monomials			not **monomials**		

---

multiple
múltiplo
(27)

A product of a counting number and another number.

The ***multiples*** *of 3 include 3, 6, 9, and 12.*

---

multiplicative identity
identidad multiplicativa
(2)

The number 1. *See also* **Identity Property of Multiplication.**

$$-2 \times 1 = -2$$

↑ **multiplicative identity**

The number 1 is called the ***multiplicative identity*** *because multiplying any number by 1 does not change the number.*

N

natural numbers
números naturales
(1)

See **counting numbers.**

---

negative numbers
números negativos
(4)

Numbers less than zero.

−15 and −2.86 are ***negative numbers.***
19 and 0.74 are not ***negative numbers.***

---

nonlinear equations
ecuaciones no lineales
(120)

An equation whose graph does not lie on a line.

---

number line
recta numérica
(4)

A line for representing and graphing numbers. Each point on the line corresponds to a number.

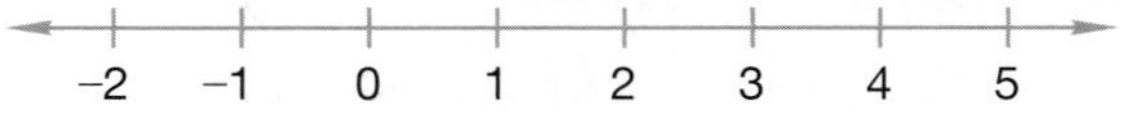

number line

---

numeral
número
(maintenance)

A symbol or group of symbols that represents a number.

4, 72, and $\frac{1}{2}$ are examples of ***numerals.*** *"Four," "seventy-two," and "one-half" are words that name numbers but are not* ***numerals.***

---

numerator
numerador
(8)

The top term of a fraction.

$\frac{9}{10}$ 9 ⟵ **numerator**; 10 ⟵ denominator

O

oblique line(s)
línea(s) oblicua(s)
(7)

1. A line that is neither horizontal nor vertical.

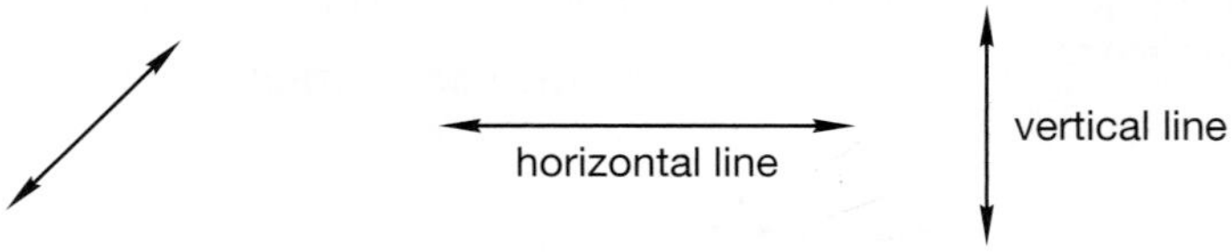

oblique line not **oblique lines**

2. Lines in the same plane that are neither parallel nor perpendicular.

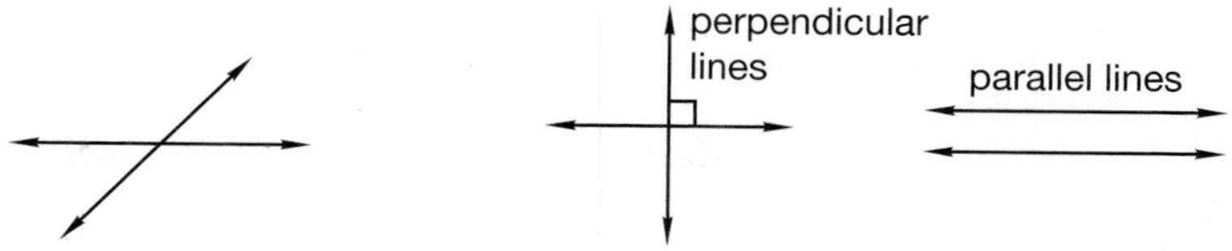

oblique lines — not **oblique lines**

obtuse angle
ángulo obtuso
(7)

An angle whose measure is between 90° and 180°.

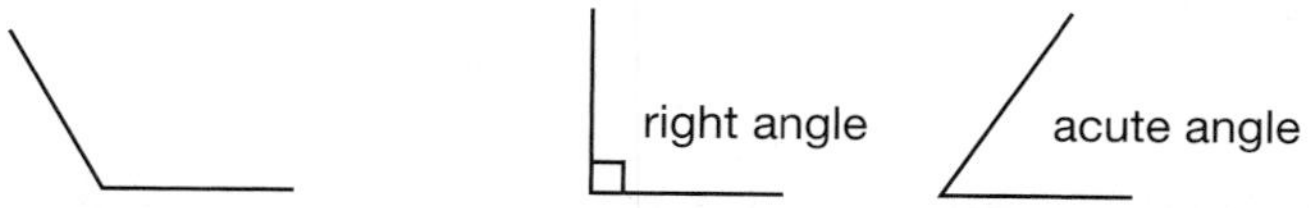

obtuse angle — not **obtuse angles**

*An **obtuse angle** is larger than both a right angle and an acute angle.*

obtuse triangle
triángulo obtusángulo
(62)

A triangle whose largest angle measures between 90° and 180°.

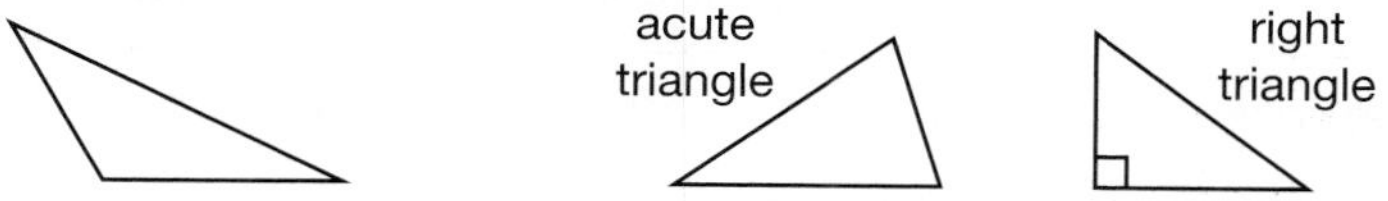

obtuse triangle — not **obtuse triangles**

odds
posibilidad
(Inv. 8)

A way of describing the likelihood of an event; the ratio of favorable outcomes to unfavorable outcomes.

*If you roll a number cube, the **odds** of getting a 3 are 1 to 5.*

operations of arithmetic
operaciones aritméticas
(1)

The four basic mathematical operations: addition, subtraction, multiplication, and division.

$1 + 9$ $\quad 21 - 8$ $\quad 6 \times 22$ $\quad 3 \div 1$

the **operations of arithmetic**

opposites
opuestos
(4)

Two numbers whose sum is zero; a positive number and a negative number whose absolute values are equal.

$(-3) + (+3) = 0$

*The numbers +3 and −3 are **opposites.***

origin
origen
(4, Inv. 3)

1. The location of the number 0 on a number line.

origin on a number line

2. The point (0, 0) on a coordinate plane.

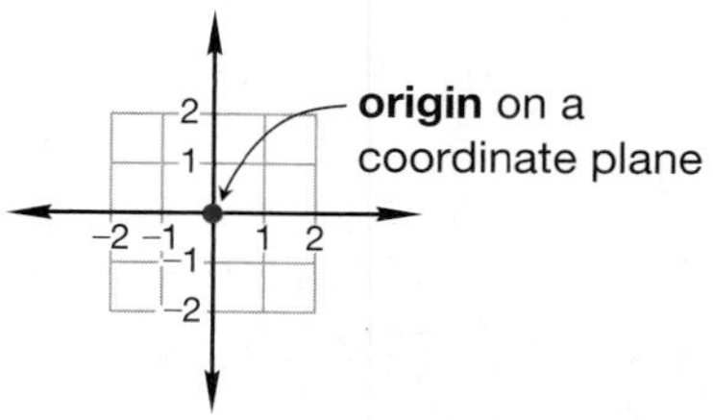

outlier
valor alejado
(Inv. 4)

A number in a list of data that is distant from the other numbers in the list.

1, 5, 4, 3, 6, 28, 7, 2

*In this list, the number 28 is an **outlier** because it is distant from the other numbers in the list.*

P

parallel lines
líneas paralelas
(7)

Lines in the same plane that do not intersect.

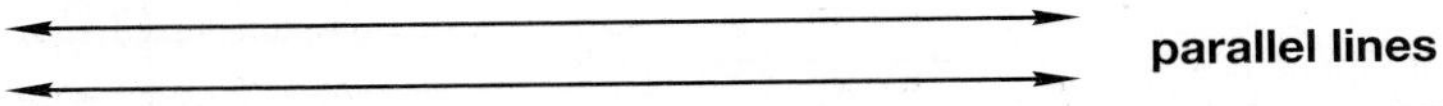

parallelogram
paralelogramo
(Inv. 6)

A quadrilateral that has two pairs of parallel sides.

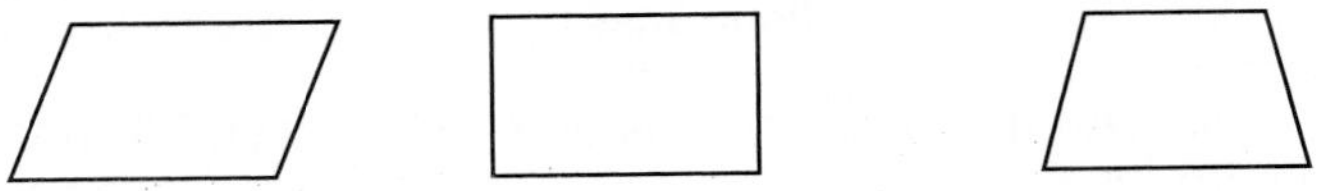

parallelograms — not a **parallelogram**

percent
por ciento
(8)

A fraction whose denominator of 100 is expressed as a percent sign (%).

$\frac{99}{100} = 99\% = 99$ ***percent***

perfect square
cuadrado perfecto
(4)

The product when a whole number is multiplied by itself.

*The number 9 is a **perfect square** because $9 = 3^2$.*

perimeter
perímetro
(19)

The distance around a closed, flat shape.

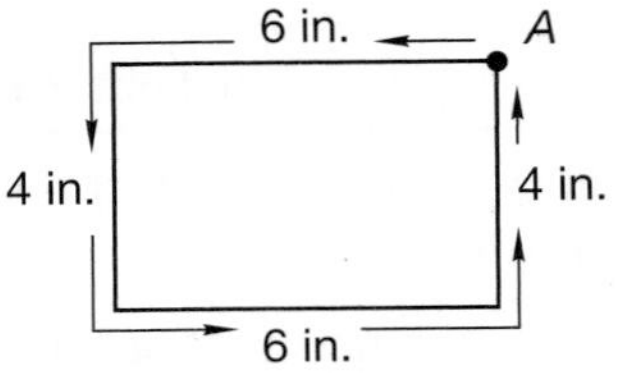

*The **perimeter** of this rectangle (from point A around to point A) is 20 inches.*

permutation
permutación
(8)

One possible arrangement of a set of objects.

2 4 3 1

*The arrangement above is one possible **permutation** of the numbers 1, 2, 3, and 4.*

perpendicular bisector
mediatriz
(Inv. 10)

A perpendicular line, ray, or segment that intersects another segment at its midpoint.

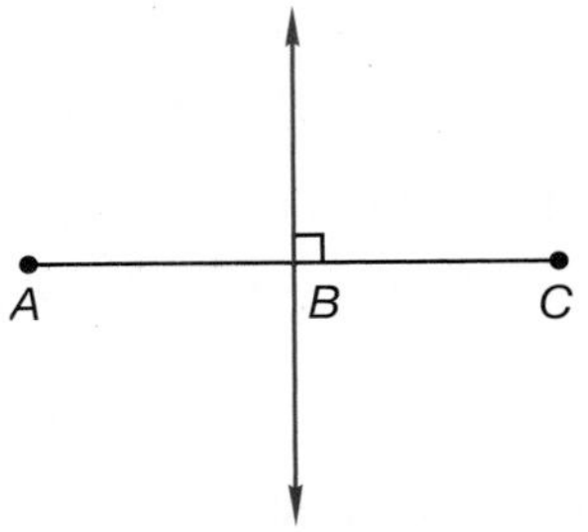

*This vertical line is a **perpendicular bisector** of $\overline{AC}$.*

perpendicular lines
líneas perpendiculares
(7)

Two lines that intersect at right angles.

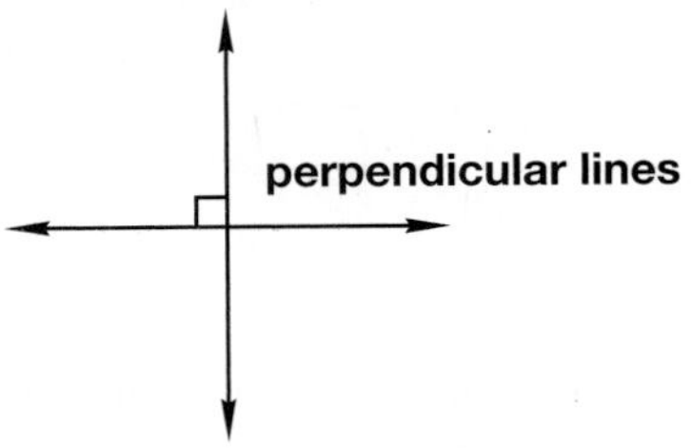

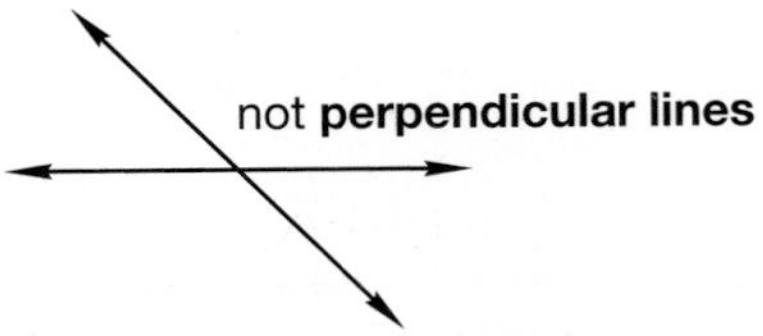

pi (π)
pi (π)
(65)

The number of diameters equal to the circumference of a circle.

*Approximate values of **pi** are 3.14 and $\frac{22}{7}$.*

place value
valor posicional
(5)

The value of a digit based on its position within a number.

$$\begin{array}{r} 341 \\ 23 \\ +\quad 7 \\ \hline 371 \end{array}$$

Place value** tells us that the 4 in 341 is worth "4 tens." In addition problems, we align digits with the same **place value.

---

plane
plano
(7)

In geometry, a flat surface that has no boundaries.

*The flat surface of a desk is part of a **plane.***

---

point
punto
(7)

An exact position on a line, on a plane, or in space.

• *A*   *This dot represents **point** A.*

---

point of symmetry
punto de simetría
(Inv. 6)

A type of rotational symmetry in which the image of the figure reappears after a 180° turn, because every point on the figure has a corresponding point on the figure on the opposite side of and equally distant from a central point called the point of symmetry.

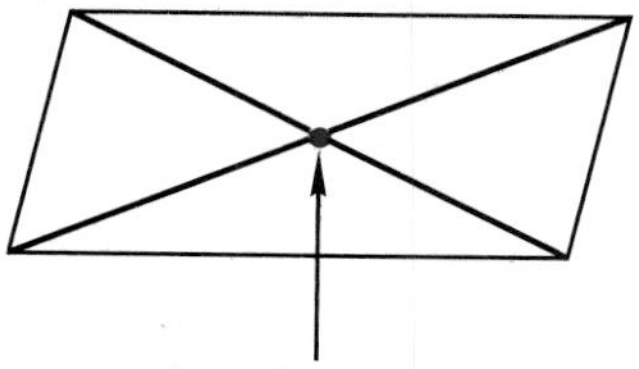

---

polygon
polígono
(18)

A closed, flat shape with straight sides.

---

polyhedron
poliedro
(67)

A geometric solid whose faces are polygons.

---

polynomial
polinomio
(84)

An algebraic expression that has one or more terms.

*The expression $3x^2 + 13x + 12y$ is a **polynomial.***

---

positive numbers
números positivos
(4)

Numbers greater than zero.

*0.25 and 157 are **positive numbers.***
*−40 and 0 are not **positive numbers.***

---

power
potencia
(20)

1. The value of an exponential expression.

*16 is the fourth **power** of 2 because $2^4 = 16$.*

2. An exponent.

*The expression 2^4 is read "two to the fourth **power.**"*

---

prime factorization
factorización prima
(21)

The expression of a composite number as a product of its prime factors.

*The **prime factorization** of 60 is $2 \times 2 \times 3 \times 5$.*

prime factors The factors of a number that are prime numbers.
factores primos
(21)

*The factors of 45 are 1, 3, 5, 9, 15, and 45. Its **prime factors** are 3 and 5.*

---

prime number A counting number greater than 1 whose only two factors are the number 1 and itself.
número primo
(21)

*7 is a **prime number.** Its only factors are 1 and 7.*

*10 is not a **prime number.** Its factors are 1, 2, 5, and 10.*

---

principal The amount of money borrowed in a loan, deposited in an account that earns interest, or invested in a fund.
capital
(110)

*If we borrow $750.00, our **principal** is $750.00.*

---

prism A polyhedron with two congruent parallel bases.
prisma
(67)

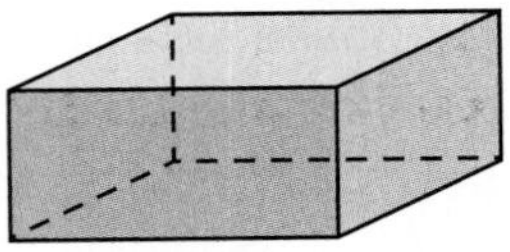

rectangular **prism**

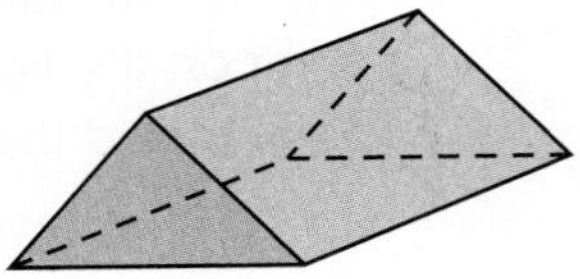

triangular **prism**

---

probability A way of describing the likelihood of an event; the ratio of favorable outcomes to all possible outcomes.
probabilidad
(14)

*The **probability** of rolling a 3 with a standard number cube is* $\frac{1}{6}$.

---

product The result of multiplication.
producto
(1)

$5 \times 4 = 20$ *The **product** of 5 and 4 is 20.*

---

proportion A statement that shows two ratios are equal.
proporción
(39)

$\frac{6}{10} = \frac{9}{15}$ *These two ratios are equal, so this is a **proportion.***

---

protractor A tool that is used to measure and draw angles.
transportador
(17)

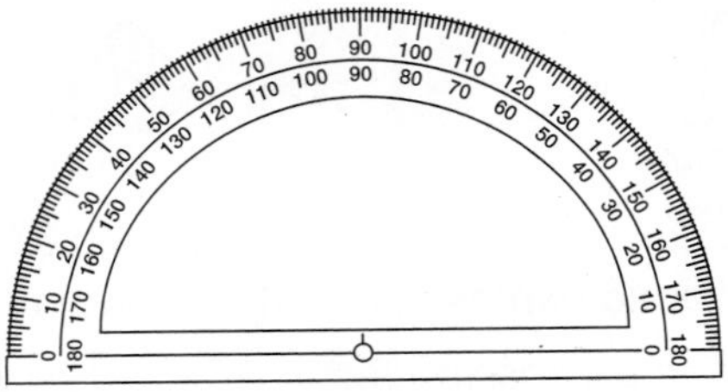

protractor

Pythagorean theorem
teorema de Pitágoras
(99)

The area of a square constructed on the hypotenuse of a right triangle is equal to the sum of the areas of squares constructed on the legs of the right triangle.

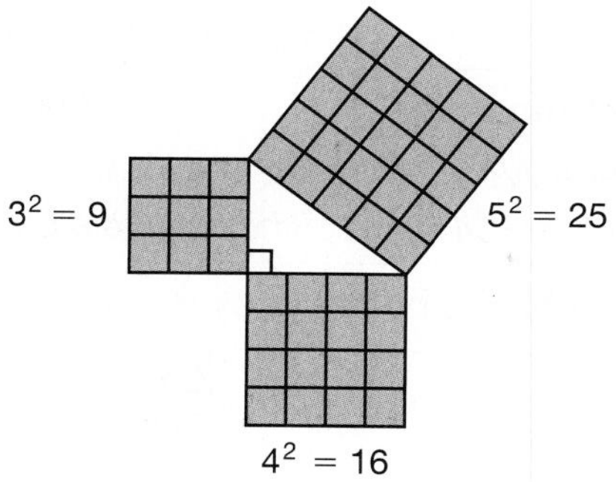

$5^2 = 4^2 + 3^2$
$25 = 16 + 9$
$25 = 15$

Q

quadrant
cuadrante
(Inv. 3)

A region of a coordinate plane formed when two perpendicular number lines intersect at their origins.

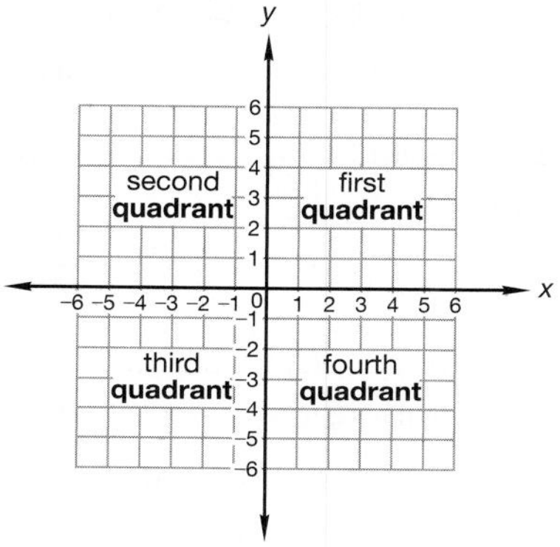

quotient
cociente
(1)

The result of division.

$12 \div 3 = 4$ $\quad \begin{array}{r}4\\3\overline{)12}\end{array}$ $\quad \dfrac{12}{3} = 4$

*The **quotient** is 4 in each of these problems.*

R

radical expression
expresión con radical
(20)

An expression that indicates the root of a number. A radical expression contains a radical sign, $\sqrt{\ }$.

radical expressions		not radical expressions	
$\sqrt{15^2}$	$\sqrt{9}$	$2 + 4$	16
$\sqrt{x}$	$2 + \sqrt{13}$	xy	4133

radius
radio
(Inv. 2)

(Plural: *radii*) The distance from the center of a circle or sphere to a point on the circle or sphere.

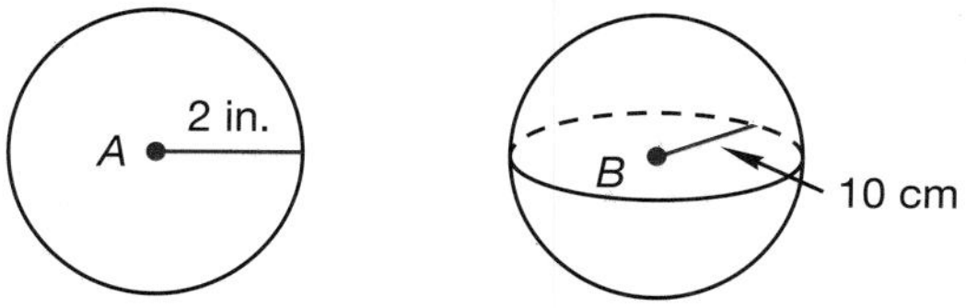

*The **radius** of circle A is 2 inches.*
*The **radius** of sphere B is 10 centimeters.*

range
intervalo
(Inv. 4)

The difference between the largest number and smallest number in a list.

5, 17, 12, 34, 29, 13

To calculate the ***range*** *of this list, we subtract the smallest number from the largest number. The* ***range*** *of this list is 29.*

---

rate
tasa
(46)

A ratio of two measures.

If a car travels 240 miles in 4 hours, its average ***rate*** *is 240 miles ÷ 4 hours, which equals 60 miles per hour (mph).*

---

ratio
razón
(36)

A comparison of two numbers by division.

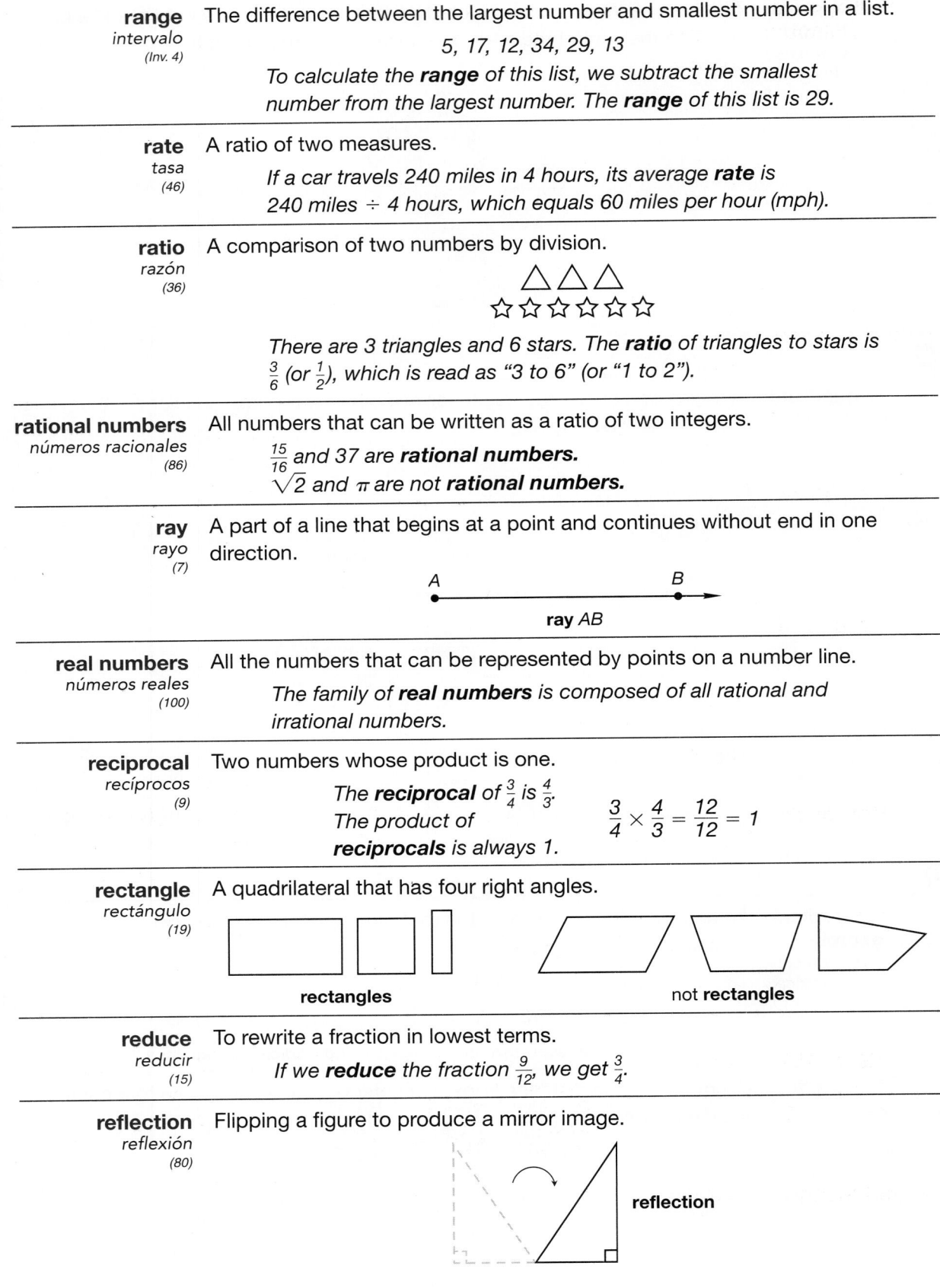

There are 3 triangles and 6 stars. The ***ratio*** *of triangles to stars is* $\frac{3}{6}$ *(or* $\frac{1}{2}$*), which is read as "3 to 6" (or "1 to 2").*

---

rational numbers
números racionales
(86)

All numbers that can be written as a ratio of two integers.

$\frac{15}{16}$ *and 37 are* ***rational numbers.***
$\sqrt{2}$ *and* π *are not* ***rational numbers.***

---

ray
rayo
(7)

A part of a line that begins at a point and continues without end in one direction.

ray *AB*

---

real numbers
números reales
(100)

All the numbers that can be represented by points on a number line.

The family of ***real numbers*** *is composed of all rational and irrational numbers.*

---

reciprocal
recíprocos
(9)

Two numbers whose product is one.

The ***reciprocal*** *of* $\frac{3}{4}$ *is* $\frac{4}{3}$*.*
The product of ***reciprocals*** *is always 1.*

$$\frac{3}{4} \times \frac{4}{3} = \frac{12}{12} = 1$$

---

rectangle
rectángulo
(19)

A quadrilateral that has four right angles.

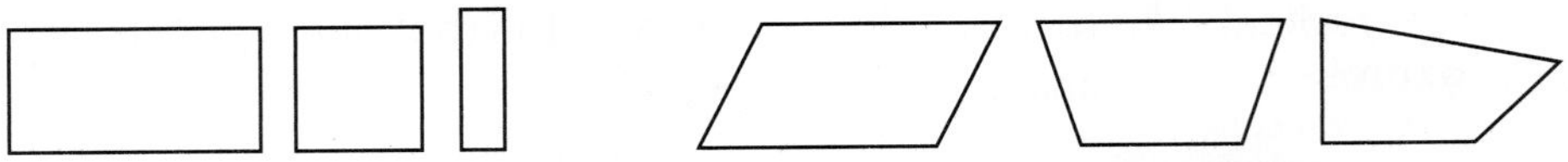

rectangles — not **rectangles**

---

reduce
reducir
(15)

To rewrite a fraction in lowest terms.

If we ***reduce*** *the fraction* $\frac{9}{12}$*, we get* $\frac{3}{4}$*.*

---

reflection
reflexión
(80)

Flipping a figure to produce a mirror image.

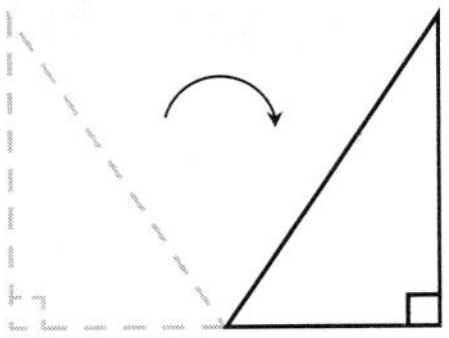

reflection

reflective symmetry
simetría de reflexión
(58)

A figure has reflective symmetry if it can be divided into two mirror images along a line; also known as line symmetry.

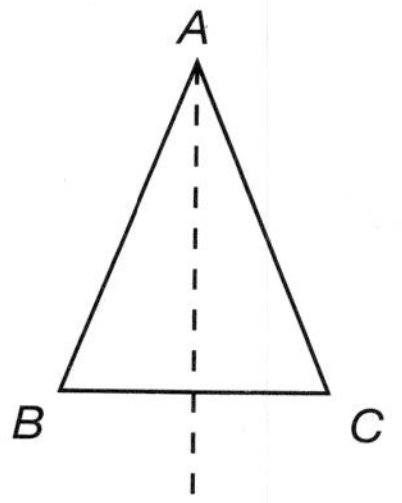

*Isosceles triangle ABC has **reflective symmetry.** A line of symmetry divides the triangle into two mirror images.*

---

regular polygon
polígono regular
(18)

A polygon in which all sides have equal lengths and all angles have equal measures.

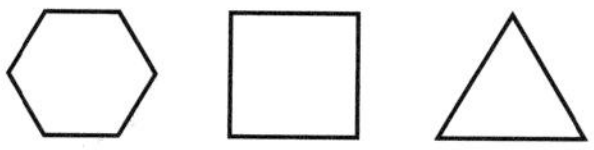

regular polygons

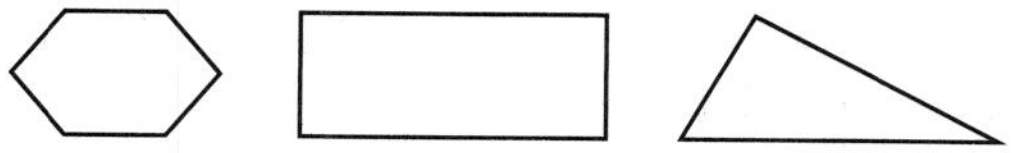

not **regular polygons**

---

repetend
término que se repite
(42)

The repeating digits of a decimal number. The symbol for a repetend is an overbar.

$$0.83333333... = 0.8\overline{3}$$

*In the number above, 3 is the **repetend.***

---

rhombus
rombo
(Inv. 6)

A parallelogram with all four sides of equal length.

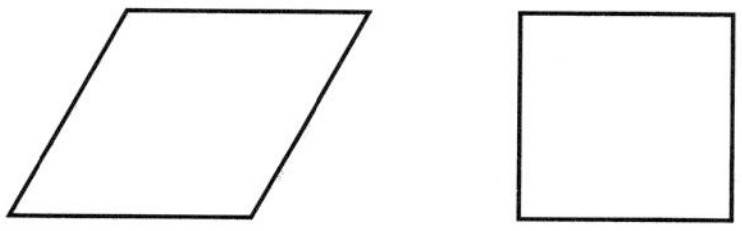

rhombuses

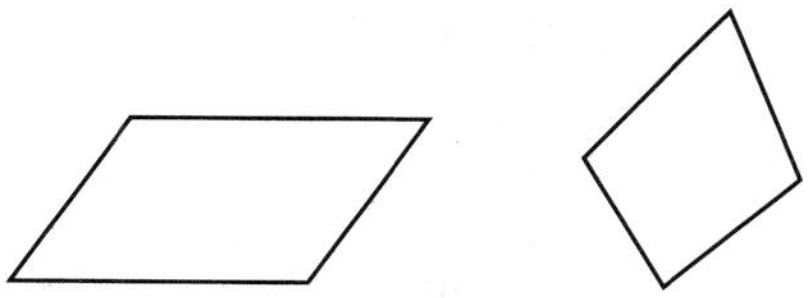

not **rhombuses**

---

right angle
ángulo recto
(7)

An angle that forms a square corner and measures 90°. It is often marked with a small square.

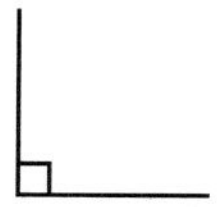

right angle

not **right angles**

*A **right angle** is larger than an acute angle and smaller than an obtuse angle.*

---

right solid
sólido rectangular
(95)

A geometric solid whose sides are perpendicular to its base.

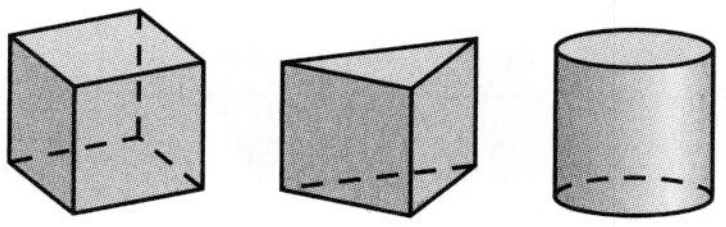

right solids

---

right triangle
triángulo rectángulo
(62)

A triangle whose largest angle measures 90°.

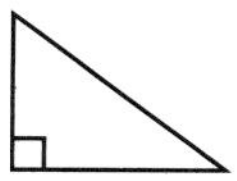

right triangle

not **right triangles**

root *raíz* *(106)*

A value of a radical expression.

$$\sqrt{16} = 4$$

*4 is a **root** of this radical expression.*

rotation *rotación* *(80)*

Turning a figure about a specified point called the *center of rotation.*

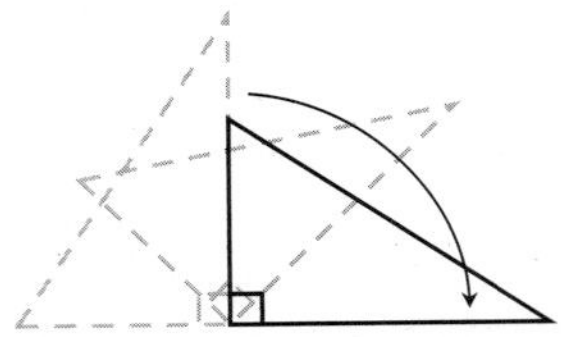

rotational symmetry *simetría rotacional* *(58)*

A figure has rotational symmetry if the figure matches itself two or more times in a full turn.

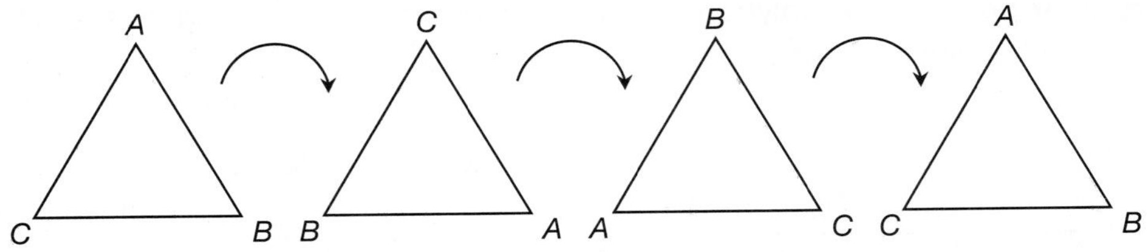

*An equilateral triangle has **rotational symmetry;** its original image reappears three times in a full turn.*

S

sample space *espacio muestral* *(36)*

A list of all the possible outcomes of an event.

*The **sample space** of outcomes when flipping a coin consists of* heads *and* tails.

scale *escala* *(98)*

A ratio that shows the relationship between a scale model and the actual object.

*If a model airplane is $\frac{1}{24}$ the size of the actual airplane, the **scale** of the model is 1 to 24.*

scale factor *factor de escala* *(98)*

The number that relates corresponding sides of similar geometric figures.

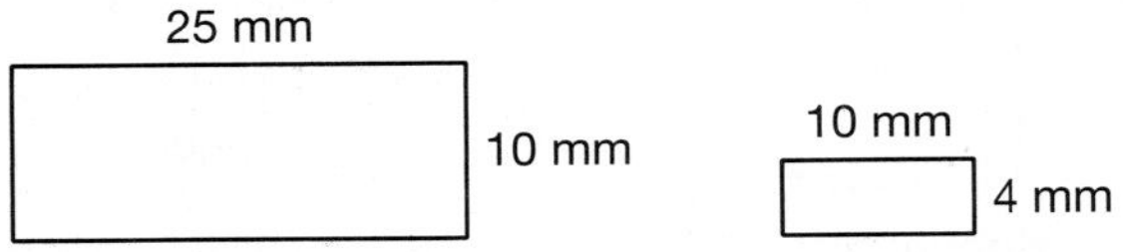

*The **scale factor** from the smaller rectangle to the larger rectangle is 2.5.*

scalene triangle *triángulo escaleno* *(62)*

A triangle with three sides of different lengths.

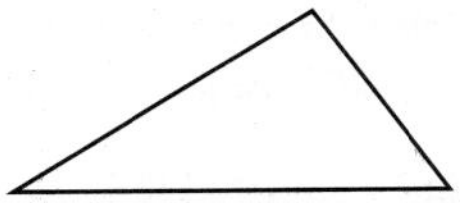

*All three sides of this **scalene triangle** have different lengths.*

scientific notation *notación científica* *(51)*

A method of writing a number as a product of a decimal number and a power of 10.

*In **scientific notation,** 34,000 is written 3.4×10^4.*

sector A region that is bordered by an arc and two radii of a circle.
sector
(Inv. 2)

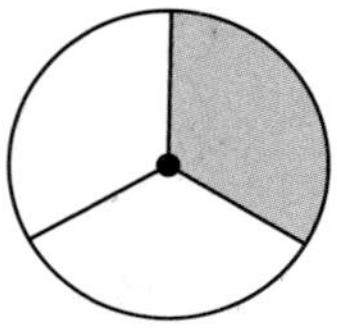

This circle is divided into 3 ***sectors.***

---

segment A part of a line with two distinct endpoints.
segmento
(7)

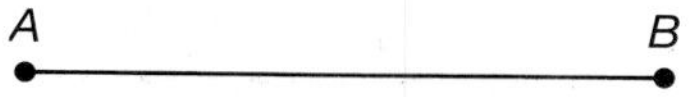

segment *AB* or **segment** *BA*

---

semicircle A half circle.
semicírculo
(Inv. 2)

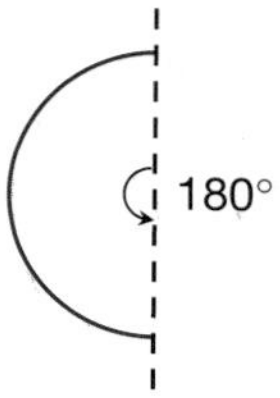

semicircle

A ***semicircle*** *is an arc whose measure is 180°.*

---

sequence A list of numbers arranged according to a certain rule.
secuencia
(4)

The numbers 2, 4, 6, 8, ... form a ***sequence.*** *The rule is "count up by twos."*

---

similar Having the same shape but not necessarily the same size. Corresponding parts of similar figures are proportional.
semejante
(18)

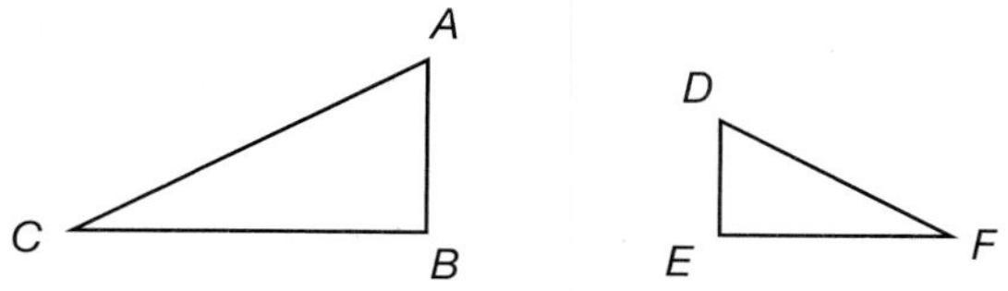

△ABC and △DEF are ***similar.*** *They have the same shape.*

---

simple interest Interest that does not pay on previously earned interest.
interés simple
(110)

Simple Interest

$100.00	*principal*
$6.00	*first-year interest*
+ $6.00	*second-year interest*
$112.00	*total after two years*

Compound Interest

$100.00	*principal*
+ $6.00	*first-year interest (6% of $100.00)*
$106.00	*total after one year*
+ $6.36	*second-year interest (6% of $106.00)*
$112.36	*total after two years*

---

skew lines In three-dimensional space, lines that do not intersect and are not in the same plane.
lineas sesgadas
(7)

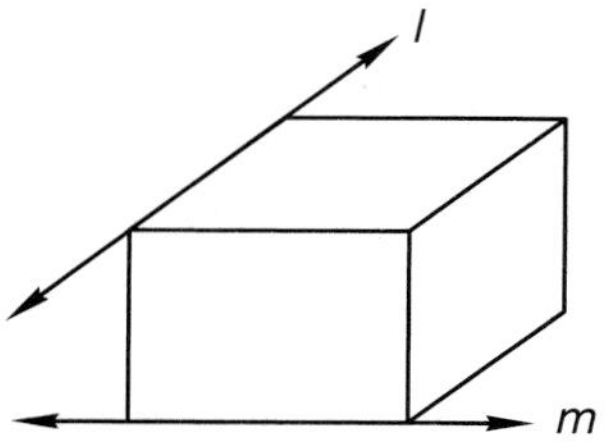

Lines l and m are ***skew lines*** *because they do not intersect but they are not parallel because they do not lie in the same plane.*

slope
pendiente
(107)

The number that represents the slant of the graph of a linear equation.

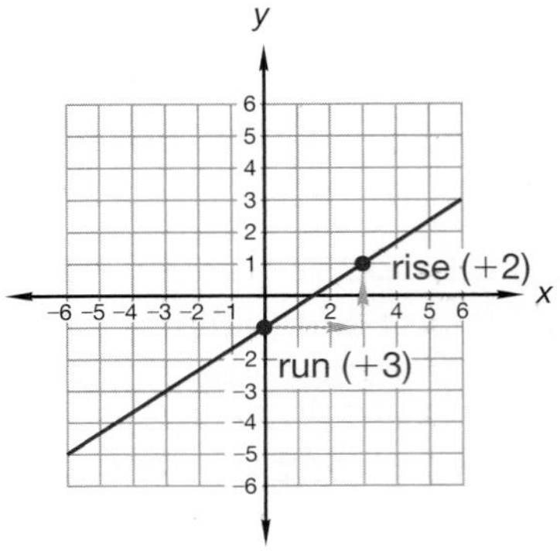

*Every vertical increase (rise) of 2 units leads to a horizontal increase (run) of 3 units, so the **slope** of this line is $\frac{2}{3}$.*

---

slope-intercept form
forma pendiente-intersección
(116)

The form $y = mx + b$ for a linear equation. In this form m is the slope of the graph of the equation, and b is the y-intercept.

In the equation $y = 2x + 3$, *the **slope** is 2 and the **y-intercept** is 3.*

---

solid
sólido
(67)

See **geometric solid.**

---

space
espacio
(7)

The dimensional space known in the physical world. In geometry, we label one-dimensional space as length, two-dimensional space as area, and three-dimensional space as volume.

---

sphere
esfera
(67)

A round geometric surface whose points are an equal distance from its center.

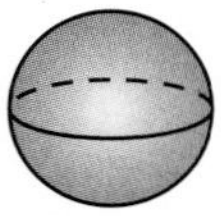

sphere

---

square
cuadrado
(18, 20)

1. A rectangle with all four sides of equal length.

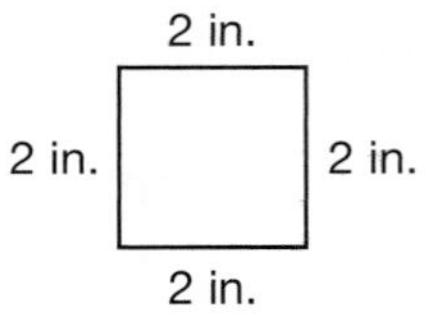

*All four sides of this **square** are 2 inches long.*

2. The product of a number and itself.

*The **square** of 4 is 16.*

---

square root
raíz cuadrada
(20)

One of two equal factors of a number. The symbol for the principal, or positive, square root of a number is $\sqrt{\ }$.

*A **square root** of 49 is 7 because* $7 \times 7 = 49$.

---

stem-and-leaf plot
diagrama de tallo y hojas
(Inv. 4)

A method of graphing a collection of numbers by placing the "stem" digits (or initial digits) in one column and the "leaf" digits (or remaining digits) out to the right.

Stem	Leaf
2	1 3 5 6 6 8
3	0 0 2 2 4 5 6 6 8 9
4	0 0 1 1 1 2 3 3 5 7 7 8
5	0 1 1 2 3 5 8

*In this **stem-and-leaf plot,** 3|2 represents 32.*

straight angle
ángulo llano
(7)

An angle that measures 180° and thus forms a straight line.

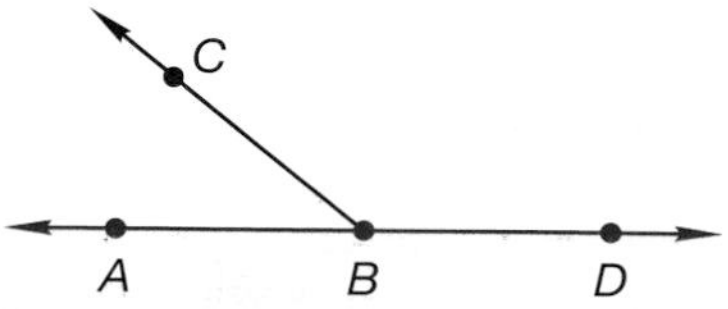

Angle ABD is a ***straight angle.*** *Angles ABC and CBD are not* ***straight angles.***

subtrahend
sustraendo
(1)

A number that is subtracted.

$12 - 8 = 4$ *The* ***subtrahend*** *in this problem is 8.*

sum
suma
(1)

The result of addition.

$7 + 6 = 13$ *The* ***sum*** *of 7 and 6 is 13.*

supplementary angles
ángulos suplementarios
(40)

Two angles whose sum is 180°.

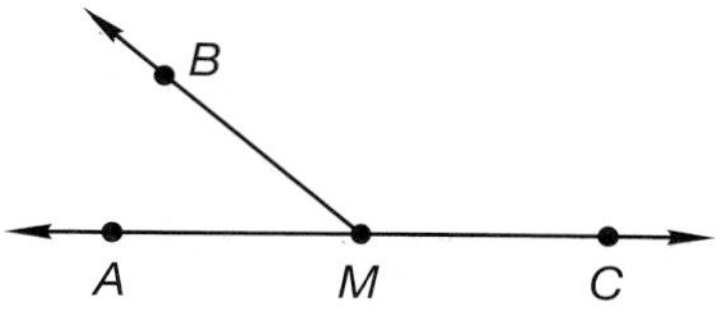

∠AMB and ∠CMB are ***supplementary.***

surface area
área superficial
(67)

The total area of the surface of a geometric solid.

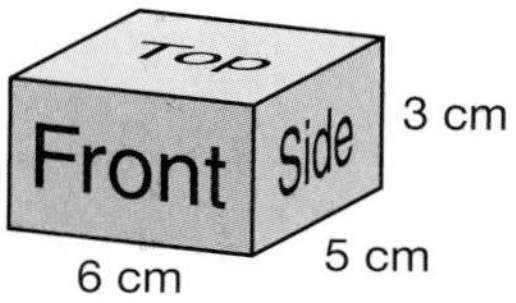

Area of top	= 5 cm × 6 cm =		$30\ cm^2$
Area of bottom	= 5 cm × 6 cm =		$30\ cm^2$
Area of front	= 3 cm × 6 cm =		$18\ cm^2$
Area of back	= 3 cm × 6 cm =		$18\ cm^2$
Area of side	= 3 cm × 5 cm =		$15\ cm^2$
+ *Area of side*	= 3 cm × 5 cm =		$15\ cm^2$
Total ***surface area***		=	$126\ cm^2$

symbols of inclusion
símbolos de inclusión
(52)

Symbols that are used to set apart portions of an expression so that they may be evaluated first: ( ), [ ], { }, and the division bar in a fraction.

In the statement $(8 - 4) \div 2$*, the* ***symbols of inclusion*** *indicate that* $8 - 4$ *should be calculated before dividing by 2.*

T

term
término
(4, 15, 84)

1. A number that serves as a numerator or denominator of a fraction.

$\frac{5}{6}$ → **terms**

2. One of the numbers in a sequence.

1, 3, 5, 7, 9, 11, ...

Each number in this sequence is a ***term.***

3. A constant or variable expression composed of one or more factors in an algebraic expression.

The expression $2x + 3xyz$ *has two* ***terms.***

theoretical probability
probabilidad teórica
(Inv. 8)

The probability that an event will occur as determined by analysis rather than by experimentation.

The ***theoretical probability*** *of rolling a 3 with a standard number cube is* $\frac{1}{6}$.

third quartile
tercer cuartil
(Inv. 4)

In a set of data, the middle number of the upper half; the number below which 75% of the data lies; the 75th percentile; also called the **upper quartile.**

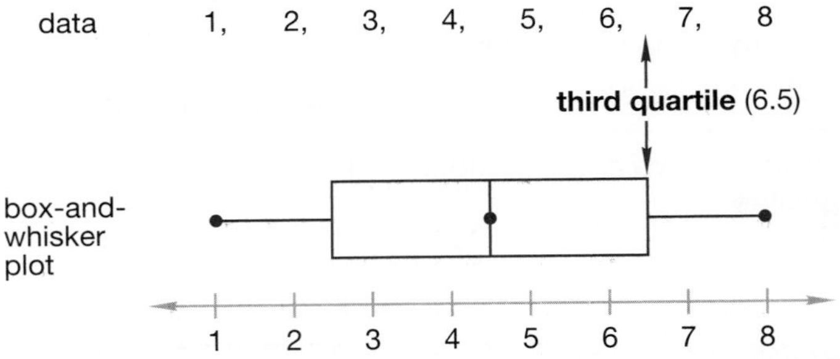

transformation
transformación
(80)

The changing of a figure's position through rotation, reflection, or translation.

Transformations

Movement	Name
flip	reflection
slide	translation
turn	rotation

translation
traslación
(80)

Sliding a figure from one position to another without turning or flipping the figure.

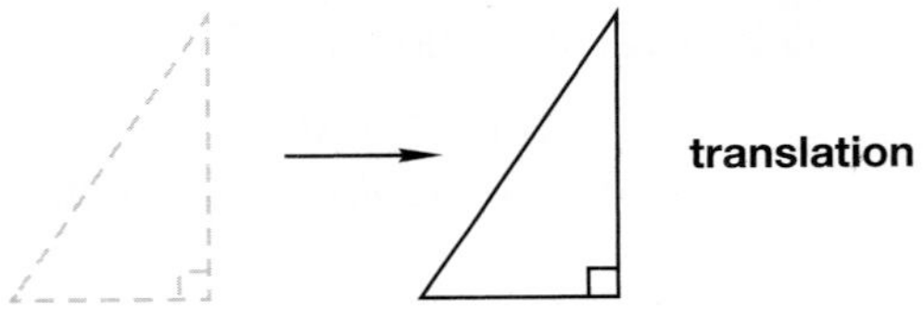

transversal
transversal
(102)

A line that intersects one or more other lines in a plane.

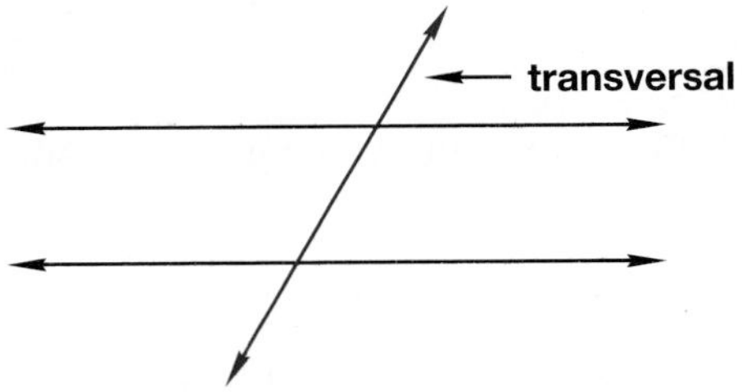

trapezoid
trapecio
(Inv. 6)

A quadrilateral with exactly one pair of parallel sides.

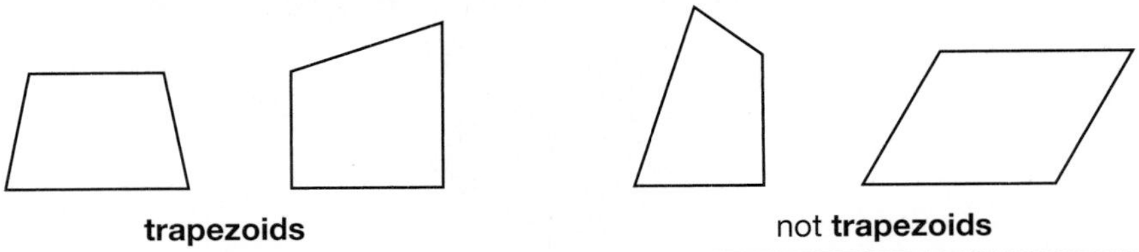

triangular prism
prisma triangular
(67)

See **prism.**

U

unit conversion
conversión de unidades
(50)

The process of changing a measure to an equivalent measure that has different units.

*Through **unit conversion,** we can write 2 feet as 24 inches.*

unit multiplier
factor de conversión
(50)

A ratio equal to 1 that is composed of two equivalent measures.

$$\frac{12 \text{ inches}}{1 \text{ foot}} = 1$$

*We can use this **unit multiplier** to convert feet to inches.*

unit price
precio unitario
(46)

The price of one unit of measure of a product.

*The **unit price** of bananas is $1.19 per pound.*

upper quartile
cuartil superior
(Inv. 4)

See **third quartile.**

U.S. Customary System
Sistema usual de EE.UU.
(16)

A system of measurement used almost exclusively in the United States.

*Pounds, quarts, and feet are units in the **U.S. Customary System.***

V

variable
variable
(1)

A quantity that can change or assume different values. Also, a letter used to represent an unknown in an expression or equation.

*In the statement $x + 7 = y$, the letters x and y are **variables.***

vertex
vértice
(7)

(Plural: *vertices*) A point of an angle, polygon, or polyhedron where two or more lines, rays, or segments meet.

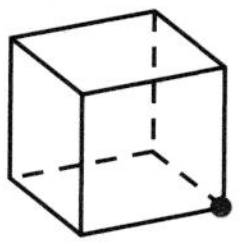

*One **vertex** of this cube is colored. A cube has eight **vertices.***

vertical angles
ángulos verticales
(40)

A pair of nonadjacent angles formed by a pair of intersecting lines. Vertical angles have the same measure.

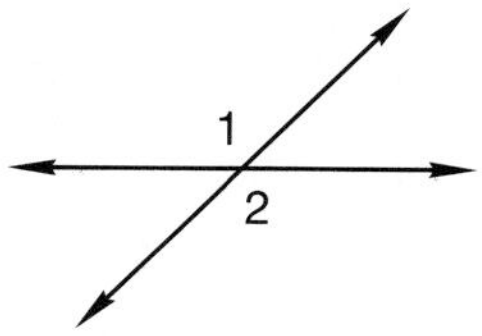

*Angles 1 and 2 are **vertical angles.***

volume
volumen
(70)

The amount of space a solid shape occupies. Volume is measured in cubic units.

*This rectangular prism is 3 units wide, 3 units high, and 4 units deep. Its **volume** is $3 \cdot 3 \cdot 4 = 36$ cubic units.*

W

whole numbers The members of the set {0, 1, 2, 3, 4, ...}.
números enteros
(1)

0, 25, and 134 are ***whole numbers.***
−3, 0.56, and $100\frac{3}{4}$ *are not* ***whole numbers.***

X

x-axis The horizontal number line of a coordinate plane.
eje de las x
(Inv. 3)

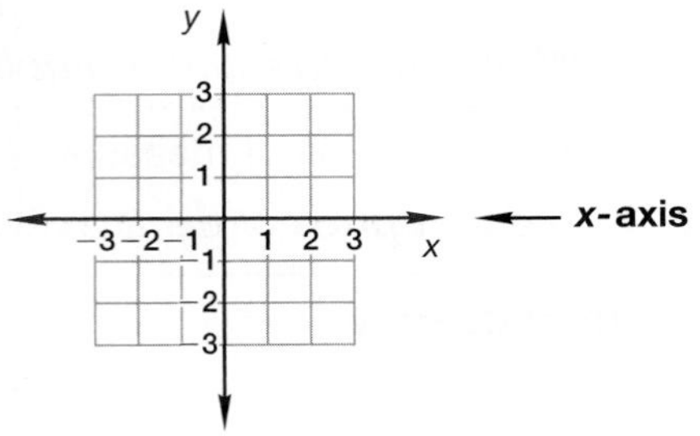

Y

y-axis The vertical number line of a coordinate plane.
eje de las y
(Inv. 3)

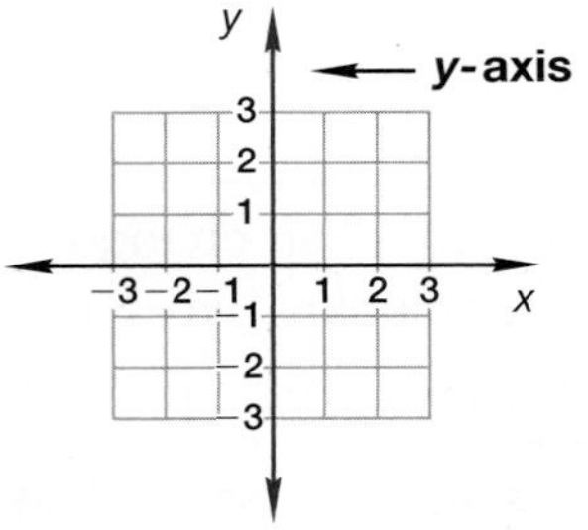

y-intercept The point on a coordinate plane where the graph of an equation intersects the *y*-axis.
intersección en y
(116)

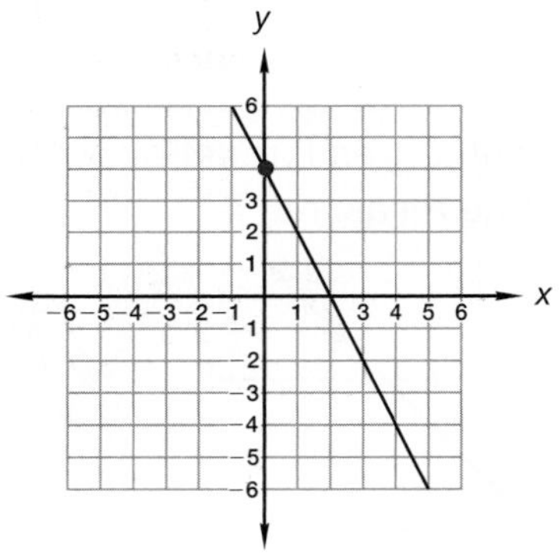

This line crosses the y-axis at y = 4, so the ***y-intercept*** *of this line is 4.*

Z

Zero Property of Multiplication Zero times any number is zero. In symbolic form, $0 \times a = 0$.
propiedad del cero en la multiplicación
(2)

The ***Zero Property of Multiplication*** *tells us that* $89 \times 0 = 0$.

A

B

C

D

INDEX

H

I

K

L

M

N

INDEX

O

P

Q

R

S

T

U

V

W

X

Y

Z